# HANDBOOK OF
# ENUMERATIVE COMBINATORICS

# DISCRETE MATHEMATICS AND ITS APPLICATIONS

*R. B. J. T. Allenby and Alan Slomson*, How to Count: An Introduction to Combinatorics, Third Edition

*Craig P. Bauer*, Secret History: The Story of Cryptology

*Juergen Bierbrauer*, Introduction to Coding Theory

*Katalin Bimbó*, Combinatory Logic: Pure, Applied and Typed

*Katalin Bimbó*, Proof Theory: Sequent Calculi and Related Formalisms

*Donald Bindner and Martin Erickson*, A Student's Guide to the Study, Practice, and Tools of Modern Mathematics

*Francine Blanchet-Sadri*, Algorithmic Combinatorics on Partial Words

*Miklós Bóna*, Combinatorics of Permutations, Second Edition

*Miklós Bóna*, Handbook of Enumerative Combinatorics

*Jason I. Brown*, Discrete Structures and Their Interactions

*Richard A. Brualdi and Dragoš Cvetković*, A Combinatorial Approach to Matrix Theory and Its Applications

*Kun-Mao Chao and Bang Ye Wu*, Spanning Trees and Optimization Problems

*Charalambos A. Charalambides*, Enumerative Combinatorics

*Gary Chartrand and Ping Zhang*, Chromatic Graph Theory

*Henri Cohen, Gerhard Frey, et al.*, Handbook of Elliptic and Hyperelliptic Curve Cryptography

*Charles J. Colbourn and Jeffrey H. Dinitz*, Handbook of Combinatorial Designs, Second Edition

*Abhijit Das*, Computational Number Theory

*Matthias Dehmer and Frank Emmert-Streib*, Quantitative Graph Theory: Mathematical Foundations and Applications

*Martin Erickson*, Pearls of Discrete Mathematics

*Martin Erickson and Anthony Vazzana*, Introduction to Number Theory

**Titles (continued)**

*Steven Furino, Ying Miao, and Jianxing Yin,* Frames and Resolvable Designs: Uses, Constructions, and Existence

*Mark S. Gockenbach,* Finite-Dimensional Linear Algebra

*Randy Goldberg and Lance Riek,* A Practical Handbook of Speech Coders

*Jacob E. Goodman and Joseph O'Rourke,* Handbook of Discrete and Computational Geometry, Second Edition

*Jonathan L. Gross,* Combinatorial Methods with Computer Applications

*Jonathan L. Gross and Jay Yellen,* Graph Theory and Its Applications, Second Edition

*Jonathan L. Gross, Jay Yellen, and Ping Zhang* Handbook of Graph Theory, Second Edition

*David S. Gunderson,* Handbook of Mathematical Induction: Theory and Applications

*Richard Hammack, Wilfried Imrich, and Sandi Klavžar,* Handbook of Product Graphs, Second Edition

*Darrel R. Hankerson, Greg A. Harris, and Peter D. Johnson,* Introduction to Information Theory and Data Compression, Second Edition

*Darel W. Hardy, Fred Richman, and Carol L. Walker,* Applied Algebra: Codes, Ciphers, and Discrete Algorithms, Second Edition

*Daryl D. Harms, Miroslav Kraetzl, Charles J. Colbourn, and John S. Devitt,* Network Reliability: Experiments with a Symbolic Algebra Environment

*Silvia Heubach and Toufik Mansour,* Combinatorics of Compositions and Words

*Leslie Hogben,* Handbook of Linear Algebra, Second Edition

*Derek F. Holt with Bettina Eick and Eamonn A. O'Brien,* Handbook of Computational Group Theory

*David M. Jackson and Terry I. Visentin,* An Atlas of Smaller Maps in Orientable and Nonorientable Surfaces

*Richard E. Klima, Neil P. Sigmon, and Ernest L. Stitzinger,* Applications of Abstract Algebra with Maple™ and MATLAB®, Second Edition

*Richard E. Klima and Neil P. Sigmon,* Cryptology: Classical and Modern with Maplets

*Patrick Knupp and Kambiz Salari,* Verification of Computer Codes in Computational Science and Engineering

*William Kocay and Donald L. Kreher,* Graphs, Algorithms, and Optimization

*Donald L. Kreher and Douglas R. Stinson,* Combinatorial Algorithms: Generation Enumeration and Search

*Hang T. Lau,* A Java Library of Graph Algorithms and Optimization

*C. C. Lindner and C. A. Rodger,* Design Theory, Second Edition

*San Ling, Huaxiong Wang, and Chaoping Xing,* Algebraic Curves in Cryptography

*Nicholas A. Loehr,* Bijective Combinatorics

*Toufik Mansour,* Combinatorics of Set Partitions

***Titles (continued)***

*Alasdair McAndrew,* Introduction to Cryptography with Open-Source Software

*Elliott Mendelson,* Introduction to Mathematical Logic, Fifth Edition

*Alfred J. Menezes, Paul C. van Oorschot, and Scott A. Vanstone,* Handbook of Applied Cryptography

*Stig F. Mjølsnes,* A Multidisciplinary Introduction to Information Security

*Jason J. Molitierno,* Applications of Combinatorial Matrix Theory to Laplacian Matrices of Graphs

*Richard A. Mollin,* Advanced Number Theory with Applications

*Richard A. Mollin,* Algebraic Number Theory, Second Edition

*Richard A. Mollin,* Codes: The Guide to Secrecy from Ancient to Modern Times

*Richard A. Mollin,* Fundamental Number Theory with Applications, Second Edition

*Richard A. Mollin,* An Introduction to Cryptography, Second Edition

*Richard A. Mollin,* Quadratics

*Richard A. Mollin,* RSA and Public-Key Cryptography

*Carlos J. Moreno and Samuel S. Wagstaff, Jr.,* Sums of Squares of Integers

*Gary L. Mullen and Daniel Panario,* Handbook of Finite Fields

*Goutam Paul and Subhamoy Maitra,* RC4 Stream Cipher and Its Variants

*Dingyi Pei,* Authentication Codes and Combinatorial Designs

*Kenneth H. Rosen,* Handbook of Discrete and Combinatorial Mathematics

*Douglas R. Shier and K.T. Wallenius,* Applied Mathematical Modeling: A Multidisciplinary Approach

*Alexander Stanoyevitch,* Introduction to Cryptography with Mathematical Foundations and Computer Implementations

*Jörn Steuding,* Diophantine Analysis

*Douglas R. Stinson,* Cryptography: Theory and Practice, Third Edition

*Roberto Tamassia,* Handbook of Graph Drawing and Visualization

*Roberto Togneri and Christopher J. deSilva,* Fundamentals of Information Theory and Coding Design

*W. D. Wallis,* Introduction to Combinatorial Designs, Second Edition

*W. D. Wallis and J. C. George,* Introduction to Combinatorics

*Jiacun Wang,* Handbook of Finite State Based Models and Applications

*Lawrence C. Washington,* Elliptic Curves: Number Theory and Cryptography, Second Edition

DISCRETE MATHEMATICS AND ITS APPLICATIONS

# HANDBOOK OF
# ENUMERATIVE COMBINATORICS

**Edited by**

**Miklós Bóna**

University of Florida
Gainesville, Florida, USA

CRC Press
Taylor & Francis Group
Boca Raton London New York

CRC Press is an imprint of the
Taylor & Francis Group, an **informa** business
A CHAPMAN & HALL BOOK

First published in paperback 2024

First published 2015 by CRC Press
2385 NW Executive Center Drive, Suite 320, Boca Raton FL 33431

and by CRC Press
4 Park Square, Milton Park, Abingdon, Oxon, OX14 4RN

*CRC Press is an imprint of Taylor & Francis Group, LLC*



Publisher's Note
The publisher has gone to great lengths to ensure the quality of this reprint but points out that some imperfections in the original copies may be apparent.

**Library of Congress Cataloging-in-Publication Data**

Bóna, Miklós.
Handbook of enumerative combinatorics / Miklós Bóna, University of Florida, Gainesville, USA.
pages cm. -- (Discrete mathematics and its applications)
"A Chapman & Hall book."
Includes bibliographical references and index.
ISBN 978-1-4822-2085-8 (alk. paper)
1. Combinatorial analysis. 2. Mathematical analysis. 3. Numeration. I. Title. II. Title: Enumerative combinatorics.

QA164.B656 2015
511'.62--dc23 2014041669

ISBN: 978-1-4822-2085-8 (hbk)
ISBN: 978-1-03-291731-3 (pbk)
ISBN: 978-0-429-17031-7 (ebk)

DOI: 10.1201/b18255

**Visit the Taylor & Francis Web site at**
**http://www.taylorandfrancis.com**

**and the CRC Press Web site at**
**http://www.crcpress.com**

# Contents

# Foreword

When I first became seriously interested in enumerative combinatorics around 1967, the subject did not really exist *per se*. There were numerous results and methods, beginning with Euler (with some hints of even earlier work), scattered throughout the literature, but there was little systematic attempt to bring some order to this chaos. It is remarkable that enumerative combinatorics has progressed so far that we now need over 1000 pages just to present a basic overview of techniques and results. Numerous topics such as pattern avoidance and parking functions existed in only very rudimentary form 40–50 years ago but are now flourishing subjects in their own right. The seventeen authors of the present volume, in addition to being world leaders in the area of their contribution, are also superb writers. Their fifteen chapters are a combination of broad surveys of major areas and techniques, together with more specialized expositions of many of the most active research topics in enumerative combinatorics today.

A prominent reason why practitioners of enumerative combinatorics find it so appealing is its unexpected connections with other areas of mathematics. These connections have grown increasingly sophisticated over the years. It is no longer sufficient to know some rudimentary algebraic topology, say, to give a significant connection with enumerative combinatorics. The papers in this handbook do an exemplary job of explaining deep connections with such areas as complex analysis, probability theory, linear algebra, commutative algebra, representation theory, algebraic geometry, algebraic topology, and computer science. Just this list of topics gives an idea of the breadth and depth of modern enumerative combinatorics. Readers from neophytes to experts have much to look forward to when they peruse the riches that follow.

Richard Stanley
Cambridge, MA
December 2014

# *Preface*

When designing a handbook of a large and rapidly developing fields like Enumerative Combinatorics, one faces several questions: What subjects to cover? How to organize the subjects? What audience to target?

We have decided to include both chapters that focus on *methods* of enumeration of various objects and chapters that focus on specific kinds of *objects* that need to be counted, by any method available. The chapters are organized so that we advance from the more general ones, namely enumeration methods, towards the more specialized ones that focus on the counting of specific objects. These objects become more and more specialized as we proceed.

As far as our preferred audience goes, we believe that each chapter can benefit at least three different kinds of readers as listed below.

- The experts, who are familiar with most of the information in the chapter, but are interested in its presentation, and some of the finer points.
- The "relative outsiders," that is, readers who have already seen a few results here and there, a proof or two here and there, but nothing systematic, and who are interested to see an organized treatment of the topic.
- The novices, who are new to the field, and have no background information past the first year of graduate school. These readers will hopefully see that the subject is interesting, accessible, and challenging.

We do hope that all three groups of our targeted readership will find the book as useful and enjoyable as we do.

Miklós Bóna
Gainesville, FL
February 2015

# Acknowledgments

At the completion of this book, I want to say thanks to Robert Ross at CRC Press, who initiated the project and selected me to lead it. The most important part of this endeavor were certainly the authors, who agreed to the difficult and time-consuming task of contributing a chapter to the book, and fulfilled that task, sometimes on time. My gratitude is extended to the mathematicians who have reviewed the various chapters. Turning 15 chapters from 17 different authors into one consistent LaTeX file was often too difficult for me, so I am grateful to Shashi Kumar at cenveo.com, who often helped me with technical matters. My gratitude is extended to my colleague Vincent Vatter, who was my local expert.

A non-negligible part of my editing work was done while I was on vacation. I owe a lot to my brother Péter who made sure that I had high-speed internet access when most people around us could not even make a phone call. Last, but not least, I am thankful to my wife Linda, and my sons Miki, Benny, and Vinnie for putting up with me when I worked at the book at highly unexpected times.

# Acknowledgements

[illegible]

[illegible]

# Part I

# Methods

# Chapter 1

# *Algebraic and Geometric Methods in Enumerative Combinatorics*

**Federico Ardila**
*San Francisco State University, San Francisco, CA, USA; Universidad de los Andes, Bogotá, Colombia*

## CONTENTS

## 1.1 Introduction

Enumerative combinatorics is about counting. The typical question is to find the number of objects with a given set of properties.

However, enumerative combinatorics is not just about counting. In "real life," when we talk about *counting*, we imagine lining up a set of objects and counting them off: $1, 2, 3, \ldots$. However, families of combinatorial objects do not come to us in a natural linear order. To give a very simple example: We do not count the squares in an $m \times n$ rectangular grid linearly. Instead, we use the rectangular structure to understand that the number of squares is $m \cdot n$. Similarly, to enumerate a more complicated combinatorial set, we usually spend most of our efforts understanding the underlying structure of the individual objects, or of the set itself.

Many combinatorial objects of interest have a rich and interesting **algebraic** or **geometric** structure, which often becomes a very powerful tool toward their enumeration. In fact, there are many families of objects that we **only** know how to count using these tools. Our goal in this chapter is to highlight some key aspects of the rich interplay between algebra, discrete geometry, and combinatorics, with an eye toward enumeration.

**About this chapter.** Over the last fifty years, combinatorics has undergone a radical transformation. Not too long ago, combinatorics mostly consisted of ad hoc methods and clever solutions to problems that were fairly isolated from the rest of mathematics. It has since grown to be a central area of mathematics, largely thanks to the discovery of deep connections to other fields. Combinatorics has become an essential tool in many disciplines. Conversely, even though ingenious methods and clever new ideas still abound, there is now a powerful, extensive toolkit of algebraic, geometric, topological, and analytic techniques that can be applied to combinatorial problems.

It is impossible to give a meaningful summary of the many facets of algebraic and geometric combinatorics in a writeup of this length. I found it very difficult but necessary to omit several beautiful, important directions. In the spirit of a *Handbook of Enumerative Combinatorics*, my guiding principle was to focus on algebraic and geometric techniques that are useful toward the solution of enumerative problems. My main goal was to state clearly and concisely some of the most useful tools in algebraic and geometric enumeration, and to give many examples that quickly and concretely illustrate how to put these tools to use.

# PART 1. ALGEBRAIC METHODS

The first part of this chapter focuses on algebraic methods in enumeration. In Section 1.2 we discuss the question, "What is a good answer to an enumerative problem." Generating functions are the most powerful tool to unify the different kinds of answers that interest us: explicit formulas, recurrences, asymptotic formulas, and generating functions. In Section 1.3 we develop the algebraic theory of generating functions. Various natural operations on combinatorial families of objects correspond to simple algebraic operations on their generating functions, and this allows us to count many families of interest. In Section 1.4 we show how many problems in combinatorics can be rephrased in terms of linear algebra, and reduced to the problem of computing determinants. Finally, Section 1.5 is devoted to the theory of posets. Many combinatorial sets have a natural poset structure, and this general theory is very helpful in enumerating such sets.

## 1.2 What is a good answer?

The main goal of enumerative combinatorics is to count the elements of a finite set. Most frequently, we encounter a family of sets $T_0, T_1, T_2, T_3, \ldots$ and we need to find the number $t_n = |T_n|$ for $n = 1, 2, \ldots$. What constitutes a good answer?

Some answers are obviously good. For example, the number of subsets of $\{1, 2, \ldots, n\}$ is $2^n$, and it seems clear that this is the simplest possible answer to this question. Sometimes an answer "is so messy and long, and so full of factorials and sign alternations and whatnot, that we may feel that the disease was preferable to the cure" [218]. Usually, the situation is somewhere in between, and it takes some experience to recognize a good answer.

A combinatorial problem often has several kinds of answers. Which answer is better depends on what one is trying to accomplish. Perhaps this is best illustrated with an example. Let us count the number $a_n$ of **domino tilings** of a $2 \times n$ rectangle into $2 \times 1$ rectangles. There are several different ways of answering this question.

**Explicit formula 1.** We first look for an explicit combinatorial formula for $a_n$. To do that, we play with a few examples, and quickly notice that these tilings are structurally very simple: They are just a sequence of $2 \times 1$ vertical tiles, and $2 \times 2$ blocks covered by two horizontal tiles. Therefore constructing a tiling is the same as writing $n$ as an ordered sum of 1s and 2s. For example, the tilings of Figure 1.1 correspond, respectively, to $1+1+1+1, 1+1+2, 1+2+1, 2+1+1, 2+2$. These sums are

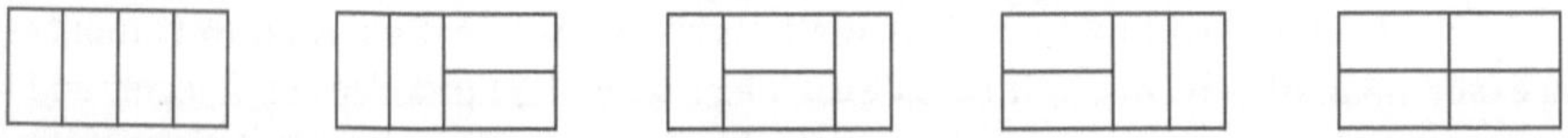

**Figure 1.1**
The five domino tilings of a $2\times 4$ rectangle.

easy to count. If there are $k$ summands equal to 2 there must be $n-2k$ summands equal to 1, and there are $\binom{n-2k+k}{k}=\binom{n-k}{k}$ ways of ordering the summands. Therefore

$$a_n=\sum_{k=0}^{\lfloor n/2\rfloor}\binom{n-k}{k}=\binom{n}{0}+\binom{n-1}{1}+\binom{n-2}{2}+\cdots. \qquad (1.1)$$

This is a pretty good answer. It is certainly an explicit formula, and it may be used to compute $a_n$ directly for small values of $n$. It does have two drawbacks. Aesthetically, it is certainly not as satisfactory as "$2^n$." In practice, it is also not as useful as it seems; after computing a few examples, we will soon notice that computing binomial coefficients is a non-trivial task. In fact there is a more efficient method of computing $a_n$.

**Recurrence.** Let $n\geq 2$. In a domino tiling, the leftmost column of a $2\times n$ can be covered by a vertical domino or by two horizontal dominoes. If the leftmost domino is vertical, the rest of the dominoes tile a $2\times(n-1)$ rectangle, so there are $a_{n-1}$ such tilings. On the other hand, if the two leftmost dominoes are horizontal, the rest of the dominoes tile a $2\times(n-2)$ rectangle, so there are $a_{n-2}$ such tilings. We obtain the **recurrence relation**

$$a_0=1,\quad a_1=1,\quad a_n=a_{n-1}+a_{n-2}\ \text{for } n\geq 2, \qquad (1.2)$$

which allows us to compute each term in the sequence in terms of the previous ones. We see that $a_n=F_{n+1}$ is the $(n+1)$th **Fibonacci number.**

This recursive answer is not as nice as "$2^n$" either; it is not even an explicit formula for $a_n$. If we want to use it to compute $a_n$, we need to compute all the first $n$ terms of the sequence $1,1,2,3,5,8,13,21,34,55,89,144,\ldots$. However, we can compute those very quickly; we only need to perform $n-1$ additions. This is an extremely efficient method for computing $a_n$.

**Explicit formula 2.** There is a well-established method that turns linear recurrence relations with constant coefficients, such as (1.2), into explicit formulas. We will review it in Theorem 1.3.5. In this case, the method gives

$$a_n=\frac{1}{\sqrt{5}}\left(\left(\frac{1+\sqrt{5}}{2}\right)^{n+1}-\left(\frac{1-\sqrt{5}}{2}\right)^{n+1}\right). \qquad (1.3)$$

This is clearly the simplest possible explicit formula for $a_n$; in that sense it is a great formula.

A drawback is that this formula is really not very useful if we want to compute the exact value of, say, $a_{1000}$. It is not even clear why (1.3) produces an integer, and to get it to produce the correct integer would require arithmetic calculations with extremely high precision.

An advantage is that, unlike (1.1), (1.3) tells us very precisely how $a_n$ grows with $n$.

**Asymptotic formula.** It follows immediately from (1.3) that

$$a_n \sim c \cdot \varphi^n, \tag{1.4}$$

where $c = \frac{1+\sqrt{5}}{2\sqrt{5}}$ and $\varphi = \frac{1+\sqrt{5}}{2} \approx 1.6179\ldots$ is the golden ratio. This notation means that $\lim_{n\to\infty} a_n/(c \cdot \varphi^n) = 1$. In fact, since $|\frac{1-\sqrt{5}}{2}| < 1$, $a_n$ is the closest integer to $c \cdot \varphi^n$.

**Generating function.** The last kind of answer we discuss is the generating function. This is perhaps the strangest kind of answer, but it is often the most powerful one.

Consider the infinite power series $A(x) = a_0 + a_1x + a_2x^2 + \cdots$. We call this the **generating function** of the sequence $a_0, a_1, a_2, \ldots$. * We now compute this power series: From (1.2) we obtain that $A(x) = 1 + x + \sum_{n\geq 2}(a_{n-1} + a_{n-2})x^n = 1 + x + x(A(x) - 1) + x^2A(x)$, which implies

$$A(x) = a_0 + a_1x + a_2x^2 + \cdots = \frac{1}{1-x-x^2}. \tag{1.5}$$

With a bit of theory and some practice, we will be able to write the equation (1.5) immediately, with no further computations (Example 18 in Section 1.3.2). To show this is an **excellent** answer, let us use it to derive all our other answers, and more.

- Generating functions help us obtain explicit formulas. For instance, rewriting

  $$A(x) = \frac{1}{1-(x+x^2)} = \sum_{k\geq 0}(x+x^2)^k$$

  we recover (1.1). If, instead, we use the method of partial fractions, we get

  $$A(x) = \left(\frac{1/\sqrt{5}}{1-\frac{1+\sqrt{5}}{2}x}\right) - \left(\frac{1/\sqrt{5}}{1-\frac{1-\sqrt{5}}{2}x}\right)$$

  which brings us to our second explicit formula (1.3).

- Generating functions help us obtain recursive formulas. In this example, we simply compare the coefficients of $x^n$ on both sides of the equation $A(x)(1-x-x^2) = 1$, and we get the recurrence relation (1.2).

*For the moment, let us not worry about where this series converges. The issue of convergence can be easily avoided (as combinatorialists often do, in a way that will be explained in Section 1.3.1) or resolved and exploited to our advantage; let us postpone that discussion for the moment.

- Generating functions help us obtain asymptotic formulas. In this example, (1.5) leads to (1.3), which gives (1.4). In general, almost everything that we know about the rate of growth of combinatorial sequences comes from their generating functions, because analysis tells us that the asymptotic behavior of $a_n$ is intimately tied to the singularities of the function $A(x)$.
- Generating functions help us enumerate our combinatorial objects in more detail, and understand some of their statistical properties. For instance, say we want to compute the number $a_{m,n}$ of domino tilings of a $2 \times n$ rectangle that use exactly $m$ vertical tiles. Once we really understand (1.5) in Section 1.3.2, we will get the answer immediately:

$$\frac{1}{1-vx-x^2} = \sum_{m,n\geq 0} a_{m,n} v^m x^n.$$

Now suppose we wish to know what fraction of the tiles is vertical in a large random tiling. Among all the $a_n$ domino tilings of the $2 \times n$ rectangle, there are $\sum_{m\geq 0} m a_{m,n}$ vertical dominoes. We compute

$$\sum_{n\geq 0}\left(\sum_{m\geq 0} m a_{m,n}\right) x^n = \left[\frac{\partial}{\partial v}\left(\frac{1}{1-vx-x^2}\right)\right]_{v=1} = \frac{x}{(1-x-x^2)^2}.$$

Partial fractions then tell us that $\sum_{m\geq 0} m a_{m,n} \sim \frac{n}{5}\left(\frac{1+\sqrt{5}}{2}\right)^{n+1} \sim \frac{1}{\sqrt{5}} n a_n$. Hence the fraction of vertical tiles in a random domino tiling of a $2 \times n$ rectangle converges to $1/\sqrt{5}$ as $n \to \infty$.

**So what is a good answer to an enumerative problem?** Not surprisingly, there is no definitive answer to this question. When we count a family of combinatorial objects, we look for explicit formulas, recursive formulas, asymptotic formulas, and generating functions. They are all useful. Generating functions are the most powerful framework we have to relate these different kinds of answers and, ideally, find them all.

## 1.3 Generating functions

In combinatorics, one of the most useful ways of "determining" a sequence of numbers $a_0, a_1, a_2, \ldots$ is to compute its **ordinary generating function**

$$A(x) = \sum_{n\geq 0} a_n x^n = a_0 + a_1 x + a_2 x^2 + a_3 x^3 + a_4 x^4 + \cdots$$

or its **exponential generating function**

$$A_{\exp}(x) = \sum_{n\geq 0} a_n \frac{x^n}{n!} = a_0 + a_1 x + a_2 \frac{x^2}{2} + a_3 \frac{x^3}{6} + a_4 \frac{x^4}{24} + \cdots.$$

This simple idea is extremely powerful because some of the most common algebraic operations on ordinary and exponential generating functions correspond to some of the most common operations on combinatorial objects. This allows us to count many interesting families of objects; this is the content of Section 1.3.2 (for ordinary generating functions) and Section 1.3.3 (for exponential generating functions). In Section 1.3.4 we see how nice generating functions can be turned into explicit, recursive, and asymptotic formulas for the corresponding sequences.

Before we get to this interesting theory, we have to understand what we mean by power series. Section 1.3.1 provides a detailed discussion, which is probably best skipped the first time one encounters power series. In the meantime, let us summarize it in one paragraph:

There are two main attitudes toward power series in combinatorics: the analytic attitude and the algebraic attitude. To harness the full power of power series, one should really understand both. Chapter 2 of this *Handbook of Enumerative Combinatorics* is devoted to the analytic approach, which treats $A(x)$ as an honest analytic function of $x$, and uses analytic properties of $A(x)$ to derive combinatorial properties of $a_n$. In this chapter we follow the algebraic approach, which treats $A(x)$ as a formal algebraic expression, and manipulates it using the usual laws of algebra, without having to worry about any convergence issues.

### 1.3.1 The ring of formal power series

Enumerative combinatorics is full of intricate algebraic computations with power series, where justifying convergence is cumbersome, and usually unnecessary. In fact, many natural power series in combinatorics, such as $\sum_{n\geq 0} n!x^n$, only converge at 0, so analytic methods are not available to study them. For these reasons we often prefer to carry out our computations algebraically in terms of **formal** power series. We will see that even in this approach, analytic considerations are often useful.

In this section we review the definition and basic properties of the **ring of formal power series** $\mathbb{C}[[x]]$. For a more in-depth discussion, including the (mostly straightforward) proofs of the statements we make here, see [149].

**Formal power series.** A **formal power series** is an expression of the form

$$A(x) = a_0 + a_1x + a_2x^2 + \cdots, \qquad a_0, a_1, a_2, \ldots \in \mathbb{C}.$$

Formally, this series is just a sequence of complex numbers $a_0, a_1, a_2, \ldots$. We will see that it is convenient to denote it $A(x)$, but we do not consider it to be a function of $x$.

Let $\mathbb{C}[[x]]$ be the **ring of formal power series**, where the sum and the product of $A(x) = \sum_{n\geq 0} a_nx^n$ and $B(x) = \sum_{n\geq 0} b_nx^n$ are

$$A(x)+B(x) = \sum_{n\geq 0}(a_n+b_n)x^n, \qquad A(x)B(x) = \sum_{n\geq 0}\left(\sum_{k=0}^{n} a_kb_{n-k}\right)x^n.$$

It is implicit in the definition, but worth mentioning, that $\sum_{n\geq 0} a_n x^n = \sum_{n\geq 0} b_n x^n$ if and only if $a_n = b_n$ for all $n \geq 0$.

We define the **degree** of $A(x) = \sum_{n\geq 0} a_n x^n$ to be the **smallest** $n$ such that $a_n \neq 0$. We also write

$$[x^n]A(x) := a_n, \qquad A(0) := [x^0]A(x) = a_0.$$

We also define formal power series inspired by series from analysis, such as

$$e^x := \sum_{n\geq 0} \frac{x^n}{n!}, \qquad -\log(1-x) := \sum_{n\geq 1} \frac{x^n}{n}, \qquad (1+x)^r := \sum_{n\geq 0} \binom{r}{n} x^n,$$

for any complex number $r$, where $\binom{r}{n} := r(r-1)\cdots(r-n+1)/n!$.

The ring $\mathbb{C}[[x]]$ is commutative with $0 = 0 + 0x + \cdots$ and $1 = 1 + 0x + \cdots$. It is an integral domain; that is, $A(x)B(x) = 0$ implies that $A(x) = 0$ or $B(x) = 0$. It is easy to describe the units:

$$\sum_{n\geq 0} a_n x^n \text{ is invertible} \iff a_0 \neq 0.$$

For example, $\frac{1}{1-x} = 1 + x + x^2 + \cdots$ because $(1-x)(1+x+x^2+\cdots) = 1 + 0x + 0x^2 + \cdots$.

**Convergence.** When working in $\mathbb{C}[[x]]$, we will not consider convergence of sequences or series of complex numbers. In particular, we will never substitute a complex number $x$ into a formal power series $A(x)$.

However, we do need a notion of convergence for sequences in $\mathbb{C}[[x]]$. We say that a sequence $A_0(x), A_1(x), A_2(x), \ldots$ of formal power series **converges** to $A(x) = \sum_{n\geq 0} a_n x^n$ if $\lim_{n\to\infty} \deg(A_n(x) - A(x)) = \infty$; that is, if for any $n \in \mathbb{N}$, the coefficient of $x^n$ in $A_m(x)$ **equals** $a_n$ for all sufficiently large $m$. This gives us a useful criterion for convergence of infinite sums and products in $\mathbb{C}[[x]]$:

$$\sum_{j=0}^{\infty} A_j(x) \text{ converges} \iff \lim_{j\to\infty} \deg A_j(x) = \infty$$

$$\prod_{j=0}^{\infty} (1 + A_j(x)) \text{ converges} \iff \lim_{j\to\infty} \deg A_j(x) = \infty \qquad (A_j(0) = 0)$$

For example, the infinite sum $\sum_{n\geq 0} (x+1)^n/2^n$ does not converge in this topology. Notice that the coefficient of $x^0$ in this sum cannot be obtained through a finite computation; it would require interpreting the infinite sum $\sum_{n\geq 0} 1/2^n$. On the other hand, the following infinite sum converges:

$$\sum_{n\geq 0} \frac{1}{n!} \left( -\sum_{m\geq 1} \frac{x^m}{m} \right)^n = 1 - x. \tag{1.6}$$

It is clear from the criterion above that this series converges; but why does it equal $1 - x$?

**Borrowing from analysis.** In $\mathbb{C}[[x]]$, (1.6) is an algebraic identity which says that the coefficients of $x^k$ in the left-hand side, for which we can give an ugly but finite formula, equal $1,-1,0,0,0,\ldots$. If we were to follow a purist algebraic attitude, we would give an algebraic or combinatorial proof of this identity. This is probably possible, but intricate and rather dogmatic. A much simpler approach is to shift toward an analytic attitude, at least momentarily, and recognize that (1.6) is the Taylor series expansion of

$$e^{-\log(1-x)} = 1-x$$

for $|x|<1$. Then we can just invoke the following simple fact from analysis.

**Theorem 1.3.1** *If two analytic functions are equal in an open neighborhood of* 0, *then their Taylor series at* 0 *are equal coefficient-by-coefficient; that is, they are equal as formal power series.*

**Composition.** The **composition** of two series $A(x) = \sum_{n\geq 0} a_n x^n$ and $B(x) = \sum_{n\geq 0} b_n x^n$ with $b_0 = 0$ is naturally defined to be

$$A(B(x)) = \sum_{n\geq 0} a_n \left( \sum_{m\geq 0} b_m x^m \right)^n .$$

Note that this sum converges if and only if $b_0 = 0$. Two very important special cases in combinatorics are the series $\frac{1}{1-B(x)}$ and $e^{B(x)}$.

**"Calculus."** We define the **derivative** of $A(x) = \sum_{n\geq 0} a_n x^n$ to be

$$A'(x) = \sum_{n\geq 0} (n+1) a_{n+1} x^n .$$

This formal derivative satisfies the usual laws of derivatives, such as

$$(A+B)' = A' + B', \qquad (AB)' = A'B + AB', \qquad [A(B(x))]' = A'(B(x))B'(x).$$

We can still solve differential equations formally. For example, if we know that $F'(x) = F(x)$ and $F(0) = 1$, then $(\log F(x))' = F'(x)/F(x) = 1$, which gives $\log F(x) = x$ and $F(x) = e^x$.

This concludes our discussion on the formal properties of power series. Now let us return to combinatorics.

### 1.3.2 Ordinary generating functions

Suppose we are interested in enumerating a family $\mathscr{A} = \mathscr{A}_0 \sqcup \mathscr{A}_1 \sqcup \mathscr{A}_2 \sqcup \cdots$ of combinatorial structures, where $\mathscr{A}_n$ is a finite set consisting of the objects of "size" $n$. Denote by $|a|$ the size of $a \in \mathscr{A}$. The **ordinary generating function** of $\mathscr{A}$ is

$$A(x) = \sum_{a\in\mathscr{A}} x^{|a|} = a_0 + a_1 x + a_2 x^2 + \cdots$$

where $a_n$ is the number of elements of size $n$.

We are not interested in the philosophical question of determining what it means for $\mathscr{A}$ to be "combinatorial"; we are willing to call $\mathscr{A}$ a combinatorial structure as long as $a_n$ is finite for all $n$. We consider two structures $\mathscr{A}$ and $\mathscr{B}$ combinatorially equivalent, and write $\mathscr{A} \cong \mathscr{B}$, if $A(x) = B(x)$.

More generally, we may consider a family $\mathscr{A}$ where each element $a$ is given a weight $\text{wt}(a)$, often a constant multiple of $x^{|a|}$, or a monomial in one or more variables $x_1, \ldots, x_n$. Again, we require that there are finitely many objects of any given weight. Then we define the **weighted** ordinary generating function of $\mathscr{A}$ to be the formal power series

$$A_{\text{wt}}(x_1, \ldots, x_n) = \sum_{a \in \mathscr{A}} \text{wt}(a).$$

Examples of combinatorial structures (with their respective size functions) are words on the alphabet $\{0,1\}$ (length), domino tilings of rectangles of height 2 (width), or Dyck paths (length). We may weight these objects by $t^k$ where $k$ is, respectively, the number of 1s, the number of vertical tiles, or the number of returns to the $x$ axis.

#### 1.3.2.1 Operations on combinatorial structures and their generating functions

There are a few simple but very powerful operations on combinatorial structures, all of which have nice counterparts at the level of ordinary generating functions. Many combinatorial objects of interest may be built up from basic structures using these operations.

**Theorem 1.3.2** *Let $\mathscr{A}$ and $\mathscr{B}$ be combinatorial structures.*

1. *($\mathscr{C} = \mathscr{A} + \mathscr{B}$: Disjoint union) If a $\mathscr{C}$-structure of size $n$ is obtained by choosing an $\mathscr{A}$-structure of size $n$* **or** *a $\mathscr{B}$-structure of size $n$, then*

$$C(x) = A(x) + B(x).$$

   *This result also holds for weighted structures if the weight of a $\mathscr{C}$-structure is the same as the weight of the respective $\mathscr{A}$- or $\mathscr{B}$-structure.*

2. *($\mathscr{C} = \mathscr{A} \times \mathscr{B}$: Product) If a $\mathscr{C}$-structure of size $n$ is obtained by choosing an $\mathscr{A}$-structure of size $k$* **and** *a $\mathscr{B}$-structure of size $n-k$ for some $k$, then*

$$C(x) = A(x)B(x).$$

   *This result also holds for weighted structures if the weight of a $\mathscr{C}$-structure is the product of the weights of the respective $\mathscr{A}$- and $\mathscr{B}$-structures.*

3. *($\mathscr{C} = \mathsf{Seq}(\mathscr{B})$: Sequence) Assume $|\mathscr{B}_0| = 0$. If a $\mathscr{C}$-structure of size $n$ is obtained by choosing a sequence of $\mathscr{B}$-structures of total size $n$, then*

$$C(x) = \frac{1}{1 - B(x)}.$$

   *This result also holds for weighted structures if the weight of a $\mathscr{C}$-structure is the product of the weights of the respective $\mathscr{B}$-structures.*

4. *($\mathscr{C} = \mathscr{A} \circ \mathscr{B}$: Composition)* ***Compositional formula.*** *Assume that $|\mathscr{B}_0| = 0$. If a $\mathscr{C}$-structure of size n is obtained by choosing a sequence of (say, k) $\mathscr{B}$-structures of total size n and placing an $\mathscr{A}$-structure of size k on this sequence of $\mathscr{B}$-structures, then*

$$C(x) = A(B(x)).$$

*This result also holds for weighted structures if the weight of a $\mathscr{C}$-structure is the product of the weights of the $\mathscr{A}$-structure on its blocks and the weights of the $\mathscr{B}$-structures on the individual blocks.*

5. *($\mathscr{C} = \mathscr{A}^{-1}$: Inversion)* ***Lagrange inversion formula.***

   *(a)* ***Algebraic version.*** *If $A^{<-1>}(x)$ is the compositional inverse of $A(x)$ then*

$$n[x^n]A^{<-1>}(x) = [x^{n-1}]\left(\frac{x}{A(x)}\right)^n.$$

   *(b)* ***Combinatorial version.*** *Assume $|\mathscr{A}_0| = 0$, $|\mathscr{A}_1| = 1$, and let*

$$A(x) = x - a_2x^2 - a_3x^3 - a_4x^4 - \cdots$$

   *where $a_n$ is the number of $\mathscr{A}$-structures of size n for $n \geq 2$.* *

*Let an* ***$\mathscr{A}$-decorated plane rooted tree*** *(or simply $\mathscr{A}$-tree) be a rooted tree T where every internal vertex v has an ordered set $D_v$ of at least two "children," and each set $D_v$ is given an $\mathscr{A}$-structure. The* ***size*** *of T is the number of leaves. Let $C(x)$ be the generating function for $\mathscr{A}$-decorated plane rooted trees. Then*

$$C(x) = A^{<-1>}(x).$$

*This result also holds for weighted structures if the weight of a tree is the product of the weights of the $\mathscr{A}$-structures at its vertices.*

*Proof.* 1. is clear. The identity in 2. is equivalent to $c_n = \sum_k a_k b_{n-k}$, which corresponds to the given combinatorial description. Iterating 2., the generating function for $k$-sequences of $\mathscr{B}$-structures is $B(x)^k$, so in 3. we have $C(x) = \sum_k B(x)^k = 1/(1-B(x))$ and in 4. we have $C(x) = \sum_k a_k B(x)^k$. The weighted statements follow similarly.

5(b). Observe that, by the Compositional Formula, an $\mathscr{A} \circ \mathscr{C}$ structure is either (i) an $\mathscr{A}$-tree, or (ii) a sequence of $k \geq 2$ $\mathscr{A}$-trees $T_1, \ldots, T_k$ with an $\mathscr{A}$-structure on $\{T_1, \ldots, T_k\}$.

The structure in (ii) is equivalent to an $\mathscr{A}$-tree $T$, obtained by grafting $T_1, \ldots, T_k$ at a new root and placing the $\mathscr{A}$-structure on its offspring (which contributes a negative

*Here, to simplify matters, we introduced signs into $A(x)$. Instead we could let $A(x)$ be the ordinary generating function for $\mathscr{A}$-structures, but we would need to give each $\mathscr{A}$-tree the sign $(-1)^m$, where $m$ is the number of internal vertices. Similarly, we could allow $a_1 \neq 1$ at the cost of some factors of $a_1$ on the $\mathscr{A}$-trees.

sign). This $T$ also arises in (i) with a positive sign. These two appearances of $T$ cancel each other out in $A(C(x))$, and the only surviving tree is the trivial tree • with one vertex, which only arises once and has weight $x$.

5(a). Let a **sprig** be a rooted plane tree consisting of a path $r = v_1 v_2 \cdots v_k = l$ starting at the root $r$ and ending at the leaf $l$, and at least one leaf hanging from each $v_i$ and to the right of $v_{i+1}$ for $1 \le i \le k-1$. The trivial tree • is an allowable sprig with $k = 1$.

An $\mathscr{A}$**-sprig** is a sprig where the children of $v_i$ are given an $\mathscr{A}$-structure for $1 \le i \le k-1$; its **size** is the number of leaves other than $l$ minus 1. * The right panel of Figure 1.2 shows several $\mathscr{A}$-sprigs. An $\mathscr{A}$-sprig is equivalent to a sequence of $\mathscr{A}$-structures, with weights shifted by $-1$, so Theorem 1.3.2.3 tells us that

$$\frac{1}{A(x)} = \frac{1}{x} \cdot \frac{1}{1-(a_2x+a_3x^2+\cdots)} = \sum_{n \ge -1} (\text{\# of } \mathscr{A}\text{-sprigs of size } n)\, x^n$$

Hence $[x^{n-1}](x/A(x))^n = [x^{-1}](1/A(x))^n$ is the number of sequences of $n$ $\mathscr{A}$-sprigs of total size $-1$ by Theorem 1.3.2.2. We need to show that

$$n \cdot (\text{\# of } \mathscr{A}\text{-trees with } n \text{ leaves}) = (\text{\# of sequences of } n\ \mathscr{A}\text{-sprigs of total size } -1)$$

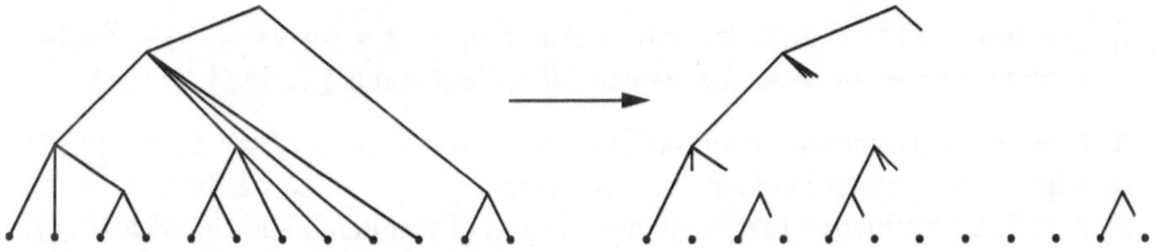

**Figure 1.2**
The map from $\mathscr{A}$-trees to sequences of $\mathscr{A}$-sprigs.

An $\mathscr{A}$-tree $T$ can be trimmed into a sequence of $n$ $\mathscr{A}$-sprigs $S_1, \ldots, S_n$ as follows. At each step, look at the leftmost leaf and the path $P$ to its highest remaining ancestor. Remove $P$ and all the branches hanging directly from $P$ (which form an $\mathscr{A}$-sprig), but do not remove any other vertices. Repeat this until the tree is completely decomposed into $\mathscr{A}$-sprigs. The total size of these sprigs is $-1$. Figure 1.2 shows a tree of weight $x^{5+(-1)+0+(-1)+2+(-1)+(-1)+(-1)+(-1)+(-1)+0+(-1)} = x^{-1}$. Notice that all the partial sums of the sum $5+(-1)+0+(-1)+2+(-1)+(-1)+(-1)+(-1)+(-1)+0+(-1) = -1$ are non-negative.

Conversely, suppose we wish to recover the $\mathscr{A}$-tree corresponding to a sequence of sprigs $S_1, \ldots, S_n$ with $|S_1| + \cdots + |S_n| = -1$. We must reverse the process, adding

*We momentarily allow negative sizes, since the trivial $\mathscr{A}$-sprig • has size $-1$. Thus we need to compute with Laurent series, which are power series with finitely many negative exponents.

$S_1, \dots, S_n$ to $T$ one at a time; at each step we must graft the new sprig at the leftmost free branch. Note that after grafting $S_1, \dots, S_k$ we are left with $1 + |S_1| + \cdots + |S_k|$ free branches, so a sequence of sprigs corresponds to a tree if and only if the partial sums $|S_1| + \cdots + |S_k|$ are non-negative for $k = 1, \dots, n-1$. Finally, it remains to observe that any sequence $a_1, \dots, a_n$ of integers adding to $-1$ has a unique cyclic shift $a_i, \dots, a_n, a_1, \dots, a_{i-1}$ whose partial sums are all non-negative. Therefore, out of the $n$ cyclic shifts of $S_1, \dots, S_n$, exactly one of them corresponds to an $\mathscr{A}$-tree. The desired result follows. ■

The last step of the proof above is a special case of the **cycle lemma** of Dvoretsky and Motzkin [192, Lemma 5.3.7], which is worth stating explicitly. Suppose $a_1, \dots, a_n$ is a string of 1s and $-1$s with $a_1 + \cdots + a_n = k > 0$. Then there are exactly $k$ cyclic shifts $a_i, a_{i+1}, \dots, a_n, a_1, \dots, a_{i-1}$ whose partial sums are all non-negative.

#### 1.3.2.2 Examples

**Classical applications.** With practice, these simple ideas give very easy solutions to many classical enumeration problems.

1. (Trivial classes) It is useful to introduce the trivial class $\circ$ having only one element of size 0, and the trivial class $\bullet$ having only one element of size 1. Their generating functions are 1 and $x$, respectively.

2. (Sequences) The slightly less trivial class $\mathsf{Seq} = \{\emptyset, \bullet, \bullet\bullet, \bullet\bullet\bullet, \dots\} = \mathsf{Seq}(\bullet)$ contains one set of each size. Its generating function is $\sum_n x^n = 1/(1-x)$.

3. (Subsets and binomial coefficients) Let Subset consist of the pairs $([n], A)$ where $n$ is a natural number and $A$ is a subset of $[n]$. Let the size of that pair be $n$. A Subset-structure is equivalent to a word of length $n$ in the alphabet $\{0,1\}$, so $\mathsf{Subset} \cong \mathsf{Seq}(\{0,1\})$ where $|0| = |1| = 1$, and

$$\mathsf{Subset}(x) = \frac{1}{1-(x^1+x^1)} = \sum_{n\geq 0} 2^n x^n.$$

We can use the extra variable $y$ to keep track of the size of the subset $A$, by giving $([n], A)$ the weight $x^n y^{|A|}$. This corresponds to giving the letters 0 and 1 weights $x$ and $xy$ respectively, so we get the generating function

$$\mathsf{Subset}_{\mathrm{wt}}(x) = \frac{1}{1-(x+xy)} = \sum_{n\geq k\geq 0} \binom{n}{k} x^n y^k$$

for the **binomial coefficients** $\binom{n}{k} = \frac{n!}{k!(n-k)!}$, which count the $k$-subsets of $[n]$.

From this generating function, we can easily obtain the main results about binomial coefficients. Computing the coefficient of $x^n y^k$ in $\left(\sum \binom{n}{k} x^n y^k\right)(1-x-xy) = 1$ gives **Pascal's recurrence**

$$\binom{n}{k} = \binom{n-1}{k} + \binom{n-1}{k-1}, \qquad n \geq k \geq 1$$

with initial values $\binom{n}{0} = \binom{n}{n} = 1$. Expanding $\mathsf{Subset}_{\mathrm{wt}}(x) = \frac{1}{1-x(1+y)} = \sum_{n\geq 0} x^n(1+y)^n$ gives the **Binomial theorem**

$$(1+y)^n = \sum_{k=0}^{n} \binom{n}{k} y^k.$$

4. (Multinomial coefficients) Let $\mathsf{Words}^k \cong \mathsf{Seq}(\{1,\ldots,k\})$ consist of the words in the alphabet $\{1,2,\ldots,k\}$. The words of length $n$ are in bijection with the ways of putting $n$ numbered balls into $k$ numbered boxes. The placements having $a_i$ balls in box $i$, where $a_1+\cdots+a_k = n$, are enumerated by the **multinomial coefficient** $\binom{n}{a_1,\ldots,a_k} = \frac{n!}{a_1!\cdots a_k!}$.
   Giving the letter $i$ weight $x_i$, we obtain the generating function

$$\mathsf{Words}^k(x_1,\ldots,x_k) = \sum_{a_1,\ldots,a_k\geq 0} \binom{a_1+\cdots+a_k}{a_1,\ldots,a_k} x_1^{a_1}\cdots x_k^{a_k} = \frac{1}{1-x_1-\cdots-x_k}$$

   from which we obtain the recurrence

$$\binom{n}{a_1,\ldots,a_k} = \binom{n-1}{a_1-1,a_2,\ldots,a_k} + \cdots + \binom{n-1}{a_1,\ldots a_{k-1},a_k-1}$$

   and the multinomial theorem

$$(x_1+\cdots+x_k)^n = \sum_{\substack{a_1,\ldots,a_k\geq 0\\ a_1+\cdots+a_k=n}} \binom{n}{a_1,\ldots,a_k} x_1^{a_1}\cdots x_k^{a_k}.$$

5. (Compositions) A **composition** of $n$ is a way of writing $n = a_1+\cdots+a_k$ as an **ordered** sum of positive integers $a_1,\ldots,a_k$. For example, 523212 is a composition of 15. A composition is just a sequence of natural numbers, so $\mathsf{Comp} \cong \mathsf{Seq}(\mathbb{N})$ where $|a| = a$. Therefore

$$\mathsf{Comp}(x) = \frac{1}{1-(x+x^2+x^3+\cdots)} = \frac{1-x}{1-2x} = \sum_{n\geq 1} 2^{n-1}x^n.$$

   and there are $2^{n-1}$ compositions of $n$.
   If we give a composition of $n$ with $k$ summands the weight $x^n y^k$, the weighted generating function is

$$\mathsf{Comp}_{\mathrm{wt}}(x) = \frac{1}{1-(xy+x^2y+x^3y+\cdots)} = \frac{1-x}{1-x(1+y)} = \sum_{n\geq 1} \binom{n-1}{k-1} x^n y^k$$

   so there are $\binom{n-1}{k-1}$ compositions of $n$ with $k$ summands.

6. (Compositions into restricted parts) Given a subset $A \subseteq \mathbb{N}$, an **$A$-composition** of $n$ is a way of writing $n$ as an ordered sum $n = a_1+\cdots+a_k$ where $a_1,\ldots,a_k \in$

$A$. The corresponding combinatorial structure is $A\text{-Comp} \cong \mathsf{Seq}(A)$ where $|a| = a$, so

$$A\text{-Comp}(x) = \frac{1}{1-(\sum_{a\in A} x^a)}.$$

For example, the number of compositions of $n$ into odd parts is the Fibonacci number $F_{n-1}$, because the corresponding generating function is

$$\mathsf{OddComp}(x) = \frac{1}{1-(x+x^3+x^5+\cdots)} = \frac{1-x^2}{1-x-x^2} = 1+\sum_{n\geq 1} F_{n-1}x^n.$$

7. (Multisubsets) Let $\mathsf{Multiset}^m$ be the collection of multisets consisting of possibly repeated elements of $[m]$. The size of a multiset is the number of elements, counted with repetition. For example, $\{1,2,2,2,3,5\}$ is a multisubset of $[7]$ of size 6. Then $\mathsf{Multiset}^m \cong \mathsf{Seq}(\{1\}) \times \cdots \times \mathsf{Seq}(\{m\})$, where $|i| = 1$ for $i = 1,\ldots,m$, so the corresponding generating function is

$$\mathsf{Multiset}^m(x) = \left(\frac{1}{1-x}\right)^m = \sum_{n\geq 0}\binom{-m}{n}(-x)^n,$$

and the number of multisubsets of $[m]$ of size $n$ is $\left(\!\binom{m}{n}\!\right) := (-1)^n\binom{-m}{n} = \binom{m+n-1}{n}$.

8. (Partitions) A **partition** of $n$ is a way of writing $n = a_1 + \cdots + a_k$ as an **unordered** sum of positive integers $a_1,\ldots,a_k$. We usually write the parts in weakly decreasing order. For example, 532221 is a partition of 15 into 6 parts. Let $\mathsf{Partition}$ be the family of partitions weighted by $x^ny^k$ where $n$ is the sum of the parts and $k$ is the number of parts. Then $\mathsf{Partition} \cong \mathsf{Seq}(\{1\}) \times \mathsf{Seq}(\{2\}) \times \cdots$, where $\mathrm{wt}(i) = x^iy$ for $i = 1,2,\ldots$, so the corresponding generating function is

$$\mathsf{Partition}(x,y) = \left(\frac{1}{1-xy}\right)\left(\frac{1}{1-x^2y}\right)\left(\frac{1}{1-x^3y}\right)\cdots$$

There is no simple explicit formula for the number $p(n)$ of partitions of $n$, although there is a very elegant and efficient recursive formula. Setting $y = -1$ in the previous identity, and invoking Euler's pentagonal theorem [2]

$$\prod_{n\geq 0}(1-x^n) = 1+\sum_{j\geq 1}(-1)^j\left(x^{j(3j-1)/2}+x^{j(3j+1)/2}\right) \tag{1.7}$$

we obtain

$$p(n) = p(n-1)+p(n-2)-p(n-5)-p(n-7)+p(n-12)+p(n-15)-\cdots$$

where $1,2,5,7,12,15,22,26,\ldots$ are the pentagonal numbers.

9. (Partitions into distinct parts) Let DistPartition be the family of partitions into distinct parts, weighted by $x^n y^k$ where $n$ is the sum of the parts and $k$ is the number of parts. Then DistPartition $\cong \{1,\overline{1}\} \times \{2,\overline{2}\} \times \cdots$, where $\mathrm{wt}(i) = x^i y$ and $\mathrm{wt}(\overline{i}) = 1$ for $i = 1, 2, \ldots$, so the corresponding generating function is

$$\mathsf{DistPartition}(x,y) = (1+xy)(1+x^2y)(1+x^3y)\cdots$$

10. (Partitions into restricted parts) It is clear how to adapt the previous generating functions to partitions where the parts are restricted. For example, the identity

$$\frac{1}{1-x} = (1+x)(1+x^2)(1+x^4)(1+x^8)(1+x^{16})\cdots$$

expresses the fact that every positive integer can be written uniquely in binary notation, as a sum of distinct powers of 2. The identity

$$(1+x)(1+x^2)(1+x^3)(1+x^4)\cdots = \frac{1}{1-x}\cdot\frac{1}{1-x^3}\cdot\frac{1}{1-x^5}\cdot\frac{1}{1-x^7}\cdots,$$

which may be proved by writing $1+x^k = (1-x^{2k})/(1-x^k)$, expresses that the number of partitions of $n$ into distinct parts equals the number of partitions of $n$ into odd parts.

11. (Partitions with restrictions on the size and the number of parts) Let $p_{\leq k}(n)$ be the number of partitions of $n$ into at most $k$ parts. This is also the number of partitions of $n$ into parts of size at most $k$. To see this, represent a partition $n = a_1 + \cdots + a_j$ as a left-justified array of squares, where the $i$th row has $a_i$ squares. Each partition $\lambda$ has a conjugate partition $\lambda'$ obtained by exchanging the rows and the columns of the Ferrers diagram. Figure 1.3 shows the Ferrers diagram of 431 and its conjugate partition 3221. It is clear that $\lambda$ has at most $k$ parts if and only if $\lambda'$ has parts of size at most $k$.

From the previous discussion it is clear that

$$\sum_{n\geq 0} p_{\leq k}(n)x^n = \frac{1}{1-x}\cdot\frac{1}{1-x^2}\cdots\frac{1}{1-x^k}.$$

Now let $p_{\leq j,\leq k}(n)$ be the number of partitions of $n$ into at most $j$ parts of size at most $k$. Then

$$\sum_{n\geq 0} p_{\leq j,\leq k}(n)x^n = \frac{(1-x)(1-x^2)\cdots(1-x^{j+k})}{(1-x)(1-x^2)\cdots(1-x^j)\cdot(1-x)(1-x^2)\cdots(1-x^k)}.$$

This is easily proved by induction, using that $p_{\leq j,\leq k}(n) = p_{\leq j,\leq k-1}(n) + p_{\leq j,\leq k}(n-k)$.

12. (Even and odd partitions) Setting $y = -1$ into the generating function for partitions into distinct parts of Example 9, we get

$$\sum_{n\geq 0}(\mathsf{edp}(n) - \mathsf{odp}(n))x^n = \prod_{j\geq 1}(1-x^j) = 1 - x - x^2 + x^5 + x^7 - x^{12} - x^{15} + \cdots$$

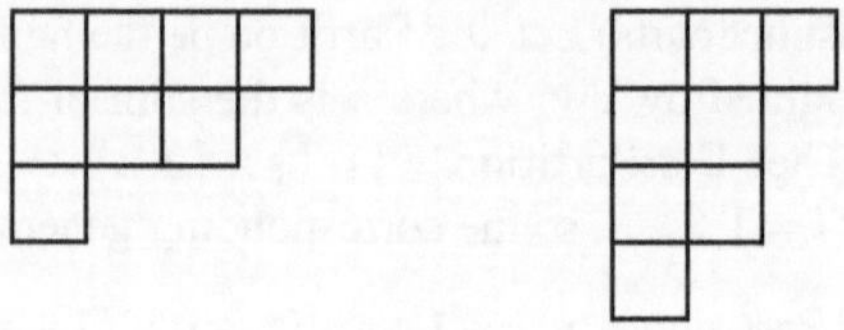

**Figure 1.3**
The Ferrers diagrams of the conjugate partitions 431 and 3221.

where $\mathsf{edp}(n)$ (respectively, $\mathsf{odp}(n)$) counts the partitions of $n$ into an even (respectively, odd) number of distinct parts. Euler's pentagonal formula (1.7) says that $\mathsf{edp}(n) - \mathsf{odp}(n)$ equals 0 for all $n$ except for the pentagonal numbers, for which it equals 1 or $-1$.

There are similar results for partitions into distinct parts coming from a given set $S$.

- When $S$ is the set of Fibonacci numbers, the coefficients of the generating function
$$\prod_{n \geq 1}(1 - x^{F_n}) = 1 - x - x^2 + x^4 + x^7 - x^8 + x^{11} - x^{12} - x^{13} + x^{14} + \cdots$$
are also equal to $0, 1$, or $-1$. [5, 168]
- This is also true for any "$k$-Fibonacci sequence" $S = \{a_1, a_2, \ldots\}$ given by $a_n = a_{n-1} + \cdots + a_{n-k}$ for $n > k$ and $a_j > a_{j-1} + \cdots + a_1$ for $1 \leq j \leq k$. [67]
- The result also holds trivially for $S = \{2^j : j \in \mathbb{N}\}$ since there is a unique partition of any $n$ into distinct powers of 2.

These three results seem qualitatively different from (and increasingly less surprising than) Euler's result, as these sequences $S$ grow much faster than $\{1, 2, 3, \ldots\}$, and $S$-partitions are sparser. Can more be said about the sets $S$ of positive integers for which the coefficients of $\prod_{n \in S}(1 - x^n)$ are all $1, 0$ or $-1$?

13. (Set partitions) A **set partition** of a set $S$ is an unordered collection of pairwise disjoint sets $S_1, \ldots, S_k$ whose union is $S$. The family of set partitions with $k$ parts is $\mathsf{SetPartition}^k \cong \bullet \times \mathsf{Seq}(\{1\}) \times \bullet \times \mathsf{Seq}(\{1,2\}) \times \cdots \times \bullet \times \mathsf{Seq}(\{1,2,\ldots,k\})$, where the singleton $\bullet$ and all numbers $i$ have size 1. To see this, we regard a word such as $w = \bullet 11 \bullet 1221 \bullet 31$ as an instruction manual to build a set partition $S_1, \ldots, S_k$. The $j$th symbol $w_j$ tells us where to put the number $j$: if $w_j$ is a number $h$, we add $j$ to the part $S_h$; if $w_j$ is the $i$th $\bullet$, then we add $j$ to a new part $S_i$. The sample word above leads to the partition $\{1,2,3,5,8,11\}, \{4,6,7\}, \{9,10\}$. This process is easily reversible. It follows

that

$$\sum_{n\geq 0} S(n,k)x^n = \frac{x}{1-x}\cdot\frac{x}{1-2x}\cdot\cdots\cdot\frac{x}{1-kx},$$

where $S(n,k)$ is the number of set partitions of $[n]$ into $k$ parts. These numbers are called the **Stirling numbers of the second kind.**

The equation $(1-kx)\sum_{n\geq 0} S(n,k)x^n = x\sum_{n\geq 0} S(n,k-1)x^n$ gives the recurrence

$$S(n,k) = kS(n-1,k)+S(n-1,k-1), \qquad 1\leq k\leq n,$$

with initial values $S(n,0)=S(n,n)=1$. Note the great similarity with Pascal's recurrence.

14. (Catalan structures) It is often said that if you encounter a new family of mathematical objects, and you have to guess how many objects of size $n$ there are, you should guess "the Catalan number $C_n = \frac{1}{n+1}\binom{2n}{n}$." The Catalan family has more than 200 incarnations in combinatorics and other fields [185, 192]; let us see three important ones.

    (a) (Plane binary trees) A **plane binary tree** is a rooted tree where every internal vertex has a left and a right child. Let PBTree be the family of plane binary trees, where a tree with $n$ internal vertices (and necessarily $n+1$ leaves) has size $n$. A tree is either the trivial tree $\circ$ of size 0, or the grafting of a left subtree and a right subtree at the root $\bullet$, so PBTree $\cong$ $\circ +$ (PBTree $\times \bullet \times$ PBTree). It follows that the generating function for plane binary trees satisfies

    $$T(x) = 1+T(x)xT(x).$$

    We may use the quadratic formula * and the binomial theorem to get

    $$T(x) = \frac{1-\sqrt{1-4x}}{2x} = \sum_{n\geq 0}\frac{1}{n+1}\binom{2n}{n}x^n.$$

    It follows that the number of plane binary trees with $n$ internal vertices (and $n+1$ leaves) is the **Catalan number** $C_n = \frac{1}{n+1}\binom{2n}{n}$.

    (b) (Triangulations) A **triangulation** of a convex polygon is a subdivision into triangles using only the diagonals of $P$. A triangulation of an $(n+2)$-gon has $n$ triangles; we say it has size $n$. If we fix an edge $e$ of $P$, then a triangulation of $P$ is obtained by choosing the triangle $T$ that will cover $e$, and then choosing a triangulation of the two polygons to the left and to the right of $T$. Therefore Triang $\cong \circ +$ (Triang $\times \bullet \times$ Triang) and the number of triangulations of an $(n+2)$-gon is also the Catalan number $C_n$.

*Since this is the first time we are using the quadratic formula, let us do it carefully. Rewrite the equation as $(1-2xT(x))^2 = 1-4x$ , or $(1-2xT(x)-\sqrt{1-4x})(1-2xT(x)+\sqrt{1-4x}) = 0$. Since $\mathbb{C}[[x]]$ is an integral domain, one of the factors must be 0. From the constant coefficients we see that it must be the first factor.

(c) (Dyck paths) A **Dyck path** $P$ of length $n$ is a path from $(0,0)$ to $(2n,0)$ that uses the steps $(1,1)$ and $(1,-1)$ and never goes below the $x$-axis. Say $P$ is **irreducible** if it touches the $x$-axis exactly twice, at the beginning and at the end. Let $D(x)$ and $I(x)$ be the generating functions for Dyck paths and irreducible Dyck paths.

A Dyck path is equivalent to a sequence of **irreducible** Dyck paths. Also, an irreducible path of length $n$ is the same as a Dyck path of length $n-1$ with an additional initial and final step. Therefore

$$D(x) = \frac{1}{1-I(x)}, \qquad I(x) = xD(x)$$

from which it follows that $D(x) = \frac{1-\sqrt{1-4x}}{2x}$ as well, and the number of Dyck paths of length $n$ is also the Catalan number.

Generatingfunctionology gives us fairly easy algebraic proofs that these three families are enumerated by the Catalan numbers. Once we have discovered this fact, the temptation to search for nice bijections is hard to resist.

Our algebraic analysis suggests a bijection $\phi$ from (b) to (a). The families of plane binary trees and triangulations grow under the same recursive recipe, and so we can let the bijection grow with them, mapping a triangulation $T \times \bullet \times T'$ to the tree $\phi(T) \times \bullet \times \phi(T')$. A non-recursive description of the bijection is the following. Consider a triangulation $T$ of the polygon $P$, and fix an edge $e$. Put a vertex inside each triangle of $T$, and a vertex outside $P$ next to each edge other than $e$. Then connect each pair of vertices separated by an edge. Finally, root the resulting tree at the vertex adjacent to $e$. This bijection is illustrated in Figure 1.4.

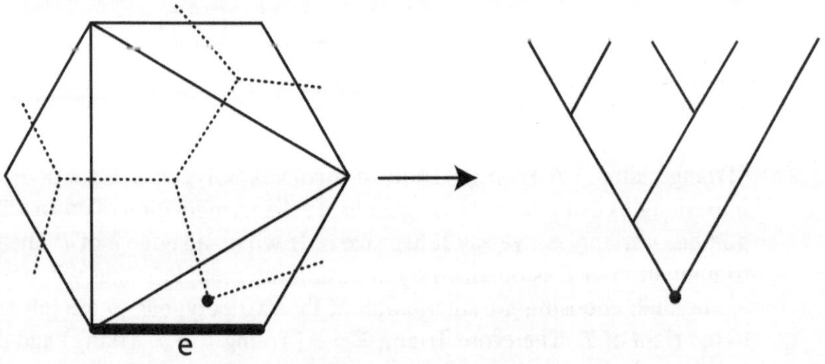


**Figure 1.4**
The bijection from triangulations to plane binary trees.

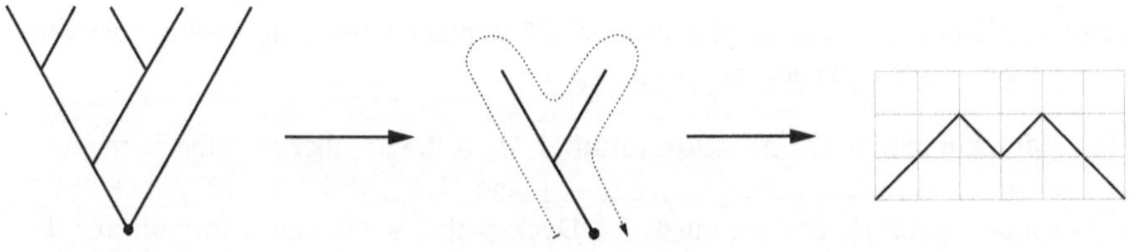

**Figure 1.5**
The bijection from plane binary trees to Dyck paths.

A bijection from (a) to (c) is less obvious from our algebraic computations, but is still not difficult to obtain. Given a plane binary tree $T$ of size $n$, prune all the leaves to get a tree $T'$ with $n$ vertices. Now walk around the periphery of the tree, starting on the left side from the root, and continuing until we traverse the whole tree. Record the walk in a Dyck path $D(T)$ : every time we walk up (respectively, down) a branch we take a step up (respectively, down) in $D(T)$. One easily checks that this is a bijection. See Figure 1.5 for an illustration.

Even if it may be familiar, it is striking that two different (and straightforward) algebraic computations show us that two families of objects that look quite different are in fact equivalent combinatorially. Although a simple, elegant bijection can often explain the connection between two families more transparently, the algebraic approach is sometimes simpler, and better at discovering such connections.

15. ($k$-Catalan structures) Let $\mathsf{PTree}_k$ be the class of **plane $k$-ary trees**, where every vertex that is not a leaf has $k$ ordered children; let the size of such a tree be its number of leaves. In the sense of Theorem 1.3.2.5(a), this is precisely an $\mathscr{A}$-tree, where $\mathscr{A} = \{\bullet, \bullet^k\}$ consisting of one structure of size 1 and one of size $k$. Therefore $\mathsf{PTree}_k = (x - x^k)^{<-1>}$. Lagrange inversion then gives

$$m[x^m]A^{<-1>}(x) = [x^{m-1}]\left(\frac{1}{1-x^{k-1}}\right)^m = [x^{m-1}]\sum_{n\geq 0}\binom{m+n-1}{n}x^{(k-1)n}.$$

It follows that a plane $k$-ary tree must have $m = (k-1)n+1$ leaves for some integer $n$, and the number of such trees is the **$k$-Catalan number**

$$C_n^k = \frac{1}{(k-1)n+1}\binom{kn}{n}.$$

This is an alternative way to compute the ordinary Catalan numbers $C_n = C_n^2$.

The $k$-Catalan number $C_n^k$ also has many different interpretations [100]; we mention two more. It counts the subdivisions of an $(n(k-1)+2)$-gon $P$ into (necessarily $n$) $(k+1)$-gons using diagonals of $P$, and the paths from $(0,0)$ to $(n,(k-1)n)$ with steps $(0,1)$ and $(1,0)$ that never rise above the line $y = (k-1)x$.

**Other applications.** Let us now discuss a few other interesting applications that illustrate the power of Theorem 1.3.2.

16. (Motzkin paths) The **Motzkin number** $M_n$ is the number of paths from $(0,0)$ to $(n,0)$ using the steps $(1,1)$, $(1,-1)$, and $(1,0)$, which never go below the $x$-axis. Imitating our argument for Dyck paths, we obtain a formula for the generating function

$$M(x) = \frac{1}{1-(x+xM(x)x)} \qquad \Longrightarrow \qquad M(x) = \frac{1-x-\sqrt{1-2x-3x^2}}{2x^2}.$$

The quadratic equation $x^2M^2+(x-1)M+1=0$ gives rise to the quadratic recurrence $M_n = M_{n-1} + \sum_i M_i M_{n-2-i}$. The fact that $M(x)$ satisfies a polynomial equation leads to a more efficient recurrence

$$(n+2)M_n = (2n+1)M_{n-1} + (3n-3)M_{n-2}.$$

We will see this in Section 1.3.4.2.

17. (Schröder paths) The (large) **Schröder number** $r_n$ is the number of paths from $(0,0)$ to $(2n,0)$ using steps $NE=(1,1), SE=(1,-1)$, and $E=(2,0)$ which stays above the $x$-axis. Their generating function satisfies $R(x) = 1/(1-x-xR(x))$, and therefore

$$R(x) = \frac{1-x-\sqrt{1-6x+x^2}}{2x}.$$

Theorem 1.3.2.3 is useful when we are counting combinatorial objects that "factor" uniquely into an ordered "product" of "irreducible" objects. It tells us that we can count all objects if and only if we can count the irreducible ones. We have already used that idea several times; let us see it in action in some other examples.

18. (Domino tilings of rectangles) In Section 1.2 we let $a_n$ be the number of domino tilings of a $2 \times n$ rectangle. Such a tiling is uniquely a sequence of blocks, where each block is either a vertical domino (of width 1) or two horizontal dominoes (of width 2). This truly explains the formula:

$$A(x) = \frac{1}{1-(x+x^2)}.$$

Similarly, if $a_{m,n}$ is the number of domino tilings of a $2 \times n$ rectangle using $v$ vertical tiles, we immediately obtain

$$\sum_{m,n \geq 0} a_{m,n} v^m x^n = \frac{1}{1-(vx+x^2)}.$$

Sometimes the enumeration of irreducible structures is not immediate, but still tractable.

19. (Monomer-dimer tilings of rectangles) Let $T(2,n)$ be the number of tilings of a $2 \times n$ rectangle with dominoes and unit squares. Say a tiling is irreducible if it does not contain an internal vertical line from top to bottom. Then $\mathsf{Tilings} \cong \mathsf{Seq}(\mathsf{IrredTilings})$. It now takes some thought to recognize the irreducible tilings in Figure 1.6.

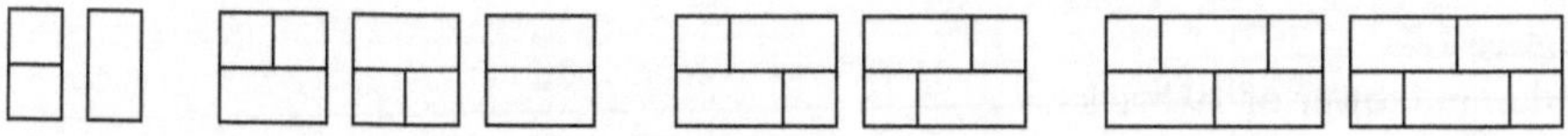

**Figure 1.6**
The irreducible tilings of $2 \times n$ rectangles into dominoes and unit squares.

There are three irreducible tilings of length 2, and two of every other length greater than or equal to 1. Therefore

$$\sum_{n \geq 0} T(2,n)x^n = \frac{1}{1-(2x+3x^2+2x^3+2x^4+\cdots)} = \frac{1-x}{1-3x-x^2+x^3}.$$

We will see in Theorem 1.3.5.2 that this gives $T(2,n) \sim c \cdot \alpha^n$ where $\alpha \approx 3.214\ldots$ is the inverse of the smallest positive root of the denominator.

Sometimes the enumeration of **all** objects is easier than the enumeration of the irreducible ones. In that case we can use Theorem 1.3.2.3 in the opposite direction.

20. (Irreducible permutations) A permutation $\pi$ of $[n]$ is **irreducible** if it does not factor as a permutation of $\{1,\ldots,m\}$ and a permutation of $\{m+1,\ldots,n\}$ for $1 \leq m < n$; that is, if $\pi([m]) \neq [m]$ for all $1 \leq m < n$. Clearly every permutation factors uniquely into irreducibles, so

$$\sum_{n \geq 0} n!x^n = \frac{1}{1-\mathsf{IrredPerm}(x)}.$$

This gives the series for $\mathsf{IrredPerm}$.

There are many interesting situations where it is possible, but not at all trivial, to decompose the objects that interest us into simpler structures. To a combinatorialist this is good news, the techniques of this section are useful tools, but are not enough. There is no shortage of interesting work to do. Here is a great example.

21. (Domino towers) [91, 31, 221] A **domino tower** is a stack of horizontal $2 \times 1$ bricks in a brickwork pattern, so that no brick is directly above another brick, such that the bricks on the bottom level are contiguous, and every higher brick is (half) supported on at least one brick in the row below it. Let the size of a domino tower be the number of bricks. See Figure 1.7 for an illustration.

**Figure 1.7**
A domino tower of 19 bricks.

Remarkably, there are $3^{n-1}$ domino towers consisting of $n$ bricks. Equally remarkably, no simple bijection is known. The nicest argument known is as follows.

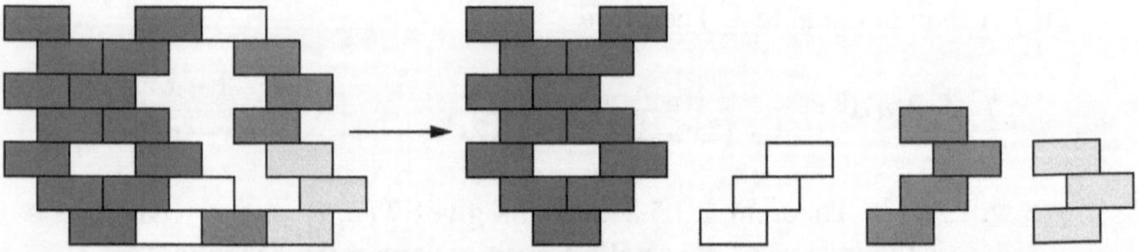

**Figure 1.8**
The decomposition of a domino tower into a pyramid and three half-pyramids.

We decompose a domino tower $x$ into smaller pieces, as illustrated in Figure 1.8. Each new piece is obtained by pushing up the leftmost remaining brick in the bottom row, dragging with it all the bricks encountered along the way. The first piece $p$ will be a **pyramid**, which we define to be a domino tower with only one brick in the bottom row. All subsequent pieces $h_1, \ldots, h_k$ are **half-pyramids**, which are pyramids containing no bricks to the left of the bottom brick. This decomposition is reversible. To recover $x$, we drop $h_k, h_{k-1}, \ldots, h_1, p$ from the top in that order; each piece is dropped in its correct horizontal position, and some of its bricks may get stuck on the previous pieces. This shows that the corresponding combinatorial classes satisfy $X \cong P \times \mathsf{Seq}(H)$.

Similarly, we may decompose a pyramid $p$ into half-pyramids, as shown in Figure 1.9. Each new half-pyramid is obtained by pushing up the leftmost remaining brick (which is not necessarily in the bottom row), dragging with it all the bricks that it encounters along the way. This shows that $P \cong \mathsf{Seq}_{\geq 1}(H) := H + (H \times H) + \cdots$.

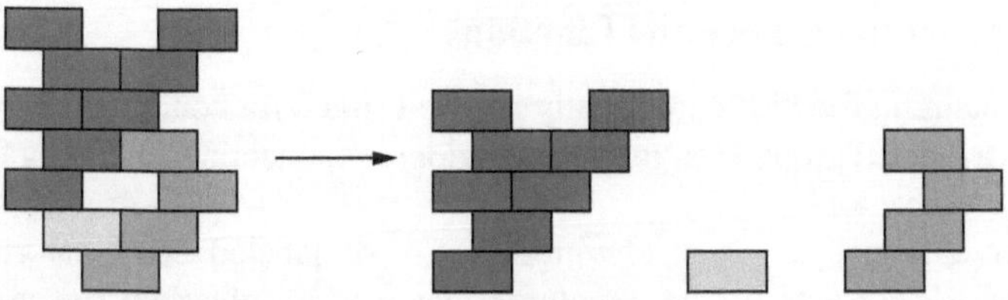

**Figure 1.9**
A pyramid and its decomposition into half-pyramids.

Finally consider a half-pyramid $h$; there are two cases. If there are other bricks on the same horizontal position as the bottom brick, consider the lowest such brick, and push it up, dragging with it all the bricks it encounters along the way, obtaining a half-pyramid $h_1$. Now remove the bottom brick; what remains is a half-pyramid $h_2$. This is shown in Figure 1.10. As before, we can recover $h$ from $h_1$ and $h_2$. On the other hand, if there are no bricks above the bottom brick, removing the bottom brick leaves either a half-pyramid or the empty set. Therefore $H \cong (H \times \bullet \times H) + (\bullet \times H) + \bullet$.

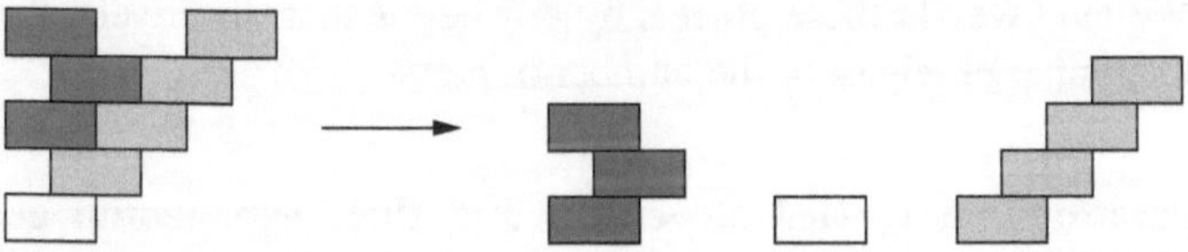

**Figure 1.10**
A (non-half-pyramid) pyramid and its decomposition into two half-pyramids and a bottom brick.

The above relations correspond to the following identities for the corresponding generating functions:

$$X = \frac{P}{1-H}, \qquad P = \frac{H}{1-H}, \qquad H = xH^2 + xH + x.$$

Surprisingly cleanly, we obtain $X(x) = x/(1-3x) = \sum_{n\geq 1} 3^{n-1}x^n$. This proves that there are $3^{n-1}$ domino towers of size $n$.

Although we do not need this here, it is worth noting that half-pyramids are enumerated by Motzkin numbers; their generating functions are related by $H(x) = xM(x)$.

### 1.3.3 Exponential generating functions

Ordinary generating functions are usually not well suited for counting combinatorial objects with a labeled ground set. In such situations, exponential generating functions are a more effective tool.

Consider a family $\mathscr{A} = \mathscr{A}_0 \sqcup \mathscr{A}_1 \sqcup A_2 \sqcup \cdots$ of labeled combinatorial structures, where $\mathscr{A}_n$ consists of the structures that we can place on the ground set $[n] = \{1, \ldots, n\}$ (or, equivalently, on any other labeled ground set of size $n$). If $a \in \mathscr{A}_n$, we let $|a| = n$ be the **size** of $a$. We also let $a_n$ be the number of elements of size $n$. The **exponential generating function** of $\mathscr{A}$ is

$$A(x) = \sum_{a \in \mathscr{A}} \frac{x^{|a|}}{|a|!} = a_0 \frac{x^0}{0!} + a_1 \frac{x^1}{1!} + a_2 \frac{x^2}{2!} + a_3 \frac{x^3}{3!} + \cdots.$$

We may again assign a weight $\mathrm{wt}(a)$ to each object $a$, usually a monomial in variables $x_1, \ldots, x_n$, and consider the **weighted exponential generating function** of $\mathscr{A}$ to be the formal power series

$$A_{\mathrm{wt}}(x_1, \ldots, x_n, x) = \sum_{a \in \mathscr{A}} \mathrm{wt}(a) \frac{x^{|a|}}{|a|!}.$$

Examples of combinatorial structures (with their respective size functions) are permutations (number of elements), graphs (number of vertices), or set partitions (size of the set). We may weight these objects by $t^k$ where $k$ is, respectively, the number or cycles, the number of edges, or the number of parts.

#### 1.3.3.1 Operations on labeled structures and their exponential generating functions

Again, there are some simple operations on *labeled* combinatorial structures, which correspond to simple algebraic operations on the exponential generating functions. Starting with a few simple structures, these operations are sufficient to generate many interesting combinatorial structures. This will allow us to compute the exponential generating functions for those structures.

**Theorem 1.3.3** *Let $\mathscr{A}$ and $\mathscr{B}$ be labeled combinatorial structures.*

1. *($\mathscr{C} = \mathscr{A} + \mathscr{B}$: Disjoint union) If a $\mathscr{C}$-structure on a finite set $S$ is obtained by choosing an $\mathscr{A}$-structure on $S$* **or** *a $\mathscr{B}$-structure on $S$, then*

$$C(x) = A(x) + B(x).$$

*This result also holds for weighted structures if the weight of a $\mathscr{C}$-structure is the same as the weight of the respective $\mathscr{A}$- or $\mathscr{B}$-structure.*

2. *($\mathscr{C} = \mathscr{A} * \mathscr{B}$: Labeled Product) If a $\mathscr{C}$-structure on a finite set $S$ is obtained by partitioning $S$ into disjoint sets $S_1$ and $S_2$ and putting an $\mathscr{A}$-structure on $S_1$* **and** *a $\mathscr{B}$-structure on $S_2$, then*

$$C(x) = A(x)B(x).$$

*This result also holds for weighted structures if the weight of a $\mathscr{C}$-structure is the product of the weights of the respective $\mathscr{A}$- and $\mathscr{B}$-structures.*

3. *($\mathscr{C} = \mathrm{Seq}_*(\mathscr{B})$: Labeled Sequence) If a $\mathscr{C}$-structure on a finite set S is obtained by choosing an* **ordered** *partition of S into a sequence of blocks and putting a $\mathscr{B}$-structure on each block, then*

$$C(x) = \frac{1}{1-B(x)}.$$

*This result also holds for weighted structures if the weight of a $\mathscr{C}$-structure is the product of the weights of the respective $\mathscr{B}$-structures.*

4. *($\mathscr{C} = \mathrm{Set}(\mathscr{B})$: Set)* **Exponential formula.** *If a $\mathscr{C}$-structure on a finite set S is obtained by choosing an* **unordered** *partition of S into a set of blocks and putting a $\mathscr{B}$-structure on each block, then*

$$C(x) = e^{B(x)}.$$

*This result also holds for weighted structures if the weight of a $\mathscr{C}$-structure is the product of the weights of the respective $\mathscr{B}$-structures.*

*In particular, if $c_k(n)$ is the number of $\mathscr{C}$-structures of an n-set that decompose into k components ($\mathscr{B}$-structures), we have*

$$\sum_{n,k,\geq 0} c_k(n)\frac{x^n}{n!}y^k = e^{yB(x)} = C(x)^y$$

5. *($\mathscr{C} = \mathscr{A} \circ \mathscr{B}$: Composition)* **Compositional formula.** *If a $\mathscr{C}$-structure on a finite set S is obtained by choosing an* unordered *partition of S into a set of blocks, putting a $\mathscr{B}$-structure on each block, and putting an $\mathscr{A}$-structure on the set of blocks, then*

$$C(x) = A(B(x)).$$

*This result also holds for weighted structures if the weight of a $\mathscr{C}$-structure is the product of the weights of the $\mathscr{A}$-structure on its set of blocks and the weights of the $\mathscr{B}$-structures on the individual blocks.*

Again, Theorem 1.3.4.4 is useful when we are counting combinatorial objects that "decompose" uniquely as a set of "indecomposable" objects. It tells us that we can count all objects if and only if we can count the indecomposable ones. Amazingly, we also obtain for free the finer enumeration of the objects by their number of components.

*Proof.* 1. is clear. The identity in 2. is equivalent to $c_n = \sum_k \binom{n}{k} a_k b_{n-k}$, which corresponds to the given combinatorial description. Iterating 2., we see that the exponential generating functions for $k$-sequences of $\mathscr{B}$-structures is $B(x)^k$, and hence the one for $k$-sets of $\mathscr{B}$-structures is $B(x)^k/k!$. This readily implies 3, 4, and 5. The weighted statements follow similarly. ■

The following statements are perhaps less fundamental, but also useful.

**Theorem 1.3.4** *Let $\mathscr{A}$ be a labeled combinatorial structure.*

1. *($\mathscr{C} = \mathscr{A}_+$: Shifting) If a $\mathscr{C}$-structure on S is obtained by adding a new element t to S and choosing an $\mathscr{A}$-structure on $S \cup \{t\}$, then*
$$C(x) = A'(x).$$

2. *($\mathscr{C} = \mathscr{A}_\bullet$: Rooting) If a $\mathscr{C}$-structure on S is a* rooted *$\mathscr{A}$*-structure, *obtained by choosing an $\mathscr{A}$-structure on S and an element of S called the* **root**, *then*
$$C(x) = xA(x).$$

3. *(Sieving by parity of size) If the $\mathscr{C}$-structures are precisely the $\mathscr{A}$-structures of even size,*
$$C(x) = \frac{A(x)+A(-x)}{2}.$$

4. *(Sieving by parity of components) Suppose $\mathscr{A}$-structures decompose uniquely into components, so $\mathscr{A} = \mathsf{Set}(\mathscr{B})$ for some $\mathscr{B}$. If the $\mathscr{C}$-structures are the $\mathscr{A}$-structures having only components of even size,*
$$C(x) = \sqrt{A(x)A(-x)}.$$

5. *(Sieving by parity of number of components) Suppose $\mathscr{A}$-structures decompose uniquely into components, so $\mathscr{A} = \mathsf{Set}(\mathscr{C})$ for some $\mathscr{C}$. If the $\mathscr{C}$-structures are precisely the $\mathscr{A}$-structures having an even number of components,*
$$C(x) = \frac{1}{2}\left(A(x) + \frac{1}{A(x)}\right).$$

*Similar sieving formulas hold modulo k for any $k \in \mathbb{N}$.*

*Proof.* We have $c_n = a_{n+1}$ in 1., $c_n = na_n$ in 2., and $c_n = \frac{1}{2}(a_n + (-1)^n a_n)$ in 3., from which the generating function formulas follow. Combining 3. with the Exponential Formula we obtain 4. and 5.

Similarly we see that the generating function for $\mathscr{A}$-structures whose size is a multiple of $k$ is $\frac{1}{k}\left(A(x) + A(\omega x) + \cdots + A(\omega^{k-1}x)\right)$ where $\omega$ is a primitive $k$th root of unity. If we wish to count elements of size $i$ mod $k$, we use 1. to shift this generating function $i$ times. ■

#### 1.3.3.2 Examples

**Classical applications.** Once again, these simple ideas give very easy solutions to many classical enumeration problems.

1. (Trivial classes) Again we consider the trivial classes $\circ$ with only one element of size 0, and $\bullet$ with only one element of size 1. Their exponential generating functions are 1 and $x$, respectively.

2. (Sets) A slightly less trivial class of Set contains one set of each size. We also let $\mathsf{Set}_{\geq 1}$ denote the class of non-empty sets, with generating function $e^x - 1$. The exponential generating functions are

$$\mathsf{Set}(x) = e^x, \qquad \mathsf{Set}_{\geq 1}(x) = e^x - 1.$$

3. (Set partitions) In Section 1.3.2.2 we found the ordinary generating function for Stirling numbers $S(n,k)$ for a given $k$; but in fact it is easier to use exponential generating functions. Simply notice that $\mathsf{SetPartition} \cong \mathsf{Set}(\mathsf{Set}_{\geq 1})$, and the Weighted Exponential Formula then gives

$$\mathsf{SetPartition}(x,y) = \sum_{n,k \geq 0} S(n,k) \frac{x^n}{n!} y^k = e^{y(e^x - 1)}.$$

4. (Permutations) Let $\mathsf{Perm}_n$ consist of the $n!$ permutations of $[n]$. A permutation is a labeled sequence of singletons, so $\mathsf{Perm} = \mathsf{Seq}_*(\bullet)$, and the generating function for permutations is

$$\mathsf{Perm}(x) = \sum_{n \geq 0} n! \frac{x^n}{n!} = \frac{1}{1-x}.$$

5. (Cycles) Let $\mathsf{Cycle}_n$ consist of the cyclic orders of $[n]$. These are the ways of arranging $1, \ldots, n$ around a circle, where two orders are the same if they differ by a rotation of the circle. There is an $n$-to-1 mapping from permutations to cyclic orders obtained by wrapping a permutation around a circle, so

$$\mathsf{Cycle}(x) = \sum_n (n-1)! x^n / n! = -\log(1-x).$$

   There is a more indirect argument that will be useful to us later. Recall that a permutation $\pi$ can be written uniquely as a (commutative) product of disjoint cycles of the form $(i, \pi(i), \pi^2(i), \ldots, \pi^{k-1}(i))$, where $k$ is the smallest index such that $\pi^k(i) = i$. For instance, the permutation 835629741 can be written in cycle notation as (18469)(235)(7). Then $\mathsf{Perm} = \mathsf{Set}(\mathsf{Cycle})$ so $1/(1-x) = e^{\mathsf{Cycle}(x)}$.

6. (Permutations by number of cycles) The (signless) **Stirling number of the first kind** $c(n,k)$ is the number of permutations of $n$ having $k$ cycles. The Weighted Exponential Formula gives

$$\sum_{n,k \geq 0} c(n,k) \frac{x^n}{n!} y^k = e^{y\mathsf{Cycle}(x)} = \left(\frac{1}{1-x}\right)^y = \sum_{n \geq 0} y(y+1)\cdots(y+n-1) \frac{x^n}{n!}$$

   so the Stirling numbers $c(n,k)$ of the first kind are the coefficients of the polynomial $y(y+1)\cdots(y+n-1)$.

**Other applications.** The applications of these techniques are countless; let us consider a few more applications, old and recent.

7. (Permutations by cycle type) The **type** of a permutation $\pi \in S_n$ is type$(w) = (c_1, \ldots, c_n)$ where $c_i$ is the number of cycles of length $i$. For indeterminates $\mathbf{t} = (t_1, \ldots, t_n)$, let $\mathbf{t}^{\text{type}(w)} = t_1^{c_1} \cdots t_n^{c_n}$. The **cycle indicator** of the symmetric group $S_n$ is $Z_n = \frac{1}{n!} \sum_{w \in S_n} \mathbf{t}^{\text{type}(w)}$. The Weighted Exponential Formula immediately gives
$$\sum_{n \geq 0} Z_n x^n = e^{t_1 x + t_2 x^2/2 + t_3 x^3/3 + \cdots}.$$
Let us discuss two special cases of interest.

8. (Derangements) A **derangement** of $[n]$ is a permutation such that $\pi(i) \neq i$ for all $i \in [n]$. Equivalently, a derangement is a permutation with no cycles of length 1. It follows that Derangement $=$ Set(Cycle$_{\geq 2}$), so the number $d_n$ of derangements of $[n]$ is given by
$$\mathsf{Derangement}(x) = \sum_{n \geq 0} d_n \frac{x^n}{n!} = e^{-\log(1-x)-x} = e^{-x} + xe^{-x} + x^2 e^{-x} + \cdots,$$
which leads to the explicit formula
$$d_n = n! \left(1 - \frac{1}{1!} + \frac{1}{2!} - \frac{1}{3!} + \cdots \pm \frac{1}{n!}\right) \sim \frac{n!}{e}.$$

9. (Involutions) An **involution** of $[n]$ is a permutation $w$ such that $w^2$ is the identity. Equivalently, an involution is a permutation with cycles of length 1 and 2, so the number $i_n$ of involutions of $[n]$ is given by
$$\mathsf{Inv}(x) = \sum_{n \geq 0} i_n \frac{x^n}{n!} = e^{x + \frac{x^2}{2}}.$$
Note that $\mathsf{Inv}'(x) = (x+1)\mathsf{Inv}(x)$, which gives $i_n = i_{n-1} + (n-1)i_{n-2}$. In Section 1.3.4.2 we will explain the theory of D-finite power series, which turns differential equations for power series into recurrences for the corresponding sequences.

10. (Trees) A **tree** is a connected graph with no cycles. Consider a "birooted" tree $(T, a, b)$ on $[n]$ with two (possibly equal) root vertices $a$ and $b$. Regard the unique path $a = v_0, v_1, \ldots, v_k = b$ as a "spine" for $T$; the rest of the tree consists of rooted trees hanging from the $v_i$s; direct their edges toward the spine. Now regard $v_1 \ldots v_k$ as a permutation in one-line notation, and rewrite it in cycle notation, while continuing to hang the rooted trees from the respective $v_i$'s. This transforms $(T, a, b)$ into a directed graph consisting of a disjoint collection of cycles with trees directed toward them. Every vertex has outdegree 1, so this defines a function $f : [n] \to [n]$. A moment's thought will convince us that this is a bijection. Therefore there are $n^n$ birooted trees on $[n]$, and hence there are $n^{n-2}$ trees on $[n]$. See Figure 1.11 for an illustration.

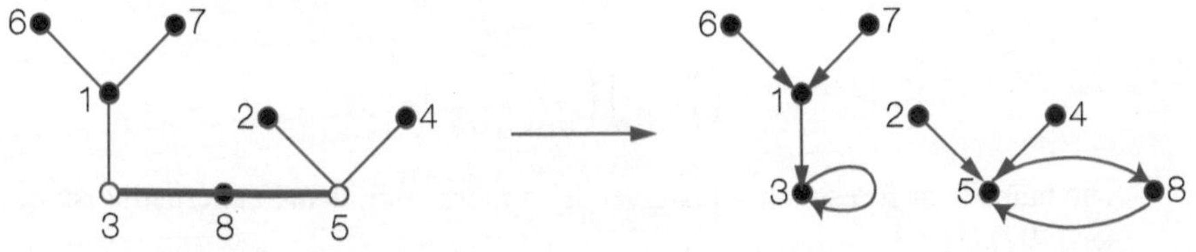


**Figure 1.11**
A tree on [8] birooted at $a = 3$ and $b = 5$, and the corresponding function $f : [8] \to [8]$.

11. (Trees, revisited) Let us count trees in a different way. Let a **rooted tree** be a tree with a special vertex called the root, and a **planted forest** be a graph with no cycles where each connected component has a root. Let $t_n, r_n, f_n$ and $T(x), R(x), F(x)$ be the sequences and exponential generating functions enumerating trees, rooted trees, and planted forests, respectively.

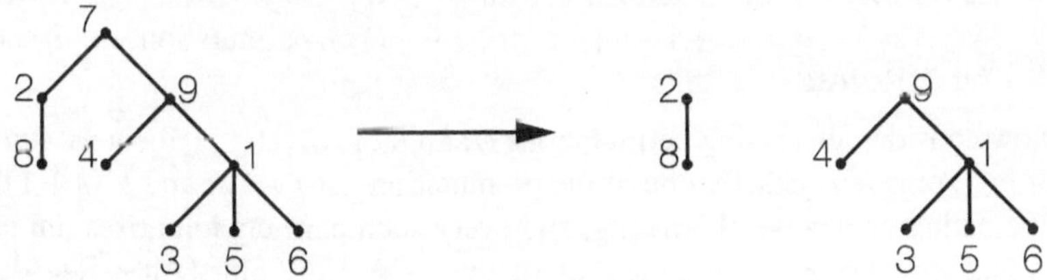


**Figure 1.12**
A rooted tree seen as a root attached to the roots of a planted forest.

Planted forests are vertex-disjoint unions of rooted trees, so $F(x) = e^{R(x)}$. Also, as illustrated in Figure 1.12, a rooted tree $T$ consists of a root attached to the roots of a planted forest, so $R(x) = xF(x)$. It follows that $x = R(x)e^{-R(x)}$, so

$$R(x) = (xe^{-x})^{<-1>}.$$

Lagrange inversion gives $n \cdot \frac{r_n}{n!} = [x^{n-1}]e^{nx} = \frac{n^{n-1}}{(n-1)!}$, so

$$r_n = n^{n-1}, \qquad f_n = (n+1)^{n-1}, \qquad t_n = n^{n-2}.$$

We state a finer enumeration; see [192, Theorem 5.3.4] for a proof. The **degree sequence** of a rooted forest on $[n]$ is $(\deg 1, \ldots, \deg n)$ where $\deg i$ is the number of children of $i$. For example the degree sequence of the rooted tree in Figure 1.12 is $(3,1,0,0,0,0,2,0,2)$. Then the number of planted forests with a given degree sequence $(d_1, \ldots, d_n)$ and (necessarily) $k = n - (d_1 + \cdots + d_n)$

components is

$$\binom{n-1}{k-1}\binom{n-k}{d_1,\ldots,d_n}.$$

The number of forests on $[n]$ is given by a more complicated alternating sum; see [197].

12. (Permutations, revisited) Here is an unnecessarily complicated way of proving there are $n!$ permutations of $[n]$. A permutation $\pi$ of $[n+1]$ decomposes uniquely as a concatenation $\pi = L(n+1)R$ for permutations $L$ and $R$ of two complementary subsets of $[n]$. Therefore $\mathsf{Shift}(\mathsf{Perm}) = (\mathsf{Perm}) * (\mathsf{Perm})$, and the generating function $P(x)$ for permutations satisfies $P'(x) = P(x)^2$ with $P(0) = 1$. Solving this differential equation gives $P(x) = \frac{1}{1-x} = \sum_{n\geq 0} n!\frac{x^n}{n!}$.

13. (Alternating permutations) The previous argument was gratuitous for permutations, but it will now help us to enumerate the class $\mathsf{Alt}$ of **alternating** permutations $w$, which satisfy $w_1 < w_2 > w_3 < w_4 > \cdots$. The **Euler numbers** are $E_n = |\mathsf{Alt}_n|$; let $E(x)$ be their exponential generating function. We will need the class $\mathsf{RevAlt}$ of permutations $w$ with $w_1 > w_2 < w_3 > w_4 < \cdots$. The map $w = w_1 \ldots w_n \mapsto w' = (n+1-w_1)\ldots(n+1-w_n)$ on permutations of $[n]$ shows that $\mathsf{Alt} \cong \mathsf{RevAlt}$.

    Now consider alternating permutations $L$ and $R$ of two complementary subsets of $[n]$. For $n \geq 1$, exactly one of the permutations $L(n+1)R$ and $L'(n+1)R$ is alternating or reverse alternating, and every such permutation arises uniquely in that way. For $n = 0$ both are alternating. Therefore $\mathsf{Shift}(\mathsf{Alt} + \mathsf{RevAlt}) = \mathsf{Alt} * \mathsf{Alt} + \circ$, so $2E'(x) = E(x)^2 + 1$ with $E(0) = 1$. Solving this differential equation we get

$$E(x) = \sum_{n\geq 0} E_n \frac{x^n}{n!} = \sec x + \tan x.$$

    Therefore $\sec x$ and $\tan x$ enumerate the alternating permutations of even and odd length, respectively. The Euler numbers are also called secant and tangent numbers. This surprising connection allows us to give combinatorial interpretations of various trigonometric identities, such as $1 + \tan^2 x = \sec^2 x$.

14. (Graphs) Let $g(v)$ and $g_{\text{conn}}(v)$ be the number of simple graphs and connected graphs on $[v]$, respectively. (A graph is simple if it contains no loops and no multiple edges.) The Exponential Formula tells us that their exponential generating functions are related by $G(x) = e^{G_{\text{conn}}(x)}$. In this case it is hard to count the connected graphs directly, but it is easy to count all graphs: to choose a graph we just have to decide whether each edge is present or not, so $g(v) = 2^{\binom{v}{2}}$. This gives us

$$\sum_{v\geq 0} g_{\text{conn}}(v)\frac{x^v}{v!} = \log\left(\sum_{v\geq 0} 2^{\binom{v}{2}}\frac{x^v}{v!}\right).$$

    We may easily adjust this computation to account for edges and components.

There are $\binom{v(v-1)/2}{e}$ graphs on $[v]$ with $e$ edges; say $g(v,c,e)$ of them have $c$ components, and give them weight $y^c z^e$. Then

$$\sum_{v,c,e\geq 0} g(v,c,e)\frac{x^v}{v!}y^c z^e = \left(\sum_{v,e\geq 0}\binom{\binom{v}{2}}{e}\frac{x^v}{v!}z^e\right)^y = F(x,1+z)^y$$

where

$$F(\alpha,\beta) = \sum_{n\geq 0}\frac{\alpha^n \beta^{\binom{n}{2}}}{n!}$$

is the **deformed exponential function** of [179].

15. (Signed graphs) A **signed graph** $G$ is a set of vertices, with at most one "positive" edge and one "negative" edge connecting each pair of vertices. We say $G$ is *connected* if and only if its underlying graph $\overline{G}$ (ignoring signs) is connected. A *cycle* in $G$ corresponds to a cycle of $\overline{G}$; we call it *balanced* if it contains an even number of negative edges, and *unbalanced* otherwise. We say that $G$ is *balanced* if all its cycles are balanced. Let $s(v,c_+,c_-,e)$ be the number of signed graphs with $v$ vertices, $e$ edges, $c_+$ balanced components, and $c_-$ unbalanced components; we will need the generating function

$$S(x,y_+,y_-,z) = \sum_{G \text{ signed graph}} s(v,c_+,c_-,e)\frac{x^v}{v!}y_+^{c_+}y_-^{c_-}z^e$$

in order to carry out a computation in Section 1.8.9; we follow [11].

Let $S(x,y_+,y_-,z)$, $B(x,y_+,z)$, $C_+(x,z)$, and $C_-(x,z)$ be the generating functions for signed, balanced, connected balanced, and connected unbalanced graphs, respectively. The Weighted Exponential Formula gives

$$B = e^{y_+C_+}, \qquad S = e^{y_+C_+ + y_-C_-},$$

so if we can compute $C_+$ and $C_-$ we will obtain $B$ and $S$. In turn, these equations give

$$C_+(x,z) = \frac{1}{2}\log B(x,2,z), \qquad C_+(x,z)+C_-(x,z) = \log S(x,1,1,z),$$

and we now compute the right-hand side of these two equations. (In the first equation, we set $t_+ = 2$ because, surprisingly, $B(x,2,z)$ is easier to compute than $B(x,1,z)$.) One is easy:

$$S(x,1,1,z) = \sum_{e,v\geq 0}\binom{v(v-1)}{e}\frac{x^v}{v!}z^e = F(x,(1+z)^2).$$

For the other one, we count balanced signed graphs by relating them with **marked graphs**, which are simple graphs with a sign + or − on each **vertex**. [95] A marked graph $M$ gives rise to a balanced signed graph $G$ by assigning to each edge the product of its vertex labels. Furthermore, if $G$ has $c$

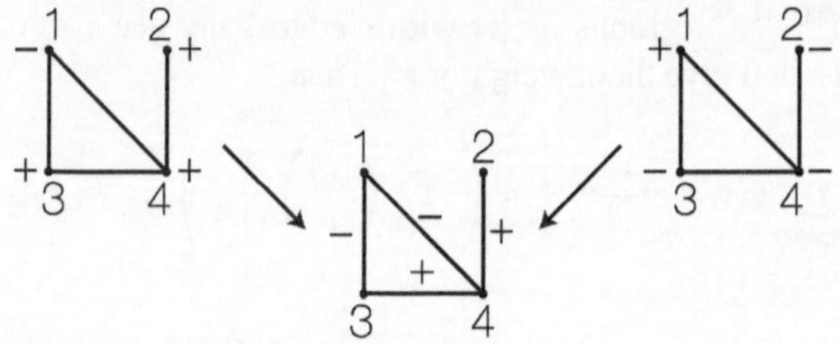


**Figure 1.13**
The two marked graphs that give rise to one balanced signed graph.

components, then it arises from precisely $2^c$ different marked graphs, obtained from $M$ by choosing some connected components and changing their signs. This correspondence is illustrated in Figure 1.13. It follows that $B(x,2y,z) = \sum_{B \text{ balanced}} 2^c b(v,c,e)\frac{x^v}{v!}y^c z^e = \sum_{M \text{ marked}} m(v,c,e)\frac{x^v}{v!}y^c z^e$ is the generating function for marked graphs, and hence $B(x,2,z)$ may be computed easily:

$$B(x,2,z) = \sum_{e,v} \binom{\binom{v}{2}}{e} 2^v \frac{x^v}{v!} z^e = F(2x, 1+z).$$

Putting these equations together yields

$$S(x,y_+,y_-,z) = F(2x,1+z)^{(y_+-y_-)/2} F(x,(1+z)^2)^{y_-}.$$

### 1.3.4 Nice families of generating functions

In this section we discuss three nice properties that a generating function can have: being rational, algebraic, or D-finite. Each one of these properties gives rise to useful properties for the corresponding sequence of coefficients.

#### 1.3.4.1 Rational generating functions

Many sequences in combinatorics and other fields satisfy three equivalent properties: They satisfy a recursive formula with constant coefficients, they are given by an explicit formula in terms of polynomials and exponentials, and their generating functions are rational. We understand these sequences very well. The following theorem tells us how to translate any one of these formulas into the others.

**Theorem 1.3.5** [194, Theorem 4.1.1] *Let $a_0, a_1, a_2, \ldots$ be a sequence of complex numbers and let $A(x) = \sum_{n \geq 0} a_n x^n$ be its ordinary generating function. Let $q(x) = 1 + c_1 x + \cdots + c_d x^d = (1-r_1 x)^{d_1} \cdots (1-r_k x)^{d_k}$ be a complex polynomial of degree d. The following are equivalent:*

1. *The sequence satisfies the linear recurrence with constant coefficients*

$$a_n + c_1 a_{n-1} + \cdots + c_d a_{n-d} = 0 \qquad (n \geq d).$$

2. *There exist polynomials* $f_1(x),\ldots,f_k(x)$ *with* $\deg f_i(x) < d_i$ *for* $1 \le i \le n$ *such that*

$$a_n = f_1(n)\, r_1^n + \cdots + f_k(n)\, r_k^n.$$

3. *There exists a polynomial* $p(x)$ *with* $\deg p(x) < d$ *such that* $A(x) = p(x)/q(x)$.

Notice that Theorem 1.3.5.2 gives us the asymptotic growth of $a_n$ immediately. Let us provide more explicit recipes.

($1 \Rightarrow 2$) Extract the inverses $r_i$ of the roots of $q(x) = 1 + c_1x + \cdots + c_dx^d$ and their multiplicities $d_i$. The $d_1 + \cdots + d_k = d$ coefficients of the $f_i$s are the unknowns in the system of $d$ linear equations $a_n = f_1(n)\, r_1^n + \cdots + f_k(n) r_k^n$ ($n = 0, 1, \ldots, d-1$), which has a unique solution.

($1 \Rightarrow 3$) Read off $q(x) = 1 + c_1x + \cdots + c_dx^d$ from the recurrence; the coefficients of $p(x)$ are $[x^k]p(x) = a_k + c_1a_{k-1} + \cdots + c_da_{k-d}$ for $0 \le k < d$, where $a_i = 0$ for $i < 0$.

($2 \Rightarrow 1$) Compute the $c_i$s using $q(x) = \prod_i^k (1 - r_ix)^{\deg f_i + 1}$.

($2 \Rightarrow 3$) Let $q(x) = \prod_i (1 - r_ix)^{\deg f_i + 1}$, and compute the first $k$ terms of $p(x) = A(x)q(x)$; the others are 0.

($3 \Rightarrow 1$) Extract the $c_i$s from the denominator $q(x)$.

($3 \Rightarrow 2$) Compute the partial fraction decomposition $p(x)/q(x) = \sum_{i=1}^k p_i(x)/(1 - r_ix)^{d_i}$ where $\deg p_i(x) < d_i$ and use $(1 - r_ix)^{-d_i} = \sum_n \binom{d_i+n-1}{d_i-1} r_i^n x^n$ to extract $a_n = [x^n]p(x)/q(x)$.

**Characterizing polynomials.** As a special case of Theorem 1.3.5, we obtain a useful characterization of sequences given by a polynomial. The **difference operator** $\Delta$ acts on sequences, sending the sequence $\{a_n : n \in \mathbb{N}\}$ to the sequence $\{\Delta a_n : n \in \mathbb{N}\}$ where $\Delta a_n = a_{n+1} - a_n$.

**Theorem 1.3.6** [194, Theorem 4.1.1] *Let* $a_0, a_1, a_2, \ldots$ *be a sequence of complex numbers and let* $A(x) = \sum_{n \ge 0} a_n x^n$ *be its ordinary generating function. Let* $d$ *be a positive integer. The following are equivalent:*

1. *We have* $\Delta^{d+1} a_n = 0$ *for all* $n \in \mathbb{N}$.

2. *There exists a polynomial* $f(x)$ *with* $\deg f \le d$ *such that* $a_n = f(n)$ *for all* $n \in \mathbb{N}$.

3. *There exists a polynomial* $p(x)$ *with* $\deg p(x) \le d$ *such that* $A(x) = p(x)/(1-x)^{d+1}$.

We have already seen some combinatorial polynomials and generating functions whose denominator is a power of $1 - x$; we will see many more examples in the following sections.

#### 1.3.4.2 Algebraic and D-finite generating functions

When the generating function $A(x) = \sum_n a_n x^n$ we are studying is not rational, the next natural question to ask is whether $A(x)$ is algebraic. If it is, then just as in the rational case, the sequence $a_n$ still satisfies a linear recurrence, although now the coefficients are polynomial in $n$. This general phenomenon is best explained by introducing the wider family of "D-finite" (also known as "differentially finite" or "holonomic") power series. Let us discuss a quick example before we proceed to the general theory.

We saw that the ordinary generating function for the Motzkin numbers satisfies the quadratic equation

$$x^2M^2 + (x-1)M + 1 = 0 \tag{1.8}$$

which gives rise to the quadratic recurrence $M_n = M_{n-1} + \sum_i M_i M_{n-2-i}$ with $M_0 = 1$. This is not a bad recurrence, but we can find a better one. Differentiating (1.8) we can express $M'$ in terms of $M$. Our likely first attempt leads us to $M' = -(2xM^2 + M)/(2x^2M + x - 1)$, which is not terribly enlightening. However, using (1.8) and a bit of purposeful algebraic manipulation, we can rewrite this as a linear equation with polynomial coefficients:

$$(x - 2x^2 - 3x^3)M' + (2 - 3x - 3x^2)M - 2 = 0.$$

Extracting the coefficient of $x^n$ we obtain the much more efficient recurrence relation

$$(n+2)M_n - (2n+1)M_{n-1} - (3n-3)M_{n-2} = 0. \qquad (n \geq 2)$$

We now explain the theoretical framework behind this example.

**Rational, algebraic, and D-finite series.** Consider a formal power series $A(x)$ over the complex numbers. We make the following definitions.

| | |
|---|---|
| $A(x)$ is **rational** | There exist polynomials $p(x)$ and $q(x) \neq 0$ such that $$q(x)A(x) = p(x).$$ |
| $A(x)$ is **algebraic** | There exist polynomials $p_0(x), \ldots, p_d(x)$ such that $$p_0(x) + p_1(x)A(x) + p_2(x)A(x)^2 + \cdots + p_d(x)A(x)^d = 0.$$ |
| $A(x)$ is **D-finite** | There exist polynomials $q_0(x), \ldots, q_d(x), q(x)$ such that $$q_0(x)A(x) + q_1(x)A'(x) + q_2(x)A''(x) + \cdots + q_d(x)A^{(d)}(x) = q(x).$$ |

Now consider the corresponding sequence $a_0, a_1, a_2 \ldots$ and make the following definitions.

| | |
|---|---|
| $\{a_0, a_1, \ldots\}$ is **c-recursive** | There are constants $c_0, \ldots, c_d \in \mathbb{C}$ such that for all $n \geq d$ $$c_0 a_n + c_1 a_{n-1} + \cdots + c_d a_{n-d} = 0$$ |
| $\{a_0, a_1, \ldots\}$ is **P-recursive** | There are complex polynomials $c_0(x), \ldots, c_d(x)$ such that for all $n \geq d$ $$c_0(n) a_n + c_1(n) a_{n-1} + \cdots + c_d(n) a_{n-d} = 0$$ |

These families contain most (but certainly not all) series and sequences that we encounter in combinatorics. They are related as follows.

**Theorem 1.3.7** *Let $A(x) = a_0 + a_1 x + a_2 x^2 + \cdots$ be a formal power series. The following implications hold.*

| | | |
|---|---|---|
| *$A(x)$ is rational* | $\Longrightarrow$ *$A(x)$ is algebraic* $\Longrightarrow$ | *$A(x)$ is D-finite* |
| $\Updownarrow$ | | $\Updownarrow$ |
| *$\{a_0, a_1, \ldots\}$ is c-recursive* | $\Longrightarrow$ | *$\{a_0, a_1, \ldots\}$ is P-recursive* |

*Proof.* We already discussed the correspondence between rational series and c-recursive functions, and rational series are trivially algebraic. Let us prove the remaining statements.

*(Algebraic $\Rightarrow$ D-finite)* Suppose $A(x)$ satisfies an algebraic equation of degree $d$. Then $A$ is algebraic over the field $\mathbb{C}(x)$, and the field extension $\mathbb{C}(x, A)$ is a vector space over $\mathbb{C}(x)$ having dimension of at most $d$.

Taking the derivative of the polynomial equation satisfied by $A$, we get an expression for $A'$ as a rational function of $A$ and $x$. Taking derivatives repeatedly, we find that all derivatives of $A$ are in $\mathbb{C}(x, A)$. It follows that $1, A, A', A'', \ldots, A^{(d)}$ are linearly dependent over $\mathbb{C}(x)$, and a linear relation between them is a certificate for the D-finiteness of $A$.

*(P-recursive $\Leftrightarrow$ D-finite)* If $q_0(x)A(x) + q_1(x)A'(x) + \cdots + q_d(x)A^{(d)}(x) = q(x)$, comparing the coefficients of $x^n$ gives a $P$-recursion for the $a_i$s. In the other direction, given a P-recursion for the $a_i$s of the form $c_0(n)a_n + \cdots + c_d(n)a_{n-d} = 0$, it is easy to obtain the corresponding differential equation after writing $c_i(x)$ in terms of the basis $\{(x+i)_k : k \in \mathbb{N}\}$ of $\mathbb{C}[x]$, where $(y)_k = y(y-1)\cdots(y-k+1)$. ■

The converses are not true. For instance, $\sqrt{1+x}$ is algebraic but not rational, and $e^x$ and $\log(1-x)$ are D-finite but not algebraic.

**Corollary 1.3.8** *The ordinary generating function $\sum_n a_n x^n$ is D-finite if and only if the exponential generating function $\sum_n a_n \frac{x^n}{n!}$ is D-finite.*

*Proof.* This follows from the observation that $\{a_n : n \in \mathbb{N}\}$ is P-recursive if and only if $\{a_n/n! : n \in \mathbb{N}\}$ is P-recursive. ■

**A few examples.** Before we discuss general tools, we collect some examples. We will prove all of the following statements later in this section.

The power series for subsets, Fibonacci numbers, and Stirling numbers are rational:

$$\sum_{n\geq 0} 2^n x^n = \frac{1}{1-2x}, \quad \sum_{n\geq 0} F_n x^n = \frac{x}{1-x-x^2}, \quad \sum_{n\geq k} S(n,k)x^n = \frac{x}{1-x}\cdot\frac{x}{1-2x}\cdots\frac{x}{1-kx}.$$

The "diagonal binomial," $k$-Catalan, and Motzkin series are algebraic but not rational:

$$\sum_{n\geq 0}\binom{2n}{n}x^n = \frac{1}{\sqrt{1-4x}}, \quad \sum_{n\geq 0}\frac{1}{(k-1)n+1}\binom{kn}{n}x^n,$$

and

$$\sum_{n\geq 0} M_n x^n = \frac{1-x-\sqrt{1-2x-3x^2}}{2x^2}.$$

The following series are D-finite but not algebraic:

$$e^x, \quad \log(1+x), \quad \sin x, \quad \cos x, \quad \arctan x, \quad \sum_{n\geq 0}\binom{2n}{n}^2 x^n, \quad \sum_{n\geq 0}\binom{3n}{n,n,n}x^n$$

The following series are not D-finite:

$$\sqrt{1+\log(1+x^2)}, \quad \sec x, \quad \tan x, \quad \sum_{n\geq 0} p(n)x^n = \prod_{k\geq 0}\frac{1}{1-x^k}.$$

**Recognizing algebraic and D-finite series.** It is not always obvious whether a given power series is algebraic or D-finite, but there are some tools available. Fortunately, algebraic functions behave well under a few operations, and D-finite functions behave even better. This explains why these families contain most examples arising in combinatorics.

The following table summarizes the properties of formal power series that are preserved under various key operations. For example, the fifth entry on the bottom row says that if $A(x)$ and $B(x)$ are D-finite, then the composition $A(B(x))$ is not necessarily D-finite.

| | $cA$ | $A+B$ | $AB$ | $1/A$ | $A\circ B$ | $A\star B$ | $A'$ | $\int A$ | $A^{\langle -1\rangle}$ |
|---|---|---|---|---|---|---|---|---|---|
| rational | Y | Y | Y | Y | Y | Y | Y | N | N |
| algebraic | Y | Y | Y | Y | Y | N | Y | N | Y |
| D-finite | Y | Y | Y | N | N | Y | Y | Y | N |

Here $A \star B(x) := \sum_{n \geq 0} a_n b_n x^n$ denotes the **Hadamard product** of $A(x)$ and $B(x)$, $\int A(x) := \sum_{n \geq 1} \frac{a_{n-1}}{n} x^n$ is the **formal integral** of $A(x)$, and $A^{\langle -1 \rangle}(x)$ is the **compositional inverse** of $A(x)$. In the fourth column we are assuming that $A(0) \neq 0$ so that $1/A(x)$ is well-defined, in the fifth column we are assuming that $B(0) = 0$ so that $A(B(x))$ is well-defined, and in the last column we are assuming that $A(0) = 0$ and $A'(0) \neq 0$, so that $A^{\langle -1 \rangle}(x)$ is well-defined.

For proofs of the "Yes" entries, see [183], [192], and [79]. For the "No" entries, we momentarily assume the statements of the previous subsection. Then we have the following counterexamples:

- $\cos x$ is D-finite but $1/\cos x = \sec x$ is not.
- $\sqrt{1+x}$ and $\log(1+x^2)$ are D-finite but their composition $\sqrt{1+\log(1+x^2)}$ is not.
- $A(x) = \sum_{n \geq 0} \binom{2n}{n} x^n$ is algebraic but $A \star A(x) = \sum_{n \geq 0} \binom{2n}{n}^2 x^n$ is not.
- $1/(1+x)$ is rational and algebraic but its integral $\log(1+x)$ is neither.
- $x + x^2$ is rational but its compositional inverse $(-1+\sqrt{1+4x})/2$ is not.
- $\arctan x$ is D-finite but its compositional inverse $\tan x$ is not.

Some of these negative results have weaker positive counterparts:

- If $A(x)$ is algebraic and $B(x)$ is rational, then $A(x) \star B(x)$ is algebraic.
- If $A(x)$ is D-finite and $A(0) \neq 0$, $1/A(x)$ is D-finite if and only if $A'(x)/A(x)$ is algebraic.
- If $A(x)$ is D-finite and $B(x)$ is algebraic with $B(0) = 0$, then $A(B(x))$ is D-finite.

See [192, Proposition 6.1.11], [96], and [192, Theorem 6.4.10] for the respective proofs.

The following result is also useful.

**Theorem 1.3.9** [192, Section 6.3] *Consider a multivariate formal power series $F(x_1, \ldots, x_d)$ that is rational in $x_1, \ldots, x_d$ and its* ***diagonal****:*

$$F(x_1, \ldots, x_d) = \sum_{n_1, \ldots, n_d \geq 0} a_{n_1, \ldots, n_d} x_1^{n_1} \cdots x_d^{n_d}, \qquad \text{diag } F(x) = \sum_{n \geq 0} a_{n, \ldots, n} x^n.$$

1. *If $d = 2$, then* diag $F(x)$ *is algebraic.*
2. *If $d > 2$, then* diag $F(x)$ *is D-finite but not necessarily algebraic.*

Now we are ready to prove our positive claims about the series at the beginning of this section. The first three expressions are visibly rational, and the diagonal binomial and Motzkin series are visibly algebraic. We proved that the $k$-Catalan series is algebraic in Section 1.3.2.2. The functions $e^x$, $\log(1+x)$, $\sin x$, $\cos x$, $\arctan x$ satisfy the differential equations $y' = y$, $(1+x)y' = 1$, $y'' = -y$, $y'' = -y$, $(1+x^2)y' = 1$, respectively. The series $\sum_{n\geq 0}\binom{2n}{n}^2 x^n$ is the Hadamard product of $(1-4x)^{-1/2}$ with itself, and hence D-finite. Finally, $\sum_{n\geq 0}\binom{3n}{n,n,n}x^n$ is the diagonal of the rational function $\frac{1}{1-x-y-z} = \sum_{a,b,c\geq 0}\binom{a+b+c}{a,b,c}x^a y^b z^c$, and hence D-finite.

Proving the negative claims requires more effort and, often, a bit of analytic machinery. We briefly outline some key results.

**Recognizing series that are not algebraic.** There are a few methods available to prove that a series is **not** algebraic. The simplest algebraic and analytic criteria are the following.

**Theorem 1.3.10 (Eisenstein's theorem [155])** *If a series $A(x) = \sum_{n\geq 0} a_n x^n$ with rational coefficients is algebraic, then there exists a positive integer $m$ such that $a_n m^n$ is an integer for all $n > 0$.*

This shows that $e^x, \log(1+x), \sin x, \cos x$, and $\arctan x$ are not algebraic.

**Theorem 1.3.11** [111] *If the coefficients of an algebraic power series $A(x) = \sum_{n\geq 0} a_n x^n$ satisfy $a_n \sim cn^r\alpha^n$ for nonzero $c, \alpha \in \mathbb{C}$ and $r < 0$, then $r$ cannot be a negative integer.*

Stirling's approximation $n! \sim \sqrt{2\pi n}\,(n/e)^n$ gives $\binom{2n}{n}^2 \sim c\cdot 16^n/n$ and $\binom{3n}{n,n,n} \sim c\cdot 27^n/n$, so the corresponding series are not algebraic.

Another useful analytic criterion is that an algebraic series $A(x)$ must have a Newton-Puiseux expansion at any of its singularities. See [79, Theorem VII.7] and [78] for details.

**Recognizing series that are not D-finite.** The most effective methods to show that a function is **not** D-finite are analytic.

**Theorem 1.3.12** [97, Theorem 9.1] *Suppose that $A(x)$ is analytic at $x = 0$, and it is D-finite, satisfying the equation $q_0(x)A(x) + q_1(x)A'(x) + \cdots + q_d(x)A^{(d)}(x) = q(x)$ with $q_d(x) \neq 0$. Then $A(x)$ can be extended to an analytic function in any simply connected region of the complex plane not containing the (finitely many) zeroes of $q_d(x)$.*

Since $\sec x$ and $\tan x$ have a pole at every odd multiple of $\pi$, they are not D-finite. Similarly, $\sum_n p(n)x^n = \prod_{k=1}^{\infty}\frac{1}{1-x^k}$ is not D-finite because it has the circle $|x| = 1$ as a natural boundary of analyticity.

There are other powerful analytic criteria to prove a series is not D-finite. See [78, Theorem VII.7] for details and further examples.

Sometimes it is possible to give *ad hoc* proofs that series are D-finite. For instance, consider $y = \sqrt{1+\log(1+x^2)}$. By induction, for any $k \in \mathbb{N}$ there exist polynomials $r_1(x), \ldots, r_k(x)$ such that $y^{(k)} = r_1/y + r_2/y^3 + \cdots + r_k/y^{2k-1}$. An equation of the form $\sum_{i=0}^{d} q_i(x) y^{(i)} = q(x)$ would then give rise to a polynomial equation satisfied by $y$. This would also make $y^2 - 1 = \log(1+x^2)$ algebraic; but this contradicts Theorem 1.3.10.

## 1.4 Linear algebra methods

There are several important theorems in enumerative combinatorics that express a combinatorial quantity in terms of a determinant. Of course, evaluating a determinant is not always straightforward, but there is a wide array of tools at our disposal.

The goal of Section 1.4.1 is to reduce many combinatorial problems to "just computing a determinant"; examples include walks in a graph, spanning trees, Eulerian cycles, matchings, and routings. In particular, we discuss the transfer matrix method, which allows us to encode many combinatorial objects as walks in graphs, so that these linear algebraic tools apply. These problems lead us to many beautiful, mysterious, and highly non-trivial determinantal evaluations. We will postpone the proofs of the evaluations until Section 1.4.2, which is an exposition of some of the main techniques in the subtle science of computing combinatorial determinants.

### 1.4.1 Determinants in combinatorics

#### 1.4.1.1 Preliminaries: Graph matrices

An **undirected graph**, or simply a **graph** $G = (V, E)$ consists of a set $V$ of vertices and a set $E$ of edges $\{u, v\}$ where $u, v \in V$ and $u \neq v$. In an undirected graph, we write $uv$ for the edge $\{u, v\}$. The **degree** of a vertex is the number of edges incident to it. A **walk** is a set of edges of the form $v_1v_2, v_2v_3, \ldots, v_{k-1}v_k$. This walk is **closed** if $v_k = v_1$.

A **directed graph** or **digraph** $G = (V, E)$ consists of a set $V$ of vertices and a set $E$ of oriented edges $(u, v)$ where $u, v \in V$ and $u \neq v$. In an undirected graph, we write $uv$ for the directed edge $(u, v)$. The **outdegree** (respectively, **indegree**) of a vertex is the number of edges coming out of it (respectively, coming into it). A **walk** is a set of directed edges of the form $v_1v_2, v_2v_3, \ldots, v_{k-1}v_k$. This walk is **closed** if $v_k = v_1$.

We will see in this section that many graph theory problems can be solved using tools from linear algebra. There are several matrices associated to graphs that play a crucial role; we review them here.

**Directed graphs.** Let $G = (V,E)$ be a directed graph.

- The **adjacency matrix** $A = A(G)$ is the $V \times V$ matrix whose entries are
$$a_{uv} = \text{number of edges from } u \text{ to } v.$$
- The **incidence matrix** $M = M(G)$ is the $V \times E$ matrix with
$$m_{ve} = \begin{cases} 1 & \text{if } v \text{ is the final vertex of edge } e, \\ -1 & \text{if } v \text{ is the initial vertex of edge } e, \\ 0 & \text{otherwise.} \end{cases}$$
- The **directed Laplacian matrix** $\overrightarrow{L} = \overrightarrow{L}(G)$ is the $V \times V$ matrix whose entries are
$$\overrightarrow{l}_{uv} = \begin{cases} -(\text{number of edges from } u \text{ to } v) & \text{if } u \neq v, \\ \text{outdeg}(u) & \text{if } u = v. \end{cases}$$

**Undirected graphs.** Let $G = (V,E)$ be an undirected graph.

- The (undirected) **adjacency matrix** $A = A(G)$ is the $V \times V$ matrix whose entries are
$$a_{uv} = \text{number of edges connecting } u \text{ and } v.$$
This is the directed adjacency matrix of the directed graph on $V$ containing edges $u \to v$ and $v \to u$ for every edge $uv$ of $G$.
- The (undirected) **Laplacian matrix** $L = L(G)$ is the $V \times V$ matrix with entries
$$l_{uv} = \begin{cases} -(\text{number of edges connecting } u \text{ and } v) & \text{if } u \neq v \\ \deg u & \text{if } u = v \end{cases}$$
If $M$ is the incidence matrix of any orientation of the edges of $G$, then $L = MM^T$.

#### 1.4.1.2 Walks: the transfer matrix method

Counting walks in a graph is a fundamental problem, which (often in disguise) includes many important enumerative problems. The transfer matrix method addresses this problem by expressing the number of walks in a graph $G$ in terms of its adjacency matrix $A(G)$, and then uses linear algebra to count those walks.

**Directed or undirected graphs.** The transfer matrix method is based on the following simple, powerful observation, which applies to directed and undirected graphs:

**Theorem 1.4.1** *Let $G = (V,E)$ be a graph and let $A = A(G)$ be the $V \times V$ adjacency matrix of $G$, where $a_{uv}$ is the number of edges from $u$ to $v$. Then*

$$(A^n)_{uv} = \text{number of walks of length } n \text{ in } G \text{ from } u \text{ to } v.$$

*Proof.* Observe that

$$(A^n)_{uv} = \sum_{w_1,\ldots,w_{n-1}\in V} a_{uw_1}a_{w_1w_2}\cdots a_{w_{n-1}v}$$

and there are $a_{uw_1}a_{w_1w_2}\cdots a_{w_{n-1}v}$ walks of length $n$ from $u$ to $v$ visiting vertices $u, w_1, \ldots, w_{n-1}, v$ in that order. ■

**Corollary 1.4.2** *The generating function $\sum_{n\geq 0}(A^n)_{uv}x^n$ for walks of length $n$ from $u$ to $v$ in $G$ is a rational function.*

*Proof.* Using Cramer's formula, we have

$$\sum_{n\geq 0}(A^n)_{uv}x^n = ((I-xA)^{-1})_{uv} = (-1)^{u+v}\frac{\det(I-xA : v,u)}{\det(I-xA)}$$

where $(M : v,u)$ is the cofactor of $M$ obtained by removing row $v$ and column $u$. ■

**Corollary 1.4.3** *If $C_G(n)$ is the number of closed walks of length $n$ in $G$, then*

$$C_G(n) = \lambda_1^n + \cdots + \lambda_k^n, \qquad \sum_{n\geq 1} C_G(n)x^n = \frac{-xQ'(x)}{Q(x)}$$

*where $\lambda_1, \ldots, \lambda_k$ are the eigenvalues of adjacency matrix $A$ and $Q(x) = \det(I-xA)$.*

*Proof.* Theorem 1.4.1 implies that $C_G(n) = \operatorname{tr}(A^n) = \lambda_1^n + \cdots + \lambda_k^n$. The second equation then follows from $Q(x) = (1-\lambda_1 x)\cdots(1-\lambda_k x)$. ■

In view of Theorem 1.4.1, we want to be able to compute powers of the adjacency matrix $A$. As we learn in linear algebra, this is very easy to do if we are able to diagonalize $A$. This is not always possible, but we can do it when $A$ is undirected.

**Undirected graphs.** When our graph $G$ is undirected, the adjacency matrix $A(G)$ is symmetric, and hence diagonalizable.

**Theorem 1.4.4** *Let $G = (V,E)$ be an* ***undirected*** *graph and let $\lambda_1, \ldots, \lambda_k$ be the eigenvalues of the adjacency matrix $A = A(G)$. Then for any vertices $u$ and $v$ there exist constants $c_1, \ldots, c_k$ such that*

$$\text{number of walks of length } n \text{ from } u \text{ to } v = c_1\lambda_1^n + \cdots + c_k\lambda_k^n.$$

*Proof.* The key fact is that a real symmetric $k \times k$ matrix $A$ has $k$ real orthonormal eigenvectors $q_1, \dots, q_k$ with real eigenvalues $\lambda_1, \dots, \lambda_k$. Equivalently, the $k \times k$ matrix $Q$ with columns $q_1, \dots, q_k$ is orthogonal (so $Q^T = Q^{-1}$) and diagonalizes $A$:

$$Q^{-1}AQ = D = \operatorname{diag}(\lambda_1, \dots, \lambda_k)$$

where $D = \operatorname{diag}(\lambda_1, \dots, \lambda_k)$ is the diagonal matrix with diagonal entries $\lambda_1, \dots, \lambda_k$. The result then follows from $A^n = QD^nQ^{-1} = Q\operatorname{diag}(\lambda_1^n, \dots, \lambda_k^n)\,Q^T$, with $c_t = q_{it}q_{jt}$. ■

**Applications.** Many families of combinatorial objects can be enumerated by first recasting the objects as walks in a "transfer graph" and then applying the transfer matrix method. We illustrate this technique with a few examples.

1. (Colored necklaces) Let $f(n,k)$ be the number of ways of coloring the beads of a necklace of length $n$ with $k$ colors so that no two adjacent beads have the same color. (Different rotations and reflections of a coloring are considered different.) There are several ways to compute this number, but a very efficient one is to notice that such a coloring is a graph walk in disguise. If we label the beads $1, \dots, n$ in clockwise order and let $a_i$ be the color of the $i$th bead, then the coloring corresponds to the closed walk $a_1, a_2, \dots, a_n, a_1$ in the complete graph $K_n$. The adjacency graph of $K_n$ is $A = J - I$ where $J$ is the matrix all of whose entries equal 1, and $I$ is the identity. Since $J$ has rank 1, it has $n-1$ eigenvalues equal to 0. Since the trace is $n$, the last eigenvalue is $n$. It follows that the eigenvalues of $A = J - I$ are $-1, -1, \dots, -1, n-1$. Then Corollary 1.4.3 tells us that

$$f(n,k) = (n-1)^k + (n-1)(-1)^k.$$

It is possible to give a bijective proof of this formula, but this algebraic proof is much simpler.

2. (Words with forbidden subwords, 1) Let $h_n$ be the number of words of length $n$ in the alphabet $\{a, b\}$ that do not contain $aa$ as a consecutive subword. This is the same as a walk of length $n-1$ in the transfer graph with vertices $a$ and $b$ and edges $a \to b$, $b \to a$ and $b \to b$. The absence of the edge $a \to a$ guarantees that these walks produce only the valid words we wish to count. The adjacency matrix and its powers are

$$A = \begin{pmatrix} 0 & 1 \\ 1 & 1 \end{pmatrix}, \qquad A^n = \begin{pmatrix} F_{n-1} & F_n \\ F_n & F_{n+1} \end{pmatrix},$$

where the Fibonacci numbers $F_0, F_1, \dots$ are defined recursively by $F_0 = 0, F_1 = 1$, and $F_k = F_{k-1} + F_{k-2}$ for $k \geq 2$.

Since $h_n$ is the sum of the entries of $A^{n-1}$, we get that $h_n = F_{n+2}$, and $g_n \sim c \cdot \alpha^n$ where $\alpha = \frac{1}{2}(1+\sqrt{5}) \approx 1.6179\dots$ is the golden ratio. Of course there are easier proofs of this fact, but this approach works for any problem of enumerating words in a given alphabet with given forbidden consecutive subwords. Let us study a slightly more intricate example, which should make it clear how to proceed in general.

3. (Words with forbidden subwords, 2) Let $g_n$ be the number of cyclic words of length $n$ in the alphabet $\{a,b\}$ that do not contain *aa* or *abba* as a consecutive subword. We wish to model these words as walks in a directed graph. At first this may seem impossible because, as we construct the word sequentially, the validity of a new letter depends on more than just the previous letter. However, a simple trick resolves this difficulty: We can introduce more memory into the vertices of the transfer graph. In this case, since the validity of a new letter depends on the previous three letters, we let the vertices of the transfer graph be $aba, abb, bab, bba, bbb$ (the allowable "windows" of length 3) and put an edge $wxy \to xyz$ in the graph if the window $wxy$ is allowed to precede the window $xyz$; that is, if $wxyz$ is an allowed subword. The result is the graph of Figure 1.14, whose adjacency matrix $A$ satisfies

$$\det(I-xA) = \det\begin{pmatrix} 1 & 0 & -x & 0 & 0 \\ 0 & 1 & 0 & 0 & -x \\ -x & -x & 1 & 0 & 0 \\ 0 & 0 & -x & 1 & 0 \\ 0 & 0 & 0 & -x & 1-x \end{pmatrix} = -x^4+x^3-x^2-x+1.$$

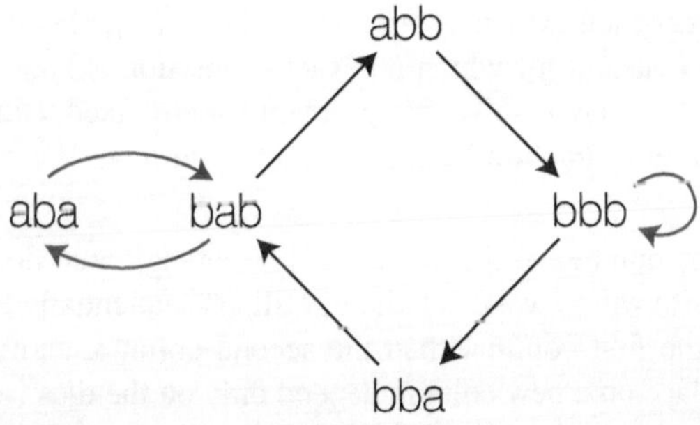


**Figure 1.14**
The transfer graph for words on the alphabet $\{a,b\}$ avoiding *aa* and *abba* as consecutive subwords.

The valid cyclic words of length $n$ correspond to the closed walks of length $n$ in the transfer graph, so Corollary 1.4.3 tells us that the generating function for $g_n$ is

$$\sum_{n\geq 0} g_n x^n = \frac{x+2x^2-3x^3+4x^4}{1-x-x^2+x^3-x^4}$$
$$= x+3x^2+x^3+7x^4+6x^5+15x^6+15x^7+31x^8+37x^9+\cdots.$$

Theorem 1.3.5.2 then tells us that $g_n \approx c\cdot\alpha^n$ where $\alpha \approx 1.5129$ is the inverse of the smallest positive root of $1-x-x^2+x^3-x^4=0$. The values of $g_1, g_2, g_3$ may surprise us. Note that the generating function does something counterintuitive: it does not count the words *a* (because *aa* is forbidden), *aba* (because *aa* is forbidden), or *abb* (because *abba* is forbidden).

This example serves as a word of caution: When we use the transfer matrix method to enumerate "cyclic" objects using Corollary 1.4.3, the initial values of the generating function may not be the ones we expect. In a particular problem of interest, it will be straightforward to adjust those values accordingly.

To illustrate the wide applicability of this method, we conclude this section with a problem where the transfer graph is less apparent.

4. (Monomer-dimer problem) An important open problem in statistical mechanics is the **monomer-dimer problem** of computing the number of tilings $T(m,n)$ of an $m\times n$ rectangle into dominoes ($2\times 1$ rectangles) and unit squares. Equivalently, $T(m,n)$ is the number of partial matchings of an $m\times n$ grid, where each node is matched to at most one of its neighbors.

   There is experimental evidence, but no proof, that $T(n,n)\sim c\cdot\alpha^{n^2}$ where $\alpha\approx$ 1.9402... is a constant for which no exact expression is known. The transfer-matrix method is able to solve this problem for any fixed value of $m$, proving that the generating function $\sum_{n\geq 0} T(m,n)x^n$ is rational. We carry this out for $m=3$.

   Let $t(n)$ be the number of tilings of a $3\times n$ rectangle into dominoes and unit squares. As with words, we can build our tilings sequentially from left to right by covering the first column, then the second column, and so on. The tiles that we can place on a new column depend only on the tiles from the previous column that are sticking out, and this can be modeled by a transfer graph.

   More specifically, let $T$ be a tiling of a $3\times n$ rectangle. We define $n+1$ triples $v_0,\ldots,v_n$ which record how $T$ interacts with the $n+1$ vertical grid lines of the rectangle. The $i$th grid line consists of three unit segments, and each coordinate of $v_i$ is 0 or 1 depending on whether these three segments are edges of the tiling or not. For example, Figure 1.15 corresponds to the triples $111, 110, 011, 101, 010, 111$.

   The choice of $v_i$ is restricted only by $v_{i-1}$. The only restriction is that $v_{i-1}$ and $v_i$ cannot both have a 0 in the same position, because this would force us to put

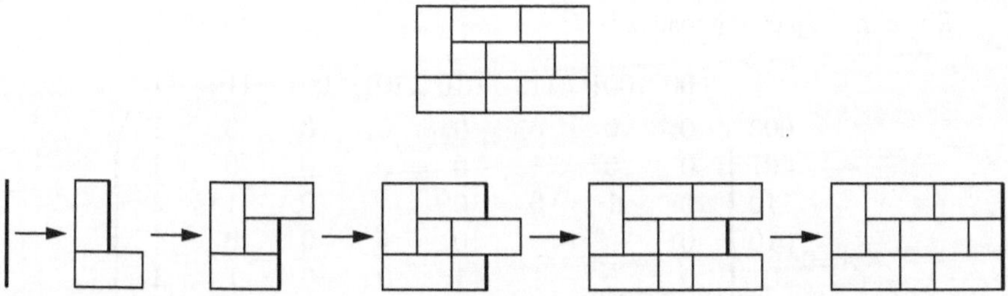

**Figure 1.15**
A tiling of a $3 \times 5$ rectangle into dominoes and unit squares.

two overlapping horizontal dominoes in $T$. These compatibility conditions are recorded in the transfer graph of Figure 1.16. When $v_{i-1} = v_i = 111$, there are three ways of covering column $i$. If $v_{i-1}$ and $v_i$ share two 1s in consecutive positions, there are two ways. In all other cases, there is a unique way. It follows that the tilings of a $3 \times n$ rectangle are in bijection with the walks of length $n$ from 111 to 111 in the transfer graph.

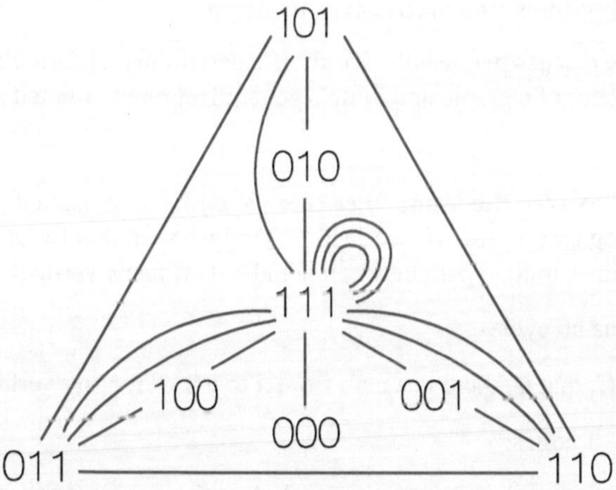


**Figure 1.16**
The transfer graph for tilings of $3 \times n$ rectangles into dominoes and unit squares.

Since the adjacency matrix is

$$A = \begin{array}{c} \\ 000 \\ 001 \\ 110 \\ 010 \\ 101 \\ 001 \\ 110 \\ 111 \end{array} \begin{array}{c} \begin{array}{cccccccc} 000 & 001 & 110 & 010 & 101 & 001 & 110 & 111 \end{array} \\ \begin{pmatrix} 0 & 0 & 0 & 0 & 0 & 0 & 0 & 1 \\ 0 & 0 & 1 & 0 & 0 & 0 & 0 & 1 \\ 0 & 1 & 0 & 0 & 1 & 0 & 1 & 2 \\ 0 & 0 & 0 & 0 & 1 & 0 & 0 & 1 \\ 0 & 0 & 1 & 1 & 0 & 0 & 1 & 1 \\ 0 & 0 & 0 & 0 & 0 & 0 & 1 & 1 \\ 0 & 0 & 1 & 0 & 1 & 1 & 0 & 2 \\ 1 & 1 & 2 & 1 & 1 & 1 & 2 & 3 \end{pmatrix} \end{array}$$

Theorem 1.4.1 tells us that

$$\begin{aligned} \sum_{n\geq 0} t(n)x^n &= \frac{\det(I - xA : 111, 111)}{\det(I - xA)} \\ &= \frac{(1+x-x^2)(1-2x-x^2)}{(1+x)(1-5x-9x^2+9x^3+x^4-x^5)} \\ &= 1+3x+22x^2+131x^3+823x^4+5096x^5+31687x^6+\cdots \end{aligned}$$

By Theorem 1.3.5.2, $t_n \sim c \cdot \alpha^n$ where $\alpha \approx 6.21207\ldots$ is the inverse of the smallest positive root of the denominator $(1+x)(1-5x-9x^2+9x^3+x^4-x^5)$.

#### 1.4.1.3 Spanning trees: the matrix-tree theorem

In this section we discuss two results: Kirkhoff's determinantal formula for the number of spanning trees of a graph, and Tutte's generalization to oriented spanning trees of directed graphs.

**Undirected matrix-tree theorem.** Let $G = (V, E)$ be a connected graph with no loops. A **spanning tree** $T$ of $G$ is a collection of edges such that for any two vertices $u$ and $v$, $T$ contains a unique path between $u$ and $v$. If $G$ has $n$ vertices, then

- $T$ contains no cycles,
- $T$ spans $G$; that is, there is a path from $u$ to $v$ in $T$ for any vertices $u \neq v$, and
- $T$ has $n-1$ edges.

Furthermore, any two of these properties imply that $T$ is a spanning tree. Our goal in this section is to compute the number $c(G)$ of spanning trees of $G$.

Orient the edges of $G$ arbitrarily. Recall that the incidence matrix $M$ of $G$ is the $V \times E$ matrix whose $e$th column is $\mathbf{e}_v - \mathbf{e}_u$ if $e = u \to v$, where $\mathbf{e}_i$ is the $i$th basis vector. The Laplacian $L = MM^T$ has entries

$$l_{uv} = \begin{cases} -(\text{number of edges connecting } u \text{ and } v) & \text{if } u \neq v \\ \deg u & \text{if } u = v \end{cases}$$

Note that $L(G)$ is singular because all its row sums are 0. A **principal cofactor** $L_v(G)$ is obtained from $L(G)$ by removing the $v$th row and $v$th column for some vertex $v$.

**Theorem 1.4.5 (Kirkhoff's Matrix-Tree Theorem)** *The number $c(G)$ of spanning trees of a connected graph $G$ is*

$$c(G) = \det L_v(G) = \frac{1}{n}\lambda_1 \cdots \lambda_{n-1},$$

*where $L_v(G)$ is any principal cofactor of the Laplacian $L(G)$, and $\lambda_1, \ldots, \lambda_{n-1}, \lambda_n = 0$ are the eigenvalues of $L(G)$.*

*Proof.* We use the Binet-Cauchy formula, which states that if $A$ and $B$ are $m \times n$ and $n \times m$ matrices, respectively, with $m < n$, then

$$\det AB = \sum_{S \subseteq [n]\,:\,|S|=m} \det A[S] \det B[S]$$

where $A[S]$ (respectively, $B[S]$) is the $n \times n$ matrix obtained by considering only the columns of $A$ (respectively, the rows of $B$) indexed by $S$.

We also use the following observation: If $M_v$ is the "reduced" adjacency matrix $M$ with the $v$th row removed, and $S$ is a set of $n-1$ edges of $E$, then

$$\det M_v[S] = \begin{cases} \pm 1 & \text{if } S \text{ is a spanning tree,} \\ 0 & \text{otherwise.} \end{cases}$$

This observation is easily proved: If $S$ is not a spanning tree, then it contains a cycle $C$, which gives a linear dependence among the columns indexed by the edges of $C$. Otherwise, if $S$ is a spanning tree, think of $v$ as its root, and "prune" it by repeatedly removing a leaf $v_i \neq v$ and its only incident edge $e_i$ for $1 \leq i \leq n-1$. Then if we list the rows and columns of $M[S]$ in the orders $v_1, \ldots, v_{n-1}$ and $e_1, \ldots, e_{n-1}$, respectively, the matrix will be lower triangular with 1s and $-1$s in the diagonal.

Combining these two equations, we obtain the first statement:

$$\det L_v(G) = \sum_{S \subseteq [n]\,:\,|S|=m} \det M[S] \det M^T[S] = \sum_{S \subseteq [n]\,:\,|S|=m} \det M[S]^2 = c(G).$$

To prove the second one, observe that the coefficient of $-x^1$ in the characteristic polynomial $\det(L - xI) = (\lambda_1 - x) \cdots (\lambda_{n-1} - x)(0 - x)$ is the sum of the $n$ principal cofactors, which are all equal to $c(G)$. ■

The matrix-tree theorem is a very powerful tool for computing the number of spanning trees of a graph. Let us state a few examples.

The **complete graph** $K_n$ has $n$ vertices and an edge joining each pair of vertices. The **complete bipartite graph** $K_{m,n}$ has $m$ "top" vertices and $n$ "bottom" vertices, and $mn$ edges joining each top vertex to each bottom vertex. The **hyperoctahedral graph** $\diamondsuit_n$ has vertices $\{1, 1', 2, 2', \ldots, n, n'\}$ and its only missing edges are $ii'$ for $1 \leq$

$i \leq n$. The **$n$-cube graph** $C_n$ has vertices $(\varepsilon_1, \ldots, \varepsilon_n)$ where $\varepsilon_i \in \{0,1\}$, and an edge connecting any two vertices that differ in exactly one coordinate. The **$n$-dimensional grid of size** $m$, denoted $mC_n$, has vertices $(\varepsilon_1, \ldots, \varepsilon_n)$ where $\varepsilon_i \in \{1, \ldots, m\}$, and an edge connecting any two vertices that differ in exactly one coordinate $i$, where they differ by 1.

**Theorem 1.4.6** *The number of spanning trees of some interesting graphs are as follows.*

1. *(Complete graph)* $c(K_n) = n^{n-2}$
2. *(Complete bipartite graph)* $c(K_{m,n}) = m^{n-1}n^{m-1}$
3. *(Hyperoctahedral graph)* $c(\Diamond_n) = 2^{2n-2}(n-1)^n n^{n-2}$
4. *(n-cube)* $c(C_n) = 2^{2^n-n-1}\prod_{k=1}^{n} k^{\binom{n}{k}}$
5. *(n-dimensional grid of size m)* $c(mC_n) = m^{m^n-n-1}\prod_{k=1}^{n} k^{\binom{n}{k}(m-1)^k}$

We will see proofs of the first and third example in Section 1.4.2. For the others, and many additional examples, see [55].

**Directed matrix-tree theorem.** Now let $G = (V,E)$ be a **directed** graph containing no loops. An **oriented spanning tree rooted at** $v$ is a collection of edges $T$ such that for any vertex $u$ there is a unique path from $u$ to $v$. The underlying unoriented graph $\underline{T}$ is a spanning tree of the unoriented graph $\underline{G}$. Let $c(G,v)$ be the number of spanning trees rooted at $G$.

Recall that the **directed Laplacian matrix** $\overrightarrow{L}$ has entries

$$\overrightarrow{l}_{uv} = \begin{cases} -(\text{number of edges from } u \text{ to } v) & \text{if } u \neq v \\ \text{outdeg } u & \text{if } u = v \end{cases}$$

Now the matrix $\overrightarrow{L}(G)$ is not necessarily symmetric, but it is still singular.

**Theorem 1.4.7 (Tutte's Directed Matrix-Tree Theorem)** *Let $G$ be a directed graph and $v$ be a vertex. The number $c(G,v)$ of oriented spanning trees rooted at $v$ is*

$$c(G,v) = \det \overrightarrow{L}_v(G)$$

*where $\overrightarrow{L}_v(G)$ is obtained from $L(G)$ by removing the $v$th row and column. Furthermore, if $G$ is* **balanced**, *so* indeg $v$ = outdeg $v$ *for all vertices $v$, then*

$$c(G,v) = \frac{1}{n}\lambda_1 \cdots \lambda_{n-1}$$

*where $\lambda_1, \ldots, \lambda_{n-1}, \lambda_n = 0$ are the eigenvalues of $L(G)$.*

*Proof.* Proceed by induction. Consider a vertex $w \neq v$ and an edge $e$ starting at $w$. Let $G' = G - e$ be obtained from $G$ by removing $e$. If $e$ is the only edge starting at $w$, then every spanning tree must use it, and we have

$$c(G,v) = c(G',v) = \det \overrightarrow{L}_v(G') = \det \overrightarrow{L}_v(G).$$

Otherwise, let $G''$ be obtained from $G$ by removing all edges starting at $w$ other than $e$. There are $c(G',v)$ oriented spanning trees rooted at $v$ that do not contain $e$, and $c(G'',v)$ that do contain $e$, so we have

$$c(G,v) = c(G',v) + c(G'',v) = \det \overrightarrow{L}_v(G') + \det \overrightarrow{L}_v(G'') = \det \overrightarrow{L}_v(G)$$

where the last equality holds since determinants are multilinear.

We postpone the proof of the second statement to the next section, where it will be an immediate consequence of Theorem 1.4.9. ■

#### 1.4.1.4 Eulerian cycles: the BEST theorem

One of the earliest combinatorial questions is the problem of the Seven Bridges of Königsberg. In the early 1700s, the Prussian city of Königsberg was separated by the Pregel river into four regions, connected to each other by seven bridges. In the map of Figure 1.17 we have labeled the regions $N, S, E$, and $I$; there are two bridges between $N$ and $I$, two between $S$ and $I$, and three bridges connecting $E$ to each of $N, I$, and $S$. The problem was to find a walk through the city that crossed each bridge exactly once. Euler proved in 1735 that it was impossible to find such a walk; this is considered to be the first paper in graph theory.

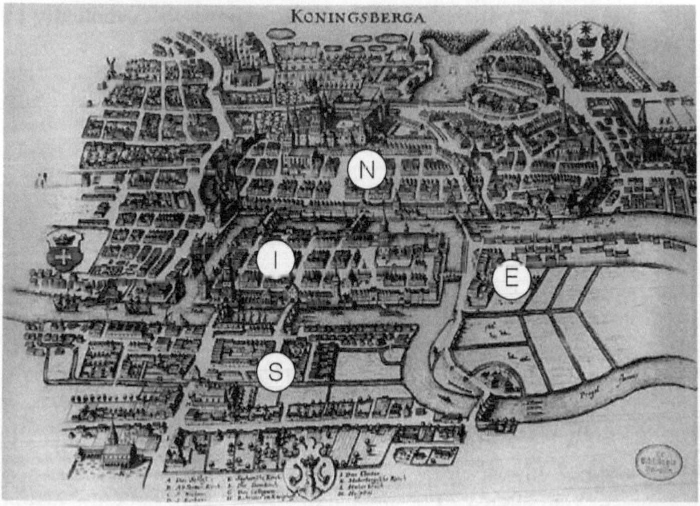


**Figure 1.17**
The seven bridges of Königsberg. Public domain map by Merian-Erben, 1652.

Euler's argument is simple, and relies on the fact that every region of Königsberg is adjacent to an odd number of bridges. Suppose there existed such a walk, starting at region $A$ and ending at region $B$. Now consider a region $C$ other than $A$ and $B$. Then our path would enter and leave $C$ the same number of times; but then it would not use all the bridges adjacent to $C$, because there is an odd number of such bridges.

In modern terminology, each region of the city is represented by a vertex, and each bridge is represented by an edge connecting two vertices. We will be more interested in the directed case, where every edge has an assigned direction. An **Eulerian path** is a path in the graph that visits every edge exactly once. If the path starts and ends at the same vertex, then it is called an **Eulerian cycle**. We say $G$ is an **Eulerian graph** if it has an Eulerian cycle.

**Theorem 1.4.8** *A directed graph is Eulerian if and only if it is connected and every vertex $v$ satisfies $indeg(v) = outdeg(v)$.*

*Proof.* If a graph has an Eulerian cycle $C$, then $C$ enters and leaves each vertex $v$ the same number of times. Therefore $\text{indeg}(v) = \text{outdeg}(v)$.

To prove the converse, let us start by arbitrarily "walking around $G$ until we get stuck." More specifically, we start at any vertex $v_0$, and at each step, we exit the current vertex by walking along any outgoing edge we have not used yet. If there is no available outgoing edge, we stop.

Whenever we enter a vertex $v \neq v_0$, we will also be able to exit it since $\text{indeg}(v) = \text{outdeg}(v)$; so the walk can only get stuck at $v_0$. Hence the resulting walk $C$ is a cycle. If $C$ uses all edges of the graph, we are done. If not, then since $G$ is connected we can find a vertex $v'$ of $C$ with an unused outgoing edge, and we use this edge to start walking around the graph $G - C$ until we get stuck, necessarily at $v'$. The result will be a cycle $C'$. Starting at $v'$ we can traverse $C$ and then $C'$, thus obtaining a cycle $C \cup C'$ that is longer than $C$. Repeating this procedure, we will eventually construct an Eulerian cycle. ∎

There is a remarkable formula for the number of Eulerian cycles, due to de Bruijn, van Ardenne-Ehrenfest, Smith, and Tutte.

**Theorem 1.4.9 (BEST Theorem)** *If $G$ is an Eulerian directed graph, then the number of Eulerian cycles of $G$ is*

$$c(G,v) \cdot \prod_{w \in V} (outdeg(w) - 1)!$$

*for any vertex $v$, where $c(G,v)$ is the number of oriented spanning trees rooted at $v$.*

*Proof.* We fix an edge $e$ starting at $v$, and let each Eulerian cycle start at $e$. For each vertex $w$ let $E_w$ be the set of outgoing edges from $w$.

Consider an Eulerian cycle $C$. For each vertex $w \neq v$, let $e_w$ be the last outgoing edge from $w$ that $C$ visits, and let $\pi_w$ (respectively, $\pi_v$) be the ordered set $E_w - e_w$ (respectively, $E_v - e$) of the other outgoing edges from $w$ (respectively, $v$), listed in the order that $C$ traverses them. It is easy to see that $T = \{e_w : w \neq v\}$ is an oriented spanning tree rooted at $v$.

Conversely, an oriented tree $T$ and permutations $\{\pi_w : w \in V\}$ serve as directions to tour $G$. We start with edge $e$. Each time we arrive at vertex $w$, we exit it by using the first unused edge according to $\pi_w$. If we have used all the edges $E_w - e_w$ of $\pi_w$, then we use $e_w \in T$. It is not hard to check that this is a bijection. This completes the proof. ■

**Corollary 1.4.10** *In an Eulerian directed graph, the number of oriented spanning trees rooted at v is the same for all vertices v; it equals*

$$c(G,v) = \frac{1}{n}\lambda_1 \cdots \lambda_{n-1}$$

*where* $\lambda_1, \ldots, \lambda_{n-1}, \lambda_n = 0$ *are the eigenvalues of* $\overrightarrow{L}(G)$.

*Proof.* The BEST theorem implies that $c(G,v)$ is independent of $v$, and then the argument in the proof of Theorem 1.4.5 applies to give the desired formula. ■

The BEST theorem can be used beautifully to enumerate a very classical, and highly nontrivial, family of objects. A $k$-ary **de Bruijn sequence** of order $n$ is a cyclic word $W$ of length $k^n$ in the alphabet $\{1,\ldots,k\}$ such that the $k^n$ consecutive subwords of $W$ of length $n$ are the $k^n$ distinct words of length $n$. For example, the 2-ary deBruijn sequences of order 3 are 11121222 and 22212111; these "memory wheels" were described in Sanskrit poetry several centuries ago [112]. Their existence and enumeration was proved by Flye Saint-Marie in 1894 for $k = 2$ and by van Aardenne-Ehrenfest and de Bruijn in 1951 in general.

**Theorem 1.4.11** [80, 58] *The number of k-ary de Bruijn sequences of order n is* $(k!)^{k^{n-1}}/k^n$.

*Proof.* Consider the **de Bruijn graph** whose vertices are the $k^{n-1}$ sequences of length $n-1$ in the alphabet $\{1,\ldots,k\}$, and where there is an edge from $a_1a_2\ldots a_{n-1}$ to the word $a_2a_3\ldots a_n$ for all $a_1,\ldots,a_n$. It is natural to label this edge $a_1a_2\ldots a_n$. It then becomes apparent that $k$-ary de Bruijn sequences are in bijection with the Eulerian cycles of the de Bruijn graph. Since $\text{indeg}(v) = \text{outdeg}(v) = k$ for all vertices $v$, this graph is indeed Eulerian, and we proceed to count its Eulerian cycles. Notice that for any vertices $u$ and $v$ there is a unique path of length $n$ from $u$ to $v$. Therefore the $k^{n-1} \times k^{n-1}$ adjacency matrix $A$ satisfies $A^n = J$, where $J$ is the matrix whose entries are all equal to 1. We already saw that the eigenvalues of $J$ are $0,\ldots,0,k^n$. Since the trace of $A$ is $k$, the eigenvalues of $A$ must be $0,\ldots,0,k$. Therefore the Laplacian $L = kI - A$ has eigenvalues $k,\ldots,k,0$. It follows from Corollary 1.4.10 that the de Bruijn graph has $c(G,v) = \frac{1}{k^{n-1}}k^{(k^{n-1}-1)} = k^{k^{n-1}-n}$ oriented spanning trees rooted at any vertex $v$, and

$$c(G,v) \cdot \prod_{w \in V} (\text{outdeg}(w) - 1)! = k^{k^{n-1}-n} \cdot (k-1)!^{k^{n-1}}$$

Eulerian cycles, as desired. ■

#### 1.4.1.5 Perfect matchings: the Pfaffian method

A **perfect matching** of a graph $G = (V,E)$ is a set $M$ of edges such that every vertex of $G$ is on exactly one edge from $M$. We are interested in computing the number $m(G)$ of perfect matchings of a graph $G$. We cannot expect to be able to do this in general; in fact, even for bipartite graphs $G$, the problem of computing $m(G)$ is #P-complete. However, for many graphs of interest, including all planar graphs, there is a beautiful technique that produces a determinantal formula for $m(G)$.

**Determinants and Pfaffians.** Let $A$ be a **skew-symmetric** matrix of size $2m \times 2m$, so $A^T = -A$. The Pfaffian is a polynomial encoding the matchings of the complete graph $K_{2m}$. A perfect matching $M$ of the complete graph $K_{2m}$ is a partition $M$ of $[2m]$ into disjoint pairs $\{i_1, j_1\}, \ldots, \{i_m, j_m\}$, where $i_k < j_k$ for $1 \leq k \leq m$. Draw the points $1, \ldots, 2m$ in order on a line and connect each $i_k$ to $j_k$ by a semicircle above the line. Let $\mathrm{cr}(M)$ be the number of crossings in this drawing, and let $\mathrm{sign}(M) = (-1)^{\mathrm{cr}(M)}$. Let $a_M = a_{i_1 j_1} \cdots a_{i_m j_m}$. The **Pfaffian** of $A$ is

$$\mathrm{Pf}(A) = \sum_M \mathrm{sign}(M) a_M$$

summing over all perfect matchings $M$ of the complete graph $K_m$.

**Theorem 1.4.12** *If $A$ is a skew-symmetric matrix, so $A^T = -A$, then*

$$\det(A) = \mathrm{Pf}(A)^2.$$

*Sketch of Proof.* The first step is to show that the skew symmetry of $A$ causes many cancellations in the determinant, and

$$\det A = \sum_{\pi \in ECS_n} \mathrm{sign}(\pi) a_\pi$$

where $ECS_n \subset S_n$ is the set of permutations of $[n]$ having only cycles of even length. Then, to prove that this equals $(\sum_M \mathrm{sign}(M) a_M)^2$, we need a bijection between ordered pairs $(M_1, M_2)$ of matchings and permutations $\pi$ in $ECS_n$ such that $a_{M_1} a_{M_2} = a_\pi$ and $\mathrm{sign}(M_1)\mathrm{sign}(M_2) = \mathrm{sign}(\pi)$. We now describe such a bijection.

Draw the matchings $M_1$ and $M_2$ above and below the points $1, \ldots, n$ on a line, respectively. Let $\pi$ be the permutation given by the cycles of the resulting graph, where each cycle is oriented following the direction of $M_1$ at its smallest element. This is illustrated in Figure 1.18. It is clear that $a_{M_1} a_{M_2} = a_\pi$, while some care is required to show that $\mathrm{sign}(M_1)\mathrm{sign}(M_2) = \mathrm{sign}(\pi)$. For details, see [2]. ■

**Counting perfect matchings via Pfaffians.** Suppose we wish to compute the number $m(G)$ of perfect matchings of a graph $G = (V,E)$ with no loops. After choosing an orientation of the edges, we define the $V \times V$ **signed adjacency matrix** $S(G)$ whose entries are

$$s_{ij} = \begin{cases} 1 & \text{if } i \to j \text{ is an edge of } G \\ -1 & \text{if } j \to i \text{ is an edge of } G \\ 0 & \text{otherwise.} \end{cases}$$

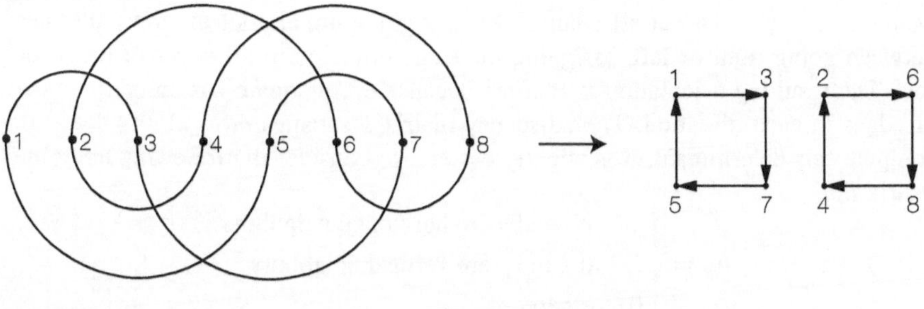


**Figure 1.18**
The pair of matchings $\{1,3\},\{2,6\},\{4,8\},\{5,7\}$ and $\{1,5\},\{2,4\},\{3,7\},\{6,8\}$ gives the permutation $(1375)(2684)$ with $(-a_{13}a_{26}a_{48}a_{57})(-a_{15}a_{24}a_{37}a_{68}) = a_{13}a_{37}a_{75}a_{51}a_{26}a_{68}a_{84}a_{42}$.

Then, for $\{i_1 j_1 \ldots i_m j_m\} = \{1,\ldots,2m\}$, $s_M = s_{i_1 j_1} \cdots s_{i_m j_m}$ is nonzero if and only if $\{i_1, j_1\},\ldots,\{i_m, j_m\}$ is a perfect matching of $G$.

We say that our edge orientation is **Pfaffian** if all the perfect matchings of $G$ have the same sign. At the moment there is no efficient test to determine whether a graph admits a Pfaffian orientation. There is a simple combinatorial restatement: An orientation is Pfaffian if and only if every even cycle $C$ for which $G\backslash V(C)$ has a perfect matching has an odd number of edges in each direction.

Fortunately, we have the following result of Kasteleyn [115, 131]:

Every planar graph has a Pfaffian orientation.

This is very desirable, because Theorem 1.4.12 implies the following:

For a Pfaffian orientation of $G$, $m(G) = \sqrt{\det S(G)}$.

Therefore the number of matchings of a planar graph is reduced to the evaluation of a combinatorial determinant. We will see in Section 1.4.2 that there are many techniques at our disposal to carry out this evaluation.

Let us illustrate this method with an important example, due to Kasteleyn [115] and Temperley–Fisher [198].

**Theorem 1.4.13** *The number $m(R_{a,b})$ of matchings of the $a \times b$ rectangular grid $R_{a,b}$ (where we assume $b$ is even) is*

$$m(R_{a,b}) = 4^{\lfloor a/2 \rfloor (b/2)} \prod_{j=1}^{\lfloor a/2 \rfloor} \prod_{k=1}^{b/2} \left( \cos^2 \frac{\pi j}{a+1} + \cos^2 \frac{\pi k}{b+1} \right) \sim c \cdot e^{\frac{G}{\pi} ab} \sim c \cdot 1.3385^{ab},$$

*where $G = 1 - \frac{1}{9} + \frac{1}{25} - \frac{1}{49} + \cdots$ is Catalan's constant.*

Clearly this is also the number of domino tilings of an $a \times b$ rectangle.

*Sketch of Proof.* Orient all columns of $R_{a,b}$ going up, and let the rows alternate between going right or left, assigning the same direction to all edges of the same row. The resulting orientation is Pfaffian because every square has an odd number of edges in each direction. The adjacency matrix $S$ satisfies $m(R_{a,b}) = \sqrt{\det S}$. To compute this determinant, it is slightly easier * to consider the following $mn \times mn$ matrix $B$:

$$b_{ij} = \begin{cases} 1 & \text{if } i \text{ and } j \text{ are horizontal neighbors} \\ i & \text{if } i \text{ and } j \text{ are vertical neighbors} \\ 0 & \text{otherwise.} \end{cases}$$

We can obtain $B$ from the $S$ by scaling the rows and columns by suitable powers of $i$, so we still have $m(R_{a,b}) = \sqrt{|\det B|}$. We will prove the product rule for this determinant in Section 1.4.2.

We then use this product formula to give an asymptotic formula for $m(R_{a,b})$. Note that $\log m(R_{a,b})/ab$ may be regarded as a Riemann sum; as $m,n \to \infty$ it converges to

$$c = \frac{1}{\pi^2}\int_0^{\pi/2}\int_0^{\pi/2} \log(4\cos^2 x + 4\cos^2 y)\,dx\,dy = \frac{G}{\pi}$$

where $G$ is Catalan's constant. Therefore

$$m(R_{a,b}) \approx e^{\frac{G}{\pi}ab} \approx 1.3385^{ab}.$$

Loosely speaking, this means that in a matching of the rectangular grid there are about 1.3385 degrees of freedom per vertex. ■

Obviously, this beautiful formula is not an efficient method of computing the exact value of $m(R_{a,b})$ for particular values of $a$ and $b$; it is not even clear why it gives an integer! There are alternative determinantal formulas for this quantity that are more tractable; see for example [2, Section 10.1].

#### 1.4.1.6 Routings: the Lindström–Gessel–Viennot lemma

Let $G$ be a directed graph with no directed cycles, which has a weight $\mathrm{wt}(e)$ on each edge $e$. We are most often interested in the unweighted case, where all weights are 1. Let $S = \{s_1, \ldots, s_n\}$ and $T = \{t_1, \ldots, t_n\}$ be two (not necessarily disjoint) sets of vertices, which we call sources and sinks, respectively. A **routing** from $S$ to $T$ is a set of paths $P_1, \ldots, P_n$ from the $n$ sources $s_1, \ldots, s_n$ to the $n$ sinks $t_1, \ldots, t_n$ such that no two paths share a vertex. Let $\pi$ be the permutation of $[n]$ such that $P_i$ starts at source $s_i$ and ends at sink $t_{\pi(i)}$, and define $\mathrm{sign}(R) = \mathrm{sign}(\pi)$.

Let the weight of a path or a routing be the product of the weights of the edges it contains. Consider the $n \times n$ **path matrix** $Q$ whose $(i, j)$ entry is

$$q_{ij} = \sum_{P \text{ path from } s_i \text{ to } t_j} \mathrm{wt}(P).$$

*In fact, this is the matrix that Kasteleyn uses in his computation.

**Theorem 1.4.14 (Lindström–Gessel–Viennot lemma)** *Let $G$ be a directed acyclic graph with edge weights, and let $S = \{s_1, \ldots, s_n\}$ and $T = \{t_1, \ldots, t_n\}$ be sets of vertices in $G$. Then the determinant of the $n \times n$ path matrix $Q$ is*

$$\det Q = \sum_{R \text{ routing from } S \text{ to } T} \operatorname{sign}(R)\operatorname{wt}(R).$$

*In particular, if all edge weights are 1 and if every routing takes $s_i$ to $t_i$ for all $i$, then*

$$\det Q = \textit{number of routings from } S \textit{ to } T.$$

*Proof.* We have $\det A = \sum_P \operatorname{sign}(P)\operatorname{wt}(P)$ summing over **all** path systems $P = \{P_1, \ldots, P_n\}$ from $S$ to $T$; we need to cancel out the path systems that are not routings. For each such $P$, consider the lexicographically first pair of paths $P_i$ and $P_j$ that intersect, and let $v$ be their first vertex of intersection. Now exchange the subpath of $P_i$ from $s_i$ to $v$ and the subpath of $P_j$ from $s_j$ to $v$, to obtain new paths $P_i'$ and $P_j'$. Replacing $\{P_i, P_j\}$ with $\{P_i', P_j'\}$, we obtain a new path system $\varphi(P)$ from $S$ to $T$. Notice that $\varphi(\varphi(P)) = P$, and $\operatorname{sign}(\varphi(P))\operatorname{wt}(\varphi(P)) + \operatorname{sign}(P)\operatorname{wt}(P) = 0$; so for all non-routings $P$, the path systems $P$ and $\varphi(P)$ cancel each other out. ■

This theorem was also anticipated by Karlin and McGregor [114] in the context of birth-and-death Markov processes.

**Determinants via routings.** The Lindström-Gessel-Viennot lemma is also a useful combinatorial tool for computing determinants of interest, usually by enumerating routings in a lattice. We illustrate this with several examples.

1. (Binomial determinants) Consider the **binomial determinant**

$$\binom{a_1, \ldots, a_n}{b_1, \ldots, b_n} = \det\left[\binom{a_i}{b_j}\right]_{1 \leq i,j \leq n}$$

where $0 \leq a_1 < \cdots < a_n$ and $0 \leq b_1 < \cdots < b_n$ are integers. These determinants arise as coefficients of the Chern class of the tensor product of two vector bundles. [129] This algebro-geometric interpretation implies these numbers are positive integers; as combinatorialists, we would like to know what they count.

A **SE path** is a lattice path in the square lattice $\mathbb{N}^2$ consisting of unit steps south and east. Consider the sets of points $A = \{A_1, \ldots, A_n\}$ and $B = \{B_1, \ldots, B_n\}$ where $A_i = (0, a_i)$ and $B_i = (b_i, b_i)$ for $1 \leq i \leq n$. Since there are $\binom{a_i}{b_j}$ SE paths from $A_i$ to $B_j$, and since every SE routing from $A$ to $B$ takes $A_i$ to $B_i$ for all $i$, we have

$$\binom{a_1, \ldots, a_n}{b_1, \ldots, b_n} = \text{number of SE routings from } A \text{ to } B.$$

This is the setting in which Gessel and Viennot discovered Theorem 1.4.14; they also evaluated these determinants in several special cases. [88] We now discuss one particularly interesting special case.

2. (Counting permutations by descent set) The **descent set** of a permutation $\pi$ is the set of indices $i$ such that $\pi_i > \pi_{i+1}$. We now prove that

$$\binom{c_1,\ldots,c_k,n}{0,c_1,\ldots,c_k} = \text{number of permutations of } [n] \text{ with descent set } \{c_1,\ldots,c_k\},$$

for any $0 < c_1 < \cdots < c_k < n$. It is useful to define $c_0 = 0, c_{k+1} = n$.

Encode such a permutation $\pi$ by a routing as follows. For each $i$ let $f_i$ be the number of indices $j \leq i$ such that $\pi_j \leq \pi_i$. Note that the descents $c_1,\ldots,c_k$ of $\pi$ are the positions where $f$ does not increase. Splitting $f$ at these positions, we are left with $k+1$ increasing subwords $f^1,\ldots,f^{k+1}$. Now, for $1 \leq i \leq n+1$ let $P_i$ be the NW path from $(c_{i-1},c_{i-1})$ to $(0,c_i)$ taking steps north precisely at the steps listed in $f^i$. These paths give one of the routings enumerated by the binomial determinant in question, and this is a bijection. See Figure 1.19 for an illustration. [88]

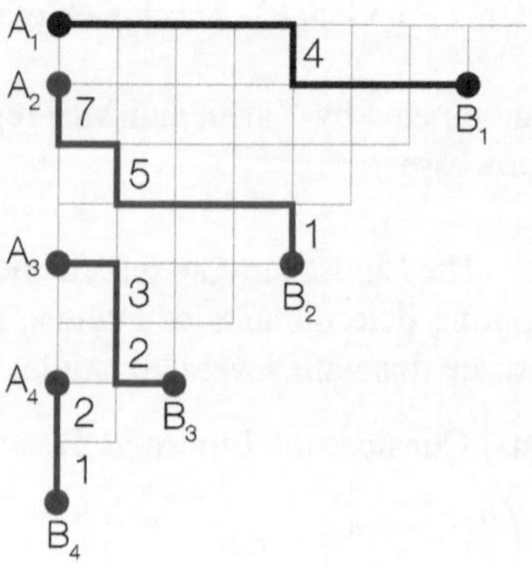


**Figure 1.19**
The routing corresponding to $\pi = 28351674$ and $f(\pi) = 12.23.157.4$.

3. (Rhombus tilings and plane partitions) Let $R_n$ be the number of tilings of a regular hexagon of side length $n$ using unit rhombi with angles 60° and 120°. Their enumeration is due to MacMahon [133]. There are several equivalent combinatorial models for this problem, illustrated in Figure 1.20, which we now discuss.

Firstly, it is almost inevitable to view these tilings as three-dimensional pictures. This shows that $R_n$ is also the number of ways of stacking unit cubes into the corner of a cubical box of side length $n$. Incidentally, this three-dimensional view makes it apparent that there are exactly $n^2$ rhombi of each one of the three possible orientations.

Secondly, we may consider the triangular grid inside our hexagon, and place a dot on the center of each triangle. These dots form a hexagonal grid, where

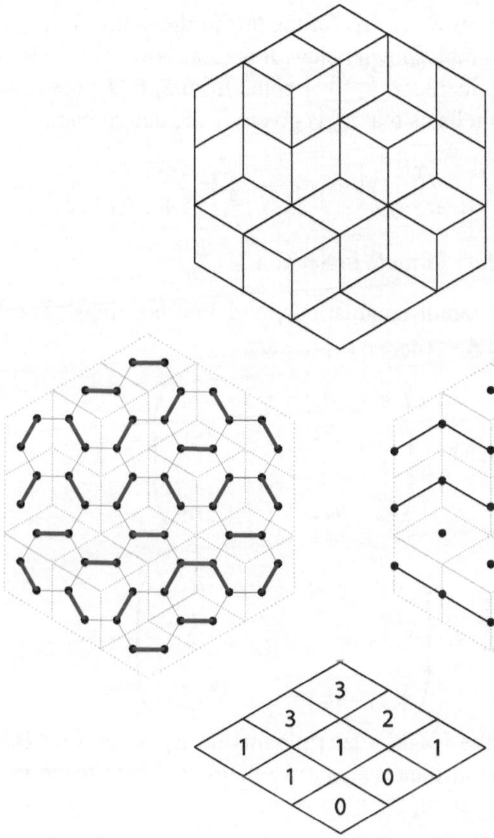


**Figure 1.20**
Four models for the rhombus tilings of a hexagon.

two dots are neighbors if they are at distance 1 from each other. Finally, join two neighboring dots when the corresponding triangles are covered by a tile. The result is a perfect matching of the hexagonal grid.

Next, on each one of the $n^2$ squares of the floor of the box, write down the number of cubes above it. The result is a **plane partition**: an array of non-negative integers (finitely many of which are non-zero) that is weakly decreasing in each row and column. We conclude that $R_n$ is also the number of plane partitions whose non-zero entries are at most $n$, and fit inside an $n \times n$ square.

Finally, given such a rhombus tiling, construct $n$ paths as follows. Each path starts at the center of one of the vertical edges on the western border of the hexagon, and successively crosses each tile splitting it into equal halves. It eventually comes out at the southeast side of the diamond, at the same height where it started (as is apparent from the 3-D picture). The final result is a routing from the $n$ sources $S_1,\ldots,S_n$ on the left to the sinks $T_1,\ldots,T_n$ on the right in the "rhombus" graph shown below. It is clear how to recover the tiling from the routing. Since there are $\binom{2n}{n+i-j}$ paths from $S_i$ to $T_j$, the Lindström–Gessel–Viennot lemma tells us that $R_n$ is given by the determinant

$$R_n = \det\left[\binom{2n}{n+i-j}\right]_{1\leq i,j\leq n} = \prod_{i,j,k=1}^{n} \frac{i+j+k-1}{i+j+k-2}.$$

We will prove this product formula in Section 1.4.2.

4. (Catalan determinants, multitriangulations, and Pfaffian rings) The **Hankel matrices** of a sequence $A = (a_0, a_1, a_2, \ldots)$ are

$$H_n(A) = \begin{pmatrix} a_0 & a_1 & \cdots & a_n \\ a_1 & a_2 & \cdots & a_{n+1} \\ \vdots & \vdots & \ddots & \vdots \\ a_n & a_{n+1} & \cdots & a_{2n} \end{pmatrix},$$

and

$$H'_n(A) = \begin{pmatrix} a_1 & a_2 & \cdots & a_{n+1} \\ a_2 & a_3 & \cdots & a_{n+2} \\ \vdots & \vdots & \ddots & \vdots \\ a_{n+1} & a_{n+2} & \cdots & a_{2n+1} \end{pmatrix}.$$

Note that if we know the Hankel determinants $\det H_n(A)$ and $\det H'_n(A)$ and they are nonzero for all $n$, then we can use them as a recurrence relation to recover each $a_k$ from $a_0, \ldots, a_{k-1}$.

There is a natural interpretation of the Hankel matrices of the Catalan sequence $C = (C_0, C_1, C_2, \ldots)$. Consider the "diagonal" grid on the upper half plane with steps $(1,1)$ and $(1,-1)$. Let $A_i = (-2i, 0)$ and $B_i = (2i, 0)$. Then there are $C_{i+j}$ paths from $A_i$ to $B_j$, and there is clearly a unique routing from $(A_0, \ldots, A_n)$ to $(B_0, \ldots, B_n)$. See Figure 1.21 for an illustration. This proves that $\det H_n(C) = 1$, and an analogous argument proves that $\det H'_n(C) = 1$. Therefore

$$\det H_n(A) = \det H'_n(A) = 1 \text{ for all } n \geq 0 \iff A \text{ is the Catalan sequence.}$$

The Hankel determinants of the shifted Catalan sequences also arise naturally in several contexts; they are given by:

$$\det \begin{pmatrix} C_{n-2k} & C_{n-2k+1} & \cdots & C_{n-k-1} \\ C_{n-2k+1} & C_{n-2k+2} & \cdots & C_{n-k} \\ \vdots & \vdots & \ddots & \vdots \\ C_{n-k-1} & C_{n-k} & \cdots & C_{n-2} \end{pmatrix} = \prod_{i+j\leq n-2k-1} \frac{i+j+2(k-1)}{i+j}. \tag{1.9}$$

There are several ways of proving this equality; for instance, it is a consequence of [122, Theorem 26]. We describe three appearances of this determinant.

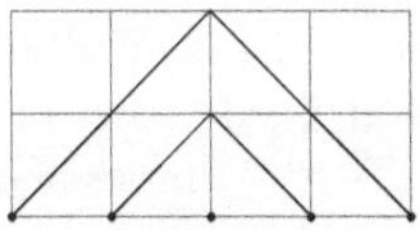

**Figure 1.21**
Routing interpretation of the Hankel determinant $H_n(C)$.

(a) A **$k$-fan of Dyck paths of length** $2n$ is a collection of $k$ Dyck paths from $(-n,0)$ to $(n,0)$ that do not cross (although they necessarily share some edges). Shifting the $(i+1)$th path $i$ units up and adding $i$ upsteps at the beginning and $i$ downsteps at the end, we obtain a routing of $k$ Dyck paths starting at the points $A = \{-(n+k-1),\ldots,-(n+1),-n\}$ and ending at the points $B = \{n, n+1,\ldots,n+k-1\}$ on the $x$-axis. See Figure 1.22 for an illustration. It follows that the number of $k$-fans of Dyck paths of length $2(n-2k)$ is given by (1.9).

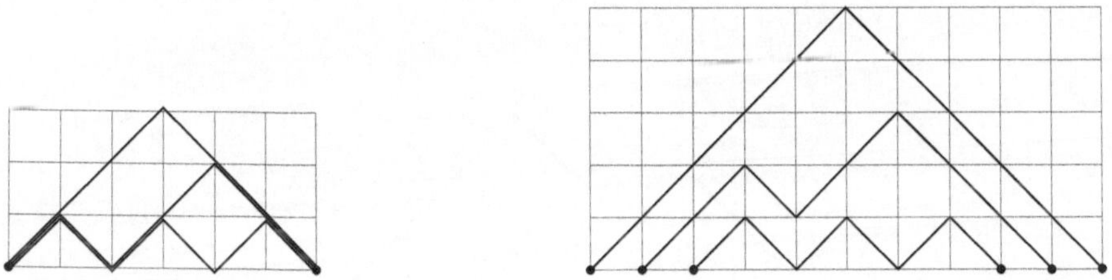

**Figure 1.22**
A $k$-fan of Dyck paths.

(b) There is also an extension of the classical one-to-one correspondence between Dyck paths and triangulations of a polygon. Define a **$k$-crossing** in an $n$-gon to be a set of $k$ diagonals that cross pairwise. A **$k$-triangulation** is a maximal set of diagonals with no $(k+1)$-crossings. The main enumerative result, due to Jonsson [110], is that the number of $k$-triangulations of an $n$-gon is also given by (1.9). A subtle bijection with fans of Dyck paths is given in [177].

Several properties of triangulations extend non-trivially to this context. For example, every $k$-triangulation has exactly $k(2n-2k-1)$ diagonals [148, 69]. The $k$-triangulations are naturally the facets of a simplicial complex called the **multiassociahedron**, which is topologically a sphere [109]; it is not currently known whether it is polytopal. There is a further generalization in the context of Coxeter groups, with connections to cluster algebras [49].

(c) These determinants also arise naturally in the commutative algebraic properties of Pfaffians, defined earlier in this section. Let $A$ be a skew-symmetric $n \times n$ matrix whose entries above the diagonal are indeterminates $\{a_{ij} : 1 \leq i < j \leq n\}$ over a field . Consider the **Pfaffian ideal** $I_k(A)$ generated by the $\binom{n}{2k}$ Pfaffian minors of $A$ of size $2k \times 2k$, and the **Pfaffian ring** $R_k(A) =\ [a_{ij}]/I_k(A)$. Then the multiplicity of the Pfaffian ring $R_k(X)$ is also given by (1.9). [99, 89]

5. (Schröder determinants and Aztec diamonds) Recall from Section 1.3.2.2 that a **Schröder path** of length $n$ is a path from $(0,0)$ to $(2n,0)$ using steps $NE = (1,1), SE = (1,-1)$, and $E = (2,0)$ that stays above the $x$-axis. The Hankel determinant $\det H_n(R)$ counts the routings of Schröder paths from the points $A = \{0,-2,\ldots,-(2n)\}$ to the points $B = \{0,2,\ldots,2n\}$ on the $x$-axis.

   These Hankel determinants have a natural interpretation in terms of tilings. Consider the **Aztec diamond** * $AD_n$ consisting of $2n$ rows centered horizontally, consisting successively of $2,4,\ldots,2n,2n,\ldots,4,2$ squares. We are interested in counting the tilings of the Aztec diamond into dominoes.

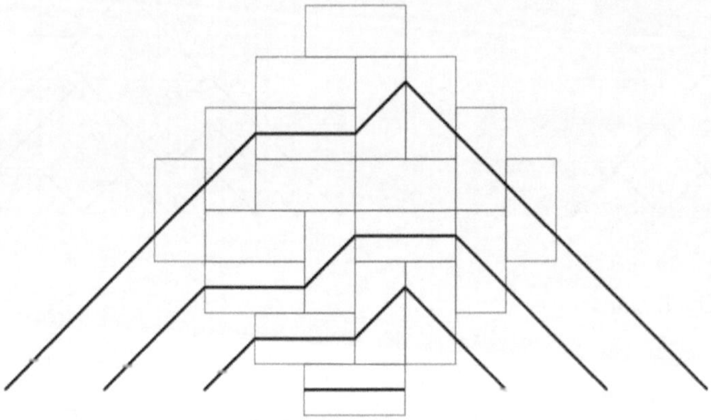

**Figure 1.23**
A tiling of the Aztec diamond and the corresponding routing.

Given a domino tiling of $AD_n$, construct $n$ paths as follows. Each path starts at the center of one of the vertical unit edges on the southwest border of the diamond, and successively crosses each tile that it encounters following a straight line through the center of the tile. It eventually comes out at the

*This shape is called the **Aztec diamond** because it is reminiscent of designs of several Native American groups. Perhaps the closest similarity is with Mayan pyramids, such as the Temple of Kukulcán in Chichén Itzá; the name **Mayan diamond** would have been more appropriate.

southeast side of the diamond, at the same height where it started. See Figure 1.23 for an illustration. If we add $i$ initial $NE$ steps and $i$ final $SE$ steps to the $(i+1)$th path for each $i$, the result will be a routing of Schröder paths from $A = \{-(2n), \ldots, -2, 0\}$ to $B = \{0, 2, \ldots, 2n\}$. In fact this correspondence is a bijection [76].

We will prove in Section 1.4.2 that

$$\begin{cases} \det H_n(A) = 2^{n(n-1)/2} \\ \det H_n'(A) = 2^{n(n+1)/2} \end{cases} \text{for all } n \geq 0 \quad \Longleftrightarrow \quad A \text{ is the Schröder sequence.}$$

It will follow that

$$\text{number of domino tilings of the Aztec diamond } AD_n = 2^{n(n+1)/2}.$$

This elegant result is originally due to Elkies, Kuperberg, Larsen, and Propp. For several other proofs, see [73, 74].

## 1.4.2 Computing determinants

In light of Section 1.4.1, it is no surprise that combinatorialists have become talented at computing determinants. Fortunately, this is a very classical topic with connections to many branches of mathematics and physics, and by now there are numerous general techniques and guiding examples available to us. Krattenthaler's surveys [122] and [123] are excellent references that have clearly influenced the exposition in this section. We now highlight some of the key tools and examples.

### 1.4.2.1 Is it known?

Of course, when we wish to evaluate a new determinant, one first step is to check whether it is a special case of some known determinantal evaluation. Starting with classical evaluations such as the Vandermonde determinant

$$\det(x_i^{j-1})_{1 \leq i,j \leq n} = \prod_{1 \leq i < j \leq n} (x_j - x_i), \tag{1.10}$$

there is now a wide collection of powerful results at our disposal. A particularly useful one [122, Lemma 3] states that for any $x_1, \ldots, x_n, a_2, \ldots, a_n, b_2, \ldots, b_n$ we have:

$$\det\left[(x_i + b_2) \cdots (x_i + b_j)(x_i + a_{j+1}) \cdots (x_i + a_n)\right]_{1 \leq i,j \leq n} = \prod_{1 \leq i < j \leq n} (x_j - x_i) \prod_{1 \leq i < j \leq n} (b_i - a_j). \tag{1.11}$$

For instance, as pointed out in [89] and [122], the Catalan determinant (1.9) is a special case of this formula. Recognizing it as such is not immediate, but the product formula for Catalan numbers gives an indication of why this is feasible.

In fact, here is a counterintuitive principle: Often the easiest way to prove a determinantal identity is to generalize it. It is very useful to introduce as many parameters as possible into a determinant, while making sure that the more general determinant still evaluates nicely. We will see this principle in action several times in what follows.

#### 1.4.2.2 Row and column operations

A second step is to check whether the standard methods of computing determinants are useful: Laplace expansion by minors, or performing row and column operations until we get a matrix whose determinant we can compute easily. For example, recall the determinant $L_0(K_n)$ of the $(n-1)\times(n-1)$ reduced Laplacian of the complete graph $K_n$, discussed in Section 1.4.1.3. We can compute it by first adding all rows to the first row, and then adding the first row to all rows:

$$\begin{aligned}\det L_0(K_n) &= \begin{vmatrix} n-1 & -1 & \cdots & -1 \\ -1 & n-1 & \cdots & -1 \\ \vdots & \vdots & \ddots & \vdots \\ -1 & -1 & \cdots & n-1 \end{vmatrix} = \begin{vmatrix} 1 & 1 & \cdots & 1 \\ -1 & n-1 & \cdots & -1 \\ \vdots & \vdots & \ddots & \vdots \\ -1 & -1 & \cdots & n-1 \end{vmatrix} \\ &= \begin{vmatrix} 1 & 1 & \cdots & 1 \\ 0 & n & \cdots & 0 \\ \vdots & \vdots & \ddots & \vdots \\ 0 & 0 & \cdots & n \end{vmatrix} = n^{n-2},\end{aligned}$$

reproving Theorem 1.4.6.1.

#### 1.4.2.3 Identifying linear factors

Many $n \times n$ determinants of interest have formulas of the form $\det M(\mathbf{x}) = cL_1(\mathbf{x})\cdots L_n(\mathbf{x})$ where $c$ is a constant and the $L_i(\mathbf{x})$ are linear functions in the variables $\mathbf{x} = (x_1,\ldots,x_k)$. We may prove such a formula by first checking that each $L_i(\mathbf{x})$ is indeed a factor of $M$, and then computing the constant $c$.

The best known application of this technique is the proof of the formula (1.10) for Vandermonde's determinant $V(x_1,\ldots,x_n)$. If $x_i = x_j$ for $i \neq j$, then rows $i$ and $j$ are equal, and the determinant is 0. It follows that $x_i - x_j$ must be a factor of the polynomial $\det V(x_1,\ldots,x_n)$. Since this polynomial is homogeneous of degree $\binom{n}{2}$, it must equal a constant times $\prod_{i<j}(x_i - x_j)$. Comparing the coefficients of $x_1^0 x_2^1 \cdots x_n^{n-1}$ we see that the constant equals 1.

A similar argument may be used to prove the more general formula (1.11).

To use this technique, it is sometimes necessary to introduce new variables into our determinant. For example, the formula $\det(i^{j-1})_{1\leq i,j\leq n} = 1^{n-1}2^{n-2}\cdots(n-1)^1$ cannot immediately be treated with this technique. However, the factorization of the answer suggests that this may be a special case of a more general result where this method does apply; in this case, Vandermonde's determinant.

### 1.4.2.4 Computing the eigenvalues

Sometimes we can compute explicitly the eigenvalues of our matrix, and multiply them to get the determinant. One common technique is to produce a complete set of eigenvectors.

1. (The Laplacian of the complete graph $K_n$) Revisiting the example above, the Laplacian of the complete graph is $L(K_n) = nI - J$ where $I$ is the identity matrix and $J$ is the matrix all of whose entries equal 1. We first find the eigenvalues of $J$: 0 is an eigenvalue of multiplicity $n-1$, as evidenced by the linearly independent eigenvectors $\mathbf{e}_1 - \mathbf{e}_2, \ldots, \mathbf{e}_{n-1} - \mathbf{e}_n$. Since the sum of the eigenvalues is $\mathrm{tr}(J) = n$, the last eigenvalue is $n$; an eigenvector is $\mathbf{e}_1 + \cdots + \mathbf{e}_{n-1}$. Now, if $v$ is an eigenvector for $J$ with eigenvalue $\lambda$, then it is an eigenvector for $nI - J$ with eigenvalue $n - \lambda$. Therefore the eigenvalues of $nI - J$ are $n, n, \ldots, n, 0$. Using Theorem 1.4.5, we have reproved yet again that $\det L_0(K_n) = \frac{1}{n}(n^{n-1}) = n^{n-2}$.

2. (The Laplacian of the $n$-cube $C_n$) A more interesting example is the reduced Laplacian $L_0(C_n)$ of the graph of the $n$-dimensional cube, from Theorem 1.4.6.4. By producing explicit eigenvectors, one may prove that if the Laplacians $L(G)$ and $L(H)$ have eigenvalues $\{\lambda_i : 1 \le i \le a\}$ and $\{\mu_j : i \le j \le b\}$ then the Laplacian of the product graph $L(G \times H)$ has eigenvalues $\{\lambda_i + \mu_j : 1 \le i \le a, 1 \le j \le b\}$. Since $C_1$ has eigenvalues 0 and 2, this implies that $C_n = C_1 \times \cdots \times C_1$ has eigenvalues $0, 2, 4, \ldots, 2n$ with multiplicities $\binom{n}{0}, \binom{n}{1}, \ldots, \binom{n}{n}$, respectively. Therefore the number of spanning trees of the cube $C_n$ is

$$\det L_0(C_n) = \frac{1}{2^n} 2^{\binom{n}{1}} 4^{\binom{n}{2}} \cdots (2n)^{\binom{n}{n}} = 2^{2^n - n - 1} 1^{\binom{n}{1}} 2^{\binom{n}{2}} \cdots n^{\binom{n}{n}}.$$

3. (The perfect matchings of a rectangle) An even more interesting example comes from the perfect matchings of the $a \times b$ rectangle, which we discussed in Section 1.4.1.5. Let $V$ be the $4mn$-dimensional vector space of functions $f : [2m] \times [2n] \to \mathbb{C}$, and consider the linear transformation $L : V \to V$ given by

$$(Lf)(x,y) = f(x-1,y) + f(x+1,y) + if(x,y-1) + if(x,y+1),$$

where $f(x,y) = 0$ when $x \in \{0, a+1\}$ or $y \in \{0, b+1\}$. The matrix of this linear transformation is precisely the one we are interested in. A straightforward computation shows that the following are eigenfunctions and eigenvalues of $L$:

$$g_{k,l}(x,y) = \sin \frac{k\pi x}{a+1} \sin \frac{l\pi y}{b+1}, \qquad \lambda_{k,l} = 2\cos \frac{k\pi}{a+1} + 2i \cos \frac{l\pi}{b+1}$$

for $1 \le k \le a$ and $1 \le l \le b$. (Note that $g_{k,l}(x,y) = 0$ for $x \in \{0, a\}$ or $y \in \{0, b\}$.) This is then the complete list of eigenvalues for $L$, so

$$\det L = 2^{ab} \prod_{k=1}^{a} \prod_{l=1}^{b} \left( \cos \frac{k\pi}{a+1} + i \cos \frac{l\pi}{b+1} \right),$$

which is easily seen to equal the expression in Theorem 1.4.13.

#### 1.4.2.5 LU factorizations

A classic result in linear algebra states that, under mild hypotheses, a square matrix $M$ has a unique factorization

$$M = LU$$

where $L$ is a lower triangular matrix and $U$ is an upper triangular matrix with all diagonal entries equal to 1. Computer algebra systems can compute the LU-factorization of a matrix, and if we can guess and prove such a factorization it will follow immediately that $\det M$ equals the product of the diagonal entries of $L$.

An interesting application of this technique is the determinant

$$\det(\gcd(i,j))_{1\leq i,j\leq n} = \prod_{i=1}^{n} \varphi(i), \tag{1.12}$$

where $\varphi(k) = \{i \in \mathbb{N} : (\gcd(i,k) = 1 \text{ and } 1 \leq i \leq k\}$ is Euler's totient function. This is a special case of a more general formula for semilattices that is easier to prove. For this brief computation, we assume familiarity with the Möbius function $\mu$ and the zeta function $\zeta$ of a poset; these will be treated in detail in Section 1.5.5.3.

Let $P$ be a finite meet semilattice and consider any function $F : P \times P \to$ . We will prove the **Lindström–Wilf determinantal formula**:

$$\det F(p \vee q, p)_{p,q \in P} = \prod_{p \in P} \left( \sum_{r \geq p} \mu(p,r) F(r,p) \right). \tag{1.13}$$

Computing some examples will suggest that the LU factorization of $F$ is $F = MZ$ where

$$M_{pq} = \begin{cases} \sum_{r \geq q} \mu(q,r) F(r,p) & \text{if } p \leq q, \\ 0 & \text{otherwise,} \end{cases} \qquad Z_{pq} = \begin{cases} 1 & \text{if } p \geq q, \\ 0 & \text{otherwise.} \end{cases}$$

This guess is easy to prove, and it immediately implies (1.13). In turn, applying the Lindström–Wilf to the poset of integers $\{1,\ldots,n\}$ ordered by reverse divisibility and the function $F(x,y) = x$, we obtain (1.12).

Another interesting special case is the determinant

$$\det(x^{\text{rank}(p \vee q)})_{p,q \in P} = \prod_{p \in P} \left( x^{\text{rank}(p)} \chi_{[p,\hat{1}]}(1/x) \right),$$

where $\chi_{[p,\hat{1}]}(x)$ is the characteristic polynomial of the interval $[x,\hat{1}]$. When $P$ is the partition lattice $\Pi_n$, this determinant arises in Tutte's work on the Birkhoff-Lewis equations [203].

#### 1.4.2.6 Hankel determinants and continued fractions

For Hankel determinants, the following connection with continued fractions [209] is extremely useful. If the expansion of the generating function for a sequence $f_0, f_1, \ldots$

as a **J-fraction** is

$$\sum_{n=0}^{\infty} f_n x^n = \cfrac{f_0}{1 + a_0 x - \cfrac{b_1 x^2}{1 + a_1 x - \cfrac{b_2 x^2}{1 + a_2 x - \cdots}}},$$

then the Hankel determinants of $f_0, f_1, \ldots$ equal

$$\det H_n(A) = f_0^n b_1^{n-1} b_2^{n-2} \cdots b_{n-2}^2 b_{n-1}$$

For instance, using the generating function for the Schröder numbers $r_n$, it is easy to prove that

$$\sum_{n=0}^{\infty} r_n x^n = \cfrac{1}{1 - 2x - \cfrac{2x^2}{1 - 3x - \cfrac{2x^2}{1 - 3x - \cdots}}}, \qquad \sum_{n=0}^{\infty} r_{n+1} x^n = \cfrac{2}{1 - 3x - \cfrac{2x^2}{1 - 3x - \cfrac{2x^2}{1 - 3x - \cdots}}}.$$

Therefore

$$\det H_n(R) = 2^{n(n-1)/2}, \qquad \det H_n'(R) = 2^{n(n+1)/2},$$

as stated in Example 5 of Section 1.4.1.6.

By computer calculation, it is often easy to guess J-fractions experimentally. With a good guess in place, there is an established procedure for proving their correctness, rooted in the theory of orthogonal polynomials; see [122, Section 2.7].

**Dodgson condensation.** It is often repeated that Lewis Carroll, author of *Alice in Wonderland*, was also an Anglican deacon and a mathematician, publishing under his real name, Rev. Charles L. Dodgson. His contributions to mathematics are discussed less often, and one of them is an elegant method for computing determinants.

To compute an $n \times n$ determinant $A$, we create a square pyramid of numbers, consisting of $n+1$ levels of size $n+1, n \ldots, 1$, respectively. On the bottom level we place an $(n+1) \times (n+1)$ array of 1s, and on the next level we place the $n \times n$ matrix $A$. Each subsequent floor is obtained from the previous two by the following rule: Each new entry is given by $f = (ad - bc)/e$ where $f$ is directly above the entries $\begin{pmatrix} a & b \\ c & d \end{pmatrix}$ and two floors above the entry $e$. * The top entry of the pyramid is the determinant. For example, the computation

$$\begin{pmatrix} 1 & 1 & 1 & 1 & 1 \\ 1 & 1 & 1 & 1 & 1 \\ 1 & 1 & 1 & 1 & 1 \\ 1 & 1 & 1 & 1 & 1 \\ 1 & 1 & 1 & 1 & 1 \end{pmatrix} \to \begin{pmatrix} 2 & 7 & 5 & 4 \\ 1 & 9 & 7 & 7 \\ 2 & 3 & 2 & 1 \\ 5 & 7 & 6 & 3 \end{pmatrix} \to \begin{pmatrix} 11 & 4 & 7 \\ -15 & -3 & -7 \\ -1 & 4 & 0 \end{pmatrix} \to \begin{pmatrix} 3 & -1 \\ -21 & 28 \end{pmatrix} \to (21)$$

shows that the determinant of the $4 \times 4$ determinant is 21.

Dodgson's condensation method relies on the following fact, due to Jacobi. If $A$ is an $n \times n$ matrix and $A_{i_1,\ldots,i_k; j_1,\ldots,j_k}$ denotes the matrix $A$ with rows $i_1, \ldots, i_k$ and columns $j_1, \ldots, j_k$ removed, then

$$\det A \cdot \det A_{1,n;1,n} = \det A_{1;1} \cdot \det A_{n;n} - \det A_{1;n} \cdot \det A_{n;1}. \tag{1.14}$$

*Special care is required when 0s appear in the interior of the pyramid.

This proves that the numbers appearing in the pyramid are precisely the determinants of the "contiguous" submatrices of $A$, consisting of consecutive rows and columns.

If we have a guess for the determinant of $A$, as well as the determinants of its contiguous submatrices, Dodgson condensation is an extremely efficient method to prove it. All we need to do is to verify that our guess satisfies (1.14).

To see how this works in an example, let us use Dodgson condensation to prove the formula in Section 1.4.1.6 for $R_n = \det\binom{2n}{n+i-j}_{1\le i,j\le n}$, the number of stacks of unit cubes in the corner of an $n\times n\times n$ box. The first step is to guess the determinant of the matrix in question, as well as all its contiguous submatrices; they are all of the form $R(a,b,c) = \det\binom{a+b}{a+i-j}_{1\le i,j\le c}$, where $a+b=2n$. This more general determinant is equally interesting combinatorially: it counts the stacks of unit cubes in the corner of an $a\times b\times c$ box. By computer experimentation, it is not too difficult to arrive at the following guess:

$$R(a,b,c) = \det\left[\binom{a+b}{a+i-j}\right]_{1\le i,j\le c} = \prod_{i=1}^{a}\prod_{j=1}^{b}\prod_{k=1}^{c}\frac{i+j+k-1}{i+j+k-2}.$$

Proving this formula by Dodgson condensation is then straightforward; we just need to check that our conjectural product formula holds for $c=0,1$ and that it satisfies (1.14); that is,

$$R(a,b,c+1)R(a,b,c-1) = R(a,b,c)^2 - R(a+1,b-1,c)R(a-1,b+1,c).$$

For more applications of Dodgson condensation, see for example [3].

There is a wonderful connection between Dodgson condensation, Aztec diamonds, and **alternating sign matrices**, which we now describe. Let us construct a square pyramid of numbers where levels $n+2$ and $n+1$ are given by two matrices $\mathbf{y} = (y_{ij})_{1\le i,j\le n+2}$ and $\mathbf{x} = (x_{ij})_{1\le i,j\le n+1}$, respectively, and levels $n-1,\ldots,2,1$ are computed in terms of the lower rows using Dodgson's recurrence $f = (ad-bc)/e$. Let $f_n(\mathbf{x},\mathbf{y})$ be the entry at the top of the pyramid.

Remarkably, all the entries of the resulting pyramid will be Laurent monomials in the $x_{ij}$s and $y_{ij}$s; that is, their denominators are always monomials. This is obvious for the first few levels, but it becomes more and more surprising as we divide by more and more intricate expressions.

The combinatorial explanation for this fact is that each entry in the $(n-k)$th level of the pyramid encodes the domino tilings of an Aztec diamond $AD_k$. For instance, if $n=3$, the entry at the top of the pyramid is

$$\begin{aligned} f_2(\mathbf{x},\mathbf{y}) = {} & \frac{x_{11}x_{22}x_{33}}{y_{22}y_{33}} - \frac{x_{11}x_{23}x_{32}}{y_{22}y_{33}} - \frac{x_{12}x_{21}x_{33}}{y_{22}y_{33}} + \frac{x_{12}x_{21}x_{23}x_{32}}{x_{22}y_{22}y_{33}} \\ & - \frac{x_{12}x_{21}x_{23}x_{32}}{x_{22}y_{23}y_{32}} + \frac{x_{12}x_{23}x_{31}}{y_{23}y_{32}} + \frac{x_{13}x_{21}x_{32}}{y_{23}y_{32}} - \frac{x_{13}x_{31}x_{32}}{y_{23}y_{32}} \end{aligned}$$

There is a simple bijection between the eight terms of $f_3$ and the eight domino tilings of $AD_2$. Regard a tiling of $AD_2$ as a graph with vertices on the underlying lattice, and add a vertical edge above and below the tiling, and a horizontal edge to the

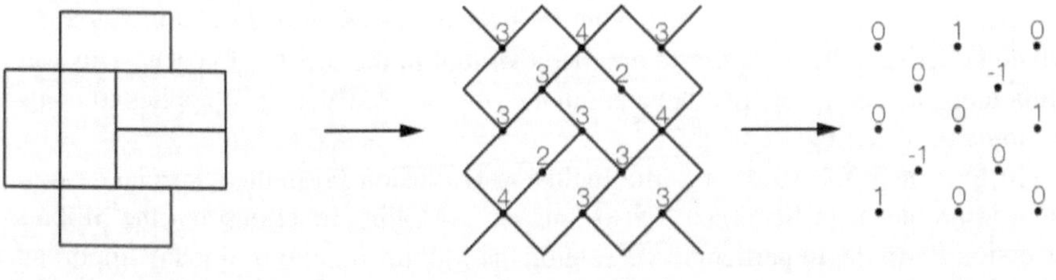


**Figure 1.24**
A domino tiling of $AD_2$ and the corresponding monomial in $f_2(\mathbf{x},\mathbf{y})$.

left and to the right of $T$. Now rotate the tiling 45°. Record the degree of each vertex, ignoring the outside corners on the boundary of the diamond, and subtract 3 from each vertex. This leaves us with an $n \times n$ grid of integers within an $(n+1) \times (n+1)$ grid of integers. Assign to it the monomial whose $x$ exponents are given by the outer grid and whose $y$ exponents are given by the inner grid. For example, the tiling in Figure 1.24 corresponds to the monomial $(x_{12}x_{23}x_{31})/(y_{23}y_{32})$.

In general, this gives a bijection between the terms of $f_n(\mathbf{x},\mathbf{y})$ and the domino tilings of the Aztec diamond $AD_n$. One may also check that there are no cancellations, so Dodgson condensation tells us that the number $m_n$ of terms in $f_n$ satisfies $m_{n-1}m_{n+1} = 2m_n^2$. This gives an alternative proof that the Aztec diamond $AD_n$ has $2^{n(n+1)/2}$ domino tilings.

We may also consider the patterns formed by the $x_{ij}$s by themselves (or of the $y_{ij}$s by themselves). In each individual monomial of $f_n(\mathbf{x},\mathbf{y})$, the exponents of the $x_{ij}$s form an $n \times n$ **alternating sign matrix** (ASM): a matrix of 1s, 0s, and $-1$s such that the nonzero entries in any row or column alternate $1,-1,\ldots,-1,1$. Similarly, the negatives of the exponents of the $y_{ij}$s form an ASM of size $n-1$.

Alternating sign matrices are fascinating objects in their own right, with connections to representation theory, statistical mechanics, and other fields. The number of alternating sign matrices of size $n$ is

$$\frac{1!\,4!\,7!\cdots(3n-2)!}{n!\,(n+1)!\,(n+2)!\cdots(2n-1)!}.$$

For details on the history and solution of this difficult enumeration problem see [44, 167, 222].

## 1.5 Posets

This section is devoted to the enumerative aspects of the theory of partially ordered sets (posets). Section 1.5.1 introduces key definitions and examples. Section 1.5.2 discusses some families of lattices that are of special importance. In Section 1.5.3 we count chains and linear extensions of posets.

The remaining sections are centered around the *Möbius Inversion Formula*, which is perhaps the most useful enumerative tool in the theory of posets. This formula helps us count sets that have an underlying poset structure; it applies to many combinatorial settings of interest.

In Section 1.5.4 we discuss the Inclusion-Exclusion Formula, a special case of great importance. In Section 1.5.5 we introduce Möbius functions and the Möbius Inversion Formula. In particular, we catalog the Möbius functions of many important posets. The incidence algebra, a nice algebraic framework for understanding and working with the Möbius function, is discussed in Section 1.5.5.3. In Section 1.5.5.4 we discuss methods for computing Möbius functions of posets, and sketch proofs for the posets of Section 1.5.5. Finally, in Section 1.5.6, we discuss Eulerian posets and the enumeration of their flags, which gives rise to the **ab**-index and **cd**-index.

### 1.5.1 Basic definitions and examples

A **partially ordered set** or **poset** $(P, \leq)$ is a set $P$ together with a binary relation $\leq$, called a **partial order**, such that

- For all $p \in P$, we have $p \leq p$.
- For all $p, q \in P$, if $p \leq q$ and $q \leq p$ then $p = q$.
- For all $p, q, r \in P$, if $p \leq q$ and $q \leq r$ then $p \leq r$.

We say that $p < q$ if $p \leq q$ and $p \neq q$. We say that $p$ and $q$ are **comparable** if $p < q$ or $p > q$, and they are **incomparable** otherwise. We say that $q$ **covers** $p$ if $q > p$ and there is no $r \in P$ such that $q > r > p$. When $q$ covers $p$ we write $q \gtrdot p$.

**Example 1.5.1** *Many sets in combinatorics come with a natural partial order, and often the resulting poset structure is very useful for enumerative purposes. Some of the most important examples are the following:*

1. (Chain) *The poset* $\mathbf{n} = \{1, 2, \ldots, n\}$ *with the usual total order.* $(n \geq 1)$
2. (Boolean lattice) *The poset* $2^A$ *of subsets of a set* $A$*, where* $S \leq T$ *if* $S \subseteq T$.
3. (Divisor lattice) *The poset* $D_n$ *of divisors of* $n$*, where* $c \leq d$ *if* $c$ *divides* $d$. $(n \geq 1)$
4. (Young's lattice) *The poset* $Y$ *of integer partitions, where* $\lambda \leq \mu$ *if* $\lambda_i \leq \mu_i$ *for all* $i$.
5. (Partition lattice) *The poset* $\Pi_n$ *of set partitions of* $[n]$*, where* $\pi \leq \rho$ *if* $\pi$ refines $\rho$*; that is, if every block of* $\rho$ *is a union of blocks of* $\pi$. $(n \geq 1)$
6. (Non-crossing partition lattice) *The subposet* $NC_n$ *of* $\Pi_n$ *consisting of the non-crossing set partitions of* $[n]$*, where there are no elements* $a < b < c < d$ *such that* $a, c$ *are together in one block and* $b, d$ *are together in a different block.* $(n \geq 1)$

7. (Bruhat order on permutations) *The poset $S_n$ of permutations of $[n]$, where $\pi$ covers $\rho$ if $\pi$ is obtained from $\rho$ by choosing two adjacent numbers $\rho_i = a < b = \rho_{i+1}$ in $\rho$ and exchanging their positions.* $(n \geq 1)$

8. (Subspace lattice) *The poset $L(\mathbb{F}_q^n)$ of subspaces of a finite dimensional vector space $\mathbb{F}_q^n$, where $U \leq V$ if $U$ is a subspace of $V$.* $(n \geq 1,\ q$ *a prime power*$)$

9. (Distributive lattice) *The poset $J(P)$ of order ideals of a poset $P$ (subsets $I \subseteq P$ such that $j \in P$ and $i < j$ imply $i \in P$) ordered by containment.*

10. (Face poset of a polytope) *The poset $F(P)$ of faces of a polytope $P$, ordered by inclusion.*

11. (Face poset of a subdivision of a polytope) *The poset $\widehat{\mathcal{T}}$ of faces of a subdivision $\mathcal{T}$ of a polytope $P$ ordered by inclusion, with an additional maximum element.*

12. (Subgroup lattice of a group) *The poset $L(G)$ of subgroups of a group $G$, ordered by containment.*

The **Hasse diagram** of a finite poset $P$ is obtained by drawing a dot for each element of $P$ and an edge going down from $p$ to $q$ if $p$ covers $q$. Figure 1.25 shows the Hasse diagrams of some of the posets above. In particular, the Hasse diagram of $2^{[n]}$ is the 1-skeleton of the $n$-dimensional cube.

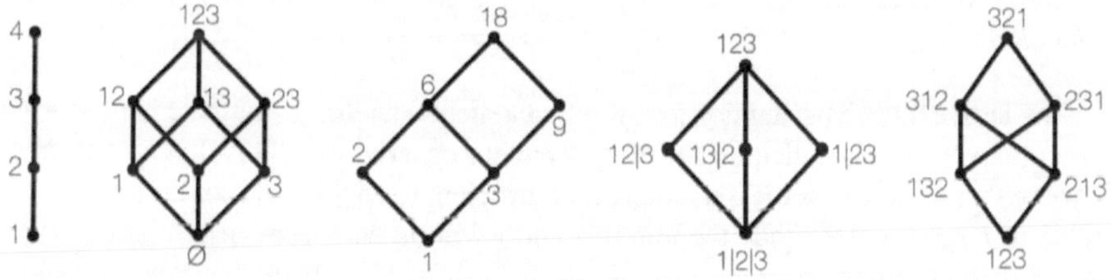


**Figure 1.25**
The Hasse diagrams of the chain **4**, Boolean lattice $2^{[3]}$, divisor lattice $D_{18}$, partition lattice $\Pi_3$, and Bruhat order $S_3$.

A subset $Q$ of $P$ is a **chain** if every pair of elements is comparable, and it is an **antichain** if every pair of elements is incomparable. The length of a chain $C$ is $|C| - 1$. If there is a **rank function** $r : P \to \mathbb{N}$ such that $r(x) = 0$ for any minimal element $x$ and $r(y) = r(x) + 1$ whenever $y \gtrdot x$, then $P$ is called **graded** or **ranked**. The largest rank is called the **rank** or **height** of $P$. The **rank-generating function** of a finite graded poset is

$$R(P;x) = \sum_{p \in P} x^{r(p)}.$$

All the posets of Example 1.5.1 are graded except for subgroup lattices.

A poset $P$ induces a poset structure on any subset $Q \subseteq P$; a special case of interest is the **interval** $[p,q] = \{r \in P : p \leq r \leq q\}$. We call a poset **locally finite** if all its intervals are finite. Given posets $P$ and $Q$ on disjoint sets, the **direct sum** $P+Q$ is the poset on $P \cup Q$ inheriting the order relations from $P$ and $Q$, and containing no additional order relations between elements of $P$ and $Q$. The **direct product** $P \times Q$ is the poset on $P \times Q$ where $(p,q) \leq (p',q')$ if $p \leq p'$ and $q \leq q'$.

We have already seen examples of product posets. The Boolean lattice is $2^A \cong \mathbf{2} \times \cdots \times \mathbf{2}$. Also, if $n = p_1^{t_1} \cdots p_k^{t_k}$ is the prime factorization of $n$, then $D_n \cong (\mathbf{t_1}+\mathbf{1}) \times \cdots \times (\mathbf{t_k}+\mathbf{1})$.

### 1.5.2 Lattices

A poset is a **lattice** if every two elements $p$ and $q$ have a least upper bound $p \vee q$ and a greatest lower bound $p \wedge q$, called their **meet** and **join**, respectively. We will see this additional algebraic structure can be quite beneficial for enumerative purposes.

**Example 1.5.2** *All the posets in Example 1.5.1 are lattices, except for the Bruhat order. In most cases, the meet and join have easy descriptions. In* $\mathbf{n}$*, the meet and join are the minimum and maximum, respectively. In* $2^A$ *they are the intersection and union. In* $D_n$ *they are the greatest common divisor and least common multiple. In* $Y$ *they are the componentwise minimum and maximum. In* $\Pi_n$ *and in* $NC_n$ *the meet of two partitions* $\pi$ *and* $\rho$ *is the collection of intersections of a block of* $\pi$ *and a block of* $\rho$*. In* $L(\mathbb{F}_q^n)$ *the meet and join are the intersection and the span. In* $J(P)$ *they are the intersection and the union. In* $F(P)$ *the meet is the intersection. In* $L(G)$ *the meet is the intersection.*

Any lattice must have a unique minimum element $\widehat{0}$ and maximum element $\widehat{1}$. An element covering $\widehat{0}$ is called an **atom**; an element covered by $\widehat{1}$ is called a **coatom**. To prove that a finite poset $P$ is a lattice, it is sufficient to check that it has a $\widehat{1}$ and that any $x,y \in P$ have a meet; then the join of $x$ and $y$ will be the (necessarily non-empty) meet of their common upper bounds. Similarly, it suffices to check that $P$ has a $\widehat{0}$ and that any $x,y \in P$ have a join.

**Distributive lattices.** A lattice $L$ is **distributive** if the join and meet operations satisfy the distributive properties:

$$x \vee (y \wedge z) = (x \vee y) \wedge (x \vee z), \qquad x \wedge (y \vee z) = (x \wedge y) \vee (x \wedge z) \tag{1.15}$$

for all $x,y,z \in L$. To prove that $L$ is distributive, it is sufficient to verify that one of the equations in (1.15) holds for all $x,y,z \in L$.

**Example 1.5.3** *There are several distributive lattices in Example 1.5.1: the chains* $\mathbf{n}$*, the Boolean lattices* $2^A$*, the divisor lattices* $D_n$*, and Young's lattice* $Y$*. This follows from the fact that the pairs of operations* $(\min,\max)$*,* $(\gcd,\mathrm{lcm})$ *and* $(\cap,\cup)$ *satisfy the distributive laws. The others are not necessarily distributive; for example,* $\Pi_3$ *and* $S_3$*.*

The most important, and in fact, the only, source of finite distributive lattices is the construction of Example 1.5.1.9: Given a poset $P$, a **downset** or **order ideal** $I$ is a subset of $P$ such that if $i \in I$ and $j < i$ then $j \in I$. A **principal** order ideal is one of the form $P_{\leq p} = \{q \in P : q \leq p\}$. Let $J(P)$ be the **poset of order ideals** of $P$, ordered by inclusion.

**Theorem 1.5.4 (Fundamental Theorem for Finite Distributive Lattices)** *A poset $L$ is a distributive lattice if and only if there exists a poset $P$ such that $L \cong J(P)$.*

*Sketch of Proof.* Since the collection of order ideals of a poset $P$ is closed under union and intersection, $J(P)$ is a sublattice of $2^P$. The distributivity of $2^P$ then implies that $J(P)$ is a distributive lattice.

For the converse, let $L$ be a distributive lattice, and let $P$ be the set of join-irreducible elements of $L$; that is, the elements $p > \widehat{0}$ that cannot be written as $p = q \vee r$ for $q, r < p$. These are precisely the elements of $L$ that cover exactly one element. The set $P$ inherits a partial order from $L$, and this is the poset such that $L \cong J(P)$. The isomorphism is given by

$$\begin{array}{rcl} \phi : J(P) & \longrightarrow & L \\ I & \longmapsto & \bigvee_{p \in I} p \end{array}$$

and the inverse map is given by $\phi^{-1}(l) = \{p \in P : p \leq l\}$. ■

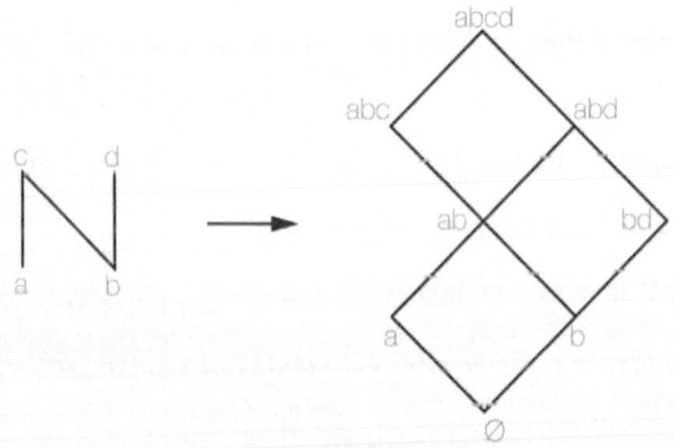


**Figure 1.26**
A poset and the corresponding distributive lattice.

Theorem 1.5.4 extends to some infinite posets with minor modifications. Let $J_f(P)$ be the set of finite order ideals of a poset $P$. Then the map $P \mapsto J_f(P)$ is a bijection between the posets whose principal order ideals are finite and the locally finite distributive lattices with $\widehat{0}$.

**Example 1.5.5** *The posets $P$ of join-irreducibles of the distributive lattices $L \cong J(P)$ of Example 1.5.3 are as follows. For $L = \mathbf{n}$, $P = \mathbf{n-1}$ is a chain. For $L = 2^A$, $P = \mathbf{1} + \cdots + \mathbf{1}$ is an antichain. For $L = D_n$, where $n = p_1^{t_1} \cdots p_k^{t_k}$, $P = \mathbf{t_1} + \cdots + \mathbf{t_k}$ is the disjoint sum of $k$ chains. For $L = Y$, $P = \mathbb{N} \times \mathbb{N}$ is a "quadrant."*

Theorem 1.5.4 explains the abundance of cubes in the Hasse diagram of a distributive lattice $L$. For any element $l \in L$ covered by $n$ elements $l_1, \ldots, l_n$ of $L$, the joins of the $2^n$ subsets of $\{l_1, \ldots, l_n\}$ are distinct, and form a copy of the Boolean lattice $2^{[n]}$ inside $L$. The dual result holds as well.

The **width** of a poset $P$ is the size of the largest antichain of $P$. **Dilworth's theorem** [68] states that this is the smallest integer $w$ such that $P$ can be written as the disjoint union of $w$ chains.

**Theorem 1.5.6** *The distributive lattice $J(P)$ can be embedded as an induced subposet of the poset $\mathbb{N}^w$, where $w$ is the width of $P$.*

*Proof.* Decompose $P$ as the disjoint union of $w$ chains $C_1, \ldots, C_c$. The map

$$\begin{aligned} \phi : J(P) &\longrightarrow \mathbb{N}^w \\ I &\longmapsto (|I \cap C_1|, \ldots, |I \cap C_w|) \end{aligned}$$

gives the desired inclusion. ■

**Geometric lattices.** Now we introduce another family of lattices of great importance in combinatorics. We say that a lattice $L$ is:

- **semimodular** if the following two equivalent conditions hold:
  - $L$ is graded and $r(p) + r(q) \geq r(p \wedge q) + r(p \vee q)$ for all $p, q \in L$.
  - If $p$ and $q$ both cover $p \wedge q$, then $p \vee q$ covers both $p$ and $q$.
- **atomic** if every element is a join of atoms.
- **geometric** if it is semimodular and atomic.

**Example 1.5.7** *In Figure 1.25, the posets $2^{[3]}$ and $\Pi_3$ are geometric, while the posets $\mathbf{4}$, $D_{18}$, and $S_3$ are not.*

Not surprisingly, the prototypical example of a geometric lattice comes from a natural geometric construction, illustrated in Figure 1.27. Let $A = \{v_1, \ldots, v_n\}$ be a set of vectors in a vector space $V$. A **flat** is a subspace of $V$ generated by a subset of $A$. We identify a flat with the set of $v_i$s that it contains. Let $L_A$ be the set of flats of $A$, ordered by inclusion. Then $L_A$ is a geometric lattice.

The theory of geometric lattices is equivalent to the rich theory of *matroids*, which is the subject of Section 1.8.

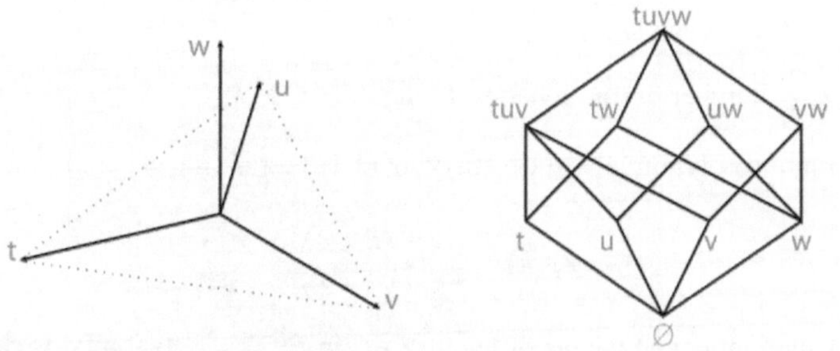


**Figure 1.27**
A vector configuration $\mathbf{t}, \mathbf{u}, \mathbf{v}, \mathbf{w}$ (where $\mathbf{t}, \mathbf{u}, \mathbf{v}$ are coplanar in $\mathbb{R}^3$) and the corresponding geometric lattice.

**Supersolvable lattices.** A lattice $L$ is **supersolvable** if there exists a maximal chain, called an M-chain, such that the sublattice generated by $C$ and any other chain of $L$ is distributive. [182]

Again unsurprisingly, an important example comes from supersolvable groups, but there are several other interesting examples. Here is a list of supersolvable lattices, and an M-chain in each case.

1. Distributive lattices: every maximal chain is an M-chain.
2. Partition lattice $\Pi_n$: $1|2|\cdots|n < 12|3|\cdots|n < 123|4|\cdots|n < \cdots < 123\cdots n$.
3. Noncrossing partition lattice $NC_n$: the same chain as above.
4. Lattice of subspaces $L(\mathbb{F}_q^n)$ of the vector space $\mathbb{F}_q^n$ over a finite field $\mathbb{F}_q$: every maximal chain is an M-chain.
5. Subgroup lattices of finite supersolvable groups $G$: an M-chain is given by any normal series $1 = H_0 \triangleleft H_1 \triangleleft \cdots \triangleleft H_k = G$ where each $H_i$ is normal and $H_i/H_{i-1}$ is cyclic of prime order.

Fortunately, there is a simple criterion to verify semimodularity. An $R$**-labeling** of a poset $P$ is a labeling of the edges of the Hasse diagram of $P$ with integers such that for any $s \leq t$ there exists a unique maximal chain $C$ from $s$ to $t$.

**Theorem 1.5.8** [139] *A finite graded lattice of rank $n$ is supersolvable if and only if it has an R-labeling for which the labels on every maximal chain are a permutation of* $\{1,\ldots,n\}$.

### 1.5.3 Zeta polynomials and order polynomials

The **zeta polynomial** of a finite poset $P$ counts the multichains of various lengths in $P$. A **multichain** of length $k$ in $P$ is a sequence of possibly repeated elements

$t_0, t_1, \ldots, t_k \in P$ such that $t_0 \le t_1 \le \cdots \le t_k$. Let

$$Z_P(k) = \text{number of multichains of length } k-2 \text{ in } P \qquad (k \ge 2). \tag{1.16}$$

There is a unique polynomial $Z_P(k)$ satisfying (1.16) for all integers $k \ge 2$; it is given by

$$Z_P(k) = \sum_{i \ge 2} b_i \binom{k-2}{i-2}, \tag{1.17}$$

where $b_i$ is the number of chains of length $i-2$ in $P$. This polynomial is called the **zeta polynomial** of $P$.

**Example 1.5.9** *The following posets have particularly nice zeta polynomials:*

1. $P = \mathbf{n}$*:*
$$Z(k) = \binom{n+k-2}{n-1}$$
2. $P = B_n$*:*
$$Z(k) = k^n$$
3. $P = NC_n$*: (Kreweras, [124])*
$$Z(k) = \frac{1}{n}\binom{kn}{n-1}$$

The **order polynomial** of $P$ counts the order-preserving labelings of $P$; it is defined by

$$\Omega_P(k) = \text{number of maps } f : P \to [k] \text{ such that } p < q \text{ implies } f(p) \le f(q)$$

for $k \in \mathbb{N}$. The next proposition shows that, once again, there is a unique polynomial taking these values at the natural numbers.

**Proposition 1.5.10** *For any poset $P$, $\Omega_P(k) = Z_{J(P)}(k)$.*

*Proof.* An order-preserving map $f : P \to [k]$ gives rise to a sequence of order ideals $f^{-1}(\{1\}) \subseteq f^{-1}(\{1,2\}) \subseteq \cdots f^{-1}(\{1,\ldots,k\})$, which is a multichain in $J(P)$. Conversely, every sequence arises uniquely in this way. ■

A **linear extension** of $P$ is an order-preserving labeling of the elements of $P$ with the labels $1, \ldots, n = |P|$, which extends the order of $P$; that is, a bijection $f : P \to [n]$ such that $p < q$ implies $f(p) < f(q)$. Let

$$e(P) = \text{ number of linear extensions of } P.$$

It follows from Proposition 1.5.10 and (1.17) that the order polynomial $\Omega_P$ has degree $|P|$, and leading coefficient $e(P)/|P|!$.

The following is a method for computing $e(P)$ recursively.

**Proposition 1.5.11** *Define $e : J(P) \to \mathbb{N}$ recursively by*

$$e(I) = \begin{cases} 1 & \text{if } I = \widehat{0}, \\ \sum_{J \lessdot I} e(J) & \text{otherwise.} \end{cases}$$

*Then $e(\widehat{1})$ is the number $e(P)$ of linear extensions of $P$.*

*Proof.* Let $p_1, \ldots, p_k$ be the maximal elements of $P$. In a linear extension of $P$, one of $p_1, \ldots, p_k$ has to be labeled $n$, and therefore

$$e(P) = e(P \backslash \{p_1\}) + \cdots + e(P \backslash \{p_k\}).$$

This is equivalent to the desired recurrence. ■

It is useful to keep in mind that $J(P)$ is a subposet of $\mathbb{N}^w$ for $w = w(P)$. The recurrence of Proposition 1.5.11 generalizes Pascal's triangle, which corresponds to the case $P = \mathbb{N} + \mathbb{N}$. When we apply it to $P = \mathbb{N} + \cdots + \mathbb{N}$, we get the recursive formula for multinomial coefficients.

**Example 1.5.12** *In some special cases, the problem of enumerating linear extensions is of fundamental importance.*

- $e(\mathbf{n}_1 + \cdots + \mathbf{n}_k) = \binom{n_1 + \cdots + n_k}{n_1, \ldots, n_k}$
- $e(\mathbf{2} \times \mathbf{n}) = C_n = \frac{1}{n+1}\binom{2n}{n}$.
- *Let $T$ be a* tree poset *of $n$ elements, such that the Hasse diagram is a tree rooted at $\widehat{0}$. For each vertex $v$ let $t_v = |T_{\geq v}| = |\{w \in T : w \geq v\}|$. Then*

$$e(T) = \frac{n!}{\prod_{v \in T} t_v}.$$

- *Let $\lambda$ be a Ferrers diagram of $n$ cells, partially ordered by decreeing that each cell is covered by the cell directly below and the cell directly to the right, if they are in $\lambda$. The* **hook** *$H_c$ of a cell $c$ consists of cells on the same row and to the right of $c$, those on the same column and below $c$, and $c$ itself. Let $h_c = |H_c|$. Then*

$$e(\lambda) = \frac{n!}{\prod_{c \in D} h_c}.$$

*This is the dimension of the irreducible representation of the symmetric group $S_n$ corresponding to $\lambda$. [173]*

### 1.5.4 The inclusion-exclusion formula

Our next goal is to discuss one of the most useful enumerative tools for posets: Möbius functions and the Möbius inversion theorem. Before we do that, we devote this section to a special case that preceded and motivated them: the inclusion-exclusion formula.

**Theorem 1.5.13 (Inclusion-Exclusion Formula)** *For any finite sets $A_1, \dots, A_n \subseteq X$, we have*

1. $|A_1 \cup \cdots \cup A_n| = \sum_i |A_i| - \sum_{i<j} |A_i \cap A_j| + \sum_{i<j<k} |A_i \cap A_j \cap A_k| - \cdots \pm |A_1 \cap \cdots \cap A_n|$.

2. $|\overline{A_1} \cap \cdots \cap \overline{A_n}| = |X| - \sum_i |A_i| + \sum_{i<j} |A_i \cap A_j| - \cdots \pm |A_1 \cap \cdots \cap A_n|$.

*Proof.* It suffices to prove one of these two equivalent equations. To prove the first one, consider an element $x$ appearing in $k \geq 1$ of the given sets. The number of times that $x$ is counted in the right-hand side is $k - \binom{k}{2} + \cdots \pm \binom{k}{k} = 1$. ■

We now present a slightly more general formulation.

**Theorem 1.5.14 (Inclusion-Exclusion Formula)** *Let $A$ be a set and consider two functions $f_=, f_\geq : 2^A \longrightarrow$ from $2^A$ to a field . Then*

1. $f_\geq(S) = \sum_{T \supseteq S} f_=(T)$ *for* $S \subseteq A \iff f_=(S) = \sum_{T \supseteq S} (-1)^{|T-S|} f_\geq(T)$ *for* $S \subseteq A$.

2. $f_\leq(S) = \sum_{T \subseteq S} f_=(T)$ *for* $S \subseteq A \iff f_=(S) = \sum_{T \subseteq S} (-1)^{|S-T|} f_\leq(T)$ *for* $S \subseteq A$.

The most common interpretation is the following. Suppose we have a set $U$ of objects and a set $A$ of properties that each object in $U$ may or may not satisfy. If we know, for each $S \subseteq A$, the number $f_\geq(S)$ of elements having *at least* the properties in $S$ (or the number $f_\leq(S)$ of elements having *at most* the properties in $S$), then we obtain, for each $S \subseteq A$, the number $f_=(S)$ of elements having *exactly* the properties in $S$. We are often interested in the number $f_=(\emptyset)$ or $f_=(A)$ of elements satisfying none or all of the given properties.

Theorem 1.5.14 has a simple linear algebraic interpretation. Consider the two $2^A \times 2^A$ matrices $C, D$ whose non-zero entries are $C_{S,T} = 1$ for $S \subseteq T$, and $D_{S,T} = (-1)^{|T-S|}$ for $S \subseteq T$. Then the inclusion-exclusion formula is equivalent to the assertion that $C$ and $D$ are inverse matrices. This can be proved directly, but we prefer to deduce it as a special case of the Möbius inversion formula (Theorem 1.5.16). We now present two applications.

**Derangements.** One of the classic applications of the inclusion-exclusion formula is the enumeration of the **derangements** of $[n]$, which are the permutations $\pi \in S_n$ such that $\pi(i) \neq i$ for all $i$. Let $A = \{A_1, \dots, A_n\}$ where $A_i$ is the property that $\pi(i) = i$. Then $f_\geq(T) = (n - |T|)!$, so the number $D_n$ of derangements of $[n]$ is

$$\begin{aligned} D_n &= f_=(\emptyset) = \sum_T (-1)^{|T|} f_\geq(T) = \sum_{k=0}^{n} \binom{n}{k} (-1)^k (n-k)! \\ &= n! \left( \frac{1}{0!} - \frac{1}{1!} + \frac{1}{2!} - \cdots \pm \frac{1}{n!} \right). \end{aligned}$$

It follows that $D_n$ is the integer closest to $n!/e$.

**Discrete derivatives.** Consider the -vector space $\Gamma$ of functions $f : \mathbb{Z} \to$ . The **discrete derivative** of $f$ is the function $\Delta f$ given by $\Delta f(n) = f(n+1) - f(n)$. We now wish to show that, just as with ordinary derivatives,

$$\Delta^{d+1} f = 0 \text{ if and only if } f \text{ is a polynomial of degree at most } d.$$

This was part of Theorem 1.3.6. Regarding $\Delta$ as a linear operator on $\Gamma$, we have $\Delta = E - 1$ where $Ef(n) = f(n+1)$ and 1 is the identity. Then $\Delta^k = (E-1)^k = \sum_{i=0}^{k} \binom{k}{i} E^i (-1)^{k-i}$, so the $k$th discrete derivative is $\Delta^k f(n) = \sum_{i=0}^{k} (-1)^{k-i} \binom{k}{i} f(n+i)$.

The functions $f_{\leq}(S) = f(n+|S|)$ and $f_{=}(S) = \Delta^k f(|S|)$ satisfy Theorem 1.5.14.2, so we have $f(n+k) = \sum_{i=0}^{k} \binom{k}{i} \Delta^i f(n)$. (This is equivalent to $E^k = (\Delta+1)^k$.) If $\Delta^{d+1} f = 0$, this gives $f(k) = \sum_{i=0}^{d} \binom{k}{i} \Delta^i f(0)$, which is a polynomial in $k$ of degree at most $d$. The converse follows from the observation that $\Delta$ lowers the degree of a polynomial by 1.

### 1.5.5 Möbius functions and Möbius inversion

#### 1.5.5.1 The Möbius function

Given a locally finite poset $P$, let $\mathrm{Int}(P) = \{[x,y] : x,y \in P, x \leq y\}$ be the set of intervals of $P$. The (two-variable) Möbius function of a poset $P$ is the function $\mu : \mathrm{Int}(P) \to \mathbb{Z}$ defined by

$$\sum_{p \leq r \leq q} \mu(p,r) = \begin{cases} 1 & \text{if } p = q, \\ 0 & \text{otherwise.} \end{cases} \tag{1.18}$$

Here we are denoting $\mu(p,q) = \mu([p,q])$. We will later see that the Möbius function can be defined equivalently by the equations:

$$\sum_{p \leq r \leq q} \mu(r,q) = \begin{cases} 1 & \text{if } p = q, \\ 0 & \text{otherwise.} \end{cases} \tag{1.19}$$

When $P$ has a minimum element $\widehat{0}$, the (one-variable) **Möbius function** $\mu : P \to \mathbb{Z}$ is $\mu(x) = \mu(\widehat{0}, x)$. If $P$ also has a $\widehat{1}$, the **Möbius number** of $P$ is $\mu(P) = \mu(\widehat{0}, \widehat{1})$.

Computing the Möbius function is a very important problem, because the Möbius function is the poset analog of a derivative; and as such, it is a fundamental invariant of a poset. This problem often leads to very interesting enumerative combinatorics, as can be gleaned from the following gallery of Möbius functions.

**Theorem 1.5.15** *The Möbius functions of some key posets are as follows.*

1. *(Chain)* $P = \mathbf{n}$*:*

$$\mu_{\mathbf{n}}(i,j) = \begin{cases} 1 & \text{if } j = i \\ -1 & \text{if } j = i+1 \\ 0 & \text{otherwise.} \end{cases}$$

2. *(Boolean lattice)* $P = 2^A$:

$$\mu_{2^A}(S,T) = (-1)^{T-S}.$$

3. *(Divisor lattice)* $P = D_n$: *We have* $\mu_{D_n}(k,l) = \mu(l/k)$ *where*

$$\mu(m) = \begin{cases} (-1)^t & \text{if } m \text{ is a product of } t \text{ distinct primes, and} \\ 0 & \text{otherwise.} \end{cases}$$

*is the classical* ***Möbius function*** *from number theory.*

4. *(Young's lattice)* $P = Y$:

$$\mu(\lambda,\mu) = \begin{cases} (-1)^{|\mu-\lambda|} & \text{if } \mu-\lambda \text{ has no two adjacent squares, and} \\ 0 & \text{otherwise.} \end{cases}$$

5. *(Partition lattice)* $P = \Pi_n$: *The Möbius number of* $\Pi_n$ *is*

$$\mu(\Pi_n) = (-1)^{n-1}(n-1)!,$$

*from which a (less elegant) formula for the complete Möbius function can be derived.*

6. *(Non-crossing partition lattice)* $P = NC_n$: *The Möbius number of* $NC_n$ *is*

$$\mu(NC_n) = (-1)^{n-1}C_{n-1},$$

*where* $C_{n-1}$ *is the* $(n-1)$*th Catalan number. This gives a (less elegant) formula for the complete Möbius function.*

7. *(Bruhat order)* $P = S_n$:

$$\mu(u,v) = (-1)^{\ell(v)-\ell(u)},$$

*where the* ***length*** $\ell(w)$ *of a permutation* $w \in S_n$ *is the number of* ***inversions*** $(i,j)$ *where* $1 \leq i < j \leq n$ *and* $w_i > w_j$. *(There is a generalization of this result to the Bruhat order on any Coxeter group* $W$*, or even on a parabolic subgroup* $W^J$*; see Section 1.5.5.4.)*

8. *(Subspace lattice)* $P = L(\mathbb{F}_q^n)$:

$$\mu(U,V) = (-1)^d q^{\binom{d}{2}},$$

*where* $d = \dim V - \dim U$.

9. *(Distributive lattice)* $L = J(P)$:

$$\mu(I,J) = \begin{cases} (-1)^{|J-I|} & \text{if } J-I \text{ is an antichain in } P\text{, and} \\ 0 & \text{otherwise.} \end{cases}$$

10. *(Face poset of a polytope)* $L = F(P)$:

$$\mu(F,G) = (-1)^{\dim G - \dim F}.$$

11. *(Face poset of a subdivision $\mathcal{T}$ of a polytope P)* $L = \widehat{T}$:

$$\mu(F,G) = \begin{cases} (-1)^{\dim G - \dim F} & \text{if } G < \widehat{1} \\ (-1)^{\dim P - \dim F + 1} & \text{if } G = \widehat{1} \text{ and } F \text{ is not on the boundary of } P \\ 0 & \text{if } G = \widehat{1} \text{ and } F \text{ is on the boundary of } P. \end{cases}$$

12. *(Subgroup lattice of a finite p-group) If* $|G| = p^n$ *for p prime,* $n \in \mathbb{N}$, *then in* $L = L(G)$:

$$\mu(A,B) = \begin{cases} (-1)^k p^{\binom{k}{2}} & \text{if } A \text{ is a normal subgroup of } B \text{ and } B/A \cong \mathbb{Z}_p^k, \text{ and} \\ 0 & \text{otherwise.} \end{cases}$$

Some of the formulas above follow easily from the definitions, while others require more sophisticated methods. In the following sections we will develop some of the basic theory of Möbius functions and discuss the most common methods for computing them. Along the way, we will sketch proofs of all the formulas above.

It is worth remarking that a version of item 11 of Theorem 1.5.15 holds more generally for the face poset of any finite regular cell complexes $\Gamma$ such that the underlying space $|\Gamma|$ is a manifold with or without boundary; see [194, Prop. 3.8.9] for details.

#### 1.5.5.2 Möbius inversion

In enumerative combinatorics, there are many situations where we have a set $U$ of objects, and a natural way of assigning to each object $u$ of $U$ an element $f(u)$ of a poset $P$. We are interested in counting the objects in $U$ that map to a particular element $p \in P$. Often we find that it is much easier to count the objects in $s$ that map to an element **less than or equal to** * $p$ in $P$. The following theorem tells us that this easier enumeration is sufficient for our purposes, as long as we can compute the Möbius function of $P$.

**Theorem 1.5.16 (Möbius Inversion formula)** *Let P be a poset and let* $f,g : P \to$ *be functions from P to a field . Then*

1. $\forall p \in P \ g(p) = \sum_{q \geq p} f(q) \iff \forall p \in P \ f(p) = \sum_{q \geq p} \mu(p,q) g(q)$ *and*
2. $\forall p \in P \ g(p) = \sum_{q \leq p} f(q) \iff \forall p \in P \ f(p) = \sum_{q \leq p} \mu(q,p) g(q)$.

In his paper [169], which pioneered the use of the Möbius inversion formula as a tool for counting in combinatorics, Rota described this enumerative philosophy as follows:

---

*or greater than or equal to

> It often happens that a set of objects to be counted possesses a natural ordering, in general only a partial order. It may be unnatural to fit the enumeration of such a set into a linear order such as the integers: instead, it turns out in a great many cases, that a more effective technique is to work with the natural order of the set. One is led in this way to set up a "difference calculus" relative to an arbitrary partially ordered set.

Indeed, one may think of the Möbius function as a poset-theoretic analog of the Fundamental Theorem of Calculus: $g$ is analogous to the integral of $f$, as it stores the cumulative value of this function. Like the Fundamental Theorem of Calculus, the Möbius inversion formula tells us how to recover the function $f$ from its cumulative values.

It is not difficult to prove Theorem 1.5.16 directly, but we will soon discuss an algebraic framework that really explains it. In the meantime, we discuss two key applications. We will see several other applications later on.

**Example 1.5.17** *Möbius inversion is particularly important for chains, Boolean and divisor lattices.*

1. *For $P = \mathbb{N}$, Theorem 1.5.16.1 is a simple but important result for partial sums:*

$$g(n) = \sum_{i=0}^{n} f(i) \text{ for } n \in \mathbb{N} \quad \text{if and only if} \quad f(n) = g(n) - g(n-1) \text{ for } n \in \mathbb{N}$$

*where $g(-1) = 0$.*

2. *For $P = 2^A$, Theorem 1.5.16.1 is the inclusion-exclusion formula:*

$$g(S) = \sum_{T \supseteq S} f(T) \text{ for all } S \subseteq A \quad \text{if and only if}$$

$$f(S) = \sum_{T \supseteq S} (-1)^{|T-S|} g(T) \text{ for all } S \subseteq A.$$

3. *For $P = D_n$, Theorem 1.5.16.2 is the Möbius inversion formula from number theory:*

$$g(m) = \sum_{m|d|n} f(d) \text{ for } m|n \quad \text{if and only if} \quad f(m) = \sum_{m|d|n} \mu(d/m) g(d) \text{ for } m|n.$$

*A typical application is the computation of Euler's totient function*

$$\varphi(n) = |\{u : 1 \leq u \leq n, \ \gcd(u,n) = 1\}|,$$

*or more generally $f(m) = |\{u : 1 \leq u \leq n, \ \gcd(u,n) = m\}|$ for $m|n$. Here $U = [n]$, and we assign to each $u \in U$ the divisor $\gcd(n,u)$ of $n$. There are $g(m) = n/m$ multiples of $m$ in $U$, so*

$$\varphi(n) = f(1) = \sum_{d|n} \mu(d) \frac{n}{d} = n\left(1 - \frac{1}{p_1}\right) \cdots \left(1 - \frac{1}{p_k}\right).$$

We will be interested in applying the Möbius inversion formula in many other contexts, and for that reason it is important that we gain a deeper understanding of Möbius functions, and develop techniques to compute them. This is one of the main goals of the following sections.

#### 1.5.5.3 The incidence algebra

The Möbius function has a very natural algebraic interpretation, which we discuss in this section. Given a field , recall that a -algebra $A$ is a vector space over equipped with a bilinear product.

The **incidence algebra** $I(P)$ of a locally finite poset $P$ is the -algebra of functions $f : \mathrm{Int}(P) \to$ from the intervals of $P$ to , equipped with the **convolution product** $f \cdot g$ given by

$$f \cdot g(p,r) = \sum_{p \leq q \leq r} f(p,q)g(q,r) \qquad \text{for } p \leq r.$$

Alternatively, let $\mathrm{Mat}(P)$ be the set of $P \times P$ matrices $A$ with entries in whose only nonzero entries $a_{pq} \neq 0$ occur in positions where $p \leq q$ in $P$. There is no canonical way of listing the rows and columns of the matrix $P$ in a linear order. We normally list them in the order given by a linear extension of $P$, so that the matrices in $\mathrm{Mat}(P)$ will be upper triangular. Then $\mathrm{Mat}(P)$ is a -algebra under matrix multiplication, and it is clear from the definitions that

$$I(P) \cong \mathrm{Mat}(P).$$

**The unit and inverses.** The product in $I(P)$ is clearly associative, and has a **unit**

$$\mathbf{1}(p,q) = \begin{cases} 1 & \text{if } p = q, \\ 0 & \text{if } p < q. \end{cases}$$

which is a (two-sided) multiplicative identity. An element $f \subset I(P)$ has a (necessarily unique) left and right multiplicative inverse $f^{-1}$ if and only if $f(p,p) \neq 0$ for all $p \in P$.

**The zeta function and counting chains.** An important element of $I(P)$ is the **zeta function**

$$\zeta(p,q) = 1 \qquad \text{for all } p \leq q.$$

Notice that $\zeta^k(p,q)$ is the number of multichains of length $k$ from $p$ to $q$.

If $P$ has a $\widehat{0}$ and $\widehat{1}$, then $\zeta_P^k(\widehat{0},\widehat{1})$ counts all multichains of length $k-2$ in $P$, so the zeta polynomial and zeta function are related by

$$Z_P(k) = \zeta_P^k(\widehat{0},\widehat{1}) \qquad \text{for } k \geq 2. \tag{1.20}$$

Similarly, $(\zeta - \mathbf{1})^k(p,q)$ is the number of chains of length $k$ from $p$ to $q$. The sum $\mathbf{1} + (\zeta - \mathbf{1}) + (\zeta - \mathbf{1})^2 + \cdots$ has finitely many non-zero terms, and it equals $(\mathbf{2} - \zeta)^{-1}$, where $\mathbf{2} = 2 \cdot \mathbf{1}$; so

$$(\mathbf{2} - \zeta)^{-1}(p,q) = \text{ total number of chains in } P \text{ from } p \text{ to } q.$$

**The Möbius function and Möbius inversion.** The (equivalent) equations (1.18) and (1.19) defining the Möbius function can be rewritten as $\mu\zeta = \mathbf{1}$ and $\zeta\mu = \mathbf{1}$, respectively. This explains why these two equations are equivalent: they say that

$$\mu = \zeta^{-1}.$$

*Proof of the Möbius inversion formula.* Consider the left action of the incidence algebra $I(P)$ on the vector space $^P$ of functions $f : P \to$ , given by

$$a \cdot f(p) = \sum_{q \geq p} a(p,q) f(q)$$

for $a \in I(P)$ and $f : P \to$ . The Möbius inversion formula then states that $g = \zeta \cdot f$ if and only if $\mu \cdot g = f$; this follows immediately from $\mu = \zeta^{-1}$. ■

#### 1.5.5.4 Computing Möbius functions

In this section, we discuss some of the main tools for computing Möbius functions. Along the way, we prove the formulas for the Möbius functions of the posets of Theorem 1.5.15.

1. (Chain $\mathbf{n}$) We can check the formula for $\mu_{\mathbf{n}}$ manually.

**Möbius functions of products.** A simple but important fact is that Möbius functions behave well under poset multiplication:

$$\mu_{P\times Q}((p,q),(p',q')) = \mu_P(p,p')\mu_Q(q,q').$$

This is easily verified directly, and also follows from the fact that $I(P\times Q) \cong I(P) \otimes_{\mathbb{k}} I(Q)$.

2. (Boolean lattice $2^A$) Since $2^A \cong \mathbf{2} \times \cdots \times \mathbf{2}$, the Möbius function of $\mathbf{2}$ tells us that the Möbius function of $2^A$ is $\mu(S,T) = (-1)^{|T-S|}$.

3. (Divisor lattice $D_n$) Since $D_n \cong (\mathbf{t_1}+\mathbf{1}) \times \cdots \times (\mathbf{t_k}+\mathbf{1})$ for $n = p_1^{t_1} \cdots p_k^{t_k}$, we get the formula for $\mu_{D_n}$ from the formula for the Möbius function of a chain.

**Möbius functions through Möbius inversion.** So far we have thought of the Möbius function as a tool to apply Möbius inversion. Somewhat counterintuitively, it is possible to use Möbius inversion in the other direction, as a tool to compute Möbius functions. We carry out this approach to compute the Möbius function of the partition lattice $\Pi_n$.

4. (Partition lattice $\Pi_n$) Let $\Pi_A$ denote the lattice of set partitions of a set $A$ ordered by refinement. Every interval $[\pi,\rho]$ in $\Pi_A$ is a product of partition lattices, as illustrated by the following example: In $\Pi_9$ we have $[18|2|37|4|569, 13478|2569] \cong \Pi_{\{18,37,4\}} \times \Pi_{\{2,569\}} \cong \Pi_3 \times \Pi_2$. Since the

Möbius function is multiplicative, to compute the Möbius function of the partition lattices it suffices to show that $\mu_{\Pi_n}(\widehat{0},\widehat{1}) = (-1)^{n-1}(n-1)!$.

Let $W$ be the set of words $w = w_1 \ldots w_n$ of length $n$ in the alphabet $\{0,1,\ldots,q-1\}$. Classify the words according to the equalities among their coordinates; namely, to each word $w$, associate the partition of $[n]$ where $i$ and $j$ are in the same block when $w_i = w_j$.

Let $f(\pi)$ be the number of words whose partition is $\pi$, and let $g(\pi)$ be the number of words whose partition is a coarsening of $\pi$. This is a situation where both $f(\pi)$ and $g(\pi)$ are easily computed: If $\pi$ has $b$ blocks then we have

$$f(\pi) = q(q-1)\cdots(q-b+1), \qquad g(\pi) = q^b.$$

Since $g(\pi) = \sum_{\rho\geq\pi} f(\rho)$ we get $f(\pi) = \sum_{\rho\geq\pi} \mu(\pi,\rho)g(\rho)$. For $\pi = \widehat{0}$ this says that

$$q(q-1)\cdots(q-n+1) = \sum_{\rho\in\Pi_n} \mu(\widehat{0},\rho)q^{|\rho|}.$$

Equating the coefficients of $q^1$ we get the desired result.

**Möbius functions through closures.** A function $\overline{\cdot} : P \to P$ is a **closure operator** if

- $p \leq \overline{p}$ for all $p \in P$,
- $p \leq q$ implies $\overline{p} \leq \overline{q}$ for all $p,q \in P$, and
- $\overline{\overline{p}} = \overline{p}$ for all $p \in P$.

An element of $p$ is **closed** if $\overline{p} = p$; let $\mathrm{Cl}(P)$ be the subposet of closed elements of $P$.

**Proposition 1.5.18** *If $\overline{\cdot} : P \to P$ is a closure operator, then for any $p \leq q$ in $P$,*

$$\sum_{r:\overline{r}=q} \mu(p,r) = \begin{cases} \mu_{Cl(P)}(p,q) & \text{if } p \text{ and } q \text{ are closed} \\ 0 & \text{otherwise.} \end{cases}$$

*Proof.* We have an inclusion of incidence algebras $I(Cl(P)) \to I(P)$ given by $f \mapsto \overline{f}$ where

$$\overline{f}(p,q) = \begin{cases} f(p,q) & \text{if } p \text{ and } q \text{ are closed} \\ 0 & \text{otherwise.} \end{cases}$$

Let $\overline{1}, \overline{\zeta}$, and $\overline{\mu}$ be the image of the unit, zeta, and Möbius functions of $Cl(P)$ in $I(P)$. Also consider the "closure function" $c \in I(P)$ whose non-zero values are $c(p,\overline{p}) = 1$ for all $p$. Note that $c\overline{\zeta} = \zeta\overline{1}$ because when $q$ is closed, $\overline{p} \leq q$ if and only if $p \leq q$. Now we compute in $I(P)$:

$$\mu c \overline{1} = \mu c \overline{\zeta} \overline{\mu} = \mu \zeta \overline{1} \overline{\mu} = 1 \overline{\mu} = \overline{\mu}.$$

This is equivalent to the desired equality. ■

**Möbius functions of lattices.** When $L$ is a lattice, there are two other methods for computing Möbius functions. The first one gives an alternative to the defining recursion for $\mu(p,q)$. This new recurrence is usually much shorter, at the expense of requiring some understanding of the join operation.

**Proposition 1.5.19 (Weisner's Theorem)** *For any $p < a \leq q$ in a lattice $L$ we have*

$$\sum_{p \leq r \leq q : r \vee a = q} \mu(p,r) = 0.$$

Weisner's theorem follows from Proposition 1.5.18 because $r \mapsto r \vee a$ is a closure operation, whose closed sets are the elements greater than or equal to $a$. There is also a dual version, obtained by applying the result to the reverse lattice $L^{\mathrm{op}}$ obtained from $L$ by reversing all order relations.

Let us apply Weisner's theorem to three additional examples of computing Möbius functions.

5. (Subspace lattice $L(\mathbb{F}_q^n)$) Since an interval of height $r$ in $L(\mathbb{F}_q^n)$ is isomorphic to $L(\mathbb{F}_q^r)$, it suffices to compute $\mu(L(\mathbb{F}_q^n)) =: \mu_n$. Let $p = \widehat{0}$, $q = \widehat{1}$, and let $a$ be any line. The only subspaces $r \neq \widehat{1}$ with $r \vee a = \widehat{1}$ are the $q^{n-1}$ hyperplanes not containing $a$. Each one of these hyperplanes satisfies $[\widehat{0}, r] \cong L(\mathbb{F}_q^{n-1})$, so Weisner's theorem gives $\mu_n + q^{n-1}\mu_{n-1} = 0$, from which $\mu_n = (-1)^n q^{n(n-1)/2}$.

6. (Partition lattice $\Pi_n$, revisited) A similar (and easier) argument may be used to compute $\mu_n = \mu(\Pi_n)$, though it is now easier to use the dual to Weisner's theorem. Let $p = \widehat{0}$, $q = \widehat{1}$, and let $a$ be the coatom $12\ldots n-1|n$. The only partitions $r \neq \widehat{0}$ with $r \wedge a = \widehat{0}$ are those with only one non-singleton block of the form $\{i,n\}$ for $i \neq n$. Each such partition $\pi$ has $[\pi, \widehat{1}] \cong \Pi_{n-1}$. The dual to Weisner's theorem tells us that $\mu_n + (n-1)\mu_{n-1} = 0$, from which $\mu_n = (-1)^n(n-1)!$.

7. (Subgroup lattice $L(G)$ of a finite $p$-group $G$) A similar argument may be used for the subgroup lattice $L(G)$ of a $p$-group $G$. However, now one needs to invoke some facts about $p$-groups; see [210].

**Theorem 1.5.20 (Crosscut Theorem)** *Let $L$ be a lattice and let $X$ be the set of atoms of $L$. Then*

$$\mu(\widehat{0}, \widehat{1}) = \sum_k (-1)^k N_k,$$

*where $N_k$ is the number of $k$-subsets of $X$ whose join is $\widehat{1}$.*

We will sketch a proof at the end of this section. Meanwhile, we point out a simple corollary of the Crosscut theorem:

If the join of the atoms is not $\widehat{1}$ then $\mu(\widehat{0}, \widehat{1}) = 0$.

8. (Distributive lattice $L = J(P)$) If $J - I$ is an antichain in $P$, then $[I,J]$ is a Boolean lattice in $J(P)$ and $\mu_L(I,J) = (-1)^{J-I}$. Otherwise, the join of the atoms of $[I,J]$ is $I \cup \min(J-I) \neq J$, and hence $\mu_L(I,J) = 0$.

9. (Young's lattice $Y$) We obtain this Möbius function for free since $Y$ is distributive.

Naturally, there are dual formulations to the previous two propositions, obtained by reversing the order of $L$. Also, there are many different versions of the crosscut theorem; see for example [169].

**Möbius functions through multichains.** If we know the zeta polynomial of a poset $P$ with $\widehat{0}$ and $\widehat{1}$, we can obtain its Möbius number $\mu(P)$ immediately.

**Proposition 1.5.21** *If $P$ is a poset with $\widehat{0}$ and $\widehat{1}$,*

$$Z_P(-1) = \mu_P(\widehat{0},\widehat{1}). \tag{1.21}$$

*Proof.* We saw in (1.20) that $Z_P(k) = \zeta_P^k(\widehat{0},\widehat{1})$ for all integers $k \geq 2$. It would be irresponsible to just set $k=-1$, but it is very tempting, since $\zeta^{-1} = \mu$.

In the spirit of **combinatorial reciprocity**, this is an instance where such irresponsible behavior pays off, with a bit of extra care. We know that $Z_P(k)$ is polynomial for $k \in \mathbb{Z}$, and we leave it as an exercise to show that $\zeta^k(\widehat{0},\widehat{1})$ is also polynomial for $k \in \mathbb{Z}$. Since these two polynomials agree on infinitely many values, they also agree for $k=-1$, and the result follows. ∎

In light of (1.21), the zeta polynomial of $P$ will give us the Möbius number $\mu(P)$ automatically. This is advantageous because sometimes the zeta polynomial is easier to compute than the Möbius function, as it is the answer to an explicit enumerative question.

10. (Non-crossing partition lattice $NC_n$) From the zeta polynomial of $NC_n$ in Example 1.5.9 we immediately obtain that $\mu(NC_n) = (-1)^{n-1}C_{n-1}$ where $C_{n-1}$ is the $(n-1)$th Catalan number. This gives a formula for the full Möbius function, since every interval in $NC_n$ is a product of smaller non-crossing partition lattices.

**Möbius functions through topology.** Equation (1.21) has a topological interpretation that is an extremely powerful method for computing Möbius functions.

**Proposition 1.5.22 (Phillip Hall's Theorem)** *Let $P$ be a finite poset with a $\widehat{0}$ and $\widehat{1}$, and let $c_i$ be the number of chains $\widehat{0} = p_0 < p_1 < \cdots < p_i = \widehat{1}$ of length $i$ from $\widehat{0}$ to $\widehat{1}$ in $P$. Then*

$$\mu_P(\widehat{0},\widehat{1}) = c_0 - c_1 + c_2 - \cdots$$

This formula is equivalent to (1.21) in light of (1.17) and the relations $b_i = c_i + 2c_{i-1} + c_{i-2}$; it may also be proved directly in the incidence algebra of $P$.

Let us now interpret this result topologically. The **order complex** $\Delta(P)$ of a poset $P$ is the simplicial complex whose vertices are the elements of $P$, and whose faces are the chains of $P$.

**Theorem 1.5.23** *Let $P$ be a finite poset with a $\widehat{0}$ and $\widehat{1}$, and let $\overline{P} = P - \{\widehat{0}, \widehat{1}\}$. Then*

$$\mu_P(\widehat{0}, \widehat{1}) = \widetilde{\chi}(\Delta(\overline{P}))$$

*is the reduced Euler characteristic of $\Delta(\overline{P})$.*

*Proof.* In light of Proposition 1.5.22, this follows immediately from the combinatorial formula

$$\widetilde{\chi}(\Delta) = \sum_{k=-1}^{d} (-1)^k f_k$$

for the Euler characteristic of $|\Delta|$, where $f_k$ is the number of $k$-dimensional faces of $\Delta$. ■

Let us use this topological description to sketch proofs of the remaining Möbius functions of Theorem 1.5.15.

11. (Face lattice of a polytope $L(P)$) The **barycentric subdivision** $\mathrm{sd}(P)$ is a simplicial complex with a vertex $v(F)$ at the barycenter of each proper face $F$. It has a simplex connecting vertices $v(F_1), \ldots, v(F_k)$ whenever $F_1 \subset \cdots \subset F_k$. As abstract simplicial complexes, $\mathrm{sd}(P)$ equals $\Delta(\overline{F(P)})$. Geometrically, $\mathrm{sd}(P)$ is a subdivision of the boundary of the polytope $P$, and hence is homeomorphic to the sphere $\mathbb{S}^{\dim P-1}$. Therefore

$$\mu(\widehat{0}, \widehat{1}) = \widetilde{\chi}(\Delta(\overline{P})) = \widetilde{\chi}(\mathbb{S}^{\dim P-1}) = (-1)^{\dim P}.$$

We will see in Section 1.6 that every interval of $F(P)$ is itself the face poset of a polytope. It then follows that $\mu(F,G) = (-1)^{\dim G - \dim F}$.

12. (Face lattice of a subdivision $\mathscr{T}$ of a polytope $P$) A similar argument holds, though the details are slightly more subtle; see [194, Prop. 3.8.9].

13. (Bruhat order in permutations $S_n$) There are several known proofs of the fact that $\mu(u,v) = (-1)^{l(u)-l(v)}$, none of which is easy. The first proof was an ad hoc combinatorial argument due to Verma [205]. Later Kazhdan and Lusztig [116] and Stembridge [195] gave algebraic proofs using Kazhdan-Lusztig polynomials and Hecke algebras, respectively. Björner and Wachs [40] gave a topological proof, based on Theorem 1.5.23 and further tools from topological combinatorics. As we remarked earlier, there are similar formulas for the Möbius function of an arbitrary parabolic quotient of an arbitrary Coxeter group; see [38] or [195].

Theorem 1.5.23 tells us that in order to compute Möbius functions of posets of interest, it can be very useful to understand the topology of their underlying order complexes. Conversely, combinatorial facts about Möbius functions often lead to the discovery of topological properties of these complexes. This is the motivation for the very rich study of **poset topology**. We refer the reader to the survey [206] for further information on this topic.

### 1.5.6 Eulerian posets, flag $f$-vectors, and flag $h$-vectors

Let $P$ be a graded poset of height $r$. The **flag $f$-vector** $(f_S : S \subseteq [0,r])$ and the **flag $h$-vector** $(h_S : S \subseteq [0,r])$ are defined by

$$f_S = \text{number of chains } p_1 < p_2 < \cdots < p_k \text{ with } \{r(p_1), \ldots r(p_k)\} = S, \quad \text{and}$$

$$h_S = \sum_{T \subseteq S} (-1)^{|S|-|T|} f_T, \qquad f_S = \sum_{T \subseteq S} h_T$$

for $S \subseteq [0,r]$.

**Example 1.5.24** *Let $H$ be a hexagonal prism of Figure 1.28, and let $P = L(H) - \{\widehat{0}, \widehat{1}\}$ be its face lattice, with the top and bottom element (the empty face and the full face $H$) removed. It is not so enlightening to draw the poset, but we can still compute its flag $f$ and $h$-vectors. For example, $f_{\{0,2\}} = 36$ counts the pairs $(v, f)$ of a vertex $v$ contained in a 2-face $f$. We get:*

| | $\emptyset$ | 0 | 1 | 2 | 01 | 02 | 12 | 012 |
|---|---|---|---|---|---|---|---|---|
| $f_S$ | 1 | 12 | 18 | 8 | 36 | 36 | 36 | 72 |
| $h_S$ | 1 | 11 | 17 | 7 | 7 | 17 | 11 | 1 |

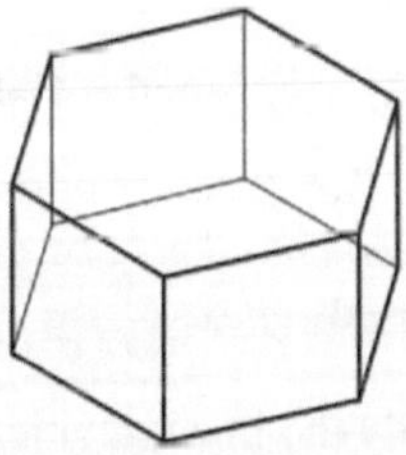

**Figure 1.28**
A hexagonal prism.

This example suggests that there may be additional structure in the flag $h$-vector; for instance in this example we have $h_S = h_{[0,d]-S}$ for all $S$. This equation is not true in general, but it does hold for an important class of posets, which we now discuss.

**Eulerian posets.** Say a graded poset $P$ is **Eulerian** if $\mu(x,y) = (-1)^{r(y)-r(x)}$ for all $x \leq y$.

In Theorem 1.5.15 we saw three important families of Eulerian posets: Boolean lattices, the Bruhat order, and face posets of polytopes. In particular, the poset $P$ of Example 1.5.24 is Eulerian.

For an Eulerian poset, there are many linear relations among the $f_S$s, which are easier to describe in terms of the $h_S$s; for example, $h_S = h_{[0,d]-S}$ for all $S$. To describe them all, we further encode the flag $f$-vector in a polynomial in non-commuting variables **a** and **b** called the **ab-index**, which is defined to be

$$\Phi_P[\mathbf{a},\mathbf{b}] = \sum_{S \subseteq [d]} h_S u_S$$

where $u_S = u_1 \dots u_d$ and $u_i = \mathbf{a}$ if $i \notin S$ and $u_i = \mathbf{b}$ if $i \in S$. This is an element of the ring $\mathbb{Z}\langle \mathbf{a},\mathbf{b}\rangle$ of integer polynomials in the non-commutative variables **a** and **b**.

**Theorem 1.5.25** [28, 29, 191] *The* **ab**-*index of an Eulerian poset $P$ can be expressed uniquely as a polynomial in*

$$\mathbf{c} = \mathbf{a} + \mathbf{b}, \qquad \mathbf{d} = \mathbf{ab} + \mathbf{ba}$$

*called the* **cd**-*index* $\Psi_P(\mathbf{c},\mathbf{d})$ *of $P$. Furthermore, if $P$ is the face poset of a polytope, then the coefficients of the* **cd**-*index are non-negative.*

**Example 1.5.26** *The* **ab**-*index of the hexagonal prism in Example 1.5.24 is*

$$\begin{aligned} \Phi(\mathbf{a},\mathbf{b}) &= \mathbf{aaa} + 11\mathbf{baa} + 17\mathbf{aba} + 7\mathbf{aab} + 7\mathbf{bba} + 17\mathbf{bab} + 11\mathbf{abb} + \mathbf{bbb} \\ &= (\mathbf{a}+\mathbf{b})^3 + 6(\mathbf{a}+\mathbf{b})(\mathbf{ab}+\mathbf{ba}) + 10(\mathbf{ab}+\mathbf{ba})(\mathbf{a}+\mathbf{b}) \end{aligned}$$

*so the* **cd**-*index is*

$$\Psi(\mathbf{c},\mathbf{d}) = \mathbf{c}^3 + 6\mathbf{cd} + 10\mathbf{dc}.$$

*In general, the eight entries of the flag $f$-vector of a* 3*-polytope (or any other Eulerian poset of rank* 3*) are determined completely by only three numbers, namely, the coefficients of* $\mathbf{c}^3$, **cd**, *and* **dc** *in the* **cd**-*index.*

The **cd**-index encodes optimally the linear relations among the entries of flag $f$-vectors of polytopes. Since the number of monomials in **c** and **d** of **ab**-degree $d$ is the Fibonacci number $F_{d+1}$, this is the smallest number of entries in $(f_S)$ from which we can recover the whole flag $f$-vector—this is much smaller than $2^d$.

**Corollary 1.5.27** [28, 29] *The subspace of* $\mathbb{R}^{2^d}$ *spanned by the flag $f$-vectors of Eulerian posets of rank $d$ (or by the flag $f$-vectors of $d$-polytopes) has dimension equal to the Fibonacci number* $F_{d+1}$.

# PART 2. DISCRETE GEOMETRIC METHODS

Part 2 is devoted to discrete geometry, which studies the connections between combinatorics and the geometry of subspaces (points, lines, planes, ..., hyperplanes) in a Euclidean space. Configurations in the plane and 3-space have received great attention, and feature many interesting results and open questions; however, very few of them involve exact enumeration in a meaningful way. Instead, we will focus on the discrete geometry of higher dimensions, where

- studying "general" geometric configurations leads to interesting enumerative questions.
- we have enough room to construct "special" geometric configurations that model various combinatorial structures of interest.

This second part is divided into three sections on closely interrelated topics. In Section 1.6 we discuss polytopes, which are the higher dimensional generalization of polygons. Section 1.7 discusses arrangements of hyperplanes in a vector space. Finally Section 1.8 is devoted to matroids, which are combinatorial objects that simultaneously abstract arrangements of vectors, graphs, and matching problems, among others.

## 1.6 Polytopes

The theory of polytopes is a vast area of study, with deep connections to pure (algebraic geometry, commutative algebra, representation theory) and applied mathematics (optimization). Again, we focus on aspects related to enumeration. For a general introduction to polytopes, see [93, 225].

After discussing the basic definitions and facts in Section 1.6.1 and some important examples in Section 1.6.2, we turn to enumerative questions. Section 1.6.3 is devoted to the enumeration of faces of various dimensions. Section 1.6.4 is on *Ehrhart theory*, which measures polytopes by counting the lattice points that they (and their dilations) contain.

### 1.6.1 Basic definitions and constructions

Recall that a set $S$ in Euclidean space $\mathbb{R}^d$ is **convex** if for every pair of points $u, v$ in $S$, the line segment $uv$ is in $S$. The **convex hull** $\text{conv}(S)$ of a set $S \subseteq \mathbb{R}^d$ is the minimal convex set containing $S$. If $S = \{\mathbf{v}_1, \ldots, \mathbf{v}_n\}$ then

$$\begin{aligned} \text{conv}\{\mathbf{v}_1, \ldots, \mathbf{v}_n\} &= \{\lambda_1 \mathbf{v}_1 + \cdots + \lambda_n \mathbf{v}_n : \lambda_1, \ldots, \lambda_n \geq 0, \lambda_1 + \cdots + \lambda_n = 1\} \\ &= \text{intersection of all convex sets containing } \mathbf{v}_1, \ldots, \mathbf{v}_n. \end{aligned}$$

We will only be interested in convex polytopes, and when we talk about polytopes, it will be assumed that they are convex.

A **hyperplane** $H$ in Euclidean space $\mathbb{R}^d$ is an affine subspace of dimension $d-1$; it is given by a linear equation

$$H = \{\mathbf{x} \in \mathbb{R}^d : \mathbf{a} \cdot \mathbf{x} = b\} \qquad \text{for some } \mathbf{a} \in \mathbb{R}^d - \{\mathbf{0}\} \text{ and } b \in \mathbb{R}.$$

It separates $\mathbb{R}^d$ into the two halfspaces given by the inequalities $\mathbf{a} \cdot \mathbf{x} \leq b$ and $\mathbf{a} \cdot \mathbf{x} \geq b$, respectively.

There are two equivalent ways of defining convex polytopes: The *V-description* gives a polytope in terms of its vertices, and the *H-description* gives it in terms of its defining inequalities.

**The V-description.** A **convex polytope** $P$ is the convex hull of finitely many points $\mathbf{v}_1, \ldots, \mathbf{v}_n$ in a Euclidean space $\mathbb{R}^d$:

$$P = \text{conv}\{\mathbf{v}_1, \ldots, \mathbf{v}_n\}$$

**The H-description.** A **convex polytope** $P$ is a bounded intersection of finitely many halfspaces in a Euclidean space $\mathbb{R}^d$:

$$\begin{aligned} P &= \left\{\mathbf{x} \in \mathbb{R}^d : \mathbf{a}_1 \cdot \mathbf{x} \leq b_1, \ldots, \mathbf{a}_m \cdot \mathbf{x} \leq b_m\right\} \\ &= \left\{x \in \mathbb{R}^d : A\mathbf{x} \leq \mathbf{b}\right\} \end{aligned}$$

where $\mathbf{a}_1, \ldots, \mathbf{a}_m \in \mathbb{R}^d$ and $b_1, \ldots, b_m \in \mathbb{R}$. In the second expression we think of $\mathbf{x}$ as a column vector, $A$ is the $m \times d$ matrix with rows $\mathbf{a}_1, \ldots \mathbf{a}_m$ and $\mathbf{b}$ is the column vector with entries $b_1, \ldots, b_m$.

**Theorem 1.6.1** *A subset $P \subseteq \mathbb{R}^d$ is the convex hull of a finite set of points if and only if it is a bounded intersection of finitely many halfspaces.*

For theoretical and practical purposes, it is useful to have **both** the V-description and the H-description of a polytope. For example, it is clear from the H-description (but not at all from the V-description) that the intersection of two polytopes is a polytope, and it is clear from the V-description (but not at all from the H-description) that a projection of a polytope is a polytope. It is a nontrivial task to translate one description into the other; see [225, Notes to Chapter 1] for a discussion and references on this problem. Here are some simple examples.

**Example 1.6.2** *The* ***standard simplex*** *$\Delta_{d-1}$, the* ***cube*** *$\square_d$, and the* ***crosspolytope*** *$\diamondsuit_d$ are:*

1. $\begin{aligned} \Delta_{d-1} &= \text{conv}\{\mathbf{e}_1, \ldots, \mathbf{e}_d\} \\ &= \{\mathbf{x} \in \mathbb{R}^d : x_1 + \cdots + x_d = 1 \textit{ and } x_i \geq 0 \textit{ for } i = 1, \ldots, d\}, \end{aligned}$
2. $\begin{aligned} \square_d &= \text{conv}\{\pm\mathbf{e}_1 \pm \cdots \pm \mathbf{e}_d \textit{ for any choice of signs}\} \\ &= \{\mathbf{x} \in \mathbb{R}^d : -1 \leq x_i \leq 1 \textit{ for } i = 1, \ldots, d\}, \end{aligned}$
3. $\begin{aligned} \diamondsuit_d &= \text{conv}\{-\mathbf{e}_1, \mathbf{e}_1, \ldots, -\mathbf{e}_d, \mathbf{e}_d\} \\ &= \{\mathbf{x} \in \mathbb{R}^d : \pm x_1 \pm \cdots \pm x_d \leq 1 \textit{ for any choice of signs}\}, \end{aligned}$

*respectively, where* $\mathbf{e}_1, \ldots, \mathbf{e}_d$ *are the standard basis in* $\mathbb{R}^d$*. These polytopes are illustrated in Figure 1.29 for* $d = 3$.

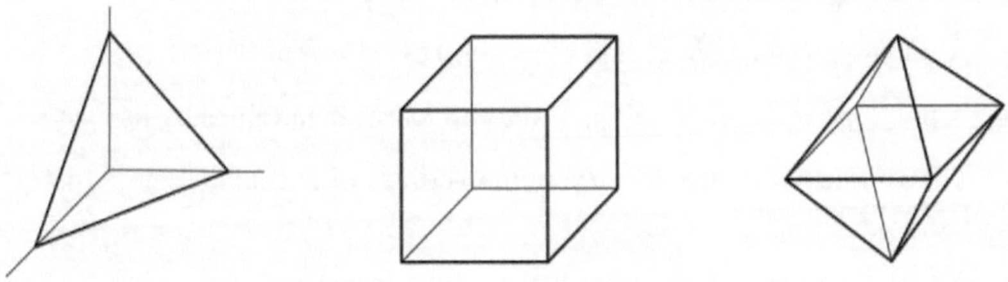

**Figure 1.29**
The triangle $\Delta_2$, the cube $\square_3$, and the octahedron $\lozenge_3$.

The **dimension** of a polytope $P$ is the dimension of the affine subspace spanned by $P$:

$$\text{aff}(P) = \left\{\lambda_1 \mathbf{v}_1 + \cdots + \lambda_k \mathbf{v}_k : \mathbf{v}_1, \ldots, \mathbf{v}_k \in \mathbb{R}^d, \lambda_1, \ldots, \lambda_k \in \mathbb{R}, \lambda_1 + \cdots + \lambda_k = 0\right\}.$$

For example, even though $\Delta_{d-1}$ lives in $\mathbb{R}^d$, its dimension is $d-1$ because it lies on the hyperplane $x_1 + \cdots + x_d = 1$.

The **interior** $\text{int}(P)$ of a polytope $P \subseteq \mathbb{R}^d$ is the topological interior of $P$. It is often more useful to consider the **relative interior** $\text{relint}(P)$, which is its topological interior as a subset of the affine space $\text{aff}(P)$.

**Polar polytopes.** The similarity of our descriptions of $\square_d$ and $\lozenge_d$ is a manifestation of a general notion of duality between V-descriptions and H-descriptions of polytopes.

Let $P$ be a polytope in $\mathbb{R}^d$ such that $\mathbf{0} \in \text{int}(P)$. * The **polar polytope** $P^\triangle$ of $P$ is

$$P^\triangle = \left\{\mathbf{a} \in \mathbb{R}^d : \mathbf{a} \cdot \mathbf{x} \leq 1 \text{ for all } \mathbf{x} \in P\right\}.$$

If $P$ has vertices $\mathbf{v}_1, \ldots \mathbf{v}_n$ and facets $\mathbf{a}_1 \cdot \mathbf{x} \leq 1, \ldots, \mathbf{a}_m \cdot \mathbf{x} \leq 1$ (and hence contains $\mathbf{0}$), then the polar $P^\triangle$ has vertices $\mathbf{a}_1, \ldots \mathbf{a}_m$ and facets $\mathbf{v}_1 \cdot \mathbf{x} \leq 1, \ldots, \mathbf{v}_n \cdot \mathbf{x} \leq 1$. It follows that if $\mathbf{0} \in P$, then

$$(P^\triangle)^\triangle = P.$$

**Faces.** For each vector $\mathbf{a} \in \mathbb{R}^d$, let

$$P_{\mathbf{a}} = \{\mathbf{x} \in P : \mathbf{a} \cdot \mathbf{x} \geq \mathbf{a} \cdot \mathbf{y} \text{ for all } \mathbf{y} \in P\}$$

*We can apply an affine transformation to any full-dimensional polytope to make its interior contain the origin.

be the subset of $P$ where the linear function $\mathbf{a}\cdot\mathbf{x}$ is maximized. Such a set is called a **face** of $P$. It is customary to allow the empty set to be a face of $P$ as well. If $\dim(P) = d$, then the faces of dimension $0, 1, d-2, d-1$ are called **vertices, edges, ridges**, and **facets**, respectively. We let $V(P)$ be the set of vertices of $P$. Faces other than $P$ are called **proper faces**. We collect a few basic facts about faces.

- A polytope is the convex hull of its vertices: $P = \mathrm{conv}(V(P))$.
- A polytope is the intersection of the halfspaces determined by its facets.
- The vertices of a face $F$ of $P$ are the vertices of $P$ contained in $F$: $V(F) = V(P) \cap F$.
- A face $F$ of $P$ equals the intersection of the facets of $P$ containing $F$.
- If $F$ is a face of $P$, then any face of $F$ is a face of $P$.
- The intersection of two faces of $P$ is a face of $P$.
- A polytope is the disjoint union of the relative interiors of its faces: $P = \bigcup_{F \text{ face}} \mathrm{relint}(F)$.

**The face lattice.** The **face lattice** $L(P)$ is the poset of faces of $P$, ordered by containment. It is indeed a lattice with $F \wedge G = F \cap G$ and $F \vee G = \mathrm{aff}(F \cup G) \cap P$. It is graded with $\mathbf{rk}(F) = \dim F + 1$. We say that polytopes $P$ and $Q$ are **combinatorially isomorphic** if $L(P) \cong L(Q)$. We collect some basic properties of face lattices:

- For each face $F$ of $P$, the interval $[\widehat{0}, F]$ of $L(P)$ is isomorphic to the face lattice of $F$.
- For each face $F$ of $P$, the interval $[F, \widehat{1}]$ of $L(P)$ is isomorphic to the face lattice of a polytope, called the **face figure** $P/F$ of $F$.
- Every interval $[F, G]$ of $L(P)$ is isomorphic to the face lattice of some polytope.
- The face lattice $L(P^{\triangle})$ of the polar polytope $P^{\triangle}$ is isomorphic to the opposite poset $L(P)^{\mathrm{op}}$, obtained by reversing the order relations of $L(P)$.

There are different constructions of the face figure $P/F$, giving rise to combinatorially isomorphic polytopes. We discuss one such construction. Let $F^{\diamond}$ be the face of $P^{\triangle}$ corresponding to the face $F$ of $P$ under the isomorphism $L(P^{\triangle}) \cong L(P)^{\mathrm{op}}$. Then we can define the face figure to be the polar polytope of $F^{\diamond}$; that is, $P/F = (F^{\diamond})^{\triangle}$. When $F = \mathbf{v}$ is a vertex, there is a more direct construction: let $P/\mathbf{v} = P \cap H$, where $H$ is a hyperplane separating $v$ from all other vertices of $P$.

We say a $d$-polytope $P$ is **simplicial** if every face is a simplex. We say $P$ is **simple** if every vertex is on exactly $d$ facets (or, equivalently, on $d$ edges). Note that the convex hull of generically chosen points is a simplicial polytope. Similarly, a bounded intersection of generically chosen half-spaces is a simple polytope. Also note that $P$ is simplicial if and only if its polar $P^{\triangle}$ is simple.

**Triangulations and subdivisions.** In many contexts, it is useful to subdivide a polytope into simpler polytopes (most often simplices). It is most convenient to do it in such a way that the pieces of the subdivision meet face to face.

A **subdivision** of a polytope $P$ is a finite collection $\mathcal{T}$ of polytopes such that

- $P$ is the union of the polytopes in $\mathcal{T}$,
- if $P \in \mathcal{T}$ then every face of $P$ is in $\mathcal{T}$, and
- if $P, Q \in \mathcal{T}$ then $P \cap Q \in \mathcal{T}$.

The elements of $\mathcal{T}$ are called the **faces** of $\mathcal{T}$; the full-dimensional ones are called **facets**. If all the faces are simplices, then $\mathcal{T}$ is called a **triangulation** of $P$.

In some situations, it is useful to assume that a subdivision does not introduce new vertices; that is, that the only points in $\mathcal{T}$ are the vertices of $P$. We will assume that throughout the rest of this chapter.

**Theorem 1.6.3** *Every convex polytope has a triangulation.*

*Sketch of Proof.* Let $V$ be the set of vertices of our polytope $P \in \mathbb{R}^d$. For each **height function** $h : V \to \mathbb{R}$ consider the set of lifted points $P^h = \{(v, h(v)), v \in V\}$ in $\mathbb{R}^{d+1}$. Let $Q$ be the convex hull of $P^h$, and consider the set $\mathcal{F}$ of "lower facets" of $Q$ that are visible from below, that is, the facets maximizing some linear function $\mathbf{a} \cdot \mathbf{x}$ with $a_{d+1} = -1$. For each such facet $F$, let $\pi(F)$ be its projection back down to $\mathbb{R}^d$. One may check that $\{\pi(F) : F \text{ is a lower facet of } Q\}$ is a subdivision of $P$. If the height function is chosen generically, then this subdivision is actually a triangulation. See Figure 1.30 for an illustration. ∎

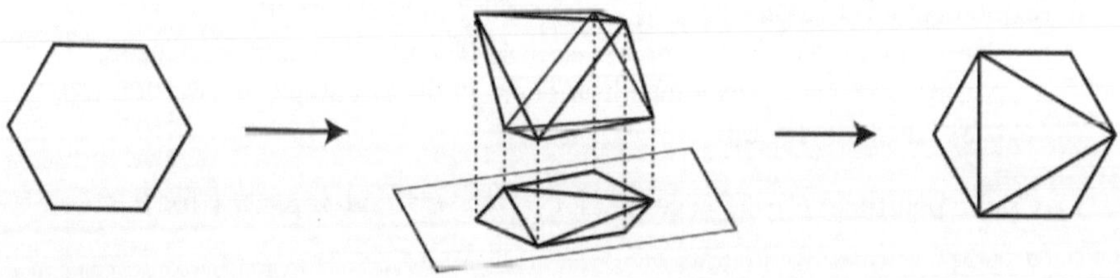

**Figure 1.30**
A regular subdivision of a hexagon.

The subdivisions that may be obtained from the lifting construction above are called **regular** or **coherent**. In general, not every subdivision is regular, and the question of distinguishing regular and non-regular triangulations is subtle. In the simplest case, when $P$ is a convex polygon, every subdivision is regular, and the number of subdivisions is a Catalan number, as we saw in Section 1.3.2.2. Few other exact enumerative results of this sort are known.

In any case, regular subdivisions are of great importance for several reasons. One reason is that they are easy to define and construct; for example, choosing the heights $h(v) = v_1^2 + \cdots v_d^2$ leads to the **Delaunay triangulation**, which is very easy to compute and has several desirable properties. Secondly, they have a very elegant structure. Every polytope $P$ gives rise to a **secondary polytope**, whose faces are in bijection with the regular subdivisions of $P$; and faces $F$ and $G$ satisfy that $F \subset G$ if and only if the subdivision of $F$ refines the subdivision of $G$. Thirdly, they are widely applicable, in particular, because they play a key role in the theory of Gröbner bases in commutative algebra. For our enumerative purposes the regularity question will not be too important. For readers interested in this and other aspects of triangulations, we recommend [63, 87, 196].

### 1.6.2 Examples

The following polytopes have particularly nice enumerative properties. We give references to some relevant results at the end of Section 1.6.4.

1. (Product of two simplices) $\Delta_{c-1} \times \Delta_{d-1} = \text{conv}\{(\mathbf{e}_i, \mathbf{f}_j) : 1 \leq i \leq c, 1 \leq j \leq d\}$ in $\mathbb{R}^c \times \mathbb{R}^d$ where $\{\mathbf{e}_1, \ldots, \mathbf{e}_c\}$ and $\{\mathbf{f}_1, \ldots, \mathbf{f}_d\}$ are the standard bases for $\mathbb{R}^c$ and $\mathbb{R}^d$, respectively.

2. (Hypersimplex) $\Delta(r,d) = \text{conv}\{\mathbf{e}_{i_1} + \cdots + \mathbf{e}_{i_r} : 1 \leq i_1 < \cdots < i_r \leq d\}$ in $\mathbb{R}^d$.

3. (Permutahedron)
$$\Pi_{d-1} = \text{conv}\{(a_1, \ldots, a_d) : \{a_1, \ldots, a_d\} \text{ is a permutation of } [d]\}.$$

4. (Zonotopes) The zonotope of a vector configuration $A = \{\mathbf{a}_1, \ldots, \mathbf{a}_k\} \subset \mathbb{R}^d$ is the *Minkowski sum* $Z(A) = [\mathbf{0}, \mathbf{a}_1] + \cdots + [\mathbf{0}, \mathbf{a}_k] := \{\lambda_1 \mathbf{a}_1 + \cdots + \lambda_k \mathbf{a}_k : 0 \leq \lambda_1, \ldots, \lambda_k \leq 1\}$. The permutahedron $\Pi_{d-1}$ is the zonotope of the *root system* $\{\mathbf{e}_i - \mathbf{e}_j : 1 \leq i < j \leq d\}$.

5. (Cyclic polytope) $C_d(m) = \text{conv}\{(1, t_i, t_i^2, \ldots, t_i^{d-1}) \mid 1 \leq i \leq m\}$ for $t_1 < \cdots < t_m$.

6. (Order polytope of a poset $P$) $\mathscr{O}(P) = \{\mathbf{x} \in \mathbb{R}^P : x_i \geq 0 \text{ for all } i \text{ and } x_i \leq x_j \text{ if } i < j \text{ in } P\}$.

7. (Chain polytope of a poset $P$)
$$\mathscr{C}(P) = \{\mathbf{x} \in \mathbb{R}^P : x_i \geq 0 \text{ for all } i \text{ and}$$
$$x_{i_1} + \cdots + x_{i_k} \leq 1 \text{ for each chain } i_1 < \cdots < i_k \text{ in } P\}.$$

8. (Type A root polytope) $A_{d-1} = \text{conv}\{\mathbf{e}_i - \mathbf{e}_j : 1 \leq i \neq j \leq d\}$ in $\mathbb{R}^d$.

9. (Type A positive root polytope) $A^+_{d-1} = \text{conv}\{\mathbf{e}_i - \mathbf{e}_j : 1 \leq i < j \leq d\}$ in $\mathbb{R}^d$.

10. (Flow polytope) Given a directed graph $G = (V,E)$ and a vector $\mathbf{b} \in \mathbb{R}^V$, the *flow polytope* is $F_G(\mathbf{b}) = \{\mathbf{f} \in \mathbb{R}^E : f_e \geq 0 \text{ for all } e \in E \text{ and } \sum_{vw \in E} f_{vw} - \sum_{uv \in E} f_{uv} = b_v \text{ for all } v \in V\}$. We think of $f_e$ as a *flow* on edge $e$, so that the *excess flow* or *leak* at each vertex $v$ equals $b_v$.

11. (CRY polytope) $\text{CRY}_n = F_{K_{n+1}}(1,0,\ldots,0,-1)$ where $K_{n+1}$ is the complete graph on $[n+1]$ with edges directed $i \to j$ for $i < j$.

12. (Associahedron) $\text{Assoc}_{d-1}$ is a polytope whose faces are in bijection with the ways of subdividing a convex $(d+3)$-gon into polygons without introducing new vertices. Faces $F$ and $G$ satisfy that $F \subset G$ if and only if the subdivision of $F$ refines the subdivision of $G$. There are several different polytopal realizations of $\text{Assoc}_{d-1}$; see [50] for a survey.

13. (Matroid polytope) If $M = \{v_1, \ldots, v_d\}$ is a set of vectors spanning a vector space $^r$, the *matroid (basis) polytope* is $P_M = \text{conv}\{\mathbf{e}_{i_1} + \cdots + \mathbf{e}_{i_r} : \{v_{i_1}, \ldots, v_{i_r}\} \text{ is a basis of } ^r\}$ in $\mathbb{R}^d$. When $M$ is generic we get $P_M = \Delta(r,d)$. This construction is better understood in the context of matroids; see Section 1.8.

14. (Generalized permutahedra / polymatroids) Many interesting polytopes are deformations of the permutahedron, obtained by moving the vertices of $\Pi_{d-1}$ while respecting all the edge directions. Examples include the polytopes $\Delta_{d-1}$, $\Delta_{e-1} \times \Delta_{d-e-1}$, $\Delta(r,d)$, $\Pi_{d-1}$, $A_{d-1}$, $P_M$, and $\text{Assoc}_{d-1}$ above. Such polytopes are called *polymatroid base polytopes* or *generalized permutahedra*; see [82, 158].

Many of these polytopes are related to the permutation group $S_d$ and the corresponding *type A root system* $\text{conv}\{\mathbf{e}_i - \mathbf{e}_j : 1 \leq i \neq j \leq d\}$. Many have generalizations to the wider context of finite Coxeter groups and root systems; see for example [8, 42, 81, 83, 143] and the references therein.

The faces of most of these polytopes can be described combinatorially. This should help us appreciate these polytopes, because there are not many interesting families of polytopes for which we can do this. We describe the most interesting ones.

3. (Permutahedron) The H-description of $\Pi_{d-1}$ is

$$x_1 + \cdots + x_d = d(d+1)/2,$$

$$x_{i_1} + \cdots + x_{i_k} \geq k(k+1)/2 \quad \text{for } \emptyset \subsetneq \{i_1, \ldots, i_k\} \subsetneq [d].$$

There is a bijection $\mathscr{S} \leftrightarrow F_{\mathscr{S}}$ between the ordered set partitions of $[d]$ and the faces of the permutahedron $\Pi_{d-1}$.

4. (Zonotope) The face enumeration of $Z(A)$ is the subject of the upcoming Theorem 1.7.15.

5. (Cyclic polytope) Let $\mathbf{t}_i = (1, t_i, \ldots, t_i^{d-1})$. The facets of the cyclic polytope $C_d(m)$ are the simplices $\mathrm{conv}\{\mathbf{t}_s : s \in S\}$ for the $d$-subsets $S \subseteq [m]$ satisfying **Gale's evenness condition**: Between any $i < j$ not in $S$ there is an even number of elements of $S$. There is a similar description for all faces of the cyclic polytope. In particular, the combinatorics of $C_d(m)$ is independent of $t_1, \ldots, t_m$. Two other remarkable facts are the following.

   - Every subset of at most $d/2$ vertices forms a face of $C_d(m)$.
   - (McMullen's Upper Bound Theorem [138]) Among all $d$-polytopes with $m$ vertices, $C_d(m)$ maximizes the number of faces of dimension $k$ for all $2 \leq k \leq d-1$.

   The rich theory of *positroids* can be seen as a generalization of the study of cyclic polytopes; see Section 1.8 and [16, 157] for this connection.

6. (Order polytope) The vertices of $\mathscr{O}(P)$ are $\sum_{i \notin I} \mathbf{e}_i$ for the order ideals $I \subseteq P$.

7. (Chain polytope) The vertices of $\mathscr{C}(P)$ are $\sum_{i \in A} \mathbf{e}_i$ for the antichains $A \subseteq P$.

13. (Matroid polytope) The matroid polytope is cut out (non-minimally) by the inequalities

$$\sum_{e \in E} x_e = r(E), \qquad \sum_{e \in S} x_e \leq r(S) \text{ for } E \subset S.$$

   The facets are characterized in [77].

14. (Generalized permutahedra) There are several interesting results on various classes of generalized permutahedra; see [156, 158].

### 1.6.3 Counting faces

The $f$-**vector** of a $d$-polytope $P$ is

$$f_P = (f_0, f_1, \ldots, f_{d-1}, f_d)$$

where $f_i$ is the number of $(i-1)$-dimensional faces of $P$ for $0 \leq i \leq d$. Note that we include the empty face and omit the full-dimensional face $P$ in this enumeration. The problem of characterizing the $f$-vectors of various kinds of polytopes (or more general polyhedral complexes) is a central one in combinatorics. We offer a very brief discussion; for more detailed accounts, see [33, 146, 225]. A one sentence summary is this: We completely understand the $f$-vectors of simplicial (or equivalently, of simple) polytopes, but we are far from understanding the $f$-vectors of general polytopes. There are many interesting results and (mostly) open questions in between.

***f*-vectors.** The most important result about $f$-vectors of arbitrary polytopes is the following.

**Theorem 1.6.4 (McMullen's Upper Bound Theorem [138])** *For any polytope $P$ of dimension $d$ with $m$ vertices, we have*

$$f_i(P) \leq f_i(C_d(m)) \qquad \text{for } i = 0, 1, \ldots, d-1,$$

*where $C_d(m)$ is the cyclic polytope.*

***h*-vectors.** If $P$ is a simplicial $d$-polytope (or more generally, any simplicial complex of dimension $d-1$), we define the **$h$-vector** $h_P = (h_0, h_1, \ldots, h_d)$ by the equivalent equations:

$$h_0 x^d + h_1 x^{d-1} + \cdots + h_d x^0 = f_0(x-1)^d + f_1(x-1)^{d-1} + \cdots + f_d(x-1)^0, \quad \text{or}$$

$$h_i = \sum_{j=0}^{i} (-1)^{i-j} \binom{d-j}{i-j} f_j \qquad \text{and} \qquad f_i = \sum_{j=0}^{i} \binom{d-j}{i-j} h_j \qquad \text{for } 0 \leq i \leq d.$$

The $h$-vector is a more economical way of storing the $f$-vector, due to the *Dehn-Sommerville relations*: $h_i = h_{d-i}$ for $0 \leq i \leq d$.

In fact, the $g$-**theorem** characterizes completely the $f$-vectors of simplicial polytopes [36, 188]! This spectacular result (which McMullen conjectured, and described as "*even more intriguing, if rather less plausible*" [138]) is one of the most important achievements of algebraic and geometric combinatorics to date. To state it, we need some definitions.

We say a sequence of nonnegative integers $(m_0, m_1, \ldots, m_d)$ is an **M-sequence** if there exists a set $S$ of monomials in $x_1, \ldots, x_n$, containing exactly $m_i$ monomials of degree $i$ for $i = 0, 1, \ldots, d$, such that $m' \in S$ and $m | m'$ implies $m \in S$. Macaulay gave a numerical characterization, as follows.

For any positive integers $a$ and $i$, there is a unique representation $a = \binom{a_i}{i} + \binom{a_{i-1}}{i-1} + \cdots + \binom{a_j}{j}$ for some $a_i > a_{i-1} > \cdots > a_j \geq j \geq 1$. We then define $a^{\langle i \rangle} = \binom{a_i}{i+1} + \binom{a_{i-1}}{i} + \cdots + \binom{a_j}{j+1}$. Then $(m_0, m_1, \ldots, m_d)$ is an M-sequence if and only if $m_0 = 1$ and $m_{i+1} \leq m_i^{\langle i \rangle}$ for $i = 1, 2, \ldots, d-1$.

**Theorem 1.6.5 (Billera-Lee-Stanley's $g$-theorem [36, 188])** *A sequence $(h_0, \ldots, h_d)$ of positive integers is the $h$-vector of a simplicial $d$-polytope if and only if*

1. *$h_i = h_{d-i}$ for $i = 0, 1, \ldots, \lfloor d/2 \rfloor$ (Dehn-Somerville equations), and*
2. *$(g_0, g_1, \ldots, g_{\lfloor d/2 \rfloor})$ is an M-sequence, where $g_0 = h_0$ and $g_i = h_i - h_{i-1}$ for $1 \leq i \leq \lfloor d/2 \rfloor$.*

For general polytopes, the situation is much less clear, even in dimension 4, a case that has been studied extensively. [226] Stanley [184] defined a more subtle **toric $h$-vector**. It coincides with the $h$-vector when $P$ is simplicial, and it also satisfies the Dehn-Sommerville equations. His definition may be seen as a combinatorial formula for the dimensions of the intersection cohomology groups of the corresponding projective toric variety (if $P$ is a rational polytope). In a different direction, for "cubical" polytopes $P$ whose proper faces are all cubes, Adin [1] defined the **cubical $h$-vector**, which also satisfies the Dehn-Sommerville relations. Characterizing these toric and cubical $h$-vectors is an important open problem.

**Flag $f$-vectors and $h$-vectors.** The *flag $f$-vector* $(f_D)_{D\subseteq[0,d-1]}$ of a $d$-polytope enumerates the flags $F_1 \subset \cdots \subset F_k$ of given dimensions $D = \{d_1 < \cdots < d_k\}$ for all $D \subseteq [0, d-1]$. As we described in Section 1.5.6 (in the wider context of Eulerian posets), this information can be more economically stored in the **cd**-index of $P$. This encoding incorporates all linear relations among the flag $f$-vector.

As we saw in Theorem 1.5.25, the **cd**-index of a polytope $P$ is non-negative; this is not true for general Eulerian posets. Another very interesting question, which is wide open, is to classify the **cd**-indices of polytopes.

### 1.6.4 Counting lattice points: Ehrhart theory

In this section we are interested in "measuring" a polytope $P$ by counting the lattice points in its integer dilations $P, 2P, 3P, \ldots$. We limit our attention to **lattice polytopes**, whose vertices are lattice points, and to **rational polytopes**, whose vertices have rational coordinates; at the moment there is no good theory for general polytopes.

**Theorem 1.6.6** *Let $P \subset \mathbb{R}^d$ be a lattice polytope. There are polynomials $L_P(x)$ and $L_{P^o}(x)$ of degree $\dim P$, called the **Ehrhart polynomial** and **interior Ehrhart polynomial** of $P$, such that*

1. *The number of lattice points in the nth dilation of $P$ and its interior $P^o$ are*

$$L_P(n) = |nP \cap \mathbb{Z}^d| \qquad L_{P^o}(n) = |nP^o \cap \mathbb{Z}^d| \qquad \text{for all } n \in \mathbb{N}.$$

2. *The **Ehrhart reciprocity law** holds:*

$$(-1)^{\dim P} L_P(-x) = L_{P^o}(x).$$

**Note.** If $P$ is a rational polytope, then $L_P(n)$ is instead given by a **quasipolynomial**; that is, there exist an integer $m$ and polynomials $L_1(x), \ldots, L_m(x)$ such that we have $|nP \cap \mathbb{Z}^d| = L_k(n)$ whenever $n \equiv k \pmod{m}$.

*Sketch of Proof of Theorem 1.6.6.* By working in one dimension higher, we can consider the various dilations of $P$ all at once. We embed $\mathbb{R}^d$ into $\mathbb{R}^{d+1}$ by mapping $\mathbf{v}$ to $(\mathbf{v}, 1)$, and consider the cone

$$\text{cone}(P) = \{\lambda_1(\mathbf{v}_1, 1) + \cdots + \lambda_n(\mathbf{v}_n, 1) : \lambda_1, \ldots, \lambda_n \geq 0\}.$$

Then for each $n \in \mathbb{N}$, the slice $x_{d+1} = n$ of cone$(P)$ is a copy of the dilation $nP$, as shown in Figure 1.31. Two key ingredients of the proof will be the **lattice point**

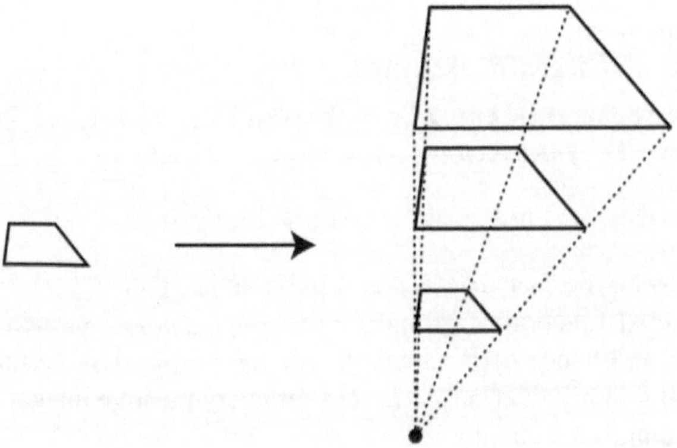

**Figure 1.31**
The cone of a polytope and its slices at height $0, 1, 2, 3$.

**enumerator** of cone$(P)$ and the **Ehrhart series** of $P$:

$$\sigma_{\text{cone}(P)}(\mathbf{z}) = \sum_{\mathbf{p}\in\text{cone}(P)} \mathbf{z}^{\mathbf{p}}, \qquad \text{Ehr}_P(z) = \sum_{n\geq 0} L_P(n) z^n = \sigma_{\text{cone}(P)}(1,\dots,1,z), \tag{1.22}$$

where $\mathbf{z}^{\mathbf{p}} = z_1^{p_1}\cdots z_{d+1}^{p_{d+1}}$.

We first prove the theorem for simplices, and then use triangulations to prove the general case.

*Step 1. (Simplices)* First we prove the result when $P$ is a simplex, which we may assume is full-dimensional. In this case cone$(P)$ is an orthant generated by $d+1$ linearly independent vectors $\mathbf{w}_1,\dots,\mathbf{w}_{d+1}$, where $\mathbf{w}_i = (\mathbf{v}_i, 1)$.

1. Define the **lower and upper fundamental (half-open) parallelepipeds**

$$\begin{aligned} \Pi &= \{\lambda_1\mathbf{w}_1+\cdots+\lambda_{d+1}\mathbf{w}_{d+1} : 0\leq\lambda_1,\dots,\lambda_{d+1}<1\} \\ \Pi^+ &= \{\lambda_1\mathbf{w}_1+\cdots+\lambda_{d+1}\mathbf{w}_{d+1} : 0<\lambda_1,\dots,\lambda_{d+1}\leq 1\}. \end{aligned}$$

Then cone$(P)$ is tiled by the various non-negative $\mathbf{w}$-integer translates of $\Pi$, namely, the parallelepipeds $\Pi + k_1\mathbf{w}_1+\cdots+k_{d+1}\mathbf{w}_{d+1}$ for $\mathbf{k}=(k_1,\dots,k_{d+1})\in\mathbb{N}^{d+1}$. Similarly, the interior int(cone$(P)$) is tiled by the various non-negative $\mathbf{w}$-integer translates of $\Pi^+$. Therefore

$$\begin{aligned} \sigma_{\text{cone}(P)}(\mathbf{z}) &= \sigma_{\Pi}(\mathbf{z})\left(\frac{1}{1-\mathbf{z}^{\mathbf{w}_1}}\right)\cdots\left(\frac{1}{1-\mathbf{z}^{\mathbf{w}_{d+1}}}\right) \\ \sigma_{\text{int}(\text{cone}(P))}(\mathbf{z}) &= \sigma_{\Pi^+}(\mathbf{z})\left(\frac{1}{1-\mathbf{z}^{\mathbf{w}_1}}\right)\cdots\left(\frac{1}{1-\mathbf{z}^{\mathbf{w}_{d+1}}}\right). \end{aligned}$$

Using (1.22) we get

$$\text{Ehr}_P(z) = \frac{\sigma_\Pi(1,\dots,1,z)}{(1-z)^{d+1}}, \qquad \text{Ehr}_{P^\circ}(z) = \frac{\sigma_{\Pi^+}(1,\dots,1,z)}{(1-z)^{d+1}},$$

which are rational functions. Then, by Theorem 1.3.6, $L_P(n)$ and $L_{P^o}(n)$ are polynomial functions of $n$ of degree $d$, as desired.

2. Now observe that $\sigma_{\Pi^+}(\mathbf{z}) = \mathbf{z}^{\mathbf{w}_1+\cdots+\mathbf{w}_{d+1}}\sigma_\Pi(1/z_1,\ldots,1/z_{d+1})$, because $\Pi^+$ is the translation of $-\Pi$ by the vector $\mathbf{w}_1+\cdots+\mathbf{w}_{d+1}$. This gives

$$\mathrm{Ehr}_{P^o}(z)+(-1)^d\mathrm{Ehr}_P(1/z)=0.$$

It remains to invoke the fact that if $f$ is a polynomial, then $F^+(z)=\sum_{n\geq 0} f(n)z^n$ (which is a rational function of $z$) and $F^-(z)=\sum_{n<0} f(n)z^n$ (which is a rational function of $1/z$, and hence of $z$) satisfy $F^+(z)+F^-(z)=0$ as rational functions. This implies that $(-1)^{\dim P}L_P(-n)=L_{P^o}(n)$ for every positive integer $n$, and hence these two polynomials are equal.

*Step 2. (The general case)* Now let $P$ be a general lattice polytope, and let $\mathcal{T}$ be a triangulation of $P$. Recall that $\mathcal{T}^o$ is the set of non-boundary faces of $\mathcal{T}$.

1. We have

$$L_P(n)=\sum_{F\in\mathcal{T}} L_{F^o}(n), \qquad L_{P^o}(n)=\sum_{F\in\mathcal{T}^o} L_{F^o}(n)$$

which implies that $L_P$ and $L_{P^o}$ are polynomials of degree $\dim P$, by Step 1.

2. Let $\widehat{\mathcal{T}}$ be the face poset of $\mathcal{T}$, with an additional maximum element $\hat{1}$. We have the Möbius dual relations

$$L_F=\sum_{G\leq F} L_{G^o}, \qquad L_{F^o}=\sum_{G\leq F}\mu(G,F)L_G \qquad \text{for all } F\in\mathcal{T}$$

omitting the arguments of the polynomials in question. This gives us

$$L_P=\sum_{F\in\mathcal{T}} L_{F^o}=\sum_{F\in\mathcal{T}}\sum_{G\leq F}\mu(G,F)L_G=\sum_{G\in\mathcal{T}} -\mu(G,\hat{1})L_G=\sum_{G\in\mathcal{T}^o}(-1)^{\dim P-\dim G}L_G,$$

where in the last step we are invoking Theorem 1.5.15.11. Now, using that the simplices $G\in\mathcal{T}^o$ satisfy Ehrhart reciprocity, we obtain

$$L_P(-n)=\sum_{G\in\mathcal{T}^o}(-1)^{\dim P-\dim G}L_G(-n)=(-1)^{\dim P}\sum_{G\in\mathcal{T}^o}L_{G^o}(n)=(-1)^{\dim P}L_{P^o}(n),$$

as desired. ■

Next we observe that the volume of a lattice polytope can be recovered from its Ehrhart polynomial. If $\dim P<d$, there is a small subtlety: Let $V$ be the affine span of $P$, and $\Lambda=V\cup\mathbb{Z}^d$. Then we need to normalize the volumes along $V$, so that any "unit" cube, generated by a $\mathbb{Z}$-basis of $\Lambda$, has normalized volume 1. We let $\mathrm{vol}(P)$ denote the **normalized volume** or **lattice volume** of $P$. When $P$ is full-dimensional, this is the usual volume.

For example, the segment from $(1,1)$ to $(4,7)$ has normalized volume (lattice length) 3, because in this case the lattice $\Lambda$ is generated by the primitive vector $(1,2)$, and $(4,7)-(1,1)=\mathbf{3}(1,2)$.

**Proposition 1.6.7** *If $L_P(t) = c_d t^d + \cdots + c_1 t + c_0$ is the Ehrhart polynomial of a lattice polytope $P$, then $c_d = \mathrm{vol}(P)$ and $c_0 = 1$.*

*Sketch of Proof.* We obtain better and better approximations of the volume of $P$ by choosing finer and finer grids $(\frac{1}{n}\mathbb{Z})^d$, and placing a cube of volume $\frac{1}{n^d}$ centered at each lattice point in $P \cap (\frac{1}{n}\mathbb{Z})^d$. Therefore

$$\mathrm{vol}(P) = \lim_{n\to\infty} \frac{1}{n^d}\left|P \cap \left(\frac{1}{n}\mathbb{Z}\right)^d\right| = \lim_{n\to\infty} \frac{1}{n^d}\left|nP \cap \mathbb{Z}^d\right| = \lim_{n\to\infty} \frac{L_P(n)}{n^d} = c_d.$$

Since the only lattice point in $0P$ is the origin, $c_0 = L_P(0) = 1$. ■

A lattice simplex $S$ in $\mathbb{R}^d$ with vertices $\mathbf{v}_1, \ldots, \mathbf{v}_k$ is **unimodular** if the vectors $\mathbf{v}_2 - \mathbf{v}_1, \ldots, \mathbf{v}_d - \mathbf{v}_1$ are a lattice basis for the lattice $\mathrm{aff}\, S \cap \mathbb{Z}^d$. A triangulation is **unimodular** if all its simplices are unimodular. The following result shows that unimodular triangulations are particularly useful for enumerative purposes.

**Proposition 1.6.8** *1. For every lattice polytope $P \subset \mathbb{R}^d$ the* ***Ehrhart $h^*$-polynomial*** *$h_P^*(x) = h_0^* + h_1^* x + \cdots + h_d^* x^d$, which is defined by*

$$\mathrm{Ehr}_P(z) = \frac{h_P^*(z)}{(1-z)^{d+1}},$$

*has nonnegative coefficients: $h_k^* \geq 0$ for all $k$.*
*2. If $P$ has a unimodular triangulation $\mathcal{T}$, then*

$$h_P^*(z) = h_{\mathcal{T}}(z),$$

*where the* ***h-polynomial*** *of $\mathcal{T}$ is the generating function $h_{\mathcal{T}}(z) = h_0 + h_1 z + \cdots + h_d z^d$ for the $h$-vector $(h_0, \ldots, h_d)$ of $\mathcal{T}$, which is defined in terms of the $f$-vector $(f_0, \ldots, f_d)$ of $\mathcal{T}$ as in Section 1.6.3.*

*Proof.* 1. is due to Stanley [187]; for a short proof see [30, Theorem 3.12]. For 2., we have

$$\mathrm{Ehr}_P(z) = \sum_{F\in\mathcal{T}} \mathrm{Ehr}_{F^\circ}(z) = \sum_{F\in\mathcal{T}} \frac{\sigma_{\Pi_F^+}(1,\ldots,1,z)}{(1-z)^{\dim F+1}}$$

where $\Pi_+(F)$ is the upper fundamental parallelepiped of $F$. Since $F$ is unimodular, the only lattice point in $\Pi_F^+$ is the sum of its generators, which is at height $\dim F + 1$, so

$$\mathrm{Ehr}_P(z) = \sum_{F\in\mathcal{T}} \left(\frac{z}{1-z}\right)^{\dim F+1} = \sum_{k=0}^{d+1} f_k \left(\frac{z}{1-z}\right)^k = \frac{\sum_{k=0}^{d+1} h_k z^k}{(1-z)^{d+1}}$$

as desired. ■

**Examples.** The following polytopes have particularly nice Ehrhart polynomials and Ehrhart series.

1. (Polygon) $P$ a lattice polygon in $\mathbb{R}^2$:

$$L_P(n) = \left(I + \frac{B}{2} - 1\right) n^2 + \frac{B}{2} n + 1, \qquad \mathrm{Ehr}_P(z) = \frac{Iz^2 + (I+B-3)z + 1}{(1-z)^3},$$

where $P$ contains $I$ lattice points in its interior and $B$ lattice points on its boundary. This is equivalent to Pick's formula for the area of a lattice polygon.

2. (Simplex) $\Delta_{d-1}$:

$$L_{\Delta_{d-1}}(n) = \binom{n+d-1}{d-1}, \qquad \mathrm{Ehr}_{\Delta_{d-1}}(z) = \frac{1}{(1-z)^d}.$$

3. (Cube) $\square_d$:

$$L_{\square_d}(n) = (n+1)^d, \qquad \mathrm{Ehr}_{\square_d}(z) = \frac{\sum_{k=0}^{d} A(d,k) z^k}{(1-z)^{d+1}},$$

where the **Eulerian number** $A(d,k)$ is the number of permutations $\pi$ of $[d]$ with $k-1$ descents; that is, positions $i$ with $\pi(i) > \pi(i+1)$.

4. (Crosspolytope) $\lozenge_d$:

$$L_{\lozenge_d}(n) = \sum_{k=0}^{d} 2^k \binom{d}{k}\binom{n}{k}, \qquad \mathrm{Ehr}_{\lozenge_d}(z) = \frac{(1+z)^d}{(1-z)^{d+1}}.$$

5. (Product of two simplices) $P = \Delta_{c-1} \times \Delta_{d-1}$:

$$L_P(n) = \binom{n+c-1}{c-1}\binom{n+d-1}{d-1}, \qquad \mathrm{Ehr}_P(z) = \frac{\sum_{k=0}^{\min(c-1,d-1)} \binom{c-1}{k}\binom{d-1}{k} z^k}{(1-z)^{c+d-1}}.$$

Every triangulation of $\Delta_{c-1} \times \Delta_{d-1}$ is unimodular, with $h$-vector given by the $h^*$-vector above. These triangulations are very interesting combinatorially, and play an important role in tropical geometry and other contexts; see [10, 12, 66, 175] and the references therein.

6. (Hypersimplex) $\Delta(r,d)$:

$$L_{\Delta(r,d)}(n) = [z^{rn}] \left(\frac{1-z^{n+1}}{1-z}\right)^d, \qquad \mathrm{vol}(\Delta_{r,d}) = \frac{A(d-1,r)}{(d-1)!}$$

where, again, $A(d-1,r)$ denotes the Eulerian numbers. No simple formula is known for the Ehrhart series of $\Delta(r,d)$. There is a nice formula for the "half-open hypersimplex"; see [102]. An elegant triangulation of $\Delta(r,d)$ was given (using four different descriptions) by Lam-Postnikov, Stanley, Sturmfels, and Ziegler; see [127].

7. (Permutahedron) $P = \Pi_{d-1}$:

$$L_{\Pi_{d-1}}(n) = \sum_{i=0}^{d-1} f_i n^i, \qquad \mathrm{vol}(\Pi_{d-1}) = d^{d-2}$$

where $f_i$ is the number of forests on $[d]$ with $i$ vertices. [190]

8. (Zonotope) $P = Z(A)$:

$$L_{Z(A)}(n) = n^r M\left(1+\frac{1}{n}, 1\right)$$

where $r$ is the rank of $A$ and $M(x,y)$ is the *arithmetic Tutte polynomial* of Section 1.8.9. [190] This polynomial is difficult to compute in general; when $A$ is a root system, explicit formulas are given in [11].

9. (Cyclic polytope) $P = C_d(t_1,\ldots,t_m)$:

$$L_P(n) = \sum_{k=0}^{d} \mathrm{vol}\, C_k(t_1,\ldots,t_m) n^k.$$

See [130]. The triangulations of the cyclic polytope are unusually well behaved; see [63, 163] and the references therein. In particular, when $m = d+4$, this is one of the few polytopes whose triangulations have been enumerated exactly (and non-trivially) [23].

10. (Order polytope and chain polytope) $\mathcal{O}(P), \mathcal{C}(P)$:

$$L_{\mathcal{O}(P)} = L_{\mathcal{C}(P)} = \Omega_P(n+1), \qquad \mathrm{vol}(\mathcal{O}(P)) = \mathrm{vol}(\mathcal{C}(P)) = e(P)/|P|!,$$

where $\Omega_P$ is the order polynomial of $P$ and $e(P)$ is the number of linear extensions, as discussed in Section 1.5.3. [189] Remarkably, $\mathcal{O}(P)$ and $\mathcal{C}(P)$ have the same Ehrhart polynomial, even though they are not metrically, or even combinatorially, equivalent in general. There is a nice characterization of the posets $P$ such that $\mathcal{O}(P)$ and $\mathcal{C}(P)$ may be obtained from one another by a unimodular change of basis [101].

11. (Root polytope) $A_{d-1}$:

$$\mathrm{Ehr}_{A_d}(z) = \frac{\sum_{k=0}^{d} \binom{d}{k}^2 z^k}{(1-x)^d}, \qquad \mathrm{vol}(A_d) = \frac{\binom{2d}{d}}{d!}.$$

There are similar formulas for the other classical root polytopes $B_d, C_d, D_d$, as well as for the positive root polytopes. For example, $\mathrm{vol}(A_d^+) = C_d/d!$ where $C_d$ is the $d$th Catalan number. Explicit unimodular triangulations were constructed in [8].

12. (CRY polytope / Flow polytope) $\mathrm{CRY}_n = F_{K_{n+1}}(1,0,\ldots,0,-1)$:

$$\mathrm{vol}(\mathrm{CRY}_n) = C_0C_1C_2\cdots C_{n-2},$$

a product of Catalan numbers. [223]. No combinatorial proof of this fact is known. Flow polytopes are of great importance due to their close connection with the Kostant partition function, which Gelfand described as "*the transcendental element which accounts for many of the subtleties of the Cartan-Weyl theory*" of representations of semisimple Lie algebras. [165]. There are many other interesting combinatorial results; see for example [25, 143].

13. (Matroid polytopes) A combinatorial formula for the volume of a matroid polytope is given in [9].

14. (Generalized permutahedra / polymatroids) There are many interesting results about the volumes and lattice points of various families of generalized permutahedra; see [158].

15. (Cayley polytope) $\mathbf{C}_n = \{\mathbf{x} \in \mathbb{R}^n : 1 \leq x_1 \leq 2 \text{ and } 1 \leq x_i \leq 2x_{i-1} \text{ for } 2 \leq i \leq n\}$:

$$\mathrm{vol}(\mathbf{C}_n) = \frac{c_{n+1}}{n!}$$

where $c_n$ is the number of connected graphs on $[n]$. The related *Tutte polytope* has volume given by an evaluation of the Tutte polynomial (see Section 1.8.6) of the complete graph. [120]

## 1.7 Hyperplane arrangements

We now discuss arrangements of hyperplanes in a vector space. The questions that we ask depend on whether the underlying field is $\mathbb{R}, \mathbb{C}$, or a finite field $\mathbb{F}_q$; but in every case, the underlying combinatorics plays an important role. The presentation of this section is heavily influenced by [193]. See [153] for a great introduction to more algebraic and topological aspects of the theory of hyperplane arrangements.

After developing the basics in Section 1.7.1, in Section 1.7.2 we introduce the *characteristic polynomial*, which plays a crucial role in this theory. In Section 1.7.3 we discuss some of its important properties, and in Section 1.7.4 we develop several techniques for computing it. We illustrate these techniques by computing the characteristic polynomials of many arrangements of interest. Finally, in Section 1.7.5, we give a remarkable formula for the **cd**-index of an arrangement, which enumerates the flags of faces of given dimensions.

### 1.7.1 Basic definitions

Let  be a field and $V = {}^d$. A **hyperplane arrangement** $\mathscr{A} = \{H_1, \ldots, H_n\}$ is a collection of affine hyperplanes in $V$, say,

$$H_i = \{x \in V : v_i \cdot x = b_i\}$$

for nonzero normal vectors $v_1, \ldots, v_n \in V$ and constants $b_1, \ldots, b_n \in$ . We say $\mathscr{A}$ is **central** if all hyperplanes have a common point; in the most natural examples, the origin is a common point. Figure 1.32 shows a central arrangement of four hyperplanes in $\mathbb{R}^3$.

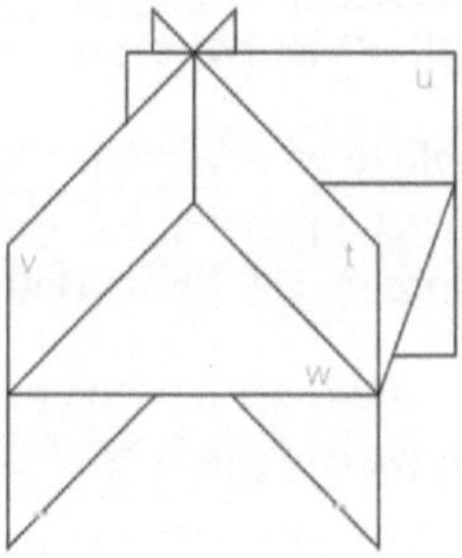


**Figure 1.32**
A hyperplane arrangement.

In some ways, central arrangements are slightly better behaved than affine arrangements. We can **centralize** an affine arrangement $\mathscr{A}$ in ${}^n$ to obtain the **cone** of $\mathscr{A}$, an arrangement $c\mathscr{A}$ in ${}^{n+1}$, by turning the hyperplane $a_1x_1 + \cdots + a_nx_n = a$ in ${}^n$ into the hyperplane $a_1x_1 + \cdots + a_nx_n = ax_{n+1}$ in ${}^{n+1}$, and adding the hyperplane $x_{n+1} = 0$.

Sometimes arrangements are "too central," in the sense that their intersection is a subspace $L$ of positive dimension. In that case, there is little harm in intersecting our arrangement with the orthogonal complement $L^\perp$. We define the **essentialization** of $\mathscr{A}$ to be the arrangement $\mathrm{ess}(\mathscr{A}) = \{H \cap L^\perp : H \in \mathscr{A}\}$ in $L^\perp$. The result is an **essential** arrangement, where the intersection of the hyperplanes is the origin. In most problems of interest, there is no important difference between $\mathscr{A}$ and $\mathrm{ess}(\mathscr{A})$.

A key object is the **complement**

$$V(\mathscr{A}) = V \setminus \left( \bigcup_{H \in \mathscr{A}} H \right),$$

and we now introduce a polynomial that is a fundamental tool in the study of $V(\mathscr{A})$.

### 1.7.2 The characteristic polynomial

There is a combinatorial polynomial that knows a tremendous amount about the complement $V(\mathscr{A})$ of an arrangement $\mathscr{A}$. The kinds of questions that we ask about $V(\mathscr{A})$ depend on the underlying field.

- If $\quad = \mathbb{R}$ then every hyperplane $v_i \cdot x = b_i$ divides $V$ into two half-spaces, where $v_i \cdot x < b_i$ and $v_i \cdot x > b_i$, respectively. Therefore an arrangement $\mathscr{A}$ divides $\mathbb{R}^d$ into $a(\mathscr{A})$ **regions**, which are the connected components of the complement $\mathbb{R}^d \backslash \mathscr{A}$. Let $b(\mathscr{A})$ be the number of those regions that are bounded. If $\mathscr{A}$ is not essential, we let $b(\mathscr{A})$ be the number of **relatively bounded** regions. These are the regions that become bounded in the essentialization $\text{ess}(\mathscr{A})$. A central question about real hyperplane arrangements is to compute the numbers $a(\mathscr{A})$ and $b(\mathscr{A})$ of regions and bounded regions.
- If $\quad = \mathbb{C}$, then it is possible to walk around a hyperplane without crossing it; this produces a loop in the complement $V(\mathscr{A})$. Therefore $V(\mathscr{A})$ has nontrivial topology, and it is natural to ask for its Betti numbers.
- If $\quad = \mathbb{F}_q$ is a finite field, where $q$ is a prime power, then $V(\mathscr{A})$ is a finite set, and the simplest question we can ask is how many points it contains.

Amazingly, the **characteristic polynomial** $\chi_{\mathscr{A}}(q)$ can answer these questions immediately. Let us define it.

Define a **flat** of $\mathscr{A}$ to be an affine subspace obtained as an intersection of hyperplanes in $\mathscr{A}$. The **intersection poset** $L_{\mathscr{A}}$ is the set of flats partially ordered by reverse inclusion. If $\mathscr{A}$ is central, then $L_{\mathscr{A}}$ is a **geometric lattice**, as discussed in Section 1.5.2. If $\mathscr{A}$ is not central, then $L_{\mathscr{A}}$ is only a **geometric meet semilattice** [207]. The **rank** $r = r(\mathscr{A})$ of $\mathscr{A}$ is the height of $L_{\mathscr{A}}$.

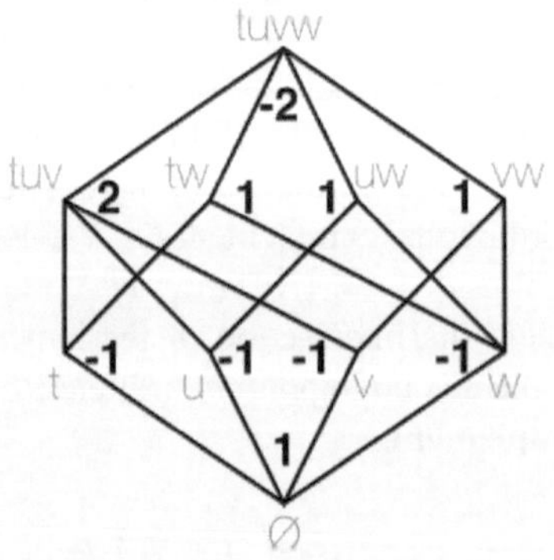


**Figure 1.33**
The intersection poset of $\mathscr{A}$ and its Möbius function.

The **characteristic polynomial** of $\mathscr{A}$ is

$$\chi_A(q) = \sum_{F \in L_{\mathscr{A}}} \mu(\widehat{0}, F) q^{\dim F}.$$

Figure 1.33 shows the intersection poset of the arrangement $\mathscr{A}$ in Figure 1.32; its characteristic polynomial $\chi_{\mathscr{A}}(q) = q^3 - 4q^2 + 5q - 2$ is easily computed by adding the Möbius numbers on each level of $L_{\mathscr{A}}$.

We will see some general techniques to compute characteristic polynomials in Section 1.7.4.

**Theorem 1.7.1** *The characteristic polynomial $\chi_{\mathscr{A}}(x)$ contains the following information about the complement $V(\mathscr{A})$ of a hyperplane arrangement $\mathscr{A}$.*

1. *( $= \mathbb{R}$, Zaslavsky's Theorem) [220] Let $\mathscr{A}(A)$ be a* **real** *hyperplane arrangement in $\mathbb{R}^n$. The number of regions and relatively bounded regions of the complement $V(\mathscr{A})$ are*

$$a(\mathscr{A}) = (-1)^n \chi_{\mathscr{A}}(-1), \qquad b(\mathscr{A}) = (-1)^{r(\mathscr{A})} \chi_{\mathscr{A}}(1).$$

2. *( $= \mathbb{C}$) [90, 152] Let $\mathscr{A}(A)$ be a* **complex** *hyperplane arrangement in $\mathbb{C}^n$. The complement $V(\mathscr{A})$ has Poincaré polynomial*

$$\sum_{k \geq 0} \operatorname{rank} H^k(V(A), \mathbb{Z}) q^k = (-q)^n \chi_{\mathscr{A}}\left(\frac{-1}{q}\right).$$

3. *( $= \mathbb{F}_q$) [20, 54] Let $\mathscr{A}(A)$ be a hyperplane arrangement in $\mathbb{F}_q^n$ where $\mathbb{F}_q$ is the finite field of $q$ elements for a prime power $q$. The complement $V(\mathscr{A})$ has size*

$$|V(A)| = \chi_{\mathscr{A}}(q).$$

*Proof.* 3. This is a typical enumerative problem where our set of objects (the points of $\mathbb{F}_q^n$) are stratified by a partial order, according to how special their position is with respect to $\mathscr{A}$. This is a natural setting to apply Möbius inversion.

For each flat $F$, let $f_=(F)$ be the number of points of $\mathbb{F}_q^n$ which are on $F$, and on no smaller flat. Since there are $f_{\geq}(F) = q^{\dim F}$ points on $F$, we have

$$q^{\dim F} = \sum_{G \geq F} f_=(G)$$

which inverts to

$$f_=(F) = \sum_{G \geq F} \mu(F, G) q^{\dim G}.$$

Setting $F = \widehat{0}$ gives the desired result.

1. For each flat $F$ let $\mathscr{A}/F$ be the arrangement inside $F$ obtained by intersecting the hyperplanes $\mathscr{A} - F$ with $F$. The arrangement subdivides each flat $F$ into (relatively) open faces, namely, the $(\dim G)$-dimensional regions of $\mathscr{A}/G$ for each flat

$G \geq F$. Since the Euler characteristic of $F \cong \mathbb{R}^{\dim F}$ is $(-1)^{\dim F}$, we get

$$(-1)^{\dim F} = \sum_{G \geq F} (-1)^{\dim G} a(\mathscr{A}/G),$$

which inverts to

$$(-1)^{\dim F} a(\mathscr{A}/F) = \sum_{G \geq F} \mu(F,G)(-1)^{\dim G}.$$

Setting $F = \widehat{0}$ gives the desired result. The same strategy works for $b(\mathscr{A})$, using the result that the union of the bounded faces of $\mathscr{A}$ is contractible, and hence has Euler characteristic equal to 1. [39, Theorem 4.5.7(b)]

2. This proof is beyond the scope of this writeup; see [90, 153]. ■

### 1.7.3 Properties of the characteristic polynomial

**Whitney's formula and the Tutte polynomial.** Since the intersection poset $\mathscr{A}$ is constructed from its atoms (the hyperplanes of $\mathscr{A}$), it is natural to invoke the Crosscut Theorem 1.5.20 and obtain the following result.

**Theorem 1.7.2 (Whitney's Theorem)** *The characteristic polynomial of an arrangement $\mathscr{A}$ is given by*

$$\chi_{\mathscr{A}}(q) = \sum_{\substack{\mathscr{B} \subseteq \mathscr{A} \\ \mathscr{B} \text{ central}}} (-1)^{|\mathscr{B}|} q^{n-r(\mathscr{B})}$$

*where $r(\mathscr{B}) = n - \dim \cap_{H \in \mathscr{B}} H$.*

The **Tutte polynomial** is another important polynomial associated to an arrangement:

$$T_{\mathscr{A}}(x,y) = \sum_{\substack{\mathscr{B} \subseteq \mathscr{A} \\ \mathscr{B} \text{ central}}} (x-1)^{r-r(\mathscr{B})} (y-1)^{|\mathscr{B}|-r(\mathscr{B})}.$$

Whitney's Theorem can then be rephrased as:

$$\chi_{\mathscr{A}}(q) = (-1)^r q^{n-r} T_{\mathscr{A}}(1-q,0).$$

#### 1.7.3.1 Deletion and contraction

A common technique for inductive arguments in hyperplane arrangement $\mathscr{A}$ is to choose a hyperplane $H$ and study how $\mathscr{A}$ behaves without $H$ (in the deletion $\mathscr{A}\backslash H$) and how $H$ interacts with the rest of $\mathscr{A}$ (in the contraction $\mathscr{A}/H$).

For a hyperplane $H$ of an arrangement $\mathscr{A}$ in $V$, the **deletion**

$$\mathscr{A}\backslash H = \{A \in \mathscr{A} : A \neq H\}$$

is the arrangement in $V$ consisting of the hyperplanes other than $H$, and the **contraction**

$$\mathscr{A}/H = \{A \cap H : A \in \mathscr{A}, A \neq H\}$$

is the arrangement in $H$ consisting of the intersections of the other hyperplanes with $H$. Figure 1.34 shows a hyperplane arrangement $\mathscr{A} = \{t,u,v,w\}$ in $\mathbb{R}^3$ and the deletion $\mathscr{A}\backslash w$ and contraction $\mathscr{A}/w$.

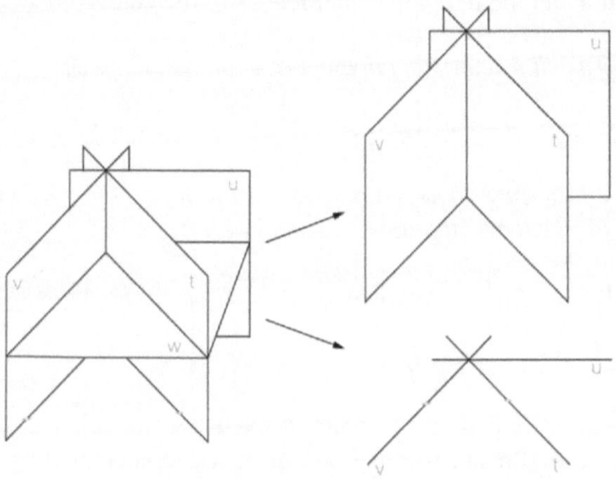


**Figure 1.34**
An arrangement $\mathscr{A}$ and its deletion $\mathscr{A}\backslash w$ and contraction $\mathscr{A}/w$.

**Proposition 1.7.3 (Deletion-Contraction)** *If $\mathscr{A}$ is an arrangement and $H$ is a hyperplane in $\mathscr{A}$ then*

$$\chi_{\mathscr{A}}(q) = \chi_{\mathscr{A}\backslash H}(q) - \chi_{\mathscr{A}/H}(q).$$

*Proof.* Whitney's formula gives

$$\chi_{\mathscr{A}}(q) = \sum_{\substack{H \notin \mathscr{B} \subseteq \mathscr{A} \\ \mathscr{B} \text{ central}}} (-1)^{|\mathscr{B}|} q^{n-r(\mathscr{B})} + \sum_{\substack{H \in \mathscr{B} \subseteq \mathscr{A} \\ \mathscr{B} \text{ central}}} (-1)^{|\mathscr{B}|} q^{n-r(\mathscr{B})} = \chi_{\mathscr{A}\backslash H}(q) - \chi_{\mathscr{A}/H}(q),$$

where we use the fact that if $H \in \mathscr{B}$ then $r(\mathscr{B}) = r_{\mathscr{A}/H}(\mathscr{B}\backslash H) + 1$. ■

Unfortunately hyperplane arrangements are not closed under contraction; in the example of Figure 1.34, the image of $t$ in $(\mathscr{A}/u)/v$ is not a hyperplane. Strictly speaking, this deletion-contraction formula applies only when there is only one copy of $H$ in $\mathscr{A}$.

These small but annoying difficulties are elegantly solved by working in the wider class of *matroids* (for central arrangements; see Section 1.8) or *semimatroids* (for affine arrangements; see [7]).

The characteristic polynomial of the cone $c\mathscr{A}$ can be expressed as follows.

**Proposition 1.7.4** *If $c\mathscr{A}$ is the cone of arrangement $\mathscr{A}$ then*

$$\chi_{c\mathscr{A}}(q) = (q-1)\chi_{\mathscr{A}}(q).$$

#### 1.7.3.2 Sign alternation and unimodality

More can be said about the individual coefficients of the characteristic polynomial.

**Theorem 1.7.5** *The characteristic polynomial of an arrangement*

$$\chi_{\mathscr{A}}(q) = q^n - a_{n-1}q^{n-1} + a_{n-2}q^{n-2} - \cdots + (-1)^n a_n q^0$$

*has coefficients alternating signs, so $a_i \geq 0$. Furthermore, the coefficients are unimodal and even log-concave, that is,*

$$a_1 \leq a_2 \leq \cdots \leq a_{i-1} \leq a_i \geq a_{i+1} \geq \cdots \geq a_n \qquad \text{for some } i, \text{ and}$$

$$a_{j-1}a_{j+1} \leq a_j^2 \qquad \text{for all } j.$$

*Proof.* The sign alternation is easily proved by induction using deletion-contraction. The second result is much deeper. It was conjectured by Rota in 1970 [170], and recently proved by Huh [104] for fields of characteristic 0 and by Huh and Katz [105] for arbitrary fields, drawing from toric geometry, tropical geometry, and matroid theory. This result is conjectured to be true for any geometric lattice; this is still open. ■

#### 1.7.3.3 Graphs and proper colorings

There is a special case of interest, corresponding to arrangements coming from graphs. To each graph $G$ on vertex set $[n]$ we associate the **graphical arrangement** in $^n$, consisting of the hyperplanes $x_i = x_j$ for all edges $ij$ of $G$.

Given $q$ colors, a **proper $q$-coloring** of $G$ assigns a color to each vertex of $G$ so that two vertices $i$ and $j$ that share an edge $ij$ in $G$ must have different colors. The number of proper $q$-colorings of $G$ is given by the **chromatic polynomial** $\chi_G(q)$.

Say an orientation of the edges of $G$ is **acyclic** if it creates no directed cycles.

**Theorem 1.7.6** *The chromatic polynomial of a graph $G$ equals the characteristic polynomial of its graphical arrangement $\mathscr{A}_G$*

$$\chi_{\mathscr{A}_G(q)} = \chi_G(q),$$

*and the number of acyclic orientations of $G$ equals the number of regions of $\mathscr{A}_G$ in $\mathbb{R}^n$.*

*Proof.* A proper $q$-coloring of $G$ is equivalent to a point $\mathbf{a} \in \mathbb{F}_q^n$ which is on none of the hyperplanes $x_i = x_j$ for $ij \in G$. Also, the arrangement $\mathscr{A}_G$ in $^n$ has the same intersection lattice for any field . (This is not true for every arrangement.) The first equality then follows from Theorem 1.7.1.3.

Given a region $R$ of $\mathscr{A}_G$ we give each edge $ij$ the orientation $i \to j$ if $x_i > x_j$ in $R$, and $i \leftarrow j$ if $x_i < x_j$ in $R$. This is a bijection between the regions of $\mathscr{A}_G$ and the acyclic orientations of $G$. ∎

In particular, this theorem proves that the chromatic polynomial $\chi_G(q)$ is indeed given by a polynomial in $q$ when $q$ is a positive integer. It also gives us a reciprocity theorem, telling us what happens when we irreverently substitute the negative integer $q = -1$ into this polynomial:

**Corollary 1.7.7** *The graph G has* $|\chi_G(-1)|$ *acyclic orientations of G.*

#### 1.7.3.4 Free arrangements

Far more often than we might expect, characteristic polynomials of hyperplane arrangements factor as products of linear forms. [172] Such factorizations are often a manifestation of the underlying algebraic structure. The theory of free arrangements gives one possible explanation for this phenomenon.

Let $\mathscr{A}$ be a real central arrangement and let $R = \mathbb{R}[x_1, \ldots, x_n]$ be the polynomial ring in $n$ variables, graded by total degree. A **derivation** is a linear map $D$ satisfying Leibniz's law

$$D(fg) = f(Dg) + (Df)g \qquad \text{for all } f, g \in R.$$

The set Der of derivations is an $R$-module; that is, if $p \in R$ and $D \in \text{Der}$ then $pD \in \text{Der}$. It is a graded module, where $D$ is homogeneous of degree $d$ if it takes polynomials of degree $k$ to polynomials of degree $k+d$. It is a free module with basis $\frac{\partial}{\partial x_1}, \ldots \frac{\partial}{\partial x_n}$; that is,

$$\text{Der} = \left\{ p_1 \frac{\partial}{\partial x_1} + \cdots + p_n \frac{\partial}{\partial x_n} : p_i \in R \right\}.$$

Indeed, if $D$ is a derivation with $Dx_i = p_i \in R$ for $1 \leq i \leq n$, then $D = p_1 \frac{\partial}{\partial x_1} + \cdots + p_n \frac{\partial}{\partial x_n}$ by linearity.

Now consider the submodule of $\mathscr{A}$**-derivations**:

$$\text{Der}(\mathscr{A}) - \{D \in \text{Der} : \alpha_H \text{ divides } D(\alpha_H) \text{ for all } H \in A\}$$

where $\alpha_H$ is a linear form defining hyperplane $H$, so $H = \{v \in {}^n : \alpha_H(v) = 0\}$. Let $Q_{\mathscr{A}} = \prod_{H \in \mathscr{A}} \alpha_H$ be the **defining polynomial** of $\mathscr{A}$.

For example, if $\mathscr{H}$ is the arrangement of coordinate hyperplanes with defining polynomial $Q_{\mathscr{H}} = x_1 \cdots x_n$, then $E_i = x_i \frac{\partial}{\partial x_i}$ is an $\mathscr{H}$-derivation for $1 \leq i \leq n$ since $E_i(x_i) = x_i$ and $E_i(x_j) = 0$ for $j \neq i$.

We say the arrangement $\mathscr{A}$ is **free** if $\text{Der}(\mathscr{A})$ is a free $R$-module. This notion is not well understood at the moment; it is not even known if it is a combinatorial condition. For example, Ziegler gave an example of a free arrangement over $\mathbb{F}_2$ and a non-free arrangement over $\mathbb{F}_3$ with isomorphic intersection posets.

**Conjecture 1.7.8** [199] *If two real arrangements $\mathscr{A}_1$ and $\mathscr{A}_2$ have isomorphic intersection posets and $\mathscr{A}_1$ is free, then $\mathscr{A}_2$ is free.*

Mysterious as it is, freeness is a very useful property.

**Theorem 1.7.9 (Terao's Factorization Theorem [199])** *If $\mathscr{A}$ is free then Der($\mathscr{A}$) has a homogeneous basis $D_1, \ldots, D_n$ whose degrees $d_1, \ldots, d_n$ only depend on $\mathscr{A}$, and the characteristic polynomial of $\mathscr{A}$ is*

$$\chi_{\mathscr{A}}(q) = (q - d_1 - 1) \cdots (q - d_n - 1).$$

Freeness is made more tractable thanks to the following two useful criteria.

**Theorem 1.7.10 (Saito Criterion [174])** *Let $D_1, \ldots, D_n$ be $\mathscr{A}$-derivations and let $Q_{\mathscr{A}}$ be the defining polynomial of $\mathscr{A}$. Then Der($\mathscr{A}$) is free with basis $D_1, \ldots, D_n$ if and only if*

$$\det(D_i(x_j))_{1 \leq i,j \leq n} = c \cdot Q_{\mathscr{A}}$$

*for some constant c.*

**Theorem 1.7.11** [199] *Let $\mathscr{A}$ be an arrangement and H be a hyperplane of $\mathscr{A}$. Any two of the following statements imply the third:*

- *$\mathscr{A}/H$ is free with exponents $b_1, \ldots, b_{n-1}$.*
- *$\mathscr{A} \backslash H$ is free with exponents $b_1, \ldots, b_{n-1}, b_n - 1$.*
- *$\mathscr{A}$ is free with exponents $b_1, \ldots, b_{n-1}, b_n$*

Theorem 1.7.10 can be very easy to use if we have the right candidate for a basis. For example, we saw that $E_i = x_i \frac{\partial}{\partial x_i}$ is an $\mathscr{H}_n$-derivation for $1 \leq i \leq n$. The matrix $E_i(x_j)$ is diagonal with determinant $x_1 \cdots x_n$, so this must in fact be a basis for Der($\mathscr{H}$) with exponents $0, \ldots, 0$, and

$$\chi_{\mathscr{H}_n}(q) = (q-1)^n.$$

To use Theorem 1.7.11 to prove that an arrangement $\mathscr{A}$ is free, we need $\mathscr{A}$ to belong to a larger family of free arrangements with predictable exponents, which behaves well under deletion and contraction. For example, to prove inductively that $\mathscr{H}_n$ is free, we need the stronger statement that an arrangement of $k$ coordinate hyperplanes in $^n$ (where $k \leq n$) has exponents 0 ($k$ times) and $-1$ ($n-k$ times). A more interesting example is given in Section 1.7.4.

#### 1.7.3.5 Supersolvability

There is a combinatorial counterpart to the notion of freeness that produces similar results. Recall that a lattice $L$ is **supersolvable** if it has an **M-chain** $C$ such that the sublattice generated by $C$ and any other chain is distributive.

In this section we are interested in arrangements whose intersection poset is supersolvable. Since this poset is a geometric lattice, it is semimodular, so the following theorem applies.

**Theorem 1.7.12** [182] *Let L be a finite supersolvable semimodular lattice, and suppose that* $\widehat{0} = t_0 < t_1 < \cdots < t_n = \widehat{1}$ *is an M-chain. Let* $a_i$ *be the number of atoms s such that* $s \leq t_i$ *but* $s \not\leq t_{i-1}$*. Then*

$$\chi_L(q) = (q - a_1) \cdots (q - a_n).$$

Graphical arrangements are an interesting special case. Say a graph $G$ is **chordal** if there exists an ordering of the vertices $v_1, \ldots, v_n$ such that for each $i$, the vertices among $\{v_1, \ldots, v_{i-1}\}$ that are connected to $v_i$ form a complete subgraph. An equivalent characterization is that $G$ has no induced cycles of length greater than 3.

It is very easy to compute the chromatic polynomial of a chordal graph. Suppose we wish to assign colors from $[q]$ to $v_1, \ldots, v_n$ in order, to get a proper $q$-coloring. If $v_i$ is connected to $a_i$ vertices among $\{v_1, \ldots, v_{i-1}\}$, since these are all connected pairwise, they must have different colors, so there are exactly $q - a_i$ colors available for $b_i$. It follows that $\chi_G(q) = (q - a_1) \cdots (q - a_n)$. The similarity in these formulas is not a coincidence.

**Theorem 1.7.13** [182] *The intersection lattice of the graphical arrangement* $\mathscr{A}_G$ *is supersolvable if and only if the graph G is chordal.*

### 1.7.4 Computing the characteristic polynomial

The results of the previous section, and Theorem 1.7.1 in particular, show the importance of computing $\chi_{\mathscr{A}}(q)$ for arrangements of interest. In this section we discuss the most common techniques for doing this.

**Computing the Möbius function directly.** In Section 1.5.5.4 we saw many techniques for computing Möbius functions, and we can use them to compute $\chi_{\mathscr{A}}(x)$.

1. (Generic arrangement) $\mathscr{A}_{n,r}$: $n$ generic hyperplanes in $\,^r$.

   Consider a **generic** arrangement of $n$ hyperplanes in $\,^r$, where any $k \leq r$ hyperplanes have an intersection of codimension $k$. There are $\binom{n}{m}$ flats of rank $m$, and for each flat $F$ we have $[\widehat{0}, F] \cong 2^{[m]}$, so $\mu(\widehat{0}, F) = (-1)^m$. Therefore the characteristic polynomial is

   $$\chi_{\mathscr{A}_{n,r}}(x) = \sum_{m=0}^{r} (-1)^m \binom{n}{m} x^m$$

   and the number of regions and bounded regions are

   $$\begin{aligned} a(\mathscr{A}_{n,r}) &= \binom{n}{r} + \binom{n}{r-1} + \cdots + \binom{n}{0}, \\ b(\mathscr{A}_{n,r}) &= \binom{n}{r} - \binom{n}{r-1} + \cdots \pm \binom{n}{0} = \binom{n-1}{r}. \end{aligned}$$

This method works in some examples, but when there is a nice formula for the characteristic polynomial, this is usually not the most efficient technique.

**The finite field method.** Theorem 1.7.1.3 is about arrangements over finite fields, but it may also be used as a powerful technique for computing $\chi_{\mathscr{A}}(x)$ for real or complex arrangements. The idea is simple: Most arrangements $\mathscr{A}$ we encounter "in nature" (that is, in mathematics) are given by equations with integer coefficients.

We can use the equations of $\mathscr{A}$ to determine an arrangement $\mathscr{A}_q$ over a prime $q$. For large enough $q$, the arrangements $\mathscr{A}$ and $\mathscr{A}_q$ will have the same intersection poset, so

$$\chi_{\mathscr{A}}(q) = |\mathbb{F}_q^n \setminus \bigcup_{H \in \mathscr{A}_q} H|.$$

We have thus reduced the computation of $\chi_{\mathscr{A}}$ to an enumerative problem in a finite field. By now we are pretty good at counting, and we can solve these problems for many arrangements of interest.

2. (Coordinate arrangement) $\mathscr{H}_n: \quad x_i = 0 \qquad (1 \leq i \leq n)$

   Here $\chi_{\mathscr{H}_n}(q)$ is the number of $n$-tuples $(a_1, \ldots, a_n) \in \mathbb{F}_q^n$ with $a_i \neq 0$ for all $i$, so

   $$\chi_{\mathscr{H}_n}(q) = (q-1)^n, \qquad a(\mathscr{H}_n) = 2^n.$$

   There is an easy bijective proof for the number of regions: each region $R$ of $\mathscr{H}_n$ is determined by whether $x_i < 0$ or $x_i > 0$ in $R$.

3. (Braid arrangement) $\mathscr{A}_{n-1}: \quad x_i = x_j \qquad (1 \leq i < j \leq n)$.

   Here $\chi_{\mathscr{A}_{n-1}}(q)$ is the number of $n$-tuples $(a_1, \ldots, a_n) \in \mathbb{F}_q^n$ such that $a_i \neq a_j$ for $i \neq j$. Selecting them in order, $a_i$ can be any element of $\mathbb{F}_q$ other than $a_1, \ldots, a_{i-1}$, so

   $$\chi_{\mathscr{A}_{n-1}}(q) = q(q-1)(q-2)\cdots(q-n+1), \qquad a(\mathscr{A}_{n-1}) = n!$$

   A region is determined by whether $x_i > x_j$ or $x_i < x_j$ for all $i \neq j$; that is, by the relative linear order of $x_1, \ldots, x_n$. This explains why the braid arrangement has $n!$ regions.

4. (Threshold arrangement) $\mathscr{T}_n: \quad x_i + x_j = 0, \qquad (1 \leq i < j \leq n)$.

   Let $q$ be an odd prime. We need to count the points $\mathbf{a} \in \mathbb{F}_q^n$ on none of the hyperplanes, that is, those satisfying $a_i + a_j \neq 0$ for all $i \neq j$. To specify one such point, we may first choose the partition $[n] = S_0 \sqcup S_1 \sqcup \cdots \sqcup S_{(q-1)/2}$ where $S_i$ is the set of positions $j$ such that $a_j \in \{i, -i\}$. Note that $S_0$ can have at most one element. Then, for each non-empty block $S_i$ with $i > 0$ we need to decide whether $a_j = i$ or $a_j = -i$ for all $j \in S_i$. The techniques of Section 1.3.3 then give

   $$\sum_{n \geq 0} \chi_{\mathscr{T}_n}(x) \frac{z^n}{n!} = (1+z)(2e^z - 1)^{(x-1)/2}.$$

   The regions of $\mathscr{T}_n$ are in bijection with the **threshold graphs** on $[n]$. These are the graphs for which there exist vertex weights $w(i)$ for $1 \leq i \leq n$ and a "threshold" $w$ such that edge $ij$ is present in the graph if and only if $w(i) + w(j) > w$. Threshold graphs have many interesting properties and applications; see [135].

There are many interesting deformations of the braid arrangement, obtained by considering hyperplanes of the form $x_i - x_j = a$ for various constants $a$. The left panel of Figure 1.35 shows the braid arrangement $\mathscr{A}_2$. This is really an arrangement in $\mathbb{R}^3$, but since all hyperplanes contain the line $x = y = z$, we draw its essentialization by intersecting it with the plane $x + y + z = 0$. Similarly, the other panels show the Catalan, Shi, Ish, and Linial arrangements, which we now discuss.

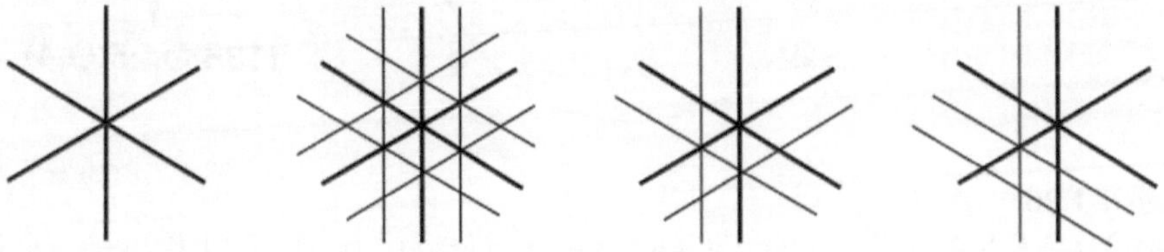

**Figure 1.35**
The arrangements $\mathscr{A}_2, \mathrm{Cat}_2, \mathrm{Shi}_2$, and $\mathrm{Ish}_2$.

5. (Catalan arrangement) $\mathrm{Cat}_{n-1}: \quad x_i - x_j = -1, 0, 1 \qquad (1 \leq i < j \leq n)$.

   To compute $\chi_{\mathrm{Cat}_{n-1}}(q)$, we need to count the $n$-tuples $\mathbf{a} = (a_1, \ldots, a_n) \in \mathbb{F}_q^n$ where $a_i$ and $a_j$ are never equal or adjacent modulo $q$. There are $q$ choices for $a_1$, and once we have chosen $a_1$ we can "unwrap" $\mathbb{F}_q - \{a_1\}$ into a linear sequence of $q-1$ dots. The set $A = \{a_2, \ldots, a_n\}$ consists of $n-1$ non-adjacent dots, and choosing them is equivalent to choosing a partition of $q-n$ into $n$ parts, corresponding to the gaps between the dots; there are $\binom{q-n-1}{n-1}$ choices. Finally there are $(n-1)!$ ways to place $A$ in a linear order in $\mathbf{a}$. Therefore

   $$\chi_{\mathrm{Cat}_{n-1}}(q) = q(q-n-1)(q-n-2)\cdots(q-2n+1)$$

   $$a(\mathrm{Cat}_{n-1}) = n!C_n, \qquad b(\mathrm{Cat}_{n-1}) = n!C_{n-1},$$

   where $C_n$ is the $n$th Catalan number. It is not too difficult to show bijectively that each region of the braid arrangement $\mathscr{A}_{n-1}$ contains $C_n$ regions of the Catalan arrangement, $C_{n-1}$ of which are bounded. [193, Section 5.4]

6. (Shi arrangement) $\mathrm{Shi}_{n-1}: \quad x_i - x_j = 0, 1 \qquad (1 \leq i < j \leq n)$.

   Consider a point $\mathbf{a} = (a_1, \ldots, a_n) \in \mathbb{F}_q^n$ not on any of the Shi hyperplanes. Consider dots $0, 1, \ldots, q-1$ around a circle, and mark dot $a_i$ with the number $i$ for $1 \leq i \leq n$. Mark the remaining dots $\bullet$. Now let $w$ be the word of length $q$ obtained by reading the labels clockwise, starting at the label 1. Note that each block of consecutive numbers must be listed in increasing order. By recording the sets between adjacent $\bullet$s, and dropping the initial 1, we obtained an ordered partition $\Pi$ of $\{2, 3, \ldots, n\}$ into $q-n$ parts. There are $(q-n)^{n-1}$ such partitions. Figure 1.36 shows an example of this construction.

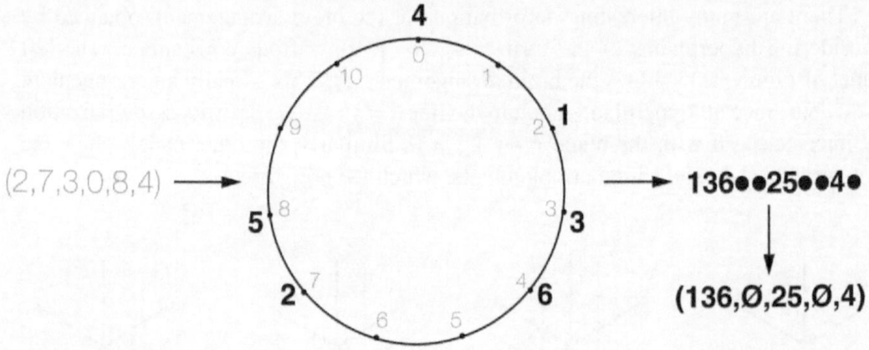


**Figure 1.36**
$\mathbf{a} = (2,7,3,0,8,4) \in \mathbb{F}_{11}^6$, $w = 136\bullet\bullet25\bullet\bullet4\bullet$, $\Pi = (\{3,6\},\emptyset,\{2,5\},\emptyset,\{4\})$.

To recover $a$ from this partition, we only need to know where to put marking 1 on the circle; that is, we need to know $a_1$. It follows that

$$\chi_{\mathrm{Shi}_{n-1}}(q) = q(q-n)^{n-1}$$

$$a(\mathrm{Shi}_{n-1}) = (n+1)^{n-1}, \qquad b(\mathrm{Shi}_{n-1}) = (n-1)^{n-1}.$$

7. (Ish arrangement) $\mathrm{Ish}_{n-1}:\quad x_i = x_j, \quad x_1 - x_j = i \qquad (1 \le i < j \le n).$

   To choose a point $\mathbf{a}$ in the complement of the Ish arrangement over $\mathbb{F}_q$, choose $a_1 \in \mathbb{F}_q$, and then choose $a_n, a_{n-1}, \ldots, a_2$ subsequently. At each step we need to choose $a_i \notin \{a_1, a_1 - 1, \ldots, a_1 - i + 1, a_n, a_{n-1}, \ldots, a_{i+1}\}$. These $n$ forbidden values are distinct, so

$$\chi_{\mathrm{Ish}_{n-1}}(q) = q(q-n)^{n-1},$$

$$a(\mathrm{Ish}_{n-1}) = (n+1)^{n-1}, \qquad b(\mathrm{Ish}_{n-1}) = (n-1)^{n-1}.$$

   The Shi and Ish arrangements share several features, which are generally easier to verify for the Ish arrangement. Together, they give a nice description of the "$q,t$-Catalan numbers" [18].

8. (Linial arrangement) $\mathscr{L}_{n-1}:\quad x_i - x_j = 1, \qquad (1 \le i < j \le n).$

   The Linial arrangement has characteristic polynomial

$$\chi_{\mathscr{L}_{n-1}}(q) = \frac{q}{2^n} \sum_{k=0}^{n} \binom{n}{k} (q-k)^{n-1}.$$

   The number of regions of $\mathscr{L}_{n-1}$ equals the number of **alternating trees** with vertex set $[n+1]$, where each vertex is either smaller than all its neighbors or greater than all its neighbors. There are two different proofs, using Whitney's Theorem [161] and the finite field method [22] respectively. Both are somewhat indirect, combining combinatorial tricks and algebraic manipulations. To date, there is no known bijection between the regions of the Linial arrangement and alternating trees.

Here are a few other nice examples.

9. (Coxeter arrangement) $\mathscr{BC}_n: \quad x_i \pm x_j = 0, \quad x_i = 0 \qquad (1 \leq i < j \leq n).$
To specify an $n$-tuples $(a_1, \ldots, a_n) \in \mathbb{F}_q^n \backslash \mathscr{BC}_n$, we can choose $a_1, \ldots, a_n$ successively. We find there are $q - 2i + 1$ choices for $a_i$, namely, any number other than $0, \pm a_1, \ldots, \pm a_{i-1}$ (that are all distinct). Therefore
$$\chi_{\mathscr{BC}_n}(q) = (q-1)(q-3)\cdots(q-2n+3)(q-2n+1), \qquad r(\mathscr{BC}_n) = 2^n \cdot n!.$$

10. (Coxeter arrangement) $\mathscr{D}_n: \quad x_i \pm x_j = 0 \qquad (1 \leq i < j \leq n).$
Here there are $(q-1)(q-3)(q-5)\cdots(1-2n+3)(q-2n+1)$ $n$-tuples in $\mathbb{F}_q^n \backslash \mathscr{D}_n$ with no $a_i$ equal to 0, and $n(q-1)(q-3)\cdots(q-2n+3))$ $n$-tuples with one $a_i$ equal to 0. Therefore
$$\chi_{\mathscr{D}_n}(q) = (q-1)(q-3)\cdots(q-2n+3)(q-n+1), \qquad r(\mathscr{D}_n) = 2^{n-1} \cdot n!$$

11. (Finite projective space) $\mathscr{A}(p,n)$: all linear hyperplanes in $\mathbb{F}_p^n$
The equations $\sum_i c_i x_i = 0$ (where $c_i \in \mathbb{F}_p$) define an arrangement over $\mathbb{F}_q$ where $q = p^k$, which has the same intersection poset. Recall that $\mathbb{F}_q$ is a $k$-dimensional vector space over $\mathbb{F}_p$. Now let us count the points $(a_1, \ldots, a_n)$ in $\mathbb{F}_q$ that are not on any hyperplane. We choose $a_1, \ldots, a_n$ subsequently and at each step we need $a_i \notin \mathrm{span}_{\mathbb{F}_p}(a_1, \ldots, a_{n-1})$. Therefore
$$\chi_{\mathscr{A}(p,n)}(q) = (q-1)(q-p)(q-p^2)\cdots(q-p^{n-1}).$$
See [27, 145].

12. (All-subset arrangement) $\mathrm{All}_n: \quad \sum_{i \in A} x_i = 0 \qquad (A \subseteq [n],\ A \neq \emptyset, [n]).$
This arrangement appears naturally in combinatorics, representation theory, algebraic geometry, and physics, among many other contexts. To date, we do not know a simple formula for the characteristic polynomial, though we do have a nice bound. [32].

The reduction of $\mathrm{All}_n$ modulo 2 is the arrangement $\mathscr{A}(2,n)$ in $\mathbb{F}_2^n$ considered above. The map from $\mathrm{All}_n$ to $\mathscr{A}(2,n)$ changes the combinatorics; for example, the subset of hyperplanes $x_1 + x_2 = 0, x_2 + x_3 = 0, x_3 + x_1 = 0$ decreases from rank 3 to rank 2. However, this map is a rank-preserving *weak map*, in the sense that the rank of a subset never increases, and the total rank stays the same. This implies [125, Cor. 9.3.7] that the coefficients of $\chi_{\mathrm{All}_n}(q)$ are greater than the respective coefficients of $\chi_{\mathscr{A}(2,n)}(q)$ in absolute value, so

$$r(\mathrm{All}_n) = |\chi_{\mathrm{All}_n}(-1)| > |\chi_{\mathscr{A}(2,n)}(-1)| = \prod_{i=0}^{n-2}(2^i+1) > 2^{\binom{n-1}{2}}$$

It is also known that $r(\mathrm{All}_n) < 2^{(n-1)^2}$; see [32].

**Whitney's formula.** We can sometimes identify combinatorially the terms in Whitney's Theorem 1.7.2 to obtain a useful formula for the characteristic polynomial.

13. (Coordinate arrangement) $\mathscr{H}_n : \quad x_i = 0 \qquad (1 \le i \le n)$

    Whitney's formula gives

$$\chi_{\mathscr{H}_n}(q) = \sum_{A \subseteq [n]} (-1)^{|A|} q^{n-|A|} = (q-1)^n.$$

14. (Generic deformation of $\mathscr{A}_n$) $\mathscr{G}_n : \quad x_i - x_j = a_{ij} \qquad (1 \le i < j \le n, \quad a_{ij}$ generic )

    Let $H_{ij}$ represent the hyperplane $x_i - x_j = a_{ij}$. By the genericity of the $a_{ij}$s, a subarrangement $H_{i_1 j_1}, \ldots, H_{i_k j_k}$ is central if and only if the graph with edges $i_1 j_1, \ldots, i_n j_n$ has no cycles; that is, it is a forest. Therefore

$$\chi_{\mathscr{G}_n}(q) = \sum_{F \text{ forest on } [n]} (-1)^{|F|} q^{n-|F|}, \qquad r(\mathscr{G}_n) = \mathrm{forests}(n)$$

    where forests$(n)$ is the number of forests on vertex set $[n]$. There is no simple formula for this number, though we can use the techniques of Section 1.3.3 to compute the exponential generating function for $\chi_{\mathscr{G}_n}(q)$.

15. (Other deformations of $\mathscr{A}_n$) The same approach can work for any arrangement consisting of hyperplanes of the form $x_i - x_j = a_{ij}$, which correspond to edges marked $a_{ij}$. If we have enough control over the $a_{ij}$s to describe combinatorially which subarrangements are central, we will obtain a combinatorial formula for the characteristic polynomial.

    For simple arrangements like the Shi and Catalan arrangement, the finite field method gives slicker proofs than Whitney's formula. However, this unified approach is also very powerful; for many interesting examples, see [161]. [6]

**Freeness.** Terao's Factorization Theorem 1.7.9 is a powerful algebraic technique for computing characteristic polynomials that factor completely into linear terms.

16. (Braid arrangement, revisited) It is not difficult to describe the $\mathscr{A}_{n-1}$-derivations for the braid arrangement $\mathscr{A}_{n-1}$. Note that $F_d = x_1^d \frac{\partial}{\partial x_1} + \cdots + x_n^d \frac{\partial}{\partial x_n}$

is an $\mathscr{A}_{n-1}$-derivation for $d = 0, 1, \ldots, n-1$ because $F_d(x_i - x_j) = x_i^d - x_j^d$. By Saito's criterion, since

$$\det(F_i(x_j))_{0\le i,j\le n-1} = \det(x_j^i)_{1\le i,j\le n} = \prod_{1\le i<j\le n} (x_i - x_j) = Q_{\mathscr{A}},$$

$\mathrm{Der}(\mathscr{A})$ is free with basis $F_0, \ldots, F_{n-1}$ of degrees $-1, 0, 1, \ldots, n-2$. Therefore

$$\chi_{\mathscr{A}_n}(q) = q(q-1)\cdots(q-n+1).$$

17. (Coxeter arrangements $\mathscr{BC}_n$ and $\mathscr{D}_n$, revisited) For $\mathscr{D}_n$, the function $F_d$ above satisfies $F_d(x_i + x_j) = x_i^d + x_j^d$, so it is a $\mathscr{D}_n$-derivation only for $d$ odd. Fortunately,

$$\det(F_{2i-1}(x_j))_{1\le i,j,\le n} = \prod_{1\le i<j\le n} (x_i - x_j)(x_i + x_j),$$

as can be seen by identifying linear factors as in Section 1.4.2. It follows that $\mathrm{Der}(\mathscr{D}_n)$ is free with basis $F_1, F_3, F_5, \ldots, F_{2n-1}$ and degrees $0, 2, 4, \ldots, 2n-2$.

For $\mathscr{BC}_n$ we need the additional derivation $F = x_1 \cdots x_n \sum_i \frac{1}{x_i}\frac{\partial}{\partial x_i}$. This is indeed a $\mathscr{BC}_n$-derivation because $F(x_i \pm x_j) = (x_j \pm x_i)x_1 \cdots x_n / x_i x_j$. Once again it is easy to check Saito's criterion to see that $\mathrm{Der}(\mathscr{BC}_n)$ is free with basis $F_1, F_3, \ldots, F_{2n-3}, F$ and degrees $0, 2, 4, \cdots, 2n-4, n-2$.

18. (Coxeter arrangements, in general) A **finite reflection group** is a finite group generated by reflections through a family of hyperplanes. For example the reflections across the hyperplanes $x_i = x_j$ of the braid arrangement $\mathscr{A}_{n-1}$ correspond to the transpositions $(ij)$, and generate the symmetric group $S_n$. Every finite reflection group is a direct product of irreducible ones. Here we focus on the **crystallographic** ones, which can be written with integer coordinates. The irreducible crystallographic finite reflection groups $\Phi$ come in three infinite families $A_n, BC_n, D_n$ and five exceptional groups $G_2, F_4, E_6, E_7, E_8$. The corresponding hyperplane arrangements $\mathscr{A}_\Phi$ are the following.

| $\Phi$ | $n$ | equations |
|---|---|---|
| $\mathscr{A}_{n-1}$ | $n$ | $x_i = x_j$ |
| $\mathscr{BC}_n$ | $n$ | $x_i = \pm x_j, \quad x_i = 0$ |
| $\mathscr{D}_n$ | $n$ | $x_i = \pm x_j$ |
| $\mathscr{G}_2$ | 3 | $x_i = \pm x_j, \quad 2x_i = x_j + x_k$ |
| $\mathscr{F}_4$ | 4 | $x_i = \pm x_j, \quad x_i = 0, \quad \pm x_1 \pm x_2 \pm x_3 \pm x_4 = 0.$ |
| $\mathscr{E}_6$ | 9 | $x_i = x_j, \quad \sum_{i\in\{a,b,c\}} 2x_i = \sum_{i\notin\{a,b,c\}} x_i \quad \left(\begin{smallmatrix}1\le a\le 3\\ 4\le b\le 6\\ 7\le c\le 9\end{smallmatrix}\right).$ |
| $\mathscr{E}_7$ | 8 | $x_i = x_j, \quad \sum_{i\in A} x_i = \sum_{i\notin A} x_i \quad \left(\begin{smallmatrix}A\subseteq[8]\\ \lvert A\rvert=4\end{smallmatrix}\right)$ |
| $\mathscr{E}_8$ | 8 | $x_i = \pm x_j, \quad \sum_{i=1}^{8} \varepsilon_i x_i = 0 \quad \left(\begin{smallmatrix}\varepsilon_i=\pm 1,\\ \varepsilon_1\cdots\varepsilon_8=1\end{smallmatrix}\right)$ |

Note that arrangement $\Phi_r$ has rank $r$, but we have embedded some of them in ${}^d$ for higher $d$ in order to obtain nicer equations. Here the indices range over the corresponding ${}^n$; for instance, $\mathcal{G}_2$ includes the hyperplanes $2x_1 = x_2+x_3$, $2x_2 = x_1+x_3$, $2x_3 = x_1+x_2$. We now list the characteristic polynomials and number of regions for these arrangements.

| $\Phi$ | characteristic polynomial | regions |
|---|---|---|
| $\mathcal{BC}_n$ | $(q-1)(q-3)(q-5)\cdots(q-2n+3)(q-2n+1)$ | $2^n n!$ |
| $\mathcal{D}_n$ | $(q-1)(q-3)(q-5)\cdots(q-2n+3)(q-n+1)$ | $2^{n-1}n!$ |
| $\mathcal{G}_2$ | $(q-1)(q-5)$ | 12 |
| $\mathcal{F}_4$ | $(q-1)(q-5)(q-7)(q-11)$ | 1,152 |
| $\mathcal{E}_6$ | $(q-1)(q-4)(q-5)(q-7)(q-8)(q-11)$ | 51,840 |
| $\mathcal{E}_7$ | $(q-1)(q-5)(q-7)(q-9)(q-11)(q-13)(q-17)$ | 2,903,040 |
| $\mathcal{E}_8$ | $(q-1)(q-7)(q-11)(q-13)(q-17)(q-19)$ $\times(q-23)(q-29)$ | 696,729,600 |

It follows from the theory of reflection groups that the number of regions equals the order of the reflection group:

$$r(\mathcal{A}_\Phi) = |\Phi|.$$

We already computed in two ways the characteristic polynomials of the classical Coxeter groups $\mathcal{A}_n, \mathcal{BC}_n$, and $\mathcal{D}_n$. We first gave simple finite field proofs; it would be interesting to do this for the other ("exceptional") groups. We also computed explicit bases for the modules of $\mathcal{A}$-derivations; doing this for the other Coxeter groups is not straightforward, but it can be done. [153, Appendix B]

However, these beautiful formulas clearly illustrate that there are deeper things at play, and they make it very desirable to have a unified explanation. Indeed, invariant theory produces an elegant case-free proof that all Coxeter arrangement are free; for details, see [153, Theorem 6.60].

19. (Shi arrangement) Given its characteristic polynomial, it is natural to guess that the Shi arrangement $\mathcal{S}_{n-1}$ is free. Since we only defined freeness for central arrangements, we consider the cone $c\mathcal{S}_{n-1}$ instead. Following [21], we may prove that this is a free arrangement inductively using Theorem 1.7.11. The strategy is to remove the hyperplanes of $\mathcal{S}_{n-1}$ one at a time, by finding in each step a hyperplane $H$ such that $\mathcal{A}\backslash H$ and $A/H$ are free with predictable exponents. This leads us to consider a more general family of arrangements.

    We claim that for any $m \geq 0$ and $2 \leq k \leq n+1$, the arrangement $\mathcal{S}_{n-1}^{m,k}$

$$\begin{aligned} x_i - x_j &= 0,1 \qquad \text{for } 2 \leq i < j \leq n \\ x_1 - x_j &= 0,1,\ldots,m \qquad \text{for } 2 \leq j < k \\ x_1 - x_j &= 0,1,\ldots,m,m+1 \qquad \text{for } k \leq j \leq n \end{aligned}$$

is free with exponents $n+m-1,\ldots,n+m-1$ $(n-k+1$ times$)$ , $n+m-2,\ldots,n+m-2$ $(k-2$ times$)$, and $-1$.

Obviously it takes some care to assemble this family of free arrangements and their exponents. However, once we have done this, it is straightforward to prove this more general statement by induction using Theorem 1.7.11, by removing one hyperplane at a time while staying within the family of arrangements $\mathscr{S}_{n-1}^{m,k}$. Setting $k=2$ and $m=0$ we get that the Shi arrangement is free.

Incidentally, Athanasiadis also characterized all the arrangements $\mathscr{A}$ with $\mathscr{A}_{n-1} \subseteq \mathscr{A} \subseteq \mathscr{S}_{n-1}$ such that $c\mathscr{A}$ is free [21]. These arrangements have Ish counterparts that have the same characteristic polynomial, although their cones are not free. [19, Corollary 3.3]

### 1.7.5 The cd-index of an arrangement

We now discuss a vast strengthening of Zaslavsky's Theorem 1.7.1.1 due to Billera, Ehrenborg, and Readdy. [34] They showed that the enumeration of chains of faces of a real hyperplane arrangement $\mathscr{A}$ (or, equivalently, of a zonotope) depends only on the intersection poset of $L_{\mathscr{A}}$. The equivalence with zonotopes is explained by the following result.

**Proposition 1.7.14** *Let A be an arrangement of non-zero real vectors, and let $\mathscr{A}$ be the arrangement of hyperplanes perpendicular to the vectors of A. Then there is an order-reversing bijection between the faces of the zonotope Z(A) and the faces of the real arrangement $\mathscr{A}$.*

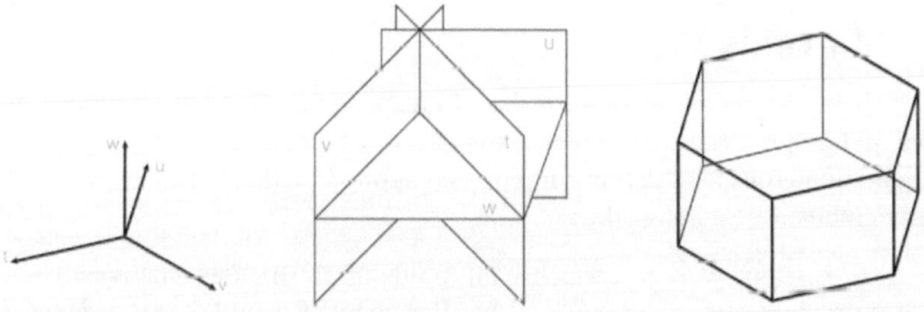


**Figure 1.37**
A vector arrangement, the normal hyperplane arrangement, and its dual zonotope.

From this result, it follows easily that the **cd**-indices of $Z(A)$ and $F(\mathscr{A})$ are reverses of each other; that is, they are related by the linear map $^*: \mathbb{Z}\langle \mathbf{c},\mathbf{d}\rangle \to \mathbb{Z}\langle \mathbf{c},\mathbf{d}\rangle$ that reverses any word in **c** and **d**, so that $(v_1\ldots v_k)^* = v_k \ldots v_1$. To describe these **cd**-indices, we introduce the linear map $\omega : \mathbb{Z}\langle \mathbf{a},\mathbf{b}\rangle \to \mathbb{Z}\langle \mathbf{c},\mathbf{d}\rangle$ obtained by first replacing each occurrence of **ab** with 2**d**, and then replacing every remaining letter with a **c**.

**Theorem 1.7.15** [34] *The flag f-vector of a real hyperplane arrangement (or equivalently, of the dual zonotope) depends only on its intersection poset.*

*More explicitly, let $\mathscr{A}$ be a real hyperplane arrangement. Let $F(\mathscr{A})$ be its face poset and $L_{\mathscr{A}}$ be its intersection poset. The* **cd***-indices of the face poset $F(\mathscr{A})$ and the dual zonotope $Z(\mathscr{A})$ are given in terms of the* **ab***-index of the intersection lattice $L_{\mathscr{A}}$ by the formula:*

$$\Psi_{F(\mathscr{A})}(\mathbf{c},\mathbf{d}) = \left[\omega(\mathbf{a}\Phi_{L_{\mathscr{A}}}(\mathbf{a},\mathbf{b}))\right]^*, \qquad \Psi_{Z(\mathscr{A})}(\mathbf{c},\mathbf{d}) = \omega(\mathbf{a}\Phi_{L_{\mathscr{A}}}(\mathbf{a},\mathbf{b})).$$

Billera, Ehrenborg, and Readdy [34] extended this result to *orientable matroids*, and asked for an interesting interpretation of $\omega(\mathbf{a}\Phi_L(\mathbf{a},\mathbf{b}))$ for an arbitrary geometric lattice $L$.

To illustrate this result we revisit Example 1.5.24, where we computed the **cd**-index of a hexagonal prism. This is the dual zonotope to the arrangement of Figure 1.37, whose intersection poset $L_{\mathscr{A}}$ is Figure 1.33. The flag $f$ and $h$-vectors of $L_{\mathscr{A}}$ are $(f_{\emptyset}, f_{\{1\}}, f_{\{2\}}, f_{\{1,2\}}) = (1,4,4,9)$ and $(h_{\emptyset}, h_{\{1\}}, h_{\{2\}}, h_{\{1,2\}}) = (1,3,3,2)$, so its **ab**-index is $\Phi_{L_{\mathscr{A}}}(\mathbf{a},\mathbf{b})) = \mathbf{aa} + 3\mathbf{ab} + 3\mathbf{ba} + 2\mathbf{bb}$. Therefore

$$\begin{aligned}\Psi_{Z(\mathscr{A})}(\mathbf{c},\mathbf{d}) &= \omega(\mathbf{aaa} + 3\mathbf{aab} + 3\mathbf{aba} + 2\mathbf{abb}) = \mathbf{ccc} + 3\mathbf{c}(2\mathbf{d}) + 3(2\mathbf{d})c + 2(2\mathbf{d})\mathbf{c} \\ &= \mathbf{c}^3 + 6\mathbf{cd} + 10\mathbf{dc}\end{aligned}$$

and

$$\Psi_{F(\mathscr{A})}(\mathbf{c},\mathbf{d}) = \mathbf{c}^3 + 6\mathbf{dc} + 10\mathbf{cd}$$

in agreement with the computation in Section 1.5.6.

## 1.8 Matroids

Matroid theory is a combinatorial theory of independence that has its roots in linear algebra and graph theory, but which turns out to have deep connections with many fields, and numerous applications in pure and applied mathematics. Rota was a particularly enthusiastic ambassador:

> "It is as if one were to condense all trends of present day mathematics onto a single finite structure, a feat that anyone would a priori deem impossible, were it not for the fact that matroids do exist." [171]

There are natural notions of independence in linear algebra, graph theory, matching theory, the theory of field extensions, and the theory of routings, among others. Matroids capture the combinatorial essence that those notions share.

In this section we will mostly focus on enumerative aspects of matroid theory. For a more complete introduction, see [154, 211] and the three volume series [214, 215, 216].

In Section 1.8.1 we discuss some combinatorial aspects of vector configurations and linear independence, and introduce some terminology. This example and terminology strongly motivates the definition(s) of a matroid, which we give in Section

1.8.2. Section 1.8.3 discusses many important families of matroids arising in algebra, combinatorics, and geometry. Section 1.8.4 introduces some basic constructions, and Section 1.8.5 gives a very brief overview of structural matroid theory. The Tutte polynomial is our main enumerative tool; it is introduced in Section 1.8.6. In Section 1.8.7 we answer many enumeration problems in terms of Tutte polynomials, and in Section 1.8.8 we see how we can actually compute the polynomial in some examples of interest. Section 1.8.9 is devoted to two generalizations: the multivariate and the arithmetic Tutte polynomials. Finally, we give a brief overview of matroid subdivisions in Section 1.8.10, including a discussion on matroid valuations and the Derksen–Fink invariant.

### 1.8.1 Main example: Vector configurations and linear matroids

Let $E \subset {}^d$ be a finite set of vectors. For simplicity, we assume that $E$ does not contain $\mathbf{0}$, does not contain repeated vectors, and spans ${}^d$. The **matroid** of $E$ is given by any of the following definitions:

- The **independent sets** of $E$ are the subsets of $E$ that are linearly independent over .
- The **bases** of $E$ are the subsets of $E$ that are bases of ${}^d$.
- The **circuits** of $E$ are the minimal linearly dependent subsets of $E$.
- The **rank function** $r : 2^E \to \mathbb{N}$ is $r(A) = \dim(\operatorname{span} A)$ for $A \subseteq E$.
- The **flats** $F$ of $E$ are the subspaces of ${}^d$ spanned by subsets of $E$. We identify a flat with the set of vectors of $E$ that it contains.
- The **lattice of flats** $L_M$ is the poset of flats ordered by containment.
- The **matroid (basis) polytope** is

$$P_M = \operatorname{conv}\{\mathbf{e}_{b_1} + \cdots + \mathbf{e}_{b_r} : \{b_1, \ldots, b_r\} \text{ is a basis}\} \subset \mathbb{R}^E.$$

More concretely, if we know one of the following objects, we can determine them all: the list $\mathscr{I}$ of independent sets, the list $\mathscr{B}$ of bases, the list $\mathscr{C}$ of circuits, the rank function $r$, the list $\mathscr{F}$ of flats, the lattice $L_M$ of flats, or the matroid polytope $P_M$. For this reason, we define the **matroid** of $E$ to be any one (and all) of these objects. A matroid that arises in this way from a set of vectors in a vector space (over ) is called a **linear** or **representable matroid** (over ).

**Example 1.8.1** *Let* $\mathbf{t} = (1,-1,0), \mathbf{u} = (0,1,-1), \mathbf{v} = (-1,0,1), \mathbf{w}, = (1,1,1)$ *be the vector configuration shown in the left panel of Figure 1.37. Then*

$$\begin{aligned}
\mathscr{I} &= \{\emptyset, \mathbf{t}, \mathbf{u}, \mathbf{v}, \mathbf{w}, \mathbf{tu}, \mathbf{tv}, \mathbf{tw}, \mathbf{uv}, \mathbf{uw}, \mathbf{vw}, \mathbf{tuw}, \mathbf{tvw}, \mathbf{uvw}\},\\
\mathscr{B} &= \{\mathbf{tuw}, \mathbf{tvw}, \mathbf{uvw}\},\\
\mathscr{C} &= \{\mathbf{tuv}\}\\
\mathscr{F} &= \{\emptyset, \mathbf{t}, \mathbf{u}, \mathbf{v}, \mathbf{w}, \mathbf{tuv}, \mathbf{tw}, \mathbf{uw}, \mathbf{vw}, \mathbf{tuvw}\}.
\end{aligned}$$

*(We are omitting brackets for clarity, so for example we write* **tuv** *for* $\{\mathbf{t},\mathbf{u},\mathbf{v}\}$*.) The rank function is* $r(\mathbf{tuvw}) = 3, r(\mathbf{tuv}) = 2$*, and* $r(A) = |A|$ *for all other sets A. The matroid polytope is the equilateral triangle with vertices* $(1,1,0,1)$, $(1,0,1,1)$, *and* $(0,1,1,1)$ *in* $\mathbb{R}^4$.

A vector arrangement $E \subset {}^d$ determines a hyperplane arrangement $\mathscr{A}$ in ${}^d$ consisting of the normal hyperplanes $\{\mathbf{x} \in {}^d : \mathbf{a} \cdot \mathbf{x} = 0\}$ for $\mathbf{a} \in E$. The rank function and lattice of flats of $E$ are the same as the rank function and intersection poset of $\mathscr{A}$, as defined in Section 1.7.2. For instance, the lattice of flats of the example above is the same as the intersection poset of the corresponding arrangement, which is illustrated in Figure 1.33.

### 1.8.2 Basic definitions

A simple but important insight of matroid theory is that many of the properties of linear independence (and many other notions of independence in mathematics) are inherently combinatorial. That combinatorial structure is unexpectedly rich, and matroids provide a general combinatorial framework that is ideally suited for exploring it.

A **matroid** $M = (E, \mathscr{I})$ consists of a finite set $E$ and a collection $\mathscr{I}$ of subsets of $E$ such that

(I1) $\emptyset \in \mathscr{I}$

(I2) If $J \in \mathscr{I}$ and $I \subseteq J$ then $I \in \mathscr{I}$.

(I3) If $I, J \in \mathscr{I}$ and $|I| < |J|$ then there exists $j \in J - I$ such that $I \cup j \in \mathscr{I}$.

The sets in $\mathscr{I}$ are called **independent**.

The following result, which is an immediate consequence of (I3), shows that this simple definition already captures the notion of dimension.

**Proposition 1.8.2** *In a matroid* $M = (E, \mathscr{I})$*, all the* **bases** *(the maximal elements of* $\mathscr{I}$*) have the same size, called the* **rank** *of the matroid.*

Now we develop the combinatorial versions of the definitions of the previous section. Let $M = (E, \mathscr{I})$ be a matroid.

- An **independent set** is a set in $\mathscr{I}$.
- A **basis** is a maximal independent set.
- The **rank function** $r : 2^E \to \mathbb{N}$ is $r(A) =$(size of the largest independent subset of $A$).
- A **circuit** is a minimal dependent set; i.e., a minimal set not in $\mathscr{I}$.
- A **flat** is a set $F$ such that $r(F \cup e) > r(F)$ for all $e \notin F$.
- The **lattice of flats** $L_M$ is the poset of flats ordered by containment.
- The **matroid (basis) polytope** $P_M = \text{conv}\{\mathbf{e}_{b_1} + \cdots + \mathbf{e}_{b_r} : \{b_1, \ldots, b_r\}$ is a basis$\}$.

Again, we let $\mathscr{B}$, $r$, $\mathscr{C}$ and $\mathscr{F}$ denote the set of bases, the rank function, the set of circuits, and the set of flats of $M$. If we know the ground set $E$ and any one of $\mathscr{B}$, $r$, $\mathscr{C}$, $\mathscr{F}$, $L_M$, or $P_M$ (assuming that we know the labels of the elements of $L_M$ and the embedding of the polytope $P_M$ in $\mathbb{R}^E$), we know them all. Knowing $E$, we can recover the collection $\mathscr{I}$ of independent sets from $\mathscr{B}$, $r$, $\mathscr{C}$, or $\mathscr{F}$. For that reason, we call any one (and all) of $(E,\mathscr{I}),(E,\mathscr{B}),(E,r),(E,\mathscr{C})$ and $(E,\mathscr{F})$ "the matroid $M$." Each one of these points of view has its own axiomatization.

**Proposition 1.8.3** *We have the following characterizations of the possible bases $\mathscr{B}$, rank function $r$, circuits $\mathscr{C}$, lattice of flats $L$, independent sets $\mathscr{I}$, and matroid polytopes $P_M$ of a matroid.*

1. *The collection $\mathscr{B} \subseteq 2^E$ is the set of bases of a matroid if and only if*

   *(B1) $\mathscr{B}$ is nonempty.*

   *(B2) If $B_1, B_2 \in \mathscr{B}$ and $b_1 \in B_1 - B_2$, then there exists $b_2 \in B_2 - B_1$ such that $(B_1 - b_1) \cup b_2 \in \mathscr{B}$.*

2. *The function $r : 2^E \to \mathbb{N}$ is the rank function of a matroid if and only if*

   *(R1) $0 \leq r(A) \leq |A|$ for all $A \subseteq E$.*

   *(R2) If $A \subseteq B \subseteq E$, then $r(A) \leq r(B)$.*

   *(R3) $r(A)+r(B) \geq r(A \cup B)+r(A \cap B)$ for all $A,B \subseteq E$.*

3. *The collection $\mathscr{C} \subseteq 2^E$ is the set of circuits of a matroid if and only if*

   *(C1) $\emptyset \notin \mathscr{C}$.*

   *(C2) If $C \in \mathscr{C}$ and $C \subsetneq D$, then $D \notin \mathscr{C}$.*

   *(C3) If $C_1, C_2 \in \mathscr{C}$ and $e \in C_1 \cap C_2$, then there exists $C \in \mathscr{C}$ with $C \subseteq C_1 \cup C_2 - e$.*

4. *The poset $L$ is the lattice of flats of a matroid if and only if it is a geometric lattice.*

5. *The collection $\mathscr{I} \subseteq 2^E$ is the set of independent sets of a matroid if and only if*

   *(I1) $\emptyset \in \mathscr{I}$.*

   *(I2) If $J \in \mathscr{I}$ and $I \subseteq J$ then $I \in \mathscr{I}$.*

   *(I3′) For every weight function $w : E \to \mathbb{R}$, the following* greedy algorithm *produces a maximal element $B$ of $\mathscr{I}$ of minimum weight $w(B) = \sum_{b \in B} w(b)$:*

   > *Start with $B = \emptyset$. Then, at each step, add to $B$ an element $e \notin B$ of minimum weight $w(e)$ such that $B \cup e \in \mathscr{I}$. Stop when $B$ is maximal in $\mathscr{I}$.*

6. *[86] The collection $\mathscr{B} \subseteq 2^E$ is the set of bases of a matroid if and only if every edge of the polytope* $\operatorname{conv}\{\mathbf{e}_{b_1} + \cdots + \mathbf{e}_{b_r} : \{b_1,\ldots,b_r\} \in \mathscr{B}\}$ *in $\mathbb{R}^E$ is a translate of $\mathbf{e}_i - \mathbf{e}_j$ for some $i, j \in E$.*

This proposition allows us to give six other definitions of matroids in terms of bases, ranks, circuits, flats, independent sets, and polytopes. The first three axiom systems are "not too far" from each other, but it is useful to have them all. For instance, to prove that linear matroids are indeed matroids, it is easier to check the circuit axioms (C1)–(C3) than the independence axioms (I1)–(I3). The last three axiom systems suggest deeper, fruitful connections to posets, optimization, and polytopes.

In fact, there are several other equivalent (and useful) definitions of matroids. It is no coincidence that matroid theory gave birth to the notion of a **cryptomorphism**, which refers to the equivalence of different axiomatizations of the same object. This feature of matroid theory can be frustrating at first; but as one goes deeper, it becomes indispensable and extremely powerful. A very valuable resource is [46], a multilingual cryptomorphism dictionary that translates between these different points of view.

We say $(E, \mathscr{B})$ and $(E', \mathscr{B}')$ are **isomorphic** if there is a bijection between $E$ and $E'$ that induces a bijection between $\mathscr{B}$ and $\mathscr{B}'$. A **loop** of $M$ is an element $e \in E$ of rank 0, so $\{e\}$ is dependent. A **coloop** is an element $e$ that is independent of $E - e$, so $r(E - e) = r - 1$. Non-loops **e** and **f** are **parallel** if $r(\mathbf{ef}) = 1$; they are indistinguishable inside the matroid. A matroid is **simple** if it contains no loops or parallel elements. The map $M \mapsto L_M$ is a bijection between simple matroids and geometric lattices.

### 1.8.3 Examples

We now describe a few contexts where matroids arise naturally. These statements are nice (and not always trivial) exercises in their respective fields. Most proofs may be found, for example, in [154]. We give references for the rest.

0. (Uniform matroids) For $n \geq k \geq 1$, the **uniform matroid** $U_{k,n}$ is the matroid on $[n]$ where every $k$-subset is a basis. It is the linear matroid of $n$ generic vectors in a $k$-dimensional space.

1a. (Vector configurations) Let $E$ be a set of vectors in ${}^d$ and let $\mathscr{I}$ be the set of linearly independent subsets of $E$. Then $(E, \mathscr{I})$ is a matroid. Such a matroid is called **linear** or **representable** (over ). [132, 217] There are at least three other ways of describing the same family of matroids, listed in 1b, 1c, and 1d below.

1b. (Point configurations) Let $E$ be a set of points in ${}^d$ and let $\mathscr{I}$ be the set of affinely independent subsets of $E$; that is, the subsets $\{\mathbf{a}_1, \ldots, \mathbf{a}_k\}$ whose affine span is $k$-dimensional. Then $(E, \mathscr{I})$ is a matroid.

1c. (Hyperplane arrangements) Let $E$ be a hyperplane arrangement in ${}^d$ and let $r(\{H_1, \ldots, H_k\}) = d - \dim(H_1 \cap \cdots \cap H_k)$ for $\{H_1, \ldots, H_k\} \subseteq E$. Then $(E, r)$ is a matroid.

1d. (Subspaces) Let $V \cong {}^E$ be a vector space with a chosen basis, and let $U \subseteq V$ be a subspace of codimension $r$. Consider the coordinate subspaces $V_S = \{\mathbf{v} \in$

$V : v_s = 0$ for $s \in S\}$. Say an $r$-subset $B \subseteq [d]$ is a basis if $U \cap V_B = \{0\}$. If $\mathscr{B}$ is the set of bases, $(E, \mathscr{B})$ is a matroid.

Given a matrix $A$ over , the matroid on the columns of $A$ (in the sense of 1a) is the same as the matroid of the rowspace of $A$ (in the sense of 1d).

2. (Graphs) Let $E$ be the set of edges of a connected graph $G$ and let $\mathscr{C}$ be the set of cycles of $G$ (where each cycle is regarded as the list of its edges). Then $(E, \mathscr{C})$ is a matroid. Such a matroid is called **graphical**. The bases are the spanning trees of $G$.

3. (Field extensions) Let $\mathbb{F} \subseteq \mathbb{K}$ be two fields, and let $E$ be a finite set of elements in the extension field $\mathbb{K}$. Say $I = \{i_1, \ldots, i_k\} \subseteq E$ is independent if it is algebraically independent over $\mathbb{F}$; that is, if there does not exist a non-trivial polynomial $P(x_1, \ldots, x_k)$ with coefficients in $\mathbb{F}$ such that $P(i_1, \ldots, i_k) = 0$. If $\mathscr{I}$ is the collection of independent sets, $(E, \mathscr{I})$ is a matroid. Such a matroid is called **algebraic**. [204]

4. (Matchings) Let $G = (S \cup T, E)$ be a bipartite graph, so every edge in $E$ joins a vertex of $S$ and a vertex of $T$. Recall that a **partial matching** of $G$ is a set of edges, no two of which have a common vertex. Let $\mathscr{I}$ be the collection of subsets $I \subseteq T$ that can be matched to $S$; that is, those for which there exists a partial matching whose edges contain the vertices in $I$. Then $(T, \mathscr{I})$ is a matroid. Such a matroid is called **transversal**. [71]

5. (Routings) Let $G = (V, E)$ be a directed graph and $B_0$ be an $r$-subset of the vertices. Let $\mathscr{B}$ be the collection of $r$-subsets of $V$ for which there exists a routing (a collection of $r$ vertex-disjoint paths) starting at $B$ and ending at $B_0$. Then $(V, \mathscr{B})$ is a matroid. Such a matroid is called **cotransversal**. [136]

6. (Lattice paths) For each Dyck path $P$ of length $n$, let $B \subseteq [2n]$ be the upstep set of $P$; that is, the set of integers $i$ for which the $i$th step of $P$ is an upstep. Let $\mathscr{B}$ be the collection of upstep sets of Dyck paths of length $n$. Then $([2n], \mathscr{B})$ is the set of basis of a matroid. This matroid is called a **Catalan matroid** $\mathbf{C}_n$. More generally, we may consider the lattice paths of $2n$ steps $(1,1)$ and $(1,-1)$ which stay between an upper and a lower border; their upstep sets form the bases of a **lattice path matroid**. [4, 41]

7. (Schubert matroids) Given $n \in \mathbb{N}$ and a set of positive integers $I = \{i_1 < \cdots < i_k\}$, the sets $\{j_1 < \cdots < j_k\} \subseteq [n]$ such that $j_1 \geq i_1, \ldots, j_k \geq i_k$ are the bases of a matroid. These matroids are called **Schubert matroids** due to their connection to the Schubert cells in the Grassmannian $\mathrm{Gr}(k,n)$. This simple but important family of matroids has been rediscovered many times under names such as freedom matroids [52] and shifted matroids [4, 118].

8. (Positroids) A **positroid** is a matroid on $[n]$ that can be represented by the columns of a full rank $d \times n$ matrix such that all its maximal minors are nonnegative. [157] These matroids arise in the study of the totally nonnegative

part of the Grassmannian, and have recently found applications in the physics of scattering amplitudes. Remarkably, these matroids can be described combinatorially, and they are in bijection with several interesting classes of objects with elegant enumerative properties. [17, 150, 157, 219]

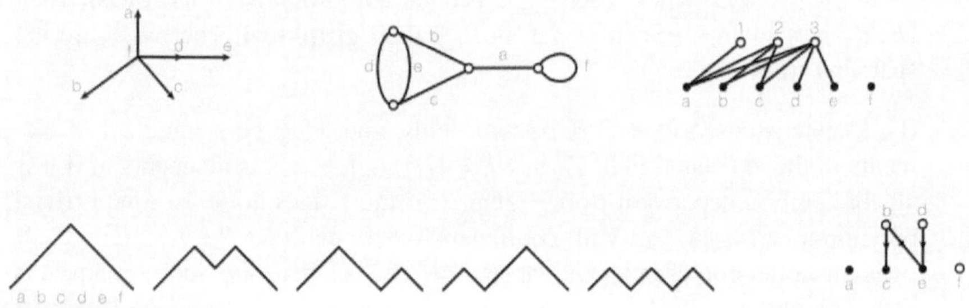


**Figure 1.38**
Many manifestations of the same matroid.

**Example 1.8.4** *The list above gives eight different manifestations of the matroid on* $\{a,b,c,d,e,f\}$ *with bases* $\mathscr{B} = \{abc, abd, abe, acd, ace\}$. *Figure 1.38 shows a vector configuration in* $\mathbb{R}^3$ *(with* $f = \mathbf{0}$*) that realizes* $M$ *as a linear matroid, a graph that realizes it as a graphical matroid, a bipartite graph that realizes it as a transversal matroid, the five Dyck paths of length 3 that show that* $M$ *is the Catalan matroid* $\mathbf{C}_3$, *and a directed graph with sinks* $B_0 = \{a,c,e\}$ *that realizes* $M$ *as a cotransversal matroid. It can be realized as an algebraic matroid in the field extension* $\mathbb{R} \subset \mathbb{R}(x,y,z)$ *with* $a = z^3, b = x+y, c = x-y, d = xy, e = x^2y^2, f = 1$. *It is also isomorphic to the Schubert matroid on* $[6]$ *for* $I = \{2,4,6\}$. *Finally, it is a positroid, since in the realization shown, all five bases are positive: They all satisfy the "right-hand rule" for vectors in* $\mathbb{R}^3$ *when listed in alphabetical order.*

To give a small illustration of the power of matroids, let us revisit the almost trivial observation that all bases of a matroid have the same size. This result has many interesting consequences.

- All bases of a vector space have the same size, called its "dimension."
- All spanning trees of a connected graph have the same number $(v-1)$ of edges, where $v$ is the number of vertices of the graph.
- In a bipartite graph $G = (S \cup T, E)$, all the maximal subsets of $T$ that can be matched to $S$ have the same size.
- All transcendence bases of a field extension have the same size, called the "transcendence degree" of the extension.

Naturally, once we work harder to obtain more interesting results about matroids, we will obtain more impressive results in all of these different fields.

### 1.8.4 Basic constructions

We now discuss the notions of duality, minors, direct sums, and connected components in matroids; in Section 1.8.5 we explain how they generalize similar notions in linear algebra and graph theory.

**Proposition/Definition 1.8.5 (Duality)** *Let $M = (E, \mathscr{B})$ be a matroid, and let $\mathscr{B}^* = \{E - B : B \in \mathscr{B}\}$. Then $M^* = (E, \mathscr{B}^*)$ is a matroid, called the* **dual matroid** *of $M$.*

**Proposition 1.8.6** *For any matroid $M$ we have $(M^*)^* = M$.*

Note that the loops of a matroid $M$ are the coloops of $M^*$.

**Proposition/Definition 1.8.7 (Deletion, contraction)** *Let $M = (E, \mathscr{I})$ be a matroid and $s \in E$. Let*

$$\mathscr{I}' = \{I \in \mathscr{I} : s \notin I\}, \qquad \mathscr{I}'' = \begin{cases} \{I \subseteq E - s : I \cup s \in I\}, & \text{if } s \text{ is not a loop,} \\ \mathscr{I}, & \text{if } s \text{ is a loop.} \end{cases}$$

*Then $M \backslash s = (E - s, \mathscr{I}')$ and $M/s = (E - s, \mathscr{I}'')$ are also matroids. They are called the* **deletion** *of $s$ from $M$ (or the* **restriction** *of $M$ to $E - s$) and the* **contraction** *of $s$ from $M$.*

**Proposition/Definition 1.8.8 (Minors)** *Let $M$ be a matroid on $E$.*

1. *Deletion and contraction commute; that is, for all $s \neq t$ in $E$,*

$$(M\backslash s)\backslash t = (M\backslash t)\backslash s, \quad (M/s)/t = (M/t)/s, \quad (M/s)\backslash t = (M\backslash t)/s.$$

2. *Let the* **deletion** *$M \backslash S$ (respectively,* **contraction** *$M/S$) of a subset $S \subseteq E$ from $M$ be the successive deletion (respectively, contraction) of the individual elements of $S$. Their rank functions are*

$$r_{M\backslash S}(A) = r(A), \qquad r_{M/S}(A) = r(A \cup S) - r(S) \qquad (A \subseteq E - S).$$

   *Let a* **minor** *of $M$ be a matroid of the form $M/S\backslash T$ for disjoint $S, T \subseteq E$.*

3. *Deletion and contraction are dual operations; that is, for all $S \subseteq E$,*

$$(M\backslash S)^* = M^*/S.$$

**Proposition/Definition 1.8.9 (Direct sum, connected components)**

1. *If $M = (E_1, \mathscr{I}_1)$ and $M_2 = (E_2, \mathscr{I}_2)$ are matroids on disjoint ground sets, then $\mathscr{I} = \{I_1 \cup I_2 : I_1 \in \mathscr{I}_1, I_2 \in \mathscr{I}_2\}$ is the collection of independent sets of a matroid $M_1 \oplus M_2 = (E_1 \cup E_2, \mathscr{I})$, called the* **direct sum** *of $M_1$ and $M_2$.*

2. *Any matroid $M$ decomposes uniquely as a direct sum $M = M_1 \oplus \cdots \oplus M_c$. The ground sets of $M_1, \ldots, M_c$ are called the* **connected components** *of $M$. If $c = 1$, $M$ is* **connected.**

The matroid polytope of $M$ has $\dim(P_M) = n - c$ where $n$ is the number of elements of $M$ and $c$ is the number of connected components.

### 1.8.5 A few structural results

Although they will not be strictly necessary in our discussion of enumerative aspects of matroids, it is worthwhile to mention a few important structural results. In particular, in this section we explain the meaning of duality and minors for various families of matroids.

**Duality.**

- The dual of a linear matroid is linear. When $M$ is the matroid of a subspace $V$ of a vector space $W$, $M^*$ is the matroid of its orthogonal complement $V^{\perp}$. For that reason, $M^*$ is sometimes called the **orthogonal matroid** of $M$.

- The dual of a graphical matroid is not necessarily graphical; however, matroid duality is a generalization of graph duality. We say a graph $G$ is **planar** if it can be drawn on the plane so that its edges intersect only at the endpoints. Such a drawing is called a **plane** graph. The dual graph $G^*$ is obtained by putting a vertex $v^*$ inside each region of $G$ (including the "outside region"), and joining $v^*$ and $w^*$ by an edge labeled $e^*$ if the corresponding regions of $G$ are separated by an edge $e$ in $G$. This construction is exemplified in Figure 1.39. The graph $G^*$ depends on the drawing of $G$, but its matroid does not: if $G$ is planar, we have $M(G^*) = M(G)^*$.

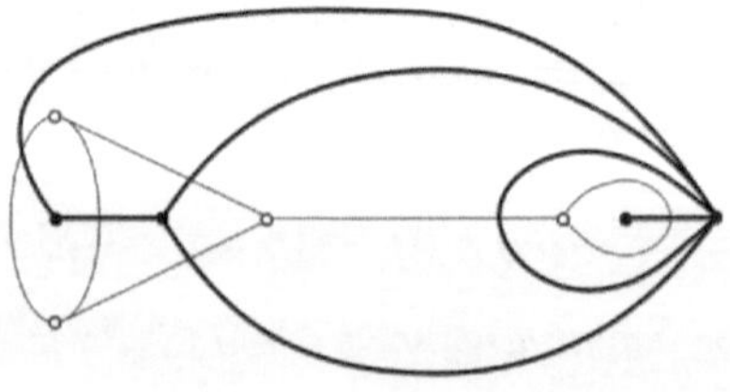

**Figure 1.39**
A plane graph $G$ and its dual $G^*$.

- Embarrassingly, no one knows whether the dual of an algebraic matroid is always algebraic. This illustrates how poorly we currently understand this family of matroids.

- Cotransversal matroids are precisely the duals of transversal matroids.

- The Catalan matroid is self-dual. The dual of a lattice path matroid is a lattice path matroid.

- Linear programming duality can be framed and generalized in the context of *oriented matroids*; see [39] for details.

**Corollary 1.8.10** *If a connected planar graph has $v$ vertices, $e$ edges, and $f$ faces, then $v - e + f = 2$.*

*Proof.* A basis of our graph $G$ has $v-1$ elements and a basis of its dual graph $G^*$ has $f-1$ elements; so if $M$ is the matroid of $G$, we have $(v-1)+(f-1) = r(M) + r(M^*) = e$. ■

**Minors.**

- The minors of a linear matroid are linear. If $M$ is the matroid of a vector configuration $E \subset V$ and $\mathbf{v} \in E$, then $M \backslash \mathbf{v}$ is the matroid of $E \backslash \mathbf{v}$, and $M/\mathbf{v}$ is the matroid of $E \backslash \mathbf{v}$ modulo $\mathbf{v}$; that is, the matroid of $\pi(E \backslash \mathbf{v})$, where $\pi : V \to V/(\operatorname{span} \mathbf{v})$ is the canonical projection map.

- The minors of a graphical matroid are graphical. If $M$ is the matroid of a graph $G$ and $e = uv$ is an edge, then $M \backslash e$ is the matroid of the graph $G \backslash e$ obtained by removing the edge $e$, and $M/e$ is the matroid of the graph $G/e$ obtained by removing the edge $e$ and identifying vertices $u$ and $v$.

- [154, Cor. 6.7.14] The minors of an algebraic matroid are algebraic.

- Transversal matroids are closed under contraction, but not deletion. Cotransversal matroids are closed under deletion, but not contraction. A **gammoid** is a deletion of a transversal matroid or, equivalently, a contraction of a cotransversal matroid. Gammoids are the smallest minor-closed class of matroids containing transversal (or cotransversal) matroids.

- [41] Catalan matroids are not closed under deletion or contraction. Lattice path matroids are closed under deletion and contraction.

**Direct sums and connectivity.**

- If $M_1$ and $M_2$ are the linear matroids of configurations $E_1 \subset V_1$ and $E_2 \subset V_2$ over the same field , then $M_1 \oplus M_2$ is the matroid of $\{(\mathbf{v}_1, \mathbf{0}) : \mathbf{v}_1 \in E_1\} \cup \{(\mathbf{0}, \mathbf{v}_2) : \mathbf{v}_2 \in E_2\}$ in $V_1 \oplus V_2$.

- If $M_1$ and $M_2$ are the matroids of graphs $G_1$ and $G_2$, then $M_1 \oplus M_2$ is the matroid of the disjoint union of $G_1$ and $G_2$. The matroid of a loopless graph $G$ is connected if and only if $G$ is 2-connected; that is, $G$ is connected, and there does not exist a vertex of $G$ whose removal disconnects it.

**Non-representability.** Although it is conjectured that almost all matroids are not linear and not algebraic, it takes some effort to construct examples of non-linear and non-algebraic matroids. We describe the simplest examples.

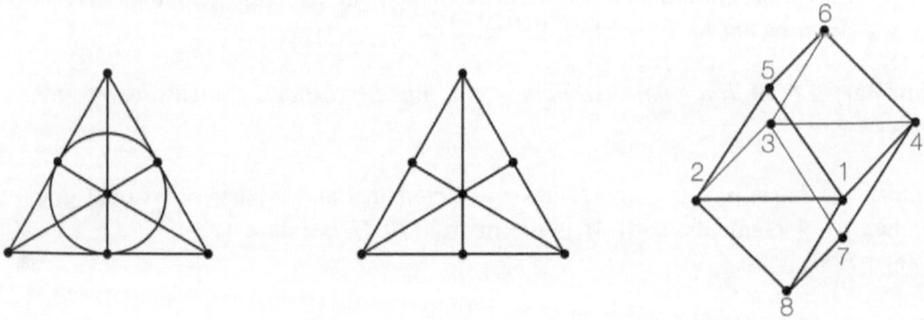


**Figure 1.40**
The Fano matroid, the non-Fano matroid $F_7^-$, and the Vámos matroid $V_8$.

- The **Fano matroid** $F_7$ is the linear matroid determined by the seven non-zero vectors in $\mathbb{F}_2^3$. This matroid is illustrated in the left panel of Figure 1.40, where an element is denoted by a vertex, and a circuit is denoted by a line or circle joining the points in question. The Fano matroid is representable over   if and only if char( ) = 2.
- The **non-Fano matroid** $F_7^-$ is the linear matroid determined by the seven non-zero 0-1 vectors $(0,0,1),(0,1,0),(1,0,0),(0,1,1),(1,0,1),(1,1,0)$, and $(1,1,1)$ in $\mathbb{R}^3$. It is representable over   if and only if char( ) $\neq 2$.
- The direct sum $F_7 \oplus F_7^-$ is not representable over any field.
- The **Vámos matroid** $V_8$ is a matroid of rank 4 whose only non-trivial circuits are 1234, 1456, 1478, 2356, 2378, as illustrated in the right panel of Figure 1.40. This matroid is not representable over any field, because in any affine configuration of points with these coplanarities, 5678 would also be coplanar. The Vámos matroid is also not algebraic over any field.

**Relating various classes.** Most matroids that we encounter naturally in combinatorics are linear, although this is usually not clear from their definition.

- All the examples of Section 1.8.3 are linear, with the possible exception of algebraic matroids.
- Graphical, transversal, cotransversal matroids, and positroids are linear.
- Linear matroids over   are algebraic over  .

- Algebraic matroids over of characteristic 0 are linear over a field extension of .
- Catalan matroids are Schubert matroids.
- Schubert matroids are lattice path matroids.
- Lattice path matroids are transversal.
- Positroids are gammoids.

**Obstructions to representability.** A general question in structural matroid theory is to try to describe a family $\mathscr{F}$ of matroids by understanding the "obstructions" belonging to $\mathscr{F}$. These results are inspired by Kuratowski's Theorem in graph theory, which asserts that a graph is planar if and only if it does not contain the complete graph $K_5$ or the complete bipartite graph $K_{3,3}$ as a minor. Similarly, one seeks a (hopefully finite) list of minors that a matroid must avoid to be in $\mathscr{F}$.

The question that has received the most attention has been that of characterizing the matroids that are representable over a particular field. Here are some results:

- $M$ is representable over $\mathbb{F}_2$ if and only if it has no minor isomorphic to $U_{2,4}$.
- $M$ is representable over $\mathbb{F}_3$ if and only if it has no minor isomorphic to $U_{2,5}, U_{3,5}, F_7$, or $F_7^*$.
- $M$ is representable over every field if and only if it has no minor isomorphic to $U_{2,4}, F_7$, or $F_7^*$.
- $M$ is graphical if and only if it has no minor isomorphic to $U_{2,4}, F_7, F_7^*, M(K_5)^*$, or $M(K_{3,3})^*$.
- For any field $\mathbb{F}$ of characteristic zero, there are infinitely many minor-minimal matroids that are not representable over $\mathbb{F}$.
- (Rota's Conjecture) For any finite field $\mathbb{F}_q$, there is a finite set $\mathscr{S}_q$ of matroids such that $M$ is representable over $\mathbb{F}_q$ if and only if it contains no minor isomorphic to a matroid in $\mathscr{S}_q$.

In 2013, Geelen, Gerards, and Whittle announced a proof of Rota's conjecture. In [85] they write: "*We are now immersed in the lengthy task of writing up our results. Since that process will take a few years, we have written this article offering a high-level preview of the proof.*"

**Asymptotic enumeration.** The asymptotic growth of various classes of matroids is not well understood. Let $P$ be a matroid property, and let $p(n)$ be the number of matroids on $[n]$ having property $P$. Let $m(n)$ be the number of matroids on $[n]$. If $\lim_{n\to\infty} p(n)/m(n) = 1$, we say that asymptotically almost every matroid has property $P$. We list some conjectures. [137]

- Asymptotically almost every matroid is connected.
- Asymptotically almost every matroid $M$ is **paving**; that is, all its circuits have size $r$ or $r+1$.
- Asymptotically almost every matroid is asymmetric, that is, it has no non-trivial automorphism.
- Asymptotically almost every matroid is not linear.
- Asymptotically almost every matroid of $n$ elements on rank $r$ satisfies the inequality $(n-1)/2 \leq r \leq (n+1)/2$.

We also list some results.

- [26, 119] The number $m(n)$ of matroids on $[n]$ satisfies

$$n - \frac{3}{2}\log n + \frac{1}{2}\log\frac{2}{\pi} - o(1) \leq \log\log m(n) \leq n - \frac{3}{2}\log n + \frac{1}{2}\log\frac{2}{\pi} + 1 + o(1).$$

- [137] Asymptotically almost every matroid is loopless and coloopless.
- [137] Asymptotically, the proportion of matroids that are connected is at least $1/2$. (It is conjectured to equal 1.)
- [17] Asymptotically, the proportion of **positroids** that are connected is $1/e^2$.

### 1.8.6 The Tutte polynomial

Throughout this section, let $R$ be an arbitrary commutative ring. A function $f$ : Matroids $\to R$ is a **matroid invariant** if $f(M) = f(N)$ whenever $M \cong N$. Let $L$ and $C$ be the matroids consisting of a single loop and a single coloop, respectively.

#### 1.8.6.1 Explicit definition

The invariant that appears most often in enumerative questions is the **Tutte polynomial**

$$T_M(x,y) = \sum_{A \subseteq M} (x-1)^{r-r(A)}(y-1)^{|A|-r(A)}. \tag{1.23}$$

When we meet a new matroid invariant, the first question we usually ask is whether it is an evaluation of $T_M(x,y)$.

#### 1.8.6.2 Recursive definition and universality property

The ubiquity of the Tutte polynomial is not an accident: it is *universal* in a large, important family of matroid invariants. Let us make this precise. Let $R$ be a ring.

Say $f$ : Matroids $\to R$ is a **generalized Tutte-Grothendieck invariant** if for every matroid $M$ and every element $e \in M$,

$$f(M) = \begin{cases} af(M\backslash e) + bf(M/e) & \text{if } e \text{ is neither a loop nor a coloop,} \\ f(M\backslash e)f(L) & \text{if } e \text{ is a loop,} \\ f(M/e)f(C) & \text{if } e \text{ is a coloop,} \end{cases} \tag{1.24}$$

for some non-zero constants $a, b \in R$. It is a **Tutte-Grothendieck invariant** if $a = b = 1$.

**Theorem 1.8.11** *The Tutte polynomial is a universal Tutte-Grothendieck invariant; namely:*

1. *For every matroid M and every element $e \in M$*

$$T_M(x,y) = \begin{cases} T_{M\backslash e}(x,y) + T_{M/e}(x,y) & \text{if } e \text{ is neither a loop nor a coloop,} \\ yT_{M\backslash e}(x,y) & \text{if } e \text{ is a loop,} \\ xT_{M/e}(x,y) & \text{if } e \text{ is a coloop.} \end{cases}$$

2. *Any generalized Tutte-Grothendieck invariant is a function of the Tutte polynomial. Explicitly, if f satisfies (1.24), then*

$$f(M) = a^{n-r} b^r T_M\left(\frac{f(C)}{b}, \frac{f(L)}{a}\right)$$

*where n is the number of elements and r is the rank of M. We do not need to assume a and b are invertible; when we multiply by $a^{n-r} b^r$, we cancel all denominators.*

*Sketch of Proof.* This is a very powerful result with a very simple proof. The first part is a straightforward computation from the definitions. The second statement then follows easily by induction on the number of elements of $M$. ■

More generally, write $T(x,y) = \sum_{i,j} t_{ij} x^i y^j$. If $f$ satisfies $f(M) = f(M\backslash e) + f(M/e)$ when $e$ is neither a loop nor a coloop, then $f(M) = \sum_{i,j} t_{ij} f(C^i \oplus L^j)$, where $C^i \oplus L^j$ denotes the matroid consisting of $i$ coloops and $j$ loops.

**Proposition 1.8.12** *The Tutte polynomial behaves well with respect to duality and direct sums:*

1. *For any matroid M, $T_{M^*}(x,y) = T_M(y,x)$.*

2. *For any matroids M and N on disjoint ground sets, $T_{M\oplus N}(x,y) = T_M(x,y)T_N(x,y)$.*

#### 1.8.6.3 Activity interpretation

An unexpected consequence of Theorem 1.8.11 is that the Tutte polynomial has non-negative coefficients; this is not at all apparent from the explicit formula (1.23). As usual, the natural question for a combinatorialist is: What do these coefficients count? The natural question for an algebraist is: What vector spaces have these coefficients as their dimensions? At the moment, the first question has a nice answer, while the second one does not.

Fix a linear order $<$ on the elements of $E$. Say an element $i \in B$ is **internally active** if there is no basis $B - i \cup j$ with $j < i$. Say an element $j \notin B$ is **externally active** if there is no basis $B - i \cup j$ with $i < j$. These are dual notions: $i$ is internally active with respect to basis $B$ in $M$ if and only if it is externally active with respect to basis $E - B$ in $M^*$.

**Theorem 1.8.13** [53, 201] *For any linear order $<$ on the ground set of a matroid, let $I(B)$ and $E(B)$ be, respectively, the set of internally active and externally active elements with respect to $B$. Then*

$$T_M(x,y) = \sum_{B \text{ basis}} x^{|I(B)|} y^{|E(B)|}.$$

*Sketch of Proof.* It is possible to give a standard deletion-contraction proof, by showing that the right-hand side is a Tutte-Grothendieck invariant, and applying Theorem 1.8.11. However, there is a much more enlightening explanation. One may prove [53] that the intervals $[B \backslash I(B), B \cup E(B)]$ partition the Boolean lattice $2^E$. In other words, every subset of the ground set can be written uniquely in the form $(B \backslash J) \cup F$ where $B$ is a basis, $J \subseteq I(B)$, and $F \subseteq E(B)$. Furthermore, under these assumptions $r((B \backslash J) \cup F) = r - |J|$. Then (1.23) becomes

$$T_M(x,y) = \sum_{B \text{ basis}} \sum_{J \subseteq I(B)} \sum_{F \subseteq E(B)} (x-1)^{|J|}(y-1)^{|F|} = \sum_{B \text{ basis}} x^{|I(B)|} y^{|E(B)|}$$

as desired. ∎

This combinatorial interpretation of the non-negative coefficients of the Tutte polynomial is very interesting, but it would be even nicer to find an interpretation that does not depend on choosing a linear order on the ground set.

In a different direction, Procesi [60] asked the following question: Given a matroid $M$, is there a natural bigraded algebra whose bigraded Hilbert polynomial is $T_M(x,y)$? This question is still open. It may be more tractable (and still very interesting) when $M$ is representable; we will see some approximations in Section 1.8.7.

#### 1.8.6.4 Finite field interpretation

The **coboundary polynomial** $\overline{\chi}_M(X,Y)$ is the following simple transformation of the Tutte polynomial:

$$\overline{\chi}_M(X,Y) = (Y-1)^r T_M\left(\frac{X+Y-1}{Y-1}, Y\right).$$

It is clear how to recover $T(x,y)$ from $\overline{\chi}(X,Y)$. We have the following interpretation of the coboundary (and hence the Tutte) polynomial.

**Theorem 1.8.14 (Finite Field Method [6, 54, 213])** *Let $\mathscr{A}$ be a hyperplane arrangement of rank $r$ in $\mathbb{F}_q^d$. For each point $p \in \mathbb{F}_q^d$ let $h(p)$ be the number of hyperplanes of $\mathscr{A}$ containing $p$. Then*

$$\sum_{p \in \mathbb{F}_q^d} t^{h(p)} = q^{d-r}\overline{\chi}(q,t).$$

*Proof.* For each $p \in \mathbb{F}_q^n$, let $H(p)$ be the set of hyperplanes of $\mathscr{A}$ that $p$ lies on. Then

$$\begin{aligned} q^{n-r}\overline{\chi}_{\mathscr{A}}(q,t) &= \sum_{\mathscr{B} \subseteq \mathscr{A}} q^{n-r(\mathscr{B})}(t-1)^{|\mathscr{B}|} = \sum_{\mathscr{B} \subseteq \mathscr{A}} q^{\dim \cap \mathscr{B}}(t-1)^{|\mathscr{B}|} \\ &= \sum_{\mathscr{B} \subseteq \mathscr{A}} |\cap \mathscr{B}|(t-1)^{|\mathscr{B}|} = \sum_{\mathscr{B} \subseteq \mathscr{A}} \sum_{p \in \cap \mathscr{B}} (t-1)^{|\mathscr{B}|} \\ &= \sum_{p \in \mathbb{F}_q^n} \sum_{\mathscr{B} \subseteq H(p)} (t-1)^{|\mathscr{B}|} = \sum_{p \in \mathbb{F}_q^n} (1+(t-1))^{h(p)}, \end{aligned}$$

as desired. ∎

Computing Tutte polynomials is extremely difficult (#P-complete [212]) for general matroids, but it is still possible in some cases of interest. Theorem 1.8.14 is one of the most effective methods for computing Tutte polynomials of (a few) particular arrangements $\mathscr{A}$ in $^d$, as follows.

If the hyperplanes of $\mathscr{A}$ have integer coefficients (as most arrangements of interest do), we may use the same equations to define an arrangement $\mathscr{A}_q$ over $\mathbb{F}_q^d$. If $q$ is a power of a large enough prime, then $\mathscr{A}$ and $\mathscr{A}_q$ have isomorphic matroids, and hence have the same Tutte polynomial. Then Theorem 1.8.14 reduces the computation of $T_{\mathscr{A}}(x,y)$ to an enumerative problem over $\mathbb{F}_q^d$, which can sometimes be solved. [6]

### 1.8.7 Tutte polynomial evaluations

Many important invariants of a matroid are generalized Tutte-Grothendieck invariants, and hence are evaluations of the Tutte polynomial. In fact, many results outside of matroid theory fit naturally into this framework. In this section we collect, without proofs, results of this sort in many different areas of mathematics and applications.

One can probably prove every statement in this section by proving that the quantities in question satisfy a deletion–contraction recursion; many of the results also have more interesting and enlightening explanations. The wonderful surveys [47], [75], and [212] include most of the results mentioned here; we provide references for the other ones.

#### 1.8.7.1 General evaluations

Let $M$ be any matroid of rank $r$.

- The number of independent sets is $T(2,1)$.
- The number of spanning sets is $T(1,2)$.
- The number of bases is $T(1,1)$.
- The number of elements of the ground set is $\log_2 T(2,2)$.
- The **Möbius number** is $\mu(M) = \mu_{L_M}(\widehat{0},\widehat{1}) = (-1)^r T(1,0)$.
- We have $\chi_M(q) = (-1)^{r(M)} T(1-q,0)$ for the **characteristic polynomial**:

$$\chi_M(q) = \sum_{F \in L_M} \mu(\widehat{0},F) q^{r-r(F)}.$$

  (Note that if $\mathscr{A}$ is a hyperplane arrangement of rank $r$ in $^d$ and $M$ is its matroid, then the characteristic polynomial $\chi_{\mathscr{A}}(q)$ defined in Section 1.7.2 is given by $\chi_{\mathscr{A}}(q) = q^{d-r}\chi_M(q)$.)

- The **beta invariant** of $M$ is defined to be $\beta(M) = [x^1y^0]T(x,y) = [x^0y^1]T(x,y)$. It has some useful properties. We always have $\beta(M) \geq 0$. We have $\beta(M) = 0$ if and only if $M$ is disconnected. We have $\beta(M) = 1$ if and only if $M$ is the graphical matroid of a **series-parallel** graph; that is, a graph obtained from a single edge by repeatedly applying series extensions (convert an edge $uv$ into two edges $uw$ and $wv$ for a new vertex $w$) and parallel extensions (convert one edge $uv$ into two edges joining $u$ and $v$). There are similar characterizations of the matroids with $\beta(M) \leq 4$.

The **independence complex** of $M$ is the simplicial complex consisting of the independent sets of $M$. Recall that the $f$-vector $(f_0,\ldots,f_r)$ and $h$-vector $(h_0,\ldots,h_r)$ are defined so that $f_i$ is the number of sets of size $i$ and $\sum h_i x^{r-i} = \sum f_i (x-1)^{r-i}$. Then the following hold.

- The reduced Euler characteristic of the independence complex is $T(0,1)$.
- The $f$-polynomial of the independence complex is $\sum_i f_i x^i = x^r T(1+\frac{1}{x},1)$.
- The $h$-polynomial of the independence complex is $\sum_i h_i x^i = x^r T(\frac{1}{x},1)$.

This polynomial $\sum_i h_i x^{r-i} = T(x,1)$ is also known as the **shelling polynomial** of $M$.

There are many other interesting geometric/topological objects associated to a matroid; we mention two. Björner [37] described the **order complex of the (proper part of the) lattice of flats** $\Delta(L_M - \{\widehat{0},\widehat{1}\})$, showing it is a wedge of $T(1,0)$ spheres. The **Bergman fan** $\mathrm{Trop}(M) = \{w \in \mathbb{R}^E :$ for every circuit $C$, $\max_{c\in C} w_c$ is achieved at least twice$\}$ is the tropical geometric analog of a linear space. The **Bergman complex** is its intersection with the hyperplane $\sum_i x_i = 0$ and the unit sphere $\sum_i x_i^2 = 1$. Ardila and Klivans showed $\mathscr{B}(M)$ and $\Delta(L_M - \{\widehat{0},\widehat{1}\})$ are homeomorphic. [14] Then the following hold.

- [37] The reduced Euler characteristic of the order complex $\Delta(L_M - \{\widehat{0}, \widehat{1}\})$ is $T(1,0)$.
- [14] The reduced Euler characteristic of the Bergman complex $\mathscr{B}(M)$ is $T(1,0)$.
- [224] The Poincaré polynomial of the *tropical cohomology* of Trop$(M)$ is $q^r T(1+1/q, 0)$.

One of the most intriguing conjectures in matroid theory is **Stanley's $h$-vector conjecture** [186], which states that for any matroid $M$, there exists a set $X$ of monomials such that

- if $m$ and $m'$ are monomials such that $m \in X$ and $m'|m$, then $m' \in X$,
- all the maximal monomials in $X$ have the same degree,
- there are exactly $h_i$ monomials of degree $i$ in $X$.

This conjecture has been proved, using rather different methods, for several families: duals of graphic matroids, [140], lattice path matroids [176] cotransversal matroids [151], paving matroids [141], and matroids up to rank 4 or corank 2 [62, 117]. The general case remains open.

#### 1.8.7.2 Graphs

Let $G = (V, E)$ be a graph with $v$ vertices, $e$ edges, and $c$ connected components. Let $M(G)$ be the matroid of $G$ and $T(x,y)$ be its Tutte polynomial.

**Colorings.** Recall that the **chromatic polynomial** $\chi_G(q)$ of a graph, when evaluated at a positive integer $q$, counts the proper colorings of $G$ with $q$ colors.

- We have
$$\chi_G(q) = (-1)^{v-c} q^c T(1-q, 0).$$
- More generally, for every $q$-coloring $\chi$ of the vertices of $G$, let $h(\chi)$ be the number of improperly colored edges, that is, those whose endpoints have the same color. Then
$$\sum_{\chi : V \to [q]} t^{h(\chi)} = (t-1)^{v-c} q^c \, T\left(\frac{q+t-1}{t-1}, t\right). \qquad (1.25)$$

**Flows.** Fix an orientation of the edges of $G$ and a finite Abelian group $H$ of $t$ elements, such as $\mathbb{Z}/t\mathbb{Z}$, for some $t \in \mathbb{N}$. An **$H$-flow** is an assignment $f : E \to H$ of a "flow" (an element of $H$) to each edge such that at every vertex, the total inflow equals the outflow as elements of $H$. We say $f$ is **nowhere zero** if $f(e) \neq 0$ for all edges $e$.

- The number of nowhere zero $t$-flows is given by the **flow polynomial**

$$\chi_G^*(t) = (-1)^{e-v+c}T(0, 1-t).$$

In particular, this number is independent of the orientation of $G$. It also does not depend on the particular group $H$, but only on its size.

- More generally, for every $H$-flow $f$ on the edges of $G$, let $h(v)$ be the number of edges having flow equal to 0. Then

$$\sum_{f:E\to H} t^{h(f)} = (t-1)^{e-v+c}T\left(t, \frac{q+t-1}{t-1}\right).$$

**Acyclic orientations.** An **acyclic orientation** of $G$ is an orientation of the edges that creates no directed cycles.

- The number of acyclic orientations is $T(2,0)$.
- A **source** (respectively, a **sink**) of an orientation is a vertex with no incoming (respectively, outgoing) edges. For any fixed vertex $w$, the number of acyclic orientations whose unique source is $w$ is $(-1)^{v-c}\mu(M(G)) = T(1,0)$. In particular, it does not depend on the choice of $w$.
- For any edge $e = uv$, the number of acyclic orientations whose unique source is $u$ and whose unique sink is $v$ is the beta invariant $\beta(M(G))$. In particular, it is independent of the choice of $e$.

**Totally cyclic orientations.** A **totally cyclic orientation** of $G$ is an orientation of the edges such that every edge is contained in a cycle. If $G$ is connected, this is the same as requiring that there are directed paths in both directions between any two vertices.

- The number of totally cyclic orientations is $T(0,2)$.
- Given an orientation $o$ of $G$, the outdegree of a vertex is the number of outgoing edges, and the outdegree sequence of $o$ is $(\text{outdeg}(v) : v \in V)$. The number of distinct outdegree sequences among the orientations of $G$ is $T(2,1)$.
- For any edge $e$, the number of totally cyclic orientations of $G$ such that every directed cycle contains $e$ is equal to $2\beta(M(G))$. In particular, it is independent of $e$.

**Eulerian orientations.** Recall that an orientation of $G$ is **Eulerian** if each vertex has the same number of incoming and outgoing edges. Say $G$ is 4-regular if every vertex has degree 4.

- A 4-regular graph $G$ has $(-1)^{v+c}T_G(0,-2)$ Eulerian orientations or **ice configurations** . (They are in easy bijection with the nowhere-zero $\mathbb{Z}_3$-flows.)

**Chip firing.** Let $G = (V, E)$ be a graph and let $q \in V$ be a vertex called the **bank**. A **chip configuration** is a map $\theta : V \to \mathbb{Z}$ with $\theta(v) \geq 0$ for $v \neq q$ and $\theta(q) = -\sum_{v \neq q} \theta(v)$. We think of vertex $v$ as having $\theta(v)$ chips. We say $v \neq q$ is **ready** if $\theta(v) \geq \deg(v)$, and $q$ is **ready** if no other vertex is ready. In each step of the **chip firing game**, a ready vertex $v$ gives away $\deg(v)$ chips, giving $e$ chips to $w$ if there are $e$ edges connecting $v$ and $w$. We say a configuration $\theta$ is **stable** if the only vertex that is ready is $q$. We say $\theta$ is **recurrent** if it is possible to fire a (non-empty) sequence of vertices subsequently, and return to $\theta$. We say $\theta$ is **critical** if it is stable and recurrent. One can check that for any starting chip configuration, the chip firing game leads us to a unique critical configuration. The following result is the key to Merino's proof of Stanley's $h$-vector conjecture in the special case of cographic matroids:

- [140] The generating function for critical configurations is

$$\sum_{\theta \text{ critical}} y^{-\theta(q)} = y^{|E| - \deg(q)} T_G(1, y).$$

Chip-firing games have recently gained prominence as they are intimately connected to the study of divisors on tropical algebraic curves. [24, 84]

**Plane graphs and Eulerian partitions.** Let $G$ be a connected plane graph. The **medial graph** $H$ has a vertex on each edge of $G$, and two vertices of $H$ are connected if the corresponding edges of $G$ are neighbors in some face of $G$. This construction is illustrated in Figure 1.41.

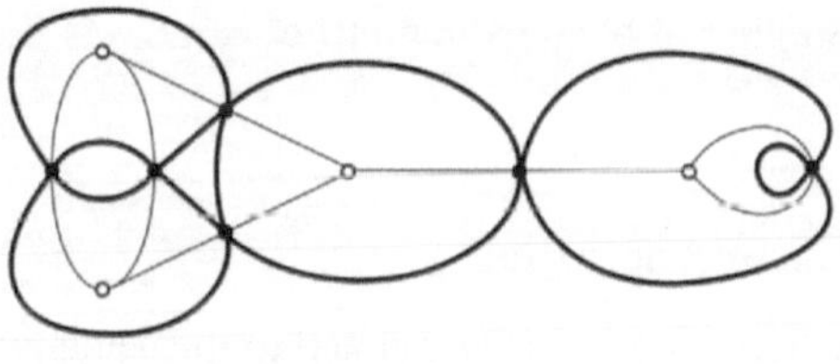

**Figure 1.41**
A connected graph $G$ and its medial graph $H$ in bold.

An **Eulerian partition** $\pi$ of $H$ is a partition of the edges into closed paths. A closed path may revisit vertices, but it may not revisit edges; and different starting points and orientations are considered to be the same path. Let $\gamma(\pi)$ be the number of paths in $\pi$. If the edges of a vertex $v$ of $H$ are $e_1, e_2, e_3, e_4$ in clockwise order, say $\pi$ has a **crossing** at $v$ if one of its paths uses edges $e_1$ and $e_3$ consecutively.

- [128] We have $T_G(-1,-1) = (-1)^e(-2)^{\gamma(\pi_c)-1}$, where $\pi_c$ is the unique **fully crossing** Eulerian partition of $H$ with a crossing at every vertex.

- [128] If $\Pi_{nc}(H)$ is the set of Eulerian partitions of $H$ with no crossings, then

$$T_G(x,x) = \sum_{P \in \Pi_{nc}(H)} (x-1)^{\gamma(P)-1}.$$

In the example above, note that the fully crossing Eulerian partition of $G$ consists of one Eulerian walk, in agreement with $T_G(-1,-1) = 1 = (-1)^6(-2)^{1-1}$. We have $T_G(x,x) = 3x^4 + 2x^3$.

#### 1.8.7.3 Hyperplane arrangements

Let $\mathscr{A}$ be an arrangement of $n$ hyperplanes in $^d$ of rank $r$. Let $V(\mathscr{A})$ be the complement of $\mathscr{A}$ in $^d$. We restate some theorems from Section 1.7.2 and present new ones.

- If $= \mathbb{R}$, the number of regions of $V(\mathscr{A})$ is $T(2,0)$.
- If $= \mathbb{C}$, the Poincaré polynomial of $V(\mathscr{A})$ is $q^r T(1+\frac{1}{q}, 0)$.
- If $= \mathbb{F}_q$, the number of elements of $V(\mathscr{A})$ is $(-1)^r q^{n-r} T(1-q, 0)$.
- Suppose $= \mathbb{R}$, and consider an affine hyperplane $H$ that is in general position with respect to $\mathscr{A}$. Then the number of regions of $\mathscr{A}$ that have a bounded (and non-empty) intersection with $H$ is $(-1)^r|\mu(M)| = T(1,0)$. In particular, it is independent of $H$.
- Suppose $= \mathbb{R}$, and add to $\mathscr{A}$ an affine hyperplane $H'$ that is a translation of $H \in \mathscr{A}$. The number of bounded regions of $\mathscr{A} \cup H'$ is $\beta(M)$. In particular, it is independent of $H$.

#### 1.8.7.4 Algebras from arrangements

There are several natural algebraic spaces related to the Tutte polynomial arising in commutative algebra, hyperplane arrangements, box splines, and index theory; we discuss a few. For each hyperplane $H$ in a hyperplane arrangement $\mathscr{A}$ in $^d$ let $l_H$ be a linear function such that $H$ is given by the equation $l_H(x) = 0$.

- [208] Let $C_{\mathscr{A},0} = \text{span}\{\prod_{H \in \mathscr{B}} l_H \,:\, \mathscr{B} \subseteq \mathscr{A}\}$. This is a subspace of a polynomial ring in $d$ variables, graded by degree. Its dimension is $T(2,1)$ and its Hilbert series is

$$\text{Hilb}(C_{\mathscr{A},0}; q) = \sum_{j \geq 0} \dim(C_{\mathscr{A},0})_j q^j = q^{n-r} T\left(1+q, \frac{1}{q}\right).$$

- [15, 57, 103, 159, 160] More generally, let $C_{\mathscr{A},k}$ be the vector space of polynomial functions such that the restriction of $f$ to any line $l$ has degree at most $\rho_{\mathscr{A}}(h) + k$, where $\rho_{\mathscr{A}}(h)$ is the number of hyperplanes of $\mathscr{A}$ not containing

$h$. It is not obvious, but true, that this definition of $C_{\mathscr{A},0}$ matches the one above. Then

$$\text{Hilb}(C_{\mathscr{A},-1};q) = q^{n-r}T\left(1,\frac{1}{q}\right), \qquad \text{Hilb}(C_{\mathscr{A},-2};q) = q^{n-r}T\left(0,\frac{1}{q}\right)$$

and similar formulas hold for any $k \geq -2$.

- [45, 162, 200] Let $R(\mathscr{A})$ be the vector space of rational functions whose poles are in $\mathscr{A}$. It may be described as the -algebra generated by the rational functions $\{1/l_H : H \in \mathscr{A}\}$; we grade it so that $\deg(1/l_H) = 1$. Then

$$\text{Hilb}(R(\mathscr{A});q) = \frac{q^d}{(1-q)^d}T\left(\frac{1}{q},0\right).$$

#### 1.8.7.5 Error-correcting codes

Suppose we wish to transmit a message over a noisy channel. We might then encode our message in a redundant way, so that we can correct small errors introduced during transmission. An **error-correcting code** is a set $C \subset A^n$ of **codewords** of length $n$ over an alphabet $A$. The sender encodes each word into a redundant codeword, which is transmitted. If the channel is not too noisy and the codewords in $C$ are sufficiently different from each other, the recipient will succeed in recovering the original message. Of course it is useful to have many codewords available under these constraints.

A common kind of error-correcting code is a **linear code** $C$, which is a $k$-dimensional vector space $\mathbb{F}_q^n$. The codewords have length $n$ and alphabet $\mathbb{F}_q$. The **support** of a word is the set of non-zero entries. The **distance** $d(\mathbf{u},\mathbf{v})$ between two words $\mathbf{u}$ and $\mathbf{v}$ is the number of coordinates where $\mathbf{u}$ and $\mathbf{v}$ differ. Since $\mathbf{u}-\mathbf{v} \in \mathscr{C}$, we have $d(\mathbf{u},\mathbf{v}) = |\text{supp}(\mathbf{u}-\mathbf{v})|$. The minimum **distance** $d$ in the code is $d = \min_{c \in C} |\text{supp}(c)|$. The code $C$ is said to be of type $[n,k,d]$.

- The **weight enumerator** of a linear code over $\mathbb{F}_q$ is $W_C(q,z) = \sum_{c \in C} z^{|\text{supp}(c)|}$. Curtis Greene [92] discovered that it can be expressed in terms of the Tutte polynomial; in fact, Theorem 1.8.14 is equivalent to the equation

$$W_C(q,z) = (1-z)^k z^{n-k} T_C\left(\frac{1+(q-1)z}{1-z},\frac{1}{z}\right). \tag{1.26}$$

The **dual code** of $C$ is $C^\perp = \{\mathbf{x} \in \mathbb{F}_q^n : (\mathbf{x},\mathbf{c}) = 0 \text{ for all } \mathbf{c} \in C\}$. An important example is the **Hamming code** $H$. Let $\mathscr{A}(2,n)$ be the rowspace of the $n \times (2^n - 1)$ matrix whose columns are all the nonzero vectors in $\mathbb{F}_2^n$, and let $H = \mathscr{A}(2,n)^\perp$. This is a largest possible code (consisting of $2^{2^n-n-1}$ words) of length $2^n - 1$ and distance 3.

- Florence MacWilliams [134] gave the following remarkable identity relating the weight enumerators of $C$ and $C^{\perp}$:

$$W_{C^{\perp}}(q,z) = \frac{(1+(q-1)z)^n}{q^k} W_C\left(q, \frac{1-z}{1+(q-1)z}\right).$$

In light of (1.26), this is equivalent to Tutte polynomial duality: $T_{C^{\perp}}(x,y) = T_C(y,x)$. It is easy to show that the weight enumerator of $\mathscr{A}(2,n)$ is $1+(q^n-1)z^{q^{n-1}}$. [92, Example 3.4] MacWilliams's Identity then gives us the weight enumerator of the Hamming code.

- Another nice result [107] is that if $C$ is a linear code in $\mathbb{F}_2^n$ then $T_C(-1,-1) = (-1)^n |C \cap C^{\perp}|$.

#### 1.8.7.6 Probability and statistical mechanics

The Tutte polynomial arises in several standard processes in probability and statistical physics. We consider three related models. [47, 178, 212]

- Suppose each edge of a connected graph $G$ is white with probability $p$ and black with probability $1-p$ (for fixed $0 \leq p \leq 1$), independently of the other edges. The probability that the white graph is connected is given by the **reliability polynomial**

$$R_G(p) = \sum_{\substack{A \subseteq E \\ A \text{ spanning}}} p^{|A|}(1-p)^{|E-A|} = (1-p)^{e-v+1} p^{v-1} T_G\left(1, \frac{1}{1-p}\right).$$

- The **random cluster model** depends on parameters $0 \leq p \leq 1$ and $q > 0$. Now we choose a white set of edges at random, and the probability of choosing $A \subset E$ is $p^{|A|}(1-p)^{|E-A|} q^{c(A)}/Z$, where $c(A) = |V| - r(A)$ is the number of components of $A$, and $Z$ is a scaling constant. To know the probability of a particular state, it is fundamental to know the scaling constant $Z$, which is the **partition function**

$$\begin{aligned} Z(p,q) &= \sum_{A \subseteq E} p^{|A|}(1-p)^{|E-A|} q^{c(A)} \\ &= p^{v-c}(1-p)^{e-v+c} q^c T_G\left(1 + \frac{q(1-p)}{p}, \frac{1}{1-p}\right). \end{aligned}$$

- In the **$q$-state Potts model**, a graph $G = (V,E)$ models a set $V$ of "atoms" and a set $E$ of "bonds" between them. (When $q = 2$, this is the **Ising model**.) Each atom can exist in one of $q$ states or "spins." Each edge $e = uv$ has an associated *interaction energy* $J_e$ between $u$ and $v$. The energy (or Hamiltonian) $H$ of a configuration is the sum of $-J_e$ over all edges $e$ whose vertices have the same spin. (The case $J_e \geq 0$ is called ferromagnetic, as it favors adjacent spins being equal. The case $J_e \leq 0$ is antiferromagnetic. Here we are assuming that there is no external magnetic field. If there were such a field, it would contribute an additional term to the Hamiltonian, and the direct connection with the Tutte polynomial is no longer valid.) A configuration of energy $H$ has Boltzmann weight $e^{-\beta H}$ where $\beta = 1/kT > 0$, $T$ is the temperature, and $k$ is Boltzmann's constant.

The **partition function** is the sum of the Boltzmann weights of all configurations:

$$Z_G(q,\mathbf{w}) = \sum_{\sigma:V\to[q]} \prod_{\substack{e=ij\in E \\ \sigma(i)=\sigma(j)}} e^{\beta J_e} = \sum_{\sigma:V\to[q]} \prod_{\substack{e=ij\in E \\ \sigma(i)=\sigma(j)}} (1+w_e)$$

where $w_e = e^{\beta J_e} - 1$. If all $J_e$s are equal to $J$ and $w = e^{\beta J} - 1$, then (1.25) gives

$$Z_G(q,w) = w^{v-c} q^c \, T\left(\frac{q}{w}+1, w+1\right).$$

For general $J_e$, a simple computation shows that

$$q^{-v} Z_G(q,\mathbf{w}) = \sum_{A\subseteq E} q^{-r(A)} \prod_{e\in A} w_e = \widetilde{Z}_G(q,\mathbf{w})$$

where $\widetilde{Z}_G(q,\mathbf{w})$ is the **multivariate Tutte polynomial** of Section 1.8.9.

#### 1.8.7.7 Other applications

- [108, 212] A **knot** is an embedding of a circle in $\mathbb{R}^3$. It is a difficult, important question to determine whether a knot can be deformed smoothly (without cutting or crossing segments) to obtain another knot; or even to determine whether a given knot is actually knotted or not. Let $\mathbf{O}$ be the unknotted circle in $\mathbb{R}^3$.

  One common approach is to construct a function $f(K)$ of a knot $K$ that does not change under smooth deformation. If $f(K) \neq f(\mathbf{O})$, then $K$ is actually knotted. One such function is the **Jones polynomial** $V(K)$. If $K$ is an **alternating knot**, there is a graph $G$ associated to it such that the Jones polynomial of $K$ is an evaluation of the Tutte polynomial of $G$.

- [164] There is a more symmetric finite field interpretation for the Tutte polynomial. Let $M$ be an integer matrix, and let $p$ and $q$ be prime powers such that the matroid of $M$ does not change when $M$ is considered as a matrix over $\mathbb{F}_p$ or over $\mathbb{F}_q$. Then

$$T_M(1-p, 1-q) = (-1)^{r(M)} \sum_{\substack{\mathbf{x}\in \mathrm{row}(M), \mathbf{y}\in\ker(M) \\ \mathrm{supp}(\mathbf{x})\sqcup \mathrm{supp}(\mathbf{y})=E}} (-1)^{|\mathrm{supp}(\mathbf{y})|},$$

  where the rowspace $\mathrm{row}(M)$ is considered as a subspace of $\mathbb{F}_p^n$ and the kernel $\ker(M)$ is considered as a subspace of $\mathbb{F}_q^n$, and $\sqcup$ denotes a disjoint union.

- [4, 41] The Tutte polynomial of the Catalan matroid is

$$T_{\mathbf{C}_n}(x,y) = \sum_{P \text{ Dyck}} x^{a(P)}y^{b(P)},$$

  where $a(P)$ is the number of upsteps before the first downstep of $P$, and $b(P)$ is the number of times that $P$ returns to the $x$-axis. Since $\mathbf{C}_n$ is self-dual, this polynomial is symmetric in $x$ and $y$.

  In fact, the coefficient of $x^iy^j$ in $T_{\mathbf{C}_n}(x,y)$ depends only on $i+j$. It would be interesting to find other families of matroids with this unusual property.

- [41] More generally, the Tutte polynomial of a lattice path matroid is the sum over the bases $P$ of $x^{a(P)}y^{b(P)}$, where $a(P)$ and $b(P)$ are the number of returns to the upper and lower boundary paths, respectively.

- [121] A T-tetromino is a T-shape made of four unit squares. Figure 1.42 shows a T-tetromino tiling of an $8 \times 8$ square. The number of **T-tetromino tilings** of a $4m \times 4n$ rectangle equals $2T_{L_{m,n}}(3,3)$, where $L_{m,n}$ is the $m \times n$ grid graph. This result extends to T-tetromino tilings of many other shapes and, more generally, to coverings of many graphs with copies of the "claw" graph $K_{1,3}$.

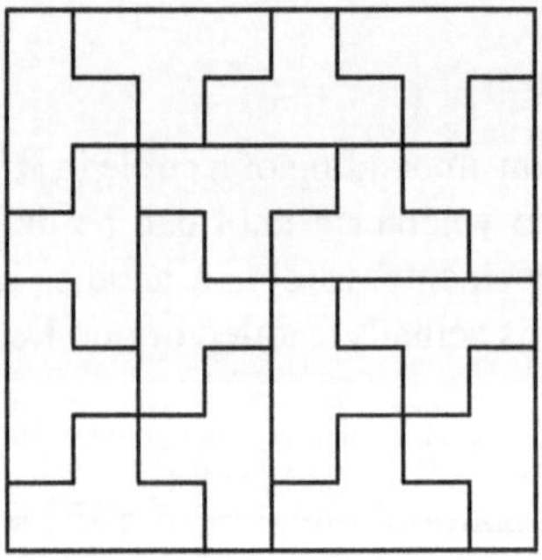

**Figure 1.42**
A tiling of an $8 \times 8$ rectangle by T-tetrominoes.

### 1.8.8 Computing the Tutte polynomial

As we already mentioned, computing the Tutte polynomial of an arbitrary graph or matroid is not a tractable problem. In the language of complexity theory, this is a #P-complete problem. [212] However, the results of Section 1.8.6 allow us to compute the Tutte polynomial of **some** matroids of interest. We now survey some of the most interesting examples; see [142] for others. Some of these formulas are best expressed in terms of the **coboundary polynomial**

$$\overline{\chi}_{\mathscr{A}}(X,Y) = (y-1)^{r(\mathscr{A})} T_{\mathscr{A}}(x,y), \qquad \text{where } x = \frac{X+Y-1}{Y-1}, \quad y = Y.$$

Almost all of them are most easily proved using the Finite Field Method (Theorem 1.8.14) or its graph version (1.25).

- For the uniform matroid $U_{k,n}$ we have $T_{U_{k,n}}(x,y) = \sum_{i=1}^{r} \binom{n-i-1}{n-r-1} x^i + \sum_{j=1}^{n-r} \binom{n-j-1}{r-1} y^j$.
- If $M^{(k)}$ is the matroid obtained from $M$ by replacing each element by $k$ copies of itself, then

$$T_{M^{(k)}}(x,y) = (y^{k-1} + y^{k-2} + \cdots + y + 1)^r T_M \left( \frac{y^{k-1} + y^{k-2} + \cdots + y + x}{y^{k-1} + y^{k-2} + \cdots + y + 1}, y^k \right).$$

  This formula is straightforward in terms of coboundary polynomials: $\overline{\chi}_{M^{(k)}}(X,Y) = \overline{\chi}_M(X, Y^K)$. For an extensive generalization, see Section 1.8.9.
- [6, 145] Root systems are arguably the most important vector configurations; these highly symmetric arrangements play a fundamental role in many branches of mathematics. For the general definition and properties, see for example [106]; we focus on the four infinite families of **classical root systems**:

$$\begin{aligned} A_{n-1} &= \{e_i - e_j, : 1 \leq i < j \leq n\} \\ B_n &= \{e_i - e_j, e_i + e_j : 1 \leq i < j \leq n\} \cup \{e_i : 1 \leq i \leq n\} \\ C_n &= \{e_i - e_j, e_i + e_j : 1 \leq i < j \leq n\} \cup \{2e_i : 1 \leq i \leq n\} \\ D_n &= \{e_i - e_j, e_i + e_j : 1 \leq i < j \leq n\}. \end{aligned}$$

  See Figure 1.43 for an illustration.

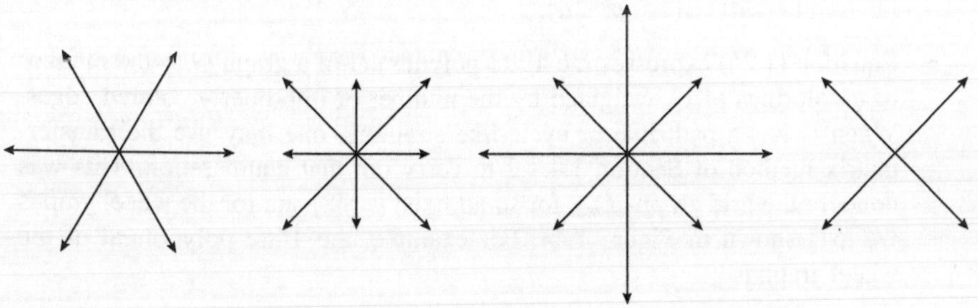

**Figure 1.43**
The root systems $A_2, B_2, C_2$, and $D_2$, respectively.

Let the **deformed exponential function** be $F(\alpha,\beta) = \sum_{n\geq 0}(\alpha^n \beta^{\binom{n}{2}})/n!$. Then the **Tutte generating functions** of the infinite families $A$ and $\Phi = B, C, D$:

$$T_A(X,Y,Z) = 1 + X\sum_{n\geq 1}\overline{\chi}_{A_{n-1}}(X,Y)\frac{Z^n}{n!}, \quad T_\Phi(X,Y,Z) = \sum_{n\geq 0}\overline{\chi}_{\Phi_n}(X,Y)\frac{Z^n}{n!}$$

are given by

$$\begin{aligned}
T_A &= F(Z,Y)^X, \\
T_B &= F(2Z,Y)^{(X-1)/2}F(YZ,Y^2), \\
T_C &= F(2Z,Y)^{(X-1)/2}F(YZ,Y^2), \\
T_D &= F(2Z,Y)^{(X-1)/2}F(Z,Y^2).
\end{aligned}$$

Aside from the four infinite families, there are a small number of exceptional root systems, which are also very interesting objects. Their Tutte polynomials are computed in [61].

- Since the matroid of the complete graph $K_n$ is isomorphic to the matroid of the vector configuration $A_{n-1}$, the first formula above is a formula for the Tutte polynomials of the complete graphs, proved originally by Tutte [202]. Similarly, the coboundary polynomials of the complete graphs $K_{m,n}$ are given by

$$1 + X\sum_{\substack{m,n\geq 0 \\ (m,n)\neq(0,0)}}\overline{\chi}_{K_{m,n}}(X,Y)\frac{Z_1^m Z_2^n}{m!\;n!} = \left(\sum_{m,n\geq 0} Y^{mn}\frac{Z_1^m Z_2^n}{m!\;n!}\right)^X.$$

- [27, 145]. The Tutte polynomial of the arrangement $\mathscr{A}(p,n)$ of all linear hyperplanes in $\mathbb{F}_p^n$ is best expressed in terms of a "$p$-exponential generating function":

$$\sum_{n\geq 0}\overline{\chi}_{\mathscr{A}(p,n)}(X,Y)\frac{u^n}{(p;p)_n} = \frac{(u;p)_\infty}{(Xu;p)_\infty}\sum_{n\geq 0} Y^{1+p+\cdots+p^{n-1}}\frac{u^n}{(p;p)_n}$$

where we define $(a;p)_\infty = (1-a)(1-pa)(1-p^2a)\cdots$ and $(a;p)_n = (1-a)(1-pa)\cdots(1-p^{n-1}a)$.

- Equation (1.25) expresses the Tutte polynomial of a graph $G$ as the number of $q$-colorings of $G$, weighted by the number of improperly colored edges. When $G$ has a path-like or cycle-like structure, one may use the transfer-matrix method of Section 1.4.1.2 to carry out that enumeration. This was done for the grid graphs $L_{m,n}$ for small fixed $m$ [48] and for the wheel graphs $W_n$ [51] shown in Figure 1.44. For example, the Tutte polynomial of the wheel graph is

$$T_{W_n}(x,y) = \frac{1}{2^n}\left(b + \sqrt{b^2-4a}\right)^n + \frac{1}{2^n}\left(b - \sqrt{b^2-4a}\right)^n + a - b$$

where $b = 1 + x + y$ and $a = xy$.

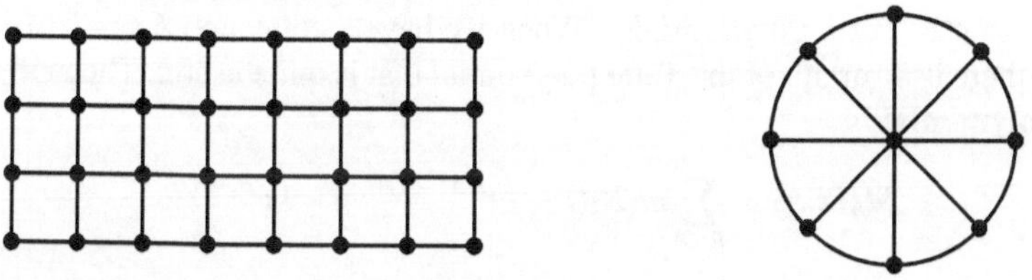

**Figure 1.44**
The grid graph $L_{4,8}$ and the wheel graph $W_8$.

### 1.8.9 Generalizations of the Tutte polynomial

**The multivariate Tutte polynomial.** The **multivariate Tutte polynomial** of a matroid $M$ is

$$\widetilde{Z}_M(q;\mathbf{w}) = \sum_{A\subseteq E} q^{-r(A)} \prod_{e\in A} w_e$$

where $q$ and $(w_e)_{e\in E}$ are indeterminates. When all $w_e = w$, we get $Z(q,w) = w^{v-c}q^cT(\frac{q}{w}+1, w+1)$.

Notice that $\widetilde{Z}_M(q;\mathbf{w})$ determines $E$ and $r$, and hence it determines $M$. In fact, among the many definitions of a matroid, we could define $M$ to be $\widetilde{Z}_M(q;\mathbf{w})$. This is a useful encoding of the matroid.

- We saw in Section 1.8.7 that for a graph $G$ and a positive integer $q$, the multivariate Tutte polynomial is equal to the partition function of the $q$-state Potts model on $G$.

- For a vector $\mathbf{a} \in \mathbb{N}^n$, let $M(\mathbf{a})$ be the matroid $M$ where each element $e$ is replaced by $a_e$ copies of $e$. It is natural to ask for the Tutte polynomials of the various matroids $M(\mathbf{a})$. For each $\mathbf{a}$,

$$T_{M(\mathbf{a})}(x,y) = (x-1)^{r(\text{supp}(\mathbf{a}))}\widetilde{Z}_M\left((x-1)(y-1); y^{a_1-1}, \ldots y^{a_n-1}\right).$$

The generating function for the Tutte polynomials of **all** the matroids $M(\mathbf{a})$ turns out to be equivalent to the multivariate Tutte polynomial, disguised under a change of variable:

$$\sum_{\mathbf{a} in \mathbb{N}^n} \frac{T_{M(\mathbf{a})}(x,y)}{(x-1)^{r(\text{supp}(\mathbf{a}))}} w_1^{a_1} \cdots w_n^{a_n}$$

$$= \frac{1}{\prod_{i=1}^n (1-w_i)} \widetilde{Z}_M\left((x-1)(y-1); \frac{(y-1)w_1}{1-yw_1}, \ldots, \frac{(y-1)w_n}{1-yw_n}\right).$$

- For an algebraic interpretation of the multivariate Tutte polynomial, see [15].

**The arithmetic Tutte polynomial.** When we have a collection $A \subseteq \mathbb{Z}^n$ of integer vectors, there is a variant of the Tutte polynomial that is quite useful. The **arithmetic Tutte polynomial** is

$$M_A(x,y) = \sum_{B \subseteq A} m(B)(x-1)^{r(A)-r(B)}(y-1)^{|B|-r(B)}$$

where, for each $B \subseteq A$, the *multiplicity* $m(B)$ is the index of $\mathbb{Z}B$ as a sublattice of $(\operatorname{span} B) \cap \mathbb{Z}^n$. Using the vectors in $B$ as the columns of a matrix, $m(B)$ is the greatest common divisor of the minors of full rank. This polynomial is related to the zonotope of $A$ [190, 56]:

- The volume of the zonotope $Z(A)$ is $M_A(1,1)$.
- The zonotope $Z(A)$ contains $M_A(2,1)$ lattice points, $M_A(0,1)$ of which are in its interior.
- The Ehrhart polynomial of the zonotope $Z(A)$ is $q^r M(1+\frac{1}{q},1)$.

Let $T = \operatorname{Hom}(\mathbb{Z}^n, G)$ be the group of homomorphisms from $\mathbb{Z}^n$ to a multiplicative group $G$, such as the unit circle $\mathbb{S}^1$ or $\mathbb{F}^* = \mathbb{F} \backslash \{0\}$ for a field $\mathbb{F}$. Each element $a \in A$ determines a (hyper)torus $T_a = \{t \in T : t(a) = 1\}$ in $T$. For instance $a = (2,-3,5)$ gives the torus $x^2y^{-3}z^5 = 1$. Let

$$\mathscr{T}(A) = \{T_a : a \in A\}, \qquad R(A) = T \setminus \bigcup_{a \in \mathscr{T}(A)} T_a$$

be the *toric arrangement* of $A$ and its complement, respectively.

- [72, 144] If $G = \mathbb{S}^1$, the number of regions of $R(A)$ in the torus $(\mathbb{S}^1)^r$ is $M_A(1,0)$.
- [59, 144] If $G = \mathbb{C}^*$, the Poincaré polynomial of $R(A)$ is $q^r M_A(2+\frac{1}{q},0)$.
- [43, 11] If $G = \mathbb{F}^*_{q+1}$ where $q+1$ is a prime power, the number of elements of $R(A)$ is $(-1)^r q^{n-r} M_A(1-q,0)$. Furthermore,

$$\sum_{p \in \mathbb{F}^*_{q+1}} t^{h(p)} = (t-1)^r q^{n-r} M_A\left(\frac{q+t-1}{t-1}, t\right)$$

  where $h(p)$ is the number of tori of $\mathscr{T}(A)$ that $p$ lies on.

As with ordinary Tutte polynomials, this last result may be used as a **finite field method** to compute arithmetic Tutte polynomials for some vector configurations $A$. However, at the moment there are very few results along these lines. One exception is the case of root systems; the study of their geometric properties motivates much of the theory of arithmetic Tutte polynomials. Explicit formulas for the arithmetic Tutte polynomials of the classical root systems $A_n, B_n, C_n$, and $D_n$ are given in [11], based on the computation of Example 15 in Section 1.3.3.

### 1.8.10 Matroid subdivisions, valuations, and the Derksen-Fink invariant

A very interesting recent development in matroid theory has been the study of matroid subdivisions and valuative matroid invariants. We offer a brief account of some key results and some pointers to the relevant bibliography.

A **matroid subdivision** is a polyhedral subdivision $\mathscr{P}$ of a matroid polytope $P_M$ where every polytope $P \in \mathscr{P}$ is itself a matroid polytope. Equivalently, it is a subdivision of $P_M$ whose only edges are the edges of $P_M$. See Figure 1.45 for an illustration. In the most important case, $M$ is the uniform matroid $U_{d,n}$ and $P_M$ is the hypersimplex $\Delta(d,n)$.

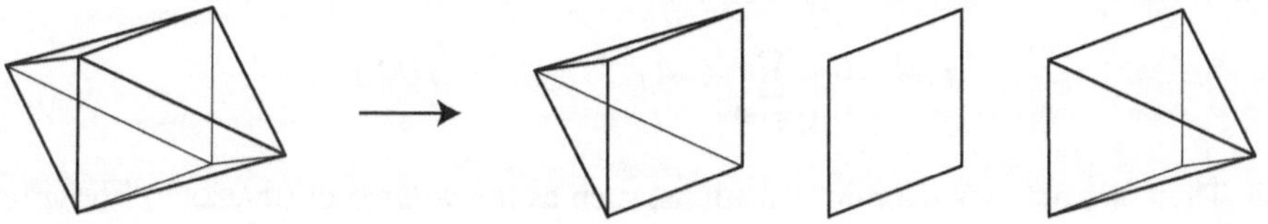

**Figure 1.45**
A matroid subdivision of $\Delta(2,4)$.

Matroid subdivisions arose in algebraic geometry [94, 113, 126], in the theory of valuated matroids [70, 147], and in tropical geometry [180]. For instance, Lafforgue showed that if a matroid polytope $P_M$ has no nontrivial matroid subdivisions, then the matroid $M$ has (up to trivial transformations) only finitely many realizations over a fixed field $\mathbb{F}$. This is one of very few results about realizability of matroids over arbitrary fields.

The connection with tropical geometry is a rich source of examples. Let $K$ be the field of Puiseux series $a_0x^{m/N} + a_1x^{(m+1)/N} + \cdots$ (where $a_0, a_1, \ldots \in \mathbb{C}$, $m, N \in \mathbb{Z}$ and $N > 0$). Roughly speaking, every subspace $L$ of $K^n$ may be *tropicalized*, and there is a canonical way of decomposing the resulting tropical linear space into Bergman fans of various matroids. These matroids give a matroid subdivision of $\Delta(d,n)$. For some nice choices of $L$, the corresponding matroid subdivisions can be described explicitly. For details, see for example [14, 166, 180].

There is also a useful connection with the subdivisions of the product of simplices $\Delta_{d-1} \times \Delta_{n-d-1}$, which are much better understood, as we discussed in Section 1.6.4. Note that the vertex figure of any vertex of $\Delta(d,n)$ is $\Delta_{d-1} \times \Delta_{n-d-1}$. Every matroid subdivision of $\Delta(d,n)$ then induces a "local" polyhedral subdivision of $\Delta_{d-1} \times \Delta_{n-d-1}$ at each one of its vertices. Conversely, we can "cone" any subdivision $\mathscr{S}$ of $\Delta_{d-1} \times \Delta_{n-d-1}$ to get a matroid subdivision that looks like $\mathscr{S}$ at a given vertex. [98, 166]

**Enumerative aspects of matroid subdivisions.** Say a face $P$ of a subdivision $\mathcal{P}$ of $P_M$ is *internal* if it is not on the boundary of $P_M$. Let $\mathcal{P}^{\text{int}}$ be the set of internal faces of $\mathcal{P}$. Currently, the most interesting enumerative question on matroid subdivisions is the following.

**Conjecture 1.8.15 (Speyer's $f$-vector conjecture [180])** *A matroid subdivision of the hypersimplex $\Delta(d,n)$ has at most $\frac{(n-c-1)!}{(d-c)!(n-d-c)!(c-1)!}$ interior faces of dimension $n-c$, with equality if and only if all facets correspond to series-parallel matroids.*

Speyer constructed a subdivision simultaneously achieving the conjectural maximum number of interior faces for all $c$. In attempting to prove this conjecture, he pioneered the study of valuative matroid invariants. We say that a matroid invariant $f$ is **valuative** if for any matroid subdivision $\mathcal{P}$ of any matroid polytope $P_M$ we have

$$f(M) = \sum_{P_{M_i} \in \mathcal{P}^{\text{int}}} (-1)^{\dim P_M - \dim P_{M_i}} f(M_i).$$

There are obvious matroid valuations, such as the volume or (thanks to Ehrhart reciprocity) the Ehrhart polynomial of $P_M$. Much more remarkably, we have the following result.

**Theorem 1.8.16** [180] *The Tutte polynomial $T_M(x,y)$ is a matroid valuation.*

**Corollary 1.8.17** *The $f$-vector conjecture is true for $c = 1$.*

*Proof.* By Theorem 1.8.16 the beta invariant $\beta(M) = [x^1y^0]T_M(x,y)$ is also a matroid valuation. Recall that $\beta(N) = 0$ if and only if $N$ is not connected (or, equivalently, if $P_N$ is not a facet of $\mathcal{P}$), $\beta(N) = 1$ if and only if $N$ is series-parallel, and $\beta(N) \geq 2$ otherwise. Also $\beta(U_{d,n}) = \binom{n-2}{d-1}$. Therefore, in a matroid subdivision of $\Delta(d,n)$ we have

$$\binom{n-2}{d-1} = \beta(U_{d,n}) = \sum_{\substack{P_N \in \mathcal{P} \\ P \text{ facet}}} \beta(N) \geq (\text{number of facets of } \mathcal{P})$$

with equality if and only if every facet is a series-parallel matroid. ∎

For matroids $M$ realizable over some field of characteristic 0, Speyer [181] used the K-theory of the Grassmannian to construct another polynomial invariant $g_M(t)$, which he used to prove the f-vector conjecture for matroid subdivisions whose matroids are realizable in characteristic 0.

**Theorem 1.8.18** [181] *The $f$-vector conjecture is true for matroid subdivisions consisting of matroids that are representable over a field of characteristic 0.*

Speyer's proof of this result relies on the nonnegativity of $g_M(t)$, which is proved geometrically for matroids realizable in characteristic 0. The nonnegativity of $g_M(t)$ for all $M$, which would prove Conjecture 1.8.15 in full generality, remains open.

**The Derksen-Fink invariant.** Many other natural matroid functions were later discovered to be valuative. [13, 35, 64]. An example of a very general valuation on matroid polytopes from [13] is the formal sum $R(M) = \sum_{A \subseteq E} R_{A,r(A)}$ of symbols of the form $R_{S,k}$ where $S$ is a subset of $E$ and $k$ is an integer. Many other nice valuations can be obtained from this one. In fact, the matroid $M$ can clearly be recovered from $R(M)$. (However, note that $R$ is not a matroid isomorphism invariant.)

Eventually, Derksen [64] constructed a valuative matroid invariant, which he and Fink proved to be universal. [65] We call it the Derksen–Fink invariant. They gave several versions; we present one that is particularly simple to define.

Let $\text{Matroids}_{n,r}$ be the set of matroids on $[n]$ of rank $r$. Let $\mathscr{C}_n$ be the set of $n!$ complete chains $\mathscr{S} : \emptyset = S_0 \subsetneq S_1 \subsetneq \cdots \subsetneq S_{n-1} \subsetneq S_n = [n]$. For a complete chain $\mathscr{S}$ and a matroid $M = ([n], r)$, let $r[S_{i-1}, S_i] = r(S_i) - r(S_{i-1})$ for $1 \leq i \leq n$; this is always 0 or 1. The Derksen–Fink invariant of $M$ is

$$\mathscr{G}(M) = \sum_{\mathscr{S} \in \mathscr{C}_n} U_{(r[S_0,S_1], r[S_1,S_2], \ldots, r[S_{n-1},S_n])}$$

in the vector space $\mathbf{U}$ spanned by the $\binom{n}{r}$ formal symbols $U_{w_1 \cdots w_n}$ where $w_1, \ldots, w_n \in \{0,1\}$ and $w_1 + \cdots + w_n = r$.

**Theorem 1.8.19** [65] *The Derksen–Fink invariant $\mathscr{G}$ : $\mathit{Matroids}_{n,r} \to \mathbf{U}$ is a universal valuative matroid invariant; that is, for any valuative matroid invariant $f : \mathit{Matroids}_{n,r} \to V$ there is a linear map $\pi : \mathbf{U} \to V$ such that $f = \pi \circ \mathscr{G}$.*

**Acknowledgments.** I am extremely grateful to my combinatorics teachers, Richard Stanley, Gian-Carlo Rota, Sergey Fomin, Sara Billey, and Bernd Sturmfels (in chronological order); their influence on the way that I understand combinatorics is apparent in these pages. I've also learned greatly from my wonderful collaborators and students.

Several superb surveys have been instrumental in the preparation of this manuscript; they are mentioned throughout the text, but it is worth acknowledging a few that I found especially helpful: Aigner's *A Course in Enumeration*, Beck and Robins's *Computing the continuous discretely*, Brylawski and Oxley's *The Tutte polynomial and its applications*, Flajolet and Sedgewick's *Analytic Combinatorics*, Krattenthaler's *Advanced Determinantal Calculus*, and Stanley's *Enumerative Combinatorics* and *Lectures on hyperplane arrangements*.

I am very grateful to Miklós Bóna for the invitation to contribute to this project. Writing this survey has not been an easy task, but I have enjoyed it immensely, and have learned an enormous amount. I also thank Alex Fink, Alicia Dickenstein, Alin Bostan, David Speyer, Federico Castillo, Mark Wildon, Richard Stanley, and the anonymous referee for their helpful comments. Finally, and most importantly, I am so very thankful to May-Li for her patience, help, and support during my months of hard work on this project.

# References

[1] Ron M. Adin. A new cubical $h$-vector. In *Proceedings of the 6th Conference on Formal Power Series and Algebraic Combinatorics (New Brunswick, NJ, 1994)*, volume 157, pages 3–14, 1996.

[2] Martin Aigner. *A course in enumeration*, volume 238 of *Graduate Texts in Mathematics*. Springer, Berlin, 2007.

[3] Tewodros Amdeberhan and Doron Zeilberger. Determinants through the looking glass. *Adv. in Appl. Math.*, 27(2-3):225–230, 2001. Special issue in honor of Dominique Foata's 65th birthday (Philadelphia, PA, 2000).

[4] Federico Ardila. The Catalan matroid. *J. Combin. Theory Ser. A*, 104(1):49–62, 2003.

[5] Federico Ardila. The coefficients of a Fibonacci power series. *Fibonacci Quart.*, 42(3):202–204, 2004.

[6] Federico Ardila. Computing the Tutte polynomial of a hyperplane arrangement. *Pacific J. Math.*, 230(1):1–26, 2007.

[7] Federico Ardila. Semimatroids and their Tutte polynomials. *Rev. Colombiana Mat.*, 41(1): 39–66, 2007.

[8] Federico Ardila, Matthias Beck, Serkan Hosten, Julian Pfeifle, and Kim Seashore. Root polytopes and growth series of root lattices. *SIAM J. Discrete Math.*, 25(1):360–378, 2011.

[9] Federico Ardila, Carolina Benedetti, and Jeffrey Doker. Matroid polytopes and their volumes. *Discrete Comput. Geom.*, 43(4):841–854, 2010.

[10] Federico Ardila and Sara Billey. Flag arrangements and triangulations of products of simplices. *Adv. Math.*, 214(2):495–524, 2007.

[11] Federico Ardila, Federico Castillo, and Michael Henley. The arithmetic Tutte polynomials of the classical root systems. arXiv.org:1305.6621, 2013.

[12] Federico Ardila and Mike Develin. Tropical hyperplane arrangements and oriented matroids. *Math. Z.*, 262(4):795–816, 2009.

[13] Federico Ardila, Alex Fink, and Felipe Rincón. Valuations for matroid polytope subdivisions. *Canad. J. Math.*, 62(6):1228–1245, 2010.

[14] Federico Ardila and Caroline J. Klivans. The Bergman complex of a matroid and phylogenetic trees. *J. Combin. Theory Ser. B*, 96(1):38–49, 2006.

[15] Federico Ardila and Alexander Postnikov. Combinatorics and geometry of power ideals. *Trans. Amer. Math. Soc.*, 362(8):4357–4384, 2010.

[16] Federico Ardila, Felipe Rincón, and Lauren Williams. Positively oriented matroids are realizable. arXiv:1310.4159, 2013.

[17] Federico Ardila, Felipe Rincón, and Lauren Williams. Positroids and non-crossing partitions. arXiv:1308.2698, 2013.

[18] Drew Armstrong. Hyperplane arrangements and diagonal harmonics. *J. Comb.*, 4(2):157–190, 2013.

[19] Drew Armstrong and Brendon Rhoades. The Shi arrangement and the Ish arrangement. *Trans. Amer. Math. Soc.*, 364(3):1509–1528, 2012.

[20] Christos A. Athanasiadis. Characteristic polynomials of subspace arrangements and finite fields. *Adv. Math.*, 122(2):193–233, 1996.

[21] Christos A. Athanasiadis. On free deformations of the braid arrangement. *European J. Combin.*, 19(1):7–18, 1998.

[22] Christos A. Athanasiadis. Extended Linial hyperplane arrangements for root systems and a conjecture of Postnikov and Stanley. *J. Algebraic Combin.*, 10(3):207–225, 1999.

[23] M. Azaola and F. Santos. The number of triangulations of the cyclic polytope $C(n, n-4)$. *Discrete Comput. Geom.*, 27(1):29–48, 2002. Geometric combinatorics (San Francisco, CA/Davis, CA, 2000).

[24] Matthew Baker and Serguei Norine. Riemann-Roch and Abel-Jacobi theory on a finite graph. *Adv. Math.*, 215(2):766–788, 2007.

[25] Welleda Baldoni and Michèle Vergne. Kostant partitions functions and flow polytopes. *Transform. Groups*, 13(3-4):447–469, 2008.

[26] N. Bansal, R. A. Pendavingh, and J. G. van der Pol. On the number of matroids. *Proceedings of the Twenty-Fourth Annual ACM-SIAM Symposium on Discrete Algorithms*, pages 675–694, 2012.

[27] Michael Barany and Victor Reiner. The Tutte polynomial of a finite projective space. `http://www.math.umn.edu/~reiner/Papers/Hamming.pdf`, 2005.

[28] Margaret M. Bayer and Louis J. Billera. Generalized Dehn-Sommerville relations for polytopes, spheres and Eulerian partially ordered sets. *Invent. Math.*, 79(1):143–157, 1985.

[29] Margaret M. Bayer and Andrew Klapper. A new index for polytopes. *Discrete Comput. Geom.*, 6(1):33–47, 1991.

[30] Matthias Beck and Sinai Robins. *Computing the continuous discretely*. Undergraduate Texts in Mathematics. Springer, New York, 2007. Integer-point enumeration in polyhedra.

[31] J. Bétréma and J.-G. Penaud. Animaux et arbres guingois. *Theoret. Comput. Sci.*, 117(1-2):67–89, 1993. Conference on Formal Power Series and Algebraic Combinatorics (Bordeaux, 1991).

[32] L. J. Billera, J. Tatch Moore, C. Dufort Moraites, Y. Wang, and K. Williams. Maximal unbalanced families, 2012.

[33] Louis J. Billera and Anders Björner. Face numbers of polytopes and complexes. In *Handbook of discrete and computational geometry*, CRC Press Ser. Discrete Math. Appl., pages 291–310. CRC, Boca Raton, FL, 1997.

[34] Louis J. Billera, Richard Ehrenborg, and Margaret Readdy. The $c$-$2d$-index of oriented matroids. *J. Combin. Theory Ser. A*, 80(1):79–105, 1997.

[35] Louis J. Billera, Ning Jia, and Victor Reiner. A quasisymmetric function for matroids. *European J. Combin.*, 30(8):1727–1757, 2009.

[36] Louis J. Billera and Carl W. Lee. A proof of the sufficiency of McMullen's conditions for $f$-vectors of simplicial convex polytopes. *J. Combin. Theory Ser. A*, 31(3):237–255, 1981.

[37] Anders Björner. On the homology of geometric lattices. *Algebra Universalis*, 14(1):107–128, 1982.

[38] Anders Björner and Francesco Brenti. *Combinatorics of Coxeter Groups*, volume 231 of *Graduate Texts in Mathematics*. Springer, April 2005.

[39] Anders Björner, Michel Las Vergnas, Bernd Sturmfels, Neil White, and Günter M. Ziegler. *Oriented matroids*, volume 46 of *Encyclopedia of Mathematics and its Applications*. Cambridge University Press, Cambridge, second edition, 1999.

[40] Anders Björner and Michelle Wachs. Bruhat order of Coxeter groups and shellability. *Adv. in Math.*, 43(1):87–100, 1982.

[41] Joseph Bonin, Anna de Mier, and Marc Noy. Lattice path matroids: enumerative aspects and Tutte polynomials. *J. Combin. Theory Ser. A*, 104(1):63–94, 2003.

[42] Alexandre V. Borovik, I. M. Gelfand, and Neil White. *Coxeter matroids*, volume 216 of *Progress in Mathematics*. Birkhäuser Boston, Inc., Boston, MA, 2003.

[43] Petter Brändén and Luca Moci. The multivariate arithmetic Tutte polynomial. arXiv:1207.3629, 2013.

[44] David Bressoud and James Propp. How the alternating sign matrix conjecture was solved. *Notices Amer. Math. Soc.*, 46(6):637–646, 1999.

[45] Michel Brion and Michèle Vergne. Arrangement of hyperplanes. I. Rational functions and Jeffrey-Kirwan residue. *Ann. Sci. École Norm. Sup. (4)*, 32(5):715–741, 1999.

[46] Thomas Brylawski. Appendix of matroid cryptomorphisms. In *Theory of matroids*, volume 26 of *Encyclopedia Math. Appl.*, pages 298–316. Cambridge Univ. Press, Cambridge, 1986.

[47] Thomas Brylawski and James Oxley. The Tutte polynomial and its applications. In *Matroid applications*, volume 40 of *Encyclopedia Math. Appl.*, pages 123–225. Cambridge Univ. Press, Cambridge, 1992.

[48] N. Calkin, C. Merino, S. Noble, and M. Noy. Improved bounds for the number of forests and acyclic orientations in the square lattice. *Electron. J. Combin.*, 10:Research Paper 4, 18 pp. (electronic), 2003.

[49] Cesar Ceballos, Jean-Philippe Labbé, and Christian Stump. Subword complexes, cluster complexes, and generalized multi-associahedra. *J. Algebraic Combin.*, 39(1):17–51, 2014.

[50] Cesar Ceballos, Francisco Santos, and Günter M. Ziegler. Many non-equivalent realizations of the associahedron. arXiv:1109.5544, 2013.

[51] S.-C. Chang and R. Shrock. Zeros of Jones polynomials for families of knots and links. *Phys. A*, 301(1-4):196–218, 2001.

[52] Henry Crapo and William Schmitt. A free subalgebra of the algebra of matroids. *European J. Combin.*, 26(7):1066–1085, 2005.

[53] Henry H. Crapo. The Tutte polynomial. *Aequationes Math.*, 3:211–229, 1969.

[54] Henry H. Crapo and Gian-Carlo Rota. *On the foundations of combinatorial theory: Combinatorial geometries*. The M.I.T. Press, Cambridge, Mass.-London, preliminary edition, 1970.

[55] Dragos M. Cvetković, Michael Doob, and Horst Sachs. *Spectra of graphs*. Johann Ambrosius Barth, Heidelberg, third edition, 1995. Theory and applications.

[56] Michele D'Adderio and Luca Moci. Ehrhart polynomial and arithmetic Tutte polynomial. *European J. Combin.*, 33(7):1479–1483, 2012.

[57] Wolfgang Dahmen and Charles A. Micchelli. On the local linear independence of translates of a box spline. *Studia Math.*, 82(3):243–263, 1985.

[58] N. G. de Bruijn. A combinatorial problem. *Nederl. Akad. Wetensch., Proc.*, 49:758–764 = Indagationes Math. 8, 461–467 (1946), 1946.

[59] C. De Concini and C. Procesi. On the geometry of toric arrangements. *Transform. Groups*, 10(3-4):387–422, 2005.

[60] C. De Concini and C. Procesi. Hyperplane arrangements and box splines. *Michigan Math. J.*, 57:201–225, 2008. With an appendix by A. Björner, Special volume in honor of Melvin Hochster.

[61] C. De Concini and C. Procesi. The zonotope of a root system. *Transform. Groups*, 13(3-4):507–526, 2008.

[62] Jesús A. De Loera, Yvonne Kemper, and Steven Klee. $h$-vectors of small matroid complexes. *Electron. J. Combin.*, 19(1):Paper 14, 11, 2012.

[63] Jesús A. De Loera, Jörg Rambau, and Francisco Santos. *Triangulations*, volume 25 of *Algorithms and Computation in Mathematics*. Springer-Verlag, Berlin, 2010. Structures for algorithms and applications.

[64] Harm Derksen. Symmetric and quasi-symmetric functions associated to polymatroids. *J. Algebraic Combin.*, 30(1):43–86, 2009.

[65] Harm Derksen and Alex Fink. Valuative invariants for polymatroids. *Adv. Math.*, 225(4):1840–1892, 2010.

[66] Mike Develin and Bernd Sturmfels. Tropical convexity. *Doc. Math.*, 9:1–27 (electronic), 2004.

[67] Hansheng Diao. A poset structure on quasi-Fibonacci partitions. *J. Combin. Theory Ser. A*, 116(8):1361–1373, 2009.

[68] R. P. Dilworth. A decomposition theorem for partially ordered sets. *Ann. of Math. (2)*, 51:161–166, 1950.

[69] A. Dress, J. H. Koolen, and V. Moulton. On line arrangements in the hyperbolic plane. *European J. Combin.*, 23(5):549–557, 2002.

[70] Andreas W. M. Dress and Walter Wenzel. Valuated matroids. *Adv. Math.*, 93(2):214–250, 1992.

[71] Jack Edmonds and D. R. Fulkerson. Transversals and matroid partition. *J. Res. Nat. Bur. Standards Sect. B*, 69B:147–153, 1965.

[72] Richard Ehrenborg, Margaret Readdy, and Michael Slone. Affine and toric hyperplane arrangements. *Discrete Comput. Geom.*, 41(4):481–512, 2009.

[73] Noam Elkies, Greg Kuperberg, Michael Larsen, and James Propp. Alternating-sign matrices and domino tilings. I. *J. Algebraic Combin.*, 1(2):111–132, 1992.

[74] Noam Elkies, Greg Kuperberg, Michael Larsen, and James Propp. Alternating-sign matrices and domino tilings. II. *J. Algebraic Combin.*, 1(3):219–234, 1992.

[75] Joanna A. Ellis-Monaghan and Criel Merino. Graph polynomials and their applications I: The Tutte polynomial. In *Structural analysis of complex networks*, pages 219–255. Birkhäuser/Springer, New York, 2011.

[76] Sen-Peng Eu and Tung-Shan Fu. A simple proof of the Aztec diamond theorem. *Electron. J. Combin.*, 12:Research Paper 18, 8 pp. (electronic), 2005.

[77] Eva Maria Feichtner and Bernd Sturmfels. Matroid polytopes, nested sets and Bergman fans. *Port. Math. (N.S.)*, 62(4):437–468, 2005.

[78] Philippe Flajolet, Stefan Gerhold, and Bruno Salvy. On the non-holonomic character of logarithms, powers, and the *n*th prime function. *Electron. J. Combin.*, 11(2):Article 2, 16, 2004/06.

[79] Philippe Flajolet and Robert Sedgewick. *Analytic combinatorics*. Cambridge University Press, Cambridge, 2009.

[80] C. Flye Sainte-Marie. Solution to question nr. 48. *L'intermédiaire des Mathématiciens*, pages 107–110, 1894.

[81] Sergey Fomin and Nathan Reading. Root systems and generalized associahedra. In *Geometric combinatorics*, volume 13 of *IAS/Park City Math. Ser.*, pages 63–131. Amer. Math. Soc., Providence, RI, 2007.

[82] Satoru Fujishige. *Submodular functions and optimization*, volume 58 of *Annals of Discrete Mathematics*. Elsevier B. V., Amsterdam, second edition, 2005.

[83] William Fulton and Joe Harris. *Representation theory*, volume 129 of *Graduate Texts in Mathematics*. Springer-Verlag, New York, 1991. A first course, Readings in Mathematics.

[84] Andreas Gathmann and Michael Kerber. A Riemann-Roch theorem in tropical geometry. *Math. Z.*, 259(1):217–230, 2008.

[85] Jim Geelen, Bert Gerards, and Geoff Whittle. Solving Rota's conjecture. *Notices of the Amer. Math. Soc.*, 61(7):736–743, 2014.

[86] I. M. Gel′fand, R. M. Goresky, R. D. MacPherson, and V. V. Serganova. Combinatorial geometries, convex polyhedra, and Schubert cells. *Adv. in Math.*, 63(3):301–316, 1987.

[87] I. M. Gelfand, M. M. Kapranov, and A. V. Zelevinsky. *Discriminants, resultants and multidimensional determinants*. Modern Birkhäuser Classics. Birkhäuser Boston, Inc., Boston, MA, 2008. Reprint of the 1994 edition.

[88] Ira Gessel and Gérard Viennot. Binomial determinants, paths, and hook length formulae. *Adv. in Math.*, 58(3):300–321, 1985.

[89] Sudhir R. Ghorpade and Christian Krattenthaler. The Hilbert series of Pfaffian rings. In *Algebra, arithmetic and geometry with applications (West Lafayette, IN, 2000)*, pages 337–356. Springer, Berlin, 2004.

[90] Mark Goresky and Robert MacPherson. *Stratified Morse theory*, volume 14 of *Ergebnisse der Mathematik und ihrer Grenzgebiete (3) [Results in Mathematics and Related Areas (3)]*. Springer-Verlag, Berlin, 1988.

[91] D. Gouyou-Beauchamps and G. Viennot. Equivalence of the two-dimensional directed animal problem to a one-dimensional path problem. *Adv. in Appl. Math.*, 9(3):334–357, 1988.

[92] Curtis Greene. Weight enumeration and the geometry of linear codes. *Studies in Appl. Math.*, 55(2):119–128, 1976.

[93] Branko Grünbaum. *Convex polytopes*, volume 221 of *Graduate Texts in Mathematics*. Springer-Verlag, New York, second edition, 2003. Prepared and with a preface by Volker Kaibel, Victor Klee and Günter M. Ziegler.

[94] Paul Hacking, Sean Keel, and Jenia Tevelev. Compactification of the moduli space of hyperplane arrangements. *J. Algebraic Geom.*, 15(4):657–680, 2006.

[95] Frank Harary and Jerald A. Kabell. Counting balanced signed graphs using marked graphs. *Proc. Edinburgh Math. Soc. (2)*, 24(2):99–104, 1981.

[96] William A. Harris, Jr. and Yasutaka Sibuya. The reciprocals of solutions of linear ordinary differential equations. *Adv. in Math.*, 58(2):119–132, 1985.

[97] Peter Henrici. *Applied and computational complex analysis. Vol. 2.* Wiley Classics Library. John Wiley & Sons, Inc., New York, 1991. Special functions—integral transforms—asymptotics—continued fractions, Reprint of the 1977 original, A Wiley-Interscience Publication.

[98] Sven Herrmann, Michael Joswig, and David Speyer. Dressians, tropical Grassmannians, and their rays. arXiv:1112.1278, 2012.

[99] Jürgen Herzog and Ngô Viêt Trung. Gröbner bases and multiplicity of determinantal and Pfaffian ideals. *Adv. Math.*, 96(1):1–37, 1992.

[100] Silvia Heubach, Nelson Y. Li, and Toufik Mansour. Staircase tilings and $k$-Catalan structures. *Discrete Math.*, 308(24):5954–5964, 2008.

[101] Takayuki Hibi and Nan Li. Unimodular equivalence of order and chain polytopes. arXiv:1208.4029, 2012.

[102] Takayuki Hibi, Nan Li, and Hidefumi Ohsugi. The face vector of a half-open hypersimplex. arXiv:1309.5155, 2014.

[103] Olga Holtz and Amos Ron. Zonotopal algebra. *Adv. Math.*, 227(2):847–894, 2011.

[104] June Huh. Milnor numbers of projective hypersurfaces and the chromatic polynomial of graphs. *J. Amer. Math. Soc.*, 25(3):907–927, 2012.

[105] June Huh and Eric Katz. Log-concavity of characteristic polynomials and the Bergman fan of matroids. *Math. Ann.*, 354(3):1103–1116, 2012.

[106] James E. Humphreys. *Reflection groups and Coxeter groups*, volume 29 of *Cambridge Studies in Advanced Mathematics*. Cambridge University Press, Cambridge, 1990.

[107] Francois Jaeger. On Tutte polynomials of matroids representable over GF$(q)$. *European J. Combin.*, 10(3):247–255, 1989.

[108] Vaughan F. R. Jones. A polynomial invariant for knots via von Neumann algebras. *Bull. Amer. Math. Soc. (N.S.)*, 12(1):103–111, 1985.

[109] Jakob Jonsson. Generalized triangulations of the n-gon, 2003.

[110] Jakob Jonsson. Generalized triangulations and diagonal-free subsets of stack polyominoes. *J. Combin. Theory Ser. A*, 112(1):117–142, 2005.

[111] R. Jungen. Sur les séries de Taylor n'ayant que des singularités algébrico-logarithmiques sur leur cercle de convergence. *Comment. Math. Helv.*, 3(1):266–306, 1931.

[112] Subhash Kak. Yamātārājabhānasalagām: an interesting combinatoric sūtra. *Indian J. Hist. Sci.*, 35(2):123–127, 2000.

[113] M. M. Kapranov. Chow quotients of Grassmannians. I. In *I. M. Gel'fand Seminar*, volume 16 of *Adv. Soviet Math.*, pages 29–110. Amer. Math. Soc., Providence, RI, 1993.

[114] Samuel Karlin and James McGregor. Coincidence probabilities. *Pacific J. Math.*, 9:1141–1164, 1959.

[115] P. W. Kasteleyn. Dimer statistics and phase transitions. *J. Mathematical Phys.*, 4:287–293, 1963.

[116] David Kazhdan and George Lusztig. Representations of Coxeter groups and Hecke algebras. *Invent. Math.*, 53(2):165–184, 1979.

[117] Steven Klee and Jose Alejandro Samper. Lexicographic shellability, matroids and pure order ideals. arXiv:1406.1951, 2014.

[118] Caroline J. Klivans. Threshold graphs, shifted complexes, and graphical complexes. *Discrete Math.*, 307(21):2591–2597, 2007.

[119] Donald E. Knuth. The asymptotic number of geometries. *J. Combinatorial Theory Ser. A*, 16:398–400, 1974.

[120] Matjaž Konvalinka and Igor Pak. Triangulations of Cayley and Tutte polytopes. *Adv. Math.*, 245:1–33, 2013.

[121] Michael Korn and Igor Pak. Tilings of rectangles with T-tetrominoes. *Theoret. Comput. Sci.*, 319(1-3):3–27, 2004.

[122] C. Krattenthaler. Advanced determinant calculus. *Sém. Lothar. Combin.*, 42:Art. B42q, 67 pp. (electronic), 1999. The Andrews Festschrift (Maratea, 1998).

[123] C. Krattenthaler. Advanced determinant calculus: a complement. *Linear Algebra Appl.*, 411:68–166, 2005.

[124] G. Kreweras. Sur les partitions non croisées d'un cycle. *Discrete Math.*, 1(4):333–350, 1972.

[125] Joseph P. S. Kung and Hien Q. Nguyen. Weak maps. In *Theory of matroids*, volume 26 of *Encyclopedia Math. Appl.*, pages 254–271. Cambridge Univ. Press, Cambridge, 1986.

[126] L. Lafforgue. *Chirurgie des Grassmanniennes*, volume 19 of *CRM Monograph Series*. American Mathematical Society, Providence, RI, 2003.

[127] Thomas Lam and Alexander Postnikov. Alcoved polytopes. I. *Discrete Comput. Geom.*, 38(3):453–478, 2007.

[128] Michel Las Vergnas. On the evaluation at $(3,3)$ of the Tutte polynomial of a graph. *J. Combin. Theory Ser. B*, 45(3):367–372, 1988.

[129] Alain Lascoux. Classes de Chern d'un produit tensoriel. *C. R. Acad. Sci. Paris Sér. A-B*, 286(8):A385–A387, 1978.

[130] Fu Liu. Ehrhart polynomials of cyclic polytopes. *J. Combin. Theory Ser. A*, 111(1):111–127, 2005.

[131] László Lovász and Michael D. Plummer. *Matching theory*. AMS Chelsea Publishing, Providence, RI, 2009. Corrected reprint of the 1986 original [MR0859549].

[132] Saunders MacLane. Some Interpretations of Abstract Linear Dependence in Terms of Projective Geometry. *Amer. J. Math.*, 58(1):236–240, 1936.

[133] Percy A. MacMahon. *Combinatory analysis*. Vol. I, II (bound in one volume). Dover Phoenix Editions. Dover Publications, Inc., Mineola, NY, 2004. Reprint of *An introduction to combinatory analysis* (1920) and *Combinatory analysis*, Vol. I, II (1915, 1916).

[134] Jessie MacWilliams. A theorem on the distribution of weights in a systematic code. *Bell System Tech. J.*, 42:79–94, 1963.

[135] N. V. R. Mahadev and U. N. Peled. *Threshold graphs and related topics*, volume 56 of *Annals of Discrete Mathematics*. North-Holland Publishing Co., Amsterdam, 1995.

[136] J. H. Mason. On a class of matroids arising from paths in graphs. *Proc. London Math. Soc. (3)*, 25:55–74, 1972.

[137] Dillon Mayhew, Mike Newman, Dominic Welsh, and Geoff Whittle. On the asymptotic proportion of connected matroids. *European J. Combin.*, 32(6):882–890, 2011.

[138] P. McMullen and G. C. Shephard. *Convex polytopes and the upper bound conjecture*. Cambridge University Press, London-New York, 1971. Prepared in collaboration with J. E. Reeve and A. A. Ball, London Mathematical Society Lecture Note Series, 3.

[139] Peter McNamara. EL-labelings, supersolvability and 0-Hecke algebra actions on posets. *J. Combin. Theory Ser. A*, 101(1):69–89, 2003.

[140] Criel Merino. The chip firing game and matroid complexes. In *Discrete models: combinatorics, computation, and geometry (Paris, 2001)*, Discrete Math. Theor. Comput. Sci. Proc., AA, pages 245–255 (electronic). Maison Inform. Math. Discrèt. (MIMD), Paris, 2001.

[141] Criel Merino, Steven D. Noble, Marcelino Ramírez-Ibáñez, and Rafael Villarroel-Flores. On the structure of the $h$-vector of a paving matroid. *European J. Combin.*, 33(8):1787–1799, 2012.

[142] Criel Merino, Marcelino Ramirez-Ibanez, and Guadalupe Rodriguez Sanchez. The Tutte polynomial of some matroids. arXiv:1203.0090, 2012.

[143] Karola Mészáros and Alejandro H. Morales. Flow polytopes of signed graphs and the Kostant partition function. arXiv:1208.0140, 2012.

[144] Luca Moci. A Tutte polynomial for toric arrangements. *Trans. Amer. Math. Soc.*, 364(2):1067–1088, 2012.

[145] Eunice Gogo Mphako. Tutte polynomials of perfect matroid designs. *Combin. Probab. Comput.*, 9(4):363–367, 2000.

[146] Satoshi Murai and Eran Nevo. On the generalized lower bound conjecture for polytopes and spheres. *Acta Math.*, 210(1):185–202, 2013.

[147] Kazuo Murota. Convexity and Steinitz's exchange property. *Adv. Math.*, 124(2):272–311, 1996.

[148] Tomoki Nakamigawa. A generalization of diagonal flips in a convex polygon. *Theoret. Comput. Sci.*, 235(2):271–282, 2000. Combinatorics and optimization (Okinawa, 1996).

[149] Ivan Niven. Formal power series. *Amer. Math. Monthly*, 76:871–889, 1969.

[150] Suho Oh. Combinatorics of positroids. In *21st International Conference on Formal Power Series and Algebraic Combinatorics (FPSAC 2009)*, Discrete Math. Theor. Comput. Sci. Proc., AK, pages 721–732. Assoc. Discrete Math. Theor. Comput. Sci., Nancy, 2009.

[151] Suho Oh. Generalized permutohedra, $h$-vectors of cotransversal matroids and pure O-sequences. *Electron. J. Combin.*, 20(3):Paper 14, 14, 2013.

[152] Peter Orlik and Louis Solomon. Combinatorics and topology of complements of hyperplanes. *Invent. Math.*, 56(2):167–189, 1980.

[153] Peter Orlik and Hiroaki Terao. *Arrangements of hyperplanes*, volume 300 of *Grundlehren der Mathematischen Wissenschaften [Fundamental Principles of Mathematical Sciences]*. Springer-Verlag, Berlin, 1992.

[154] James Oxley. *Matroid theory*, volume 21 of *Oxford Graduate Texts in Mathematics*. Oxford University Press, Oxford, second edition, 2011.

[155] George Pólya and Gabor Szegő. *Problems and theorems in analysis. II.* Classics in Mathematics. Springer-Verlag, Berlin, 1998. Theory of functions, zeros, polynomials, determinants, number theory, geometry, Translated from the German by C. E. Billigheimer, Reprint of the 1976 English translation.

[156] Alex Postnikov, Victor Reiner, and Lauren Williams. Faces of generalized permutohedra. *Doc. Math.*, 13:207–273, 2008.

[157] Alexander Postnikov. Total positivity, Grassmannians, and networks. arXiv:math/0609764, 2006.

[158] Alexander Postnikov. Permutohedra, associahedra, and beyond. *Int. Math. Res. Not. IMRN*, (6):1026–1106, 2009.

[159] Alexander Postnikov and Boris Shapiro. Trees, parking functions, syzygies, and deformations of monomial ideals. *Trans. Amer. Math. Soc.*, 356(8):3109–3142 (electronic), 2004.

[160] Alexander Postnikov, Boris Shapiro, and Mikhail Shapiro. Algebras of curvature forms on homogeneous manifolds. In *Differential topology, infinite-dimensional Lie algebras, and applications*, volume 194 of *Amer. Math. Soc. Transl. Ser. 2*, pages 227–235. Amer. Math. Soc., Providence, RI, 1999.

[161] Alexander Postnikov and Richard P. Stanley. Deformations of Coxeter hyperplane arrangements. *J. Combin. Theory Ser. A*, 91(1-2):544–597, 2000. In memory of Gian-Carlo Rota.

[162] Nicholas Proudfoot and David Speyer. A broken circuit ring. *Beiträge Algebra Geom.*, 47(1):161–166, 2006.

[163] Jörg Rambau. Triangulations of cyclic polytopes and higher Bruhat orders. *Mathematika*, 44(1):162–194, 1997.

[164] Victor Reiner. An interpretation for the Tutte polynomial. *European J. Combin.*, 20(2):149–161, 1999.

[165] Vladimir Retakh. Israel Moiseevich Gelfand, Part I. *Notices of the Amer. Math. Soc*, 60(1):24–49, 2013.

[166] Felipe Rincón. Local tropical linear spaces. *Discrete Comput. Geom.*, 50(3):700–713, 2013.

[167] David P. Robbins. The story of $1,2,7,42,429,7436,\cdots$. *Math. Intelligencer*, 13(2):12–19, 1991.

[168] Neville Robbins. Fibonacci partitions. *Fibonacci Quart.*, 34(4):306–313, 1996.

[169] Gian-Carlo Rota. On the foundations of combinatorial theory. I. Theory of Möbius functions. *Z. Wahrscheinlichkeitstheorie und Verw. Gebiete*, 2:340–368 (1964), 1964.

[170] Gian-Carlo Rota. Combinatorial theory, old and new. In *Actes du Congrès International des Mathématiciens (Nice, 1970), Tome 3*, pages 229–233. Gauthier-Villars, Paris, 1971.

[171] Gian-Carlo Rota. *Indiscrete thoughts*. Birkhäuser Boston, Inc., Boston, MA, 1997. With forewords by Reuben Hersh and Robert Sokolowski, Edited with notes and an epilogue by Fabrizio Palombi.

[172] Bruce E. Sagan. Why the characteristic polynomial factors. *Bull. Amer. Math. Soc. (N.S.)*, 36(2):113–133, 1999.

[173] Bruce E. Sagan. *The symmetric group*, volume 203 of *Graduate Texts in Mathematics*. Springer-Verlag, New York, second edition, 2001. Representations, combinatorial algorithms, and symmetric functions.

[174] Kyoji Saito. Theory of logarithmic differential forms and logarithmic vector fields. *J. Fac. Sci. Univ. Tokyo Sect. IA Math.*, 27(2):265–291, 1980.

[175] Francisco Santos. The Cayley trick and triangulations of products of simplices. In *Integer points in polyhedra—geometry, number theory, algebra, optimization*, volume 374 of *Contemp. Math.*, pages 151–177. Amer. Math. Soc., Providence, RI, 2005.

[176] Jay Schweig. On the $h$-vector of a lattice path matroid. *Electron. J. Combin.*, 17(1):Note 3, 6, 2010.

[177] Luis Serrano and Christian Stump. Maximal fillings of moon polyominoes, simplicial complexes, and Schubert polynomials. *Electron. J. Combin.*, 19(1):Paper 16, 18, 2012.

[178] Alan D. Sokal. The multivariate Tutte polynomial (alias Potts model) for graphs and matroids. In *Surveys in combinatorics 2005*, volume 327 of *London Math. Soc. Lecture Note Ser.*, pages 173–226. Cambridge Univ. Press, Cambridge, 2005.

[179] Alan D. Sokal. The leading root of the partial theta function. *Adv. Math.*, 229(5):2603–2621, 2012.

[180] David E. Speyer. Tropical linear spaces. *SIAM J. Discrete Math.*, 22(4):1527–1558, 2008.

[181] David E. Speyer. A matroid invariant via the $K$-theory of the Grassmannian. *Adv. Math.*, 221(3):882–913, 2009.

[182] R. P. Stanley. Supersolvable lattices. *Algebra Universalis*, 2:197–217, 1972.

[183] R. P. Stanley. Differentiably finite power series. *European J. Combin.*, 1(2):175–188, 1980.

[184] Richard Stanley. Generalized $H$-vectors, intersection cohomology of toric varieties, and related results. In *Commutative algebra and combinatorics (Kyoto, 1985)*, volume 11 of *Adv. Stud. Pure Math.*, pages 187–213. North-Holland, Amsterdam, 1987.

[185] Richard P. Stanley. Catalan addendum.

[186] Richard P. Stanley. Cohen-Macaulay complexes. In *Higher combinatorics (Proc. NATO Advanced Study Inst., Berlin, 1976)*, pages 51–62. NATO Adv. Study Inst. Ser., Ser. C: Math. and Phys. Sci., 31. Reidel, Dordrecht, 1977.

[187] Richard P. Stanley. Decompositions of rational convex polytopes. *Ann. Discrete Math.*, 6:333–342, 1980.

[188] Richard P. Stanley. The number of faces of a simplicial convex polytope. *Adv. in Math.*, 35(3):236–238, 1980.

[189] Richard P. Stanley. Two poset polytopes. *Discrete Comput. Geom.*, 1(1):9–23, 1986.

[190] Richard P. Stanley. A zonotope associated with graphical degree sequences. In *Applied geometry and discrete mathematics*, volume 4 of *DIMACS Ser. Discrete Math. Theoret. Comput. Sci.*, pages 555–570. Amer. Math. Soc., Providence, RI, 1991.

[191] Richard P. Stanley. Flag $f$-vectors and the $cd$-index. *Math. Z.*, 216(3):483–499, 1994.

[192] Richard P. Stanley. *Enumerative combinatorics. Vol. 2*, volume 62 of *Cambridge Studies in Advanced Mathematics*. Cambridge University Press, Cambridge, 1999. With a foreword by Gian-Carlo Rota and appendix 1 by Sergey Fomin.

[193] Richard P. Stanley. An introduction to hyperplane arrangements. In *Geometric combinatorics*, volume 13 of *IAS/Park City Math. Ser.*, pages 389–496. Amer. Math. Soc., Providence, RI, 2007.

[194] Richard P. Stanley. *Enumerative combinatorics. Volume 1*, volume 49 of *Cambridge Studies in Advanced Mathematics*. Cambridge University Press, Cambridge, second edition, 2012.

[195] John R. Stembridge. A short derivation of the Möbius function for the Bruhat order. *J. Algebraic Combin.*, 25(2):141–148, 2007.

[196] Bernd Sturmfels. *Gröbner bases and convex polytopes*, volume 8 of *University Lecture Series*. American Mathematical Society, Providence, RI, 1996.

[197] Lajos Takács. On the number of distinct forests. *SIAM J. Discrete Math.*, 3(4):574–581, 1990.

[198] H. N. V. Temperley and Michael E. Fisher. Dimer problem in statistical mechanics—an exact result. *Philos. Mag. (8)*, 6:1061–1063, 1961.

[199] Hiroaki Terao. Arrangements of hyperplanes and their freeness. I. *J. Fac. Sci. Univ. Tokyo Sect. IA Math.*, 27(2):293–312, 1980.

[200] Hiroaki Terao. Algebras generated by reciprocals of linear forms. *J. Algebra*, 250(2):549–558, 2002.

[201] W. T. Tutte. A contribution to the theory of chromatic polynomials. *Canadian J. Math.*, 6:80–91, 1954.

[202] W. T. Tutte. On dichromatic polynominals. *J. Combinatorial Theory*, 2:301–320, 1967.

[203] W. T. Tutte. On the Birkhoff-Lewis equations. *Discrete Math.*, 92(1-3):417–425, 1991.

[204] B. L. van der Waerden. Zur algebraischen Geometrie. *Math. Ann.*, 113(1):36–39, 1937.

[205] Daya-Nand Verma. Möbius inversion for the Bruhat ordering on a Weyl group. *Ann. Sci. École Norm. Sup. (4)*, 4:393–398, 1971.

[206] Michelle L. Wachs. Poset topology: tools and applications. In *Geometric combinatorics*, volume 13 of *IAS/Park City Math. Ser.*, pages 497–615. Amer. Math. Soc., Providence, RI, 2007.

[207] Michelle L. Wachs and James W. Walker. On geometric semilattices. *Order*, 2(4):367–385, 1986.

[208] David G. Wagner. Algebras related to matroids represented in characteristic zero. *European J. Combin.*, 20(7):701–711, 1999.

[209] H. S. Wall. *Analytic Theory of Continued Fractions*. D. Van Nostrand Company, Inc., New York, N. Y., 1948.

[210] Louis Weisner. Some properties of prime-power groups. *Trans. Amer. Math. Soc.*, 38(3):485–492, 1935.

[211] D. J. A. Welsh. *Matroid theory*. Academic Press [Harcourt Brace Jovanovich, Publishers], London-New York, 1976. L. M. S. Monographs, No. 8.

[212] D. J. A. Welsh. *Complexity: knots, colourings and counting*, volume 186 of *London Mathematical Society Lecture Note Series*. Cambridge University Press, Cambridge, 1993.

[213] Dominic J. A. Welsh and Geoffrey P. Whittle. Arrangements, channel assignments, and associated polynomials. *Adv. in Appl. Math.*, 23(4):375–406, 1999.

[214] Neil White, editor. *Theory of matroids*, volume 26 of *Encyclopedia of Mathematics and its Applications*. Cambridge University Press, Cambridge, 1986.

[215] Neil White, editor. *Combinatorial geometries*, volume 29 of *Encyclopedia of Mathematics and its Applications*. Cambridge University Press, Cambridge, 1987.

[216] Neil White, editor. *Matroid applications*, volume 40 of *Encyclopedia of Mathematics and its Applications*. Cambridge University Press, Cambridge, 1992.

[217] Hassler Whitney. On the Abstract Properties of Linear Dependence. *Amer. J. Math.*, 57(3):509–533, 1935.

[218] Herbert S. Wilf. *generatingfunctionology*. A K Peters, Ltd., Wellesley, MA, third edition, 2006.

[219] Lauren K. Williams. Enumeration of totally positive Grassmann cells. *Adv. Math.*, 190(2):319–342, 2005.

[220] Thomas Zaslavsky. Facing up to arrangements: face-count formulas for partitions of space by hyperplanes. *Mem. Amer. Math. Soc.*, 1(issue 1, 154):vii+102, 1975.

[221] Doron Zeilberger. The amazing $3^n$ theorem and its even more amazing proof. arXiv:1208.2258, 2012.

[222] Doron Zeilberger. Proof of the refined alternating sign matrix conjecture. *New York J. Math.*, 2:59–68, electronic, 1996.

[223] Doron Zeilberger. Proof of a conjecture of Chan, Robbins, and Yuen. *Electron. Trans. Numer. Anal.*, 9:147–148 (electronic), 1999. Orthogonal polynomials: numerical and symbolic algorithms (Leganés, 1998).

[224] Ilia Zharkov. The Orlik-Solomon algebra and the Bergman fan of a matroid. *J. Gökova Geom. Topol. GGT*, 7:25–31, 2013.

[225] Günter M. Ziegler. *Lectures on polytopes*, volume 152 of *Graduate Texts in Mathematics*. Springer-Verlag, New York, 1995.

[226] Günter M. Ziegler. Face numbers of 4-polytopes and 3-spheres. In *Proceedings of the International Congress of Mathematicians (Beijing, 2002)*, volume III, pages 625–634. Higher Ed. Press, Beijing, 2002.

# Chapter 2

## *Analytic Methods*

**Helmut Prodinger**
*Department of Mathematics*
*University of Stellenbosch*
*Stellenbosch 7602, South Africa*

**CONTENTS**

## 2.1 Introduction

This chapter should have really been written by Philippe Flajolet (1948–2011), who could not make it this time. **He** coined the name **analytic combinatorics**. The present author knew Flajolet since 1979 and followed all the developments closely since then, also being a coauthor on various occasions. Flajolet and his followers started out in **analysis of algorithms**, a subject founded by Knuth in his series of books **The art of computer programming**; it became clear over the years that many techniques from classical mathematics had to be unearthed and many new ones had to be discovered. Flajolet was a pioneer in this direction; Doron Zeilberger called him a combinatorialist who became an analyst, and it was his understanding that combinatorics should have an analytic component, like number theory has analytic number theory. Flajolet, apart from being an exceptional problem solver, had a strong desire to be clear and systematic. Eventually, with coauthor Robert Sedgewick, after many years of preparations, the book **Analytic Combinatorics** [24] was published. It has 810 pages, and only a fraction of it can be represented here.

When analyzing algorithms, there is often an algebraic (combinatorial) part, followed by an asymptotic (analytic) part. And, indeed, analytic combinatorics, as understood by Flajolet, follows the same pattern. The central objects are generating functions. First, through combinatorial constructions (a bit reminiscent of grammars for formal languages), symbolic equations for the objects are obtained. There are rules for how to translate the symbolic equations into equations for the associated generating functions. From here, there are a few typical scenarios. Ideally, one can write explicit expressions for the generating functions, and then get from them explicit expressions for the coefficients, which are the numbers of interest. But often, these expressions are involved, and one needs asymptotic techniques. It is often better to derive asymptotic equivalents directly from the generating functions. Sometimes, one cannot solve equations and thus has no explicit form for the generating functions. But even then there might still be hope. Asymptotic methods (in combinatorics) have been known for decades, and we might cite De Bruijn's book [5] and Odlyzko's treatise [40]. One of Flajolet's favorite methods was the Mellin transform. He wrote a series of survey papers about it, and a draft of some 100+ pages about it existed, but eventually did not make it into the book [24].

We follow these guidelines and include material here that, in one way or another, can be traced back to [24]. Whatever we cite from this great book we do with due respect. Even in places where it is not explicitly said, the concept and notations are probably taken from this authoritative text, since we firmly believe that it is vain to try to improve on the masters themselves.

The choice is perhaps a bit personal; during some 35 years as a working mathematician, I came across many things that are useful and also necessary to know. Whatever is included here, was interesting to myself over the years. I hope that the selection presented here will be useful for the readers of this handbook as well.

**Plan of this chapter.** We start with combinatorial constructions, for both unlabeled and labeled objects. On the way, we discuss various classical combinatorial objects, like compositions, partitions, trees, set partitions, etc. Then we elaborate on techniques related to generating functions. Before we move to asymptotic considerations, we need a few preparations, like the Gamma function, the zeta function, the harmonic numbers, etc. Then we move to the Mellin transform. Here, we start with so-called harmonic sums, move to digital sums, and eventually to divide-and-conquer type recursions. Important here is the Mellin inversion formula and the Mellin-Perron summation formula. Then we discuss Rice's method, which is based on the Cauchy integral formula, and a running example that is both instructive and important: approximate counting. Then we sketch singularity analysis of generating functions, which is a toolkit allowing us to translate from the local behavior of the generating functions around their (dominant) singularities to the asymptotic behavior of its (Taylor-)coefficients. Another interesting running example is the one of longest runs in strings consisting of two symbols. Another important asymptotic technique, the saddle point method, is only sketched. The last section deals with Gaussian limiting distributions and how they can be obtained in an important family of special instances; these developments are due to H.-K. Hwang. This section must be seen as an appetizer; it will increase the desire to read more comprehensive texts about limiting distributions in combinatorial analysis.

Let us collect a few conventions and notations and facts that will be encountered later:

**Iverson's notation.** We write $[P] = 1$ when condition $P$ is true, $[P] = 0$ otherwise. This notation is more flexible, say, than Kronecker's $\delta_{i,j} = [i = j]$.

**Probability generating functions and moments.** Assume that the power series $f(z)$ has non-negative coefficients and $f(1) = 1$ (a probability generating function). Then expectation and variance (when they exist) are given by

$$\mathbb{E} = f'(1), \qquad \mathbb{V} = f''(1) + f'(1) - \left(f'(1)\right)^2.$$

**Coefficient extraction of power series.** If $f(z) = \sum_{n \geq 0} a_n z^n$, then $[z^n]f(z) = a_n$.

## 2.2 Combinatorial constructions and associated ordinary generating functions

Let $\mathscr{A}$ be a denumerable set with a **size** (a non-negative number) associated to each of its elements. We often write $|a|$ for the size of $a \in \mathscr{A}$. Furthermore, we assume that, for each $n$, there is only a finite number of elements of size $n$ in $\mathscr{A}$; we call it

$a_n$. Then we define the associated (ordinary) generating function

$$A(z) = \sum_{n\geq 0} a_n z^n.$$

For the time being, this is just a formal construction, but later on, we will interpret $A(z)$ as a function of a complex variable $z$. This will be particularly useful when we discuss how to get asymptotic equivalents for the numbers $a_n$. At the moment, we are just interested in combinatorial constructions, and how they are reflected by their associated generating functions. Such constructions allow us to start from basic objects and create more complicated ones. Another, but equivalent point of view is decomposition, where a certain combinatorial class of objects is (uniquely) decomposed into simpler ingredients. Combinatorial constructions are also relevant in symbolic algebra systems.

The most common constructions are now discussed.

**Union.** Let $\mathscr{B}$ and $\mathscr{C}$ be such combinatorial classes with associated generating functions $B(z)$ and $C(z)$ and assume that the classes are mutually disjoint. Then the (disjoint) union $\mathscr{A} = \mathscr{B} + \mathscr{C}$, also written as $\mathscr{A} = \mathscr{B} \cup \mathscr{C}$, has associated generating function $A(z) = B(z) + C(z)$. This follows from the elementary $a_n = b_n + c_n$.

**Product.** Let $\mathscr{B}$ and $\mathscr{C}$ be such combinatorial classes with associated generating functions $B(z)$ and $C(z)$. Then we form the (Cartesian) product $\mathscr{A} = \mathscr{B} \times \mathscr{C}$. The size of an object $(b,c)$ is defined to be $|b| + |c|$. Then the associated generating function is $A(z) = B(z)C(z)$. This follows from the fact that

$$a_n = \sum_{k=0}^{n} b_k c_{n-k},$$

and this is just the Cauchy product of two series:

$$B(z)C(z) = \sum_{m\geq 0} b_m z^m \cdot \sum_{n\geq 0} c_n z^n = \sum_{n\geq 0} \left( \sum_{k=0}^{n} b_k c_{n-k} \right) z^n = A(z).$$

This idea immediately extends to several factors, not just two. In particular, for a fixed number $k$, we can consider $k$-tuples $(x_1, \ldots, x_k)$, where all $x_i \in \mathscr{A}$, and $|(x_1, \ldots, x_k)| = |x_1| + \cdots + |x_k|$. Then the generating function associated to $\mathscr{A}^k$ is $A^k(z)$.

**Sequence.** Let $\mathscr{B}$ be a combinatorial class, that does not contain elements of size 0. Then we form

$$\mathscr{A} = \mathscr{B}^0 + \mathscr{B}^1 + \mathscr{B}^2 + \cdots,$$

which describes sequences of elements of $\mathscr{B}$. The (unique) sequence of zero elements is traditionally written as $\varepsilon$. We write $\mathscr{A} = \textsc{Seq}(\mathscr{B})$, and the associated generating function is

$$A(z) = 1 + B(z) + B^2(z) + \cdots = \frac{1}{1 - B(z)}.$$

Especially when dealing with languages (sets of words), the notion $\mathscr{B}^*$ instead of $\textsc{Seq}(\mathscr{B})$ is common; then also $\mathscr{B}^+ = \mathscr{B}^1 + \mathscr{B}^2 + \cdots$.

**Power set.** For a given $\mathscr{B}$, we form finite sets of elements taken from $\mathscr{B}$; the result is $\mathscr{A} = \text{PSET}(\mathscr{B})$, and the size of such a set is defined to be the sum of the sizes of its elements. We must assume that $\mathscr{B}$ does not contain an element of size 0. We have

$$\mathscr{A} \equiv (\varepsilon + \{\beta_1\}) \times (\varepsilon + \{\beta_2\}) \times \cdots$$

for an enumeration $(\beta_1, \beta_2, \ldots)$ of the class $\mathscr{B}$. Now let

$$B(z) = \sum_{n \geq 1} B_n z^n,$$

then we can compute

$$\begin{aligned} A(z) &= \prod_{\beta \in \mathscr{B}} (1 + z^{|\beta|}) = \prod_{n \geq 1} (1 + z^n)^{B_n} \\ &= \exp\Big(\sum_{n \geq 1} B_n \log(1 + z^n)\Big) = \exp\Big(\sum_{n \geq 1} B_n \sum_{k \geq 1} \frac{(-1)^{k-1} z^{nk}}{k}\Big) \\ &= \exp\Big(\frac{B(z)}{1} - \frac{B(z^2)}{2} + \frac{B(z^3)}{3} - \cdots\Big). \end{aligned}$$

The operations with infinite series are justified, since, for given $n$, only a finite number of $B_j$'s contribute to $A_n$.

**Multiset.** This is very similar to forming sets, but now repeated elements are allowed, leading to multisets. The computation is similar:

$$\begin{aligned} A(z) &= \prod_{n \geq 1} (1 + z^n + z^{2n} + z^{3n} + \cdots)^{B_n} \\ &= \exp\Big(\sum_{n \geq 1} B_n \log \frac{1}{1 - z^n}\Big) = \exp\Big(\sum_{n \geq 1} B_n \sum_{k \geq 1} \frac{z^{nk}}{k}\Big) \\ &= \exp\Big(\frac{B(z)}{1} + \frac{B(z^2)}{2} + \frac{B(z^3)}{3} + \cdots\Big). \end{aligned}$$

**Cycles.** For a given $\mathscr{B}$, we form cycles of elements taken from $\mathscr{B}$, where again $B_0 = 0$. A cycle is $(b_1, \ldots, b_n)$ with $b_i \in \mathscr{B}$. It is identified with all cyclic rotations. So, for example, $(a,b,a,b) = (b,a,b,a)$, but $(a,a,b,b)$ is a different cycle. Again, the size of a cycle is the sum of the sizes of its elements. Then, for the associated generating functions,

$$A(z) = \sum_{k \geq 1} \frac{\phi(k)}{k} \log \frac{1}{1 - B(z^k)},$$

where $\phi(k)$ is Euler's totient function; in other words, the number of $i$'s less than $k$ that are relatively prime to $k$.

The proof of this relation will not be given; it is usually done using Polya's enumeration theory (enumeration under group action, here just the cyclic group).

Constructions under restrictions are also considered, for example $\text{SEQ}_k(\mathscr{B})$, $\text{SEQ}_{\geq k}(\mathscr{B})$ of sequences with exactly $k$ or $\geq k$ elements, and various others.

Further constructions will be introduced in this text when they occur.

Now we turn to a few examples.

**Compositions.** A composition of a positive integer $n$ is a representation $n = i_1 + \cdots + i_k$ with positive integers $i_j$; the number $k$ is referred to as the number of parts. We can interpret the integers $\mathscr{I} = \{1,2,\dots\} = \{\bullet, \bullet\bullet, \bullet\bullet\bullet, \dots\} = \mathrm{SEQ}_{\geq 1}\{\bullet\}$. The size of integer $i \cong \bullet^i$ is just $i$, and so

$$I(z) = \frac{z}{1-z}.$$

Further, compositions are described by $\mathscr{C} = \mathrm{SEQ}_{\geq 1}(\mathscr{I})$, whence

$$C(z) = \frac{\frac{z}{1-z}}{1 - \frac{z}{1-z}} = \frac{z}{1-2z},$$

and $I_n$, the number of compositions of $n$, is given by $I_n = 2^{n-1}$, which is also easy to see directly.

**Partitions.** They are defined like compositions, except that the order of the terms ("parts") is irrelevant. They can be seen as multisets of $\mathscr{I}$; the multiset construction then gives us

$$P(z) = \exp\left(I(z) + \frac{I(z^2)}{2} + \frac{I(z^3)}{3} + \cdots\right).$$

This form is, however, not very useful; the more natural way to write this is

$$P(z) = \prod_{n \geq 1} \frac{1}{1 - z^n}.$$

On expanding the product, a typical term is $z^{i_1 + 2i_2 + 3i_3 + \cdots}$, which just describes a partition ($i_1$ ones, $i_2$ twos, $i_3$ threes, etc.) The size is the sum of the parts. There will be a separate section providing the very basic elements of the extremely rich and useful theory of partitions.

**Some families of trees.** The class of binary trees $\mathscr{B}$ is either an external node or a root (an internal node) followed by a left and a right subtree, both again binary trees. This recursive definition can be stated as $\mathscr{B} = \square + \circ \cdot \mathscr{B} \cdot \mathscr{B}$. External nodes are not counted when one speaks about size, and sometimes not drawn. The equation

$$B(z) = 1 + zB^2(z)$$

is immediate, leading to

$$B(z) = \frac{1 - \sqrt{1-4z}}{2z} = \sum_{n \geq 0} \frac{1}{n+1}\binom{2n}{n} z^n,$$

so that binary trees with $n$ (internal) nodes are enumerated by Catalan numbers $\frac{1}{n+1}\binom{2n}{n}$.

The generalization to $t$-ary trees is immediate: The root has an ordered list of $t$ subtrees, and $t = 2$ means binary trees.

Another important family is $\mathscr{P}$, the family of planar trees. They are known under several names: planted plane trees, plane trees, ordered trees, ... There is a root node and a sequence of $r$ planar subtrees ($r \geq 0$). Thus $\mathscr{P} = \circ + \circ \cdot \mathscr{P} + \circ \cdot \mathscr{P} \cdot \mathscr{P} + \circ \cdot \mathscr{P} \cdot \mathscr{P} \cdot \mathscr{P} + \cdots$ Consequently

$$P(z) = zP(z) + zP^2(z) + zP^3(z) + \cdots = \frac{z}{1 - P(z)}$$
$$= \frac{1 - \sqrt{1 - 4z}}{2} = \sum_{n \geq 1} \frac{1}{n} \binom{2n-2}{n-1} z^n.$$

There are two standard bijections: Between planar trees and binary trees the rotation correspondence. Take a planar tree, let only the leftmost edge survive, connect siblings instead, cut off the root and turn the tree by 45° to obtain a binary tree with one node less. This is reversible. This construction is described in more detail in many textbooks. Figure 2.1 shows an example.

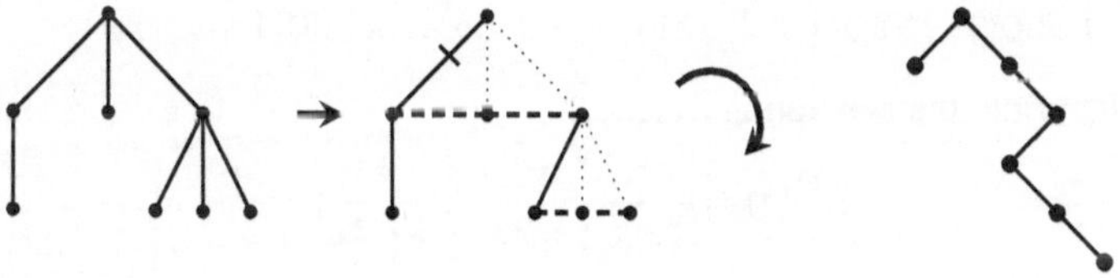

**Figure 2.1**
A planar tree with eight nodes (seven edges) and the corresponding binary tree with seven internal nodes.

Planar trees are also in bijection with non-negative lattice paths (Dyck paths), which are described a little later in this text. See Figure 2.2 for an example.

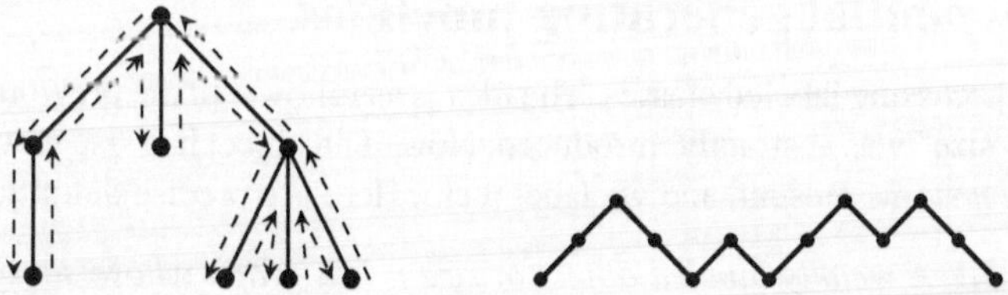

**Figure 2.2**
A planar tree with eight nodes (seven edges) and the corresponding Dyck path of length 12 (=semi-length 6).

As this example shows, one just walks around the tree and records the steps (up or down) in a diagram.

Much more about these and other families of trees can be found in this handbook in various chapters that specialize on trees.

**Set partitions.** A partition of the set $\{1,2,\dots,n\}$ into $k$ blocks consists of $k$ nonempty subsets that are mutally disjoint and whose union is the full set. The number of them is denoted by ${n \brace k}$, a Stirling subset number (in the older literature Stirling numbers of the second kind). In order to describe these set partitions symbolically, we call the smallest element in each block the leader. Then we order the blocks according to the block leaders in ascending order. In this way, it makes sense to talk about block 1, block 2, ..., block $k$. And now we write a string $b_1b_2\dots b_n$, where $b_i$ is the number of the block in which $i$ lies (its address). Such a string has the properties that $b_1 = 1$, and if a new number appears for the first time (scanning the string from left to right), it is one higher than the previous highest number, and altogether all the numbers $1,\dots,k$ appear. Example. The string 112122313241 codes the set partition $\{1,2,4,8,12\},\{3,5,6,10\},\{7,9\},\{11\}$. The set of admissible strings admits the following representation:

$$1\,\mathrm{SEQ}(1)\,2\,\mathrm{SEQ}(1+2)\,3\,\mathrm{SEQ}(1+2+3)\dots k\,\mathrm{SEQ}(1+\dots+k).$$

This translates into the generating function

$$S^{(k)}(z) = \frac{z}{1-z}\frac{z}{1-2z}\cdots\frac{z}{1-kz};$$

partial fraction decomposition gives

$$S^{(k)}(z) = \frac{1}{k!}\sum_{j=0}^{k}\binom{k}{j}(-1)^{k-j}\frac{1}{1-jz}, \text{ so } {n \brace k} = \frac{1}{k!}\sum_{j=1}^{k}\binom{k}{j}(-1)^{k-j}j^n.$$

## 2.3 Combinatorial constructions and associated exponential generating functions

We are now discussing labeled classes. The idea is as follows. In the previous section, the function "size" was abstractly introduced. Now, if an object has size $n$, we assume that there are $n$ atoms present, and we label them. Here is a precise definition.

**Definition 2.3.1** *A weakly labeled object of size n is a graph whose set of vertices is a subset of the integers. Equivalently, we say that the vertices bear labels, with the implied condition that labels are distinct integers from $\mathbb{Z}$. An object of size n is said to be well-labeled, or, simply, labeled, if it is weakly labeled and, in addition, its collection of labels is the complete integer interval $[1..n]$. A labeled class is a combinatorial class that consists of well-labeled objects.*

We will also use reduction of labels: For a weakly labeled structure of size $n$, this operation reduces its labels to the standard interval $[1..n]$ while preserving the relative order of labels. For instance, the sequence $\langle 7,3,9,2\rangle$ reduces to $\langle 3,2,4,1\rangle$. We use $\rho(\alpha)$ to denote this canonical reduction of the structure $\alpha$.

In order to count labeled objects, we appeal to exponential generating functions. The exponential generating function of a sequence $(A_n)$ is the formal power series

$$A(z) = \sum_{n\geq 0} A_n \frac{z^n}{n!}.$$

The exponential generating function of a class $\mathscr{A}$ is the exponential generating function of the numbers $A_n = \text{card}(\mathscr{A}_n)$. It is also said that the variable $z$ marks the size in the generating function. With the standard notation for coefficients of series, the coefficient $A_n$ in an exponential generating function is then recovered by

$$A_n = n![z^n]A(z),$$

since $[z^n]A(z) = A_n/n!$.

**Neutral and atomic classes.** It proves useful to introduce a neutral (empty, null) object $\varepsilon$ that has size 0 and bears no label at all, and consider it as a special labeled object; a neutral class $\mathscr{E}$ is then by definition $\mathscr{E} = \varepsilon$ and is also denoted by 1. The (labeled) atomic class $\mathscr{Z} = ①$ is formed of a unique object of size 1 that, being well-labeled, bears the integer label ①. The exponential generating functions of the neutral class and the atomic class are, respectively, $E(z) = 1$, $Z(z) = z$.

**Labeled product.** The labeled product of $\mathscr{B}$ and $\mathscr{C}$, denoted $\mathscr{B} \star \mathscr{C}$, is obtained by forming ordered pairs from $\mathscr{B} \times \mathscr{C}$ and performing all possible order-consistent relabelings. When $\mathscr{A} = \mathscr{B} \star \mathscr{C}$, the corresponding counting sequences satisfy

$$A_n = \sum_{k=0}^{n} \binom{n}{k} B_k C_{n-k};$$

the binomial coefficients count the relabelings. The new object has size $k + (n-k) = n$; $k$ of the numbers $\{1,\ldots,n\}$ are selected for the labels of the first component, $n-k$ for the second. But this is just the way exponential generating functions are multiplied:

$$\sum_{n\geq 0} A_n \frac{z^n}{n!} = \sum_{n\geq 0} B_n \frac{z^n}{n!} \cdot \sum_{n\geq 0} C_n \frac{z^n}{n!}.$$

**Sequences.** General sequences and sequences with $k$ factors can be formed as before. Here and in the following we assume again that $\varepsilon \notin \mathscr{B}$.

$$\text{SEQ}(\mathscr{B}) = \{\varepsilon\} + \mathscr{B} + \mathscr{B} \star \mathscr{B} + \mathscr{B} \star \mathscr{B} \star \mathscr{B} + \cdots = \bigcup_{k\geq 0} \text{SEQ}_k(\mathscr{B}).$$

The translations into exponential generating functions are

$$A(z) = B(z)^k \quad \text{and} \quad A(z) = \frac{1}{1-B(z)},$$

respectively.

**Sets.** We denote by $\text{SET}_k(\mathscr{B})$ the class of $k$-sets formed from $\mathscr{B}$. The set class is defined formally as the quotient $\text{SET}_k(\mathscr{B}) := \text{SEQ}_k(\mathscr{B})/R$, where the equivalence relation $R$ identifies two sequences when the components of one are a permutation of the components of the other. A "set" is like a sequence, but the order between components is immaterial. The (labeled) set construction applied to $\mathscr{B}$, denoted $\text{SET}(\mathscr{B})$, is then defined by

$$\text{SET}(\mathscr{B}) = \bigcup_{k\geq 0} \text{SET}_k(\mathscr{B}).$$

The translations into exponential generating functions are

$$A(z) = \frac{B(z)^k}{k!} \quad \text{and} \quad A(z) = \sum_{k\geq 0} \frac{B(z)^k}{k!} = \exp(B(z)),$$

respectively.

**Cycles.** We start with $k$-cycles. The class of $k$-cycles, $\text{CYC}_k(\mathscr{B})$ is formally defined to be the quotient $\text{CYC}_k(\mathscr{B}) := \text{SET}_k(\mathscr{B})/S$, where the equivalence relation $S$ identifies two sequences when the components are one cyclic permutation of the components of each other. A cycle is like a sequence whose components can be cyclically shifted, so that there is now a uniform $k$-to-one correspondence between $k$-sequences and $k$-cycles. We assume that $\mathscr{B} \neq \emptyset$ and $k \geq 1$. Then

$$\mathscr{A} = \text{CYC}_k(\mathscr{B}) \implies A(z) = \frac{1}{k} B(z)^k,$$

$$\mathscr{A} = \text{CYC}(\mathscr{B}) \implies A(z) = \sum_{k\geq 1} \frac{1}{k} B(z)^k = \log \frac{1}{1 - B(z)}.$$

In the sequel we describe a few important combinatorial objects as families of labeled objects.

**Surjections.** Fix some integer $r \geq 1$ and let $\mathscr{R}_n^{(r)}$ denote the class of all surjections from the set $\{1,\ldots,n\}$ onto $\{1,\ldots,r\}$ whose elements are called $r$-surjections. We set $\mathscr{R}^{(r)} = \bigcup_{n\geq 1} \mathscr{R}_n^{(r)}$ and compute the corresponding exponential generating function, $R^{(r)}(z)$. We observe that an $r$-surjection $\phi \in \mathscr{R}_n^{(r)}$ is determined by the ordered $r$-tuple formed from the collection of all preimage sets, $(\phi^{-1}(1), \phi^{-1}(2), \ldots, \phi^{-1}(r))$; they are disjoint non-empty sets of integers that cover the interval $[1..n]$. One has the combinatorial specification

$$\mathscr{R}^{(r)} = \text{SEQ}_r(\mathscr{V}), \quad \mathscr{V} = \text{SET}_{\geq 1}(\mathscr{Z}),$$

(a surjection is a sequence of non-empty sets), from which we conclude $R^{(r)}(z) = (e^z - 1)^r$. From this, we find also

$$\begin{aligned}\mathscr{R}_n^{(r)} &= n![z^n](e^z - 1)^r = n![z^n] \sum_{j=0}^{r} \binom{r}{j} (-1)^{r-j} e^{jz} \\ &= \sum_{j=0}^{r} \binom{r}{j} (-1)^{r-j} j^n = r! \left\{ {n \atop r} \right\}.\end{aligned}$$

**Set partitions into $r$ blocks.** Let $\mathcal{S}_n^{(r)}$ denote the number of ways of partitioning the set $\{1,\dots,n\}$ into $r$ disjoint and non-empty equivalence classes (blocks). We set $\mathcal{S}^{(r)} = \bigcup_{n\geq 1} \mathcal{S}_n^{(r)}$; the corresponding objects are called set partitions, as defined earlier. The enumeration problem for set partitions is closely related to that of surjections:

$$\mathcal{S}^{(r)} = \textsc{Set}_r(\mathcal{V}), \quad \mathcal{V} = \textsc{Set}_{\geq 1}(\mathcal{Z}) \quad \Longrightarrow \quad S^{(r)}(z) = \frac{(e^z-1)^r}{r!}.$$

Thus we find again the formula for the Stirling set partition numbers:

$$\left\{ n \atop r \right\} = \frac{1}{r!}\sum_{j=1}^{r}\binom{r}{j}(-1)^{r-j}j^n.$$

If we want to consider **all** surjections (respectively, set partitions) instead of only those that consist of $r$ blocks, then we simply have to sum over $r$. This leads to

$$R(z) = \sum_{r\geq 0}(e^z-1)^r = \frac{1}{2-e^z}$$

and

$$S(z) = \sum_{r\geq 0}\frac{(e^z-1)^r}{r!} = e^{e^z-1}.$$

The numbers $R_n = n![z^n]R(z)$ and $S_n = n![z^n]S(z)$ are called surjection numbers and Bell numbers, respectively. Clearly,

$$R_n = \sum_{r\geq 0} r!\left\{ n \atop r \right\} \quad \text{and} \quad S_n = \sum_{r\geq 0}\left\{ n \atop r \right\}.$$

We have

$$R(z) = \frac{1}{2}\frac{1}{1-\frac{e^z}{2}} = \sum_{l\geq 0}\frac{e^{lz}}{2^{l+1}},$$

therefore

$$R_n = \sum_{l\geq 0}\frac{l^n}{2^{l+1}}.$$

Similarly,

$$S(z) = \frac{1}{e}e^{e^z} = \frac{1}{e}\sum_{j\geq 0}\frac{e^{jz}}{j!},$$

whence

$$S_n = n![z^n]\frac{1}{e}\sum_{j\geq 0}\frac{e^{jz}}{j!} = \frac{1}{e}\sum_{j\geq 0}\frac{j^n}{j!}.$$

This is known as Dobinski's formula.

The present approach is also flexible with respect to restrictions. For example, $\exp(e_b(z)-1)$, with the truncated exponential series

$$e_b(z) := 1+z+\frac{z^2}{2!}+\cdots+\frac{z^b}{b!}$$

corresponds to partitions with all blocks of size $\leq b$, $e^{e^z-1-z}$ corresponds to partitions with no singletons, and $\cosh(e^z-1)$ to partitions with an even number of blocks.

**Restricted words and random allocation.** Consider an alphabet with $r$ letters, say, $\{a_1,\ldots,a_r\}$. For a word of length $n$, the sequence {set of indices of letter $a_1$}, …, {set of indices of letter $a_r$} is forming an "ordered" partition of the sets of labels $\{1,\ldots,n\}$; without restrictions, this yields the exponential generating function $(e^z)^r$.

Now we want to determine the exponential generating function of all words where all letters appear at least $b$ times. For that, we again use the truncated exponential series $e_b(z)$. Then we get $(e^z-e_{b-1}(z))^r$ as an answer. Observe that this is clear if the alphabet has just one letter, and the concept of exponential generating functions automatically takes care of the mixing of letters, whence the $r$th power. Variations of this also work. For instance, if we require that all letters appear at most $b$ times, then we get $e_b(z)^r$, and all kinds of restrictions can be handled.

Now we consider a balls-in-bins model. Throw at random $n$ distinguishable balls into $m$ distinguishable bins. We might think of the balls numbered from 1 to $n$. Each bin corresponds to one of $m$ letters, and each realization of the experiment is coded by a word of length $n$. Let MIN and MAX represent the size of the least filled and most filled bins, respectively. Then

$$\mathbb{P}(\text{MAX} \leq b) = \frac{n!}{m^n}[z^n]e_b(z)^m = n![z^n]e_b(\tfrac{z}{m})^m$$

and

$$\mathbb{P}(\text{MIN} > b) = n![z^n]\left(e^{z/m}-e_b(\tfrac{z}{m})\right)^m.$$

**Birthday paradox.** This is a classical example: Assume that there is a line of persons entering a large room one by one. Each person is let in and declares her birthday upon entering the room. How many people must enter in order to find two that have the same birthday? The birthday paradox is the counterintuitive fact that on average a birthday collision is likely to take place as early as at time $n \approx 24$. Let $B$ be the time of the first collision, which is a random variable ranging between 2 and $r+1$ (where the upper bound is derived from the pigeonhole principle; we assume that the year has $r$ days; $\mathscr{X}$ denotes an alphabet with $r$ letters). A collision has not yet occurred at time $n$, if the sequence of birthdays $\beta_1,\ldots,\beta_n$ has no repetition. In other words, the function $\beta$ from $[1..n]$ to $\mathscr{X}$ must be injective; equivalently, $\beta_1,\ldots,\beta_n$ is an $n$-arrangement of $r$ objects (=$r$ ordered objects). Thus, we have the fundamental relation

$$\mathbb{P}(B>n) = \frac{r(r-1)\ldots(r-n+1)}{r^n} = \frac{n!}{r^n}[z^n](1+z)^r = n![z^n]\left(1+\frac{z}{r}\right)^r.$$

The expectation of the random variable $B$ is

$$\mathbb{E}(B) = \sum_{n\geq 0} \mathbb{P}(B > n) = 1 + \sum_{n=1}^{r} \frac{r(r-1)\ldots(r-n+1)}{r^n}.$$

An alternative form of the expectation is now derived, which easily leads to generalizations. Let $f(z) = \sum_n f_n z^n$ be an entire function with non-negative coefficients. Then

$$\sum_n f_n n! = \sum_n f_n \int_0^\infty e^{-t} t^n dt = \int_0^\infty f(t) e^{-t} dt$$

(a Laplace transform). Therefore

$$\mathbb{E}(B) = \int_0^\infty e^{-t} \left(1 + \frac{t}{r}\right)^r dt.$$

Exactly the same reasoning leads to the following: The expected time necessary for the first occurrence of the event "$b$ persons have the same birthday" has expectation given by the integral

$$\int_0^\infty e^{-t} \left(e_{b-1}\left(\frac{t}{r}\right)\right)^r dt,$$

where the classical case means $b = 2$.

**Coupon collector.** This problem is dual to the birthday paradox. We ask for the first time $C$ when $\beta_1, \ldots, \beta_C$ contains all the elements of $\mathscr{X}$; that is, all the possible birthdays have been "collected." In other words, the event $\{C \leq n\}$ means the equality between sets, $\{\beta_1, \ldots, \beta_n\} = \mathscr{X}$. Thus, the probabilities satisfy

$$\mathbb{P}\{C \leq n\} = \frac{r!\left\{ {n \atop r} \right\}}{r^n} = \frac{n!}{r^n}[z^n](e^z - 1)^r = n![z^n](e^{z/r} - 1)^r.$$

The complementary probabilities are then

$$\mathbb{P}\{C > n\} = n![z^n]\left(e^z - (e^{z/r} - 1)^r\right).$$

In the same style as before we get

$$\mathbb{E}(C) = \int_0^\infty \left(1 - (1 - e^{-t/r})^r\right) dt = r \sum_{j=1}^{r} \binom{r}{j} \frac{(-1)^{j-1}}{j}.$$

Alternatively, we might substitute $v = 1 - e^{-t/r}$, then expand and integrate termwise; this process provides the answer in the form $rH_r$, with harmonic numbers $H_n = 1 + \frac{1}{2} + \cdots + \frac{1}{n}$. More on these numbers will appear later. This answer can also be obtained in an elementary fashion: To get the first copy, you need on average one drawing, to get a second one needs $r/(r-1)$, a third one needs $r/(r-2)$, and so on.

The symbolic approach (leading to an integral) has the advantage of straightforward generalizations. For instance, the expected time till each coupon is obtained $b$ times is

$$\int_0^\infty \left(1 - (1 - e_{b-1}(\tfrac{t}{r}) e^{-t/r})^r\right) dt.$$

**Permutations and cycles.** It is known that a permutation admits a unique decomposition into cycles: Let $\sigma$ be a permutation. Start with any element, say 1, and draw a directed edge from 1 to $\sigma(1)$, then continue connecting to $\sigma^2(1)$, $\sigma^3(1)$, and so on. A cycle containing 1 is obtained after at most $n$ steps. If one repeats the construction, taking at each stage an element not yet connected to earlier ones, the cycle decomposition of the permutation $\sigma$ is obtained. This argument shows that the class of sets-of-cycles is isomorphic to the class of permutations:

$$\mathscr{P} \cong \textsc{Set}(\textsc{Cyc}(\mathscr{Z})) \cong \textsc{Seq}(\mathscr{Z}).$$

This combinatorial isomorphism is reflected by the obvious series identity

$$P(z) = \exp\left(\log \frac{1}{1-z}\right) = \frac{1}{1-z}.$$

The advantage of it is that restrictions are handled in an almost automatic fashion.

The class $\mathscr{P}^{(A,B)}$ of permutations with cycle lengths in $A \subseteq \mathbb{N}$ and with cycle numbers that belongs to $B \subseteq \mathbb{N}_0$ has exponential generating function

$$P^{(A,B)}(z) = \beta(\alpha(z)) \quad \text{with} \quad \alpha(z) = \sum_{a \in A} \frac{z^a}{a}, \quad \beta(z) = \sum_{b \in B} \frac{z^b}{b!}.$$

A popular instance is derangements (fix-point free permutations). The restriction is that no cycles of length one are allowed, therefore

$$\alpha(z) = \sum_{a \geq 2} \frac{z^a}{a} = \log \frac{1}{1-z} - z, \quad \beta(z) = \sum_{b \geq 0} \frac{z^b}{b!} = e^z,$$

leading to

$$\exp\left(\log \frac{1}{1-z} - z\right) = \frac{e^{-z}}{1-z};$$

this produces the number of derangements as

$$n![z^n] \frac{e^{-z}}{1-z} = n! \sum_{k=0}^{n} \frac{(-1)^k}{k!}.$$

Notice that the probability that a random permutation of $n$ elements has no fix-points is very close (for large $n$) to $\frac{1}{e}$, which is a popular result.

**Stirling cycle numbers.** The class $\mathscr{P}^{(r)}$ of permutations that decompose into $r$ cycles can be represented as

$$\mathscr{P}^{(r)} = \textsc{Set}_r(\textsc{Cyc}(\mathscr{Z})),$$

which leads to

$$P^{(r)}(z) = \frac{1}{r!}\left(\log \frac{1}{1-z}\right)^r.$$

Therefore we get

$$P_n^{(r)} = \begin{bmatrix} n \\ r \end{bmatrix} = \frac{n!}{r!}[z^n]\left(\log\frac{1}{1-z}\right)^r.$$

These numbers are called Stirling cycle numbers; in the older literature, they are often called sign-less Stirling numbers of the first kind. This is a somewhat strange name, so the new name should be favored as it makes much more sense.

## 2.4 Partitions and $q$-series

Partitions have already appeared briefly before. Here, we want to describe them in more detail.

A partition of a positive integer $n$ is a representation $n = i_1 + i_2 + \cdots + i_k$ with integers $1 \le i_1 \le i_2 \le \cdots$. The $i_j$'s are called parts and $k$ is the number of parts. So partitions can be described by the formal expression

$$\mathbf{1}^*\mathbf{2}^*\mathbf{3}^*\ldots,$$

since $n = 1 \cdot j_1 + 2 \cdot j_2 + \cdots$ with integers $j_s \ge 0$. If we denote $p(n)$ the number of partitions of $n$ and

$$P(q) = 1 + \sum_{n\ge1} p(n)q^n$$

the (ordinary) generating function of partitions, then we get immediately from the formal expression that

$$P(q) = \prod_{i\ge1}\frac{1}{1-q^i}.$$

Recall that the star $*$ is a handy alternative for the construction SEQ introduced earlier. Note [1] that in the context of partitions it is customary to use the variable $q$ instead of $z$ in generating functions. Andrews' encyclopedic book contains all of this, and much more. We also introduce $p(0) = 1$ to have smoother expressions. Let us fix some notation (we assume $|q| < 1$):

$$(q)_n = (q;q)_n = (1-q)(1-q^2)\ldots(1-q^n),$$
$$(q)_\infty = (q;q)_\infty = (1-q)(1-q^2)\ldots,$$

and, more generally,

$$(x)_n = (x;q)_n = (1-x)(1-xq)\ldots(1-xq^{n-1}),$$
$$(x)_\infty = (x;q)_\infty = (1-x)(1-xq)\ldots.$$

So we have $P(q) = 1/(q)_\infty$, and also the number of partitions of $n$ into $k$ parts is given as

$$[q^nt^k]\prod_{i\ge1}\frac{1}{1-tq^i} = [q^nt^k]\frac{1}{(tq;q)_\infty}.$$

There is a graphical representation of a partition, called a Ferrers diagram. One simply codes a part $k$ as a row of $k$ unit squares and arranges them in decreasing order. Upon reflecting the diagram at the diagonal, we get another partition, called the conjugate partition. See Figure 2.3 for an example.

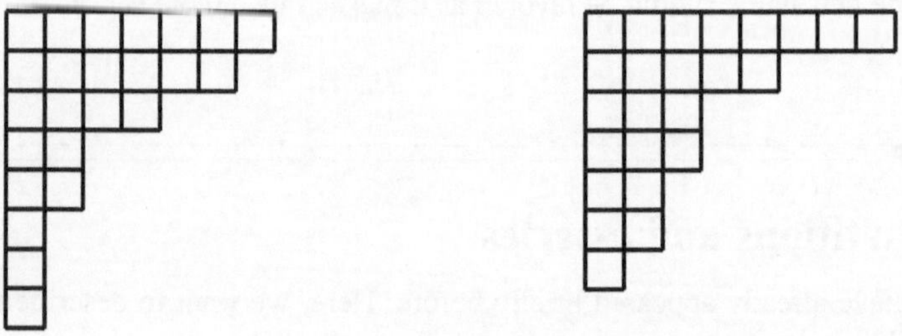

**Figure 2.3**
The partition $(7,6,4,2,2,1,1,1)$ and its conjugate $(8,5,3,3,2,2,1)$.

The number of partitions of $n$ where the parts are of size $\leq k$ is

$$[q^n]\frac{1}{(q)_k},$$

and using the concept of conjugate partitions, this number also equals the number of partitions of $n$ where the number of parts are $\leq k$.

The following theorem is very basic for the manipulation of "$q$-series."

**Theorem 2.4.1** *We have*

$$F(t) = \frac{(at)_\infty}{(t)_\infty} = \sum_{n\geq 0} \frac{(a)_n}{(q)_n} t^n.$$

This theorem is attributed to Cauchy in [1] and often called the $q$-binomial theorem.

*Proof.* Splitting off the first factors in the product, we get

$$F(t) = \frac{1-at}{1-t} F(qt), \quad \text{or} \quad (1-t)F(t) = (1-at)F(qt).$$

Writing $F(t) = \sum_{n\geq 0} A_n t^n$ and comparing coefficients of $t^n$, we get

$$A_n - A_{n-1} = q^n A_n - aq^{n-1} A_{n-1}, \quad \text{or} \quad A_n = \frac{1-aq^{n-1}}{1-q^n} A_{n-1}.$$

Since $A_0 = F(0) = 1$, iteration of this recursion results in

$$A_n = \frac{(a)_n}{(q)_n},$$

as claimed. ∎

The first important special case arises from setting $a = 0$:

$$\frac{1}{(t)_\infty} = \prod_{n\geq 0} \frac{1}{1-tq^n} = \sum_{n\geq 0} \frac{t^n}{(q)_n}.$$

It is called Euler's partition identity, because replacing $t$ by $tq$ results in

$$\frac{1}{(qt)_\infty} = \sum_{n\geq 0} \frac{q^n t^n}{(q)_n},$$

which is the generating function of partitions of $n$ where the number of parts is labeled by $t$. Thus, comparing coefficients, we find that

$$\frac{q^k}{(q)_k}$$

is the generating function of partitions with $k$ parts. The other important special case is obtained by replacing $a$ by $a/b$ and $t$ by $bt$. Then

$$(a/b)_n (bt)^n = (b-a)(b-aq)\dots(b-aq^{n-1})t^n.$$

Setting now $a = -1$ and $b = 0$ results in

$$(-t)_\infty = \prod_{n\geq 0}(1+tq^n) = \sum_{n\geq 0} \frac{q^{\binom{n}{2}}}{(q)_n} t^n.$$

This is also called Euler's partition identity. For this we consider the number of partitions into distinct parts $p_{\mathscr{D}}(n)$. Since each number $i$ can be a part either 0 or 1 times, the formal equation

$$(\varepsilon + \mathbf{1})(\varepsilon + \mathbf{2})(\varepsilon + \mathbf{3})\dots$$

describes these objects. Thus

$$\sum_{n\geq 0} p_{\mathscr{D}}(n) q^n = \prod_{k\geq 1}(1+q^k) = (-q)_\infty,$$

and $(-qt)_\infty$ is the generating function of partitions of $n$ into distinct parts and $k$ parts. Now we replace $t$ by $tq$ in Euler's partition identity:

$$(-tq)_\infty = \prod_{n\geq 1}(1+tq^n) = \sum_{n\geq 0} \frac{q^{\binom{n}{2}+n}}{(q)_n} t^n.$$

Comparing coefficients of $t^k$, we find that the generating function of partitions into distinct parts and $k$ parts is given by

$$\frac{q^{\binom{k+1}{2}}}{(q)_k}.$$

Euler's partition identities appear frequently in Analytic Combinatorics and Analysis of Algorithms.

Now we want to compute the generating function of partitions where the number of parts is bounded by $M$ and the parts are bounded by $N$. There is only a finite number of possibilities, whence this generating function is actually a polynomial, call it $G(M,N)$. Now $G(M,N)-G(M-1,N)$ is the generating function where the number of parts is exactly equal to $M$; removing one from each part shows that this equals $q^M G(M,N-1)$. Together with $G(0,N)=G(M,0)=1$, the solution of

$$G(M,N)-G(M-1,N)=q^M G(M,N-1)$$

is given by

$$G(M,N)=\frac{(q)_{M+N}}{(q)_M(q)_N},$$

which is called a Gaussian $q$-binomial coefficient $\begin{bmatrix} M+N \\ M \end{bmatrix}$.

Once again, we get

$$G(M,\infty)=G(\infty,M)=\frac{1}{(q)_M}.$$

Here is another technique of interest, nicknamed "adding a new slice" by Flajolet. Define

$$F(q,u)=\sum_{n\geq 1}\sum_{i\geq 1}[\text{number of partitions of } n \text{ with last part } i]q^n u^i.$$

Now, to create a new slice, i.e., a new part $j\geq i$, means to replace $u^i$ by

$$\sum_{j\geq i}(qu)^j=\frac{(qu)^i}{1-qu}.$$

So, taking the partitions with just one part separately into account,

$$F(q,u)=\frac{qu}{1-qu}+\frac{1}{1-qu}F(q,qu).$$

This recursion can be iterated:

$$\begin{aligned} F(q,u)&=\frac{qu}{1-qu}+\frac{1}{1-qu}\left[\frac{q^2u}{1-q^2u}+\frac{1}{1-q^2u}\left[\frac{q^3u}{1-q^3u}+\frac{1}{1-q^3u}\left[\cdots\right.\right.\right. \\ &=\sum_{k\geq 1}\frac{uq^k}{(qu)_k}. \end{aligned}$$

Forgetting what the last part is means setting $u=1$; adding 1 for the empty partition results in

$$\sum_{k\geq 0}\frac{q^k}{(q)_k}=\frac{1}{(q)_\infty},$$

which follows directly from Euler's partition identity by setting $t = 1$. This describes the generating function of all partitions in terms of those with exactly $k$ parts, as discussed earlier.

In various applications (see for instance [22]) it is important to expand

$$Q(x) = (qx)_\infty = \prod_{k\geq 1}(1 - xq^k)$$

around $x = 1$, viz.

$$Q(x) = Q(1) + Q'(1)(x-1) + \frac{Q''(1)}{2}(x-1)^2 + \cdots.$$

We have

$$Q'(x) = Q(x)\sum_{k\geq 1}\frac{-q^k}{1-xq^k}, \quad Q''(x) = 2Q(x)\sum_{1\leq j<k}\frac{q^{j+k}}{(1-xq^j)(1-xq^k)}$$

and therefore

$$\frac{Q'(1)}{Q(1)} = -\sum_{k\geq 1}\frac{1}{q^{-k}-1}, \quad \frac{Q''(1)}{Q(1)} = \left(\sum_{k\geq 1}\frac{1}{q^{-k}-1}\right)^2 - \sum_{k\geq 1}\left(\frac{1}{q^{-k}-1}\right)^2.$$

## 2.5 Some applications of the adding a slice technique

A restricted composition of a natural number $n$ in the sense of Carlitz [6] (a **Carlitz composition**) is defined to be a composition

$$n = a_1 + a_2 + \cdots + a_k \quad \text{such that} \quad a_i \neq a_{i+1} \text{ for } i = 1, \ldots, k-1.$$

We refer to $n$ as the size and to $k$ as the number of parts of the composition.

Observe that there are $2^{n-1}$ unrestricted compositions of the integer $n$ with generating function $z/(1-2z)$.

Let $c(n)$ denote the number of Carlitz compositions of $n$. In [6], Carlitz found the generating function

$$C(z) := \sum_{n\geq 0} c(n)z^n.$$

We will rederive this here with the method called "adding a new slice." This appears in [30] and is also described in the book [24].

We proceed from a Carlitz composition with $k$ parts to one with $k+1$ parts by allowing $a_{k+1}$ to be any number and then subtracting the forbidden case $a_{k+1} = a_k$. In terms of generating functions this reads as follows. Let $f_k(z,u)$ be the generating function of those Carlitz compositions with $k$ parts where the coefficient of $z^n u^j$ refers to size $n$ and last part $a_k = j$. Then

$$f_{k+1}(z,u) = f_k(z,1)\frac{zu}{1-zu} - f_k(z,zu) + [k=0] \quad \text{for } k \geq 0, \; f_0(z,u) = 1.$$

The first term means that we forget the labeling of the last part ($u := 1$) and add any term, together with a labeling by $u$, and the second one means that we subtract the forbidden term, which is a repetition of the previous last part. Introducing $F(z,u) := \sum_{k\geq 1} f_k(z,u)$ and summing on $k \geq 0$, we get

$$F(z,u) = F(z,1)\frac{zu}{1-zu} + \frac{zu}{1-zu} - F(z,zu).$$

This functional equation can now be iterated:

$$F(z,u) = (1+F(z,1))\frac{zu}{1-zu} - (1+F(z,1))\frac{z^2u}{1-z^2u} + (1+F(z,1))\frac{z^3u}{1-z^3u} - + \cdots;$$

now setting $u := 1$ and abbreviating

$$\sigma(z) = \sum_{j\geq 1} \frac{z^j(-1)^{j-1}}{1-z^j},$$

we get

$$F(z,1) = \sigma(z) + F(z,1)\,\sigma(z).$$

Since $C(z) = 1+F(z,1)$, we find the formula of Carlitz,

$$C(z) = \frac{1}{1-\sigma(z)}.$$

The next example is about **level number sequences of trees** [20]. However, the enumeration of these is equivalent to many other objects; the paper [12] describes the somewhat erratic history. The objects are sequences $(a_1, a_2, \ldots, a_k)$ with positive integers $a_i$ such that always $1 \leq a_{i+1} \leq 2a_i$. The number $k$ is arbitrary, and $n := a_1 + \cdots + a_k$. The starting value is $a_1 = 1$. The interest is in the number $H_n$ of sequences satisfying these conditions. Let $H_{n,j}^{[k]}$ be the number of such sequences relative to a fixed number $k$, and let the last element be fixed as well: $a_k = j$. Furthermore, set

$$H^{[k]}(q,u) = \sum_{n,j\geq 1} H_{n,j}^{[k]} q^n u^j \quad \text{and}$$

$$H(q,u) = \sum_{k\geq 1} H^{[k]}(q,u).$$

Now we describe how $H^{[k+1]}(q,u)$ can be obtained from $H^{[k]}(q,u)$. This is done using the substitution

$$u^j \to (uq) + (uq)^2 + \cdots + (uq)^{2j} = \frac{uq}{1-uq}(1-(uq)^{2j}).$$

This means

$$H^{[k+1]}(q,u) = \frac{uq}{1-uq}\left[H^{[k]}(q,1) - H^{[k]}(q,u^2q^2)\right]$$

and by summing on $k$ (noticing that $H^{[1]}(q,u) = uq$),

$$H(q,u) = qu + \frac{uq}{1-uq}\big[H(q,1) - H(q,u^2q^2)\big].$$

This can again be iterated: set $G(u) = qu + \frac{uq}{1-uq}H(q,1)$, then

$$H(q,u) = G(u) - \frac{uq}{1-uq}\Big[G(u^2q^2) - \frac{u^2q^3}{1-u^2q^3}\Big[G(u^4q^6) - \frac{u^4q^7}{1-u^4q^7}\Big[G(u^8q^{14}) - \cdots;$$

now it is possible to set $u = 1$, which means that we do not care about the value of the last element in the sequence anymore, and get

$$H(q,1) = G(1) - \frac{q}{1-q}\Big[G(q^2) - \frac{q^3}{1-q^3}\Big[G(q^6) - \frac{q^7}{1-q^7}\Big[G(q^{14}) - \cdots.$$

Now this equation can be solved for $H(q,1)$, with the answer

$$H(q,1) = \frac{\sum_{j\ge1} \frac{(-1)^{j-1}q^{2^{j+1}-j-2}}{(1-q)(1-q^3)\dots(1-q^{2^{j-1}-1})}}{\sum_{j\ge0} \frac{(-1)^j q^{2^{j+1}-j-2}}{(1-q)(1-q^3)\dots(1-q^{2^j-1})}}.$$

The last example is about **words** $a_1a_2\dots a_{2n+1}$ of odd length where a letter $k \in \mathbb{N}$ is weighted by a geometric probability $pq^{k-1}$ ($p+q=1$), i. e., $\mathbf{P}\{a_j = k\} = pq^{k-1}$, $k \ge 1$, and the letters **obey the pattern** $a_1 \ge a_2 \le a_3 \ge a_4 \le \cdots$. Let $T_{2n+1}(u)$ be the generating function such that the coefficient of $u^i$ in it is the mass of correct words and last letter $i$. Then we have for $n \ge 1$

$$T_{2n+1}(u) = \frac{p^2u}{(1-qu)(1-q^2u)}T_{2n-1}(1) - \frac{p^2u}{(1-qu)(1-q^2u)}T_{2n-1}(q^2u),$$
$$T_1(u) = \frac{pu}{1-qu}.$$

Adding a new slice means adding a pair $(k,j)$ with $1 \le k \le i$, $j \ge k$, replacing $u^i$ by 1 and providing the factor $u^j$. But

$$\sum_{k=1}^{i} pq^{k-1} \sum_{j\ge k} pq^{j-1}u^j = \frac{p^2u}{(1-qu)(1-q^2u)} - \frac{p^2u}{(1-qu)(1-q^2u)}(q^2u)^i,$$

which explains the recursion. The starting value is just

$$\sum_{j\ge1} pq^{j-1}u^j = \frac{pu}{1-qu}.$$

We introduce the generating functions

$$F(z,u) = \sum_{n\geq 0} T_{2n+1}(u) z^{2n+1} \quad \text{and} \quad f(z) = F(z,1).$$

Summing up we find

$$F(z,u) = \frac{puz}{1-qu} + \frac{p^2uz^2}{(1-qu)(1-q^2u)} F(z,1) - \frac{p^2uz^2}{(1-qu)(1-q^2u)} F(z,q^2u).$$

Iterating that we find

$$f(z) = \frac{pz}{1-q} + \frac{p^2z^2}{(1-q)(1-q^2)} f(z) - \frac{p^2q^2z^3}{(1-q)(1-q^2)(1-q^3)} - \frac{p^4q^2z^4}{(1-q)(1-q^2)(1-q^3)(1-q^4)} f(z) + \cdots$$

and eventually

$$f(z) = \sum_{n\geq 0} \frac{(-1)^n (pz)^{2n+1}}{(q)_{2n+1}} q^{n(n+1)} \Big/ \sum_{n\geq 0} \frac{(-1)^n (pz)^{2n}}{(q)_{2n}} q^{n(n-1)}.$$

This example was taken from [44]; in this paper it is also explained why the limit for $q \to 1$ of $f(z)$ is the tangent function $\tan z$. It is a classical result that $\tan z$ is the exponential generating function of up-down (or down-up) alternating permutations of odd length.

## 2.6 Lagrange inversion formula

Let $y = x\Phi(y)$, where $\Phi(y)$ is a power series such that $\Phi(0) \neq 0$. It is obvious to expand $x$ as a power series in $y$, but we want just the opposite. This is the celebrated inversion formula of Lagrange. We give three versions of it.

$$[x^n]y = \frac{1}{n}[y^{n-1}](\Phi(y))^n, \qquad (n \geq 1).$$

Slightly more general (for $n \geq 1$, $p \geq 0$):

$$[x^n]y^p = \frac{p}{n}[y^{n-p}](\Phi(y))^n.$$

Even more general (for $n \geq 1$ and a power series $g(y)$):

$$[x^n]g(y) = \frac{1}{n}[y^{n-1}]g'(y)(\Phi(y))^n.$$

*Proof.* We use Cauchy's integral formula. In case this is not justified analytically, it can be done on a purely formal level as explained in [25]. The present approach is also easy to remember. Observe that

$$dx = dy\frac{\Phi(y) - y\Phi'(y)}{\Phi^2(y)}.$$

Now

$$\begin{aligned}
[x^n]y^p &= \frac{1}{2\pi i}\oint \frac{dx}{x^{n+1}}y^p \\
&= \frac{1}{2\pi i}\oint dy\frac{\Phi(y) - y\Phi'(y)}{\Phi^2(y)}\frac{\Phi^{n+1}(y)}{y^{n+1}}y^p \\
&= \frac{1}{2\pi i}\oint dy\Big(\Phi(y) - y\Phi'(y)\Big)\frac{\Phi^{n-1}(y)}{y^{n+1-p}} \\
&= [y^{n-p}]\Big(\Phi(y) - y\Phi'(y)\Big)\Phi^{n-1}(y) \\
&= [y^{n-p}]\Phi^n(y) - [y^{n-p-1}]\Phi'(y)\Phi^{n-1}(y) \\
&= [y^{n-p}]\Phi^n(y) - \frac{1}{n}[y^{n-p-1}](\Phi^n(y))' \\
&= [y^{n-p}]\Phi^n(y) - \frac{n-p}{n}[y^{n-p}]\Phi^n(y) \\
&= \frac{p}{n}[y^{n-p}]\Phi^n(y).
\end{aligned}$$

Now we set

$$g(y) = \sum_{p\geq 1} c_p y^p$$

and get

$$\begin{aligned}
[x^n]g(y) &= \sum_{p\geq 1} c_p[x^n]y^p = \sum_{p\geq 1} c_p\frac{p}{n}[y^{n-p}](\Phi(y))^n \\
&= \frac{1}{n}[y^{n-1}]\sum_{p\geq 1} pc_p y^{p-1}(\Phi(y))^n = \frac{1}{n}[y^{n-1}]g'(y)(\Phi(y))^n.
\end{aligned}$$

■

**Applications.** (The following tree structures have been introduced in Section 2.2.)

***t*-ary trees.** These objects are recursively built from a root and $t$ successors, which are themselves $t$-ary trees. A tree might be empty as well. For $t = 2$, we get the important special case of binary trees. The equation $B = 1 + xB^t$ for the generating function, counting trees according to the number of vertices, is immediate. Set $B = 1 + y$, then $y = x(1+y)^t$, in order to make the Lagrange inversion formula applicable. Then $\Phi(y) = (1+y)^t$, and thus

$$b_n = [x^n]B(x) = [x^n]y(x) = \frac{1}{n}[y^{n-1}](1+y)^{tn} = \frac{1}{n}\binom{tn}{n-1}.$$

**Number of leaves in planar trees.** The recursive description of planar trees immediately translates into a bivariate generating function:

$$G(z,u) = zu + \frac{zG(z,u)}{1-G(z,u)};$$

the variable $u$ counts leaves, $z$ nodes (in planar trees). Note that for $u=1$ and $y = G(z,1)$, we have $y = z\Phi(y)$ with $\Phi(y) = \frac{1}{1-y}$.

$$\begin{aligned} G_{n,k} &= [u^k][z^n]G(z,u) = [u^k]\frac{1}{n}[y^{n-1}]\left(u + \frac{y}{1-y}\right)^n \\ &= \frac{1}{n}[y^{n-1}]\binom{n}{k}\frac{y^{n-k}}{(1-y)^{n-k}} \\ &= \frac{1}{n}\binom{n}{k}[y^{k-1}]\frac{1}{(1-y)^{n-k}} \\ &= \frac{1}{n}\binom{n}{k}\binom{n-2}{k-1}. \end{aligned}$$

These numbers are sometimes called **Narayana** numbers.

**Planar trees according to degree of the root.** Let $P_{n,k}$ be the number of planar trees with $n$ nodes and root degree $k$. Then, again, $y = z\Phi(y)$ with $\Phi(y) = y/(1-y)$ enumerates planar trees. Let us assume that $p \geq 1$ and $n \geq 2$, since $P_{n,0} = [n=1]$. We find

$$P_{n,k} = [z^{n-1}]y^k = \frac{k}{n-1}[y^{n-1-k}](1-y)^{-n+1} = \frac{k}{n-1}\binom{2n-3-k}{n-2}.$$

## 2.7 Lattice path enumeration: the continued fraction theorem

We follow here [24] and [13].

**Definition 2.7.1 (Lattice path)** *A Motzkin path $\upsilon = (U_0, U_1, \ldots, U_n)$ is a sequence of points in the discrete quarter-plane $\mathbb{N}_0 \times \mathbb{N}_0$, such that $U_j = (j, y_j)$ and the jump condition $|y_{j+1} - y_j| \leq 1$ is satisfied. An edge $\langle U_j, U_{j+1}\rangle$ is called an ascent if $y_{j+1} - y_j = 1$, a descent if $y_{j+1} - y_j = -1$, and a level step if $y_{j+1} - y_j = 0$. A path that has no level steps is called a Dyck path. The quantity n is the length of the path, $ini(\upsilon) := y_0$ is the initial altitude, $fin(\upsilon) := y_n$ is the final altitude. A path is called an excursion if both its initial and final altitudes are zero. The extremal quantities $\sup \upsilon := \max_j y_j$ and $\inf \upsilon := \min_j y_j$ are called the height and depth of the path.*

A path can always be encoded by a word with $a$, $b$, $c$ representing ascents, descents, and level steps, respectively. What we call the standard encoding is such a word in which each step $a$, $b$, $c$ is (redundantly) subscripted by the value of the $y$-coordinate of its initial point. For instance,

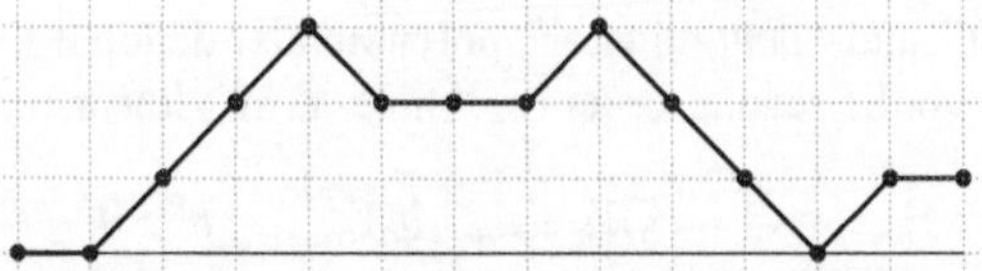

$$c_0\ a_0\ a_1\ a_2\ b_3\ c_2\ c_2\ a_2\ b_3\ b_2\ b_1\ a_0\ c_1$$

encodes a path that connects the initial point $(0,0)$ to the point $(13,1)$.

Let us examine the description of the class written $\mathscr{H}_{0,0}^{[<1]}$ of Motzkin excursions of height $< 1$. We have

$$\mathscr{H}_{0,0}^{[<1]} \cong (c_0)^* \quad \Longrightarrow \quad H_{0,0}^{[<1]} = \frac{1}{1-c_0}.$$

The class of excursions of height $< 2$ is obtained from here by a substitution

$$c_0 \mapsto c_0 + a_0(c_1)^* b_1,$$

whence

$$H_{0,0}^{[<2]} = \cfrac{1}{1-c_0-\cfrac{a_0b_1}{1-c_1}}.$$

Iteration of this simple mechanism yields the finite version of the continued fraction theorem of Flajolet [13].

**Theorem 2.7.2 (Continued fraction theorem, finite version)**

$$H_{0,0}^{[<h]} = \cfrac{1}{1-c_0-\cfrac{a_0b_1}{1-c_1-\cfrac{a_1b_2}{\ddots\,\overline{\,1-c_{h-1}\,}}}} = \frac{P_h}{Q_h}.$$

The unrestricted version leads to an infinite continued fraction.

**Theorem 2.7.3 (Continued fraction theorem, infinite version)**

$$H_{0,0} = \cfrac{1}{1-c_0-\cfrac{a_0b_1}{1-c_1-\cfrac{a_1b_2}{\ddots}}}.$$

Generating functions written in this way are nothing but a concise description of usual counting generating functions: for instance if individual weights $\alpha_j, \beta_j, \gamma_j$ are assigned to the letters $a_j, b_j, c_j$, respectively, then the ordinary generating function of multiplicatively weighted paths with $z$ marking length is obtained by setting $a_j = \alpha_j z$, $b_j = \beta_j z$, $c_j = \gamma_j z$.

The "numerator" and "denominator" polynomials, denoted by $P_h$ and $Q_h$, are defined as solutions to the second-order (or "three-term") linear recurrence equation

$$Y_{h+1} = (1-c_h)Y_h - a_{h-1}b_hY_{h-1}, \quad h \geq 0,$$

together with the initial conditions $(P_{-1}, Q_{-1}) = (-1, 0)$, $(P_0, Q_0) = (0, 1)$, and with the convention $a_{-1} = b_0 = 1$. These recursions are easy to obtain by replacing $1 - c_{n-1}$ by $1 - c_{n-1} - \frac{a_{n-1}b_n}{1-c_n}$ and comparing numerators and denominators separately.

These polynomials are also known as continuant polynomials [26, 48]. For the computation of $H_{0,0}^{[<h]}$ and $P_h$, $Q_h$, one classically introduces the linear fractional transformations

$$g_j(y) = \frac{1}{1 - c_j - a_j b_{j+1} y}$$

so that

$$H_{0,0}^{[<h]} = g_0 \circ g_1 \circ \cdots \circ g_{h-1}(0).$$

Linear fractional transformations are representable by $2 \times 2$ matrices

$$\frac{ay+b}{cy+d} \mapsto \begin{pmatrix} a & b \\ c & d \end{pmatrix}$$

in such a way that the composition corresponds to the matrix product. By induction on the compositions that build up $H_{0,0}^{[<h]}$, there follows the equality

$$g_0 \circ g_1 \circ \cdots \circ g_{h-1}(y) = \frac{P_h - P_{h-1}a_{h-1}b_h y}{Q_h - Q_{h-1}a_{h-1}b_h y}.$$

Eventually, one sets $y := 0$. The polynomials $P_h$ and $Q_h$ both satisfy the recursion for $Y_h$ as just given.

Now we come to applications. In order to count Dyck paths, it is sufficient to substitute

$$a_j \mapsto z, \qquad b_j \mapsto z, \qquad c_j \mapsto 0.$$

Because of the natural bijection (sometimes called the glove bijection), described earlier, a Dyck path of length $2n$ and height $h$ translates into a planar tree with $n+1$ nodes and height $h+1$, so that the results translate directly; compare [10]. In order to avoid misunderstandings, we state explicitly that the height of a planar tree is the length of the longest path from the root to a leaf; the length of a path is counted in terms of the number of nodes on it. This is the original definition used in [10]; sometimes people count the number of edges, which is then one less than what is considered here. The height of a Dyck path is of course the maximal vertical level reached.

The families of polynomials $P_h$, $Q_h$ are in this case determined by a recurrence with constant coefficients. Define the Fibonacci polynomials by the recurrence

$$F_{h+2}(z) = F_{h+1}(z) - zF_h(z), \quad F_0(z) = 0,\ F_1(z) = 1,$$

then it turns out that $Q_h(z) = F_{h+1}(z^2)$ and $P_h(z) = F_h(z^2)$. The Fibonacci polynomials admit an explicit form

$$F_h(z) = \frac{1}{\sqrt{1-4z}}\left[\left(\frac{1+\sqrt{1-4z}}{2}\right)^h - \left(\frac{1-\sqrt{1-4z}}{2}\right)^h\right].$$

If we take the limit $h \to \infty$ in

$$\frac{F_h(z)}{F_{h+1}(z)},$$

then we get

$$D(z) = \frac{1-\sqrt{1-4z}}{2z} = \sum_{n\geq 0} \frac{1}{n+1}\binom{2n}{n} z^n,$$

the generating function of Dyck paths of halflength $n$; the coefficients are Catalan numbers. The limit is taken in the sense of the "discrete topology," but it works as well as the "pointwise" limit for analytic functions. As demonstrated in [10], all our expressions become easier with the substitution

$$z = \frac{u}{(1+u)^2} \quad \Longrightarrow \quad dz = \frac{1-u}{(1+u)^3}du.$$

Then, by solving the second order recursion and substituting,

$$F_h = \frac{1}{1-u}\frac{1-u^h}{(1+u)^{h-1}}, \qquad \frac{F_h}{F_{h+1}} = (1+u)\frac{1-u^h}{1-u^{h+1}}.$$

The limit of the last expression for $h \to \infty$ is $1+u$. The function $D(z) - F_h(z)/F_{h+1}(z) = \frac{1-u^2}{u}\frac{u^{h+1}}{1-u^{h+1}}$ is easier and describes the Dyck paths with height $\geq h$, according to halflength; call it $H^{[\geq h]}(z)$.

The following method to extract coefficients is in [10]; it is the Lagrange inversion formula in disguise and uses the Cauchy integral formula:

$$\begin{aligned}
[z^n]H^{[\geq h-1]}(z) &= \frac{1}{2\pi i}\oint \frac{dz}{z^{n+1}}\frac{1-u^2}{u}\frac{u^h}{1-u^h} \\
&= \frac{1}{2\pi i}\oint \frac{du(1+u)^{2n-1}(1-u)}{u^{n+1}}\frac{1-u^2}{u}\frac{u^h}{1-u^h} \\
&= [u^{n+1}](1-u)^2(1+u)^{2n}\sum_{k\geq 1} u^{hk} \\
&= \sum_{k\geq 1}[u^{n+1-hk}](1-2u+u^2)(1+u)^{2n} \\
&= \sum_{k\geq 1}\left[\binom{2n}{n+1-hk} - 2\binom{2n}{n-hk} + \binom{2n}{n-1-hk}\right].
\end{aligned} \tag{2.1}$$

We would like to introduce an alternative method to compute the generating function of Dyck paths of height $< h$; it appears for instance in [41]. Define $\varphi_i(z)$ the generating function (according to length) of non-negative lattice paths starting at $(0,0)$, ending at $(n,i)$, and height $< h$, for $i = 0,\dots,h-1$. Then

$$\varphi_0(z) = 1 + z\varphi_1(z), \quad \varphi_{h-1}(z) = z\varphi_{h-2}(z),$$
$$\varphi_i(z) = z\varphi_{i-1}(z) + z\varphi_{i+1}(z) \quad \text{for } 1 \le i \le h-2.$$

This system is best written as a matrix equation:

$$\begin{pmatrix} \varphi_0 \\ \varphi_1 \\ \vdots \\ \varphi_{h-1} \end{pmatrix} = \begin{pmatrix} 1 & -z & 0 & \dots \\ -z & 1 & -z & \dots \\ & & \ddots & \\ & & -z & 1 \end{pmatrix} \begin{pmatrix} \varphi_0 \\ \varphi_1 \\ \vdots \\ \varphi_{h-1} \end{pmatrix} + \begin{pmatrix} 1 \\ 0 \\ \vdots \\ 0 \end{pmatrix}.$$

It can be solved using Cramer's rule. Denote by $a_h$ the determinant of the matrix

$$\begin{pmatrix} 1 & -z & 0 & \dots \\ -z & 1 & -z & \dots \\ & & \ddots & \\ & & -z & 1 \end{pmatrix}$$

with $h$ rows and columns. Expanding with respect to the first row yields the recursion

$$a_h = a_{h-1} - z^2 a_{h-2}, \qquad \text{for } h \ge 2,\ a_0 = a_1 = 1.$$

The solution is

$$a_h = \frac{1}{\sqrt{1-4z^2}} \left[ \left( \frac{1+\sqrt{1-4z^2}}{2} \right)^{h+1} - \left( \frac{1-\sqrt{1-4z^2}}{2} \right)^{h+1} \right],$$

and thus

$$\varphi_i(z) = \frac{z^i a_{h-1-i}}{a_h}.$$

The good substitution in this case is $z = \frac{u}{1+u^2}$, because then

$$a_h = \frac{1}{1-u^2} \frac{1-u^{2h+2}}{(1+u^2)^h}.$$

This approach works also in the Motzkin case (level steps allowed), see [42]; the equation is then

$$\begin{pmatrix} \varphi_0 \\ \varphi_1 \\ \vdots \\ \varphi_{h-1} \end{pmatrix} = \begin{pmatrix} 1-z & -z & 0 & \dots \\ -z & 1-z & -z & \dots \\ & & \ddots & \\ & & -z & 1-z \end{pmatrix} \begin{pmatrix} \varphi_0 \\ \varphi_1 \\ \vdots \\ \varphi_{h-1} \end{pmatrix} + \begin{pmatrix} 1 \\ 0 \\ \vdots \\ 0 \end{pmatrix};$$

the recursion for the determinants is

$$a_h = (1-z)a_{h-1} - z^2 a_{h-2},$$

and

$$a_h = \frac{1}{\sqrt{1-2z-3z^2}}\left[\left(\frac{1-z+\sqrt{1-2z-3z^2}}{2}\right)^{h+1} - \left(\frac{1-z-\sqrt{1-2z-3z^2}}{2}\right)^{h+1}\right],$$

and the good substitution is $z = \frac{v}{1+v+v^2}$. Explicit expressions are also possible; they involve trinomial coefficients $\binom{n,3}{k} = [v^k](1+v+v^2)^n$ (notation from [9]).

It is worthwhile to notice that Motzkin paths can be obtained from Dyck paths by squeezing in arbitrary sequences of level steps between consecutive steps. That amounts to replacing $z$ by $z(1+z+\cdots) = \frac{z}{1-z}$, if we assume that an up/down-step is followed by a sequence of level steps. We must provide for an arbitrary sequence of level steps in the beginning. So we get

$$\frac{1}{1-z}D\left(\left(\frac{z}{1-z}\right)^2\right) = \frac{1-z-\sqrt{1-2z-3z^2}}{2z^2},$$

as expected. Since this procedure does not affect the height, functions like $H_{0,0}^{[<h]}$ also translate.

## 2.8 Lattice path enumeration: the kernel method

This is taken from the survey paper [45]; I am confident that the present Handbook will have much more about the subject in different chapters.

In the present author's view, the **kernel method** originated in Knuth's book [31], where it was presented as an innocent exercise (Exercise 2.2.1.4). Later, it was turned into a method; see [4] and the literature cited therein. It was probably rediscovered independently by many people; I recommend to follow the references in [4].

I feel that I cannot do anything better as an introduction than to reproduce Knuth's original exercise. One starts at the origin, and can advance from $(n,i)$ to both $(n+1,i\pm 1)$, except in the case when $i=0$, when one can only go to $(n+1,1)$. In this way, one models non-negative lattice paths (or random walks). The Dyck paths of the previous section are the case where one ends at level 0 after $n$ steps. We want to know how many paths lead from the origin to $(n,0)$, and, more generally, to $(n,i)$. (Clearly, this is a very classical subject, but the derivation that Knuth presented is the subject of this presentation.) One uses generating functions $f_i(z)$, describing walks leading to $(n,i)$; the coefficient of $z^n$ is the number of walks from the origin to $(n,i)$. The following recursions are immediate:

$$f_i(z) = zf_{i-1}(z) + zf_{i+1}(z), \qquad i \geq 1,$$
$$f_0(z) = 1 + zf_1(z).$$

Now one introduces $F(z,x) = \sum_{n\geq 0} f_n(z)x^n$, multiplies the recursion by $x^i$, and sums:

$$F(z,x) - f_0(z) = zxF(z,x) + \frac{z}{x}\left[F(z,x) - f_0(z) - xf_1(z)\right],$$

or

$$F(z,x) = zxF(z,x) + \frac{z}{x}\left[F(z,x) - F(z,0)\right] + 1,$$

whence

$$F(z,x) = \frac{zF(z,0) - x}{zx^2 - x + z}.$$

Plugging in $x = 0$ leads to nothing, but the denominator factors as $z(x - r_1(z))(x - r_2(z))$, with

$$r_{1,2}(z) = \frac{1 \mp \sqrt{1-4z^2}}{2z}.$$

Note that $x - r_1(z) \sim x - z$ as $x, z \to 0$. Therefore the factor $1/(x - r_1(z))$ has no power series expansion around $(0,0)$, but $F(z,x)$ **has,** so this "bad" factor must actually disappear, i.e., $(x - r_1(z))$ must be a factor of the **numerator** as well, which leads to the equation $zF(z,0) = r_1(z)$, from which $F(z,0)$ can be computed. Consequently, $F(z,x)$ is then also explicitly computed, and the factor $(x - r_1(z))$ can be cancelled from both numerator and denominator.

From this, one finds for instance that $[z^{2n}]F(z,0) = \frac{1}{n+1}\binom{2n}{n}$, a well-known **Catalan number**, and similar expressions for $[z^n x^i]F(z,x)$, for $n \equiv i \bmod 2$.

The next example revisits the **toilet paper problem**, a popular subject introduced by Knuth [34]. He considers two rolls of tissues, with $m$ and $n$ units, respectively, and random users, who are with probability $p$ **big-choosers** (taking one unit from the larger roll) and with probability $q = 1 - p$ **little-choosers** (taking one unit from the smaller roll), respectively. The parameter of interest is the (average) number of units remaining on the larger roll, when the smaller one became empty.

Let $m$ be the number of units on the larger, and $n$ on the smaller roll; $M_{m,n}$ is the expected number of units left on the larger roll, when the smaller one becomes empty.

The recursions are

$$\begin{aligned} M_{m,0} &= m, \\ M_{m,m} &= M_{m,m-1}, \qquad m \geq 1, \\ M_{m,n} &= pM_{m-1,n} + qM_{m,n-1}, \qquad m > n > 0. \end{aligned}$$

Define

$$F_0(z) = \sum_{m\geq 0} M_{m,m} z^m, \qquad F_1(z) = \sum_{m\geq 1} M_{m,m-1} z^m.$$

Note that

$$F_0(z) = \sum_{m\geq 0} M_{m,m} z^m = \sum_{m\geq 1} M_{m,m-1} z^m = F_1(z).$$

Define

$$F(z,x) = \sum_{m \geq n \geq 0} M_{m,n} z^m x^{m-n}.$$

Then, by summing up,

$$\begin{aligned}
F(z,x) &= \sum_{m>n>0} M_{m,n} z^m x^{m-n} + \sum_{m>0} M_{m,0} z^m x^m + \sum_{m \geq 0} M_{m,m} z^m \\
&= \sum_{m>n>0} [pM_{m-1,n} + qM_{m,n-1}] z^m x^{m-n} + \frac{zx}{(1-zx)^2} + F_0(z) \\
&= pzx \sum_{m-1 \geq n>0} M_{m-1,n} z^{m-1} x^{(m-1)-n} \\
&\quad + \frac{q}{x} \sum_{m>n>0} M_{m,n-1} z^m x^{m-(n-1)} + \frac{zx}{(1-zx)^2} + F_0(z) \\
&= pzx \Big[ F(z,x) - \sum_{m \geq 0} M_{m,0} z^m x^m \Big] \\
&\quad + \frac{q}{x} \sum_{m \gg n \geq 0} M_{m,n} z^m x^{m-n} + \frac{zx}{(1-zx)^2} + F_0(z) \\
&= pzx \Big[ F(z,x) - \frac{zx}{(1-zx)^2} \Big] \\
&\quad + \frac{q}{x} \Big[ F(z,x) - x \sum_{n \geq 0} M_{n+1,n} z^{n+1} - \sum_{n \geq 0} M_{n,n} z^n \Big] \\
&\quad + \frac{zx}{(1-zx)^2} + F_0(z) \\
&= pzxF(z,x) + \frac{q}{x} \Big[ F(z,x) - xF_1(z) - F_1(z) \Big] \\
&\quad + \frac{zx(1-pzx)}{(1-zx)^2} + F_1(z).
\end{aligned}$$

(Here, we used the ad hoc notation $a \gg b :\Leftrightarrow a - b \geq 2$.) Solving,

$$\begin{aligned}
F(z,x) &= \frac{\frac{zx(1-pzx)}{(1-zx)^2} + F_1(z)[1-q-q/x]}{1-pzx-q/x} = \frac{F_1(z)[q-px] - \frac{zx^2(1-pzx)}{(1-zx)^2}}{pzx^2 - x + q} \\
&= \frac{F_1(z)[q-px] - \frac{zx^2(1-pzx)}{(1-zx)^2}}{pz(x-r_1(z))(x-r_2(z))},
\end{aligned}$$

with

$$r_{1,2}(z) = \frac{1 \mp \sqrt{1-4pqz}}{2pz}.$$

Therefore, for $x = r_1(z)$, the numerator must vanish, yielding

$$F_1(z)[q - pr_1(z)] - \frac{zr_1^2(z)(1-pzr_1(z))}{(1-zr_1(z))^2} = 0,$$

or

$$F_1(z) = \frac{zr_1^2(z)(1-pzr_1(z))}{(q-pr_1(z))(1-zr_1(z))^2} = \frac{z}{q(1-z)^2}(q - C(pqz)),$$

with

$$C(z) = \frac{1-\sqrt{1-4z}}{2}.$$

Note that $r_1(z) = \frac{C(pqz)}{pz}$ and that $1/r_2(z) = r_1(z)pz/q$. The argument that $1/(x - r_1(z))$ has no power series expansion around $(x,z) \sim (0,0)$ must be replaced here by something else, for instance, that $F(z,x)/z \sim 1$ for $(x,z) \sim (0,0)$, which follows from the combinatorial description of the problem.

The expression for $F(z,x)$ is ugly, but we can extend Knuth's asymptotic analysis to $M_{m,m-n}$ for $m \to \infty$ and fixed $n$; the instance $n = 0$ was given in [34]. This asymptotic analysis is perhaps best understood after consulting Section 2.17 first. However, we decided to put it in here, so that we avoid having to repeat the description of the problem later.

For $q < p$, Knuth has shown that the local expansion of $C(pqz)$ around $z = 1$ starts like

$$q + \tfrac{pq}{p-q}(z-1) + \tfrac{(pq)^2}{(p-q)^3}(z-1)^2 + \cdots.$$

Hence the local expansion of $F(z,x)$ around $z = 1$ is given by

$$\frac{1}{1-z} \cdot \frac{p}{(2p-1)(1-x)} + Q(z),$$

where $Q(z)$ has a radius of convergence $> 1$. So $M_{m,m-n} = p/(2p-1) + O(r^m)$ (for a suitable $0 < r < 1$), and the $n$ plays no role here. This is intuitive, the big-choosers dominate, so it does not really make a difference whether the second roll is slightly smaller. Now let us assume that $p < q$. Then $F(z,x)$ starts like

$$\frac{1}{(1-z)^2} \cdot \frac{2p-1}{(p-1)(1-x)} + \frac{1}{1-z}\left[-\frac{1}{1-x} + \frac{p}{q(1-x)^2} - \frac{p(1-p)}{(2p-1)(q-px)}\right] + \cdots.$$

The coefficient of $z^m$ is asymptotic to

$$(m+1) \cdot \frac{1-2p}{(1-p)(1-x)} + \left[-\frac{1}{1-x} + \frac{p}{q(1-x)^2} - \frac{p(1-p)}{(2p-1)(q-px)}\right].$$

And the coefficient of $x^n$ ($n$ fixed) in this is

$$(m+1) \cdot \frac{1-2p}{1-p} + \left[-1 + \frac{p}{q}(n+1) - \frac{p}{2p-1}\left(\frac{p}{q}\right)^n\right],$$

or

$$m \cdot \frac{1-2p}{q} + \frac{pn}{q} + \frac{p}{1-2p}\left(\frac{p}{q}\right)^n.$$

For $n=0$, we find again Knuth's value $\frac{q-p}{q}m+\frac{p}{q-p}$. Perhaps it is not very intuitive at first glance why this **grows** with $n$. However, for larger $n$, the process tends to be over more quickly, and so more will be left on the large roll.

Now let us discuss the case $p=q$. Then

$$C(pqz)=\frac{1}{2}-\frac{1}{2}\sqrt{1-z},$$

and *

$$F(z,x)\sim(1-z)^{-3/2}\cdot\frac{1}{1-x}-(1-z)^{-1/2}\cdot\frac{1-2x}{(1-x)^3},$$

and the coefficient of $z^m$ behaves like

$$\left(2\sqrt{\frac{m}{\pi}}+\frac{3}{4\sqrt{\pi m}}\right)\cdot\frac{1}{1-x}-\frac{1}{\sqrt{\pi m}}\cdot\frac{1-2x}{(1-x)^3}.$$

Furthermore the coefficient of $x^n$ ($n$ fixed) in this is

$$2\sqrt{\frac{m}{\pi}}+\frac{3}{4\sqrt{\pi m}}+\frac{1}{\sqrt{\pi m}}\cdot\frac{(n+1)(n-2)}{2}.$$

For $n=0$ we find again

$$2\sqrt{\frac{m}{\pi}}-\frac{1}{4\sqrt{\pi m}}.$$

Again, since everybody takes at random, the process tends to be over more quickly, leaving more on the larger roll.

## 2.9 Gamma and zeta function

These two special functions appear in many contexts; therefore we only collect here a few basic facts. General references are [49] and [24].

Euler defined

$$\Gamma(z)=\int_0^\infty e^{-t}t^{z-1}dt \quad \text{for } \Re z>0.$$

It is easy to show via integration by parts that $\Gamma(z+1)=z\Gamma(z)$ and $\Gamma(1)=1$, whence $\Gamma(n+1)=n!$ for positive integers $n$. There is another definition due to Gauss,

$$\Gamma(z)=\lim_{n\to\infty}\frac{n!n^z}{z(z+1)\dots(z+n)}.$$

*We use the symbol $\sim$ in the sense that a full **asymptotic series** in powers $1-z$ would be available, at least in principle, to which the method of **singularity analysis of generating functions (transfer theorems)** [19] is applicable. See also Section 2.17.

The meaning of $n^z$ is here $e^{z\log n}$, with $\log n$ being real and positive. Now

$$\frac{z(z+1)\dots(z+n)}{n!e^{z\log n}} = e^{-z\log n}z\left(1+\frac{z}{1}\right)\left(1+\frac{z}{2}\right)\dots\left(1+\frac{z}{n}\right)$$
$$= e^{zH_n - z\log n}z\prod_{k=1}^{n}\left(1+\frac{z}{k}\right)e^{-z/k}.$$

Here, $H_n = 1+\frac{1}{2}+\cdots+\frac{1}{n}$ is a **harmonic number.** Thus, the limit can be made explicit by writing

$$\frac{1}{\Gamma(z)} = ze^{\gamma z}\prod_{n\geq 1}\left[\left(1+\frac{z}{n}\right)e^{-z/n}\right],$$

with Euler's constant

$$\gamma = \lim_{n\to\infty}(H_n - \log n) = \sum_{n\geq 1}\left[\frac{1}{n} - \log\left(1+\frac{1}{n}\right)\right] = 0.5772156649.$$

Gauss' definition is more general, but coincides with Euler's when both make sense. From it, it is not hard to derive Legendre's **duplication** formula

$$\Gamma(z)\Gamma(z+\tfrac{1}{2}) = \Gamma(2z)\sqrt{\pi}\,2^{1-2z}$$

as the **reflection** formula

$$\Gamma(z)\Gamma(1-z) = \frac{\pi}{\sin \pi z}.$$

From this, the special value $\Gamma(\frac{1}{2}) = \sqrt{\pi}$ follows.

The logarithmic derivate of the Gamma function is

$$\psi(z) := \frac{\Gamma'(z)}{\Gamma(z)};$$

using the product form it can be written as

$$\psi(z+1) = -\gamma - \sum_{n\geq 1}\left(\frac{1}{n+z} - \frac{1}{n}\right) = -\gamma - \sum_{n\geq 2}(-1)^n\zeta(n)z^{n-1},$$

with the Riemann zeta function $\zeta(s) = \sum_{n\geq 1} n^{-s}$ for $\Re s > 1$. This also expresses harmonic numbers: $H_n = \psi(n+1)+\gamma$. From the last expansion of $\psi(z+1)$, the expansion of $\Gamma(z)$ around any integer can be reconstructed; for example $\Gamma(z+1) \sim 1-\gamma z$ for $z\to 0$; the Gamma function has simple poles at $z=-k$, with residue $(-1)^k/k!$, $k = 0,1,2,\dots$. This can be seen directly from Gauss' definition as well.

For purely imaginary values, we have

$$|\Gamma(iz)| = \sqrt{\frac{\pi}{z\sinh(\pi z)}},$$

which also shows the rapid decay when $z$ becomes large.

The Riemann zeta function is for $\Re s > 1$ defined by

$$\zeta(s) = \sum_{n\ge 1} \frac{1}{n^s}.$$

It can be continued to the whole of $\mathbb{C}$ and has only one simple pole at $s = 1$ with the Laurent expansion

$$\zeta(s) = \frac{1}{s-1} + \gamma + \cdots.$$

The following special values are useful:

$$\zeta(2k) = \frac{2^{2k-1}B_{2k}(-1)^{k-1}\pi^{2k}}{(2k)!} \quad \text{for } k \in \mathbb{N},$$
$$\zeta(-2k+1) = -\frac{B_{2k}}{2k}, \quad \zeta(-2k) = 0,$$
$$\zeta(s) \sim -\frac{1}{2} - \log\sqrt{2\pi}\cdot s + \cdots, \qquad (s \to 0),$$

with $B_n$ the Bernoulli numbers.

The following functional equation due to Riemann is very famous:

$$\Gamma\Big(\frac{s}{2}\Big)\zeta(s) = \pi^{s-\frac{1}{2}}\Gamma\Big(\frac{1-s}{2}\Big)\zeta(1-s).$$

Some properties of the Bernoulli numbers (taken from [26]) are now collected: They are defined by their exponential generating function

$$\frac{z}{e^z - 1} = \sum_{n\ge 0} B_n \frac{z^n}{n!};$$

from this the recursion

$$\sum_{k=0}^{n} \binom{n+1}{k} B_k = [n=0] \quad \text{for all } n \ge 0$$

follows. Bernoulli numbers with odd indices, except $B_1 = -\frac{1}{2}$, are zero. The coefficients of the tangent may be expressed by them:

$$\tan z = \sum_{n\ge 0} (-1)^{n-1} 4^n (4^n - 1) B_{2n} \frac{z^{2n-1}}{(2n)!}.$$

And the celebrated sum of the first $m$th powers is

$$1^m + 2^m + \cdots + (n-1)^m = \frac{1}{m+1}\big(B_{m+1}(n) - B_{m+1}\big).$$

Here we have Bernoulli polynomials defined by

$$B_m(x) = \sum_k \binom{m}{k} B_k x^{m-k}.$$

They satisfy $B_m(0) = B_m$ and have the exponential generating function

$$\sum_{m\geq 0} B_m(x)\frac{z^m}{m!} = \frac{ze^{xz}}{e^z - 1}.$$

## 2.10 Harmonic numbers and their generating functions

We start with

$$[z^n]\frac{1}{1-z}\log\frac{1}{1-z} = \sum_{k=1}^{n}[z^k]\log\frac{1}{1-z} = \sum_{k=1}^{n}\frac{1}{k} = H_n.$$

For many applications it is necessary to know the coefficients of

$$\frac{1}{(1-z)^{m+1}}\log^k\frac{1}{1-z}$$

for integers $m, k \geq 0$. The following derivation is based on [50]. We start from a bivariate generating function

$$\begin{aligned}
\sum_{k\geq 0}\frac{1}{(1-z)^{m+1}}\log^k\frac{1}{1-z}\frac{t^k}{k!} &= \frac{1}{(1-z)^{m+1}}\exp\left(t\log\frac{1}{1-z}\right) \\
&= \frac{1}{(1-z)^{m+1+t}} = \sum_{n\geq 0}\binom{m+t+n}{n}z^n \\
&= \sum_{n\geq 0}\binom{m+n}{n}z^n\prod_{j=1}^{n}\left(1+\frac{t}{m+j}\right) \\
&= \sum_{n\geq 0}\binom{m+n}{n}z^n\exp\left(\sum_{j=1}^{n}\log\left(1+\frac{t}{m+j}\right)\right) \\
&= \sum_{n\geq 0}\binom{m+n}{n}z^n\exp\left(\sum_{i\geq 1}\frac{(-1)^{i-1}t^i}{i}\sum_{j=1}^{n}\frac{1}{(m+j)^i}\right) \\
&= \sum_{n\geq 0}\binom{m+n}{n}z^n\exp\left(\sum_{i\geq 1}\frac{(-1)^{i-1}t^i}{i}\left(H_{m+n}^{(i)} - H_m^{(i)}\right)\right),
\end{aligned}$$

with harmonic numbers of order $i$,

$$H_n^{(i)} = \sum_{k=1}^{n}\frac{1}{k^i}.$$

Now we can read off coefficients:

$$k![z^n t^k]\frac{1}{(1-z)^{m+1}}\log^k\frac{1}{1-z}\frac{t^k}{k!} = \binom{m+n}{n}k!(-1)^k$$
$$\times \sum_{1j_1+2j_2+\cdots=k} \frac{(-1)^{j_1+j_2+\cdots}(H^{(1)}_{m+n}-H^{(1)}_m)^{j_1}(H^{(2)}_{m+n}-H^{(2)}_m)^{j_2}\cdots}{1^{j_1}2^{j_2}\cdots j_1!j_2!\cdots}.$$

For $k=1$ and $k=2$ we get the important special cases

$$\frac{1}{(1-z)^{m+1}}\log\frac{1}{1-z} = \sum_{n\ge 0}\binom{m+n}{n}(H_{m+n}-H_m),$$
$$\frac{1}{(1-z)^{m+1}}\log^2\frac{1}{1-z} = \sum_{n\ge 0}\binom{m+n}{n}\Big[(H_{m+n}-H_m)^2-(H^{(2)}_{m+n}-H^{(2)}_m)\Big].$$

The general expression can be written using Bell polynomials [50].

## 2.11 Approximation of binomial coefficients

One often needs to approximate binomial coefficients $\binom{2n}{n+k}$ in a central region (e.g. $|k| \le \sqrt{n}\log n$). We obtain, for instance by Stirling's formula (2.4),

$$\frac{\binom{2n}{n+k}}{\binom{2n}{n}} \sim e^{-k^2/n}\cdot\left(1+\frac{k^2}{2n^2}-\frac{k^4}{6n^3}+\cdots\right). \tag{2.2}$$

A derivation of Stirling's formula will be given later (2.4). Quite often $r$th differences of binomial coefficients appear, like (for $k=2$)

$$\frac{\binom{2n}{n+k+1}-2\binom{2n}{n+k}+\binom{2n}{n+k-1}}{\binom{2n}{n}}.$$

Let us recall the difference operator $\Delta$ (operating on $k$):

$$\Delta^r\binom{2n}{n+k} = \sum_{l=0}^{r}\binom{r}{l}(-1)^{r-l}\binom{2n}{n+k+l}.$$

Let us consider $\Delta^r e^{-k^2/n}$. It is not difficult to see that one can apply this difference operator term by term to the expansion (2.2).

For this, we need the Hermite polynomials. Here are the necessary properties [2]:

$$e^{2zt-t^2} = \sum_{n\ge 0} H_n(z)\frac{t^n}{n!},$$
$$H_0 = 1,\quad H_1 = 2z,\quad H_{n+1} = 2zH_n - 2nH_{n-1},$$

$$H_n(z) = \sum_{k\geq 0} \frac{n!}{k!(n-2k)!}(-1)^k(2z)^{n-2k},$$
$$H_n(-z) = (-1)^n H_n(z).$$

Thus

$$\begin{aligned}\Delta^r e^{-k^2/n} &= \sum_{l=0}^{r}(-1)^{r-l}\binom{r}{l}e^{-(k+l)^2/n}\\ &= e^{-k^2/n}\sum_{l=0}^{r}(-1)^{r-l}\binom{r}{l}e^{-\frac{2kl}{n}-\frac{l^2}{n}}\\ &= e^{-k^2/n}\sum_{l=0}^{r}(-1)^{r-l}\binom{r}{l}\sum_{m\geq 0}H_m\left(-\frac{k}{\sqrt{n}}\right)\left(\frac{l}{\sqrt{n}}\right)^m\frac{1}{m!}.\end{aligned}$$

We know (with Stirling subset numbers) that

$$\sum_{l=0}^{r}(-1)^{r-l}\binom{r}{l}l^m = r!\left\{ {m \atop r} \right\}.$$

Hence

$$(-1)^r\Delta^r e^{-k^2/n} \sim e^{-k^2/n}\left[H_r\left(\frac{k}{\sqrt{n}}\right)\frac{1}{n^{r/2}} - H_{r+1}\left(\frac{k}{\sqrt{n}}\right)\frac{1}{n^{(r+1)/2}}\frac{r}{2}+\cdots\right].$$

Therefore we have

$$\sum_{l=0}^{r}(-1)^l\binom{r}{l}\frac{\binom{2n}{n+k+r}}{\binom{2n}{n}} \sim e^{-k^2/n}H_r\left(\frac{k}{\sqrt{n}}\right)\frac{1}{n^{r/2}},$$

which was announced in [21].

To obtain more terms we have to consider $\Delta^r k^t e^{-k^2/n}$. For this, we can use the general formula (of Leibniz type)

$$\Delta^r\big(f(k)g(k)\big) = \sum_{l=0}^{r}\binom{r}{l}\big(\Delta^{r-l}f(k+l)\big)\big(\Delta^l g(k)\big)$$

with $f(k) = k^t$ and $g(k) = e^{-k^2/n}$.

A very general answer that includes the approximation of binomial coefficients as a special case is presented in [27], which we sketch here; compare the original text for some technical conditions.

Let

$$g(z) = \sum_{k\geq 0} p_k z^k$$

be a probability generating function, and its Thiele expansion given as

$$g(e^t) = \exp\left(\mu t + \frac{\sigma^2 t^2}{2!} + \frac{\kappa_3 t^3}{3!} + \frac{\kappa_4 t^4}{4!} + \cdots\right),$$

where $\mu = g'(1)$ and $\sigma^2 = g''(1) + g'(1) - (g'(1))^2$ are expectation and variance. The constants $\kappa_n$ are called **cumulants**. Further, let

$$A_{n,k} = [z^{\mu n+k}]g(z)^n,$$

where $k$ is chosen to make the exponent an integer, which is "not too far away" from the expected value $\mu n$.

For our application, we choose $g(z) = \frac{1+z}{2}$, so that $\mu = \frac{1}{2}$ and $\sigma^2 = \frac{1}{4}$. Then

$$A_{2n,k} = \frac{\binom{2n}{n+k}}{2^{2n}};$$

it does not matter whether we normalize by $2^{2n}$ or $\binom{2n}{n}$, since, by Stirling's formula (2.4),

$$\binom{2n}{n} 2^{-2n} = \frac{1}{\sqrt{\pi n}} - \frac{1}{8\sqrt{\pi} n^{3/2}} + O(n^{-5/2}),$$

where any number of terms would be available.

The answer in [27] is

$$A_{n,k} = \frac{1}{\sigma\sqrt{2\pi n}} \exp\left(-\frac{k^2}{2\sigma^2 n}\right)\left[1 - \frac{\kappa_3}{2\sigma^4}\frac{k}{n} + \frac{\kappa_3}{6\sigma^6}\frac{k^3}{n^2}\right] + O(n^{-3/2});$$

there is also a general version given, including terms of the form $k^K/n^N$, which we do not reproduce here.

## 2.12 Mellin transform and asymptotics of harmonic sums

This section is based on the survey [16]. For details and proofs we refer to this classical paper.

The Mellin transform associates to a function $f(x)$ defined over the positive reals the complex function $f^*(s)$ where

$$f^*(s) = \int_0^\infty f(x)x^{s-1}dx;$$

it is a close relative to the Laplace and Fourier transform.

The major use of the Mellin transform examined here is for the asymptotic analysis of sums obeying the general pattern

$$F(x) = \sum_k \lambda_k f(\mu_k x),$$

either as $x \to 0$ or as $x \to \infty$. Sums of this type are called **harmonic sums**; $f(x)$ is called the base function.

Harmonic sums surface at many places in combinatorial mathematics as well as in the analysis of algorithms and data structures. De Bruijn and Knuth are responsible in an essential way for introducing the Mellin transform in this range of problems, as attested by Knuth's account in [32] and the classic paper [10], which have been the basis of many later combinatorial applications.

By a simple change of variable, we find

$$F^*(s) = \sum_k \lambda_k \mu_k^{-s} \cdot f^*(s).$$

It is this factorization property that makes the Mellin transform useful for harmonic sums. The Mellin transform exists in a fundamental strip $\langle -u, -v \rangle := \{z \in \mathbb{C} \mid -u < \Re z < -v\}$, if

$$f(x) = O(x^u) \quad \text{as } x \to 0, \qquad f(x) = O(x^v) \quad \text{as } x \to \infty.$$

As the integral defining $f^*(s)$ depends analytically on the complex parameter $s$, a Mellin transform is in addition analytic in its fundamental strip.

For instance, the function $f(x) = (1+x)^{-1}$ is $O(x^0)$ at 0 and $O(x^{-1})$ at infinity, hence a guaranteed existence strip for $f^*(s)$ is $(0,1)$, which here coincides with the fundamental strip. In this case, the Mellin transform may be found from the classical Beta integral to be $f^*(s) = \frac{\pi}{\sin \pi s}$, which is analytic in $\langle 0,1 \rangle$.

The function $f(x) = e^{-x}$ satisfies

$$e^{-x} \sim 1 \quad \text{as } x \to 0, \qquad e^{-x} = O(x^{-b}) \quad \text{for any } b \text{ as } x \to \infty.$$

Therefore the Mellin transform

$$f^*(s) = \int_0^\infty e^{-x} x^{s-1} dx = \Gamma(s)$$

exists in $\langle 0, \infty \rangle$ and is analytic there.

Let $H(x)$ be the step function defined by $H(x) = [0 \le x < 1]$. Then $H^*(s) = \frac{1}{s}$ in $\langle 0, \infty \rangle$.

The following list covers the essential functional properties.

| $f(x)$ | $f^*(s)$ | $\langle \alpha, \beta \rangle$ |
|---|---|---|
| $x^\nu f(x)$ | $f^*(s+\nu)$ | $\langle \alpha - \nu, \beta - \nu \rangle$ |
| $f(x^\rho)$ | $\frac{1}{\rho} f^*(\frac{s}{\rho})$ | $\langle \alpha\rho, \beta\rho \rangle$ $(\rho > 0)$ |
| $f(\frac{1}{x})$ | $-f^*(-s)$ | $\langle -\beta, -\alpha \rangle$ |
| $f(\mu x)$ | $\mu^{-s} f^*(s)$ | $\langle \alpha, \beta \rangle$, $(\mu > 0)$ |
| $\sum_k \lambda_k f(\mu_k x)$ | $\sum_k \lambda_k \mu_k^{-s} \cdot f^*(s)$ | |
| $f(x) \log x$ | $\frac{d}{ds} f^*(s)$ | $\langle \alpha, \beta \rangle$ |
| $x \frac{d}{dx} f(x)$ | $-s f^*(s)$ | |
| $\frac{d}{dx} f(x)$ | $-(s-1) f^*(s-1)$ | |
| $\int_0^\infty f(t) dt$ | $-\frac{1}{s} f^*(s+1)$ | |

The empty ranges depend on the situation. They are usually the intersection of the strips of the ingredients, like the Dirichlet-like series and the strip for the base function. For instance, let

$$F(x) = \frac{1}{e^x - 1} = e^{-x} + e^{-2x} + e^{-3x} + \cdots,$$

then

$$F^*(s) = (1 + 2^{-s} + 3^{-s} + \cdots)\Gamma(s) = \zeta(s)\Gamma(s).$$

It is valid in the intersection of $\langle 1, \infty \rangle$ ($\zeta$-function) and $\langle 0, \infty \rangle$ ($\Gamma$-function), so it is $\langle 1, \infty \rangle$.

Here is a little list of some common Mellin transforms.

| | | |
|---|---|---|
| $e^{-x}$ | $\Gamma(s)$ | $\langle 0, \infty \rangle$ |
| $e^{-x} - 1$ | $\Gamma(s)$ | $\langle -1, 0 \rangle$ |
| $e^{-x} - 1 + x$ | $\Gamma(s)$ | $\langle -2, -1 \rangle$ |
| $e^{-x^2}$ | $\frac{1}{2}\Gamma(\frac{s}{2})$ | $\langle 0, \infty \rangle$ |
| $\frac{1}{1+x}$ | $\frac{\pi}{\sin \pi s}$ | $\langle 0, 1 \rangle$ |
| $\log(1+x)$ | $\frac{\pi}{s \sin \pi s}$ | $\langle -1, 0 \rangle$ |
| $H(x) \equiv [0 \le x < 1]$ | $\frac{1}{s}$ | $\langle 0, \infty \rangle$ |
| $x^\alpha (\log x)^k H(x)$ | $\frac{(-1)^k k!}{(s+\alpha)^{k+1}}$ | $\langle -\alpha, \infty \rangle$ |

**Theorem 2.12.1 (Mellin's inversion theorem)**

(i) *Let $f(x)$ be integrable with fundamental strip $\langle \alpha, \beta \rangle$. If $c$ is such that $\alpha < c < \beta$ and $f^*(c + it)$ is integrable, then the equality*

$$\frac{1}{2\pi i} \int_{c-i\infty}^{c+i\infty} f^*(s) x^{-s} ds = f(x)$$

*holds almost everywhere. Moreover, if $f(x)$ is continuous, then equality holds everywhere on $(0, \infty)$.*

(ii) *Let $f(x)$ be locally integrable with fundamental strip $\langle \alpha, \beta \rangle$ and be of bounded variation in a neighborhood of $x_0$. Then, for any $c$ in the interval $(\alpha, \beta)$,*

$$\lim_{T \to \infty} \frac{1}{2\pi i} \int_{c-iT}^{c+iT} f^*(s) x^{-s} ds = \frac{f(x_0^+) + f(x_0^-)}{2}.$$

*Recall that a function of bounded variation is a real-valued function whose total variation is bounded; the total variation of a real-valued function $f(x)$, defined on an interval $[a, b]$, is the quantity*

$$\sup_P \sum_{0 \le i < n_P} |f(x_{i+1}) - f(x_i)|,$$

*taken over all partitions $P = (x_0, \ldots, x_{n_P})$ of the interval $[a, b]$.*

We need the notion of a singular expansion. Let $\phi(s)$ be meromorphic in $\Omega$, and let $\mathscr{S}$ include all the poles of $\phi(s)$ in $\Omega$. A singular expansion of $\phi(s)$ in $\Omega$ is a formal sum of singular elements of $\phi(s)$ at all points of $\mathscr{S}$. When $E$ is a singular expansion of $\phi(s)$ in $\Omega$, we write $\phi(s) \asymp E$, $s \in \Omega$. This formal expansion is a concise way of combining information contained in the Laurent expansions of the function $\phi(s)$ at various points. For example,

$$\Gamma(s) \asymp \sum_{k\geq 0} \frac{(-1)^k}{k!} \frac{1}{s+k}$$

is the singular expansion in the whole of $\mathbb{C}$.

**Theorem 2.12.2 (Direct mapping)** *Let $f(x)$ have a transform $f^*(s)$ with nonempty fundamental strip $\langle \alpha, \beta \rangle$.*

*(i) Assume that $f(x)$ admits as $x \to 0^+$ a finite asymptotic expansion of the form*

$$f(z) = \sum_{(\xi,k)\in A} c_{\xi,k} x^{\xi} (\log x)^k + O(x^{\gamma}), \tag{2.3}$$

*where the $\xi$ satisfy $-\gamma < -\xi \leq \alpha$ and the $k$ are non-negative. Then $f^*(s)$ is continuable to a meromorphic function in the strip $\langle -\gamma, \beta \rangle$ where it admits the singular expansion*

$$f(z) \asymp \sum_{(\xi,k)\in A} c_{\xi,k} \frac{(-1)^k k!}{(s+\xi)^{k+1}} + O(x^{\gamma}), \qquad s \in \langle -\gamma, \beta \rangle.$$

*(ii) Similarly, assume that $f(x)$ admits as $x \to \infty$ a finite asymptotic expansion of the form (2.3) where now $\beta \leq -\xi < -\gamma$. Then $f^*(s)$ is continuable to a meromorphic function in the strip $\langle \alpha, -\gamma \rangle$ where*

$$f(z) \asymp - \sum_{(\xi,k)\in A} c_{\xi,k} \frac{(-1)^k k!}{(s+\xi)^{k+1}} + O(x^{\gamma}), \qquad s \in \langle \alpha, -\gamma \rangle.$$

Thus terms in the asymptotic expansion of $f(x)$ at 0 induce poles of $f^*(s)$ in a strip to the left of the fundamental strip; terms in the expansion at $\infty$ induce poles in a strip to the right.

This principle is general: Subtracting from a function a truncated form of its asymptotic expansion at either 0 or $\infty$ does not alter its Mellin transform and only shifts the fundamental strip. An instance is provided by the functions $e^{-x}$, $e^{-x}-1$, $e^{-x}-1+x$, all having the Mellin transform $\Gamma(s)$, in different strips.

Under a set of mild conditions, a converse to the Direct Mapping theorem also holds: The singularities of a Mellin transform that is small enough toward $i\infty$ encode the asymptotic properties of the original function. See [16] for a precise statement.

The Mellin summation formula is

$$\sum_k \lambda_k f(\mu_k x) \sim \pm \sum_{s\in H} \operatorname{Res}(f^*(s)\Lambda(s)x^{-s}),$$

with $\Lambda(s) = \sum_k \lambda_k \mu_k^{-s}$.

For an expansion near 0, the sum is over the set $H$ of poles to the left of the fundamental strip, and the sign is $+$.

For an expansion near $\infty$, the sum is over the set $H$ of poles to the right of the fundamental strip, and the sign is $-$.

**Example 2.12.3 (Harmonic Numbers)** *We refer here to Section 2.9 about properties of the zeta function and the Bernoulli numbers. We write*

$$h(x) = \sum_{k\geq 1}\left(\frac{1}{k} - \frac{1}{k+x}\right) = \sum_{k\geq 1}\frac{1}{k}\frac{x/k}{1+x/k}$$

*and notice that* $H_n = h(n)$ *and that* $h(x)$ *is a harmonic sum, with* $\lambda_k = \mu_k = \frac{1}{k}$. *We have*

$$h^*(s) = -\frac{\pi}{\sin \pi s}\zeta(s),$$

*and the fundamental strip is* $\langle -1, 0\rangle$. *The singular expansion to the right of this fundamental strip is*

$$h^*(s) \asymp \left(\frac{1}{s^2} - \frac{\gamma}{s}\right) - \sum_{k\geq 1}(-1)^k\frac{\zeta(1-k)}{s-k}.$$

*Hence*

$$H_n \sim \log n + \gamma + \frac{1}{2n} + \sum_{k\geq 2}\frac{(-1)^k B_k}{k}\frac{1}{n^k}.$$

*The dominant terms come from the expansion at* 0. *We have to take the residue (with a negative sign) of*

$$-\frac{\pi}{\sin \pi s}\zeta(s)x^{-s},$$

*which is*

$$[s^{-1}]\left(-s^{-2} + (\log x + \gamma)s^{-1} + \cdots\right) = \log x + \gamma.$$

*The full expansion for the harmonic numbers is*

$$H_n \sim \log n + \gamma + \frac{1}{2n} - \sum_{k\geq 1}\frac{B_{2k}}{2kn^{2k}}.$$

**Example 2.12.4 (Stirling's formula for the Γ-function)** *From the product decomposition of the Gamma function, one has*

$$l(x) = \Gamma(x+1) - \gamma x = \sum_{n\geq 1}\left[\frac{x}{n} - \log\left(1 + \frac{x}{n}\right)\right].$$

*One computes*

$$l^*(s) = -\zeta(-s)\frac{\pi}{s\sin(\pi s)},$$

*with fundamental strip $\langle -2,-1\rangle$. There are double poles at $s=-1$, $s=0$ and simple poles at the positive integers. The main contribution is*

$$[(s+1)^{-1}]\zeta(-s)\frac{\pi}{s\sin(\pi s)}x^{-s} = x\log x + x(\gamma-1).$$

*The full expansion is*

$$\log(x!) \sim \log\left(x^x e^{-x}\sqrt{2\pi x}\right) + \sum_{n\geq 1}\frac{B_{2n}}{2n(2n-1)}\frac{1}{x^{2n-1}}. \tag{2.4}$$

**Example 2.12.5 (A divisor sum)** *Consider*

$$D(x) = \sum_{k\geq 1} d(k)e^{-kx},$$

*where $d(k)$ is the number of (positive) divisors of $k$. Since*

$$\sum_{k\geq 1}\frac{d(k)}{k^s} = \zeta^2(s),$$

*we find*

$$D^*(s) = \Gamma(s)\zeta^2(s),$$

*and the main contribution of the asymptotic expansion at $x=0$ is given by*

$$[(s-1)^{-1}]\Gamma(s)\zeta^2(s)x^{-s} = \frac{-\log x+\gamma}{x}.$$

**Example 2.12.6 (Height of planar trees)** *We have seen (2.1) that the probability that a Dyck path of length $2n$ has height $\geq h-1$ is given by*

$$\sum_{k\geq 1}\frac{\binom{2n}{n+1-hk} - 2\binom{2n}{n-hk} + \binom{2n}{n-1-hk}}{\frac{1}{n+1}\binom{2n}{n}}.$$

*Because of the fundamental correspondence between Dyck paths and planar trees this enumerates as well the family of planar trees with $n+1$ nodes and height $\geq h$. In order to compute the expectation, one has to sum this on $h\geq 1$, which leads to*

$$E_{n+1} = (n+1)\sum_{k\geq 1} d(k)\frac{\binom{2n}{n+1-k} - 2\binom{2n}{n-k} + \binom{2n}{n-1-k}}{\binom{2n}{n}}.$$

*Here, again, $d(k)$ is the number of (positive) divisors of the integer $k$. Approximation of the second difference of binomial coefficients as described earlier (Section 2.11) leads to the approximation*

$$E_{n+1} \sim n\sum_{k\geq 1} d(k)e^{-k^2/n}\left(\frac{4k^2}{n}-2\right).$$

*The asymptotic evaluation of this series is a typical application of the Mellin transform. With $x = 1/\sqrt{n}$, we need to estimate*

$$\sum_{k\geq 1} d(k)e^{-k^2x^2}(4k^2x^2-2),$$

*which is a harmonic sum, and its Mellin transform is*

$$\sum_{k\geq 1} d(k)k^{-s}\cdot\int_0^\infty e^{-x^2}(4x^2-2)x^{s-1}dx = \zeta^2(s)(s-1)\Gamma\Big(\frac{s}{2}\Big).$$

*Using the inversion formula, this sum can be written as*

$$\frac{1}{2\pi i}\int_{2-i\infty}^{2+i\infty} \zeta^2(s)(s-1)\Gamma\Big(\frac{s}{2}\Big)x^{-s}ds.$$

*The residue at $s = 1$ (after shifting the line of integration to the left) produces $\frac{\sqrt{\pi}}{x}$, which is $\sqrt{\pi n}$, which is the asymptotic equivalent of the average height of a planar trees with n nodes. For more details, compare [24] and the literature cited therein.*

**Example 2.12.7 (A doubly exponential sum and periodicities)** *The prototype of harmonic sums with a fluctuating behavior is the function*

$$F(x) = \sum_{k\geq 0} e^{-x2^k},$$

*whose behavior is sought as $x \to 0$. The Mellin transform is*

$$\sum_{k\geq 0} 2^{-ks}\Gamma(s) = \frac{1}{1-2^{-s}}\Gamma(s),$$

*and the fundamental strip is $\langle 0,\infty\rangle$. The poles to the left of the strip are at $s = 0$ (double) and at $s = \chi_k := \frac{2\pi ik}{\log 2}$ (simple). Now, as $s \sim 0$,*

$$\frac{1}{1-2^{-s}}\Gamma(s)x^{-s} \sim \frac{1}{(\log 2)^2 s^2} + \frac{1}{s}\Big(\frac{1}{2} - \log_2 x - \frac{\gamma}{\log 2}\Big) + \cdots,$$

*and thus, as $x \to 0$,*

$$F(x) \sim \frac{1}{2} - \log_2 x - \frac{\gamma}{\log 2} + \frac{1}{\log 2}\sum_{k\neq 0}\Gamma(-\chi_k)e^{2\pi ik\log_2 x}.$$

*The series is a Fourier series and represents a periodic function. Since the Gamma function is small along the imaginary axis [49], as mentioned earlier in Section 2.9, the amplitude of this function is small.*

## 2.13 The Mellin-Perron formula

The following treatment is borrowed from [18]. We start from the Mellin inversion formula

$$\sum_k \lambda_k f(\mu_k x) = \frac{1}{2\pi i}\int_{c-i\infty}^{c+i\infty} \Big(\sum_k \lambda_k \mu^{-s}\Big) f^*(s) x^{-s} ds, \tag{2.5}$$

where $c$ is in the fundamental strip. Introduce the step function

$$H_0(x) = \begin{cases} 1 & \text{if } x \in [0,1], \\ 0 & \text{otherwise,} \end{cases}$$

also written as $H_0(x) = [0 \le x \le 1]$, together with $H_m(x) = H_0(x)(1-x)^m$. Then we get the following theorem.

**Theorem 2.13.1** *Let $c > 0$ lie in the half-plane of absolute convergence of $\sum_k \lambda_k \mu_k^{-s}$. Then, for any $m \ge 1$, we have*

$$\frac{1}{m!}\sum_{1\le k<n} \lambda_k \Big(1-\frac{k}{n}\Big)^m = \frac{1}{2\pi i}\int_{c-i\infty}^{c+i\infty} \sum_k \frac{\lambda_k}{\mu_k^s}\cdot n^s \frac{ds}{s(s+1)\dots(s+m)}.$$

*For $m = 0$,*

$$\sum_{1\le k<n} \lambda_k + \frac{\lambda_n}{2} = \frac{1}{2\pi i}\int_{c-i\infty}^{c+i\infty} \sum_k \frac{\lambda_k}{\mu_k^s}\cdot n^s \frac{ds}{s}.$$

This formula is obtained by setting $x := 1/n$, $f(x) := H_m(x)$ in the Mellin inversion formula (2.5) and noticing that

$$H_m^*(s) = \frac{m!}{s(s+1)\dots(s+m)}.$$

For $m = 0$, the formula has to be modified slightly by taking a principal value, since $H_0(x)$ is discontinuous at $x = 1$. See [3] for a direct proof of this instance.

For example, for $\lambda_k \equiv 1$, $\mu_k \equiv k$ and $m = 1$, we get

$$\sum_{1\le k<n} \Big(1-\frac{k}{n}\Big) = \frac{n-1}{2} = \frac{1}{2\pi i}\int_{2-i\infty}^{2+i\infty} \zeta(s) n^s \frac{ds}{s(s+1)}.$$

Shifting the line of integration to the left and taking the poles at $s = 1$ and $s = 0$ into account (note that $\zeta(0) = -\frac{1}{2}$), we get

$$0 = \frac{1}{2\pi i}\int_{-\frac{1}{4}-i\infty}^{-\frac{1}{4}+i\infty} \zeta(s) n^s \frac{ds}{s(s+1)}. \tag{2.6}$$

This formula is the basis of some **exact** formulæ.

We apply the Mellin-Perron machinery now to the binary sum of digits function. Let $\nu(k)$ be the number of digits 1 in the binary expansion of the integer $k$.

Furthermore, let $v_2(k)$ be the exponent of 2 in the prime decomposition of $k$ (so, if $k = 2^i(2j+1)$, then $v_2(k) = i$). Then we notice that the binary representation looks like $**{*}10^i$ (for $k$), and $**{*}01^i$ (for $k-1$). Taking differences, we see that $\nu(k) - \nu(k-1) = 1 - v_2(k)$. Summing this on $k$, we see that

$$\nu(k) = k - \sum_{j\le k} v_2(j).$$

We will study the summatory function

$$S(n) := \sum_{k<n} \nu(k) = \frac{n(n-1)}{2} - \sum_{j\le k<n} v_2(j) = \binom{n}{2} - \sum_{j\le n}(n-j)v_2(j).$$

Now we compute

$$V(s) = \sum_{k\ge1} \frac{v_2(k)}{k^s} = \sum_{k\ge0} \frac{v_2(2k+1)}{(2k+1)^s} + \sum_{k\ge1} \frac{v_2(2k)}{(2k)^s} = \frac{1}{2^s} \sum_{k\ge1} \frac{1+v_2(k)}{k^s},$$

or

$$V(s) = \frac{1}{2^s}\zeta(s) + \frac{1}{2^s}V(s) \implies V(s) = \frac{\zeta(s)}{2^s-1}.$$

Thus

$$S(n) = \frac{n(n-1)}{2} - \frac{n}{2\pi i}\int_{2-i\infty}^{2+i\infty} \frac{\zeta(s)}{2^s-1} n^s \frac{ds}{s(s+1)}.$$

The integrand has a simple pole at $s = 1$, a double pole at $s = 0$ and simple poles at $s = \chi_k := \frac{2\pi i k}{\log 2}$. Shifting the line of integration* to $\Re(s) = -\frac{1}{4}$ and taking residues into account, we get

$$S(n) = \frac{1}{2} n \log_2 n + nF_0(\log_2 n) - nR(n),$$

where the Fourier series akin to $F_0$ occurs as the sum of residues of the integrand at the imaginary poles $s = \chi_k$. The remainder term is

$$R(n) = \frac{1}{2\pi i}\int_{-\frac{1}{4}-i\infty}^{-\frac{1}{4}+i\infty} \frac{\zeta(s)}{2^s-1} n^s \frac{ds}{s(s+1)}.$$

We are going to prove that $R(n) = 0$ whenever $n$ is an integer. The integral converges since $|\zeta(-\frac{1}{4}+it)| \ll |t|^{3/4}$ (cf. [49]). Using the expansion

$$\frac{1}{2^s-1} = -1 - 2^s - 2^{2s} - 2^{3s} - \cdots$$

*Technically, we integrate along a rectangle with upper and lower sides passing through $(2N+1)\pi i/\log 2$, respectively, and let $N \to \infty$. Because of growth properties of the zeta function, the contribution along the horizontal segments vanishes. This also proves directly that the sum of residues at the complex points (which gives the Fourier series) converges.

in the integral, which is legitimate since now $\Re(s) < 0$, we find that $R(n)$ is a sum of terms of the form

$$\frac{1}{2\pi i}\int_{-\frac{1}{4}-i\infty}^{-\frac{1}{4}+i\infty} \zeta(s)(2^k n)^s \frac{ds}{s(s+1)},$$

and each of these terms is 0 by virtue of (2.6).

This proves a result originally due to Delange [11].

**Theorem 2.13.2** *The sum-of-digits function $S(n)$ satisfies*

$$S(n) = \frac{1}{2} n \log_2 n + nF_0(\log_2 n),$$

*where $F_0(x)$ is representable as a Fourier series $F_0(x) = \sum_{k\in\mathbb{Z}} f_k e^{2\pi ikx}$ and*

$$f_0 = \frac{\log_2 \pi}{2} - \frac{1}{2\log 2} - \frac{3}{4},$$

$$f_k = -\frac{1}{\log 2}\frac{\zeta(\chi_k)}{\chi_k(\chi_k+1)} \quad \textit{for} \quad \chi_k = \frac{2\pi ik}{\log 2},\ k \neq 0.$$

It is interesting to compare this derivation with the elementary and pretty arguments provided by Delange [11]: We use the notation $x = \lfloor x \rfloor + \{x\}$ with integer part $\in \mathbb{Z}$ and fractional part $0 \le \{x\} < 1$ and write $m$ in binary as $(\ldots a_2 a_1 a_0)_2$; it is not hard to see that

$$a_k = \left\lfloor \frac{m}{2^k} \right\rfloor - 2\left\lfloor \frac{m}{2^{k+1}} \right\rfloor = \int_m^{m+1} \left( \left\lfloor \frac{t}{2^k} \right\rfloor - 2\left\lfloor \frac{t}{2^{k+1}} \right\rfloor \right) dt.$$

Therefore, with $l = \lfloor \log_2 n \rfloor$,

$$S(n) = \sum_{k=0}^{l} \int_0^n \left( \left\lfloor \frac{t}{2^k} \right\rfloor - 2\left\lfloor \frac{t}{2^{k+1}} \right\rfloor \right) dt = \sum_{k=0}^{l} 2^{k+1} \int_0^{n/2^{k+1}} (\lfloor 2u \rfloor - 2\lfloor u \rfloor) du.$$

Introducing $g(u) = \lfloor 2u \rfloor - 2\lfloor u \rfloor - \frac{1}{2}$, then

$$\begin{aligned} S(n) &= \frac{n(l+1)}{2} + \sum_{k=0}^{l} 2^{k+1} g\left(\frac{n}{2^{k+1}}\right) \\ &= \frac{n\log_2 n}{2} + n\frac{1-\{\log_2 n\}}{2} + n2^{1-\{\log_2 n\}} \sum_{k\ge 0} 2^{-k} g(2^{\{\log_2 n\}-1} \cdot 2^k); \end{aligned}$$

the last step was the change $k := l - k$ and noticing that $g(x) = 0$ whenever $x$ is an integer. Introducing

$$h(x) = \sum_{k\ge 0} 2^{-k} g(x \cdot 2^k),$$

this reads as

$$S(n) = \frac{n\log_2 n}{2} + n\frac{1-\{\log_2 n\}}{2} + n2^{1-\{\log_2 n\}} h(2^{\{\log_2 n\}-1}).$$

The final step is to introduce

$$F_0(x) = \frac{1-\{x\}}{2} + 2^{1-\{x\}}h(2^{\{x\}-1}),$$

then

$$S(n) = \frac{n\log_2 n}{2} + nF_0(\log_2 n).$$

It can be shown that $F_0(x)$ is periodic and continuous, and the Fourier coefficients can be computed as well.

We want to finish the discussion of digital properties by considering the Gray code and the sum-of-digits function of it, again using the Mellin-Perron technique.

The Gray code representation of the integers starts like

$$0, 1, 11, 10, 110, 111, 101, 100, 1100, 1101, \ldots;$$

its characteristic is that the representations of $n$ and $n+1$ differ in exactly one binary position, and it is constructed in a simple manner by reflections based on powers of two. Let $\gamma(k)$ be the number of 1-digits in the Gray code representation of $k$, and $\delta_k = \gamma(k) - \gamma(k-1)$. It is easy to see that $\delta_{2k} = \delta_k$, and the pattern for odd values is $\delta_{2k+1} = (-1)^k$. Thus the Dirichlet series $\delta(s)$ relative to the sequence $\delta_k$ is given by

$$\delta(s) = \sum_{k\geq 1} \frac{\delta_k}{k^s} = \frac{2^s}{2^s-1} \sum_{k\geq 0} \frac{(-1)^k}{(2k+1)^s}.$$

So the summatory function $G(n) = \sum_{k<n} \gamma(k)$ of the sum-of-digits function of the Gray code representation can be expressed via

$$\frac{n}{2\pi i} \int_{2-i\infty}^{2+i\infty} \delta(s) n^s \frac{ds}{s(s+1)};$$

and the rest of the analysis is very similar to before, and leads to the explicit result

$$G(n) = \frac{1}{2} n \log_2 n + nF_1(\log_2 n),$$

with explicit Fourier coefficients of $F(x)$. We refer for details to the fundamental paper [18].

We would like to mention the Hurwitz zeta function [49]

$$\zeta(s,a) := \sum_{n\geq 0} \frac{1}{(n+a)^s} \qquad \text{for } 0 < a \leq 1 \text{ and } \Re s > 1.$$

Then

$$\sum_{k\geq 0} \frac{(-1)^k}{(2k+1)^s} = \sum_{k\geq 0} \frac{1}{(4k+1)^s} - \sum_{k\geq 0} \frac{1}{(4k+3)^s} = \frac{1}{4^s}\zeta(s,\tfrac{1}{4}) - \frac{1}{4^s}\zeta(s,\tfrac{3}{4}).$$

This shows that $\delta(s)$ can be expressed via $\zeta(z,a)$.

## 2.14 Mellin-Perron formula: divide-and-conquer recursions

The aim of this section, as described in [15], is to solve the divide-and-conquer recursion

$$f_n = f_{\lfloor n/2 \rfloor} + f_{\lceil n/2 \rceil} + e_n,$$

where $e_n$ is a given sequence. The initial conditions are chosen to be $e_0 = e_1 = f_0 = 0$. The tool will be a special case ($m = 2$) of the Mellin-Perron formula as discussed before.

**Lemma 2.14.1** *Let $w_n$ be a sequence and $W(s) = \sum_{n\geq 1} w_n/n^s$ its generating Dirichlet series. Assume that $W(s)$ converges absolutely for $\Re(s) > 2$. Then*

$$\frac{n}{2\pi i}\int_{3-i\infty}^{3+i\infty} W(s)n^s \frac{ds}{s(s+1)} = \sum_{k=1}^{n}(n-k)w_k.$$

The recursion is now written for even and odd indices separately:

$$f_{2m} = 2f_m + e_{2m},$$
$$f_{2m+1} = f_m + f_{m+1} + e_{2m+1},$$

which holds for $m \geq 1$. Taking backward differences with $\nabla f_n = f_n - f_{n-1}$ and $\nabla e_n = e_n - e_{n-1}$ yields

$$\nabla f_{2m} = \nabla f_m + \nabla e_{2m},$$
$$\nabla f_{2m+1} = \nabla f_{m+1} + \nabla e_{2m+1}$$

for $m \geq 1$. Now we take forward differences. Note that $\Delta\nabla f_m = \Delta(f_m - f_{m-1}) = f_{m+1} - 2f_m + f_{m-1}$. So

$$\Delta\nabla f_{2m} = \Delta\nabla f_m + \Delta\nabla e_{2m},$$
$$\Delta\nabla f_{2m+1} = \Delta\nabla e_{2m+1}$$

for $m \geq 1$, with $\Delta\nabla f_1 = f_2 - 2f_1 = e_2 = \Delta\nabla e_1$. Now set $w_n = \Delta\nabla f_n$ and its Dirichlet generating function $W(s) = \sum_{n\geq 1} w_n n^{-s}$. From the recursion we get by summing

$$W(s) = \sum_{m\geq 1}\frac{\Delta\nabla f_m}{(2m)^s} + \Delta\nabla f_1 + \sum_{m\geq 2}\frac{\Delta\nabla e_m}{m^s} = \frac{W(s)}{2^s} + \sum_{m\geq 1}\frac{\Delta\nabla e_m}{m^s},$$

or

$$W(s) = \frac{1}{1-2^{-s}}\sum_{m\geq 1}\frac{\Delta\nabla e_m}{m^s}.$$

It is easy to check that

$$\sum_{k=1}^{n}(n-k)\Delta\nabla f_k = f_n - nf_1.$$

We assume that $e_n = O(n)$. The Mellin-Perron formula thus gives us

$$f_n = nf_1 + \frac{n}{2\pi i}\int_{3-i\infty}^{3+i\infty} \frac{\Xi(s)n^s}{1-2^{-s}} \frac{ds}{s(s+1)},$$

where

$$\Xi(s) = \sum_{n\geq 1} \frac{\Delta\nabla e_n}{n^s}.$$

The growth condition on $e_n$ ensures that this Dirichlet series converges for $\Re(s) > 2$, as required.

This is an exact formula. Asymptotics can be derived from it as in previous instances, by shifting the line of integration to the left, and taking residues into account.

**Example 2.14.2** *We study the "worst case of the number of comparisons in merge-sort," which, according to [15] and the literature cited therein is given by our recursion, for $e_n = n-1$ and $f_1 = 0$, which implies $\Delta\nabla f_1 = e_2 = 1$ and $\Delta\nabla e_n = 0$ for $n \geq 2$. Then*

$$\frac{f_n}{n} = \frac{1}{2\pi i}\int_{3-i\infty}^{3+i\infty} \frac{n^s}{1-2^{-s}} \frac{ds}{s(s+1)}.$$

*The residue calculations involve a double pole at $s = 0$ and simple poles at $s = \chi_k = 2\pi ik/\log 2$. Then one gets*

$$f_n = n\log_2 n + nA(\log_2 n) + O(\sqrt{n}),$$

*with*

$$A(x) = \sum_{k\in\mathbb{Z}} a_k e^{2\pi ikx}$$

*and*

$$a_0 = \frac{1}{2} - \frac{1}{\log 2} \quad \text{and} \quad a_k = \frac{1}{\log 2}\frac{1}{\chi_k(\chi_k+1)} \text{ for } k \neq 0.$$

For more details and more examples we refer to [15].

## 2.15 Rice's method

Rice's method made its first appearence as Exercise 5.2.2–54 in [32]: It allows us to write the alternating sum

$$\sum_{k=0}^{n} \binom{n}{k} (-1)^k f_k$$

as a contour integral

$$\frac{1}{2\pi i}\int_{\mathscr{C}} \frac{(-1)^n n!}{z(z-1)\dots(z-n)} f(z)dz,$$

where $f(z)$ is an analytic function that extrapolates the sequence $f_k$ ($f(k) = f_k$), and $\mathscr{C}$ encircles the interval $[0..n]$. As was pointed out in [23], such integrals can be traced back to Nörlund [39]. The advantage of this representation is that, by extending the contour of integration, asymptotic equivalents can be derived, usually by collecting additional residues. In many cases, the integral disappears when the contour goes to infinity, leading to identities. Furthermore, the alternating sum usually involves heavy cancellations, and is not always easy to analyze in a direct way.

Consider the difference operator $\Delta$, defined by $\Delta f_k = f_{k+1} - f_k$. Then it is a standard exercise by induction to prove that

$$\Delta^n f_0 = \sum_{k=0}^{n} \binom{n}{k} (-1)^{n-k} f_k.$$

The transformation of sequences

$$f_n \mapsto g_n = \sum_{k=0}^{n} \binom{n}{k} (-1)^k f_k$$

is an involution, called Euler transform; if $F(z)$ and $G(z)$ are the ordinary generating functions of the sequences $f_n$ and $g_n$, then

$$\begin{aligned} G(z) &= \sum_{0 \le k \le n} z^n \binom{n}{k} (-1)^k f_k = \sum_{0 \le k} (-1)^k f_k \sum_{n \ge k} \binom{n}{k} z^n \\ &= \sum_{0 \le k} (-1)^k f_k \frac{z^k}{(1-z)^{k+1}} = \frac{1}{1-z} F\left(\frac{z}{z-1}\right). \end{aligned}$$

If $f(z)$ and $g(z)$ are the exponential generating functions, then

$$g(z) = \sum_{0<k<n} \frac{z^n}{n!} \binom{n}{k} (-1)^k f_k = \sum_{k \ge 0} \frac{(-z)^k f_k}{k!} \sum_{n-k \ge 0} \frac{z^{n-k}}{(n-k)!} = e^z f(-z).$$

We state the integral formula as a lemma.

**Lemma 2.15.1** *Let $f(z)$ be analytic in a domain that contains the half-line $[n_0, \infty)$. Then, the differences of the sequence $(f_n)$ admit the integral representation*

$$\sum_{k=n_0}^{n} (-1)^k \binom{n}{k} f_k = \frac{(-1)^n}{2\pi i} \int_{\mathscr{C}} \frac{n!}{z(z-1)\dots(z-n)} f(z) dz, \qquad (2.7)$$

*where $\mathscr{C}$ is a positively oriented curve that lies in the domain of analyticity of $f(z)$, encircles $[n_0, n]$, and does not include any of the integers $0, 1, \dots, n_0 - 1$.*

*Proof.* This is a direct application of residue calculus:

$$\begin{aligned} \operatorname{Res}_{z=k} \frac{n!}{z(z-1)\dots(z-n)} f(z) &= \frac{n!}{k(k-1)\dots1(-1)(-2)\dots(k-n)} f_k \\ &= (-1)^{n-k} \binom{n}{k} f_k. \end{aligned}$$

■

The kernel in (2.7) can be expressed by Gamma functions as follows

$$\frac{n!}{z(z-1)\dots(z-n)} = \frac{\Gamma(n+1)\Gamma(z-n)}{\Gamma(z+1)} = \frac{(-1)^{n+1}\Gamma(n+1)\Gamma(-z)}{\Gamma(n+1-z)}.$$

We will deal now with rational and meromorphic functions $f(z)$. It turns out that the differences of the sequence $(f_n)$ can be expressed via collecting residues.

**Theorem 2.15.2 (Rational functions)** *Let $f(z)$ be a rational function analytic on $[n_0,\infty)$ Then, except for a finite number of values of n, one has*

$$\sum_{k=n_0}^{n}(-1)^k\binom{n}{k}f_k = (-1)^{n-1}\mathrm{Res}_z\frac{n!f(z)}{z(z-1)\dots(z-n)}$$

*where the sum is extended to all poles z of $f(z)/(z(z-1)\dots(z-n))$ not on $[n_0,\infty)$.*

*Proof.* The idea is to integrate along a large circle with radius $R$ and use trivial bounds. The details are in [23]. ■

The first example that we treat is

$$S_n(m) = \sum_{k=1}^{n}\binom{n}{k}\frac{(-1)^k}{k^m}$$

for $m$ a positive integer. Note that for negative $m$, this is basically a Stirling subset number. In this example, $f(z) = z^{-m}$ is a rational function, and there is only one additional pole at $z = 0$. So, we get (compare again [23])

$$\begin{aligned}
S_n(m) &= (-1)^{n-1}\mathrm{Res}_{z=0}\frac{n!}{z^{m+1}(z-1)\dots(z-n)} \\
&= (-1)^{n-1}[z^{-1}]\frac{n!}{z^{m+1}(z-1)\dots(z-n)} \\
&= (-1)^{n-1}[z^m]\frac{n!}{(z-1)\dots(z-n)} = -[z^m]\frac{1}{(1-\frac{z}{1})\dots(1-\frac{z}{n})} \\
&= -[z^m]\exp\left\{\sum_{k=1}^{n}\log\left(\frac{1}{1-\frac{z}{k}}\right)\right\} = -[z^m]\exp\left\{\sum_{k=1}^{n}\sum_{j\ge1}\frac{1}{j}\left(\frac{z}{k}\right)^j\right\} \\
&= -[z^m]\exp\left\{\sum_{j\ge1}\frac{1}{j}z^jH_n^{(j)}\right\} = -[z^m]\prod_{j\ge1}\exp\left\{\frac{1}{j}z^jH_n^{(j)}\right\} \\
&= -\sum_{1l_1+2l_2+3l_3+\dots=m}\frac{1}{l_1!l_2!l_3!\dots}\left(\frac{H_n^{(1)}}{1}\right)^{l_1}\left(\frac{H_n^{(2)}}{2}\right)^{l_2}\left(\frac{H_n^{(3)}}{3}\right)^{l_3}\dots
\end{aligned}$$

The quantities that arise here are harmonic numbers:

$$H_n^{(m)} = \sum_{k=1}^{n}\frac{1}{k^m}.$$

In particular, we get

$$\sum_{k=1}^{n} \binom{n}{k} \frac{(-1)^{k-1}}{k} = H_n,$$

which is a classic formula.

**Theorem 2.15.3 (Meromorphic functions)** *Let $f(z)$ be a function that is analytic on $[n_0, \infty)$.*

*(i) Assume that $f(z)$ is meromorphic in the whole of $\mathbb{C}$ and analytic on $\Omega = \bigcup_{j=1}^{\infty} \gamma_j$ where the $\gamma_j$ are positively oriented concentric circles whose radii tend to infinity. Let $f(z)$ be of polynomial growth on $\Omega$. Then, for n large enough,*

$$\sum_{k=n_0}^{n} (-1)^k \binom{n}{k} f_k = (-1)^{n-1} \mathrm{Res}_z \frac{n! f(z)}{z(z-1)\dots(z-n)}$$

*where the sum is extended to all poles z not on $[n_0, \infty)$.*

*(ii) Assume that $f(z)$ is meromorphic in the half-plane $\Omega$ defined by $\Re z \geq d$ for some $d < n_0$. Let $f(z)$ be of polynomial growth in the complement in $\Omega$ of some compact set. Then, for n large enough,*

$$\sum_{k=n_0}^{n} (-1)^k \binom{n}{k} f_k = (-1)^{n-1} \mathrm{Res}_z \frac{n! f(z)}{z(z-1)\dots(z-n)} + O(n^d)$$

*where the sum is extended to all poles z in $\Re z > d$ and not on $[n_0, \infty)$.*

The proof can be found in [23].

Now we study the next example, which is about trie sums; see [32, 22, 23]. They originate from the solution of the divide-and-conquer recursion

$$f_n = a_n + 2^{-n} \sum_{k=0}^{n} \binom{n}{k} (f_k + f_{n-k}),$$

for a given (toll) sequence $a_n$. A prototype is the sequence

$$U_n = \sum_{k=2}^{n} \binom{n}{k} \frac{(-1)^k}{2^{k-1} - 1}.$$

It arises when we set $a_n = n - 1$ for $n \geq 2$ and $f_0 = f_1 = 0$: Translating the recursion into an equation for the exponential generating function $F(z) = \sum_{n \geq 0} f_n z^n / n!$, we get

$$F(z) = (z-1)e^z + 1 + 2e^{z/2} F\left(\frac{z}{2}\right).$$

Setting $G(z) = e^{-z} F(z) = \sum_{n \geq 0} g_n z^n / n!$, this means

$$G(z) = z - 1 + e^{-z} + 2G\left(\frac{z}{2}\right),$$

and so

$$g_n = \frac{(-1)^n}{1-2^{1-n}}, \quad n \geq 2, \quad g_0 = g_1 = 0.$$

Therefore, with $F(z) = e^z G(z)$,

$$f_n = \sum_{k=2}^{n} \binom{n}{k} \frac{(-1)^k}{1-2^{k-1}} = \sum_{k=2}^{n} \binom{n}{k} \frac{(-1)^k}{2^{k-1}-1} + U_n = n-1+U_n.$$

The analysis of $U_n$ is a direct application of Theorem 2.15.3 when taking as integration contours large circles that go in between the poles of the function $(2^{s-1}-1)^{-1}$. The poles are at

$$\chi_k = 1 + \frac{2\pi i k}{\log 2};$$

each of these induces a contribution of the form

$$n^{\chi_k} = n e^{2\pi i \cdot \log_2 n}.$$

The asymptotic formula follows now from the residues at $s = 0$ (double pole) and $s = \chi_k$, $k \neq 0$ (simple poles):

$$\begin{aligned} U_n &= \frac{n}{\log 2}(H_{n-1} - 1) - \frac{n}{2} + 2 + \frac{1}{\log 2} \sum_{k \neq 0} \frac{\Gamma(n+1)\Gamma(-1+\chi_k)}{\Gamma(n+\chi_k)} \\ &= n \log_2 n + n\left(\frac{\gamma - 1}{\log 2} - \frac{1}{2}\right) + \frac{n}{\log 2} \sum_{k \neq 0} \Gamma(-\chi_k) e^{2\pi i k \cdot \log_2 n} + O(\sqrt{n}). \end{aligned}$$

The error term $O(\sqrt{n})$ is rather arbitrary here and could be replaced by $O(n^a)$ for any $0 < a < 1$.

Here, we used the approximation

$$\frac{\Gamma(n+1)}{\Gamma(n+1-\chi_k)} = n^{\chi_k}\left(1 + O\left(\frac{|\chi_k|^2}{n}\right)\right),$$

which is uniform in $k$ and follows from Stirling's formula.

## 2.16 Approximate counting

Consider the following graph.

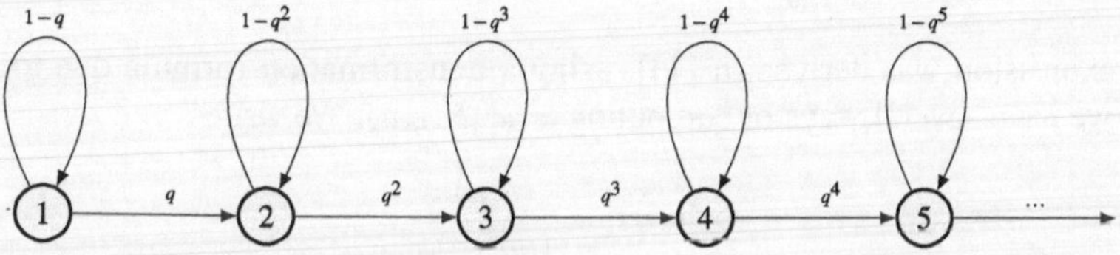

Assume that a random walk starts at state 1. Random steps are done as indicated: if we are in state $i$, the probability to advance to state $i+1$ is given by $q^i$, and with the complementary probability, we stay in state $i$. The question is where are we after $n$ random steps? In [24], this is called a **walk of the pure-birth type**, but in the older literature [14], it is called **approximate counting**, and that is where the motivation comes from. The states represent a counter, and the final state $k$ is considered to be an approximate count for—not $n$ of course, but $\log_Q n$, with $Q = \frac{1}{q}$. We are not discussing the applications for that idea, but since the example is quite instructive and appears often, also in disguised form, and was also rediscovered several times, we decided to include it here. A relatively new application is in [8, 46].

Let $p(n,k)$ be the probability that, starting in state 1, we end up in state $k$ after $n$ random steps.

Flajolet [14] proves the following.

**Theorem 2.16.1**

$$p(n,k) = \sum_{t=0}^{k-1} \frac{(-1)^t q^{\binom{t}{2}}}{(q)_t (q)_{k-1-t}} (1-q^{k-t})^n.$$

The following derivation is from [36]. We obtain by a direct translation from the graph the recursion

$$p(n,k) = q^{k-1} p(n-1,k-1) + (1-q^k) p(n-1,k), \quad p(0,1) = 1.$$

We will use a bivariate **generating function.** If we set

$$F(z,u) := \sum_{n=0}^{\infty} \sum_{k=0}^{\infty} z^n u^k p(n,k),$$

we derive

$$F(z,u) - u = zuF(z,qu) + zF(z,u) - zF(z,qu),$$

or

$$F(z,u) = \frac{u}{1-z} + \frac{z(u-1)}{1-z} F(z,qu).$$

Iterating, this gives

$$\begin{aligned} F(z,u) &= \frac{u}{1-z} + \frac{z(u-1)}{1-z}\frac{uq}{1-z} + \frac{z(u-1)}{1-z}\frac{z(qu-1)}{(1-z)^2}uq^2 \\ &\quad + \frac{z(u-1)}{1-z}\frac{z(qu-1)}{1-z}\frac{z(q^2u-1)}{(1-z)^2}uq^3 + \cdots \qquad (2.8) \\ &= \sum_{j=0}^{\infty} \frac{(-1)^j z^j (u;q)_j u q^j}{(1-z)^{j+1}}. \end{aligned}$$

This expression was derived in [43], using a transformation formula due to Heine. Now we have several ways of computing $[z^n u^k]F(z,u)$. We write

$$(u;q)_j = \frac{(u;q)_\infty}{(uq^j;q)_\infty},$$

and with Euler's partition identity (see Section 2.4), we have

$$p(n,k) = \sum_{t=0}^{k-1} \frac{(-1)^t q^{\binom{t}{2}}}{(q)_t (q)_{k-1-t}} \sum_{j=0}^{n} (-1)^j q^{(k-1-t)j} q^j [z^{n-j}](1-z)^{-(j+1)}$$
$$= \sum_{t=0}^{k-1} \frac{(-1)^t q^{\binom{t}{2}}}{(q)_t (q)_{k-1-t}} (1-q^{k-t})^n,$$

which is exactly Flajolet's formula.

There is a second expression (given by Charalambides [7]),

$$p_C(n,k) = \frac{q^{\binom{k}{2}}}{(q)_k} \sum_{j=k}^{n} (-1)^{j-k} \frac{(q)_j}{(q)_{j-k}} \binom{n}{j}, \quad p_C(0,0) = 1.$$

This is equivalent to Flajolet's formula for $p(n-1,k)$; we give an independent proof of this fact.

$$p(n-1,k) = \sum_{j=0}^{k-1} \frac{(-1)^j q^{\binom{j}{2}}}{(q)_j (q)_{k-1-j}} (1-q^{j-k})^{n-1} = \sum_{j=0}^{k} \frac{(-1)^j q^{\binom{j}{2}}}{(q)_j (q)_{k-j}} (1-q^{k-j})^n.$$

Let us consider the generating function

$$S = \sum_{k\geq 0} x^k p(n-1,k)$$
$$= \sum_{k\geq 0} x^k \sum_{j=0}^{k} \frac{(-1)^j q^{\binom{j}{2}}}{(q)_j (q)_{k-j}} (1-q^{k-j})^n$$
$$= \sum_{l=0}^{n} \binom{n}{l} (-1)^l \sum_{k\geq 0} x^k \sum_{j=0}^{k} \frac{(-1)^j q^{\binom{j}{2}}}{(q)_j (q)_{k-j}} q^{(k-j)l}$$
$$= \sum_{l=0}^{n} \binom{n}{l} (-1)^l \sum_{j\geq 0} \frac{(-1)^j q^{\binom{j}{2}}}{(q)_j} \sum_{k\geq j} x^k \frac{1}{(q)_{k-j}} q^{(k-j)l}$$
$$= \sum_{l=0}^{n} \binom{n}{l} (-1)^l \sum_{j\geq 0} \frac{(-1)^j x^j q^{\binom{j}{2}}}{(q)_j} \sum_{k\geq 0} (xq^l)^k \frac{1}{(q)_k}$$
$$= \sum_{l=0}^{n} \binom{n}{l} (-1)^l \frac{1}{(xq^l;q)_\infty} \sum_{j\geq 0} \frac{(-1)^j x^j q^{\binom{j}{2}}}{(q)_j}$$
$$= \sum_{l=0}^{n} \binom{n}{l} (-1)^l \frac{1}{(xq^l;q)_\infty} (x;q)_\infty$$
$$= \sum_{l=0}^{n} \binom{n}{l} (-1)^l (x;q)_l.$$

On the other hand, let us consider the generating function

$$\begin{aligned} T = \sum_{k\geq 0} x^k p_C(n,k) &= \sum_{k\geq 0} x^k q^{\binom{k}{2}} \sum_{j=k}^{n} (-1)^{j-k} \begin{bmatrix} j \\ k \end{bmatrix}_q \binom{n}{j} \\ &= \sum_{j=0}^{n} \binom{n}{j} (-1)^j \sum_{k=0}^{j} x^k q^{\binom{k}{2}} (-1)^k \begin{bmatrix} j \\ k \end{bmatrix}_q \\ &= \sum_{j=0}^{n} \binom{n}{j} (-1)^j \prod_{l=0}^{j-1} (1-q^l x) \\ &= \sum_{j=0}^{n} \binom{n}{j} (-1)^j (x;q)_j. \end{aligned}$$

So $S = T$, which ends the proof.

A third expression for Flajolet's formula consists in using a $q$-binomial in (2.8) to extract $[u^{k-1}]$. First,

$$\begin{aligned} [u^k]F(z,u) &= \sum_{j\geq 0} \frac{(-1)^j z^j q^j}{(1-z)^{j+1}} [u^{k-1}](u;q)_j \\ &= \sum_{j\geq 0} \frac{(-1)^j z^j q^j}{(1-z)^{j+1}} q^{\binom{k-1}{2}} \begin{bmatrix} j \\ k-1 \end{bmatrix}_q (-1)^{k-1}, \end{aligned}$$

and consequently:

$$\begin{aligned} p(n,k) &= [z^n] \sum_{j\geq 0} \frac{(-1)^j z^j q^j}{(1-z)^{j+1}} q^{\binom{k-1}{2}} \begin{bmatrix} j \\ k-1 \end{bmatrix}_q (-1)^{k-1} \\ &= q^{\binom{k-1}{2}} (-1)^{k-1} \sum_{j=k-1}^{n} (-1)^j \begin{bmatrix} j \\ k-1 \end{bmatrix}_q \binom{n}{j} q^j. \end{aligned}$$

For the asymptotics of the expected value $C_n$ and the variance, Flajolet [14] first approximated the probabilities $p(n,k)$ using real analysis, and then continued with the approximate values, using the Mellin transform. This provides additional information about the probability distribution in the limit. To compute $C_n$, there is, however a more direct way, starting from the generating function $F(z,u)$, differentiating it w.r.t. $u$, followed by $u := 1$. It should be noted that this operation, applied to $(u;q)_j$ for $j \geq 1$, simply results in zero. Therefore

$$\begin{aligned} C_n &= [z^n] \frac{\partial}{\partial u} F(z,u) \Big|_{u=1} \\ &= -[z^n] \sum_{j\geq 1} \frac{(-1)^j z^j (q)_{j-1} q^j}{(1-z)^{j+1}} + [z^n] \frac{1}{1-z} \\ &= -\sum_{j\geq 1} (-1)^j \binom{n}{j} (q)_{j-1} q^j + 1. \end{aligned}$$

The asymptotic evaluation of the sum is a typical application of Rice's method, see Section 2.15. We extend $(q)_{j-1}q^j$ by

$$\frac{(q)_\infty}{(1-q^z)(q^z)_\infty}q^z.$$

Then we have to compute residues of

$$\frac{(-1)^n n!}{z(z-1)\dots(z-n)}\frac{(q)_\infty}{(1-q^z)(q^z)_\infty}q^z$$

at $z=0$ (double pole) and at $z=\chi_k$ (simple poles for $k\neq 0$); we write $\chi_k = 2\pi ik/L$, with $L=\log Q$ and $Q=1/q$. The double pole requires more work, since all the factors have to be expanded to two terms. We collect the expansions:

$$\frac{(-1)^n n!}{z(z-1)\dots(z-n)} \sim \frac{1}{z}(1+H_n), \qquad \frac{1}{1-q^z} \sim \frac{1}{Lz}\Big(1+\frac{Lz}{2}\Big),$$

$$\frac{(q)_\infty}{(q^z)_\infty} \sim 1-\alpha Lz, \qquad q^z \sim 1-Lz,$$

with $\alpha = \sum_{k\geq 1}\frac{1}{Q^k-1}$. Using the asymptotics for $H_n$ and adding the extra term 1, we get $\log_Q n + \frac{\gamma}{L} + \frac{1}{2} - \alpha$. The residue at $\chi_k$ is simpler:

$$-\frac{1}{L}\frac{\Gamma(n+1)\Gamma(-\chi_k)}{\Gamma(n+1-\chi_k)} \sim \frac{1}{L}n^{\chi_k}\Gamma(-\chi_k).$$

Altogether we have a theorem, first proved by Flajolet [14].

**Theorem 2.16.2** *The following equality holds:*

$$C_n = \log_Q n + \frac{\gamma}{L} + \frac{1}{2} - \alpha + \delta(\log_Q n) + O(n^{1/2}),$$

*where the (small) periodic function $\delta(x)$ is given by its Fourier expansion*

$$\delta(x) = -\frac{1}{L}\sum_{k\neq 0}\Gamma(-\chi_k)e^{2\pi ikx}.$$

For more results on the subject we refer to the original papers and [24].

## 2.17 Singularity analysis of generating functions

Our goal is to say something about $[z^n]f(z)$, using information about the singularities of $f(z)$. Let us start with the simple case of a rational function $f(z)$. Using partial fraction decomposition, it can be written as a polynomial plus a finite number of terms of the form

$$\frac{A}{(1-z/\rho)^k},$$

for positive integers $k$. Now, the polynomial part can only influence a finite number of terms, is thus irrelevant for asymptotics, and

$$[z^n]\frac{A}{(1-z/\rho)^k} = A\rho^{-n}\binom{n+k-1}{n} \sim A\rho^{-n}\frac{n^{k-1}}{(k-1)!}.$$

We see that the singularities $\rho$ that are closest to the origin are poles in this instance. They produce the largest exponential growth $\rho^{-n}$. For example, let $f(z) = \frac{8}{(1-z^2)(1-z/3)}$. Then there are two poles at $\pm 1$ of smallest modulus 1, and one at 3, which will produce an exponentially small error term. We find the local expansions

$$f(z) \sim \frac{6}{1-z} \quad (z \to 1), \qquad f(z) \sim \frac{3}{1+z} \quad (z \to -1),$$

whence the asymptotic formula

$$[z^n]f(z) = 6 + 3(-1)^n + O(3^{-n}).$$

Now this instance was extremely simple, but we can deduce an important principle: The location of the singularity is responsible for the exponential growth, and a pole of order $k$ produces a (leading) term of order $n^{k-1}$.

Now let us consider Catalan numbers, given by

$$f(z) = \frac{1-\sqrt{1-4z}}{2z} = \sum_{n\geq 0} \frac{1}{n+1}\binom{2n}{n} z^n.$$

Stirling's formula gives us

$$\frac{1}{n+1}\binom{2n}{n} \sim \frac{4^n}{\sqrt{\pi}n^{3/2}}.$$

Can we see this directly from the generating function $f(z)$? Well, the singularity is at $z = \frac{1}{4}$, and the local expansion is

$$f(z) = 2 - 2\sqrt{1-4z} + \cdots,$$

and

$$[z^n](-2\sqrt{1-4z}) \sim \frac{-2 4^n}{\Gamma(-\frac{1}{2})n^{3/2}} = \frac{4^n}{\sqrt{\pi}n^{3/2}}.$$

The rest of this section is devoted to the study of why and how this works. The main references are the highly cited article [19] and of course [24].

For a precise notion of exponential growth, we write

$$a_n \bowtie K^n \qquad \text{if and only if} \qquad \limsup |a_n|^{1/n} = K$$

(read $a_n$ is of exponential growth $K^n$).

**Theorem 2.17.1 (Exponential growth formula)** *If $f(z)$ is analytic at 0 and $R$ is the modulus of the singularity nearest to the origin, then the coefficient $f_n = [z^n]f(z)$ satisfies*

$$f_n \bowtie \left(\frac{1}{R}\right)^n.$$

This allows us to deal with meromorphic functions (the only singularities are poles). We can consider all poles nearest to the origin; there can only be a finite number of them, usually it is just one pole. Then one can subtract these poles (the principal parts). The resulting function has a larger radius of convergence, and thus what remains is exponentially small compared to the contribution from the dominant poles. Let us consider the example

$$R(z) = \frac{1}{2-e^z},$$

which is the exponential generating function of surjections. The poles are the solutions of $e^z = 2$, or the points $\chi_k = \log 2 + 2\pi ik$, $k \in \mathbb{Z}$. The closest pole to the origin is at $\log 2$, so

$$[z^n]R(z) \bowtie \left(\frac{1}{\log 2}\right)^n.$$

But we can write

$$R(z) = -\frac{1}{2}\frac{1}{z-\log 2} + S(z),$$

and $S(z)$ has radius of convergence $|\log 2 + 2\pi i| = 6.321302922$. Consequently

$$[z^n]R(z) = \frac{1}{2}\frac{1}{(\log 2)^{n+1}} + O(6^{-n}).$$

In general, if $\rho$ is a dominant pole of order $k$, there will be a contribution $p(n)\rho^{-n}$, where the polynomial $p(n)$ has degree $k-1$.

The process can be iterated by including more poles, and obtaining a smaller (exponential) error term. In a variety of cases, the formal process that includes all the poles will lead to a series expansion that is exact and asymptotic. In general, there is a theorem due to Mittag-Leffler that explains how a series expansion can be obtained. For asymptotic purposes, collecting the contributions from dominant poles will be sufficient.

The process called singularity analysis of generating functions is about transfer theorems, from information of the generating function around the singularity to the asymptotics of the coefficients. Sufficient conditions will be provided for the implications

$$\begin{array}{llll} (T1) & f(z) = O(g(z)) & \Longrightarrow & f_n = O(n), \\ (T2) & f(z) = o(g(z)) & \Longrightarrow & f_n = o(n), \\ (T3) & f(z) \sim g(z) & \Longrightarrow & f_n \sim g_n. \end{array}$$

We will also speak about $O$-transfers, $o$-transfers, and $\sim$-transfers. $O$-transfers are the most basic; refinements usually lead to $o$-transfers, and $\sim$-transfers follow from these, since

$$f(z) \sim g(z) \quad \text{is equivalent to} \quad f(z) = g(z) + o(g(z)).$$

**Basic transfer.** We know from basic principles that

$$[z^n](1-z)^{-\alpha} = \binom{n+\alpha-1}{n} = \frac{\Gamma(n+\alpha)}{\Gamma(\alpha)\Gamma(n+1)}$$

for $\alpha \notin \{0,-1,-2,\dots\}$; from Stirling's formula this gives us

$$[z^n](1-z)^{-\alpha} \sim \frac{n^{\alpha-1}}{\Gamma(\alpha)}\left[1+\frac{\alpha(\alpha-1)}{2n}+\frac{\alpha(\alpha-1)(\alpha-2)(3\alpha-1)}{24n^2}+\cdots\right].$$

The full expansion is given in [19].

Without loss of generality, when analyzing a singularity at $\rho$, we can assume that it is as 1, because of the simple transformation $g(z) := f(z/\rho)$, and

$$[z^n]g(z) = \rho^{-n}[z^n]f(z).$$

From a technical point of view, the following domains are important:

$$\Delta(\phi,\eta) = \{z \mid |z| \leq 1+\eta, |\arg(z-1)| \geq \phi\},$$

where we take $\eta > 0$ and $0 < \phi < \pi/2$. This domain has the form of an indented disk as shown in Figure 2.4.

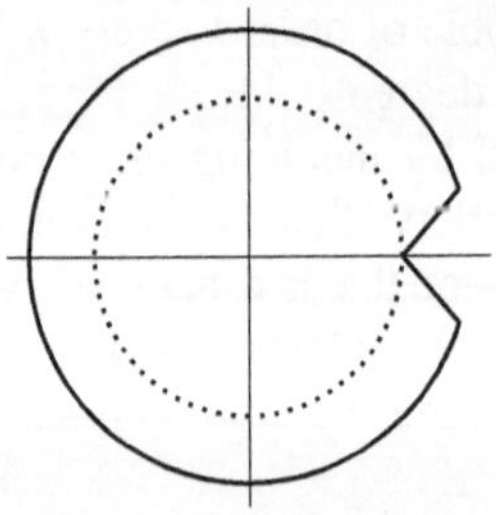

**Figure 2.4**
Domain $\Delta(\phi,\eta)$.

Extraction of coefficients is then done using Cauchy's integral formula, with a contour that stays as close to the boundary of the domain $\Delta(\phi,\eta)$ as possible. See Figure 2.5 for an illustration. We must refer to the original paper for details.

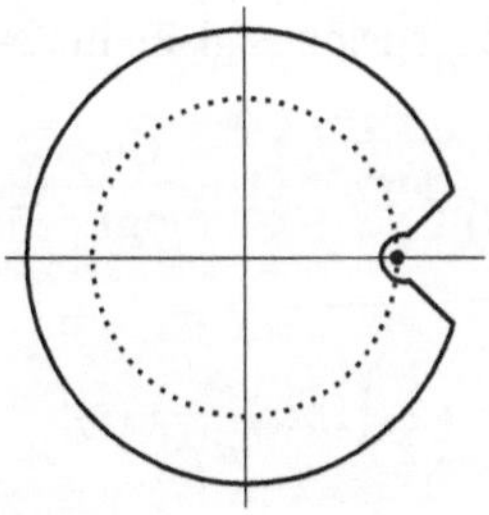

**Figure 2.5**
A typical path of integration.

**Theorem 2.17.2** *Assume that, with the sole exception of the singularity $z = 1$, $f(z)$ is analytic in the domain $\Delta = \Delta(\phi, \eta)$, where $\eta > 0$ and $0 < \phi < \pi/2$. Assume further that as $z$ tends to 1 in $\Delta$,*

$$f(z) = O(|1-z|^{-\alpha}).$$

*for some real number $\alpha$. Then the n-th Taylor coefficient of $f(z)$ satisfies*

$$f_n = [z^n]f(z) = O(n^{\alpha-1}).$$

**Corollary 2.17.3** *Assume that $f(z)$ is analytic in $\Delta \setminus \{1\}$, and that as $z$ tends to 1 in $\Delta$,*

$$f(z) = o(|1-z|^{-\alpha}),$$

*Then, as $n \to \infty$,*

$$f_n = o(n^{\alpha-1}).$$

**Corollary 2.17.4** *Assume that $f(z)$ is analytic in $\Delta \setminus \{1\}$, and that as $z$ tends to 1 in $\Delta$,*

$$f(z) \sim K(1-z)^{-\alpha},$$

*for $\alpha \notin \{0, -1, -2, \dots\}$. Then, as $n \to \infty$,*

$$f_n \sim \frac{K}{\Gamma(\alpha)} n^{\alpha-1}.$$

This settles the instance about the Catalan numbers from before. Since $\sqrt{1-4z}$ can be extended to the complex plane except for a cut from $\frac{1}{4}$ to infinity on the real axis, this also includes a fortiori a $\Delta$-domain.

Now we also allow logarithmic factors. In other words we consider

$$f(z) = (1-z)^{-\alpha} \left( \frac{1}{z} \log \left( \frac{1}{1-z} \right) \right)^{\beta}.$$

The factor $1/z$ is introduced for convenience: Since $\log 1/(1-z) = z + O(z^2)$, dividing by $z$ lets this expansion start with 1, and then raising it to the power $\beta$ results in a power series expansion. The expansion is given in [24]:

$$[z^n]f(z) = \frac{n^{\alpha-1}}{\Gamma(\alpha)}(\log n)^{\beta}\left[1+\frac{C_1}{\log n}+\frac{C_2}{\log^2 n}+\cdots\right],$$

with

$$C_k = \binom{\beta}{k}\Gamma(\alpha)\frac{d^k}{ds^k}\Gamma(s)\Big|_{s=\alpha}.$$

Note that this expansion comprises terms that have the factor $n^{\alpha-1}$. Terms including $n^{\alpha-2}$ go into the remainder term.

Here is an example:

$$[z^n]\frac{1}{\sqrt{1-z}}\frac{1}{\frac{1}{z}\log\frac{1}{1-z}} = \frac{1}{\sqrt{\pi n}\log n}\left(1-\frac{\gamma+2\log 2}{\log n}+O\Big(\frac{1}{\log^2 n}\Big)\right).$$

It is also possible to introduce an additional factor of the form $(\log\log n)^{\delta}$; we are not citing the details here.

The instance of $\alpha$ being a negative integer had to be excluded before, since it resulted in a polynomial. With the presence of logarithmic factors, however, this still makes sense. Here is one particular expansion: $(n > m \geq 0)$

$$[z^n](1-z)^m\log\frac{1}{1-z} = \sum_{k=0}^{m}\binom{m}{k}(-1)^k\frac{1}{n-k} = \frac{m!(-1)^m}{n(n-1)\dots(n-m)}.$$

The following proof is quite instructive: Consider

$$f(z) = \frac{m!}{z(z-1)\dots(z-m)}\frac{1}{z-n}$$

and perform partial fraction decomposition:

$$f(z) = \sum_{k=0}^{m}\binom{m}{k}(-1)^{m-k}\frac{1}{k-n}\frac{1}{z-k}+\frac{m!}{n(n-1)\dots(n-m)}\frac{1}{z-n}.$$

Now multiply this by $z$ and let $z\to\infty$. The result is

$$0 = \sum_{k=0}^{m}\binom{m}{k}(-1)^{m-k}\frac{1}{k-n}+\frac{m!}{n(n-1)\dots(n-m)}.$$

## 2.18 Longest runs in words

We consider a binary alphabet $\{0,1\}$, and are interested in the length of the longest run of ones in a random word of length $n$. Much more general scenarios are discussed in [24]. For a given parameter $k$, we want to find the generating functions of words where all runs of ones have length $< k$; call it $W^{<k}(z)$. There is a natural decomposition of the words with this property:

$$\mathbf{1}^{<k}(\mathbf{01}^{<k})^*,$$

where $\mathbf{1}^{<k} = \{\varepsilon, \mathbf{1}, \mathbf{11}, \ldots, \mathbf{1}^{k-1}\}$. From this,

$$W^{<k}(z) = \frac{1-z^k}{1-z}\frac{1}{1-z\frac{1-z^k}{1-z}} = \frac{1-z^k}{1-2z+z^{k+1}}.$$

The first step is the location of the dominant pole $\rho_k$, which we expect to be close to $\frac{1}{2}$. Set $Q_k(z) = 1-2z+z^{k+1}$. As $z$ traverses the circle $|z| = 1$ in the complex plane, the value of $Q_k(z)$ winds around the origin exactly once, hence the polynomial $Q_k$ has exactly one root in $|z| < 1$. We call this root $\rho_k = \frac{1}{2} + \varepsilon_k$. It satisfies the equation

$$z = \frac{1}{2} + \frac{1}{2}z^{k+1}.$$

One can start with a crude bound $\varepsilon_k = O(\frac{1}{k})$ and plug this in:

$$\frac{1}{2} + \varepsilon_k = \frac{1}{2} + \frac{1}{2}\Big(\frac{1}{2} + O\Big(\frac{1}{k}\Big)\Big)^{k+1}.$$

This leads to the approximation

$$\rho_k = \frac{1}{2} + \frac{1}{2^{k+2}} + O\Big(\frac{k}{2^{2k}}\Big),$$

which is enough for practical purposes. We can do better, however, using the Lagrange inversion formula. We write $y = z - \frac{1}{2}$ and $x = \frac{1}{2}$, then

$$y = x\Big(\frac{1}{2} + y\Big)^{k+1} = x\Phi(y).$$

Hence

$$[x^n]y = \frac{1}{n}[y^{n-1}]\Big(\frac{1}{2} + y\Big)^{n(k+1)} = \frac{1}{n}\binom{n(k+1)}{n-1}\frac{1}{2^{nk+1}},$$

and

$$y = \sum_{n\ge 1}\frac{1}{n}\binom{n(k+1)}{n-1}\frac{1}{2^{nk+1}}x^n.$$

Plugging in the special value for $x$, we get

$$\rho_k = \frac{1}{2} + \sum_{n\ge 1}\frac{1}{n}\binom{n(k+1)}{n-1}\frac{1}{2^{nk+n+1}} = \frac{1}{2} + \frac{1}{2^{k+2}} + \cdots.$$

The iterative process to get better and better approximations for $\rho_k$ is called **bootstrapping** and appears as early as in [33]. The next step is to expand $W^{<k}(z)$ around the dominant pole. We have

$$W^{<k}(z) \sim \frac{A}{1 - z/\rho_k},$$

and

$$A = \lim_{z\to\rho_k}(1-z/\rho_k)\frac{1-z^k}{Q_k(z)} = \frac{1-\rho_k^k}{-\rho_k Q'(\rho_k)} = \frac{1-\rho_k^k}{\rho_k(2-(k+1)\rho_k^k)}.$$

Therefore

$$[z^n]W^{<k}(z) = \frac{1-\rho_k^k}{2-(k+1)\rho_k^k}\rho_k^{-n-1} + O(1).$$

The error comes from the fact that the remaining poles are larger than 1 in absolute value. This derivation, as it stands, works for $k$ fixed and $n\to\infty$, but it is not too hard to extend the range of validity of it. In particular, one replaces $A$ by the easier $\frac{1}{2}$ and

$$[z^n]W^{<k}(z) \sim \frac{1}{2}\Big(\frac{1}{2}+\frac{1}{2^{k+2}}\Big)^{-n-1} \sim 2^n\Big(1-\frac{1}{2^{k+1}}\Big)^{n+1} \sim 2^n e^{-n/2^{k+1}},$$

when $k$ is close to $\log_2 n$ where the main contribution comes from. We must refer to [24] for this series of approximations.

It is interesting to sketch the enumeration problem that Knuth [33] encountered in his carry propagation problem. Words of length $n$ over the alphabet $\{0,1,2\}$ are studied where letter 1 appears with probability $\frac{1}{2}$, and 0 and 2 with probability $\frac{1}{4}$ each. For given $k$, the (contiguous) substring $1^k2$ is forbidden. The allowed words may be described as

$$\Big((\mathbf{1}^{<k}\mathbf{2})^*\mathbf{1}^*\mathbf{0}\Big)^*(\mathbf{1}^{<k}\mathbf{2})^*\mathbf{1}^*.$$

Translating that into a generating function ($\mathbf{0}\mapsto z/4$, $\mathbf{1}\mapsto z/2$, $\mathbf{2}\mapsto z/4$), we get

$$\frac{1}{1-z+\frac{1}{2}(\frac{z}{2})^{k+1}}.$$

The rest of the analysis is very similar to the previous discussion; the dominant pole is here close to 1.

## 2.19 Inversions in permutations and pumping moments

We consider permutations $\pi = p_1p_2\dots p_n$ of $\{1,2,\dots,n\}$ (written in one-line notation) and assume that all $n!$ of them are equally likely. An inversion in $\pi$ is a pair $i<j$ with $p_i > p_j$, and $I(\pi)$ is the total number of inversions of $\pi$. There is a convenient way to study the statistics of this parameter, namely the inversion table $b_1b_2\dots b_n$. The meaning is that $b_i$ counts the number of elements larger than $i$ that stand to the left of $i$. Clearly, $I(\pi) = b_1+\cdots+b_n$. We have the natural restrictions $0\le b_i\le n-i$, since there are only $n-i$ elements larger than $i$ altogether. The cute observation is that, given an inversion table, the permutation itself can be reconstructed: The value $b_1$ tells us that number 1 will be in position $b_1+1$, then $b_2$ tells us the position of number 2 ($b_2$ slots must be left open for later numbers), and so on. There is thus a bijection between permutations and inversion tables; the latter ones are easier to handle when it comes to generating functions, because of independence of the entries. If we consider the product

$$1(1+q)(1+q+q^2)\cdots(1+q+\cdots+q^{n-1}) = \prod_{j=1}^{n} \frac{1-q^j}{1-q},$$

then the coefficient of $q^k$ in it is the number of permutations of $n$ elements with $k$ inversions. Now let

$$F_n(q) = \prod_{j=1}^{n} \frac{1-q^j}{j(1-q)} = F_{n-1}(q)\frac{1-q^n}{n(1-q)}$$

be the probability generating function of the parameter $I(\pi)$. Our goal is to get information about the $s$th factorial moment, $\frac{d^s}{dq^s}F_n(q)\big|_{q=1}$. We form a bivariate generating function

$$H(z,q) = \sum_{n\geq 0} z^n F_n(q).$$

The recurrence relation then translates into

$$(1-q)\frac{\partial}{\partial z}H(z,q) = H(z,q) - qH(zq,q), \qquad H(z,1) = \frac{1}{1-z}.$$

It is impossible to solve this functional differential equation explicitly, but it is possible to derive information about

$$g_s(z) = \sum_{n\geq 0} z^n \frac{d^s}{dq^s}F_n(q)\big|_{q=1} = \frac{d^s}{dq^s}H(z,q)\big|_{q=1},$$

by differentiating the functional equation several times, and expressing $g_s(z)$ by the known values $g_i(z)$ with $i < s$. This procedure is nicknamed **pumping moments** in [24]; the following computations are taken from [29].

With a bit of patience, this program results in

$$g_s'(z) - \frac{1}{1-z}g_s(z) = h_s(z), \qquad g_0(z) = \frac{1}{1-z},$$

where

$$h_s(z) = \frac{1}{1-z}\sum_{j=1}^{s}\binom{s}{j}g_{s-j}^{j}z^j + \frac{1}{(s+1)(1-z)}\sum_{j=2}^{s+1}\binom{s+1}{j}g_{s+1-j}^{(j)}z^j.$$

Note that $h_s(z)$ is a "known" function since it involes only $g_i$'s and its derivatives that were already computed. This differential equation is easy to solve:

$$g_s(z) = \frac{1}{1-z}\int_0^z h_s(t)(1-t)dt.$$

By inspection, one "sees" (and then proves by induction) that

$$h_s(z) = \frac{(2s)!}{4^s(1-z)^{s+1}} + \frac{c_s(2s-1)!}{(1-z)^{2s}} + \text{lower order terms.}$$

The constants $c_s$ satisfy

$$c_s = c_{s-1}\frac{s}{2(2s-1)} + \frac{s(1-4s)}{3(2s-1)4^{s-1}}, \quad \text{with} \quad c_0 = 0.$$

It is easy to prove that

$$c_s = -\frac{s(4s+5)}{9 \cdot 4^{s-1}}.$$

This leads by direct translation to the formula

$$\frac{d^s}{dq^s}F_n(q)\big|_{q=1} = \frac{1}{4^s}n^{2s} + \frac{s(2s-11)}{9 \cdot 4^s}n^{2s-1} + O(n^{2s-2}).$$

For $s = 1$, we get the expected value $\frac{n(n-1)}{4}$, which is exact.

If one applies the pumping moments method to a suitably shifted random variable, one can derive limiting distribution results. See [24] for more information.

**Area under Dyck paths.** In this example, we consider Dyck paths (counted by Catalan numbers) and their area; if the path is $x_0x_1 \ldots x_{2n}$ with $x_0 = x_{2n} = 0$, $x_i \geq 0$, $x_i - x_{i+1} = \pm 1$, then the area is defined to be $x_0 + x_1 + \cdots + x_{2n}$. Let $F(z,q)$ be the generating function according to half-length and area. Mapping $a_j \mapsto q^j z$, $b_j \mapsto q^j$, $c_j \mapsto 0$, the continued fraction theorem (Theorem 2.7.3, Section 2.7) leads directly to

$$F(z,q) = \cfrac{1}{1 - \cfrac{zq}{1 - \cfrac{zq^2}{\ddots}}} = \frac{1}{1 - zqF(zq,q)}.$$

It is natural to set $F(z,q) = A(z)/B(z)$, then

$$\frac{A(z)}{B(z)} = \frac{1}{1 - zq\frac{A(zq)}{B(zq)}} = \frac{B(zq)}{B(zq) - zqA(zq)}.$$

Comparing numerator and denominator,

$$A(z) = B(zq), \qquad B(z) = 1 - zqA(zq) = 1 - zqB(zq^2).$$

Setting $B(z) = \sum_n b_n z^n$, this leads to

$$b_n = q^n b_n - q^{2n-1} b_{n-1}, \qquad b_0 = 1.$$

This recurrence can be iterated, with the result

$$b_n = \frac{(-1)^n q^{n^2}}{(1-q)(1-q^2)\dots(1-q^n)}.$$

Defining

$$E(z,q) = \sum_{n\geq 0} \frac{(-z)^n q^{n^2}}{(q)_n},$$

we may express the sought generating function as

$$F(z,q) = \frac{E(zq,q)}{E(z,q)}.$$

Now we look at the generating functions $\mu_r(z) = \frac{\partial^r}{\partial q^r} F(z,q)\big|_{q=1}$; they are (apart from normalization) the generating functions of the $r$th factorial moments. Pumping the moments can now be done as follows: First we rewrite the functional equation as $F(z,q) = 1 + zF(z,q)F(zq,q)$. Then this will be differentiated $r$ times according to the Leibniz rule, resulting in

$$\mu_r(z) = z\sum_{j=0}^{r} \binom{r}{j} \mu_{r-j}(z) \sum_{k=0}^{j} z^k \mu_{j-k}^{(k)}(z).$$

One sees by inspection that

$$\mu_r(z) = \frac{K_r}{(1-4z)^{(3r-1)/2}} + O((1-4z)^{-(3r-2)/2}).$$

The constants $K_r$ follow a recurrence, which cannot be solved explicitly, but this information is sufficient to characterize its probability distribution [24, 47]. This is also true for our previous (simpler) example of inversions, which leads to the normal distribution.

Many similar problems exist in the literature for which such an approach works.

## 2.20 The tree function

The objects we consider are labeled rooted non-planar trees. There is a symbolic description of this family:

$$\mathcal{T} = \mathcal{Z} \star \text{SET}(\mathcal{T}).$$

This is a recursive description, which uses the labeled product that conveniently does the relabeling for us once we use exponential generating functions:

$$T(z) = ze^{T(z)}.$$

Reading off coefficients is a typical application of the Lagrange inversion formula:

$$[z^n]T = \frac{1}{n}[T^{n-1}]e^{Tn} = \frac{1}{n}\frac{n^{n-1}}{(n-1)!}.$$

Therefore the number of labeled rooted non-planar trees with $n$ nodes is given by $n![z^n]T = n^{n-1}$.

This is a famous formula, and many direct proofs exist for it. We refer to [24, 31] for pointers to the literature.

For comparison, let us also study the unlabeled counterparts, defined by

$$\mathscr{A} = \mathscr{Z} \cdot \mathrm{MSET}(\mathscr{A}),$$

which leads to

$$A(z) = z\exp\left(A(z) + \frac{1}{2}A(z^2) + \frac{1}{3}A(z^3) + \cdots\right);$$

this equation for the ordinary generating function $A(z) = \sum_{n\geq 1} a_n z^n$ was first derived by Polya. There is an equivalent formula,

$$A(z) = \frac{z}{(1-z)^{a_1}(1-z^2)^{a_2}(1-z^3)^{a_3}\dots},$$

which is also implicit, but allows us to compute the numbers $a_n$ in a recursive fashion. The formula follows directly from the combinatorial description: What follows the root is a multiset of already existing trees of all possible sizes.

The implicitly defined function $y = ze^y$ is known as the **tree function**. This is an important function that, although implicitly defined, we should add to our arsenal of known functions. Whenever one sees quantities like $n^n$, one should think about this tree function. We know that

$$y = \sum_{n\geq 1} n^{n-1}\frac{z^n}{n!},$$

and then

$$\sum_{n\geq 0} n^n \frac{z^n}{n!} = 1 + zy'(z) = 1 + \frac{y}{1-y} = \frac{1}{1-y}.$$

Many similar quantities may be expressed via the tree function. A celebrated example is Ramanujan's $Q$-function [17]:

$$Q(n) = 1 + \frac{n-1}{n} + \frac{(n-1)(n-2)}{n^2} + \cdots.$$

It follows immediately that

$$\frac{n^n}{n!}Q(n) = \frac{n^{n-1}}{(n-1)!} + \frac{n^{n-2}}{(n-2)!} + \cdots + 1.$$

Now consider $\log \frac{1}{1-y(z)}$ and its coefficients. This is a typical application of the third version of the Lagrange inversion formula, with $g(y) = \log \frac{1}{1-y}$. We compute

$$[z^n] \log \frac{1}{1-y} = \frac{1}{n}[y^{n-1}]g'(y)e^{yn} = \frac{1}{n}[y^{n-1}]\frac{1}{1-y}e^{yn} = \frac{1}{n}\sum_{k=0}^{n-1}\frac{n^k}{k!},$$

so we see that $\log \frac{1}{1-y}$ is essentially the generating function of Ramanujan's $Q$-function:

$$\log \frac{1}{1-y(z)} = \sum_{n\geq 1} \frac{n^{n-1}}{n!} Q(n) z^n.$$

Next, we look again at the Lagrange inversion scenario, $y = z\Phi(y)$. In combinatorial contexts, it is natural to assume that $\Phi(y)$ is given as a power series with non-negative coefficients $\Phi(y) = \sum_k \phi_k y^k$ and that $\phi_0 > 0$. To avoid trivialities, we exclude the linear function $\Phi(y) = \phi_0 + \phi_1 y$. This describes planar trees, and the number $\phi_k$ can be interpreted as a weight when branching with $k$ successors occurs; in particular, if $\phi_k = 0$, $k$-way branching is not allowed. The family of such trees is often called **simply generated family of trees**, introduced by Meir and Moon [38]. The tree function $y = ze^y$ is the instance $\Phi(y) = e^y$. The radius of convergence of $y(z)$ (and thus the exponential growth of the coefficients) can be determined by a theorem that we cite from [24].

**Theorem 2.20.1** *Let $\Phi$ be a function analytic at 0, having non-negative Taylor coefficients, and such that $\Phi(0) \neq 0$. Let $R \leq \infty$ be the radius of convergence of the series representing $\Phi$ at 0. Under the condition*

$$\lim_{z\to R^-} \frac{z\Phi'(z)}{\Phi(z)} > 1,$$

*there exists a unique solution $\tau \in (0, R)$ of the characteristic equation*

$$\frac{\tau\Phi'(\tau)}{\Phi(\tau)} = 1.$$

*Then, the formal solution $y(z)$ of the equation $y(z) = z\Phi(y(z))$ is analytic at 0 and its coefficients satisfy the exponential growth formula:*

$$[z^n]y(z) \bowtie \rho^{-n} \quad \textit{where} \quad \rho = \frac{\tau}{\Phi(\tau)} = \frac{1}{\Phi'(\tau)}.$$

We just give the remark that the equation $\frac{\tau\Phi'(\tau)}{\Phi(\tau)} = 1$ follows from the implicit function theorem.

Now we apply this to the tree function. It is plain to see that $\tau = 1$ and hence $\rho = \frac{1}{e}$. We want to expand $y(z)$ around its singularity $\rho = \frac{1}{e}$. In order to do so, it is easier to look at $z = z(y) = ye^{-y}$, expand this around $\tau = 1$, and then invert the expansion. We easily get

$$ez = 1 - \frac{1}{2}(1-y)^2 - \frac{1}{3}(1-y)^3 + \cdots.$$

Note that the linear term is not present, which will lead us to a singularity of the square-root type. We get

$$(1-y)^2 = 2(1-ez)+\cdots \Longrightarrow y = 1-\sqrt{2}\sqrt{1-ez}+\cdots.$$

The conditions of singularity analysis are satisfied, since the function $y(z)$ is analytic except for a cut from $\frac{1}{e}$ to $\infty$ on the real axis. Therefore

$$[z^n]y(z) \sim -\sqrt{2}\frac{1}{\Gamma(-\frac{1}{2})n^{3/2}}e^n = \frac{e^n}{\sqrt{2\pi}\,n^{3/2}}.$$

But we know the exact answer $[z^n]y(z) = \frac{n^{n-1}}{n!}$. Hence

$$n! \sim \sqrt{2\pi n}\left(\frac{n}{e}\right)^n.$$

Note that in principle as many terms as desired could be obtained. Singularity analysis has proved Stirling's formula via the tree function!

## 2.21 The saddle point method

Our main source here is Odlyzko's chapter [40].

The saddle point method is the most useful method for obtaining asymptotic information about rapidly growing functions. It is based on the freedom to shift contours of integration when estimating integrals of analytic functions. We assume that $f(z)$ is analytic in $|z| < R \leq \infty$. We will also make the assumption that for some $R_0$, if $R_0 < r < R$, then

$$\max_{|z|=r} |f(z)| = f(r).$$

This assumption is clearly satisfied by all functions with real non-negative coefficients, which are the most common ones in combinatorial enumeration. We will also suppose that $|z| = r$ is the unique point with $|z| = r$ where the maximum value is assumed. The first step in estimating $[z^n]f(z)$ by the saddle point method is to find the saddle point. Under our assumptions, that will be a point $r \in (R_0, R)$ that minimizes $r^{-n}f(r)$. The minimizing $r = r_0$ will usually be unique, at least for large $n$. Cauchy's integral formula is then applied with the contour $|z| = r_0$. The reason for this choice is that for many functions, on this contour the integrand is large only near $z = r_0$, the contributions from the region near $z = r_0$ do not cancel each other, and remaining regions contribute little. By Cauchy's theorem, any simple closed contour enclosing the origin gives the correct answer. However, on most of them the integrand is large, and there is so much cancellation that it is hard to derive any estimates. The circle going through the saddle point, on the other hand, yields an integral that can be controlled well.

**Example 2.21.1 (Stirling's formula)** *We compute the reciprocal of $n!$ as $[z^n]e^z$. The saddle point is that real $r$ that minimizes $r^{-n}e^r$, which is $r = n$. Consider the contour $|z| = n$, and set $z = n\exp(i\theta)$, $-\pi \leq \theta \leq \pi$. Then*

$$[z^n]e^z = \frac{1}{2\pi i}\int_{|z|=n} e^z \frac{dz}{z^{n+1}} = \frac{1}{2\pi}\int_{-\pi}^{\pi} n^{-n}\exp(ne^{i\theta} - ni\theta)\,d\theta.$$

*Since $|\exp(z)| = \exp(\Re(z))$, the absolute value of the integrand is $n^{-n}\exp(n\cos\theta)$, which is maximized for $\theta = 0$. Now*

$$e^{i\theta} = \cos\theta + i\sin\theta = 1 - \frac{\theta^2}{2} + i\theta + O(|\theta|^3),$$

*so for any $\theta_0 \in (0;\pi)$,*

$$\int_{-\theta_0}^{\theta_0} n^{-n}\exp(ne^{i\theta} - ni\theta)\,d\theta = \int_{-\theta_0}^{\theta_0} n^{-n}\exp\Big(n - \frac{n\theta^2}{2} + O(n|\theta|^3)\Big)\,d\theta.$$

*Note the cancellation of the $ni\theta$ term. We select $\theta_0 = n^{-2/5}$, so that $n|\theta|^3 \leq n^{-1/5}$, and therefore*

$$\exp\left(n - \frac{n\theta^2}{2} + O(n|\theta|^3)\right) = \exp\left(\left(n - \frac{n\theta^2}{2}\right)\left(1 + O(n^{-1/5})\right)\right).$$

*Hence*

$$\int_{-\theta_0}^{\theta_0} n^{-n}\exp(ne^{i\theta} - ni\theta)\,d\theta = (1 + O(n^{-1/5}))n^{-n}e^n\int_{-\theta_0}^{\theta_0}\exp(-\frac{n\theta^2}{2})\,d\theta.$$

*But*

$$\begin{aligned}\int_{-\theta_0}^{\theta_0}\exp(-\frac{n\theta^2}{2})\,d\theta &= \int_{-\infty}^{\infty}\exp(-\frac{n\theta^2}{2})\,d\theta - 2\int_{\theta_0}^{\infty}\exp(-\frac{n\theta^2}{2})\,d\theta \\ &= (2\pi/n)^{1/2} + O(\exp(-n^{1/5}/2)),\end{aligned}$$

*so*

$$\int_{-\theta_0}^{\theta_0} n^{-n}\exp(ne^{i\theta} - ni\theta)\,d\theta = (1 + O(n^{-1/5}))n^{-n}e^n(2\pi/n)^{1/2}.$$

*On the other hand, for $\theta_0 < |\theta| \leq \pi$,*

$$\cos\theta \leq \cos\theta_0 = 1 - \frac{\theta_0^2}{2} + O(\theta_0^4)$$

*so*

$$n\cos\theta \leq n - n^{1/5}/2 + O(n^{3/5})$$

*and therefore for large $n$*

$$\left|\int_{\theta_0}^{\pi} n^{-n}\exp(ne^{i\theta} - ni\theta)\,d\theta\right| \leq n^{-n}\exp(n - n^{1/5}/2).$$

*Combining all these estimates we find that*

$$\frac{1}{n!} = [z^n]e^z = (1+O(n^{-1/5}))(2\pi/n)^{1/2}n^{-n}e^n,$$

*which is a weak form of Stirling's approximation.*

Let us summarize: We decompose the contour $\mathscr{C} = \mathscr{C}^{(0)} \cup \mathscr{C}^{(1)}$, where $\mathscr{C}^{(0)}$ (the "central part") contains the saddle point (or passes very near to it) and $\mathscr{C}^{(0)}$ is formed of the two remaining "tails." This splitting has to be determined in each case in accordance with the growth of the integrand. The basic principle rests on two major conditions: the contributions of the two tails should be asymptotically negligible; in the central region, the quantity $f(z)$ in the integrand (which is written as $\exp(f(z))$) should be asymptotically well approximated by a quadratic function. Under these conditions, the integral is asymptotically equivalent to an incomplete Gaussian integral. It then suffices to verify that tails can be completed back, introducing only negligible error terms. By this sequence of steps, the original integral is asymptotically reduced to a complete Gaussian integral, which evaluates in closed form.

Let us consider another example where the outcome is less predictable. It goes back to van Lint [35], but compare with [37] for further developments.

**Representations of 0 as a weighted sum.** Let $A(N)$ be the number of solutions of the equation

$$\sum_{k=-N}^{N} \varepsilon_k k = 0, \quad \text{where} \quad \varepsilon_k \in \{0,1\}.$$

Now

$$A(N) = [z^0] \prod_{k=-N}^{N} (1+z^k) = \frac{1}{2\pi i} \oint \prod_{k=-N}^{N} (1+z^k) \frac{dz}{z}.$$

In order to find the saddle point, it is convenient to take the logarithm of the integrand, differentiate, and solve the equation

$$1 = \sum_{k=1}^{N} \frac{k(z^k-1)}{1+z^k}.$$

The solution (approximate saddle point) is close to one; if one wants a closer approximation, one can use bootstrapping, as explained earlier. We thus choose the unit circle as the path of integration:

$$A(N) = \frac{2^{2N+2}}{\pi} \int_0^{\frac{\pi}{2}} \prod_{k=1}^{N} \cos^2 kx\, dx.$$

Now we sketch how to choose appropriate ranges and approximations. For the range $\frac{\pi}{2N} \le x \le \frac{\pi}{2}$, the integrand is exponentially small, and can be ignored. For $0 \le x < N^{-4/3}$, the Gaussian approximation is valid:

$$\prod_{k=1}^{N} \cos^2 kx = \exp\Big(-x^2 \frac{N(N+1)(2N+1)}{6} + O(N^{-1/3})\Big);$$

the integral over the remaining range $N^{-4/3} \leq x < \frac{\pi}{2N}$ is not neglible, but smaller than the contribution from the central range. All error terms can be made explicit, and since

$$\int_0^{N^{-4/3}} \exp\Big(-x^2\frac{N(N+1)(2N+1)}{6}\Big)\,dx \sim \frac{1}{2}(3\pi)^{1/2}N^{-3/2},$$

there is the result

$$A(N) \sim \Big(\frac{3}{\pi}\Big)^{1/2} 2^{2N+1} N^{-3/2}.$$

## 2.22 Hwang's quasi-power theorem

Here, we want to discuss a method that turns out to be very useful in combinatorial contexts to prove Gaussian limit laws. This section is by no means a full description of the rich interplay between combinatorics and probability theory. The source of our short treatment is once again [24].

A real random variable $Y$ is specified by its distribution function,

$$\mathbb{P}\{Y \leq x\} = F(x).$$

It is said to be continuous if $F(x)$ is continuous. In that case, $F(x)$ has no jump, and there is no single value in the range of $Y$ that carries a non-zero probability mass. If in addition $F(x)$ is differentiable, the random variable $Y$ is said to have a density, $g(x) = F'(x)$, so that

$$\mathbb{P}\{Y \leq x\} = \int_{\infty}^{x} g(x)dx.$$

A particularly important case is the standard Gaussian or normal $\mathcal{N}(0,1)$ distribution function,

$$\Phi(x) = \frac{1}{\sqrt{2\pi}}\int_{\infty}^{x} e^{-w^2/2}dw;$$

the corresponding density is

$$\Phi'(x) = \frac{1}{\sqrt{2\pi}}e^{-x^2/2}.$$

Let $Y$ be a continuous random variable with distribution function $F_Y(x)$. A sequence of random variables $Y_n$ with distribution functions $F_{Y_n}(x)$ is said to converge **in distribution** to $Y$ if, pointwise, for each $x$,

$$\lim_{n\to\infty} F_{Y_n}(x) = F_Y(x).$$

In that case, one writes $Y_n \Rightarrow Y$ and $F_{Y_n} \Rightarrow F_Y$.

For the readers' convenience, we cite a classic theorem.

**Theorem 2.22.1 (Basic central limit theorem)** *Let $T_j$ be independent random variables supported by $\mathbb{R}$ with a common distribution of (finite) mean $\mu$ and (finite) standard deviation $\sigma$. Let $S_n := T_1 + \cdots + T_n$. Then the standardized sum $S_n^*$ converges in distribution to the standard normal distribution,*

$$S_n^* = \frac{S_n - \mu n}{\sigma\sqrt{n}} \Rightarrow \mathcal{N}(0,1).$$

Short proofs use the concept of characteristic functions, which we do not discuss here.

A particularly simple application is when $T \equiv T_j$ takes the values 0 and 1, both with probability $\frac{1}{2}$. Then we can consider the probability generating function

$$p_n(u) = \left(\frac{1}{2} + \frac{u}{2}\right)^n,$$

so that $[u^k]p_n(u) = 2^{-n}\binom{n}{k}$ is the probability that $S_n = k$. We see here that the probability generating function is a large power of a fixed function, here $\frac{1+u}{2}$.

We will see that it suffices that the probability generating function of a combinatorial parameter behaves nearly like a large power of a fixed function to ensure convergence to a Gaussian limit. This is the quasi-powers framework, a concept that is largely due to Hwang [28].

Before we state a theorem, we discuss the cycle distribution. Assume that we have a random permutation of $n$ elements, and the random variable $X_n$ counts the number of cycles. Since a permutation can be seen as a set of cycles, it is easy to write down the relevant generating functions:

$$\mathscr{P} = \textsc{Set}(\textsc{Cyc}(\mathscr{Z})) \quad \Longrightarrow \quad P(z,u) = \exp\left(u \log \frac{1}{1-z}\right) = (1-z)^{-u}.$$

So, $n![z^n u^k]P(z,u)$ is the probability that a random permutation of $n$ elements has $k$ cycles. Extracting the coefficient of $z^n$, we find the probability generating function $p_n(u)$ related to $X_n$:

$$p_n(u) = \binom{n+u-1}{n} = \frac{\Gamma(u+n)}{\Gamma(u)\Gamma(n+1)};$$

near $u = 1$:

$$p_n(u) = \frac{n^{u-1}}{\Gamma(u)}\left(1 + O\left(\frac{1}{n}\right)\right) = \frac{\left(e^{u-1}\right)^{\log n}}{\Gamma(u)}\left(1 + O\left(\frac{1}{n}\right)\right).$$

Thus, as $n \to \infty$, the probability generating function $p_n(u)$ approximately equals a large power of $e^{u-1}$, taken with exponent $\log n$ and multiplied by the fixed function, $\Gamma(u)^{-1}$. This is enough to ensure a Gaussian limit law, as we will see.

The following notations will be convenient: Given a function $f(u)$ analytic at $u = 1$ and assumed to satisfy $f(1) \neq 0$, we set

$$\mathsf{mean}(f) = \frac{f'(1)}{f(1)}, \qquad \mathsf{var}(f) = \frac{f''(1)}{f(1)} + \frac{f'(1)}{f(1)} - \left(\frac{f'(1)}{f(1)}\right)^2.$$

**Theorem 2.22.2 (Hwang's quasi-power theorem)** *Let the $X_n$ be non-negative discrete random variables (supported by $\mathbb{N}_0$), with probability generating functions $p_n(u)$. Assume that, uniformly in a fixed complex neighborhood of $u = 1$, for sequences $\beta_n, \kappa_n \to \infty$, there holds*

$$p_n(u) = A(u)B(u)^{\beta_n}\left(1 + O\left(\frac{1}{\kappa_n}\right)\right),$$

*where $A(u)$, $B(u)$ are analytic at $u = 1$ and $A(1) = B(1) = 1$. Assume finally that $B(u)$ satisfies the so-called "variability condition,"*

$$\mathit{var}(B(u)) = B''(1) + B'(1) - (B'(1))^2 \neq 0.$$

*Under these conditions, the mean and variance of $X_n$ satisfy*

$$\mu_n = \mathbb{E}X_n = \beta_n \mathit{mean}(B(u)) + \mathit{mean}(A(u)) + O(\kappa_n^{-1}),$$
$$\sigma_n^2 = \mathbb{V}X_n = \beta_n \mathit{var}(B(u)) + \mathit{var}(A(u)) + O(\kappa_n^{-1}).$$

*The distribution of $X_n$ is, after standardization, asymptotically Gaussian, and the speed of convergence to the Gaussian limit is $O(\kappa_n^{-1} + \beta_n^{-1/2})$. That is,*

$$\mathbb{P}\left\{\frac{X_n - \mathbb{E}X_n}{\sqrt{\mathbb{V}X_n}} \leq x\right\} = \Phi(x) + O\left(\frac{1}{\kappa_n} + \frac{1}{\sqrt{\beta_n}}\right).$$

This theorem is a direct application of a lemma, also due to Hwang, that applies more generally to arbitrary discrete or continuous distributions.

## References


[1] G. Andrews. *The Theory of Partitions*, volume 2 of *Encyclopedia of Mathematics and its Applications*. Addison–Wesley, 1976.

[2] G. Andrews, R. Askey, and R. Roy. *Special Functions*, volume 71 of *Encyclopedia of Mathematics and its Applications*. Cambridge University Press, 1999.

[3] T. Apostol. *Introduction to Analytic Number Theory*. Springer, 1976.

[4] C. Banderier, M. Bousquet-Mélou, A. Denise, P. Flajolet, D. Gardy, and D. Gouyou-Beauchamps. Generating functions of generating trees. *Discrete Mathematics*, 246:29–55, 2002.

[5] N. G. De Bruijn. *Asymptotic Methods in Analysis*. Cover, 1970.

[6] L. Carlitz. Restricted compositions. *FQ*, 14:254–264, 1976.

[7] Ch. A. Charalambides. Moments of a class of discrete $q$-distributions. *Journal of Statistical Planning and Inference*, 135:64–76, 2005.

[8] J. Cichoń and W. Macyna. Approximate counters for flash memory. *17th IEEE International Conference on Embedded and Real-Time Computing Systems and Applications (RTCSA), Toyama, Japan*, 2011.

[9] L. Comtet. *Advanced Combinatorics*. Reidel, Dordrecht, 1974.

[10] N. G. De Bruijn, D. E. Knuth, and S. O. Rice. The average height of planted plane trees. In R. C. Read, editor, *Graph Theory and Computing*, pages 15–22. Academic Press, 1972.

[11] H. Delange. Sur la fonction sommatoire de la fonction somme des chiffres. *Enseignement Mathématique*, 21:31–47, 1975.

[12] C. Elsholtz, C. Heuberger, and H. Prodinger. The number of huffman codes, compact trees, and sums of unit fractions. *IEEE Transactions of Information Theory*, 59:1065–1075, 2013.

[13] P. Flajolet. Combinatorial aspects of continued fractions. *Discrete Mathematics*, 32:125–161, 1980.

[14] P. Flajolet. Approximate counting: a detailed analysis. *BIT*, 25:113–134, 1985.

[15] P. Flajolet and M. Golin. Mellin transforms and asymptotics. The mergesort recurrence. *Acta Inform.*, 31(7):673–696, 1994.

[16] P. Flajolet, X. Gourdon, and P. Dumas. Mellin transforms and asymptotics: Harmonic sums. *Theoretical Computer Science*, 144:3–58, 1995.

[17] P. Flajolet, P. Grabner, P. Kirschenhofer, and H. Prodinger. On Ramanujan's $Q(n)$–function. *Journal of Computational and Applied Mathematics*, 58:103–116, 1995.

[18] P. Flajolet, P. Grabner, P. Kirschenhofer, H. Prodinger, and R. Tichy. Mellin transforms and asymptotics: Digital sums. *Theoretical Computer Science*, 123(2):291–314, 1994.

[19] P. Flajolet and A. Odlyzko. Singularity analysis of generating functions. *SIAM Journal on Discrete Mathematics*, 3:216–240, 1990.

[20] P. Flajolet and H. Prodinger. Level number sequences for trees. *Discrete Mathematics*, 65:149–156, 1987.

[21] P. Flajolet, J.-C. Raoult, and J. Vuillemin. The number of registers required for evaluating arithmetic expressions. *Theoretical Computer Science*, 9(1):99–125, 1979.

[22] P. Flajolet and R. Sedgewick. Digital search trees revisited. *SIAM Journal on Computing*, 15:748–767, 1986.

[23] P. Flajolet and R. Sedgewick. Mellin transforms and asymptotics: Finite differences and Rice's integrals. *Theoretical Computer Science*, 144:101–124, 1995.

[24] P. Flajolet and R. Sedgewick. *Analytic Combinatorics*. Cambridge University Press, Cambridge, 2009.

[25] I. Goulden and D. Jackson. *Combinatorial Enumeration*. John Wiley, 1983.

[26] R. L. Graham, D. E. Knuth, and O. Patashnik. *Concrete Mathematics (Second Edition)*. Addison Wesley, 1994.

[27] D. H. Greene and D. E. Knuth. *Mathematics for the analysis of algorithms*. Birkhäuser, Boston, second edition, 1982.

[28] H.-K. Hwang. On convergence rates in the central limit theorems for combinatorial structures. *European Journal of Combinatorics*, 19:329–343, 1998.

[29] P. Kirschenhofer, H. Prodinger, and R. F. Tichy. A contribution to the analysis of in situ permutation. *Glasnik Mathematicki*, 22(42):269–278, 1987.

[30] A. Knopfmacher and H. Prodinger. On Carlitz compositions. *European Journal of Combinatorics*, 19:579–589, 1998.

[31] D. E. Knuth. *The Art of Computer Programming*, volume 1: Fundamental Algorithms. Addison-Wesley, 1973. Third edition, 1997.

[32] D. E. Knuth. *The Art of Computer Programming*, volume 3: Sorting and Searching. Addison-Wesley, 1973.

[33] D. E. Knuth. The average time for carry propagation. *Indagationes mathematicae*, 40:238–242, 1978.

[34] D. E. Knuth. The toilet paper problem. *Amer. Math. Monthly*, 91:465–470, 1984.

[35] J. H. Van Lint. Representation of 0 as $\sum_{k=-N}^{n} \varepsilon_k k$. *Proc. Amer. Math. Soc.*, 18:182–184, 1967.

[36] G. Louchard and H. Prodinger. Generalized approximate counting revisited. *Theoretical Computer Science*, 391:109–125, 2008.

[37] G. Louchard and H. Prodinger. Representations of numbers as $\sum_{k=-n}^{n} \varepsilon_k k$: A saddle point approach. In M. Archibald, V. Brattka, V. F. Goranko, and B. Löwe, editors, *LNCS 5489 (Infinity in Logic and Computation)*, pages 87–96. Springer Verlag, 2010.

[38] A. Meir and J. W. Moon. On the altitude of nodes in random trees. *Canadian Journal of Mathematics*, 30:997–1015, 1978.

[39] N. F. Nörlund. *Vorlesungen über Differenzenrechnung*. Chelsea, New York, 1954.

[40] A. Odlyzko. *Asymptotic Enumeration Methods (in: Handbook of Combinatorics).* Eds. R. Graham, M. Grötschel, L. Lovász. Elsevier Science, 1995.

[41] W. Panny and H. Prodinger. The expected height of paths for several notions of height. *Studia Scientiarum Mathematicarum Hungarica*, 20:119–132, 1985.

[42] H. Prodinger. The average height of a stack where three operations are allowed and some related problems. *Journal of Combinatorics, Information and System Sciences*, 5:287–304, 1980.

[43] H. Prodinger. Approximate counting via Euler transform. *Mathematica Slovaka*, 44:569–574, 1994.

[44] H. Prodinger. Combinatorics of geometrically distributed random variables: New $q$–tangent and $q$–secant numbers. *International Journal of Mathematical Sciences*, 24:825–838, 2000.

[45] H. Prodinger. The kernel method: a collection of examples. *Séminaire Lotharingien de Combinatoire*, B50f:19 pages, 2004.

[46] H. Prodinger. Approximate counting with $m$ counters: a detailed analysis. *Theoretical Computer Science*, 439:58–68, 2012.

[47] L. Takács. Conditional limit theorems for branching processes. *Journal of Applied Mathematics and Stochastic Analysis*, 4:263–292, 1991.

[48] H. S. Wall. *Analytic Theory of Continued Fractions*. Chelsea Publishing Company, 1948.

[49] E. T. Whittaker and G. N. Watson. *A Course of Modern Analysis*. Cambridge University Press, fourth edition, 1927. Reprinted 1973.

[50] D. A. Zave. A series expansion involving the harmonic numbers. *Information Processing Letters*, 5(3):75–77, 1976.

# Part II

# Topics

# Chapter 3

# *Asymptotic Normality in Enumeration*

**E. Rodney Canfield**

*Department of Computer Science*
*University of Georgia*

**CONTENTS**



## 3.1 Introduction

The focus of this chapter is the frequent appearance of the normal distribution in the context of combinatorial enumeration. The notion of **asymptotic normality** is defined, and four methods for establishing its presence are stated and illustrated. The connection of asymptotic normality with other methods of approximate enumeration is briefly explored.

## 3.2 The normal distribution

A random variable $X$ is said to be **normally distributed** when

$$\text{Prob}\{X \le x\} = \frac{1}{\sqrt{2\pi}} \int_{-\infty}^{x} e^{-t^2/2} dt.$$

We shall denote the function on the right by $\mathcal{N}(x)$. Even though their work is separated by about one hundred years [24, page 55], de Moivre and Laplace share credit for the following fundamental theorem concerning independent Bernoulli trials.

**Theorem 3.2.1** *Let $0 < p < 1$, let $X_i$, $1 \le i \le n$, be a sequence of independent random variables each of which takes the value 1 with probability $p$ and the value 0 with probability $q = 1 - p$, and let $S_n = \sum_{i=1}^{n} X_i$. Then, for any fixed $x$,*

$$Prob\,(S_n < np + x\sqrt{npq}) \;\to\; \mathcal{N}(x), \quad \text{as } n \to \infty.$$

We remark that $np$ and $npq$ are the mean and variance, respectively, of $S_n$, and that the assertion of the theorem may be written explicitly in terms of binomial coefficients as

$$\sum_{k<np+x\sqrt{npq}} \binom{n}{k} p^k q^{n-k} \to \mathcal{N}(x), \quad \text{as } n \to \infty. \tag{3.1}$$

Many textbooks on probability and statistics introduce the normal distribution as an approximation to the binomial distribution, as in the previous equation. Theorem 3.2.1 may be proved by estimating the individual terms via Stirling's formula and approximating the sum with an integral. Exactly this approach is followed in Section VII.2 of Feller [17].

The normal distribution is also known as the **Gaussian distribution**. Having achieved fame with his calculation of the orbit of asteroid Ceres (discovered in 1801), Gauss later published a treatise on the subject of celestial orbits. In this work he propounded the method of least squares, and provided theoretical justification by assuming that observational measurements are normally distributed about their mean.

The focus of this article is the frequent appearance of the normal distribution in the context of combinatorial enumeration. The typical situation is that for each positive integer $n$ we have a family of combinatorial objects, such as all permutations of the integers $[n] = \{1,2,\ldots,n\}$. It is desired to count these with respect to a parameter, for instance how many permutations of $[n]$ have $k$ cycles ? This leads to a doubly-indexed array of counts, $a(n,k)$ say. When the objects are given the uniform probability, the parameter can be viewed as a random variable $X_n$. Denoting the mean of $X_n$ as $\mu_n$ and the variance as $\sigma_n^2$, it is often found that the probability of the event $X_n < \mu_n + x\sigma_n$ approaches $\mathcal{N}(x)$ as $n \to \infty$. In this situation, we say the counts $a(n,k)$ are **asymptotically normal**. We repeat the definition.

**Definition 3.2.2** *Let $a(n,k)$ be a doubly-indexed sequence of nonnegative numbers. We say that the sequence $a(n,k)$ is* **asymptotically normal with mean $\mu_n$ and variance $\sigma_n^2$** *provided that for the normalized probabilities*

$$p(n,k) = \frac{a(n,k)}{\sum_k a(n,k)}$$

*we have, for each x,*

$$\sum_{k<\mu_n+x\sigma_n} p(n,k) \to \mathcal{N}(x), \text{ as } n \to \infty.$$

In the next few sections we give, with illustrative examples, four methods of proving asymptotic normality: direct, negative roots, moments, and singularity analysis. In following sections we consider passing from asymptotic normality to asymptotic estimates for the underlying counts $a(n,k)$, multivariate normality, and the role of asymptotic normality in other methods of approximate enumeration.

The de Moivre-Laplace formulation given above in Theorem 3.2.1 is not always adequate for our task, and so we have ready the following generalization.

**Theorem 3.2.3** *Let $X_i$, $i \geq 1$, be a sequence of independent* 0, 1 *random variables with $p_i$ denoting the probability that $X_i = 1$, and let $S_n$ be the nth partial sum $\sum_{i=1}^n X_i$. Let $\mu_n, \sigma_n^2$ be the mean and variance of $S_n$:*

$$\mu_n = \sum_{i=1}^n p_i, \quad \sigma_n^2 = \sum_{i=1}^n p_i(1-p_i).$$

*Then, provided $\sigma_n \to \infty$,*

$$Prob(S_n < \mu_n + x\sigma_n) \to \mathcal{N}(x),$$

*for all x.*

Theorem 3.2.1 above is the original, primordial so to speak, central limit theorem. One may dismiss it as a simple application of Stirling's formula. Consider Theorem 3.2.3, on the other hand. There are some very "combinatorial looking" expressions involved. Namely, one has exactly $2^n$ distinct products

$$A_1A_2\cdots A_n, \quad A_i \in \{p_i, 1-p_i\}. \tag{3.2}$$

Let $a(n,k)$ be the sum of the $\binom{n}{k}$ products in which the first alternative ( $A_i = p_i$ ) is exercised exactly $k$ times. Theorem 3.2.3 asserts that the numbers $a(n,k)$ are asymptotically normal, *provided* the variance becomes infinite. The binomial coefficients have appeared as the number of products making up $a(n,k)$, but not so prominently as in Equation (3.1). It is not clear that Stirling's formula can be brought to bear in proving Theorem 3.2.3.

Through the ages mathematicians have sought ways to weaken the hypotheses of Theorem 3.2.1 to fit the latest problem at hand. In Theorem 3.2.3 we still employ independent 0,1 variates, but no longer are they assumed to be identically distributed. One may consider other improvements such as variables that assume more than two values, or variables that are weakly dependent, or error estimates for the convergence to normality. The following refinement addresses the first and last of these three.

**Theorem 3.2.4** [4, 15] *Let $X_i$, $1 \le i \le n$, be independent random variables with means $\mu_i$, variances $\sigma_i^2$, and absolute third central moments $\rho_i = \mathbb{E}|X_i - \mu_i|^3$. With $\mu = \sum_i \mu_i$, $\sigma^2 = \sum_i \sigma_i^2$ denoting the mean and variance of the sum $S_n = \sum_{i=1}^n X_i$, we have*

$$|Prob\,(S_n < \mu + x\sigma) \;-\; \mathcal{N}(x)| \;\le\; \frac{C\sum_{i=1}^n \rho_i}{\sigma^3},$$

*where $C$ is a universal constant.*

This is the Berry-Esseen theorem. It can be used to prove asymptotic normality, and at the same time provide a bound on the error of the approximation. Improved estimates of the universal constant $C$ have been the topic of many papers, even recently. From a purely analytical point of view, requiring the existence of the third moment may be regarded as an excessive assumption. But in combinatorial applications we typically are looking at the limit of finite distributions and the availability of third (and higher) moments is not an issue. Thus, Theorem 3.2.4 can be a widely applicable tool for the combinatorial domain. It is left as an exercise for the reader to deduce Theorem 3.2.3 from 3.2.4.

## 3.3 Method 1: direct approach

Sometimes one can realize the enumeration problem under consideration, suitably normalized, as the distribution of a sum of independent 0,1 variables. Then, Theorem 3.2.3 is directly applicable. We illustrate this with three examples.

**Example 3.3.1** *Let*

$$\pi_1\,\pi_2\,\cdots\,\pi_n$$

*be a permutation of the integers $[n]$. A* **left-to-right-minimum** *of $\pi$ is a value $\pi_i$, $1 \le i \le n$, satisfying*

$$\pi_i = \min\{\pi_j : j \le i\}.$$

*Let $X_n$ be the number of left-to-right minima in a permutation selected uniformly at random.*

**Theorem 3.3.2** [16, 20] *With $X_n$ the number of left-to-right minima in a random permutation,*

$$Prob\left(X_n < \log n + x(\log n)^{1/2}\right) \;\to\; \mathcal{N}(x),$$

*for each fixed value of $x$. Moreover, the same conclusion holds for the random variable $Y_n$ equal to the number of cycles in a random permutation.*

*Proof.* The key to seeing that Theorem 3.2.3 is applicable is to build up the permutation $\pi$ by placing the numbers $1,2,\ldots,n$ one at a time. When 2 is placed, there are 2 possible positions relative to the previously placed element 1; when 3 is placed, there are 3 possible positions relative to the previously placed elements 1 and 2; etc. Letting the $i$ possible places for the element $i$ be equally likely means that the final permutation $\pi$ will have been formed in accordance with the uniform distribution. In the process, we can see that the probability that $i$ is a left-to-right minimum is $1/i$, and that these events are independent. The last assertion holds because whether or not $i$ will be a left-to-right minimum is unaffected by the order of the first $i-1$ placements; it depends only on whether or not $i$ is placed at the extreme left of the latter. The desired conclusion is now an immediate consequence of Theorem 3.2.3, with $p_i$ taken to be $1/i$, and the well-known formula

$$\sum_{n=1}^{n} \frac{1}{i} = \log n + O(1).$$

A bit more combinatorial thinking will allow us to extend the conclusion regarding $X_n$ to the random variable $Y_n$, the number of cycles. Suppose a certain permutation $\pi$ is presented as an unordered set of cycles. Let the smallest element of each cycle serve as the cycle's representative. Taking the cycles in the order of decreasing representatives, list the elements of each cycle starting with the smallest (the representative) and then following the cyclic order. In this manner, we go from $\pi$ given as an unordered set of cycles to $\pi'$ in one line notation, with the number of cycles of $\pi$ corresponding to the number of left-to-right minima of $\pi'$. Thus, $X_n$ and $Y_n$ have exactly the same distribution. ∎

**Example 3.3.3** *This example concerns another permutation statistic, the number of inversions. An* **inversion** *of a permutation $\pi$ of $[n]$*

$$\pi_1\,\pi_2\,\cdots\,\pi_n$$

*is a pair $(i,j)$ such that $i<j$ but $\pi_i > \pi_j$. Again taking the uniform probability distribution on permutations, we define random variable $Z_n$ by letting $Z_n(\pi)$ be the number of inversions in permutation $\pi$.*

**Theorem 3.3.4** [16, 20] *With $Z_n(\pi)$ the number of inversions of random permutation $\pi$,*

$$Prob\left(Z_n < n^2/4 + xn^{3/2}/6\right) \;\to\; \mathcal{N}(x).$$

*Proof.* The proof is only a slight embellishment of that given for Theorem 3.3.2. Let the permutation $\pi$ be built by placing the numbers $1,2,\ldots,n$ as before. When $i$ is placed, it has a $1/i$ probability of creating $j$ inversions with the previously placed numbers, where $0 \le j < i$. Thus,

$$Z_n = \sum_{i=1}^{n} Z_n^{(i)},$$

where the $Z_n^{(i)}$ are independent, and $Z_n^{(i)}$ assumes each of the values $0, 1, \ldots, i-1$ with probability $1/i$. Clearly,

$$\mathbb{E}(Z_n^{(i)}) = \frac{i-1}{2}, \quad \text{and } \mathbb{E}(Z_n) = \frac{n(n-1)}{4}.$$

Moreover,

$$\mathbb{E}\left((Z_n^{(i)})^2\right) = \frac{1}{i}\sum_{j=0}^{i-1} j^2 = i^2/3 - i/2 + 1/6,$$

whence

$$\operatorname{Var}(Z_n^{(i)}) \sim i^2/12, \quad \text{and } \sigma_n^2 \stackrel{\text{def}}{=} \operatorname{Var}(Z_n) \sim n^3/36.$$

To use Theorem 3.2.4, we need $\rho_i$, the expectation of $\left|Z_n^{(i)} - \frac{i-1}{2}\right|^3$. Clearly, for some $c$,

$$\rho_i \sim ci^3, \quad \text{and } \sum_i \rho_i \sim cn^4/4,$$

whence

$$\frac{\sum_i \rho_i}{\sigma_n^3} \sim 54cn^{4-9/2} \to 0.$$

By Theorem 3.2.4 the proof is complete. ■

**Example 3.3.5** *Both the previous examples are well explained in [17], where citations to the original papers [20] and [16] are given. Our third example is chronologically the earliest of the three, but is presented last due to its greater difficulty. Mark Kac gave a lecture shortly after his arrival in America about independence. One of his topics was divisibility by distinct primes, and he told the audience that there surely must be a central limit theorem awaiting demonstration. A member of the audience, Paul Erdős, "perked up and asked me to explain once again what the difficulty was. Within the next few minutes, even before the lecture was over, he interrupted to announce that he had the solution!" [24], page 90. The Erdős–Kac theorem marks the beginning of probabilistic number theory.*

**Theorem 3.3.6** [13] *Let $x$ be a fixed real number, and let $f(m)$ be the number of distinct prime divisors of $m$. Define*

$$K_n \stackrel{\text{def}}{=} \#\{m \le n : f(m) < \log\log n + x(\log\log n)^{1/2}\}.$$

*Then, as $n \to \infty$,*

$$\frac{K_n}{n} \to \mathcal{N}(x).$$

*Sketch of proof.* Consider the product $L = p_1 \cdots p_h$ of all the primes less than some fixed bound $H$, and treat the integers from 1 to $L$ as a (uniform) probability space. (Keep in mind throughout this proof sketch the $p_i$ are prime numbers, not probabilities.) Then the event "$m$ is divisible by $p_i$" occurs with probability $1/p_i$, and these events are (precisely) independent. So, the probability (let us call it $\alpha_k$) that an integer in $[1,L]$ has $k$ distinct prime divisors equals exactly the probability that the sum $S_h = \sum_{i=1}^{h} X_i$ of independent 0,1 random variables $X_i$, $1 \le i \le h$,

$$\text{Prob}(X_i = 1) = \frac{1}{p_i},$$

be $k$. It is seen that the same equality holds if instead of $L$ we consider any integral multiple $qL$ of $L$. If instead of $qL$ we consider a very large but arbitrary number $n$, then the equality becomes not exact, but approximate, and the error can be proven negligible by writing $n = qL + r$, with $r$ always limited to $0 \le r < L$ no matter how large $n$. In short, a certain density exists:

$$\frac{\#\{m \le n : f_H(m) = k\}}{n} \to \alpha_k,$$

where $f_H(m)$ equals by definition the number of distinct prime divisors of $m$ that are less than or equal to $H$. It is known (a weak form of Mertens' Second Theorem)

$$\mathbb{E}(S_h) = \sum_{p \le H} \frac{1}{p} \sim \log\log H,$$

and that $\text{Var}(S_h)$ is asymptotically the same. Directly by Theorem 3.2.3 above, which Erdős and Kac call the "central limit theorem of the calculus of probability,"

$$\lim_{H\to\infty} \lim_{n\to\infty} \frac{\#\{m \le n : f_H(m) < \log\log H + x(\log\log H)^{1/2}\}}{n} = \mathcal{N}(x). \tag{3.3}$$

The hitch of getting from here to the final result, Kac tells the not necessarily mathematically trained readers of his autobiography, is "one of a common variety having to do with the difficulty of justifying the interchange of taking limits." The number theoretic tool which Kac was lacking at the time of his lecture, and which Erdős supplied, is the theorem of Brun [6]: provided

$$\frac{\log n}{\log H_n} \to \infty,$$

we have

$$\#\{m \le n : m \text{ is not divisible by any prime less than } H_n\} = \left(e^{-\gamma} + o(1)\right) \frac{n}{H_n}.$$

The rest of the proof consists in defining $H = H_n$ as a function of $n$, and in lieu of the iterated limit in (3.3) considering instead

$$\lim_{n\to\infty} \frac{\#\{m \le n : f_{H_n}(m) < \log\log H_n + x(\log\log H_n)^{1/2}\}}{n}.$$

Provided $H_n$ grows sufficiently slowly with $n$, the limiting behavior seen in (3.3) is preserved; provided $H_n$ grows sufficiently quickly with $n$, the two sets

$$\{m \le n : f_{H_n}(m) < \log\log H_n + x(\log\log H_n)^{1/2}\}$$

and

$$\{m \le n : f(m) < \log\log n + x(\log\log n)^{1/2}\}$$

will be sufficiently close in size. Brun's theorem given above, plus two other asymptotic formulas due to Mertens [28],

$$\sum_{p<n} \frac{\log p}{p} = \log n + O(1),$$

and

$$\prod_{p<n}\left(1 - \frac{1}{p}\right) \sim \frac{e^{-\gamma}}{\log n},$$

constitute the necessary tools. The reader may enjoy attempting the proof. Fair warning, despite Erdős' having found the solution in "minutes," Kac reports (as the one designated to write up the proof) spending considerably more time in "assimilating Erdős' proof." ∎

## 3.4 Method 2: negative roots

By the fundamental theorem of algebra a monic polynomial $P(x)$ can be factored

$$P(x) = \prod_{r\in\mathscr{R}} (x + r),$$

where $\mathscr{R}$ is the multiset of the negatives of the (complex) roots. The coefficient of $x^k$ in $P(x)$ is given by the familiar formula

$$[x^k]P(x) = \sum_{\substack{\mathscr{S}\subseteq\mathscr{R} \\ |\mathscr{S}|=k}} \prod_{r\notin\mathscr{S}} r,$$

the sum of the products of the elements of $\mathscr{R}$ taken $|\mathscr{R}| - k$ at the time. Comparing this with the earlier appearing expression (3.2) leads to an interesting conclusion: if $r_i \ge 0$, $1 \le i \le d$, are nonnegative real numbers then

$$[x^k]\prod_{i=1}^{d} \frac{x + r_i}{1 + r_i} = \text{Prob}\left(\sum_{i=1}^{d} X_i = k\right),$$

where $X_i$, $1 \le i \le d$, are independent 0,1 random variables with

$$\text{Prob}\,(X_i = 1) = \frac{1}{1 + r_i}.$$

(The product of $(1 + r_i)$ on the left is needed to normalize the coefficients into probabilities.) In brief, when a polynomial $P(x)$ has real and nonpositive roots, we have a probabilistic interpretation of its coefficients. Theorem 3.2.4 is applicable. A bit of algebra shows that the mean and variance of $\sum_i X_i$ are given by the following "root-free" expressions

$$\mu = \frac{P'(1)}{P(1)}, \quad \sigma^2 = \frac{P(1)(P'(1)+P''(1)) - P'(1)^2}{P(1)^2}.$$

Adapting Theorem 3.2.4 to this situation yields the following.

**Theorem 3.4.1** [21] *Let*

$$P_n(x) = \sum_k c(n,k)x^k$$

*be a sequence of polynomials with real nonpositive roots, and let $\mu_n, \sigma_n^2$ be the associated means and variances. Then, provided $\sigma_n \to \infty$,*

$$\frac{1}{P_n(1)} \sum_{k<\mu_n + x\sigma_n} c(n,k) \ \to \ \mathcal{N}(x),$$

*as $n \to \infty$.*

**Example 3.4.2** *This technique was exploited by Lévy [26], and has been independently discovered at least one other time: Combinatorialists trace it to [21]. In that work, Harper studies the Stirling numbers of the second kind, $S(n,k)$, which is the number of partitions of an n-set into k nonempty and pairwise disjoint subsets. Harper's proof bears repeating.*

**Theorem 3.4.3** [21] *The polynomials $P_n(x) = \sum_k S(n,k)x^k$ have all real and nonpositive roots, and the Stirling numbers of the second kind are asymptotically normal.*

*Proof.* By a simple combinatorial argument

$$S(n+1,k) = kS(n,k) + S(n,k-1). \tag{3.4}$$

(The two terms on the right count the partitions in which $n+1$ is not, and is, respectively, in a block by itself.) Multiplying both sides by $x^k$ and summing on $k$, this translates into

$$P_{n+1}(x) \ = \ x\left(P_n'(x) + P_n(x)\right). \tag{3.5}$$

From here, one finds that the roots of these polynomials, (other than the common root $x=0$ which they all have), exhibit a nice **interlacing property**. Assume, inductively, that $P_n(x)$ has the roots

$$r_{n-1} < r_{n-2} < \cdots < r_1 < r_0 = 0.$$

It is clear that $P_n'(x)$ will be alternately positive, then negative, at the points $0, r_1, \ldots$ . Then by (3.5) $P_{n+1}(x)$ is alternately positive, then negative, at the points $r_1, r_2, \ldots$ . We have found roots for $P_{n+1}(x)$ at $x=0$, and in each of the $n-1$ intervals $(r_1,r_0),(r_2,r_1),\ldots,(r_{n-1},r_{n-2})$. That's $1+(n-1) = n$ real and nonpositive roots so far. The sign of $P_{n+1}(x)$ at $r_{n-1}$ is opposite that of $P_n'(x)$, that is, $(-1)\times(-1)^{n-1} = (-1)^n$. But $P_{n+1}(x)$, being of degree $n+1$, has sign $(-1)^{n+1}$ as $x \to -\infty$. So, there is an $(n+1)$th root in the interval $(-\infty, r_{n-1})$. By induction then, each polynomial $P_n(x)$ has $n$ roots in the interval $(-\infty, 0]$, and these roots do indeed exhibit an interlacing phenomenon.

If only we knew $\sigma_n \to \infty$, the Stirling numbers of the second kind, $S(n,k)$, would be proven asymptotically normal. Let $B_n$, the $n$th Bell number, equal $\sum_k S(n,k)$, the total number of partitions of $[n]$. Using (3.4), as found by Harper,

$$\mu_n = B_{n+1}/B_n - 1, \quad \sigma_n^2 = B_{n+2}/B_n - (B_{n+1}/B_n)^2 - 1. \tag{3.6}$$

Some careful asymptotic estimates for $B_n$ complete Harper's proof. ■

In colorful language Harper says that traditionally proofs of asymptotic normality proceed by "torturing the characteristic function until it converges to $e^{-t^2/2}$." (See Section 3.5). But in the present proof, one needs a "hat from which to pull a rabbit," and that hat is the central limit theorem. With no intention to diminish the magical aura of the proof, might not one nevertheless seek a definite combinatorial meaning for the probabilities $1/(1+r_i)$? That is, perhaps we can define some independent random variables $Y_i$ with $\sum_i Y_i$ having the desired distribution, and apply Theorem 3.2.3 directly. Think back to the first example of the direct method: the number of cycles in a random permutation. There we let each cycle be represented by its smallest element, and produced a listing of numbers whose left-to-right minima corresponded to the cycles. A similar process suggests itself for partitions of $[n]$. Let the partition $\pi$ be given as an unordered collection of subsets, usually referred to as blocks. Represent each block by its smallest element, and in this way the blocks may be numbered $1,2,\ldots,k$, $k$ being the number of blocks. Now define a function $f_\pi : [n] \to [k]$ by letting all members of the $i$th block be mapped by $f_\pi$ to $i$. Make a list of the values of this function

$$f_\pi(1)\, f_\pi(2)\, \cdots\, f_\pi(n).$$

It will be seen that $f_\pi$ is a **restricted growth function**, meaning

$$f_\pi(i) \le 1 + \max\{f_\pi(j) : j < i\},$$

and that the number of blocks of partition $\pi$ is exactly the number of left-to-right maxima of the associated restricted growth function $f_\pi$. Let the set of all restricted growth functions on $[n]$ have the uniform probability, and define the 0,1 random variables $Y_i$ to be the indicators of the event "a left-to-right maximum for $f$ occurs at position $i$." We have just seen that

$$\text{Prob}\left(\sum_i Y_i - k\right) = \frac{S(n,k)}{B_n},$$

so the sum $\sum_i Y_i$ has the desired distribution. Unfortunately, the $Y_i$ are not independent, and the theorems available to us do not apply. (The one-line notation for set partitions, widely used now, appears first in [32, page 96].)

**Example 3.4.4** *We close this section with an example taken from the theory of matchings in graphs. Let $G = (V, E)$ be a simple graph; a subset $M \subseteq E$ is called a* **matching** *in $G$ provided no two distinct edges $e, e' \in M$ have a vertex in common. Let $\mathscr{M}(G)$ be the set of all matchings in graph $G$, and define the matching polynomial $P_G(x)$ by*

$$P_G(x) \stackrel{\text{def}}{=} \sum_{M \in \mathscr{M}(G)} x^{|M|}.$$

*Thus, $[x^k]P_G(x)$ equals the number of matchings in $G$ of size $k$. In addition to its graph theoretical origins, this concept has been a long-time interest in statistical physics where the terminology is that of* **monomer-dimer coverings**. *The remarkable fact that $P_G(x)$ always has all its roots real and negative is due to this latter community. See the seminal paper of Heilmann and Lieb [23]. In that work Heilmann and Lieb prove a more general assertion using different polynomials, namely*

$$Q_G(x) \stackrel{\text{def}}{=} \sum_{M \in \mathscr{M}(G)} (-1)^{|M|} W(M) x^{n-2|M|},$$

*where whenever used as a subscript of the symbol "$Q$" $G$ is a complete graph on $n$ vertices whose edges carry positive weights, and $W(M)$ is the product of the weights of the edges comprising $M$. For a vertex $i$ in the edge-weighted complete graph $G$, partition $\mathscr{M}(G)$ into those $M$ which, respectively, do not and do cover vertex $i$; one is led to the recursion*

$$Q_G(x) = xQ_{G-i}(x) - \sum_{j \in [n]-i} w_{ij} Q_{G-i-j}(x).$$

*($w_{ij}$ is the weight on edge $ij$.) From here an argument resembling that given earlier for $S(n,k)$ establishes by induction that the roots of $Q_{G-i}$ are real, distinct, and interlace those of $Q_G$. By taking a limit in which all edge weights are 1 and 0, one finds that $x^n P_G(-1/x^2)$ ($G$ now arbitrary) has all real, albeit not necessarily distinct, roots. Whence the matching polynomial $P_G(x)$ has all roots real and nonpositive.*

Godsil [18] took up this topic and, building on Harper's work, found the following version of Theorem 3.4.1 for the matching polynomials.

**Theorem 3.4.5** [18] *Let $G_n$ be a sequence of graphs, $X_n$ be the size of a matching $M$ drawn uniformly at random from $\mathscr{M}(G_n)$, $\mu_n = \mathbb{E}(X_n)$, and $\sigma_n^2 = Var(X_n)$. Then, provided $\sigma_n^2 \to \infty$,*

$$Prob\,(X_n < \mu_n + x\sigma_n) \;\to\; \mathscr{N}(x).$$

Two applications are noted: $G_n$ a sequence of regular graphs of degree $r$, with the number of vertices becoming infinite; and $G_n$ equal to the complete graph, $K_n$. Later, Kahn [25] continued the study of this topic, and gave, among other things, a formulation of the condition $\sigma_n \to \infty$ in purely graph theoretical notions.

**Theorem 3.4.6** [25] *Let $G_n$ be a sequence of graphs, $X_n$ be the size of a matching $M$ drawn uniformly at random from $\mathscr{M}(G_n)$, $\mu_n = \mathbb{E}(X_n)$, and $\nu_n$ the size of the largest matching in $\mathscr{M}(G_n)$. Then*

$$Prob\,(X_n < \mu_n + x\sigma_n) \;\to\; \mathscr{N}(x)$$

*if and only if*

$$\nu_n - \mu_n \;\to\; \infty.$$

## 3.5 Method 3: moments

For a random variable $X$, the associated **moment generating function** is defined by

$$M_X(\tau) \;\stackrel{\text{def}}{=}\; \mathbb{E}e^{\tau X}.$$

No question of convergence arises when $\tau$ is pure imaginary, but $\tau = 0$ could be the only real value for which $M_X(\tau)$ is defined. For $X$ distributed as standard normal, $M_X(\tau)$ is the entire function $e^{\tau^2/2}$. When $M_X(\tau)$ is analytic near $\tau = 0$, it has the power series expansion

$$M_X(\tau) \;=\; \sum_{k=0}^{\infty} \mathbb{E}X^k \frac{\tau^k}{k!},$$

hence the nomenclature "moment generating function." The following theorem finds frequent combinatorial application.

**Theorem 3.5.1** [11] *Suppose $X_n$ is a sequence of random variables with distribution functions $F_n(x)$ such that*

$$\mathbb{E}e^{\tau X_n} \to e^{\tau^2/2}$$

*for all $\tau$ in a nonempty real interval $(-a, +a)$. Then, for all $x$,*

$$F_n(x) \to \mathscr{N}(x).$$

Similar limit theorems are known for the characteristic function $\mathbb{E}e^{i\tau X}$ (Lévy's Continuity Theorem, [5], Theorem 26.3) and for moments (the Fréchet-Shohat Theorem, [5], Theorem 30.2). The theorem for moments states that if each member of sequence $X_n$ has moments of all orders, and for each $k$ the $k$th moments of $X_n$ converge to the $k$th moment of the standard normal, then the $X_n$ are asymptotically normal with mean 0 and variance 1.

**Example 3.5.2** *Let $q(n,k)$ be the number of partitions of $n$ into exactly $k$* **distinct** *parts. For example, $q(8,3) = 2$, the two partitions being $5+2+1$ and $4+3+1$. The following theorem is stated without proof in [14], excluding the explicit formula for the variance.*

**Theorem 3.5.3** [14] *Define*

$$\mu_n = \frac{2\sqrt{3}\ln(2)}{\pi} n^{1/2}$$
$$\sigma_n^2 = \frac{\sqrt{3}}{\pi}\left(1 - \left(\frac{2\sqrt{3}\ln(2)}{\pi}\right)^2\right) n^{1/2}.$$

*Then the numbers $q(n,k)$ are asymptotically normal with mean $\mu_n$ and variance $\sigma_n^2$.*

*Proof.* Define the polynomials $Q_n(y) = \sum_k q(n,k)y^k$, and the generating function

$$f(x,y) = \sum_{n=0}^{\infty} x^n Q_n(y) = \prod_{m=1}^{\infty}(1+yx^m).$$

By Theorem E it suffices to show

$$e^{-\tau\mu_n/\sigma_n}\frac{Q_n(e^{\tau/\sigma_n})}{Q_n(1)} \to e^{\tau^2/2}. \tag{3.7}$$

Using contour integration around a circle of radius $e^{-t}$, $t > 0$, we have

$$e^{-nt}Q_n(e^{\tau}) = \frac{1}{2\pi}\int_{-\pi}^{+\pi} f(e^{-t+i\theta}, e^{\tau})e^{-ni\theta}\,d\theta.$$

Throughout the remainder of the proof, asymptotic formulas hold uniformly for $n \to \infty$ and $t, \tau \to 0$ with $t = (c_2/n)^{1/2}$ and $\tau = O(t^{1/2})$. (The constants $c_0, c_1, c_2$ are respectively $1/2, \ln 2, \pi^2/12$.)

Using

$$f(e^{-t+i\theta}, e^{\tau})e^{-ni\theta} \sim f(e^{-t}, e^{\tau})\exp\left(c_1\tau t^{-2}i\theta - c_2 t^{-3}\theta^2\right), \; |\theta| \le t^{7/5}, \tag{3.8}$$

we find

$$\int_{-t^{7/5}}^{+t^{7/5}} f(e^{-t+i\theta}, e^{\tau})e^{-ni\theta}\,d\theta \sim \sqrt{\frac{\pi}{c_2}}f(e^{-t}, e^{\tau})t^{3/2}\exp\left(-\frac{c_1^2}{4c_2}\tau^2 t^{-1}\right).$$

Using

$$|f(e^{-t+i\theta}, e^{\tau})| \le f(e^{-t}, e^{\tau})e^{-c/t^{1/7}}, \quad c > 0,\ t^{7/5} \le |\theta| \le \pi, \tag{3.9}$$

we find that the integral over $t^{7/5} \le |\theta| \le \pi$ is negligible by comparison. This gives an asymptotic for $e^{-nt}Q_n(e^{\tau})$, which we use twice to find

$$\frac{Q_n(e^{\tau})}{Q_n(1)} \sim \frac{f(e^{-t}, e^{\tau})}{f(e^{-t}, 1)}\exp\left(-\frac{c_1^2}{4c_2}\tau^2 t^{-1}\right).$$

Finally, using

$$\frac{f(e^{-t}, e^{\tau})}{f(e^{-t}, 1} \sim \exp\left(c_1\tau t^{-1} + \frac{c_0}{4}\tau^2 t^{-1}\right), \tag{3.10}$$

we reach the desired conclusion (3.7). The three key formulas (3.8), (3.9), and (3.10) can be proven either by Euler-Maclaurin summation, or by the Cahier-Mellin formula; see [8], for a similar calculation. ∎

The value for $\sigma_n^2$ appears in [31]. In the latter work, Szekeres proves a formula equivalent to

$$q(n,k) \sim \frac{e^{-x^2/2}}{\sigma_n(2\pi)^{1/2}} Q_n(1), \quad k = \mu_n + x\sigma_n,$$

for $k - \mu_n = o(n^{1/3})$. That is, he proves a local limit theorem; see Section 3.7. In [14] Erdős and Lehner concentrate on $p(n,k)$, the analogous quantity for unrestricted partitions. These numbers are **not** asymptotically normal, and it would be interesting to follow the methodology of Theorem 6 to see where the difference arises. Erdős and Lehner's derivation of the limiting distribution of $p(n,k)$ is a brilliant application of inclusion/exclusion—reminiscent of Brun's sieve in Example 3. Another theorem about partitions, similar in conclusion to Theorem 6, has been given by Goh and Schmutz [19]: The number of distinct parts in a random unrestricted partition is asymptotically normal with mean $(\sqrt{6}/\pi)n^{1/2}$ and variance $\left(\sqrt{6}/(2\pi) - \sqrt{54}/\pi^3\right) n^{1/2}$.

**Example 3.5.4** *Let $g_i$, $i \geq 1$, be a sequence of nonnegative integers, and define a doubly-indexed array $a(n,k)$ by the equation*

$$a(n,k) \stackrel{\text{def}}{=} \frac{1}{k!} \sum_{\substack{(i_1,\dots,i_k) \\ \sum i\alpha = n}} \binom{n}{i_1 \cdots i_k} g_{i_1} \cdots g_{i_k}. \tag{3.11}$$

*Combinatorially, $a(n,k)$ is the number of ways to partition $[n]$ into $k$ blocks, and to build a g-object on each block, $g_i$ being the number of g-objects on an i-set. If, for instance, we let $g_i$ equal 1 for all i, then the above becomes a familiar formula for $S(n,k)$, the number of partitions of $[n]$ into k blocks. As another illustration, if $g_i = (i-1)!$, the number of full cycles on an i-set, then the above $a(n,k)$ is the number of permutations of $[n]$ having k cycles. Letting $g(u)$ be the exponential generating function of the sequence $g_i$,*

$$g(u) = \sum_{i \geq 1} g_i \frac{u^i}{i!},$$

*the generating function equivalent of (3.11) is*

$$\sum_{n=0}^{\infty} \frac{u^n}{n!} \left( \sum_k a(n,k) y^k \right) = \exp(yg(u)).$$

*The polynomials $P_n(y) = \sum_k a(n,k)y^k$ are* **polynomials of binomial type** *[30].*

**Theorem 3.5.5** [7] *Let*

$$g(u) = \sum_{j=1}^{d} g_j u^j / j! = \sum_{j=1}^{d} c_j u^j$$

*be a real polynomial with nonnegative coefficients, at least two of which are nonzero, and such that*

$$gcd\{j : c_j \neq 0\} \;=\; 1.$$

*Let the doubly-indexed array $a(n,k)$ be defined by (3.11). Then the numbers $a(n,k)$ are asymptotically normal.*

*Sketch of proof.* As in the prior example, relying on Theorem 3.5.1, it suffices to define $\mu_n, \sigma_n^2$ that satisfy

$$e^{-\tau\mu_n/\sigma_n}\frac{P_n(e^{\tau/\sigma_n})}{P_n(1)} \;\to\; e^{\tau^2/2}. \tag{3.12}$$

Again, we begin with contour integration around a circle to extract $P_n(y)/n!$ as the coefficient of $u^n$ in $\exp(yg(u))$, and, as before, the integral is dominated by a small arc near the positive real axis, provided the radius $\rho$ is chosen by the saddlepoint equation

$$\rho g'(\rho) \;=\; n/y.$$

Let $r(X)$ denote the inverse function of $Xg'(X)$, so the latter radius is $\rho = r(n/y)$. The estimate resulting from the contour integration is

$$r(n/y)^n \frac{P_n(y)}{n!} \;\sim\; \frac{\exp(yg(r(n/y)))}{(2\pi B)^{1/2}}, \quad n \to \infty,\; y \to 1,$$

with $B = d^2 c_d r(n)^d$ (recall $d$ is the degree of $g(u)$). In Example 3.5.2, the next step was to form the quotient, but here a nice feature of Example 3.5.2 is missing: Even for $y \to 1$ we must keep up with both $r(n)$ and $r(n/y)$, and we cannot use the same radius for both $P_n(1)$ and $P_n(y)$. However, the common terms, $B$ and $n!$, do drop out of the quotient leaving

$$\frac{P_n(y)}{P_n(1)} \;\sim\; \frac{\exp(yg(r(n/y)))}{\exp(g(r(n)))} \cdot \left(\frac{r(n)}{r(n/y)}\right)^n, \quad n \to \infty,\; y \to 1.$$

The next step is to recognize the right side of this formula as

$$\exp\left(\int_1^y g(r(n/\beta))\,d\beta\right).$$

Looking back at (3.12), we need to define $\mu_n, \sigma_n^2$ so that

$$-\frac{\tau\mu_n}{\sigma_n} + \int_1^{e^{\tau/\sigma_n}} g(r(n/\beta))\,d\beta \;\to\; \frac{\tau^2}{2}.$$

The expansion of $g(r(n/\beta))$ about $\beta = 1$ begins

$$g(r(n/\beta)) \;=\; g(r(n)) - \frac{n^2}{B}(\beta - 1) + \cdots$$

and so we are led to define

$$\mu_n = g(r(n))$$
$$\sigma_n^2 = g(r(n)) - \frac{n^2}{B}.$$

The rest of the proof involves looking carefully at the higher-order derivatives of the function $\beta \mapsto g(r(n/\beta))$. The number of derivatives that come into play, call it $K$, must satisfy $KJ/2 > d$, where $J$ is the second highest index (after $d$) such that $c_J \neq 0$. Details can be found in [7]. ∎

Applications of Theorem 3.5.5 include partitions into blocks of bounded size, permutations with cycles of bounded size, and permutations whose order divides a given $d > 1$.

## 3.6 Method 4: singularity analysis

The next theorem furnishes an example of a simply stated criterion on the bivariate generating function $f(x,y)$, which implies asymptotic normality for the coefficients $a(n,k)$. In this case, the criterion is that the function have only one singularity on the circle of convergence, and that it be well behaved.

**Theorem 3.6.1** [3] *Let $f(x,y) = \sum_{k,n} a(n,k)x^n y^k$, with $a(n,k) \geq 0$. Suppose there exist*

*(i) a function $A(s)$ continuous and nonzero near 0,*

*(ii) a function $r(s)$ with bounded third derivative near 0,*

*(iii) a nonnegative integer $m$, and*

*(iv) positive numbers $\varepsilon$ and $\delta$ such that*

$$\left(1 - \frac{x}{r(s)}\right)^m f(x,e^s) - \frac{A(s)}{1 - x/r(s)}$$

*is analytic and bounded for $|s| < \varepsilon$, $|x| < r(0) + \delta$.*

*Put $\mu = -r'(0)/r(0)$ and $\sigma^2 = \mu^2 - r''(0)/r(0)$. If $\sigma^2 \neq 0$, then the numbers $a(n,k)$ are asymptotically normal with mean $\mu n$ and variance $\sigma^2 n$.*

*Sketch of proof.* Let

$$f(z,e^s) = \sum \phi_n(s) z^n.$$

If a power series $\sum a_n z^n$ has finite radius of convergence $r$, and $f(z) - A/(1-z/r)^m$ is analytic in a larger circle, then $a_n \sim A n^{m-1} r^{-n}/(m-1)!$. By this sort of analysis it is shown

$$\phi_n(s) \sim A(s) n^m r(s)^{-n}/m!,$$

uniformly in $s$. The characteristic function is

$$e^{-it\mu_n/\sigma_n} \frac{\phi(it/\sigma_n)}{\phi(1)} \sim \left(\frac{r(it/\sigma_n)}{r(0)}\right)^{-n},$$

and by expanding

$$\frac{r(s)}{r(0)} = \exp\left(\frac{r'(0)}{r(0)} s + \left(\frac{r''(0)}{r(0)} - \left(\frac{r'(0)}{r(0)}\right)^2\right)(s^2/2) + \cdots\right)$$

the proof is completed. ∎

**Example 3.6.2** *Say that a permutation $\pi$ of $[n]$ has a rise at position $i$, $1 \le i < n$, if $\pi_i < \pi_{i+1}$. For $n \ge 1$ we follow the (fairly) standard practice of declaring there also to be a rise at position $i = 0$. Let $A(n,k)$, $1 \le k \le n$, be the number of permutations of $[n]$ having $k$ rises. These are called the* **Eulerian numbers**.

**Theorem 3.6.3** [3, 12] *The Eulerian numbers $A(n,k)$ are asymptotically normal with mean $\mu_n = n/2$ and variance $\sigma_n^2 = n/12$.*

*Proof.* We start with the two-variable exponential generating function as given by David and Barton [12]

$$f(x,y) \stackrel{\text{def}}{=} \sum_{n,k \ge 0} E_n(y) \frac{x^n}{n!} = \frac{1-y}{1 - ye^{-x(y-1)}}. \tag{3.13}$$

By solving for the vanishing of the denominator, we find that the radius of convergence for $f(x, e^s)$ is

$$r(s) = \frac{s}{e^s - 1} = 1 - s/2 + s^2/12 + \cdots.$$

Theorem 3.6.1 is applicable, and $r(0), r'(0), r''(0)$ equal $1, -1/2, 1/6$ respectively. Hence, $\mu = 1/2$ and $\sigma^2 = (1/2)^2 - 1/6 = 1/12$. ∎

**Remarks.** The asymptotic normality of the Eulerian numbers can be, and has been, obtained in other ways. Let $E_n(y) = \sum_k A(n,k) y^k$ be the associated polynomials. With the convention that $E_0 = 1$, the following recursion can be deduced by combinatorial reasoning:

$$A(n,k) = kA(n-1,k) + (n-k+1)A(n-1,k-1), \quad n \ge 1.$$

From here we find

$$E_n(y) = y(1-y)E'_{n-1}(y) + nyE_{n-1}, \quad n \ge 1. \tag{3.14}$$

By taking derivatives with respect to $y$, evaluating at $y = 1$, and solving the resulting recursions, one can obtain the two exact formulas

$$\mu_n = \frac{n+1}{2}$$
$$\sigma_n^2 - \frac{n+1}{2}$$

for the mean and variance. Using the technique illustrated in Section 3.4 one can prove from (3.14) that the polynomials $E_n(y)$ have interlacing real negative roots. Hence, by Theorem 3.4.1, the Eulerian numbers $A(n,k)$ are asymptotically normal. This is the approach taken in [9].

Another approach is that taken by David and Barton [12]. Working with the two-variable exponential generating function (3.13), they prove asymptotic normality using a variant of the method of moments, the method of cumulants.

## 3.7 Local limit theorems

Suppose a doubly-indexed array $a(n,k)$ is asymptotically normal with mean $\mu_n$ and variance $\sigma_n^2$. This may tell us nothing about an individual number $a(n,k)$. For instance, in Example 3, concerning the number, $f(m)$, of prime divisors of a random integer $m$, we dare not hope for an asymptotic formula for the number of prime divisors of $m$. Indeed, we suspect that integers with one prime divisor may very well be adjacent to an integer with an extraordinary number of prime divisors. Just knowing the **frequency** with which $f(m)$ assumes various values, the very content of asymptotic normality, is a surprising and pleasing amount of information for such a noisy function.

On the other hand, suppose for a given $n$ the numbers $a(n,k)$ do not vary chaotically like the divisor counting function; suppose instead they vary smoothly. Let $\mathscr{D}_x$ be the discrete set of real $t$ in the interval $(-\infty, x]$ with the property that $\mu_n + t\sigma_n$ is an integer. The $t \in \mathscr{D}_x$ are spaced at a distance $1/\sigma_n$, and if $\sigma_n \to \infty$,

$$\frac{1}{\sigma_n} \sum_{t \in \mathscr{D}_x} \frac{e^{-t^2/2}}{\sqrt{2\pi}} \to \mathscr{N}(x).$$

But by asymptotic normality

$$\sum_{t \in \mathscr{D}_x} \frac{a(n, \mu_n + t\sigma_n)}{\sum_k a(n,k)} \to \mathscr{N}(x).$$

Here we have two smoothly varying sequences of numbers whose partial sums are asymptotic; optimistically, one hopes

$$\frac{a(n,k)}{\sum_k a(n,k)} \sim \frac{e^{-t^2/2}}{\sigma_n \sqrt{2\pi}} \Big|_{t=(k-\mu_n)/\sigma_n}.$$

This prompts the next definition.

**Definition 3.7.1** *We say the doubly indexed sequence $a(n,k)$ satisfies a* **local limit theorem on the set** $S$ **of real numbers** *provided*

$$\sup_{x\in S}\left|\frac{\sigma_n a(n,\lfloor \mu_n + x\sigma_n\rfloor)}{\sum_k a(n,k)} - \frac{e^{-x^2/2}}{\sqrt{2\pi}}\right| \to 0.$$

**Theorem 3.7.2** [3] *Suppose that $a(n,k)$ are asymptotically normal, and $\sigma_n^2 \to \infty$. If for each $n$ the sequence $a(n,k)$ is unimodal in $k$, then $a(n,k)$ satisfy a local limit theorem on the set $\{x : |x| \geq \varepsilon\}$, any $\varepsilon > 0$. If for each $n$ the sequence $a(n,k)$ is log concave in $k$, then $a(n,k)$ satisfy a local limit theorem on the set $\mathbb{R}$.*

Unimodality,

$$a(n,1) \leq \cdots \leq a(n,K) \geq a(n,K+1) \geq \cdots,$$

and log concavity,

$$a(n,k)^2 \geq a(n,k+1)a(n,k-1),$$

are features that arise often in combinatorics. Both of these constitute "smoothness" adequately enough to pass from a central to local limit theorem, as stated in Theorem 3.7.2.

**Example 3.7.3** *If $P_n(y) = \sum_k a(n,k)y^k$ has all its roots real and nonpositive, then its coefficients are log concave. Thus, from Examples 3.4.2 and 3.6.2, we derive genuine asymptotic formulas for Stirling numbers of the second kind $S(n,k)$ and Eulerian numbers $A(n,k)$ when $(k-\mu_n)/\sigma_n$ is bounded.*

**Theorem 3.7.4** [21, 29] *Let $\mu_n = n/\ln n$ and $\sigma_n^2 = n/(\ln n)^2$. Then, uniformly for $k - \mu_n = O(\sigma_n)$,*

$$\frac{S(n,k)}{B_n} \sim \frac{e^{-x_n^2/2}}{\sigma_n(2\pi)^{1/2}}, \quad x_n = (k-\mu_n)/\sigma_n.$$

*Proof.* To obtain the formulas for $\mu_n, \sigma_n^2$ we use Equation (3.6) and an asymptotic estimate of the Bell number $B_n$. Then, Theorem 3.7.2 is applicable. ∎

**Theorem 3.7.5** [9] *Let $\mu_n = n/2$ and $\sigma_n^2 = n/12$. Then, uniformly for $k - \mu_n = O(\sigma_n)$,*

$$\frac{A(n,k)}{n!} \sim \frac{e^{-x_n^2/2}}{\sigma_n(2\pi)^{1/2}}, \quad x_n = (k-\mu_n)/\sigma_n.$$

*Proof.* Derivation of the formulas for $\mu_n, \sigma_n^2$ has been indicated earlier. Then, Theorem 3.7.2 is applicable. ∎

**Remark 3.7.6** Harper [21] states Theorem 9.1 as a corollary to his central limit theorem for $S(n,k)$, giving a quite succinct and informal justification. The first asymptotic formula for $S(n,k)$ covering all $k$ was provided earlier by Moser and Wyman [29]. In Moser and Wyman's formula the connection to the central and local limit theorems is not at all apparent unless one specializes $k$ to be near $\mu_n$ and does the necessary algebra. In their work [9] Carlitz et al. acknowledge their debt to Harper, and utilize the Berry-Esseen theorem in deriving an asymptotic formula for $A(n,k)$ with explicit error bound. Their proof, which Riordan complements in his math review, can be read as a careful fleshing out of Harper's succinct proof. Theorem G, due to Bender, can be read as an adaptation of the local limit theorem in [9] that does not use the Berry-Esseen theorem and that holds in perfect generality.

## 3.8 Multivariate asymptotic normality

With the intent of providing the reader with the basic and most widely used principles of our topic, the discussion thus far has been totally one-dimensional. Of course, there is such a thing as a $d$-dimensional Gaussian distribution, and multivariate distributions are ubiquitous in combinatorial settings.

**Definition 3.8.1** *Let* $\mathbf{k},\mathbf{i}$ *be $d$-dimensional vectors of integers, and* $a(n,\mathbf{k})$ *an array of nonnegative numbers. We say that* $a(n,\mathbf{k})$ **is asymptotically normal with mean** $\mathbf{m}_n$ **and covariance matrix** $B_n$ *if*

$$\sup_{\mathbf{u}}\left|\sum_{\mathbf{i}\le\mathbf{m}_n+\mathbf{u}}\frac{a(n,\mathbf{i})}{\sum a(n,\mathbf{k})}-\frac{1}{(2\pi)^{d/2}|B_n|^{1/2}}\int_{\mathbf{x}\le\mathbf{u}}\exp\left(-\frac{1}{2}\mathbf{x}B_n^{-1}\mathbf{x}^T\right)d\mathbf{x}\right|\to 0.$$

For a $d$-dimensional normal $Z$ with mean $\mathbf{0}$ and covariance matrix $B$ (the probability density in the second operand above), the moment generating function is found to be

$$\mathbb{E}\left(e^{\tau\cdot Z}\right) = \exp\left(\frac{1}{2}\tau B\tau^T\right).$$

As in the one-dimensional case, one may prove convergence to normality by the method of moments. For instance, let each of $\xi_1,\xi_2,\ldots,\xi_\ell$ be a $d$-dimensional real vector, and consider the random variable $X=(x_1,\ldots,x_d)$ given by

$$X=\xi_i \text{ with probability } c_i,\quad 1\le i\le\ell,\quad \sum_i c_i=1.$$

Form the $\ell\times d$ matrix $M$ using $(c_i)^{1/2}\xi_i$ as the $i$th row:

$$(M)_{ij}=(c_i)^{1/2}(\xi_i)_j;$$

then, the $(j,k)$ entry of the product $M^T M$ is $\mathbb{E}(x_i x_j)$. With $\mathbf{m}$ the vector of means, we find

$$\mathbb{E}e^{\tau \cdot X} = \exp\left( \mathbf{m} \cdot \tau + \frac{1}{2}\tau\left(M^T M - N\right)\tau^T + O(\sum_j |\tau_j|^3) \right),$$

where $(N)_{jk} = (\mathbb{E}x_j)(\mathbb{E}x_k)$. Now $B = M^T M - N$ is the covariance matrix for $X$, and it follows that the sums $\sum_{i=1}^n X_i$ (with the $X_i$ independent and each distributed like $X$) are asymptotically normal with mean $n\mathbf{m}$ and covariance matrix $nB$. So, we have a $d$-dimensional analog of Theorem 3.2.3 courtesy of the method of moments, and we might reasonably anticipate products $g(\mathbf{x})^n$ to be likely sources of normality.

In [1] it is shown that if the coefficients $\phi_n(\mathbf{x})$ in the ordinary generating function

$$\sum_n a(n,\mathbf{k}) z^n \mathbf{x}^{\mathbf{k}} = \sum_n \phi_n(\mathbf{x}) z^n$$

behave asymptotically like a product $\alpha_n h(\mathbf{x}) g(\mathbf{x})^n$, then under suitable general conditions the $a(n,\mathbf{k})$ will satisfy a central limit theorem. Generalizing Theorem 3.6.1, Bender and Richmond prove in [1] that this principle applies to generating functions $f(\mathbf{x},z)$ for which

$$\left(1 - \frac{z}{r(\mathbf{s})}\right)^q f(e^{\mathbf{s}}, z) - \frac{A(\mathbf{s})}{1 - z/r(\mathbf{s})}$$

is analytic and bounded for $\| \mathbf{s} \| < \varepsilon$, $|z| < |r(\mathbf{0})| + \delta$, with $A(\mathbf{s}), r(\mathbf{s})$ satisfying conditions much as in Theorem F. As an example, they apply this theorem to the generating function

$$f(x_1,x_2,z) = \sum a(n,k_1,k_2) x_1^{k_1} x_2^{k_2} \frac{z^n}{n!} = \frac{x_2 S + x_2(x_1+1)C}{1 - (x_0 + x_1)C},$$

in which $a(n,k_1,k_2)$ is the number of permutations of $[n]$ having $k_1$ rises in odd positions and $k_2$ falls in even positions [10]. Here,

$$C = \cosh(\beta z) - 1)/\beta^2, \;\; S = \sinh(\beta z)/\beta, \;\; \beta = \sqrt{(x_2-1)(x_1-1)}.$$

They deduce that $a(n,k_1,k_2)$ are asymptotically normal with mean equal to $n\mathbf{m}$ and covariance matrix equal to $nB$ where

$$\mathbf{m} = (1/4, 1/4), \;\; B = \begin{bmatrix} 1/8 & -1/12 \\ -1/12 & 1/8 \end{bmatrix}.$$

As another example, consider $a(n,k_1,k_2)$ equal to the number of matchings in the product graph $[n] \times [2]$ consisting of $k_1$ horizontal and $k_2$ vertical edges. Then the generating function is given by

$$\sum_{n,k_1,k_2} a(n,k_1,k_2) z^n x_1^{k_1} x_2^{k_2} = \left( \sum_{n \geq 0} \begin{bmatrix} 1+x_1 & 1 & 1 & 1 \\ x_2 & 0 & x_2 & 0 \\ x_2 & x_2 & 0 & 0 \\ x_2^2 & 0 & 0 & 0 \end{bmatrix}^n \right)_{(1,1)}$$

(indicating the extraction of element $(1,1)$ from the matrix) [2]. For a great variety of such matrix recursions, the authors of [2] show that both central and local limit theorems hold. The mean and variance are again proportional to $n$, but the details of how to calculate these are not easy and the interested reader is asked to consult the original paper.

## 3.9 Normality in service to approximate enumeration

Let $g(u) = \sum g_j u^j$ be a power series with real, nonnegative coefficients having radius of convergence $R$, $0 < R \leq \infty$. For each $r < R$ we may consider the random variable $X_r$ which assumes the value $j$ with probability $g_j r^j / g(r)$. Define

$$a(r) \stackrel{\text{def}}{=} rg'(r)/g(r), \quad b(r) \stackrel{\text{def}}{=} ra'(r);$$

it will be seen that these are the mean and variance of $X_r$. Assuming $X_r$ to be asymptotically normal gives

$$\frac{1}{g(r)} \sum_{j \leq a(r) + xb(r)^{1/2}} g_j r^j \to \mathcal{N}(x), \quad r \to R. \tag{3.15}$$

Let the sequence $r_n$ be defined by the equation

$$a(r_n) = n.$$

If $r_n \to R$, and if passage from central to local limit theorem is permitted, we have (after isolating $g_n$ on one side)

$$g_n \sim \frac{g(r_n)}{r_n^n (2\pi b(r_n))^{1/2}}, \tag{3.16}$$

$n$ being $a(r_n) + 0 \times b(r_n)^{1/2}$ and $\mathcal{N}(0)$ being $(2\pi)^{-1/2}$. In [22] Hayman has axiomatized, so to speak, a large class of power series $g(u)$ for which the conclusions (3.15) and (3.16) hold true. Hayman calls these functions $g(u)$ **admissible**. We present the exact definition in a moment, but first the prototypical example: $g(u) = e^u$. Then, $a(r) = r$, $b(r) = r$, $r_n = n$, and from (3.16)

$$\frac{1}{n!} \sim \frac{e^n}{n^n (2\pi n)^{1/2}},$$

Stirling's formula (upside down). Hence the title of [22]. Now for the definition.

Observe that the characteristic function of $X_r$ is $g(re^{i\theta})/g(r)$; by Section 3.5 a necessary and sufficient condition for $X_r$ to be asymptotically normal is

$$e^{-i\theta a(r)/b(r)^{1/2}} g(re^{i\theta/b(r)^{1/2}})/g(r) \;\to\; e^{-\theta^2/2}.$$

**Definition 3.9.1** *A power series $g(z)$ convergent for $|z| < R$, with $0 < R \leq \infty$, is* **admissible** *provided we have a function $\delta(r)$, $0 < \delta(r) < \pi$, defined for $R_0 < r < R$ such that*

$$g(re^{i\theta}) \sim g(r)\exp(i\theta a(r) - (1/2)\theta^2 b(r)), \quad \text{as } r \to R,$$

*uniformly for $|\theta| \leq \delta(r)$, while uniformly for $\delta(r) \leq |\theta| \leq \pi$,*

$$g(re^{i\theta}) = \frac{o(g(r))}{b(r)^{1/2}}, \quad \text{as } r \to R.$$

*It is also required that*

$$b(r) \to +\infty, \quad \text{as } r \to R.$$

Deriving (3.15) and (3.16) for an admissible function is a simple application of the circle method. The value of Hayman's paper lies in its characterization of a large class of power series to which the circle method is applicable. Numerous properties of admissible functions are proven, as well as closure of the class under a number of operations. The paper provides a template of how to use the circle method effectively. For example, the proofs in Section 3.5 are found by making simple modifications of this template. The later and deeper sections of the paper have not been explored by combinatorialists whose immediate intent is applications to enumeration.

In later years McKay, jointly with a variety of co-authors, has developed a new method of asymptotic enumeration that follows this same general pattern. The seminal example is the generating function

$$f(\mathbf{x}) = \prod_{1\leq i<j\leq n} (1 + x_i x_j),$$

treated by McKay and Wormald in [27]. The coefficient

$$[x_1^{d_1} \cdots x_n^{d_n}]\, f(\mathbf{x})$$

is the number of simple, labeled graphs on $n$ vertices having degree sequence $d_1, \ldots, d_n$. For certain degree sequences the coefficient can be estimated by the multi-dimensional circle method. For the multi-dimensional integral a set of radii $r_1, \ldots, r_n$ is needed, and these are found by solving a system of saddlepoint equations. The main contribution to the integral might arise from $x_i = r_i e^{\theta_i}$ with $|\theta_i|$ all small, but in some instances a host of small contributing regions is identified. Around the points that do contribute, the shape of the integrand is found to be asymptotically normal, now a multivariate normal, and the estimation of the primary integral involves diagonalizing the Hermitian covariance matrix. Complications arise due to the fact that the number of variables is growing along with $n$. This means that the Taylor series used in the primary integrals cannot stop with a linear and quadratic, but must go at least, typically, to quintic. Estimating the integral of these quintic terms is a new challenge. The necessary argument that the contribution to the integral from outside the primary region is negligible becomes quite complicated and combinatorial in nature. Although similar issues recur in each new application of this method, no useful general systematization has yet been proposed. The intermediate algebra and analysis can become too large to handle without computer-based methods, yet the final result is typically a succinct and elegant asymptotic formula. These formulas strikingly agree with formulas found for other ranges of the parameters and proven by totally different methods.

# References


[1] E. A. Bender and L. B. Richmond. Central and local limit theorems applied to asymptotic enumeration ii. multivariate generating functions. *J. Combin. Theory, Series A*, 34:255–265, 1983.

[2] E. A. Bender, L .B. Richmond, and S. G. Williamson. Central and local limit theorems applied to asymptotic enumeration iii. matrix recursions. *J. Combin. Theory, Series A*, 34:255–265, 1983.

[3] Edward A. Bender. Central and local limit theorems applied to asymptotic enumeration. *J. Combinatorial Theory*, 15:91–111, 1973.

[4] Andrew C. Berry. The accuracy of the Gaussian approximation to the sum of independent variates. *Transactions of the American Mathematical Society*, 49:122–136, 1941.

[5] Patrick Billingsley. *Probability and Measure*. Wiley, 1995.

[6] Viggo Brun. Le crible d'erasosthène et le théorème de goldbach, 1920. Skrifter Videns, Kristiania.

[7] E. Rodney Canfield. Central and local limit theorems for the coefficients of polynomials of binomial type. *Journal of Combinatorial Theory, Series A*, 23:275–290, 1977.

[8] E. Rodney Canfield. Durfee and dilogarithmic deductions. *Adv. in Appl. Math.*, 34:768–797, 2005.

[9] Leonard Carlitz, David C. Kurtz, Richard Scoville, and O. P. Stackelberg. Asymptotic properties of Eulerian numbers. *Z. Wahrscheinlichkeitstheorie und Verw. Gebiete*, 23:47–54, 1972.

[10] Leonard Carlitz and Richard Scoville. Enumeration of rises and falls by position. *Discrete Math.*, 5:45–59, 1973.

[11] J. H. Curtiss. A note on the theory of moment generating functions. *Ann. Math. Statist.*, 13:371–453, 1942.

[12] F. N. David and E. E. Barton. *Combinatorial Chance*. Griffin, 1962.

[13] Paul Erdős and Mark Kac. The Gaussian law of errors in the theory of additive number theoretic functions. *American Journal of Mathematics*, 62:738–742, 1940.

[14] Paul Erdős and Joseph Lehner. The distribution of the number of summands in the partitions of a positive integer. *Duke Math. J.*, 8:335–345, 1941.

[15] Carl-Gustav Esseen. On the liapunoff limit of error in the theory of probability. *Arkiv for Matematik, Astronomi och Fysik*, 28A:1–19, 1942.

[16] William Feller. The fundamental limit theorems in probability. *Bull. Amer. Math. Soc.*, 51:800–832, 1945.

[17] William Feller. *An Introduction to Probability Theory and Its Applications, volume I.* Wiley, 1957.

[18] Christopher D. Godsil. Matching behaviour is asymptotically normal. *Combinatorica*, 1:369–376, 1981.

[19] William M. Y. Goh and Eric Schmutz. The number of distinct part sizes in a random partition. *J. Comb. Theory Ser A.*, 69:149–158, 1995.

[20] V. L. Goncharov. From the realm of combinatorics. *Izv. Akad. Nauk SSSR*, 8:3–48, 1944.

[21] Larry H. Harper. Stirling behaviour is asymptotically normal. *Ann. Math. Statist.*, 38:410–414, 1967.

[22] Walter K. Hayman. A generalisation of stirling's formula. *J. Reine Angew. Math.*, 196:67–95, 1956.

[23] O. J. Heilmann and E. H. Lieb. Theory of monomer-dimer systems. *Comm. Math. Phys.*, 25:190–232, 1972.

[24] Mark Kac. *Enigmas of Chance: An Autobiography*. Harper and Row, 1985.

[25] Jeff Kahn. A normal law for matchings. *Combinatorica*, 20:339–391, 2000.

[26] Paul Lévy. Sur une propriété de la loi de poisson relative aux petites probabilités. *Soc. Math. de France, Comptes rendus des scéances de l'année 1936*, page 29, 1937.

[27] Brendan D. McKay and Nicholas C. Wormald. Asymptotic enumeration by degree sequence of graphs of high degree. *European J. Combin.*, 11:565–580, 1990.

[28] Franz Mertens. Ein beitrag zur analytischen zahlentheorie. *J. reine angew. Math.*, 78:46–62, 1874.

[29] Leo Moser and Max Wyman. Stirling numbers of the second kind. *Duke Math. J.*, 25:29–43, 1957.

[30] Ron Mullin and Gian-Carlo Rota. On the foundations of combinatorial theory iii: Theory of binomial enumeration. In Bernard Harris, editor, *Graph Theory and Its Applications*. Academic Press, 1970.

[31] George Szekeres. Asymptotic distribution of the number and size of parts in unequal partitions. *Bull. Austral. Math. Soc*, 36:89–97, 1987.

[32] S. Gill Williamson. *Combinatorics for Computer Science*. Computer Science Press, 1985.

# Chapter 4

# *Trees*

**Michael Drmota**
*Institute of Discrete Mathematics and Geometry, TU Wien*

**CONTENTS**



## 4.1 Introduction

Trees are a fundamental object in graph theory and combinatorics as well as a basic object for data structures and algorithms in computer science. This is certainly the reason why (several types of) trees and tree-like structures are of constant interest in research.

In this chapter we will survey and explain several methods for the enumeration of trees. The main focus is, however, the method of generating functions that is a very appropriate and powerful tool for tree enumeration. This is mainly due to the recursive structure of trees so that combinatorial decompositions can be easily translated in corresponding relations for generating functions. We, thus, decided to use the generating functions terminology throughout the chapter.

After an introductory section we concentrate on the enumeration problem of unlabeled trees and that of labeled trees separately. A concluding section is devoted to three selected topics that are related to trees and tree enumeration.

The literature on trees is vast, in particular on algorithmic questions and on probabilistic properties of trees. There are also several books that are specialized on tree enumeration. Probably, the first book is due to Moon [21] from 1970 that specializes on labeled trees. A few years later the classical book by Harary and Palmer [18] on *Graphical Enumeration* was published. Trees are an an important part of that book, but the book goes much beyond them, and generating functions are widely used. Another **milestone** is the book by Bergeron, Labele and Leroux [3] on *Combinatorial Species and Tree-Like Structures* (that was first published in French in 1994 and then in an English version in 1998). It develops a theory for the enumeration of unlabeled tree-like structures by the use of Pólya's theory of counting. Finally the author wants to mention two more recent books. The book by Flajolet and Sedgewick [13] on *Analytic Combinatorics* systematically uses the relation between combinatorial constructions and generating functions (we adopt this philosophy in this chapter) and trees and tree-like structures are discussed there in various aspects. Another focus in [13] is the asymptotic analysis of combinatorial objects based on generating functions. Finally we also refer to the book [11] on *Random Trees* by the author, where the approach of [13] is extended to a systematic analysis on the stochastic behavior of tree parameters under various random tree models. Again generating functions (in particular in several variables) are used as a basic tool. Some parts of the present chapter follow the presentation of [11].

In the following text there are several references to the aforementioned books (and, of course, also to other sources) since the limited space of this chapter can only provide an introduction to tree enumeration. The choice of the presented results reflects the authors's taste, too. So the author encourages the interested reader to consult in particular these references in order to obtain more detailed information on more refined enumeration problems on trees.

## 4.2 Basic notions

Trees are finite simple connected graphs without cycles. If we just require that a finite simple graph has no cycles then it is called **forest**. The connected components of a forest are then trees.

The degree $d(v)$ of a node $v$ in a tree is the number of nodes that are adjacent to $v$ or the number of neighbors of $v$.

Nodes of degree $\leq 1$ are usually called **leaves** or **external nodes** and the remaining ones are **internal nodes**. If a leaf has degree 0 then the tree consists just of this vertex and has no edges. In all other cases leaves $v$ are characterized by $d(v) = 1$. It is easy to see that leaves always exist.

**Lemma 4.2.1** *Every tree with at least two vertices has at least two leaves.*

*Proof.* We fix an arbitrary vertex $v_0$ of the tree $T$ to start a path at $v$ that uses an edge of $T$ at most once. Since $T$ has no cycles no vertex will be visited twice. Furthermore the path can be extended as long as the degree of present vertex is $\geq 2$. However, since a tree is a finite graph such a path cannot be infinite. Thus it has to terminate at a vertex with degree 1, which is a leaf. If the starting point $v_0$ was already a leaf then we are done. In the other case we just start a second path at $v_0$ and will find a second leaf. ∎

Lemma 4.2.1 can be used to deconstruct a tree by removing edges. Since a leaf has degree 1 there is precisely one edge attached to it. So if you remove the leaf and the corresponding edge the remaining graph is still a connected and cyle-free graph. In this way it follows by induction that we construct any tree by starting with a single vertex and by adding step by step an edge (together with a new leaf). This also implies the following property of trees.

**Lemma 4.2.2** *Let $\alpha_0$ denote the number of vertices of a tree $T$ and $\alpha_1$ the number of edges of $T$. Then we have*

$$\alpha_1 = \alpha_0 - 1.$$

*Conversely if a simple connected graph has the property that the number of edges is 1 less than the number of vertices then the graph is a tree.*

*Proof.* As already mentioned the relation $\alpha_1 = \alpha_0 - 1$ follows by induction. Next we observe that every connected graph $G$ has a spanning tree $T$, that is, a maximal subgraph $G$ that is a tree. It is easy to see that $T$ contains all vertices of $G$. Since $T$ is a tree the number of edges of $T$ equals the number of vertices minus 1. Hence, $T$ contains all edges of $G$, too. Thus $T$ and $G$ coincide. ∎

Another characterization of a tree is the property that for every pair of vertices there exists a unique path that connects these two vertices.

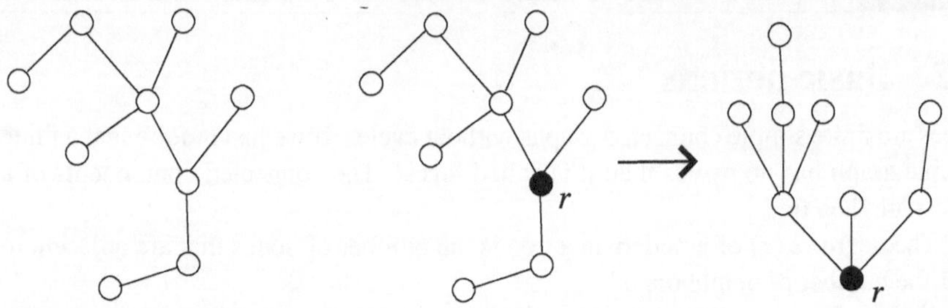


**Figure 4.1**
Tree and rooted tree.

**Rooted versus unrooted trees.** If we mark a specific node $r$ in a tree $T$, which we denote the **root** of $T$, we call the resulting tree a **rooted tree**. See Figure 4.1 for an illustration. A root induces also a kind of hierarchy in terms of **generations** or **levels**. The root is the zeroeth generation. The neighbors of the root constitute the first generation, and in general the nodes at distance $k$ from the root form the $k$th generation (or level). If a node of level $k$ has neighbors of level $k+1$ then these neighbors are also called **successors**. The number of successors of a node $v$ is also called the **out-degree** $d^+(v)$. For all nodes $v$ different from the root we have $d(v) = d^+(v)+1$.

Furthermore, if $v$ is a node in a rooted tree $T$ then $v$ may be considered as the root of a subtree $T_v$ of $T$ that consists of all iterated successors of $v$. This means that rooted trees can be constructed in a recursive way. Due to that property counting problems on rooted trees are usually easier than on unrooted trees.

Rooted trees also have various applications in computer science. They naturally appear as data structures, e.g. the recursive structure of folders in any computer is just a rooted tree. Furthermore, fundamental algorithms such as Quicksort or the Lempel-Ziv data compression algorithm are closely related to rooted trees, namely to binary and digital search trees which are also used to store (and search for) data. Rooted trees even occur in information theory. For example, prefix free codes on an alphabet of order $m$ are encoded as the set of leaves in $m$-ary trees.

**Plane versus non-plane trees.** Trees are planar graphs since they can be embedded into the plane without crossings. Nevertheless, a tree may have different embeddings (compare with Figure 4.2). This makes a difference in counting problems. When we say that we are counting **planar trees** we mean that we are counting all possible different embeddings into the plane.

In the context of rooted trees it is common to use the notion **plane tree** or **ordered tree** when successors of the root and recursively the successors of each node are equipped with a **left-to-right-order**. Alternatively one can give the successors a rank so that one can speak of the $j$th successor ($j \geq 1$). Of course, this induces a natural

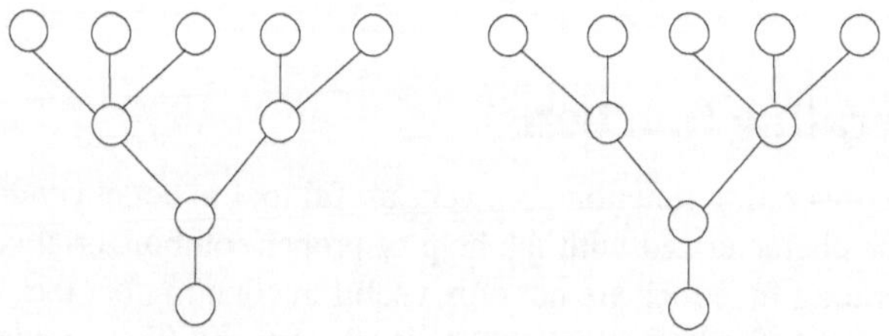

**Figure 4.2**
Two different embeddings of a tree.

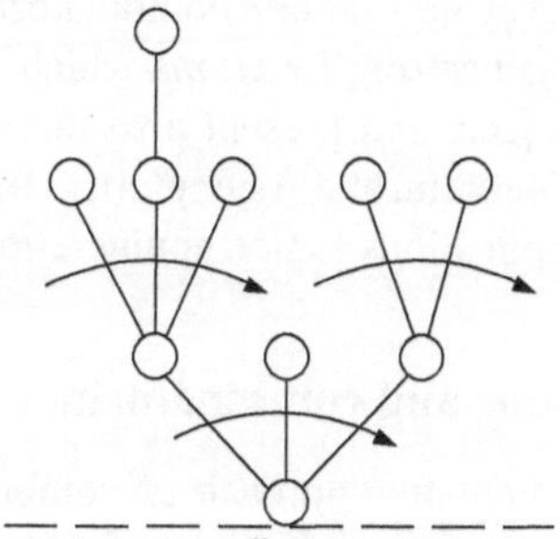


**Figure 4.3**
Plane rooted tree.

embedding into the half-plane (compare with Figure 4.3). Note that this notion is different from considering all embeddings into the plane, since it is not allowed to rotate the subtrees of the root cyclically around the root.

**Labeled versus unlabeled trees.** We also distinguish between labeled trees, where the nodes are labeled by different numbers, and unlabeled trees, where the nodes are indistinguishable. This is particularly important for the counting problem. For example, there is only one unlabeled tree with three nodes whereas there are three different labeled trees of size 3 with labels $1, 2, 3$.

There is much latitude in choosing labels on trees. The simplest model is to assume that the nodes of a tree of size $n$ are labeled by the numbers $1, 2, \ldots, n$, but there are many other ways to do so. For so-called embedded trees one only assumes that the labels of adjacent vertices differ (at most) by 1. Another possibility is to put labels consistently with the structure of the tree. For example, recursive trees have the property that the root is labeled by 1 and the labels on all paths away from the root are strictly increasing.

## 4.3 Generating functions

The concept of generating functions is a very useful tool to count combinatorial objects that can be characterized with the help of proper combinatorial constructions. However, generating functions are not only useful in counting but they can be used, too, to obtain asymptotics by evaluating the Cauchy integral (that computes the coefficient of a power series) asymptotically. Finally, but this will be not covered in this chapter, they can be used to encode the distribution of random variables that are related to counting problems and, hence, asymptotic methods can be applied to obtain probabilistic limit theorems like central limit theorems.

In this chapter we first survey how certain combinatorial constructions have their counterparts in relations for generating functions which can then be used to count the original combinatorial objects and present also three special topics (Lagrange inversion, the Dissymmetry Theorem, and Asymptotics) that are related to generating functions and have several applications to tree enumeration.

### 4.3.1 Generating functions and combinatorial constructions

We present very briefly a systematic approach of combinatorial constructions and generating functions. The presentation is inspired by the work of Flajolet and his co-authors (see in particular the monograph [13] where this concept is described in much more detail.) *

**Definition 4.3.1** *The ordinary generating function (ogf) of a sequence $(A_n)_{n\geq 0}$ (of complex numbers) is the formal power series*

$$A(x) = \sum_{n\geq 0} A_n x^n.$$

*Similarly the exponential generating function (egf) of the sequence $(A_n)_{n\geq 0}$ is given by*

$$\hat{A}(x) = \sum_{n\geq 0} A_n \frac{x^n}{n!}.$$

We use the notation

$$[x^n]A(x) = A_n$$

to **read off** the coefficient of $x^n$ in a generating function.

It is clear that certain algebraic operations on the sequence $A_n$ have their counterpart on the level of generating functions. The two tables in Figure 4.4 collect some of them.

*More detailed descriptions can be found in Chapter 1 on Algebraic and Geometric Methods in Enumerative Combinatorics and in Chapter 2 on Analytic Methods. In order to keep the chapter self-contained we keep this short overview of notions that we will need for the enumeration problems on trees.

| | sequence | ogf |
|---|---|---|
| sum | $C_n = A_n + B_n$ | $C(x) = A(x) + B(x)$ |
| product | $C_n = \sum_{k=0}^{n} A_k B_{n-k}$ | $C(x) = A(x)B(x)$ |
| partial sums | $C_n = \sum_{k=0}^{n} A_k$ | $C(x) = \frac{1}{1-x}A(x)$ |
| marking | $C_n = nA_n$ | $C(x) = xA'(x)$ |
| scaling | $C_n = c^n A_n$ | $C(x) = A(cx)$ |

| | sequence | egf |
|---|---|---|
| sum | $C_n = A_n + B_n$ | $\hat{C}(x) = \hat{A}(x) + \hat{B}(x)$ |
| product | $C_n = \sum_{k=0}^{n} \binom{n}{k} A_k B_{n-k}$ | $\hat{C}(x) = \hat{A}(x)\hat{B}(x)$ |
| binomial sums | $C_n = \sum_{k=0}^{n} \binom{n}{k} A_k$ | $\hat{C}(x) = e^x \hat{A}(x)$ |
| marking | $C_n = nA_n$ | $\hat{C}(x) = x\hat{A}'(x)$ |
| scaling | $C_n = c^n A_n$ | $\hat{C}(x) = \hat{A}(cx)$ |

**Figure 4.4**
Basic relations between sequences and their generating functions.

A generating function $A(x)$ represents an analytic function for $|x| < R$, where

$$R = \left( \limsup_{n\to\infty} |A_n|^{\frac{1}{n}} \right)^{-1}$$

denotes the radius of convergence. Thus, if $R > 0$ then we can either use a differentiation to represent the sequence

$$A_n = \frac{A^{(n)}(0)}{n!},$$

or we use Cauchy's formula

$$A_n = \frac{1}{2\pi i} \int_C A(x) \frac{dx}{x^{n+1}},$$

where $C$ is a closed curve inside the region of analyticity of $A(x)$ with winding number $+1$ around the origin.

Another point of view on generating functions is that they can be considered as a power series generated by certain **combinatorial objects**. Let $\mathscr{C}$ be a (countable) set of objects, for example, a set of graphs and

$$|\cdot|_{\mathscr{C}} : \mathscr{C} \to \mathbb{N}$$

a **weight function** that assigns to every element $\gamma \in \mathscr{C}$ a weight or size $|\gamma|_{\mathscr{C}}$. We assume that the sets

$$\mathscr{C}_n := |\cdot|_{\mathscr{C}}^{-1}(\{n\}) = \{\gamma \in \mathscr{C} : |\gamma|_{\mathscr{C}} = n\} \qquad (n \in \mathbb{N})$$

are all finite. Set $C_n = |\mathscr{C}_n|$. Then the ordinary generating function $C(x)$ of the pair $(\mathscr{C}, |\cdot|_{\mathscr{C}})$ that we also call **combinatorial structure** is given by

$$C(x) = \sum_{\gamma \in \mathscr{C}} x^{|\gamma|_{\mathscr{C}}} = \sum_{n \geq 0} C_n x^n,$$

and the exponential generating function $\hat{C}(x)$ by

$$\hat{C}(x) = \sum_{\gamma \in \mathscr{C}} \frac{x^{|\gamma|}}{|\gamma|_{\mathscr{C}}!} = \sum_{n \geq 0} C_n \frac{x^n}{n!}.$$

The choice of ordinary generating functions or exponential generating functions depends on the kind of problem. As a rule unlabeled (or unordered) structures should be counted with the help of ordinary generating functions and labeled (or ordered) structures with exponential generating functions.

For example, the ogf of binary sequences, where the weight denotes the length, is given by

$$B(x) = \sum_{n \geq 0} 2^n x^n = \frac{1}{1-2x}.$$

An example of an egf is the egf of permutations of finite sets, where the weight is the size of the finite set:

$$\hat{P}(x) = \sum_{n \geq 0} n! \frac{x^n}{n!} = \frac{1}{1-x}.$$

One major aspect in the use of generating functions is that certain combinatorial constructions have their counterparts in relations for the corresponding generating functions.

We again use the notation $\mathscr{A}, \mathscr{B}, \mathscr{C}, \ldots$ for sets of combinatorial objects with corresponding size functions $|\cdot|_{\mathscr{A}}, |\cdot|_{\mathscr{B}}, |\cdot|_{\mathscr{C}}, \ldots$. First suppose that the objects that we consider are (in some sense) unlabeled or unordered. We have the following basic operations:

**Disjoint union.** If $\mathscr{A}$ and $\mathscr{B}$ are disjoint then $\mathscr{C} = \mathscr{A} + \mathscr{B} = \mathscr{A} \cup \mathscr{B}$ denotes the union of $\mathscr{A}$ and $\mathscr{B}$.

**Cartesian product.** $\mathscr{C} = \mathscr{A} \times \mathscr{B}$ denotes the Cartesian product. The size function of a pair $\gamma = (\alpha, \beta)$ is given by $|\gamma|_{\mathscr{C}} = |\alpha|_{\mathscr{A}} + |\beta|_{\mathscr{B}}$.

**Sequence.** Suppose that $\mathscr{A}$ contains no object of size 0 and that the sets $\mathscr{A}$, $\mathscr{A} \times \mathscr{A}$, $\mathscr{A} \times \mathscr{A} \times \mathscr{A}, \ldots$ are disjoint. Then

$$\mathscr{C} = \mathrm{Seq}(\mathscr{A}) := \{\varepsilon\} + \mathscr{A} + \mathscr{A} \times \mathscr{A} + \mathscr{A} \times \mathscr{A} \times \mathscr{A} + \cdots$$

is the set of (finite) sequences of elements of $\mathscr{A}$ ($\varepsilon$ denotes the empty object of size zero). Alternatively $\mathrm{Seq}(\mathscr{A})$ is denoted by $\mathscr{A}^*$.

**Powerset.** Let $\mathscr{C} = \mathrm{PSet}(\mathscr{A})$ denote the set of all finite subsets of $\mathscr{A}$. The size of a subset $\{\alpha_1, \alpha_2, \ldots, \alpha_k\} \subseteq \mathscr{A}$ is given by $|\{\alpha_1, \alpha_2, \ldots, \alpha_k\}|_{\mathrm{PSet}(\mathscr{A})} = |\alpha_1|_{\mathscr{A}} + |\alpha_2|_{\mathscr{A}} + \cdots + |\alpha_k|_{\mathscr{A}}$.

**Multiset.** Let $\mathscr{C} = \mathrm{MSet}(\mathscr{A})$ denote the set of all finite multisets $\{\alpha_1^{j_1}, \alpha_2^{j_2}, \ldots, \alpha_k^{j_k}\}$ of $\mathscr{A}$, that is, the element $\alpha_i$ is taken $j_i$ times $(1 \le i \le k)$. Its size is given by $|\{\alpha_1^{j_1}, \alpha_2^{j_2}, \ldots, \alpha_k^{j_k}\}|_{\mathrm{MSet}(\mathscr{A})} = j_1|\alpha_1|_{\mathscr{A}} + j_2|\alpha_2|_{\mathscr{A}} + \cdots + j_k|\alpha_k|_{\mathscr{A}}$.

**Cycle.** The cycle construction $\mathscr{C} = \mathrm{Cyc}(\mathscr{A})$ can be seen as $(\mathrm{Seq}(\mathscr{A}) \setminus \{\varepsilon\})/S$, where $S$ is the equivalence relation that identifies two finite sequences $(\alpha_1, \ldots, \alpha_r)$ and $(\alpha_1', \ldots, \alpha_r')$ if a cyclic shift transfers them into each other.

As already indicated these combinatorial constructions have counterparts in relations of generating functions. Compare also with the relations in Figure 4.4:

**Theorem 4.3.2** *The above mentioned combinatorial constructions imply the following relations for the corresponding generating functions:*

| *combinat. constr.* | *ogf* |
|---|---|
| $\mathscr{C} = \mathscr{A} + \mathscr{B}$ | $C(x) = A(x) + B(x)$ |
| $\mathscr{C} = \mathscr{A} \times \mathscr{B}$ | $C(x) = A(x)B(x)$ |
| $\mathscr{C} = \mathrm{Seq}(\mathscr{A})$ | $C(x) = \dfrac{1}{1 - A(x)}$ |
| $\mathscr{C} = \mathrm{PSet}(\mathscr{A})$ | $C(x) = \exp\left(A(x) - \frac{1}{2}A(x^2) + \frac{1}{3}A(x^3) \mp \cdots\right)$ |
| $\mathscr{C} = \mathrm{MSet}(\mathscr{A})$ | $C(x) = \exp\left(A(x) + \frac{1}{2}A(x^2) + \frac{1}{3}A(x^3) + \cdots\right)$ |
| $\mathscr{C} = \mathrm{Cyc}(\mathscr{A})$ | $C(x) = \sum_{k \ge 1} \dfrac{\varphi(k)}{k} \log \dfrac{1}{1 - A(x^k)}$ |

*($\varphi(k)$ denotes the Euler totient function.)*

*Proof.* We just sketch some parts of the proof. For example, the ogf of the Cartesian product $\mathscr{C} = \mathscr{A} \times \mathscr{B}$ is given by

$$C(x) = \sum_{(a,b) \in \mathscr{A} \times \mathscr{B}} x^{|(a,b)|} = \sum_{(a,b) \in \mathscr{A} \times \mathscr{B}} x^{|a|+|b|} = \sum_{a \in \mathscr{A}} x^{|a|} \cdot \sum_{b \in \mathscr{B}} x^{|b|} = A(x) \cdot B(x).$$

This implies that $1/(1 - A(x))$ represents the ogf of $\mathrm{Seq}(\mathscr{A})$.

The generating function of the finite subsets of $\mathscr{A}$ is given by

$$C(x) = \prod_{a \in \mathscr{A}} \left(1 + x^{|a|}\right) = \prod_{n \ge 1} (1 + x^n)^{A_n}.$$

By taking logarithms and expanding the series $\log(1 + x^n)$ one obtains the proposed representation $C(x) = \exp\left(A(x) - \frac{1}{2}A(x^2) + \frac{1}{3}A(x^3) \mp \cdots\right)$. Similarly the ogf of finite multisets is given by

$$C(x) = \prod_{a \in \mathscr{A}} \left( \frac{1}{1 - x^{|a|}} \right) = \prod_{n \geq 1} \left( \frac{1}{1 - x^n} \right)^{A_n}.$$

Finally we mention that the relation for the cycle construction relies on Pólya's theory of counting (see [13, p. 729]). ■

There are similar constructions of so-called labeled or ordered combinatorial objects with corresponding relations for their exponential generating functions. We call a combinatorial object $\gamma$ of size $n$ labeled if it is formally of the form $\gamma = \tilde{\gamma} \times \pi$, where $\pi \in \mathfrak{S}_n$ is a permutation. For example, we can think of a graph $\tilde{\gamma}$ of $n$ vertices and the permutation $\pi$ represents a labeling of its vertices. We list some of the combinatorial constructions where we have to take care of the involved permutations.

**Disjoint union.** If $\mathscr{A}$ and $\mathscr{B}$ are disjoint then $\mathscr{C} = \mathscr{A} + \mathscr{B} = \mathscr{A} \cup \mathscr{B}$ denotes the union of $\mathscr{A}$ and $\mathscr{B}$.

**Labeled product.** The **labeled product** $\mathscr{C} = \mathscr{A} * \mathscr{B}$ of two labeled structures is defined as follows. Suppose that $\alpha = \tilde{\alpha} \times \pi \in \mathscr{A}$ has size $|\alpha| = k$ and $\beta = \tilde{\beta} \times \sigma \in \mathscr{B}$ has size $|b| = m$. Then we define $\alpha * \beta$ as the set of objects $((\tilde{\alpha}, \tilde{\beta}), \tau)$, where $\tau \in \mathfrak{S}_{k+m}$ runs over all permutations that are consistent with $\pi$ and $\sigma$ in the following way: There is a partition $\{j_1, j_2, \ldots, j_k\}, \{\ell_1, \ell_2, \ldots, \ell_m\}$ of $\{1, 2, \ldots, k+m\}$ with $j_1 < j_2 < \cdots < j_k$ and $\ell_1 < \ell_2 < \cdots < \ell_m$ such that

$$\tau(1) = j_{\pi(1)},\ \tau(2) = j_{\pi(2)}, \ldots, \tau(k) = j_{\pi(k)} \quad \text{and}$$
$$\tau(k+1) = \ell_{\sigma(1)},\ \tau(k+2) = \ell_{\sigma(2)}, \ldots, \tau(k+m) = \ell_{\sigma(m)}.$$

Finally we set

$$\mathscr{A} * \mathscr{B} = \bigcup_{\alpha \in \mathscr{A},\ \beta \in \mathscr{B}} \alpha * \beta.$$

The size of $((\tilde{\alpha}, \tilde{\beta}), \tau)$ is given by $|((\tilde{\alpha}, \tilde{\beta}), \tau)| = |\alpha| + |\beta|$.

Figure 4.5 shows a small example for the labeled product.

**Figure 4.5**
Labeled product.

**Sequence.** Suppose that $\mathscr{A}$ contains no object of size 0 and that the sets $\mathscr{A}$, $\mathscr{A}*\mathscr{A}$, $\mathscr{A}*\mathscr{A}*\mathscr{A}, \ldots$ are disjoint. Then

$$\mathscr{C} = \mathrm{Seq}(\mathscr{A}) := \{\varepsilon\} + \mathscr{A} + \mathscr{A}*\mathscr{A} + \mathscr{A}*\mathscr{A}*+\cdots$$

is the set of (finite labeled) sequences of elements of $\mathscr{A}$.

**Set.** Similarly we define unordered labeled sequences (or a set) by

$$\mathscr{C} = \mathrm{Set}(\mathscr{A}) = \{\varepsilon\} + \mathscr{A} + \frac{1}{2!}\mathscr{A}*\mathscr{A} + \frac{1}{3!}\mathscr{A}*\mathscr{A}*\mathscr{A} + \cdots,$$

where the shorthand notation $\frac{1}{n!}\mathscr{A}*\mathscr{A}*\cdots*\mathscr{A}$ means that we do not take care of the order of the $n$ elements in the sequence $\mathscr{A}*\mathscr{A}*\cdots*\mathscr{A}$.

**Cycle.** We again define the cycle construction by $\mathscr{C} = \mathrm{Cyc}(\mathscr{A}) = (\mathrm{Seq}(\mathscr{A})\setminus\{\varepsilon\})/S$, where $S$ is the equivalence relation that identifies two finite sequences when they can be transferred into each other by a cyclic shift.

**Theorem 4.3.3** *The above mentioned combinatorial constructions for labeled objects imply the following relations for the corresponding exponential generating functions:*

| *combinat. constr.* | *egf* |
|---|---|
| $\mathscr{C} = \mathscr{A} + \mathscr{B}$ | $\hat{C}(x) = \hat{A}(x) + \hat{B}(x)$ |
| $\mathscr{C} = \mathscr{A} * \mathscr{B}$ | $\hat{C}(x) = \hat{A}(x)\hat{B}(x)$ |
| $\mathscr{C} = \mathrm{Seq}(\mathscr{A})$ | $\hat{C}(x) = \dfrac{1}{1-\hat{A}(x)}$ |
| $\mathscr{C} = \mathrm{Set}(\mathscr{A})$ | $\hat{C}(x) = \exp\left(\hat{A}(x)\right)$ |
| $\mathscr{C} = \mathrm{Cyc}(\mathscr{A})$ | $\hat{C}(x) = \log\dfrac{1}{1-\hat{A}(x)}$ |

*Proof.* The only case that has to be explained is the labeled product $\mathscr{C} = \mathscr{A}*\mathscr{B}$. If $|\alpha| = k$ and $|\beta| = m$ then there are exactly $\binom{k+m}{k}$ possible ways to partition $\{1,2,\ldots,k+m\}$ into two sets of size $k$ and $m$. Thus $\alpha*\beta$ has size $\binom{k+m}{k}$ and consequently

$$\hat{C}(x) = \sum_{\alpha\in\mathscr{A},\beta\in\mathscr{B}} \binom{|\alpha|+|\beta|}{|\alpha|} \frac{x^{|\alpha|+|\beta|}}{(|\alpha|+|\beta|)!} = \sum_{\alpha\in\mathscr{A}} \frac{x^{|\alpha|}}{|\alpha|!} \cdot \sum_{\beta\in\mathscr{B}} \frac{x^{|\beta|}}{|\beta|!} = \hat{A}(x)\cdot\hat{B}(x).$$

■

Finally we mention that several notions and properties of the above discussed univariate generating functions can be extended to the multivariate case. For example, the underlying idea in the binary case is to consider (in addition to $|\cdot|_{\mathscr{C}}$) a second

**size function** that we denote by $p_{\mathscr{C}}(\cdot)$. Then the corresponding bivariate generating function is given by

$$C(x,u) = \sum_{\gamma\in\mathscr{C}} x^{|\gamma|_{\mathscr{C}}} u^{p_{\mathscr{C}}(\gamma)} = \sum_{n,k\geq 0} C_{n,k} x^n u^k,$$

where $C_{n,k} = |\{\gamma \in \mathscr{C} : |\gamma|_{\mathscr{C}} = n, p_{\mathscr{C}}(\gamma) = k\}|$.

We just give here a simple example how the above mentioned combinatorial constructions can be adapted. Let us consider $\mathscr{C} = \mathrm{Seq}(\mathscr{A})$ and let $p_{\mathscr{C}}(\gamma)$ denote the number of $\mathscr{A}$-parts of $\gamma \in \mathscr{C}$. Then we have

$$C(x,u) = 1 + uA(x) + u^2A(x)^2 + \cdots = \frac{1}{1-uA(x)}.$$

We will use this kind of extension in several enumeration problems below.

### 4.3.2 The Lagrange inversion formula

Let $A(x) = \sum_{n\geq 0} A_n x^n$ be a power series with $A_0 = 0$ and $A_1 \neq 0$. The Lagrange inversion formula provides an explicit representation of the coefficients of the inverse power series $A^{[-1]}(x)$, which is defined by $A(A^{[-1]}(x)) = A^{[-1]}(A(x)) = x$.

**Theorem 4.3.4** *Let $A(x) = \sum_{n\geq 0} A_n x^n$ be a formal power series with $A_0 = 0$ and $A_1 \neq 0$. Let $A^{[-1]}(x)$ be the inverse power series and $g(x)$ an arbitrary power series. Then the n-th coefficient of $g(A^{[-1]}(x))$ is given by*

$$[x^n]g(A^{[-1]}(x)) = \frac{1}{n}[u^{n-1}]g'(u)\left(\frac{u}{A(u)}\right)^n \qquad (n \geq 1).$$

In tree enumeration problems the following variant is more appropriate.

**Theorem 4.3.5** *Let $\Phi(x)$ be a power series with $\Phi(0) \neq 0$ and $Y(x)$ the (unique) power series solution of the equation*

$$Y(x) = x\Phi(Y(x)).$$

*Then $Y(x)$ is invertible and the n-th coefficient of $g(Y(x))$ (where $g(x)$ is an arbitrary power series) is given by*

$$[x^n]g(Y(x)) = \frac{1}{n}[u^{n-1}]g'(u)\Phi(u)^n \qquad (n \geq 1).$$

Theorems 4.3.4 and 4.3.5 are equivalent. If $A(x) = x/\Phi(x)$ then $A^{[-1]}(x) = Y(x)$, where $Y(x)$ satisfies the equation $Y(x) = x\Phi(Y(x))$. *

As an example we consider the equation $Y(x) = x(1+Y(x))^2$ that is related to the enumeration of binary trees (see Section 4.4.1). Here we have $\Phi(x) = (1+x)^2$ and, thus, we obtain

$$Y_n = [x^n]Y(x) = \frac{1}{n}[u^{n-1}](1+u)^{2n} = \frac{1}{n}\binom{2n}{n-1} = \frac{1}{n+1}\binom{2n}{n}.$$

*A proof of Theorem 4.3.5 can be found in Chapter 2 on Analytic Methods.

### 4.3.3 The dissymmetry theorem

The enumeration of rooted objects (like rooted trees) is usually easier than the enumeration of unrooted objects. Nevertheless there is a general method to relate unrooted and rooted objects by the so-called **Dissymmetry theorem**. It was originally observed by Otter [22] and generalized by Bergeron, Labele and Leroux [3].

**Theorem 4.3.6** *Let $\mathscr{A}$ be a class of trees, $\mathscr{A}^{\circ}$ the corresponding class of trees, where a node is marked, $\mathscr{A}^{\circ-\circ}$ the class, where an edge is marked, and $\mathscr{A}^{\circ\to\circ}$ the class, where an edge is marked and is given a direction. Then there is a bijection between*

$$\mathscr{A}+\mathscr{A}^{\circ\to\circ} \quad \textit{and} \quad \mathscr{A}^{\circ}+\mathscr{A}^{\circ-\circ}$$

*Proof.* Recall that the **center** of a tree is either a single node, the **central node**, or a pair of adjacent vertices, the **central edge**. (The center is defined as the set of nodes of a tree, where the maximal distance to other nodes is minimal.) This means that an unrooted tree has either a distinguished root or a distinguished edge. Consequently we can see the class

$$\mathscr{A}^{nc} := (\mathscr{A}^{\circ}+\mathscr{A}^{\circ-\circ})\setminus\mathscr{A}$$

as the class of trees that are either rooted at a non-central node or at a non-central edge. Therefore it is sufficient to formulate a bijection between $\mathscr{A}^{nc}$ and $\mathscr{A}^{\circ\to\circ}$.

We distinguish between two cases. If a tree in $\mathscr{A}^{nc}$ is marked at a vertex $v_0$ that is different from its central node $v_c$ (when the center is a node) let $v_1$ be the vertex adjacent to $v_0$ on the unique path from $v_0$ to the center (that contains the central edge, too, if the center is an edge). Now we mark the tree at the directed edge from $v_0$ to $v_1$ and obtain a tree in $\mathscr{A}^{\circ\to\circ}$. Second, if a tree in $\mathscr{A}^{nc}$ is marked at an edge $e_0$ that is different from its central edge $e_c$ (when the center is an edge) then orient $e_0$ away from the center and we obtain a tree in $\mathscr{A}^{\circ\to\circ}$.

It is easy to see that this mapping is actually a bijection. ∎

As an example we apply the Dissymmetry theorem for the class of unlabeled trees $\mathscr{A}=\tilde{\mathscr{T}}$ (see Section 4.4.4). Then $\mathscr{A}^{\circ}$ is in the class of rooted unlabeled trees $\mathscr{T}$, which has generating function $T(x)$. The class $\mathscr{A}^{\circ\to\circ}$ can be identified with $\mathscr{T}\times\mathscr{T}$, the class of ordered pairs of rooted trees; we can join two rooted trees by a directed edge from the root of the first tree to the root of the second tree. Clearly the generating function of $\mathscr{A}^{\circ\to\circ}$ is $T(x)^2$. Finally, the class $\mathscr{A}^{\circ-\circ}$ can be seen, too, as a pair of rooted trees, however, as unordered pairs. In the corresponding counting problem we have to distinguish between the case, where the two rooted trees are different and where they are the same. In the first case we can identify an unordered pair $\{T_1,T_2\}$ with the two (different) ordered pairs $(T_1,T_2),(T_2,T_1)$, whereas in the second case we just have one ordered pair $(T_1,T_1)$. This implies that the generating functions corresponding to $\mathscr{A}^{\circ-\circ}$ equals $\frac{1}{2}\left(T(x)^2+T(x^2)\right)$ since $T(x^2)$ counts pairs of the form $(T_1,T_1)$. Summing up it follows that

$$\tilde{T}(x)+T(x)^2=T(x)+\frac{1}{2}\left(T(x)^2+T(x^2)\right)$$

or

$$\tilde{T}(x) = T(x) - \frac{1}{2}\left(T(x)^2 - T(x^2)\right)$$

which is Otter's theorem [22].

### 4.3.4 Asymptotics

We mention here only a theorem (by Meir and Moon [20]) that is related to the Lagrange inversion and, of course, to tree enumeration problems. *

**Theorem 4.3.7** *Let R denote the radius of convergence of a power series* $\Phi(x) = \sum_{k\ge 0}\varphi_k x^k$ *with non-negative coefficients and let* $Y(x) = \sum_{n\ge 1} Y_n x^n$ *be the solution of the equation* $Y(x) = x\Phi(Y(x))$. *Suppose that there exists* $\tau$ *with* $0 < \tau < R$ *that satisfies* $\tau\Phi'(\tau) = \Phi(\tau)$. *Set* $d = \gcd\{j > 0 : \varphi_j > 0\}$. *Then*

$$Y_n = d\sqrt{\frac{\Phi(\tau)}{2\pi\Phi''(\tau)}}\,\frac{\Phi'(\tau)^n}{n^{3/2}}\left(1 + O(n^{-1})\right) \qquad (n \equiv 1 \bmod d)$$

*and* $Y_n = 0$ *if* $n \not\equiv 1 \bmod d$.

Note that in the most frequent case $d = 1$ the condition $n \equiv 1 \bmod 1$ is satisfied for all integers $n$ (and the case $n \not\equiv 1 \bmod 1$ does not appear).

This theorem applies to several classes of trees, in particular to simply generated trees that will be discussed in Section 4.4.5. So it is quite **universal**.

As an example let us discuss the equation $Y(x) = x/(1 - Y(x))$ that is related to the counting problem of planted plane trees (see Section 4.4.2). Here $\Phi(x) = 1/(1-x)$ and the equation $\tau\Phi'(\tau) = \Phi(\tau)$ has the solution $\tau = 1/2$. We have $d = 1$, $\Phi'(\tau) = 4$, $\Phi''(\tau) = 16$, and consequently $Y_n = [x^n]Y(x)$ is asymptotically given by

$$Y_n \sim \frac{4^{n-1}}{\sqrt{\pi}\,n^{3/2}}.$$

In Section 4.4.2 we will see that $Y_n = \frac{1}{n}\binom{2n-2}{n-1}$ and it is easy to check by Stirling's formula that this asymptotic expansion is correct.

It should be mentioned, too, that Theorem 4.3.7 can be generalized (or adjusted) in severals ways that are relevant for tree counting problems. Therefore we discuss the background of Theorem 4.3.7 a little. The main observation from the analytic point of view is that the solution $Y(x)$ of the implicit equation $Y = x\Phi(Y)$ has a so-called squareroot singularity at $x_0 = \tau/\Phi(\tau)$, which means that there is a local expansion of the form

$$Y(x) = a_0 + a_1\sqrt{x_0 - x} + a_2(x_0 - x) + a_3(x_0 - x)^{3/2} + \cdots,$$

that converges for $|x - x_0| < \eta$ (for some $\eta > 0$) and where $a_0 = \tau$ and $a_1 = -\sqrt{2\Phi(\tau)/(x_0\Phi''(\tau))} < 0$. If $x_0$ is the only singularity on the circle of convergence

*It is also discussed in Chapter 2 on Analytic Methods.

$|x| = x_0$, which is easy to check, then it follows by the transfer theorem by Flajolet and Odlyzko [12] that $Y_n$ is asymptotically given by

$$Y_n = -\frac{a_1\sqrt{x_0}}{2\sqrt{\pi}} x_0^{-n} n^{-3/2} + O\left(x_0^{-n} n^{-5/2}\right).$$

A similar procedure applies if we consider a function of the form $C(x) = g(Y(x))$, where $g(x)$ is regular for $x = \tau = Y(x_0)$. Then a local expansion shows that $C(x)$ can be represented similarly:

$$C(x) = c_0 + c_1\sqrt{x_0 - x} + c_2(x_0 - x) + c_3(x_0 - x)^{3/2} + \cdots,$$

where $c_0 = g(\tau)$ and $c_1 = g'(\tau)a_1$. If $c_1 \neq 0$ then we are in precisely the same situation as above. However, if $c_1 = 0$ and $c_3 \neq 0$ then the singular term $(x_0 - x)^{3/2}$ is dominating and we obtain an asymptotic expansion of the form

$$C_n = [x^n]C(x) = \frac{3c_3 x_0^{3/2}}{4\sqrt{\pi}} x_0^{-n} n^{-5/2} + O\left(x_0^{-n} n^{-7/2}\right).$$

## 4.4 Unlabeled trees

As mentioned above, generating functions are quite natural in the context of tree counting since (rooted) trees have a recursive structure that usually translates into recurrence relations for corresponding counting problems. In what follows we will count several classes of unlabeled trees with the help of generating functions. We start with binary trees, which are probably the most basic ones.

### 4.4.1 Binary trees

Binary trees are planted rooted trees, where each node is either a leaf or it has two distinguishable successors: the left successor and the right successor. The leaves of a binary tree are usually called **external** nodes and those nodes with two successors **internal** nodes. It is an easy exercise to show that a binary tree with $n$ internal nodes has $n+1$ external nodes. Thus, the total number of nodes is always odd.

**Theorem 4.4.1** *The number $B_n$ of binary trees with n internal nodes is given by the Catalan number*

$$B_n = \frac{1}{n+1}\binom{2n}{n}.$$

*Proof.* Since this is the first example, we will provide several variants of the proof. However, all of them all related to generating functions.

The first variant is the most **basic** one that sets up a recurrence that is then solved with the help of generating functions. Suppose that a binary tree has $n+1$ internal

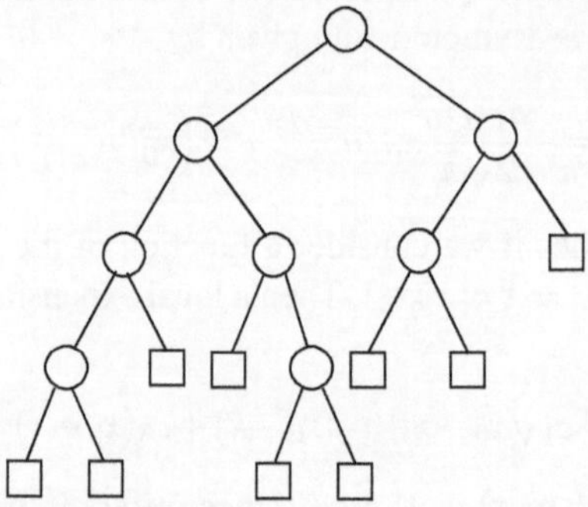

**Figure 4.6**
A binary tree.

nodes. Then the left and right subtrees are also binary trees (with $k$ and $n-k$ internal nodes, where $0 \le k \le n$). Thus, one gets directly a recurrence for the corresponding numbers:

$$B_{n+1} = \sum_{k=0}^{n} B_k B_{n-k}. \tag{4.1}$$

The initial value is $B_0 = 1$ (where the tree consists just of the root that is an external node).

This recurrence can be solved using the generating function

$$B(x) = \sum_{n\ge 0} B_n x^n.$$

By the properties of generating functions we find the relation

$$B(x) = 1 + xB(x)^2 \tag{4.2}$$

and consequently an explicit representation of the form

$$B(x) = \frac{1-\sqrt{1-4x}}{2x}.$$

(Note that the other solution $(1+\sqrt{1-4x}/(2x)$ of the quadratic equation does not represent a power series, since it is singular at $x = 0$.) Hence we obtain

$$B_n = [x^n]\frac{1-\sqrt{1-4x}}{2x} = -\frac{1}{2}[x^{n+1}](1-4x)^{\frac{1}{2}}$$
$$= -\frac{1}{2}\binom{\frac{1}{2}}{n+1}(-4)^{n+1} = \frac{1}{n+1}\binom{2n}{n}.$$

By inspecting this first variant of the proof of Theorem 4.4.1 one observes that the recurrence relation (4.1), together with its initial condition, is exactly a translation of a recursive description of binary trees:

A binary tree $\mathscr{B}$ is either just an external node or an internal node (the root) with two subtrees that are again binary trees.

Of course this can be rewritten in terms of combinatorial constructions:

$$\mathscr{B} = \square + \circ \times \mathscr{B} \times \mathscr{B} = \square + \circ \times \mathscr{B}^2, \tag{4.3}$$

where $\mathscr{B}$ denotes the system of binary trees; $\square$ represents an external and $\circ$ an internal node. This leads to a corresponding relation (4.2) for the generating function (see Theorem 4.3.2):

$$B(x) = 1 + xB(x)^2,$$

where the size of a binary tree is the number of internal nodes. See also the schematic Figure 4.7. Thus, setting up the equation (4.2) for the generating is just a translation of the combinatorial construction.

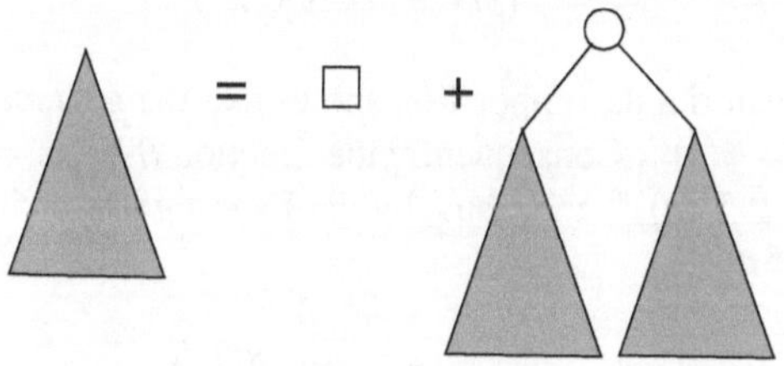

**Figure 4.7**
Recursive structure of binary trees.

In order to obtain a formula for the coefficient $B_n = [x^n]B(x)$ we can either use the above procedure (explicit solution of the quadratic equation and using the binomial series) or we just apply the Lagrange inversion formula.

By setting $\tilde{B}(x) = B(x) - 1$ the equation (4.2) rewrites to

$$\tilde{B}(x) = x(1 + \tilde{B}(x))^2.$$

Now we can apply Lagrange's inversion formula (Theorem 4.3.5) with $\Phi(x) = (1 + x)^2$ and obtain (for $n \geq 1$)

$$\begin{aligned} B_n = [x^n]\tilde{B}(x) &= \frac{1}{n}[u^{n-1}](1+u)^{2n} \\ &= \frac{1}{n}\binom{2n}{n-1} = \frac{1}{n+1}\binom{2n}{n}, \end{aligned}$$

see also Section 4.3.2. ■

The explicit formula for $B_n = \frac{1}{n+1}\binom{2n}{n}$ can also be used to obtain an asymptotic expansion. By applying Stirling's formula $n! \sim \sqrt{2\pi n} e^{-n} n^n$ we obtain

$$B_n \sim \frac{4^n}{\sqrt{\pi} n^{3/2}}.$$

This is, of course, consistent with Theorem 4.3.7 applied to $\Phi(x) = (1+x)^2$.

A direct generalization of binary trees are ***m*-ary rooted trees,** where $m \geq 2$ is a fixed integer. As in the binary case ($m = 2$) we just take into account the number $n$ of internal nodes. The number of leaves is then given by $(m-1)n+1$ and the total number of nodes by $mn+1$.

**Theorem 4.4.2** *The number $B_n^{(m)}$ of m-ary trees ($m \geq 2$) with n internal nodes is given by*

$$B_n^{(m)} = \frac{1}{(m-1)n+1}\binom{mn}{n}.$$

*Proof.* The combinatorial decomposition shows that the generating function satisfies $B^{(m)}(x) = 1 + xB^{(m)}(x)^m$. Consequently the function $\tilde{B}^{(m)}(x) = B^{(m)}(x) - 1$ satisfies $\tilde{B}^{(m)}(x) = x(1+\tilde{B}^{(m)}(x))^m$ and, thus, by the Lagrange inversion formula (Theorem 4.3.5) we get for $n \geq 1$

$$B_n^{(m)} = [x^n]\tilde{B}^{(m)}(x) = \frac{1}{n}[u^{n-1}](1+u)^{mn} = \frac{1}{n}\binom{mn}{n-1} = \frac{1}{(m-1)n+1}\binom{mn}{n}.$$

∎

Another variant of binary trees are **Motzkin trees** that are ordered rooted trees, where each node has either no, or one, or two successors (recall that **ordered** means that successors of each node are equipped with a left-to-right-order). In this context it is common to take the number or edges as the size of a Motzkin tree. Then the generating function $M(x)$ of Motzkin trees satisfies the equation

$$M(x) = 1 + xM(x) + x^2 M(x)^2.$$

By using the substitution $T(x) = xM(x)$ we get $T(x) = x(1 + T(x) + T(x)^2)$. Hence, the **Motzkin numbers** are given by

$$M_n = [x^n]M(x) = [x^{n+1}]T(x) = \frac{1}{n+1}[u^n](1+u+u^2)^{n+1}.$$

There is no *simple formula* for $M_n$. Nevertheless, they can be expressed as hypergeometric sums. For example we have

$$M_n = {}_2F_1\left(\frac{1}{2}(1-n), -\frac{1}{2}n; 2; 4\right),$$

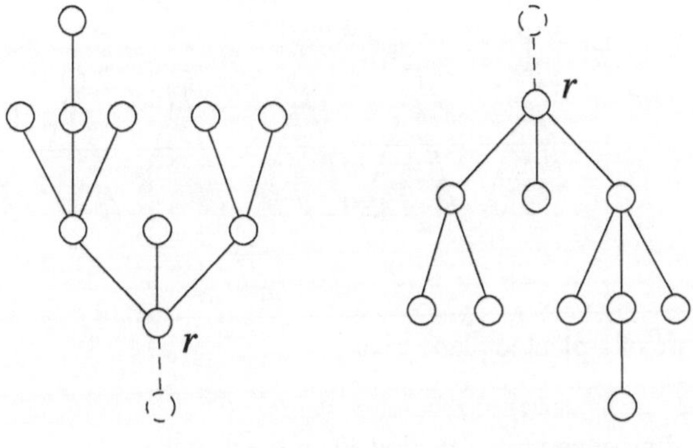

**Figure 4.8**
Planted plane tree.

where ${}_2F_1(a,b;c;z) = \sum_{n\ge 0} \frac{(a)_n (b)_n}{(c)_n} \frac{z^n}{n!}$ is the usual hypergeometric function. When we apply Theorem 4.3.7 to the function $U(x) = xM(x)$ and to $\Phi(x) = 1+x+x^2$ we also obtain the asymptotic formula

$$M_n \sim \sqrt{\frac{3}{4\pi n^3}}\, 3^{n+1}.$$

### 4.4.2 Planted plane trees

Another interesting class of trees are **planted plane trees**. Sometimes they are also called **Catalan trees**. Planted plane trees are again rooted trees, where each node has an arbitrary number of successors with a natural left-to-right-order (this again means that we are considering plane or oriented trees). The term **planted** comes from the interpretation that the root is connected (or planted) to an additional **phantom** node that is not taken into account (see Figure 4.8). Usually we will not even depict this additional node when we deal with planted trees. However, it is quite useful to define the degree of the root $r$ by $d(r) = d^+(r) + 1$, which means that the additional (planted) node is considered a neighbor node. This has the advantage that in this case all nodes have the property $d(v) = d^+(v) + 1$.

**Theorem 4.4.3** *The numbers $P_n$ of planted plane trees with $n \ge 1$ nodes are given by*

$$P_n = \frac{1}{n}\binom{2n-2}{n-1}.$$

*Proof.* Let $\mathscr{P}$ denote the set of planted plane trees, where the size is the number of nodes. The combinatorial decomposition $\mathscr{P} = \{\circ\} \times \mathrm{Seq}(\mathscr{P})$ implies

$$P(x) = \frac{x}{1-P(x)}$$

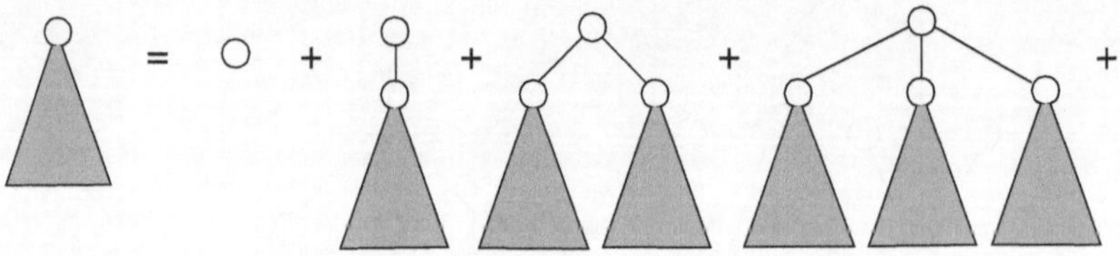

**Figure 4.9**
Recursive structure of a planted plane tree.

for the corresponding generating function $P(x)$, see also the schematic Figure 4.9.

Consequently

$$P_n = [x^n]P(x) = \frac{1}{n}[u^{n-1}](1-u)^{-n} = \frac{(-1)^{n-1}}{n}\binom{-n}{n-1} = \frac{1}{n}\binom{2n-2}{n-1}.$$

Alternatively we have $P(x) = x + P(x)^2$ and consequently $P(x) = xB(x)$, which is equivalent to $P_n = B_{n-1}$. ■

The property that the number of planted plane trees of size $n$ equals the $(n-1)$th Catalan number explains the term **Catalan tree.**

It is worth mentioning that the relation $P_n = B_{n-1}$ has a natural interpretation. First there is a **natural bijection** between planted plane trees with $n$ nodes and binary trees with $n-1$ internal nodes, the so-called **rotation correspondence.** Let us start with a planted plane tree with $n$ nodes and apply the following procedure:

1. Delete the root and all edges going to the root.
2. If a node has successors delete all edges to these successors despite one edge to the leftmost one.
3. Join all these (previous) successors with a path (by **horizontal edges**).
4. Rotate all these new (horizontal) edges by the angle $\pi/4$ below.
5. The remaining $n-1$ nodes are now considered as internal nodes of a binary tree. Append the (missing) $n+1$ external leaves.

The result is a binary tree with $n-1$ internal nodes. It is easy to verify that this procedure is bijective (compare with the example given in Figure 4.10).

A second interpretation of the relation $P_n = B_{n-1}$ comes from an alternate recursive description of planted plane trees. If a planted plane tree has more than one node then we can delete the left-most edge of the root and obtain two planted plane trees, the original one minus the left-most subtree of the root and the left-most subtree of the root (see Figure 4.11). Obviously this description leads to the recursive description

$$\mathcal{P} = \circ + \mathcal{P}^2,$$

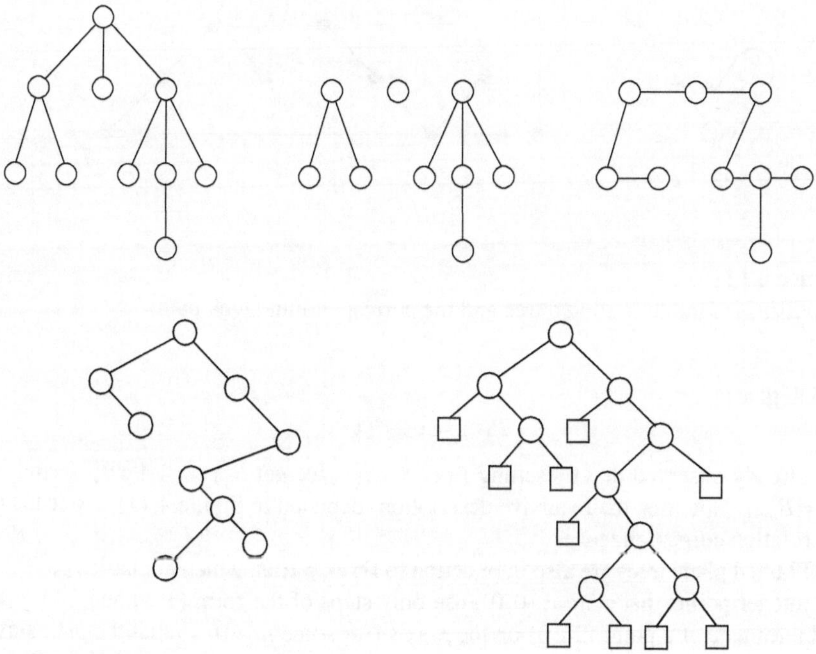

**Figure 4.10**
Rotation correspondence.

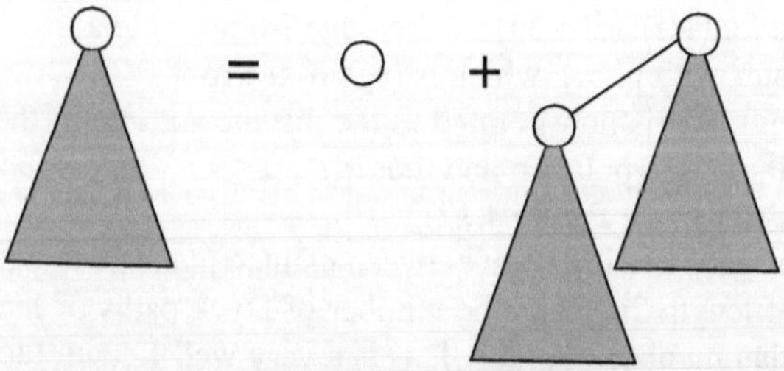

**Figure 4.11**
Alternative recurrence for planted plane trees.

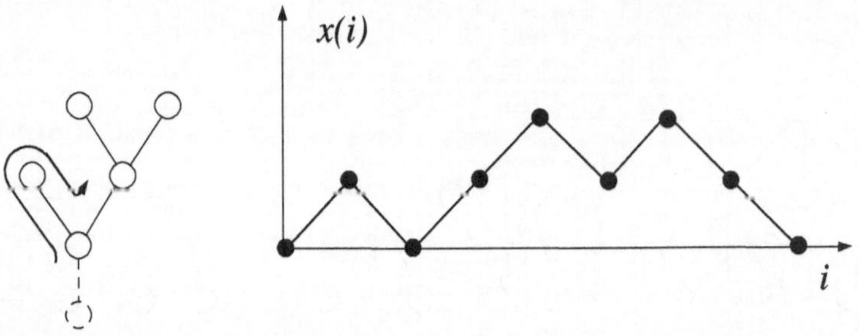


**Figure 4.12**
Depth-first search of a rooted tree and the corresponding Dyck path.

which gives

$$P(x) = x + P(x)^2.$$

We already observed that by setting $P(x) = xB(x)$ we get $B(x) = 1 + xB(x)^2$ and so $P_n = B_{n-1}$. Note that the recursive description (depicted in Figure 4.11) also leads to the rotation correspondence.

Planted plane trees are also in bijection to **Dyck paths**, which are lattice paths on the integer points that start at $(0,0)$, use only steps of the form $(1,1)$ and $(-1,-1)$ and terminate at a point $(2n,0)$ on the $x$-axis (for some $n \geq 0$) so that the path stays always above the $x$-axis. The right part of Figure 4.12 shows an example for $n = 4$.

The bijection can be stated in terms of the so-called **depth-first search**. For the sake of consistency we assume that the planted plane tree has $n$ edges (or $n+1$ vertices; the planted vertex is not taken into account). The depth-first search can be described as a walk $(v(i), 0 \leq i \leq 2n)$ **around the vertices**, where $v(0) = v(2n)$ is the root. We proceed inductively. Given $v(i)$ choose (if possible) the first edge (in the natural order) incident to $v(i)$ leading away from the root that has not already been used, and let $(v(i), v(i+1))$ be that edge. If all edges leading away from $v(i)$ have been already used let $(v(i), v(i+1))$ be the edge from $v(i)$ leading toward the root. This walk terminates with $v(2n)$, which is (again) the root.

The search depth $x(i)$ is now defined as the distance of $v(i)$ to the root. For non-integer $i$ we use linear interpolation and thus $x(t)$, $0 \leq t \leq 2n$, can be identified with a Dyck path of $2n$ steps (see Figure 4.12).

Obviously this leads to a bijection between planted plane trees with $n+1$ vertices and Dyck paths of length $2n$. Thus the number of Dyck paths of length $2n$ is again given by the Catalan number $\frac{1}{n+1}\binom{2n}{n}$. This is a very well-known fact (and there are several different proofs for this formula).

In contrast to binary (or $m$-ary trees) the number of leaves in a planted plane tree is not related to the total number of leaves. Actually we can refine the counting problem and also get an explicit formula for the number of planted plane trees with a given number of leaves.

**Theorem 4.4.4** *The numbers $P_{n,k}$ of planted plane trees of size n with k leaves are given by the Narayana numbers*

$$P_{n,k} = \frac{1}{n}\binom{n}{k}\binom{n-1}{k}.$$

*Proof.* Let $P(x,u) = \sum_{n,k} P_{n,k} x^n u^k$ denote the bivariate generating function of the numbers $P_{n,k}$. Then, following the recursive description of planted plane trees, one gets

$$P(x,u) = xu + x \sum_{k \geq 1} p(x,u)^k = xu + \frac{xP(x,u)}{1 - P(x,u)}.$$

For instance, let $x$ be considered as a *parameter*. Then we have

$$P(x,u) = \frac{ux}{\left(1 - \frac{x}{1-p(x,u)}\right)},$$

and consequently by applying Lagrange's inversion formula for the variable $u$ we obtain

$$[u^k]P(x,u) = \frac{1}{k}[v^{k-1}]\left(\frac{x}{1 - \frac{x}{1-v}}\right)^k.$$

Finally this implies

$$\begin{aligned}
P_{n,k} &= [x^n u^k]P(x,u) \\
&= \frac{1}{k}[x^n v^{k-1}]\left(\frac{x}{1 - \frac{x}{1-v}}\right)^k \\
&= \frac{1}{k}\binom{n-1}{k-1}[v^{k-1}](1-v)^{-n+k} \\
&= \frac{1}{k}\binom{n-1}{k-1}\binom{n-1}{k} \\
&= \frac{1}{n}\binom{n}{k}\binom{n-1}{k}.
\end{aligned}$$

■

It is interesting to discuss the asymptotic behavior of $P_{n,k}$. By using Stirling's formula we obtain a bivariate asymptotic expansion for $P_{n,k}$ of the form

$$\begin{aligned}
P_{n,k} &= \frac{1}{2\pi kn}\left(\frac{n}{k}\right)^{2k}\left(\frac{n}{n-k}\right)^{2(n-k)}\left(1 + O\left(\frac{1}{k}\right) + O\left(\frac{1}{n-k}\right)\right) \\
&= \frac{1}{2\pi n^2}\frac{n}{k}\left(\frac{1-\frac{k}{n}}{\frac{k}{n}}\right)^{2k}\left(\frac{1}{1-\frac{k}{n}}\right)^{2n}\left(1 + O\left(\frac{1}{k}\right) + O\left(\frac{1}{n-k}\right)\right),
\end{aligned}$$

where we assume that $k \to \infty$ and $n-k \to \infty$. In particular, if we fix $n$ then $P_{n,k}$ is maximal if $k \approx n/2$ and we locally get a behavior of the kind

$$P_{n,k} \sim \frac{4^n}{\pi n^2} \exp\left(-\frac{(n-2k)^2}{n}\right), \tag{4.4}$$

if $|k-n/2| \le C\sqrt{n}$ (for an arbitrary constant $C > 0$). This approximation has several implications. First, it shows that it is most likely that a typical tree of size $n$ has approximately $n/2$ leaves and the distribution of the number of leaves around $n/2$ looks like a Gaussian distribution.

We can make this observation more precise. Let $n$ be given and assume that each of the $P_n$ planted plane trees of size $n$ is equally likely. Then the number of leaves is a random variable on this set of trees, which we will denote by $X_n$. More precisely, we have

$$\mathbb{P}\{X_n = k\} = \frac{P_{n,k}}{P_n}.$$

Then $\mathbb{E}X_n \sim n/2$ and $\mathbb{V}\mathrm{ar}\, X_n \sim n/8$, and (4.4) can be restated as a weak limit theorem:

$$\frac{X_n - \mathbb{E}X_n}{\sqrt{\mathbb{V}\mathrm{ar}\, X_n}} \to N(0,1).$$

Actually there are lots of statements of this kind and the method of generating functions can be used to prove them directly (for more details see Chapter 3 on Asymptotic Normality in Enumeration or [13, 11]).

### 4.4.3 Unlabeled plane trees

A tree usually has several different embeddings into the plane. Planted plane trees, in particular, take into account all possible planar embeddings of rooted trees into the half plane, where the root is at the boundary of the half plane. With the help of this observation and by using the cycle construction and the Dissymmetry theorem it is possible to derive from planted plane trees the structure of plane trees $\tilde{\mathscr{P}}$, that is, all different embeddings of (unlabeled) trees into the plane.

**Theorem 4.4.5** *The generating function $\tilde{P}(x)$ of plane trees $\tilde{\mathscr{P}}$ is given by*

$$\tilde{P}(x) = x + x\sum_{k\ge1} \frac{\varphi(k)}{k} \log\frac{1}{1-P(x^k)} - \frac{1}{2}P(x)^2 + \frac{1}{2}P(x^2),$$

*where $P(x) = (1-\sqrt{1-4x})/2$ denotes the generating function of planted plane trees. The numbers $\tilde{P}_n = [x^n]\tilde{P}(x)$ (that count different embeddings of (unrooted) trees of size $n$ into the plane) are asymptotically given by*

$$\tilde{P}_n = \frac{1}{8\sqrt{\pi}} 4^n n^{-5/2}\left(1 + O(n^{-1})\right). \tag{4.5}$$

*Proof.* First, rooted plane trees $\mathscr{R}$ can be described as $\mathscr{R} = \circ + \circ \times \mathrm{Cyc}(\mathscr{P})$: A rooted plane tree it is either just the root (when it is of size 1) or it is the root attached with a cyclically ordered sequence of planted plane trees.

Hence, the generating function $R(x)$ is given by

$$R(x) = x + x \sum_{k \geq 1} \frac{\varphi(k)}{k} \log \frac{1}{1 - P(x^k)}.$$

Second, the Dissymmetry theorem (Theorem 4.3.6)) implies

$$\tilde{P}(x) = R(x) - \frac{1}{2}\left(P(x)^2 - P(x^2)\right).$$

We just have to mimic the arguments for unlabeled trees following the proof of Theorem 4.3.6.

In order to obtain the asymptotics we observe that $P(x^k)$ is analytic at $x_0 = 1/4$ for all $k \geq 2$, and thus, only the leading term $1/(1 - P(x))$ (corresponding to $k = 1$) has to be taken into account in order to obtain the local behavior of $\tilde{P}(x)$ around $x_0 = 1/4$. By using the explicit form of $P(x) = (1 - \sqrt{1 - 4x})/2$ it follows that $\tilde{P}(x)$ has a local expansion of the form

$$\tilde{P}(x) = c_0 + c_2(1 - 4x) + \frac{1}{6}(1 - 4x)^{3/2} + \cdots$$

which means that the term corresponding to $\sqrt{1 - 4x}$ cancels. Hence we obtain (4.5); compare with Section 4.3.4. ■

### 4.4.4 General unlabeled trees

The counting problem of unlabeled unrooted trees $\tilde{\mathscr{T}}$ is actually a quite involved one. Here we do not care about the possible embeddings into the plane. We just think of trees in the graph-theoretical sense. In addition to unrooted trees we consider—as usual—also the class of rooted unlabeled trees $\mathscr{T}$.

Let us denote by $\tilde{T}_n$ and $T_n$ the corresponding numbers of those trees of size $n$, for example we have

$$\tilde{T}_1 = 1,\ \tilde{T}_2 = 1,\ \tilde{T}_3 = 1,\ \tilde{T}_4 = 2 \quad \text{and} \quad T_1 = 1,\ T_2 = 1,\ T_3 = 2,\ T_4 = 4.$$

We already observed in Section 4.3.3 that the Dissymmetry theorem implies a relation between the generating function of rooted and unrooted trees:

$$\tilde{T}(x) = T(x) - \frac{1}{2}T(x)^2 + \frac{1}{2}T(x^2). \tag{4.6}$$

Thus we only have to get some relation for rooted trees. Of course, they can be recursively defined, however, instead of a sequence of trees we have a multiset of trees attached to the root:

$$\mathscr{T} = \circ \times \mathrm{MSet}(\mathscr{T}).$$

Thus, the generating function $T(x)$ satisfies

$$T(x) = x\exp\left(T(x) + \frac{1}{2}T(x^2) + \frac{1}{3}T(x^3) + \cdots\right). \tag{4.7}$$

Of course, the equation (4.7) can be transformed into a recurrence

$$T_{n+1} = \frac{1}{n}\sum_{j=1}^{n}\sum_{d|j} dT_d T_{n-j+1},$$

and an even simpler relation holds between $\tilde{T}_n$ and $T_n$. However, it seems that there is no proper explicit formula for $T_n$ and $\tilde{T}_n$. Nevertheless there are asymptotic expansions for them and by using extensions of the mentioned counting procedure it is also possible to study several shape characteristics of these kinds of trees.

Historically the relation (4.7) was already known to Pólya as an application of **Pólya's theory of counting** [23]. The amazing relation (4.6) is due to Otter [22]. More precisely Pólya [23] already discussed analytic properties of the generating function $T(x)$ and showed that the radius of convergence $\rho$ satisfies $0 < \rho < 1$ and that $x = \rho$ is the only singularity on the circle of convergence $|x| = \rho$. Later Otter [22] showed that $T(\rho) = 1$ and used the representation (4.8) to deduce the asymptotics for $T_n$. He also calculated $\rho \approx 0.338219$. However, his main contribution was to show (4.6). Consequently he derived (4.9) and (4.11).

**Theorem 4.4.6** *The generating functions $T(x)$ and $\tilde{T}(x)$ satisfy the functional equations (4.7) and (4.6) and have a common radius of convergence $\rho \approx 0.338219$, which is given by the relation $T(\rho) = 1$. Moreover they have a local expansion of the form*

$$T(x) = 1 - b(\rho - x)^{1/2} + c(\rho - x) + d(\rho - x)^{3/2} + O((\rho - x)^2)) \tag{4.8}$$

*and*

$$\tilde{T}(x) = \frac{1 + T(\rho^2)}{2} + \frac{b^2 - \rho t'(\rho^2)}{2}(\rho - x) + bc(\rho - x)^{3/2} + O((\rho - x)^2), \tag{4.9}$$

*where $b \approx 2.6811266$ and $c = b^2/3 \approx 2.3961466$ and $x = \rho$ is the only singularity on the circle of convergence $|x| = \rho$. Finally, $T_n$ and $\tilde{T}_n$ are asymptotically given by*

$$T_n = \frac{b\sqrt{\rho}}{2\sqrt{\pi}} n^{-3/2} \rho^{-n} \left(1 + O(n^{-1})\right) \tag{4.10}$$

*and*

$$\tilde{T}_n = \frac{b^3 \rho^{3/2}}{4\sqrt{\pi}} n^{-5/2} \rho^{-n} \left(1 + O(n^{-1})\right). \tag{4.11}$$

We do not give the proof here. We just mention that it follows the same lines as the proof of Theorem 4.4.5. In a first step it is shown that $T(x)$ has a local expansion of the form (4.8), which follows from a proper extension of Theorem 4.3.7. This provides the asymptotic expansion (4.10) for $T_n$. Finally by using (4.8) and (4.6) the local expansion (4.9) for $\tilde{T}(x)$ follows and the asymptotic expansion for $\tilde{T}_n$ is an immediate consequence.

In a similar way we can deal with unlabeled binary trees $\mathscr{B}$, where we do not care about the embedding in the plane. We only require that every node has outdegree 0 or 2 and do not care about the order of the subtrees. The corresponding unrooted version $\tilde{\mathscr{B}}$ is the set of unlabeled trees, where every node has either degree 1 or 3 (if it is not only a single vertex). Let the corresponding cardinalities of these trees (of size $n$) be denoted by $\tilde{B}_n$ and $B_n$, and the generating functions by

$$\tilde{B}(x) = \sum_{n\geq 1} \tilde{B}_n x^n \quad \text{and} \quad B(x) = \sum_{n\geq 1} B_n x^n.$$

Then we have (similarly to the above, compare also with [7]):

**Theorem 4.4.7** *The generating functions $B(x)$ and $\tilde{B}(x)$ satisfy the functional equations*

$$B(x) = x + \frac{x}{2}\left(B(x)^2 + B(x^2)\right) \tag{4.12}$$

*and*

$$\tilde{B}(x) = x + \frac{x}{6}\left(B(x)^3 + 3B(x)B(x^2) + 2B(x^3)\right) - \frac{1}{2}B(x)^2 + \frac{1}{2}B(x^2). \tag{4.13}$$

*They have a common radius of convergence $\rho_2 \approx 0.6345845127$ which is given by $\rho_2 B(\rho_2) = 1$ and singular expansions corresponding to (4.8) and (4.9). Furthermore, $B_n$ and $\tilde{B}_n$ are asymptotically given by*

$$B_n = c_1 n^{-3/2} \rho_2^{-n}\left(1 + O(n^{-1})\right) \quad (n \equiv 1 \bmod 2)$$

*and*

$$\tilde{B}_n = c_2 n^{-5/2} \rho_2^{-n}\left(1 + O(n^{-1})\right)$$

*with certain positive constants $c_1, c_2$.*

We just comment on the combinatorial part, that is, on (4.12) and (4.13). The first equation comes from the fact that rooted binary trees of this kind can be recursively defined as a single (root) vertex or the root followed by an unordered pair of binary trees.

With the help of $B(x)$ we obtain the generating function $R(x)$ of rooted (unlabeled) trees, where all vertices (including the root) have either degree 1 or 3, that is, a root followed by an unordered triple of binary trees:

$$R(x) = x + \frac{x}{6}\left(B(x)^3 + 3B(x)B(x^2) + 2B(x^3)\right).$$

Finally, we apply the Dissymmetry theorem and obtain

$$\tilde{B}(x) = R(x) - \frac{1}{2}B(x)^2 + \frac{1}{2}B(x^2).$$

This completes the proof of (4.13).

The proof of the analytic part runs along the same lines as in the proof of Theorem 4.4.5.

As mentioned in the Introduction, the book by Bergeron, Labele, and Leroux [3] provides a systematic treatment of the counting problem of unlabeled trees and tree-like structures that is based on Pólya's theory of counting, and many other things. The above two examples are only the **tip of the iceberg**.

### 4.4.5 Simply generated trees and Galton-Watson trees

Simply generated trees are weighted versions of rooted trees and have been introduced by Meir and Moon [20]. The idea is to put a weight to a rooted tree according to its degree distribution.

Let $\varphi_j$, $j \ge 0$, be a sequence of non-negative real numbers, called the **weight sequence**. Usually one assumes that $\varphi_0 > 0$ and $\varphi_j > 0$ for some $j \ge 2$. We then define the weight $\omega(T)$ of a finite rooted ordered tree $T$ by

$$\omega(T) = \prod_{v \in V(T)} \varphi_{d^+(v)} = \prod_{j \ge 0} \varphi_j^{D_j(T)},$$

where $d^+(v)$ denotes the out-degree of the vertex $v$ (or the number of successors) and $D_j(T)$ the number of nodes in $T$ with $j$ successors. The numbers

$$Y_n = \sum_{|T|=n} \omega(T)$$

are then the weighted numbers of trees of size $n$.

It is convenient to introduce the generating series

$$\Phi(x) = \varphi_0 + \varphi_1 x + \varphi_2 x^2 + \cdots = \sum_{j \ge 0} \varphi_j x^j.$$

**Theorem 4.4.8** *The generating function $Y(x) = \sum_{n \ge 1} Y_n x^n$ of a simply generated family of trees satisfies*

$$Y(x) = x\Phi(Y(x))$$

*so that $Y_n$ is given by $Y_n = \frac{1}{n}[u^{n-1}]\Phi(u)^n$.*

*Proof.* Similarly to binary trees or planted plane trees we can write a relation for the weighted combinatorial objects of simply generated trees:

$$\mathscr{Y} = \varphi_0 \cdot \circ + \varphi_1 \cdot \circ \times \mathscr{Y} + \varphi_2 \cdot \circ \times \mathscr{Y}^2 + \cdots.$$

In terms of the generating function we, thus, obtain

$$Y(x) = x\varphi_0 + x\varphi_1 Y(x) + x\varphi_2 Y(x)^2 + \cdots = x\Phi(Y(x)).$$

■

Simply generated trees generalize several of the above examples of combinatorial trees.

If $\varphi_j = 1$ for all $j \geq 0$, that is, $\Phi(x) = 1/(1-x)$, then all planted plane trees have weight $\omega(T) = 1$ and $Y_n$ is the number of planted plane trees.

Binary trees (counted according to their internal nodes) are also covered by this approach. If we set $\varphi_0 = 1$, $\varphi_1 = 2$, $\varphi_2 = 1$, and $\varphi_j = 0$ for $j \geq 3$, that is, $\Phi(x) = (1+x)^2$, then nodes with one successor get weight 2. This takes into account that binary trees (where external nodes are disregarded) have two kinds of nodes with one successor, namely those with a left branch but no right branch and those with a right branch but no left branch. Similarly, $m$-ary trees are covered with the help of the weights $\varphi_j = \binom{m}{j}$ or with $\Phi(x) = (1+x)^m$.

If $\varphi_0 = \varphi_1 = \varphi_2 = 1$ and $\varphi_j = 0$ for $j \geq 3$ or $\Phi(x) = 1 + x + x^2$, then we recover **Motzkin trees**. Here only rooted trees, where all nodes have less than 3 successors, get (a non-zero) weight $\omega(T) = 1$: $Y_n$ is then the number $M_{n-1}$ of Motzkin trees with $n$ nodes (or $n-1$ edges).

If we set $\varphi_j = 1/j!$ then

$$n! \cdot Y_n = n^{n-1}$$

denotes precisely the number of labeled rooted non-plane trees, see Section 4.5.1. The weight $\varphi_j = 1/j!$ disregards all possible orderings of the successors of a vertex of out-degree $j$ and the factor $n!$ corresponds to all possible labelings of $n$ nodes. Hence, every labeled tree gets the same weight.

We also note that Theorem 4.3.7 provides an asymptotic formula for $Y_n$ for a wide class of simply generated families of trees (we only have to ensure the existence of $\tau < R$ that satisfies $\tau\Phi'(\tau) = \Phi(\tau)$).

There is an intimate relation between simply generated trees and **Galton-Watson branching processes**. Let $\xi$ be a non-negative integer-valued random variable, the so-called offspring distribution. The Galton-Watson branching process starts with a single individual (generation 0); each individual has a number of children distributed as independent copies of $\xi$.

It is clear that Galton-Watson branching processes can be represented by ordered (finite or infinite) rooted trees $T$. We denote by $\nu(T)$ the probability that a specific tree $T$ occurs. We assume that $\mathbb{P}\{\xi = 0\} > 0$ so that there is positive probability that the resulting tree $T$ is finite. We also mention an important parameter for Galton-Watson processes, namely the value $\mathbb{E}\,\xi$, which is the expected offspring and governs the overall behavior of the process. The most interesting case is the case $\mathbb{E}\,\xi = 1$ which means that the average number of children is 1. This case is called **critical**.

The generating function $Y(x) = \sum_{n\geq 1} Y_n x^n$ of the numbers

$$Y_n = \mathbb{P}\{|T| = n\} = \sum_{|T|=n} \nu(T)$$

satisfies the functional equation

$$Y(x) = x\Phi(Y(x)),$$

where

$$\Phi(t) = \mathbb{E}t^{\xi} = \sum_{j\ge 0} \varphi_j t^j$$

with $\varphi_j = \mathbb{P}\{\xi = j\}$. Observe that

$$\nu(T) = \prod_{j\ge 0} \varphi_j^{D_j(T)} = \omega(T).$$

The **weight** of $T$ is now the **probability** of $T$. Thus Galton-Watson trees (conditioned on the size $n$) can be considered as special cases of simply generated trees.

The converse is almost true. It is possible to scale a general simply generated family properly so that it mimics a Galton-Watson process with the property that (conditioned on $n$) they are equivalent. The reason is the following one. If we replace $\varphi_j$ by $\tilde{\varphi}_j = ab^j\varphi_j$, which is the same as replacing $\Phi(x)$ by $\tilde{\Phi}(x) = a\Phi(bx)$ for two numbers $a,b > 0$, then $\omega(T)$ is replaced by

$$\tilde{\omega}(T) = \prod_{j\ge 0} \left(ab^j\varphi_j\right)^{D_j(T)} = a^{|T|}b^{|T|-1}\omega(T).$$

Note that $\sum_j jD_j(T) = |T| - 1$. Hence, $\tilde{Y}_n = a^n b^{n-1} Y_n$. In particular by setting $b = \tau$, where $\tau$ is the solution of the equation $\tau\Phi'(\tau) = \Phi(\tau)$ (if it exists) and $a = 1/\Phi(b)$ then we have

$$\tilde{\Phi}(1) = 1 \quad \text{and} \quad \tilde{\Phi}'(1) = 1.$$

This means that $\tilde{\Phi}(x)$ can be interpreted as the probability distribution of a discrete random variable with expected value 1, that is, the corresponding Galton-Watson process is critical.

If we consider just trees $T$ of size $n$ then all weights satisfy $\tilde{\omega}(T) = a^n b^{n-1}\omega(T)$ which means that the **distribution** of the weights is the same. More precisely if we define the probability $\pi_n(T)$ of a tree of size $n$ by $\omega(T)/Y_n$ then these probabilities do not change by introducing these scaling constants $a,b$.

## 4.5 Labeled trees

We call a tree **labeled** if the vertices $v$ are mapped to integer (or other) values $\ell(v)$, the labels, and we distinguish two trees to be different if they have different labels even if the underlying unlabeled trees are the same. Of course the counting problem depends on the kind of labels, the conditions on the labels and on other properties of the trees (if we consider planted trees or not, etc.). In what follows we will consider some of the most interesting versions of labeled trees (partly taken from Moon's book [21], where many more results can be found). We start with the famous formula by Cayley.

### 4.5.1 Cayley trees and related trees

A Cayley tree of size $n$ is a labeled tree, where the $n$ nodes are labeled by $1, 2, \ldots, n$ and where we do not take care on embeddings into the plane. In labeled trees there is an easy relation between rooted and unrooted trees. Since there are exactly $n$ different ways to make an unrooted tree to a rooted one by choosing one of the labeled nodes, the number of rooted labeled trees of size $n$ equals the number of unrooted labeled trees exactly $n$ times. Consequently it is sufficient to consider rooted labeled trees which has the advantage that one can use the recursive structure.

**Theorem 4.5.1** *The number $T_n$ of rooted labeled trees of size n is given by*

$$T_n = n^{n-1}.$$

*Consequently the number of unrooted labeled trees of size n equals $\tilde{T}_n = n^{n-2}$.*

*Proof.* Let $\mathcal{T}$ denote the set of labeled rooted trees. Then $\mathcal{T}$ can be recursively described as a root followed by an unordered $k$-tuple of labeled rooted trees for some $k \geq 0$:

$$\mathcal{T} = \circ * \text{Set}(\mathcal{T}).$$

Thus, it is appropriate to use the exponential generating function $T(x) = \sum_{n\geq 0} T_n x^n/n!$ of $T_n$. The above recursive description is then translated into

$$T(x) = xe^{T(x)}.$$

Finally, by Lagrange's inversion formula (Theorem 4.3.4)

$$T_n = n!\frac{1}{n}[u^{n-1}]e^{un} = n^{n-1}.$$

We remark that the generating function $\tilde{T}(x)$ of the number $\tilde{T}_n = n^{n-2}$ can be expressed as

$$\tilde{T}(x) = T(x) - \frac{T(x)^2}{2}.$$

This follows either by observing that the derivative of $T(x) - T(x)^2/2$ is given by $T'(x) - T(x)T'(x) = T(x)/x$ (since $T(x) = xe^{T(x)}$ implies $T'(x) = T(x)/(x(1 - T(x)))$). Hence

$$T(x) - \frac{T(x)^2}{2} = \int_0^x \sum_{n\geq 1} n^{n-1}\frac{t^{n-1}}{n!}\,dt = \sum_{n\geq 1} n^{n-2}\frac{x^n}{n!} = \tilde{T}(x).$$

Alternatively we can use the Dissymmetry theorem (Theorem 4.3.6) to express $\tilde{T}(x)$ in terms of $T(x)$. ∎

There are several other proofs of Cayley's formula. We just mention two further ones. One is based on the interesting fact that the number of trees of size $n$, where the vertex with label $j$ has degree $d_j$, $1 \leq j \leq n$, is given by the number (see [21])

$$\binom{n-2}{d_1 - 1, \ldots, d_n - 1}.$$

Equivalently we have

$$\sum_{T:|T|=n} x_1^{d(v_1(T))-1} x_2^{d(v_2(T))-1} \cdots x_n^{d(v_n(T))-1} = (x_1+x_2+\cdots+x_n)^{n-2}, \qquad (4.14)$$

where the sum is taken over all labeled trees of size $n$ and where $v_j(T)$ denotes the vertex with label $j$. Of course by setting $x_1 = x_2 = \cdots = x_n = 1$ we recover Cayley's formula $\tilde{T}_n = n^{n-2}$.

Let $C_n(x_1,\ldots,x_n)$ denote the left-hand side of (4.14). Then $C_n$ is a symmetric and homogeneous polynomial of degree $n-2$. The specialized polynomial $C_n(x_1,\ldots,x_{n-1},0)$ corresponds to those trees with $d(v_n(T)) = 1$. This means that $v_n$ is a leaf and is linked to a vertex $v_j$ with $1 \le j < n$. By removing $v_n$ and the edge to $v_j$ we are left with a labeled tree of size $n-1$. By rephrasing this procedure in term of $C_n$ we get

$$C_n(x_1,\ldots,x_{n-1},0) = (x_1+\cdots+x_{n-1})C_{n-1}(x_1,\ldots,x_{n-1})$$

and, thus, it follows by induction that

$$C_n(x_1,\ldots,x_{n-1},0) = (x_1+\cdots+x_{n-1})^{n-2}.$$

Since the degree of $C_n$ equals $n-2$, which is smaller than $n$, it is possible to reconstruct $C_n(x_1,\ldots,x_n)$ from $T(x_1,\ldots,x_{n-1},0)$. We just have to **complete** the appearing monomials in $C_n(x_1,\ldots,x_{n-1},0)$ to make them symmetric in $x_1,\ldots,x_n$. Thus, it follows that $C_n(x_1,\ldots,x_n) = (x_1+\cdots+x_n)^{n-2}$.

A further and completely different way for computing $\tilde{T}_n = n^{n-2}$ is to observe that an unrooted labeled tree can be interpreted as a spanning tree of the complete graph $K_n$ with nodes $1,2,\ldots,n$ (see Figure 4.13). Thus, the matrix-tree theorem (Theorem 4.6.1) can be applied. (We will discuss spanning trees of a graph and how to compute their number in more detail in Section 4.6.1.)

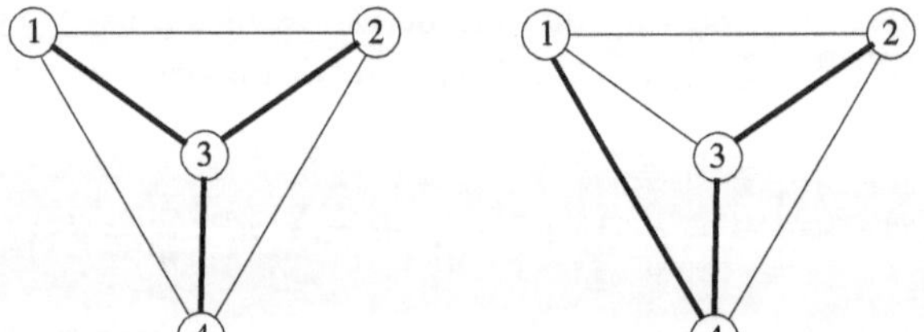


**Figure 4.13**
Two of 16 possible spanning trees of $K_4$.

If we distinguish also the number of leaves in rooted labeled trees we obtain a formula in terms of the Stirling numbers of the second kind $S_{n,k}$, which denote the number of partitions of a set of size $n$ into $k$ parts. Alternatively, these Stirling numbers can be defined by

$$\sum_{n,k\ge 0} S_{n,k} u^k \frac{x^n}{n!} = e^{u(e^x-1)}$$

or explicitly by

$$S_{n,k} = \frac{n!}{k!}[x^n](e^x-1)^k = \frac{1}{k!}\sum_{j=0}^{k}(-1)^{k-j}\binom{k}{j} j^n.$$

**Theorem 4.5.2** *The number $T_{n,k}$ of rooted labeled trees of size $n$ with $k$ leaves is given by*

$$T_{n,k} = \frac{n!}{k!} S_{n-1,n-k}.$$

*Proof.* By adapting the counting procedure in the proof of Theorem 4.5.1 we obtain that the function $T(x,w) = \sum_{n,k\ge 0} T_{n,k} w^k \frac{x^n}{n!}$ satisfies

$$T(x,w) = x\left(e^{T(x,w)} - 1 + w\right).$$

Consequently

$$T_{n,k} = n!\frac{1}{n}[u^{n-1}w^k](e^u - 1 + w)^n = (n-1)!\binom{n}{k}[u^{n-1}](e^u-1)^{n-k} = \frac{n!}{k!}S_{n-1,n-k}.$$

■

It is also of interest to consider the degree of the root.

**Theorem 4.5.3** *The number $L_{n,k}^{(r)}$ of rooted labeled trees of size $n$, where the root has degree $k$, is given by*

$$L_{n,k}^{(r)} = \binom{n-2}{k-1} n(n-1)^{n-k-1}.$$

*In particular, the number of rooted labeled trees, of size $n$, where the vertex with a specified label (e.g. label $n$) has degree $k$ is given by*

$$\binom{n-2}{k-1}(n-1)^{n-k-1}.$$

*Proof.* With the help of the generating function $T(x)$ from the proof of Theorem 4.5.1 and by the Lagrange inversion formula (Theorem 4.3.4 applied with $g(x) = x^k$) we have

$$\begin{aligned} L_{n,k}^{(r)} &= n![x^n]x\frac{T(x)^k}{k!} = \frac{n!}{k!}[x^{n-1}]T(x)^k = \frac{n!}{k!}\frac{1}{n-1}[u^{n-2}]ku^{k-1}e^{(n-1)u} \\ &= \frac{n!}{k!}\frac{k}{n-1}[u^{n-k-1}]e^{(n-1)u} = \binom{n-2}{k-1}n(n-1)^{n-k-1}. \end{aligned}$$

■

It is also possible to consider sets of trees (or forests).

**Theorem 4.5.4** *The number $F_{n,k}$ of vertex labeled forests with k rooted tree components is given by*

$$F_{n,k} = \binom{n}{k} kn^{n-k-1}$$

*whereas the number $\tilde{F}_{n,k}$ of vertex labeled forests with k unrooted tree components is given by*

$$\tilde{F}_{n,k} = \binom{n}{k} n^{n-k-1} \sum_{i=0}^{k} (-1)^i 2^{-i} \binom{k}{i} (k+i) \frac{(n-k)_i}{n^i}.$$

*Proof.* The generating function for forests that consist of $k$ rooted trees is given by $T(x)^k/k!$. Thus we can proceed as in the proof of Theorem 4.5.3.

Similarly the generating function for forests of unrooted trees is given by

$$\frac{\left(T(x) - \frac{T(x)^2}{2}\right)^k}{k!}.$$

Again we can apply the Lagrange inversion formula and obtain an explicit formula for $\tilde{F}_{n,k}$ (after a few lines of computation). ∎

Trees are bipartite graphs, that is, we can partition the vertices of a tree into two sets $V_1$ and $V_2$ so that there are only edges between vertices of $V_1$ and $V_2$. For example if we fix a vertex $v_0$ in a tree then we can define $V_1$ as the set of vertices with even distance to $v_0$ and $V_2$ as the set of vertices with odd distance to $v_0$. Actually the partition $\{V_1, V_2\}$ is uniquely defined. It is therefore possible to consider labeled bipartite trees $\tilde{\mathcal{T}}_{m,n}$, where $V_1$ and $V_2$ have given sizes $m$ and $n$, respectively, and where we label the vertices of $V_1$ by $1, \ldots, m$ and those of $V_2$ by $1, \ldots, n$.

**Theorem 4.5.5** *The number $\tilde{T}_{m,n} = |\tilde{\mathcal{T}}_{m,n}|$ of labeled bipartite trees of size $m+n$ with labels $1, \ldots, m$ and $1, \ldots, n$, respectively, is given by*

$$\tilde{T}_{m,n} = m^{n-1} n^{m-1}.$$

*Proof.* In order to simplify the presentation of the proof we use properly vertex colored trees with the colors black and white, where **properly colored** means that adjacent vertices have different colors. (Of course, the black vertices can be considered as the set $V_1$ and the white ones as the set $V_2$.) Furthermore we label the black and white vertices independently. Now let $\mathcal{T}_1$ denote the set of these trees, where we root at a black vertex and $\mathcal{T}_2$ the set of trees, where we root at a white vertex. The corresponding generating functions are

$$T_1(x,y) = \sum_{m,n \geq 0} T_{1,m,n} \frac{x^m}{m!} \frac{y^n}{n!} \quad \text{and} \quad T_2(x,y) = \sum_{m,n \geq 0} T_{2,m,n} \frac{x^m}{m!} \frac{y^n}{n!},$$

where $T_{j,m,n}$, $j = 1, 2$, is the number of trees in $\mathcal{T}_j$ with $m$ black vertices and $n$ white vertices.

Similarly to the usual labeled trees there is a corresponding recursive description for bipartite trees that leads to the following relations for $T_1(x,y)$ and $T_2(x,y)$:

$$T_1(x,y) = xe^{T_2(x,y)} \quad \text{and} \quad T_2(x,y) = ye^{T_1(x,y)}.$$

Now Lagrange inversion applied to $T_1(x,y) = x\exp\left(ye^{T_1(x,y)}\right)$ yields:

$$\begin{aligned}
[x^m y^n]T_1(x,y) &= [y^n]\frac{1}{m}[u^{m-1}]\left(e^{ye^u}\right)^m \\
&= [y^n]\frac{1}{m}[u^{m-1}]\sum_{k\ge 0}\sum_{l\ge 0}\frac{m^k y^k}{k!}\frac{u^l k^l}{l!} \\
&= [y^n]\frac{1}{m}\sum_{k\ge 0}\frac{m^k y^k}{k!}\frac{k^{m-1}}{(m-1)!} = \frac{m^n n^{m-1}}{m!n!}.
\end{aligned}$$

Furthermore, $\tilde{T}_{m,n} = T_{1,m,n}/m = T_{2,m,n}/n$ since there are exactly $m$ ways to choose the root of type 1 in an unrooted tree with $m$ vertices of type 1. ∎

Finally we also consider labeled trees together with their embedding into the plane. Interestingly there is also an explicit formula for the number of different planar embeddings.

**Theorem 4.5.6** *The number $\hat{T}_n$ of different planar embeddings of rooted labeled trees of size n is given by*

$$\hat{T}_1 = 1 \quad \textit{and} \quad \hat{T}_n = n\frac{(2n-3)!}{(n-1)!} \quad (n \ge 2).$$

*Consequently the number of different planar embeddings of unrooted labeled trees of size n equals $(2n-3)!/(n-1)!$ (for $n \ge 2$).*

*Proof.* Let $\hat{P}(x)$ denote the exponential generating function of labeled rooted plane trees. Then due to the recursive structure of these kinds of trees we have (compare with the proof of Theorem 4.4.3)

$$\hat{P}(x) = \frac{x}{1-\hat{P}(x)}.$$

Consequently, the exponential generating function $\hat{T}(x)$ for the numbers $\hat{T}_n$ of different embeddings for labeled rooted trees $\mathscr{T} = \circ + \circ * \mathrm{Cyc}(\mathscr{P})$ is given by

$$\hat{T}(x) = x + x\log\frac{1}{1-\hat{P}(x)}.$$

Hence, by Lagrange's inversion formula (Theorem 4.3.5) we obtain (for $n \ge 2$)

$$\hat{T}_n = n![x^{n-1}]\log\frac{1}{1-\hat{P}(x)}$$

$$= n!\frac{1}{n-1}[u^{n-2}]\frac{1}{1-u}\frac{1}{(1-u)^{n-1}}$$
$$= n\frac{(2n-3)!}{(n-1)!}.$$

■

### 4.5.2 Recursive trees

Recursive trees are rooted labeled trees, where the root is labeled by 1 and the labels of all successors of any node $v$ are larger than the label of $v$ (see Figure 4.14).

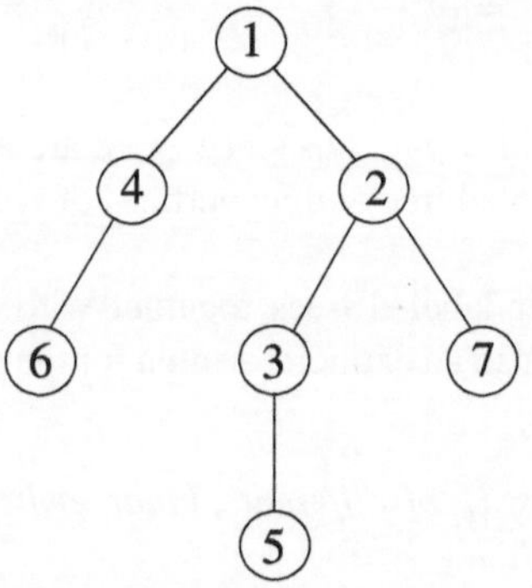


**Figure 4.14**
Recursive tree.

Usually one does not take care of the possible embeddings of a recursive tree into the plane. In this sense recursive trees can be seen as the result of the following evolution process. Suppose that the process starts with a node carrying the label 1. This node will be the root of the tree. Then attach a node with label 2 to the root. The next step is to attach a node with label 3. However, there are two possibilities: either to attach it to the root or to the node with label 2. Similarly one proceeds further. After having attached the nodes with labels $1,2,\ldots,k$, attach the node with label $k+1$ to one of the existing nodes.

This procedure is actually the basis of the proof of the following enumeration result.

**Theorem 4.5.7** *The number $R_n$ of recursive trees of size $n$ is given by*

$$R_n = (n-1)!$$

*Proof.* Every recursive tree of size $n$ is obtained by the above evolution process in a unique way. (Actually the labels represent the history of the evolution process.)

Since there are exactly $k$ ways to attach the node with label $k+1$, there are exactly $1 \cdot 2 \cdots (n-1) = (n-1)!$ possible trees of size $n$. ∎

Note that the left-to-right-order of the successors of the nodes in a recursive tree was not relevant in the above counting procedure. It is, however, relatively easy to consider all possible embeddings as plane rooted trees. These kind of trees are usually called plane oriented recursive trees (PORTs).

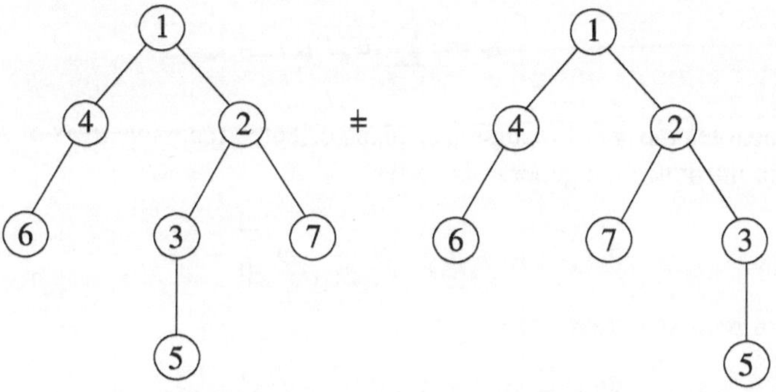


**Figure 4.15**
Two different plane oriented trees.

**Theorem 4.5.8** *The number $P_n$ of plane oriented recursive trees of size n is given by*

$$P_n = (2n-3)!! = 1 \cdot 3 \cdot \ldots \cdot (2n-3).$$

*Proof.* Plane oriented recursive trees can again be seen as the result of an evolution process, where the left-to-right-order of the successors is taken into account. More precisely, if a node $v$ has out-degree $d$, then there are $d+1$ possible ways to attach a new node to $v$. Furthermore we have for a rooted tree

$$\sum_{v \in V(T)} (d^+(v)+1) = 2|V(T)| - 1.$$

Hence, the number of different plane oriented recursive trees with $n$ nodes equals

$$1 \cdot 3 \cdot \ldots \cdot (2n-3) = (2n-3)!! = \frac{1}{2^{n-1}} \frac{(2(n-1))!}{(n-1)!}.$$

∎

The model of simply generated trees was to define a weight for a tree that reflects the degree distribution of rooted trees. The same idea can be applied to recursive and to plane oriented recursive trees. The resulting classes of trees are called **increasing trees**. They have been first introduced by Bergeron, Flajolet, and Salvy [2].

As for simply generated grees we define the weight $\omega(T)$ of a recursive or a plane oriented recursive tree $T$ by

$$\omega(T) = \prod_{v \in V(T)} \varphi_{d^+(v)} = \prod_{j \geq 0} \varphi_j^{D_j(T)},$$

where $d^+(v)$ denotes the out-degree of the vertex $v$ (or the number of successors) and $D_j(T)$ the number of nodes in $T$ with $j$ successors. Then we set

$$Y_n = \sum_{T \in \mathcal{I}_n} \omega(T),$$

where $\mathcal{I}_n$ denotes the set of recursive or plane oriented recursive trees of size $n$.

We again introduce the generating series

$$\Phi(x) = \varphi_0 + \varphi_1 x + \varphi_2 \frac{x^2}{2!} + \varphi_2 \frac{x^3}{3!} + \cdots$$

in the case of recursive trees and

$$\Phi(x) = \varphi_0 + \varphi_1 x + \varphi_2 x^2 + \varphi_3 x^3 + \cdots$$

in the case of plane oriented recursive trees. The generating function

$$Y(x) = \sum_{n \geq 0} Y_n \frac{x^n}{n!}$$

satisfies the differential equation

$$Y'(x) = \Phi(Y(x)), \quad Y(0) = 0.$$

In the interest of clarity we state how the general concept specializes.

Recursive trees (that is, every non-planar recursive tree gets weight 1) are given by $\Phi(x) = e^x$. Here $Y_n = R_n = (n-1)!$ and $R(x) = \log(1/(1-x))$.

Plane oriented recursive trees are given by $\Phi(x) = 1/(1-x)$. This means that every planar recursive tree gets weight 1. Here $Y_n = P_n = (2n-3)!! = 1 \cdot 3 \cdot 5 \cdots (2n-3)$ and $Y(x) = 1 - \sqrt{1-2x}$.

Binary recursive trees are defined by $\Phi(x) = (1+x)^2$. We have $Y_n = n!$ and $Y(x) = 1/(1-x)$. The model that is induced by this (planar) binary increasing trees is exactly the standard permutation model of **binary search trees** that is discussed in Section 4.6.3.

### 4.5.3 Well-labeled trees

Well-labeled trees are planted plane trees, where the vertices $v$ are labeled with integer numbers $\ell(v) \geq 1$ such that the root gets label $\ell(\text{root}) = 1$ and labels $\ell(v), \ell(w)$ of adjacent vertices satisfy

$$|\ell(v) - \ell(w)| \leq 1.$$

See Figure 4.16 for an example.

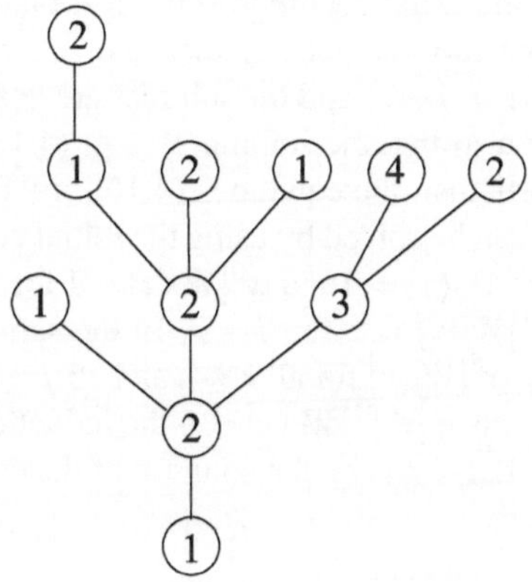

**Figure 4.16**
Well-labeled tree.

Well-labeled trees (and variations of them) have been discussed quite intensively during the last few years. This comes from the fact that there is a remarkable bijection between (rooted) quadrangulations with $n$ faces and well-labeled trees with $n$ edges that is due to Gilles Schaeffer [24]. One of the **main features** of this bijection is that the labels of the well-labeled tree correspond to the graph distance to the root of the (corresponding) quadrangulations. This observation made it possible to study the distance profile of quadrangulations and to prove that the diameter of a randomly chosen quadrangulation is of order $n^{1/4}$, see [8].

At first sight it is by no means obvious how these kinds of trees can be counted. Of course, the resulting number has to coincide with the number of quadrangulations. Nevertheless there is a direct proof that uses an ingenious method due to Bouttier, Di Francesco, and Guitter [5, 6]; see also Bousquet-Mélou [4].

**Theorem 4.5.9** *The number $W_n$ of well-labeled trees (with increments* 0 *and* $\pm 1$*) with n edges, is given by the formula*

$$W_n = \frac{2 \cdot 3^n}{(n+1)(n+2)} \binom{2n}{n}. \tag{4.15}$$

*Proof.* We consider variants of well-labeled trees, where the root has a label not necessarily equal to 1 but again all vertices are labeled by positive integers and labels of adjacent vertices differ at most by 1. For $j \geq 1$, let $W_j(x)$ denote the (ordinary) generating function of those generalized well-balanced trees where the root has label $j$ and where the exponent of $x$ counts the number of edges. Then, by using the convention $W_0(x) = 0$, we immediately get the relation

$$W_j(x) = \frac{1}{1 - x(W_{j-1}(x) + W_j(x) + W_{j+1}(x))}, \quad (j \geq 1). \tag{4.16}$$

This relation comes from the usual decomposition of a planted plane tree into the root and its subtrees. If the root has label $j$ then the successors of the root must have a label in the set $\{j-1, j, j+1\}$ and the subtrees are again well-balanced trees. This justifies (4.16). Note also that the infinite system (4.16) uniquely determines the functions $W_j(x)$. In particular, the equations (4.16) provide a recurrence for the coefficients $[x^n]\,W_j(x)$ that can be solved by using the initial condition $W_0(x) = 0$. For example we have indeed $[x^0]\,W_0(x) = 0$ and $[x^0]\,W_j(x) = 1$ for $j \ge 1$. Hence, it follows that $[x^1]\,W_1(x) = 2$ and $[x^1]\,W_j(x) = 3$ for $j \ge 2$. In the same way we can proceed further. If we already know $[x^k]\,W_j(x)$ for all $k \le n$ and all $j \ge 0$ then $[x^{n+1}]\,W_j(x)$ can be computed for all $j \ge 1$. Since $[x^{n+1}]\,W_0(x) = 0$ the induction works.

Let $W(x) = (1-\sqrt{1-12x})/(6x)$ be the solution of the equation

$$W(x) = \frac{1}{1-3xW(x)}, \tag{4.17}$$

and let $Z(x)$ be defined by

$$Z(x) + \frac{1}{Z(x)} + 1 = \frac{1}{xW(x)^2}. \tag{4.18}$$

Note that one of the two possible choices of $Z(x)$ represents a power series in $x$ with constant term 0 (and non-negative coefficients) which we will use in the sequel.

By using the ansatz

$$W_j(x) = W(x)\frac{u_j u_{j+3}}{u_{j+1}u_{j+2}}$$

with unknown functions $u_j$, the recurrence (4.16) is equivalent to

$$\begin{aligned} u_j u_{j+1} u_{j+2} u_{j+3} = {} & \frac{1}{W(x)} u_{j+1}^2 u_{j+2}^2 \\ & + xW(x)\left(u_{j-1}u_{j+2}^2 u_{j+3} + u_j^2 u_{j+3}^2 + u_j u_{j+1}^2 u_{j+4}\right). \end{aligned} \tag{4.19}$$

By using (4.17) and (4.18) it is easy to check that

$$u_j = 1 - Z(x)^j$$

satisfies (4.19). Hence, the functions

$$W_j(x) = W(x)\frac{(1-Z(x)^j)(1-Z(x)^{j+3})}{(1-Z(x)^{j+1})(1-Z(x)^{j+2})} \tag{4.20}$$

satisfy the system of equations (4.16) and we also have $W_0(x) = 0$. Since all these functions $W_j(x)$ are power series in $x$, we have found the solution of interest of the system (4.16).

In particular, we have

$$W_1(x) = W(x)\frac{(1-Z(x))(1-Z(x)^4)}{(1-Z(x)^2)(1-Z(x)^3)}$$

$$= W(x)\frac{1+Z(x)^2}{1+Z(x)+Z(x)^2}$$
$$= W(x)(1-xW(x)^2),$$

and it is an easy exercise (by using Lagrange's inversion formula) to show that

$$q_n = [x^n]W_1(x) = \frac{2\cdot 3^n}{(n+1)(n+2)}\binom{2n}{n}.$$

This completes the proof of the theorem. ∎

The **unexpected** part of the preceding proof is formula (4.20). The resulting Formula (4.15) for $W_n$ has, however, a quite natural interpretation using the bijection between quadrangulations with $n$ faces and well-labeled trees with $n$ edges (the Schaeffer bijection [24]).

There are many more results on well-labeled trees (see [4]). We just mention that there is, for example, a similar explicit formula for the number of well-labeled trees, where adjacent vertices $v, w$ satisfy $|\ell(v)-\ell(w)| = 1$. Here we have

$$\tilde{W}_n = \frac{3\cdot 2^{n-1}}{(n+1)(n+2)}\binom{2n}{n}$$

for these kinds of trees with $n$ edges.

## 4.6 Selected topics on trees

### 4.6.1 Spanning trees

Let $G$ be a connected graph (without loops or multiple edges). Then a tree $T$ is a **spanning tree** of $G$ if it has the same vertex set as $G$ and if the edge set of $T$ is a subset of the edge set of $G$. Alternatively $T$ is the maximal subgraph of $G$ that is a tree. There always exists a spanning tree. We just have to **delete** edges that are within a cycle till we are left with a tree.

The enumeration problem for spanning trees is solved by Kirchhoff's matrix tree theorem (see for example [21]).

**Theorem 4.6.1** *Let $G$ be a connected graph without multiple edges and loops. Let $\{v_1, v_2, \ldots, v_n\}$ denote its vertex set, $d(v_1), d(v_2), \ldots, d(v_n)$ the degree sequence and $D = \operatorname{diag}(d(v_1), d(v_2), \ldots, d(v_n))$ the corresponding diagonal matrix. Furthermore let $A = (a_{ij})_{1\le i,j\le n}$ denote the adjacency matrix of $A$, that is, $a_{ij} = 1$ if $v_i$ and $v_j$ are linked by an edge and $a_{ij} = 0$ else.*

*Then the number of spanning subgraphs of $G$ equals the absolute value of the determinant of any $(n-1)\times(n-1)$ submatrix of $D-A$.*

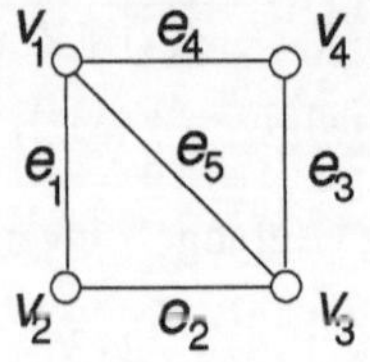


**Figure 4.17**
An illustration of the use of the matrix tree theorem.

Actually much more can be said. For example, let us consider the graph shown in Figure 4.17 with four vertices, where we also label the edges by $e_1,\ldots,e_5$.

We then consider a generalized version $\tilde{D}$ of the diagonal matrix $D$, where the degrees are replaced by the **sum** of the edges that are incident with the corresponding vertices. Similarly we define $\tilde{A}$ as a refined adjacency matrix, where $a_{ij}$ is replaced by the actual edge that connects $v_i$ and $v_j$ (if it exists):

$$\tilde{D}=\begin{pmatrix} e_1+e_4+e_5 & 0 & 0 & 0 \\ 0 & e_1+e_2 & 0 & 0 \\ 0 & 0 & e_2+e_3+e_5 & 0 \\ 0 & 0 & 0 & e_3+e_4 \end{pmatrix}, \quad \tilde{A}=\begin{pmatrix} 0 & e_1 & e_5 & e_4 \\ e_1 & 0 & e_2 & 0 \\ e_5 & e_2 & 0 & e_3 \\ e_4 & 0 & e_3 & 0 \end{pmatrix}.$$

Of course, if we set $e_j=1$, $1\le j\le 5$, then we obtain the original matrices $D$ and $A$. If we now delete the first row and column of $\tilde{D}-\tilde{A}$ and compute the determinant, we get

$$\begin{vmatrix} e_1+e_2 & -e_2 & 0 \\ -e_2 & e_2+e_3+e_5 & -e_3 \\ 0 & -e_3 & e_3+e_4 \end{vmatrix}$$
$$=e_1e_2e_3+e_1e_2e_4+e_1e_3e_4+e_1e_3e_5+e_1e_4e_5+e_2e_3e_4+e_2e_3e_5+e_2e_4e_5.$$

By inspecting this sum we observe that each of these eight monomials corresponds to precisely one of the eight possible spanning trees, where the edges are just the factors of the monomials. This means that this determinant is the **generating series** of all spanning trees that can be used to do a more refined analysis of spanning trees (for example we can count how many spanning trees use the edge $e_1$, etc.).

As mentioned in Section 4.5.1 a labeled tree can be considered as a spanning tree of the complete graph $K_n$. By applying Theorem 4.6.1 we are led to compute the $(n-1)\times(n-1)$-determinant

$$d_n=\begin{vmatrix} n-1 & -1 & \cdots & -1 \\ -1 & n-1 & \cdots & -1 \\ \vdots & \vdots & \ddots & \vdots \\ -1 & -1 & \cdots & n-1 \end{vmatrix}.$$

Now we do the following row computations. First we add all but the first row to the first row so that the first row consists just of 1's. Then we add this new first row to all other rows and obtain the following determinant:

$$d_n = \begin{vmatrix} 1 & 1 & \cdots & 1 \\ -1 & n-1 & \cdots & -1 \\ \vdots & \vdots & \ddots & \vdots \\ -1 & -1 & \cdots & n-1 \end{vmatrix} = \begin{vmatrix} 1 & 1 & \cdots & 1 \\ 0 & n & \cdots & 0 \\ \vdots & \vdots & \ddots & \vdots \\ 0 & 0 & \cdots & n \end{vmatrix}.$$

Now it is clear that $d_n = n^{n-2}$ since we just have to take the product of the diagonal elements.

There are several other applications of this method and also an extension to directed graphs (for more details see Moon's book [21]).

Most of the proofs of Theorem 4.6.1 are quite involved and use multi linear algebra (see Chapter 1 or [21]). However, there is a nice recent analytic proof by Amitai Zernik [25] that we can present here.

*Proof.* Let $L = (L_{ij}) = \tilde{A} - \tilde{D}$. Then $L$ is symmetric and has the property that for all $i$ the sums $\sum_j L_{ij} = 0$. Let $C_{ij}$ denote the cofactor of $L$ corresponding to row $i$ and column $j$, that is, $(-1)^{i+j} \det M_{ij}(L)$, where $M_{ij}(L)$ denotes the $(n-1) \times (n-1)$-matrix that is obtained from $L$ by deleting the $i$th row and the $j$th column. It is easy to see that in our case (where $\sum_j L_{ij} = 0$ for all $i$) all cofactors have a common value that we denote by $C(L)$.

For every tree $T$ with vertex set $\{1,2,\ldots,n\}$ we set

$$A_T(L) = \prod_{i<j,\ \{i,j\} \in E(T)} L_{ij}.$$

The main observation is that

$$C(L) = \sum_T A_T(L).$$

This is precisely the generalized version of Theorem 4.6.1 (note that the diagonal is not used to compute $A_T(L)$), which specializes to Theorem 4.6.1 if we set $L = A - D$.

We denote by $\mathscr{L}$ the linear space of all symmetric matrices $L = (L_{ij})$ with the property that for all $i$ the sums $\sum_j L_{ij} = 0$. It is easy to see that $\mathscr{L}$ is path connected and that a (continuously differentiable) function $f$ on $\mathscr{L}$ is constant if and only if $D_{ij} f = 0$ for all $i < j$, where $D_{ij}$ is defined by

$$D_{ij} = \frac{\partial}{\partial L_{ij}} + \frac{\partial}{\partial L_{ji}} - \frac{\partial}{\partial L_{ii}} - \frac{\partial}{\partial L_{jj}}.$$

(Formally $D_{ij}$, $i < j$, span the tangent space of $\mathscr{L}$ at every point of $\mathscr{L}$.) Next set $B(L) = \sum_T A_T(L)$. The idea of the proof is to compare the derivatives $D_{ij}$ of both sides of the proposed equality $C(L) = B(L)$. More precisely we show that $D_{ij}(C(L)) = D_{ij}(B(L))$ for all $i < j$. Since $C(\mathbf{0}) = B(\mathbf{0}) = 0$ the equality $C(L) = B(L)$ follows for all $L \in \mathscr{L}$.

Let $L' = M_{jj}(L_+)$, where $L_+$ is an $(n-1)\times(n-1)$-matrix obtained from $L$ by adding the $j$th row to the $i$th row and the $j$th column to the $i$th column. Note that $L'$ has the same properties as $L$, that is, it is symmetric and $\sum_{j'} L'_{i'j'} = 0$ for all $i'$. The main step of the proof is to show that $D_{ij}(B(L)) = B(L')$ and $D_{ij}(C(L)) = -C(L')$. The result then follows by induction on the size of the matrix.

To see that $D_{ij}(B(L)) = B(L')$, note that $D_{ij}(A_T(L)) = \frac{\partial}{\partial L_{ij}} A_T(L)$ is nonzero only if $i, j$ are connected by an edge in $T$, in which case we may contract the edge to produce a new spanning tree $T' = \text{contract}_{ij}(T)$ with vertices $\{1,2,\ldots,n-1\}$. Explicitly, we erase the vertex $j$ and the edge $\{i,j\}$ and reconnect all the other neighbors of $j$ to the vertex $i$. We also relabel the vertices $j+1,\ldots,n$ by $j,\ldots,n-1$, respectively. Fixing some $T'$ it is then easy to see that

$$\sum_{T:\text{contract}_{ij}(T)=T'} \frac{\partial}{\partial L_{ij}} A_T(L) = A_{T'}(L')$$

which implies $D_{ij}(B(L)) = B(L')$.

Finally we show $-D_{ij}(C(L)) = C(L')$. Indeed,

$$\begin{aligned}
-D_{ij}(C(L)) &= -D_{ij}(C_{ii}(L)) = \frac{\partial}{\partial L_{jj}} C_{ii}(L) = \frac{\partial}{\partial L_{jj}} C_{ii}(L_+) \\
&= \frac{\partial}{\partial L_{jj}} \det(M_{ii}(L_+)) = \det(M_{j-1,i-1}(M_{ii}(L_+))) = \det M_{ii}(M_{jj}(L_+)) \\
&= \det M_{ii}(L') = C(L'),
\end{aligned}$$

which completes the proof. ■

### 4.6.2 $k$-Trees

A $k$-tree is a graph that can be considered as a generalized tree and can be defined recursively: A $k$-tree is either a complete graph on $k$ vertices or a graph obtained from a smaller $k$-tree by adjoining a new vertex together with $k$ edges connecting it to a $k$-clique of the smaller $k$-tree and, thus, producing a $(k+1)$-clique. (A $k$-clique is a subgraph that is isomorphic to a complete graph on $k$ vertices.) In particular, a 1-tree is a usual tree. Figure 4.18 shows an example of a 2-tree.

The notion of a $k$-tree originates from the parameter *tree-width* $\text{tw}(G)$ of a graph $G$, which is the minimum width among all possible tree decompositions of $G$, or equivalently, $\text{tw}(G)$ is the minimum $k$ such that $G$ is a subgraph of a $k$-tree. A $k$-tree, as a bounded tree-width graph, is exactly the maximal graph with a fixed tree-width $k$ such that no more edges can be added without increasing its tree width.

Labeled $k$-trees have been already counted by Beineke, Pippert, Moon and Foata [1, 21, 14] four decades ago.

**Theorem 4.6.2** *Let $k \geq 1$. Then the number $B_n^{(k)}$ of $k$-trees having $n$ labeled vertices is given by*

$$B_n^{(k)} = \binom{n}{k}(k(n-k)+1)^{n-k-2}. \tag{4.21}$$

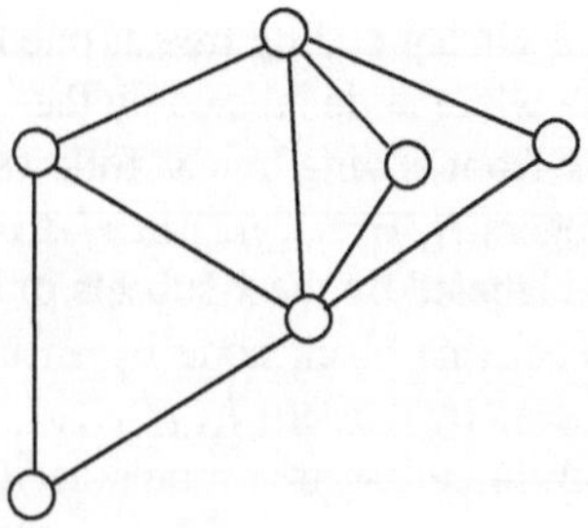

**Figure 4.18**
A $k$-tree.

We present here a proof that is related to generating functions (see [9]).

*Proof.* We call a $k$-clique of a $k$-tree **front** and a $(k+1)$-clique **hedron**. It is clear that $k$-tree with $n$ vertices contains precisely $n-k$ hedra. In the following proof we will work with $k$-trees with $n$ hedra (and thus with $n+k$ vertices).

First we introduce a bijection between labeled $k$-trees and so-called $k$-front coding trees. A $k$-front coding tree has unlabeled black nodes and labeled white nodes. To construct a $k$-front coding tree from a labeled $k$-tree, an unlabeled black node in a $k$-front coding tree represents a hedron in the labeled $k$-tree. A white node labeled by the set $\{i_1, i_2, \ldots, i_k\}$ represents a front with vertices that are labeled by the integers $i_1, i_2, \ldots, i_k$ in the labeled $k$-tree. A black node connects with a white node if the corresponding hedron contains the corresponding front. It is clear that labeled $k$-trees having $n$ hedra are in bijection with $k$-front coding trees with $n$ black nodes and $kn+1$ labeled white nodes. See Figure 4.19 for an example.

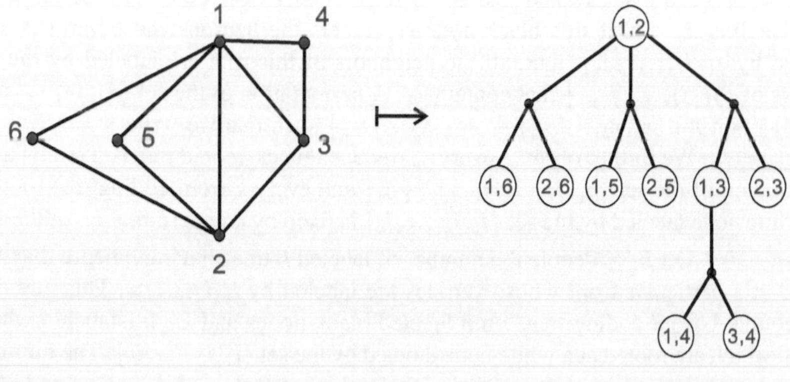


**Figure 4.19**
Bijection between a labeled 2-tree (left) and a 2-front coding tree (right).

We will next reduce the problem of counting unrooted trees into rooted trees. This can be done by rooting a $k$-front coding tree at one of its fronts. Suppose a $k$-front coding tree is rooted at a white node labeled by the set $\{i_1, i_2, \ldots, i_k\}$. Then we label the black nodes in this $k$-front coding tree as follows. For every hedron whose vertices are labeled by the integers $j_1, j_2, \ldots, j_{k+1}$ in a $k$-tree, the corresponding black node connects to white nodes labeled by the $k$ subsets of $\{j_1, j_2, \ldots, j_{k+1}\}$ in the $k$-front coding tree. We will label this black node by an integer $j_m$ if among all its neighbors, the white node labeled by the set $\{j_1, j_2, \ldots, j_{k+1}\} \setminus \{j_m\}$ is closest to the root. In this way we get a $k$-front coding tree rooted at $\{i_1, i_2, \ldots, i_k\}$ with all black nodes labeled. So a labeled $k$-tree rooted at a front can be identified with a $k$-front coding tree rooted at a white node. See Figure 4.20 for an example.

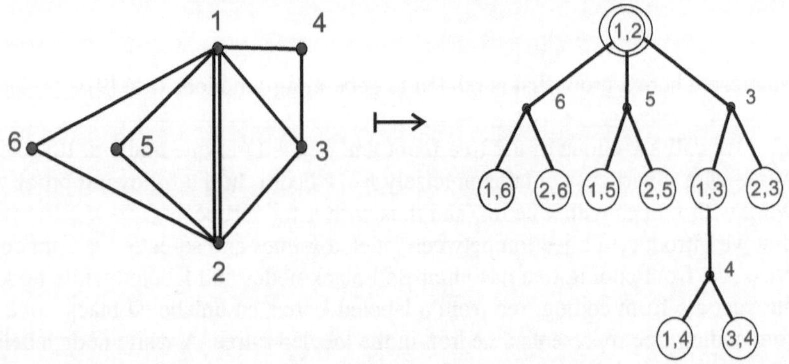


**Figure 4.20**
Bijection between a labeled 2-tree rooted at a front whose vertices are labeled by 1, 2 (left) and a 2-front coding tree rooted at a white node labeled by $\{1,2\}$ (right).

A $j_m$-reduced black-rooted tree is a $k$-front coding tree rooted at a black node with labeling $j_m$ and if this black node represents the hedron whose vertices are labeled by $j_1, j_2, \ldots, j_{k+1}$, then all the neighbors of this root are labeled by the $k$-subsets of $\{j_1, j_2, \ldots, j_{k+1}\}$ that contain $j_m$. A permutation of the set $\{x_1, x_2, \ldots, x_k\}$ is a bijection $\pi : \{x_1, x_2, \ldots, x_k\} \to \{x_1, x_2, \ldots, x_k\}$ such that we write $\pi = i_1 i_2 \cdots i_k$ if $\pi(x_m) = i_m$. We further observe every $j_m$-reduced black-rooted tree is fixed by any permutation of the set $\{j_1, j_2, \ldots, j_{k+1}\} \setminus \{j_m\}$ and every $k$-front coding tree rooted at a white node labeled by the set $\{i_1, i_2, \ldots, i_k\}$ is fixed by any permutation of the set $\{i_1, i_2, \ldots, i_k\}$. Let $R^{(k)}_{n+k}$ denote the number of labeled $k$-trees having $n$ hedra that are rooted at a particular front whose vertices are labeled by $i_1, i_2, \ldots, i_k$. This number also counts $k$-front coding trees on $n$ labeled black nodes and $kn+1$ labeled white nodes, which are rooted at a white node labeled by the set $\{i_1, i_2, \ldots, i_k\}$. The number $R_n(n+k)$ is independent of the choice of the particular root $\{i_1, i_2, \ldots, i_k\}$. Let $C_k(x)$ be the exponential generating function for $\{i_1, i_2, \ldots, i_k\}$-rooted $k$-front coding trees where the size is the number of black nodes, that is,

$$C_k(x) = \sum_{n=0}^{\infty} R_k(n+k)\frac{x^n}{n!}.$$

Let $B_k(x)$ be the exponential generating function for $j_m$-reduced black-rooted $k$-front coding trees. Since every $k$-front coding tree rooted at a white node labeled by $\{i_1, i_2, \ldots, i_k\}$ can be identified as a set of reduced black-rooted trees that are fixed by any permutation of $\{i_1, i_2, \ldots, i_k\}$. Thus we have

$$C_k(x) = \exp(B_k(x)). \tag{4.22}$$

We observe that the black root $j_m$ in a reduced black-rooted tree connects with $k$ white-rooted coding trees. All these $k$ white-rooted coding trees are respectively rooted at the white nodes labeled by the $k$-subsets of $\{i_1, i_2, \ldots, i_k, j_m\}$ except $\{i_1, i_2, \ldots, i_k\}$. This yields

$$B_k(x) = x(C_k(x))^k$$

and combining it with (4.22)

$$B_k(x) = x\exp(kB_k(x)). \tag{4.23}$$

By applying Lagrange's inversion formula to (4.23), we obtain

$$[x^n]B_k(x) = \frac{1}{n}[u^{n-1}]e^{knu} = \frac{(kn)^{n-1}}{n!}$$

which implies that the number of $j_m$-reduced black rooted $k$-front coding trees having $n$ black nodes is $(kn)^{n-1}$. Similarly we can derive the coefficients of $C_k(x)$ from equation (4.22):

$$[x^n]C_k(x) = \frac{1}{n}[u^{n-1}]e^{(kn+1)u} = \frac{(kn+1)^{n-1}}{n!}.$$

Equivalently, $R_{n+k}^{(k)} = (kn+1)^{n-1}$ counts the number of labeled $k$-trees having $n$ hedra that are rooted at a particular front whose vertices are labeled by $i_1, i_2, \ldots, i_k$. Since there are $\binom{n+k}{k}$ ways to choose the set $\{i_1, i_2, \ldots, i_k\}$, the number of labeled $k$-trees having $n$ hedra that are rooted at a front is

$$(kn+1)B_{n+k}^{(k)} = \binom{n+k}{k}R_{n+k}^{(k)}$$

and the explicit formula for $B_n^{(k)}$ given in (4.21) follows. ■

The counting problem of unlabeled $k$-trees is much more difficult. Only the case of 2-trees was already solved by Harary and Palmer [18] and Fowler *et al* [15] by using the dissimilarity theorem. The general case was a long-standing open problem and was solved just recently by Gainer-Dewar [16]; see also [17].

In order to state the result (as given in [17]) we have to introduce some notation. Let $g \in \mathfrak{S}_m$ be a permutation of $\{1,2,\cdots,m\}$ that has $\ell_i$ cycles of size $i$, $1 \le i \le k$, in its cyclic decomposition. Then its cycle type $\lambda = (1^{\ell_1} 2^{\ell_2} \cdots k^{\ell_k})$ is a partition of $m$ where $m = \ell_1 + 2\ell_2 + \cdots + k\ell_k$. We will denote by $\lambda \vdash m$ that $\lambda$ is a partition of $m$. Furthermore we set $z_\lambda = 1^{\ell_1}\ell_1! 2^{\ell_2}\ell_2! \cdots k^{\ell_k}\ell_k!$ and $\frac{m!}{z_\lambda}$ is the number of permutations in $\mathfrak{S}_m$ of cycle type $\lambda$.

The result says that the generating function $U(x) = \sum_{n\ge k} U_n^{(k)} x^n$ of unlabeled $k$-trees is given by

$$U(x) = B(x) + C(x) - E(x),$$

where

$$B(x) = \sum_{\lambda\vdash k+1} \frac{B_\lambda(x)}{z_\lambda} \qquad B_\lambda(x) = x\prod_i C_{\lambda^i}(x^i)$$

$$C(x) = \sum_{\mu\vdash k} \frac{C_\mu(x)}{z_\mu} \qquad \bar{B}_\mu(x) = x\prod_i C_{\mu^i}(x^i)$$

$$E(x) = \sum_{\mu\vdash k} \frac{\bar{B}_\mu(x) C_\mu(x)}{z_\mu} \qquad C_\mu(x) = \exp\left[\sum_{m=1}^{\infty} \frac{\bar{B}_{\mu^m}(x^m)}{m}\right].$$

Here $\lambda$ is always the cycle type of a permutation in $\mathfrak{S}_{k+1}$ and $\mu$ the cycle type of a permutation in $\mathfrak{S}_k$. Furthermore $\lambda^i$ denotes the cycle type of permutation $\pi^i$ where $\pi \in \mathfrak{S}_{k+1}$ has cycle type $\lambda$ and $i$ is a part of $\lambda$. The above products over $i$ range over all parts $i$ of $\lambda$ or $\mu$, respectively, that is, if $i$ is contained $m$ times in $\lambda$ then it appears $m$ times in the product. Finally we implicitly use the relation $\bar{B}_\mu(x) = \bar{B}_\lambda(x)$, $C_\mu(x) = C_\lambda(x)$ if $\lambda$ is obtained from $\mu$ by adding a part 1. So starting from $\mu = 1^k$, where we get an implicit equation for $\bar{B}_{1^k}(x)$, we can compute step by step all involved generating functions.

Recently the asymptotic behavior of $U_n^{(k)}$ was determined in [10]:

$$U_n^{(k)} \sim c_k \gamma_k^n n^{-5/2},$$

where $c_k$ and $\gamma_k$ are certain positive constants.

Similar to recursive (and increasing) trees we can consider all possible **evolution processes** of $k$-trees, when we start with a specified complete graph (on $k$ vertices) and add step by step a vertex and its corresponding edges. This means that we have $k$ unlabeled vertices from the starting complete graph and $n$ labeled vertices that encode the evolution. In each step we have $ik+1$ choices for the new vertex with label $i+1$. Thus the resulting number of increasing $k$-trees (with $k+n$ vertices) is

$$R_n^{(k)} = \prod_{i=0}^{n-1} (ik+1).$$

The corresponding (exponential) generating function $R(x) = \sum_{n\ge 0} R_n^{(k)} x^n/n!$ satisfies the differential equation

$$R'(x) = R(x)^k, \qquad T(0) = 1$$

and is explicitly given by $R(x) = (1-kx)^{-1/k}$.

### 4.6.3 Search trees

Search trees are used in computer science for storing and searching data. We describe here a few standard models, where the underlying tree is binary. The counting problem is not the main focus here, since this has been already done in Section 4.4.1. However, the combinatorial and algorithmic background are of considerable interest. Furthermore these kinds of trees give rise to very interesting probabilistic tree models, however, we will not go into these details. We refer to the monographs [11, 19].

We start with **binary search trees**. The origin of binary search trees dates to a fundamental problem in computer science: the dictionary problem. In this problem a set of records is given where each can be addressed by a key and the purpose is to maintain this set under insertion, deletion and membership queries. The binary search tree is a data structure used for storing the records in its internal nodes (in a way that will explained below). It will then also be clear how the operations **insert**, **delete** and **search** can be implemented then.

In principle binary search trees are plane binary trees generated by a permutation (or list) $\pi(1), \pi(2), \ldots, \pi(n)$ of the numbers $\{1, 2, \ldots, n\}$. The elements $\pi(1), \pi(2), \ldots, \pi(n)$ serve as keys and will be stored in the internal nodes of the resulting tree. Equivalently we could think of a list of keys, where each pair of keys can be compared. Of course, this is the situation in practice.

Starting with one of the keys (for example with the first on the list $\pi(1)$) one first compares $\pi(1)$ with $\pi(2)$. If $\pi(2) < \pi(1)$, then $\pi(2)$ becomes root of the left subtree; otherwise, $\pi(2)$ becomes root of the right subtree. When having constructed a tree with nodes $\pi(1), \ldots, \pi(k)$, the next node $\pi(k+1)$ is inserted by comparison with the existing nodes in the following way: Start with the root as current node. If $\pi(k+1)$ is less than the current node, then descend into the left subtree, otherwise into the right subtree. Now continue with the root of the chosen subtree as current, according to the same rule. Finally, attach $n+1$ external nodes (= leaves) at the possible places. Figure 4.21 shows an example of a binary search tree (without and with external nodes) for the input keys $(4, 6, 3, 5, 1, 8, 2, 7)$.

Alternatively one can describe the construction of the binary search tree recursively in the following way. If $n > 1$, we select (as above) a pivot (for example $\pi(1)$) and subdivide the remaining keys into two sublists $I_1, I_2$:

$$I_1 = (x \in \{\pi(2), \ldots, \pi(n)\} : x < \pi(1)) \quad \text{and}$$
$$I_2 = (x \in \{\pi(2), \ldots, \pi(n)\} : x > \pi(1)).$$

The pivot $\pi(1)$ is put to the root and by recursively applying the same procedure, the elements of $I_1$ constitute the left subtree of the root and the elements of $I_2$ the right subtree. This is precisely the standard Quicksort algorithm.

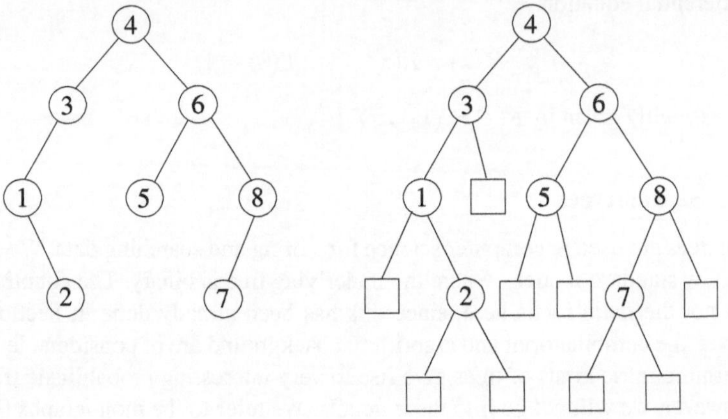


**Figure 4.21**
Binary search tree.

Now the operations **insert** and **search** are easy to implement. If we want to insert a new record we just compare its key with the key of the root. If it is smaller then we go to the left subtree and if it is larger we go to the right subtree. Then we iterate this procedure till we find an external node, where we can store this record. The searching procedure is almost the same (the only difference is that we stop if the keys are the same). Finally if we want to **delete** a record $r$, we first have to find it and then we have to distinguish between two cases. First if the corresponding internal node $i(r)$ has no or just one internal node as a successor we just delete the internal node $i(r)$ (and thus $r$) and contract the tree. Second, if $i(r)$ has two internal nodes as successors replace the value at $i(r)$ with the largest value in its left subtree and then delete the node with the largest value from its left subtree. (Note that largest value in the left subtree will never have two subtrees so this node can be safely deleted as in the first case.)

The standard probabilistic model for binary search trees (and their variants) is the so-called permutation model, where it is assumed that every permutation of the input data is equally likely. We again refer to the books [11, 19], where several tree parameters of interest are studied under this model.

**Digital search trees** are intended for the same kind of problems as binary search trees, that is to **insert**, to **delete**, and to **search**. However, they are not constructed from the total order structure of the keys for the data stored in the internal nodes of the tree but from digital representations (or binary sequences) that serve as keys.

Consider a set of records, numbered from 1 to $n$ and let $x_1, \ldots, x_n$ be binary sequences for each item (that represent the keys). We construct a binary tree, the digital search tree, from such a sequence as follows. First, the root is left empty, we can say that it stores the empty word. Then the first item occupies the right or left child of the root depending on its first symbol. If the first symbol is 1 then it occupies the right one, otherwise the left one. After having inserted the first $k$ items, we insert item $k+1$: Choose the root as current node and look at the binary key $x_{k+1}$. If the first digit is 1, descend into the right subtree, otherwise into the left one. If the root of the subtree is occupied, continue by looking at the next digit of the key. This procedure terminates at the first unoccupied node where the $(k+1)$th item is stored.

For example, consider the items

$$x_1 = 110011\cdots \qquad x_5 = 000110\cdots$$
$$x_2 = 100110\cdots \qquad x_6 = 010111\cdots$$
$$x_3 = 010010\cdots \qquad x_7 = 000100\cdots$$
$$x_4 = 101110\cdots \qquad x_8 = 100101\cdots.$$

If we apply the above described procedure we end up with the binary tree depicted in Figure 4.22. As in the case of binary search trees we can append external nodes to make it a complete binary tree.

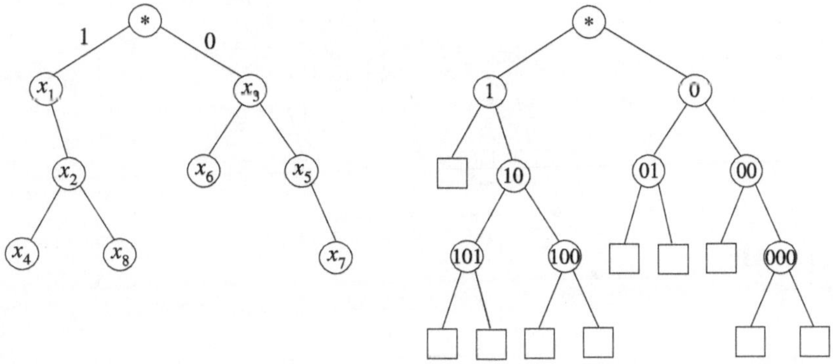


**Figure 4.22**
Digital search tree.

The standard probabilistic model, the Bernoulli model, is to assume that the keys $x_1,\ldots,x_n$ are binary sequences, where the digits 0 and 1 are drawn independently and identically distributed with probability $p$ for 1 and probability $q = 1-p$ for 0. The case $p = q = \frac{1}{2}$ is called symmetric.

There are several natural generalizations of this basic model. Instead of a binary alphabet one can use an $m$-ary one leading to $m$-ary digital search trees. One can also change the probabilistic model by using, for example, discrete Markov processes to generate the key sequences.

The construction idea of **tries** is similar to that of digital search trees except that the records are stored in the leaves rather than in the internal nodes. Again a 1 indicates a descent into the right subtree, and 0 indicates a descent into the left subtree. Insertion causes some rearrangement of the tree, since a leaf becomes an internal node. In contrast to binary search trees and digital search trees, the shape of the trie is independent of the actual order of insertion. The position of each item is determined by the shortest unique prefix of its key. If we use the same input data as for the example of a digital search tree, then we obtain the trie that is depicted in Figure 4.23.

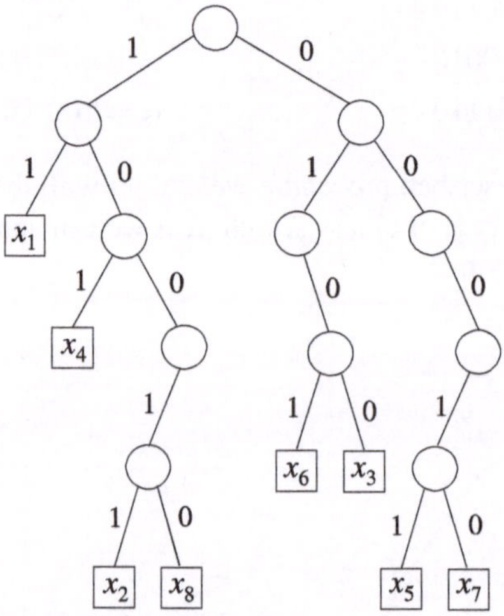


**Figure 4.23**
Trie.

An alternative description runs: Given a set $\mathscr{X}$ of strings, we partition $\mathscr{X}$ into two parts, $\mathscr{X}_L$ and $\mathscr{X}_R$, such that $x_j \in \mathscr{X}_L$ (respectively $x_j \in \mathscr{X}_R$) if the first symbol of $x_j$ is 0 (respectively 1). The rest of the trie is defined recursively in the same way, except that the splitting at the $k$th level depends on the $k$th symbol of each string. The first time that a branch contains exactly one string, a leaf is placed in the trie at that location (denoting the placement of the string into the trie), and no further branching takes place from such a portion of the trie.

This description implies also a recursive definition of tries. As above consider a sequence of $n$ binary strings. If $n = 0$, then the trie is empty. If $n = 1$, then a single (external) node holding this item is allocated. If $n \geq 1$, then the trie consists of a root (internal) node directing strings to the two subtrees according to the first letter of each string, and string directed to the same subtree are themselves tries, however, constructed from the second letter on.

As in the case of digital search trees we can construct $m$-ary tries by using strings over an $m$-ary alphabet leading to $m$-ary trees.

**Acknowledgment.** The author is grateful to an anonymous referee for his careful reading of a previous version of this chapter and for his many valuable comments for improving the presentation.

## References

[1] L. W. Beineke and R. E. Pippert. The number of labeled $k$-dimensional trees. *J. Combinatorial Theory*, 6:200–205, 1969.

[2] F. Bergeron, P. Flajolet, and B. Salvy. Varieties of increasing trees. In *CAAP '92 (Rennes, 1992)*, volume 581 of *Lecture Notes in Comput. Sci.*, pages 24–48. Springer, Berlin, 1992.

[3] F. Bergeron, G. Labelle, and P. Leroux. *Combinatorial species and tree-like structures*, volume 67 of *Encyclopedia of Mathematics and its Applications*. Cambridge University Press, Cambridge, 1998. Translated from the 1994 French original by Margaret Readdy, with a foreword by Gian-Carlo Rota.

[4] M. Bousquet-Mélou. Limit laws for embedded trees: applications to the integrated super-Brownian excursion. *Random Structures Algorithms*, 29(4):475–523, 2006.

[5] J. Bouttier, P. Di Francesco, and E. Guitter. Geodesic distance in planar graphs. *Nucl. Phys., B*, 663(3):535–567, 2003.

[6] J. Bouttier, P. Di Francesco, and E. Guitter. Statistics of planar graphs viewed from a vertex: A study via labeled trees. *Nucl. Phys., B*, 675(3):631–660, 2003.

[7] N. Broutin and P. Flajolet. The height of random binary unlabelled trees. In *Fifth Colloquium on Mathematics and Computer Science*, Discrete Math. Theor. Comput. Sci. Proc., AC. 2008.

[8] P. Chassaing and G. Schaeffer. Random planar lattices and integrated super-Brownian excursion. *Probab. Theory Relat. Fields*, 128(2):161–212, 2004.

[9] M. Drmota and E. Y. Jin. An asymptotic analysis of labeled and unlabeled $k$-trees. *Algorithmica, submitted.*

[10] M. Drmota and E. Y. Jin. An asymptotic analysis of unlabeled $k$-trees. *DMTCS proc.*, pages 85–96, 2014.

[11] Michael Drmota. *Random trees: An interplay between combinatorics and probability*. Springer, Wien–New York, 2009.

[12] P. Flajolet and A. Odlyzko. Singularity analysis of generating functions. *SIAM J. Discrete Math.*, 3(2):216–240, 1990.

[13] P. Flajolet and R. Sedgewick. *Analytic Combinatorics*. Cambridge University Press, Cambridge, 2008.

[14] Dominique Foata. Enumerating $k$-trees. *Discrete Math.*, 1(2):181–186, 1971/72.

[15] Tom Fowler, Ira Gessel, Gilbert Labelle, and Pierre Leroux. The specification of 2-trees. *Adv. in Appl. Math.*, 28(2):145–168, 2002.

[16] A. Gainer Dewar. $\gamma$-species and the enumeration of $k$-trees. *Electron. J. Combin.*, 19(4):P45, 2012.

[17] I. M. Gessel and A. Gainer-Dewar. Counting unlabeled $k$-trees. *J. Combinatorial Theory Ser. A*, 126:177–193, 2014.

[18] F. Harary and E. M. Palmer. *Graphical enumeration*. Academic Press, New York, 1973.

[19] H. M. Mahmoud. *Evolution of random search trees*. John Wiley & Sons Inc., New York, 1992.

[20] A. Meir and J. W. Moon. On the altitude of nodes in random trees. *Canad. J. Math.*, 30(5):997–1015, 1978.

[21] J. W. Moon. *Counting labelled trees*, volume 1969 of *From lectures delivered to the Twelfth Biennial Seminar of the Canadian Mathematical Congress (Vancouver)*. Canadian Mathematical Congress, Montreal, Que., 1970.

[22] R. Otter. The number of trees. *Ann. of Math. (2)*, 49:583–599, 1948.

[23] G. Pólya. Kombinatorische Anzahlbestimmungen für Gruppen, Graphen und chemische Verbindungen. *Acta Math.*, 68:145–254, 1937.

[24] G. Schaeffer. *Conjugaison darbres et cartes combinatoires aléatoires*. PhD Thesis, Université Bordeaux I, 1998.

[25] A. Zernik. Taylor expansion proof of the matrix tree theorem – part i. arXiv:1308.2633, 2013.

# Chapter 5

# *Planar Maps*

**Gilles Schaeffer**
*CNRS, Laboratoire d'informatique de l'École Polytechnique, France*

## CONTENTS

## 5.1 Introduction

The enumerative theory of planar maps is born in the early sixties with the seminal work of William T. Tutte on the enumeration of planar triangulations. Over 50 years it has led several parallel lives in combinatorics, statistical physics, quantum gravity, enumerative topology and probability theory, that have started to interact intensely only in the last ten to fifteen years. Writing a fair survey of this whole story appears to be a great challenge. We concentrate instead in this text on the combinatorial point of view on map enumeration, and only review very tangentially the physics, topology and probability literature.

## 5.2 What is a map?

### 5.2.1 A few definitions

Combinatorial maps usually arise from one of two settings: either the study of some planar graph drawings, or the construction of surfaces via polygon gluings. Accordingly we give two definitions of maps and discuss how duality reconciles them.

**The graph drawing point of view.** An **embedding** (or **drawing**) of a graph $G = (V,E)$ on the oriented sphere $\mathbb{S}$ is **proper** if the vertices are represented by distinct points and the edges are represented by arcs that only intersect at their endpoints and in agreements with the incidence relation of $G$.

**Definition 5.2.1** *A* **planar map** *$M$ is a proper embedding of a connected graph $G$ in the sphere $\mathbb{S}$, considered up to orientation preserving homeomorphisms of $\mathbb{S}$.*

Loops and multiple edges are allowed, and the map is instead said to be **simple** if it contains neither multiple edges nor loops. A **face** is a connected component of $\mathbb{S} \setminus G$. A **corner** is the angular sector delimited by two consecutive edges around a vertex. Each corner $c$ is incident to a vertex $v(c)$, to a face $f(c)$, and to two edges: In counterclockwise direction around $v(c)$, let $cw(c)$ denote the edge preceeding $c$ and $ccw(c)$ denote the edge following $c$. The **degree** of a vertex or face is the number of incident corners. A map is **Eulerian** if all its vertices have even degrees. It is $m$-**valent** if all its vertices have degree $m$, it is a $m$-**angulation** if all its faces have degree $m$. In the special cases $m = 3,4$ we use the standard terminology trivalent maps and tetravalent maps, triangulations and quadrangulations. Observe that with these definitions, triangulations and quadrangulations are allowed to have multiple edges or loops. Some examples are given in Figure 5.1. Observe that the leftmost map is not simple because the edge $e'$ has a parallel edge.

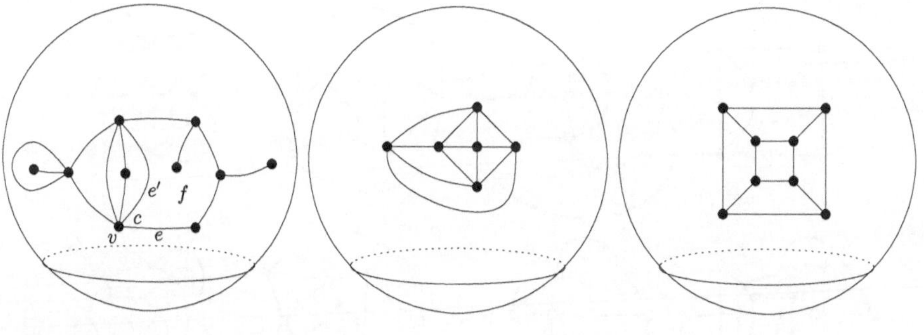


**Figure 5.1**
Three planar maps. The first has 10 vertices, 14 edges and 6 faces. The second is a tetravalent triangulation with 8 triangles. The third is a trivalent quadrangulation with 6 quadrangles; it is usually called the **cube**. On the first map the corner $c$ is incident to the vertex $v = v(c)$, to the face $f = f(c)$, and the edges $e = cw(c)$ and $e' = ccw(c)$.

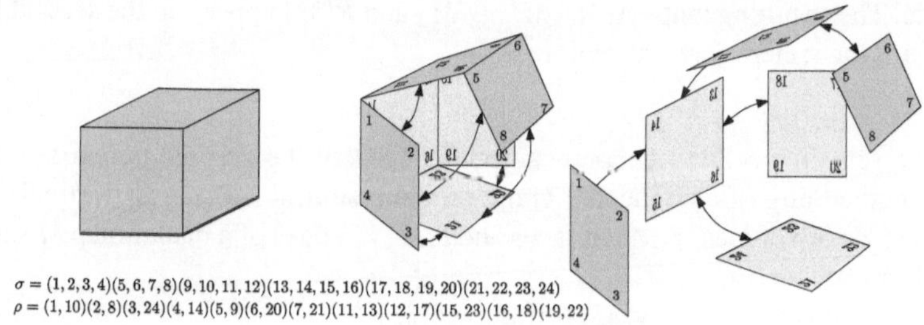


**Figure 5.2**
The cube as a cellular decomposition of the sphere. To make the picture more readable, the side identifications are performed in two stages.

**The polygon gluing point of view.** A **cellular decomposition** of the sphere is a collection of oriented polygons with labeled corners, and a complete set of coherent side identifications such that the resulting surface is the sphere. An example of cellular decomposition of the sphere is given by Figure 5.2. More precisely, a cellular decomposition can be given by the associated **rotation system** $(\sigma, \rho)$, consisting of a permutation $\sigma$ whose cycles describe the clockwise arrangement of corner labels around polygons and a matching (or fix point free involution) $\rho$ describing side identifications: If $\rho(i) = j$ then the polygon side $(i, \sigma(i))$ is identified with the polygon side $(j, \sigma(j))$.

**Definition 5.2.2** *A* **planar map** *is a cellular decomposition of the sphere considered up to relabeling of the corners of the polygons.*

Equivalence between Definitions 5.2.1 and 5.2.2 follows from standard results in surface topology: The faces of a proper embedding of a connected graph $G$ in $\mathbb{S}$ are simply connected, so that the components of $\mathbb{S} \setminus G$ can be identified with polygons,

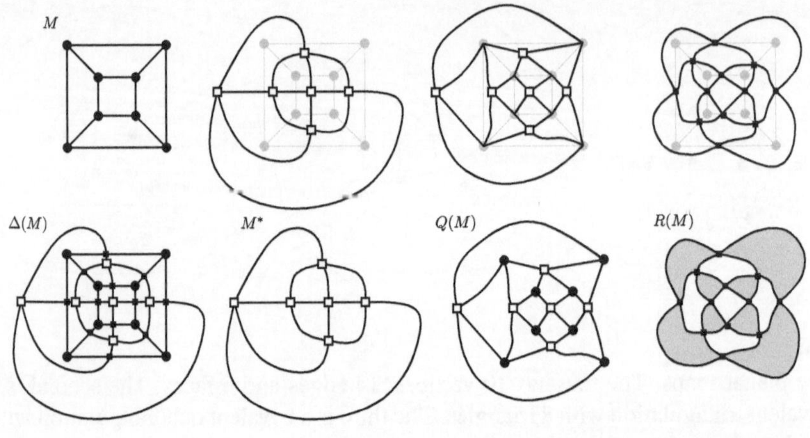


**Figure 5.3**
The cube map $M$, and in the first line, the construction of its dual, incidence map and edge map. The resulting maps $\Delta(M)$, $M^*$, $Q(M)$ and $R(M)$ appear on the second line. The underlying spheres are omitted.

and conversely, in a cellular decomposition the sides of the original polygons define a proper embedding of a connected graph on the resulting surface. In particular the numbers $v(M)$ of vertices, $f(M)$ of faces and $e(M)$ of edges of a planar map $M$ satisfy Euler's formula:

$$\mathrm{v}(M)+\mathrm{f}(M)=\mathrm{e}(M)+2. \tag{5.1}$$

**The dual of a map.** There is a natural symmetry between the role of vertices in Definition 5.2.1 and the role of polygons in Definition 5.2.2. This observation directly leads to the fundamental idea of **duality**, illustrated by Figure 5.3. The **dual** of a map $M$, denoted $M^*$, is the map obtained by drawing a vertex $f^*$ of $M^*$ in each face $f$ of $M$ and an edge $e^*$ of $M^*$ across each edge $e$ of $M$ (see Figure 5.3). By construction each face of $M^*$ then contains exactly one vertex of $M$. The superimposition of a map $M$ and its dual $M^*$ (with tetravalent vertices created at the intersection of dual edges) is a quadrangulation $\Delta(M)$, which is called the **derived map** of $M$. Observe that faces of $\Delta(M)$ are in one-to-one correspondence with corners of $M$.

**Theorem 5.2.3** *Duality is an involution on the set of planar maps. It preserves the number of edges, and exchanges the numbers of vertices and faces:* $M^{**}=M$, $\mathrm{e}(M^*)=\mathrm{e}(M)$, *and* $\mathrm{v}(M^*)=\mathrm{f}(M)$.

Let $M$ be a map with vertex set $V$ and edge set $E$, and let $M^*$ be its dual with vertex set $V^*$ and edge set $E^*$. The **incidence map** (or quadrangulation) $Q(M)$ of the map $M$ is the map whose vertex set is $V \cup V^*$ and with one edge per corner $c$ connecting $v(c)$ and $f(c)^*$. The **edge map** $R(M)$ of $M$ is instead the map with vertex set $E$ and with one edge per corner $c$ connecting $cw(c)$ and $ccw(c)$. The mapping $Q$ and $R$ are bijections from maps with $n$ edges respectively onto vertex-bicolored

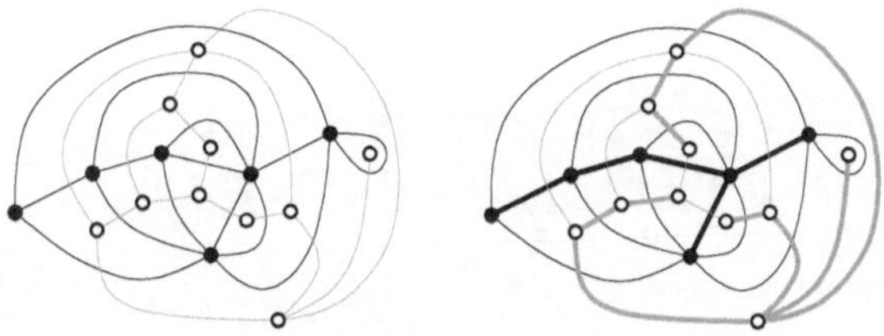

**Figure 5.4**
The superimposition of a map $M$ and its dual $M^*$, and two dual spanning trees.

quadrangulations with $n$ faces, and onto face-bicolored tetravalent maps with $n$ vertices. Moreover $Q(M) = \overline{Q(M^*)}$, $R(M) = \overline{R(M^*)}$ and $Q(M) = R(M)^*$, where the bar denotes the exchange of colors.

The transformation between $M$, $M^*$, $\Delta(M)$, $Q(M)$ and $R(M)$ should be viewed as mere changes of representations for a same underlying object: In the language of computer science, one would say that they represent different data structures that can be used to encode a same cellular decomposition of the sphere. Alternatively this statement can be made precise using the language of topology, and in particular branched covers of the sphere (see [125, Chapter 1] for an exposition, which requires too many definitions to be reproduced here).

**Duality, spanning trees and Euler's formula revisited.** A **spanning tree** of a planar map $M$ is a subset $T$ of the set of edges of $M$ that forms a tree and that is incident to every vertex of $M$. The picture on the right of Figure 5.4 illustrates the following fundamental property of duality and spanning trees for planar maps:

**Theorem 5.2.4** *Let $(T_1, T_2)$ be a partition of the edges of a planar map $M$. Then $T_1$ is a spanning tree of $M$ if and only if $T_2^*$ is a spanning tree of $M^*$.*

The trees $T_1$ and $T_2^*$ are usually called **dual spanning trees**, although this terminology is somewhat improper since the edges of $T_2^*$ are not the duals of the edges of $T_1$, but rather the duals of the edges **not** in $T_1$. The proposition can be viewed as a consequence of the characterization of the sphere by Jordan's lemma ($\mathbb{S} \setminus T_1$ is connected if and only if there is no simple cycle in $T_1$), together with the fact that $\mathbb{S} \setminus T_1$ is connected if and only if $T_2^*$ is connected.

Observe that the above proposition, together with the facts that every map admits a spanning tree, and that any tree with $v$ vertices has $v-1$ edges, gives an interpretation of Euler's formula (5.1): The $e(M)$ edges of a map are partitioned into the $v(M)-1$ edges of a spanning tree and the $f(M)-1$ edges of the dual spanning tree.

### 5.2.2 Plane maps, rooted maps and orientations

Rather than drawing maps on the sphere, we usually draw maps on the plane. This naturally leads to the notion of rooted maps, and to the discussion of orientations.

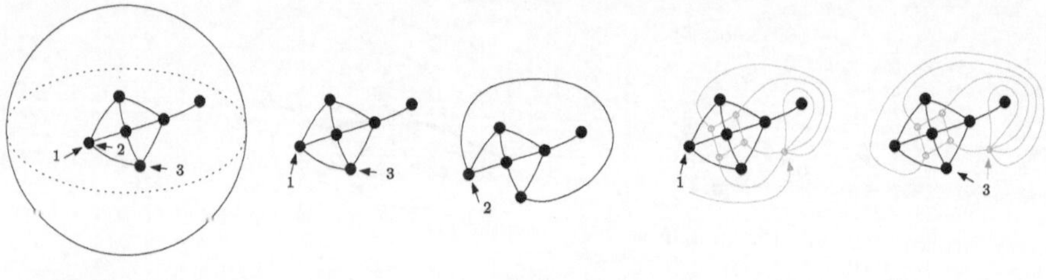


**Figure 5.5**
Three roots for a planar map: the corresponding plane maps, and the plane embedding of the derived map of the first and third one.

**Plane maps, rooted planar maps.** In order to represent a planar map $M$ in the plane, we choose a point $x_0$ of $\mathbb{S}$ in a face of $M$ and identify the punctured sphere $\mathbb{S}^2 \setminus \{x_0\}$ with the plane, sending $x_0$ at infinity. In such a representation, all faces are homeomorphic to discs, except for the face containing $x_0$, which is usually called the **exterior** or **outer face**. Depending on the choice of $x_0$ we **a priori** get different drawings, but up to homeomorphisms of the plane only the choice of the face in which $x_0$ is taken matters. Accordingly, let a **plane map** $(M, f)$ be a planar map $M$ with a marked face $f$, so that plane maps are in one-to-one correspondence with equivalence classes of proper embeddings of connected graphs in the plane up to homeomorphisms of the oriented plane.

Given a planar map $M$, the choice of a marked face does however not fix in general the embedding of the dual map $M^*$ and of the derived map $\Delta M$ up to homeomorphisms of the plane. As illustrated by Figure 5.5, what is needed for this is the choice of a face of $\Delta M$, or equivalently of a corner of $M$. Accordingly, let a **rooted planar map** $(M, c)$ be a planar map with a marked corner $c$. The **root face**, **root vertex** and **root edge** of $(M, c)$ are then defined to be respectively $f(c)$ and $v(c)$ and $ccw(c)$. (In the literature, a rooted map is sometimes defined as a map with a marked and oriented edge, or with a marked half-edge. This definition is equivalent to ours: to each oriented edge $\vec{e}$ is associated the unique corner $c$ such that $v(c)$ is the origin of $\vec{e}$ and $ccw(c) = e$.)

Upon setting $(M, c)^* = (M^*, c)$, duality extends into an involution on rooted planar maps. The derived map of a rooted map $(M, c)$ is instead naturally endowed with a marked face (the face of $\Delta(M)$ that corresponds to $c$). The incidence map of a rooted map $(M, c)$ is a bicolored quadrangulation with a marked edge (the edge $e$ of $Q(M)$ that corresponds to $c$), or equivalently a rooted quadrangulation (taking as root the unique corner $c'$ incident to the root vertex of $M$ and such that $ccw(c') = e$). Similarly the edge map of a rooted map can be considered as a rooted tetravalent map.

**Proposition 5.2.5** *The incidence map $Q$ and edge map $R$ are one-to-one correspondences between rooted planar maps with $n$ edges and respectively rooted planar quadrangulations with $n$ faces, and rooted tetravalent planar maps with $n$ vertices.*

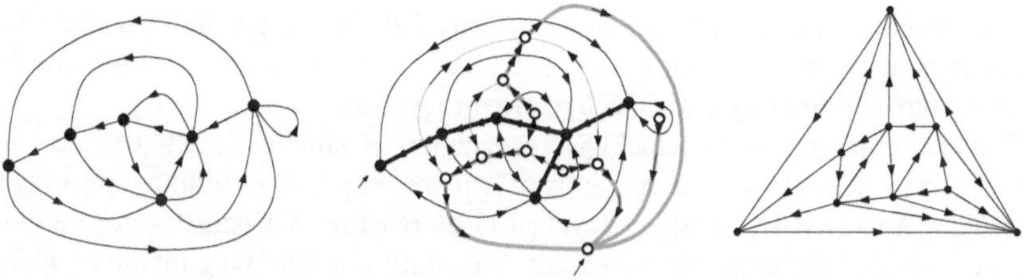

**Figure 5.6**
An Eulerian orientation, the $t$-orientation induced by the dual spanning trees of Figure 5.4, and a $s$-orientation of a triangulation.

**Orientations.** Let $(M,c)$ be a rooted map, and $\mathcal{O}$ an orientation of the edges of $M$. A **circuit** is a directed cycle of oriented edges, it is **simple** if it visits each vertex at most once. Any simple circuit $C$ divides the sphere into a **left component**, which borders the left- hand side of every edge of $C$, and a **right component**, which borders the right-hand side of every edge of $C$.

Let us say that a simple circuit $C$ in a rooted planar map is **clockwise** if the root corner lies in its left component, **counterclockwise** otherwise. With the convention that the outer face of a rooted plane map is its root face, these notions of clockwise and counterclockwise circuits coincide with the usual definitions: A simple circuit is clockwise if the unbounded component it defines is on its left-hand side, counterclockwise otherwise.

A function $\alpha : V \to \mathbb{N}$ is **feasible on the planar map** $M$ if there exists an orientation $\mathcal{O}$ of $M$ such that for every vertex $v$ of $M$, $\alpha(v)$ is the outdegree of $v$. Such an orientation is then called an $\alpha$-orientation.

A classical example of feasible function is the half-degree function $h(v) = \deg(v)/2$ on an Eulerian map (recall that a map is Eulerian if all its vertices have even degree). The $h$-orientations are precisely **Eulerian orientations**, that is, orientations such that the in- and out-degree are equal on each vertex. Another example is given by the orientation induced by spanning trees on derived maps: let $(M,c)$ be a rooted planar map and for each vertex $v$ of the derived map $\Delta(M)$ let $t(v) = 0$ if $v$ is the root vertex of $M$ or $M^*$, $t(v) = 1$ if $v$ is a non-root vertex of $M$ or $M^*$, and $t(v) = 3$ if $v$ is a dual edge intersection vertex. Then $t$ is feasible since each pair of dual spanning trees on $M$ and $M^*$ induces a $t$-orientation of $\Delta(M)$ by selecting as unique out-going edge on each non-root vertex of $M$ or $M'$ the edge going toward its father in the tree it belongs to. Finally a third example is Schnyder's orientations of triangulations [158]: Let $T$ be a plane triangulation and $s(v) = 1$ if $v$ is incident to the outerface, $s(v) = 3$ otherwise. Then $s$ is feasible and the $s$-orientations are called **3-orientations**. See Figure 5.6 for an illustration.

Let $\alpha$ be a feasible function on a plane map $(M, f)$, and let $\mathcal{O}$ be an $\alpha$-orientation of $M$. Observe that simultaneously changing the orientation of all the edges of a circuit in $(M, \mathcal{O})$ yields another orientation $\mathcal{O}'$ that is still an $\alpha$-orientation. Moreover two $\alpha$-orientations $\mathcal{O}$ and $\mathcal{O}'$ of a planar map $M$ always differ on a set of edges that form a collection of Eulerian submaps: since any Eulerian graph admits an Eulerian tour (that is, a circuit that visit every edge once), it is possible to go from one $\alpha$-orientation to any other by a sequence of circuit reversals.

The clockwise or counterclockwise orientation of simple circuits can now be used to endow the set of $\alpha$-orientations of a plane map $(M,f)$ with an even nicer structure: Let us say that a circuit reversal is **increasing** if it consists in returning a ccw-circuit into the opposite cw-circuit. We admit the following theorem, which gives a first illustration of the rich combinatorial properties enjoyed by planar maps.

**Theorem 5.2.6 (Felsner [100])** *Let $\alpha$ be a feasible function on a planar map $(M,f)$. Then increasing circuit reversals endow the set of $\alpha$-orientation of a plane map $(M,f)$ with a lattice structure. In particular there exists a unique minimal $\alpha$-orientation in this lattice, which is the unique $\alpha$-orientation without cw-circuit.*

### 5.2.3 Which maps shall we count?

Let $\mathcal{M}^u$, $\mathcal{M}^r$ and $\mathcal{M}^\ell$ denote the sets of (unrooted) planar maps, rooted planar maps and corner labeled planar maps respectively. Then the following four counting problems are the most commonly considered:

1. Count rooted planar maps with $n$ edges, or compute the ordinary generating function (gf)
$$M^r(z) = \sum_{M\in\mathcal{M}^r} z^{e(M)} = 1+2z+9z^2+O(z^3).$$

2. Count planar maps with $2n$ labeled corners, or compute the exponential generating function
$$M^\ell(z) = \sum_{M\in\mathcal{M}^\ell} \frac{z^{e(M)}}{(2e(M))!} = 1+2\frac{z}{2!}+9\cdot 3!\frac{z^2}{4!}+O(z^3).$$

3. Count unrooted planar maps with $n$ edges with a weight $1/|\mathrm{aut}(M)|$ per map $M$ with automorphism group $\mathrm{aut}(M)$, or compute
$$M^a(z) = \sum_{M\in\mathcal{M}^u} \frac{z^{e(M)}}{|\mathrm{aut}(M)|} = 1+(\tfrac{1}{2}+\tfrac{1}{2})z+(\tfrac{1}{2}+\tfrac{1}{1}+\tfrac{1}{4}+\tfrac{1}{2})z^2+O(z^3).$$

4. Count unrooted planar maps with $n$ edges, or compute
$$M^u(z) = \sum_{M\in\mathcal{M}^u} z^{e(M)} = 1+2z+4z^2+O(z^3).$$

Observe that we classify maps according to their number of edges (or corners, equivalently): In view of duality this is the most natural size parameter to consider, but we

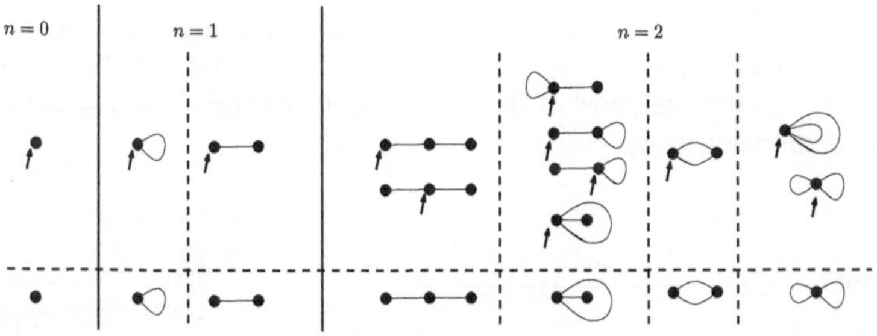


**Figure 5.7**
The 1, 2 and 9 rooted planar maps with 0, 1 and 2 edges, and the corresponding 1, 2 and 4 unrooted planar maps.

shall see that most results can later be refined to take into account the numbers of vertices and faces.

The first three problems are in fact essentially equivalent: On the one hand, rooted maps have no nontrivial automorphisms (see below), so that each rooted map admits $(2e(M)-1)!$ distinct labelings such that the root corner has label 1. On the other hand, each unrooted map corresponds by definition of aut$(M)$ to $\frac{(2e(M))!}{|\text{aut}(M)|}$ labeled maps. Therefore:

$$M^r(z) = \frac{2zd}{dz}M^\ell(z) = \frac{2zd}{dz}M^a(z).$$

**Automorphisms of maps.** Let $(M,r)$ be a rooted map and assume once again that its corners are labeled with $\{1,\ldots,2n\}$, with $(\sigma,\rho)$ the corresponding rotation system. By definition of rotation systems, two corners are neighboring in $M$ if their labels can be mapped one onto the other by $\sigma$ or $\rho$: as a consequence, the connexity of $M$ implies that for any two corners $(c,c')$ there exists a sequence $(\tau_1,\ldots,\tau_k)$ of elements of $\{\sigma,\rho\}$ such that the labels $i$ and $i'$ of $c$ and $c'$ satisfy $i' = \tau_1 \circ \tau_2 \circ \cdots \circ \tau_k(i)$. By definition an **automorphism** of the map $M$ is a relabeling $\pi : \{1,\ldots,2n\} \to \{1,\ldots,2n\}$ of its corners that preserves the map $M$, or equivalently, that commutes with $\sigma$ and $\rho$: $\sigma \circ \pi = \pi \circ \sigma$, and $\rho \circ \pi = \pi \circ \rho$. An automorphism $\pi$ of the rooted map $(M,c)$ is an automorphism of $M$ that also preserves the label $i_0$ of the root corner $c$: $\pi(i_0) = i_0$. Now according to the previous discussion, $\pi(i) = \pi \circ \tau_1 \circ \cdots \circ \tau_k(i_0)$ for some sequence $(\tau_1,\ldots,\tau_k)$ in $\{\sigma,\rho\}^*$. Using the commutation relations between $\pi$ and $\sigma$ or $\rho$, and the equality $\pi(i_0) = i_0$, this shows that $\pi(i) = i$: The only automorphism of a rooted map is the trivial one. An important consequence of this discussion is that the number of different labelings of the corners of a rooted map with $2n$ edges is always $(2n)!$, and the number of such labelings where the root corner has label 1 is $(2n-1)!$. A more detailed discussion of map automorphisms can be found in [88], or [143].

As illustrated by Figure 5.7 the fourth problem is not equivalent to the other three. Observe that we could have considered a fifth problem, namely to count plane maps (**i.e.** unrooted planar maps with a marked face). This problem is less advertised in the literature but it arises as an intermediary step while counting classes of unrooted maps. The literature overwhelmingly concentrates on the rooted (or labeled) case: It is technically simpler, and most of the results for rooted maps can *a posteriori* be transfered to unrooted maps.

## 5.3 Counting tree-rooted maps

We start this exposition with rooted planar maps with a marked spanning tree, called tree-rooted maps: these maps are the main characters of some simple extensions of classical bijections between rooted plane trees, balanced parenthesis words and non-crossing (**aka** planar) arch diagrams. There are at least two reasons to discuss at once these bijections: First it is a gentle way to get the reader used to manipulating planar maps, by building on the more standard combinatorics of plane trees. Second the bijections for tree-rooted planar maps are useful tools to simplify the later description of bijections for rooted planar maps. This section is largely inspired by the seminal work [28].

### 5.3.1 Mullin's decomposition

Our first series of results dates back to the work of Tutte and Mullin in the 60's.

**Walking around a spanning tree.** Let a **tree-rooted planar map** be a rooted planar map $(M,c)$ with $n$ edges, endowed with a spanning tree $T_1$. The **counterclockwise walk** around $(T_1,c)$ induces a total order on the $2n$ corners of $M$ given by the order in which these corners are visited by a 2d little ant traveling on the border of the tree $T_1$ in counterclockwise direction. This process is illustrated by Figure 5.8.

Each edge is visited twice during the walk and four symbols can be used to record the four types of moves of the ant during the walk: $\bullet$ and $\bar{\bullet}$ for the first and second time it goes across an edge, and $\circ$ and $\bar{\circ}$ for the first and second time it goes along an edge. The **counterclockwise contour code** is then the word $w$ on the alphabet $\{\circ,\bar{\circ},\bullet,\bar{\bullet}\}$ whose $i$th letter is the type $w_i$ of the $i$th move of the ant.

The restriction of the ccw contour code $w$ to the letters $\{\circ,\bar{\circ}\}$ is the standard counterclockwise contour code of the rooted plane tree $(T_1,c)$ (see Chapter 4 of the present book): In particular it is a **balanced parenthesis word**, that is, $|w|_\circ = |w|_{\bar{\circ}}$ and for all prefix $w'$ of $w$, $|w'|_\circ \geq |w'|_{\bar{\circ}}$, where $|w|_x$ denotes the number of letters $x$ in the word $w$. The restriction of the ccw contour code $w$ to the letters $\{\bullet,\bar{\bullet}\}$ is instead a direct encoding of the arch diagram formed around $T_1$ by the edges not in $T_1$: It is in particular again a balanced parenthesis word, $|w|_\bullet = |w|_{\bar{\bullet}}$ and for all prefix $w'$ of $w$, $|w'|_\bullet \geq |w'|_{\bar{\bullet}}$.

The word $w$ is thus composed of the shuffle of two balanced parenthesis words on the alphabet $\{\circ,\bar{\circ}\}$ and $\{\bullet,\bar{\bullet}\}$ respectively. Mullin's result is essentially that this encoding is one-to-one.

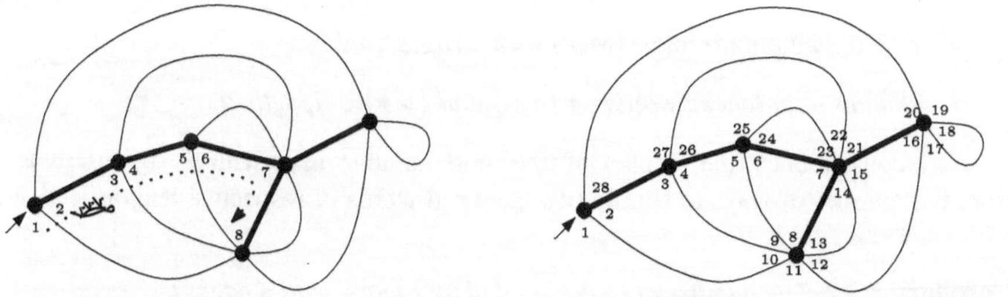

**Figure 5.8**
The eight first corners visited during a ccw walk around a spanning tree, and the full induced numbering of corners. The transition from one corner to the next is

$$1 \xrightarrow{\bullet} 2 \xrightarrow{\circ} 3 \xrightarrow{\bullet} 4 \xrightarrow{\circ} 5 \xrightarrow{\bullet} 6 \xrightarrow{\circ} 7 \xrightarrow{\circ} 8 \xrightarrow{\bar{\bullet}} 9 \xrightarrow{\bar{\bullet}} 10 \xrightarrow{\bar{\bullet}} 11 \xrightarrow{\bullet} 12 \xrightarrow{\bullet} 13 \xrightarrow{\bar{\circ}} 14 \xrightarrow{\bar{\bullet}} 15 \xrightarrow{\circ} 16$$
$$\xrightarrow{\bar{\bullet}} 17 \xrightarrow{\bullet} 18 \xrightarrow{\bar{\bullet}} 19 \xrightarrow{\bullet} 20 \xrightarrow{\bar{\circ}} 21 \xrightarrow{\bullet} 22 \xrightarrow{\bullet} 23 \xrightarrow{\bar{\circ}} 24 \xrightarrow{\bar{\bullet}} 25 \xrightarrow{\bar{\circ}} 26 \xrightarrow{\bar{\bullet}} 27 \xrightarrow{\bar{\circ}} 28 \xrightarrow{\bar{\bullet}}$$

so that the contour code is: $\bullet\circ\bullet\circ\bullet\circ\circ\bar{\bullet}\bar{\bullet}\bar{\bullet}\bullet\bullet\bar{\circ}\bar{\bullet}\circ\bar{\bullet}\bullet\bar{\bullet}\bullet\bar{\circ}\bullet\bullet\bar{\circ}\bar{\bullet}\bar{\circ}\bar{\bullet}\bar{\circ}\bar{\bullet}$.

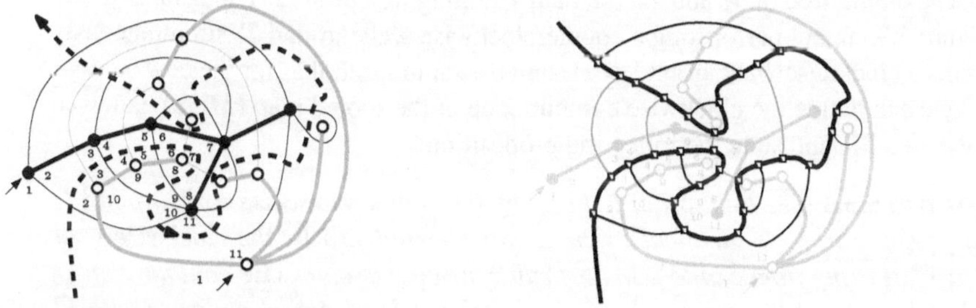

**Figure 5.9**
(i) The contour walk between a spanning tree and its dual. (ii) The associated cubic map with a rooted Hamiltonian cycle.

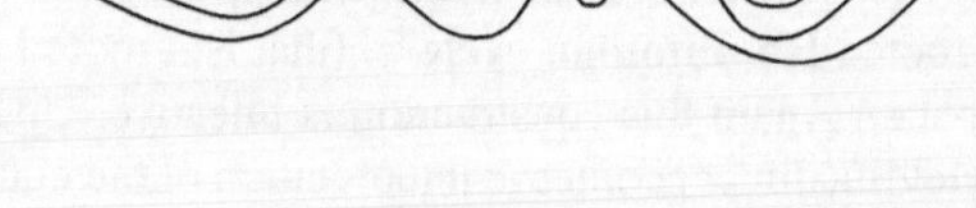

**Figure 5.10**
A shuffle of two arc diagrams.

**Theorem 5.3.1 (Mullin's encoding [144])** *The contour code is a bijection between*

- *tree-rooted planar maps $(M,c)$ with n edges, and*
- *shuffles of balanced words on $\{\circ,\bar{\circ}\}$ and $\{\bullet,\bar{\bullet}\}$ of length 2n.*

As a consequence, the number of tree-rooted planar maps with $n$ edges derives from the number of ways to shuffle two balanced words of respective length $i$ and $j$ with $i+j=n$.

**Corollary 5.3.2** *The number of tree-rooted planar maps with n edges is*

$$\sum_{\substack{i+j=n\\ i,j\geq 0}} \binom{2n}{2i} C_i C_j = C_n C_{n+1}$$

*where $C_n = \frac{1}{n+1}\binom{2n}{n}$ denotes the nth Catalan number, and the second expression follows from the Chu-Vandermonde identity [15]*

**Duality and cubic maps with a rooted Hamiltonian cycle.** It is worth observing that the contour code is compatible with duality. Let $(M,c)$ be a rooted planar map, $T_1$ be a spanning tree of $M$ and $T_2^*$ the dual spanning tree of $M^*$. Then, as illustrated by Figure 5.9, an ant performing a counterclockwise walk around $T_1$ simultaneously performs a clockwise walk around $T_2^*$. Using the same symbol as for the ccw contour code, one can define the clockwise contour code of the rooted map $(M^*,c)$ endowed with the tree $T_2^*$, and state the following proposition:

**Proposition 5.3.3** *The total order induced by the counterclockwise walk around $T_1$ starting from c in M is identical to the total order induced by the clockwise walk around $T_2^*$ starting from c in $M^*$. In particular the counterclockwise contour code of $(M,c)$ endowed with $T_1$ and the clockwise contour code of $(M^*,c)$ endowed with $T_2^*$ are mapped one onto the other by the exchanges $\circ \leftrightarrow \bullet$, $\bar{\circ} \leftrightarrow \bar{\bullet}$.*

From this primal/dual point of view, it is natural to draw the contour walk as a curve $\mathscr{C}$ traveling between the spanning tree $T_1$ and the dual spanning tree $T_2^*$ in the superimposition of $M$ and $M^*$, as illustrated by Figure 5.9. The intersections of the curve $\mathscr{C}$ with edges of $M \setminus T_1$ and $M^* \setminus T_2^*$ create tetravalent vertices in the middle of every half-edge of $M$ and $M^*$ that is not in $T_1$ or $T_2^*$. In the superimposition of $M \setminus T_1$, $M^* \setminus T_2^*$ and $\mathscr{C}$, each of these new vertices is adjacent to three other new vertices (two along $\mathscr{C}$ and one along an edge of $M \setminus T$ or $M^* \setminus T_2^*$) and to one vertex of $M$ or $M^*$. Upon keeping only the new vertices and the induced map, one obtains a cubic planar map endowed with a rooted **Hamiltonian cycle** $\mathscr{C}$ (that is a rooted cycle that visits every vertex exactly once). Again this construction is bijective. Upon straightening the rooted cycle $\mathscr{C}$ it yields a direct geometric intepretation of the contour code as the shuffle of two arch diagrams above and below $\mathscr{C}$. See Figure 5.10 for an illustration.

**Theorem 5.3.4 (Mullin [144])** *There is a bijection between tree-rooted planar maps with n edges and cubic planar maps endowed with a rooted Hamiltonian cycle of length 2n.*

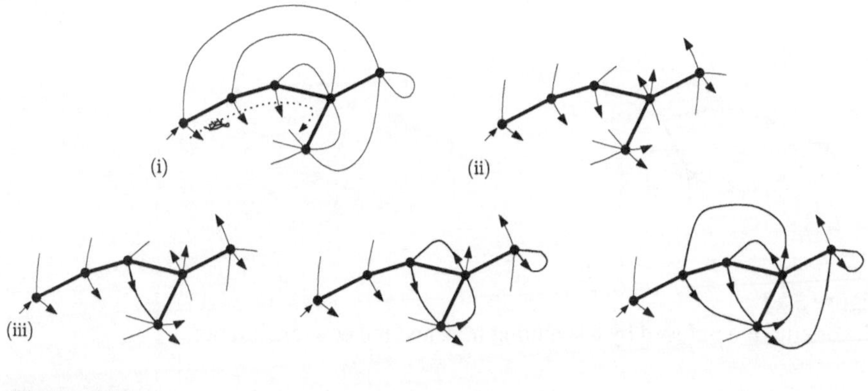


**Figure 5.11**
(i) The opening of the tree-rooted map of Figure 5.8. (ii) The resulting balanced tree. (iii) The first few steps of the closure of the previous balanced tree.

**Corollary 5.3.5 (Tutte [160])** *The number of cubic planar maps endowed with a rooted Hamiltonian cycle of length 2n is* $\sum_{i+j=n}\binom{2n}{2i}C_iC_j = C_nC_{n+1}$.

**Breaking the symmetry, balanced trees.** The previous constructions were symmetric with respect to duality. Let us now consider an alternative construction that breaks this symmetry. The idea is to break during the ccw walk the edges that are crossed into pairs of half-edges. More precisely, in view of the previous discussion, it is natural to distinguish for each edge the half-edge that corresponds to the first visit and the half-edge that corresponds to the second visit: This is done by using outgoing half-edges for the first ones, and incoming half-edges for the second ones. The resulting operation, called the **opening**, turns a tree-rooted map $M$ into a **rooted decorated plane tree**, that is, a rooted plane tree with two types of dangling half-edges, as illustrated by Figure 5.11. In view of the previous discussion, the decorated trees that are produced in this way are characterized by the fact that the number of outgoing and incoming half-edges are equal, and that the sequence of outgoing and incoming half-edges that are met during a counterclockwise walk around the tree forms a balanced parenthesis word. These rooted decorated trees are called **balanced trees**.

Conversely the fact that the outgoing and incoming half-edges of a balanced tree form by definition a balanced parenthesis word ensures that there is a unique way to **close** any balanced tree into a tree-rooted planar map such that the root corner stays in the outer face. In particular this can be done by repeating the following local closure rule: Starting from the root corner and walking counterclockwise around the current map, let $h$ be the first outgoing half-edge that is followed by an incoming half-edge $h'$, and glue the pair $(h,h')$ into an edge in the unique way that leaves the root corner in the outer face.

**Theorem 5.3.6 (Walsh and Lehman [169])** *Opening and closure are inverse bijections between tree-rooted planar maps with n edges, v vertices and f faces, and balanced trees with v vertices and* $f+1$ *outgoing half edges.*

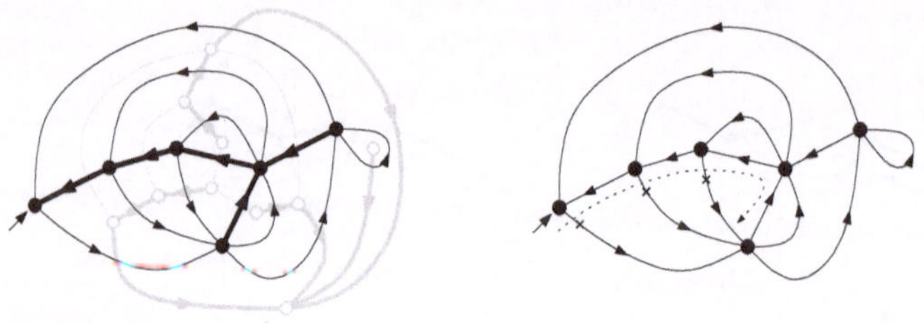

**Figure 5.12**
The orientation induced by a spanning tree, and the ccw-exploration.

A balanced tree should be viewed as a rooted plane tree that carries on its border the contour code of another tree. We will reuse this idea several times.

### 5.3.2 Spanning trees and orientations

Mullin's encoding explains his convolution formula for the number of tree-rooted planar maps, but not the simple product form $C_nC_{n+1}$ arising from the Chu-Vandermonde formula. In preparation of an explanation of this formula, we first "forget about the spanning trees" and reformulate the result in terms of orientations.

**The orientation induced by a spanning tree.** Let $(M,c)$ be a rooted map, and let $T$ be a spanning tree of $M$. The **orientation of** $(M,c)$ **induced by** $T$ is the orientation $\mathcal{O}_T$ such that each edge of $T$ is oriented toward the root, and each edge $e$ of $M \setminus T$ **turns counterclockwise around** $T$, that is, the unique simple circuit formed by $e$ and edges of $T$ is counterclockwise (see Figure 5.12). Equivalently, the map $(M,c)$ endowed with $\mathcal{O}_T$ can be constructed using the closure of Theorem 5.3.6, upon orienting the edges of the balanced tree toward the root and keeping the orientation of the matched half-edges.

It is instructive to perform the closure incrementally: Starting from the oriented balanced tree, close one edge at a time, and observe that at each step no cw-circuit can be created. Finally in view of the orientation of the edges of $T$, the orientation $\mathcal{O}_T$ is **root-accessible**, i.e. there is an oriented path from each vertex to the root vertex.

**Proposition 5.3.7** *Let $T$ be a spanning tree of a rooted map $(M,c)$. Then the orientation $\mathcal{O}_T$ induced by $T$ is root-accessible and has no cw-circuit.*

**The ccw-exploration of an oriented map.** A key observation now is that on the rooted oriented map $(M,c,\mathcal{O}_T)$, the ccw walk around $T$ can be performed without knowing $T$. Indeed in the ccw walk, each edge is reached for the first time from its endpoint if it is an edge of $T$, and from its origin if is not an edge of $T$. This implies that the orientation alone allows us to decide at the first visit of each edge whether it belongs to $T$ or not, and whether the walk should follow the edge or cross it. The resulting ccw walk around the initialy unknown tree $T$ is called the **ccw-exploration** of the oriented map $M$.

Conversely an elementary case analysis allows us to characterize the rooted oriented maps on which ccw-exploration produces a spanning tree.

**Proposition 5.3.8** *The ccw exploration of a rooted planar map $(M,c)$ endowed with an orientation $\mathscr{O}$ without cw-circuit outputs a tree $T$, and this tree is a spanning tree of $M$ if and only if the orientation $\mathscr{O}$ is root-accessible.*

Summarizing these results yields the following theorem, first explicitely stated in this general form in Bernardi's PhD thesis.

**Theorem 5.3.9 (Bernardi [29])** *Let $(M,c)$ be a rooted planar map. The ccw-exploration and induced orientation are inverse bijections between accessible orientations without cw-circuit of $(M,c)$ and spanning trees of $(M,c)$.*

Earlier instances of the result in the special case of triangulations and 3-connected planar maps appears in [148, 105], extensions were proposed in [29, 6]. Observe that ccw-exploration is just leftmost oriented depth first search; it might thus be the case that this result is known already in other contexts.

**Corollary 5.3.10** *The number of rooted planar maps with n edges endowed with a root-accessible orientation without cw-circuit is $\sum_{i+j=n}\binom{2n}{2i}C_iC_j = C_nC_{n+1}$.*

Again in this corollary only the convolution formula is obtained bijectively.

In view of Theorem 5.2.6, it is tempting to define a **root-accessible feasible function** on the set $V$ of vertices of a rooted planar map $(M,c)$ as a feasible function $\alpha : V \to \mathbb{N}$ such that $\alpha$-orientations are root-accessible (observe that if one $\alpha$-orientation is accessible then all are since circuit reversal preserves accessibility).

**Corollary 5.3.11** *The number of rooted planar maps with n edges endowed with a root-accessible feasible function is $\sum_{i+j=n}\binom{2n}{2i}C_iC_j = C_nC_{n+1}$.*

Further reformulations and extensions of these results in terms for instance of recurrent sand pile configurations, and relations to Tutte polynomials are given in [29].

**Ccw-exploration and opening.** already observed, the orientation induced by a spanning tree of a map coincides with the orientation induced by the closure of Theorem 5.3.6. As a consequence, Theorems 5.3.6 and 5.3.9 can be combined to obtain the following corollary, which will be useful when we turn to the problem of enumerating rooted planar maps.

**Corollary 5.3.12** *Ccw-exploration followed by opening is a bijection between rooted planar maps with i vertices and j faces endowed with an accessible orientation without cw-circuit and balanced trees with i vertices and j outgoing half-edges. Moreover, this bijection preserves in- and out-degrees of every vertex.*

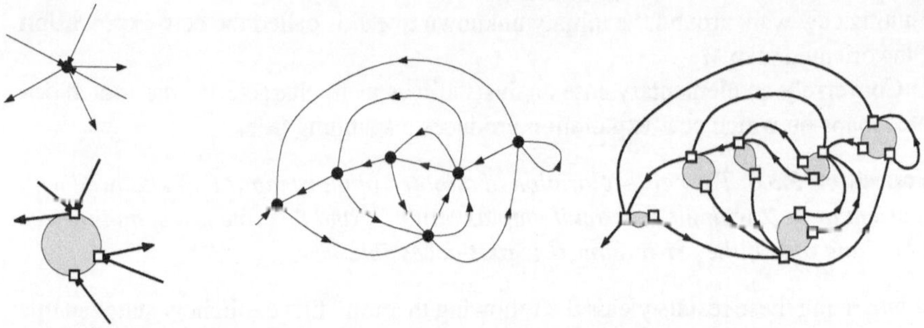

**Figure 5.13**
The blowing of a vertex, a map $M$, its complete blowing with the split tree.

### 5.3.3 Vertex blowing and polyhedral nets

The last result we shall present in this section is a construction that was first used in a particular case by Cori and Vauquelin in the 80's [90] and that was shown by Bernardi 30 years later to finally explain the product formula $C_nC_{n+1}$ [28].

**Vertex blowing and split trees.** Let $(M,c)$ be a rooted planar map and $\mathcal{O}$ an orientation of $M$. The **blowing** of a vertex $v$ with out-degree $k$ is the operation replacing $v$ by a white polygon with $k$ nodes each carrying one outgoing edge and the incoming edges that precede it the cw direction, as illustrated by Figure 5.13. The **complete blowing** $\Sigma(M,\mathcal{O})$ of an oriented rooted map $(M,\mathcal{O})$ consists in blowing each vertex of $M$ independently (the blowing of the root vertex is performed as if there was an outgoing half-edge at the root corner). The map $\Sigma(M,\mathcal{O})$ naturally inherits from $\mathcal{O}$ a partial orientation that we continue to denote $\mathcal{O}$: the edges inherited from $M$ are oriented while the edges of the white polygons created by blowings are not.

A key observation here is that, if $(M,\mathcal{O})$ is accessible, then the oriented edges of $\Sigma(M,\mathcal{O})$ cannot form ccw circuits: Indeed if they would form a ccw circuit then this circuit would also be a ccw circuit in $M$. In view of the blowing rule, the vertices of $M$ on this circuit would only have outgoing edges on the left-hand side of the circuit. But in a ccw circuit the root lies in the right-hand side of the circuit, so this contradicts accessibility.

Assuming moreover that $\mathcal{O}$ is an orientation without cw-circuit, then $\Sigma(M,\mathcal{O})$ is an accessible oriented map without circuit, that is a tree.

**Proposition 5.3.13** *Let $(M,c)$ be a rooted planar map and $\mathcal{O}$ a root-accessible orientation without cw-circuit. Then the oriented edges of the complete blowing $\Sigma = \Sigma(M,\mathcal{O})$ form a spanning tree of $\Sigma$ oriented toward c, called the* **split tree** *of $M$.*

**Polyhedral nets.** Let us use the split tree to cut the sphere on which the map $\Sigma$ is drawn. As illustrated by Figure 5.14, this **splitting** construction yields a polyhedral net. In order to be able to reconstruct the map $\Sigma$ from the polyhedral net one needs

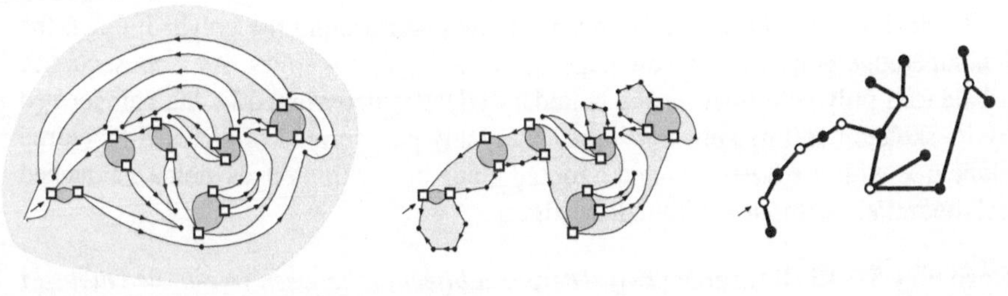

**Figure 5.14**
Cutting along the split tree to get a polyhedral net, another representation of the same polyhedral net, and its skeleton.

to specify the way edges should be glued together. This can be done by explicitely recording a planar arch diagram as in Figure 5.2, but it turns out to be more convenient to just keep track of the orientation of the edges of the split tree on the sides of the polyhedral net, as shown in Figure 5.14. The **folding** of the polyhedral net then just consists in iteratively gluing pairs of oriented border edges that originate from a same vertex until all boundary edges have been matched.

Let us describe more precisely the oriented polyhedral nets that arise from this construction: They consist of an outerface and a collection of black and white simple faces called **polygons**, such that the sides of white polygons are incident only to black polygons, and the boundary of each black polygon consists of an alternating sequence of sides incident to white polygons and **superedges** formed of a ccw edge incident to the outerface, followed by a (possibly empty) sequence of cw edges incident to the outerface.

An oriented polyhedral net is **rooted** if a corner incident to a white node in the outerface is marked, and **balanced** if moreover, starting from the root corner and turning clockwise around the polyhedral net, its boundary edges form the contour code of a planted planar tree (equivalently if there are always strictly more already seen ccw edges than cw edges during the tour).

**Theorem 5.3.14** *Splitting and folding are inverse bijections between rooted planar maps with $i$ vertices and $j$ faces and $n = i + j - 2$ edges endowed with an accessible orientation without cw-circuit, and balanced polyhedral nets with $i$ white bounded faces and $j$ black bounded faces and $2n+2$ boundary edges.*

Let the **skeleton** of a balanced polyhedral net be the bicolored tree obtained by putting a (black or white) vertex in each (black or white) polygon and joining vertices corresponding to adjacent polygons by an edge. In particular the degree of a black (respectively white) vertex in the skeleton corresponds to the out-degree of the associated vertex of the initial map (respectively to the number of ccw oriented edges around the associated face). Now observe that in a balanced polyhedral net with $2n+2$ boundary edges, the $2n+2$ symbols forming the contour code of the planted split tree are written sequentially on the $2n+2$ boundary edges of the polyhedral net in a deterministic way: The $k$th opening parenthesis of the code is preceeded by $\ell_k$ closing parenthesis if and only if, in cw order around the polyhedral net, the $k$th superedge consists of a ccw edge preceded by $\ell_k$ cw edges. As a consequence a balanced polyhedral net $2n+2$ boundary edges is determined by the pair formed by its skeleton and its split tree. Conversely any pair formed of a rooted bicolored plane tree with $n+2$ vertices and a rooted plane tree with $n$ edges yields a balanced polyhedral net with $2n+2$ boundary edges.

**Corollary 5.3.15 (Bernardi [28])** *There is a bijection between tree-rooted maps $M$ with $i$ vertices and $j$ faces and $n = i + j - 2$ edges, and pairs $(t_1, t_2)$ where $t_1$ is a (bipartite) rooted plane tree with $i$ black and $j$ white vertices and $n+1$ edges, and $t_2$ is a rooted plane tree with $n$ edges. In particular the number of tree-rooted maps with $n$ edges is $C_n C_{n+1}$.*

A nice feature of the above result is that it not only explains the univariate product formula $C_{n+1}C_n$, but also give an alternate proof of the bivariate formula as $N_{i,j}C_n$, where $N_{i,j} = \frac{n!(n+1)!}{i!(i+1)!j!(j+1)!}$ is the Narayana number of rooted bipartite trees with $i$ black and $j$ white vertices. Or conversely, combined with the formula $\binom{2n}{2i}C_iC_j$ obtained by Mullin's encoding, it allows us to recover Naranaya's formula.

### 5.3.4 A summary and some observations

In this section we have successively shown that tree-rooted maps are in one-to-one correspondence with shuffle of parenthesis words, cubic maps endowed a rooted Hamiltonian cycle, balanced trees, root-accessible oriented maps without cw-circuit, balanced polyhedral nets and finally pairs of rooted plane trees.

On the one hand, these results form a coherent and nice chapter of bijective combinatorics that relies on and extends several extremely classical results of the **Catalan garden** [165]. These constructions may seem intrinsically planar: For instance on a surface of genus $g$, the dual of the edges that are not in a spanning tree of a map do not form a spanning tree of the dual, and the combinatorics of higher genus tree-rooted maps is not very satisfying. Instead a remarkable observation of Bernardi and Chapuy is the fact that many of these constructions extend to higher genus maps upon considering **covered maps** rather than tree-rooted maps: A **covering** of a map is a spanning subset of edges $C$ such that the local cyclic arrangements of edges around $T$ forms a map with one face (recall that a **planar** map is a tree if and only if it has one face). This approach leads to various elegant enumerative consequences, as explained in [35].

On the other hand, these bijections are some of the most fundamental ingredients in the search of bijections for rooted planar maps: Indeed a natural approach to the enumeration of rooted planar maps is to endow each map with a canonical spanning tree such that the family of balanced trees arising from Theorem 5.3.6 or the family of balanced polyhedral net arising from Theorem 5.3.14 are simple to describe and enumerate.

## 5.4 Counting planar maps

### 5.4.1 The exact number of rooted planar maps

**Rooted planar maps.** The most striking result about planar maps is certainly the fact that the number of rooted planar maps with $n$ edges has a simple closed formula

$$|\mathcal{M}_n^r| = \frac{2 \cdot 3^n \cdot (2n)!}{n!(n+2)!} = \frac{2}{n+2} \cdot \frac{3^n}{n+1}\binom{2n}{n}, \tag{5.2}$$

and in particular satisfies the following linear recurrence

$$(n+2)|\mathcal{M}_n^r| = 6(2n-1)|\mathcal{M}_{n-1}^r|, \quad \text{with} \quad |\mathcal{M}_0^r| = 1. \tag{5.3}$$

Equivalently, the ordinary generating function (gf) of rooted planar maps with respect to the number of edges $M^r(z) = \sum_{M \in \mathcal{M}^r} z^{e(M)}$ satisfies

$$M^r(z) = T(z) - zT(z)^3 \tag{5.4}$$

where $T(z)$ is the unique power series solution of

$$T(z) = 1 + 3zT(z)^2. \tag{5.5}$$

(The equivalence of Formula (5.2) and Equations (5.4)-(5.5) follows from a direct application of Lagrange inversion formula [infra, Chapter on Tree].) Another way to write the relation between $M^r(z)$ and $T(z)$ is

$$\frac{\partial}{\partial z}(z^2 M^r(z)) = 2T(z). \tag{5.6}$$

More blandly, $M^r(z)$ is the unique power series root of the polynomial

$$P(M,z) = 1 - 16z + 18zM - 27z^2M^2. \tag{5.7}$$

In particular $M^r(z)$ is an algebraic function over the field $\mathbb{Q}(z)$.

Formula (5.2) was first discovered by W. T. Tutte in a series of papers written between January 1961 and February 1962 where he deals with the enumeration of various subfamilies of planar maps. Using little more than the simple idea of root-edge deletion, Tutte established in [161] a recurrence for the number $E(d_1, d_2, \ldots)$ of rooted Eulerian planar maps with $d_i$ vertices of degree $2i$ and $e = \sum_{i \geq 1} id_i$ edges, then guessed that these numbers admit the following simple form

$$E(d_1, d_2, \ldots) = \frac{2(e!)}{(e - \sum_{i \geq 1} d_i + 2)!} \prod_{i \geq 1} \frac{1}{d_i!}\binom{2i-1}{i}^{d_i}. \tag{5.8}$$

He then checked that this formula satisfies his recursion by an incredible computational **tour de force**. Quickly after this, Tutte observed in [162] that the special case

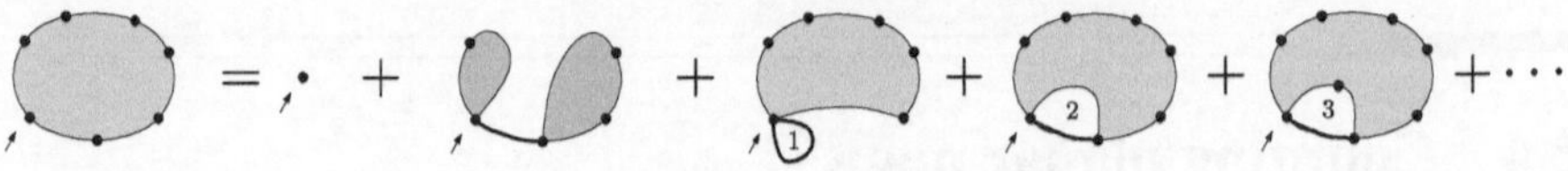


**Figure 5.15**
Tutte's root edge deletion.

$d_2 = n$ and $d_i = 0$ for all $i \neq 2$ of Formula (5.8), that counts rooted tetravalent planar maps with $n$ vertices, gives Formula (5.2) for the number of rooted planar maps with $n$ edges in view of the edge map transformation (see page 338).

A few years later, in [163], Tutte proposed a streamlined version of the root edge deletion method he uses to establish recursive decompositions and to translate them into functional equation for gfs: He is literally applying there what is now known as the symbolic method for combinatorial enumeration [102, Chapter 1]. Let $M_j(z,\mathbf{y})$ denote the gf of rooted planar maps with a root face of degree $j$, counted by number of edges (variable $z$) and by number of non-root faces with degree $i$ (variable $y_i$) for all $i \geq 1$, and let $M(z,\mathbf{y};u) = \sum_{j\geq 0} u^j M_j(z,\mathbf{y})$. Then Tutte's equation reads

$$M(z,\mathbf{y};u) = 1 + zu^2 M(z,\mathbf{y};u)^2 + z\sum_{i\geq 1} y_i \frac{M(z,\mathbf{y};u) - \sum_{j=0}^{i-2} u^j M_j(z,\mathbf{y})}{u^{i-2}}, \tag{5.9}$$

where each term has a clear interpretation in terms of a decomposition by deletion of the root edge, as illustrated by Figure 5.15.

- The quadratic term accounts for planar maps whose root edge is a separating edge: Such a map can be uniquely obtained from a pair of rooted planar maps by joining their root corners with a new root edge.
- The $i$th term in the sum instead corresponds to planar maps whose root edge has a face of degree $i$ on its left-hand side: Such a map can be uniquely obtained from a rooted planar map whose outerface has degree at least $i-1$. Hence the truncated $M$ series.

Iterating Equation (5.9) as a fixed point equation in the space of formal power series, one obtains the first coefficients:

$$\begin{aligned}
M(z,\mathbf{y},u) &= 1 + zu^2(1+O(z))^2 + zy_1\tfrac{1+O(z)}{u^{-1}} = 1 + z(u^2+uy_1) + O(z^2) \\
&= 1 + zu^2(1+z(u^2+uy_1))^2 + zy_1\tfrac{1+z(u^2+uy_1)}{u^{-1}} \\
&\quad + zy_2\frac{1+z(u^2+uy_1)-1}{u^0} + zy_3\frac{1+z(u^2+uy_1)-(1+zuy_1)}{u} + O(z^3) \\
&= 1 + z(u^2+uy_1) + z^2(2u^4+3u^3y_1+u^2y_1^2+u^2y_2+uy_2y_1+uy_3) \\
&\quad + O(z^3).
\end{aligned}$$

One can check that these first terms agree with what can be seen on Figure 5.7, and for $u = 1$ and $y_i = 1$ for all $i$, with the first values of Formula (5.2).

In the case $y_i = 0$ for all odd $i$, Equation (5.9) is essentially equivalent to the recurrence used by Tutte to conjecture and prove Formula (5.8), but no such simple formula is known to enumerate generic (**i.e.** non necessarily Eulerian) rooted planar maps with a fixed vertex degree distribution, and Tutte was unable in the 70's to guess the general solution. Instead he observed that for $y_i = y$, his equation rewrites as

$$\begin{aligned} M(z,y;u) &= 1 + zu^2 M(z,y;u)^2 + zyu \sum_{j \geq 0} \left( \sum_{i=0}^{j} u^{j-i} \right) M_j(z,y) \\ &= 1 + zu^2 M(z,y;u)^2 + zyu \frac{M(z,y;1) - uM(z,y;u)}{1-u}, \end{aligned} \tag{5.10}$$

and using again a guess and check approach, he was able to solve this equation to count maps by number of edges and faces. The resulting number of rooted planar maps with $k$ vertices and $\ell$ faces and $n = k + \ell - 2$ edges is a not particularly appealing triple summation of binomial coefficients [13] but the bivariate gf $M^r(x,y) = \sum_{M \in \mathcal{M}^r} x^{v(M)} y^{f(M)} = xy(1 + M(x, y/x; 1))$ satisfies an elegant variant of Formula (5.4):

$$M^r(x,y) = N(x,y)P(x,y)(1 - 2N(x,y) - 2P(x,y)) \tag{5.11}$$

where $N(x,y)$ and $P(x,y)$ are given by the bivariate analog of (5.5)

$$\begin{cases} N(x,y) &= x + N(x,y)^2 + 2N(x,y)P(x,y), \\ P(x,y) &= y + P(x,y)^2 + 2N(x,y)P(x,y). \end{cases} \tag{5.12}$$

Equivalently, the bivariate analog of Equation (5.6) reads

$$\frac{\partial}{\partial x} M^r(x,y) = P(x,y).$$

**Solving Tutte's equations.** The difficulty in deriving the explicit expressions (5.2)-(5.5) or their refinement (5.11)-(5.12) from Equation (5.10) lies in the particular role played by variable $u$: Observe indeed that variable $y$ can be considered as an optional parameter (setting $y = 1$ yields a valid equation that allows us to compute $M(z,1;u)$ order by order in $z$ by iteration as above), whereas variable $u$ appears to be necessary to get a non trivial equation (setting $u = 1$ yields instead an **a priori** useless equation involving $M(z,y,u)$ and $\frac{d}{du} M(z,y;u)|_{u=1}$ because of the discrete derivative appearing in the last summand). Following Zeilberger, the variable $u$ is called a **catalytic variable** (in analogy with the **catalytic** ingredients that are sometime added to allow for a chemical reaction to take place), and Equations (5.9) and (5.10) are **equations with one catalytic variable.**

Starting with Brown [74], methods to solve quadratic equation with one catalytic variable without guessing were developed, that allowed in particular Bender and Canfield [21], almost 30 years later, to give an essentially complete solution to Tutte's equation. Brown's method was later turned into a systematic approach to polynomial equations with one catalytic variable by Bousquet-Mélou and Jehanne [55]. Let us quote here their main general statement.

Let $\mathbb{K}$ be a field of characteristic 0. Let $F(u) \equiv F(z;u)$ be an element of $\mathbb{K}[u][[z]]$, that is, a power series in $z$ with coefficients that are polynomial in $u$ over $\mathbb{K}$. Then the following divided difference is well-defined:

$$\Delta F(u) = \frac{F(u) - F(0)}{u},$$

and $\lim_{u\to 0} \Delta F(u) = F'(0)$ (from now on in this section derivatives are taken with respect to the variable $u$ unless explicitely mentioned). More generally the iterated application of $\Delta$ yields

$$\Delta^{(i)} F(u) = \frac{F(u) - F(0) - uF'(0) - \ldots - \frac{u^{i-1}}{(i-1)!} F^{(i-1)}(0)}{u^i}.$$

Then the main result of [55] is the following theorem (see also [89] for an earlier derivation of the case $k = 1$).

**Theorem 5.4.1** *Let $Q(w_0, w_1, w_2, \ldots, w_k; z)$ be a polynomial in u, z and the $w_i$ with coefficients in a field $\mathbb{K}$ with characteristic 0. Then the unique formal power series solution $F(u)$ of the equation*

$$F(u) \equiv F(z,u) = zQ\left(F(u), \Delta F(u), \Delta^{(2)} F(u), \ldots, \Delta^{(k)} F(u), u; z\right) \qquad (5.13)$$

*is algebraic over $\mathbb{K}(z,u)$.*

For any $i_0 \geq 2$, this theorem directly applies to the specialization $y_i = 0$, $i > i_0$ of Equation (5.9) and proves that the corresponding series are algebraic.

The proof of the theorem is in fact constructive, and in principle gives a method to derive a system of algebraic equations that determines $F(u)$ and all the $\Delta^i F(u)$, as well as the specializations $F^{(i-1)}(0)$. We content ourselves here with the case $k = 1$, which is sufficient to derive Equation (5.7) from Equation (5.10), and refer to [53] for a gentle introduction to the general case and to [55] for the full details (see also [34, Section 9] for a slight extention allowing $F$ and $Q$ to be rational in $u$).

**A derivation of Formula (5.2).** The first observation is that up to a slight change of variable $F(u) = (1+u)(M(1+u) - 1)$, the simplified Tutte's equation (5.10) can be recasted into the form (5.13) with $k = 1$, and

$$Q(w_0, w_1, w_2; z) = (1 + w_2)(1 + w_2 + w_0)^2 + (1 + w_2)^2(1 + w_1).$$

The initial trick is to differentiate Equation (5.13) with respect to $u$ to get

$$\begin{aligned} F'(u) \;&=\; zF'(u)Q_0'(F(u), \Delta F(u), u; z) + \frac{zF'(u)}{u} Q_1'(F(u), \Delta F(u), u; z) \\ &\quad - \frac{z\Delta F(u)}{u^2} Q_1'(F(u), \Delta F(u), u; z) + zQ_2'(F(u), \Delta(F(u), u; z). \end{aligned} \qquad (5.14)$$

Now observe that since $Q(w_0, w_1, w_2; z)$ is a polynomial and $F(u)$ is a power series in $z$ with polynomial coefficients in $u$, the equation

$$U = zUQ_0'(F(U), \Delta F(U), U; z) + zQ_1'(F(U), \Delta F(U), U; z) \tag{5.15}$$

is contractant on the space of power series in $z$ and has a unique power series solution $U(z) = zQ_1'(0,0,0;0) + O(z^2)$. Upon multiplying by $u$ and substituting $u = U(z)$ in Equation (5.14), all terms in the first line cancel and we obtain a second equation

$$\Delta F(U)Q_1'(F(U), \Delta F(U), U; z) = U^2 Q_2'(F(U), \Delta F(U), U; z)$$

which can be rewriten, using Equation (5.15) to eliminate $Q_1'$, as

$$\Delta F(U)(1 - zQ_0'(F(U), \Delta F(U), U; z)) = zUQ_2'(F(U), \Delta F(U), U; z).$$

Summarizing, the triple of unknown functions $(F(U(z)), \Delta F(U(z)), U(z))$ is a solution of the system of algebraic equations in the variables $w_0$, $w_1$ and $w_2$:

$$\left\{ \begin{array}{lcl} w_0 & = & zQ(w_0, w_1, w_2; z) \\ w_1 & = & z(w_1 Q_0'(w_0, w_1, w_2; z) + Q_2'(w_0, w_1, w_2; z)) \\ w_2 & = & z(w_2 Q_0'(w_0, w_1, w_2; z) + Q_1'(w_0, w_1, w_2; z)). \end{array} \right. \tag{5.16}$$

In particular the series $F(U(z))$, $\Delta F(U(z))$ and $U(z)$ are algebraic series over $\mathbb{K}(z)$, and so is

$$F(0) = F(U(z)) - U(z)\Delta F(U(z)). \tag{5.17}$$

Finally, returning to Equation (5.13), $F(u)$ is seen to be algebraic on $\mathbb{K}(z,u)$. Observe moreover that if $Q$ has positive coefficients then the system (5.16) is $\mathbb{N}$-algebraic in the sense of [53].

Eliminating the unknown functions $F(V(z))$, $\Delta F(V(z))$ and $V(z)$ in the system of equations (5.16)-(5.17) results in an algebraic equation for $F(0)$ that directly yields Equation 5.7 for $M(z,0) = 1 + F(0)$.

**The complete solution of Tutte's equation.** Let us finally state the complete solution of Equation (5.9), as taken from [21, 59, 55]: the gf $M(z, \mathbf{y}) = M(z, \mathbf{y}, 1)$ of rooted planar maps with respect to the number of edges (variable $z$), and the number of faces of degree $i$ (variable $y_i$) for all $i \geq 1$ satisfies

$$\frac{\partial}{\partial z}(zM(z, \mathbf{y})) = \frac{1}{z^2} S_2(9S_2 - 4S_1) \tag{5.18}$$

where $S_1$ and $S_2$ are the unique formal power series in $z$ with coefficients in $\mathbb{Q}[y_1, y_2, \ldots]$ solutions of the system

$$\left\{ \begin{array}{lcl} S_1 & = & t[v^0] \sum_{i \geq 1} y_i (v + S_1 + S_2/v)^{i-1} \\ S_2 & = & t + t[v^{-1}] \sum_{i \geq 1} y_i (v + S_1 + S_2/v)^{i-1}. \end{array} \right. \tag{5.19}$$

These expressions may look complicated at first sight, and maybe one could be tempted to argue that Tutte's equation (5.9) is more compact. However Equations (5.18)-(5.19) should be considered as much more explicit: They have a clear tree-gf structure, directly amenable for instance to bivariate Lagrange inversion, or to singularity analysis, and they imply that for any non-degenerate finite set $\mathscr{D}$ of allowed degrees, the gf of rooted planar maps with vertex degrees in $\mathscr{D}$ with respect to the number of edges (variable $z$) are given by the specialization $y_i = \delta_{i\in\mathscr{D}}$ of the above system, and in particular this gf is algebraic over $\mathbb{Q}(z)$.

**Tutte equations and matrix integrals.** Remarkably Tutte's results were reproduced in the physics literature [73, 41] using an ingenious representation of the all-genus map gf using Hermitian matrix integrals. We shall not discuss here this alternative approach because various accessible texts already exist. We only stress the fact that there are (at least) two point of views regarding matrix integral representations.

On the one hand they can be used as a convenient shorthand notation for the all genus gf of maps viewed as a formal power series, and this point of view leads to the derivation of **loop equations** or **Schwinger-Dyson equations** that are essentially equivalent to Tutte equations (see [96]). These equations are dealt with using clever changes of variables and more or less detailed discussions of the possible singularities of the resulting expressions. These approaches appear to be analytic variants of Brown's more algebraic approach (a rigorous discussion of this is still missing in the literature however). From there many explicit formulas have appeared in the physics literature, culminating with Eynard and Orantin's topological recurrence [95, 99]. All these results are closely related to the present enumerative approach and we refer to [125, Chapter Matrix] for an elementary introduction, and [97, 58] for more detailed discussions.

On the other hand one can try to give a *bona fide* matrix integral representation to the all genus gf viewed as an analytic function. This is the original point of view of [73, 41], which allows them to use more analytic tools to derive explicit expressions, like saddle point approximations and orthogonal polynomials. However the validity of these representations requires careful discussion [94, 118, 151, 121] and involves deep relations with matrix integral theory, which makes this approach less elementary to follow from an enumerative point of view. Finally let us refer to older but very complete surveys [8, 92] for discussions of the motivations for the study of maps in physics.

### 5.4.2 Unrooted planar maps

For unrooted maps on the oriented sphere, the formulas are still explicit, although somewhat more involved. In particular the number of unrooted planar maps with $n$ edges is

$$|\mathscr{M}_n^u| = \frac{1}{2n}|\mathscr{M}_n^r| + \sum_{\substack{d<n\\ d|n}} \frac{\phi(\frac{n}{d})}{2n}\binom{d+2}{2}|\mathscr{M}_d^r| + \left(\frac{n+1}{4} - \frac{(-1)^n}{2}\right)|\mathscr{M}_{\lfloor\frac{n}{2}\rfloor}^r| \quad (5.20)$$

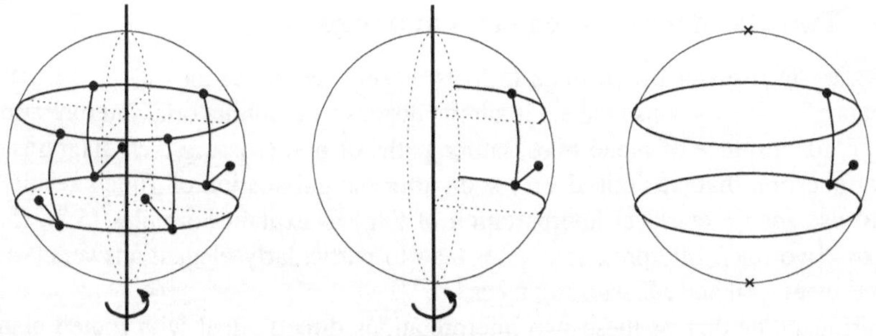

**Figure 5.16**
A symmetric map and the construction of the quotient map associated with the rotation of order 4 with vertical axis.

where $\phi(x)$ is Euler's totient function, giving the number of non-trivial divisors of $x$.

The form of this formula admits a nice explanation. Generically there are $2n$ choices of root corner for an unrooted map, hence the first term in the right-hand side. But in this first attempt, maps with non trivial automorphisms are undercounted since they correspond to less than $2n$ rooted maps each: According to Burnside lemma, to obtain the correct formula, one has to introduce a correction term for each pair $(M, \omega)$ where $\omega$ is an automorphism of $M$.

The exact correction terms to be used arise from a classification of the possible automorphisms of planar maps. This classification is best understood in geometric terms, using the (non-trivial, but very appealing) fact that any symmetric map admits a drawing on standard unit sphere in $\mathbb{R}^3$ that realizes its automorphisms as isometries of the oriented sphere, **i.e.** rotations. Rotations acting on maps are then simply classified according to their order, and to the type of cells (vertex, edge, or face) that they leave invariant (a rotation fixes exactly the two cells intersecting its axis).

Once the classification is established, the idea is that each pair $(M, \omega)$ with $\omega$ a rotation of order $n/d$ can be constructed from its **quotient map** $M/\omega$, which is identified as a map with $d$ edges and 2 marked cells as illustrated by Figure 5.16. The construction slightly differs depending whether the rotation fixes an edge or not: If not, the two fixed cells are chosen among the $v(M) + f(M) = e(M) + 2$ vertices or faces of $M$, and this yields the first correcting term; otherwise, depending on the parity of $n$ either one or two fixed cells are edges, and this explains the alternating second correcting term.

This elegant approach of unrooted map enumeration via the study of quotient maps was pioneered by Liskovets [132] (see also [166]) and extends to count planar maps up to sense reversing automorphisms [133] (see also the more recent [135, 137] for a streamlined presentation). An alternative approach based on multiply rooted maps is due to Wormald [171, 172], building on earlier work of Brown [75], but it yields less attractive formulas.

### 5.4.3 Two bijections between maps and trees

The series $T(z)$ given by Formula (5.5) is closely related to the classical Catalan numbers gf, and as a consequence it admits dozens of combinatorial interpretations in terms for instance of plane trees, lattice paths or non crossing arch diagrams. A natural question that was raised very soon after the publication of Tutte's results is how to use such a classical interpretation of $T(z)$ to explain Formulas (5.2), (5.4) or (5.6). Two main interpretations have lead to particularly elegant answers: **well-labeled trees** [90] and **blossoming trees** [153].

Curiously, neither of these two interpretations directly deal with rooted planar maps, but rather with quadrangulations and tetravalent maps: Recall indeed again that, thanks to the incidence map and edge map constructions of page 338, all the above formulas apply to rooted planar quadrangulations with $n$ faces, or to rooted tetravalent maps with $n$ vertices.

**The blossoming tree approach.** This first approach builds on a variant of binary trees to interpret Formula (5.5) and to arrive at a direct explanation of Formulas (5.2) and (5.4) in terms of tetravalent maps.

More specifically, let a **blossoming tree** of size $n$ be a rooted planar map with one face (**i.e.** a plane tree) with $n$ vertices of degree 4 (the **nodes**) and $2n+2$ vertices of degree one (the **leaves**) that are colored in black and white in such a way that every node is incident to exactly one black leaf. As illustrated by Figure 5.17, the standard decomposition of rooted plane trees applied to blossoming trees rooted at a white leaf matches Equation (5.5), so that the gf of these **white trees** is $T(z)$. Similarly, the gf of **black trees** (blossoming trees rooted at a black leaf) is $zT(z)^3$.

Now the right-hand side $T(z) - zT(z)^3$ of Formula (5.4) can be interpreted as the gf of **balanced blossoming trees**: Select for each unrooted blossoming tree with $n$ nodes a canonical way to match its $n$ black leaves to $n$ of its $n+2$ white leaves, and declare **balanced** the white trees that are rooted on an unmatched leaf of the underlying unrooted blossoming tree. The gf of balanced blossoming trees is then the difference of the gf of white and black trees. Equivalently the right-hand side of Formula (5.2) can be interpreted as follows: A fraction $\frac{2}{n+2}$ of the $\frac{3^n}{n+1}\binom{2n}{n}$ white trees with $n$ nodes are balanced because among the $n+2$ leaves of any unrooted blossoming tree, exactly two are unmatched.

The bottom line of this approach is that the requested canonical matching can be performed in a greedy iterative way that preserves planarity and maintains unmatched leaves in the outer face, as illustrated by Figure 5.17: in particular the closure of Theorem 5.3.6 describes how to do this. In view of Corollary 5.3.12, closure yields in fact a one-to-one correspondence between balanced blossoming trees and some oriented almost tetravalent maps with two vertices of degree 1 in the outer face. Upon gluing these two vertices to form a root edge, the correspondence is easily recasted as a bijection between balanced blossoming trees with $n$ nodes and rooted tetravalent planar maps with $n$ vertices endowed with an Eulerian orientation without cw-circuit.

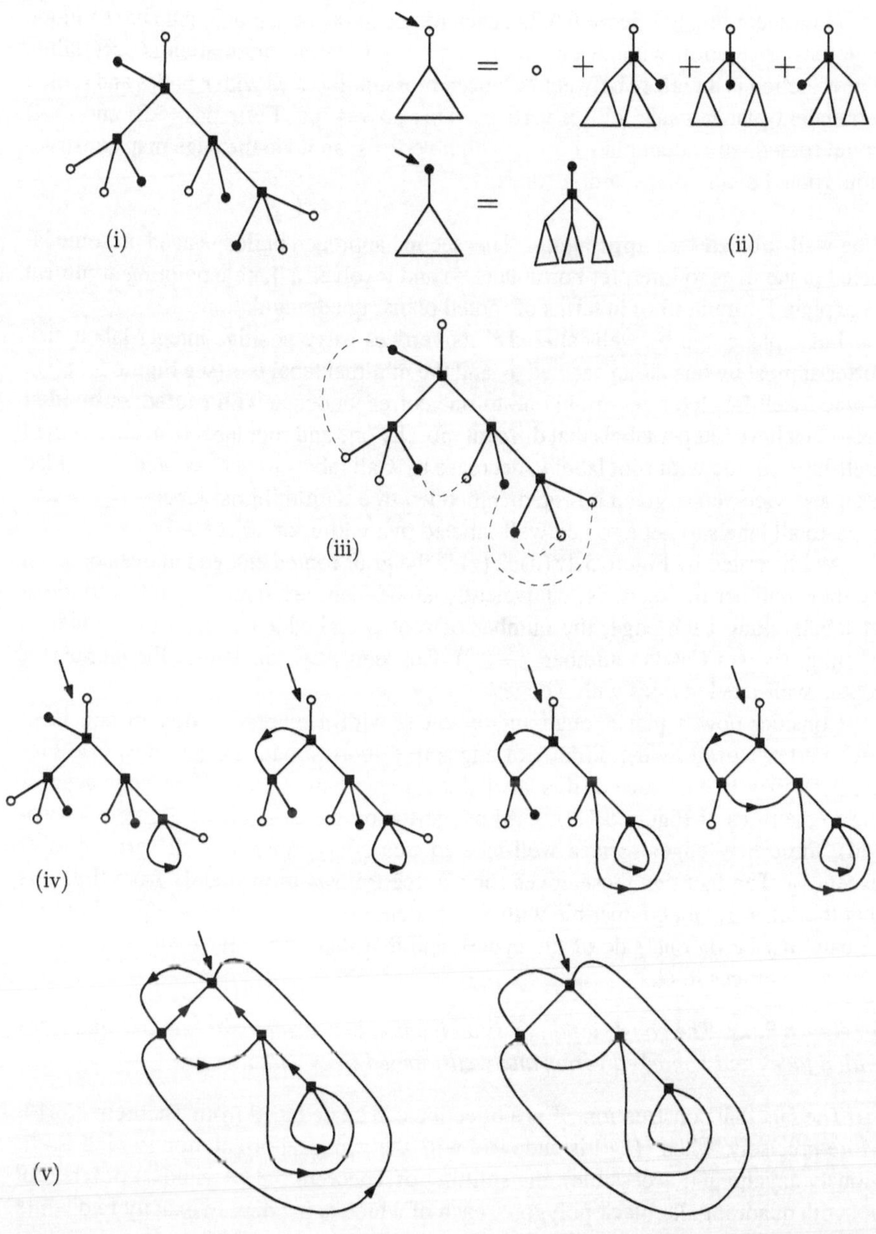


**Figure 5.17**
(i) An unrooted blossoming tree $T$. (ii) The decomposition of white and black trees. (iii) The canonical matching of the black and white leaves of $T$. (iv) The iterative closure of a balanced blossoming tree into a 2-leg tetravalent map. (v) The associated 2-oriented tetravalent maps, and the underlying non-oriented rooted tetravalent map.

Now according to Theorem 5.2.6, each rooted tetravalent planar map has a unique Eulerian orientation without cw-circuit, and any Eulerian orientation is accessible. The bijection is therefore between balanced blossoming trees with $n$ nodes and rooted tetravalent planar maps with $n$ vertices. This proves that Formulas (5.2) and (5.4) count rooted tetravalent planar maps with $n$ vertices, and, via the edge map construction, rooted planar maps with $n$ edges.

**The well-labeled tree approach.** This second approach builds instead on some labeled plane trees to interpret Formula (5.5) and involves a double pointing argument to explain Formula (5.6) in terms of rooted planar quadrangulations.

Let a plane tree be **well-labeled** if its vertices carry positive integer labels that differ at most by one along each edge, and the minimal label is 1 (see Figure 5.18(i)). Rooted well-labeled trees are in one-to-one correspondence with **rooted embedded trees** that have integer labels that differ at most by one and root label 0: given a rooted well-labeled tree with root label $k$, decrease by $k$ all labels to get a rooted embedded tree, and vice-versa, given a rooted embedded tree with minimal label $-\ell \leq 0$, add $\ell+1$ to all labels to get a rooted well-labeled tree with root label $\ell+1$.

As illustrated by Figure 5.18(ii), $T(z)$ is the gf of rooted embedded trees counted by their number of edges, or, equivalently, since there are three possible variations of labels along each edge, the number of rooted embedded trees with $n$ edges is $3^n$ times the $n$th Catalan number $\frac{3^n}{n+1}\binom{2n}{n}$. Consequently, this is also the number of rooted well-labeled trees with $n$ edges.

Consider now a planar quadrangulation $Q$ with a marked vertex $v_0$ and label each vertex $v$ of $Q$ by the number of edges in a shortest path from $v$ to $v_0$ (see Figure 5.18(iii)). In each face of this labeled quadrangulation, draw a new edge according to the rules of Figure 5.18(iv). As suggested by the example in Figure 5.18(vi)-(vii), these new edges form a well-labeled tree $\mathscr{T}(Q)$ spanning all vertices of $Q$ except $v_0$: The fact that these edges form a tree follows immediately from the facts that the rules are not compatible with the existence of a cycle of new edges (because $v_0$ can only be on one side of the cycle), and that there are $n$ new edges and $n+1$ non-root vertices in $Q$.

**Theorem 5.4.2** *The construction $\mathscr{T}$ is a bijection between planar quadrangulations with $n$ faces and a marked vertex and well-labeled trees with $n$ edges.*

The fact that construction $\mathscr{T}$ is a bijection can be deduced from Theorem 5.3.14. More precisely, when $(Q,c)$ is endowed with the geodesic orientation (which is obviously acyclic and accessible), the splitting of Theorem 5.3.14 yields a polyhedral net with quadrangular black polygons, each of which is incident to exactly two white polygons (see Figure 5.18(viii)). The local rules of Figure 5.18(iv) are seen to describe the two possible local configurations in the construction of the polyhedral net, as shown in Figure 5.18(v). These observations are sufficient to prove that $\mathscr{T}$, as a special case of splitting, is injective. To conclude the proof of Theorem 5.4.2, one needs to check that folding the polyhedral net described by any well-labeled tree yields a quadrangulation endowed with the geodesic orientation from one of its vertices. This follows from the fact that the labels of the vertices of the well-labeled

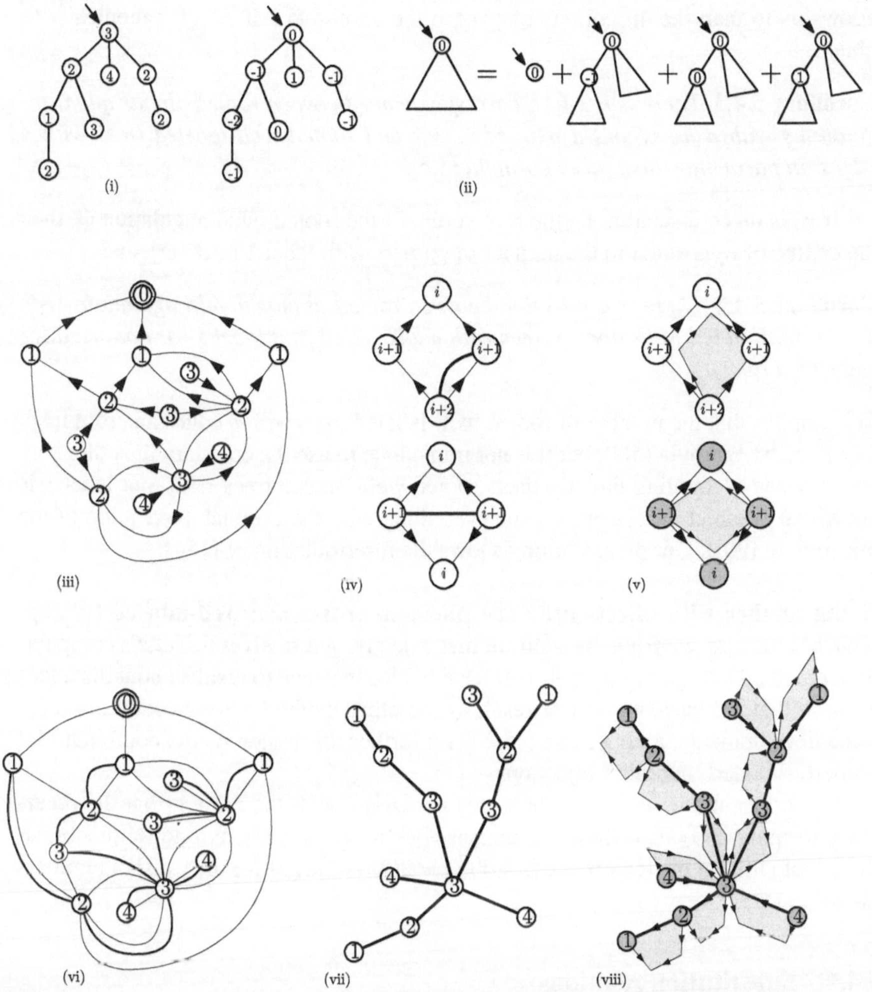

**Figure 5.18**
(i) A rooted well-labeled tree and the associated rooted embedded tree. (ii) The decomposition of rooted embedded trees. (iii) A planar quadrangulation $Q$ with a marked vertex endowed with distance labels, and the associated geodesic orientation. (iv) The local rules defining $\mathscr{T}$. (v) Local configurations in the geodesic splitting. (vi) The construction of $\mathscr{T}(Q, v_0)$. (vii) The resulting tree $\mathscr{T}(Q, v_0)$. (viii) The polyhedral net encoded by $\mathscr{T}(Q, v_0)$.

trees coincide with the natural geodesic labels on the nodes of the folding tree, as shown by Figure 5.18(vi)-(v).

Since each edge of $\mathscr{T}(Q)$ is drawn in a different face of $Q$, any local convention allows us to map the $4n$ corners of $Q$ onto the $2n$ corners of $\mathscr{T}(Q)$ canonically to obtain:

**Corollary 5.4.3** *There is a 2-to-1 correspondence between rooted planar quadrangulations with n faces and a marked vertex and rooted well-labeled trees with n edges. In particular this proves Formula (5.6).*

If $v_0$ is taken instead to be the root vertex of the rooted quadrangulation $Q$, then the degree of $v_0$ is equal to the number of corners with label 1 in $\mathscr{T}(Q)$, and

**Corollary 5.4.4** *There is a bijection between rooted planar quadrangulations with n faces and rooted well-labeled trees with n edges and root label 1 (or non-negative embedded trees).*

This implies that the number of rooted well-labeled trees with $n$ edges and root label 1 is given by Formula (5.2), but it is not immediate to use the corollary the other way round because counting directly these rooted well-labeled trees with root label 1 is not trivial (a direct proof appears however along with the original description of the bijection in [90]). Our presentation follows the reformulation of [154].

**Going further with bijections.** The **blossoming tree** and **well-labeled tree** approaches have grown from the status of nice bijective alternatives to Tutte's computational method into powerful tools that have led for instance to results about distances in maps that are currently out of reach of the other methods: See Section 5.4.6 for these developments, and Section 5.5.2 for a further discussion of the combinatorial properties underlying these bijections.

In the meantime, let us conclude this paragraph with the remark that the recurrence formula (5.3) also admits an elegant bijective proof [43], akin to Rémi's bijective proof [149] of the recurrence $(n+1)C_n = 2(2n-1)C_{n-1}$ for the Catalan numbers $C_n = \frac{1}{n+1}\binom{2n}{n}$.

### 5.4.4 Substitution relations

**The non-separable core of a map and 2-connected planar maps.** Following Tutte [162], let us call a planar map **non-separable** if it is either reduced to a single edge (which can be a bridge or a loop), or if it is loopless and 2-connected. Let $\mathscr{C}$ denote the set of rooted 2-connected planar maps. The **non-separable core** of a rooted planar map is the largest non-separable submap containing the root. As illustrated by Figure 5.19, any rooted planar map decomposes bijectively into its non-separable core and a collection of rooted submaps attached to each corner of the core. This immediately yields the substitution equation

$$M^r(z) = 1 + C(zM^r(z)^2), \tag{5.21}$$

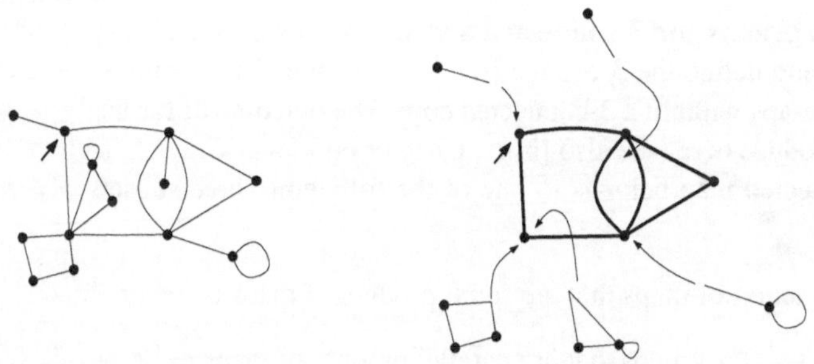

**Figure 5.19**
A rooted planar map and its decomposition into its non-separable core and submaps attached to corners.

where $C(t)$ is the gf of rooted non-separable planar maps counted by edges with variable $t$ (recall that the number of corners in a map is twice the number of edges, hence the substitution $t \to zM^r(z)^2$).

Equation (5.21) allows us to determine $C(t)$ from our knowledge of $M(z)$: Recall that $M(z)$ is an algebraic function, and consider the polynomial $P$ such that $P(M(z),z) = 0$ as given by Formula (5.7). Then the series $H(z) = zM(z)^2$ is a root of the polynomial $Q(t,z) = \text{Resultant}_x(P(x,z), t - zx^2)$. Moreover, $H(z) = z + O(z^2)$ clearly admits a compositional inverse $Y(t)$ in the space of formal power series (that is, the unique formal power series $Y(t)$ such that $H(Y(t)) = t$). This series is algebraic as well as satisfying $Q(t,Y(t)) = 0$. Finally letting $z = Y(t)$ in $P(M(z),z) = 0$ yields $P(C(t),Y(t)) = 0$, from which we deduce that $C(t)$ is a root of the polynomial $R(x,t) = \text{Resultant}_z(P(x,z), Q(t,z), z)$: explicit computations give $R(C(t),t) = 0$ with

$$R(C,t) = C^3 + 2C^2 + (1 - 18t)C + 27t^2 - 2t.$$

Alternatively the computation can be made using the parametrization (5.4)-(5.5) to obtain directly a parametrization of $C(t)$ as

$$C(t) = 2B(t) - 3B(t)^2 \quad \text{where } B(t) = 1/(1 - B(t))^2. \tag{5.22}$$

The number of rooted non-separable planar maps with $n$ edges follows using Lagrange inversion formula,

$$|\mathscr{C}_n| = \frac{4}{2n} \cdot \frac{1}{2n-1} \binom{3n-3}{n-1}.$$

The previous discussion is easily adapted to take into account the number of vertices, and this yields the refined formula

$$|\mathscr{C}_{i+1,j+1}| = \frac{(2i+j-2)!(i+2j-2)!}{(2i-1)!(2j-1)!i!j!},$$

for the number of rooted non-separable planar maps with $i+1$ vertices and $j+1$ faces.

**Polyhedral graphs and 3-connected maps.** Following Tutte's steps [162], one can go further and define the 3-connected core of a map: This requires some extra care to classify maps without a 3-connected core. The outcome of his analysis, which we do not reproduce here (see also [infra, Chapter on Planar Graphs, Section 2]), is that any 2-connected map belongs to one of the following three subsets $\mathscr{S}$, $\mathscr{P}$, or $\mathscr{H}$, where

- the set $\mathscr{S}$ of maps that are serial product of maps of $\mathscr{P}$ or $\mathscr{H}$,
- the set $\mathscr{P}$ of maps that are parallel product of maps of $\mathscr{S}$ or $\mathscr{H}$,
- the set $\mathscr{H}$ of maps that have a non-trivial 3-connected core.

Then, by definition of $\mathscr{S}$ and $\mathscr{P}$, the gfs $S(t)$, $P(t)$ and $H(t)$ are determined as rational functions of $C(t)$ by the system of equations

$$\left\{\begin{array}{rcl} C(t) & = & S(t)+P(t)+H(t) \\ S(t) & = & \frac{P(t)+H(t)}{1-P(t)-H(t)} \\ P(t) & = & \frac{S(t)+H(t)}{1-S(t)-H(t)}. \end{array}\right. \tag{5.23}$$

Finally, rooted non-separable maps that have a non-trivial 3-connected core can be related to rooted 3-connected maps by a substitution scheme: Each such map is indeed uniquely obtained from a rooted 3-connected map in which each non-root edge is replaced by a rooted 2-connected map (the replacement operation of an oriented edge $e$ by a rooted map $N$ in a map $M$ consists in identifying the endpoints of $e$ and of the root of $N$ and removing these two edges). If $G(z)$ denotes the gf of rooted planar 3-connected maps counted by non-root edges (variable $z$), then the resulting substitution equation reads

$$H(t) = G(C^{\dagger}(t))$$

where $C^{\dagger}(t) = \frac{1}{t}C(t) - 2$ is the generating function of rooted planar 2-connected maps counted by non-root edges, $C(t)$ given by Formula (5.22) and $H(t)$ by Equations (5.23). As for Equation (5.21), this substitution equation determines $G(z)$ in terms of $C(z)$. Again, as shown by Mullin and Schellenberg [145], the system of equations can be refined to take into account the number of vertices, and we directly state the bivariate result as

**Theorem 5.4.5 (Mullin-Schellenberg [145])** *Let $U(z,x)$ and $V(z,x)$ denote the unique pair of formal power series solutions of the system*

$$U = x(1+V)^2 \quad \textit{and} \quad V = y(1+U)^2.$$

*Then*

$$G(x,y) = x^2y^2\left(\frac{1}{1+x}+\frac{1}{1+y}-1-\frac{(1+U)^2(1+V)^2}{(1+U+V)^3}\right). \tag{5.24}$$

**Ten steps to planar graphs.** By substitution we went from 1-connected to 2-connected and to 3-connected maps. The enumeration of 3-connected planar maps is a great achievement from the point of view of enumerative graph theory because of Whitney's theorem: A vertex labeled 3-connected planar graph essentially has only one embedding as a vertex labeled 3-connected planar map. We already argued that the exponential generating series of edge labeled planar maps is, up to a derivative, the ordinary generating series of rooted planar maps. In the 3-connected case, a similar relation can be devised for vertex-labeled planar maps so that Equation (5.24) essentially yields an exponential generating series for labeled 3-connected planar graphs. As shown in Chapter 6 of this book, substitution relations, now written for planar graphs instead of planar maps, can be used the other way round to obtain successively from (5.24) the generating series of labeled 2-connected planar graphs, of labeled connected planar graphs, and of planar graphs.

This up and down approach to the enumeration of planar graphs was first explicitly proposed by Liskovets and Walsh [136] in the form of a ten step program to count unlabeled planar graphs.

### 5.4.5 Asymptotic enumeration and uniform random planar maps

**Asymptotic formulas.** From the asymptotic counting perspective, the exact results of Sections 5.4.1 and 5.4.4 have direct simple consequences: When $n$ goes to infinity, the number of rooted planar maps with $n$ edges satisfies

$$|\mathcal{M}_n| = c \cdot \rho_0^n n^{-5/2} \cdot (1 + O(1/n)) \tag{5.25}$$

where $\rho_0 = 12$ and $c$ is an explicit positive constant, and for any fixed $\delta \in (-\frac{1}{2}, \frac{1}{2})$, the number of rooted planar maps with $n$ edges and $k$ vertices satisfies

$$|\mathcal{M}_{n,k}| = c_\delta \rho_\delta^n n^{-3}(1 + o(1)), \quad \text{for} \quad k = \lceil n(\tfrac{1}{2} - \delta) \rceil \tag{5.26}$$

where again $c_\delta$ and $\rho_\delta$ are positive constants that have explicit expressions in terms of $\delta$ [22], with $\rho_\delta < \rho_0 = 12$ for all $\delta \neq 0$. Asymptotic enumerative results on rooted planar maps are intimately related to the study of the random variable $\mathbf{M}_n$ with uniform distribution on the set of rooted planar maps with $n$ edges. From this point of view Formula (5.26) can be restated as a Gaussian local limit law with linear variance for the number of vertices of $\mathbf{M}_n$, using the expansion $\ln \rho_\delta / \rho_0 = -\alpha \delta^2 (1 + o(1))$ near $\delta = 0$, where $\alpha$ is a positive constant (see [infra, Chapter on Planar Graphs, Section 3] for similar results in the case of planar graphs).

**Random maps and face degrees.** The root face degree is a natural parameter of $\mathbf{M}_n$ to study: As shown by Equations (5.9)-(5.10), the root face degree $d_f(M)$ plays a distinguised role in Tutte's decomposition, and it follows from his approach that the series

$$M^r(z,u) = \sum_{M \in \mathcal{M}^r} z^{e(M)} u^{d_f(M)},$$

is algebraic and in fact fairly explicit. As a consequence $d_f(\mathbf{M}_n)$ and by duality $d_v(\mathbf{M}_n)$ are quite well known: As $n$ goes to infinity, the degree has a discrete limit law,

$D_f(k) = \lim_{n\to\infty} \mathbf{P}(d_f(\mathbf{M}_n) = k) > 0$, with exponential decay with $k$. More precisely, as $k$ goes to infinity,

$$D_f(k) \underset{k\to\infty}{\sim} c'(k/\pi)^{1/2}(5/6)^k$$

where $c' = \sqrt{10}/20$ [108]. Upon selecting uniformly a second root corner at random in $\mathbf{M}_n$, one realizes that the root face degree distribution is the distribution of the degree of the face bordering any random edge in $\mathbf{M}_n$. This rerooting trick makes it possible to derive results about the general local structure of large uniform random maps: For instance [111], the maximum face degree satisfies for large $n$,

$$\mathbb{E}(\text{max\_deg}(\mathbf{M}_n)) = \frac{\ln n - \frac{1}{2}\ln\ln n}{\ln(6/5)} + O(1)$$

with a variance of order $O(\ln n)$. More precisely, for values of $k$ close to the above expected value, the numbers of faces of degree $k$ behave like independent Poisson random variables.

**Submaps and asymmetry.** Another example is the number $n_H(M)$ of induced copies of a plane submap $H$ in $\mathbf{M}_n$ (by an induced copy of $H$ in $M$ we mean a simple cycle $C$ such that the vertices, faces, and edges inside and on $C$ form exactly a copy of $H$). In general for a given $H$ there exists a constant $c$ such that the probability that $n_H(\mathbf{M}_n) < cn$ is exponentially small: In other terms, the number of copies of $H$ in $\mathbf{M}_n$ is almost surely linear. Under further technical restrictions on $H$, the random variable $n_H(\mathbf{M}_n)$ is conjectured to have a mean, a variance, and a Gaussian fluctuation (but this is only proved for restricted classes of maps like 3-connected triangulations [112]). The almost sure linearity property implies in turn by a remarkably elegant argument [150] that among rooted planar maps with $n$ edges, the proportion of maps having a non trivial automorphism (of unrooted planar map) is exponentially small (essentially because an automorphism must fix the relative orientations of a large number of copies of any given asymmetric submap $H$). In particular this allows us to recover independently of Formula (5.20) the asymptotic number of unrooted planar maps as

$$|\mathcal{M}_n^u| = |\mathcal{M}_n^r|/(2n)\cdot(1+o(\varepsilon^n))$$

for some positive constant $\varepsilon < 1$. It implies furthermore that the random variable $\mathbf{U}_n$ with uniform distribution on unrooted planar maps with $n$ edges behaves asymptotically like $\mathbf{M}_n$ at least for parameters that are independent of the root and polynomially bounded. More precisely if $p$ is an integer-valued parameter of unrooted planar maps such that $p(M) < c\cdot e(M)^\alpha$ for some $\alpha > 0$ and $c > 0$, then the total variation distance between the distribution of $p(\mathbf{U}_n)$ and $p(\mathbf{M}_n)$ is small:

$$\sum_k |\mathbf{P}(p(\mathbf{U}_n) = k) - \mathbf{P}(p(\mathbf{M}_n) = k)| \underset{n\to\infty}{\longrightarrow} 0.$$

This gives an *a posteriori* further justification of the fact that the literature focuses on the slightly easier problem of enumeration of rooted maps.

Another interesting consequence of the submap density results is the possibility to prove a 0-1 law for the first order logic on planar maps, as pioneered by Bender, Compton and Richmond [23]: Roughly speaking one proves that any property expressible in this framework has a probability to be true on $\mathbf{M}_n$, which tends to 0 or to 1 as $n$ goes to infinity.

**Separation properties.** Continuing with properties that are accessible via explicit bivariate enumeration, another fundamental problem is that of the existence of small cuts in random maps. This problem was first raised in mathematical physics, where $\mathbf{M}_n$ appears as a relevant model of 2d quantum geometry: In this context, the random map is viewed as a random discretized surface, a **2-dimensional quantum universe** [8], and a natural question is whether this random universe is expected to branch into several almost independent parts. One way to formalize the question (with two variants) is to ask about the existence of a vertex $v$ in $\mathbf{M}_n$ whose removal produces at least two components and **(i)** one of the components is a tree of size at least $k$, or **(ii)** both components have size at least $k$.

In the first variant **(i)** the probability that there is such a tree-cutting vertex is bounded by the expected number of tree-cutting vertices, itself bounded by

$$\sum_{i=k}^{n-1} \frac{2nT_iM_{n-i}}{M_n} \approx \sum_{i=k}^{n-1} \frac{n^{7/2}}{i^{5/2}(n-i)^{5/2}} \frac{4^i 12^{n-i}}{12^i} \approx 3^{-k},$$

where we have used the asymptotic formula (5.25) and the classic approximation $C_n \approx 4^n n^{-3/2}$ for the number of rooted plane trees with $n$ edges. In particular the above bound is exponentially decreasing with $k$ and one can prove that the size of the largest induced subtree in $\mathbf{M}_n$ is $\Theta(\ln n)$. The precise analysis is made possible by the existence of a substitution scheme analog to those of Section 5.4.4: Removing all induced subtrees from a map $M$ (or equivalently removing iteratively all vertices of degree one that are not incident to the root edge) yields its **1-core** $C^1(M)$, which is a rooted planar map without non-root vertex of degree 1. This decomposition yields the gf of rooted planar maps counted by number of edges (variable $z$) and by number of edges in the 1-core (variable $u$)

$$M_T(z,u) = 1 + C^1(uzT(z)^2), \tag{5.27}$$

where $C^1(z)$ is the gf of rooted planar maps without non-root vertex of degree 1, and this equation for $u=1$ determines the gf $C^1(z)$. Then the singular analysis of the bivariate substitution scheme (5.27) in the sense of [102] makes it possible to show that $C^1(\mathbf{M}_n)$ has expected size $\alpha n$ for some constant $\alpha$ with Gaussian fluctuations in the range $n^{1/2}$. In the quantum gravity literature, this situation received the evocative description of that of a **unique mother quantum universe** from which logarithmic size **tree-like baby universes** emerge.

In the second variant **(ii)** the probability that the separation is possible is bounded by

$$\sum_{i=k}^{n-k} \frac{2nM_iM_{n-i}}{M_n} \approx \sum_{i=k}^{n/2} \frac{n^{7/2}}{i^{5/2}(n-i)^{5/2}} \approx n \cdot k^{-3/2}.$$

This suggests, and it can actually be proved [110], that at most one 2-connected component of $\mathbf{M}_n$ has linear size (the **mother universe**) and the second largest component (the **largest baby universe**) has size $k = \Theta(n^{2/3})$. The substitution Equation (5.21) refines into the bivariate equation

$$M(z,u) = 1 + C^{\text{ns}}(uzM(z)^2)$$

whose analysis shows more precisely [16] that the largest non-separable component $C^{\text{ns}}(\mathbf{M}_n)$ has expected size $n/3$ with fluctuations in the range $n^{2/3}$ given by a stable law of index $2/3$.

The qualitative difference between variants **(i)** and **(ii)** is directly explained by the fact that plane trees have a strictly lower growth constant than planar maps. Equivalently, at the technical level, the difference is between subcritical and critical compositions of gfs (see [102, Chapter VI.9]). In both cases, the existence of a bivariate substitution scheme to describe the decomposition of $\mathbf{M}_n$ into a mother universe and baby universes implies that the mother universe, conditionally to its size, is itself uniformly chosen among the corresponding set of maps: $C^{\text{ns}}(\mathbf{M}_n)$ is a uniform random 2-connected map with $\mathbf{m}$ edges, where the number $\mathbf{m}$ of edges is a concentrated random variable.

The above discussion concentrates on submaps that can be separated by a unique cut vertex, but it can be extended to consider 2-vertex separators, and, qualitatively, the results are expected to hold for fixed separating cycles of fixed length (although in that case there is some ambiguity in the definition of the **mother universe**). A natural next step is to ask about the length of a shortest cycle allowing the separation of a random planar map into two big regions (**i.e.** so that there are at least $\alpha n$ edges on each side of the cycle). The answer to this question requires a detour via the study of the intrinsic geometry of random planar maps.

### 5.4.6 Distances in planar maps

**The intrinsic geometries of planar quadrangulations and planar maps.** Given two vertices $x$ and $y$ in a map $M$, let $d_M(x,y)$ denote their distance, that is the minimal number of edges in a path from $x$ to $y$. The **intrinsic geometry** of $M$ is the metric space $(V_M, d_M)$, where $V_M$ denotes the set of vertices of $M$. In order to study the intrinsic geometry of $\mathbf{M}_n$ a natural first step is to consider the distances between two random vertices, or between the root vertex and a random vertex.

This can be done thanks to the two above mentioned bijections between maps and trees. In particular our presentation of the bijection $\mathscr{T}$ of Theorem 5.4.2 clearly shows that all the distances to the marked point in a rooted quadrangulation with a marked vertex are encoded in the labels of the associated rooted well-labeled tree. This gives access to the intrinsic geometry of $\mathbf{Q}_n$, the r.v. with uniform distribution on the set of rooted planar quadrangulations with $n$ faces. As a consequence most of the results on distances in the literature are stated for this related but **a priori** different model of random map. Let us temporarily ignore this problem and discuss the results in terms of $\mathbf{Q}_n$.

From a probabilistic point of view the sole fact that labels of rooted well-labeled trees can be interpreted in terms of distances in quadrangulations is already extremely fruitful: Let $\mathbf{T}_n$ denote a random variable taken uniformly at random in the set of rooted well-labeled trees with $n$ edges. The random well-labeled tree $\mathbf{T}_n$ is obtained by shifting the labels of a random rooted embedded tree $\mathbf{E}_n$ by their minimum. The tree $\mathbf{E}_n$ can be constructed in two steps, first taking a uniform random plane tree with $n$ edges, and then choosing the label increment on each edge in $\{-1,0,+1\}$ uniformly and independently. As a consequence a uniform random vertex $\mathbf{v}$ in $\mathbf{E}_n$ is typically at distance $\ell = \Theta(n^{1/2})$ of the root in the tree (by standard results on the height profile of trees) and its label $\lambda_{\mathbf{T}_n}(\mathbf{v})$ is a sum of $\ell$ i.i.d. random variables taken uniformly in $\{-1,0,1\}$: it is thus almost surely of order $n^{1/4}$, as $n$ goes to infinity. It can be proved that this is also almost surely the case for the minimal label, so that the statement remains true for $\mathbf{T}_n$ [85]. In terms of distances in $\mathbf{Q}_n$, we conclude that the typical distance between two random vertices is of order $n^{1/4}$. This combinatorial derivation of the typical distance in $\mathbf{Q}_n$, proposed in [85], confirms earlier semi-rigourous prediction of the physics literature (see [9] and reference therein).

**Exact counting results for labeled trees and distances in quadrangulations.** At the enumerative level, results can be made even much more precise. For all $i \geq 0$, let $T_i$ be the gf of rooted embedded trees with label strictly larger than $-i$. Then the standard recursive decomposition of rooted plane trees can be refined to write the following system of equations for the infinite family of gfs $(T_i)_{i\geq 0}$:

$$T_i = 1 + zT_i(T_{i-1} + T_i + T_{i+1}) \text{ for } i \geq 0, \quad \text{and} \quad T_0 = 0. \tag{5.28}$$

A true miracle is that the system of equations (5.28) can actually be solved exactly. As first shown by Bouttier, Di Francesco, and Guitter in [62],

$$T_i(z) = T\frac{(1-Y^i)(1-Y^{i+3})}{(1-Y^{i+1})(1-Y^{i+2})} \tag{5.29}$$

where $Y \equiv Y(z)$ and $T \equiv T(z)$ are the unique power series solutions of

$$Y = zT^2(1+Y+Y^2) \quad \text{and} \quad T = 1+3zT^2.$$

Observe that the series $T$ in this expression is the same as in Equation (5.5): accordingly the limit when $i$ goes to infinity of $T_i$ is just $T$. While Formula (5.29) was guessed and checked in [62], it was recently observed by Eynard and Guitter that the system of equations (5.28) belongs to a family of equations known to admit explicit computable solution in a different context [98].

The bijection between rooted well-labeled trees with root label 1 (or non negative rooted embedded trees) and rooted quadrangulations implies that $M(z) = T_1(z)$ and

$$M(z) = T\frac{(1-Y)(1-Y^4)}{(1-Y^2)(1-Y^3)} = T - \frac{Y}{1+Y+Y^2}T = T - zT^3$$

in agreement with Formula (5.4). The general explicit expression (5.29) has far

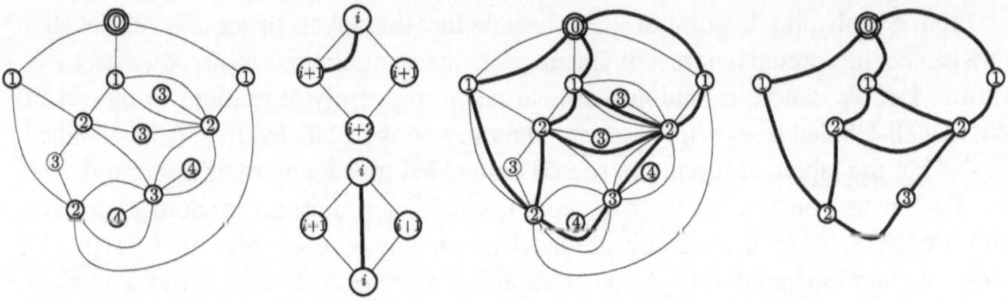

**Figure 5.20**
A quadrangulation $Q$ with distances to a marked $v_0$, the Ambjørn-Budd rules, and their applications to $(Q, v_0)$.

reaching implications for distance statistics on quadrangulations, as discussed in [62, 63, 66, 68, 67, 120] (see also the results mentioned below for distances in $\mathbf{M}_n$).

Finally, let us mention that the combinatorial structure of Formula (5.29) is still not completely clear: from the point of view of quadrangulations, an appealing interpretation of $Y$ and $T_i(z)$ was given in [70] (see also [3]) in terms of **slices** of quadrangulations, from which the explicit expressions can be understood. However we still miss a direct explanation of Formula (5.29) in terms of the natural interpretation of $Y/z$ as gf of well-labeled trees with a marked branch whose labels form an excursion.

**Exact results for distances in planar maps.** Quite unexpectedly, many of the exact results for quadrangulations can be easily transfered to planar maps: As already indicated above, the classical incidence map transformation does not help much for this, but the solution is provided by another bijection between marked quadrangulations and marked planar maps that was recently discovered by Ambjørn and Budd [7] (see also [65] for a broader discussion of the relation between this bijection and well-labeled tree approach). As illustrated by Figure 5.20 this bijection consists in applying to a quadrangulation $Q$ with a marked vertex $v_0$ the rules opposite to those of Figure 5.18(iv).

The two key properties of this bijection for our purpose are that (i) it preserves the distance to $v_0$ of the common vertices of the two maps, namely if $(M, \bar{v}_0) = AB(Q, v_0)$ then $d_Q(v, v_0) = d_M(\bar{v}, \bar{v}_0)$, and (ii) the vertices of $Q$ that do not appear in $M$ are local maxima of the distance to $v_0$ in $Q$, and they correspond to vertices whose labels are local maxima in the associated well-labeled tree. As a consequence of these two properties the labels of a random well-labeled tree that are not local maxima describe the distances to a random marked vertex in a random planar map.

Let us illustrate by an example the very explicit results on the intrinsic geometry of random planar maps that are derived in [65]. Let us say that the root edge of a vertex-pointed rooted planar map $M$ has type $(i, j)$ if the root vertex is at distance $i$ of the pointed vertex, and the other extremity of the root edge is at distance $j$ of the pointed vertex (by definition $|i - j| \leq 1$). Let then $R_i(z)$ denote the gf of pointed rooted maps with a root edge of type $(j-1, j)$ for $j \leq i$. Upon reverting the root,

$R_{j+1}(z)$ is the gf of such maps with a root edge of type $(j+1, j)$ for $j \geq i$, and let $S_j^2(z)$ denote the gf of such maps with a root edge of type $(j, j)$ for $j \leq i$ (for consistency with the literature this gf is written as a square). Then the combination of the well-labeled tree approach and the AB bijection is that $R_i$ and $S_i^2$ can be written for all $i \geq 0$ as

$$R_i = zT_iT_{i+1} \quad \text{and} \quad S_i^2 = tT_i^2,$$

where the $T_i$ are given by the system of equation (5.29). For completeness let us state the explicit expressions for the $R_i$ and $S_i$,

$$R_i(z) = (1+z(1+T)^2)\frac{(1-Y^{i+1})(1-Y^{i+3})}{(1-Y^{i+2})^2} \tag{5.30}$$

$$S_i^2(z) = z(1+T)^2\left(\frac{(1-Y^i)(1-Y^{i+3})}{(1-Y^{i+1})(1-Y^{i+2})}\right)^2. \tag{5.31}$$

In particular these explicit expressions allow to compute distance statistics for $\mathbf{M}_n$: for instance, for any fixed $i \geq 1$, the expected number of vertices at distance $i$ of a uniformly chosen vertex in $\mathbf{M}_n$ is

$$\lim_{n\to\infty} \mathbb{E}(|\{v \mid d_{\mathbf{M}_n}(v, \mathbf{v_0})\} = i|) = \frac{3}{280}(2i+3)(10i^2+30i+9).$$

Similarly the limit for large $n$ of the expected number of vertices in the ball of radius $i$ around a uniform random vertex of $\mathbf{M}_n$ grows like $i^4$.

### 5.4.7 Local limit, continuum limit

**Local limits.** A nice way to subsum the above results on the local structure of random planar maps is the following statement: Large uniform random rooted planar maps have a local limit, the **uniform infinite planar map**, which has Hausdorf dimension 4. The first part of this statement means that, as $n$ goes to infinity, for any fixed $k$, the submap of radius $k$ around the root of $\mathbf{M}_n$ converges in distribution to a random planar map $\mathbf{P_k}$ with radius $k$, and the family of random planar map $(R_k)_{k\geq 1}$ coherently defines a random infinite (but locally finite) planar map $\mathbf{P}$. The second part means that in the random infinite planar map $\mathbf{P}$ the number of vertices at distance at most $i$ of the root roughly grows like $i^4$. This statement was first stated and proved in the case of random triangulations and quadrangulations [11, 10, 84], leading to the definition of the Uniform Infinite Planar Triangulation and Quadrangulation (UIPT, UIPQ), but again, the Ambjørn-Budd bijection makes it possible to transfer the results to general planar maps to define the UIPM.

A natural way to continue the study of the local properties of large planar maps is to concentrate on these limit uniform infinite planar maps. We will however not discuss further this direction for at least two reasons: Many of the methods involved there escape from the strict range of enumerative combinatorics, and the topic is currently evolving at a very fast pace. We refer the reader to the elegant presentations of Nicolas Curien.

**The continuum limit.** Instead of concentrating on local properties, one can return to the observation that the distance between the root and a random vertex of $\mathbf{M}_n$ is of order $n^{1/4}$. The correspondence between quadrangulations and well-labeled trees makes it possible to show more precisely that the **rescaled profile** (average number of vertices at distance $k$ of the pointed vertex) and **radius** (maximal distance to the pointed vertex) of uniform random rooted planar quadrangulation with a random marked vertex converge upon renormalizing the distances by a factor $n^{-1/4}$ to functionals of a well-studied continuum random process called the **Integrated SuperBrownian Excursion (ISE)** or of its variant the **Brownian snake** [85].

The example of the convergence of rescaled simple random walks to the Brownian motion and that of rescaled simple trees to the Continuum Random Tree then suggests that after rescaling distances by such a factor $n^{-1/4}$, one should look for a continuum limit of random planar quadrangulations. This question has attracted a lot of attention in the last few years, that has culminated with the proof of the existence and uniqueness of such a continuum limit, the **Brownian planar map** [129, 140]. A key ingredient underlying these results is the fact that the bijective correspondence between well-labeled trees and pointed rooted maps is robust enough to go through the process of taking a continuum limit: The only currently known explicit construction of the **Brownian planar map** consists indeed in starting from the continuum limit of rescaled uniform random well-labeled trees of size $n$ when $n \to \infty$, the above mentioned **Brownian snake**, and applying a continuum analog of the discrete bijection to define a metric on this embedded continuum tree. Again the initial constructions of [129, 140] were done for quadrangulations, but the Ambjørn-Budd correspondence makes it possible to transfer most of these results to the intrinsic geometry of uniform random planar maps with $n$ edges [44].

Remarkably it has been proved that the Brownian planar map is a random metric space with the topology of the sphere [131, 138]: A consequence of this last statement is the fact that for any positive $\alpha < 1/2$, the shortest cycle on $\mathbf{M}_n$ with at least $\alpha n$ edges on each side has almost surely length $\Theta(n^{1/4})$ when $n$ goes to infinity. Although, as discussed above, the enumerative results and in particular the bijections between maps and trees are fundamental ingredients of construction of the Brownian planar map, these consequences about shortest $\alpha$-separating cycles currently appear to be out of reach with purely enumerative methods (see however [67] for a tractable variant of the problem). Further discussion of these results is thus largely out of scope here and we refer to the [130, 141] for a survey of these developments, or to [127, 128] for short reviews.

## 5.5 Beyond planar maps, an even shorter account

At this point we are more or less done with general planar maps, and there are (at least) four ways to go further: universality, master theorems, maps on surfaces, and decorated maps.

### 5.5.1 Patterns and universality

A first direction to explore is the observation that literally all of the above results admit variants for various natural subfamilies of planar maps. In the combinatorial literature, **patterns** in the asymptotic behavior have been observed, and equivalently, in the physics and probabilistic literature some **critical exponents** are expected to be **universal**.

**Pattern in the asymptotic behavior and universality.** A first example is the fact that for a large collection of subfamilies $\mathscr{F}$ of maps for which the asymptotic number of rooted planar maps with $n$ edges in $\mathscr{F}$ is known, the polynomial growth exponent takes the same value $\gamma = -\frac{5}{2}$: let $\mathscr{M}_n^{\mathscr{F}}$ denote the set of rooted planar maps with $n$ edges in the subfamily $\mathscr{F}$, then

$$|\mathscr{M}_n^{\mathscr{F}}| \underset{n\to\infty}{\sim} c_{\mathscr{F}} \rho_{\mathscr{F}}^n n^{-5/2} \tag{5.32}$$

for all admissible values of $n$, where $c_{\mathscr{F}}$ and $\rho_{\mathscr{F}}$ are constants depending on $\mathscr{F}$. This asymptotic behavior holds in particular for various families of planar maps defined by combinations of a finite restriction on the allowed vertex or face degrees (but not restrictions on both), a finite restriction on the girth (length of the shortest simple cycle), or an irreducibility condition (a map is irreducible if the length of the shortest non-facial cycle is strictly larger than its girth), and for some of these families, with a further bipartiteness constraint [26, 27, 18, 107, 79, 31]. Similar patterns or universal exponents hold for each of the asymptotic/probabilistic statements of the previous section about degree distribution [134], absence of non-trivial automorphisms [24], separation properties [152, 110, 16], and distances [69].

A quite natural way to unify these results is to state the universality of the Brownian planar map as a continuum limit: For all "reasonable" family $\mathscr{F}$ of planar maps, there should exist a constant $\alpha_{\mathscr{F}}$ and a definition of the size such that uniform random planar maps of size $n$ in $\mathscr{F}$ with distances rescaled by a factor $\alpha_{\mathscr{F}} n^{-1/4}$ converge to the Brownian map in the sense of [129, 140]. The collection of families of "reasonable" maps for which this statement has actually been proved is more restricted than for the previous ones but it has recently grown quickly to include planar quadrangulations and more generally $2p$-angulations, general, simple, and bipartite planar maps, and simple triangulations [1, 2, 4, 44], and all the above mentioned families are expected to belong to the same **universality class**.

A common point of these families is that they are defined by constraints that can be checked locally (at finite distance in the derived map). From this point of view examples of "non-reasonable" subfamilies are outerplanar maps because the existence of a large outerface is not a local constraint, series-parallel maps or stack triangulations whose characterization needs to be checked recursively. And indeed these families lead to other continuum limits [5, 76]. A different example is that of planar maps with a global rotation or reflexion symmetry whose continuum limit is expected to conserve the initial symmetry: They do not even satisfy the initial asymptotic pattern because, in view of Section 5.4.2, they are in bijection quotient maps, that are multiply rooted planar maps. Finally let us observe that too stringent restrictions on both vertex and face degrees cannot either lead to the same universality class due to rigidity constraints: For instance triangulations with too many vertices of degree 6 have to be constituted of patches of regular triangular lattices. The question of whether a family of maps belongs or not to some universality class is a key issue in the physics literature on maps, and in particular on decorated maps: Accordingly we shall return briefly to this question in Section 5.5.4.

**Parametric families and meta theorems.** A different kind of recurrent statements in the theory of maps is illustrated by the observation that many families of rooted planar maps defined by local restrictions have algebraic gfs. Here the universality is more questionable and it is in fact not difficult to design "reasonable" families of rooted planar maps that most likely do not have algebraic gfs and yet will satisfy most other universality properties: The family of rooted planar maps with only prime vertex degrees seems an obvious candidate, and several families of non-critical decorated maps, as discussed in Section 5.5.4, provide other examples.

One way to circumvent the difficulty of defining universality is to obtain general parametric results of the form of Equations (5.18) and (5.19). Tutte's equation can be generalized to bicolored maps counted by number of faces of degree $i$ of each color [55]. Similar parametric equations have also been written for families of rooted non-separable planar maps [170]. However for 3-connected maps and other families of maps defined by girth restrictions, equations with catalytic variables are harder to establish from the root edge deletion approach (see for instance [45] for the already quite intricate case of 3-connected maps counted by edges and root face degree).

General parametric results have instead been obtained by Bouttier and Guitter via a far reaching and beautiful extension of the substitution schemes of Section 5.4.4 to the gfs of maps with girth $g$ and $d$-irreductible maps with respect to the distribution of face degrees [72, 71]. Through standard transformations like duality, incidence map, edge map, and a few others, their theorem encompasses all the known critical substitution schemes summarized in [16] and allows us to recover easily the enumeration of rooted planar 5-connected triangulations of [109].

Another way to assert some form of universality could be to resort to logic and prove a **meta theorem**: Let $f$ be a formula of the first order logic on maps (with quantifiers on vertices, faces, edges and corner, and adjacency/incidence predicates). It is tempting to conjecture that the family $\mathcal{M}^f$ of rooted planar maps $M$ such that $f(M)$ is true has an algebraic gf, and that the number of such maps satisfies the universal asymptotic pattern above. (The second part of the statement is actually more likely to be true than the first, but probably hard to prove independently.) This approach is reminiscent of the 0-1 laws briefly discussed in Page 369, but appears to requires new technical ingredients.

### 5.5.2 The bijective canvas and master bijections

The initial motivation for looking for bijective proofs for the formulas of Tutte was closely related to Schützenberger's methodology according to which combinatorial structures with algebraic gfs should admit natural encodings by words of context free languages (or equivalently, they should be in natural correspondence with simple families of trees). Accounts of early attempts in this directions are [86, 87]. A recent discussion (and partial refutation) of the statement that algebraic gfs should be the trace of such natural encodings can be found in [52].

In the case of planar maps, most of the known gf algebraicity results do admit derivations by bijections with some trees. Rather than universality results, master bijections have been proposed [6, 36, 38, 39], that allow us to derive these bijections in a unified way. A beautiful outcome of these results is that they make explicit the structural properties of maps that make the existence of such bijections possible.

**Blossoming trees and well-labeled trees made parametric.** The first extensions of the original two bijections of Section 5.4.3 that were proposed were rather ad-hoc variants for specific subclasses like bipartite maps [12, 154], loopless, simple or irreducible triangulations or quadrangulations [154, 147, 148, 105, 104, 103] (see also [91] for an independantly found recursive variant for non-separable maps). In all these cases, known formulas were instrumental to help guessing an adequate family of balanced trees, which was then used to design a bijection.

A first parametric extension of the blossoming tree approach was to the case of planar constellations in [56]: Although no previous enumerative results were available for this case, conjecturing a formula was again a preliminary to the design of the right family of trees. It is yet a first example of a result that is actually much harder to derive from Tutte's root deletion approach [55] than from bijections.

Similarly, while basic blossoming trees and well-labeled trees are natural combinatorial interpretations of Equation (5.5), parametric extensions were build as interpretations of the system (5.19). This approach was then extended further to bicolored maps counted by number of faces of degree $i$ of each color [57], building on the enumerative indications given by partial enumerative results for trivalent and tetravalent maps [60]. With this exemple the paradigm started to change and the bijective approach yields new formulas that were not even conjectured before.

**A structural result about orientations.** This shift of paradigm lead to the question of understanding under what conditions bijections between maps and trees can work. In view of Section 5.3, the bijections of Section 5.4.3 are seen to rely on identifying an easy to enumerate family of decorated trees whose closure yields the expected family of maps by a specialization of the bijections of Theorems 5.3.6 and 5.3.14. Conversely, given a family of planar maps, these bijections rely on identifying the right notion of canonical spanning tree for which these theorems lead to a simple family of decorated trees. A first step is Theorem 5.3.9, which suggests that in order to find the right canonical spanning trees, one should look for accessible orientations without cw-circuit. This in turn is made easier by Theorem 5.2.6, which asserts that it is sufficient to look for an accessible feasible function on our family of maps.

The key structural result is now the fact that the existence of feasible and accessible $\alpha$-functions is a natural graph theoretic property, expressible in terms of girth conditions. A most general result of this type is given by Bernardi and Fusy in [39] for what they call **fittingly charged hypermaps**. We only quote here a subcase of their classification taken from [36, 38]:

**Theorem 5.5.1** *A planar d-angulation has girth d if and only if it admits an accessible* $(d+2)/d$*-fractional orientation.*

The exact definition of fractional orientation is out of the scope of this text, but the key point is that the theory of $\alpha$-orientations extends to these fractional orientations, and in particular the fact that there is a unique such orientation without cw-circuit. It suggests that, at least for those classes of maps, the quest is over: Instead of guessing the right family of tree from enumerative formulas, one can start from the naturally associated feasible $\alpha$ function and work the other way round.

**The bijective canvas.** The resulting canvas for bijection between maps and trees is the following: Start with a subfamily $\mathscr{F}$ of rooted planar maps defined by degree and girth conditions.

1. Reformulate the degree and girth conditions in terms of the existence of some generalized $\alpha$-orientations (using **e.g.** Theorem 5.5.1).
2. For a given $\alpha$, the set of generalized $\alpha$-orientations on a given map has a lattice structure, and in particular there is a unique such orientation without cw-circuit (Theorem 5.2.6).

At this point there are two options: The first option generalizes the blossoming tree approach and was formalized in [6] (see also [155]).

3a. Use ccw-exploration as in Section 5.3.2 to obtain a canonical spanning tree $T$ of the map $M$, and check that external edges can be encoded as decorations of this tree $T$ in such a way that the resulting set of decorated trees can be described by local rules, directly inherited from the degree and $\alpha$-constraints.

In the case of tetravalent maps, we have already seen that this approach directly yields the blossoming trees of Section 5.4.3: A planar map is tetravalent if and only if it admits a 2-orientation (a particular case of Euler characterization of Eulerian maps); in particular a tetravalent map admits a unique 2-orientation without cw-circuit; the cw-exploration of this orientation yields a spanning tree whose opening has vertices of degree 4 with in and out-degree 2; by construction, apart from the root vertex, each vertex has one outgoing edge toward the root, so that the other outgoing edge must be a dangling half-edge, and each vertex has two incoming edges or half-edges: This provides, without guessing, the definition of blossoming trees of Section 5.4.3.

Another example is the case of simple triangulations (that is, triangulations with girth three): As first shown by Schnyder [158], a planar map is a simple triangulation if and only if it admits a 3-orientation; in particular a simple triangulation admits a unique 3-orientation without cw-circuit; the cw-exploration of this orientation yields a spanning tree whose opening has two dangling outgoing half-edges per vertices; the face degree condition implies that incoming half-edges are not necessary to perform the closure; the resulting family of blossoming trees is exactly the family of plane trees such that each inner vertex is incident to two leaves and this allows us to recover the bijection first proposed in [146].

The second option generalizes the well-labeled approach and was formalized in [36, 38, 39].

3b. Use vertex blowing and splitting as in Section 5.3.3 to obtain a balanced polyhedral net $M$, and again, check that, apart from the fact of being balanced, all the constraint on the resulting set of polyhedral nets can be described by local rules inherited from the degree and $\alpha$-constraints.

In the case of quadrangulations, we have already seen that using the geodesic distance to orient the map directly yields the bijection of Section 5.4.3 with well-labeled trees. The extension of the bijection to general bicolored maps, known as Bouttier, Di Francesco, Guitter's bijection [59] is recovered as well using the orientation induced by oriented geodesics and a similar local analysis of the possible configurations around black and white polygons: the mobiles of [59] naturally arise as simplified descriptions of the resulting polyhedral nets.

This second approach yields the currently most general master theorem [39]: It extends in particular to hypermaps with **ingirth** $d$ (that naturally generalize maps with girth $d$) and in this context gives a very general set of parametric equations for hypermaps with ingirth $d$ and outerface of degree $d$ counted by degrees of black and white faces, that extends Equations (5.18)-(5.19). As already mentioned the most general statement involves some **fittingly charged hypermaps** that arise as the natural setting to which the technics of proof underlying Theorem 5.5.1 apply.

Finally we should mention that a third bijective canvas arises from the extension of the substitution scheme approach proposed in [72]: This last approach has the advantage that it allows to perform non-trivial but directly ℕ-algebraic decompositions for planar maps with a boundary, as first proposed in [70, 3].

**Bijections with other combinatorial structures.** Tutte's formula (5.2) for the number of rooted planar maps with $n$ edges, and its variants for rooted non-separable planar or triangulations have such a simple closed form that one should expect **a priori** unrelated combinatorial structures to be counted by the same numbers. This is indeed the case and bijections have been devised to explain some of these coincidences, in particular in the study of permutations with forbidden pattern (see [117] and references therein). A higher level "explanation" is that pattern avoiding permutation tends to admit natural decompositions leading to polynomial equations with catalytic variables, and we have seen that the simplest of those equations count maps. Exhibiting structurally identical recursive decompositions for two equinumerous families of combinatorial structures is a natural way to get a (recursive) bijection and this approach has been fruitful in this context (see [46] and the references therein). Another remarkable example is that of Baxter permutations and intervals in the Catalan and Tamari orders (see [106] and reference therein).

### 5.5.3 Maps on surfaces

The third way to go is to move to **maps on surfaces**, that is, map on the torus, or more generally on an oriented or non oriented surface of genus $g$. The main idea is that, *mutatis mutandis*, many results have analogs for maps on oriented and also on non-oriented surfaces but actual statements are more complicated, and proofs much more technical.

**Exact and asymptotic counting.** A particularly nice result is that Formula (5.25) is replaced for rooted map on orientable surfaces of genus $g$ by the more general

$$|\mathcal{M}_n^g| = c \cdot \tau_g \cdot 12^n n^{\frac{5}{2}(g-1)} \cdot (1 + O(1/\sqrt{n})), \tag{5.33}$$

where the limit is to be taken at $g$ fixed and $n$ going to infinity [19] (for similar results in the physics literature see references in the survey [92]). This result is a consequence of a more technical statement giving the general form of the gf of these maps [19, 20, 14]: There exists a family of polynomials $P_g(x)$ such that

$$M^g(z) = \frac{P_g(T)}{(2 - T(z))^{3g}}, \quad \text{where} \quad T(z) = 1 + 3zT(z)^2.$$

As usual, this statement can be refined to take into account the number of vertices [22]. Earlier statements for small values of $g$ were already given in [73, 41], and probably the general form of the result was known to the physics community in the early 80's as well but we could not trace the reference. Recently, Eynard and Orantin have proposed a more explicit expression of the series $M^g(z)$ as a finite sum of residues of rational functions in $T(z)$ indexed by some simple trivalent graphs, which they derive from an explicit **topological recurrence** [95, 99]. The statement of this result requires too many notations to be reproduced here.

But maybe the most surprising result is the following simple quadratic recurrence recently discovered by Carrell and Chapuy [78]:

$$\begin{aligned}(n+1)M_n^g = 4(2n-1)M_{n-1}^g &+ \frac{(2n-3)(2n-2)(2n-1)}{2} M_{n-2}^{g-1} \\ &+3 \sum_{\substack{k+\ell=n \\ k,\ell\geq 1}} \sum_{\substack{i+j=g \\ i,j\geq 0}} (2k-1)(2\ell-1) M_{k-1}^i M_{\ell-1}^j,\end{aligned} \tag{5.34}$$

for $n \geq 0$, $g \geq 0$, with initial condition $M_0^0 = 1$, and $M_n^g = 0$ for $n < 0$ or $g < 0$. This recurrence even admits a barely more complicated refinement for the number $M_{i,j}^g$ of rooted planar maps with $i$ vertices and $j$ faces on a surface of genus $g$:

$$\begin{aligned}(n+1)M_{i,j}^g = 2(2n-1)(M_{i-1,j}^g + M_{i,j-1}^g) &+ \frac{(2n-3)(2n-2)(2n-1)}{2} M_{i,j}^{g-1} \\ &+3 \sum_{\substack{i_1+i_2=i \\ i_1,i_2\geq 1}} \sum_{\substack{j_1+j_2=j, \\ j_1,j_2\geq 1}} \sum_{\substack{g_1+g_2=g \\ g_1,g_2\geq 0}} (2n_1-1)(2n_2-1) M_{i_1,j_1}^{g_1} M_{i_2,j_2}^{g_2},\end{aligned} \tag{5.35}$$

for $i, j \geq 1$, with the initial conditions that $M_{i,j}^g = 0$ if $i + j + 2g < 2$, that if $i + j +$

$2g=2$ then $M^g_{i,j} = \mathbf{1}_{\{(i,j)=(1,1)\}}$, and where we use the notation $n=i+j+2g-2$, $n_1 = i_1+j_1+2g_1-1$, and $n_2 = i_2+j_2+2g_2-1$.

While Formula (5.33) arises from (non-trivial) refinements of the gf technics traditionally used to study planar maps, it is worth observing that the proof of Formula (5.34) requires more algebraic ingredients based on the encoding of maps in terms of permutations briefly used in formalizing Definition 5.2.2 (see [114, 78], the earliest reference we could find is [123] in the combinatorial literature and [41, Appendix 6] in the physics literature, both in the case of unicellular maps). Another remarkable outcome of the approach of [114, 78] is the possibility to derive from it detailed asymptotics for the constant $\tau_g$ [25]. Analogous results for non-orientable surfaces have also appeared [19, 107, 77].

Unrooted maps in higher genus surfaces have also been considered, see [167] and reference therein for exact results, and asymptotic asymmetry results like Formula (5.4.5) are in fact valid on higher genus surfaces as well [24].

**Universality again.** As in the planar case, the asymptotic results are valid not only for the family of all rooted planar maps but the pattern holds for various subfamilies:

$$|\mathcal{M}^{\mathcal{F}}_{g;n}| = c_{\mathcal{F}} \cdot \tau_g \cdot (\rho_{\mathcal{F}})^n \cdot (\alpha_{\mathcal{F}} n)^{\frac{5}{2}(g-1)} \cdot (1+O(1/n)) \qquad (5.36)$$

where $c_{\mathcal{F}}$, $\rho_{\mathcal{F}}$ and $\alpha_{\mathcal{F}}$ depend on the family, but the dependency in the genus is entirely controlled by the constants $\tau_g$ and $\gamma_g = \frac{5}{2}(g-1)$ that already appear in Formula (5.33). All the families of maps satisfying the pattern (5.32) are expected to satisfy (5.36) as well on higher genus surfaces, but this is currently proved only for fewer families (in particular for those defined by one of the aforementioned restrictions: degrees, girth, or irreducibility) [107, 79].

Another general observation is the claim by Eynard, that map gfs should in general satisfy a **topological recurrence** parameterized by the so-called **spectral curve** [97]. This statement is more general than the algebraicity and asymptotic pattern (5.36) since the topological recurrence can be written for gfs of critical decorated maps as well as for series arising from algebraic geometry problems. For the numerous developments around this recurrence we refer to the forthcoming book [97].

**Bijection for maps on surfaces.** Regarding bijective proofs, only the approach via well-labeled trees currently extends to higher genus maps: it yields in particular a correspondence between rooted maps on a surface of genus $g$ and well-labeled unicellular maps of genus $g$ (a map is **unicellular** if it has only one face). Combined with decompositions **à la Wright**, it yields a combinatorial explanation [83] of the occurence of the exponent $\frac{5}{2}(g-1)$ in Formula (5.33). A closer relation with well-labeled trees with several marked points was given by Chapuy in [80] using a remarkable bijective decomposition of unicellular maps [81]. Another remarkable extension to higher genus is Miermont's result for maps with several marked points [139]: In particular as discussed in [65], this result unifies the well-labeled tree and the Ambjørn-Budd results.

As in the planar case, the bijective approach allows us to keep track of distances in quadrangulations. Until now exact distance enumeration results like Formulas (5.30)-

(5.31) have been obtained only in the case $g = 1$, [119], but the approach has allowed us to show that in general typical distances remain in the order $n^{1/4}$, [83, 80], and it has opened the way to the construction of continuum limits of random quadrangulations on surfaces [42].

It is instead an open problem to give combinatorial interpretations of the Eynard-Orantin topological recurrence, or of the Carrell-Chapuy recurrences (5.34) or (5.35). The later problem is particularly intriguing: Each term in the recurrence has a clear combinatorial interpretation, and the special case $j = 1$ is the celebrated Harer-Zagier recurrence for the number $\varepsilon_g(n) = M^g_{n+1-2g,1}$ of rooted unicellular maps with $n$ edges and genus $g$, for which an elegant bijective proof was recently proposed by Chapuy, Feray and Fusy in [82], building on the earlier construction of Chapuy [81].

It is worth mentioning the fact that there is a parallel story of unicellular maps, starting with the root deletion method [168, 169, 170], continuing with characters of the symmetric group [123, 116, 146] and matrix integrals [122] to arrive to more combinatorial methods [126, 115, 156, 40, 32].

### 5.5.4 Decorated maps

Finally the fourth direction to discuss is the study of decorated maps. The decoration of a map can be a coloring of its vertices, a spanning tree, a spin or loop configuration, an orientation, etc. Of interest is then the gf of maps endowed with such a structure, or more generally the weighted gf where each decoration is given a weight and the weights are summed over all pairs (map,decoration). We briefly discuss at the end of this section some motivations for this study arising from physics and probability, but we concentrate on counting results that have been obtained in the combinatorial literature, either in the form of explicit counting formula or of equations for gfs.

**Exact counting results by direct decompositions.** As discussed in Section 5.3, probably the first model to have been dealt with is that of cubic maps endowed with a Hamiltonian cycle, which were enumerated by Tutte in [160] even before counting planar maps. Spanning trees on general maps were counted later by Mullin [144] who realized that there is a simple correspondence between these two models (see also Theorem 5.3.4). Other models like bipolar orientations [17, 48, 106], realizers [47, 33, 106], or Schnyder decompositions [37] share with spanning trees and Hamiltonian cycles the property that they enjoy quite explicit formulas and bijective proofs, even though their generating series are non algebraic. We include in this list the bijection of [30] although it deals **a priori** with the family of cubic planar map without decoration but the proof here uses the fact that each such a map admits $2^n$ spanning trees and then deals with tree-rooted maps. We believe that all these results await for a unifying **master bijection** mapping rooted planar maps with an $\alpha$-orientation onto some bidimensional walks, maybe extending the master bijection between rooted planar maps endowed with a **minimal** $\alpha$-orientation onto decorated trees as discussed in Section 5.5.2.

**Trees and Tutte's equations for decorated maps.** Another family of results is that of dimer models, hard particles, and Ising models on planar maps that admit bijective proofs based on the blossoming or well-labeled approach [61, 57, 64], and enjoy algebraic generating series. Most of these results were first obtained via the matrix model formalism, and several have been rederived via Tutte root-deletion method before getting bijective proof with trees.

The enumeration of planar maps weighted by the Tutte polynomial (also known as the Potts model on random lattices) plays a particular role in this list, as it contains many of the above previously solved models: As first shown by Tutte for the special case of well $q$-colored triangulations in a long series of 10 papers spanning 20 years of research ([164] and ref. therein), this model is still amenable to the root-deletion approach but the equation is amazingly difficult to solve. The resulting gf is non-algebraic but it satisfies an explicit differential algebraic equation of order three. Tutte's solution remained as an isolated artifact for almost 25 years until Bernardi and Bousquet-Mélou were able to distillate and extend his approach to the general Tutte polynomial on triangulations and to deal with non-trivial variants of the problem on rooted planar maps (see [34, 53] and reference therein). As opposed to the now well-understood polynomial equations with one catalytic variable that is associated to standard Tutte equation (see Section 5.4.1), much is still to be understood on the class of polynomial equations with two catalytic variables that are involved in these decorated models.

**Exact results from substitution schemes.** Other models have been dealt with by variants and generalizations of Tutte's substitution schemes that we discussed in Section 5.4.4. Using a refinement of this approach Sundberg and Thistlewaite obtained the gf of diagrams of open prime alternating links [159] and the asymptotic number of prime alternating links then follows from an asymmetry analysis [124]. Zinn-Justin and Zuber [173] extended the approach to 2-colored links, in an attempt to attach the problem of enumerating prime alternating knots (see also [157]). Tutte's substitution scheme allows us also to deal with the Ising model on non-separable maps [34] or with orientations on maps with girth or connectivity constraints [101]. In all those examples, the main idea is that the decomposition follows the structures of the underlying non-decorated map, and the interactions between the decoration and the decomposition is local and relatively well controlled.

In the case of the so-called $O(n)$ of colored loops on maps, Borot, Bouttier, and Guitter [51, 50, 49] apply instead the idea of decomposing along the maximal loops of the model. By a careful analysis of the possible interface between the core (here renamed **gasket**) and the submaps along these maximal loops, they obtain a functional equation that they are not able to solve exactly in general but which is sufficient to derive remarkably precise non-trivial asymptotic results.

Finally let us quote the recent work of Bousquet-Mélou and Courtiel [54], which starts this time from a direct, non-recursive, substitution scheme: Rooted planar map with a spanning forest can indeed be obtained from non-decorated maps upon inserting spanning trees in vertices with proper weights, as already observed in [64]. This approach yields almost directly an equation, but its study is quite difficult and leads to remarkable developments involving differentially algebraic gf and to non standard asymptotic expansions for a particular interpretation of the model in terms of forested maps weighted by the external activity of the spanning forest.

**Decorated maps in physics and probability.** To conclude, let us give a beotian's motivation for the study of decorated map in physics. From the physics point of view, we are considering the **annealed** partition function of a toy model of statistical physics coupled to a random lattice. Toy models of statistical physics are usually defined on a fixed regular lattice (typically the triangular, square or hexagonal lattice) and are viewed as microscopic models of matter. The idea to couple such a toy model with an irregular lattice arises from quantum geometry: The regular **classical euclidean geometry** is to be replaced by a **quantum geometry**, that is a distribution of probability on a set of irregular geometries: In the pair (map, decoration), the map plays the role of the random lattice and it is decorated by the configuration of the model. Accordingly, non-decorated maps are considered as a model of **pure 2d quantum geometry**, while decorated maps correspond to **2d quantum geometry with matter**. We refer to the survey [92] and the book [8] for an introduction to the motivations for the study of these toy models on random lattices in physics.

The physics community has developed a remarkable expertise in classifying the toy models into universality classes, and in predicting which universality classes and critical exponents are possible. From this point of view the pattern and universality statements of Section 5.5.1 are just the tip of the iceberg, corresponding as already mentioned to the simplest case of pure 2d quantum geometry without coupling to matter. The more general situation of quantum geometry with matter is quite comparable to that of the study of statistical physics conformally invariant toy models on regular 2d lattices, and in fact a tight relation exists between these two areas, first discovered by Knitchik, Polyakov and Zamolochikov and now refered to as the KPZ relation. Again the interested reader is directed to [92], and to the work of Duplantier and Sheffield [93] and Miller and Sheffield [142] for recent developments on these interactions, see [113] for a survey.

# References

[1] C. Abraham. Rescaled bipartite planar maps converge to the brownian map. arXiv:1312.5959, 2013.

[2] L. Addario-Berry and M. Albenque. The scaling limit of random simple triangulations and random simple quadrangulations. arXiv:1306.5227, 2013.

[3] M. Albenque and J. Bouttier. Constellations and multicontinued fractions: application to Eulerian triangulations. In *24th International Conference on Formal Power Series and Algebraic Combinatorics (FPSAC 2012)*, Discrete Math. Theor. Comput. Sci. Proc., AR, pages 805–816. Assoc. Discrete Math. Theor. Comput. Sci., Nancy, 2012.

[4] M. Albenque, G. Collet, and E. Fusy. Simple planar maps (also) converge to the brownian planar map. Talk in Paris, July 2014.

[5] M. Albenque and J.-F. Marckert. Some families of increasing planar maps. *Electronic journal of probability*, 13(56):1624–1671, 2008.

[6] M. Albenque and D. Poulalhon. Generic method for bijections between blossoming trees and planar maps. arXiv:1305.1312, 2013.

[7] J. Ambjørn and T. G. Budd. Trees and spatial topology change in causal dynamical triangulations. *J. Phys. A: Math. Theor.*, 46(31):315201, 2013.

[8] J. Ambjørn, B. Durhuus, and T. Jonsson. *Quantum geometry*. Cambridge Monographs on Mathematical Physics. Cambridge University Press, Cambridge, 1997. A statistical field theory approach.

[9] J. Ambjørn and Y. Watabiki. Scaling in quantum gravity. *Nuclear Phys. B*, 445(1):129–142, 1995.

[10] O. Angel. Growth and percolation on the infinite planar triangulation. *Geom. Funct. Anal.*, 13(5):935–974, 2003.

[11] O. Angel and O. Schramm. Uniform infinite planar triangulations. *Comm. Math. Phys.*, 241(2-3):191–213, 2003.

[12] D. Arquès. Les hypercartes planaires sont des arbres très bien étiquetés. *Discrete Math.*, 58:11–24, 1986.

[13] D. Arquès. Relations fonctionnelles et dénombrement des cartes pointées sur le tore. *Journal of Combinatorial Theory, Series B*, 43(3):253 – 274, 1987.

[14] D. Arquès and A. Giorgetti. Counting rooted maps on a surface. *Theoret. Comput. Sci.*, 234:255–272, 2000.

[15] R. Askey. *Orthogonal polynomials and special functions*, volume 21 of *Regional Conference Series in Applied Mathematics*. SIAM, Philadelphia, 1975.

[16] C. Banderier, P. Flajolet, G. Schaeffer, and M. Soria. Random maps, coalescing saddles, singularity analysis, and Airy phenomena. *Random Structures Algorithms*, 19(3/4):194–246, 2001.

[17] R. J. Baxter. Dichromatic polynomials and Potts models summed over rooted maps. *Ann. Comb.*, 5:17, 2001.

[18] E. A. Bender. The number of three-dimensional convex polyhedra. *Amer. Math. Monthly*, 94(1):7–21, 1987.

[19] E. A. Bender and E. R. Canfield. The asymptotic number of rooted maps on a surface. *J. Combin. Theory Ser. A*, 43:244–257, 1986.

[20] E. A. Bender and E. R. Canfield. The number of rooted maps on a surface. *J. Combin. Theory Ser. B*, 53:293–299, 1991.

[21] E. A. Bender and E. R. Canfield. The number of degree-restricted rooted maps on the sphere. *SIAM J. Discrete Math.*, 7(1):9–15, 1994.

[22] E. A. Bender, E. R. Canfield, and L. B. Richmond. The asymptotic number of rooted maps on surfaces. ii. enumeration by vertices and faces. *J. Combin. Theory Ser. A*, 63(2):318–329, 1993.

[23] E. A. Bender, K. J. Compton, and L. B. Richmond. 0 1 laws for maps. *Random Structures & Algorithms*, 14(3):215–237, 1999.

[24] E. A. Bender, Z. Gao, and L. B. Richmond. Almost all rooted maps have large representativity. *J. Graph Theory*, 18(6):545–555, October 1994.

[25] E. A. Bender, Z. Gao, and L. B. Richmond. The map asymptotics constant $t_g$. *Electron. J. Combin.*, 15(1):Research paper 51, 8, 2008.

[26] E. A. Bender and L. B. Richmond. The asymptotic enumeration of rooted convex polyhedra. *J. Combin. Theory Ser. B*, 36(3):276–283, 1984.

[27] E. A. Bender and N. C. Wormald. The number of rooted convex polyhedra. *Canad. Math. Bull.*, 31(1):99–102, 1988.

[28] O. Bernardi. Bijective counting of tree-rooted maps and shuffles of parenthesis systems. *Electron. J. Combin.*, 14(9), 2006.

[29] O. Bernardi. *Combinatoire des cartes et polynôme de Tutte*. PhD thesis, Université de Bordeaux I, 2006.

[30] O. Bernardi. Bijective counting of Kreweras walks and loopless triangulations. *Journal of Combinatorial Theory, Series A*, 114(5):931–956, 2007.

[31] O. Bernardi. On triangulations with high vertex degree. *Ann. Comb.*, 12(1):17–44, 2008.

[32] O. Bernardi. An analogue of the Harer-Zagier formula for unicellular maps on general surfaces. *Adv. in Appl. Math.*, 48(1):164–180, 2012.

[33] O. Bernardi and N. Bonichon. Intervals in catalan lattices and realizers of triangulations. *J. Combin. Theory Ser. A*, 116(1):55–75, 2009.

[34] O. Bernardi and M. Bousquet-Mélou. Counting colored planar maps: Algebraicity results. *J. Comb. Theory Ser. B*, 101(5):315–377, 2011.

[35] O. Bernardi and G. Chapuy. A bijection for covered maps, or a shortcut between harer-zagier's and jackson's formulas. *J. Comb. Theory Ser. A*, 118(6):1718–1748, August 2011.

[36] O. Bernardi and É. Fusy. A bijection for triangulations, quadrangulations, pentagulations, etc. *J. Comb. Theory, Ser. A*, 119(1):218–244, 2012.

[37] O. Bernardi and É. Fusy. Schnyder decompositions for regular plane graphs and application to drawing. *Algorithmica*, 62(3-4):1159–1197, 2012.

[38] O. Bernardi and É. Fusy. Unified bijections for maps with prescribed degrees and girth. *J. Combin. Theory Ser. A*, 119(6):1352–1387, 2012.

[39] O. Bernardi and É. Fusy. Unified bijections for planar hypermaps with general cycle-length constraints. arXiv:1403.5371, 2014.

[40] O. Bernardi and A. H. Morales. Bijections and symmetries for the factorizations of the long cycle. *Adv. in Appl. Math.*, 50(5):702–722, 2013.

[41] D. Bessis, C. Itzykson, and J.-B Zuber. Quantum field theory techniques in graphical enumeration. *Advances in Applied Mathematics*, 1(2):109 – 157, 1980.

[42] J. Bettinelli. The topology of scaling limits of positive genus random quadrangulations. *Ann. Probab.*, 40(5):1897–1944, 2012.

[43] J. Bettinelli. Increasing forests and quadrangulations via a bijective approach. *J. Combin. Theory Ser. A*, 122(0):107–125, 2014.

[44] J. Bettinelli, E. Jacob, and G. Miermont. The scaling limit of uniform random plane maps, via the Ambjørn Budd bijection. arXiv:1312.5842, 2013.

[45] M. Bodirsky, C. Gröpl, D. Johannsen, and M. Kang. A direct decomposition of 3-connected planar graphs. *Séminaire Lotharingien de Combinatoire*, 54A(B54Ak):15, 2007.

[46] M. Bóna. Exact enumeration of 1342-avoiding permutations; a close link with labeled trees and planar maps. *J. Combin. Theory Ser. A*, 80:257–272, 1997.

[47] N. Bonichon. A bijection between realizers of maximal plane graphs and pairs of non-crossing dyck paths. *Discrete Math.*, 298(1-3):104–114, 2005.

[48] N. Bonichon, M. Bousquet-Mélou, and É. Fusy. Baxter permutations and plane bipolar orientations. arXiv:0805.4180, 2008.

[49] G. Borot, J. Bouttier, and E. Guitter. Loop models on random maps via nested loops: the case of domain symmetry breaking and application to the Potts model. *J. Phys. A*, 45(49):494017, 35, 2012.

[50] G. Borot, J. Bouttier, and E. Guitter. More on the $O(n)$ model on random maps via nested loops: loops with bending energy. *J. Phys. A*, 45(27):275206, 32, 2012.

[51] G. Borot, J. Bouttier, and E. Guitter. A recursive approach to the $O(n)$ model on random maps via nested loops. *J. Phys. A*, 45(4):045002, 38, 2012.

[52] M. Bousquet-Mélou. Rational and algebraic series in combinatorial enumeration. In *Proceedings of the Internatinal Congress of Mathematicians*, 2006.

[53] M. Bousquet-Mélou. Counting planar maps, coloured or uncoloured. In Robin Chapman, editor, *Surveys in Combinatorics 2011*, pages 1–50. Cambridge University Press, 2011. Cambridge Books Online.

[54] M. Bousquet-Mélou and J. Courtiel. Spanning forests in regular planar maps. *DMTCS Proceedings*, 0(01), 2013.

[55] M. Bousquet-Mélou and A. Jehanne. Polynomial equations with one catalytic variable, algebraic series, and map enumeration. *J. Combin. Theory Ser. B*, to appear, 2005.

[56] M. Bousquet-Mélou and G. Schaeffer. Enumeration of planar constellations. *Adv. in Appl. Math.*, 24(337-368), 2000.

[57] M. Bousquet-Mélou and G. Schaeffer. The degree distribution in bipartite planar maps: applications to the ising model, 2002. 32pp.

[58] J. Bouttier. Matrix integrals and the enumeration of maps. In J. Baik G. Akemann and P. Di Francesco, editors, *The Oxford Handbook of Random Matrix Theory*. Oxford university press, 2013.

[59] J. Bouttier, P. Di Francesco, and E. Guitter. Census of planar maps: from the one-matrix model solution to a combinatorial proof. *Nuclear Phys.*, B 645:477–499, 2002.

[60] J. Bouttier, P. Di Francesco, and E. Guitter. Critical and tricritical hard objects on bicolorable random lattices: exact solutions. *J. Phys. A*, 35:3821–3854, 2002.

[61] J. Bouttier, P. Di Francesco, and E. Guitter. Combinatorics of hard particles on planar graphs. *Nuclear Phys. B*, 655(3):313–341, 2003.

[62] J. Bouttier, P. Di Francesco, and E. Guitter. Geodesic distance in planar graphs. *Nucl. Phys.*, B663:535–567, 2003.

[63] J. Bouttier, P. Di Francesco, and E. Guitter. Statistics of planar graphs viewed from a vertex: a study via labeled trees. *Nuclear Phys. B*, 675(3):631–660, 2003.

[64] J. Bouttier, P. Di Francesco, and E. Guitter. Blocked edges on Eulerian maps and mobiles: Application to spanning trees, hard particles and the Ising model. *J. Phys. A*, 40:7411–7440, 2007.

[65] J. Bouttier, É. Fusy, and E. Guitter. On the two-point function of general planar maps and hypermaps. arXiv:1312.0502, 2013.

[66] J. Bouttier and E. Guitter. The three-point function of planar quadrangulations. *J. Stat. Mech. Theory Exp.*, 39(7):P07020, 2008.

[67] J. Bouttier and E. Guitter. Confluence of geodesic paths and separating loops in large planar quadrangulations. *Journal of Statistical Mechanics: Theory and Experiment*, 2009(03):P03001, 2009.

[68] J. Bouttier and E. Guitter. Distance statistics in quadrangulations with a boundary, or with a self-avoiding loop. *Journal of Physics A: Mathematical and Theoretical*, 42(46):465208, 2009.

[69] J. Bouttier and E. Guitter. Distance statistics in quadrangulations with no multiple edges and the geometry of minbus. *J. Phys. A*, A 43:313–341, 2010.

[70] J. Bouttier and E. Guitter. Planar maps and continued fractions. *Comm. Math. Phys.*, 309(3):623–662, 2012.

[71] J. Bouttier and E. Guitter. A note on irreducible maps with several boundaries. arXiv:1305.4816, 2013.

[72] J. Bouttier and E. Guitter. On irreducible maps and slices. arXiv:1303.3728, 2013.

[73] E. Brézin, C. Itzykson, G. Parisi, and J. B. Zuber. Planar diagrams. *Communications in Mathematical Physics*, 59(1):35–51, 1978.

[74] W. G. Brown. Enumeration of triangulations of the disk. *Proc. Lond. Math. Soc.*, pages 746–768, 1964.

[75] W. G. Brown. Enumeration of quadrangular dissections of the disk. *Canad. J. Math.*, 17:302–317, 1965.

[76] A. Caraceni. The scaling limit of random outerplanar maps. arXiv:1405.1971, 2014.

[77] S. R. Carrell. The non-orientable map asymptotics constant $p_g$. arXiv:1406.1760, 2014.

[78] S. R. Carrell and G. Chapuy. Simple recurrence formulas to count maps on orientable surfaces. arXiv:1402.6300, 2014.

[79] G. Chapuy. Asymptotic enumeration of constellations and related families of maps on orientable surfaces. *Combinatorics, Probability, and Computing*, 18(4):477–516, 2009.

[80] G. Chapuy. The structure of unicellular maps, and a connection between maps of positive genus and planar labelled trees. *Probability Theory and Related Fields*, 147(3-4):415–447, 2010.

[81] G. Chapuy. A new combinatorial identity for unicellular maps, via a direct bijective approach. *Advances in Applied Mathematics*, 47(4):874–893, 2011.

[82] G. Chapuy, V. Féray, and É. Fusy. A simple model of trees for unicellular maps. *Journal of Combinatorial Theory, Series A*, 120(8):2064 – 2092, 2013.

[83] G. Chapuy, M. Marcus, and G. Schaeffer. A bijection for rooted maps on orientable surfaces. *SIAM J. Discrete Math.*, 23(3):1587–1611, 2009.

[84] P. Chassaing and B. Durhuus. Local limit of labeled trees and expected volume growth in a random quadrangulation. *The Annals of Probability*, 34(3):879–917, 05 2006.

[85] P. Chassaing and G. Schaeffer. Random planar lattices and integrated super-Brownian excursion. *Probab. Theory Related Fields*, 128(2):161–212, 2004.

[86] R. Cori. Un code pour les graphes planaires et ses applications. *Astérisque*, 27:1–169, 1975.

[87] R. Cori. Planarité et algébricité. *Astérisque*, 38-39:33–44, 1976.

[88] R. Cori and A. Machi. Maps, hypermaps and their automorphisms: a survey. *Expo. Math.*, 10:403–467, 1992.

[89] R. Cori and J. Richard. Énumération des graphes planaires à l'aide des séries formelles en variables non commutatives. *Discrete Mathematics*, 2(2):115–162, 1972.

[90] R. Cori and B. Vauquelin. Planar maps are well labeled trees. *Canad. J. Math.*, 33(5):1023–1042, 1981.

[91] A. Del Lungo, F. Del Ristoro, and J.-G. Penaud. Left ternary trees and non-separable rooted planar maps. *Theoretical computer science*, 233(1–2):201–215, 2000.

[92] P. Di Francesco, P. Ginsparg, and J. Zinn-Justin. 2d gravity and random matrices. *Physics Reports*, 254(12):1 – 133, 1995.

[93] B. Duplantier and S. Sheffield. Liouville quantum gravity and KPZ. *Invent. Math.*, 185(2):333–393, 2011.

[94] N. M. Ercolani and K. D. T.-R. McLaughlin. Asymptotics of the partition function for random matrices via Riemann-Hilbert techniques and applications to graphical enumeration. *Int. Math. Res. Not.*, 14:755–820, 2003.

[95] B. Eynard. Topological expansion for the 1-hermitian matrix model correlation functions. *Journal of High Energy Physics*, 2004(11):031, 2004.

[96] B. Eynard. Formal matrix integrals and combinatorics of maps. In J. Harnad, editor, *Random Matrices, Random Processes and Integrable Systems*, CRM Series in Mathematical Physics, pages 415–442. Springer, 2011.

[97] B. Eynard. *Counting surfaces: Combinatorics, Matrix Models and Algebraic Geometry*. Springer, to appear.

[98] B. Eynard and E. Guitter. Distances sur les mobiles et systèmes intégrables. Talk at ENS, April 2014.

[99] B. Eynard and N. Orantin. Topological expansion of the 2-matrix model correlation functions: diagrammatic rules for a residue formula. *Journal of High Energy Physics*, 2005(12):034, 2005.

[100] S. Felsner. Lattice structures for planar graphs. *Electron. J. Combin.*, 11(1):Research paper R15, 24p, 2004.

[101] S. Felsner, É. Fusy, and M. Noy. Asymptotic enumeration of orientations. *Discrete Math. Theor. Comput. Sci.*, 12(2):249–262, 2010.

[102] P. Flajolet and R. Sedgewick. *Analytic combinatorics*. Cambridge University Press, 2009.

[103] É. Fusy. Straight-line drawing of quadrangulations. In *Proceedings of Graph Drawing'06*, volume 4372 of *LNCS*, pages 234–239. Springer, 2006.

[104] É. Fusy. Transversal structures on triangulations: A combinatorial study and straight-line drawings. *Discrete Math.*, 309:1870–1894, 2009.

[105] É. Fusy, D. Poulalhon, and G. Schaeffer. Dissections, orientations, and trees, with applications to optimal mesh encoding and to random sampling. *Transactions on Algorithms*, 4(2):Art. 19, April 2008.

[106] É. Fusy, D. Poulalhon, and G. Schaeffer. Bijective counting of plane bipolar orientations and Schnyder woods. *European J. Combin.*, 30(7):1646–1658, 2009.

[107] Z. Gao. A pattern for the asymptotic number of rooted maps on surfaces. *Journal of Combinatorial Theory, Series A*, 64(2):246–264, November 1993.

[108] Z. Gao and L. B. Richmond. Root vertex valency distributions of rooted maps and rooted triangulations. *European Journal of Combinatorics*, 15(5):483–490, 1994.

[109] Z. Gao, I. M. Wanless, and N. C. Wormald. Counting 5-connected planar triangulations. *Journal of Graph Theory*, 38(1):18–35, 2001.

[110] Z. Gao and N. C. Wormald. The size of the largest components in random planar maps. *SIAM J. Discrete Math.*, 12(2):217–228, 1999.

[111] Z. Gao and N. C. Wormald. The distribution of the maximum vertex degree in random planar maps. *Journal of Combinatorial Theory, Series A*, 89(2):201–230, 2000.

[112] Z. Gao and N. C. Wormald. Asymptotic normality determined by high moments, and submap counts of random maps. *Probability Theory and Related Fields*, 130(3):368–376, 2004.

[113] C. Garban. Quantum gravity and the kpz formula. In *Bourbaki seminar*, volume 1052. Institut Henri Poincarr, 2012.

[114] I. P. Goulden and D. M. Jackson. The kp hierarchy, branched covers, and triangulations. *Advances in Mathematics*, 219:932–951, 2008.

[115] I. P. Goulden and A. Nica. A direct bijection for the Harer-Zagier formula. *J. Comb. Theory, Ser. A*, 111(2):224–238, 2005.

[116] A. Goupil and G. Schaeffer. Factoring n-cycles and counting maps of given genus. *European J. Combin.*, 19(7):819–834, 1998.

[117] O. Guibert. Stack words, standard young tableaux, permutation with forbidden subsequences, and planar maps. *Discrete Mathematics*, 210(71-85), 2000.

[118] A. Guionnet. Random matrices and enumeration of maps. In *International Congress of Mathematicians. Vol. III*, pages 623–636. Eur. Math. Soc., Zürich, 2006.

[119] E. Guitter. Distance statistics in large toroidal maps. *Journal of Statistical Mechanics: Theory and Experiment*, 2010(04):P04018, 2010.

[120] E. Guitter and E. Fusy. The three-point function of general planar maps. arXiv:1403.3514, 2014.

[121] R. Gurau. The 1/N Expansion of Tensor Models Beyond Perturbation Theory. *Comm. Math. Phys.*, 330(3):973–1019, 2014.

[122] J. Harer and D. Zagier. The Euler characteristic of the moduli space of curves. *Invent. Math.*, 85:457–486, 1986.

[123] D. M. Jackson. Some combinatorial problems associated with products of conjugacy classes of the symmetric group. *J. Comb. Theory, Ser. A*, 49(2):363–369, 1988.

[124] S. Kunz-Jacques and G. Schaeffer. The asymptotic number of prime alternating links. In H. Barcelo, editor, *FPSAC 2001*, 2001.

[125] S. K. Lando and A. K. Zvonkin. *Graphs on Surfaces and Their Applications*. Springer, 2004.

[126] B. Lass. Démonstration combinatoire de la formule de Harer-Zagier. *C. R. Acad. Sci. Paris*, 333, Série I:155–160, 2001.

[127] J.-F. Le Gall. Large random planar maps and their scaling limits. Address to the European Congress of Mathematics.

[128] J.-F. Le Gall. Random geometry on the sphere. Address to the International Congress of Mathematics.

[129] J.-F. Le Gall. Uniqueness and universality of the brownian map. *Ann. Probab.*, 41:2880–2960, 2013.

[130] J.-F. Le Gall and G. Miermont. Scaling Limits of Random Trees and Planar Maps. *Proceedings, Clay Mathematics*, pages 1–57, 2012.

[131] J.-F. Le Gall and F. Paulin. Scaling limits of bipartite planar maps are homeomorphic to the 2-sphere. *Geom. Funct. Anal.*, 18:893–918, 2008.

[132] V. A. Liskovets. A census of non-isomorphic planar maps. In *Colloq. Math. Soc. János Bolyai. 25th Algebraic Methods in Graph Theory, Vol. I, II, Szeged (Hungary)*, pages 479–484, 1978.

[133] V. A. Liskovets. A reductive technique for enumerating non-isomorphic planar maps. *Discrete Math.*, 156:197–217, 1996.

[134] V. A. Liskovets. A pattern of asymptotic vertex valency distributions in planar maps. *J. Combin. Theory Ser. B*, 75(1):116–133, 1999.

[135] V. A. Liskovets. Enumerative formulae for unrooted planar maps: a pattern. *Electron. J. Combin.*, 11(1):Research Paper 88, 14, 2004.

[136] V. A. Liskovets and T. R. Walsh. Ten steps to counting planar graphs. *Congr. Numer.*, 60:269–277, 1987. Eighteenth Southeastern International Conference on Combinatorics, Graph Theory, and Computing (Boca Raton, Fla., 1987).

[137] V. A. Liskovets and T. R. Walsh. Counting unrooted maps on the plane. *Adv. in Appl. Math.*, 36(4):364–387, 2006.

[138] G. Miermont. On the sphericity of scaling limits of random planar quadrangulations. *Electronic Communications in Probability*, 13, 2008.

[139] G. Miermont. Tessellations of random maps of arbitrary genus. *Ann. Sci. Éc. Norm. Supér*, 42(fascicule 5): 725–781, 2009.

[140] G. Miermont. The Brownian map is the scaling limit of uniform random plane quadrangulations. *Acta Math.*, 210:319–401, 2013.

[141] G. Miermont. Aspects of random maps. Cours à l'école d'été de Saint Flour, 2014.

[142] J. Miller and S. Sheffield. Quantum Loewner evolution. arXiv:1312.5745, 2013.

[143] B. Mohar and C. Thomassen. *Graphs on Surfaces*. Johns Hopkins University Press, 2001.

[144] R. C. Mullin. On the enumeration of tree-rooted maps. *Canad. J. Math.*, 19:174–183, 1967.

[145] R. C. Mullin and P. J. Schellenberg. The enumeration of c-nets via quadrangulations. *J. Combin. Theory*, 4:259–276, 1968.

[146] D. Poulalhon and G. Schaeffer. Factorisations of a $n$-cycle into $m$ permutations. *Discrete Math.*, 254(1-3):433–458, 2002.

[147] D. Poulalhon and G. Schaeffer. A bijection for triangulations of a polygon with interior points and multiple edges. *Theoret. Comput. Sci.*, 307(2):385–401, 2003.

[148] D. Poulalhon and G. Schaeffer. Optimal coding and sampling of triangulations. *Algorithmica*, 46(3-4):505–527, 2006.

[149] J.-L. Rémy. Un procédé itératif de dénombrement d'arbres binaires et son application à leur génération aléatoire. *RAIRO Inform. Théor.*, 19(2):179–195, 1985.

[150] L. B. Richmond and N. C. Wormald. Almost all maps are asymmetric. *Journal of Combinatorial Theory, Series B*, 63(1):1–7, 1995.

[151] V. Rivasseau. Constructive matrix theory. *J. High Energy Phys.*, 9:008, 13, 2007.

[152] R. W. Robinson. Enumeration of non-separable graphs. *J. Combin. Theory*, 9:327–356, 1970.

[153] G. Schaeffer. Bijective census and random generation of Eulerian planar maps with prescribed vertex degrees. *Electron. J. Combin.*, 4(1):# 20, 14 pp., 1997.

[154] G. Schaeffer. *Conjugaison d'arbres et cartes combinatoires aléatoires*. PhD thesis, Université Bordeaux I, 1998.

[155] G. Schaeffer. Combinatoire énumérative. Habilitation à diriger des recherches, Université Bordeaux 1, 2005.

[156] G. Schaeffer and E. A. Vassilieva. A bijective proof of Jackson's formula for the number of factorizations of a cycle. *J. Comb. Theory, Ser. A*, 115(6):903–924, 2008.

[157] G. Schaeffer and P. Zinn-Justin. On the asymptotic number of plane curves and alternating knots. *Experiment. Math.*, 13(4):483–494, 2003.

[158] W. Schnyder. Planar graphs and poset dimension. *Order*, 5:323–343, 1989.

[159] C. Sundberg and M. Thistlethwaite. The rate of growth of the number of prime alternating links and tangles. *Pacific J. Math.*, 182(2):329–358, 1998.

[160] W. T. Tutte. A census of hamiltonian polygons. *Canad. J. Math.*, 14:402–417, 1962.

[161] W. T. Tutte. A census of slicings. *Canad. J. Math.*, 14:708–722, 1962.

[162] W. T. Tutte. A census of planar maps. *Canad. J. Math.*, 15:249–271, 1963.

[163] W. T. Tutte. On the enumeration of planar maps. *Bulletin of the American Mathematical Society*, 74(1):64–74, 1968.

[164] W. T. Tutte. Chromatic sums revisited. *Aequationes Math.*, 50(1–2):95–134, 1995.

[165] X. Viennot. Promenade dans le jardin de catalan. Large audience talk., 2000-2014.

[166] T. R. S. Walsh. Efficient enumeration of sensed maps. *Discrete Math.*, 293(1-3):263–289, 2005.

[167] T. R. S. Walsh, A. Giorgetti, and A. Mednykh. Enumeration of unrooted orientable maps of arbitrary genus by number of edges and vertices. *Discrete Mathematics*, 312(17):2660–2671, 2012. Proceedings of the 8th French Combinatorial Conference.

[168] T. R. S. Walsh and A. B. Lehman. Counting rooted maps by genus. I. *J. Combin. Theory Ser. B*, 13:192–218, 1972.

[169] T. R. S. Walsh and A. B. Lehman. Counting rooted maps by genus. II. *J. Combin. Theory Ser. B*, 13:122–141, 1972.

[170] T. R. S. Walsh and A. B. Lehman. Counting rooted maps by genus. III: Nonseparable maps. *J. Combin. Theory Ser. B*, 18:222–259, 1975.

[171] N. C. Wormald. Counting unrooted planar maps. *Discrete Math.*, 36:205–225, 1981.

[172] N. C. Wormald. On the number of planar maps. *Canad. J. Math.*, 33(1):1–11, 1981.

[173] P. Zinn-Justin and J.-B. Zuber. Knot theory and matrix integrals. In *The Oxford handbook of random matrix theory*, pages 557–577. Oxford Univ. Press, Oxford, 2011.

# Chapter 6

# *Graphs*

**Marc Noy**
*Universitat Politècnica de Catalunya*
*Barcelona, Spain*

**CONTENTS**

## 6.1 Introduction

Many references to date on the enumeration of graphs deal with unlabeled graphs, like the monographs of Harary and Palmer [32] and Pólya and Read [50]. The main tool there is the cycle index polynomial and the goal is to obtain exact expressions for the number of graphs of a given kind, like trees or connected graphs. Our emphasis in this survey is instead on labeled graphs. One reason is that in many interesting cases the number of unlabeled graphs with $n$ vertices is asymptotically the number of labeled graphs divided by $n!$, the number of possible labelings. This is not always so, as in the case of trees, since a random tree has almost surely exponentially many automorphisms.

A **class** of graphs is a set of labeled graphs closed under isomorphism. Typical examples of interest are bipartite, acyclic, triangle-free or planar graphs. We use $n$ to denote the number of vertices in a graph. Let $\mathcal{G}$ be a class of graphs and $\mathcal{G}_n$ the graphs in $\mathcal{G}$ with $n$ vertices. The main problem we consider is to compute or to estimate $|\mathcal{G}_n|$ when $n \to \infty$. In very few cases we have exact formulas. Here are some examples:

- The number of labeled graphs is $2^{\binom{n}{2}}$. This is because each of the $\binom{n}{2}$ edges of the complete graph can be chosen independently to be or not be in a graph. Likewise, the number of graphs with $n$ vertices and $m$ edges is equal to $\binom{\binom{n}{2}}{m}$.
- The number of labeled **even** graphs (all vertices have even degree) is $2^{\binom{n-1}{2}}$. There is a very simple proof of this fact: Given a graph $G$ on $n-1$ vertices, add a new vertex labeled $n$ and connect it to all vertices in $G$ of odd degree. This gives an even graph on $n$ vertices and it is easy to check that this is in fact a bijection.
- The number of labeled **trees** is $n^{n-2}$. This is the well-known Cayley's formula, which admits many different proofs. There is no simple formula for the number of forests (acyclic graphs). However, the number of labeled forests consisting of $k$ rooted trees is $\binom{n}{k}kn^{n-k-1}$.

The main emphasis in this chapter is on asymptotic results. In many cases we use generating functions, which are of the exponential type since graphs are labeled. We start with a basic example. Let $G_n = 2^{\binom{n}{2}}$ be the number of graphs on $n$ vertices, with the convention that $G_0 = 1$. In order to compute the number $C_n$ of connected graphs, we introduce the generating functions

$$G(x) = \sum_{n\geq 0} G_n \frac{x^n}{n!}, \qquad C(x) = \sum_{n\geq 0} C_n \frac{x^n}{n!}.$$

The exponential formula $G(x) = e^{C(x)}$ implies that

$$C(x) = \log\left(1 + \sum_{n\geq 1} 2^{\binom{n}{2}} \frac{x^n}{n!}\right).$$

The sequence of coefficients is sequence A001187 in The Online Encyclopedia of Integer Sequences [64]. In the rest of this Chapter, the sequences appearing in the Online Encyclopedia of Integer Sequences are referred by their code in that database.

Using the expansion

$$\log(1+t) = t - \frac{t^2}{2} + \frac{t^3}{3} - \cdots$$

and extracting coefficients, it is easy to see that $C_n \sim G_n$. In other words, almost all graphs are connected. But this can be proved more easily using the theory of random graphs, as discussed next.

Consider the random graph binomial model $G(n,p)$ with $n$ vertices, in which every edge is drawn independently with probability $p$. The probability of a graph with $m$ edges is

$$p^m(1-p)^{\binom{n}{2}-m}.$$

If $p = 1/2$, then every graph has the same probability $2^{-\binom{n}{2}}$, and the distribution is uniform among all graphs with $n$ vertices. We say that almost all graphs in $G(n,p)$ satisfy property $\mathscr{A}$ if the probability that a graph with $n$ vertices satisfies $\mathscr{A}$ tends to 1, as $n$ tends to $\infty$. It is easy to prove (see [10]) that the following properties hold for almost all graphs in $G(n,1/2)$, hence they hold for almost all graphs under the uniform distribution. Let $R_n$ be a random labeled graph with $n$ vertices. Then almost surely, as $n \to \infty$, $R_n$ satisfies:

- $R_n$ has diameter two. In particular, it is connected.
- $R_n$ is $k$-connected for every fixed $k \geq 1$.
- For every fixed graph $H$, $R_n$ contains an induced subgraph isomorphic to $H$.
- $R_n$ is Hamiltonian. In particular, it has a perfect matching if the number of vertices is even.
- $R_n$ has no non-trivial automorphisms.

In the other direction, almost no graph is regular, or planar (since it contains $K_5$ as a subgraph), or $r$-colorable for fixed $r$ (since it contains $K_{r+1}$ as subgraph). One can prove much deeper results using the theory of random graphs. For instance, the chromatic number of a random graph with $n$ vertices is close to $n/(2\log_2 n)$, and the independence number is close to $2\log_2 n$. We will not pursue this topic here, the interested reader can find many more results in the reference textbooks [10, 36] of this fascinating topic.

Throughout the paper we use the following conventions. Given a class of labeled graphs $\mathscr{G}$, let $G_n = |\mathscr{G}_n|$ be the number of graphs in $\mathscr{G}$ with $n$ vertices. The associated generating function is

$$G(x) = \sum_n G_n \frac{x^n}{n!}.$$

As a rule we use the same letter for the class, the sequence we want to enumerate and the corresponding generating function. We use the variable $x$ for marking vertices, and variable $y$ for edges. If $G(x,y)$ is a series in two variables, $G_x(x,y)$ and $G_y(x,y)$ denote the partial derivatives. We make systematic use of the symbolic method, as in [24]. For instance, if $\mathcal{G}$ is a class of graphs closed under the operation of taking connected components, the exponential formula

$$G(x) = e^{C(x)}$$

relates the generating function $G(x)$ of graphs in $\mathcal{G}$ with $C(x)$, the one for connected graphs in $\mathcal{G}$. As another example, $xG'(x) = \sum nG_n x^n/n!$ is the generating function of **rooted** graphs, since a labeled graph with $n$ vertices can be rooted in $n$ different ways.

The notation and terminology for graph theory is standard. Given a graph $G = (V,E)$, the subgraph induced by $U \subseteq V$ is the graph $(U,A)$, where $A$ is the set of edges joining vertices of $U$. A set $U$ of vertices is independent if there are no edges joining vertices of $G$. A graph is $k$-partite if the vertex set can be partitioned as $V = V_1 \cup \cdots \cup V_k$ such that each $V_i$ is independent. The chromatic number $\chi(G)$ is the minimum number $k$ of colors in a coloring of $V$ such that adjacent vertices receive different colors. A graph $G$ is $k$-partite if and only if $\chi(G) \leq k$. A graph is bipartite if and only if it has no odd cycles.

The rest of the paper is organized as follows. In Section 6.2 we present basic decompositions of graphs and the corresponding equations for the associated generating functions. This is applied in particular to arbitrary and bipartite graphs. In Section 6.3 we discuss the enumeration of connected graphs with given excess; this is a classical problem with strong connections to the theory of random graphs. Section 6.4 is devoted to regular graphs, another classical problem. In Section 6.5 we cover monotone and hereditary classes, defined in terms of forbidden graphs. This is an active area of research closely related to extremal graph theory. In Sections 6.6 and 6.7 we discuss the enumeration of planar graphs, graphs on surfaces, and classes of graphs defined in terms of excluded minors, which are more recent topics of research. In Section 6.8 we present briefly some results on the enumeration of digraphs. We conclude in Section 6.9 with a quick overview on the enumeration of unlabeled graphs.

## 6.2 Graph decompositions

The results in this section are based on the classical decompositions of graphs into $k$-connected components, and on the decomposition of a connected graph into the 2-core and the trees attached to it. As is well-known, a graph decomposes into its connected components, and a connected graph decomposes into its 2-connected components, also called blocks: A block is a maximal 2-connected subgraph or a separating edge. The decomposition into blocks has a tree structure. Furthermore, a 2-connected graph decomposes into its so-called 3-connected components: These are not necessarily subgraphs of the host graph and the decomposition is more difficult to describe (see for instance [15, 27] for a self-contained discussion). Briefly, one decomposes a 2-connected graph $G$ along its 2-separators in a canonical way into **bricks**, which are either cycles, multi-edges or 3-connected graphs. The decomposition is tree-like and can be encoded using generating functions as discussed next.

### 6.2.1 Graphs with given connectivity

Let $\mathcal{G}$ be a class of graphs with the following properties:

**(C1)** A graph $G$ is in $\mathcal{G}$ if and only if the connected components of $G$ are in $\mathcal{G}$.

**(C2)** A connected graph $G$ is in $\mathcal{G}$ if and only if the blocks of $G$ are in $\mathcal{G}$.

This holds clearly for the class of all graphs, but also for other classes such as planar graphs.

The following derivations can be found in [32]. We denote by $\mathcal{C}$ the *connected* graphs in $\mathcal{G}$, and by $\mathcal{B}$ the *2-connected* graphs in $\mathcal{G}$. The corresponding generating functions are denoted by

$$G(x) = \sum G_n \frac{x^n}{n!}, \qquad C(x) = \sum C_n \frac{x^n}{n!}, \qquad B(x) = \sum B_n \frac{x^n}{n!}.$$

The decomposition of a graph into connected components, and the decomposition of a connected graph into 2-connected components implies the equations

$$G(x) = e^{C(x)}, \tag{6.1}$$

and

$$C'(x) = e^{B'(xC'(x))}. \tag{6.2}$$

Equation (6.1) is the exponential formula for sets of labeled combinatorial objects, and (6.2) is based on the recursive decomposition of a rooted connected graph into its blocks. The generating function $\sum nC_n x^n/n!$ enumerates connected graphs rooted at a vertex, since a labeled graph can be rooted at $n$ different vertices.

It follows that

$$C(x) = \log G(x),$$

and $R(x) = xC'(x)$ satisfies the implicit equation

$$R(x) = xe^{B'(R(x))}.$$

We can rephrase the former equation by saying that $R(x)$ is the functional inverse of $xe^{-B'(x)}$.

If $\mathcal{G}$ is the class of all graphs then $G_n = 2^{\binom{n}{2}}$. The former equations determine $C(x)$ and $G(x)$ uniquely and we obtain

$$C(x) = x + \frac{x^2}{2!} + 4\frac{x^3}{3!} + 38\frac{x^4}{4!} + 728\frac{x^5}{5!} + 26704\frac{x^6}{6!} + \cdots,$$

which is sequence A001187. Using this expansion we also obtain

$$B(x) = \frac{x^2}{2!} + \frac{x^3}{3!} + 10\frac{x^4}{4!} + 238\frac{x^5}{5!} + 11368\frac{x^6}{6!} + \cdots,$$

which is sequence A013922.

The previous analysis can be enriched by enumerating graphs also by the number of edges. Let $G_{n,k}$ be the number of graphs with $n$ vertices and $k$ edges and let

$$G(x,y) = \sum G_{n,k} y^k \frac{x^n}{n!},$$

and define analogously $C(x,y)$. Notice that $G(x,y)$ is an ordinary generating function in $y$, since edges are not labeled.

Equations (6.1) and (6.2) generalize to

$$G(x,y) = e^{C(x,y)}, \qquad C_x(x,y) = xe^{B_x(xC_x(x,y),y)},$$

where $R(x,y) = xC_x(x,y)$. These equations hold because connected components and blocks do not share edges.

If again $\mathcal{G}$ is the class of all graphs then

$$G(x,y) = \sum_{n,m} \binom{\binom{n}{2}}{m} y^m \frac{x^n}{n!} = \sum_n (1+y)^{\binom{n}{2}} \frac{x^n}{n!}.$$

We obtain

$$C(x,y) = x + y\frac{x^2}{2!} + (3y^2+y^3)\frac{x^3}{3!} + (16y^3+15y^4+6y^5+y^6)\frac{x^4}{4!} + \cdots$$

and

$$B(x,y) = y\frac{x^2}{2!} + y^3\frac{x^3}{3!} + (3y^4+6y^5+y^6)\frac{x^4}{4!} + \cdots$$

Next we turn to the enumeration of 3-connected graphs using the decomposition of 2-connected graphs into 3-connected components, for which we follow [66]. We add the hypothesis

**(C3)** A 2-connected graph $G$ is in $\mathcal{G}$ if and only if the 3-connected components of $G$ are in $\mathcal{G}$.

Define a **network** as a 2-connected graph rooted at a directed edge (which may be deleted or not) and whose endpoints, called **poles**, are not labeled. Let $D(x,y)$ be the generating functions of networks. Then

$$2(1+y)B_y(x,y) = x^2(1+D(x,y)). \tag{6.3}$$

The left-hand term corresponds to marking and directing an edge, and keeping or not the edge; the right-hand side corresponds to adding the empty network and labeling the poles. We find

$$D(x,y) = y + (y^2+y^3)x + (2y^3+7y^4+6y^5+y^6)\frac{x^2}{2!} + \cdots$$

Let $T(x,y)$ be the generating function of 3-connected graphs. The decomposition into 3-connected components gives rise to the equation

$$1+D(x,y) = (1+y)\exp\left(\frac{xD(x,y)^2}{1+xD(x,y)} + \frac{2}{x^2}T_y(x,D(x,y))\right).$$

Given $D(x,y)$, this equation determines $T(x,y)$ and

$$T(x,y) = y^6\frac{x^4}{4!} + (15y^8+10y^9+y^{10})\frac{x^5}{5!} + \cdots$$

Setting $y=1$ we obtain the counting series of 3-connected graphs

$$T(x,1) = \frac{x^4}{4!} + 26\frac{x^5}{5!} + 1768\frac{x^6}{6!} + 225096\frac{x^7}{7!} + \cdots$$

This last sequence is A005644.

As noticed in [28], it is possible to integrate (6.3) with respect to $y$ and express $B(x,y)$ as an explicit function of $D$ and $B$:

$$\begin{aligned} B(x,y) &= T(x,D(x,y)) - \frac{1}{2}xD(x,y) + \frac{1}{2}\log(1+xD(x,y)) \\ &+ \frac{x^2}{2}\left(D(x,y) + \frac{1}{2}D(x,y)^2 + (1+D(x,y))\log\left(\frac{1+y}{1+D(x,y)}\right)\right). \end{aligned}$$

It is worth noticing that all the previous equations can be proved more combinatorially using grammars and the dissymmetry theorem for trees [15]. We summarize the previous results in a single statement.

**Theorem 6.2.1** *Let $\mathcal{G}$ be a class of graphs satisfying conditions (C1), (C2) and (C3), and let $G(x,y), C(x,y), B(x,y), D(x,y)$ and $T(x,y)$ be the generating functions associated to, respectively, graphs, connected graphs, 2-connected graphs, networks and 3-connected graphs in $\mathcal{G}$. Then*

$$\begin{aligned} G(x,y) &= e^{C(x,y)} \\ C_x(x,y) &= e^{B_x(xC_x(x,y),y)} \\ B_y(x,y) &= \frac{x^2(1+D(x,y))}{2(1+y)} \\ D(x,y) &= (1+y)\exp\left(\frac{xD(x,y)^2}{1+xD(x,y)} + \frac{2}{x^2}T_y(x,D(x,y))\right) - 1 \\ B(x,y) &= T(x,D(x,y)) - \frac{1}{2}xD(x,y) + \frac{1}{2}\log(1+xD(x,y)) \\ &+ \frac{x^2}{2}\left(D(x,y) + \frac{1}{2}D(x,y)^2 + (1+D(x,y))\log\left(\frac{1+y}{1+D(x,y)}\right)\right). \end{aligned}$$

For the class of all graphs, starting from the easy expression for $G(x,y)$ we reach an equation defining $T(x,y)$. In other situations it is the opposite way; as we will see in Section 6.6, for planar graphs one starts from the known expression for $T(x,y)$ and then goes all the way up to $G(x,y)$.

**Differential equations.** In addition to the previous system of equations, it is possible to obtain directly $B(x,y)$ and $T(x,y)$ as solutions of second order partial differential equations. This can be done algebraically or more combinatorially as in [68] and [67]. Setting $B = B(x,y)$ and $T = T(x,y)$, one has

$$\left(2(1+y)B_y - x^2(1+B_{xx})\right)(1-xB_{xx}) - x^3B_{xx}^2 = 0$$

and

$$(1+y)T_y - \frac{x^2}{2}T_{xx} = \frac{x^4F_x^2}{4F_y} - \frac{x^4y^4}{4(1+xy)^2},$$

where

$$F = \log(1+y) - \frac{xy^2}{1+xy} - \frac{2}{x^2}T_y.$$

### 6.2.2 Graphs with given minimum degree

In this section we find the generating functions for graphs with minimum degree at least two and at least three. For brevity, we define a *k*-graph as a connected graph with minimum degree at least *k*.

Given a connected graph $G$, its *2-core* is the maximum subgraph $C$ with minimum degree at least two. The 2-core $C$ is obtained from $G$ by repeatedly removing vertices of degree one. Conversely, $G$ is obtained by attaching rooted trees at the vertices of $C$. Let $H_n$ be the number of 2-graphs and $H(x)$ the associated generating function. Let $T(x)$ be the generating function of rooted trees, that satisfies $T(x) = xe^{T(x)}$. The generating function of unrooted trees (see Section 4.5.1 in this book) is equal to

$$U(x) = T(x) - \frac{T(x)^2}{2}.$$

The decomposition of a graph into its core and the forest of rooted trees implies the following equation:

$$C(x) = H(T(x)) + U(x).$$

The second summand takes care of trees, which have an empty 2-core. If we change variables $z = T(x)$, the inverse function is $x = ze^{-z}$. This implies

$$H(x) = C(xe^{-x}) - x + \frac{x^2}{2} = \frac{x^3}{3!} + 10\frac{x^4}{4!} + 253\frac{x^5}{5!} + 12058\frac{x^6}{6!} + \cdots$$

which is sequence A059166. If as before $y$ marks edges, then

$$H(x,y) = C(xe^{-xy}, y) - x + \frac{x^2y}{2} = y^3\frac{x^3}{3!} + (3y^4 + 6y^5 + y^6)\frac{x^4}{4!} + \ldots$$

The **kernel** of $G$ is obtained by replacing each maximal path of vertices of degree two in the 2-core $C$ by a single edge. The kernel has minimum degree at least three, and $C$ can be recovered from $K$ by replacing edges with paths. The kernel may have loops and multiple edges, which must be taken into account since we are dealing with simple graphs. Also, when replacing loops and multiple edge by paths the same graph can be produced several times. This can be dealt with by weighting multigraphs appropriately according to the number of loops and edges of each multiplicity. Given a multigraph $M$ with $\alpha_i$ vertices incident to exactly $i$ loops, and with $\beta_i$ edges of multiplicity $i$, the associated weight is

$$w(G) = \prod_{i\geq 1}\left(\frac{1}{2^i i!}\right)^{\alpha_i}\prod_{i\geq 1}\left(\frac{1}{i!}\right)^{\beta_i}.$$

The justification is that when replacing an edge of multiplicity $i$ with $i$ different paths, the order of the paths is irrelevant, and similarly for the loops. This weight is called the compensation factor in the literature [35].

We have $0 < w(G) < 1$, and moreover $w(G) = 1$ if and only if $G$ is simple. With this definition, the sum of the weights of all 3-multigraphs with $n$ vertices is finite. Working first with multigraphs and then moving to simple graphs by setting to 0 the variables involving loops and multiedges, one shows the following [47] (this can also be deduced from the results in [34]).

Let $K(x,y)$ be the generating function of simple 3-graphs. Then

$$K(x,y) = C(A(x,y), B(x,y)) + E(x,y),$$

where

$$\begin{aligned}
A(x,y) &= x\exp\left(\frac{x^2y^3 - 2xy}{2+2xy}\right),\\
B(x,y) &= (1+y)\exp\left(\frac{-xy^2}{1+xy}\right) - 1,\\
E(x,y) &= \frac{x^2y}{2+2xy} - \frac{x^2y^2}{4} + \frac{xy}{2} - x - \frac{1}{2}\ln(1+xy).
\end{aligned}$$

In particular

$$K(x,1) = \frac{x^4}{4!} + 26\frac{x^5}{5!} + 1858\frac{x^6}{6!} + 236926\frac{x^7}{7!} + \cdots$$

This sequence is not referenced in OEIS.

Finally $e^{H(x)}$ and $e^{K(x)}$ enumerate graphs (not necessarily connected), with minimum degree at least two and three, respectively.

### 6.2.3 Bipartite graphs

In this section we compute the number of bipartite graphs. The number of bipartite graphs with parts of sizes $a$ and $b$ is $\binom{a+b}{a}2^{ab}$, but in this way a graph may be counted more than once. A 2-colored graph is a bipartite graph equipped with a fixed bipartition. A connected bipartite graph gives rise to two 2-colored graphs, and a bipartite graph with $c$ connected components to $2^c$ such 2-colored graphs. The number $g_n$ of 2-colored bipartite graphs is

$$g_n = \sum_{k=0}^{n} \binom{n}{k} 2^{k(n-k)}.$$

Since each connected bipartite graph is counted exactly twice, the generating function $C_b(x)$ for connected bipartite graphs satisfies $\exp(2C_b(x)) = \sum g_n x^n/n!$, and

$$C_b(x) = \frac{1}{2}\log\left(\sum g_n \frac{x^n}{n!}\right) = x + \frac{x^2}{2!} + 3\frac{x^3}{3!} + 19\frac{x^4}{4!} + 195\frac{x^5}{5!} + \cdots,$$

which is sequence A001832. Hence the generating function for all bipartite graphs is equal to

$$G_b(x) = e^{C_b(x)} = \left(\sum_n \sum_{k=0}^{n} \binom{n}{k} 2^{k(n-k)} \frac{x^n}{n!}\right)^{1/2}$$

$$= 1 + x + 2\frac{x^2}{2!} + 7\frac{x^3}{3!} + 41\frac{x^4}{4!} + 376\frac{x^5}{5!} + \cdots,$$

which is sequence A047864. The dominant term in the sum for $g_n$ is when $k = n/2$, giving $2^{n+n^2/4}$ as the exponential growth. As for general graphs, almost every bipartite graph is connected.

A graph is bipartite if and only if all its blocks are bipartite. Hence, according to Equation (6.2), the generating function $B_b(x)$ of 2-connected bipartite graphs satisfies

$$C_b'(x) = e^{B_b'(xC_b'(x))}.$$

This gives

$$B_b(x) = \frac{x^2}{2!} + +3\frac{x^4}{4!} + 10\frac{x^5}{5!} + 6986\frac{x^6}{6!} + \cdots,$$

which is sequence A004100. Recurrence relations for the previous numbers are derived in [33]. As far as we know, the problem of counting 3-connected bipartite graphs is open. The difficulty here is to control whether a graph is bipartite in terms of its 3-connected components.

Counting $k$-partite graphs for $k > 2$ is not so simple, since a connected $k$-colored graph can be colored in a number of ways that depends on the graph. We discuss this problem in Section 6.5.

## 6.3 Connected graphs with given excess

The excess of a connected graph $G$ is defined as $|E(G)| - |V(G)|$, the number of edges minus the number of vertices. Trees have excess $-1$, and graphs with excess

0 are precisely unicyclic graphs, that is, connected graphs with a unique cycle. For $k \geq -1$, let $C_{n,n+k}$ be the number of connected graphs with $n$ vertices and excess $k$. Computing and estimating $C_{n,n+k}$ is a classical problem that arises in the study of random graphs [21, 10, 35], and is a crucial ingredient in the analysis of the giant component phenomenon.

The number of trees is $C_{n,n-1} = n^{n-2}$, by the well-known Cayley's formula. There are several formulas for the number of unicyclic graphs. Wright [74] showed that

$$C(n,n) = \frac{1}{2}\left(\frac{h(n)}{n} - (n-1)n^{n-2}\right),$$

where

$$h(n) = \sum_{s=1}^{n-1} \binom{n}{s} s^s (n-s)^{n-s}.$$

Similar expressions were obtained for larger excess. For instance,

$$C_{n,n+1} = \frac{1}{24}\left((n-1)(5n^2+3n+2)n^{n-2} - 14h(n)\right).$$

The first values for these sequences are as follows:

| $n$ | 3 | 4 | 5 | 6 | 7 | 8 | OEIS |
|---|---|---|---|---|---|---|---|
| $C_{n,n}$ | 1 | 15 | 222 | 3660 | 68295 | 1436568 | A006351 |
| $C_{n,n+1}$ | | 6 | 205 | 5700 | 156555 | 4483360 | A058864 |

These formulas follow from the general recurrence relation found by Wright:

$$2(n+k+1)C_{n,n+k+1} = 2(n(n-1)/2 - n - k)C_{n,n+k} \tag{6.4}$$

$$+ \sum_{s=1}^{n} \binom{n}{s} s(n-s) \sum_{h=-1}^{k+1} C_{s,s+h} C_{n-s,n-s+k-h}.$$

The proof is based on analyzing the result of removing one edge from a connected graph, producing either a connected graph or a graph with two connected components. Wright also showed that, for $k \geq 1$,

$$C_{n,n+k} = (-1)^k M_k(n) h(n) + (-1)^{k-1} P_k(n)(n-1)n^{n-2}, \tag{6.5}$$

where $M_k$ and $P_k$ are polynomials in $n$ that can be computed recursively. This implies that, for $k \geq 1$, the generating function

$$W_k(x) = \sum C_{n,n+k} \frac{x^n}{n!}$$

is a rational function in $T(x) = \sum_{n \geq 1} n^{n-1} \frac{x^n}{n!}$, the generating function of rooted labeled trees, which satisfies the equation $T(x) = xe^{T(x)}$.

The first values of the $W_k(x)$ are

$$\begin{aligned} W_{-1}(x) &= T(x) - \frac{T(x)^2}{2} \\ W_0(z) &= \frac{1}{2}\left(\log\frac{1}{1-T(x)} - T(x) - \frac{T(x)^2}{2}\right) \\ W_1(z) &= \frac{T(x)^4(6-T(x))}{24(1-T(x))^3}. \end{aligned}$$

The expression for $W_{-1}(x)$ is the well-known generating function for unrooted trees, as we have already remarked. The one for $W_0(x)$ follows from viewing a unicyclic graph as an undirected cycle of rooted trees of length at least three.

From recurrence (6.5) Wright obtained an estimate for $C_{n,n+k}$ for fixed $k$. In a subsequent paper Wright [75] extended this estimate to values $k = o(n^{1/3})$. The estimate is

$$C_{n,n+k} = f_k n^{(3k-1)/2} n^n \left(1 + O(n^{-1/2})\right),$$

where

$$f_0 = \frac{\pi}{8}, \qquad f_k = \frac{\sqrt{\pi} 3^k (k-1) d_k}{2^{(5k-1)/2}\Gamma(3k/2)}, \qquad k \geq 1$$

and the $d_k$ satisfy the recurrence

$$d_1 = d_2 = \frac{5}{36}, \qquad d_{k+1} = d_k + \sum_{h=1}^{k-1} \frac{d_h d_{k-h}}{(k+1)\binom{k}{h}}, \qquad k \geq 1.$$

Later on, Bender, Canfield and McKay [2] translated recurrence (6.4) into differential equations and extended the estimates to the widest range of parameters. This result is reproved by Pittel and Wormald [49] in a different way, using the 2-core (Section 6.2) of a connected graph. They first estimate the number of connected graphs with minimum degree at least two (called 2-graphs in Section 6.2) with given number of vertices and edges, and then apply it to estimate the number of connected graphs.

A purely analytic approach to the problem was developed in [23]. The authors find an integral representation for the generating function of connected graphs

$$C(x,y) = \log\left(1 + \sum_{n\geq 1} (1+y)^{\binom{n}{2}} \frac{x^n}{n!}\cdot\right),$$

which linearizes the quadratic exponent in $y$. They reprove the fact that

$$W_k(z) = \frac{A_k(T(z))}{(1-T(z))^{3k}},$$

where the $A_k$ are polynomials, related to the Airy Ai function. For fixed $k \geq 2$ they obtain complete asymptotic estimates for $C_{n,n+k}$. It is shown that

$$C_{n,n+k} = \frac{A_k(1)\sqrt{\pi}}{2^{(3k-1)/2}} n^n n^{(3k-1)/2} \left(\frac{1}{\Gamma\left(\frac{3k}{2}\right)} + \frac{\frac{A_k'(1)}{A_k(1)} - k}{\Gamma\left(\frac{3k-1}{2}\right)}\sqrt{\frac{2}{n}} + O\left(\frac{1}{n}\right)\right),$$

and lower order terms depend on the derivatives $A_k^{(j)}(1)$.

The values $A_k(1)$, which give the first order asymptotics, can be computed directly from the following identity

$$\sum_{k=1}^{\infty} A_k(1)(-x)^k = \log\left(1+\sum_{k=1}^{\infty} c_k(-x)^k\right),$$

where

$$c_k = \frac{(6k)!}{(3k)!(2k)!3^{2k}2^{5k}}.$$

The higher derivatives can be computed in terms of the Airy function.

## 6.4 Regular graphs

A graph is $d$-regular if all the vertices have degree $d$. Let $R_{n,d}$ be the number of (labeled) $d$-regular graphs with $n$ vertices. Read [56] gave an exact formula for $R_{n,d}$ using Pólya's counting theory, but this formula is not suitable for asymptotic analysis. The key to the enumeration of regular graphs is the pairing model.

### 6.4.1 The pairing model

In the pairing model for $d$-regular graphs, there are $n$ labeled cells, each with $d$ labeled elements, corresponding to the half-edges at each vertex. A pairing of the $nd$ elements produces a $d$-regular graph by regarding the cells as vertices and the pairs as edges. Notice that loops and multiple edges may be created by choosing a pair from the same cell, or two different pairs joining elements from two distinct cells. The number of such pairings is

$$\frac{(dn)!}{(dn/2)!2^{dn/2}}.$$

One can estimate the probability that the resulting graph is simple, by considering the number $X_i$ of cycles of length $i$ created in the pairing. Using the method of moments, it can be shown that the random variables $X_1,\ldots,X_k$ are asymptotically (when $n$ goes to infinity) independent Poisson distributed with means $\lambda_i = (d-1)^i/(2i)$. Hence the probability of being simple, which is the probability of not having cycles of lengths one or two, is asymptotically $e^{-\lambda_1}e^{-\lambda_2}$. This gives

$$\mathbf{P}(\text{simple}) \sim e^{(1-d^2)/4}.$$

Since a simple graph is produced exactly $d!^n$ times (permuting the $d$ elements at each cell gives the same graph), one obtains for fixed $d$

$$R_{n,d} \sim \exp\left(\frac{1-d^2}{4}\right)\frac{(dn)!}{(dn/2)!2^{dn/2}d!^n}.$$

Using Stirling's estimate this can be rewritten as

$$R_{n,d} \sim \sqrt{2}e^{(1-d^2)/4}\left(\frac{d^d}{e^d(d!)^2}\right)^{n/2} n^{dn/2}. \tag{6.6}$$

The dominating term in the expression above is $n^{dn/2}$, corrected by an exponential term and a constant that depend on $d$. This estimate was obtained independently by several authors (see the historical discussion in [70]).

For $d$ growing with $n$, estimating the probability that the resulting graph is simple is considerably more difficult. McKay [42] introduced a technique called **switchings** and extended the range of $d$ to $d = o(n^{1/3})$ by showing

$$\mathbf{P}(\text{simple}) \sim \exp\left(\frac{1-d^2}{4} - \frac{d^3}{12n} + O\left(\frac{d^2}{n}\right)\right).$$

Then McKay and Wormald [43] used a new sort of switching to find the formula for $d = o(\sqrt{n})$.

The basic idea of a switching can be explained through the following example. Suppose a pairing has a pair $p_1p_2$ that produces a loop in the resulting multigraph. Choose another pair $p_3p_4$ and replace the two pairs by $p_1p_3$ and $p_2p_4$. The pair giving rise to a loop disappears but a new loop can be created. Using double counting one can estimate the probability of these types of events, and then obtain an estimate of the ratio $|S_a|/|S_{a-1}|$, where $S_a$ is the number of parings having exactly $a$ loops. By telescoping one can estimate $|S_0|$, the number of pairings without loops. More general switchings can involve more than two pairs [70].

The previous analysis usually extends to graphs with a given degree sequence. Let $d_1 \geq d_2 \geq \cdots \geq d_n \geq 1$ with $\sum_{1=1}^n d_i = 2m$. Let $G_{n,\mathbf{d}}$ be the number of graphs with degree sequence $\mathbf{d} = (d_1, \ldots, d_n)$. One can use again the pairing model with $n$ cells containing $d_1, \ldots, d_n$ elements. The probability that the resulting graph is simple can be estimated (for $d_1$ bounded) as

$$\mathbf{P}(\text{simple}) \sim e^{-\lambda/2-\lambda^2/4},$$

where

$$\lambda = \frac{1}{m}\sum_{i=1}^{m}\binom{d_i}{2}.$$

When all the $d_i$ are equal to $d$, one recovers the previous expression for $d$-regular graphs. This can be extended in some cases when the $d_i$ grow with $n$.

The case of 2-regular graphs is particularly simple, since they consist of a collection of cycles of lengths of at least three. If $R(x)$ is the EGF of 2-regular graphs, then

$$R_2(x) = \exp\left[\frac{1}{2}\left(\log\frac{1}{1-x} - x - \frac{x^2}{2}\right)\right] = \frac{e^{-x/2-x^2/4}}{\sqrt{1-z}} \tag{6.7}$$

$$= \frac{x^3}{3!} + 3\frac{x^4}{4!} + 12\frac{x^5}{5!} + 70\frac{x^6}{6!} + \cdots,$$

which is sequence A001205. Singularity analysis leads immediately (see [24, Example IV.1]) to the estimate for the number of 2-regular graphs

$$\frac{e^{-3/4}}{\sqrt{\pi n}} n!,$$

which agrees with (6.6) for $d = 2$.

Let us mention that the pairing model is the key for analyzing random regular graphs. Many properties, like the existence of Hamilton cycles or the value of the chromatic number, can be proved using this model [36, 70].

### 6.4.2 Differential equations

Let $R_d(x) = \sum_n R_{n,d} \frac{x^n}{n!}$ be the generating function of labeled $d$-regular graphs. Gessel [25] proved that $R_d(x)$ is D-finite for each $d \geq 2$, that is, satisfies a linear differential equation with polynomial coefficients.

**Theorem 6.4.1** *For each $d \geq 2$, the generating function of d-regular graphs is D-finite.*

For $d = 2$, the explicit expression (6.7) immediately gives

$$2(1-x)R_2'(x) - x^2 R_2(x) = 0.$$

For $d = 3$ it was first proved in [56]. In [69] one finds in addition differential equations for the generating functions of connected, 2-connected and 3-connected cubic graphs, and for $d = 4$ it was proved in [58]. In both cases the arguments were combinatorial, based on removing one edge from a $d$-regular graph and analyzing the possible resulting configurations.

Let $R_3(x) = Q(x^2)$. Then

$$6x^2(2-2x-x^2)Q''(x) - (x^5+6x^4+6x^3-32x+8)Q'(x) + \frac{x}{6}(2-2x-x^2)^2 Q(x) = 0.$$

The first terms are

$$R_3(x) = \frac{x^4}{4!} + 70\frac{x^6}{6!} + 19355\frac{x^8}{8!} + 11180820\frac{x^{10}}{10!} + \cdots,$$

which is sequence A002829.

A linear second-order differential equation with polynomials coefficients can be obtained also for 4-regular graphs but it is a bit long to write down. The initial terms are

$$R_4(x) = \frac{x^5}{5!} + 15\frac{x^6}{6!} + 465\frac{x^7}{7!} + 19355\frac{x^8}{8!} + \cdots,$$

which is sequence A005815.

The problem was revisited by Goulden, Jackson and Reilly in [30], in terms of symmetric functions. Consider labeled graphs on vertices $\{1, 2, \ldots, \}$ and let $t_i$ be a variable marking the degree of vertex $i \geq 1$. Then

$$T = \prod_{1 \leq i < j} (1 + t_i t_j)$$

is the generating function of labeled graphs marking all possible degrees, because if $\{i,j\}$ is an edge, there is a contribution of one to both the degree of $i$ and $j$. This is a symmetric function in infinitely many variables $t=(t_1,t_2,\ldots)$. The number of $d$-regular graphs on $n$ vertices is the coefficient $[t_1^d\cdots t_n^d]T(t)$. These coefficients (called regular in [30]) can be expressed in terms of the so-called $H$-series of $T$. The authors then showed that the $H$-series satisfies a system of partial differential equations. From here they proved D-finiteness of $R_d(x)$ for $d=2,3,4$, but the general case remained open.

Gessel [25] proved D-finiteness in the general case and at the same time extended it to graphs with loops (a loop contributes 2 to the degree of a vertex) and multiple edges, by considering variations of the $T$ symmetric function. For instance

$$\prod_{1\leq i\leq j}(1+t_it_j)$$

corresponds to graphs where loops are allowed (by taking $i=j$), and

$$\prod_{1\leq i\leq j}\frac{1}{1-t_it_j}$$

to graphs where loops and multiple edges are allowed.

## 6.5 Monotone and hereditary classes

In this section we discuss classes of graphs closed under the operation of taking subgraphs or under taking induced subgraphs. This is a very active area of research with close connections to extremal graph theory and related areas. Most of the results are asymptotic and only in a few cases does one have access to the counting generating functions.

### 6.5.1 Monotone classes

A class of graphs $\mathscr{G}$ is *monotone* if it is closed under taking subgraphs. Every monotone class is defined by a collection $\mathscr{F}$, possibly infinite, of minimal forbidden subgraphs. These are the graphs $G$ not in $\mathscr{G}$ but such that every proper subgraph is in $\mathscr{G}$. We denote by $\mathrm{Mon}(\mathscr{F})$ the class of graphs not containing any subgraph isomorphic to a graph in $\mathscr{F}$. Next we briefly describe the relation between monotone classes and extremal graph theory.

Denote by $\mathrm{ex}(n;F)$ the maximum number of edges in a graph with $n$ vertices not containing $F$ as a subgraph. More generally, $\mathrm{ex}(n;\mathscr{F})$ is the maximum number of edges in a graph with $n$ vertices belonging to $\mathrm{Mon}(\mathscr{F})$, that is, containing no subgraph in $\mathscr{F}$. Let $G$ be a graph in $\mathrm{Mon}(\mathscr{F})_n$ with maximal number of edges. Since every subgraph of $G$ is also in $\mathrm{Mon}(\mathscr{F})$, we have

$$|\mathrm{Mon}(\mathscr{F})_n|\geq 2^{\mathrm{ex}(n;\mathscr{F})}.$$

It turns out that in many cases this lower bound is not too far from the right answer. This is a consequence of the fundamental Erdős-Stone-Simonovits Theorem in extremal graph theory. Let $\chi(G)$ be the chromatic number of a graph $G$, and let

$$r = \min\{\chi(F) : F \in \mathscr{F}\} - 1.$$

Then the extremal function $\mathrm{ex}(n;\mathrm{Mon}(\mathscr{F}))$ satisfies

$$\mathrm{ex}(n;\mathrm{Mon}(\mathscr{F})) = \left(1 - \frac{1}{r}\right)\binom{n}{2} + o(n^2), \qquad \text{as } n \to \infty.$$

For $r = 1$ this result says that the number of edges is subquadratic, and for $r \geq 2$ it is quadratic and of order $(1 - 1/r)n^2/2$.

A fundamental result of Erdős, Frankl and Rödl [19] gives the rough order of magnitude for the number of graphs in a monotone class, when $r \geq 2$.

**Theorem 6.5.1** *Let* $\mathrm{Mon}(\mathscr{F})$ *be a monotone class with* $r \geq 2$. *Then*

$$|\mathrm{Mon}(\mathscr{F})_n| = 2^{(1-\frac{1}{r}+o(1))\binom{n}{2}}.$$

Notice that the $o(1)$ term is in the exponent, hence this estimate is very far from being precise. When $r = 1$ (at least one bipartite graph is forbidden) the situation is more complicated. A case that has been much studied is $\mathrm{Mon}(C_4)$. It is proved in [37] that the number $F_n$ of graphs not containing $C_4$ satisfies

$$2^{cn^{3/2}} \leq F_n \leq 2^{c'n^{3/2}}, \qquad 0 < c < c',$$

an estimate that is still far from being precise.

In some cases though one has very precise estimates. We start discussing the class of triangle-free graphs. Clearly every bipartite-graph is triangle-free. It was proved by Erdős, Kleitman, and Rothschild [20] that, somewhat surprisingly, almost all triangle-free graphs are bipartite. Let $B_n$ be the number of bipartite graphs.

**Theorem 6.5.2** *Almost all triangle-free graphs are bipartite. More precisely,*

$$|\mathrm{Mon}(K_3)| = B_n\left(1 + O\left(\frac{1}{n}\right)\right).$$

Since the number of bipartite graphs is known (see Section 6.2), this gives a precise estimate for the number of triangle-free graphs. This fundamental result was extended by Kolaitis, Prömel and Rothschild [38].

**Theorem 6.5.3** *Almost all* $K_{t+1}$*-free graphs are* $t$*-partite.*

The Erdős-Kleitman-Rothschild theorem was generalized in a different direction, by showing that almost all graphs in $\mathrm{Mon}(C_{2\ell+1})$ are bipartite [39]. The two previous results were widely generalized by Prömel and Steger [53]. An edge $e$ of a graph $H$ is color-critical if $\chi(H - e) < \chi(H)$. Notice that both in complete graphs and odd cycles every edge is color-critical.

**Theorem 6.5.4** *Let $H$ be a graph containing a color-critical edge that satisfies $\chi(H) = r+1 \geq 3$. Then almost all graphs not containing $H$ as a subgraph are $r$-partite.*

In fact the authors show that the previous condition is also necessary: If almost every graph in $\mathrm{Mon}(H)$ is $r$-partite then $H$ contains a color-critical edge.

We now discuss the problem of counting $k$-partite graphs. A simpler problem is to count $k$-colored graphs. A $k$-colored graph is a graph together with a particular $k$-coloring using all $k$ colors. A recursion for the number $c_{n,k}$ of $k$-colored graphs was obtained in [57]:

$$c_{n,k} = \sum_{j=0}^{n} \binom{n}{k} 2^{j(n-j)} c_{j,k-1},$$

with the initial conditions $c_{0,k} = 1$ and $c_{n,0} = 0$ for $n \geq 1$. Prömel and Steger [55] showed that almost all $k$-partite graphs are uniquely $k$-colorable, hence the number of $k$-partite graphs is asymptotically $k!c_{n,k}$. Estimates for $c_{n,k}$ are given in [71]. The dominant term is

$$2^{\left(1-\frac{1}{k}\right)\frac{n^2}{2}} k^n,$$

and the subexponential term depends on a subtle way on the value of $n$ modulo $k$.

## 6.5.2 Hereditary classes

Recall that a class of graphs is *hereditary* if it is closed under the operation of taking *induced* subgraphs. Notable examples include perfect graphs, and several subclasses such as chordal graphs, interval graphs or permutation graphs [29]. A graph $G$ is perfect if for every induced subgraph $H$, $\chi(H)$ equals the size of the largest complete subgraph in $H$. A graph is chordal if every cycle of length more than three contains a chord. The theory of perfect graphs is closely related to combinatorial optimization [29].

A hereditary class $\mathcal{G}$ is characterized by the list, not necessarily finite, of *forbidden* subgraphs, the minimal graphs not in $\mathcal{G}$ all whose proper induced subgraphs are in $\mathcal{G}$. We write $\mathrm{Forb}(H)$ for the class of graphs not containing $H$ as an induced subgraph. The class $\mathrm{Mon}(K_t)$ discussed previously is hereditary, since a $K_t$ subgraph is always induced. However the class $\mathrm{Mon}(C_4)$ is not hereditary, since a graph can contain copies of $C_4$, none of them induced.

A general asymptotic result is available for hereditary classes, similar to Theorem 6.5.1. It is based on a modified version of the extremal function $ex(n,H)$ that makes sense for induced subgraphs. The following concept was introduced in [54]. Given positive integers $s \leq r$, an $(r,s)$-coloring of a graph $G$ is a coloring $c\colon V(G) \to [r]$ such that vertices with color $i$ induce a clique for $1 \leq i \leq s$, and an independent set for $s+1 \leq i \leq r$. Notice that $(r,0)$-colorable is the same as $r$-colorable in the classical sense. The *coloring number* of a hereditary class $\mathcal{G}$ is defined as

$$r(\mathcal{G}) = \max\{r\colon \text{ there exists } s \leq r \text{ such every } (r,s)\text{-colorable graph is in } G\}.$$

The following was proved in [54] for hereditary classes excluding a single graph, and in the general case in [12].

**Theorem 6.5.5** *Let $\mathcal{G}$ be a non-trivial hereditary class of graphs with coloring number $r = r(\mathcal{G})$. Then*

$$|\mathcal{G}_n| = 2^{(1-\frac{1}{r}+o(1))\binom{n}{2}}.$$

Observe that again the $o(1)$ term in the exponent makes the estimate very far from being precise.

In a different direction, it is know that the rate of growth of the number of graphs in a hereditary class cannot be arbitrary. This was first established in [61], and later extended by several authors [11]. In the next statement $B(n)$ is the Bell number (the number of partitions of an $n$-set), that grows like $B(n) \sim ((1+o(1))n/\log n)^n$.

**Theorem 6.5.6** *Let $\mathcal{G}$ be a non-trivial hereditary class of graphs with coloring number $r = r(\mathcal{G})$. Then one of the following cases holds for sufficiently large n.*

1. *$|\mathcal{G}_n|$ is identically zero, one or two.*
2. *$|\mathcal{G}_n|$ is a polynomial in n.*
3. *$|\mathcal{G}_n|$ has exponential order of the form $\sum_{i=1}^{k} p_i(n)i^n$ for some $k > 0$ and polynomials $p_i$.*
4. *$|\mathcal{G}_n| = n^{(1-1/r+o(1))}$ for $r > 1$.*
5. *$B(n) \leq |\mathcal{G}_n| \leq 2^{o(n^2)}$.*
6. *$|\mathcal{G}_n| = 2^{(1-1/r+o(1))n^2/2}$ for $r > 1$.*

These results extend to other combinatorial structures such as oriented graphs, hypergraphs, or posets.

Next we turn to particular hereditary classes. A fundamental difference between $\mathrm{Mon}(C_4)$ and $\mathrm{Forb}(C_4)$ can be observed by looking at the extremal functions. It is well-known that $\mathrm{ex}(n;C_4) = \Theta(n^{3/2})$, but a graph in $\mathrm{Forb}(C_4)$ can have $\Theta(n^2)$ edges: consider the disjoint union of $K_{n/2}$ and the complement of $K_{n/2}$. The former is an example of a split graph. A graph is *split* if its vertex set can be partitioned into a clique and an independent set. Clearly a split graph is in $\mathrm{Forb}(C_4)$. The following was proved by Prömel and Steger [51].

**Theorem 6.5.7** *Almost all graphs in* $\mathrm{Forb}(C_4)$ *are split.*

It follows that the number of graphs in $\mathrm{Forb}(C_4)$ is of the same order as the number of bipartite graphs.

A graph $G$ is *generalized split* if either $G$ or its complement satisfy the following property: the vertex set can be partitioned into disjoint cliques $H_0, H_1, \ldots, H_k$ such that there is no edge between $H_i$ and $H_j$ for $i > j \geq 1$. It is easy to check that generalized split graphs contain no induced $C_5$. The following was proved in [52].

**Theorem 6.5.8** *Almost all graphs in* $\mathrm{Forb}(C_5)$ *are generalized split.*

The former result has a striking consequence. A graph $G$ is perfect if $\omega(H) = \chi(H)$ for every induced subgraph $H$ of $G$, where $\omega(G)$ is the size of the largest completed subgraph in $G$. A perfect graph does not contain an induced $C_5$ and it can be checked that generalized split graphs are perfect. Hence we have

$$\text{Generalized split} \subset \text{Perfect} \subset \text{Forb}(C_5).$$

It follows that

**Theorem 6.5.9** *Almost all perfect graphs are generalized split.*

This is an unexpected result. Perfect graphs are fundamental in graph theory and some difficult conjectures have been proved only recently. Let us mention the perfect graph conjecture, which says that perfect graphs are precisely those excluding odd cycles of length at least five and their complements as induced subgraphs [16]. However, most perfect graphs have a very simple structure.

**Cographs.** In the last part of this section we analyze hereditary classes that can be enumerated exactly using generating functions. A **cograph** is a graph not containing the path $P_4$ as an induced subgraph. They can be characterized alternatively as follows:

- The empty graph is a cograph and a single vertex is a cograph.
- If $G$ and $H$ are cographs so are the disjoint union $G \cup H$ and the sum $G+H$ (the sum is the disjoint union plus all edges between $G$ and $H$).

Cographs are relevant in computer science, since several hard computational problems can be solved in polynomial time when the input is a cograph [29].

Let $G(x)$ and $C(x)$ be, respectively, the generating functions of cographs and connected cographs. Since a cograph is connected if and only if its complement is disconnected, we have $G(x) - 1 + x = 2C(x)$. Together with $G(x) = e^{C(x)}$, this gives

$$e^{C(x)} - 2C(x) + x - 1 = 0.$$

It follows that the dominant singularity of $C(x)$ is at $\rho = 2\log(2) - 1$. From here it is easy to deduce the estimates

$$C_n \sim cn^{-3/2}\rho^{-n}n!, \qquad G_n \sim 2C_n.$$

This derivation can be found in several places, for instance [31].

Two subclasses of cographs that have been considered in the literature are the following.

1. A graph is **trivially perfect** if it does not contain $P_4$ or $C_4$ as induced subgraphs. A similar argument as before gives the equation

$$(1 - e^{-x})e^{C(x)} - C(x) = 0.$$

The dominant singularity is $\rho = 1 - \log(e-1)$ and

$$C_n \sim cn^{-3/2}\rho^{-n}n!, \qquad G_n \sim eC_n.$$

2. A graph is a **threshold graph** if it does not contain $P_4, C_4$ or $K_2 \cup K_2$ (two disjoint edges) as induced subgraphs. As for cographs we have $G_n = 2C_n$. In this case we have an explicit expression for $C(x)$:

$$C(x) = \frac{1-x}{2-e^x}.$$

The function $C(x)$ has a simple pole at $\log(2)$ and

$$C_n \sim c\log(2)^{-n} n!, \qquad G_n \sim 2C_n.$$

The first values for the number of cographs and their subclasses are given in the next table.

| $n$ | 1 | 2 | 3 | 4 | 5 | 6 | OEIS |
|---|---|---|---|---|---|---|---|
| Cographs | 1 | 2 | 8 | 52 | 472 | 5504 | *A006351* |
| Trivially perfect | 1 | 2 | 8 | 49 | 402 | 4144 | *A058864* |
| Threshold | 1 | 2 | 8 | 46 | 332 | 2874 | *A005840* |

## 6.6 Planar graphs

The enumeration of planar graphs started with Tutte in the 1960s, with his fundamental papers on map enumeration, where a map is a graph with a fixed embedding in the plane. Map enumeration has grown into an important area of research; see Chapter 5. In this section we are interested in enumerating planar graphs as combinatorial structures, without considering a particular embedding. As we are going to see, the enumeration of planar graphs is based on the enumeration of planar maps and the graph decompositions described in Section 6.2. Throughout this section, $T(x,y)$, $D(x,y)$, $B(x,y)$, $C(x,y)$ and $G(x,y)$ have the same meaning as in Section 6.2 restricted to planar graphs.

The starting point of the analysis is Whitney's theorem, claiming that a 3-connected planar graph has a unique embedding in the sphere up to homeomorphism. Let $M_{n,k}$ be the number of rooted 3-connected planar maps with $n$ vertices and $k$ edges, and let $T_{n,k}$ be the number of 3-connected planar graphs with $n$ vertices and $k$ edges. Then we have the relation

$$M_{n,k} n! = 4k T_{n,k}.$$

This is because there are $n!$ ways to label a rooted map with $n$ vertices (since vertices in a rooted map are all distinguishable), and there are $4k$ ways of rooting a planar graph with $k$ edges to obtain a rooted map ($2k$ choices for a directed edge and two choices for the root face). Let the generating functions associated to 3-connected maps and graphs, respectively, be

$$M(x,y) = \sum M_{n,k} y^k x^n, \qquad T(x,y) = \sum T_{n,k} y^k \frac{x^n}{n!}.$$

The former relation implies

$$M(x,y) = 4yT_y(x,y).$$

This was first made explicit in [4], using the fact (proved in [44]) that $M(x,y)$ is algebraic of degree four and is given by

$$M(x,y) = x^2y^2 \left( \frac{1}{1+xy} + \frac{1}{1+y} - 1 - \frac{(1+U)^2(1+V)^2}{(1+U+V)^3} \right),$$

where

$$U = xy(1+V)^2, \qquad V = y(1+U)^2.$$

From the previous equations one can estimate easily the number $T_n$ of 3-connected planar graphs as

$$T_n \sim c_3 n^{-7/2} \gamma_3^n n!,$$

where $\gamma_3 = (272 + 112\sqrt{7})/27 \approx 21.05$.

The series $T(x,y)$ determines the series $B(x,y)$ of 2-connected planar graphs via Theorem 6.2.1. From here, Bender, Gao and Wormald [4] obtained the asymptotic number $B_n$ of 2-connected planar graphs as

$$B_n \sim c_2 n^{-7/2} \gamma_2^n n!,$$

where $\gamma_2 \approx 26.18$. This was an important step, since it opened the way to the full enumeration of planar graphs.

In order to proceed further one needs an expression for $B(x,y)$ in terms of $D(x,y)$. This was achieved by Giménez and Noy [26].

**Lemma 6.6.1** *Let $D = D(x,y)$ be the generating function of planar networks and let $B(x,y)$ be that of 2-connected planar graphs. Let $W = z(1+U)$, where $U$ is defined in (6.6). Then*

$$B(x,y) = \frac{x^2}{2} B_1(x,y) - \frac{x}{4} B_2(x,y), \tag{6.8}$$

*where*

$$\begin{aligned} B_1 &= \frac{D(6x-2+xD)}{4x} + (1+D)\log\left(\frac{1+y}{1+D}\right) - \frac{\log(1+D)}{2} + \frac{\log(1+xD)}{2x^2}; \\ B_2 &= \frac{2(1+x)(1+W)(D+W^2)+3(W-D)}{2(1+W)^2} - \frac{1}{2x}\log(1+xD+xW+xW^2) \\ &+ \frac{1-4x}{2x}\log(1+W) + \frac{1-4x+2x^2}{4x}\log\left(\frac{1-x+xD+-xW+xW^2}{(1-x)(D+W^2+1+W)}\right). \end{aligned}$$

The former expression is obtained by integrating $T_y(x,y)$ with respect to $y$, and using Equations (6.3–6.2.1). It is remarkable that the same result can be recovered in a purely combinatorial way using grammars [15].

With this explicit expression one can check that the equation

$$C'(x) = e^{B'(xC'(x))}$$

has no branch point. This is equivalent to the fact that $B''(R) < 1/R$. It follows that the singularity $\rho$ of $C(x)$ is given by $\rho = Re^{-B'(R)}$, where $R = 1/\gamma_2$, and is of the same kind as that of $B(x)$. Finally, the singularity of $G(x) = e^{C(x)}$ is the same as that of $C(x)$. Applying singularity analysis one obtains estimates for the numbers $C_n$ and $G_n$ of the same kind as for $T_n$ and $B_n$. The results are summarized as follows.

**Theorem 6.6.2** *For $\ell \in \{0,1,2,3\}$, the numbers of $\ell$-connected planar graphs are asymptotically of the form*

$$c_\ell n^{-7/2} \gamma_\ell^n n!,$$

*where*

$$\gamma_3 \approx 21.05, \qquad \gamma_2 \approx 26.18, \qquad \gamma_1 = \gamma_0 \approx 27.23.$$

*Moreover, the asymptotic probability that a random planar graph is connected is $c_1/c_0 \approx 0.95$.*

We include a table with the numbers of small planar graphs of given connectivity, with the convention that a $k$-connected graph must have at least $k+1$ vertices.

| $n$ | 1 | 2 | 3 | 4 | 5 | 6 | 7 | OEIS |
|---|---|---|---|---|---|---|---|---|
| Planar | 1 | 2 | 8 | 64 | 1023 | 32071 | 1823707 | A066537 |
| Connected | | 1 | 4 | 38 | 727 | 26013 | 1597690 | A096332 |
| 2-connected | | | 1 | 10 | 237 | 10707 | 774924 | A096331 |
| 3-connected | | | | 1 | 25 | 1227 | 84672 | A096330 |

It is also possible to estimate the number of planar graphs with given edge density.

**Theorem 6.6.3** *For each $\alpha \in (1,3)$, the number of planar graphs with $n$ vertices and $\lfloor \alpha n \rfloor$ edges is asymptotically*

$$c(\alpha) n^{-4} \gamma(\alpha)^n n!, \tag{6.9}$$

*where $\gamma(\alpha)$ is an analytic function that achieves its maximum at $\alpha = \kappa \approx 2.21$. The expected number of edges in a random planar graph is precisely $\kappa n$.*

The proof is based on a local limit theorem. For each $\alpha \in (1,3)$ there is a $y = y(\alpha)$ such that giving a graph with $k$ edges the weight $y^k$, the weight is concentrated on graphs with $\alpha n + O(\sqrt{n})$ edges. For this value of $y$, the generating function $G(x,y)$ has radius of convergence $\rho(\alpha)$ and we have

$$[x^n]G(x,y) \sim c(y) n^{-7/2} \gamma(\alpha)^n n!, \qquad \gamma(\alpha) = \rho(\alpha)^{-1}.$$

The extra $n^{-1/2}$ factor in (6.9) comes from the local limit theorem (for details, see [26]).

The previous enumerative results opened the way to the fine analysis of random planar graphs. Basic parameters such as the number of edges and the number of connected components were already analyzed in [26]. Extremal parameters, such as the size of the largest component or the size of the largest block, the maximum degree and the diameter can be analyzed too [46].

Some special classes of planar graphs have been enumerated too, in particular cubic planar graphs [9]. The decomposition into 3-connected components can be adapted for cubic graphs. If $G$ is a 3-connected planar graph, its dual (which is unique by Whitney's theorem) is a 3-connected triangulation. Using Tutte's enumeration of triangulations [65] one can express the generating function of 3-connected cubic planar graphs in terms of the (ordinary) generating function $T(z)$ of triangulations, given by

$$T(z) = u(z)(1-2u(z)), \qquad u(z)(1-u(z))^3 = z.$$

Theorem 6.2.1 can be adapted to this situation and one can show [9] that the number of cubic planar graphs is asymptotically, for even $n \to \infty$,

$$cn^{-7/2}\delta^n n!, \qquad \delta \approx 3.13, \quad c > 0.$$

On the other hand, counting 4-regular planar graphs is yet an open problem. No decomposition into 3-connected components is known for 4-regular graphs, and the enumeration of general 4-regular graphs [58] does not seem apply when restricted to planar graphs. Another interesting open problem is to enumerate bipartite planar graphs.

## 6.7 Graphs on surfaces and graph minors

Planar graphs are graphs embeddable on the sphere, hence it is only natural to consider graphs in other surfaces. Graphs on surfaces are a special case of classes of graphs defined in terms of excluded minors.

### 6.7.1 Graphs on surfaces

The analysis from the previous section can be extended to graphs embeddable in a fixed surface. The starting point is the enumeration of maps on surfaces. After the seminal work of Tutte, the theory of map enumeration was extended to arbitrary closed surfaces. Using Tutte's technique of removing the root edge and using induction on the genus, Bender and Canfield [1] proved the following fundamental result (see Section 5.4.3 in the previous chapter for a discussion).

**Theorem 6.7.1** *Let $M_{g,n}$ be the number of rooted maps of orientable genus $g \geq 0$ with n edges, and let $N_{h,n}$ be the analogous quantity for non-orientable genus $h \geq 1$. Then*

$$M_{g,n} \sim c_g n^{5(g-1)/2} 12^n$$

*and*

$$N_{h,n} \sim \widetilde{c}_h n^{5(h-2)/4} 12^n$$

*for some positive constants $c_g, \widetilde{c}_h$.*

As for planar graphs, it took some time to go from enumeration of maps to graphs in a surface. The corresponding result for graphs was obtained independently in [3, 14].

**Theorem 6.7.2** *Let $G_{g,n}$ be the number of rooted maps of orientable genus $g \geq 0$ with n edges, and let $H_{h,n}$ be the analogous quantity for non-orientable genus $h \geq 1$.*

$$G_{g,n} \sim d_g n^{5(g-1)/2-1} \gamma^n n!$$

*and*

$$H_{g,n} \sim \widetilde{d}_g n^{5(h-2)/4-1} \gamma^n n!$$

*for some positive constants $c_g, c'_h$, and where $\gamma \approx 27.23$ is as in Theorem 6.6.2.*

The number of graphs that can be embedded in the orientable surface of genus $g$ has the same estimate, since graphs of genus less than $g$ are of a smaller order of magnitude than graphs of genus $g$. The same applies to the non-orientable case.

The proof of Theorem 6.7.2 is rather technical, here we limit ourselves to give the main ideas. Suppose one wishes, as for planar graphs, to use the enumeration of maps on the surface $S$ for counting graphs (without an embedding) on $S$. There are two main obstacles for this program:

1. No degree of connectivity guarantees a unique embedding.
2. The class of graphs embeddable in $S$ is not closed under taking connected components and blocks (genus is additive on components and blocks).

Hence the basic equations among generating functions from Theorem 6.2.1 no longer hold. The road to the solution is to consider a parameter called *face-width*. The face-width fw$(M)$ of a graph $G$ embedded in $S$ is the minimum number of intersections of $G$ with a simple non-contractible curve $C$ on $S$. It is easy to see that this minimum is achieved when $C$ meets $G$ only at vertices. Face-width is in some sense a measure of local planarity, if the face-width is large then the embedding is planar in balls of large radius. The face-width of a graph $G$ is the maximum face-width among all the embeddings of $G$. The key result is that a 3-connected graph with large enough face-width has a unique embedding. It turns out that the generating series of 3-connected graphs of any fixed face-width has a negligible contribution in the asymptotic analysis. Therefore, the enumeration of 3-connected graphs in $S$ can be reduced, up to negligible terms, to the enumeration of 3-connected maps in $S$. It is important to remark that, since maps with small face-width are discarded, one does not work with exact counting series. Instead, if $f(x)$ is the series of interest, one finds computable series $f_1(x)$ and $f_2(x)$ such that $f_1(x) \preceq f(x) \preceq f_2(x)$ (where $\preceq$ means coefficient-wise inequality) and $f_1(x)$ and $f_2(x)$ have the same leading asymptotic estimates.

For the second obstacle one can use a result from [59]: If a connected graph $G$ of genus $g$ has face-width at least two, then $G$ has a unique block of genus $g$ and the remaining blocks are planar. A similar result holds for 2-connected graphs and 3-connected components. Since for planar graphs we have exact expressions for all the generating functions involved, starting from the (asymptotic) enumeration of 3-connected graphs of genus g we can achieve the enumeration of all graphs of genus $g$. To be more precise, let $G^g(x)$ and $C^g(x)$ be the generating functions of graphs and connected graphs of genus $g$, respectively. The usual relation $G^g(x) = \exp C^g(x)$ does not hold, since the union of graphs of genus $g$ can have larger genus. Instead, we have

$$G^g(x) \sim C^g(x) e^{C^0(x)}$$

where the symbol must be understood as that the two functions have the same dominant asymptotic terms. Similarly, the relation between $C^g(x)$ and the generating function $B^g(x)$ of 2-connected graphs of genus $g$ is not an exact equation as in the planar case, since genus is also additive in blocks, but rather an approximate version. The technical details are quite involved but the essence is to discard maps and graphs with small face-width.

### 6.7.2 Graph minors

Graphs on surfaces are one of the main examples of the theory of graph minors. A graph $H$ is a minor of $G$ if $H$ can be obtained from a subgraph of $G$ by contracting edges. Equivalently, if $H$ can be obtained from $G$ by edge contractions and deletions, and deletion of isolated vertices. A class $\mathcal{G}$ of graphs is **minor-closed** if whenever $G$ is in $\mathcal{G}$ and $H$ is a minor of $G$, $H$ is also in $\mathcal{G}$. The class is **proper** if it does not contain all graphs. The class $\mathcal{G}_S$ of graphs that can be embedded in a fixed surface $S$ is minor-closed. Other examples are graphs with bounded tree-width and $\Delta Y$-reducible graphs.

Given a minor-closed class $\mathcal{G}$, an excluded minor for $\mathcal{G}$ is a graph $H$ such that $H$ is not in $\mathcal{G}$ but every proper minor of $H$ is in $\mathcal{G}$. For instance, Kuratowski's theorem says that $K_5$ and $K_{3,3}$ are the excluded minors for planar graphs. The theory of graph minors is one of the main achievements in modern combinatorics, culminating with the great theorem of Robertson and Seymour: For each minor-closed class the number of excluded minors is finite. We write $\mathcal{G} = \mathrm{ex}(H_1,\ldots,H_k)$ if the $H_i$ are the excluded minors of $\mathcal{G}$. A class $\mathcal{G}$ is said to be **decomposable** if it satisfies condition (C1) in Section 6.2. It is **addable** if it is decomposable and in addition satisfies condition (C2). It is easy to see that a class is decomposable if and only if its excluded minors are connected, and is addable if and only if its excluded minors are 2-connected.

Let $\mathcal{G}$ be a proper minor-closed class of (labeled) graphs. We say that $\mathcal{G}$ is **small** if there exists a constant $c > 0$ such that

$$|\mathcal{G}_n| \leq c^n n!.$$

Observe that this is equivalent to the fact that the exponential generating function $\sum |\mathcal{G}_n| x^n / n!$ has positive radius of convergence. A general enumerative result says that a minor-closed class is small. It was first proved in [45] and later in a more general context in [18].

**Theorem 6.7.3** *If $\mathcal{G}$ is a proper minor-closed class of graphs, then $\mathcal{G}$ is small.*

Hence the possible growth rates are bounded by $c^n n!$. A classification of the possible growth rates of minor-closed classes of graphs can be found in [7], similar to that for hereditary classes in Section 6.5, although growth rates above $c^n n!$ are excluded.

To measure the size of a class we consider $\gamma = \limsup_{n\to\infty} (G_n/n!)^{1/n}$. Because of the previous theorem, if $\mathcal{G}$ is proper minor-closed, then $\gamma < +\infty$. We say that $\mathcal{G}$ has a growth constant if the lim sup is actually a limit, that is,

$$\gamma = \lim_{n\to\infty} \left( \frac{G_n}{n!} \right)^{1/n}.$$

The class $\mathcal{G}$ is smooth if the limit $\lim_{n\to\infty} G_n/(nG_{n-1})$ exists. This is a stronger condition than having a growth constant and if this is the case then

$$\lim_{n\to\infty} \frac{G_n}{nG_{n-1}} = \gamma.$$

The following very general theorem was proved by McDiarmid [41].

**Theorem 6.7.4** *A $\mathcal{G}$ proper addable minor-closed class of graphs is smooth.*

It is an open problem whether every minor-closed class has a growth constant.

Let $\Gamma$ be the set of real numbers that are growth constants of minor-closed classes of graphs. It is known [6] that $0, 1, 2$ belong to $\Gamma$, and the only other number in $\Gamma \cap [0, 2]$ is $\xi \approx 1.76$, which is the growth constant of forests of caterpillars. Given a minor-closed class $\mathcal{G}$, one shows that either $\mathcal{G}$ contains all paths and its growth constant is at least 1, or else it excludes some path and the growth constant is 0. The proof that $\xi$ is the only growth constant in $(1, 2)$ uses a similar idea using caterpillars instead of paths. Another result is that if $\gamma \in \Gamma$, then $2\gamma \in \Gamma$. All possible values in $\Gamma \cap [2, 2.25]$ are also determined [6], where it is shown that there are infinitely many gaps.

### 6.7.3 Particular classes

There are no general results on the enumeration of minor-closed classes besides those discussed above. But there are precise results for a number of relevant classes. All the following examples satisfy conditions (C1) and (C2) from Section 6.2 (they are addable) and the equations in Theorem 6.2.1 apply. The series $G, C, B, D, T$ has the same meaning as in Section 6.2. We use throughout the equations relating these generating functions stated in Theorem 6.2.1.

**Forests.** A forest is an acyclic graph, that is, all its components are trees. Alternatively, the class of forest is $\mathrm{ex}(K_3)$. The generating function of trees is $C(x) = T(x) - T(x)^2$, where $T(x) = xe^{T(x)}$ (see Section 6.3). The generating function of forests is

$$G(x) = e^{C(x)}.$$

From here it follows easily that the number of forests is asymptotically (a result first proved by Takács [63])

$$e^{1/2}n^{n-2} \sim \left(\frac{e}{2\pi}\right)^{1/2} n^{-5/2}e^n n!.$$

**Outerplanar graphs.** A graph is outerplanar if it can be embedded in the plane so that all the vertices are in the outer face. They are also the same as the class $\mathrm{ex}(K_4, K_{3,2})$. A 2-connected outerplanar graph has a unique Hamilton cycle and it can be drawn in the plane as a polygon plus some non-intersecting diagonals; in classical enumeration these are known as polygon dissections [22], and are enumerated by the small Schröder numbers. From here it follows easily [8] that the generating function $B(x)$ of 2-connected outerplanar graphs satisfies

$$B'(x) = \frac{1+5x-\sqrt{1-6x+x^2}}{8}.$$

The singularity of $B(x)$ is at $R = 3-2\sqrt{2}$. One checks that $B''(R) > 1/R$, which implies that in this case Equation (6.2) has a branch point, as opposed to what happened for planar graphs. In other words, the singularity of $C(X)$ from a branch point and not from the singularity of $B(x)$. From here one gets the estimate for the number of outerplanar graphs with $n$ vertices

$$cn^{-5/2}\gamma^n n!, \qquad \gamma \approx 7.32.$$

Here and in the sequel $c$ is a computable positive constant, that depends only on the class under consideration.

If we compare with the numeration of forests, we see the same behavior. For forests $B(x) = x^2/2$, while for outerplanar graphs it is an algebraic function. However in both cases the singularity comes from the existence of a branch point and the behavior of $C'(x)$ near its singularity is in both cases of square-root type. This remark applies to the next example too.

**Series-parallel graphs.** A graph is series-parallel if it does not contain $K_4$ as a minor, that is, series-parallel graphs constitute the class $\mathrm{ex}(K_4)$. Equivalently, series-parallel graphs are those that can be obtained from a forest by repeatedly applying series and parallel operations. Clearly, outerplanar graphs are series-parallel. The starting point for the enumeration is the observation that a series-parallel graph always has a vertex of degree at most two, hence it cannot be 3-connected. Hence the generating function of 3-connected graphs is identically zero. The equation for $D$ in Theorem 6.2.1 reduces to

$$1+D(x,y) = (1+y)\exp\left(\frac{xD(x,y)^2}{1+xD(x,y)}\right).$$

Applying (6.2.1), this gives a direct expression for $B(x,y)$ in terms of $D(x,y)$. It is not as explicit as for outerplanar graphs, but the situation is very similar. Equation (6.2) has again a branch point and one gets the estimate for the number of series parallel graphs

$$cn^{-5/2}\gamma^n n!, \qquad \gamma \approx 9.07.$$

**Classes defined by 3-connected components.** One can define a class of graphs $\mathcal{G}$ by fixing the class $\mathcal{T}$ of 3-connected members in the class, and letting $\mathcal{G}$ be the graphs whose 3-connected components are in $\mathcal{T}$. For instance, if $\mathcal{T} = \emptyset$, then $\mathcal{G}$ is the class of series-parallel graphs. A detailed analysis of classes defined in terms of 3-connected components is done in [28]. It is shown that the asymptotic enumeration of $\mathcal{G}$ depends essentially on the analytic properties of the generating function $T(x,y)$ of 3-connected graphs at its singularity (for each fixed value of $y$). In a number of cases one shows that the behavior is similar to the class of series-parallel graphs, with a subexponential term $n^{-5/2}$. In other cases, $\mathcal{G}$ is close to the class of planar graphs and the subexponential term is $n^{-7/2}$. This difference in the asymptotic estimates implies a profound difference in the properties of random graphs from $\mathcal{G}$ with respect to the size of the largest block [28].

We reproduce part of a table from [28] for several minor-closed classes.

| Class | $\alpha$ | $\gamma$ |
|---|---|---|
| Series-parallel | $-5/2$ | 9.07 |
| $\text{ex}(W_4)$ | $-5/2$ | 11.54 |
| $\text{ex}(W_5)$ | $-5/2$ | 14.67 |
| $\text{ex}(K_5^-)$ | $-5/2$ | 15.65 |
| $\text{ex}(K_3 \times K_2)$ | $-5/2$ | 14.67 |
| Planar | $-7/2$ | 27.2269 |
| $\text{ex}(K_{3,3})$ | $-7/2$ | 27.2293 |
| $\text{ex}(K_{3,3}^+)$ | $-7/2$ | 27.2295 |

The constants $\gamma$ and $\alpha$ are such that the asymptotic estimate is of the form

$$cn^{\alpha}\gamma^n n!.$$

The classes are defined in terms of excluded minors. $W_n$ is the wheel with $n$ vertices, $K_5^-$ is the graph obtained from $K_5$ by removing one edge, and $K_{3,3}^+$ is obtained from $K_{3,3}$ by adding one edge.

The first five classes in the table are *subcritical* classes. A class is subcritical if Equation (6.2) relating the generating functions $B(x)$ and $C(x)$ has a branch point. Hence the singularity of $C(x)$ comes from this branch point and not from the singularity of $C(x)$. As shown in [17], for subcritical classes the exponent is always $-5/2$. One may notice that in these cases the excluded minors are planar, as opposed to the cases where the exponent is $-7/2$. It is conjectured in [46] that the class $\text{ex}(H)$ is subcritical precisely when $H$ is planar.

All the previous examples are from addable classes. A first investigation of non-addable classes is carried out in [13]. Here we list some examples.

- The class of graphs that contain at most one cycle. These are trees and unicyclic graphs. The estimate for the number of graphs is
$$c \cdot n^{-3/4} e^n n!, \qquad c = \frac{1}{(2e)^{1/4}\Gamma(1/4)}.$$
The unusual exponent comes from a singularity of type fourth-root.

- The class of forests of caterpillars, that is, all the components are caterpillars. The estimate in this case is
$$cn^{-3/4} e^{2\sqrt{\alpha n}} \gamma^n n!,$$
where $\gamma$ is the inverse of the only positive root $\rho$ of $xe^x = 1$, $\alpha = \frac{1-\rho)^2}{2(1+\rho)}$, and $c$ is expressed in terms of $\rho$ and $\gamma$. The unusual asymptotics are due to the fact that $C(x)$ has a simple pole, so that $G(x)$ has an essential singularity.

- The class of graphs all whose connected components have size at most $k$. The generating function for connected graphs is $C(x) = \sum_{i=1}^{k} c_i x^i / i!$, where $c_i$ is the number of connected graphs with $i$ vertices. Hence $G(x) = \exp C(x)$ is the exponential of a polynomial. Applying Hayman's admissibility theorem, one obtains the estimate
$$\frac{1}{\sqrt{2\pi kn}} \frac{A(\zeta)}{\zeta^n} n!,$$
where $\zeta = \zeta_n$ is the solution to the saddle-point equation $\zeta C'(\zeta) = n$, and satisfies $\zeta = \alpha n^{1/k} + \beta + O(n^{-1/k})$ for suitable constants $\alpha$ and $\beta$.

## 6.8 Digraphs

A digraph is a labeled directed graph without loops or multiple arcs, but possibly with both arcs $uv$ and $vu$ for a given pair of vertices. The number of digraphs on $n$ vertices is of course $2^{n(n-1)}$, since there are $n(n-1)$ possible arcs joining distinct vertices. A source in a digraph is a vertex of indegree zero. A digraph is acyclic if it has no directed cycle, and it is strongly connected if any two vertices are joined by a directed path.

### 6.8.1 Acyclic digraphs

A very classical problem is to enumerate acyclic digraphs; they are clearly loopless and given two distinct vertices $u$ and $v$, only one of the two arcs $uv$ and $vu$ can be present. Let $a_n$ be the number of acyclic digraphs on $n$ labeled vertices, and let $a_{n,m}$ be the number of those having $m$ arcs. There is a simple recurrence for computing these numbers, found independently in [60, 62] and rediscovered several times. The number of acyclic digraphs such that $k$ given vertices are sources is equal to $2^{k(n-k)}a_{n-k}$, since the remaining $n-k$ vertices induce an acyclic digraph and every arc from the sources to the complement is allowed. Given that an acyclic digraph always has at least one source, inclusion-exclusion gives

$$a_n = \sum_{t=0}^{n-1} \binom{n}{t} (-1)^{n-t-1} 2^{t(n-t)} a_t, \qquad a_0 = 1.$$

This gives the sequence

$$1, 1, 3, 19, 219, 4231, 130023, \ldots,$$

which is sequence A001035. More generally, if

$$A_n(y) = \sum_{m=0}^{\binom{n}{2}} a_{n.m} y^m$$

is the edge-enumerator of acyclic digraphs with $n$ vertices, then

$$A_n(y) = \sum_{t=0}^{n-1} \binom{n}{t} (-1)^{n-t-1} (1+x)^{t(n-t)} A_t(y).$$

In order to determine the asymptotic behavior of the $a_n$, we introduce the generating functions

$$A(x) = \sum_{n=0}^{\infty} \frac{a_n x^n}{2^{\binom{n}{2}} n!}, \qquad A(x,y) = \sum_{n=0}^{\infty} \frac{A_n(y) x^n}{(1+y)^{\binom{n}{2}} n!}.$$

Define also

$$\Psi(x) = \sum_{n=0}^{\infty} \frac{(-1)^n z^n}{2^{\binom{n}{2}} n!}, \qquad A(x,y) = \sum_{n=0}^{\infty} \frac{(-1)^n x^n}{(1+y)^{\binom{n}{2}} n!}.$$

The previous recurrence relations imply the equations

$$A(x) = \frac{1}{\Psi(x)}, \qquad A(x,y) = \frac{1}{\Psi(x,y)}.$$

By locating the smallest zero of $\Psi(x)$ one obtains the estimate

$$a_n \sim \lambda \gamma^n n! 2^{\binom{n}{2}},$$

where $\gamma \approx 0.67201$ and $\lambda \approx 1.74106$.

The asymptotic enumeration of dense acyclic digraphs (with a quadratic number of arcs) is analyzed in [5].

### 6.8.2 Strongly connected digraphs

It is well-known that almost all digraphs are strongly connected, since with high probability there is a path of length two between any two vertices. Liskovec [40] found a recurrence for the number $s_n$ of strongly connected digraphs on $n$ vertices, later simplified by Wright [72]. Let $b_n$ be defined recursively by

$$b_n = 2^{n(n-1)} - \sum_{t=1}^{n-1} \binom{n}{t} 2^{(n-1)(n-t)} b_t, \qquad b_1 = 1.$$

Then

$$s_n = b_n + \sum_{t=1}^{n-1} \binom{n-1}{t-1} s_t b_{n-t} \qquad s_1 = 1.$$

This gives the sequence

$$1, 1, 18, 1606, 565080, \ldots,$$

which is sequence A003030.

Wright [73] also studied the number $s_{n,m}$ of strongly connected digraphs with $n$ vertices and $m$ edges. He showed that

$$s_{n,n+k} = P_k(n)n!,$$

where $P_k$ is polynomial in $n$ of degree $3k-1$. Notice the similarity of the previous expression with equation (6.5) for connected graphs with excess $k$. The first values are

$$P_1(n) = \frac{1}{4}(n-2)(n+3),$$
$$P_2(n) = \frac{1}{2880}(n-2)(51n^4 + 297n^3 - 271n^2 - 1937n - 1020).$$

This was extended more recently in [48] to cover the range $k = O(n \log n)$. Beyond this range a digraph is almost surely strongly connected.

## 6.9 Unlabeled graphs

An unlabeled graph is an isomorphism class of labeled graphs. The number of ways of labeling an unlabeled graph $G$ is equal to $n!/\text{aut}(G)$, where aut(G) is the number of automorphisms of $G$. Clearly counting unlabeled graphs is more demanding than counting their labeled counterparts, since symmetries must be taken into account. In this section we review briefly the theory of counting under symmetries, and also discuss the asymptotic enumeration of unlabeled graphs.

### 6.9.1 Counting graphs under symmetries

The starting point of the theory is the orbit-counting lemma, also known as Burnside-Frobenius lemma. Let $\Gamma$ be a permutation group acting on a finite set $X = \{1,\ldots,n\}$. Then the number of orbits of the action of $\Gamma$ on $X$ is equal to

$$\frac{1}{|\Gamma|}\sum_{\sigma\in\Gamma}\operatorname{fix}(\sigma),$$

where $\operatorname{fix}(\sigma)$ is the number of points fixed by $\sigma$.

The type of a permutation $\sigma$ is $1^{s_1}2^{s_2}\cdots n^{s_n}$, where $s_i$ is the number of $i$-cycles in the cycle decomposition of $\sigma$. Its cycle index is a monomial in $n$ variables defined as

$$z(\sigma;x_1,\ldots,x_n) = x_1^{s_1}\cdots x_n^{s_n}.$$

The cycle index polynomial of $\Gamma$ is

$$Z(\Gamma;x_1,\ldots,x_n) = \frac{1}{|\Gamma|}\sum_{\sigma\in\Gamma} z(\sigma;x_1,\ldots,x_n) = \frac{1}{|\Gamma|}\sum_{\sigma\in\Gamma}\prod_{i=1}^{n} x_i^{s_i(g)}.$$

The cycle index polynomial can be used to count inequivalent colorings (in a very general sense) of combinatorial objects under symmetries. Let $C$ be a set of "colors" and $c = |C|$. A coloring is a mapping $f\colon X \to C$. Two mapping $f,g$ are equivalent if there exists $\sigma \in \Gamma$ such that $g(x) = f(\sigma(x))$. That is, they are the same mapping up to a symmetry. The basic version of Pólya theorem is that the number of inequivalent colorings of $X$ by colors in $C$ under the action of $\Gamma$ is equal to

$$Z(G;c,\ldots,c).$$

Classical applications of the previous result are, for instance, the enumeration of necklaces (bicolored cycles) up to cyclic or dihedral symmetry, or counting the number of ways of colorings the faces of a cube with a given number of colors, up to symmetries of the cube.

A useful and important generalization is the following, also known as Redfield-Pólya's theorem. Let $w\colon X \to R$ be a weight function into a commutative ring $R$, which in the cases of interest will be either the ring of integers or a polynomial ring. A **coloring** is a mapping $f\colon X \to R$. The weight of $f$ is defined as $w(f) = \prod_{x\in X} w(f(x))$. Notice that two colorings in the same orbit have the same weight. Then we have the following result.

**Theorem 6.9.1 (Refield-Pólya)** *Let $\mathcal{O}$ be the set of orbits of $X$ by $R$ under the action of $\Gamma$. Then*

$$\sum_{f\in\mathcal{O}} w(f) = Z\left(\Gamma;\sum_{c\in C} w(c),\sum_{c\in C} w(c)^2,\ldots,\sum_{c\in C} w(c)^n\right),$$

When all the weights are equal to 1, one recovers Pólya's theorem.

We now apply Pólya's theorem to counting unlabeled graphs. Let $V = \{1, \ldots, n\}$ and let $E$ be the set of all 2-subsets of $V$. A simple graph with vertex set $V$ can be seen as a mapping $f\colon E \to \{0,1\}$, such that $f(\{x,y\}) = 1$ if and only if $\{x,y\}$ is an edge. Two graphs are isomorphic if there exists a permutation of $V$ mapping edges of one graph to edges of the other one. Let $S_n$ be the symmetric group on $V$ and let $S'_n = \{\sigma' : \sigma \in S_n\}$ be the group (isomorphic to $S_n$) acting on edges as

$$\sigma' : \{x,y\} \to \{\sigma(x), \sigma(y)\}.$$

Then $f, g\colon E \to \{0,1\}$ correspond to isomorphic graphs if they are in the same orbit of the action of $S'_n$ on $E$. For instance, $\sigma \in S_5$ has type $(2^1 3^1)$, then $\sigma'$ has type $(1^1 3^1 6^1)$. If we perform all the calculations (not a pleasant task if done by hand) then we obtain

$$Z(S'_5; x_1, x_2, x_3, x_4, x_5, \ldots) =$$
$$\frac{1}{120}\left(x_1^{10} + 10x_1^4x_2^3 + 15x_1^2x_2^4 + 20x_1x_3^3 + 20x_1x_3x_6 + 30x_2x_4^2 + 24x_5^2\right).$$

The number of unlabeled graphs with five vertices is then

$$Z(S'_5; 2, 2, 2, 2, 2) = 34.$$

The counting sequence of unlabeled graphs, starting with $n = 1$, is

$$1, 2, 4, 11, 34, 156, 1044, 12346, \ldots$$

which is sequence A000088. We can also count edges; it suffices to takes as weights $w(0) = 1, w(1) = x$. Then the enumerator polynomial for edges is

$$Z(S'_5; 1+x, 1+x^2, 1+x^3, 1+x^4, 1+x^5) =$$
$$1 + x + 2x^2 + 4x^3 + 6x^4 + 6x^5 + 6x^6 + 4x^7 + 2x^8 + x^9 + x^{10}.$$

This is an example that shows the usefulness of the general weighted version. Many other examples can be found in [32]. The case of trees can be handled directly with the symbolic method (for rooted trees) and the dissymmetry theorem (for unrooted trees) that is discussed in Chapter 4 of this book.

### 6.9.2 Asymptotics

Let $\mathcal{G}$ be a class of labeled graphs and $\mathcal{U}$ the corresponding class of unlabeled graphs. Suppose we have a result saying that almost all graphs in $\mathcal{G}$ have no automorphism as $n \to \infty$. Then it follows that

$$|\mathcal{U}_n| \sim \frac{|\mathcal{G}_n|}{n!}. \tag{6.10}$$

Hence an estimate for $|\mathcal{G}_n|$ provides an estimate for $\mathcal{U}_n|$. Of course one can ask for finer asymptotics and investigate the error term.

In order to prove such a result it is enough to show that

$$\mathbf{E}(\mathrm{aut}(G)) \to 1, \qquad n \to \infty.$$

This is known in several important cases:

- *General graphs.* As we mentioned in the introduction almost every labeled graph has no automorphism. This statement can be refined by taking into account also the number of edges. Let $G_{n,m}$ and $U_{n,m}$ be, respectively, the number of labeled and unlabeled graphs with $n$ vertices and $m$ edges. Notice that $G_{n,m} = \binom{N}{m}$, where $N = \binom{n}{2}$, and trivially $U_{n,m} \geq G_{n,m}/n!$. The following holds [10, Chap. 9]:
  If $c > 1$ is a constant and
  $$cn\log n \leq m \leq N - cn\log n,$$
  then
  $$U_{n,m} \sim G_{n,m}/n!.$$
  The condition $m \geq cn\log n$ is necessary as otherwise almost surely there is more than one isolated vertex and there is non-trivial automorphism (the second inequality is also necessary by complementation).
- *Regular graphs.* The fact that $\mathbf{E}(\mathrm{aut}(G)) \to 1$ for $d$-regular graphs was first proved by Bollobás for fixed $d$, and then extended to $d = o(n^2)$ by McKay and Wormald; see the discussion in [70].

There are situations where the expected number of automorphisms is very large and the relation (6.10) does not hold. The first notable example is the case of trees. It is well-known that a random tree contains almost surely $\alpha n$ subtrees isomorphic to any fixed subtree $T_0$. If we take $T_0$ as the tree on three vertices rooted at the vertex of degree two, then $T_0$ has a non-trivial automorphism that exchanges the two vertices of degree one. This action can be applied independently to any copy of $T_0$, and this implies that almost surely there are at least $c^n$ autormophisms, where $c = 2^{\alpha}$.

This phenomenon is true more generally for addable minor-closed classes (Section 6.7), since again almost surely there are linearly many pendant copies of each connected graph in the class. For subcritical families there is a general result [17] saying the number of unlabeled graphs in the class is asymptotically

$$c_u n^{-5/2} \gamma_u^n,$$

where $\gamma_u$ is the unlabeled growth constant. Because almost surely the number of automorphisms is exponential, necessarily $\gamma_u > \gamma$, the labeled growth constant. In some cases (for example for outerplanar graphs), $\gamma_u$ has been determined. But for planar graphs this is a remarkable open problem. As we have seen in this case $\gamma \approx 27.23$, which is a lower bound for $\gamma_u$. The best upper bound for $\gamma_u$ is

$$\gamma_u < 30.06,$$

which is proved by encoding unlabeled planar graphs with $\alpha n$ bits, and $\alpha \approx 4.91$. The main obstacle for extending the techniques of [17] to planar graphs is the numeration of unlabeled planar graphs.

# References

[1] E. A. Bender and E. R. Canfield. The asymptotic number of rooted maps on a surface. *J. Combin. Theory Ser. A*, 43:244–257, 1986.

[2] E. A. Bender, E. R. Canfield, and B. D. McKay. The asymptotic number of labeled connected graphs with a given number of vertices and edges. *Random Structures Algorithms*, 2:127–169, 1990.

[3] E. A. Bender and Z. Gao. Asymptotic enumeration of labelled graphs by genus. *Electron. J. Combin.*, 18(1), 2011.

[4] E. A. Bender, Z. Gao, and N. C. Wormald. The number of 2-connected labelled planar graphs. *Electron. J. Combin*, 9, 2002.

[5] E. A. Bender, L. B. Richmond, R. W. Robinson, and N. C. Wormald. The asymptotic number of acyclic digraphs. i. *Combinatorica*, 6:15–22, 1986.

[6] O. Bernardi, M. Noy, and D. Welsh. Growth constants of minor-closed classes of graphs. *J. Combin. Theory Ser. B*, 100:468–484, 2010.

[7] O. Bernardi, M. Noy, and D. Welsh. On the growth rate of minor-closed classes of graphs. arXiv:0710.2995, 2010.

[8] M. Bodirsky, O. Giménez, M. Kang, and M. Noy. Enumeration and limit laws for series-parallel graphs. *European J. Combin*, 28:2091–2105, 2007.

[9] M. Bodirsky, M. Kang, M. Löffler, and C. McDiarmid. Random cubic planar graphs. *Random Structures Algorithms*, pages 78–94, 2007.

[10] B. Bollobás. *Random Graphs*. Academic Press, London, 1985.

[11] B. Bollobás. *Hereditary and monotone properties of combinatorial structures*, pages 1–39. Cambridge U. Press, 2007.

[12] B. Bollobás and A. Thomason. *Hereditary and monotone properties of graphs*, volume 14 of *Algorithms and Combinatorics*, pages 70–78. Springer, 1997.

[13] M. Bousquet-Mélou and K. Weller. Asymptotic properties of some minor-closed classes of graphs. *Combin. Probab. Comput*, 23(1):749–795, 2014.

[14] G. Chapuy, E. Fusy, O. Giménez, B. Mohar, and M. Noy. Asymptotic enumeration and limit laws for graphs of fixed genus. *J. Combin. Theory Ser. A*, 118:748–777, 2011.

[15] G. Chapuy, E. Fusy, M. Kang, and B. Shoilekova. A complete grammar for decomposing a family of graphs into 3-connected components. *Electron. J. Combin.*, 15(1), 2008.

[16] M. Chudnovsky, N. Robertson, P. Seymour, and R. Thomas. The strong perfect graph theorem. *Annals Math.*, pages 51–229, 2006.

[17] M. Drmota, É. Fusy, M. Kang, V. Kraus, and J. Rué. Asymptotic study of subcritical graph classes. *SIAM J. Discrete Math.*, 25:1615–1651, 2011.

[18] Z. Dvořák and S. Norine. Small graph classes and bounded expansion. *J. Combin. Theory Ser. B*, 100:171–175, 2010.

[19] P. Erdős, P. Frankl, and V. Rödl. The asymptotic number of graphs not containing a fixed subgraph and a problem for hypergraphs having no exponent. *Graphs Combin.*, 2:113–121, 1986.

[20] P. Erdős, D. J. Kleitman, and B. L. Rothschild. Asymptotic enumeration of $k_n$-free graphs. *International Colloquium on Combinatorial Theory, Atti dei Convegni Lincei*, 17(2):19–27, 1976.

[21] P. Erdős and A. Rényi. On the evolution of random graphs. *Publ. Math. Inst. Hung. Acad. Sci.*, 5:17–61, 1960.

[22] P. Flajolet and M. Noy. Analytic combinatorics of non-crossing planar configurations. *Discrete Math.*, 204:203–229, 1999.

[23] P. Flajolet, B. Salvy, and G. Schaeffer. Airy phenomena and analytic combinatorics of connected graphs. *Electron. J. Combin*, 11, 2004.

[24] P. Flajolet and R. Sedgewick. *Analytic Combinatorics*. Cambridge University Press, Cambridge, UK, 2009.

[25] I. M. Gessel. Symmetric functions and p-recursiveness. *J. Combin. Theory Ser. A*, 53:257–285, 1990.

[26] O. Giménez and M. Noy. Asymptotic enumeration and limit laws of planar graphs. *J. Amer. Math. Soc.*, 22:309–329, 2009.

[27] O. Giménez and M. Noy. *Counting planar graphs and related families of graphs*, pages 169–210. 2009.

[28] O. Giménez, M. Noy, and J. Rué. Graph classes with given 3-connected components: asymptotic enumeration and random graphs. *Random Structures Algorithms*, 42:438–479, 2013.

[29] M. C. Golumbic. *Algorithmic Graph Theory and Perfect Graphs*. Academic Press, New York, 1980.

[30] I. P. Goulden, D. M. Jackson, and J. W. Reilly. The hammond series of a symmetric function and its application to $p$-recursiveness. *SIAM J. Algebraic Discrete Methods*, 4:179–193, 1983.

[31] V. Guruswami. Enumerative aspects of certain subclasses of perfect graphs. *Discrete Math.*, 205:97–117, 1999.

[32] F. Harary and E. Palmer. *Graphical Enumeration*. Academic Press, New York, London, 1973.

[33] F. Harary and R. W. Robinson. Labeled bipartite blocks. *Canad. J. Math.*, 31:60–68, 1979.

[34] D. M. Jackson and J. W. Reilly. The enumeration of homeomorphically irreducible labelled graphs. *J. Combin. Theory Ser. B*, 19:272–286, 1975.

[35] S. Janson, D. E. Knuth, T. Luczak, and B. Pittel. The birth of the giant component. *Random Structures Algorithms*, 4:233–358, 1993.

[36] S. Janson, T. Luczak, and A. Rucinski. *Random Graphs*. John Wiley, New York, 2000.

[37] D. Kleitman and K. Winston. On the number of graphs without 4-cycles. *Discrete Math.*, 41:167–172, 1982.

[38] P. G. Kolaitis, H. J. Prömel, and B. L. Rothschild. $k_l+1$-free graphs: asymptotic structure and a 0-1 law. *Trans. Amer. Math. Soc*, 303:637–671, 1987.

[39] E. Lamken and B. Rothschild. The numbers of odd cycle-free graphs. *Finite and Infinite Sets*, pages 547–553, 1984.

[40] V. A. Liskovec. The number of strongly connected oriented graphs. *Mat. Zametki*, 8:721–732, 1970.

[41] C. McDiarmid. Random graphs from a minor-closed class. *Combin. Probab. Comput.*, 18:583–599, 2009.

[42] B. D. McKay. Asymptotics for symmetric 0-1 matrices with prescribed row sums. *Ars Combinatoria*, 19A:15–25, 1985.

[43] B. D. McKay and N. C. Wormald. Asymptotic enumeration by degree sequence of graphs with degrees $o(n^{1/2})$. *Combinatorica*, 11:369–382, 1991.

[44] R. C. Mullin and P. J. Schellenberg. The enumeration of c-nets via quadrangulations. *J. Combin. Theory*, 4, 1968.

[45] S. Norine, P. Seymour, R. Thomas, and P. Wollan. Proper minor-closed families are small. *J. Combin. Theory Ser. B*, 96:754–757, 2006.

[46] M. Noy. Random planar graphs and beyond. In *Proc. Int. Congress of Math, Seoul*, IMU, 2014.

[47] M. Noy and L. Ramos. Random planar graphs with given minimum degree. arXiv:1403.5211, 2014.

[48] X. Pérez-Giménez and N. Wormald. Asymptotic enumeration of strongly connected digraphs by vertices and edges. *Random Structures Algorithms*, 43:80–114, 2013.

[49] B. Pittel and N. C. Wormald. Counting connected graphs inside-out. *J. Combin. Theory B*, 93:127–172, 2006.

[50] G. Pólya and R. C. Read. *Combinatorial enumeration of groups*. Springer-Verlag, 1987.

[51] H. J. Prömel and A. Steger. Excluding induced subgraphs: quadrilaterals. *Random Structures Algorithms*, 2:55–71, 1991.

[52] H. J. Prömel and A. Steger. Almost all berge graphs are perfect. *Combin. Probab. Comput.*, 1:53–79, 1992.

[53] H. J. Prömel and A. Steger. The asymptotic number of graphs not containing a fixed color-critical subgraph. *Combinatorica*, 12:463–473, 1992.

[54] H. J. Prömel and A. Steger. *Excluding induced subgraphs III: a general asymptotic*, volume 3. 1992.

[55] H. J. Prömel and A. Steger. Random l-colorable graphs. *Random Structures Algorithms*, 6:21–37, 1995.

[56] R. C. Read. The enumeration of locally restricted graphs (ii). *J. London Math. Soc.*, 35:344–351, 1960.

[57] R. C. Read. The number of k-colored graphs on labelled nodes. *Canad. J. Math.*, 12:410–414, 1960.

[58] R. C. Read and N. C. Wormald. Number of labeled 4-regular graphs. *J. Graph Theory*, 4:203–212, 1980.

[59] N. Robertson and R. P. Vitray. *Representativity of surface embedding*, pages 293–328. Springer, 1990.

[60] R. W. Robinson. *Counting labeled acyclic digraphs*, pages 239–273. 1971.

[61] E. R. Scheinerman and J. Zito. On the size of hereditary classes of graphs. *J. Combin. Theory Ser. B*, 61:16–39, 1994.

[62] R. P. Stanley. Acyclic orientations of graphs. *Discrete Math.*, 5:171–178, 1973.

[63] L. Takacs. On the number of distinct forests. *SIAM J. Discrete Math.*, 3:574–581, 1990.

[64] The On-line Encyclopedia of Integer Sequences. Published electronically at `http://oeis.org/`.

[65] W. T. Tutte. A census of planar triangulations. *Canad. J. Math.*, 14:21–38, 1962.

[66] T. R. S. Walsh. Counting labelled three-connected and homeomorphically irreducible two-connected graphs. *J. Combin. Theory Ser. B*, 32:1–11, 1982.

[67] N. Wormald. Enumeration of labelled graphs i: 3-connected graphs. *J. London Math. Soc.*, 19:7–12, 1979.

[68] N. Wormald and E. M. Wright. The exponential generating function of labelled blocks. *Discrete Math.*, 25:93–96, 1979.

[69] N. C. Wormald. Enumeration of labelled graphs ii: Cubic graphs with a given connectivity. *J. London Math. Soc.*, 20:1–7, 1979.

[70] N. C. Wormald. *Models of random regular graphs*, pages 239–298. 1999.

[71] E. M. Wright. Counting coloured graphs. *Canad. J. Math*, 13, 1961.

[72] E. M. Wright. The number of strong digraphs. *Bull. London Math. Soc.*, 3:348–350, 1971.

[73] E. M. Wright. Formulae for the number of sparsely-edged strong labelled digraphs. *Quart. J. Math. Oxford Ser. (2)*, 28:363–367, 1977.

[74] E. M. Wright. The number of connected sparsely edged graphs. *J. Graph Theory*, 1:317–330, 1977.

[75] E. M. Wright. The number of connected sparsely edged graphs iii. asymptotic results. *J. Graph Theory*, 4:393–407, 1980.

# Chapter 7

## *Unimodality, Log-concavity, Real-rootedness and Beyond*

**Petter Brändén**
*Royal Institute of Technology, Stockholm, Sweden*

**CONTENTS**

## 7.1 Introduction

Many important sequences in combinatorics are known to be log-concave or unimodal, but many are only conjectured to be so although several techniques using methods from combinatorics, algebra, geometry and analysis are now available. Stanley [90] and Brenti [25] have written extensive surveys of various techniques that can be used to prove real-rootedness, log-concavity or unimodality. After a brief introduction and a short section on probabilistic consequences of real-rootedness, we will complement [25, 90] with a survey over new techniques that have been developed, and problems and conjectures that have been solved. I stress that this is not a comprehensive account of *all* work that has been done in the area since *op. cit.*. The selection is certainly colored by my taste and knowledge.

If $\mathscr{A} = \{a_k\}_{k=0}^n$ is a finite sequence of real numbers, then

- $\mathscr{A}$ is **unimodal** if there is an index $0 \le j \le n$ such that

$$a_0 \le \cdots \le a_{j-1} \le a_j \ge a_{j+1} \ge \cdots \ge a_n.$$

- $\mathscr{A}$ is **log-concave** if

$$a_j^2 \ge a_{j-1}a_{j+1}, \qquad \text{for all } 1 \le j < n.$$

- the generating polynomial, $p_{\mathscr{A}}(x) := a_0 + a_1x + \cdots + a_nx^n$, is called **real-rooted** if all its zeros are real. By convention we also consider constant polynomials to be real-rooted.

We say that the polynomial $p_{\mathscr{A}}(x) = \sum_{k=0}^n a_kx^k$ has a certain property if $\mathscr{A} = \{a_k\}_{k=0}^n$ does. The most fundamental sequence satisfying all of the properties above is the $n$th row of Pascal's triangle $\{\binom{n}{k}\}_{k=0}^n$. Log-concavity follows easily from the explicit formula $\binom{n}{k} = n!/k!(n-k)!$:

$$\frac{\binom{n}{k}^2}{\binom{n}{k-1}\binom{n}{k+1}} = \frac{(k+1)(n-k+1)}{k(n-k)} > 1.$$

The following lemma relates the three properties above.

**Lemma 7.1.1** *Let $\mathscr{A} = \{a_k\}_{k=0}^n$ be a finite sequence of nonnegative numbers.*

- *If $p_{\mathscr{A}}(x)$ is real-rooted, then the sequence $\mathscr{A}' := \{a_k/\binom{n}{k}\}_{k=0}^n$ is log-concave.*
- *If $\mathscr{A}'$ is log-concave, then so is $\mathscr{A}$.*
- *If $\mathscr{A}$ is log-concave and positive, then $\mathscr{A}$ is unimodal.*

The author is a Wallenberg Academy fellow supported by a grant from the Knut and Alice Wallenberg Foundation. The author is also supported by a grant from the Göran Gustafsson Foundation.

*Proof.* Suppose $p_{\mathscr{A}}(x)$ is real-rooted. Let $a_k = \binom{n}{k} b_k$, for $1 \le k \le n$. By the Gauss–Lucas theorem below, the polynomial

$$\frac{1}{n} p'_{\mathscr{A}}(x) = \sum_{k=0}^{n} \frac{k}{n} \binom{n}{k} b_k x^{k-1} = \sum_{k=0}^{n-1} \binom{n-1}{k} b_{k+1} x^k \tag{7.1}$$

is real-rooted. The operation

$$x^n p_{\mathscr{A}}(1/x) = \sum_{k=0}^{n} \binom{n}{k} b_{n-k} x^k, \tag{7.2}$$

preserves real-rootedness. Let $1 \le j \le n-1$. Applying the operations (7.1) and (7.2) appropriately to $p_{\mathscr{A}}(x)$, we end up with the real-rooted polynomial

$$b_{j-1} + 2b_j x + b_{j+1} x^2,$$

and thus $b_j^2 \ge b_{j-1} b_{j+1}$. This proves the first statement.

The term-wise (Hadamard) product of a positive and log-concave sequence and a log-concave sequence is again log-concave. Since $\{\binom{n}{k}\}_{k=0}^{n}$ is positive and log-concave, the second statement follows.

The third statement follows directly from the definitions. ∎

**Example 7.1.2** *Natural examples of log-concave polynomials that are not real-rooted are the q-***factorial** *polynomials,*

$$[n]_q! = [n]_q \cdot [n-1]_q \cdots [2]_q \cdot [1]_q,$$

*where* $[k]_q = 1 + q + \cdots + q^{n-1}$. *The polynomial* $[n]_q!$ *is the generating polynomial for the* **number of inversions** *over the symmetric group* $\mathfrak{S}_n$*:*

$$[n]_q! = \sum_{\pi \in \mathfrak{S}_n} q^{\mathrm{inv}(\pi)},$$

*where*

$$\mathrm{inv}(\pi) = |\{1 \le i < j \le n : \pi(i) > \pi(j)\}|,$$

*see [94]. The easiest way to see that* $[n]_q!$ *is log-concave is to observe that* $[k]_q$ *is log-concave. Log-concavity of* $[n]_q!$ *then follows from the fact that if* $A(x)$ *and* $B(x)$ *are generating polynomials of positive log-concave sequences, then so is* $A(x)B(x)$*, see [90].*

**Example 7.1.3** *Examples of unimodal sequences that are not log-concave are the q-***binomial coefficients**

$$\begin{bmatrix} n \\ k \end{bmatrix}_q = \frac{[n]_q!}{[k]_q![n-k]_q!}.$$

*These are polynomials with nonnegative coefficients*

$$\begin{bmatrix} n \\ k \end{bmatrix}_q = a_0(n,k) + a_1(n,k)q + \cdots + a_{k(n-k)}(n,k) q^{k(n-k)}, \tag{7.3}$$

*which are unimodal and symmetric. There are several proofs of this fact, see [90]. For example the Cayley–Sylvester theorem, first stated by Cayley in the 1850's and proved by Sylvester in 1878, implies unimodality of* (7.3)*, see [90]. However* $\begin{bmatrix}4\\2\end{bmatrix}_q = 1+q+2q^2+q^3+q^4$*, which is not log-concave.*

For a proof of the following fundamental theorem we refer to [82].

**Theorem 7.1.4 (The Gauss–Lucas theorem)** *Let* $f(x) \in \mathbb{C}[x]$ *be a polynomial of degree at least one. All zeros of* $f'(x)$ *lie in the convex hull of the zeros of* $f(x)$*.*

**Example 7.1.5** *Let* $\{S(n,k)\}_{k=0}^{n}$ *be the* **Stirling numbers of the second kind***, see [94]. Then* $\bar{S}(n,k) := k!S(n,k)$ *counts the number of surjections from* $[n] := \{1,2,\dots,n\}$ *to* $[k]$*. For a surjection* $f : [n+1] \to [k]$*, let* $j = f(n+1)$*. Conditioning on whether* $|f^{-1}(\{j\})| = 1$ *or* $|f^{-1}(\{j\})| > 1$*, one sees that*

$$\bar{S}(n+1,k) = k\bar{S}(n,k-1) + k\bar{S}(n,k), \quad \textit{for all } 1 \le k \le n+1. \tag{7.4}$$

*Let* $E_n(x) = \sum_{k=1}^{n} \bar{S}(n,k)x^k$*. Then* (7.4) *translates as*

$$E_{n+1}(x) = xE_n(x) + x(x+1)E_n'(x) = x\frac{d}{dx}\Big((x+1)E_n(x)\Big).$$

*By induction and the Gauss–Lucas theorem, we see that* $E_n(x)$ *is real-rooted, and that all its zeros lie in the interval* $[-1,0]$ *for all* $n \ge 1$*. Later, in Example 7.7.2 we will see that the operation of dividing the kth coefficient by k!, for each k, preserves real-rootedness. Hence also the polynomials* $\sum_{k=1}^{n} S(n,k)x^k$*,* $n \ge 1$*, are real-rooted.*

A generalization of finite nonnegative sequences with real-rooted generating polynomials is that of Pólya frequency sequences. A sequence $\{a_k\}_{k=0}^{\infty} \subseteq \mathbb{R}$ is a **Pólya frequency sequence** (PF for short) if all minors of the infinite Toeplitz matrix $(a_{i-j})_{i,j=0}^{\infty}$ are nonnegative. In particular, PF sequences are log-concave. PF sequences are characterized by the following theorem of Edrei [42], first conjectured by Schoenberg.

**Theorem 7.1.6** *A sequence* $\{a_k\}_{k=0}^{\infty} \subseteq \mathbb{R}$ *of real numbers is* PF *if and only its generating function may be expressed as*

$$\sum_{k=0}^{\infty} a_k x^k = Cx^m e^{ax} \prod_{k=0}^{\infty}(1+\alpha_k x) \Big/ \prod_{k=0}^{\infty}(1-\beta_k x),$$

*where* $C, a \ge 0$*,* $m \in \mathbb{N}$*,* $\alpha_k, \beta_k \ge 0$ *for all* $k \in \mathbb{N}$*, and* $\sum_{k=0}^{\infty}(\alpha_k + \beta_k) < \infty$*.*

Hence a finite nonnegative sequence is PF if and only its generating polynomial is real-rooted. This was first proved by Aissen, Schoenberg and Whitney [1]. Theorem 7.1.6 provides, at least in theory, a method of proving combinatorially that a combinatorial polynomial with nonnegative coefficients is real-rooted. Namely to find a combinatorial interpretation of the minors of $(a_{i-j})_{i,j=0}^{\infty}$. This method was used by e.g. Gasharov [52] to prove that the independence polynomial of a $(3+1)$-free graph is real-rooted. For more on PF sequences in combinatorics, see [24].

## 7.2 Probabilistic consequences of real-rootedness

Below we will explain two useful probabilistic consequences of real-rootedness. For further consequences, see Pitman's survey [79]. If $X$ is a random variable taking values in $\{0,\ldots,n\}$, let $a_k = \mathbb{P}[X=k]$ for $0 \le k \le n$, and let

$$p_X(t) = a_0 + a_1 t + \cdots + a_n t^n,$$

be the **partition function** of $X$. Then $X$ has mean

$$\mu = \mathbb{E}[X] = \sum_{k=0}^{n} k\mathbb{P}[X=k] = p_X'(1),$$

and variance

$$\operatorname{Var}(X) = \mathbb{E}[X^2] - \mu^2 = p_X''(1) + p_X'(1) - p_X'(1)^2.$$

The following theorem of Bender [4] has been used on numerous occasions to prove asymptotic normality of combinatorial sequences, see e.g. [4, 5, 8].

**Theorem 7.2.1** *Let $\{X_n\}_{n=1}^{\infty}$ be a sequence of random variables taking values in $\{0,1,\ldots,n\}$ such that*

1. *$p_{X_n}(t)$ is real-rooted for all $n$, and*
2. $\operatorname{Var}(X_n) \to \infty$.

*Then the distribution of the random variable*

$$\frac{X_n - \mathbb{E}[X_n]}{\sqrt{\operatorname{Var}(X_n)}}$$

*converges to the standard normal distribution $N(0,1)$ as $n \to \infty$.*

**Example 7.2.2** *Let $X_n$ be the random variable on the symmetric group $\mathfrak{S}_n$ counting the number of cycles in a uniform random permutation. Since the number of permutations in $\mathfrak{S}_n$ with exactly $k$ cycles is the* **signless Stirling number** *of the first kind $c(n,k)$ (see [94]),*

$$p_{X_n}(t) = \frac{1}{n!} x(x+1)\cdots(x+n-1).$$

*Thus $X_n$ has mean $H_n = 1 + 1/2 + \cdots + 1/n$ and variance*

$$\sigma_n^2 = H_n - \sum_{k=1}^{n} k^{-2}.$$

*Hence the distribution of the random variable*

$$\frac{X_n - H_n}{\sigma_n}$$

*converges to the standard normal distribution $N(0,1)$ as $n \to \infty$.*

For more examples using Theorem 7.2.1, see [4], and for recent examples, see [5, 8].

A simple consequence of Lemma 7.1.1 is that if a polynomial $a_0 + a_1x + \cdots + a_nx^n$ has only real and nonpositive zeros, then there is either a unique index $m$ such that $a_m = \max_k a_k$, or two consecutive indices $m \pm 1/2$ (whence $m$ is a half-integer) such that $a_{m\pm 1/2} = \max_k a_k$. The number $m = m(\{a_k\}_{k=0}^n)$ is called the mode of $\{a_k\}_{k=0}^n$. A theorem of Darroch [40] enables us to easily compute the mode.

**Theorem 7.2.3** *Suppose $\{a_k\}_{k=0}^n$ is a sequence of nonnegative numbers such that the polynomial $p(x) = a_0 + a_1x + \cdots + a_nx^n$ is real-rooted. If $m$ is the mode of $\{a_k\}_{k=0}^n$, and $\mu := p'(1)/p(1)$ its mean, then*

$$\lfloor \mu \rfloor \leq m \leq \lceil \mu \rceil.$$

Applying Theorem 7.2.3 to the signless Stirling numbers of the first kind $\{c(n,k)\}_{k=1}^n$ (Example 7.2.2), we see that

$$\lfloor H_n \rfloor \leq m(\{c(n,k)\}_{k=1}^n) \leq \lceil H_n \rceil.$$

## 7.3 Unimodality and $\gamma$-nonnegativity

We say that the sequence $\{h_k\}_{k=0}^d$ is **symmetric** with **center of symmetry** $d/2$ if $h_k = h_{d-k}$ for all $0 \leq k \leq d$. A property called **$\gamma$-nonnegativity**, which implies symmetry and unimodality, has recently been considered in topological, algebraic and enumerative combinatorics.

The linear space of polynomials $h(x) = \sum_{k=0}^d h_k x^k \in \mathbb{R}[x]$, which are symmetric with center of symmetry $d/2$, has a basis

$$B_d := \{x^k(1+x)^{d-2k}\}_{k=0}^{\lfloor d/2 \rfloor}.$$

If $h(x) = \sum_{k=0}^{\lfloor d/2 \rfloor} \gamma_k x^k (1+x)^{d-2k}$, we call $\{\gamma_k\}_{k=0}^{\lfloor d/2 \rfloor}$ the **$\gamma$-vector** of $h$. Since the binomial numbers are unimodal, having a nonnegative $\gamma$-vector implies unimodality of $\{h_k\}_{k=0}^n$. If the $\gamma$-vector of $h$ is nonnegative, then we say that $h$ is **$\gamma$-nonnegative**. Let $\Gamma_+^d$ be the convex cone of polynomials that have nonnegative coefficients when expanded in $B_d$. Clearly

$$\Gamma_+^m \cdot \Gamma_+^n := \{fg : f \in \Gamma_+^m \text{ and } g \in \Gamma_+^n\} \subseteq \Gamma_+^{m+n}. \tag{7.5}$$

**Remark 7.3.1** Suppose $h(x) = \sum_{k=0}^d h_k x^k \in \mathbb{R}[x]$ is the generating polynomial of a nonnegative and symmetric sequence with center of symmetry $d/2$. If all its zeros are real, then we may pair the negative zeros into reciprocal pairs

$$h(x) = Ax^k \prod_{i=1}^{\ell}(x+\theta_i)(x+1/\theta_i) = Ax^k \prod_{i=1}^{\ell}((1+x)^2 + (\theta_i + 1/\theta_i - 2)x),$$

where $A > 0$. Since $x$ and $(1+x)^2 + (\theta_i + 1/\theta_i - 2)x$ are polynomials in $\Gamma_+^1$, we see that $h$ is $\gamma$-nonnegative by (7.5).

### 7.3.1 An action on permutations

There is a natural $\mathbb{Z}_2^n$-action on $\mathfrak{S}_n$, first considered in a modified version by Foata and Strehl [49], which has been used to prove $\gamma$-nonnegativity. Let $\pi = a_1 a_2 \cdots a_n \in \mathfrak{S}_n$ be a permutation written as a word ($\pi(i) = a_i$), and set $a_0 = a_{n+1} = n+1$. If $k \in [n]$, then $a_k$ is a

- **valley** if $a_{k-1} > a_k < a_{k+1}$,
- **peak** if $a_{k-1} < a_k > a_{k+1}$,
- **double ascent** if $a_{k-1} < a_k < a_{k+1}$, and
- **double descent** if $a_{k-1} > a_k > a_{k+1}$.

Define functions $\varphi_x : \mathfrak{S}_n \to \mathfrak{S}_n$, $x \in [n]$, as follows:

- If $x$ is a double descent, then $\varphi_x(\pi)$ is obtained by moving $x$ into the slot between the first pair of letters $a_i, a_{i+1}$ to the right of $x$ such that $a_i < x < a_{i+1}$;
- If $x$ is a double ascent, then $\varphi_x(\pi)$ is obtained by moving $x$ to the slot between the first pair of letters $a_i, a_{i+1}$ to the left of $x$ such that $a_i > x > a_{i+1}$;
- If $x$ is a valley or a peak, then $\varphi_x(\pi) = \pi$.

There is a geometric interpretation of the functions $\varphi_x$, $x \in [n]$, first considered in [87]. Let $\pi = a_1 a_2 \cdots a_n \in \mathfrak{S}_n$ and imagine marbles at the points $(i, a_i) \in \mathbb{N} \times \mathbb{N}$, for $i = 0, 1, \ldots, n+1$. For $i = 0, 1, \ldots, n$ connect $(i, a_i)$ and $(i+1, a_{i+1})$ with a wire. Suppose gravity acts on the marbles, and that $x$ is not at an equilibrium. If $x$ is released it will slide and stop when it has reached the same height again. The resulting permutation is $\varphi_x(\pi)$, see Figure 7.1.

The functions $\varphi_x$ are commuting involutions. Hence for any subset $S \subseteq [n]$, we may define the function $\varphi_S : \mathfrak{S}_n \to \mathfrak{S}_n$ by

$$\varphi_S(\pi) = \prod_{x \in S} \varphi_x(\pi).$$

Hence the group $\mathbb{Z}_2^n$ acts on $\mathfrak{S}_n$ via the functions $\varphi_S$, $S \subseteq [n]$. For example

$$\varphi_{\{2,3,7,8\}}(573148926) = 857134926.$$

For $\pi \in \mathfrak{S}_n$, let $\mathrm{Orb}(\pi) = \{g(\pi) : g \in \mathbb{Z}_2^n\}$ be the orbit of $\pi$ under the action. There is a unique element in $\mathrm{Orb}(\pi)$ that has no double descents, which we denote by $\hat{\pi}$.

**Theorem 7.3.2** *Let $\pi = a_1 a_2 \cdots a_n \in \mathfrak{S}_n$. Then*

$$\sum_{\sigma \in \mathrm{Orb}(\pi)} x^{\mathrm{des}(\sigma)} = x^{\mathrm{des}(\hat{\pi})}(1+x)^{n-1-2\mathrm{des}(\hat{\pi})} = x^{\mathrm{peak}(\pi)}(1+x)^{n-1-2\mathrm{peak}(\pi)},$$

*where* $\mathrm{des}(\pi) = |\{i \in [n] : a_i > a_{i+1}\}|$ *and* $\mathrm{peak}(\pi) = |\{i \in [n] : a_{i-1} < a_i > a_{i+1}\}|$.

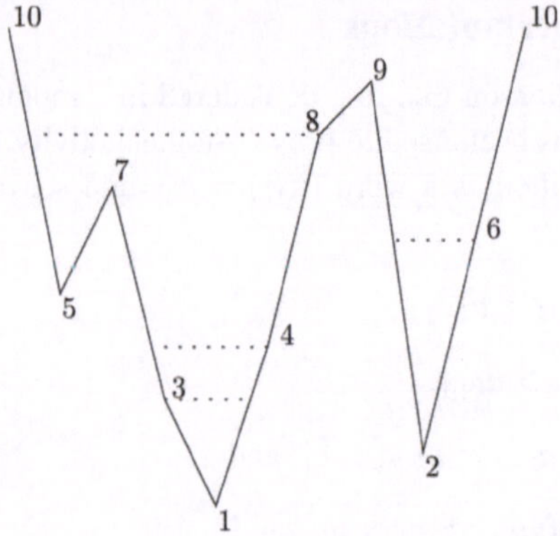


**Figure 7.1**
Graphical representation of $\pi = 573148926$. The dotted lines indicate where the double ascents/descents move to.

*Proof.* If $x$ is a double ascent in $\pi$ then $\operatorname{des}(\varphi_x(\pi)) = \operatorname{des}(\pi)+1$. It follows that

$$\sum_{\sigma\in \operatorname{Orb}(\pi)} x^{\operatorname{des}(\sigma)} = x^{\operatorname{des}(\hat{\pi})}(1+x)^a,$$

where $a$ is the number of double ascents in $\hat{\pi}$. If we delete all double ascents from $\hat{\pi}$ we get an alternating permutation

$$n+1 > b_1 < b_2 > b_3 < \cdots > b_{n-a} < n+1,$$

with the same number of descents. Hence $n-a = 2\operatorname{des}(\hat{\pi})+1$. Clearly $\operatorname{des}(\hat{\pi}) = \operatorname{peak}(\pi)$ and the theorem follows. ■

For a subset $T$ of $\mathfrak{S}_n$ let

$$A(T;x) := \sum_{\pi\in T} x^{\operatorname{des}(\pi)}.$$

**Corollary 7.3.3** *If $T \subseteq \mathfrak{S}_n$ is invariant under the $\mathbb{Z}_2^n$-action, then*

$$A(T;x) = \sum_{i=0}^{\lfloor n/2\rfloor} \gamma_i(T)x^i(1+x)^{n-1-2i},$$

*where*

$$\gamma_i(T) = 2^{-n+1+2i}|\{\pi\in T : \operatorname{peak}(\pi) = i\}|.$$

*In particular $A(T,x)$ is $\gamma$-nonnegative.*

*Proof.* It is enough to prove the theorem for an orbit of a permutation $\pi \in \mathfrak{S}_n$. Since the number of peaks is constant on $\operatorname{Orb}(\pi)$ the equality follows from Theorem 7.3.2. ■

**Example 7.3.4** *Recall that the* **Eulerian polynomials** *are defined by*

$$A_n(x) = \sum_{\pi \in \mathfrak{S}_n} x^{\operatorname{des}(\pi)+1}, \tag{7.6}$$

*see [94]. By Corollary 7.3.3,*

$$A_n(x)/x = \sum_{i=0}^{\lfloor n/2 \rfloor} \gamma_{ni} x^i (1+x)^{n-1-2i},$$

*where*

$$\gamma_{ni} = 2^{-n+1+2i} |\{\pi \in \mathfrak{S}_n : \operatorname{peak}(\pi) = i\}|.$$

**Example 7.3.5** *This example is taken from [18]. The* **stack-sorting operator** *$S$ may be defined recursively on permutations of finite subsets of $\{1,2,\ldots\}$ as follows. If $w$ is empty, then $S(w) := w$. If $w$ is nonempty, write $w$ as the concatenation $w = LmR$ where $m$ is the greatest element of $w$, and $L$ and $R$ are the subwords to the left and right of $m$, respectively. Then $S(w) := S(L)S(R)m$.*

*If $\sigma, \tau \in \mathfrak{S}_n$ are in the same orbit under the $\mathbb{Z}_2^n$-action, then it is not hard to prove that $S(\sigma) = S(\tau)$, see [18]. Let $r \in \mathbb{N}$. A permutation $\pi \in \mathfrak{S}_n$ is said to be* **r-stack sortable** *if. $S^r(\pi) = 12\cdots n$. Denote by $\mathfrak{S}_n^r$ the set of r-stack sortable permutations in $\mathfrak{S}_n$. Hence $\mathfrak{S}_n^r$ is invariant under the $\mathbb{Z}_2^n$-action for all $n, r \in \mathbb{N}$, so Corollary 7.3.3 applies to prove that for all $n, r \in \mathbb{N}$*

$$A(\mathfrak{S}_n^r; x) = \sum_{i=0}^{\lfloor n/2 \rfloor} \gamma_i(\mathfrak{S}_n^r) x^i (1+x)^{n-1-2i},$$

*where*

$$\gamma_i(\mathfrak{S}_n^r) = 2^{-n+1+2i} |\{\pi \in \mathfrak{S}_n^r : \operatorname{peak}(\pi) = i\}|.$$

*Unimodality and symmetry of $A(\mathfrak{S}_n^r; x)$ was first proved by Bona [7]. Bona conjectured that $A(\mathfrak{S}_n^r; x)$ is real-rooted for all $n, r \in \mathbb{N}$. This conjecture remains open for all $3 \le r \le n-3$, see [18].*

*More generally, if $A \subseteq \mathfrak{S}_n$, then the polynomial*

$$\sum_{\substack{\pi \in \mathfrak{S}_n \\ S(\pi) \in A}} x^{\operatorname{des}(\pi)}$$

*is $\gamma$-nonnegative.*

Postnikov, Reiner and Williams [81] modified the $\mathbb{Z}_2^n$-action to prove Gal's conjecture (see Conjecture 7.3.8) for so-called chordal nestohedra.

In [88], Shareshian and Wachs proved refinements of the $\gamma$-positivity of Eulerian polynomials. Let

$$A_n(q,p,s,t) = \sum_{k=0}^{n} A_{n,k}(q,p,t) s^k = \sum_{\sigma \in \mathfrak{S}_n} q^{\operatorname{maj}(\sigma)} p^{\operatorname{des}(\sigma)} t^{\operatorname{exc}(\sigma)} s^{\operatorname{fix}(\sigma)},$$

where

$$\begin{aligned}\operatorname{exc}(\sigma) &= |\{i : \sigma(i) > i\}|,\\ \operatorname{fix}(\sigma) &= |\{i : \sigma(i) = i\}|, \text{ and}\\ \operatorname{maj}(\sigma) &= \sum_{i:\sigma(i)>\sigma(i+1)} i.\end{aligned}$$

**Theorem 7.3.6** *Let* $B_d = \{t^k(1+t)^{d-2k}\}_{k=0}^{\lfloor d/2 \rfloor}$.

1. *The polynomial* $A_{n,0}(q,p,q^{-1}t)$ *has coefficients in* $\mathbb{N}[q,p]$ *when expanded in* $B_n$.

2. *If* $1 \leq k \leq n$, *then* $A_{n,k}(q,1,q^{-1}t)$ *has coefficients in* $\mathbb{N}[q]$ *when expanded in* $B_{n-k}$.

3. *The polynomial* $A_n(q,1,1,q^{-1}t)$ *has coefficients in* $\mathbb{N}[q]$ *when expanded in* $B_{n-1}$.

Gessel [53] has conjectured a fascinating property that resembles $\gamma$-nonnegativity for the joint distribution of descents and inverse descents:

**Conjecture 7.3.7 (Gessel [53, 18, 78])** *If n is a positive integer, then there are nonnegative numbers* $c_n(k,j)$ *for all* $k,j \in \mathbb{N}$ *such that*

$$\sum_{\pi \in \mathfrak{S}_n} x^{\operatorname{des}(\pi)} y^{\operatorname{des}(\pi^{-1})} = \sum_{\substack{k,j \in \mathbb{N} \\ k+2j \leq n-1}} c_n(k,j)(x+y)^k(xy)^j(1+xy)^{n-k-1-2j}. \tag{7.7}$$

The existence of integers $c_n(k,j)$ satisfying (7.7) follows from symmetry properties, see [78]. The open problem is nonnegativity.

### 7.3.2 $\gamma$-nonnegativity of $h$-polynomials

In topological combinatorics the $\gamma$-vectors were introduced in the context of face numbers of simplicial complexes [15, 51]. The **$f$-polynomial** of a $(d-1)$-dimensional simplicial complex $\Delta$ is

$$f_\Delta(x) = \sum_{k=0}^{d} f_{k-1}(\Delta)x^k,$$

where $f_k(\Delta)$ is the number of $k$-dimensional faces in $\Delta$, and $f_{-1}(\Delta) := 1$. The **$h$-polynomial** is defined by

$$h_\Delta(x) = \sum_{k=0}^{d} h_k(\Delta)x^k = (1-x)^d f_\Delta(x/(1-x)), \quad \text{or equivalently,} \tag{7.8}$$

$$f_\Delta(x) = (1+x)^d h_\Delta(x/(1+x)).$$

Hence $f_\Delta(x)$ and $h_\Delta(x)$ contain the same information. If $\Delta$ is a $(d-1)$-dimensional homology sphere, then the **Dehn–Sommerville relations** (see [91]) tell us that $h_\Delta(x)$ is symmetric, so we may expand it in the basis $B_d$. Recall that a simplicial complex $\Delta$ is **flag** if all minimal non-faces of $\Delta$ have cardinality two. Motivated by the Charney–Davis conjecture below, Gal made the following intriguing conjecture:

**Conjecture 7.3.8 (Gal [51])** *If $\Delta$ is a flag homology sphere, then $h_\Delta(x)$ is $\gamma$-nonnegative.*

Gal's conjecture is true for dimensions less than five, see [51]. If $h_\Delta(x)$ is symmetric with center of symmetry $d/2$, then $h_\Delta(-1) = 0$ if $d$ is odd, and $h_\Delta(-1) = (-1)^{d/2}\gamma_{d/2}(\Delta)$ if $d$ is even. Hence Gal's conjecture implies the Charney–Davis conjecture:

**Conjecture 7.3.9 (Charney–Davis [30])** *If $\Delta$ is a flag $(d-1)$-dimensional homology sphere, where $d$ is even, then $(-1)^{d/2}h_\Delta(-1)$ is nonnegative.*

Postnikov, Reiner and Williams [81] proposed a natural extension of Conjecture 7.3.8.

**Conjecture 7.3.10** *If $\Delta$ and $\Delta'$ are flag homology spheres such that $\Delta'$ geometrically subdivides $\Delta$, then the $\gamma$-vector of $\Delta'$ is entry-wise larger or equal to the $\gamma$-vector of $\Delta$.*

Conjecture 7.3.10 was proved for dimensions $\leq 4$ in a slightly stronger form by Athanasiadis [3]. In [3], Athanasiadis also proposes an analog of Gal's conjecture for local $h$-polynomials.

### 7.3.3 Barycentric subdivisions

The collection of faces of a regular cell complex $\Delta$ are naturally partially ordered by inclusion; If $F$ and $G$ are open cells in $\Delta$, then $F \leq G$ if $F$ is contained in the closure of $G$, where we assume that the empty face is contained in every other face. A **Boolean cell complex** is a regular cell complex such that each interval $[\emptyset, F] = \{G \in \Delta : G \leq F\}$ is isomorphic to a Boolean lattice. Hence simplicial complexes are Boolean. The **barycentric subdivision**, sd$(\Delta)$, of a Boolean cell complex, $\Delta$, is the simplicial complex whose $(k-1)$-dimensional faces are strictly increasing flags

$$F_1 < F_2 < \cdots < F_k,$$

where $F_j$ is a nonempty face of $\Delta$ for each $1 \leq j \leq k$. The $f$-polynomials and $h$-polynomials for cell complexes are defined just as for simplicial complexes.

Brenti and Welker [27] investigated positivity properties, such as real-rootedness and $\gamma$-positivity, of the $h$-polynomials of complexes under the operation of taking barycentric subdivisions. This was done by using analytic properties (obtained in [16, 27]) of the linear operator that takes the $f$-polynomial of a Boolean complex to the $f$-polynomial of its barycentric subdivision. These analytic properties will be discussed in Section 7.7.1. In this section we describe the topological consequences of the analytic properties.

Let $\mathscr{E} : \mathbb{R}[x] \to \mathbb{R}[x]$ be the linear operator defined by its image on the binomial basis:

$$\mathscr{E}\binom{x}{k} = x^k, \quad \text{for all } k \in \mathbb{N}, \text{ where } \binom{x}{k} = \frac{x(x-1)\cdots(x-k+1)}{k!}.$$

The operator $\mathscr{E}$ appears in several combinatorial settings. Using the binomial theorem one sees

$$\mathscr{E}(f)(x) = \sum_{n=0}^{\infty} f(n)\frac{x^n}{(1+x)^{n+1}}.$$

It follows e.g. from the theory of $P$-partitions (or from (7.4) and induction) that

$$\mathscr{E}(x^n) = E_n(x) = \sum_{k=1}^{n} k!S(n,k)x^k, \quad \text{for all } n \geq 1,$$

where $\{S(n,k)\}_{k=0}^{n}$ are the Stirling numbers of the second kind, see [94, 102].

The following lemma was proved by Brenti and Welker [27].

**Lemma 7.3.11** *For any Boolean cell complex* $\Delta$,

$$f_{\mathrm{sd}(\Delta)} = \mathscr{E}(f_\Delta).$$

*Proof.* By definition

$$f_{\mathrm{sd}(\Delta)}(x) = \sum_{F \in \Delta} W_F(x),$$

$$W_F(x) = \sum_{k=1}^{\dim F+1} x^k |\{\emptyset < F_1 < \cdots < F_k = F\}|,$$

if $F \neq \emptyset$. Since $\Delta$ is Boolean, there is a one-to-one correspondence between flags $\emptyset < F_1 < \cdots < F_k = F$, $1 \leq k \leq \dim F + 1$, and ordered set-partitions of $[n]$, where $n = \dim F + 1$. Hence $W_F(x) = E_n(x) = \mathscr{E}(x^n)$, and the lemma follows. ∎

**Lemma 7.3.12** *Let $\Delta$ be a $(d-1)$-dimensional Boolean cell complex. If $h_\Delta(x)$ is symmetric, then so is $h_{\mathrm{sd}(\Delta)}(x)$.*

*Proof.* By (7.8), $h_\Delta(x)$ is symmetric if and only if $(-1)^d f_\Delta(-1-x) = f_\Delta(x)$. Let $I : \mathbb{R}[x] \to \mathbb{R}[x]$ be the algebra automorphism defined by $I(x) = -1-x$. It was observed in [16, Lemma 4.3] that

$$I \circ \mathscr{E} = \mathscr{E} \circ I, \tag{7.9}$$

from which the lemma follows. ∎

**Corollary 7.3.13** [27] *Let $\Delta$ be a Boolean cell complex. If the h-polynomial of $\Delta$ has nonnegative coefficients, then all zeros of $h_{\mathrm{sd}(\Delta)}(x)$ are nonpositive and simple.*

*If $h_\Delta(x)$ is also symmetric, then $h_{\mathrm{sd}(\Delta)}(x)$ is $\gamma$-nonnegative.*

*Proof.* The first conclusion is immediate from Theorem 7.7.11, Lemma 7.3.11 and (7.8). The second conclusion follows from Remark 7.3.1 and Lemma 7.3.12. ∎

The second conclusion of Corollary 7.3.13 was strengthened in [71], where it was shown that with the same hypothesis, the $\gamma$-vector of $\mathrm{sd}(\Delta)$ is the $f$-vector of a balanced simplicial complex.

If $\Delta$ is a Boolean cell complex and $k$ is a positive integer, let $\mathrm{sd}^k(\Delta)$ be the simplicial complex obtained by a $k$-fold application of the subdivision operator sd. Most of the following corollary appears in [27].

**Corollary 7.3.14** *Let $\Delta$ be a $(d-1)$-dimensional Boolean cell complex with reduced Euler characteristic $\tilde{\chi}(\Delta)$, where $d \geq 2$. There exists a number $N(\Delta)$ such that*

1. *all zeros of $h_{\mathrm{sd}^n(\Delta)}(x)$ are real and simple for all $n \geq N(\Delta)$,*
2. *if $(-1)^{d-1}\tilde{\chi}(\Delta) \geq 0$, then all zeros of $h_{\mathrm{sd}^n(\Delta)}(x)$ are nonpositive and simple for all $n \geq N(\Delta)$,*
3. *if $(-1)^{d-1}\tilde{\chi}(\Delta) < 0$, then all zeros of $h_{\mathrm{sd}^n(\Delta)}(x)$ except one are nonpositive and simple for all $n \geq N(\Delta)$.*

*Moreover*

$$\lim_{n\to\infty} \frac{1}{d!^n} f_{\mathrm{sd}^n(\Delta)}(x) = f_{d-1}(\Delta)p_d(x), \tag{7.10}$$

*where $p_d(x)$ is the unique monic degree $d$ eigenpolynomial of $\mathscr{E}$ (see Theorem 7.7.12).*

*Proof.* The identity (7.10) follows from the proof of Theorem 7.7.12 by choosing $f = f_\Delta(x)/f_{d-1}(\Delta)$. By Theorem 7.7.12, all zeros of $p_d(x)$ are real, simple and lie in the interval $[-1,0]$. In view of (7.10) all zeros of $f_{\mathrm{sd}^n(\Delta)}(x)$ will be real and simple for $n$ sufficiently large. The same holds for $h_{\mathrm{sd}^n(\Delta)}(x)$ by (7.8).

Assume $(-1)^{d-1}\tilde{\chi}(\Delta) \geq 0$. By Theorem 7.7.12, $p_d(0) = p_d(-1) = 0$. Since $f_{\mathrm{sd}^n(\Delta)}(-1) = f_\Delta(-1) = -\tilde{\chi}(\Delta)$, we see by (7.10) that for all $n$ sufficiently large all zeros of $f_{\mathrm{sd}^n(\Delta)}(x)$ are simple and lie in $[-1,0)$ (since $f_{\mathrm{sd}^n(\Delta)}(x)$ has the correct sign to the left of $-1$). By (7.8) this is equivalent to (2). Statement (3) follows similary. ∎

**Corollary 7.3.15** *Let $\Delta$ be a $(d-1)$-dimensional Boolean cell complex such that $h_\Delta(x)$ is symmetric and $(-1)^{d-1}\tilde{\chi}(\Delta) \geq 0$. Then there is a number $N(\Delta)$ such that $h_{\mathrm{sd}^n(\Delta)}(x)$ is $\gamma$-nonnegative whenever $n \geq N(\Delta)$.*

*Proof.* Combine Remark 7.3.1, Lemma 7.3.12 and Corollary 7.3.14. ∎

### 7.3.4 Unimodality of $h^*$-polynomials

Let $P \subset \mathbb{R}^n$ be an $m$-dimensional **integral polytope**, i.e., all vertices have integer coordinates. Ehrhart [43, 44] proved that the function

$$i(P,r) = |rP \cap \mathbb{Z}^n|,$$

which counts the number of integer points in the $r$-fold dilate of $P$, is a polynomial in $r$ of degree $m$. It follows that we may write

$$\sum_{r=0}^{\infty} i(P,r)x^r = \frac{h_0^*(P) + h_1^*(P)x + \cdots + h_m^*(P)x^m}{(1-x)^{m+1}}. \tag{7.11}$$

Stanley [89] proved that the coefficients of the polynomial, $h_P^*(x)$, in the numerator of (7.11) are nonnegative, and Hibi [59] conjectured that $h_P^*(x)$ is unimodal whenever it is symmetric. Hibi [59] proved the conjecture for $n \leq 5$. However Payne and Mustaţă [69, 74] found counterexamples to Hibi's conjecture for each $n \geq 6$. Let us mention a weaker conjecture that is still open. An integral polytope $P$ is **Gorenstein** if $h_P^*(x)$ is symmetric, and $P$ is **integrally closed** if each integer point in $rP$ may be written as a sum of $r$ integer points in $P$, for all $r \geq 1$.

**Conjecture 7.3.16 (Ohsugi–Hibi [73])** *If $P$ is a Gorenstein and integrally closed integral polytope, then $h_P^*(x)$ is unimodal.*

Inspired by work of Reiner and Welker [84], Athanasiadis [2] provided conditions on an integral polytope $P$ that imply that $h_P^*(x)$ is the $h$-polynomial of the boundary complex of a simplicial polytope. Hence, by the $g$-theorem (see [91]), $h_P^*(x)$ is unimodal. Athanasiadis used this result to prove the following conjecture of Stanley. An **integer stochastic matrix** is a square matrix with nonnegative integer entries having all row and column sums equal to each other. Let $H_n(r)$ be the number of $n \times n$ integer stochastic matrices with row and column sums equal to $r$. The function $r \mapsto H_n(r)$ is the Ehrhart polynomial of the integral polytope $P_n$ of real doubly stochastic matrices. Stanley [91] conjectured that $h_{P_n}^*(x)$ is unimodal for all positive integers $n$, and Athanasiadis' proof of Stanley's conjecture was the main application of the techniques developed in [2]. Subsequently Bruns and Römer [28] generalized Athanasiadis results to the following general theorem.

**Theorem 7.3.17** *Let $P$ be a Gorenstein integral polytope such that $P$ has a regular unimodular triangulation. Then $h_P^*(x)$ is the $h$-polynomial of the boundary complex of a simplicial polytope. In particular, $h_P^*(x)$ is unimodal.*

## 7.4 Log-concavity and matroids

Several important sequences associated to matroids have been conjectured to be log-concave. Progress on these conjectures have been very limited until the recent breakthrough of Huh and Huh–Katz [61, 62]. Recall that the **characteristic polynomial** of a matroid $M$ is defined as

$$\chi_M(x) = \sum_{F \in L_M} \mu(\hat{0}, F)x^{r(M)-r(F)} = \sum_{k=0}^{r} (-1)^k w_k(M)x^{r(M)-k},$$

where $L_M$ is the lattice of flats, $\mu$ its Möbius function, $r$ is the rank function of $M$ and $\{(-1)^k w_k(M)\}_{k=0}^{r(M)}$ are the **Whitney numbers of the first kind**. The sequence $\{w_k(M)\}_{k=0}^{r}$ is nonnegative, and it was conjectured by Rota and Heron to be unimodal. Welsh later conjectured that $\{w_k(M)\}_{k=0}^{r(M)}$ is log-concave. It is known that $\chi_M(1) = 0$. Define the **reduced characteristic polynomial** by

$$\bar{\chi}_M(x) = \chi_M(x)/(x-1) =: \sum_{k=0}^{r-1} (-1)^k v_k(M) x^{r(M)-1-k}.$$

Note that if $\{v_k(M)\}_{k=0}^{r(M)-1}$ is log-concave, then so is $\{w_k(M)\}_{k=0}^{r(M)}$, see [90].

**Theorem 7.4.1 (Huh–Katz [62])** *If $M$ is representable over some field, then the sequence $\{v_k(M)\}_{k=0}^{r(M)-1}$ is log-concave.*

Since the **chromatic polynomial** of a graph is the characteristic polynomial of a representable matroid we have the following corollary:

**Corollary 7.4.2 (Huh [61])** *Chromatic polynomials of graphs are log-concave.*

Let

$$f_M(x) = \sum_{k=0}^{r(M)} (-1)^k f_k(M) x^{r(M)-k},$$

where $f_k(M)$ is the number of independent sets of $M$ of cardinality $k$. Hence $f_M(x)$ is the (signed) $f$-polynomial of the independence complex of $M$. Now, $f_M(x) = \bar{\chi}_{M\times e}(x)$, where $M \times e$ is the **free coextension** of $M$, see [29, 65]. Also if $M$ is representable over some field, then so is $M \times e$. Hence the following corollary is a consequence of Theorem 7.4.1.

**Corollary 7.4.3** *If $M$ is representable over some field, then $\{f_k(M)\}_{k=0}^{r(M)}$ is log-concave.*

This corollary, first noted by Lenz [65], verifies the weakest version of Mason's conjecture below for the class of representable matroids.

**Conjecture 7.4.4 (Mason)** *Let $M$ be a matroid and $n = f_1(M)$. The following sequences are log-concave:*

$$\{f_k(M)\}_{k=0}^{r(M)}, \quad \{k! f_k(M)\}_{k=0}^{r(M)}, \quad \text{and} \quad \left\{ f_k(M) \Big/ \binom{n}{k} \right\}_{k=0}^{r(M)}.$$

The proofs in [61, 62] use involved algebraic machinery that falls beyond the scope of this survey. It is unclear if the method can be extended to the case of non-representable matroids.

## 7.5 Infinite log-concavity

Consider the operator $\mathscr{L}$ on sequences $\mathscr{A} = \{a_k\}_{k=0}^{\infty} \subset \mathbb{R}$ defined by $\mathscr{L}(\mathscr{A}) = \{b_k\}_{k=0}^{\infty}$, where

$$b_0 = a_0^2 \quad \text{and} \quad b_k = a_k^2 - a_{k-1}a_{k+1}, \quad \text{for } k \geq 1.$$

This definition makes sense for finite sequences by regarding these as infinite sequences with finitely many nonzero entries. Hence a sequence $\mathscr{A}$ is log-concave if and only if $\mathscr{L}(\mathscr{A})$ is a nonnegative sequence. A sequence is **$k$-fold log-concave** if $\mathscr{L}^j(\mathscr{A})$ is a nonnegative sequence for all $0 \leq j \leq k$. A sequence is **infinitely log-concave** if it is $k$-fold log-concave for all $k \geq 1$. Although similar notions were studied by Craven and Csordas [37, 38], the following questions asked by Boros and Moll [13] spurred the interest in infinite log-concavity in the combinatorics community:

(A) For $m \in \mathbb{N}$, let $\{d_\ell(m)\}_{\ell=0}^{m}$ be defined by

$$d_\ell(m) = 4^{-m} \sum_{k=\ell}^{m} 2^k \binom{2m-2k}{m-k}\binom{m+k}{m}\binom{k}{\ell}.$$

Is the sequence $\{d_\ell(m)\}_{\ell=0}^{m}$ infinitely log-concave?

(B) For $n \in \mathbb{N}$, is the sequence $\{\binom{n}{k}\}_{k=0}^{n}$ infinitely log-concave?

Question (A) is still open. However Chen et al. [31] proved 3-fold log-concavity of $\{d_\ell(m)\}_{\ell=0}^{m}$ by proving a related conjecture of the author that implies 3-fold log-concavity of $\{d_\ell(m)\}_{\ell=0}^{m}$, for each $m \in \mathbb{N}$, by the work of Craven and Csordas [38].

In connection to (B), Fisk [47], McNamara–Sagan [68], and Stanley [95] independently conjectured the next theorem from which (B) easily follows. We may consider $\mathscr{L}$ to be an operator on the generating function of the sequence, i.e.,

$$\mathscr{L}\left(\sum_{k=0}^{\infty} a_k x^k\right) = \sum_{k=0}^{\infty} (a_k^2 - a_{k-1}a_{k+1})x^k.$$

**Theorem 7.5.1** [19] *If $f(x) = \sum_{k=0}^{n} a_k x^k$ is a polynomial with real and nonpositive zeros only, then so is $\mathscr{L}(f)$. In particular, the sequence $\{a_k\}_{k=0}^{n}$ is infinitely log-concave.*

The proof of Theorem 7.5.1 uses multivariate techniques, and will be given in Section 7.9.4.

There is a simple criterion on a nonnegative sequence $\mathscr{A} = \{a_k\}_{k=0}^{\infty}$ that guarantees infinite log-concavity [38, 68]. Namely

$$a_k^2 \geq r a_{k-1}a_{k+1}, \quad \text{for all } k \geq 1,$$

where $r \geq (3+\sqrt{5})/2$.

McNamara and Sagan [68] conjectured that the operator $\mathscr{L}$ preserves the class of PF sequences. In particular they conjectured that the columns of Pascal's triangle $\{\binom{n+k}{k}\}_{n=0}^{\infty}$, where $k \in \mathbb{N}$, are infinitely log-concave. In [20], Chasse and the author found counterexamples to the first mentioned conjecture and proved the second. They considered PF sequences that are interpolated by polynomials, i.e., PF sequences $\{p(k)\}_{k=0}^{\infty}$ where $p$ is a polynomial, and asked when classes of such sequences are preserved by $\mathscr{L}$.

Let $\mathscr{P}$ be the following class of PF sequences that are interpolated by polynomials:

$$\{\{p(k)\}_{k=0}^{\infty} \in \mathrm{PF} : p(x) \in \mathbb{R}[x] \text{ and } p(-j) = p(-j+1) = 0 \text{ for some } j \in \{0,1,2\}\}.$$

**Theorem 7.5.2** [20] *The operator $\mathscr{L}$ preserves the class $\mathscr{P}$. In particular each sequence in $\mathscr{P}$ is infinitely log-concave.*

Note that for each $k \in \mathbb{N}$, $\{\binom{n+k}{k}\}_{n=0}^{\infty} \in \mathscr{P}$. The following corollary solves the above mentioned conjecture of McNamara and Sagan.

**Corollary 7.5.3** *The columns of Pascal's triangle are infinitely log-concave, i.e., for each $k \in \mathbb{N}$, the sequence $\{\binom{n+k}{k}\}_{n=0}^{\infty}$ is infinitely log-concave.*

Let us end this section with an interesting open problem posed by Fisk [47].

**Problem 7.5.4** *Suppose all zeros of $\sum_{k=0}^{n} a_k x^k$ are nonpositive. If $d \in \mathbb{N}$, are all zeros of*

$$\sum_{k=0}^{n} \det(a_{k+i-j})_{i,j=0}^{d} \cdot x^k,$$

*where $a_i = 0$ if $i \notin \{0,\ldots,n\}$, nonpositive?*

Hence the case $d = 1$ of Problem 7.5.4 is Theorem 7.5.1.

## 7.6 The Neggers–Stanley conjecture

It is natural to ask if the real-rootedness of the Eulerian polynomials may be extended to generating polynomials of linear extensions of any poset. Define a **labeled poset** to be a poset of the form $P = ([n], \leq_P)$, where $n$ is a positive integer. The **Jordan–Hölder** set of $P$,

$$\mathfrak{L}(P) = \{\sigma \in \mathfrak{S}_n : i < j \text{ whenever } \sigma(i) <_P \sigma(j)\},$$

is the set of all linear extensions of $P$. Here $<$ denotes the usual order on the integers. The $P$-**Eulerian polynomial** is defined by

$$W_P(x) = \sum_{\sigma \in \mathfrak{L}(P)} x^{\mathrm{des}(\sigma)+1}.$$

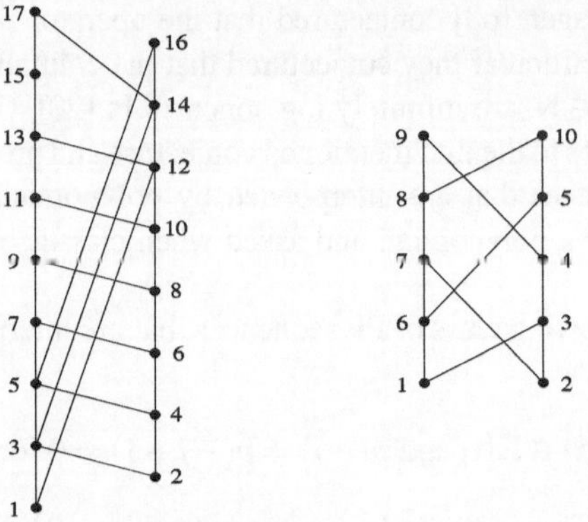


**Figure 7.2**
Counterexamples to Neggers conjecture (left) and the Neggers–Stanley conjecture (right), taken from [97].

Recall that $P$ is **naturally labeled** if $i < j$ whenever $i <_P j$. Neggers [70] conjectured in 1978 that $W_P(x)$ is real-rooted for any naturally labeled poset $P$, and Stanley extended the conjecture to all labeled posets in 1986, see [24, 25, 102]. Counterexamples to Stanley's conjecture were first found by the author in [14], and shortly thereafter naturally labeled counterexamples were found by Stembridge in [97], see Figure 7.2.

However, this does not seem to be the end of the story. Recall that a poset $P$ is **graded** if all maximal chains in $P$ have the same size.

**Theorem 7.6.1 (Reiner and Welker [84])** *If $P$ is a graded and naturally labeled poset, then $W_P(x)$ is unimodal.*

Reiner and Welker proved Theorem 7.6.1 by associating to $P$ a simplicial polytope whose $h$-polynomial is equal to $W_P(x)/x$, and then invoking the $g$-theorem for simplicial polytopes.

Theorem 7.6.1 was refined in [15, 18] to establish $\gamma$-nonnegativity for the $P$-Eulerian polynomials of a class of labeled posets that contain the graded and naturally labeled posets. Let $E(P) = \{(i, j) : j \text{ covers } i\}$ be the **Hasse diagram** of a labeled poset $P$. Define a function $\varepsilon : E(P) \to \{-1, 1\}$, by

$$\varepsilon(i, j) = \begin{cases} 1 & \text{if } i < j, \text{ and} \\ -1 & \text{if } j < i. \end{cases}$$

A labeled poset $P$ is **sign-graded** if for all maximal chains $x_0 <_P x_1 <_P \cdots <_P x_k$ in $P$, the quantity

$$r = \sum_{i=1}^{k} \varepsilon(x_{i-1}, x_i)$$

is the same, see Figure 7.3. Note that a naturally labeled poset is sign-graded if and only if it is graded.

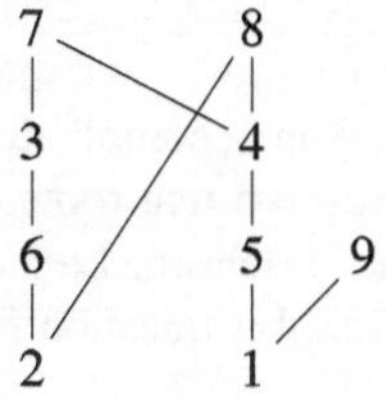


**Figure 7.3**
A sign-graded poset of rank 1.

**Theorem 7.6.2** *If $P$ is sign-graded, then $W_P(x)$ is $\gamma$-nonnegative.*

Two proofs are known for Theorem 7.6.2. The first proof [15] uses a partitioning of $\mathcal{L}(P)$ into Jordan–Hölder sets of refinements of $P$ for which $\gamma$-positivity is easy to prove. The second proof [18] uses an extension to $\mathcal{L}(P)$ of the $\mathbb{Z}_2^n$-action described in Section 7.3.1.

Here are two questions left open regarding the Neggers–Stanley conjecture.

**Question 7.6.3** *Are the coefficients of $P$-Eulerian polynomials log-concave or unimodal?*

**Question 7.6.4** *Are $P$-Eulerian polynomials of graded (or sign-graded) posets real-rooted?*

The work in [15] was generalized by Stembridge [98] to certain Coxeter cones. Let $\Phi$ be a finite root system in a real Euclidian space $V$ with inner product $\langle \cdot,\cdot \rangle$. A **Coxeter cone** is a closed convex cone of the form

$$\Delta(\Psi) = \{\mu \in V : \langle \mu, \beta \rangle \geq 0 \text{ for all } \beta \in \Psi\},$$

where $\Psi \subseteq \Phi$. This cone is a closed union of cells of the Coxeter complex defined by $\Phi$, so it forms a simplicial complex that we identify with $\Delta(\Psi)$. A **labeled Coxeter cone** is a cone of the form

$$\Delta(\Psi, \lambda) = \{\mu \in \Delta(\Psi) : \langle \mu, \beta \rangle > 0 \text{ for all } \beta \in \Psi \text{ with } \langle \lambda, \beta \rangle < 0\},$$

where $\Delta(\Psi)$ is a Coxeter cone and $\lambda \in V$. Hence $\Delta(\Psi, \lambda)$ may be identified with a **relative** complex inside $\Delta(\Psi)$. When $\Phi$ is crystallographic, Stembridge defines what it means for a (labeled) Coxeter cone to be graded. In type $A$, graded labeled Coxeter cones correspond to sign-graded posets.

**Theorem 7.6.5 (Stembridge [98])** *The h-vectors of graded labeled Coxeter cones are $\gamma$-nonnegative.*

## 7.7 Preserving real-rootedness

If a sequence of polynomials satisfies a linear recursion, then to prove that the polynomials are real-rooted it is sufficient to prove that the defining recursion "preserves" real-rootedness. Hence it is natural, from a combinatorial point of view, to ask which linear operators on polynomials preserve real-rootedness. This question has a rich history that goes back to the work of Jensen, Laguerre and Pólya, see the survey [39]. In his thesis, Brenti [24] studied this question focusing on operators occurring naturally in combinatorics.

Let us recall Pólya and Schur's [80] celebrated characterization of **diagonal** operators preserving real-tootedness. A sequence $\Lambda = \{\lambda_k\}_{k=0}^{\infty}$ of real numbers is called a **multiplier sequence (of the first kind)**, if the linear operator $T_\Lambda : \mathbb{R}[x] \to \mathbb{R}[x]$ defined by

$$T_\Lambda(x^k) = \lambda_k x^k, \quad k \in \mathbb{N},$$

preserves real-rootedness.

**Theorem 7.7.1 (Pólya and Schur [80])** *Let $\Lambda = \{\lambda_k\}_{k=0}^{\infty}$ be a sequence of real numbers, and let*

$$G_\Lambda(x) = \sum_{k=0}^{\infty} \frac{\lambda_k}{k!} x^k,$$

*be its exponential generating function. The following assertions are equivalent:*

1. *$\Lambda$ is a multiplier sequence.*

2. *For all nonnegative integers n, the polynomial*

$$T((x+1)^n) = \sum_{k=0}^{n} \binom{n}{k} \lambda_k x^k,$$

*is real-rooted, and all its zeros have the same sign.*

3. *Either $G_\Lambda(x)$ or $G_\Lambda(-x)$ defines an entire function that can be written as*

$$G_\Lambda(\pm x) = Cx^m e^{ax} \prod_{k=1}^{\infty} (1 + \alpha_k x),$$

*where $m \in \mathbb{N}$, $C \in \mathbb{R}$, $a \geq 0$, $\alpha_k \geq 0$ for all $k \in \mathbb{N}$, and $\sum_{k=1}^{\infty} \alpha_k < \infty$.*

4. *$G_\Lambda(x)$ defines an entire function that is the limit, uniform on compact subsets of $\mathbb{C}$, of real-rooted polynomials whose zeros all have the same sign.*

**Example 7.7.2** *Let $\Lambda = \{1/k!\}_{k=0}^{\infty}$. Then $T_\Lambda((x+1)^n)) = L_n(-x)$, where $L_n(x)$ is the nth* **Laguerre polynomial**. *Since orthogonal polynomials are real-rooted, we see that (2) of Theorem 7.7.1 is satisfied, and thus $\Gamma$ is a multiplier sequence.*

Only recently a complete characterization of linear operators preserving real-rootedness was obtained by Borcea and the author in [10]. This characterization is in terms of a natural extension of real-rootedness to several variables. A polynomial $P(x_1,\dots,x_m) \in \mathbb{C}[x_1,\dots,x_m]$ is called **stable** if

$$\mathrm{Im}(x_1) > 0,\dots,\mathrm{Im}(x_m) > 0 \quad \text{implies} \quad P(x_1,\dots,x_n) \neq 0.$$

By convention we also consider the identically zero polynomial to be stable. Hence a univariate real polynomial is stable if and only if it is real-rooted. Let $\alpha_1 \leq \cdots \leq \alpha_n$ and $\beta_1 \leq \cdots \leq \beta_m$ be the zeros of two real-rooted polynomials. We say that these zeros **interlace** if

$$\alpha_1 \leq \beta_1 \leq \alpha_2 \leq \beta_2 \leq \cdots \quad \text{or} \quad \beta_1 \leq \alpha_1 \leq \beta_2 \leq \alpha_2 \leq \cdots.$$

By convention, the "zeros" of any two polynomials of degree 0 or 1 interlace. Interlacing zeros is characterized by a linear condition as the following theorem, often attributed to Obreschkoff, describes:

**Theorem 7.7.3** [72, Satz 5.2] *Let $f,g \in \mathbb{R}[x]\setminus\{0\}$. Then the zeros of $f$ and $g$ interlace if and only if all polynomials in the linear space*

$$\{\alpha f + \beta g : \alpha,\beta \in \mathbb{R}\}$$

*are real-rooted.*

Let $\mathbb{R}_n[x] = \{p \in \mathbb{R}[x] : \deg p \leq n\}$. The **symbol** of a linear operator $T : \mathbb{R}_n[x] \to \mathbb{R}[x]$ is the bivariate polynomial

$$G_T(x,y) = T\left((x+y)^n\right) := \sum_{k=0}^{n} \binom{n}{k} T(x^k)y^{n-k} \in \mathbb{R}[x,y].$$

**Theorem 7.7.4** [10] *Let $T : \mathbb{R}_n[x] \to \mathbb{R}[x]$ be a linear operator. Then $T$ preserves real-rootedness if and only if (1), (2) or (3) below is satisfied.*

*(1) $T$ has rank at most two and is of the form*

$$T(p) = \alpha(p)f + \beta(p)g,$$

*where $\alpha,\beta : \mathbb{R}[x] \to \mathbb{R}$ are linear functionals and $f,g$ are real-rooted polynomials whose zeros interlace.*

*(2) $G_T(x,y)$ is stable.*

*(3) $G_T(x,-y)$ is stable.*

**Example 7.7.5** *The operators of type (1) are the ones achieved by Theorem 7.7.3. An example of an operator of type (2) is $T = d/dx$, because then $G_T(x,y) = n(x+y)^{n-1}$. An example of an operator of type (3) is the algebra automorphism, $S : \mathbb{R}[x] \to \mathbb{R}[x]$, defined by $S(x) = -x$. Indeed $T$ is of type (2) if and only if $T \circ S$ is of type (3).*

To illustrate how Theorem 7.7.4 may be used let us give a simple example from combinatorics.

**Example 7.7.6** *The Eulerian polynomials satisfy the recursion* $A_{n+1}(x) = T_n(A_n(x))$, *where*

$$T_n = x(1-x)\frac{d}{dx} + (n+1)x,$$

*see [94]. The symbol of* $T_n : \mathbb{R}_n[x] \to \mathbb{R}[x]$ *is*

$$T_n\big((x+y)^n\big) = x(x+y)^{n-1}(x+(n+1)y+n),$$

*which is trivially stable. Hence* $A_n(x)$ *is real-rooted for all* $n \in \mathbb{N}$ *by Theorem 7.7.4 and induction. This was first proved by Frobenius [50].*

A characterization of stable polynomials in two variables, and hence of the symbols of preservers of real-rootedness, follows from Helton and Vinnikov's characterization of real-zero polynomials in [58], see [11].

**Theorem 7.7.7** *Let* $P(x,y) \in \mathbb{R}[x,y]$ *be a polynomial of degree d. Then P is stable if and only if there exists three real symmetric* $d \times d$ *matrices A, B and C and a real number r such that*

$$P(x,y) = r \cdot \det(xA + yB + C),$$

*where A and B are positive semidefinite and* $A+B$ *is the identity matrix.*

For the unbounded degree analog of Theorem 7.7.4 we define the **symbol** of a linear operator $T : \mathbb{R}[x] \to \mathbb{R}[x]$ to be the formal powers series

$$\bar{G}_T(x,y) = T(e^{-xy}) := \sum_{n=0}^{\infty} (-1)^n \frac{T(x^n)}{n!} y^n \in \mathbb{R}[x][[y]].$$

The **Laguerre–Pólya class**, $\mathscr{L}\text{–}\mathscr{P}_n$, is defined to be the class of real entire functions in $n$ variables that are the uniform limits on compact subsets of $\mathbb{C}$ of real stable polynomials. For example $\exp(-x_1x_2 - x_3x_4 + 2x_5) \in \mathscr{L}\text{–}\mathscr{P}_5$ since it is the limit of the stable polynomials

$$\left(1 - \frac{x_1x_2}{n}\right)^n \left(1 - \frac{x_3x_4}{n}\right)^n \left(1 + 2\frac{x_5}{n}\right)^n.$$

**Theorem 7.7.8** [10] *Let* $T : \mathbb{R}[x] \to \mathbb{R}[x]$ *be a linear operator. Then T preserves real-rootedness if and only if (1), (2) or (3) below is satisfied.*

*(1) T has rank at most two and is of the form*

$$T(p) = \alpha(p)f + \beta(p)g,$$

*where* $\alpha, \beta : \mathbb{R}[x] \to \mathbb{R}$ *are linear functionals and* $f, g$ *are real-rooted polynomials whose zeros interlace.*

*(2)* $\bar{G}_T(x,y) \in \mathscr{L}\text{–}\mathscr{P}_2$.

*(3)* $\bar{G}_T(x,-y) \in \mathscr{L}\text{–}\mathscr{P}_2$.

There are, as of yet, no analogs of Theorems 7.7.4 and 7.7.8 for linear operators that preserve the property of having all zeros in a prescribed interval (other than $\mathbb{R}$ itself).

**Problem 7.7.9** *Let $I \subset \mathbb{R}$ be an interval. Characterize all linear operators on polynomials that preserve the property of having all zeros in $I$.*

For polynomials appearing in combinatorics the case when $I = (-\infty, 0]$ is the most important.

### 7.7.1 The subdivision operator

An example of an operator of the kind appearing in Problem 7.7.9 is the "subdivision" operator $\mathscr{E} : \mathbb{R}[x] \to \mathbb{R}[x]$ in Section 7.3.2. The following theorem by Wagner proved the Neggers–Stanley conjecture for series-parallel posets, see [102, 103].

**Theorem 7.7.10** [103] *If all zeros of $\mathscr{E}(f)$ and $\mathscr{E}(g)$ lie in the interval $[-1,0]$, then so does the zeros of $\mathscr{E}(fg)$.*

As we have seen in Section 7.3.2, the next theorem has consequences in topological combinatorics.

**Theorem 7.7.11** [16] *If*

$$f(x) = \sum_{k=0}^{d} h_k x^k (1+x)^{d-k},$$

*where $h_k \geq 0$ for all $0 \leq k \leq d$, then all zeros of $\mathscr{E}(f)$ are real, simple and located in $[-1,0]$.*

The main part of the next theorem was proved by Brenti and Welker in [27], while (2) was proved in [41] . We take the opportunity to give alternative simple proofs below.

**Theorem 7.7.12** *For each integer $n \geq 2$, $\mathscr{E}$ has a unique monic eigen-polynomial, $p_n(x)$, of degree $n$.*

*Moreover,*

*(1) all zeros of $p_n(x)$ are real, simple and lie in the interval $[-1,0]$;*

*(2) $p_n(x)$ is symmetric around $-1/2$, i.e.,*

$$(-1)^n p_n(-1-x) = p_n(x).$$

*Proof.* Let $n \geq 2$. Consider the map $\phi : [-1,0]^n \to [-1,0]^n$ defined as follows. Let $\theta = (\theta_1, \ldots, \theta_n) \in [-1,0]^n$. Since $\mathscr{E}$ preserves the property of having all zeros in $[-1,0]$ (Theorems 7.7.10 and 7.7.11), we may order the zeros of the polynomial $\mathscr{E}((x-\theta_1)\cdots(x-\theta_n))$ as $-1 \leq \alpha_1 \leq \cdots \leq \alpha_n \leq 0$. Let $\phi(\theta) := (\alpha_1, \ldots, \alpha_n)$. By Hurwitz' theorem on the continuity of zeros [82], $\phi$ is continuous. Hence by Brouwer's fixed point theorem $\phi$ has a fixed point, which then corresponds to a degree $n$ eigenpolynomial, $p_n$, of $\mathscr{E}$. It follows by examining the leading coefficients that the corresponding eigenvalue is $n!$.

Set $p_0 := 1$ and $p_1 := x + 1/2$. Let $f$ be an arbitrary monic polynomial of degree $n \geq 2$, and let $T = n!^{-1}\mathscr{E}$. Now by expanding $f$ as a linear combination of $\{p_k\}_{k=0}^{n}$,

$$f = \sum_{i=0}^{n} a_i p_i,$$

we see that

$$\lim_{k\to\infty} T^k(f) = \lim_{k\to\infty} \sum_{i=0}^{n} \left(\frac{i!}{n!}\right)^k a_i p_i = p_n,$$

since $a_n = 1$. Hence $p_n$ is unique. By choosing $f$ to be $[-1,0]$-rooted, we see that $T^k(f)$ is also $[-1,0]$-rooted for all $k$. By Hurwitz' theorem, so is $p_n$. Since $p_n$ is $[-1,0]$-rooted, it is certainly of the form displayed in Theorem 7.7.11. By Theorem 7.7.11 again, the zeros of $p_n = n!^{-1}\mathscr{E}(p_n)$ are distinct.

Property (2) follows immediately from (7.9). ■

It is easy to see that the coefficients of $p_n(x)$ are rational numbers for each $n \geq 2$.

**Question 7.7.13** *Is there a closed formula for $p_n(x)$? What are the generating functions*

$$A(x,y) = \sum_{n=0}^{\infty} p_n(x)y^n \quad and \quad B(x,y) = \sum_{n=0}^{\infty} \frac{p_n(x)}{n!}y^n?$$

*Note that $\mathscr{E}(B) = A$.*

## 7.8 Common interleavers

A powerful technique for proving that families of polynomials are real-rooted is that of **compatible polynomials**. This was employed by Chudnovsky and Seymour [35] to prove a conjecture of Hamidoune and Stanley on the zeros of independence polynomials of clawfree graphs. Subsequently an elegant alternative proof was given by Lass [63], by proving a Mehler formula for independence polynomials of clawfree graphs. An **independent set** in a finite and simple graph $G = (V,E)$ is a set of pairwise non-adjacent vertices. The **independence polynomial** of $G$ is the polynomial

$$I(G,x) = \sum_{S} x^{|S|},$$

where the sum is over all independent sets $S \subseteq V$. Recall that a **claw** is a graph isomorphic to the graph on $V = \{1,2,3,4\}$ with edges $E = \{12,13,14\}$. Note that the independence polynomial of a claw is $1+4x+3x^2+x^3$, which has two non-real zeros. A graph is clawfree if no induced subgraph is a claw. The next theorem was posed as a question by Hamidoune [57] and later as a conjecture by Stanley [92].

**Theorem 7.8.1** [35, 63] *If $G$ is a clawfree graph, then all zeros of $I(G,x)$ are real.*

Let $f,g \in \mathbb{R}[x]$ be two real-rooted polynomials with positive leading coefficients. We say that $f$ is an **interleaver** of $g$ (written $f \ll g$) if

$$\cdots \le \alpha_2 \le \beta_2 \le \alpha_1 \le \beta_1,$$

where $\{\alpha_i\}_{i=1}^n$ and $\{\beta_i\}_{i=1}^m$ are the zeros of $f$ and $g$, respectively. By convention we also write $0 \ll 0$, $0 \ll h$ and $h \ll 0$, where $h$ is any real-rooted polynomial with positive leading coefficient. If $f \ll g$ and $f \not\equiv 0$, we say that $f$ is a **proper interleaver** of $g$. The polynomials $f_1(x),\dots,f_m(x)$ are $k$-**compatible**, where $1 \le k \le m$, if

$$\sum_{j\in S} \lambda_j f_j(x)$$

is real-rooted whenever $S \subseteq [m]$, $|S| = k$ and $\lambda_j \ge 0$ for all $j \in S$. The following theorem was used in Chudnovsky and Seymour's proof of Theorem 7.8.1.

**Theorem 7.8.2 (Chudnovsky–Seymour [35])** *Suppose that the leading coefficients of $f_1(x),\dots,f_m(x) \in \mathbb{R}[x]$ are positive. The following are equivalent.*

1. *$f_1(x),\dots,f_m(x)$ are 2-compatible;*
2. *For all $1 \le i < j \le m$, $f_i(x)$ and $f_j(x)$ have a proper common interleaver;*
3. *$f_1(x),\dots,f_m(x)$ have a proper common interleaver;*
4. *$f_1(x),\dots,f_m(x)$ are $m$-compatible.*

Theorem 7.8.2 is useful in situations when the polynomials of interest may be expressed as nonnegative sums of similar polynomials. In order to prove that the polynomials of interest are real-rooted it then suffices to prove that the similar polynomials are 2-compatible.

A sequence $F_n = (f_i)_{i=1}^n$ of real-rooted polynomials is called **interlacing** if $f_i \ll f_j$ for all $1 \le i < j \le n$. Let $\mathscr{F}_n$ be the family of all interlacing sequences $(f_i)_{i=1}^n$ of polynomials, and let $\mathscr{F}_n^+$ be the family of $(f_i)_{i=1}^n \in \mathscr{F}_n$ such that $f_i$ has nonnegative coefficients for all $1 \le i \le n$. We are interested in when an $m \times n$ matrix $G = (G_{ij}(x))$ of polynomials maps $\mathscr{F}_n$ to $\mathscr{F}_m$ (or $\mathscr{F}_n^+$ to $\mathscr{F}_m^+$) by the action

$$G \cdot F_n = (g_1,\dots,g_m)^T, \text{ where } g_k = \sum_{i=0}^{n} G_{ki} f_i \text{ for all } 1 \le k \le m.$$

This problem was considered by Fisk [46, Chapter 3], who proved some preliminary results. Since this approach has been proved succesful in combinatorial situations (see [86] where it was used to prove e.g. that the type $D$ Eulerian polynomials are real-rooted), we take the opportunity to give a complete characterization for the case of nonnegative polynomials.

**Lemma 7.8.3** *If $(f_i)_{i=1}^n$ and $(g_i)_{i=1}^n$ are two interlacing sequences of polynomials, then the polynomial*

$$f_1g_n + f_2g_{n-1} + \cdots + f_ng_1$$

*is real-rooted.*

*Proof.* By Theorem 7.8.2 it suffices to prove that the sequence $(f_ig_{n+1-i})_{i=1}^n$ is 2-compatible. If $i < j$, then $f_ig_{n+1-j}$ is a common interleaver of $f_ig_{n+1-i}$ and $f_jg_{n+1-j}$. Hence the lemma follows from Theorem 7.8.2. ∎

See [86] for a proof of the following lemma.

**Lemma 7.8.4** *Let $f$ and $g$ be two polynomials with nonnegative coefficients. Then $f \ll g$ if and only if for all $\lambda, \mu > 0$, the polynomial*

$$(\lambda x + \mu)f + g$$

*is real-rooted.*

**Theorem 7.8.5** *Let $G = (G_{ij}(x))$ be an $m \times n$ matrix of polynomials. Then $G : \mathscr{F}_n^+ \to \mathscr{F}_m^+$ if and only if*

1. *$G_{ij}$ has nonnegative coefficients for all $i \in [m]$ and $j \in [n]$, and*
2. *for all $\lambda, \mu > 0$, $1 \le i < j \le n$ and $1 \le k < \ell \le m$*

$$(\lambda x + \mu)G_{kj}(x) + G_{\ell j}(x) \ll (\lambda x + \mu)G_{ki}(x) + G_{\ell i}(x). \qquad (7.12)$$

*Proof.* Let

$$g_k = \sum_{i=0}^{n} G_{ki}f_i.$$

By Lemma 7.8.4, $G : \mathscr{F}_n^+ \to \mathscr{F}_m^+$ if and only if for all $k < \ell$ and $\lambda, \mu > 0$

$$(\lambda x + \mu)g_k + g_\ell = \sum_{i=0}^{n}((\lambda x + \mu)G_{ki} + G_{\ell i})f_i =: \sum_{i=0}^{n} h_{n+1-i}f_i$$

is real-rooted and has nonnegative coefficients. The sufficiency follows from Lemma 7.8.3, since if (7.12) holds, then the sequence $(h_i)_{i=1}^n$ is interlacing. To prove the necessity, let $i < j$ and $(f_r)_{r=1}^n$ be the interlacing sequence defined by

$$f_r(x) = \begin{cases} 1 & \text{if } r = i, \\ \alpha x + \beta & \text{if } r = j, \text{ and} \\ 0 & \text{otherwise,} \end{cases}$$

where $\alpha, \beta > 0$. Hence if $G : \mathscr{F}_n^+ \to \mathscr{F}_m^+$, then $h_{n+1-i} + (\alpha x + \beta)h_{n+1-j}$ is real-rooted for all $\alpha, \beta > 0$. Thus $h_{n+1-j} \ll h_{n+1-i}$, by Lemma 7.8.4, which is inequality (7.12). ∎

**Corollary 7.8.6** *Let $G = (G_{ij})$ be an $m \times n$ matrix over $\mathbb{R}$. Then $G : \mathscr{F}_n^+ \to \mathscr{F}_m^+$ if and only if $G$ is $\mathrm{TP}_2$, i.e., all minors of $G$ of size less than three are nonnegative.*

*Proof.* By Theorem 7.8.5 we may assume that all entries of $G$ are nonnegative. Now

$$(\lambda x+\mu)G_{kj}+G_{\ell j} \ll (\lambda x+\mu)G_{ki}+G_{\ell i}$$

for all $\lambda, \mu > 0$ if and only if

$$xG_{kj}+G_{\ell j} \ll xG_{ki}+G_{\ell i},$$

which is seen to hold if and only if $G_{ki}G_{\ell j} \geq G_{\ell i}G_{kj}$. ∎

If $\lambda = (\lambda_1, \lambda_2, \ldots, \lambda_m)$ are integers such that $0 \leq \lambda_1 \leq \lambda_2 \leq \cdots \leq \lambda_m \leq n$, let $G_\lambda = (g_{ij}^\lambda(x))$ be the $m \times n$ matrix with entries

$$g_{ij}^\lambda(x) = \begin{cases} x & \text{if } 1 \leq j \leq \lambda_i \text{ and} \\ 1 & \text{otherwise.} \end{cases}$$

The following corollary was first proved in [86].

**Corollary 7.8.7** *If $\lambda = (\lambda_1, \lambda_2, \ldots, \lambda_m)$ are integers such that $0 \leq \lambda_1 \leq \lambda_2 \leq \cdots \leq \lambda_m \leq n$, then $G_\lambda : \mathscr{F}_n^+ \to \mathscr{F}_m^+$.*

*Proof.* The possible $2 \times 2$ sub-matrices of $G_\lambda$ are

$$\begin{pmatrix} x & x \\ x & x \end{pmatrix}, \begin{pmatrix} x & 1 \\ x & x \end{pmatrix}, \begin{pmatrix} x & 1 \\ x & 1 \end{pmatrix}, \begin{pmatrix} 1 & 1 \\ x & x \end{pmatrix}, \begin{pmatrix} 1 & 1 \\ x & 1 \end{pmatrix} \text{ and } \begin{pmatrix} 1 & 1 \\ 1 & 1 \end{pmatrix}.$$

By Theorem 7.8.5 we need to check (7.12) for these matrices. For example for the second matrix from the right we need to check

$$(\lambda+1)x+\mu \ll x(\lambda x+\mu+1),$$

for all $\lambda, \mu > 0$, which is equivalent to

$$-\frac{\mu+1}{\lambda} \leq -\frac{\mu}{\lambda+1},$$

which is certainly true. The other cases follows similarly. ∎

**Example 7.8.8** *Let $n$ be a positive integer and define polynomials $A_{n,i}(x)$, $i \in [n]$, by*

$$A_{n,i}(x) = \sum_{\substack{\sigma \in \mathfrak{S}_n \\ \sigma(1)=i}} x^{\mathrm{des}(\pi)}.$$

*By conditioning on $\sigma(2) = k$, where $\sigma \in \mathfrak{S}_n$ and $\sigma(1) = i$, we see that*

$$A_{n+1,i}(x) = \sum_{k<i} xA_{n,k}(x) + \sum_{k \geq i} A_{n,k}(x), \quad 1 \leq i \leq n+1.$$

*Hence if* $\mathscr{A}_n = (A_{n,i}(x))_{i=1}^n$, *then*

$$\mathscr{A}_{n+1} = G_{(0,1,2,\ldots,n)} \cdot \mathscr{A}_n.$$

*Since* $\mathscr{A}_2 = (1,x)$, *we have by induction and Corollary 7.8.7 that* $\mathscr{A}_n$ *is an interlacing sequence of polynomials for all* $n \geq 2$.

### 7.8.1 s-Eulerian polynomials

Corollary 7.8.7 was used by Savage and Visontai [86] to prove real-rootedness of a large family of $h^*$-polynomials. Let $\mathbf{s} = \{s_i\}_{i=1}^n$ be a sequence of positive integers. Define an integral polytope $P_\mathbf{s}$ by

$$P_\mathbf{s} = \left\{ (x_1,\ldots,x_n) \in \mathbb{R}^n : 0 \leq \frac{x_1}{s_1} \leq \frac{x_2}{s_2} \leq \cdots \leq \frac{x_n}{s_n} \leq 1 \right\}.$$

The **s-Eulerian polynomial** may be defined as the $h^*$-polynomial of $P_\mathbf{s}$:

$$\sum_{k=0}^{\infty} i(P_\mathbf{s},k)x^k = \frac{E_\mathbf{s}(x)}{(1-x)^{n+1}}.$$

Savage and Schuster [85] provided a combinatorial description of s-Eulerian polynomials. The **s-inversion sequences** are defined by

$$\mathscr{I}_\mathbf{s} = \{\mathscr{E} = (e_1,\ldots,e_n) \in \mathbb{N}^n : e_i/s_i < 1 \text{ for all } 1 \leq i \leq n\}.$$

The **ascent statistic** on $\mathscr{I}_\mathbf{s}$ is defined as

$$\operatorname{asc}(\mathscr{E}) = |\{i \in [n] : e_{i-1}/s_{i-1} < e_i/s_i\}|,$$

where $\mathscr{E} = (e_1,\ldots,e_n)$, $e_0 = 0$ and $s_0 = 1$.

**Theorem 7.8.9** [85]

$$E_\mathbf{s}(x) = \sum_{\mathscr{E} \in \mathscr{I}_\mathbf{s}} x^{\operatorname{asc}(\mathscr{E})}.$$

It turns out that several much studied families of polynomials in combinatorics are s-Eulerian polynomials for various s. For example the $n$th ordinary Eulerian polynomial corresponds to $\mathbf{s} = (1,2,\ldots,n)$, while the $n$th Eulerian polynomial of type $B$ corresponds to $\mathbf{s} = (2,4,\ldots,2n)$. If $\mathbf{s} = (s_1,\ldots,s_n)$, let

$$E_{\mathbf{s},i}(x) = \sum_{\substack{\mathscr{E} \subset \mathscr{I}_\mathbf{s} \\ e_n = i}} x^{\operatorname{asc}(\mathscr{E})}.$$

It is not hard to see that the polynomials $E_{\mathbf{s},i}(x)$ satisfy the following recurrences, which make them ideal for an application of Corollary 7.8.7.

**Lemma 7.8.10** [86] *If* $\mathbf{s} = (s_1, \ldots, s_n)$, $n > 1$, *is a sequence of positive integers and* $0 \leq i < n$, *then*

$$E_{\mathbf{s},i}(x) = \sum_{j=0}^{t_i - 1} x E_{\mathbf{s}',j}(x) + \sum_{j=t_i}^{s_{n-1}-1} E_{\mathbf{s}',j}(x),$$

*where* $\mathbf{s}' = (s_1, \ldots, s_{n-1})$ *and* $t_i = \lceil i s_{n-1}/s_n \rceil$.

An application of Corollary 7.8.7 proves the following theorem.

**Theorem 7.8.11** [86] *If* $\mathbf{s} = (s_1, \ldots, s_n)$ *is a sequence of positive integers, then* $E_{\mathbf{s}}(x)$ *is real-rooted. Moreover if* $n > 1$, *then the sequence* $\{E_{\mathbf{s},i}(x)\}_{i=0}^{s_n - 1}$ *is interlacing.*

### 7.8.2 Eulerian polynomials for finite Coxeter groups

For undefined terminology on Coxeter groups we refer to [6]. Let $(W, S)$ be a Coxeter system. The **length** of an element $w \in W$ is the smallest number $k$ such that

$$w = s_1 s_2 \cdots s_k, \quad \text{where } s_i \in S \text{ for all } 1 \leq i \leq n.$$

Let $\ell_W(w)$ denote the length of $w$. The (right) **descent set** of $w$ is

$$D_W(w) = \{s \in S : \ell_W(ws) < \ell_W(w)\},$$

and the **descent number** is $\operatorname{des}_W(w) = |D_W(w)|$. The $W$-**Eulerian polynomial** of a finite Coxeter group $W$ is the polynomial

$$\sum_{w \in W} x^{\operatorname{des}_W(w)}$$

which is known to be the $h$-polynomial of the Coxeter complex associated to $W$, see [26]. The type $A$ Eulerian polynomials are the common Eulerian polynomials. In [26], Brenti conjectured that the Eulerian polynomial of any finite Coxeter group is real-rooted. Brenti's conjecture is true for type $A$ and $B$ Coxeter groups [26, 50], and one may check with the aid of the computer that the conjecture holds for the exceptional groups $H_3, H_4, F_4, E_6, E_7$, and $E_8$. Moreover, the Eulerian polynomial of the direct product of two finite Coxeter groups is the product of the Eulerian polynomials of the two groups. Hence it remains to prove Brenti's conjecture for type $D$ Coxeter groups. The type $D$ case resisted many attempts, and it was not until very recently that the first sound proof was given by Savage and Visontai [86]. Their proof used compatibility arguments and ascent sequences. We will give a similar proof below that avoids the detour via ascent sequences.

Recall that a combinatorial description of a rank $n$ Coxeter group of type $B$ is the group $B_n$ of signed permutations $\sigma : [\pm n] \to [\pm n]$, where $[\pm n] = \{\pm 1, \ldots, \pm n\}$, such that $\sigma(-i) = -\sigma(i)$ for all $i \in [\pm n]$. An element $\sigma \in B_n$ is conveniently encoded by the **window notation** as a word $\sigma_1 \cdots \sigma_n$, where $\sigma_i = \sigma(i)$. The type $B$ **descent number** of $\sigma$ is then

$$\operatorname{des}_B(\sigma) = |\{i \in [n] : \sigma_{i-1} > \sigma_i\}|,$$

where $\sigma_0 := 0$, see [26]. The $n$th **type $B$ Eulerian polynomial** is thus

$$B_n(x) = \sum_{\sigma \in B_n} x^{\mathrm{des}_B(\sigma)}.$$

A combinatorial description of a rank $n$ Coxeter group of type $D$ is the group $D_n$ consisting of all elements of $B_n$ with an even number of negative entries in their window notation. The type $D$ **descent number** of $\sigma \in D_n$ is then

$$\mathrm{des}_D(\sigma) = |\{i \in [n] : \sigma_{i-1} > \sigma_i\}|,$$

where $\sigma_0 := -\sigma_2$, see [26]. The $n$th **type $D$ Eulerian polynomial** is

$$D_n(x) = \sum_{\sigma \in D_n} x^{\mathrm{des}_D(\sigma)}.$$

For $n \geq 2$ and $k \in [\pm n]$, let

$$D_{n,k}(x) = \sum_{\substack{\sigma \in D_n \\ \sigma_n = -k}} x^{\mathrm{des}_D(\sigma)}.$$

If $k \notin [\pm n]$, we set $D_{n,k}(x) := 0$. The following table is conveniently generated by the recursion in Lemma 7.8.13 below.

| $k$ | $D_{2,k}$ | $D_{3,k}$ | $D_{4,k}$ |
|---|---|---|---|
| $-4$ | $0$ | $0$ | $(x+1)(x^2+10x+1)$ |
| $-3$ | $0$ | $(x+1)^2$ | $2x(x+1)(x+5)$ |
| $-2$ | $1$ | $x(3+x)$ | $x(3x^2+14x+7)$ |
| $-1$ | $x$ | $2x(x+1)$ | $x(5x^2+14x+5)$ |
| $1$ | $x$ | $2x(x+1)$ | $x(5x^2+14x+5)$ |
| $2$ | $x^2$ | $x(3x+1)$ | $x(7x^2+14x+3)$ |
| $3$ | $0$ | $x(x+1)^2$ | $2x(x+1)(5x+1)$ |
| $4$ | $0$ | $0$ | $x(x+1)(x^2+10x+1)$ |

(7.13)

Note that the type $D$ descents make sense for any element of $B_n$, where $n \geq 2$.

**Lemma 7.8.12** *If $n \geq 2$, then*

$$D_{n,k}(x) = \frac{1}{2} \sum_{\substack{\sigma \in B_n \\ \sigma_n = -k}} x^{\mathrm{des}_D(\sigma)}. \tag{7.14}$$

*Proof.* For $k \in [n]$, let $\phi_k : B_n \to B_n$ be the involution that swaps the letters $k$ and $-k$ in the window notation of the permutation. Clearly $\phi_1$ is a bijection between $D_n$ and $B_n \setminus D_n$ that preserves the type $D$ descents for all $n \geq 2$. This proves (7.14) for $k \notin \{1, -1\}$.

For $k \in [\pm n]$, let $B_n[k]$ be the set of $\sigma \in B_n$ with $\sigma_n = k$. Then $\phi_1$ is a bijection between $B_n[1]$ and $B_n[-1]$ that preserves the type $D$ descents for all $n \geq 2$. Similarly let $D_n[k]$ be the set of $\sigma \in D_n$ with $\sigma_n = k$. Now $B_n[1] = D_n[1] \cup \phi_1(D_n[-1])$ and

$B_n[-1] = D_n[-1] \cup \phi_1(D_n[1])$, where the unions are disjoint. Hence to prove (7.14) for $k = \pm 1$, it remains to prove $D_{n,1}(x) = D_{n,-1}(x)$. We prove this by induction on $n \geq 2$, where the case $n = 2$ is easily checked.

Consider the involution $\phi_2\phi_1 : D_n[1] \to D_n[-1]$, where $n \geq 3$. Then $\phi_2\phi_1$ preserves type $D$ descents on $\sigma$ unless $\sigma_{n-1} = \pm 2$. Hence it remains to prove that the type $D$ descent generating polynomials of $D_n[2,1] \cup D_n[-2,1]$ and $D_n[2,-1] \cup D_n[-2,-1]$ agree, where $D_n[k,\ell]$ is the set of $\sigma \in D_n$ such that $\sigma_{n-1} = k$ and $\sigma_n = \ell$. This is easily seen by inspection. ∎

**Lemma 7.8.13** *If $n \geq 2$ and $i \in [\pm n]$, then*

$$D_{n+1,i}(x) = \sum_{k \leq i} xD_{n,k}(x) + \sum_{k>i} D_{n,k}(x), \quad \text{if } i < 0 \text{ and}$$

$$D_{n+1,i}(x) = \sum_{k<i} xD_{n,k}(x) + \sum_{k \geq i} D_{n,k}(x), \quad \text{if } i > 0.$$

*Proof.* The lemma follows easily by using the alternative description (7.14) of $D_{n,i}(x)$, and keeping track of $\sigma_n$, where $\sigma \in D_{n+1}[-i]$. We leave the details to the reader. ∎

**Theorem 7.8.14** *Let $n \geq 2$. The type D Eulerian polynomial $D_n(x)$ is real-rooted. Moreover for each $k \in [\pm n]$, the polynomial $D_{n,k}(x)$ is real-rooted, and if $n \geq 4$, then the sequence $\mathscr{D}_n := (D_{n,k}(x))_{k \in [\pm n]}$ is interlacing.*

*Proof.* One may easily check that $D_n(x)$ and $D_{n,k}(x)$ are real-rooted whenever $2 \leq n \leq 4$ and $k \in [\pm n]$, see (7.13). The sequence $\mathscr{D}_4$ is interlacing, see (7.13). By Lemma 7.8.13, up to a relabeling of $[\pm n]$,

$$\mathscr{D}_{n+1} = G_{\lambda^n} \mathscr{D}_n,$$

where $\lambda^n$ is a weakly increasing sequence. The matrix $G_{\lambda^n}$ is of the type appearing in Corollary 7.8.7. Hence the theorem follows from Corollary 7.8.7. ∎

**Theorem 7.8.15 (Frobenius [50], Brenti [26], Savage–Visontai [86])** *The Eulerian polynomial of any finite Coxeter group is real-rooted.*

**Remark 7.8.16** For $n \geq 1$ and $i \in [\pm n]$, define

$$B_{n,i}(x) = \sum_{\substack{\sigma \in B_n \\ \sigma_n = -i}} x^{\mathrm{des}_B(\sigma)}.$$

Then $B_{n,i}$ satisfies the same recursion as in Lemma 7.8.13, because the proof is ignorant to what happens in the far left in the window notation of an element of $B_n$. Moreover this recursion is valid for all $n \geq 1$. Since $(B_{1,-1}(x), B_{1,1}(x)) = (1, x)$ is interlacing, induction and Corollary 7.8.7 implies that the sequence $(B_{n,i}(x))_{i \in [\pm n]}$ is an interlacing sequence of polynomials for all $n \geq 1$.

## 7.9 Multivariate techniques

To prove that a family of univariate polynomials are real-rooted it is sometimes easier to work with multivariate analogs of the polynomials. As alluded to in Section 7.7, a fruitful generalization of real-rootedness for multivariate polynomials is that of (real-) stable polynomials. There are several benefits in a multivariate approach; the proofs sometimes become more transparent, several powerful inequalities are available for multivariate stable polynomials, it may give you a better understanding for the combinatorial problem at hand. An important class of stable polynomials are **determinantal** polynomials.

**Proposition 7.9.1** *Let* $A_1,\dots,A_n$ *be positive semidefinite hermitian matrices, and* $A_0$ *a hermitian matrix. Then the polynomial*

$$P(x_1,\dots,x_n) = \det(A_0 + x_1A_1 + \cdots + x_nA_n)$$

*is either stable or identically zero.*

*Proof.* By Hurwitz' theorem [33, Footnote 3, p. 96] and a standard approximation argument we may assume that $A_1$ is positive definite. Let $\mathbf{x} = (a_1+ib_1,\dots,a_n+ib_n) \in \mathbb{C}^n$ be such that $a_j \in \mathbb{R}$ and $b_j > 0$ for all $1 \le j \le n$. We need to prove that $P(\mathbf{x}) \neq 0$. Now $P(\mathbf{x}) = \det(iB - A)$, where $B = b_1A_1 + \cdots + b_nA_n$ is positive definite and $A = -A_0 - a_1A_1 - \cdots - a_nA_n$ is hermitian. Hence $B$ has a square root and thus $P(\mathbf{x}) = \det(B)\det(iI - B^{-1/2}AB^{-1/2}) \neq 0$, where $I$ is the identity matrix, since $B^{-1/2}AB^{-1/2}$ is hermitian and thus has real eigenvalues only. ∎

For $n = 2$ a converse of Proposition 7.9.1 holds, see Theorem 7.7.7. The analog of Theorem 7.7.7 for $n \ge 3$ fails to be true by a simple count of parameters. For possible partial converses of Proposition 7.9.1, see the survey [100].

Recently attempts have been made to find appropriate multivariate analogs of frequently studied real-rooted univariate polynomials in combinatorics. Let us illustrate by describing a multivariate Eulerian polynomial. For $\sigma \in \mathfrak{S}_n$ let

$$\begin{aligned} \mathrm{DB}(\sigma) &= \{\sigma(i) : \sigma(i-1) > \sigma(i)\}, \text{ and} \\ \mathrm{AB}(\sigma) &= \{\sigma(i) : \sigma(i-1) < \sigma(i)\}, \end{aligned}$$

where $\sigma(0) = \sigma(n+1) = \infty$, be the set of **descent bottoms** and **ascent bottoms** of $\sigma$, respectively. Let $A_n(\mathbf{x},\mathbf{y})$ be the polynomial in $\mathbb{R}[x_1,\dots,x_n,y_1,\dots,y_n]$ defined by

$$A_n(\mathbf{x},\mathbf{y}) = \sum_{\sigma \in \mathfrak{S}_n} w(\sigma), \quad \text{where } w(\sigma) = \prod_{i \in \mathrm{DB}(\sigma)} x_i \prod_{j \in \mathrm{AB}(\sigma)} y_i.$$

For example $w(573148926) = x_5x_3x_1x_2y_5y_1y_4y_8y_2y_6$. Generate a permutation $\sigma'$ in $\mathfrak{S}_n$ by inserting the letter 1 in a slot between two adjacent letters in a permutation $\sigma_0\sigma_1\cdots\sigma_{n-1}\sigma_n$ of $\{2,3,\dots,n\}$ (where $\sigma_0 = \sigma_n = \infty$). Note that there is an obvious one-to-one correspondence between the slots and the variables appearing in $w(\sigma')$. Thus if we insert 1 in the slot corresponding to the variable $z$, then

$$w(\sigma) = x_1 y_1 \frac{\partial}{\partial z} w(\sigma')$$

since the descent/ascent bottom corresponding to $z$ in $\sigma'$ will be destroyed, and 1 becomes an ascent and descent bottom. We have proved

$$A_n(\mathbf{x},\mathbf{y}) = x_1 y_1 \left( \sum_{j=2}^{n} \frac{\partial}{\partial x_j} + \frac{\partial}{\partial y_j} \right) A_{n-1}(\mathbf{x}^*,\mathbf{y}^*),$$

where $\mathbf{x}^* = (x_2,\dots,x_n)$ and $\mathbf{y}^* = (y_2,\dots,y_n)$. To prove that $A_n(\mathbf{x},\mathbf{y})$ is stable for all $n$ it remains to prove that the operators of the form $\sum_{j=1}^{n} \partial/\partial x_j$ preserve stability. Stability preservers were recently characterized in [9]. The following theorem is the algebraic characterization. For $\kappa \in \mathbb{N}^n$, let $\mathbb{C}_\kappa[x_1,\dots,x_n]$ be the linear space of all polynomials that have degree at most $\kappa_i$ in $x_i$ for each $1 \le i \le n$. The **symbol** of a linear operator $T : \mathbb{C}_\kappa[x_1,\dots,x_n] \to \mathbb{C}[x_1,\dots,x_m]$ is the polynomial

$$G_T(x_1,\dots,x_m,y_1,\dots,y_n) = T\left((x_1+y_1)^{\kappa_1}\cdots(x_n+y_n)^{\kappa_n}\right),$$

where $T$ acts on the $y$-variables as if they were constants.

**Theorem 7.9.2** [9] *Let $T : \mathbb{C}_\kappa[x_1,\dots,x_n] \to \mathbb{C}[x_1,\dots,x_m]$ be a linear operator of rank greater than one. Then $T$ preserves stability if and only if $G_T$ is stable.*

The symbol of the operator $T = \sum_{j=1}^{n} \partial/\partial x_j$ is

$$G_T = (x_1+y_1)^{\kappa_1}\cdots(x_n+y_n)^{\kappa_n} \sum_{j=1}^{n} \frac{\kappa_j}{x_j+y_j}.$$

Hence if $\mathrm{Im}(x_j) > 0$ and $\mathrm{Im}(y_j) > 0$ for all $1 \le j \le n$, then $\mathrm{Im}(x_j+y_j)^{-1} < 0$, and hence the symbol is non-zero. Thus $G_T$ is stable and by induction and Theorem 7.9.2, $A_n(\mathbf{x},\mathbf{y})$ is stable for all $n \ge 1$.

The multivariate Eulerian polynomials above and more general Eulerian-like polynomials were introduced in [22] and used to prove the Monotone Column Permanent Conjecture of Haglund, Ono and Wagner [55]. Suppose $A = (a_{ij})_{i,j=1}^{n}$ is a real matrix that has weakly increasing down columns. Then the Monotone Column Permanent Conjecture stated that the permanent of the matrix $(a_{ij}+x)_{i,j=1}^{n}$, where $x$ is a variable, is real-rooted. Subsequently multivariate Eulerian polynomials for colored permutations and various other models have been studied [23, 56, 101, 32].

### 7.9.1 Stable polynomials and matroids

Let $E$ be a finite set and let $\mathbf{x} = (x_e)_{e\in E}$ be independent variables. The **support** of a multiaffine polynomial

$$P(\mathbf{x}) = \sum_{S\subseteq E} a(S) \prod_{e\in S} x_e,$$

is the set system $\text{Supp}(P) = \{S \subseteq E : a(S) \neq 0\}$. Choe, Oxley, Sokal and Wagner [33] proved the following striking relationship between stable polynomials and matroids.

**Theorem 7.9.3** *The support of a homogeneous, multiaffine and stable polynomial is the set of bases of a matroid.*

Hence Theorem 7.9.3 suggests an alternative way of representing matroids. A matroid $M$, with set of bases $\mathscr{B}$, has the **half-plane property** (HPP) if its **bases generating polynomial**

$$P_M(\mathbf{x}) = \sum_{B \in \mathscr{B}} \prod_{e \in B} x_e$$

is stable, and $M$ has the **weak half-plane property** (WHPP) if there are positive numbers $a(B)$, $B \in \mathscr{B}$, such that

$$\sum_{B \in \mathscr{B}} a(B) \prod_{e \in B} x_e$$

is stable. For example, the Fano matroid $F_7$ is not WHPP, see [17]. The fact that graphic matroids are HPP is a consequence of the **Matrix Tree theorem** and Proposition 7.9.1. Suppose $V = [n]$, and let $\{\delta_i\}_{i=1}^n$ be the standard basis of $\mathbb{R}^n$. The **weighted Laplacian** of a connected graph $G = (V, E)$ is defined as

$$L_G(\mathbf{x}) = \sum_{e \in E} x_e (\delta_{e_1} - \delta_{e_2})(\delta_{e_1} - \delta_{e_2})^T,$$

where $e_1$ and $e_2$ are the vertices incident to $e \in E$. We refer to [99, Theorem VI.29] for a proof of the next classical theorem that goes back to Kirchhoff and Maxwell. Let $T_G(\mathbf{x})$ be the **spanning tree polynomial** of $G$, i.e., the bases generating polynomial of the graphical matroid associated to $G$.

**Theorem 7.9.4 (Matrix-tree theorem)** *For $i \in V$, let $L_G(\mathbf{x})_{ii}$ be the matrix obtained by deleting the column and row indexed by $i$ in $L_G(\mathbf{x})$. Then*

$$T_G(\mathbf{x}) = \det(L_G(\mathbf{x})_{ii}).$$

Clearly the matrices in the pencil $L_G(\mathbf{x})_{ii}$ are positive semidefinite. Hence that graphic matroids are HPP follows from Theorem 7.9.4 and Proposition 7.9.1. A similar reasoning proves that all regular matroids are HPP, and that all matroids representable over $\mathbb{C}$ are WHPP, see [17, 33]. On the other hand, the Vámos cube $V_8$ is not representable over any field, and still $V_8$ is HPP [105]. For further results on the relationship between stable polynomials and matroids we refer to [17, 21, 33, 105].

### 7.9.2 Strong Rayleigh measures

Stability implies several strong inequalities among the coefficients. Note that the multivariate Eulerian polynomial above is **multiaffine**, i.e., it is of degree at most one in each variable. We may view multiaffine polynomials with nonnegative coefficients as discrete probability measures. If $E$ is a finite set, $\mathbf{x} = (x_e)_{e\in E}$ are independent variables, and

$$P(\mathbf{x}) = \sum_{S\subseteq E} a(S)\prod_{e\in S} x_e,$$

is a multiaffine polynomial with nonnegative coefficients normalized so that $P(1,\dots 1) = 1$, we may define a discrete probability measure $\mu$ on $2^E$ by setting $\mu(S) = a(S)$ for each $S \in 2^E$. Then $P_\mu := P$ is the multivariate **partition function** of $\mu$. A discrete probability measure $\mu$ is called **strong Rayleigh** if $P_\mu$ is stable. Hence the measure $\mu_n$ on $2^{[2n]}$, defined by

$$\mu_n(S) = \frac{1}{n!}|\{\sigma \in \mathfrak{S}_n : \mathrm{DB}(\sigma)\cup\{i+n : i \in \mathrm{AB}(\sigma)\} = S\}|$$

is strong Rayleigh. A fundamental strong Rayleigh measure is the **uniform spanning tree measure,** $\mu_G$, associated to a connected graph $G = (V,E)$. This is the measure on $2^E$ defined by

$$\mu_G(S) = \frac{1}{t}\begin{cases} 1 & \text{if } S \text{ is a spanning tree,} \\ 0 & \text{otherwise} \end{cases},$$

where $t$ is the number of spanning trees of $G$. The uniform spanning tree measures, and more generally the uniform measure on the set of bases of any HPP matroid, is strong Rayleigh by the discussion in Section 7.9.1.

A general class of strong Rayleigh measures containing the uniform spanning tree measures is the class of **determinantal measures**, see [66]. Let $C$ be a hermitian $n\times n$ contraction matrix, i.e., a positive semidefinite matrix with all its eigenvalues located in the interval $[0,1]$. Define a probability measure on $2^{[n]}$ by

$$\mu_C(\{T : T \supseteq S\}) = \det C(S), \quad \text{for all } S \subseteq [n],$$

where $C(S)$ is the submatrix of $C$ with rows and columns indexed by $S$. Using Proposition 7.9.1, it is not hard to prove that $\mu_C$ is strong Rayleigh, see [12].

Negative dependence is an important notion in probability theory, statistics and statistical mechanics, see the survey [75]. In [12] several strong negative dependence properties of strong Rayleigh measures were established. Identify $2^E$ with $\{0,1\}^E$. A probability measure $\mu$ on $\{0,1\}^n$ is **negatively associated** if

$$\int fg d\mu \le \int f d\mu \int g d\mu,$$

whenever $f,g : \{0,1\}^n \to \mathbb{R}$ are increasing functions depending on disjoint sets of variables, i.e., $f(\eta)$ only depends on the variables $\eta_i, i\in A$, and $g(\eta)$ only depends on the variables $\eta_i, i \in B$, where $A\cap B = \emptyset$. In particular setting $f(\eta) = \eta_i$ and $g(\eta) = \eta_j$, where $i \neq j$, we see that $\mu$ is **pairwise negatively correlated** i.e.,

$$\mu(\eta : \eta_i = \eta_j = 1) \le \mu(\eta : \eta_i = 1)\mu(\eta : \eta_j = 1).$$

**Example 7.9.5** *For $n = 2$, a discrete probability measure $\mu$ defined by $\mu(\emptyset) = a, \mu(\{1\}) = b, \mu(\{2\}) = c, \mu(\{1\}) = d$, with $a+b+c+d = 1$ is pairwise negatively correlated if and only if $d(a+b+c+d) \leq (b+d)(c+d)$, i.e., if and only if $ad \leq bc$. Also, it is easy to see that a real polynomial $a+bx_1+cx_2+dx_1x_2$ is stable if and only if $ad \leq bc$. By the next theorem the notions strong Rayleigh, negative association and pairwise negative correlation agree for $n = 2$.*

**Theorem 7.9.6** [12] *If $\mu$ is a discrete probability measure that is strong Rayleigh, then it is negatively associated.*

Recently Pemantle and Peres [77] proved general concentration inequalities for strong Rayleigh measures. A function $f : \{0,1\}^n \to \mathbb{R}$ is **Lipschitz-1** if

$$|f(\eta) - f(\xi)| \leq d(\eta, \xi), \quad \text{for all } \eta, \xi \in \{0,1\}^n,$$

where $d$ is the **Hamming distance**, $d(\eta, \xi) = |\{i \in [n] : \eta_i \neq \xi_i\}|$.

**Theorem 7.9.7 (Pemantle and Peres, [77])** *Suppose $\mu$ is a probability measure on $\{0,1\}^n$ whose partition function is stable and has mean $m = \mathbb{E}(\sum_{i=1}^n \eta_i)$. If $f$ is any Lipschitz-1 function on $\{0,1\}^n$, then*

$$\mu(\eta : |f(\eta) - \mathbb{E}f| > a) \leq 5\exp\left(-\frac{a^2}{16(a+2m)}\right).$$

### 7.9.3 The symmetric exclusion process

The **symmetric exclusion process** (with creation and annihilation) is a Markov process that models particles jumping on a countable set of sites. Here we will just consider the case when we have a finite set of sites $[n]$. Given a **symmetric** matrix $Q = (q_{ij})_{i,j=1}^n$ of nonnegative numbers and vectors $b = (b_i)_{i=1}^n$ and $d = (d_i)_{i=1}^n$ of nonnegative numbers, define a continuous time Markov process on $\{0,1\}^n$ as follows. Let $\eta \in \{0,1\}^n$ represent the configuration of the particles, with $\eta_i = 1$ meaning that site $i$ is occupied, and $\eta_i = 0$ that site $i$ is vacant. Particles at occupied sites jump to vacant sites at specified rates. More precisely, these are the transitions in the Markov process, which we denote by SEP$(Q,b,d)$, see Figure 7.4:

(J) A particle jumps from site $i$ to site $j$ at rate $q_{ij}$: The configuration $\eta$ is unchanged unless $\eta_i = 1$ and $\eta_j = 0$, and then $\eta_i$ and $\eta_j$ are exchanged in $\eta$.

(B) A particle at site $i$ is created (is born) at rate $b_i$: The configuration $\eta$ is unchanged unless $\eta_i = 0$, and then $\eta_i$ is changed from a zero to a one in $\eta$.

(D) A particle at site $i$ is annihilated (dies) at rate $d_i$: The configuration $\eta$ is unchanged unless $\eta_i = 1$, and then $\eta_i$ is changed from a one to a zero in $\eta$.

It was proved in [12, 104] that SEP$(Q,b,d)$ preserves the family of strong Rayleigh measures.

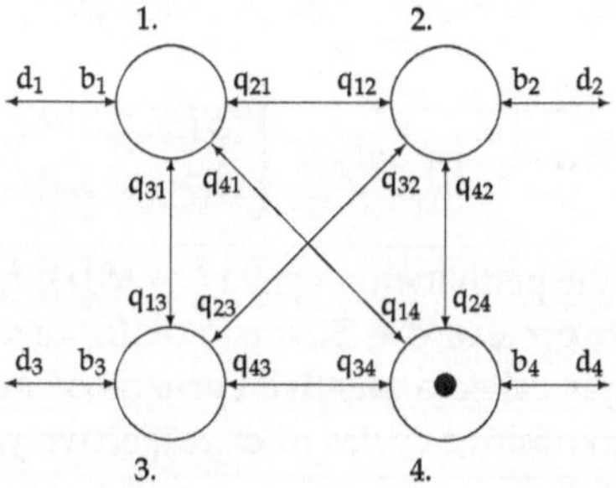


**Figure 7.4**
The transitions in SEP$(Q,b,d)$ on 4 sites, where $q_{ij} = q_{ji}$.

**Theorem 7.9.8** *If the initial distribution of a symmetric exclusion process* SEP$(Q,b,d)$ *is strong Rayleigh, then the distribution is strong Rayleigh for all positive times.*

An immediate consequence of Theorem 7.9.8 is that the stationary distribution (if unique) of the symmetric exclusion process is strong Rayleigh.

**Corollary 7.9.9** *If a symmetric exclusion process* SEP$(Q,b,d)$ *is irreducible and positive recurrent, then the unique stationary distribution is strong Rayleigh.*

*Proof.* Choose an initial distribution that is strong Rayleigh. Then the partition function, $P_t(\mathbf{x})$, of the distribution at time $t$ is stable for all $t > 0$, by Theorem 7.9.8. The partition function of the stationary distribution is given by $\lim_{t\to\infty} P_t(\mathbf{x})$. By Hurwitz' theorem [33, Footnote 3, p. 96] the partition function of the stationary distribution is stable, i.e., the stationary distribution is strong Rayleigh. ∎

In view of Corollary 7.9.9 it would be interesting to find the stationary distributions of SEP$(Q,b,d)$ for specific parameters $Q, b$, and $d$. This was achieved by Corteel and Williams [36] for the parameters

$$\begin{aligned} q_{ij} &= \begin{cases} 1 & \text{if } |j-i| = 1 \text{ and} \\ 0 & \text{if } |j-i| > 1, \end{cases} \\ b &= (\alpha, 0, \ldots, 0, \delta), \\ d &= (\gamma, 0, \ldots, 0, \beta). \end{aligned} \tag{7.15}$$

Hence the particles jump on a line, where particles are only allowed to jump to neighboring sites, and be created and annihilated at the endpoints. The description of the stationary distribution is in terms of combinatorial objects called staircase tableaux. The special case when $\delta = \gamma = 0$ is related to multivariate Eulerian polynomials. The **excedence set**, $\mathscr{X}(\sigma) \subseteq [n]$, of a signed permutation $\sigma \in B_n$ was defined by

Steingrímsson [96] as

$$i \in \mathscr{X}(\sigma) \text{ if and only if } \begin{cases} |\sigma(i)| > i, \text{ or;} \\ \sigma(i) = -i. \end{cases}$$

If $\sigma \in B_n$, let $|\sigma| \in \mathfrak{S}_n$ be the permutation where $i \mapsto |\sigma(i)|$ for all $1 \leq i \leq n$. A cycle $c$ of $|\sigma|$ is called a **negative cycle** of $\sigma \in B_n$ if $\sigma(j) < 0$, where $|\sigma(j)|$ is the maximal element of $c$. Otherwise $c$ is called a **positive cycle** of $\sigma$. Let $c_-(\sigma)$ and $c_+(\sigma)$ be the number of negative and positive cycles of $\sigma$, respectively.

**Theorem 7.9.10** [23] *The multivariate partition function of the symmetric exclusion process on* $2^{[n]}$ *with parameters as in* (7.15), *with* $\delta = \gamma = 0$, *is a constant multiple of*

$$\sum_{\sigma \in B_n} \left(\frac{2}{\alpha}\right)^{c_-(\sigma)} \left(\frac{2}{\beta}\right)^{c_+(\sigma)} \prod_{i \in \mathscr{X}(\sigma)} x_i. \tag{7.16}$$

Note that by Corollary 7.9.9, the polynomial (7.16) is stable.

**Problem 7.9.11** *Find the stationary distribution of* $\mathrm{SEP}(Q,b,d)$ *for parameters other than* (7.15).

### 7.9.4 The Grace–Walsh–Szegő theorem, and the proof of Theorem 7.5.1

The proof of Theorem 7.5.1 is an excellent example of how multivariate techniques may be used to prove statements about the zeros of univariate polynomials. The proof uses a combinatorial symmetric function identity and the Grace–Walsh–Szegő theorem, which is undoubtedly one of the most useful theorems governing the location of zeros of polynomials, see [82].

A *circular region* is a proper subset of the complex plane that is bounded by either a circle or a straight line, and is either open or closed.

**Theorem 7.9.12 (Grace–Walsh–Szegő)** *Let* $f \in \mathbb{C}[z_1,\dots,z_n]$ *be a multiaffine and symmetric polynomial, and let* $K$ *be a circular region. Assume that either* $K$ *is convex or that the degree of* $f$ *is* $n$. *For any* $\zeta_1,\dots,\zeta_n \in K$ *there is a number* $\zeta \in K$ *such that* $f(\zeta_1,\dots,\zeta_n) = f(\zeta,\dots,\zeta)$.

The second ingredient in the proof of Theorem 7.5.1 is the following symmetric function identity. Let $e_k(\mathbf{x})$ be the $k$th elementary symmetric function in the variables $\mathbf{x} = (x_1,\dots,x_n)$.

**Lemma 7.9.13** *For nonnegative integers* $n$,

$$\sum_{k=0}^{n} (e_k(\mathbf{x})^2 - e_{k-1}(\mathbf{x})e_{k+1}(\mathbf{x})) = e_n(\mathbf{x}) \sum_{k=0}^{\lfloor n/2 \rfloor} C_k e_{n-2k}\left(\mathbf{x} + \frac{1}{\mathbf{x}}\right), \tag{7.17}$$

$\mathbf{x} + 1/\mathbf{x} = (x_1 + 1/x_1,\dots,x_n + 1/x_n)$ *and* $C_k = \binom{2k}{k}/(k+1)$, $k \in \mathbb{N}$, *are the Catalan numbers.*

*Proof.* For undefined symmetric function terminology, we refer to [93, Chapter 7]. The polynomial $e_k(\mathbf{x})^2 - e_{k-1}(\mathbf{x})e_{k+1}(\mathbf{x})$ is the Schur-function $s_{2^k}(\mathbf{x})$, where $2^k = (2,2,\ldots,2)$. We may rewrite (7.17) as

$$\sum_{k=0}^{n} s_{2^k}(\mathbf{x}) = \sum_{k=0}^{\lfloor n/2 \rfloor} C_k \sum_{|S|=2k} \prod_{i\in S} x_i \prod_{j\notin S}(1+x_j^2). \tag{7.18}$$

By the combinatorial definition of the Schur-function, the left-hand side of (7.18) is the generating polynomial of all semi-standard Young tableaux with entries in $\{1,\ldots,n\}$, that are of shape $2^k$ for some $k \in \mathbb{N}$. Call this set $\mathscr{A}_n$. Given $T \in \mathscr{A}_n$, let $S$ be the set of entries that occur only once in $T$. By deleting the remaining entries we obtain a standard Young tableau of shape $2^k$, where $2k = |S|$. There are exactly $C_k$ standard Young tableaux of shape $2^k$ with set of entries $S$, see e.g. [93, Exercise 6.19.ww]. It is not hard to see that the original semi-standard Young tableau is then determined by the set of duplicates. This explains the right-hand side of (7.18). ∎

*Proof of Theorem 7.5.1.* Let $P(x) = \sum_{k=0}^{n} a_k x^k = \prod_{k=0}^{n}(1+\rho_k x)$, where $\rho_k > 0$ for all $1 \le k \le n$, and let

$$Q(x) = \sum_{k=0}^{n} (a_k^2 - a_{k-1}a_{k+1})x^k.$$

Suppose there is a number $\zeta \in \mathbb{C}$, with $\zeta \notin \{x \in \mathbb{R} : x \le 0\}$, for which $Q(\zeta) = 0$. We may write $\zeta$ as $\zeta = \xi^2$, where $\mathrm{Re}(\xi) > 0$. By (7.17),

$$0 = Q(\zeta) = a_n \xi^n \sum_{k=0}^{\lfloor n/2 \rfloor} C_k e_{n-2k}\left(\rho_1\xi + \frac{1}{\rho_1\xi}, \ldots, \rho_n\xi + \frac{1}{\rho_n\xi}\right).$$

Since $\mathrm{Re}(\rho_j\xi + 1/(\rho_j\xi)) > 0$ for all $1 \le j \le n$, the Grace–Walsh–Szegő Theorem provides a number $\eta \in \mathbb{C}$, with $\mathrm{Re}(\eta) > 0$, such that

$$0 = \sum_{k=0}^{\lfloor n/2 \rfloor} C_k e_{n-2k}(\eta,\ldots,\eta) = \sum_{k=0}^{\lfloor n/2 \rfloor} C_k \binom{n}{2k} \eta^{n-2k} =: \eta^n q_n\left(\frac{1}{\eta^2}\right).$$

Since $\mathrm{Re}(\eta) > 0$, we have $1/\eta^2 \in \mathbb{C} \setminus \{x \in \mathbb{R} : x \le 0\}$. Hence, the desired contradiction follows if we can prove that all the zeros of the polynomial $q_n(x)$ are real and negative. This follows from the identity

$$\begin{aligned}\sum_{k=0}^{\lfloor n/2 \rfloor} C_k \binom{n}{2k} x^k (1+x)^{n-2k} &= \sum_{k=0}^{n} \frac{1}{n+1}\binom{n+1}{k}\binom{n+1}{k+1} x^k \\ &= \frac{1}{n+1}(1-x)^n P_n^{(1,1)}\left(\frac{1+x}{1-x}\right),\end{aligned}$$

where $\{P_n^{(1,1)}(x)\}_n$ are **Jacobi polynomials**, see [83, p. 254]. The zeros of the Jacobi polynomials $\{P_n^{(1,1)}(x)\}_n$ are located in the interval $(-1,1)$. Note that the first identity in the equation above follows immediately from (7.17). ∎

## 7.10 Historical notes

Here are some complementary historical notes about the origin of some of the central notions of this chapter.

Although some combinatorial polynomials such as the Eulerian polynomials have been known to be $\gamma$-positive for at least 45 years [48], Gal [51] and the author [15] realized the relevance of $\gamma$-positivity to topological combinatorics and in particular to the Charney–Davis conjecture.

Multivariate stable polynomials and similar classes of polynomials have been studied in many different areas. For their importance in control theory, see [45] and the references therein. In statistical mechanics they play an important role in Lee and Yang's approach to the study of phase transitions [64, 106]. In PDE theory so-called hyperbolic polynomials play an important part in the existence of a fundamental solution to a linear PDE with constant coefficients, see [60]. The importance of stable polynomials in matroid theory was first realized in [33]. An important application of stable polynomials to a problem in combinatorics is the proof by Gurvits of a vast generalization of the Van der Waerden conjecture, [54]. A recent application is the spectacular solution to the Kadison–Singer problem by Marcus et al. [67]. See the surveys [76, 104] for further applications of stable polynomials.

The notion of HPP and WHPP matroids were introduced in Choe et al. [33]. The strong Rayleigh property was introduced for matroids by Choe and Wagner [34], and extended to discrete probability measures and studied extensively in [12].

## References

[1] Michael Aissen, I. J. Schoenberg, and A. M. Whitney. On the generating functions of totally positive sequences. I. *J. Analyse Math.*, 2:93–103, 1952.

[2] Christos A. Athanasiadis. Ehrhart polynomials, simplicial polytopes, magic squares and a conjecture of Stanley. *J. Reine Angew. Math.*, 583:163–174, 2005.

[3] Christos A. Athanasiadis. Flag subdivisions and $\gamma$-vectors. *Pacific J. Math.*, 259(2):257–278, 2012.

[4] Edward A. Bender. Central and local limit theorems applied to asymptotic enumeration. *J. Combinatorial Theory Ser. A*, 15:91–111, 1973.

[5] Moussa Benoumhani. On the modes of the independence polynomial of the centipede. *J. Integer Seq.*, 15(5):Article 12.5.1, 12, 2012.

[6] Anders Björner and Francesco Brenti. *Combinatorics of Coxeter groups*, volume 231 of *Graduate Texts in Mathematics*. Springer, New York, 2005.

[7] Miklós Bóna. Symmetry and unimodality in $t$-stack sortable permutations. *J. Combin. Theory Ser. A*, 98(1):201–209, 2002.

[8] Miklós Bóna. Real zeros and normal distribution for statistics on Stirling permutations defined by Gessel and Stanley. *SIAM J. Discrete Math.*, 23(1):401–406, 2008/09.

[9] Julius Borcea and Petter Brändén. The Lee-Yang and Pólya-Schur programs. I. Linear operators preserving stability. *Invent. Math.*, 177(3):541–569, 2009.

[10] Julius Borcea and Petter Brändén. Pólya-Schur master theorems for circular domains and their boundaries. *Ann. of Math. (2)*, 170(1):465–492, 2009.

[11] Julius Borcea and Petter Brändén. Multivariate Pólya-Schur classification problems in the Weyl algebra. *Proc. Lond. Math. Soc. (3)*, 101(1):73–104, 2010.

[12] Julius Borcea, Petter Brändén, and Thomas M. Liggett. Negative dependence and the geometry of polynomials. *J. Amer. Math. Soc.*, 22(2):521–567, 2009.

[13] George Boros and Victor Moll. *Irresistible integrals*. Cambridge University Press, Cambridge, 2004. Symbolics, analysis and experiments in the evaluation of integrals.

[14] Petter Brändén. Counterexamples to the Neggers-Stanley conjecture. *Electron. Res. Announc. Amer. Math. Soc.*, 10:155–158 (electronic), 2004.

[15] Petter Brändén. Sign-graded posets, unimodality of $W$-polynomials and the Charney-Davis conjecture. *Electron. J. Combin.*, 11(2):Research Paper 9, 15 pp. (electronic), 2004/06.

[16] Petter Brändén. On linear transformations preserving the Pólya frequency property. *Trans. Amer. Math. Soc.*, 358(8):3697–3716 (electronic), 2006.

[17] Petter Brändén. Polynomials with the half-plane property and matroid theory. *Adv. Math.*, 216(1):302–320, 2007.

[18] Petter Brändén. Actions on permutations and unimodality of descent polynomials. *European J. Combin.*, 29(2):514–531, 2008.

[19] Petter Brändén. Iterated sequences and the geometry of zeros. *J. Reine Angew. Math.*, 658:115–131, 2011.

[20] Petter Brändén and Matthew Chasse. Infinite log-concavity for polynomial Pólya frequency sequences. Preprint, arXiv:1405.6378, 2014.

[21] Petter Brändén and Rafael S. González D'León. On the half-plane property and the Tutte group of a matroid. *J. Combin. Theory Ser. B*, 100(5):485–492, 2010.

[22] Petter Brändén, James Haglund, Mirkó Visontai, and David G. Wagner. Proof of the monotone column permanent conjecture. In *Notions of positivity and*

*the geometry of polynomials*, Trends Math., pages 63–78. Birkhäuser/Springer Basel AG, Basel, 2011.

[23] Petter Brändén, Madeleine Leander, and Mirkó Visontai. Multivariate Eulerian polynomials and exclusion processes. Preprint, arXiv:1405.6919, 2014.

[24] Francesco Brenti. Unimodal, log-concave and Pólya frequency sequences in combinatorics. *Mem. Amer. Math. Soc.*, 81(413):viii+106, 1989.

[25] Francesco Brenti. Log-concave and unimodal sequences in algebra, combinatorics, and geometry: an update. In *Jerusalem combinatorics '93*, volume 178 of *Contemp. Math.*, pages 71–89. Amer. Math. Soc., Providence, RI, 1994.

[26] Francesco Brenti. $q$-Eulerian polynomials arising from Coxeter groups. *European J. Combin.*, 15(5):417–441, 1994.

[27] Francesco Brenti and Volkmar Welker. $f$-vectors of barycentric subdivisions. *Math. Z.*, 259(4):849–865, 2008.

[28] Winfried Bruns and Tim Römer. $h$-vectors of Gorenstein polytopes. *J. Combin. Theory Ser. A*, 114(1):65–76, 2007.

[29] Thomas Brylawski. Constructions. In *Theory of matroids*, volume 26 of *Encyclopedia Math. Appl.*, pages 127–223. Cambridge Univ. Press, Cambridge, 1986.

[30] Ruth Charney and Michael Davis. The Euler characteristic of a nonpositively curved, piecewise Euclidean manifold. *Pacific J. Math.*, 171(1):117–137, 1995.

[31] William Y. C. Chen, Donna Q. J. Dou, and Arthur L. B. Yang. Brändén's conjectures on the Boros-Moll polynomials. *Int. Math. Res. Not. IMRN*, (20):4819–4828, 2013.

[32] William. Y. C. Chen, Robert. X. J. Hao, and Harold. R. L. Yang. Context-free grammars and multivariate stable polynomials over Stirling permutations. arXiv:1208.1420, 2012.

[33] Young-Bin Choe, James G. Oxley, Alan D. Sokal, and David G. Wagner. Homogeneous multivariate polynomials with the half-plane property. *Adv. in Appl. Math.*, 32(1-2):88–187, 2004. Special issue on the Tutte polynomial.

[34] Youngbin Choe and David G. Wagner. Rayleigh matroids. *Combin. Probab. Comput.*, 15(5):765–781, 2006.

[35] Maria Chudnovsky and Paul Seymour. The roots of the independence polynomial of a clawfree graph. *J. Combin. Theory Ser. B*, 97(3):350–357, 2007.

[36] Sylvie Corteel and Lauren K. Williams. Tableaux combinatorics for the asymmetric exclusion process and Askey-Wilson polynomials. *Duke Math. J.*, 159(3):385–415, 2011.

[37] Thomas Craven and George Csordas. Jensen polynomials and the Turán and Laguerre inequalities. *Pacific J. Math.*, 136(2):241–260, 1989.

[38] Thomas Craven and George Csordas. Iterated Laguerre and Turán inequalities. *JIPAM. J. Inequal. Pure Appl. Math.*, 3(3):Article 39, 14 pp. (electronic), 2002.

[39] Thomas Craven and George Csordas. Composition theorems, multiplier sequences and complex zero decreasing sequences. In *Value distribution theory and related topics*, volume 3 of *Adv. Complex Anal. Appl.*, pages 131–166. Kluwer Acad. Publ., Boston, MA, 2004.

[40] J. N. Darroch. On the distribution of the number of successes in independent trials. *Ann. Math. Statist.*, 35:1317–1321, 1964.

[41] Emanuele Delucchi, Aaron Pixton, and Lucas Sabalka. Face vectors of subdivided simplicial complexes. *Discrete Math.*, 312(2):248–257, 2012.

[42] Albert Edrei. On the generating functions of totally positive sequences. II. *J. Analyse Math.*, 2:104–109, 1952.

[43] E. Ehrhart. Sur un problème de géométrie diophantienne linéaire. I. Polyèdres et réseaux. *J. Reine Angew. Math.*, 226:1–29, 1967.

[44] E. Ehrhart. Sur un problème de géométrie diophantienne linéaire. II. Systèmes diophantiens linéaires. *J. Reine Angew. Math.*, 227:25–49, 1967.

[45] Alfred Fettweis and Sankar Basu. New results on stable multidimensional polynomials. I. Continuous case. *IEEE Trans. Circuits and Systems*, 34(10):1221–1232, 1987.

[46] Steve Fisk. Polynomials, roots, and interlacing. arXiv:0612833, 2006.

[47] Steve Fisk. Questions about determinants and polynomials. arXiv:0808.1850, 2008.

[48] Dominique Foata and Marcel-P. Schützenberger. *Théorie géométrique des polynômes eulériens*. Lecture Notes in Mathematics, Vol. 138. Springer-Verlag, Berlin-New York, 1970.

[49] Dominique Foata and Volker Strehl. Euler numbers and variations of permutations. In *Colloquio Internazionale sulle Teorie Combinatorie (Roma, 1973), Tomo I*, pages 119–131. Atti dei Convegni Lincei, No. 17. Accad. Naz. Lincei, Rome, 1976.

[50] G. Frobenius. Über die Bernoullischen Zahlen und die Eulerschen Polynome. *Berl. Ber.*, 1910:809–847, 1910.

[51] Światosław R. Gal. Real root conjecture fails for five- and higher-dimensional spheres. *Discrete Comput. Geom.*, 34(2):269–284, 2005.

[52] Vesselin Gasharov. Incomparability graphs of $(3+1)$-free posets are $s$-positive. In *Proceedings of the 6th Conference on Formal Power Series*

*and Algebraic Combinatorics (New Brunswick, NJ, 1994)*, volume 157, pages 193–197, 1996.

[53] Ira M. Gessel. Personal communication, 2005.

[54] Leonid Gurvits. Van der Waerden/Schrijver-Valiant like conjectures and stable (aka hyperbolic) homogeneous polynomials: one theorem for all. *Electron. J. Combin.*, 15(1):Research Paper 66, 26, 2008. With a corrigendum.

[55] James Haglund, Ken Ono, and David G. Wagner. Theorems and conjectures involving rook polynomials with only real zeros. In *Topics in number theory (University Park, PA, 1997)*, volume 467 of *Math. Appl.*, pages 207–221. Kluwer Acad. Publ., Dordrecht, 1999.

[56] James Haglund and Mirkó Visontai. Stable multivariate Eulerian polynomials and generalized Stirling permutations. *European J. Combin.*, 33(4):477–487, 2012.

[57] Yahya Ould Hamidoune. On the numbers of independent $k$-sets in a claw free graph. *J. Combin. Theory Ser. B*, 50(2):241–244, 1990.

[58] J. William Helton and Victor Vinnikov. Linear matrix inequality representation of sets. *Comm. Pure Appl. Math.*, 60(5):654–674, 2007.

[59] Takayuki Hibi. *Algebraic combinatorics on convex polytopes.* Glebe: Carslaw Publications, 1992.

[60] Lars Hörmander. *The analysis of linear partial differential operators. II.* Classics in Mathematics. Springer-Verlag, Berlin, 2005. Differential operators with constant coefficients, Reprint of the 1983 original.

[61] June Huh. Milnor numbers of projective hypersurfaces and the chromatic polynomial of graphs. *J. Amer. Math. Soc.*, 25(3):907–927, 2012.

[62] June Huh and Eric Katz. Log-concavity of characteristic polynomials and the Bergman fan of matroids. *Math. Ann.*, 354(3):1103–1116, 2012.

[63] Bodo Lass. Mehler formulae for matching polynomials of graphs and independence polynomials of clawfree graphs. *J. Combin. Theory Ser. B*, 102(2):411–423, 2012.

[64] T. D. Lee and C. N. Yang. Statistical theory of equations of state and phase transitions. II. Lattice gas and Ising model. *Physical Rev. (2)*, 87:410–419, 1952.

[65] Matthias Lenz. The $f$-vector of a representable-matroid complex is log-concave. *Adv. in Appl. Math.*, 51(5):543–545, 2013.

[66] Russell Lyons. Determinantal probability measures. *Publ. Math. Inst. Hautes Études Sci.*, (98):167–212, 2003.

[67] Adam Marcus, Daniel A Spielman, and Nikhil Srivastava. Interlacing families ii: Mixed characteristic polynomials and the kadison-singer problem. *Ann. of Math.*, to appear.

[68] Peter R. W. McNamara and Bruce E. Sagan. Infinite log-concavity: developments and conjectures. *Adv. in Appl. Math.*, 44(1):1–15, 2010.

[69] Mircea Mustaţă and Sam Payne. Ehrhart polynomials and stringy Betti numbers. *Math. Ann.*, 333(4):787–795, 2005.

[70] Joseph Neggers. Representations of finite partially ordered sets. *J. Combin. Inform. System Sci.*, 3(3):113–133, 1978.

[71] Eran Nevo, T. Kyle Petersen, and Bridget Eileen Tenner. The $\gamma$-vector of a barycentric subdivision. *J. Combin. Theory Ser. A*, 118(4):1364–1380, 2011.

[72] N. Obreschkoff. Verteilung und Berechnung der Nullstellen reeller Polynome. Hochschulbücher für Mathematik. 55. Berlin: VEB Deutscher Verlag der Wissenschaften. VIII, 298 S. mit 2 Abb. (1963)., 1963.

[73] Hidefumi Ohsugi and Takayuki Hibi. Special simplices and Gorenstein toric rings. *J. Combin. Theory Ser. A*, 113(4):718–725, 2006.

[74] Sam Payne. Ehrhart series and lattice triangulations. *Discrete Comput. Geom.*, 40(3):365–376, 2008.

[75] Robin Pemantle. Towards a theory of negative dependence. *J. Math. Phys.*, 41(3):1371–1390, 2000. Probabilistic techniques in equilibrium and nonequilibrium statistical physics.

[76] Robin Pemantle. Hyperbolicity and stable polynomials in combinatorics and probability. In *Current developments in mathematics, 2011*, pages 57–123. Int. Press, Somerville, MA, 2012.

[77] Robin Pemantle and Yuval Peres. Concentration of Lipschitz functionals of determinantal and other strong Rayleigh measures. *Comb. Probab. Comput.*, 23(1):140–160, 2014.

[78] T. Kyle Petersen. Two-sided Eulerian numbers via balls in boxes. *Math. Mag.*, 86(3):159–176, 2013.

[79] Jim Pitman. Probabilistic bounds on the coefficients of polynomials with only real zeros. *J. Combin. Theory Ser. A*, 77(2):279–303, 1997.

[80] G. Pólya and I. Schur. Über zwei Arten von Faktorenfolgen in der Theorie der algebraischen Gleichungen. *J. Reine Angew. Math.*, 144:89–113, 1914.

[81] Alex Postnikov, Victor Reiner, and Lauren Williams. Faces of generalized permutohedra. *Doc. Math.*, 13:207–273, 2008.

[82] Q. I. Rahman and G. Schmeisser. *Analytic theory of polynomials*, volume 26 of *London Mathematical Society Monographs. New Series*. The Clarendon Press Oxford University Press, Oxford, 2002.

[83] Earl D. Rainville. *Special functions*. The Macmillan Co., New York, 1960.

[84] Victor Reiner and Volkmar Welker. On the Charney-Davis and Neggers-Stanley conjectures. *J. Combin. Theory Ser. A*, 109(2):247–280, 2005.

[85] Carla D. Savage and Michael J. Schuster. Ehrhart series of lecture hall polytopes and Eulerian polynomials for inversion sequences. *J. Combin. Theory Ser. A*, 119(4):850–870, 2012.

[86] Carla. D. Savage and Mirkó Visontai. The **s**-Eulerian polynomials have only real roots. *Trans. Amer. Math. Soc.*, 367(2):1441–1466, 2015.

[87] Louis W. Shapiro, Wen Jin Woan, and Seyoum Getu. Runs, slides and moments. *SIAM J. Algebraic Discrete Methods*, 4(4):459–466, 1983.

[88] John Shareshian and Michelle L. Wachs. Eulerian quasisymmetric functions. *Adv. Math.*, 225(6):2921–2966, 2010.

[89] Richard P. Stanley. Decompositions of rational convex polytopes. *Ann. Discrete Math.*, 6:333–342, 1980. Combinatorial mathematics, optimal designs and their applications (Proc. Sympos. Combin. Math. and Optimal Design, Colorado State Univ., Fort Collins, Colo., 1978).

[90] Richard P. Stanley. Log-concave and unimodal sequences in algebra, combinatorics, and geometry. In *Graph theory and its applications: East and West (Jinan, 1986)*, volume 576 of *Ann. New York Acad. Sci.*, pages 500–535. New York Acad. Sci., New York, 1989.

[91] Richard P. Stanley. *Combinatorics and commutative algebra*, volume 41 of *Progress in Mathematics*. Birkhäuser Boston, Inc., Boston, MA, second edition, 1996.

[92] Richard P. Stanley. Graph colorings and related symmetric functions: ideas and applications: a description of results, interesting applications, & notable open problems. *Discrete Math.*, 193(1-3):267–286, 1998. Selected papers in honor of Adriano Garsia (Taormina, 1994).

[93] Richard P. Stanley. *Enumerative combinatorics. Vol. 2*, volume 62 of *Cambridge Studies in Advanced Mathematics*. Cambridge University Press, Cambridge, 1999. With a foreword by Gian-Carlo Rota and appendix 1 by Sergey Fomin.

[94] Richard P. Stanley. *Enumerative combinatorics. Volume 1*, volume 49 of *Cambridge Studies in Advanced Mathematics*. Cambridge University Press, Cambridge, second edition, 2012.

[95] Richard P. Stanley. Personal communication, May 2008.

[96] Einar Steingrímsson. Permutation statistics of indexed permutations. *European J. Combin.*, 15(2):187–205, 1994.

[97] John R. Stembridge. Counterexamples to the poset conjectures of Neggers, Stanley, and Stembridge. *Trans. Amer. Math. Soc.*, 359(3):1115–1128 (electronic), 2007.

[98] John R. Stembridge. Coxeter cones and their $h$-vectors. *Adv. Math.*, 217(5):1935–1961, 2008.

[99] W. T. Tutte. *Graph theory*, volume 21 of *Encyclopedia of Mathematics and its Applications*. Addison-Wesley Publishing Company, Advanced Book Program, Reading, MA, 1984. With a foreword by C. St. J. A. Nash-Williams.

[100] Victor Vinnikov. LMI representations of convex semialgebraic sets and determinantal representations of algebraic hypersurfaces: past, present, and future. In *Mathematical methods in systems, optimization, and control*, volume 222 of *Oper. Theory Adv. Appl.*, pages 325–349. Birkhäuser/Springer Basel AG, Basel, 2012.

[101] Mirkó Visontai and Nathan Williams. Stable multivariate $W$-Eulerian polynomials. *J. Combin. Theory Ser. A*, 120(7):1929–1945, 2013.

[102] David G. Wagner. Enumeration of functions from posets to chains. *European J. Combin.*, 13(4):313–324, 1992.

[103] David G. Wagner. Total positivity of Hadamard products. *J. Math. Anal. Appl.*, 163(2):459–483, 1992.

[104] David G. Wagner. Multivariate stable polynomials: theory and applications. *Bull. Amer. Math. Soc. (N.S.)*, 48(1):53–84, 2011.

[105] David G. Wagner and Yehua Wei. A criterion for the half-plane property. *Discrete Math.*, 309(6):1385–1390, 2009.

[106] C. N. Yang and T. D. Lee. Statistical theory of equations of state and phase transitions. I. Theory of condensation. *Physical Rev. (2)*, 87:404–409, 1952.

# Chapter 8

# *Words*

**Dominique Perrin and Antonio Restivo**

*Université Paris-Est, Marne-la-Vallée and University of Palermo*

**CONTENTS**

## 8.1 Introduction

Combinatorics on words is a field that has both historical roots and a substantial growth. Its roots are to be found in the early results of Axel Thue on square free words and the development of combinatorial group theory (see [4] for an introduction to the early developments of combinatorics on words). The present interest in the field is pushed by its links with several connections with other topics external to pure mathematics, notably bioinformatics.

Enumerative combinatorics on words is itself a branch of enumerative combinatorics, centered on the simplest structure constructor since words are the same as finite sequences.

In this chapter, we have tried to cover a variety of aspects of enumerative combinatorics on words. We have focused on the problems of enumeration connected with conjugacy classes. This includes many interesting combinatorial aspects of words like Lyndon words and de Bruijn cycles. One of the highlights of the chapter is the connection between both of these concepts via the theorem of Fredericksen and Maiorana.

We have put aside some important aspects of enumerative combinatorics on words, which would deserve another complete chapter. This includes the enumeration of various families of words subject to a restriction. For example, the enumeration of square-free words is an important problem for which only asymptotic results are known. It is known for example that the number $s_n$ of ternary square-free words of length $n$ satisfies $\lim_{n\to\infty} s_n^{1/n} = 1.302\ldots$ (see [39] or [16]). Other examples of interest include unbordered words or words avoiding more general patterns (on this notion, see [31]).

The chapter is organized as follows.

In Section 8.2, we introduce some basic definitions concerning words used in subsequent sections. We also introduce basic notions concerning generating series and automata. Both are powerful tools for the enumeration of words.

In Section 8.3, we introduce the notion of conjugacy and the correlated notions of necklaces or circular codes. These notions play a role in almost all the remaining sections of the chapter. We review some classical formulas such as Witt's Formula or Manning's Formula for the zeta function of a set of words.

In Section 8.4, we introduce Lyndon words and prove the important factorization theorem (Theorem 8.4.3). We also discuss the problem of generating Lyndon words and present algorithms for generating them in alphabetic order.

In Section 8.5 we introduce the notion of de Bruijn cycle and their relation with Eulerian graphs. We prove the so-called BEST theorem, enumerating the spanning trees in an Eulerian graph, and apply it to the enumeration of de Bruijn cycles. We finally present the theorem of Fredericksen and Maiorana [17] which beautifully connects Lyndon words and de Bruijn cycles (Theorem 8.5.8).

In Section 8.6, we introduce unavoidable sets. We prove that, on any alphabet, there exist unavoidable sets of words of length $n$, which are a set of representatives of the conjugacy classes of words of length $n$ (Theorem 8.6.11).

In Section 8.7, we introduce a transformation on words, known as the Burrows–Wheeler transformation. This transformation is used in text compression. It is closely related with conjugacy.

We show in Section 8.8 that the Burrows–Wheeler transformation is closely related with a well-known bijection on words, known as the Gessel-Reutenauer bijection. We also prove some results due to Higgins [23], which generalize the theorem of Fredericksen and Maiorana (Theorem 8.8.9).

In Section 8.9, we show that the Burrows–Wheeler transformation is also related to a well-known concept in string processing, the so-called suffix arrays. We end the section with several results due to Schurman and Stoye [38] concerning the enumeration of suffix arrays.

## 8.2 Preliminaries

We briefly introduce the basic terminology on words. Let $A$ be a finite set usually called the **alphabet**. The elements of $A$ are called **letters**.

A word $w$ on the alphabet $A$ is denoted $w = a_1a_2\cdots a_n$ with $a_i \in A$. The integer $n$ is the length of $w$. We denote as usual by $A^*$ the set of words over $A$ and by $\varepsilon$ the empty word. For a word $w$, we denote by $|w|$ the length of $w$. We use the notation $A^+ = A^* - \{\varepsilon\}$. The set $A^*$ is a monoid. Indeed, the concatenation of words is associative, and the empty word is a neutral element for concatenation. The set $A^+$ is sometimes called the **free semigroup** over $A$, while $A^*$ is called the **free monoid**.

A word $w$ is called a **factor** (respectively, a **prefix**, respectively, a **suffix**) of a word $u$ if there exist words $x, y$ such that $u = xwy$ (respectively, $u = wy$, respectively, $u = xw$). The factor (respectively, the prefix, respectively, the suffix) is **proper** if $xy \neq \varepsilon$ (respectively, $y \neq \varepsilon$, respectively, $x \neq \varepsilon$). The prefix of length $k$ of a word $w$ is also denoted by $w[0..k-1]$.

The set of words over a finite alphabet $A$ can be conveniently seen as a tree. Figure 8.1 represents the set $\{a,b\}^*$ as a binary tree. The vertices are the elements of $A^*$. The root is the empty word $\varepsilon$. The sons of a node $x$ are the words $xa$ for $a \in A$. Every word $x$ can also be viewed as the path leading from the root to the node $x$. A word $x$ is a prefix of a word $y$ if it is an ancestor in the tree. Given two words $x$ and $y$, the longest common prefix of $x$ and $y$ is the nearest common ancestor of $x$ and $y$ in the tree.

The set of factors of a word $x$ is denoted $F(x)$. We denote by $F(X)$ the set of factors of words in a set $X \subset A^*$.

The **lexicographic order**, also called **alphabetic order**, is defined as follows. Given two words $x, y$, we have $x < y$ if $x$ is a proper prefix of $y$ or if there exist factorizations $x = uax'$ and $y = uby'$ with $a, b$ letters and $a < b$. This is the usual order in a dictionary.

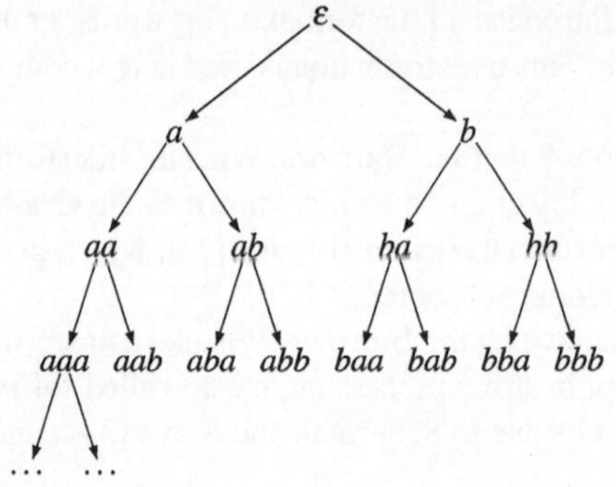


**Figure 8.1**
The tree of the free monoid on two letters.

A **border** of a word $w$ is a nonempty word that is both a prefix and a suffix of $w$. A word $w$ is **unbordered** if its only border is $w$ itself. For example, $a$ is a border of $aba$ and $aabab$ is unbordered.

### 8.2.1 Generating series

For a set $X$ of words, we denote by $f_X(z) = \sum_{n\geq 0} \mathrm{Card}(X \cap A^n)z^n$ the **generating series** of $X$.

Operations on sets can be transferred to their generating series. First, if $X, Y$ are disjoint, then

$$f_{X\cup Y}(z) = f_X(z) + f_Y(z). \tag{8.1}$$

Next, the product $XY$ of two sets $X, Y$ is defined by $XY = \{xy \mid x \in X, y \in Y\}$. We say the the product is **unambiguous** if $xy = x'y'$ for $x, x' \in X$ and $y, y' \in Y$ implies $x = x'$ and $y = y'$. Then if the product of $X, Y$ is unambiguous

$$f_{XY}(z) = f_X(z)f_Y(z). \tag{8.2}$$

A set $X \subset A^+$ is a **code** if the factorization of a word in words of $X$ is unique. Formally, $X$ is a code if $x_1x_2\cdots x_n = y_1y_2\cdots y_m$ with $x_i, y_j \in X$ and $n, m \geq 1$ implies $n = m$ and $x_i = y_i$ for $1 \leq i \leq n$.

As a particular case, a **prefix code** is a set that does not contain any proper prefix of one of its elements. The submonoid generated by a prefix code $X$ is right unitary, that is to say that $u, uv \in X^*$ implies $v \in X^*$. Conversely, any right unitary submonoid is generated by a prefix code.

If $X$ is a code, then

$$f_{X^*}(z) = \frac{1}{1 - f_X(z)}. \tag{8.3}$$

In fact, since the sets $X^n, X^m$ are disjoint for $n \neq m$, we have $f_{X^*}(z) = \sum_{n \geq 0} f_{X^n}(z)$. By unique decomposition, we also have $f_{X^n}(z) = (f_X(z))^n$. Thus $f_{X^*}(z) = \sum_{n \geq 0} f_X(z)^n$ whence the result.

**Example 8.2.1** *Let $X = \{a, ba\}$. The set $X$ is a prefix code. We have* $\mathrm{Card}(X^k \cap A^n) = \binom{k}{n-k}$*. Indeed, a word in $X^k \cap A^n$ is a product of $n-k$ words $ba$ and $2k-n$ words $a$. It is determined by the choice of the positions of the $n-k$ words $ba$ among $k$ possible ones.*

*On the other hand,* $\mathrm{Card}(X^* \cap A^n) = F_{n+1}$ *where $F_n$ is the* **Fibonacci sequence** *defined by $F_0 = 0$, $F_1 = 1$ and $F_{n+1} = F_n + F_{n-1}$ for $n \geq 1$ (the first values are given in Table 8.1).*

| $n$ | 0 | 1 | 2 | 3 | 4 | 5 | 6 | 7 | 8 | 9 | 10 | 11 | 12 | 13 |
|---|---|---|---|---|---|---|---|---|---|---|---|---|---|---|
| $F_n$ | 0 | 1 | 1 | 2 | 3 | 5 | 8 | 13 | 21 | 34 | 55 | 89 | 144 | 233 |

**Table 8.1**
The first values of the Fibonacci sequence.

*This is a consequence of the fact that $f_{X^*}(z) = \frac{1}{1-z-z^2}$ by Equation* (8.3)*. Since $f_{X^*}(z) = \sum_{k \geq 0} f_{X^k}(z)$ we obtain the well-known identity relating Fibonacci numbers and binomial coefficients*

$$F_{n+1} = \sum_{k \leq n} \binom{k}{n-k} \tag{8.4}$$

*which sums binomial coefficients along the parallels to the first diagonal in Pascal's triangle (see Table 8.2).*

| | | | | | | | | |
|---|---|---|---|---|---|---|---|---|
| | | 1 | | | | | | |
| | / | 1 | 1 | | | | | |
| 1 | / | 1 | 2 | 1 | | | | |
| 1 | / | 1 | 3 | 3 | 1 | | | |
| 2 | / | 1 | 4 | 6 | 4 | 1 | | |
| 3 | / | 1 | 5 | 10 | 10 | 5 | 1 | |
| 5 | | 1 | 6 | 15 | 20 | 15 | 6 | 1 |

**Table 8.2**
Pascal's triangle.

**Example 8.2.2** *The* **Dyck set** *is the set of words on the alphabet $\{a, b\}$ having an equal number of occurrences of $a$ and $b$. It is a right unitary submonoid and thus it* *is generated by a prefix code D called the* **Dyck code**. *Let $D_a$ (respectively, $D_b$) be the set of words of D beginning with a (respectively, b). We have*

$$D_a = aD_a^*b \quad \text{and} \quad D_b = bD_b^*a. \tag{8.5}$$

*Let us verify the first one. The second one is symmetrical. Clearly any $d \in D_a$ ends with b. Set $d = ayb$. Then y has the same number of occurrences of a and b and thus $y \in D^*$. Set $y = y_1 \cdots y_n$ with $y_i \in D$. If some $y_i$ begins with b, then $ay_1 \cdots y_{i-1}b$ is a proper prefix of d that belongs to $D^*$, a contradiction with the fact that D is a prefix code. Thus all $y_i$ are in $D_a$ and $y \in aD_a^*b$. Conversely, any word in $aD_a^*b$ is clearly in $D_a$.*

*Since all products in (8.5) are unambiguous, we obtain $f_{D_a}(z) = z^2 f_{D_a^*}(z)$. Since $D_a$ is a code, by (8.3), this implies $f_{D_a}(z) = z^2/(1 - f_{D_a}(z))$. We conclude that $f_{D_a}(z)$ is the solution of the equation*

$$y(z)^2 - y(z) + z^2 = 0. \tag{8.6}$$

*such that $y(0) = 0$. Thus, we obtain the formula*

$$f_{D_a}(z) = \frac{1 - \sqrt{1 - 4z^2}}{2} \tag{8.7}$$

*Finally, since $D = D_a \cup D_b$ and $f_{D_a}(z) = f_{D_b}(z)$ for reasons of symmetry, we obtain*

$$f_D(z) = 1 - \sqrt{1 - 4z^2} \tag{8.8}$$

*Using the binomial formula, we obtain $\mathrm{Card}(D \cap A^{2n}) = -(-4)^n \binom{1/2}{n}$. An elementary computation shows that $\binom{1/2}{n} = (2(-1)^{n-1}/n4^n)\binom{2n-2}{n-1}$. Thus*

$$\mathrm{Card}(D \cap A^{2n}) = \frac{2}{n}\binom{2n-2}{n-1} \tag{8.9}$$

*As a consequence, and since $D_a = aD_a^*b$ by (8.5), we obtain the important and well-known fact that*

$$\mathrm{Card}(D_a^* \cap A^{2n}) = \frac{1}{n+1}\binom{2n}{n} \tag{8.10}$$

*These numbers are called the* **Catalan numbers** *(see Table 8.3).*

| $n$ | 1 | 2 | 3 | 4 | 5 | 6 | 7 | 8 | 9 | 10 |
|---|---|---|---|---|---|---|---|---|---|---|
| | 1 | 1 | 2 | 5 | 14 | 42 | 132 | 429 | 1430 | 4862 |

**Table 8.3**
The first Catalan numbers.

### 8.2.2 Automata

An **automaton** on the alphabet $A$ is given by a set $Q$ of **states**, a set $E \subset Q \times A \times Q$ of edges, a set $I$ of **initial** states and a set $T$ of **terminal** states. The automaton is denoted $\mathscr{A} = (Q,E,I,T)$ or $(Q,I,T)$ if $E$ is understood.

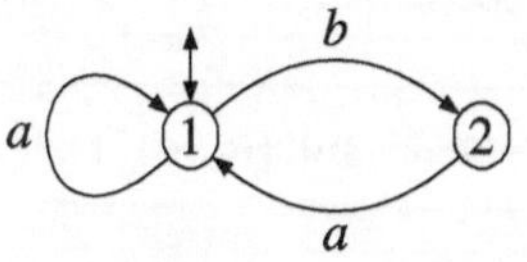


**Figure 8.2**
An automaton.

**Example 8.2.3** *Figure 8.2 represents an automaton with two states and three edges. The initial edges are indicated with an incoming edge and the terminal ones with with an outgoing edge. Here state* 1 *is both the unique initial and terminal state.*

A **path** in the automaton is a sequence of consecutive edges $(p_i, a_i, p_{i+1})$ for $1 \le i \le n$. The integer $n$ is the **length** of the path. The word $w = a_1 a_2 \cdots a_n$ is its **label**. We denote $p_1 \xrightarrow{w} p_n$ such a path. A path $i \xrightarrow{w} t$ is **successful** if $i \in I$ and $t \in T$. The set **recognized** by the automaton is the set of labels of successful paths. The automaton is said to be **unambiguous** if for each word $w$ there is at most one successful path labeled $w$. Thus, an unambiguous automaton defines a bijection between the set of successful paths and the set of their labels. As a particular case, an automaton is **deterministic** if it has at most one initial state and for each state $p$, at most one edge labeled by a given letter starting at $p$.

**Example 8.2.4** *The automaton represented in Figure 8.2 recognizes the set* $\{a, ba\}^*$ *of Example 8.2.1. It is deterministic and thus unambiguous.*

The **adjacency matrix** of the automaton $\mathscr{A} = (Q,E,I,T)$ is the $Q \times Q$-matrix with integer coefficients defined by

$$M_{p,q} = \text{Card}\{e \in E \mid e = (p,a,q) \text{ for some } a \in A\}.$$

It is clear that for each $n \ge 1$, $M^n_{p,q}$ is the number of paths of length $n$ from $p$ to $q$. Thus we have the following useful statement.

**Proposition 8.2.5** *Let* $\mathscr{A} = (Q,I,T)$ *be an unambiguous automaton, let M be its adjacency matrix and let X be the set recognized by* $\mathscr{A}$. *For each* $n \ge 1$,

$$\text{Card}(X \cap A^n) = \sum_{i \in I, t \in T} M^n_{i,t}.$$

**Example 8.2.6** *The adjacency matrix of the automaton represented in Figure 8.2 is*

$$M = \begin{bmatrix} 1 & 1 \\ 1 & 0 \end{bmatrix}.$$

*It is easy to verify that, for* $n \geq 1$,

$$M^n = \begin{bmatrix} F_{n+1} & F_n \\ F_n & F_{n-1} \end{bmatrix}.$$

*Thus, by Proposition 8.2.5, we have* $\mathrm{Card}(\{a, ba\}^* \cap A^n) = F_{n+1}$, *as already seen in Example 8.2.1.*

## 8.3 Conjugacy

We define necklaces and primitive necklaces. We enumerate first primitive necklaces (Witt's Formula, Proposition 8.3.4) and then arbitrary ones (Proposition 8.3.6). See [30] for a more detailed presentation. These notions have been extended to more general structures (see in particular the case of partial words in [6]).

### 8.3.1 Periods

An integer $p \geq 1$ is a **period** of a word $w = a_1 a_2 \cdots a_n$ where $a_i \in A$ if $a_i = a_{i+p}$ for $i = 1, \ldots, n-p$. The smallest period of $w$ is called the *minimal* period of $w$.

**Proposition 8.3.1 (Fine, Wilf)** *If* $p, q$ *are periods of a word* $w$ *of length* $\geq p+q-\gcd(p,q)$, *then* $w$ *has period* $\gcd(p,q)$.

*Proof.* Set $w = a_1 a_2 \cdots a_n$ with $a_i \in A$ and $d = \gcd(p,q)$. We may assume that $p \geq q$. Assume first that $d = 1$. Let us show that $p - q$ is a period of $w$. Let $i$ be such that $1 \leq i \leq n - p + q$. If $i \leq n - p$, we have $a_i = a_{i+p} = a_{i+p-q}$. Otherwise, we have $i > n - p$ and thus $i > q - 1$. Then $a_i = a_{i-q} = a_{i+p-q}$. Thus $w$ has period $p - q$. Since $\gcd(p,q) = \gcd(p-q, q)$ we obtain by induction on $p + q$ that $w$ has period 1.

In the general case, we consider the alphabet $B = A^d$. On this alphabet $w$ has periods $p/d$, $q/d$ and length $n/d \geq p/d + q/d$. By the first part, it has period 1 as a word on the alphabet $B$ and thus period $d$ on the alphabet $A$. ■

**Example 8.3.2** *The word* $w = abaababaaba$ *has periods* 5 *and* 8 *and length* $11 = 5 + 8 - 2$. *By Proposition 8.3.1, no word of length* 12 *can have periods* 5 *and* 8 *without having period* 1.

*More generally, let* $x_n$ *be the Fibonacci sequence of words defined by* $x_1 = b$, $x_2 = a$ *and* $x_{n+1} = x_n x_{n-1}$ *for* $n \geq 2$. *For* $n \geq 3$, *let* $y_n$ *be the word* $x_n$ *minus its two* *last letters. The word $y_7$ is the word $w$ above. Then, for $n \geq 6$, $y_{n+1}$ has periods $F_n$ and $F_{n-1}$. Indeed, $y_{n+1} = x_n y_{n-1}$ shows that $y_{n+1}$ has period $F_n$. Moreover,*

$$\begin{aligned} y_{n+1} &= x_n y_{n-1} = x_{n-1}x_{n-2}x_{n-2}y_{n-3} = x_{n-1}x_{n-2}x_{n-3}x_{n-4}y_{n-3} \\ &= x_{n-1}x_{n-1}x_{n-4}y_{n-3} \end{aligned}$$

*which shows that $F_{n-1}$ is a period since $x_{n-4}y_{n-3}$ is a prefix of $x_{n-3}$ and thus of $x_{n-1}$. Since $|y_{n+1}| = F_n + F_{n-1} - 2$, this shows that the bound of Proposition 8.3.1 is the best possible.*

A word $w \in A^+$ is **primitive** if $w = u^n$ for $u \in A^+$ implies $n = 1$.

Two words $x, y$ are **conjugate** if there exist words $u, v$ such that $x = uv$ and $y = vu$. Thus conjugate words are just cyclic shifts of one another. Conjugacy is thus an equivalence relation. The conjugacy class of a word of length $n$ and period $p$ has $p$ elements if $p$ divides $n$ and has $n$ elements otherwise. In particular, we note the following result.

**Proposition 8.3.3** *A primitive word of length n has n distinct conjugates.*

### 8.3.2 Necklaces

A class of conjugacy is often called a **necklace**, represented on a circle (read clockwise, see Figure 8.3).

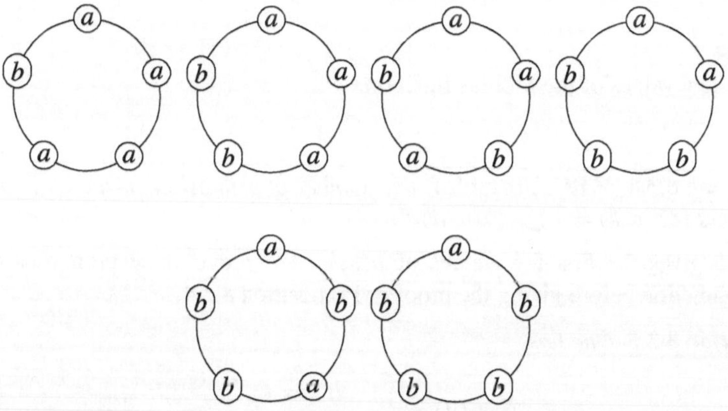


**Figure 8.3**
The six primitive necklaces of length 5 on the alphabet $\{a,b\}$.

Let $p(n,k)$ be the number of primitive necklaces of length $n$ on $k$ letters. Every word of length $n$ is in a unique way a power of a primitive word of length $d$ with $d$ dividing $n$ and such a word has $d$ distinct conjugates. Thus, for any $n \geq 1$,

$$k^n = \sum_{d|n} d\, p(d,k). \tag{8.11}$$

This can be written, using generating series, as a formula called the **Cyclotomic Identity**.

$$\frac{1}{1-kz} = \prod_{n\geq 1} \frac{1}{(1-z^n)^{p(n,k)}}. \tag{8.12}$$

Indeed, taking the logarithm of both sides in Equation (8.12), we obtain

$$\begin{aligned}\sum_{n\geq 1} \frac{k^n z^n}{n} &= \sum_{n\geq 1} -p(n,k)\log(1-z^n)\\ &= \sum_{n\geq 1} p(n,k) \sum_{m\geq 1} \frac{z^{nm}}{m} = \sum_{n\geq 1} \sum_{n=de} p(d,k)\frac{z^n}{e}\end{aligned}$$

and thus $k^n/n = \sum_{n=de} p(d,k)/e$ whence Formula (8.11).

We are going to find a converse giving an expression for the numbers $p(n,k)$. This solution of the system of linear equations (8.11) uses the following function.

The **Möbius function** is defined by $\mu(1) = 1$ and for $n > 1$

$$\mu(n) = \begin{cases} (-1)^i & \text{if } n \text{ is the product of } i \text{ distinct prime numbers} \\ 0 & \text{otherwise} \end{cases}$$

Table 8.4 gives the values of the Möbius function $\mu(n)$ for $n \leq 10$.

| $n$ | 1 | 2 | 3 | 4 | 5 | 6 | 7 | 8 | 9 | 10 |
|---|---|---|---|---|---|---|---|---|---|---|
| $\mu(n)$ | 1 | −1 | −1 | 0 | −1 | 1 | −1 | 0 | 0 | 1 |

**Table 8.4**
The first few values of the Möbius function.

**Proposition 8.3.4 (Witt's Formula)** *The number of primitive necklaces of length n on k letters is* $p(n,k) = \frac{1}{n}\sum_{d|n} \mu(n/d)k^d$.

Table 8.5 gives the first few values of $p(n,k)$. We prove some properties of the Möbius function before giving the proof of Proposition 8.3.4.

**Proposition 8.3.5** *One has*

$$\sum_{d|n} \mu(d) = \begin{cases} 1 & \text{if } n = 1 \\ 0 & \text{otherwise} \end{cases}$$

*Proof.* Indeed, for $n \geq 2$, let $n = p_1^{k_1} \cdots p_m^{k_m}$ and $d = p_1^{\ell_1} \cdots p_m^{\ell_m}$ be the prime decompositions of $n, d$. Then $\mu(d) \neq 0$ if and only if all $\ell_i$ are $0, 1$ and then $\mu(d) = (-1)^t$ with $t = \sum_{i=1}^m \ell_i$. Moreover, there are $\binom{m}{t}$ possible choices giving the same sum $t$. Thus

$$\sum_{d|n} \mu(d) = \sum_{t=0}^{m} (-1)^t \binom{m}{t} = 0$$

since, by the binomial identity, the last expression is $(1-1)^m$. ∎

| $n$ | 1 | 2 | 3 | 4 | 5 | 6 | 7 | 8 | 9 |
|---|---|---|---|---|---|---|---|---|---|
| $p(n,1)$ | 1 | 0 | 0 | 0 | 0 | 0 | 0 | 0 | 0 |
| $p(n,2)$ | 2 | 1 | 2 | 3 | 6 | 9 | 18 | 30 | |
| $p(n,3)$ | 3 | 3 | 8 | 18 | 48 | 116 | 312 | | |
| $p(n,4)$ | 4 | 6 | 20 | 60 | 204 | 670 | | | |
| $p(n,5)$ | 5 | 10 | 40 | 150 | 476 | | | | |
| $p(n,6)$ | 6 | 15 | 30 | 195 | | | | | |
| $p(n,7)$ | 7 | 21 | 27 | | | | | | |
| $p(n,8)$ | 8 | 28 | | | | | | | |
| $p(n,9)$ | 9 | | | | | | | | |

**Table 8.5**
The number $p(n,k)$ of primitive necklaces of length $n$ on $k$ letters for $2 \leq k+n \leq 10$.

For two functions $\alpha, \beta$ from $\mathbb{N} \setminus 0$ into a ring $R$, their **convolution product** is the function $\alpha * \beta : \mathbb{N} \setminus 0 \to R$ defined by

$$\alpha * \beta(n) = \sum_{de=n} \alpha(d)\beta(e).$$

This product is associative with neutral element the function $\underline{1}$ with value 1 on 1 and 0 elsewhere. By Proposition 8.3.5 the function $n \mapsto \sum_{d|n} \mu(d)$ is the function $\underline{1}$. This shows that the Möbius function is the inverse for the convolution product of the constant function equal to 1.

*Proof of Proposition 8.3.4.* Set $\alpha(n) = k^n$ and $\beta(n) = np(n,k)$. Since $k^n = \sum_{d|n} dp(d,k)$ by Equation (8.11), we have $\alpha = \beta * \gamma$ where $\gamma$ is the constant function equal to 1. Since $\gamma * \mu = \underline{1}$, the convolution product of both sides by the Möbius function gives $\alpha * \mu = \beta$, that is $np(n,k) = \sum_{n=de} \mu(d)k^e$. ■

Recall that Euler's **totient function** $\varphi$ is defined as follows. The value of $\varphi(n)$ for $n \geq 1$ is the number of integers $k$ with $1 \leq k \leq n$ such that $\gcd(n,k) = 1$. See Table 8.6 for the first few values of this function. In other words, for $n \geq 2$, $\varphi(n)$ is the number of integers $k$ for $1 \leq k < n$ that are prime to $n$. One has $n = \sum_{d|n} \varphi(d)$.

| $n$ | 1 | 2 | 3 | 4 | 5 | 6 | 7 | 8 | 9 | 10 |
|---|---|---|---|---|---|---|---|---|---|---|
| $\varphi(n)$ | 1 | 1 | 2 | 2 | 4 | 2 | 6 | 4 | 6 | 4 |

**Table 8.6**
The values of the Euler function $\varphi(n)$ for $n \leq 10$.

Indeed, for each divisor $d$ of $n$ the set $M_d$ of integers $m \leq n$ such that $\gcd(n,m) = d$ has $\varphi(n/d)$ elements. Thus $n = \sum_{d|n} \mathrm{Card}(M_d) = \sum_{d|n} \varphi(n/d) = \sum_{d|n} \varphi(d)$.

| $n$ | 1 | 2 | 3 | 4 | 5 | 6 | 7 | 8 | 9 |
|---|---|---|---|---|---|---|---|---|---|
| $c(n,1)$ | 1 | 1 | 1 | 1 | 1 | 1 | 1 | 1 | 1 |
| $c(n,2)$ | 2 | 3 | 4 | 6 | 8 | 14 | 20 | 36 | |
| $c(n,3)$ | 3 | 6 | 11 | 24 | 51 | 130 | 315 | | |
| $c(n,4)$ | 4 | 10 | 24 | 70 | 208 | 700 | | | |
| $c(n,5)$ | 5 | 15 | 45 | 165 | 481 | | | | |
| $c(n,6)$ | 6 | 21 | 36 | 216 | | | | | |
| $c(n,7)$ | 7 | 28 | 34 | | | | | | |
| $c(n,8)$ | 8 | 36 | | | | | | | |
| $c(n,9)$ | 9 | | | | | | | | |

**Table 8.7**
The values of the number $c(n,k)$ of necklaces of length $n$ on $k$ letters for $2 \leq k+n \leq 10$.

Let $c(n,k)$ be the number of necklaces of length $n$ on $k$ letters. Table 8.7 gives the first values of the numbers $c(n,k)$. The values in Table 8.7 can be easily computed from those of Table 8.5 using the fact that $c(n,k) = \sum_{d|n} p(d,k)$. The following statement gives a direct way to compute the numbers $c(n,k)$ (see [21], where it is credited to McMahon).

**Proposition 8.3.6** *The equality* $c(n,k) = \frac{1}{n}\sum_{d|n} \varphi(n/d)k^d$ *holds.*

*Proof.* Consider the multiset formed by the $n$ circular shifts of the words of length $n$ (each word of length $n$ may appear several times). The total number of the shifts in $nc(n,k)$. On the other hand, each word $w = a_0 \cdots a_{n-1}$ of length $n$ appears with a multiplicity, which is the number of integers $p$ with $0 \leq p < n$ such that $w = a_p \cdots a_{n-1}a_0 \cdots a_{p-1}$, that is which are a period of $w^2$. But $p$ is a period of $w^2$ if and only if $w$ is a power of a word of length $\gcd(n,p)$. Thus

$$nc(n,k) = \sum_{0 \leq p < n} k^{\gcd(n,p)}. \tag{8.13}$$

Since there are $\varphi(n/d)$ integers $p$ with $0 \leq p < n$ such that $d = \gcd(n,p)$, the result follows. ■

We illustrate the proof of Proposition 8.3.6 in the following example.

**Example 8.3.7** *Let* $A = \{a,b\}$. *The multiset of circular shifts of words of length* 4 *is the multiset of* $6 \times 4 = 24$ *elements represented below.*

| | | | |
|---|---|---|---|
| *aaaa* | *aaaa* | *aaaa* | *aaaa* |
| *aaab* | *aaba* | *abaa* | *baaa* |
| *aabb* | *abba* | *bbaa* | *baab* |
| *abab* | *baba* | *abab* | *baba* |
| *abbb* | *babb* | *bbab* | *bbba* |
| *bbbb* | *bbbb* | *bbbb* | *bbbb* |

*The words appearing more than once are abab, baba, which appear twice, and aaaa, bbbb, which appear four times.*

*The following array gives for each value of $p = 1,2,3$ the set of words $w$ of length 4 such that $p$ is a period of $w^2$ (for $p = 0$ it is the set of all words of length 4).*

| $p$ | | $\gcd(p,4)$ |
|---|---|---|
| 0 | **aaaa**, **aaab**, **aaba**, **aabb**, **abaa**, **abab**, **abba**, **abbb**, **baaa**, **baab**, **baba**, **babb**, **bbaa**, **bbab**, **bbba**, **bbbb** | 4 |
| 1 | **a**aaa, **b**bbb | 1 |
| 2 | **aa**aa, **ab**ab, **ba**ba, **bb**bb | 2 |
| 3 | **a**aaa, **b**bbb | 1 |

*The value of $d = \gcd(p,4)$ is indicated on the right. The corresponding prefix of length $d$ of each word is indicated in boldface. The row indexed $p$ contains $2^d$ elements coresponding to the binary words of length $d$ in boldface. In this way we have illustrated Equation (8.13) since summing the cardinalities of the sets in each row, we obtain $24 = 16+2+4+2$.*

### 8.3.3 Circular codes

A **circular code** is a set of words $X$ on the alphabet $A$ such that any necklace has a unique factorization in words of $X$. In particular, a circular code is a code.

Formally, $X$ is a circular code if for $x_1, \ldots, x_n$ and $y_1, \ldots, y_m$ in $X$ the equality $sx_2 \cdots x_n p = y_1 \cdots y_m$ with $x_1 = ps$ and $s$ nonempty implies $n = m$, $p = 1$ and $x_i = y_i$ for $1 \leq i \leq n$.

**Example 8.3.8** *The set $X = \{a, ba\}$ is a circular code. Indeed, there is at most one way to paste every occurrence of $b$ with the $a$ following it.*

**Example 8.3.9** *The set $X = \{ab, ba\}$ is not a circular code. Indeed, the necklace of $ab$ has two possible factorizations.*

It can be shown that a submonoid $M$ of $A^*$ is generated by a circular code if and only if it satisfies the following condition for any $u, v \in A^*$:

$$uv, vu \in M \Leftrightarrow u, v \in M. \tag{8.14}$$

For a proof, see [5, Chapter 7]. Note that (8.14) implies for any $u \in M$ and $n \geq 1$

$$u^n \in M \Leftrightarrow u \in M. \tag{8.15}$$

Let $S$ be a set of words on the alphabet $A$ and let $s_n = \mathrm{Card}(S \cap A^n)$ in such a way that $f_S(z) = \sum_{n \geq 0} s_n z^n$.

The **zeta function** of $S$ is the series

$$\zeta_S(z) = \exp \sum_{n \geq 1} \frac{s_n}{n} z^n.$$

The following is due to Manning (see [5, Chapter 7]). The proof given below uses an argument due to [41].

**Theorem 8.3.10** *Let $X$ be a circular code and let $S$ be the set of words having a conjugate in $X^*$. Then*

$$\zeta_S(z) = \frac{1}{1 - f_X(z)}, \tag{8.16}$$

*or equivalently*

$$f_S(z) = \frac{z f_X'(z)}{1 - f_X(z)}. \tag{8.17}$$

*Proof.* For $x \in X$, denote $g_{n,x}$ the number of words of the form $w = syp$ of length $n$ with $y \in X^*$ and $x = ps$ with $p$ nonempty. Since $X$ is circular, the triple $(s,y,p)$ is uniquely determined by $w$. Conversely, every word of $S \cap A^n$ is of this form for some $x \in X$. Thus $g_{x,n} = |x| \operatorname{Card}(X^* \cap A^{n-|x|})$ and $\operatorname{Card}(S \cap A^n) = \sum_{x \in X} g_{n,x}$. We obtain

$$\begin{aligned} \operatorname{Card}(S \cap A^n) &= \sum_{x \in X} g_{n,x} = \sum_{x \in X} |x| \operatorname{Card}(X^* \cap A^{n-|x|}) \\ &= \sum_{m=0}^{n} m \operatorname{Card}(X \cap A^m) \operatorname{Card}(X^* \cap A^{n-m}). \end{aligned}$$

This shows that $f_S(z) = z f_X'(z) f_{X^*}(z)$ whence Formula (8.17). Formula (8.16) is obtained from (8.17) by taking the derivative of the logarithm of each side. ∎

Let $u_n = \operatorname{Card}(X \cap A^n)$ in such a way that $f_X(z) = \sum_{n \geq 0} u_n z^n$. Using Formula (8.17), we obtain for any $n \geq 1$ the formula known as **Newton's formula** in the context of symmetric functions

$$s_n = n u_n + \sum_{1 \leq i \leq n-1} s_i u_{n-i}. \tag{8.18}$$

Since from Equation (8.17) we have $f_S(z) = \frac{z f_X'(z)}{1 - f_X(z)}$, we deduce that $f_S(z) = z f_X'(z) + f_S(z) f_X(z)$, whence Formula (8.18).

Let now $P$ be the set of primitive necklaces in $S$ and let $p_n = \operatorname{Card}(P \cap A^n)$. Then since a word of $S$ of length $n$ is a power of a primitive word of length $d$ with $d$ dividing $n$ and that this word has $d$ conjugates, we have the following equality, generalizing Equation (8.11)

$$s_n = \sum_{d|n} d\, p_d. \tag{8.19}$$

Like Equation (8.11), Equation (8.19) can be written as an equation relating power series and giving a generalization of the Cyclotomic Identity (8.12), namely,

$$f_{X^*}(z) = \prod_{n \geq 1} \frac{1}{(1 - z^n)^{p_n}}. \tag{8.20}$$

Let $c_n$ be the total number of necklaces in $S$, primitive or not. A word of length $n$ in $S$ is in a unique way a power of a primitive word of $S$. Thus $c_n = \sum_{d|n} p_d$ We give below two examples of computation of $s_n, p_n, c_n$.

**Example 8.3.11** *Let $S$ be the set of representatives of necklaces on $A = \{a,b\}$ without consecutive occurrences of $b$. Then $S$ is the set of words having a conjugate in $X^*$ where $X$ is the circular code $X = \{a, ba\}$. Thus, by Theorem 8.3.10, we have*

$$\zeta_S(z) = \frac{1}{1 - z - z^2}.$$

*By Newton's formula, since $u_1 = u_2 = 1$ and $u_n = 0$ for $n \geq 3$, we have $s_{n+1} = s_n + s_{n-1}$ for $n \geq 2$.*

*We obtain the values indicated in Table 8.8.*

| $n$ | 1 | 2 | 3 | 4 | 5 | 6 | 7 | 8 | 9 | 10 | 11 | 12 | 13 |
|---|---|---|---|---|---|---|---|---|---|---|---|---|---|
| $s_n$ | 1 | 3 | 4 | 7 | 11 | 18 | 29 | 47 | 76 | 123 | 199 | 322 | 521 |
| $p_n$ | 1 | 1 | 1 | 1 | 2 | 2 | 4 | 5 | 8 | 11 | 18 | 25 | 40 |
| $c_n$ | 1 | 2 | 2 | 3 | 3 | 5 | 5 | 8 | 10 | 15 | 19 | 31 | 41 |

**Table 8.8**
The values of $s_n, p_n, c_n$ for $n \leq 13$.

*The three necklaces of length 5 without bb (in agreement with $c_5 = 3$) are represented in Figure 8.4.*

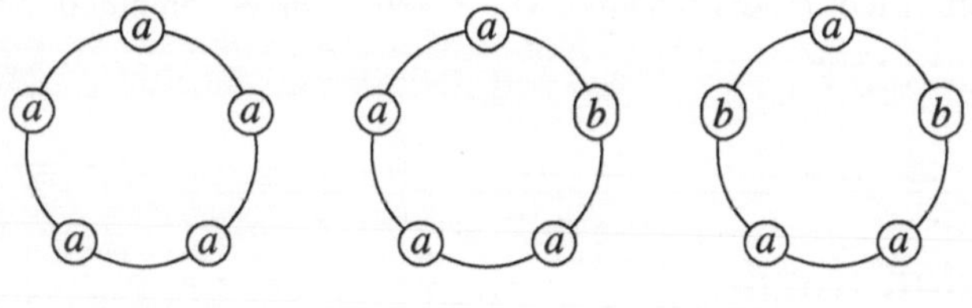


**Figure 8.4**
The three necklaces of length 5 on the alphabet $\{a,b\}$ without *bb*.

**Example 8.3.12** *Let next $S$ be the set of representatives of necklaces on $A = \{a,b\}$ without occurrence of bbb. Then $S$ is the set of words having a conjugate in $X^*$ where $X$ is the circular code $X = \{a, ba, bba\}$. Thus*

$$\zeta_S(z) = \frac{1}{1 - z - z^2 - z^3}$$

*and $s_{n+1} = s_n + s_{n-1} + s_{n-2}$ for $n \geq 3$. We obtain the values shown in Table 8.9. The four primitive necklaces of length 5 without bbb (in agreement with $p_5 = 4$) are represented in Figure 8.5.*

| $n$ | 1 | 2 | 3 | 4 | 5 | 6 | 7 | 8 | 9 | 10 | 11 | 12 | 13 |
|---|---|---|---|---|---|---|---|---|---|---|---|---|---|
| $s_n$ | 1 | 3 | 7 | 11 | 21 | 39 | 71 | 131 | 241 | 443 | 2757 | | |
| $p_n$ | 1 | 1 | 2 | 2 | 4 | 5 | 10 | 15 | 26 | 42 | 74 | 121 | 212 |
| $c_n$ | 1 | 2 | 3 | 4 | 5 | 9 | 11 | 19 | 29 | 48 | 75 | 132 | 213 |

**Table 8.9**
The values of $s_n, p_n, c_n$ for the set of necklaces without $bbb$.

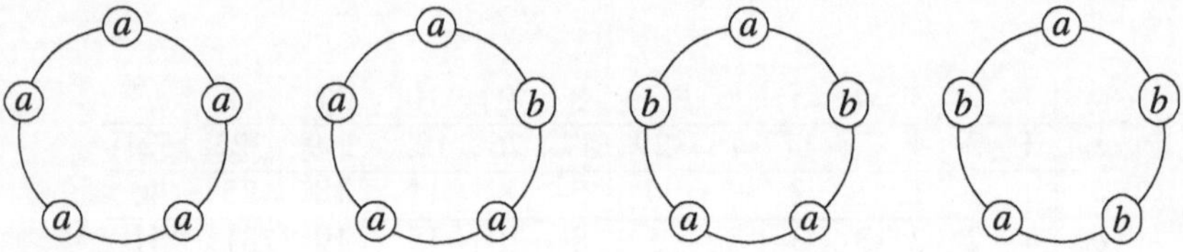


**Figure 8.5**
The four primitive necklaces of length 5 on the alphabet $\{a,b\}$ without $bbb$.

The formulae of this section generalize those of the previous one. MacMahon's Identity (8.13) also generalizes to

$$c_n = \frac{1}{n} \sum_{d|n} \varphi(n/d) s_d$$

where $\varphi$ denotes Euler totient function. This allows a direct computation of the $c_n$.

## 8.4 Lyndon words

A **Lyndon word** is a primitive word that is less than all its conjugates in the alphabetic order. We denote by $L$ the set of Lyndon words.

The first Lyndon words on $\{a,b\}$ are

$$a, b$$
$$ab$$
$$aab, abb$$
$$aaab, aabb, abbb$$
$$aaaab, aaabb, aabab, aabbb, ababb, abbbb$$

We first give the following equivalent definition.

**Proposition 8.4.1** *A word is a Lyndon word if and only if it is strictly smaller than any of its proper suffixes.*

*Proof.* The condition is sufficient. Indeed, let $w = uv$ with $u, v$ nonempty. Since $w < v$, we have $w < vu$.

It is also necessary. For $w \in L$ let $w = uv$ with $u, v$ nonempty. Assume first that $v$ is a prefix of $w$ and thus that $w = vt$. Since $w$ is a Lyndon word, $w < tv$. But $uv < tv$ implies $u < t$ and thus $vu < vt$, a contradiction. Thus $v$ is not a prefix of $w$. But then $v < w$ implies that $vu < w$, a contradiction. We conclude that $w < v$. ∎

Note that, as a consequence, a Lyndon word is unbordered. Indeed, if $u$ is both a nonempty suffix and prefix of $w$, then $u \leq w$ and thus $u = w$ by Proposition 8.4.1.

The next statement gives a recursive way to build Lyndon words.

**Proposition 8.4.2** *If $\ell, m \in L$ with $\ell < m$, then $\ell m$ is a Lyndon word.*

*Proof.* Let us first show that $\ell m < m$. If $\ell$ is a prefix of $m$, then $m = \ell m'$. Then $m < m'$ implies $\ell m < \ell m' = m$. Otherwise, $\ell < m$ implies $\ell m < m$.

Let $v$ be a nonempty proper suffix of $\ell m$. If $v$ is a suffix of $m$, then by Proposition 8.4.1, $m < v$ and thus $\ell m < m < v$. Otherwise, we have $v = v'm$. Then $\ell < v'$ and thus $\ell m < v'm = v$. By Proposition 8.4.1, we conclude that $\ell m \in L$. ∎

For example, we have $aab, ab \in L$ with $aab < ab$ and consequently $aabab \in L$.

### 8.4.1 The factorization theorem

The following result is due to Lyndon (see [30] for more references). It motivated Knuth to call Lyndon words **prime** words in [26].

**Theorem 8.4.3** *Any word factorizes uniquely as a nonincreasing product of Lyndon words.*

The proof uses the following result.

**Lemma 8.4.4** *Let $\ell_1, \ldots, \ell_m$ be a nonincreasing sequence of Lyndon words and let $w = \ell_1 \cdots \ell_m$. Then $\ell_1$ is the longest prefix of $w$, which is a Lyndon word, and $\ell_m$ is the minimal nonempty suffix of $w$.*

*Proof.* Assume that $\ell \in L$ is a prefix of $w$ longer than $\ell_1$. We have $\ell = \ell_1 \cdots \ell_i u$ with $i \geq 1$ and $u$ a nonempty prefix of $\ell_{i+1}$. Then $\ell < u \leq \ell_{i+1} \leq \ell_1 < \ell$, a contradiction.

Next, let $v$ be the minimal suffix of $w$. Then $v$ is in $L$ by Proposition 8.4.1. There is an index $j$, a nonempty suffix $s$ of $\ell_j$ and a word $t$ such that $v = st$. Then $\ell_m \leq \ell_j \leq s \leq st = v \leq \ell_m$ which implies $v = \ell_m$. ∎

*Proof of Theorem 8.4.3.* We have to show that any word $w$ can be written in a unique way $w = \ell_1 \cdots \ell_m$ with $\ell_1, \ldots, \ell_m \in L$ and $\ell_1 \geq \ldots \geq \ell_m$.

Existence: Since the letters are in $L$, any word has a factorization in Lyndon words. Consider a factorization $w = \ell_1 \cdots \ell_m$ with $m$ minimal. If $\ell_i < \ell_{i+1}$ for some $i$, then $w = \ell_1 \cdots \ell_{i-1}(\ell_i \ell_{i+1}) \cdots \ell_m$ is a factorization in Lyndon words since $\ell_i \ell_{i+1} \in L$.

Uniqueness: Assume that $\ell_1 \cdots \ell_m = \ell'_1 \cdots \ell'_{m'}$ with $\ell_i, \ell'_i \in L$, $\ell_1 \geq \ldots \geq \ell_m$ and $\ell'_1 \geq \ldots \geq \ell'_{m'}$. By Lemma 8.4.4, we have $\ell_1 = \ell'_1$, which gives the conclusion by induction on $m$. ∎

We illustrate Theorem 8.4.3 by giving below the factorization of the word *abracadabra*.

$$(abracad)(abr)(a)$$

Let $P$ be the set of prefixes of Lyndon words, also called **preprime** words in [26].

We call a word **minimal** if it is minimal for the lexicographic order in its conjugacy class. Clearly, a word is minimal if and only if it is a power of a Lyndon word.

A **sesquipower** of a word $x$ is a word $w = x^n p$ with $n \geq 1$ and $p$ a proper prefix of $x$. Set $m = |w|$. The word $w$ is determined by $x$ and $m$. It is called the ***m*-extension** of $x$.

The following result appears in Duval [13].

**Proposition 8.4.5** *The set P is the set of sesquipowers of Lyndon words distinct of the maximal letter.*

The proof uses the following lemma.

**Lemma 8.4.6** *For any word p and letter a such that pa is a prefix of a minimal word and for any letter b such that $a < b$, the word pb is in L.*

*Proof.* Let $x$ be a Lyndon word such that $pa$ is a prefix of $x^n$ for some $n \geq 1$. Then $p = x^{n-1}q$ and $x = qar$.

We first show that if $a < b$, then $qb \in L$. Indeed, this is true if $q$ is empty. Otherwise, let $t$ be a proper suffix of $q$. Then $tar$ is a proper suffix of $x$. By Proposition 8.4.1, this implies $x < tar$ and therefore $q < t$. Thus $pb < tb$. Since any proper suffix of $pb$ is of this form, this shows that $pb \in L$ by Proposition 8.4.1 again.

Now, since $x < qb$, we have $x^m qb \in L$ for any $m \geq 1$ by Proposition 8.4.2. ∎

*Proof of Proposition 8.4.5.* Let $x$ be a Lyndon word distinct of the maximal letter. Any sesquipower $w$ of $x$ is a prefix of a power $x^n$ of $x$. By hypothesis, we can write $x = paq$ with $a$ not the maximal letter. Then, by Lemma 8.4.6, for any letter $b > a$, we have $x^n pb \in L$ and thus $w$ is in $P$.

Conversely, we use an induction on the length of $w \in P$. If $|w| = 1$, then $w \in L$. Assume $|w| > 1$. Set $w = va$ with $a \in A$. By induction hypothesis, $v = y^n p$ with $y \in L$, $n \geq 1$ and $p$ proper prefix of $y$. Set $y = pbu$ with $b \in A$. Since $w$ is a prefix of a Lyndon word, we have $pb \leq pa$ and thus $b \leq a$. If $a = b$, then $w$ is is a sesquipower of $y$.

Finally if $b < a$, $w$ is a Lyndon word by Lemma 8.4.6. ∎

Observe that the Lyndon word $x$ such that $w$ is a sesquipower of $x$ is unique. Indeed, assume that $w$ is a sesquipower of $x, x' \in L$. Assuming that $|x| < |x'|$, we have $x' = x^k p$ with $p$ nonempty prefix of $x$. Then $p \leq x < x' < p$, a contradiction.

## 8.4.2 Generating Lyndon words

Proposition 8.4.5 can be used to generate Lyndon words of a given length in alphabetic order (this algorithm is due to Fredericksen and Maiorana [17], and independently to Duval [14], see [26]). The idea is to generate all preprime words of this length. This generation problem has been considered in several contexts (see [37], [34], or [26] in particular).

The algorithm SESQUIPOWERS is represented below. We use the alphabet $\{0,\ldots,k-1\}$. This algorithm visits all preprime words $a_1 \cdots a_n$ of length $n$ with an index $j$ such that $a_1 \cdots a_n$ is an extension of $a_1 \cdots a_j$ (we say equivalently that the algorithm visits $a_1a_2 \cdots a_n$ with index $j$ or that the algorithm visits $a_1a_2 \cdots a_j$).

```
SESQUIPOWERS(n,k)
 1  for i ← 1 to n do
 2        a_i ← 0
 3  j ← 1
 4  while true do
 5        ▷ Visit a_1 ··· a_n with index j
 6        j ← n
 7        while a_j = k − 1 do
 8              j ← j − 1
 9        if j = 0 then
10              return
11        a_j ← a_j + 1
12        ▷ Now a_1 ··· a_j ∈ L
13        for i ← j + 1 to n do
14              a_i ← a_{i−j}
15              ▷ Make n-extension
```

The assignment at line 11 makes $a_1 \cdots a_j$ a Lyndon word (by Lemma 8.4.6). The loop at lines 12-15 realizes the $n$ extension of the word $a_1 \cdots a_j$.

In particular, the sequence of words $a_1a_2 \cdots a_j$ visited by the algorithm is the sequence of Lyndon words of length at most $n$ in increasing order and the sequence of words $a_1a_2 \cdots a_n$ visited with index $n$ is the sequence of Lyndon words of length $n$ in increasing order.

We illustrate this on an example. Consider the list in alphabetic order of the words in $P$ of length 5 plus $b^5$ (we read the list from top to bottom and then from left to right). The letter in boldface is at index $j$.

| | | |
|---|---|---|
| **a***aaaa* | *aaba***b** | *ab***b***ab* |
| *aaaa***b** | *aab***b***a* | *abb***b***a* |
| *aaa***b***a* | *aabb***b** | *abbb***b** |
| *aaab***b** | *a***b***aba* | **b***bbbb* |
| *aa***b***aa* | *abab***b** | |

The six Lyndon words of length 5 are those with the marked letter at the last position.

A possible variant of this algorithm enumerates preprime words in decreasing order.

```
SESQUIPOWERSBIS(n,k)
1  for i ← 1 to n do
2       a_i ← k − 1
3  a_{n+1} ← −1
4  j ← 1
5  while true do
6       ▷ Visit a_1, ..., a_n with index j
7       if a_j = 0 then
8            return
9       a_j ← a_j − 1
10      for h ← j + 1 to n do
11           a_h ← k − 1
12      j ← 1
13      h ← 2
14      while a_{h−j} ≤ a_h do
15           ▷ Now a_1 ··· a_{h−1} is the (h − 1)-extension of a_1 ··· a_j
16           if a_{h−j} < a_h then
17                j ← h
18           h ← h + 1
```

At line 8, the assignement realizes the inverse of the operation at line 11 of SESQUIPOWERS. The loop at lines 13-17 implements the computation of the index $j$ such that $a_1 \cdots a_n$ is a sesquipower of $a_1 \cdots a_j$. It is guaranteed to always end by the assignment of line 3.

Recently, Kociumaka, Radoszewski and Rytter have presented a polynomial time algorithm to compute the $k$th Lyndon word [27].

## 8.5 Eulerian graphs and de Bruijn cycles

A **de Bruijn cycle** of order $n$ on $k$ letters is a necklace of length $k^n$ such that every word of length $n$ on $k$ letters appears exactly once as a factor. For example

$$aabb$$

$$aaababbb$$

$$aaaabaabbababbbb$$

$$aaaaabaaabbaababaabbbababbabbbbb$$

are de Bruijn cycles of order $2, 3, 4, 5$.

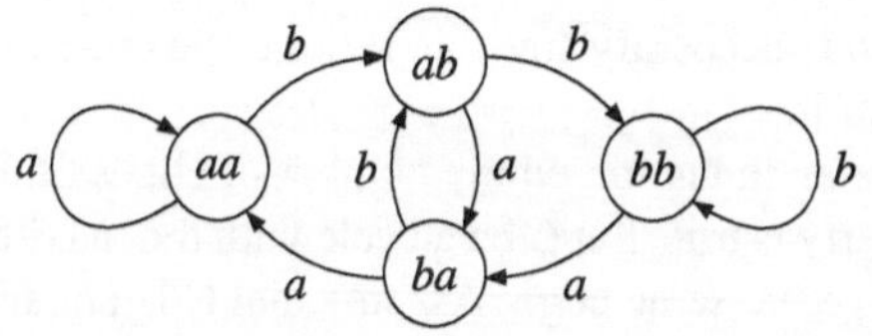


**Figure 8.6**
The de Bruijn graph of order $n = 3$.

The **de Bruijn graph** of order $n$ on an alphabet $A$ is the following labeled graph. It has $A^{n-1}$ as its set of vertices. Its edges are the pairs $(u, v)$ such that $u = aw$, $v = wb$ with $a, b \in A$. Such an edge is labeled $b$. The de Bruijn graph of orders $3, 4$ on the alphabet $\{a, b\}$ are represented in Figure 8.6 and Figure 8.7.

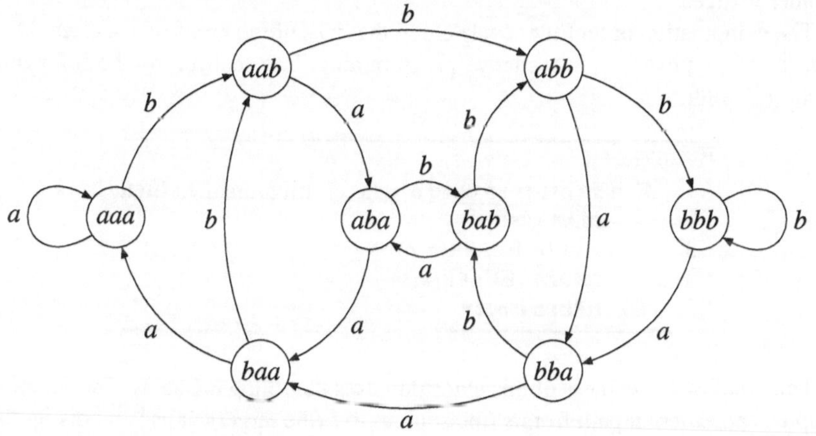


**Figure 8.7**
The de Bruin graph of order $n = 4$.

A cycle in a graph is an **Euler cycle** if it uses each edge of the graph exactly once. A finite graph is **Eulerian** if it has an Euler cycle.

It is easy to verify that the de Bruijn cycles of order $n$ are the labels of Euler cycles in the de Bruijn graph of order $n$. The following result shows the existence of de Bruijn cycles of any order.

**Theorem 8.5.1** *A strongly connected finite graph is Eulerian if and only if each vertex has an indegree equal to its outdegree.*

*Proof.* The condition is necessary since an Euler cycle enters each vertex as many times as it comes out of it.

Conversely, we use an induction on the number of edges of the graph $G$. If there are no edges, the property is true. Let $C$ be a cycle with the maximal possible number of edges not using twice the same edge. Assume that $C$ is not an Euler cycle. Then, since $G$ is strongly connected, there is a vertex $x$ which is on $C$ and in a non-trivial strongly connected component $H$ of $G \setminus C$. Every vertex of $H$ has an indegree equal to its outdegree. So, by induction hypothesis, $H$ contains an Eulerian cycle $D$. The cycles $C$ and $D$ have a vertex in common and thus can be combined to form a cycle larger than $C$, a contradiction. ■

We denote by $d^-(v)$ the indegree of $v$ (which is the number of edges entering $v$) and by $d^-(v)$ its outdegree (which is the number of edges coming out of $v$).

A variant of an Euler cycle is that of **Euler path**. It is a path using all the edges exactly once. It is easy to deduce from Theorem 8.5.1 that a graph has an Euler path from $x$ to $y$ if and only if $d^+(x) - d^-(x) = d^-(y) - d^+(y) = 1$ and $d^+(z) = d^-(z)$ for all other vertices.

The computation of an Euler cycle along the lines of the proof of Theorem 8.5.1 is an interesting exercise in recursive programming. It is realized by the following function EULER.

```
EULER(s,t)
  if there exists an edge e = (s,x) still unmarked then
        MARK(e)
        c ← (e, EULER(x,t))
        return (EULER(s,s), c)
  else return empty
```

The proof of correctness of this algorithm uses the following steps. The function computes an Eulerian path from $s$ (the source) to $t$ (the target). It uses marks on the edges of the graph, which are initially all unmarked.

It chooses an edge $e = (s,x)$ leaving $s$.

If there is an Euler path from $s$ to $t$ beginning with $e$, the solution is

$$(e, Euler(x,p)).$$

Else the solution is

$$(Euler(s,s), e, Euler(x,p)).$$

The following result is due to van Aarden-Ehrenfest and De Bruijn [1]. We are going to see a derivation of it using linear algebra.

**Theorem 8.5.2** *The number of de Bruijn cycles of order $n$ on an alphabet with $k$ letters is*

$$N(n,k) = k^{-n}(k!)^{k^{n-1}}. \tag{8.21}$$

| $n$ | 1 | 2 | 3 | 4 | 5 |
|---|---|---|---|---|---|
| $N(n,2)$ | 1 | 1 | 2 | 16 | 512 |
| $N(n,3)$ | 2 | 24 | 13824 | | |
| $N(n,4)$ | 6 | 331776 | | | |
| $N(n,5)$ | 24 | | | | |

**Table 8.10**
Some values of the number $N(n,k)$ of de Bruijn cycles of order $n$ on $k$ letters.

In particular, for $k = 2$, there are $2^{2^{n-1}-n}$ de Bruijn cycles of order $n$. Table 8.10 lists some values of the numbers $N(n,k)$.

The result for $k = 2$ was obtained as early as 1894 by Fly Sainte-Marie (see [4] for a historical survey).

Observe that $N(1,k) = (k-1)!$. This is in agreement with the fact that de Bruijn cycles of order 1 are the circular permutations of the $k$ letters.

## 8.5.1 The BEST theorem

The following result, known as the BEST theorem, is due to van Aarden-Ehrenfest and de Bruin [1], and also to Smith and Tutte [40]. For a graph $G$ on a set $V$ of vertices, denote $\pi(G) = \prod_{v\in V}(d^+(v)-1)!$. A spanning tree of $G$ oriented towards a vertex $v$ is a set of edges $T$ such that, for any vertex $w$, there is a unique path from $w$ to $v$ using the edges in $T$.

**Theorem 8.5.3** *Let $G$ be an Eulerian graph. Let $v$ be a vertex of $G$ and let $t(G)$ be the number of spanning trees oriented towards $v$. The number of Euler cycles of $G$ is $t(G)\pi(G)$.*

*Proof.* Let $\mathbb{E}$ be the set of Euler cycles and let $\mathbb{E}_v$ be the set of Euler paths from vertex $v$ to itself. Since each Euler cycle passes $d^+(v)$ times through $v$, we have $\mathrm{Card}(\mathbb{E}_v) = d^+(v)\,\mathrm{Card}(\mathbb{E})$.

Let $\mathcal{T}_v$ be the set of spanning trees of $G$ oriented towards $v$. We define a map $\varphi_v : \mathbb{E}_v \to \mathcal{T}_v$ as follows. Let $P$ be an Euler path from $v$ to $v$. We define $T = \varphi(P)$ as the set of edges of $G$ used in $P$ to leave a vertex $w \neq v$ for the last time. Let us verify that $T$ is a spanning tree oriented towards $v$.

Indeed, for each $w \neq v$, there is a unique edge in $T$ going out of $w$. Continuing in this way, we reach $v$ in a finite number of steps. Thus there is a unique path from $w$ to $v$.

Conversely, starting from a spanning tree $T$ oriented towards $v$, we build an Euler path $P$ from $v$ to $v$ as follows. We first use any edge going out of $v$. Next, from a vertex $w$, we use any edge previously unused and distinct from the edge in $T$, as long as such edge exists. There results an Euler path $P$ from $v$ to $v$, which is such that $\varphi(P) = T$. This shows that $\mathrm{Card}(\varphi^{-1}(T)) = d^+(v)!\prod_{w\neq v}(d^+(w)-1)!$. Consequently

$$\mathrm{Card}(\mathbb{E}) = \mathrm{Card}(\mathbb{E}_v)/d^+(v) = t(v)\pi(v).$$

■

We illustrate Theorem 8.5.3 on the example of the de Bruijn graph of order 3 (Figure 8.6).

**Example 8.5.4** *Figure 8.8 represents the two possible spanning trees oriented towards bb in the de Bruijn graph of order 3. Following the Eulerian path in the de Bruijn graph of order 3 (see Figure 8.6), using in turn each of these spanning trees, starting and ending at the root, we obtain the two possible de Bruijn words*

$$aaababbb,\ abaaabbb.$$

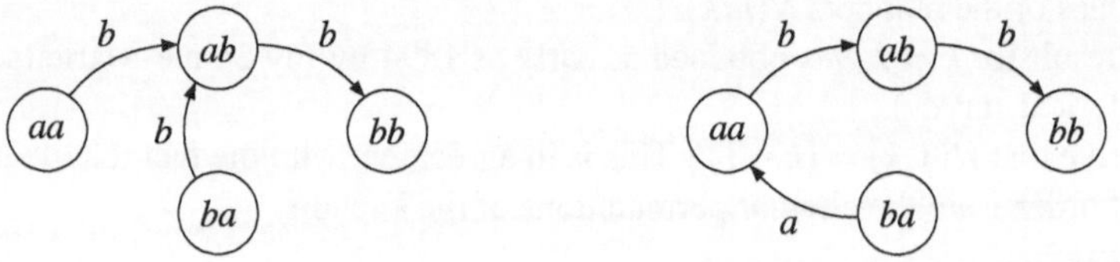


**Figure 8.8**
The two spanning trees of de Bruijn graph of order $n = 3$ oriented towards $bb$.

### 8.5.2 The Matrix-tree theorem

Let $G$ be a multigraph on a set $V$ of vertices. Let $M$ be its adjacency matrix defined by $M_{vw} = \mathrm{Card}(E_{vw})$ with $E_{vw}$ the set of edges from $v$ to $w$. Let $D$ be the diagonal matrix defined by $D_{vv} = \sum_{w \in V} M_{vw}$ and let $L = D - M$ be the **Laplacian matrix** of $G$. Note that the sum of the elements of each row of $L$ is 0. We denote by $K_v(G)$ the determinant of the matrix $C_v$ obtained by suppressing the row and the column of index $v$ in the matrix $L$.

The following result is due to Borchardt [8].

**Theorem 8.5.5 (Matrix-Tree Theorem)** *For any $v \in V$ the number of spanning trees of $G$ oriented towards $v$ is $K_v(G)$*

*Proof.* Denote by $N_v(G)$ the number of spanning trees oriented towards $v$.

We use an induction on the number of edges of $G$. The result holds if there are no edges. Indeed, if there is no edge leading to $v$, then $N_v(G) = 0$. On the other hand, since the sum of each row of $C_v$ is 0, we have $K_v(G) = 0$. Thus $N_v(G) = K_v(G)$.

Consider now an edge $e$ from $w$ to $v$. Let $G'$ be the graph obtained by deleting this edge and $G''$ the graph obtained by merging $v$ and $w$.

We have

$$N_v(G) = N_v(G') + N_v(G''). \tag{8.22}$$

Indeed, the first term of the right-hand side counts the number of spanning trees oriented towards $v$ not containing the edge $e$ and the second one the remaining spanning trees. Similarly, we have

$$K_v(G) = K_v(G') + K_v(G''). \tag{8.23}$$

Indeed, assume $v, w$ to be the first and second indices. The Laplacian matrices of the graphs $G$ and $G''$ have the form

$$L = \left[\begin{array}{c|c|c} a & b & x \\ \hline c & d & y \\ \hline & & \\ z & t & U \\ & & \end{array}\right], \quad L'' = \left[\begin{array}{c|c} a+b+c+d & x+y \\ \hline & \\ z+t & U \\ & \end{array}\right].$$

The Laplacian matrix $L'$ of $G'$ being the same as $L$ with $c+1, d-1$ instead of $c, d$. Then

$$K_v(G) = \left|\begin{array}{c|c} d & y \\ \hline & \\ t & U \\ & \end{array}\right|, \quad K_v(G') = \left|\begin{array}{c|c} d-1 & y \\ \hline & \\ t & U \\ & \end{array}\right|, \quad K_v(G'') = \det(U),$$

and thus Formula (8.23) by the linearity of determinants. By induction hypothesis, we have $K_v(G') = N_v(G')$ and $K_v(G'') = N_v(G'')$ By (8.22) and (8.23) this shows that $K_v(G) = N_v(G)$. ∎

**Example 8.5.6** *For the graph G of Figure 8.6, we have (the matrix C is obtained from L by suppressing the first row and the first column of L)*

$$L = \begin{bmatrix} 1 & -1 & 0 & 0 \\ 0 & 2 & -1 & -1 \\ -1 & -1 & 2 & 0 \\ 0 & 0 & -1 & 1 \end{bmatrix}, \quad C = \begin{bmatrix} 2 & -1 & -1 \\ -1 & 2 & 0 \\ 0 & -1 & 1 \end{bmatrix}.$$

*One has* $\det(C) = 2$ *in agreement with Theorem 8.5.5 since, by Example 8.8, the graph G has two spanning trees oriented towards bb.*

It is possible to deduce the explicit formula for the number of de Bruijn cycles of Theorem 8.5.2 from the matrix-tree theorem.

We denote by $G^*$ the **edge graph** of a graph $G$. Its set of vertices is the set $E$ of edges of $G$ and its set of edges is the set of pairs $(e, f) \in E \times E$ such that the end of $e$ is the origin of $f$. It is easy to verify that the edge graph of the de Bruijn graph $G_n$ can be identified with $G_{n+1}$.

A graph is **regular** of degree $k$ if any vertex has $k$ incoming edges and $k$ outgoing edges. If $G$ is regular, the number $t(G)$ of spanning trees oriented towards a vertex $v$ does not depend on $v$.

The following result is due to Knuth [24] (see also [25], Exercise 2.3.4.2).

**Theorem 8.5.7** *Let G be a regular graph of degree k with m vertices. Then*

$$t(G^*) = k^{m(k-1)-1} t(G).$$

The proof uses the matrix-tree theorem.

It is easy to prove Formula (8.21) by induction on $n$ using this result (and the preceding ones). Indeed, by Theorem 8.5.3, and since $G_n$ has $k^{n-1}$ vertices, we have

$$N(n,k) = (k-1)!^{k^{n-1}} t(G_n).$$

Thus (8.21) is equivalent to

$$t(G_n) = k^{-n} k^{k^{n-1}}. \tag{8.24}$$

Assuming (8.24) and using Theorem 8.5.7, we have

$$\begin{aligned} t(G_{n+1}) &= k^{k^{n-1}(k-1)-1} t(G_n) \\ &= k^{k^n - k^{n-1} - 1} k^{-n} k^{k^{n-1}} \\ &= k^{-n-1} k^{k^n} \end{aligned}$$

which proves that (8.24) holds for $n+1$.

### 8.5.3 Lyndon words and de Bruijn cycles

The following beautiful result is due to Fredericksen and Maiorana [17].

**Theorem 8.5.8** *Let $\ell_1 < \ell_2 < \ldots < \ell_m$ be the increasing sequence of Lyndon words of length dividing n. The word $\ell_1 \ell_2 \cdots \ell_m$ is a de Bruijn cycle of order n.*

The original statement contains the additional claim that the de Bruijn cycle obtained in this way is lexicographically minimal. We shall obtain this as a consequence of a variant of Theorem 8.5.8 (see Theorem 8.5.10 below).

For example, if $n = 4$ and $A = \{a, b\}$, then

$$aaaabaabbababbbb = a\ aaab\ aabb\ ab\ abbb\ b$$

is a de Bruijn cycle of order 4.

We will use the following lemma.

**Lemma 8.5.9** *Let $w$ be a prefix of length $n$ of a Lyndon word and let $\ell$ be its longest prefix in $L$. Then $w$ is the $n$-extension of $\ell$.*

*Proof.*

Set $w = \ell s$ and let $v$ be such that $wv \in L$. Set also $r = |\ell|$, $n = |w|$ and $wv = a_1 \cdots a_m$ with $a_i \in A$. By Proposition 8.4.1, we have $wv < sv$. Thus there is some index $t$ with $1 \le t \le |sv|$ such that $a_j = a_{j+r}$ for $1 \le j \le t-1$ and $a_t < a_{t+r}$. If $t \le n-r$, by Lemma 8.4.6, the word $a_1 \cdots a_{t+r}$ is a prefix of $w$, which is a Lyndon word longer than $\ell$. Thus $a_j = a_{j+r}$ for $1 \le j \le n-r$. This implies that $r$ is a period of $w$ and thus the conclusion. ∎

*Proof of Theorem 8.5.8.* Since $\ell_1 \ell_2 \cdots \ell_m$ has length $k^n$, we only need to prove that any word $w = a_1 \cdots a_n$ of length $n$ appears as a factor of $\ell_1 \cdots \ell_m \ell_1 \ell_2$. We denote by $a$ the first letter of the alphabet and by $z$ the largest one. We consider the following cases.

(a) Assume first that $w$ is primitive and that $w = uv$ with $vu = \ell_k$ and that $u$ is not a power of $z$. Set $u = pbq$ with $p \in z^*$ and $b$ a letter $b < z$. By Lemma 8.4.6, $vpz \in L$. By repeated use of Lemma 8.4.2, $vz^{|u|}$ is a Lyndon word. Thus $\ell_{k+1} \leq vz^{|u|}$. This implies that $v$ is a prefix of $\ell_{k+1}$ and thus $w$ is a factor of $\ell_k \ell_{k+1}$.

(b) Assume next that $w = uv$ is primitive, that $u \in z^*$ and that $vu \in L$. We can first rule out the case where $v \in a^*$. Indeed, $z^j a^{n-j}$ is a factor of $\ell_{m-1}\ell_m\ell_1\ell_2$. Let $k$ be the least index such that $v \leq \ell_k$ (the existence of $k$ follows from the fact that $vu = \ell_j$ for some $j$). Then $\ell_k \leq vu$ and thus $v$ is a prefix of $\ell_k$. Let $v' \leq v$ be the Lyndon word such that $v$ is a sesquipower of $v'$.

(b1) Assume first that $v' \neq \ell_{k-1}$. Let $v''$ be the word $v'$ with its last letter changed into $a$. The word visited before $v'$ by Algorithm SESQUIPOWER$(n,k)$ is, in view of Algorithm SESQUIPOWERBIS$(n,k)$, the word $v''z^{n-|v'|}$. Thus $\ell_{k-1}$ ends with $u$, $\ell_k$ begins with $v$ and thus $w = uv$ is a factor of $\ell_{k-1}\ell_k$.

(b2) Otherwise, $v' = \ell_{k-1}$. For the same reason as above, $u$ is a suffix of $\ell_{k-2}$. Since $v$ is a sesquipower of $v'$, it is a prefix of $v'v$ and thus also a prefix of $\ell_{k-1}\ell_k$. Thus $w$ is a factor of $\ell_{k-2}\ell_{k-1}\ell_k$.

(c) Assume finally that $w = (uv)^d$ with $d$ dividing $n$ and $vu = \ell_k$.

(c1) If $u \notin z^*$ then $\ell_{k+1} \leq (vu)^{d-1} vz^{|u|}$ since the latter is a Lyndon word. Thus $w$ is a factor of $\ell_k \ell_{k+1}$.

(c2) Otherwise, $\ell_{k-1}$ ends with at least $(d-1)|w|$ letters $z$ and $\ell_{k+1} \leq (vu)^{d-1} z^{|w|}$. Thus $w$ is a factor of $\ell_{k-1}\ell_k\ell_{k+1}$.

■

We illustrate the cases in the proof for $n = 6$ and $A = \{a, b\}$. Table 8.11 gives the sequence $\ell_k$.

| $k$ | 1 | 2 | 3 | 4 | 5 | 6 | 7 | 8 |
|---|---|---|---|---|---|---|---|---|
| $\ell_k$ | $a$ | $aaaaab$ | $aaaabb$ | $aaabab$ | $aaabbb$ | $aab$ | $aababb$ | $aabbab$ |

| 9 | 10 | 11 | 12 | 13 | 14 |
|---|---|---|---|---|---|
| $aabbbb$ | $ab$ | $ababbb$ | $abb$ | $abbbbb$ | $b$ |

**Table 8.11**
The Lyndon words of length dividing 6.

(a) Let $w = aabaaa$. Then $u = aab$, $v = aaa$ and $vu = \ell_2$. We find $w$ as a factor of $\ell_2\ell_3$.

(b1) Let $w = baaaab$. Then $u = b$ and $v = aaaab$. We find $k = 3$, $v' = v$ and $w$ is a factor of $\ell_2\ell_3$.

(b2) Let $w = bbabab$. Then $u = bb$, $v = abab$. We find $k = 11$. We have $v' = ab$ and we find $w$ as a factor of $\ell_9\ell_{10}\ell_{11}$.

(c1) Let $w = (ubu)^2$. Then $u = a$, $v = ba$ and $k = 6$. We find $w$ as a factor of $\ell_6\ell_7$.

(c2) Let $w = (bab)^2$. Then $u = b$, $v = ab$ and $k = 12$. We find $w$ as a factor of $\ell_{11}\ell_{12}\ell_{13}$.

Let $X$ be a set of words. A de Bruijn cycle of order $n$ relative to $X$ is a necklace such that every word of $X$ of length $n$ appears exactly once as a factor. The usual notion of de Bruijn cycle is relative to $X = A^*$.

Consider for example the set $X$ of words on $\{a,b\}$ that are representatives of necklaces without consecutive occurrences of $b$ (see Example 8.3.11). Then *aaab* is a de Bruijn cycle of order 3 relative to $X$ and *aaaabab* of order 4.

The following result, due to Moreno [34], gives a family of sets $X$ for which there are de Bruijn cycles of any order relative to $X$. Let $\ell_1 < \ell_2 < \ldots < \ell_m$ be the increasing sequence of Lyndon words of length dividing $n$. For $s < m$, we denote by $X_s$ the set of words such that no factor has a conjugate in $\{\ell_1, \ldots, \ell_s\}$.

**Theorem 8.5.10** *For any $s < m$, the sequence $\ell_s\ell_{s+1}\cdots\ell_m$ is a de Bruijn cycle of order $n$ relative to $X_s$.*

One can deduce from this result the fact that the de Bruijn cycle given by Theorem 8.5.8 is the minimal one for the alphabetic order (see [35]).

As another variant of Theorem 8.5.8, let us quote the following result due to Yu Hin Au [2]: Concatenating the Lyndon words of length $n$ in increasing order, one obtains a word that contains cyclically all primitive words of length $n$ exactly once. For example, for $n = 4$ and $A = \{a,b\}$, one obtains the word *aaab aabb abbb*, which contains cyclically all 12 primitive words of length 4.

## 8.6 Unavoidable sets

A word $t$ is said to **avoid** a word $p$ if $p$ is not a factor of $t$, i.e. if the pattern $p$ does not appear in the text $t$. For example the word `abracadabra` avoids `baba`. The set of all words avoiding a given set $X$ of words has been of interest in several contexts including the notion of a system of finite type in symbolic dynamics (see [29] for example). This notion has been extended to many other situations (see in particular the case of partial words in [7]).

Let $A$ be a finite alphabet. An **unavoidable** set on $A$ is a set $I \subset A^*$ of words on the alphabet $A$ such that any two-sided infinite word $(a_n)_{n\in\mathbb{Z}}$ on the alphabet $A$ admits at least one factor in $I$. It is of course equivalent to ask that any one-sided infinite word has a factor in $I$ or also, since the alphabet is finite, that the set of words that avoids $I$ is finite (see [31] for an exposition of the properties of unavoidable sets).

**Example 8.6.1** *Let $A = \{a,b\}$. The set $U = \{a,b^{10}\}$ is unavoidable since any word of length 10 either has a letter equal to a or is the word $b^{10}$. On the contrary, the set $V = \{aa,b^{10}\}$ is avoidable. Indeed, the infinite word $(ab)^{\omega} = ababababab\ldots$ has no factor in $V$.*

**Proposition 8.6.2** *On a finite alphabet, any unavoidable set contains a finite unavoidable set.*

*Proof.* Indeed, let $X$ be an unavoidable set and let $S$ be the set of words avoiding $X$. Since $X$ is unavoidable, $S$ is finite. Let $n$ be the maximal length of the words of $S$. Let $Z$ be the set of words of $X$ of length at most $n+1$. Every word of length $n+1$ has a factor in $X$ that is actually in $Z$. Thus $Z$ is unavoidable. ∎

The following gives an equivalent definition of unavoidable sets that holds for finite sets. It will be used below.

**Proposition 8.6.3** *Let $I \subset A^*$ be a finite set of words. The following conditions are equivalent.*

(i) *The set $I$ is unavoidable.*

(ii) *Each two-sided infinite periodic word has at least one factor in $I$.*

*Proof.* It is enough to show that (ii) $\Rightarrow$ (i). Let $(a_n)_{n\in\mathbb{Z}}$ be a two-sided infinite sequence of letters. Let $u \in A^*$ be a word longer than any word in $I$ and having an infinite number of occurrences in the sequence $(a_n)_{n\in\mathbb{Z}}$. This sequence has at least one factor of the form $uvu$. By the hypothesis, the infinite periodic word $\ldots uvuvuvuv\ldots$ has a factor $w \in I$. The word $w$ is a factor of at least one of the words $uv$ or $vu$. It is thus also a factor of the sequence $(a_n)_{n\in\mathbb{Z}}$ and thus $I$ is unavoidable. ∎

This statement is false if $I$ is infinite. For example, on a three-letter alphabet, the set of squares is avoidable but every periodic word contains obviously a square.

### 8.6.1 Algorithms

To check in practice that a given finite set $X$ is unavoidable, there are two possible algorithms.

The first one consists in computing a graph $G = (P,E)$, where $P$ is the set of prefixes of $X$ and $E$ is the set of pairs $(p,s)$ for which there is a letter $a \in A$ such that $s$ is the longest suffix of $pa$ which is in $P$.

**Proposition 8.6.4** *A finite set $X$ is unavoidable if and only if every cycle in $G$ contains a vertex in $X$.*

*Proof.* For each integer $n \geq 0$, and vertices $u, v \in P$, there is a path of length $n$ from $u$ to $v$ if and only if there exists a word $y$ of length $n$ such that $v$ is the longest suffix of $uy$ in $P$. This can be proved by induction on $n$. It follows that there is a path of length $n$ from $\varepsilon$ to a vertex in $X$ if and only if $A^*X \cap A^n \neq \emptyset$. ∎

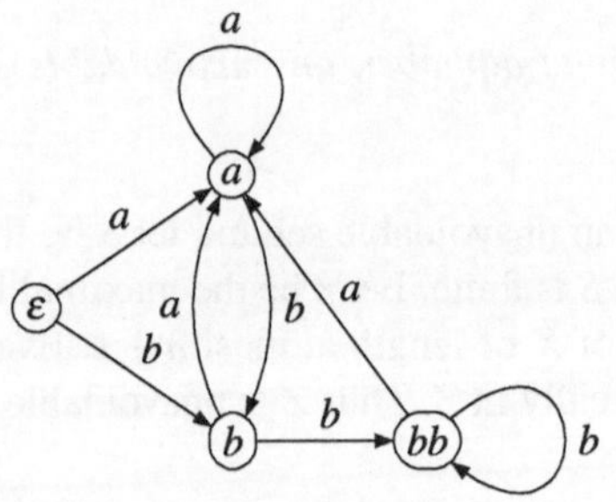


**Figure 8.9**
The graph for $X = \{a, bb\}$.

**Example 8.6.5** *For $X = \{a, bb\}$, the word graph is given in Figure 8.9. By inspection, the set $X$ is unavoidable.*

The second algorithm is sometimes easier to write down by hand. Say that a set $Y$ of words is obtained from a finite set of words $X$ by an **elementary derivation** if

(i) either there exist words $u, v \in X$ such that $u$ is a proper factor of $v$, and

$$Y = X \setminus v$$

(ii) or there exists a word $x = ya \in X$ with $a \in A$ such that, for each letter $b \in A$ there is a suffix $z$ of $y$ such that $zb \in X$, and

$$Y = (X \cup y) \setminus x.$$

A **derivation** is a sequence of elementary derivations. We say that $Y$ is derived from $X$ if $Y$ is obtained from $X$ by a derivation.

**Example 8.6.6** *Let $X = \{aaa, b\}$. Then we have the derivations*

$$X \to \{aa, b\} \to \{a, b\} \to \{\varepsilon, b\} \to \{\varepsilon\}$$

*where the first three arrows follow case (ii) and the last one case (i).*

The following result shows in particular that if $Y$ is derived from $X$, then $X$ is unavoidable if and only if $Y$ is unavoidable. We denote by $S_X$ the set of two-sided infinite words avoiding $X$.

**Proposition 8.6.7** *If $Y$ is derived from $X$, then $S_X = S_Y$.*

*Proof.* It is enough to consider the case of an elementary derivation. In the first case where $Y = X \setminus v$, where $v$ has a factor in $X$, then clearly $S_X = S_Y$. In the second case, we clearly have $S_Y \subset S_X$ since $Y$ is obtained by replacing an element of $X$ by one of its factors. Conversely, assume by contradiction the existence of some $s \in S_X \setminus S_Y$. The only possible factor of $s$ in $Y$ is $y$. Let $b$ be the letter following $y$ in $s$. Then $s$ has a factor in $X$, namely $zb$ where $z$ is the suffix of $y$ such that $zb \in X$ whose existence is granted by the definition of a derivation. This is a contradiction. ■

The notion of a derivation gives a practical method to check whether a set is unavoidable. We have indeed the following result.

**Proposition 8.6.8** *A finite set $X$ is unavoidable if and only if there is a derivation from $X$ to the set $\{\varepsilon\}$.*

*Proof.* Let $X \neq \{\varepsilon\}$ be unavoidable. We prove the existence of a derivation to $\{\varepsilon\}$ by induction on the sum $l(X)$ of the lengths of words in $X$. If $\varepsilon \in X$, we may derive $\{\varepsilon\}$ from $X$. Thus assume $\varepsilon \notin X$, and let $w$ be a word of maximal length avoiding $X$. For each $b \in A$ there is a word $x_b = zb \in X$ that is a suffix of $wb$. Let $x_a = ya$ be the longest of the words $x_b$. Then the hypotheses of case (ii) are satisfied and thus there is a derivation from $X$ to a set $Y$ with $l(Y) < l(X)$. The converse is clear by Proposition 8.6.7. ■

In practice, there is a shortcut that is useful to perform derivations. It is described in the following transformation from $X$ to $Y$.

(iii) There is a word $y$ such that $ya \in X$ for each $a \in A$ and

$$Y = (X \cup y) \setminus \{ya \mid a \in A\}.$$

It is clear that such a set $Y$ can be derived from $X$ and thus, we do not change the definition of derivations by adding case (iii) to the definition of elementary derivations. We use this new definition in the following example.

**Example 8.6.9** *Let $X = \{aaa, aba, abb, bbb\}$. We have the following sequence of derivations (with the symbol a in the word $x = ya$ underlined at each step)*

$$\begin{aligned} \{aaa, ab\underline{a}, ab\underline{b}, bbb\} &\to \{aa\underline{a}, ab, bbb\} \\ &\to \{a\underline{a}, a\underline{b}, bbb\} \to \{a, bb\underline{b}\} \to \{a, b\underline{b}\} \\ &\to \{\underline{a}, \underline{b}\} \to \{\varepsilon\} \end{aligned}$$

Derivations could of course be performed on the left rather than on the right.

### 8.6.2 Unavoidable sets of constant length

In the sequel, we will be interested in unavoidable sets made of words having all the same length $n$. The following proposition is easy to prove.

**Proposition 8.6.10** *Let $A$ be a finite alphabet and let $I$ be an unavoidable set of words of length $n$ on $A$. The cardinality of $I$ is at least equal to the number of conjugacy classes of words of length $n$ on the alphabet $A$.*

*Proof.* Let $u \in A^*$ be a word of length $n$. The factors of length $n$ of the word $u^\omega$ are the elements of the conjugacy class of $u$. Thus $I$ must contain at least one element of this class. ∎

We are going to prove the following result, which shows that the lower bound $c(n,k)$ on the size of unavoidable sets of words of length $n$ on $k$ symbols is reached for all $n,k \geq 1$.

**Theorem 8.6.11** *For all $n,k \geq 1$, there exists an unavoidable set formed of $c(n,k)$ words of length $n$ on $k$ symbols.*

This result has been obtained by J. Mykkelveit [36], solving a conjecture of Golomb. His proof uses exponential sums (see below). Later, and independently, it was solved by Champarnaud, Hansel and Perrin [11] using Lyndon words. We shall present this proof here.

It may be convenient for the reader to reformulate the statement in terms of graphs. A **feedback vertex set** in a directed graph $G$ is a set $F$ of vertices containing at least one vertex from every cycle in $G$. Consider, for $n \geq 1$, the de Bruijn graph $G_{n+1}$ of order $n+1$ on the alphabet $A$ whose vertices are the words of length $n$ on $A$ and the edges are the pairs $(au, ub)$ for all $a, b \in A$ and $u \in A^{n-1}$. It is easy to see that a set of words of length $n$ is unavoidable if the corresponding set of vertices is a feedback vertex set of the graph $G_{n+1}$. Thus, the problem of determining an unavoidable set of words of length $k$ of minimal size is the same as determining the minimal size of a feedback vertex set in $G_{n+1}$. The problem is, for general directed graphs, known to be NP-complete (see [18] for example).

As a preparation to a proof of Theorem 8.6.11, we introduce the following notions.

A **division** of a word $w$ is a pair $(\ell^i, u)$ such that $w = \ell^i u$ where $\ell \in L$, $i \geq 1$ and $u \in A^*$ with $|u| < |\ell|$.

By Proposition 8.4.5 each word in $P$ admits at least one division. We say that a Lyndon word $\ell \in L$ **meets** the word $w$ if there is a division of $w$ of the form $(\ell^i, u)$. It is clear that for any $\ell \in L$ there is at most one such division of $w$.

The **main division** of $w \in P$ is the division $(\ell^i, u)$ where $\ell$ is the shortest Lyndon word that meets $w$. The word $\ell^i$ is the **principal part** of $w$, denoted by $p(w)$, and $u$ is the **remainder**, denoted by $r(w)$.

For example, with $a < b$, the word $aabaabbba$ admits two divisions, which are $(aabaabbb, a)$ and $(aabaabb, ba)$. The first one corresponds to its decomposition as a sesquipower of a Lyndon word. The second one is its main division.

| $M_7$ | $I_7$ | | |
|---|---|---|---|
| *aaaaaaa* | *aaaaaaa* | *aabab***ab** | **ab***aabab* |
| *aaaaaab* | *aaaaaab* | *aabab***bb** | **bb***aabab* |
| *aaaaab***b** | **b***aaaaab* | *aabb***abb** | **abb***aabb* |
| *aaaab***ab** | **ab***aaaab* | *aabb***bab** | **bab***aabb* |
| *aaaab***bb** | **bb***aaaab* | *aabb***bbb** | **bbb***aabb* |
| *aaab***aab** | **aab***aaab* | *ababab***b** | **b***ababab* |
| *aaab***abb** | **abb***aaab* | *abab***bb** | **bb***ababb* |
| *aaab***bab** | **bab***aaab* | *abbabb***b** | **b***abbabb* |
| *aaab***bbb** | **bbb***aaab* | *abbb***bbb** | **bbb***abbb* |
| *aabaab***b** | **b***aabaab* | *bbbbbbb* | *bbbbbbb* |

**Table 8.12**
The sets $M_7$ and $I_7$.

Let $n \geq 1$ be an integer and let $M_n$ be the set of minimal words of length $n$. For each $m \in M_n$, let $p(m)$ be its principal part and $r(m)$ its remainder. Let $I_n$ be the set

$$I_n = \{r(m)p(m) | m \in M_n\}.$$

We remark that any minimal word that is not primitive appears in $I_k$.

**Example 8.6.12** *Table 8.12 lists the elements of $M_7$ and $I_7$ with the remainder of each word of $M_7$ in boldface.*

The object of what follows is to show that $I_n$ is an unavoidable set. By Proposition 8.6.10, the number of elements of $I_n$ is the minimal possible number of elements of an unavoidable set of words of length $n$.

Theorem 8.6.11 will be obtained as a consequence of the following one, giving a construction of the minimal unavoidable sets.

**Theorem 8.6.13** *Let $A$ be a finite alphabet and let $n \geq 1$. Let $M_n$ be the set of words on the alphabet $A$ of length $n$ and that are minimal in their conjugacy class. For every word $m \in M_n$, let $p(m)$ be the principal part of $m$ and let $r(m)$ be its remainder. Then the set*

$$I_n = \{r(m)p(m) | m \in M_n\}$$

*is an unavoidable set.*

To prove Theorem 8.6.13, we need some preliminary results.

**Proposition 8.6.14** *Let $\ell$ and $m$ be two Lyndon words, with $\ell$ a prefix of $m$. Let $s \in A^*$ be a proper suffix of $m$, with $|s| < |\ell|$. Then for all $i > 0$, the word $w = \ell^i s$ is a Lyndon word.*

*Proof.* Let $t$ be a proper suffix of $w$. Three cases may arise.

1. One has $|t| \leq |s|$. Then $t$ is a proper suffix of the Lyndon word $m$ and thus $t > m \geq \ell$ and since $|t| < |\ell|$, we have $t > \ell^n s = w$.

2. One has $|t| > |s|$ and the word $t$ factorizes as $t = \ell^j s$, with $0 \leq j < i$. Since $s$ is a proper suffix of $m$, we have $s > m \geq \ell$. Consequently $t = \ell^j s > \ell^{j+1}$ and since $|s| < |\ell|$, we have $t > \ell^i s = w$.

3. One has $|t| > |s|$ and the word $t$ factorizes as $t = s't'$, where $s'$ is a proper suffix of $\ell$. Since $\ell \in L$, one has $s' > \ell$, and consequently $t = s't' > \ell^i s = w$.

In all cases $t > w$ and thus $w$ is a Lyndon word. ■

**Proposition 8.6.15** *Let $w$ be a prefix of a minimal word and let $(\ell^i, u)$ be its main division. Let $u' \in A^*$ be a word of the same length as $u$ such that the word $w' = \ell^i u'$ is also a prefix of a minimal word. Then the main division of $w'$ is the pair $(\ell^i, u')$.*

*Proof.* Let $(m^j, v)$ be the main division of $w'$. We have $w' = m^j v$ with $|v| < |m|$. Since $(\ell^i, u')$ is a division of $w'$, the word $m$ is a prefix of $\ell$. We are going to show by contradiction that $m$ cannot be a proper prefix of $\ell$.

Suppose that $m$ is a proper prefix of $\ell$. Since the factorization of a minimal word as a power of a Lyndon word is unique, we cannot have the equality $m^j = \ell^i$. Suppose first that $|m^j| < |\ell^i|$. Since $w' = m^j v = \ell^i u'$, the word $m^j$ is a proper prefix of the word $\ell^i$. Thus there exists a non-empty word $x \in A^*$ such that $m^j x = \ell^i$ and $xu' = v$. We thus have

$$w = \ell^i u = m^j xu.$$

Since $|xu| = |xu'| = |v| < |m|$, the pair $(m^j, xu)$ is a division of $w$, which is a contradiction since $m$ is a proper prefix of $\ell$ and $(\ell^i, u)$ is the main division of $w$.

Let us now suppose that $|m^j| > |\ell^i|$. Since $w' = m^j v = \ell^i u'$, the word $\ell^i$ is a proper prefix of the word $m^j$. Since $m$ is a proper prefix of $\ell$, there exists an integer $k > 0$ and a prefix $m'$ of $m$ such that $\ell = m^k m'$. Since $\ell$ is a primitive word, $m'$ is non-empty. As a consequence, $\ell$ admits $m'$ both as a non-empty prefix and suffix, which is contradictory since $\ell$ is a Lyndon word. ■

The final property needed to prove Theorem 8.6.13 is the following.

**Proposition 8.6.16** *Let $m$ be a Lyndon word and $n$ a positive integer. Let $N \geq 1$ be the smallest integer such that $|m^N| > n$. Then the word $m^{N+1}$ has a factor in $I_n$.*

*Proof.* Let $w$ be the prefix of length $n$ of $m^N$. Let $(\ell^i, u)$ be the main division of $w$. If $u$ is the empty word, then, by construction, $w \in I_n$ and the proposition is true. Suppose that $u$ is not empty.

The word $\ell$ is a prefix of $m$ since either $|w| < |m|$ or $w$ admits a division of the form $(m^j, m')$. Let $s$ be the suffix of $m$ having the same length as $u$. By Proposition 8.6.14, the word $\ell^i s$ is a Lyndon word. Thus, by Proposition 8.6.15, the main division of $\ell^i s$ is the pair $(\ell^i, s)$. Consequently, the word $s\ell^i$ belongs to $I_n$. But this word is a factor of $m^{N+1}$. Thus $m^{N+1}$ has a factor in $I_n$. ■

We are now able to prove Theorem 8.6.13. By Proposition 8.6.3, it is enough to show that every periodic two-sided infinite word of the form $\ldots uuuuu \ldots$ has at least one factor in $I_n$. We may suppose without loss of generality that $u$ is a Lyndon word. Let $N$ be the least integer such that $N|u| > n$. Then, by Proposition 8.6.16, the word $u^{N+1}$ has a factor in $I_n$. Thus $I_n$ is unavoidable.

### 8.6.3 Conclusion

The proof of J. Mykkeltveit in [36] is based on the following principle, presented in the case of a binary alphabet. Let us associate to a word $w = a_0 a_1 \cdots a_{n-1}$ on the alphabet $\{0,1\}$ the sum $s(w) = \sum a_j \omega^j$ where $\omega = e^{2i\pi/n}$. We denote by $Is(w)$ the imaginary part of $s(w)$. It can be shown that for each conjugacy class of words, only two cases occur:

(i) either all words $w$ are such that $Is(w) = 0$ (and then, for $n > 2$ one has actually $s(w) = 0$ for each of them)

(ii) or there is, in clockwise order, one block of words $w$ such that $Is(w) > 0$ followed by one block of words $w$ such that $Is(w) < 0$ separated by at most two words $w$ such that $Is(w) = 0$.

Consider the set $S_n$ of words of length $n$ formed of

(i) a representative of each conjugacy class of words $w$ of length $n$ such that $Is(w) = 0$ for all the conjugates.

(ii) the words $w = a_0 a_1 \cdots a_{n-1}$ of length $n$ such that $Is(w) > 0$ for the first time clockwise.

It is shown in [36] that this set is unavoidable for all $n > 2$. The comparison with the previous family of minimal unavoidable set shows that the families have nothing in common. The sets obtained are indeed different. The sets defined by J. Mykkeltveit have a slight advantage in the sense that the maximal length of words avoiding the set is less. For example, for $n = 20$, there are 256 words of length 2579 that avoid $I_n$, but none of length 563 that avoid all of $S_n$ (and there is a unique way to avoid $S_n$ with length 562). This computation has been performed using D. Knuth's program UNAVOIDABLE2 (see `http://www-cs-faculty.stanford.edu/~knuth/programs.html`). Our proof has the advantage of using only elementary concepts and in particular no real or complex arithmetic.

Another proof of Theorem 8.6.11 obtained by the first two authors of [11] and presented in [10] is a construction working by stages. To explain these stages, let us consider the case of a binary alphabet $A = \{a, b\}$. Given a set $X$ of two-sided infinite words, we say that a set $Y$ of words is unavoidable in $X$ if every word of $X$ has a factor in $Y$.

For $i \geq 1$, let $X_i$ be the set of two-sided infinite words on $A$ that avoid $a^i$. Let $c_i(n,k)$ be the number of conjugacy classes of words $x$ of length $n$ on $k$ symbols such

| | 1 | 2 | 3 | 4 | 5 | 6 | 7 | 8 | 9 | 10 |
|---|---|---|---|---|---|---|---|---|---|---|
| 1 | 1 | 1 | 1 | 1 | 1 | 1 | 1 | 1 | 1 | 1 |
| 2 | 1 | 2 | 2 | 3 | 3 | 5 | 5 | 8 | 10 | 15 |
| 3 | 1 | 2 | 3 | 4 | 5 | 9 | 11 | 19 | 29 | 48 |
| 4 | 1 | 2 | 3 | 5 | 6 | 11 | 15 | 27 | 43 | 59 |
| 5 | 1 | 2 | 3 | 5 | 7 | 12 | 17 | 31 | 51 | 91 |
| 6 | 1 | 2 | 3 | 5 | 7 | 13 | 18 | 33 | 55 | 99 |
| 7 | 1 | 2 | 3 | 5 | 7 | 13 | 19 | 34 | 57 | 103 |
| 8 | 1 | 2 | 3 | 5 | 7 | 13 | 19 | 35 | 58 | 105 |
| 9 | 1 | 2 | 3 | 5 | 7 | 13 | 19 | 35 | 59 | 106 |
| 10 | 1 | 2 | 3 | 5 | 7 | 13 | 19 | 35 | 59 | 107 |

**Table 8.13**
The values of $c_i(n,2)$.

that the words of the form $x^\zeta = \cdots xxx \cdots$ are in $X_i$. It is thus also equal to the number of orbits of period $n$ in $X_i$. Table 8.13 gives the values of $c_i(n,2)$ for $1 \leq i \leq 10$ and $1 \leq n \leq 10$. The rows are indexed by $i$ and the columns by $n$. Thus the second row is the last row of Table 8.8 and the third row is the last row of Table 8.9. Moreover, subtracting 1 from the first eight entries of the last three rows, we obtain the second row of Table 8.7 (we have to subtract 1 because $a^n$ is missing).

The idea of the step-by-step construction of a minimal unavoidable set of words of length $n$ is to construct a sequence $Y_1 \subset Y_2 \subset \ldots \subset Y_n$ of sets of words of length $n$ such that for $1 \leq i \leq n$, the set $Y_n$ is unavoidable in $X_i$ with $c_i(n,k)$ elements. This can be stated as the following result.

**Theorem 8.6.17** *For each $k \geq 1$, and $n \geq i \geq 1$, there exists a set of $c_i(n,k)$ words of length $n$ on $k$ symbols that is unavoidable in $X_i$.*

The alternative proof consists in showing directly that the $c_i(n,k)$ last elements of $I_n$ form a set unavoidable in $X_i$.

It is interesting to remark that not all minimal unavoidable sets are build in this way. Indeed, there are sets that are minimal unavoidable in $X_{n+1}$ but do not contain a minimal unavoidable set in $X_n$.

For example, let $Y_1 = \{bbb\}$, $Y_2 = \{bbb, bab\}$, $Y_3 = \{bbb, bab, aab\}$. Then each $Y_i$ for $1 \leq i \leq 3$ is unavoidable in $X_i$ of size $c_i(3,2)$ and $I_3 = Y_3 \cup \{aaa\}$. In particular, the set $I_3$ contains an unavoidable set in $X_2$ with two elements, namely $Y_2$. However, the set $J_3 = \{aaa, aba, bba, bbb\}$ obtained from $I_3$ by exchanging $a$ and $b$ does not contain a two element set unavoidable in $X_2$.

A set of the form $X_i$ is a particular case of what is called a system of finite type. This is, by definition, the set of all two-sided infinite words avoiding a given finite set of words (see [29]). We do not know in general in which systems of finite type it is true that for each $n$ there exists an unavoidable set having no more elements than the number of orbits of period $n$.

## 8.7 The Burrows–Wheeler transform

The Burrows–Wheeler transform is a popular method used for text compression [9]. It produces a permutation of the characters of an input word $w$ in order to obtain a word easier to compress. The presentation given here is close to that of [12].

Suppose $w$ is a **primitive** word over a **totally ordered** alphabet $A$. Let $w_1, w_2, ..., w_n$ be the sequence of conjugates of $w$ in increasing lexicographic order. Let $M(w)$ be the matrix having $w_1, w_2, ..., w_n$ as rows. For example, if $w = aabacacb$, the matrix $M(w)$ is

$$M(aabacacb) = \begin{bmatrix} a & a & b & a & c & a & c & b \\ a & b & a & c & a & c & b & a \\ a & c & a & c & b & a & a & b \\ a & c & b & a & a & b & a & c \\ b & a & a & b & a & c & a & c \\ b & a & c & a & c & b & a & a \\ c & a & c & b & a & a & b & a \\ c & b & a & a & b & a & c & a \end{bmatrix}$$

The **Burrows–Wheeler transform** $T(w)$ of $w$ is the last column of $M(w)$, read from top to bottom. If $b_i$ denotes the last letter of the word $w_i$, for $i = 1, 2, ..., n$, then $T(w) = b_1 b_2 ... b_n$. For instance, $T(aabacacb) = babccaaa$.

It is clear that $T(w)$ depends only on the conjugacy class of $w$, i.e. $T(w) = T(w')$ if $w$ and $w'$ are conjugate. Therefore we may suppose that $w$ is a **Lyndon** word, i.e. $w = w_1$.

The matrix $M(w)$ defines a permutation $\sigma_w$ (or simply $\sigma$ when no confusion arises) of $1, 2, ..., n$ :

$$\sigma(i) = j \Longleftrightarrow w_j = a_i \cdots a_n a_1 \cdots a_{i-1}. \tag{8.25}$$

In other terms, $\sigma(i)$ is the rank in the lexicographic order of the $i$th circular shift of the word $w$. For instance, for $w = aabacacb$, we have:

$$\sigma = \begin{pmatrix} 1 & 2 & 3 & 4 & 5 & 6 & 7 & 8 \\ 1 & 2 & 6 & 3 & 7 & 4 & 8 & 5 \end{pmatrix}.$$

Let $F(w)$ denote the first column of the matrix $M(w)$. If $c_i$ denotes the first letter of the word $w_i$, for $i = 1, 2, \ldots, n$, then $F(w) = c_1 c_2 \cdots c_n$ is the nondecreasing rearrangement of $w$. By definition, we have, for each index $i$, with $1 \le i \le n$,

$$a_i = c_{\sigma(i)}. \tag{8.26}$$

The permutation $\sigma$ transforms the first column of $M(w)$ into its first row, i.e. into the word $w$. We have also the following formula expressing $T(w)$ using $\sigma$:

$$b_i = a_{\sigma^{-1}(i)-1}. \tag{8.27}$$

Indeed, $b_{\sigma(i)}$ is the last letter of $w_{\sigma(j)} = a_j \cdots a_n a_1 \cdots a_{j-1}$, hence $b_{\sigma(i)} = a_{j-1}$, which is equivalent to the above formula.

Given a primitive word $w \in A^*$, let $\pi_w$, or simply $\pi$ when no confusion arises, be the permutation defined by

$$\pi(i) = \sigma(\sigma^{-1}(i) + 1), \tag{8.28}$$

where the addition is to be taken modulo $n$.

**Remark 8.7.1** Observe that $\pi$ is just the permutation defined by writing $\sigma$ as a word and interpreting it as a $n$-cycle. Thus we have also $\sigma(i) = \pi^{i-1}(1)$ and

$$a_i = c_{\pi^{i-1}(1)}.$$

In the previous example we have, written as a cycle,

$$\pi = (1\,2\,6\,3\,7\,4\,8\,5)$$

and as an array

$$\pi = \begin{pmatrix} 1 & 2 & 3 & 4 & 5 & 6 & 7 & 8 \\ 2 & 6 & 7 & 8 & 1 & 3 & 4 & 5 \end{pmatrix}.$$

The following proposition is fundamental for defining the inverse transform.

**Proposition 8.7.2** *If $c_1 c_2 \cdots c_n$ and $b_1 b_2 ... b_n$ are the first and the last columns, respectively, of the matrix $M(w)$, then $c_i = b_{\pi(i)}$, for $i = 1, 2, \ldots, n$.*

*Proof.* Substituting in Formula (8.27) the value $a_i$ given by Formula (8.26), we obtain $b_i = c_{\sigma(\sigma^{-1}(i)-1)}$, which is equivalent to the statement of the proposition. ■

The previous proposition states that the permutation $\pi_w$ transforms the last column of the matrix $M(w)$ into the first one. Actually, it can be noted that $\pi_w$ transforms any column of the matrix $M(w)$ into the following one.

### 8.7.1 The inverse transform

We now show how the word $w$ can be recovered from $T(w)$. For this we prove a property of the matrix $M(w)$ stating that, for any letter $a \in A$, its occurrences in $F(w)$ appear in the same order as in $T(w)$, i.e. the $k$th instance of $a$ in $T(w)$ corresponds (through $\pi$) to its $k$th instance in $F(w)$. In order to formalize this property we introduce the following notation.

The **rank** of the index $i$ in the word $z = z_1 z_2 \cdots z_n$, denoted by $rank(i, z)$, is the number of occurrences of the letter $z_i$ in $z_1 z_2 \cdots z_i$. For instance, if $z = babccaaa$, then $rank(4, z) = 1$ and $rank(6, z) = 2$.

**Proposition 8.7.3** *Given the words $T(w) = b_1 b_2 \cdots b_n$ and $F(w) = c_1 c_2 \cdots c_n$, for each index $i = 1, 2, \ldots, n$, we have*

$$rank(i, F(w)) = rank(\pi(i), T(w)).$$

*Proof.* We first note that, for two words $u, v$ of the same length, and for any letter $a \in A$, one has

$$au < av \Longleftrightarrow ua < va.$$

Thus, for all indices $i, j$, $i < j$ and $c_i = c_j$ implies $\pi(i) < \pi(j)$. Hence, the number of occurrences of $c_i$ in $c_1 c_2 \cdots c_i$ is equal to the number of occurrences of $b_{\pi(i)} = c_i$ in $b_1 b_2 \cdots b_{\pi(i)}$. ∎

To obtain $w$ from $T(w) = b_1 b_2 ... b_n$, we first compute $F(w) = c_1 c_2 ... c_n$ by rearranging the letters $b_i$ in nondecreasing order. Proposition 8.7.3 shows that $\pi(i)$ is the index $j$ such that $c_i = b_j$ and $rank(j, T(w)) = rank(i, F(w))$. This defines the permutation $\pi$, from which $\sigma$ can be obtained expressing $\pi$ as a $n$-cycle, and then, by using Formula (8.26), the word $w$ can be reconstructed.

**Remark 8.7.4** Proposition 8.7.3 further shows that the permutation $\pi$ is related to the **standard** permutation of the word $T(w)$. Recall that the standard permutation of a word $v = b_1 b_2 \cdots b_n$ on a totally ordered alphabet $A$ is the permutation $\tau$ such that, for $i, j \in 1, 2, \ldots, n$, the condition $\tau(i) < \tau(j)$ is equivalent to $b_i < b_j$ or $b_i = b_j$ and $i < j$. The permutation $\tau$ may be obtained by numbering from left to right the letters of $v$, starting from the smallest letter, then the second smallest, and so on. For example, for $v = babccaaa$, we have that, written as a word, $\tau = 51678234$, and as an array:

$$\tau = \begin{pmatrix} 1 & 2 & 3 & 4 & 5 & 6 & 7 & 8 \\ 5 & 1 & 6 & 7 & 8 & 2 & 3 & 4 \end{pmatrix}.$$

It is easy to see that the permutation $\pi$ corresponds to the inverse of the standard permutation of $T(w)$.

**Remark 8.7.5** The Burrows–Wheeler transform $T(w)$ of a primitive word $w$ depends only on the conjugacy class of $w$. Therefore, $T$ defines an **injective** mapping from the primitive necklaces over an alphabet $A$ to the words of $A^*$. However such a mapping is not **surjective**. Remark 8.7.4 indeed shows that, if we consider a word $u \in A^*$ such that the standard permutation $\tau$ of $u$ (and then also the permutation $\pi = \tau^{-1}$) is not a cycle, then there does not exist any word $w$ such the $T(w) = u$. Let us, for instance, consider the word $u = bccaaab$. Its standard permutation

$$\tau = \begin{pmatrix} 1 & 2 & 3 & 4 & 5 & 6 & 7 \\ 4 & 6 & 7 & 1 & 2 & 3 & 5 \end{pmatrix}$$

is the product of two cycles

$$\tau = (1\,4)(2\,6\,3\,7\,5).$$

It follows that there does not exist any word $w$ such that $T(w) = u$.

### 8.7.2 Descents of a permutation

A **descent** of a permutation $\pi$ is an index $i$ such that $\pi(i) > \pi(i+1)$. We denote by $Des(\pi)$ the set of descents of the permutation $\pi$. Consider the permutation $\pi_w$ corresponding to a word $w$. It is clear from Proposition 8.7.3 that if $i$ is a descent of $\pi_w$, then $c_i \neq c_{i+1}$. Thus the number of descents of $\pi_w$ is at most equal to $k-1$, where $k$ is the number of distinct symbols appearing in the word $w$. For instance, for $w = aabacacb$, $\pi_w(4) > \pi_w(5)$, moreover 4 is the only descent of $\pi_w$ and so $Des(\pi_w) = \{4\}$.

Let $A = \{a_1, a_2, \ldots, a_k\}$ be a totally ordered alphabet with $a_1 < a_2 < \ldots < a_k$. If $w$ is a word of $A^*$, denote by $P(w)$ the **Parikh vector** of $w$: $P(w) = (|w|_{a_1}, |w|_{a_2}, \ldots, |w|_{a_k})$. It is clear that, if two words are conjugate, then they have the same Parikh vector, and so one can define the Parikh vector of a necklace. We say that a vector $V = (n_1, n_2, ..., n_k)$ is **positive** if $n_i > 0$ for $i = 1, 2, ..., k$. We denote by $\rho(V)$ the set of integers $\rho(V) = \{n_1, n_1 + n_2, ..., n_1 + ... + n_{k-1}\}$. When $V$ is positive, $\rho(V)$ has $k-1$ elements. Let $\pi_w$ be the permutation corresponding to word $w$ and let $P(w)$ be the Parikh vector of $w$. It is clear from Proposition 8.7.3 that we have the inclusion $Des(\pi_w) \subset \rho(P(w))$.

**Example 8.7.6** *The Parikh vector of the word $w = aabacacb$ is $V = (4,2,2)$ and $\rho(V) = \{4,6\}$. The permutation $\pi_w$ corresponding to $w$ is*

$$\pi_w = \begin{pmatrix} 1 & 2 & 3 & 4 & 5 & 6 & 7 & 8 \\ 2 & 6 & 7 & 8 & 1 & 3 & 4 & 5 \end{pmatrix}.$$

*Thus $Des(\pi_w) = \{4\} \subset \rho(V)$.*

The following statement, due to Crochemore, Désarménien and Perrin [12], results from the previous considerations and Remark 8.7.4.

**Theorem 8.7.7** *For any positive vector $V = (n_1, n_2, ..., n_k)$, with $n = n_1 + n_2 + ... + n_k$, the map $w \to \pi_w$ is one-to-one from the set of primitive necklaces of length $n$ with Parikh vector $V$ onto the set of cyclic permutations $\pi$ on $\{1, 2, ..., n\}$ such that $\rho(V)$ contains $Des(\pi)$.*

This result actually is a particular case of a result stated in [31] and closely related to the Gessel-Reutenauer bijection introduced in the next section. Since each conjugacy class of primitive words can be represented by a Lyndon word, Theorem 8.7.7 establishes a bijection between Lyndon words and cyclic permutations having special descent sets. The extension of this result (Theorem 11.6.1 of [31]) establishes a bijection between words and permutations, relating the Lyndon factorization of words and the cycle structure of the permutations.

## 8.8 The Gessel-Reutenauer bijection

We have shown that the Burrows–Wheeler transform $T(w)$ of a word $w$ depends only on the conjugacy class of $w$. Therefore, $T$ defines an injective mapping from the primitive necklaces over an alphabet $A$ to the words of $A^*$. However (cf. Remark 8.7.5) such a mapping is not surjective.

We now extend the Burrows–Wheeler transform by defining a **bijective** mapping $\Phi$ from the **multisets** of primitive necklaces over a totally ordered alphabet $A$ to the words of $A^*$. This bijection has been introduced by Gessel and Reutenauer in [20]. In order to define the mapping $\Phi$ we introduce a new order on $A^*$. For a word $z \in A^*$, $z^\omega$ denotes the infinite word $zzz\cdots$ obtained by infinitely iterating $z$. Given $u, v \in A^*$, $u \preccurlyeq_\omega v$ if and only if $u^\omega \leq v^\omega$ in the lexicographic order. Remark that the order $\preccurlyeq_\omega$ is different from the usual lexicographic order: for instance, $aba \preccurlyeq_\omega ab$.

Following [33] (see also [23]), we give here a presentation of the mapping $\Phi$ that emphasizes its relation with the Burrows–Wheeler transform.

Let $S = \{s_1, s_2, \ldots, s_m\}$ be a multiset of necklaces, represented by their Lyndon words, i.e. $s_k$ is the Lyndon word corresponding to the $k$th necklace of $S$. In some cases, it is convenient to denote by $(z_1z_2\cdots z_t)$ the necklace containing the word $z_1z_2\cdots z_t$. Denote also by $n = |s_1| + |s_2| + \ldots + |s_m|$ the **size** of $S$ and by $L$ the least common multiple of the lengths of the words in $S$.

In order to sort the elements in the necklaces of $S$ according to the order $\preccurlyeq_\omega$, consider the collection of all words of the form $u^{L\backslash|u|}$, where $u$ is an element of a necklace. All these words then have a common length $L$. We order this set of words lexicographically to yield a matrix $M(S)$ with $n$ rows and $L$ columns.

**Example 8.8.1** *Let $S = \{aab, ab, abb\}$. Then $n = 8$ and $L = 6$ and*

$$M(S) = \begin{bmatrix} a & a & b & a & a & b \\ a & b & a & a & b & a \\ a & b & a & b & a & b \\ a & b & b & a & b & b \\ b & a & a & b & a & a \\ b & a & b & a & b & a \\ b & a & b & b & a & b \\ b & b & a & b & b & a \end{bmatrix}$$

The transform $\Phi(S) = b_1b_2\cdots b_n$ corresponds to the last column of the matrix $M(S)$, read from top to bottom. In the previous example, $\Phi(S) = babbaaba$.

It is clear that if the multiset $S$ has only one necklace represented by its Lyndon word $w$, then $\Phi(S) = T(w)$, i.e. $\Phi(S)$ corresponds to the Burrows–Wheeler transform of $w$. Note also that, if there are non trivial multiplicities in the multiset $S$, then there are repeated rows in the matrix $M(S)$.

Several properties of the Burrows–Wheeler matrix $M(w)$ of a word $w$ can be easily extended to the matrix $M(S)$. In particular, if we denote by $F(S) = c_1c_2\cdots c_n$

the first column of the matrix $M(S)$, by using the same argument as in the proof of Proposition 8.7.3, it can be shown that, for any letter $a \in A$, its occurrences in $F(S)$ appear in the same order as in $\Phi(S)$. This defines a permutation $\pi$ that transforms the last column of the matrix $M(S)$ into the first one. Actually (cf. Remark 8.7.4), $\pi$ is the inverse of the standard permutation of the word $\Phi(S)$.

**Example 8.8.2** *The inverse of the standard permutation of the word babbaaba is*

$$\pi = \begin{pmatrix} 1 & 2 & 3 & 4 & 5 & 6 & 7 & 8 \\ 2 & 5 & 6 & 8 & 1 & 3 & 4 & 7 \end{pmatrix}.$$

In order to reverse the Burrows–Wheeler transform, given the word $T(w) = b_1 b_2 \cdots b_n$, we considered in Section 8.7.1 the inverse $\pi$ of its standard permutation, then we expressed it as a $n$-cycle $(j_1, j_2, \ldots, j_n)$ and we associated to this $n$-cycle the necklace $(c_{j_1} c_{j_2} \cdots c_{j_n})$.

Recall that the Burrows–Wheeler transform is injective, but not surjective. This is a consequence of the fact that, for some words $u$, the inverse of its standard permutation cannot be expressed by a single $n$-cycle, but its decomposition contains several cycles (cf. Remark 8.7.5). This remark is at the base of the surjectivity of the Gessel-Reutenauer transform.

Now we show how to reverse the transform $\Phi$, that is how the multiset of necklaces $S$ can be recovered from the word $\Phi(S) = b_1 b_2 \cdots b_n$.

As for the Burrows–Wheeler transform, first compute the first column $F(S) = c_1 c_2 \cdots c_n$ of $M(S)$ by rearranging the letters $b_i$ in nondecreasing order.

Then, consider the inverse $\pi$ of the standard permutation associated to the word $\Phi(S) = b_1 b_2 \cdots b_n$. With each cycle $(j_1, j_2, \ldots, j_i)$ of $\pi$, associate the necklace $(c_{j_1} c_{j_2} \cdots c_{j_i})$. The multiset $S$ is given by

$$S = \{(c_{j_1} c_{j_2} \cdots c_{j_i}) | (j_1, j_2, \ldots, j_i) \text{ is a cycle of } \pi\}.$$

Remark that different cycles of $\pi$ could give rise to the same necklace, and this explains the use of multisets.

**Example 8.8.3** *Let $\Psi(S) = babbaaba$ (see Example 8.8.1). Rearranging the letters in nondecreasing order, one obtains $F(S) = aaaabbbb$. Then the permutation $\pi$ is*

$$\pi = \begin{pmatrix} 1 & 2 & 3 & 4 & 5 & 6 & 7 & 8 \\ 2 & 5 & 6 & 8 & 1 & 3 & 4 & 7 \end{pmatrix}.$$

*By decomposing $\pi$ in cycles*

$$\pi = (1\,2\,5)(3\,6)(4\,8\,7),$$

*one obtains the multiset of necklaces $S = \{(aab), (ab), (abb)\}$.*

The following theorem, due to Gessel and Reutenauer [20], results from the preceding considerations.

**Theorem 8.8.4** *The map $\Phi$ defines a bijection between words over a totally ordered alphabet A and multisets of primitive necklaces over A.*

A similar, but different, bijection has been proved in [19], where, instead of the lexicographic order, is used the **alternate lexicographic** order.

### 8.8.1 Gessel-Reutenauer bijection and de Bruijn cycles

In this section we present an interesting connection, pointed out in [23], between the Gessel-Reutenauer bijection and the de Bruijn cycles.

A multiset $S = \{s_1, s_2, \ldots, s_m\}$ of necklaces is a **de Bruijn set of span** $n$ over an alphabet $A$ if $|s_1| + |s_2| + \ldots + |s_m| = \text{Card}(A)^n$ and every word $w \in A^n$ is a prefix of some power of some word in a necklace of $S$.

**Remark 8.8.5** The number of distinct prefixes of length $n$ of powers of the words in the necklaces of $S$ is at most $\text{Card}(A)^n$. So, given that $S$ is a de Bruijn set of span n, every word in $A^n$ can be read exactly once within the necklaces of $S$. It also follows, in particular, that no two necklaces in $S$ are equal, so that $S$ is indeed a set, as opposed to a multiset, of necklaces.

**Remark 8.8.6** If $S$ is a de Bruijn set of span $n$, then $S$ contains a necklace of length at least $n$. To show this, consider a Lyndon word $u$ of length $n$ (for instance, $u = ab^{n-1}$, where $a < b$). By definition, $u$ is prefix of some power of a word in a necklace of $S$. Since $u$, as a Lyndon word, is unbordered, it cannot arise as a prefix of a proper power in a necklace of $S$. It follows that $S$ contains a necklace of length at least $n$.

If $A$ is an alphabet of cardinality $k$, denote by $\Gamma$ the set of all $k!$ products of distinct elements of $A$:

$$\Gamma = \{a_1 a_2 \cdots a_k | a_i \in A \text{ for } i = 1, \ldots, k \text{ and } a_i \neq a_j \text{ for } i \neq j\}.$$

For instance, for $A = \{a, b, c\}$,

$$\Gamma = \{abc, acb, bac, bca, cab, cba\}.$$

The following result is due to Higgins [23].

**Theorem 8.8.7** *A set $S$ is a de Bruijn set of span $n$ if and only if $\Phi(S) \in \Gamma^{k^{n-1}}$.*

*Proof.* Let us first suppose that $S$ is a de Bruijn set of span $n$. Consider the matrix $M(S)$. By Remark 8.8.6, the length $L$ of the rows of $M(S)$ is at least $n$. Consider the sub-matrix consisting of the first $n$ columns of $M(S)$. Since $S$ is a de Bruijn set, the rows of this sub matrix form the set $A^n$. Each word $u \in A^{n-1}$ is prefix of $k$ successive rows of $M(S)$. We show that these successive rows of $M(S)$ end with distinct letters of $A$. Suppose, by contradiction, that two of these rows $v_1$ and $v_2$ end with the same letter $a$, i.e. $v_1 = uxa$ and $v_2 = uya$ for some $x, y \in A^*$, with $x \neq y$. Since the conjugates $aux$ and $auy$, of $v_1$ and $v_2$, respectively, correspond to distinct rows in $M(S)$, it follows that $au \in A^n$ would be a prefix of a power of distinct words in the necklaces of $S$, contrary to $S$ being a de Bruijn set of span $n$. Hence the final column $\Phi(S)$ of $M(S)$ is a product of $k^{n-1}$ elements (possibly with repetitions) taken from the set $\Gamma$.

In order to prove the converse implication, let $S$ be a multiset of necklaces such that $\Phi(S) = w \in \Gamma^{k^{n-1}}$. We first prove, by induction on the integer $r$, with $1 \leq r \leq n$, that any word $u \in A^*$ of length $r$ is the prefix of $k^{n-r}$ consecutive rows of the matrix

$M(S)$. In particular, we show that there exists an integer $j$ such that $u$ appears as a prefix in the rows of $M(S)$ ranging from the index $jk^{n-r}$ to the index $(j+1)k^{n-r}-1$. Remark that the sequence of the last letters of these rows, read from top to bottom, returns a factor of $w$, which is again a concatenation of elements of $\Gamma$.

The statement is true for $r = 1$. Indeed, since $w \in \Gamma^{k^{n-1}}$, $|w| = k^n$ and, for any letter $u \in A$, $|w|_a = k^{n-1}$. It follows that the first column $F(S)$ of $M(S)$, read from top to bottom, consists of $k^{n-1}$ occurrences of the first (in the order) letter of $A$, followed by $k^{n-1}$ occurrences of the second letter, and so on. Actually, we have also that, if $z$ is the word corresponding to an arbitrary column of $M(S)$, for each $a \in A$, $|z|_a = |z|/k$.

Let us now suppose that the statement is true for some $r < n$, and consider a word $v \in A^*$ of length $r+1$. If $a$ is the first letter of $v$, we have $v = au$, with $|u| = r$. By the inductive hypothesis, there exists an integer $j$ such that $u$ is the prefix of length $r$ of $k^{n-r}$ consecutive rows of $M(S)$ ranging from the index $jk^{n-r}$ to the index $(j+1)k^{n-r}-1$. The sequence of the last letters of these rows, read from top to bottom, forms a factor $z_u$ of $w$ (the word corresponding to the last column of $M(S)$), and moreover $z_u$ is product of elements of $\Gamma$. Thus, for any $a \in A$, $|z_u|_a = k^{n-r-1}$. It follows that, within the $k^{n-r}$ consecutive rows of $M(S)$ having $u$ as prefix, $k^{n-r-1}$ of them end with the letter $a$. By taking into account their conjugates, we have that $k^{n-r-1}$ consecutive rows of $M(S)$ have as prefix the same word $au = v$. If $b$ is the last letter of $v$, i.e. $v = u'b$, since $|u'| = r$, by the inductive hypothesis there exists an integer $i$ such that $u'$ appears as prefix of the rows of $M(S)$ ranging from the index $ik^{n-r}$ to the index $(i+1)k^{n-r}-1$. The $k$ different letters of $A$ split the interval $[ik^{n-r}, (i+1)k^{n-r}-1]$ into $k$ sub-intervals of equal length in such a way that each sub-interval contains the rows of $M(S)$ having as prefix of length $r+1$ the word $u'c$, for some $c \in A$. We conclude that there is an integer $t$ such that the $k^{n-r-1}$ consecutive rows of $M(S)$, having as prefix the word $v = u'b$, have indexes that range from $tk^{n-r-1}$ to $(t+1)k^{n-r-1}-1$. So, we have proved that, if $\Phi(S) = w \in \Gamma^{k^{n-1}}$, then, for any $r$, with $1 \le r \le n$, every word $u \in A^*$ of length $r$ is the prefix of $k^{n-r}$ consecutive rows of $M(S)$. In particular, for $r = n$, every word $u \in A^*$ of length $n$ is the prefix of exactly one row of $M(S)$. This implies that $S$ is a de Bruijn set of span $n$. ■

By Theorem 8.8.7, one can generate a de Bruijn set $S$ of span $n$, on an alphabet $A$ of cardinality $k$, by taking a word $v \in \Gamma^{k^{n-1}}$ and by computing $\Phi^{-1}(v)$.

**Example 8.8.8** *Consider the alphabet $A = \{a, b\}$ with $a < b$. Then $\Gamma = \{\alpha, \beta\}$, where $\alpha = ab$ and $\beta = ba$. Let $n = 4$, and consider the word $v = \beta\alpha\beta\beta\alpha\alpha\alpha\beta =$ baabbabaabababba $\in \Gamma^8$. Rearranging the letters of $v$ in nondecreasing order, one obtains the first column $F(S)$ of the matrix $M(S)$: $F(S) =$ aaaaaaaabbbbbbbb. The inverse $\pi$ of the standard permutation of the word $v$ is*

$$\pi = \begin{pmatrix} 1 & 2 & 3 & 4 & 5 & 6 & 7 & 8 & 9 & 10 & 11 & 12 & 13 & 14 & 15 & 16 \\ 2 & 3 & 6 & 8 & 9 & 11 & 13 & 16 & 1 & 4 & 5 & 7 & 10 & 12 & 14 & 15 \end{pmatrix}.$$

*By decomposing $\pi$ in cycles*

$$\pi = (1\,2\,3\,6\,11\,5\,9)(4\,8\,16\,15\,14\,12\,7\,13\,10),$$

*one obtains the set of necklaces*

$$S = \{(aaaabab), (aabbbbabb)\}.$$

*One can verify that any word of $A^4$ is prefix of some word in a necklace of S, i.e. S is a de Bruijn set of span 4.*

Given a totally ordered alphabet $A = \{a_1, a_2, \ldots, a_k\}$, of cardinality $k$, with $a_1 < a_2 < \ldots < a_k$, denote by $\alpha$ the element $a_1 a_2 \cdots a_k \in \Gamma$. Now we look at the special case of Theorem 8.8.7 where $v$ is a power of $\alpha$. In such a case, by specializing the arguments in the proof of Theorem 8.8.7, (cf.[23]), one can prove the following result.

**Theorem 8.8.9** *Let $v = \alpha^{k^{n-1}}$, let $S = \Phi^{-1}(v)$ and let $M = M(S)$ be the matrix corresponding to S. Then the rows of M are simply the elements of $A^n$. Moreover S is the set of necklaces of the Lyndon words of length dividing n.*

**Example 8.8.10** *Consider the alphabet $A = \{a, b\}$ with $a < b$, and the word $\alpha^{2^3} = (ab)^8$. The inverse $\pi$ of the standard permutation of the word $(ab)^8$ is*

$$\pi = \begin{pmatrix} 1 & 2 & 3 & 4 & 5 & 6 & 7 & 8 & 9 & 10 & 11 & 12 & 13 & 14 & 15 & 16 \\ 1 & 3 & 5 & 7 & 9 & 11 & 13 & 15 & 2 & 4 & 6 & 8 & 10 & 12 & 14 & 16 \end{pmatrix}.$$

*By decomposing $\pi$ in cycles*

$$\pi = (1)(2\ 3\ 5\ 9)(4\ 7\ 13\ 10)(6\ 11)(8\ 15\ 14\ 12)(16),$$

*one obtains the set of necklaces*

$$S = \{(a), (aaab), (aabb), (ab), (abbb), (b)\},$$

*which is the set of necklaces of the Lyndon words of length dividing 4. If we consider the concatenation of such Lyndon words, we obtain the word*

$$a.aaab.aabb.ab.abbb.b$$

*which is indeed the first de Bruijn word of span 4 in the lexicographic order. That this is always the case is the well-known theorem of Frederickson and Malorana (see Theorem 8.5.8).*

Actually, as a consequence of Theorem 8.8.9 and of the theorem of Frederickson and Maiorana, we obtain the following result.

**Proposition 8.8.11** *The concatenation in ascending order of the Lyndon words of the necklaces of $S = \Phi^{-1}(\alpha^{k^{n-1}})$ is the first de Bruijn word of span n in the lexicographic order.*

## 8.9 Suffix arrays

Suffix array is a widely used data structure in string algorithms (see [32] or [22]). The **suffix array** of a word $w$ of length $n$ is essentially a permutation of $\{1,2,...,n\}$ corresponding to the starting positions of all the suffixes of $w$ sorted lexicographically.

Let $A = \{a_1, a_2, \ldots, a_k\}$ be a totally ordered alphabet of size $k$, where $a_1 < a_2 < \ldots < a_k$. Given a word $w = z_1 z_2 \cdots z_n$ of length $n$ on the alphabet $A$, the suffix array of $w$ is the permutation $\vartheta_w$ (or simply $\vartheta$ when no confusion arises) of the set $\{1,2,\ldots,n\}$ such that $\vartheta(i) = j$ if the suffix $z_j z_{j+1} ... z_n$ has rank $i$ in the lexicographic ordering of all the suffixes of $w$. For instance, if $w = baaababa$, then

$$\vartheta = \begin{pmatrix} 1 & 2 & 3 & 4 & 5 & 6 & 7 & 8 \\ 8 & 2 & 3 & 6 & 4 & 7 & 1 & 5 \end{pmatrix}.$$

### 8.9.1 Suffix arrays and Burrows–Wheeler transform

We show here, following [28], the close connection between the suffix array of a word and its Burrows–Wheeler transform. For this purpose, it is convenient to introduce the **Burrows–Wheeler array** (or simply the **BW-array**) of a primitive word of $A^*$.

Given a primitive word $w = z_1 z_2 \cdots z_n$ of length $n$ on the ordered alphabet $A$, the $BW$-array of $w$ is the permutation $\varphi_w$ (or simply $\varphi$ when no confusion arises) of $\{1,2,\ldots,n\}$ such that $\varphi(i) = j$ if the conjugate $z_j \cdots z_n z_1 \cdots z_{j-1}$ has rank $i$ in the lexicographic sorting of all the conjugates of $w$. By definition, the $BW$-array of a word $w$ is just the inverse of the permutation $\sigma$ defined by the relation 8.25. i.e. $\varphi = \sigma^{-1}$.

In order to show the connection between the suffix array and the Burrows–Wheeler transform, we first introduce a **sentinel** symbol at the end of the word. Consider a symbol $\sharp \notin A$, and the ordered alphabet $A' = \{\sharp, a_1, ..., a_k\}$ where $\sharp < a_1 < \ldots < a_k$. We will examine the suffix array of the word $w' = w\sharp$. In the sequel, we denote by $S_n$ the set of permutations of $\{1,2,\ldots,n\}$. Moreover, for $\vartheta \in S_n$, $\tilde{\vartheta} \in S_{n+1}$ denotes the permutation

$$\tilde{\vartheta} = \begin{pmatrix} 1 & 2 & \ldots & n+1 \\ n+1 & \vartheta(1) & \ldots & \vartheta(n) \end{pmatrix}.$$

**Remark 8.9.1** There is a one-to-one correspondence between the suffix arrays of the words $w \in A^n$ and the suffix arrays of the words $w' \in A^n\sharp$. In particular, if the permutation $\vartheta_w \in S_n$ is the suffix array of $w \in A^n$, then the permutation $\vartheta_{w'} \subset S_{n+1}$ is the suffix array of $w' = w\sharp$ if and only if $\vartheta_{w'} = \tilde{\vartheta}$.

**Remark 8.9.2** It is easy to see that, for words in $A^*\sharp$, conjugate sorting is equivalent to suffix sorting. It follows that the suffix array of the word $w' = w\sharp$ coincides with its BW-array, i.e. $\vartheta_{w'} = \varphi_{w'}$.

The following statement follows from the previous remarks.

**Proposition 8.9.3** *A permutation $\vartheta \in S_n$ is the suffix array of a word $w \in A^n$ if and only if the permutation $\tilde{\vartheta} \in S_{n+1}$ is the BW-array of the word $w' = w\sharp$.*

Consider now the mapping $\Psi : S_n \to S_{n+1}$ defined as follows. If $\vartheta \in S_n$, $\Psi(\vartheta)$ is the permutation $\mu \in S_{n+1}$ defined by $\mu(i) = \tilde{\vartheta}^{-1}(\tilde{\vartheta}(i)+1)$, where the addition is taken modulo $n+1$. Actually, $\Psi(\vartheta)$ is just the permutation obtained by writing $\tilde{\vartheta}^{-1}$ as a word and interpreting it as a $(n+1)$-cycle.

**Example 8.9.4** *Consider the permutation*

$$\vartheta = \begin{pmatrix} 1 & 2 & 3 & 4 & 5 & 6 & 7 & 8 \\ 8 & 2 & 3 & 6 & 4 & 7 & 1 & 5 \end{pmatrix}.$$

*Then we have that*

$$\tilde{\vartheta} = \begin{pmatrix} 1 & 2 & 3 & 4 & 5 & 6 & 7 & 8 & 9 \\ 9 & 8 & 2 & 3 & 6 & 4 & 7 & 1 & 5 \end{pmatrix}$$

*and*

$$\tilde{\vartheta}^{-1} = \begin{pmatrix} 1 & 2 & 3 & 4 & 5 & 6 & 7 & 8 & 9 \\ 8 & 3 & 4 & 6 & 9 & 5 & 7 & 2 & 1 \end{pmatrix}.$$

*It follows that*

$$\Psi(\vartheta) = (8\,3\,4\,6\,9\,5\,7\,2\,1) = \begin{pmatrix} 1 & 2 & 3 & 4 & 5 & 6 & 7 & 8 & 9 \\ 8 & 1 & 4 & 6 & 7 & 9 & 2 & 3 & 5 \end{pmatrix}.$$

The following theorem, that appears in [28], gives a characterization of suffix arrays.

**Theorem 8.9.5** *Let $n_1, n_2, \ldots, n_k$ be positive integers such that $n_1 + n_2 + \ldots + n_k = n$. A permutation $\vartheta \in S_n$ is the suffix array of a word $w \in A^n$, with Parikh vector $P(w) = (n_1, n_2, \ldots, n_k)$ if and only if $Des(\Psi(\vartheta)) \subseteq \{1, 1+n_1, \ldots, 1+n_1+\ldots+n_{k-1}\}$. Moreover, in this case, $\vartheta$ is the suffix array of exactly one such word.*

*Proof.* Let us suppose that $\vartheta \in S_n$ is the suffix array of a word $w \in A^n$. Then, by Proposition 8.9.3, $\tilde{\vartheta} \in S_{n+1}$ is the $BW$-array of the word $w' = w\sharp$. By observing that the $BW$-array of $w'$ corresponds to the inverse of the permutation $\sigma_{w'}$ defined by the relation (8.25), we have that $\tilde{\vartheta} = (\sigma_{w'})^{-1}$. We show that $\Psi(\vartheta) = \pi_{w'}$, where $\pi_{w'}$ is the permutation defined by formula (8.28). Indeed, if $\mu = \Psi(\vartheta)$, we can write

$$\mu(i) = \tilde{\vartheta}^{-1}(\tilde{\vartheta}(i)+1) = \sigma_{w'}(\sigma_{w'}^{-1}(i)+1) = \pi_{w'}.$$

We now observe that, if $P(w) = (n_1, \ldots, n_k)$ is the Parikh vector of $w$, then the Parikh vector of $w' = w\sharp$ is $P(w') = (1, n_1, \ldots, n_k)$. Therefore, by Theorem 8.7.7,

$$Des(\Psi(\vartheta)) = Des(\pi_{w'}) \subseteq \rho(P(w')) = \{1, 1+n_1, \ldots, 1+n_1+\ldots+n_{k-1}\}.$$

Conversely, given a Parikh vector $V = (n_1, \ldots, n_k)$ and a permutation $\vartheta \in S_n$ such that $Des(\Psi(\vartheta)) \subseteq \{1, 1+n_1, \ldots, 1+n_1+\ldots+n_{k-1}\}$, we show that there exists a unique word $w \in A^n$ having Parikh vector $P(w) = V$ and suffix array $\vartheta_w = \vartheta$. Actually, we provide a construction of this word. Since $a_1 < a_2 < \ldots < a_k$, in the starting positions of the first $n_1$ suffixes of $w$, in the lexicographic order, there is the letter $a_1$, in the starting positions of the suffixes of $w$ having rank from $n_1+1$ to $n_1+n_2$, in the lexicographic order, there is the letter $a_2$, and so on. Therefore, for $w = z_1 z_2 \cdots z_n$, if $1 \le i \le n_1$ then $z_{\vartheta(i)} = a_1$ and, for $1 < r \le k$, if $n_1 + \ldots + n_{r-1} < i \le n_1 + \ldots + n_r$, then $z_{\vartheta(i)} = a_r$. This concludes the proof. ∎

**Example 8.9.6** *Given the permutation $\vartheta \in S_8$ in Example 8.9.4 and the vector $V = (5,3)$, we construct a word $w$ on a binary alphabet $A = \{a,b\}$, with $a < b$, having $V$ as Parikh vector and $\vartheta$ as suffix array. From Example 8.9.4, we have that*

$$\Psi(\vartheta) = \begin{pmatrix} 1 & 2 & 3 & 4 & 5 & 6 & 7 & 8 & 9 \\ 8 & 1 & 4 & 6 & 7 & 9 & 2 & 3 & 5 \end{pmatrix}.$$

*$Des(\Psi(\vartheta)) = \{1,6\}$ verifies the condition of the Theorem 8.9.5. The word $w = z_1 \ldots z_8$ having Parikh vector $(5,3)$ and suffix array $\vartheta$ is obtained as follows:*

$$z_{\vartheta(1)} = z_{\vartheta(2)} = z_{\vartheta(3)} = z_{\vartheta(4)} = z_{\vartheta(5)} = a$$

*and*

$$z_{\vartheta(6)} = z_{\vartheta(7)} = z_{\vartheta(8)} = b.$$

*Therefore $w = baaababa$.*

The following corollary of Theorem 8.9.5 will be useful in the next section.

**Proposition 8.9.7** *A permutation $\vartheta \in S_n$ is the suffix array of some word $w$ of length $n$ on an alphabet of cardinality $k$ if and only if*

$$\mathrm{Card}(Des(\Psi(\vartheta)) \setminus \{1\}) \le k-1.$$

### 8.9.2 Counting suffix arrays

The results of previous sections are used here to solve three enumeration problems concerning suffix arrays. The results are essentially due to Schurmann and Stoye [38] (see also [15], [3], and [28]).

The first problem approached here is to count the number $s(n,k)$ of distinct permutations that are suffix arrays of some word of length $n$ over an alphabet of size $k$.

The following table gives the values of $s(n,k)$ for $2 \le k \le n \le 9$.

| k \ n | 2 | 3 | 4 | 5 | 6 | 7 | 8 | 9 |
|---|---|---|---|---|---|---|---|---|
| 2 | 2 | | | | | | | |
| 3 | 5 | 6 | | | | | | |
| 4 | 12 | 23 | 24 | | | | | |
| 5 | 27 | 93 | 119 | 120 | | | | |
| 6 | 58 | 360 | 662 | 719 | 720 | | | |
| 7 | 121 | 1312 | 3728 | 4919 | 5039 | 5040 | | |
| 8 | 248 | 4541 | 20160 | 35779 | 40072 | 40319 | 40320 | |
| 9 | 503 | 15111 | 103345 | 259535 | 347769 | 362377 | 362879 | 362880 |

In the next theorem we show that the function $s(n,k)$ is related to the **Eulerian numbers** $\langle{n \atop d}\rangle$, i.e. the number of permutations of $\{1,2,...,n\}$ with exactly $d$ descents. Recall (cf.[21]) that the Eulerian numbers can be defined by the following recurrence relation

$$\left\langle{n \atop d}\right\rangle = (d+1)\left\langle{n-1 \atop d}\right\rangle + (n-d)\left\langle{n-1 \atop d-1}\right\rangle$$

with $\langle{1 \atop 0}\rangle = 1$ and $\langle{1 \atop d}\rangle = 0$ when $d \geq 1$.

**Theorem 8.9.8** *The number $s(n,k)$ of distinct permutations that are suffix arrays of some word of length $n$ over an alphabet of size $k$ is*

$$s(n,k) = \sum_{d=0}^{k-1} \left\langle{n \atop d}\right\rangle.$$

In order to prove the theorem we need a preliminary lemma. In the following it is convenient to represent a permutation $\varphi \in S_n$ by the word $\varphi(1)\varphi(2)...\varphi(n)$ on the alphabet $\{1,2,\ldots,n\}$. Now we define a mapping that, for any $\varphi \in S_n$ and for any $s \in \{2,3,\ldots,n+1\}$, gives a permutation $\psi \in S_{n+1}$. Such a mapping is described as a transformation on words performed in three steps.

For a permutation $\varphi(1)\varphi(2)\cdots\varphi(n)$ and an integer $s \in \{2,3,\ldots,n+1\}$, in the **first** step we obtain the word

$$E_s(\varphi) = \varphi_s(1)\varphi_s(2)\cdots\varphi_s(n),$$

where $\varphi_s(i) = \varphi(i)$ for $\varphi(i) < s$, and $\varphi_s(i) = \varphi(i)+1$ for $\varphi(i) \geq s$. Remark that $\varphi_s(1)\varphi_s(2)\cdots\varphi_s(n)$ is a word on the alphabet $\{1,2,\ldots,n,n+1\}$, but it does not represent a permutation, because the integer $s$ does not appear in the word. For instance, consider the permutation $\varphi \in S_6$ represented by the word 364215 and $s = 3$. Then $E_3(\varphi) = 475216$.

In the second step $I_s$, which is the most important, we move $\varphi_s(1)$ from the first position in the word to the position $s-1$. It is called the **insertion step** and it is formally defined as follows:

$$I_s(\varphi_s(1)\varphi_s(2)\cdots\varphi_s(n)) = \varphi_s(2)\cdots\varphi_s(s-1)\varphi_s(1)\varphi_s(s)\cdots\varphi_s(n).$$

For instance, $I_3(475216) = 745216$.

In the third step $C_s$ we simply insert the symbol $s$ in the first position of the word. For instance, $C_3(745216) = 3745216$.

The compositions of the above operations define the transformation $T(\varphi,s) = C_s(I_s(E_s(\varphi)))$. Remark that the word $T(\varphi,s)$ represents a permutation of $\{1,2,\ldots,n,n+1\}$. For instance, for $\varphi = 364215$ and $s = 3$, we have $T(\varphi,s) = 3745216$. Moreover, it is straightforward to check that, if $\varphi$ is **cyclic**, then $T(\varphi,s)$ is **cyclic** too. Therefore, if we denote by $S_n^c$ the set of **cyclic** permutations of $\{1,2,\ldots,n\}$, the transformation $T$ defines a mapping

$$T : S_n^c \times \{2,3,\ldots,n+1\} \to S_{n+1}^c.$$

**Lemma 8.9.9** *The mapping $T$ is a bijection from $S_n^c \times \{2,3,\ldots,n+1\}$ onto $S_{n+1}^c$.*

*Proof.* We first prove that $T$ is injective by showing that, given a permutation $\psi \in S_{n+1}^c$, one can uniquely reconstruct the pair $(\varphi,s)$, with $\varphi \in S_n^c$ and $s \in \{2,\ldots,n+1\}$, such that $T(\varphi,s) = \psi$. Let $\psi = \psi(1)\psi(2)\cdots\psi(n+1)$. Since $\psi$ is a cyclic permutation, $\psi(1) \neq 1$. By the definition of $T$, $s = \psi(1)$. We delete $\psi(1) = s$ from the word $\psi(1)\psi(2)\cdots\psi(n+1)$, and we obtain the word $\psi(2)\cdots\psi(n+1)$. Then we take the element $\psi(s)$ and move this element in the first position of the word. We obtain the word $\psi(s)\psi(2)\cdots\psi(s-1)\psi(s+1)\cdots\psi(n+1)$. Now we substitute each $\psi(j) > s$ with $\psi(j)-1$ and we obtain the permutation $\varphi \in S_n^c$ such that $T(\varphi,s) = \psi$. In order to show that the mapping $T$ is surjective, it suffices to verify that $\mathrm{Card}(S_n^c \times \{2,3,\ldots,n+1\}) = \mathrm{Card}(S_{n+1}^c)$. Indeed $\mathrm{Card}(S_n^c \times \{2,3,\ldots,n+1\}) = (n-1)!n = n! = \mathrm{Card}(S_{n+1}^c)$. ■

*Proof of Theorem 8.9.8.* According to Proposition 8.9.7, there is a bijection between the suffix arrays of words $w \in A^n$ and the cyclic permutations $\psi \in S_{n+1}^c$ such that $\mathrm{Card}(Des(\psi) \setminus \{1\}) \leq k-1$. We have then to count the number of such permutations.

Let $P(n,d)$ denote the number of permutations $\psi \in S_{n+1}^c$ such that $\mathrm{Card}(Des(\psi) \setminus \{1\}) = d$. To prove the theorem, we show that $P(n,d)$ is equal to the Eulerian number $\left\langle {n \atop d} \right\rangle$.

The proof is by induction on $n$. Trivially, $P(1,0) = 1 = \left\langle {1 \atop 0} \right\rangle$, and $P(1,d) = 0 = \left\langle {1 \atop d} \right\rangle$ when $d \geq 1$.

We now show that $P(n,d) = (d+1)P(n-1,d) + (n-d)P(n-1,d-1)$.

By Lemma 8.9.9, a permutation $\psi \in S_{n+1}^c$ can be obtained, through the transform $T$, from a permutation $\varphi \in S_n^c$ with the insertion of an element $s \in \{2,\ldots,n+1\}$. We now examine how the transform $T$ affects the number of descents of $\varphi$. Remark that the steps 1 and 3 in the definitions of the transform $T$ do not affect the number of descents. This number can be affected only in step 2 (the **insertion step** $I_s$). If $\varphi$ has $d$ descents in the interval $\{2,\ldots,n+1\}$, also $E_s(\varphi)$, the word obtained after the first step, has $d$ descents, independently from the choice of $s$. We can thus factorize $E_s(\varphi)$ in $d+1$ monotonic (increasing) runs. The second step in the transform $T$ (the insertion step $I_s$) may or may not create a new descent, depending on the position in which is inserted the first symbol $\varphi_s(1)$ of the word $E_s(\varphi)$. In each monotonic run of $E_s(\varphi)$ there is exactly one position where $\varphi_s(1)$ can be placed without creating a new descent. Otherwise one creates exactly one new descent.

How many permutations $\psi = T(\varphi, s)$ can we obtain with $\mathrm{Card}(Des(\psi) \setminus \{1\}) = d$ ? For each $\varphi \in S_n^c$ with $\mathrm{Card}(Des(\varphi) \setminus \{1\}) = d$, we have $d+1$ possibilities to choose $s$ (because in $E_s(\varphi)$ there are $d+1$ monotonic runs). For each $\varphi \in S_n^c$ with $\mathrm{Card}(Des(\varphi) \setminus \{1\}) = d-1$, we have $n-d$ possibilities to choose $s$. Since $T$ is a bijection, there is no other way to get a permutation $\psi \in S_{n+1}^c$ with $\mathrm{Card}(Des(\psi) \setminus \{1\}) = d$. It follows that

$$P(n,d) = (d+1)P(n-1,d) + (n-d)P(n-1,d-1).$$

■

We now consider the problem of counting the number of words that share the same suffix array.

**Theorem 8.9.10** *Given a permutation $\vartheta \in S_n$, the number of words of length $n$ over an alphabet of size $k$ having $\vartheta$ as their suffix array is*

$$\binom{n+k-1-d}{k-1-d},$$

*where $d = \mathrm{Card}(Des(\Psi(\vartheta)) \setminus \{1\})$.*

*Proof.* By Theorem 8.9.5, a word $w \in A^n$, with $|A| = k$, has $\vartheta$ as suffix array if and only if $w$ has a Parikh vector $P(w) = (n_1, n_2, \ldots, n_k)$ such that

$$Des(\Psi(\vartheta)) \subseteq \{1, 1+n_1, \ldots, 1+n_1+\ldots+n_{k-1}\}.$$

Therefore, given the permutation $\vartheta$, and then given the set

$$D_\vartheta = Des(\Psi(\vartheta)) \setminus \{1\} = \{m_1, m_2, \ldots, m_d\},$$

we need to count the number of tuples $(n_1, \ldots, n_k)$, with $n_1 + \ldots + n_k = n$ such that

$$D_\vartheta \subseteq \{1+n_1, 1+n_1+n_2, \ldots, 1+n_1+\ldots+n_{k-1}\}.$$

We represent the tuple $(n_1, \ldots, n_k)$ by a word $z$ on the alphabet $\{x, y\}$:

$$z = x^{n_1} y x^{n_2} y \cdots x^{n_{k-1}} y x^{n_k},$$

with $n_i \geq 0$ and $n_1 + \ldots + n_k = n$. We have that $|z| = n+k-1$. The condition $D_\vartheta = \{m_1, \ldots, m_d\} \subseteq \{1+n_1, \ldots, 1+n_1+\ldots+n_{k-1}\}$ defines the positions of $d$ occurrences of the letter $y$ in $z$. The remaining $k-1-d$ occurrences of $y$ can be placed in arbitrary positions. This can be done in

$$\binom{n+k-1-d}{k-1-d}$$

ways. ■

Note that if $k-1 < \mathrm{Card}(Des(\Psi(\vartheta)) \setminus \{1\})$, there is no word on an alphabet of size $k$ that has $\vartheta$ as its suffix array. This is confirmed by Theorem 8.9.10, since $\binom{m}{n} = 0$ for $m < n$.

In the next theorem, we require that each letter of the alphabet occurs at least once in the words that we count.

**Theorem 8.9.11** *Given a permutation $\vartheta \in S_n$, the number of words of length $n$ over an alphabet of size $k$ that have at least one occurrence of each of the $k$ letters and have $\vartheta$ as their suffix array is*

$$\binom{n-1-d}{k-1-d},$$

*where $d = \mathrm{Card}(Des(\Psi(\vartheta)) \setminus \{1\})$.*

*Proof.* The proof of Theorem 8.9.10 is modified in order to ensure that each letter occurs at least once. In the representation of the tuple $(n_1, \ldots, n_k)$ by the word $z = x^{n_1}yx^{n_2}y\cdots x^{n_{k-1}}yx^{n_k}$, we require that the $n_i$ are strictly positive, i.e. $n_i > 0$ for $i = 1, \ldots, k-1$. Then we have to distribute the occurrences of the letter $y$ among the $n-1$ possible positions. As in the proof of Theorem 8.9.10, the positions of $d$ occurrences of $y$ are determined by the permutation $\vartheta$, and the remaining $k-1-d$ are distributed among the $n-1-d$ remaining positions. ∎

From Theorem 8.9.8 and Theorem 8.9.10 we can derive a long known summation identity of Eulerian numbers. The identity

$$k^n = \sum_j \left\langle {n \atop j} \right\rangle \binom{k+j}{n},$$

as given in [21, Eq.6.37], was proven by J. Worpitzki in 1883. In order to prove it, we observe that the number of words of length $n$ over an alphabet of size $k$ can be obtained by summing the number of words for each suffix array. Thus, we have:

$$k^n = \sum_{d=0}^{k-1} \left\langle {n \atop d} \right\rangle \binom{n+k-d-1}{k-d-1}.$$

By using the symmetry rule for Eulerian and binomial numbers, from the previous equality we derive

$$k^n = \sum_{d=0}^{k-1} \left\langle {n \atop n-1-d} \right\rangle \binom{n+k-d-1}{n}.$$

By setting $j = n-d-1$, we obtain

$$k^n = \sum_{j=n-k}^{n-1} \left\langle {n \atop j} \right\rangle \binom{k+j}{n} = \sum_j \left\langle {n \atop j} \right\rangle \binom{k+j}{n},$$

where the last equality is motivated by the remark that $\left\langle {n \atop j} \right\rangle = 0$ for all $j \geq n$ and $\binom{k+j}{n} = 0$ for all $j < n-k$.

**Acknowledgments.** The authors wish to thank Nicolas Auger, Maxime Crochemore, Francesco Dolce, Gregory Kucherov, Eduardo Moreno, Giovanna Rosone and Christophe Reutenauer who have read the manuscript and made corrections. They also thank the referee, who has helped to substantially improve the presentation. The support of ANR project Eqinocs is acknowledged by the first author.

# References

[1] Tanja van Aardenne-Ehrenfest and Nicolaas Govert de Bruijn. Circuits and trees in oriented linear graphs. *Simon Stevin*, 28:203–217, 1951.

[2] Yu Hin Au. Shortest sequences containing primitive words and powers. arXiv:0904.3997, 2013.

[3] Hideo Bannai, Shunsuke Inenaga, Ayumi Shinohara, and Masayuki Takeda. Inferring strings from graphs and arrays. Volume 2747 of *Lecture Notes in Computer Science*, pages 208–217. Springer Berlin Heidelberg, 2003.

[4] Jean Berstel and Dominique Perrin. The origins of combinatorics on words. *European J. Combin.*, 28(3):996–1022, 2007.

[5] Jean Berstel, Dominique Perrin, and Christophe Reutenauer. *Codes and Automata.* Cambridge University Press, 2009.

[6] Francine Blanchet-Sadri. Algorithmic combinatorics on partial words. *Internat. J. Found. Comput. Sci.*, 23(6):1189–1206, 2012.

[7] Francine Blanchet-Sadri, N. C. Brownstein, Andy Kalcic, Justin Palumbo, and T. Weyand. Unavoidable sets of partial words. *Theory Comput. Syst.*, 45(2):381–406, 2009.

[8] Carl Wilhelm Borchardt. Ueber eine der Interpolation entsprechende Darstellung der Eliminations-Resultante. *J. reine angew. Math.*, 57:111–121, 1860.

[9] Michael Burrows and David J. Wheeler. A block sorting data compression algorithm. Technical report, DIGITAL System Research Center, 1994.

[10] Jean-Marc Champarnaud and Georges Hansel. Ensembles inévitables et classes de conjugaison. *Bull. Belg. Math. Soc. Simon Stevin*, 10(suppl.):679–691, 2003.

[11] Jean-Marc Champarnaud, Georges Hansel, and Dominique Perrin. Unavoidable sets of constant length. *Internat. J. Algebra Comput.*, 14(2):241–251, 2004.

[12] Maxime Crochemore, Jacques Désarménien, and Dominique Perrin. A note on the Burrows-Wheeler transformation. *Theoret. Comput. Sci.*, 332(1-3):567–572, 2005.

[13] Jean-Pierre Duval. Factorizing words over an ordered alphabet. *J. Algorithms*, 4(4):363–381, 1983.

[14] Jean-Pierre Duval. Génération d'une section des classes de conjugaison et arbre des mots de Lyndon de longueur bornée. *Theoret. Comput. Sci.*, 60(3):255–283, 1988.

[15] Jean-Pierre Duval and Arnaud Lefebvre. Words over an ordered alphabet and suffix permutations. *RAIRO Theor. Inform. Appl.*, 36(3):249–259, 2002.

[16] Steven R. Finch. *Mathematical constants*, volume 94 of *Encyclopedia of Mathematics and its Applications*. Cambridge University Press, Cambridge, 2003.

[17] Harold Fredricksen and James Maiorana. Necklaces of beads in $k$ colors and $k$-ary de Bruijn sequences. *Discrete Math.*, 23(3):207–210, 1978.

[18] Michael R. Garey and David S. Johnson. *Computers and intractability*. W. H. Freeman and Co., San Francisco, Calif., 1979. A guide to the theory of NP-completeness, A Series of Books in the Mathematical Sciences.

[19] Ira M. Gessel, Antonio Restivo, and Christophe Reutenauer. A bijection between words and multisets of necklaces. *European Journal of Combinatorics*, 33(7):1537–1546, 2012.

[20] Ira M. Gessel and Christophe Reutenauer. Counting permutations with given cycle structure and descent set. *J. Combin. Theory Ser. A*, 64(2):189–215, 1993.

[21] Ronald L. Graham, Donald E. Knuth, and Oren Patashnik. *Concrete mathematics*. Addison-Wesley Publishing Company, Reading, MA, second edition, 1994. A foundation for computer science.

[22] Roberto Grossi. A quick tour on suffix arrays and compressed suffix arrays. *Theoret. Comput. Sci.*, 412(27):2964 – 2973, 2011.

[23] Peter M. Higgins. Burrow-Wheeler transformations and de Bruijn words. *Theoret. Comput. Sci.*, 457(0):128 – 136, 2012.

[24] Donald E. Knuth. Oriented subtrees of an arc digraph. *J. Comb. Theory*, 3:309–314, 1967.

[25] Donald E. Knuth. *The Art of Computer Programming, volume 1, Fundamental Algorithms*. Addison Wesley, 1968. Second edition, 1973.

[26] Donald E. Knuth. *The Art of Computer Programming , Volume 4A, Combinatorial Algorithms: Part 1*. Addison Wesley, 2012.

[27] Tomasz Kociumaka, Jakub Radoszewski, and Wojciech Rytter. Computing $k$-th lyndon word and decoding lexicographically minimal de Bruijn sequence. In *Combinatorial Pattern Matching*, volume 8486 of *Lecture Notes in Computer Science*, pages 202–211, 2014.

[28] Gregory Kucherov, Lilla Tóthmérész, and Stéphane Vialette. On the combinatorics of suffix arrays. *Inform. Process. Lett.*, 113(22-24):915–920, 2013.

[29] Douglas Lind and Brian H. Marcus. *An Introduction to Symbolic Dynamics and Coding*. Cambridge, 1995.

[30] M. Lothaire. *Combinatorics on Words*. Cambridge University Press, second edition, 1997. (First edition 1983).

[31] M. Lothaire. *Algebraic Combinatorics on Words*. Cambridge University Press, 2002.

[32] Udi Manber and Gene Myers. Suffix arrays: A new method for on-line string searches. *SIAM Journal on Computing*, 22(5):935–948, 1993.

[33] Sabrina Mantaci, Antonio Restivo, Giovanna Rosone, and Marinella Sciortino. An extension of the Burrows-Wheeler Transform. *Theoret. Comput. Sci.*, 387(3):298–312, 2007.

[34] Eduardo Moreno. On the theorem of Fredricksen and Maiorana about de Bruijn sequences. *Adv. in Appl. Math.*, 33(2):413–415, 2004.

[35] Eduardo Moreno and Dominique Perrin. Corrigendum to: 'on the theorem of Fredricksen and Maiorana about de Bruijn sequences'. *Adv. in Appl. Math.*, 2014. to appear.

[36] Johannes Mykkeltveit. A proof of Golomb's conjecture for the de Bruijn graph. *J. Combinatorial Theory Ser. B*, 13:40–45, 1972.

[37] Christophe Reutenauer. *Free Lie algebras*. The Clarendon Press Oxford University Press, New York, 1993. Oxford Science Publications.

[38] Klaus-Bernd Schurmann and Jens Stoye. Counting suffix arrays and strings. *Theor. Comput. Sci.*, pages 220–234, 2008.

[39] Arseny M. Shur. Growth of power-free languages over large alphabets. *Theory Comput. Syst.*, 54(2):224–243, 2014.

[40] Cedric A. Smith and William T. Tutte. On unicursal paths in a network of degree 4. *Amer. Math. Monthly*, 48, 1941.

[41] Richard P. Stanley. *Enumerative combinatorics. Vol. 1*. Cambridge University Press, Cambridge, 1997.

# Chapter 9

# *Tilings*

**James Propp**
*University of Massachusetts at Lowell*

**CONTENTS**



## 9.1 Introduction

We begin with some examples.

**Example 9.1.1** *Figure 9.1 shows a region (a 2-by-6 rectangle) that has been tiled by* **dominos** *(1-by-2 rectangles) in one of 13 possible ways. More generally, the number of domino tilings of the 2-by-n rectangle (for $n \geq 1$) is the nth Fibonacci number (if one begins the sequence $1, 2, 3, 5, 8, \ldots$).*

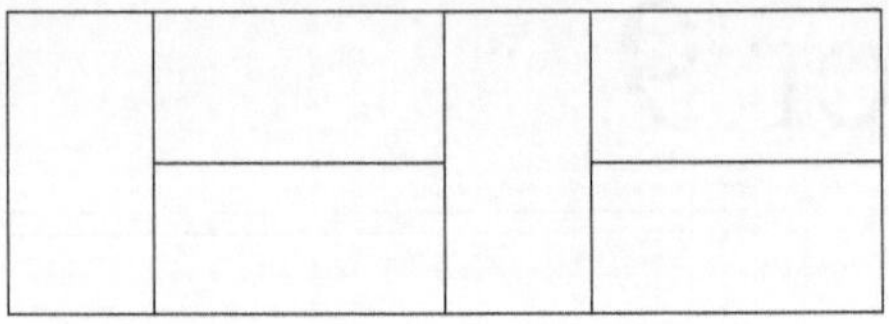

**Figure 9.1**
A domino tiling (one of 13).

**Example 9.1.2** *Figure 9.2 shows a region, composed of 24 equilateral triangles of side-length 1, that has been tiled by* **lozenges** *(unit rhombuses with internal angles of 60 and 120 degrees) in one of 14 possible ways. The region shown in Figure 9.2 is part of a one-parameter family of regions whose nth member has* $\binom{2n}{n}/(n+1)$ *tilings by lozenges. (See [47] and [51] for more on this manifestation of the Catalan numbers.)*

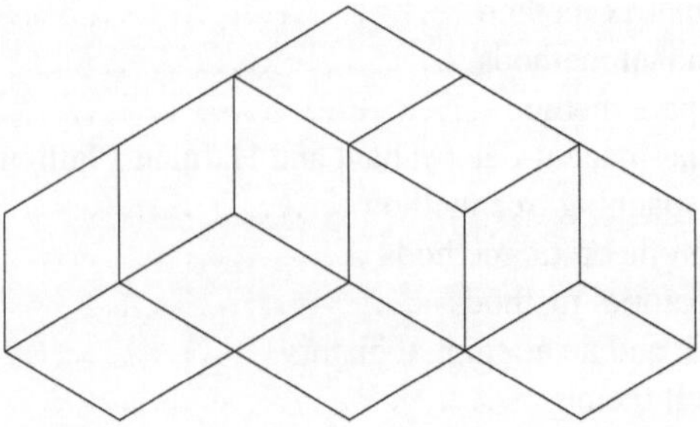

**Figure 9.2**
A lozenge tiling (one of 14).

Examples 9.1.1 and 9.1.2 are prototypes of the sorts of problems discussed in this chapter.

Specifically, we have some compact region $R$ in the plane (in Example 9.1.1, $R$ is the 2-by-6 rectangle $[0,6] \times [0,2]$) and a finite collection of other compact subsets called **prototiles** (in Example 9.1.1, the prototiles are $[0,1] \times [0,2]$ and $[0,2] \times [0,1]$), and we are looking at unordered collections $\{S_1, \ldots, S_m\}$ such that (a) each $S_i$ is a translate of a prototile, (b) the interiors of the $S_i$'s are pairwise disjoint, and (c) the union of the $S_i$'s is $R$. We call such a collection a **tiling** of $R$. Given a region to be tiled and a set of prototiles, we want to know: *In how many different ways can the region be tiled?*

What concerns us here are not individual tiling problems but infinite families of tiling problems, with each family determining not just one number but an infinite sequence of them, indexed by a size parameter we will typically call $n$. The set of prototiles remains the same, while the size of $R$ goes to infinity with $n$. In some cases, we obtain a sequence that is governed by a closed-form formula or a recurrence relation; in other cases, we settle for asymptotics. Often we find that the individual terms of the sequence exhibit number-theoretic patterns (e.g., they are perfect squares or satisfy certain congruences) that call out for a theoretical explanation, even in the absence of a general formula for the terms. The article [137] contains many such problems, as well as discussions of general issues in enumeration of tilings that overlap with the discussions contained in this chapter.

Here are two more examples of such one-parameter tiling problems:

**Example 9.1.3 (Thurston and Lagarias-Romano)** *Figure 9.3 shows a "honeycomb triangle" of order 10 with one hexagon omitted (the central hexagon), tiled by 18 regions called* **tribones**. *W. Thurston [166] conjectured that for each way of packing* $3n(n+1)/2$ *tribones into a honeycomb triangle of order* $3n+1$, *the unique uncovered cell will be the center cell. Lagarias and Romano [107] proved this conjecture and also showed that the number of such packings is* $2^n$.

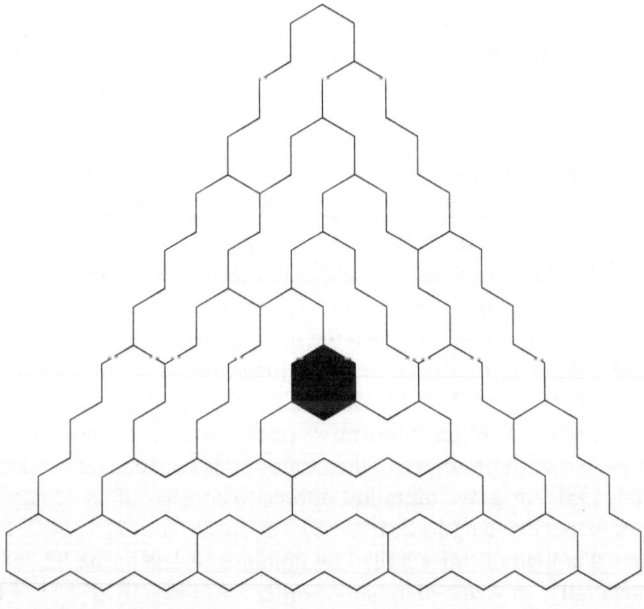

**Figure 9.3**
A tribone tiling (one of 8).

**Example 9.1.4 (Moore)** *Figure 9.4 shows a 5-by-5 square tiled by ribbon tiles of order 5, where a* **ribbon tile** *of order n is a union of n successive unit squares, each of which is either the rightward neighbor or the upward neighbor of its predecessor. It* *is an amusing puzzle to show that the number of tilings of the n-by-n square by ribbon tiles of order n is n!. (The proof is not hard, but it is not obvious.) This unpublished example was communicated to me by Cris Moore.*

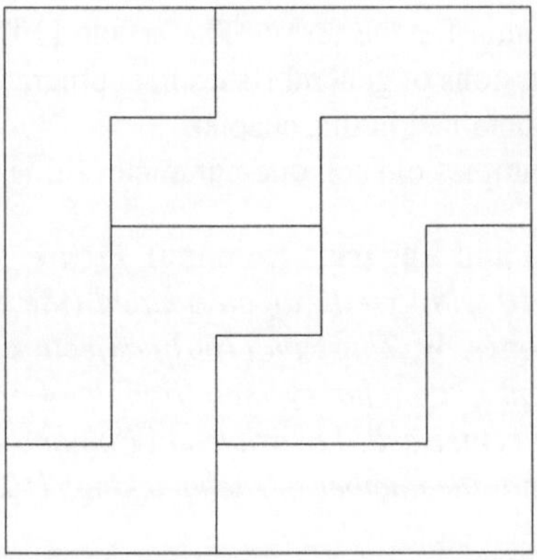

**Figure 9.4**
A ribbon tiling (one of 120).

These last two examples are pretty, but they are not typical of what you will learn about in this chapter (outside of the first part of Section 9.2). First, they do not serve as examples of a general method. Second, the sequences one gets grow too slowly to be typical of the state of the art in the modern theory of enumeration of tilings. Calling $n!$ "slow-growing" may seem surprising, since it grows faster than any exponential function. However, one thing that the $n$th Fibonacci number, the $n$th Catalan number, $n!$, and $2^n$ have in common is that their logarithms grow linearly in $n$ (at least up to a factor of $\log n$ or smaller). In contrast, the theory of enumeration of tilings shows its true strength with exact enumerative results involving functions of $n$ whose logarithms grow like $n^2$. A prototype of theorems of this kind concerns regions called Aztec diamonds [53]. An **Aztec diamond** of order $n$ consists of $2n$ centered rows of unit squares, of respective lengths $2,4,\ldots,2n-2,2n,2n,2n-2,\ldots,4,2$. Figure 9.5 shows an Aztec diamond of order 5 tiled by dominos in one of the 32,768 possible ways. More generally, an Aztec diamond of order $n$ has exactly $2^{n(n+1)/2}$ tilings by dominos. This result has now been proved a dozen different ways, though some of the differences are more cosmetic than conceptual.

Highly analogous to domino tilings of Aztec diamonds are lozenge tilings of hexagons. Call a hexagon **semiregular** if its internal angles are 120 degrees and opposite sides are of equal length (more generally, call a polygon with an even number of sides semiregular if opposite sides are parallel and of equal length). A semiregular

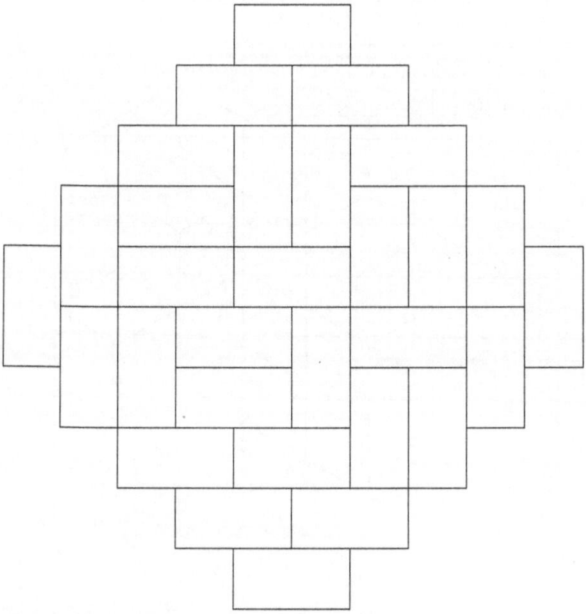

**Figure 9.5**
A domino tiling of an Aztec diamond (one of 32,768).

hexagon with side-lengths $a,b,c,a,b,c$ can be tiled by lozenges in exactly

$$\frac{H(a+b+c)H(a)H(b)H(c)}{H(a+b)H(a+c)H(b+c)}$$

ways, where $H(0) = H(1) = 1$ and $H(n) = 1!2!\cdots(n-1)!$ for $n > 1$; an equivalent expression is

$$\prod_{i=1}^{a}\prod_{j=1}^{b}\prod_{k=1}^{c}\frac{i+j+k-1}{i+j+k-2}.$$

This result goes back to [119], although MacMahon was not studying tilings but plane partitions; plane partitions whose parts fit inside an $a$-by-$b$ rectangle and with all parts less than or equal to $c$ have three-dimensional Young diagrams that are easily seen to be in bijection with lozenge tilings of the $a,b,c,a,b,c$ semiregular hexagon [156]. Figure 9.6 shows a regular hexagon of order 4, tiled by lozenges in one of the 232,848 possible ways.

Could there be higher-dimensional analogues of these tiling problems, involving sequences whose logarithms grow like $n^3$ (or faster)? Curiously, such analogues are almost entirely lacking (and I will briefly return to this issue later, in a speculative vein, at the close of the chapter). For now, I wish only to point out that any theory of exact enumeration of tilings that attempts to catch too many tiling problems in its net is likely to find that net ripped apart by the sea monsters of $\#P$-completeness. (For a

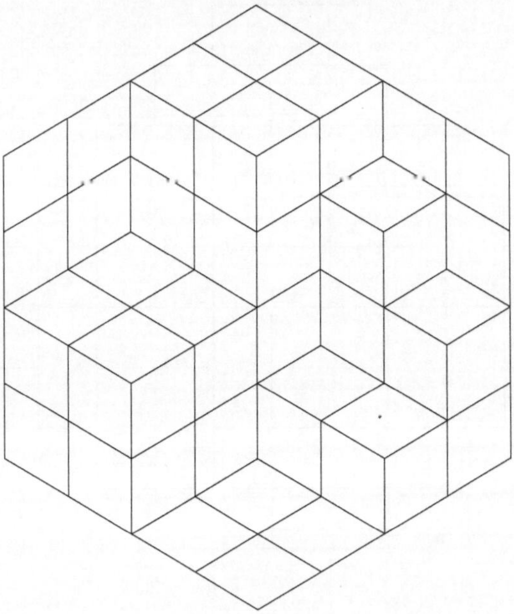

**Figure 9.6**
A lozenge tiling of a hexagon (one of 232,848).

definition of #$P$-completeness, see [125].) Hardly anyone believes that #$P$-complete problems can be solved efficiently. Beauquier, Nivat, Rémila, and Robson [6], Moore and Robson [124], and Pak and Yang [129, 130] have given examples of classes of two-dimensional tiling problems exhibiting #$P$-completeness. So there is little hope of solving the problem of counting tilings in its full generality, even in two dimensions. Still, there is much that can be done.

By far the most successful theory of enumeration of two-dimensional tilings is the theory of perfect matchings of a planar graph. A **perfect matching** of a graph $G=(V,E)$ is a subset $E'$ of the edge-set $E$ with the property that each vertex $v \in V$ is an endpoint of one and only one edge $e \in E'$. Problems involving perfect matchings of a planar graph $G$ may not immediately appear to be tiling problems, but they can easily be recast in this framework: simply take the planar dual of $G$ (the planar graph whose vertices, edges, and faces correspond respectively to the faces, edges, and vertices of $G$) and define a tile to be the union of any two regions in the dual graph that correspond to adjacent vertices in $G$. (To make this correspondence precise, one may need to limit the kinds of translation that can be applied to the prototiles.) In the other direction, some tilings correspond to perfect matchings. For example, the domino tiling of Figure 9.5 corresponds to the perfect matching in Figure 9.7. For a more careful definition of Aztec diamonds and their dual graphs, see [53]. See also Figure 1 from [135], which shows the relationship between rhombus tilings of a hexagon and dual perfect matchings, and Figure 2 from [135], which shows the relationship between domino tilings of a square and dual perfect matchings.

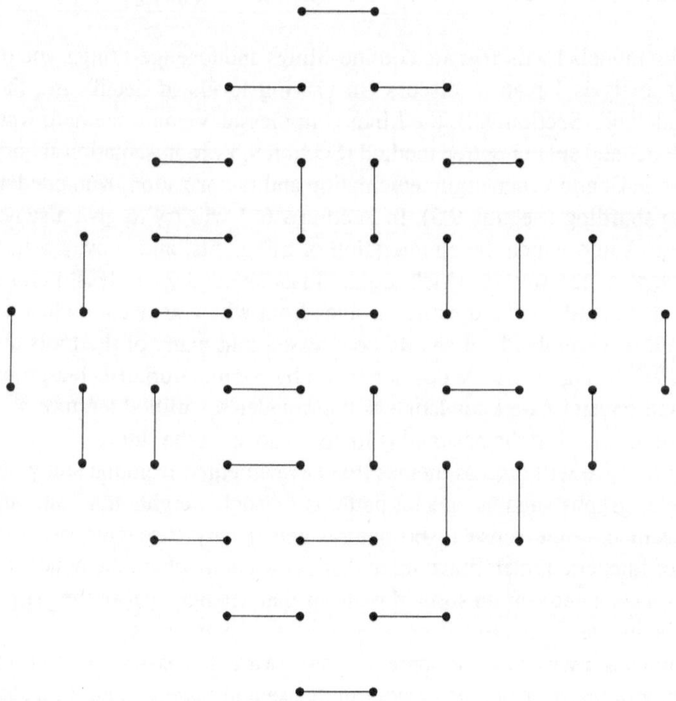

**Figure 9.7**
A matching of an Aztec diamond graph (one of 32,768).

(In the graph theory literature, a matching of a graph $G$ is a subset $E'$ of the edge-set $E$ with the property that each vertex $v \in V$ is an endpoint of *at most one* edge $e \in E'$. I will comment later on the problem of counting matchings, or as physicists call them, monomer-dimer covers. Until then, I will use the term "matching" to refer exclusively to *perfect* matchings.)

The problem of counting matchings of a general graph is #*P*-complete [168], but when the graph is planar (or not too far from being planar), the problem of counting matchings can be reduced to linear algebra, specifically, to the evaluation of determinants (and Pfaffians) of matrices. This technology was developed by the mathematical physicist Kasteleyn, who (using physicists' language) thought of his work as providing a way to evaluate "the partition function of the dimer model" (we will explain that language later). Thanks to Kasteleyn's work [78] [79] [80], much of the known body of theory relating to enumeration of tilings in the plane can be viewed as a subspecialty of the field of determinant evaluation, as developed by Krattenthaler and others (see for instance [93] and [95]). However, in many instances the matrices that arise do not belong to a general class of matrices to which established methods of determinant evaluation apply; in those instances, the only known ways of evaluating the determinant make crucial use of the combinatorics and geometry of the tiling picture.

The main models I will treat are domino-tilings and lozenge-tilings, and the main methods of analysis I plan to discuss (in varying levels of detail) are: the transfer matrix method (Section 9.2); the Lindström-Gessel-Viennot method, Kasteleyn-Percus method, and spanning tree method (Section 9.3); representation-theory methods (Section 9.4); and Ciucu complementation and factorization, Kuo condensation, and domino shuffling (Section 9.5). In Section 9.6, I will try to give a sense of the broader context into which the enumeration of tilings fits, and convey some sort of historical sense of the way in which the field has evolved (or at least point the historically minded reader toward earlier articles from which an accurate history of the subject could be assembled). It should be stressed that many of the tools discussed in Sections 9.2 through 9.5 were developed not by combinatorialists but probabilists, chemists, and physicists; a knowledge of the broader scientific literature has served combinatorialists well in the past and is likely to do so in the future.

In Section 9.7, I will discuss themes that have emerged from the study of matchings of planar graphs such as special patterns of edge-weights that, although non-periodic, seem in some sense to be natural and in any case give rise to nice $q$-analogues of integer enumerations; three distinct ways in which the concept of symmetry plays a role; and certain sorts of patterns that often appear in the prime factorization of the integers that arise from enumeration of matchings.

In Section 9.8, I will describe some of the software that exists for exploring problems in enumeration of matchings (sadly in disuse and disrepair, but that surely will change if there is sufficient new activity in this field to justify investment in infrastructure). In Section 9.9, I will give an admittedly personal and biased list of issues on which progress is needed. Unlike [137], which focused on specific problems in enumeration, the current list is about larger themes.

My aim will be not to provide definitive and general statements of results, but to give attractive examples that will make combinatorially-inclined readers want to learn more about (and perhaps work in) this area, and will help researchers starting out in this area get a better sense of what they might want to read.

For a recent article on tilings by other authors, see [2].

## 9.2 The transfer matrix method

Looking down each of the six columns in Figure 9.1, we see either a vertical tile, the left halves of two horizontal tiles, or the right halves of two horizontal tiles. Moreover, if we know, for each of the six columns, which of these three pictures we see (Vertical, Left, or Right), then we know the whole tiling. For the tiling shown in Figure 9.1, the "code" is $VLRVLR$. However, not every length-six sequence consisting of $V$'s, $L$'s, and $R$'s corresponds to an allowed tiling. The codes that correspond to tilings are precisely the ones that arise from paths of length six in the graph shown in Figure 9.8, starting at the circled node on the left and ending at the circled node on the right, where the symbols $V$, $L$, and $R$ correspond respectively to edge of slope 0, $-1$, and $+1$. For instance, the path shown in bold has successive edges of type $V$, $L$, $R$, $V$, $L$, and $R$, and therefore corresponds to the tiling of Figure 9.1.

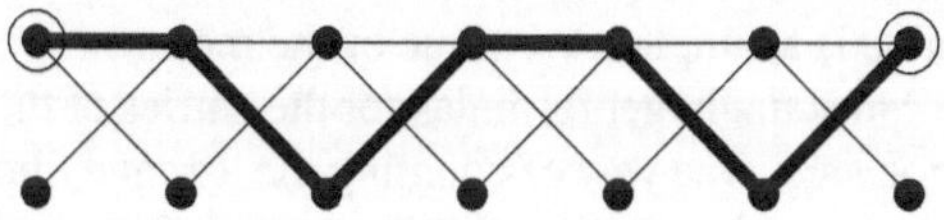

**Figure 9.8**
A trellis graph for counting domino tilings.

Figure 9.8 is an example of what I will call a **trellis graph** (though my use of the term is slightly different from the standard definition): a graph with vertex set $V_0 \cup V_2 \cup \cdots \cup V_n$ where each $V_k$ $(0 \leq k \leq n)$ is finite and all edges connect a vertex in $V_{k-1}$ to a vertex in $V_k$ for some $1 \leq k \leq n$. For each $1 \leq k \leq n$, let $M_k$ be the matrix with $|V_{k-1}|$ rows and $|V_k|$ columns whose $i,j$th entry counts the edges from the $i$th element of $V_{k-1}$ to the $j$th element of $V_k$. Then it is easy to see that the number of paths from the $i$th element of $V_0$ to the $j$th element of $V_n$ equals the $i,j$th entry in the matrix product $M_1 M_2 \cdots M_n$. We call the $M_k$'s **transfer matrices**. In our case, all the matrices are the same matrix $M = \begin{pmatrix} 1 & 1 \\ 1 & 0 \end{pmatrix}$, so we can readily compute the number of tilings as the upper-left entry of $M^6$, which is 13. More generally, we see that the number of domino-tilings of the 2-by-$n$ rectangle is the upper-left entry of $M^n$, which is the $n$th Fibonacci number.

The same method can be applied to count domino-tiling of an $m$-by-$n$ rectangle. The number of rows and columns in the matrix $M$ grows exponentially with $m$, but Theorem 4.7.2 of [159] assures us that for fixed $m$, the number of domino-tilings of an $m$-by-$n$ rectangle, viewed as a function of $n$, satisfies a linear recurrence relation with constant coefficients.

This method can be applied much more broadly to tiling problems in which the region being tiled is "essentially one-dimensional" in the sense that only one dimension is growing as $n$ increases. For example, if $a_n$ denotes the number of ways to pack an 8-by-8-by-$n$ box with 1-by-1-by-2 bricks that can be used in each of the three possible orientations, then the general theory tells us that $a_n$ must satisfy a linear recurrence. However, the span of this recurrence could be extremely long; an off-the-cuff upper bound on the span is $10^{50}$, which can probably be improved a bit but not by much without a good deal of work.

By using matrices whose entries are not just 0's and 1's, we can get more mileage out of the method. Suppose for instance we want to discriminate between tilings on the basis of how many horizontal or vertical dominos they have. Then we would look at powers of the matrix $M = \begin{pmatrix} y & x \\ x & 0 \end{pmatrix}$ (since each $V$ corresponds to a vertical domino and each $L$ or $R$ corresponds to two horizontal half-dominos). The upper-left entry of $M^6$ is $x^6 + 6x^4y^2 + 5x^2y^4 + y^6$, which tells us that the tiling of Figure 9.1 is one of exactly six tilings having four horizontal dominos and two vertical dominos.

Typically one exploits the transfer-matrix method by diagonalizing $M$ (and if one is only interested in the growth rate of the entries of $M^n$, one merely computes the dominant eigenvalue of $M$, invoking the Perron-Frobenius theorem to conclude that the dominant eigenvalue is a simple eigenvalue of the matrix).

In rare cases, one can actually get formulas for the entries of the powers of $M$ and skip the diagonalization step. One way of counting the lozenge tilings of semiregular hexagons provides an example of this: see Proposition 2.1 of [41] and its proof. One can argue that such a proof does not deserve to be called an example of the transfer-matrix method, since no linear algebra is involved. (Perhaps it should be called "enumeration by exact dynamic programming" or "the method of proving a result by finding a stronger result that can be proved by induction.")

We conclude by returning to the $x,y$-enumeration of domino tilings discussed above, putting it in the framework of statistical mechanics, and indicating how the transfer matrix method can be applied to all one-dimensional lattice models. We already mentioned in Section 9.1 that domino tilings of a region are in bijection with matchings of the graph dual to the tessellation of the region by unit squares. In physics, a (perfect) matching is called a dimer cover or dimer configuration. We assign each dimer cover a weight equal to the product of the weights of its constituent edges, where a horizontal (respectively, vertical) edge has weight $x$ (respectively, $y$). In physics, the weights have an interpretation in terms of energy (see [5]) but we will not concern ourselves with that here. The sum of the weights of the configurations is called the **partition function**, and is denoted by $Z$. The probability measure that assigns to each configuration probability equal to the weight of the configuration divided by the normalizing constant $Z$ is called the **Gibbs measure** associated with that weighting. In the case where all configurations have the same weight, and the probability distribution is correspondingly uniform, we define the **entropy** of the system as the logarithm of the number of (equally likely) states. More generally, if a system has $N$ states with respective probabilities $p_i$, we define the entropy as $\sum_{i=1}^{N} p_i \log 1/p_i$, which coincides with the former definition if $p_i = 1/N$ for $1 \le i \le N$.

These concepts, with some work, can be extended to infinite systems as well, through appropriate rescalings and limit-procedures. Some care is helpful for dealing with the boundary conditions, but these turn out not to matter so much in one dimension. For instance, in the case of the 2-by-$2n$ dimer model with edge-weights $x$ and $y$ (which we take to be positive), we might take the logarithm of the number of tilings of the 2-by-$n$ rectangle, divide by the area $2n$, and take the limit as $n \to \infty$, obtaining the normalized, or per-site, entropy of the model, a function of $x$ and $y$. Or we might look at 2-by-$n$ configurations in which dominos are allowed to protrude at the ends. The same asymptotics will prevail, since the entries of $M^n$ must all grow at the same rate.

This argument is not applicable to higher-dimensional tilings, and indeed, the per-site entropy turns out to be extremely sensitive to the boundary conditions. For example, if one modifies the Aztec diamond of order $n$ so that it has only one row of length $2n$ instead of two, the number of tilings drops from $2^{n(n+1)/2}$ down to 1.

In the case where all the states of a one-dimensional lattice model have the same probability, the transfer matrix method tells us that the entropy is just the logarithm of the dominant eigenvalue. In particular, for one-dimensional lattice models of this kind, if all the entries of the transfer matrix are 0's and 1's, the entropy must be the logarithm of an algebraic number. This, too, fails in higher dimensions (at least as far as is known); although the entropy for the square ice model introduced in Section 9.4 is the logarithm of the algebraic number $8/(3\sqrt{3})$, the entropy of the dimer model on a square grid to be discussed in Section 9.3 is $G/\pi$ (where $G$ is Catalan's constant), which is not believed to be the logarithm of an algebraic number.

# 9.3 Other determinant methods

## 9.3.1 The path method

Returning to Example 9.1.1 of Section 9.1, let us decorate each tile according to its orientation in the fashion shown in Figure 9.9, so that Figure 9.2, with all its tiles decorated, becomes Figure 9.10.

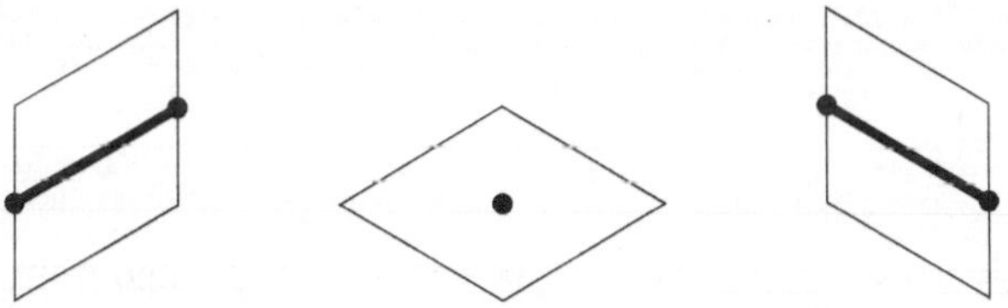

**Figure 9.9**
Decorated lozenges.

The decorations on the tiles yield a path from the leftmost node (marked $P$) to the rightmost node (marked $Q$), along with some nodes that are not visited by the path. In this way, one can see that the tilings of the region are in bijection with Dyck paths of length 6, and hence are enumerated by the Catalan number 14.

Figure 9.11 shows a domino tiling of a 3-by-4 rectangle decorated with a similar sort of path. This **Randall path** [116] is obtained by drawing a node at the midpoint of every other vertical edge (that is, when two vertical edges are neighbors in either the horizontal or vertical sense, exactly one of them has a node). Every tile then either has a single node in its interior or two nodes on its boundary, and in the latter case

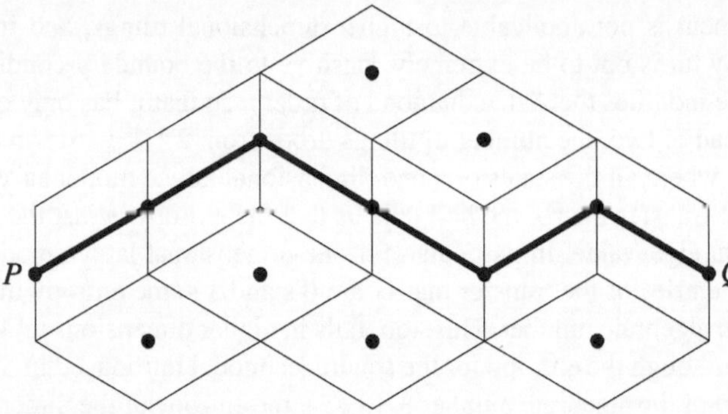


**Figure 9.10**
From tilings to paths.

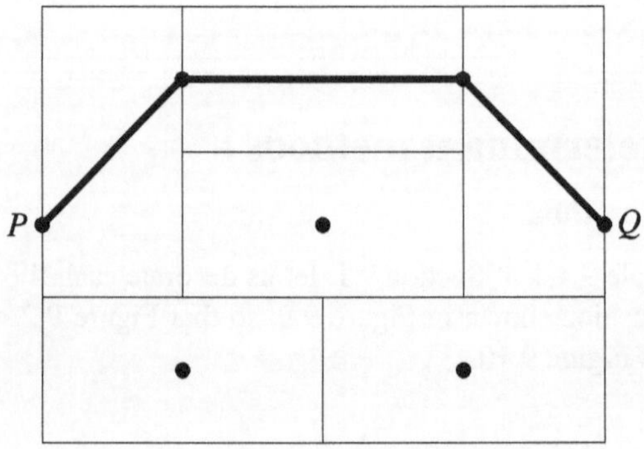


**Figure 9.11**
A tiling and a path.

we join the nodes by a line segment. All the edges in the path go from left to right, with slope 0, +1, or −1. It is easy to use dynamic programming to show that there are 11 paths from $P$ to $Q$, and hence 11 domino tilings of the rectangle, since there is a bijection between tilings and paths.

Figure 9.12 shows the other way to decorate the tiling of Figure 9.11 by drawing a node on every other vertical edge. Now we get not just one path but two nonintersecting paths, one joining $P_1$ to $Q_1$ and one joining $P_2$ to $Q_2$. There are six paths from $P_i$ to $Q_j$ when $i = j$ but only five when $i \neq j$. Lindström's lemma [113], rediscovered by Gessel and Viennot [65], shows that the number of pairs of nonintersecting paths from $P_1$ and $P_2$ to $Q_1$ and $Q_2$ respectively equals the determinant of the 2-by-2 matrix whose $i, j$th entry counts the paths from $P_i$ to $Q_j$. Thus the number of tilings is

$$\begin{vmatrix} 6 & 5 \\ 5 & 6 \end{vmatrix} = 11 \text{ (again)}$$

since there is a bijection between tilings and pairs of nonintersecting paths.

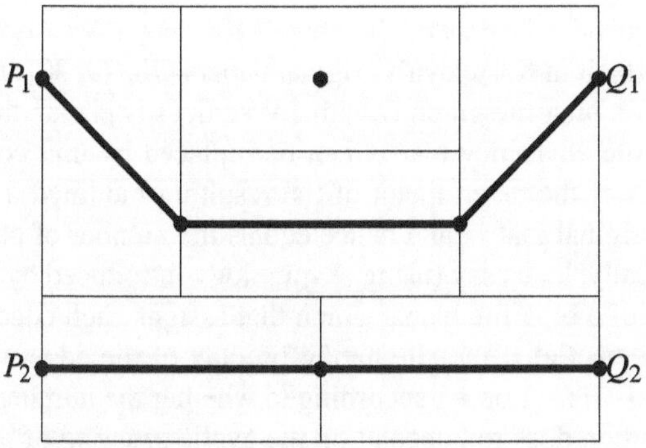


**Figure 9.12**
A tiling and two nonintersecting paths.

This method generalizes to larger collections of paths in an acyclic directed graph. As long as the only way to join up the points $P_1, \ldots, P_m$ to the points $Q_1, \ldots, Q_m$ by nonintersecting paths is to join $P_i$ to $Q_i$ for all $1 \leq i \leq m$, the number of such families of nonintersecting paths equals the determinant of the matrix whose $i, j$th entry is the number of paths from $P_i$ to $Q_j$. In many cases one can use this to count matchings of graphs. By employing this method, which was first used in [67] and was also discovered by chemists John and Sachs [77], one can write the number of lozenge tilings of a semiregular hexagon as the determinant of a matrix of binomial coefficients [65] (see also [19]); likewise one can write the number of domino tilings of an Aztec diamond as the determinant of a matrix of Schröder numbers [17, 55]. A recent (and purely bijective) approach to counting nonintersecting Schröder paths is [13]. For historical background on Lindström's lemma, see footnote 5 in [96] as well as the Notes to Chapter 2 of [159].

### 9.3.2 The permanent-determinant and Hafnian-Pfaffian method

Consider the 3-by-4 grid graph (dual to the 3-by-4 rectangle considered in the previous subsection) and two-color the vertices so that each black vertex has only white neighbors and vice versa. Number the white vertices 1 through 6 and the black vertices 1 through 6. The ordinary 12-by-12 adjacency matrix of this graph has determinant 0, as does the 6-by-6 bipartite adjacency matrix whose $i, j$th entry is 1 or 0 according to whether the $i$th white vertex and $j$th black vertex are connected or not. However, if we let $K$ be the modified bipartite adjacency matrix whose $i, j$th entry is 1, $\mathbf{i}$ $(= \sqrt{-1})$, or 0 according to whether the $i$th white vertex and $j$th black vertex are connected by a horizontal edge, a vertical edge, or no edge, then we find that the modulus of the determinant of $K$ equals the number of matchings of the graph. Likewise, if we replace the 1's in the original 12-by-12 adjacency matrix with suitable unit complex numbers (complex numbers of modulus 1), we obtain a matrix whose determinant has modulus equal to the square of the number of matchings of the graph.

This is a consequence of a general theorem of Kasteleyn. We discuss first the case of bipartite graphs (tacit in Kasteleyn's original papers [78, 79, 80] and made explicit by Percus [131]). If a bipartite graph $G$ with $2N$ vertices is planar, then the non-zero entries of its bipartite adjacency matrix can be replaced by unit complex numbers so that the modulus of the determinant of the resulting modified matrix equals the permanent of the original matrix and hence equals the number of perfect matchings of $G$. More specifically, let us say (using terminology introduced by Kuperberg) that an edge-weighting of a bipartite planar graph that assigns each edge a unit complex number is **(Kasteleyn-)flat** if the alternating product of the edge-weights around a face $e_1^{+1}e_2^{-1}e_3^{+1}e_4^{-1}\cdots$ is $-1$ or $+1$ according to whether the number of edges is 0 or 2 mod 4 (this condition does not depend on the cyclic ordering of the faces). In our case, each face has four sides, so we need the alternating product to be $-1$, and indeed we have $(\mathbf{i})^{+1}(1)^{-1}(\mathbf{i})^{+1}(1)^{-1} = -1$. (The fact that flatness holds around the exterior face is a consequence of flatness around the interior faces; one can also check directly that, in the case of the 3-by-4 grid graph, the alternating product around the 14-sided exterior face is $\mathbf{i}^3/\mathbf{i}^3 = +1$.) Let $K$ be the bipartite adjacency matrix of a bipartite planar graph $G$, with non-zero entries replaced by Kasteleyn-flat edge weights. It can be shown that if one expands the determinant of $K$ into $N!$ terms then all the non-zero terms in the expansion are unit complex numbers with the same phase. Consequently, the matchings of the graph can be counted by taking the modulus of the determinant of the modified adjacency matrix obtained from the flat edge-weighting. This is the **permanent-determinant method**.

To see this method applied to general rectangular subgraphs of the square grid, see [133]. It can be shown that the per-site entropy of the dimer model on a square grid is $G/2\pi$, where $G = \sum_{n=0}^{n-1}(-1)^n/(2n+1)^2 = 1/1 - 1/9 + 1/25 - \ldots$ (Catalan's constant).

Now suppose $G$ is a non-necessarily-bipartite planar graph with $2N$ vertices. Recall from [101] that the Pfaffian of an antisymmetric $2N$-by-$2N$ matrix $A$ equals a signed sum of terms of the form $a_{i_1,j_1}a_{i_2,j_2}\cdots a_{i_N,j_N}$ where $i_1, j_1, \ldots, i_N, j_N$ ranges over all permutations of $1, 2, \ldots, 2N$ for which $i_1 < j_1$, $i_2 < j_2$, $\ldots$, $i_N < j_N$ and $i_1 < i_2 < \cdots < i_N$; such permutations are in one-to-one correspondence with the ways of dividing $1, 2, \ldots, 2N$ into pairs. Kasteleyn showed that if a graph $G$ is planar, then there is an antisymmetric matrix $A$ whose $i, j$th entry is a unit complex number when the $i$th and $j$th vertices of $G$ are adjacent and 0 otherwise, such that the modulus of the Pfaffian of $A$ equals the number of perfect matchings of $G$. Moreover, the determinant of $A$ equals the square of the Pfaffian of $A$, and hence has modulus equal to the square of the number of perfect matchings of $G$. This is the **Hafnian-Pfaffian method** (where loosely speaking the Hafnian is to the Pfaffian as the permanent is to the determinant; see [101] for details).

Kasteleyn's method also works for graphs with (positive) edge-weights attached; one simply multiplies the weights in the weighted adjacency matrix by the required "Kasteleyn phases." Kasteleyn showed how to modify the method to cope with graphs on surfaces of small genus, but a discussion of this would take us too far afield.

We return to the bipartite case. It is known that the entries of the inverse Kasteleyn-Percus matrix $K^{-1}$ have significance of their own; for instance, when the $i$th white vertex and the $j$th black vertex are adjacent, the absolute value of the $j,i$th entry of $K^{-1}$ is the probability that the edge joining those two vertices will belong to a matching chosen uniformly at random. (For this and much else on probabilistic aspects of matchings, see [83].) A recent treatment of the inverse Kasteleyn matrix for Aztec diamond graphs is [22].

Kuperberg [105] has shown that the Gessel-Viennot matrix for a bipartite planar graph (typically of size $O(n)$) and the Kasteleyn-Percus matrix of that graph (typically of size $O(n^2)$) are not really different entities, and that it is possible to turn the latter into the former by means of combinatorially meaningful determinant-preserving matrix reductions. Extending this to the Pfaffian case for non-bipartite matching remains an open problem.

### 9.3.3 The spanning tree method

Figure 9.13 shows a domino tiling of a 5-by-5 square with a corner cell removed. The centers of certain cells have been marked, displaying a 3-by-3 grid within the 5-by-5 grid; when a domino contains a marked square, we draw an arrow from the center of its marked square to the center of its unmarked square. The dominos containing a marked square determine a spanning tree rooted at the missing corner, as shown in Figure 9.14.

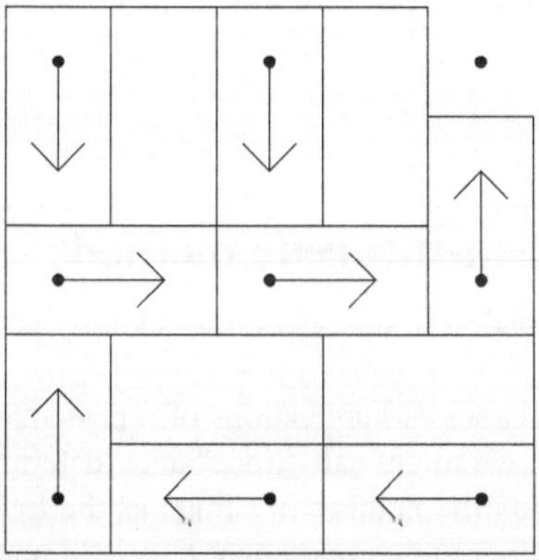

**Figure 9.13**
A tiling and a tree.

Temperley [164] gave a method for turning certain dimer problems into spanning tree problems. This method was extended [86] and later was shown to apply to all bipartite planar graphs [87].

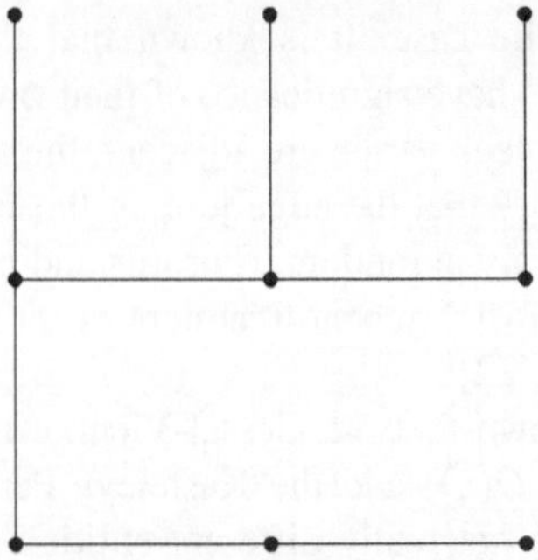

**Figure 9.14**
The tree made clearer.

The strength of the method hinges on the fact that the Matrix Tree theorem (see for example Theorem 5.6.8 of [158] or p. 57 of [10]) gives a formula for the number of spanning trees of a graph as the determinant of an associated matrix. For applications, see [29, 31, 117, 118]. To my knowledge, no one has as of yet exposed a link between the Kirchhoff matrix of a bipartite planar graph and either the Kasteleyn-Percus matrix or the Gessel-Viennot matrix.

The connection between tilings and trees yields a connection with the abelian sandpile model; see [60].

Matchings and spanning trees fit into a bigger picture involving specializations of the Tutte polynomial. See [169] as well as [154] and [21].

## 9.4 Representation-theoretic methods

(Note: Much of the information in this section was kindly provided by Greg Kuperberg.)

There are two broad classes of applications of representation theory to enumeration of matchings, which we might call "trace" and "determinant" methods. In the trace method, one interprets the number of tilings as the trace of some matrix and then uses representation theory to compute that trace. (In many cases, the matrix in question is the identity matrix so the trace is just the dimension of the vector space.) In the determinant method, one interprets the number of tilings as the determinant of some matrix and then uses representation theory to compute that determinant.

Larsen's proof of the formula for the number of domino tilings of an Aztec diamond [53] is an example of the trace method. Kuperberg [100] and Stembridge [160] (the latter building upon Proctor [132]) independently applied this approach to the study of lozenge tilings of hexagons under various symmetry constraints. In some of these results, the matrix one studies is the permutation matrix of an involution, so that the trace counts the fixed points of the involution.

Examples of the determinant method can be found in [103]. Here, one uses the representation theory of the Lie algebra sl(2). If $e$, $f$, and $h$ form a basis for sl(2) in the usual way, a block of the ladder operator $e$ goes from the $h = -1$ eigenspace to the $h = +1$ eigenspace, and the number of tilings equals the determinant of this block (up to a normalization factor). The representation theory of sl(2) lets one diagonalize this operator and thereby compute the determinant. This works for various symmetry classes of lozenge tilings of hexagons. It would be interesting to see trace methods applied to symmetry classes of domino tilings of Aztec diamonds.

Representation-theoretic arguments have also been successful in enumerative problems involving ASMs (Alternating Sign Matrices). Although the initial view of ASMs in [122] does not seem like it has anything to do with tilings, there is an attractive interpretation of ASMs as tilings that fill regions called "supergaskets" with curved tiles called "gaskets" and "baskets." See Figure 9.15 for a picture of such a tiling, and the article [138] for an explanation of the link between tilings of this kind and ASMs viewed as $n$-by-$n$ arrays of $+1$'s, $-1$'s, and 0's satisfying various constraints. It is also worth mentioning that domino tilings of Aztec diamonds were in a sense discovered by Robbins and Rumsey [149] as an outgrowth of their work on ASMs, even though the tilings picture was not made explicit. The picture of ASMs as tilings has not been helpful to researchers, but the dual picture of ASMs as states of the six-vertex model (see Figure 9.16) plays a key role in Kuperberg's proof [102] that the number of ASMs of order $n$ is equal to the product $\prod_{k=0}^{n-1}(3k+1)!/(n+k)!$ (conjectured by Mills, Robbins and Rumsey [122] and first proved by Zeilberger [176]).

In the **six-vertex model** (also called the **square ice model**), edges of a square grid are assigned orientations in such a way that at each vertex there are two arrows pointing in and two arrows pointing out. (At each vertex, there are $\binom{4}{2} = 6$ possible ways to choose orientations for the incident edges, hence the name of the model.) As in the case of domino and lozenge tilings, we find that the boundary conditions for the six-vertex model that are most relevant to physics are not the ones most conducive to exact enumeration. Just as the logarithm of the number of domino tilings of the Aztec diamond falls short of the full entropy of the dimer model on the square grid, the logarithm of the number of alternating-sign matrices falls short of the full entropy of the square ice model. Curiously, although the former entropy is not believed to be the logarithm of an algebraic number, the latter number is the logarithm of the algebraic number $8/(3\sqrt{3})$ (traditionally written as $(4/3)^{3/2}$ in the physics literature, starting with its original appearance in [112]). Although the model with free boundary conditions is exactly solvable in the sense of physics (that is, in the sense that one can find a formula for the entropy), it is not known to be exactly enumerable (in the non-asymptotic sense). For exact enumerations, it appears one must impose the "domain-wall boundary conditions" (so dubbed by Korepin; see [91]) that prevail in the particular square ice configurations that correspond to ASMs. We will use this term in a more general way to denote boundary conditions in which different "frozen" phases of the system govern the behavior of different parts of the boundary, so that the interior is subject to competing influences (and can spontaneously divide

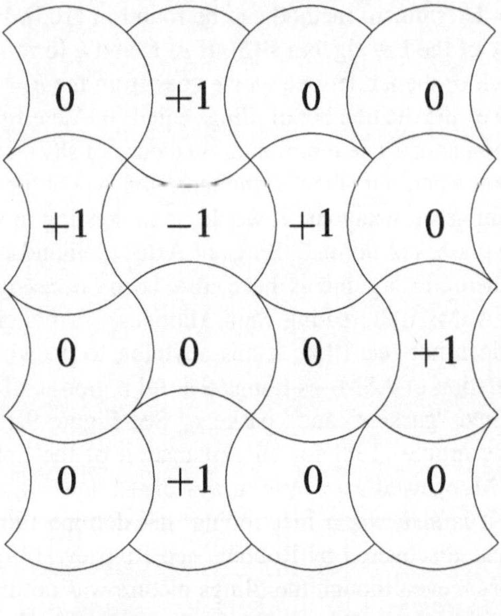


**Figure 9.15**
A tiling by baskets and gaskets (one of 42), with the associated alternating-sign matrix.

itself into multiple macroscopic domains in which the tiling displays qualitatively different microscopic behavior).

For enumeration of ASMs of order $n$, the relationship with representation theory has been quite intimate, and some of the proofs in the literature make uses of different sorts of representation theory arguments at different stages in the proof. In the first place, when one relaxes the grid-structure underlying the square ice model and looks at ice-configurations of a more general kind, one finds Reidemeister-style moves that preserve the partition function or affect it in predictable ways; this is the theory of Yang-Baxter relations, and it is tied in with the $q$-deformed representation theory of sl(2). Against this background, Izergin and Korepin were able to write the partition function in terms of the determinant of an $n$-by-$n$ matrix [73, 91] involving $2n$ variables associated with the rows and columns of the ice-grid. When one seeks to apply this formula to count ASMs, one obtains an indeterminate of type $0/0$, but by approaching the singularity along an artfully chosen path, Kuperberg was able to obtain the desired enumeration. For more details, see [102] and [106]. Okada [127] brought Lie group characters into the picture, with two distinct quantum parameters playing roles (one tending toward 1 as in Kuperberg's original proof, the other a cube root of unity). Okada showed that for all the proved or conjectural formulas in the second of Kuperberg's articles, the requisite determinant evaluations are associated with characters of Lie groups; for instance, the formula Kuperberg proved in [102] is associated with an irreducible representation of GL($2n$). Symmetry classes of alternating-sign matrices have also been enumerated by Razumov and Stroganov

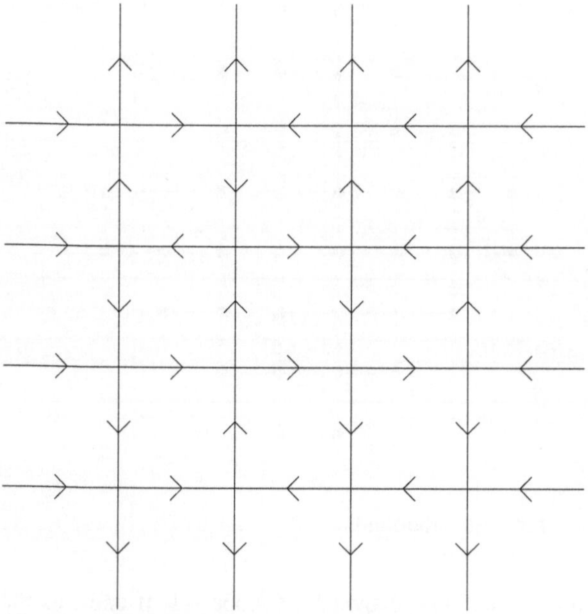

**Figure 9.16**
A state of the six-vertex model with domain-wall boundary conditions.

[144, 145, 146]. As in the case of other tiling problems with nice answers, we find that enumeration with the extra constraint of symmetry also gives rise to tiling problems with nice answers (which are usually harder to prove).

A central mystery of the subject is the remarkable fact that the number of fully symmetrical lozenge tilings of a regular hexagon of side $2n$ (which are in bijection with Totally Symmetric Self-Complementary Plane Partitions of order $2n$) are equinumerous with ASMs of order $n$. Both are governed by the product formula that appears above, but there is not at this time a conceptual explanation of why the two numbers should be the same. For instance, despite intensive work by combinatorialists over the past several decades, no bijection between TSSCCPs and ASMs is known.

## 9.5 Other combinatorial methods

A very fertile approach to enumeration of matchings, discovered by Eric Kuo while he was an undergraduate member of the Tilings Research Group at MIT, is a graphical analogue of Dodgson condensation. (For background on Dodgson condensation, see [16].) Figure 9.17 shows an Aztec diamond graph $G$ of order 4 divided by two

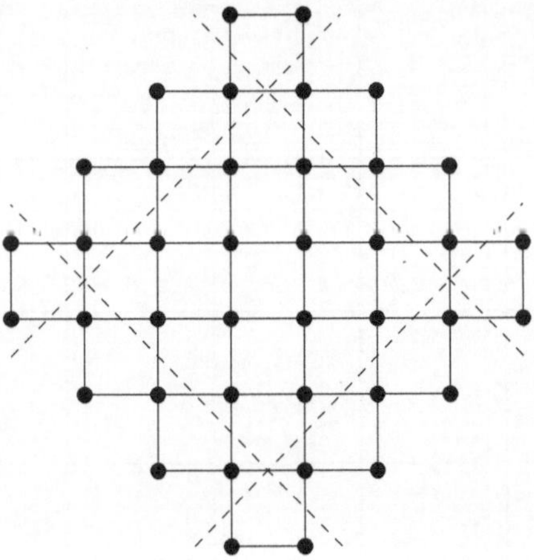

**Figure 9.17**
Kuo condensation for Aztec diamonds.

diagonals of slope $+1$ and two diagonals of slope $-1$. If one cuts the figure with one of the two lines of slope $+1$ and one of the two lines of slope $-1$, one obtains an Aztec diamond graph of order 3 (and three smaller regions). Let $G_N$, $G_S$, $G_E$, and $G_W$ respectively be the Northern, Southern, Eastern, and Western 3-by-3 Aztec diamond subgraphs obtained in this way, and let $G_C$ be the central 2-by-2 Aztec diamond subgraph. Then the original version of **Kuo condensation** [99, Theorem 2.1] tells us that $M(G)M(G_C) = M(G_N)M(G_S) + M(G_E)M(G_W)$, where $M(\cdot)$ denote the number of matchings of a graph. In this case, since the graphs $G_N$, $G_S$, $G_E$, and $G_W$ are isomorphic, we obtain the relation $M(G_4)M(G_2) = 2M(G_3)^2$ (where $G_n$ denote the Aztec diamond graph of order $n$), and more generally $M(G_{n+1})M(G_{n-1}) = 2M(G_n)^2$, which allows us to solve for $M(G_n)$ using the initial conditions $M(G_0) = 1$ and $M(G_1) = 2$. See [99] and [98]. For more details on Kuo condensation and numerous applications, see (in chronological order) [172, 171, 90, 62, 35, 36, 34, 23].

When one applies Kuo condensation repeatedly, keeping track of all the different subgraphs of the original graph that occur, one accumulates data that are mostly easily stored in a three-dimensional array in which the entries form an octahedral lattice (dual to a cubical lattice), in which the numbers $M(G)$, $M(G_C)$, $M(G_N)$, $M(G_S)$, $M(G_E)$, and $M(G_W)$ are associated with vertices that form an octahedron. That is to say, the enumerations of matchings satisfy the octahedron relation. This equation originated in the theory of discrete integrable systems, and has found numerous applications in combinatorics. Here it is worth pointing out that Kuo condensation can be applied with weightings of a fully general kind, where each edge is assigned its own weight independently of all the others.

A superficially different but probably related approach to enumerating matchings is (generalized) **domino-shuffling** [140], also known as **urban renewal**. There is some confusion in the literature regarding the intended meanings of these two terms. At least originally, urban renewal referred to a certain kind of local modification of

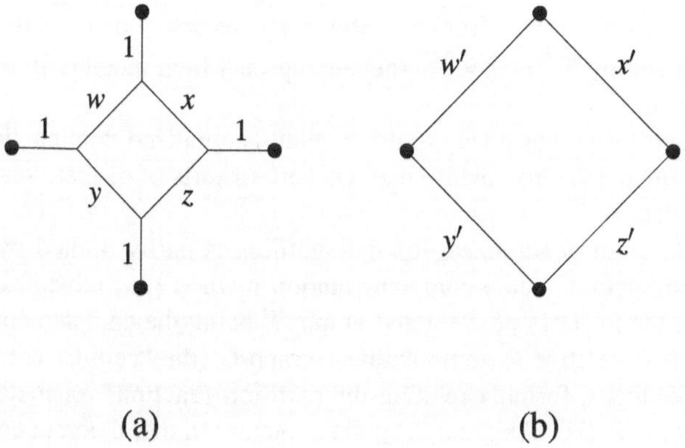


**Figure 9.18**
Urban renewal.

an edge-weighted bipartite planar graph that has a predictable effect on the partition function. In this it drew much inspiration from the Temperley-Fisher method [165] of enumerating matchings, which used similar local mutations to turn states of the dimer model into states of an Ising model on a suitable graph. (See Chapter 13 of [125], particularly Sections 13.6 and 13.7. Also see [121], in which the the dimer and Ising models are solved by use of Pfaffians in quite complete detail. Yet another worthwhile article on the subject is [123].) The inspiration for the terminology drew upon an image of small pieces of the graph as being "cities" that communicate with the outside world via certain ports (as well as being an obscure pun on a style of urban planning that had a brief vogue in the 1970s). Strictly speaking, urban renewal is an operation not on graphs but on edge-weighted graphs. By performing such operations throughout a graph in a systematic fashion (this is what was meant by "generalized domino-shuffling," on account of its relationship to the domino-shuffling algorithm introduced in [53]), one can often obtain a simpler graph of the same kind, giving rise to a recurrence relation for the number of matchings. In addition to having "local factors" associated with individual edges, it is often handy to collect together "global factors" that are associated not with individual edges but with the graph as a whole. The basic urban renewal operation is depicted in Figure 9.18 If an edge-weighted graph $G$ having a city of the kind shown in panel (a) is modified by replacing that city by a city of the kind shown in panel (b) (remaining otherwise unchanged), with $w' = z/(wz+xy)$, $x' = y/(wz+xy)$, $y' = x/(wz+xy)$, and $z' = w/(wz+xy)$, then the sum of the weights of the matchings of the first graph equals $wz+xy$ times the sum of the weights of the matchings of the second graph. Like Kuo condensation, generalized domino-shuffling can be applied with weightings of a fully general kind, where each edge is assigned its own weight independently of all the others. Using urban renewal allows one to show fairly directly that the number of matchings of the Aztec diamond graph of order $n$ equals $2^n$ times the number of matchings of the Aztec diamond of order $n-1$. Urban renewal was used in [173] to show that, for a family of finite graphs derived from the infinite square-octagon graph, the number of matchings is given by $5^{n^2}$ or $2 \times 5^{n^2}$ (depending on which member in the family is being discussed).

Di Francesco [50] gives a way to think about generalized domino-shuffling that unifies it with the octahedron recurrence, via certain sorts of discrete surfaces in the octahedron lattice.

The usefulness of generalized domino-shuffling is partly limited by the adherence to a square grid. Ciucu's complementation method [26] remedies this defect by introducing the idea of one weighted graph $H'$ being the complement of another weighted graph $H$ relative to some ambient graph $G$ (the "cellular completion" of $H$), and establishing a formula relating the partition functions for matchings of $H$ and matchings of $H'$ [26, Theorem 2.1]; the $\Delta$-factors in that theorem can be recognized as a generalization of the face-factors in generalized domino-shuffling. A similarly flexible factorization theorem [25] gives a way to write the partition function of a weighted plane graph with reflection-symmetry as the product of the partition functions of two smaller graphs (up to a predictable multiplicative factor). For applications, see [27] and [109]. It is worth mentioning that Yan and Zhang [172] used Ciucu factorization to prove a version of Kuo condensation. Thus in a sense Ciucu factorization implies Kuo condensation.

Finally, a different sort of proof of the enumeration of lozenge-tilings of a hexagon comes from Krattenthaler's bijective proof [94] of Stanley's hook-content formula for the generating function for semistandard Young tableaux of a given shape. His proof yields, as a spin-off, a way of sampling from the uniform distribution on the set of lozenge tilings of a semiregular hexagon (by way of associated semistandard tableaux). It is perhaps too isolated an achievement to be considered a method, but it deserves study and imitation.

## 9.6 Related topics, and an attempt at history

The monomer-dimer model was first considered as a simplified model of a gas [61] and later as a model of adsorption of diatomic molecules on a surface. However, the mathematical difficulties posed by the monomer-dimer model in two or three dimensions, combined with a certain amount of mismatch between physical reality and the simplified model, did not lead to successful applications of the theory.

Fortunately for combinatorialists, physicists like Fisher, Kasteleyn and Temperley were not dissuaded and, emboldened by Onsager's success [128] in exactly solving the two-dimensional Ising model, invested considerable effort and ingenuity in finding exact solution methods (in Sections 9.2–9.5) for the pure dimer model [78, 165], even though that model is even more removed from the aforementioned target applications than the monomer-dimer model. Lieb's exact solution of the square ice model [112]—that is, his calculation of the per-site entropy—and the close agreement between Lieb's answer and the experimentally determined residual entropy of actual (non-square!) three-dimensional ice gave further impetus to the study of two-dimensional lattice models in statistical mechanics.

The statistical mechanics literature (to which [5] serves as an excellent introduction) provides a unifying point of view for models that at a combinatorial level might seem to be unrelated. For instance, the dimer model on a square grid and the square ice model can both be seen as special cases of the eight-vertex model (an extension of the six-vertex model that permits vertices with all arrows pointing in or all arrows pointing out), with different sorts of weights attached to the edges of the graph; the dimer model is easier to solve than the square ice model because the former is part of the "free-fermion regime" of the eight-vertex model [56]. Statistical physicists Grensing, Carlsen, and Zapp [68] were the first researchers in any subject to look at domino tilings of Aztec diamonds, or rather (in their way of looking at things) dimer configurations in the dual graph. They went so far as to assert without proof the formula for the number of dimer configurations, but their published explorations stopped there, presumably because the model did not give rise to the known per-site entropy; they thereby missed out on discovering not just a beautiful combinatorial theorem but a rich body of theory that grew out of it, relating the dimer model to PDE and algebraic geometry [83]. This story bears out the observation that undiscovered mathematical treasures often lie in the trashbins of physicists.

A more harmonious match between theory and reality arose from work of theoretical chemists studying benzenoid hydrocarbons. These molecules, often depicted as having rings of alternating single and double bonds, actually have a more subtle quantum-mechanical nature, and can be viewed to a first approximation as being in a superposition of different valence-states or "Kekulé structures" [67]. The number of Kekulé structures on a graph has implications for its chemical properties, so chemists developed methods for counting these structures; indeed, a version of the Lindström-Gessel-Viennot method discussed above was found independently by Horst Sachs and his co-workers. (The class of regions mentioned in Example 9.1.2 of Section 9.1 was originally studied by chemists.) See [77], [47], and [151].

For a more recent application of tilings to the physical sciences, see [9], which like [61] uses the monomer-dimer model as a model of adsorption of diatomic molecules on a surface.

There has also been work on stepped surfaces (e.g., [8], [110], and [59]) of which tilings can be seen as two-dimensional projections (cf. our earlier discussion of Figure 9.6 and its interpretation as the three-dimensional Young diagram of a plane partition). Here the interests of physicists have a certain amount of overlap with the interests of probabilists and representation theorists. A phrase that has recently been associated with this line of work is "integrable probability theory" [11].

Probabilists are often interested in limit-shapes arising from tiling models when one takes the uniform distribution (or, more generally, distributions arising from a weighting) on the set of tilings of a large region and sends the size to infinity. For both random domino tilings of large Aztec diamonds and random lozenge tilings of large regular hexagons, one sees (with overwhelmingly high probability) a sharp boundary between a central circular region in which tiles of different orientations are mixed together and outer regions in which all tiles are aligned with their neighbors [39, 41].

There is a great deal of interest in the corresponding problem for alternating-sign matrices; there is a widely-believed conjecture for the asymptotic shape of the boundary between the central "liquid" domain and the four outer "solid" domains [83] of large random ASMs, but no rigorous proof is currently known. See [42], [43], and [44].

Probabilists are also interested in algorithms for randomly sampling from the uniform distribution (or, more generally, distributions arising from a weighting) on the set of tilings of a large region. In the case of matchings of planar graphs, a number of methods are available; see Section 9.8 for a discussion of this. Note that one method for random sampling is to iteratively make decisions in accordance with conditional probabilities (obtainable as ratios of tiling-enumerations); in the case of matchings of planar graphs, this is feasible because of Kasteleyn's method, and some tricks introduced by David Wilson [170] make this competitive with other methods. This method of exactly sampling from the uniform distribution cannot currently be applied to alternating-sign matrices because we lack exact formulas for the relevant conditional probabilities. It seems unlikely that the counting problems that arise are anywhere close to #$P$-complete, but we cannot currently rule out this possibility.

The broad applicability of Kasteleyn's method has inspired a whole line of research in computer science called the theory of holographic algorithms, in which one tries to see how many other counting problems one can solve using variants of Kasteleyn's trick, via computational elements called matchgates. See [70] and [18].

The Aztec diamond and its relatives have also played a role in the study of random matrices, the connecting link being fluctuations in the boundaries between phase-domains; e.g., in the Aztec diamond of order $n$, fluctuations are of order $n^{1/3}$ and converge in distribution to the Tracy-Widom distribution. For more about boundary fluctuations, see [76]. For an account more focused on lozenge tilings, see [58].

Kuo's condensation lemma is related to the octahedron relation (also called the discrete Hirota equation; see Section 12 of [97]). In this way, the study of enumeration of matchings leads to the theory of (discrete) integrable systems, and has connections to cluster algebras; [66] and [50] are good sources of more information.

Lastly, it should be mentioned that tilings, like other combinatorial objects, often arise in connection with algebraic objects. Standard examples of this phenomenon are matchings of complete bipartite graphs (which are implicit in the formula for the determinant of a matrix expanded as a sum of $n!$ monomials) and Young tableaux (which occur throughout the representation theory of the symmetric group). Matchings of Aztec diamond graphs, in disguised form, index the terms the expansion of the $\lambda$-determinant of Robbins and Rumsey [149] as a sum of Laurent monomials. Since lozenge tilings of a semi-regular hexagon can be converted to semistandard Young tableaux, they index terms in Schur polynomials associated with rectangular Young diagrams. More recently, square-triangle tilings of equilateral triangles have been used in a modern reformulation of the Littlewood-Richardson rule for expressing the product of two Schur functions as a linear combination of Schur functions [143]; see [177].

## 9.7 Some emergent themes

### 9.7.1 Recurrence relations

When we have a sequence of tiling problems, the answer is the integer sequence $a_1, a_2, \dots$ whose $n$th term counts the number of solutions to the $n$th tiling problem. What kinds of questions have nice answers? Niceness is subjective, so for present purposes we define a "nice answer" to be a sequence that appears in the mathematical literature and can be found via Sloane's On-line Encyclopedia of Integer Sequences (`http://oeis.org`) or has patterns that can be inferred empirically (e.g., satisfies a linear recurrence relation). When $a_n$ grows exponential in $n^2$ (as is usually the case for interesting two-dimensional tiling enumerations), there is no chance of finding a linear recurrence. So what else can we try?

One answer is, rational recurrences. The sequence $a_n = 2^{n(n+1)/2}$ satisfies the recurrence $a_{n+1} = 2a_n^2/a_{n-1}$, and this is not accidental; writing the recurrence in the form $a_{n-1}a_{n+1} = a_n a_n + a_n a_n$, we see it as an instance of Kuo condensation. Applying this insight in reverse, one can start with a nonlinear recurrence like $a_{n-2}\,a_{n+2} = a_{n-1}\,a_{n+1} + a_n^2$ (satisfied by the Somos-4 sequence 1,1,1,1,2,3,7,23,59,... described in [64] and [63]) and reverse-engineer an appropriate family of bipartite planar graphs whose $n$th member has $a_n$ matchings; see [14] and [155]. Just as the existence of a linear recurrence satisfied by the sequence $a_1, a_2, \dots$ is signaled by the existence of a value of $m$ for which the Hankel matrix

$$H(a_{n+1}, a_{n+2}, \dots, a_{n+2m-1}) := \begin{pmatrix} a_{n+1} & a_{n+2} & \cdots & a_{n+m} \\ a_{n+2} & a_{n+3} & \cdots & a_{n+m+1} \\ \vdots & \vdots & \ddots & \\ a_{n+m} & a_{n+m+1} & \cdots & a_{n+2m-1} \end{pmatrix}$$

appears to be singular for all $n$ in some range, the existence of a Somos-type recurrence satisfied by the sequence $a_1, a_2, \dots$ is signaled by the existence of values of $k$ and $m$ for which the Somos matrix

$$H(a_{n+1}, \dots, a_{n+2m-1}) * T(a_{n+k+1}, \dots, a_{n+k+2m-1})$$

appears to be singular for all $n$ in some range, where $T(\dots)$ denotes the Toeplitz matrix obtained by writing each row of $H(\dots)$ backwards and $*$ denotes the Kronecker product of matrices (Somos, personal communication). For example, the 3-by-3 Somos matrix

$$H(a_1, a_2, a_3, a_4, a_5) * T(a_3, a_4, a_5, a_6, a_7) = \begin{pmatrix} a_1a_5 & a_2a_4 & a_3a_3 \\ a_2a_6 & a_3a_5 & a_4a_4 \\ a_3a_7 & a_4a_6 & a_5a_5 \end{pmatrix}$$

is singular when $a_1, a_2, \dots$ satisfies the Somos-4 recurrence, and this remains true when all subscripts are shifted by adding a constant. It seems possible that some tiling-enumeration problems for which no exact formula has been discovered will turn out to governed by nonlinear recurrences, which might be discoverable by use of Somos matrices and related tools.

Inasmuch as rational recurrences like Somos-4 often arise as projections of a three-dimensional octahedron relation, an attractive research strategy for finding a nonlinear recurrence might be to imbed the sequence as a strand of a higher-dimensional array that satisfies the octahedron relation. This is one way to think about the approach to counting domino tilings of squares used in [141], although of course that construction had the benefit of hindsight (in the form of Kuo condensation).

### 9.7.2 Smoothness

In the absence of visible patterns linking the terms of a sequence, one sign of an underlying regularity is properties of the individual terms themselves, as integers. One sort of pattern can appear when one looks at how the integers factor into primes. Call a sequence $a_1, a_2, \ldots,$ **smooth** if the largest prime factor of $a_n$ is $O(n)$, and **ultra-smooth** if the largest prime factor of $a_n$ is $O(1)$, that is, bounded. (Another word that is sometimes used instead of "smooth" is "round.") For example, $H(3n)H(n)^3/H(2n)^3$ (the number of lozenge tilings of a regular hexagon of side $n$) is smooth; $2^{n(n+1)/2}$ (the number of domino tilings of an Aztec diamond of order $n$) is ultra-smooth. Quite often we find that the answers to enumerative questions are smooth sequences, and the smoothness is often a key to finding an exact product formula.

Powers of 2 arise from looking at matchings of Aztec diamond graphs [53] associated with the regular tiling of the plane by squares; powers of 5 arise from looking at matchings of "fortress graphs" [173] associated with the semiregular tessellation of the plane by squares and octagons; and powers of 13 arise from looking at matchings of "dungeon" and "Aztec dungeon" graphs [27] associated with the semiregular tessellation of the plane by squares, hexagons, and dodecagons, respectively. In all three cases, there is a highly symmetrical tessellation of the plane in the background, and the tiling problem is best thought of as the problem of counting matchings of certain finite subgraphs of the highly symmetrical infinite bipartite planar graph that arises as the dual of the tessellation. We can sharpen our inquiry by asking: what sorts of tessellations, in combination with what sorts of subgraphs of the graph dual to the tessellation, give rise to "ultra-smooth enumerations," i.e., enumerations in which only finitely many prime factors arise?

Regarding the choice of tessellation, it was observed by David Wilson that if you deform the square-hexagon-dodecagon graph so that it becomes a subgraph of the square grid, turn that into a square grid with edge-weights 0 and 1, and then perform generalized domino-shuffling (hereafter, "shuffling") three times, you get back the original edge-weights (modulo translation). This is analogous to Yang's discovery [173] that if you use urban renewal to turn the square-octagon graph into a square grid with edge-weights and apply shuffling twice, you get back what you started with (thereby obtaining exact enumeration of "diabolo tilings of fortresses"), and also analogous to the original application of urban renewal to Aztec diamonds [136].

More generally, consider an infinite (but locally finite) bipartite planar graph $G$ with two independent translation-symmetries. Suppose that it gives rise, via a suitable sort of deformation, to a doubly periodic weighting of the infinite square grid with 0's and 1's (with a caveat discussed below). Call this weighting $W_0$. Applying shuffling to $W_0$, one gets some weighting $W_1$; applying shuffling again, one gets some weighting $W_2$; etc. Suppose that $W_m = W_0$ for some $m$. Each $W_i$ is doubly periodic, and hence has only finitely many distinct "face-factors," where the face-factor associated with a face is $ac+bd$ where $a,b,c,d$ are the consecutive edge-weights around the face. Hence only finitely many primes occur in the numerators of face-factors arising from the weightings $W_0, W_1, \ldots, W_{m-1}$. These are the primes that occur in ultra-smooth enumerations of matchings of finite subgraphs of $G$.

Therefore, one question that is likely to shed light on the subject of ultra-smooth enumeration is, what doubly periodic graphs $G$ give rise to weightings of the square grid that come back to themselves after a finite number of iterations of shuffling? This is probably not the most natural setting for the question, since it gives the square grid an unjustified privileged role, and not all ultra-smooth enumerations arise in this way, but it is a place to start. (For more on iterated shuffling as an integrable system, see [66].) A related project is to classify triply periodic solutions to the octahedron recurrence (with periodicity in two "space directions" and one "time direction").

However, there is more to the story than the infinite graph $G$: One must also pick out the correct sorts of finite subgraphs. Here we face an irony that drastically augments the disappointment Grensing, Carlsen, and Zapp must have felt in their abortive study of Aztec diamond graphs: The sorts of subgraphs that are amenable to simple combinatorial methods never include the sorts that are relevant to computing the per-site entropy of the dimer model on the infinite graph. If for instance we apply shuffling to compute the number of domino tilings of a $2n$-by-$2n$ square, in the interior we pick up the face-factors equal to 2, but along the boundary we find other numbers that increase in complexity as we apply new rounds of shuffling, and we ultimately lose control over the analytical form of the answer.

In "designing" various families of bipartite planar graphs that turned out to have ultra-smooth enumerations, a heuristic that has proved to be very successful, but whose success is quite mysterious, is what one might (borrowing physicists' nomenclature) call the domain-wall heuristic. This is illustrated by the considerations that led me to (re-)invent Aztec diamonds in the 1980s. I was aware of MacMahon's beautiful product formula for the (smooth, though not ultra-smooth) enumeration of lozenge tilings of hexagons, and Kasteleyn's formula for the (un-smooth) enumeration of domino tilings of squares, and wondered what was responsible for the difference. A reading of W. Thurston's article [166], focusing on the concept of height-functions for these two sorts of tilings, led me to notice that the height-function along the boundary of the square is (modulo tiny fluctuations) flat, while the height function along the boundary of the hexagon has dramatic rises and dips. Might a region in the square grid whose boundary consisted of four zigzag paths, with dramatic rises and dips in the height function, be a better analogue of the hexagon, and have similar nice enumerative properties? The answer was a resounding "yes"; nothing is more ultrasmooth than a sequence of powers of 2.

Similar considerations, applied to other doubly periodic graphs, yielded other conjectural smooth enumerations, such as the ones later proved by Yang and by Ciucu. In every case, the starting point was a two-colored doubly periodic tessellation (often obtained from [69]) in which black cells are surrounded by white cells and vice versa. In such a tessellation one draws infinite paths that abut only white cells on one side and only black cells on the other side. A collection of such zigzags divides the plane into subregions. The finite regions that arise in this fashion turn out suspiciously often to be associated with ultra-smooth enumerations.

With these sorts of domain-walls boundaries, it is inevitable the tiling will be more tightly constrained near the boundary than in the interior, and so it is not surprising that segregation of the region into different phase-domains with interesting asymptotic shapes can occur. However, it still seems magical that domain-wall boundaries give rise to ultra-smooth enumerations.

Moreover, in the case of lozenge tilings of hexagons, the explanation in terms of face-factors does not apply, and we do not get an ultra-smooth enumeration, but we still get a smooth one, with prime factors that grow only linearly in size. The same observation applies to gaskets-and-baskets tilings: when one chooses paths through the plane that lift to maximally steep paths in the height-function picture, the regions the paths cut out are supergaskets, whose tilings (which are in bijection with ASMs) yield smooth enumerations.

Now comes the caveat that I warned you about: Kuperberg's method of turning general bipartite planar graphs into subgraphs of the square grid gives rise to weightings with lots of 0's, and these are not always amenable to shuffling when expressions of the form $0/0$ arise. The article [74] discusses some ways of circumventing this problem, but a complete solution awaits discovery. This issue has connections with a similar issue that arises when one carries out Dodgson condensation to evaluate determinants of matrices [16].

### 9.7.3 Non-periodic weights

With $q$ a formal indeterminate, define the $q$-integer $[n] = 1+q+\cdots+q^{n-1}$, the $q$-factorial $[n]! = [1][2]\cdots[n]$, and the $q$-hyperfactorial $H([n]) = [1]![2]!\cdots[n-1]!$. Many of the integer-valued enumerations associated with matchings of graphs can be obtained as the $q=1$ specializations of more refined $q$-enumerations, in which one "$q$-counts" tilings by summing the quantity $q^{h(T)}$ as $T$ varies over all the tilings, where $h$ is some integer-valued "global height" associated with $T$. More specifically, one can write $h(T)$ as a sum of local heights associated with the vertices of the tiling (or with the faces of the dual graph), as in W. Thurston's exposition [166] of Conway's approach to tilings [45]; see also [134].

One can fit $q$-enumeration into the framework of edge-weighted enumeration, but at a cost: the pattern of weights will no longer be periodic. For instance, looking at a regular procession of edges, one may find that their edge weights follow the pattern $q^1, q^{-2}, q^3, q^{-4}, \ldots$.

Factorizations in $\mathbb{Z}[q]$ are more informative than factorizations in $\mathbb{Z}$; 12 could be 2 times 6 or 3 times 4, but [2] times [6] is different from [3] times [4], so in the case of smooth enumerations that give product formulas, empirical $q$-enumerations for small regions give more clues about what the general formula might be. In this way $q$-analogues can sometimes provide information that illuminates the $q = 1$ enumerations. A recent example of this (applied to a nonsmooth enumeration) came from Adam Kalman's work on Aztec pillows, discussed in the 2015 version of [137]. Indeed, I like to think of the step of replacing $\mathbb{Z}$ by $\mathbb{Z}[q]$ as just the first step down a road that, if you can reach the end, replaces what started out as a mere number by a multivariate polynomial (or Laurent polynomial) in which every coefficient is 1. In this way, the monomials themselves are seen to encode the combinatorial objects being counted. This "very many variables" point of view, which draws its inspiration from its success in the case of Aztec diamonds, was a key ingredient in the work of Carroll and Speyer on groves [20].

It should be mentioned (apropos of Kalman's work) that there really is not a hard dichotomy between smooth and nonsmooth enumerations; in particular, a single enumeration can exhibit both smooth and nonsmooth features, with certain small primes occurring with very high exponents and larger primes occurring with much smaller exponents.

A recent and interesting way to add variables, different from the height-functions approach, is the "elliptic" approach of Borodin, Gorin, and Rains [12].

### 9.7.4 Other numerical patterns

Sometimes in examining the prime factorization of an enumeration, one sees dominance not of a particular finite set of primes but of an infinite arithmetic progression of primes. This is the case for the sequence whose terms count domino tilings of squares: one sees numbers in whose prime factorizations primes congruent to 1 mod 4 predominate.

Sometimes the patterns one sees involve not the primes but their exponents. For instance, when all the exponents in an enumeration are even (that is, when the answer is a perfect square for all $n$), surely there is something going on that demands an explanation. A variant of this phenomenon is that certain answers to tilings problems are always twice a perfect square. Typically such patterns are explained by appealing to symmetries of the graph (e.g., [75] and [25]).

Sometimes one sees patterns in the congruence classes of the enumerations themselves (not their prime factors). The prototype for this behavior is seen in the Fibonacci numbers, which are periodic mod $m$ for every $m$, and more generally any sequence whose terms satisfy a linear recurrence relation over the integers whose characteristic polynomial has first and last coefficients equal to $\pm 1$. It is less obvious that this periodicity property should be true for the Somos-4 sequence 1,1,1,1,2,3,7,23,59,... (since the defining recurrence is not linear but rational) but periodicity still holds in this case [120] and in other similar cases. In the case of tiling problems, Cohn has shown that the sequence whose $n$th term counts domino tilings of the $2n$-by-$2n$ square is periodic modulo $m$ whenever $m$ is a power of 2 [38].

Divisibility properties constitute a special case of congruence properties. For one example of a result asserting divisibility, see [72], which shows that the number of perfect matchings of any member of a certain family of graphs is always divisible by 3.

Lastly, one sometimes sees that later terms of a sequence tend to be divisible by certain earlier terms. In the case of the Fibonacci sequence (which counts domino tilings of rectangles of width 2), this observation goes back centuries. A generalization of this classical fact that applies to domino tilings of rectangles of arbitrary fixed width was proved by Strehl [162] (see also [137, Problem 21]). Quite recently, MIT undergraduate Forest Tong has found a combinatorial explanation for a broad generalization of this phenomenon [167].

### 9.7.5 Symmetry

When a region has symmetries, one can count tiling of the region that are invariant under the full symmetry group of the region or some subgroup. Choosing a larger subgroup will of course reduce the number of tilings, but curiously, even though the numbers get smaller, the problems get harder! (Though sometimes when the subgroup is too large, the number of tilings drops to 0, and then the problem is trivial.) The original example of this sort of story is Stanley's article on symmetry classes of plane partitions [156]; the smaller subgroups were the first ones to be settled (starting with the trivial subgroup, which was settled by MacMahon), culminating in the case of Totally Symmetric Self-Complementary Plane Partitions (TSSCPPs). The final piece of this ten-symmetry-classes-of-plane-partitions story was not put into place until 2011 [92].

Matchings that are invariant under a symmetry can be thought of as matchings of a quotient graph, although some care is required when the action is not free: Vertices or edges whose orbit is of anomalously small cardinality require special treatment in "orbigraph matchings" (see [104]).

More mysterious than the above is the $q = -1$ phenomenon discovered by Stembridge and significantly generalized by Reiner, Stanton, and White in their notion of the cyclic sieving phenomenon [147] (see also [152] and [148]). Consider the set $S$ of lozenge tilings of some hexagon, either unconstrained or constrained to possess certain symmetries of the hexagon. Let $f(q)$ be the $q$-enumeration of the tilings. It is clear that $f(1)$ is the number of tilings in the set $S$ we are looking at. What is surprising is that, in certain cases of this kind, $f(-1)$ turns out to have enumerative significance of its own: It is the number of tilings in $S$ that are invariant under some involution on $S$ arising from a symmetry of the hexagon. See [161].

A different symmetry concerns the size parameter $n$, rather than the region being tiled. The quantity $5^{n^2}$, which counts matchings of fortress graphs, is invariant under the substitution that replaces $n$ by $-n$. The quantity $2^{n(n+1)/2}$, which counts matchings of Aztec diamond graphs, is invariant under the substitution that replaces $n$ by $-1-n$. The quantity $(\phi^{n+1} - (-1/\phi)^{n+1})/\sqrt{5}$, which counts matchings of the 2-by-2$n$ grid, is invariant under the substitution that replaces $n$ by $-2-n$, up to a predictable factor of $\pm 1$. This phenomenon is called **reciprocity**, and it can be seen in many exact enumerations. To cite a recent example, Ciucu and Lai [108, 37] have proved a conjecture of Matt Blum (problem 25 in [137]) and shown that, for positive integers $a, b$ with $b \geq 2a$, the number of tilings of the hexagonal dungeon of type $a, 2a, b$ is $13^{2a^2} 14^{\lfloor a^2/2 \rfloor}$. This quantity is an even function of $a$; that is, it is invariant under the substitution that replaces $a$ by $-a$. Likewise one notices that the polynomials that appears in formulas (7) through (10) in [150] (as well as many other formulas in that article, both proved and conjectural) are even functions of $n$.

In some situations, reciprocity can be seen as a special case of a more general phenomenon linking two enumeration problems that yield respective integer sequences $a_1, a_2, \ldots$ and $b_1, b_2, \ldots$; the two problems are "mutually reciprocal" if (upon defining $a_0, a_{-1}, \ldots$ and $b_0, b_{-1}, \ldots$ appropriately) one has $a_n = b_{c-n}$ and $b_n = a_{c-n}$ for some constant $c$.

In the case of the 2-by-$2n$ grid, and for other 1-dimensional matching problems, the beginnings of a satisfactory explanation for reciprocity can be found in [139] and [1]. However, for 2-dimensional tiling problems, nothing has been done. Since shuffling can be done in reverse, one might guess that the octahedron recurrence would provide the scaffolding for an explanation, via an extension of Kuo condensation to signed graphs like the ones considered by [1] (featuring various sorts of negative vertices and edges). Indeed, the rich octahedral symmetries of solutions to the octahedron recurrence call out for such a point of view, in which there would not be such a stark difference between "space" and "time," let alone between the positive and negative time-directions.

Although the introduction of signed graphs might seem like a conceptual extravagance, there is a very practical application they might have for the pattern-seeking combinatorialist. If one is looking for a recurrence relation that governs a one-sided sequence $a_1, a_2, \ldots$, and if one suspects that the unknown recurrence relation would actually be satisfied by a two-sided sequence $\ldots, a_{-2}, a_{-1}, a_0, a_1, a_2, \ldots$ that extends the original one-sided sequence, and if moreover one has a principled guess for what the numbers $a_0, a_{-1}, a_{-2}, \ldots$ are, then one could apply methods like the Somos matrix singularity test to a finite sequence of the form $a_{-r}, a_{-r+1}, \ldots, a_0, \ldots, a_{s-1}, a_s$. The success of the method hinges on $r+s$ being large, so one wants to take $r$ and $s$ individually as large as possible; that is, the data will guide us toward the correct recurrence (assuming there is one!) if we give it as many terms as we can, in both the forward and reverse directions.

## 9.8 Software

The page `http://jamespropp.org/software.html` gives links to a number of resources that are likely to be useful to researchers conducting experimental work on matchings of bipartite planar graphs and associated tilings.

The program `vaxmaple` (developed by Greg Kuperberg, David Wilson and myself) allows one to count the matchings of a bipartite planar graph using the method of

Kasteleyn. One inputs a graph in VAX-format (described in [137]) and the program outputs a Maple program which, if fed into Maple, returns the number of matchings of the graph.

David Wilson's `vaxmacs` package gives the user all the power of `vaxmaple` and more in an easy-to-use interactive environment based on the popular editor `emacs`. The documentation for this package is contained in the program itself. You will need `emacs-19.30` or later in order to get full use of `vaxmacs`.

Harald Helfgott's program `ren`, written in C, implements the generalized domino-shuffling algorithm of [140].

If you want to count matchings of some non-bipartite planar graph, Matt Blum's TCL-program `graph` lets you define an arbitrary planar graph, and Ben Wieland's perl-program `planemaple` (which `graph` "knows about") lets you count the matchings of this graph efficiently, exploiting Kasteleyn's method. The program `graph` will also let you count matchings of non-planar graphs, but it does so by the brute-force method of finding them all, which will not be practicable when the graph is large. Both programs require the user to make some modifications to the program header.

Several of the above programs come equipped with compatible side-programs for printing attractive pictures of tilings, with various configurable options. Some can even be used to compute conditional probabilities of sub-configurations in a random matching, or to generate a random matching.

In addition to the above programs, which are intended as research tools, there are also programs of a more pedagogical nature.

In the 1990s, Jason Woolever created a Java applet that is now available at `http://jamespropp.org/cftp/`, and that allows one to generate random tilings of a variety of regions by means of Coupling From The Past (or CFTP). CFTP is a variant of Monte Carlo methods, and gives an approach to bias-free sampling from various sets of combinatorial objects [142]. The speed of CFTP correlates directly with the speed at which a certain Markov chain mixes, namely, the chain in which one makes random local changes in the tiling by increasing or decreasing the height at a point. It may be noticed that CFTP is considerably slower for tilings of fortresses and dungeons than for the other cases that are presented. That is because random diabolo tilings of fortresses, unlike random domino tilings of Aztec diamonds or random lozenge tilings of hexagons, exhibit not only frozen domains and a liquid domain, but also a gaseous domain (in the terminology of [83]). In height-function terms, the height function is (up to small local fluctuations) flat throughout a large expanse of the tiling far away from the boundary, so that the graph of the height-function exhibits a broad plateau. This plateau can be at any of several heights, and it is very hard for the Markov chain to change the plateau-height from one of its likely possible values to another, since this requires changing the local height at $O(n^2)$ locations, and any motion in this direction is likely to be reversed almost immediately by a kind of surface tension (entropic rather than energetic in nature). Thus there are bottlenecks in the Markov chain, and mixing is (or at least appears to be) exponentially slow. CFTP is therefore not an efficient way of generating random diabolo tilings of fortresses; other methods (such as generalized domino-shuffling or the conditional probability method) must be used instead. The same is true for dungeons.

In 2009 Ben Wieland modified the code to generate random alternating-sign matrices; see `http://nokedli.net/asm-frozen/`. It should be mentioned that for the specific case of alternating-sign matrices of large order, CFTP is the only known efficient method of generating unbiased samples from the uniform distribution.

For tilings of Aztec diamonds (or TOADs, as Hal Canary proposed calling them) there is an attractive Java applet `jamespropp.org/toadshuffle/` created by Canary that uses the original Kuperberg-Propp [53] version of domino-shuffling (with stages of annihilation, sliding, and creation) to iteratively generate uniformly random tilings of ever-larger Aztec diamonds.

## 9.9 Frontiers

Here are some directions in which further work is needed.

1. It was mentioned above that ASMs and TSSCPPs are equinumerous; specifically, both families of combinatorial objects are enumerated by the sequence $1,1,2,7,42,429,\ldots$. This remains largely a mystery, as no bijective proof is known.

   In a similar but more tractable vein, it has been noticed that the sequence $1,1,2,8,64,1024,\ldots$ has arisen in several contexts [53] [46] [126] [114]. To the extent that all of these formulas have been given bijective proofs, one can provide bijections between the different models. However, some bijections are more natural than others, and one might hope to bring some precision to bear on the question of how similar, and how different, these various models are.

2. As I mentioned in Section 9.7.1, as a step towards understanding the class of ultra-smooth enumerations of matchings, it would be good to classify the triply periodic solutions of the octahedron recurrence. A related project is to better understand all doubly periodic bipartite planar graphs. For any $k$, one can hope to classify all such graphs in which the fundamental domain contains $k$ white vertices and $k$ black vertices, up to deformations that preserve the combinatorial structure.

3. Section 9.7.5 mentions the role of symmetry. In that passage, we thought about starting with a finite graph and looking at matchings that possess invariance under some symmetry of the graph. One may take a different point of view and look at cofinite (or quasitransitive) group-actions on a finite graph, where we say an action on a graph is cofinite if there are only finitely many vertex-orbits and edge-orbits. If the infinite graph is for instance the square grid, then quotient graphs include tori, Klein bottles, and cross-caps, as well as more arcane orbifold-type quotients. The case of the torus (that is, enumeration of

matchings of a grid on a torus) is classical [78] and the case of the Klein bottle was settled by Lu and Wu [115] along with related problems in which the regions are given boundaries (ribbon graphs, Möbius strips, etc.). However, the case of the cross-cap remains unsettled.

4. Kenyon, inspired by earlier work on resistor networks and Ising models embedded in the complex plane, studied the bipartite planar dimer model on isoradial graphs [81]. In that article Kenyon showed (partly answering the final question raised in [40]) that any periodic isoradial dimer model has entropy, which is the volume of a certain 3D ideal polytope. Li recently proved a conformal invariance theorem for this class of models [111]. The connection between dimer models and hyperbolic geometry is still a mystery.

5. We have seen a host of exact enumerative results coming from the dimer model on a variety of graphs, as well as a few exact enumerative results coming from the six-vertex model on a single graph, namely, the square grid. Might there be exact enumerative results for the six-vertex model on other 4-regular graphs, such as the Kagome (or star-of-David) lattice, or other statistical mechanics models of a similar kind, such as the twenty-vertex model on the 6-regular triangular lattice?
   We might try to move the theory of enumeration of tilings beyond the framework of dimer and ice-type models. An important step in this direction has been taken in the theory of square-triangle-rhombus tilings, mentioned at the end of Section 9.6.

6. It has been hard to get traction on the problem of counting monomer-dimer configurations in two (or more) dimensions, but other variants of dimers have been amenable to algebraic methods. Successful efforts in this direction include the double dimer model [85] and the monopole-dimer model [3]. Likewise, a variant of spanning trees that can be enumerated using a variant of the matrix tree theorem is cycle-rooted spanning forests [84].

   If one is content with asymptotic results (as opposed to exact enumerations), and to work in the regime where the number of monomers is much smaller than the number of dimers, progress can be made; indeed, a great deal of interesting work has been done, much of it by Ciucu [28, 30, 32, 33, 24]. A guiding theme here is that correlational behavior of the model, in the bulk limit, mimic features of electrostatics. See also [57] and [52].

7. Ribbon tilings (see Example 9.1.4 from Section 9.1) have not played much of a role in this chapter, but there has been some extremely interesting recent work on ribbon tilings of a special kind, namely "Dyck tilings." We will not define the term or delve into the subject here, but we point the reader toward [88] and [89] as articles in this rapidly growing literature, as well as the article [153] showing the relevance of Dyck tilings to the study of Kazhdan-Lusztig polynomials.

8. As was mentioned in Section 9.1, Pak, Yang, and others have shown that tiling problems in which the allowed tiles and the region being tiled are all rectangles can be #*P*-complete. Even as we try to expand the domain of efficiently solvable tiling problems, we should try to find more and more constrained classes of tiling problems that, despite their strictures, still can be shown to be #*P*-complete.

9. By taking the $q$-analogue of the formula $H(3n)H(n)^3/H(2n)^2$ for the number of rhombus tilings of a regular hexagon of side-length $n$ and sending $n$ to infinity, one recovers MacMahon's formula for the number of unconstrained plane partitions:
$$\prod_{n=1}^{\infty} \frac{1}{(1-q^n)^n}.$$
One can think of this formula as $q$-enumerating a particular infinite set of rhombus-tilings of the plane. Let $T_0$ be a tiling of the plane that divides the plane into 120-degree sectors, with no tiles crossing between one sector and another. Suppose $T$ is a tiling of the plane that agrees with $T_0$ outside some bounded region (call such a tiling cofinitely equivalent to $T_0$). Then, interpreting the two tilings as surfaces, we can define the rank of $T$ as the volume of the portion of space bounded between the two surfaces (normalized so that the volume of a cube is 1). MacMahon's formula enumerates the tilings $T$ by rank; that is, MacMahon's product is equal to $q^{\mathrm{rank}(T)}$, summed over all tilings $T$ cofinitely equivalent to $T_0$.

   A beautiful "domino-analogue" of this result is the formula
$$\prod_{n=1}^{\infty} \frac{(1+q^{2n-1})^{2n-1}}{(1-q^{2n})^{2n}}$$
conjectured independently by Szendrői [163] and Kenyon [82] and proved by Young [174, 175]. One might ask whether other formulas of this kind await discovery.

   Recent work of Bouttier, Chapuy, and Corteel [15] creates a unifying context for results of this sort and more standard results in the theory of enumeration of tilings. Especially noteworthy is the development of the theory of vertex operators, introduced in [175].

10. It is natural to try to extend the existing theory of enumeration of tilings into higher dimensions. However, this may not be possible. For instance, considering the fact that the double product
$$\prod_{i=1}^{a}\prod_{j=1}^{b} \frac{i+j}{i+j-1}$$
counts lattice paths in a rectangle, and the fact that the triple product
$$\prod_{i=1}^{a}\prod_{j=1}^{b}\prod_{k=1}^{c} \frac{i+j+k-1}{i+j+k-2}$$

counts plane partitions in a box (and their associated discrete spanning surfaces), one might hope that the quadruple product

$$\prod_{i=1}^{a}\prod_{j=1}^{b}\prod_{k=1}^{c}\prod_{l=1}^{d}\frac{i+j+k+l-2}{i+j+k+l-3}$$

might count some sort of solid partitions (associated with some sort of discrete 3-surfaces in 4-space). However, the quadruple product is not even a whole number when $a=b=c=d=2$! (For recent work on enumerating solid partitions, see [49].)

Nonetheless, one can get some exact enumerative results relating to discrete 2-surfaces in $n$-space. (This should not be too surprising, when one considers that it is easy to count discrete 1-surfaces in $n$-space with fixed boundary using multinomial coefficients.) The first result along these lines was Elnitsky's enumeration of lozenge tilings of a semiregular octagon with side-lengths $a,b,1,1,a,b,1,1$ [54]: the number of tilings is

$$2(a+b+1)!(a+b+2)!/a!b!(a+2)!(b+2)!$$

In the case of a semiregular octagon with side-lengths $a,1,b,1,a,1,b,1$, the enumerations have much larger prime factors, so no nice product formula is possible, but there is still a double sum formula found by Elnitsky; for more details, see [48]. It should be mentioned that for both of these two-parameter families of semiregular octagons, Stembridge's $q=-1$ phenomenon has been proved to occur; in the $a,b,1,1,a,b,1,1$ case the result is due to Elnitsky [54]: in the $a,1,b,1,a,1,b,1$ case it is due to Bailey [4] (Corollary 3.3 on page 25). See also [71].

One might also seek an exact formula for the number $T_n$ of lozenge tilings of a $2n$-gon with all side-lengths equal to 1, but this appears to be a hard problem. We know from [7] that $\lim(\log T_n)/n^2 < \frac{1}{2}$, but little is known about this sequence aside from asymptotics.

Jumping off from this example into a general consideration of the growth rates of sequences that arise in enumerative combinatorics, let us say that a combinatorial sequence $a_1, a_2, \ldots$ (along with the combinatorial problem that gave rise to it) is of grade $k$ if the logarithm of $a_n$ grows like $n^k$; that is, $(\log\log a_n)/\log n$ converges to $k$. (For problems with more than one parameter, assume that all parameters are of the same order as $n$.) Combinatorics of grade 0 concerns sequences like the sequence of triangle numbers $1,3,6,10,\ldots$. Combinatorics of grade 1 concerns sequences like the Fibonacci and Catalan sequences; Examples 9.1.1 through 9.1.4 of Section 9.1 are all at grade 1. Combinatorics of grade 2 largely concerns genuinely two-dimensional tiling problems, many of which (like rhombus tilings of the $a,b,c$ hexagon) can be recast as problems involving discrete 2-surfaces in three dimensions; more generally, rhombus tilings of a semiregular polygon with $2n$ pairs of parallel sides can be recast as discrete 2-surfaces in $n$ dimensions.

In the past century, combinatorics moved from grade 1 to grade 2. However, we seem to be stuck there. Perhaps there is no such subject as grade 3 combinatorics (and perhaps, relatedly, there are no non-trivial exactly solvable 3-dimensional lattice models). Or perhaps there is such a subject, which we will begin to glimpse once we have gotten through more of the grade 2 curriculum and the teacher deems us ready for higher things.

**Acknowledgments.** This chapter would not have been possible without the assistance and encouragement of dozens of people, especially Federico Ardila, Arvind Ayyer, Art Benjamin, Jin-Yi Cai, Clara Chan, Sunil Chhita, Timothy Chow, Mihai Ciucu, Henry Cohn, Filippo Colomo, David desJardins, Nicolas Destainville, Julien Dubedat, Noam Elkies, Patrik Ferrari, Philippe di Francesco, David Gamarnik, Ira Gessel, Jan de Gier, Vadim Gorin, Tony Guttmann, Rick Kenyon, Christian Krattenthaler, Eric Kuo, Greg Kuperberg, Tri Lai, Lionel Levine, Russell Lyons, John Mangual, Barry McCoy, Roderich Moessner, Cris Moore, Igor Pak, Kyle Petersen, Dana Randall, Vic Reiner, Dan Romik, Bruce Sagan, Nicolau Saldanha, Robert Shrock, Richard Stanley, John Stembridge, Jessica Striker, Hugh Thomas, Dylan Thurston, Carlos Tomei, David Wilson, Fred Wu, and Benjamin Young. Thanks also to my wife Sandi Gubin for her encouragement and patience.

## References

[1] Nick Anzalone, John Baldwin, Ilya Bronshtein, and T. Kyle Petersen. A reciprocity theorem for monomer-dimer coverings. In *Discrete Models for Complex Systems, DMCS '03 (Lyon)*, Discrete Math. Theor. Comput. Sci. Proc., AB, pages 179–193. Assoc. Discrete Math. Theor. Comput. Sci., Nancy, 2003.

[2] Federico Ardila and Richard P. Stanley. Tilings. *Math. Intelligencer*, 32(4):32–43, 2010.

[3] A. Ayyer. A statistical model of current loops and magnetic monopoles. arXiv.org:1311.5965, 2013.

[4] Guy David Bailey. *Tilings of zonotopes: Discriminantal arrangements, oriented matroids, and enumeration.* ProQuest LLC, Ann Arbor, MI, 1997. Thesis (Ph.D.)–University of Minnesota, `www-users.math.umn.edu/~reiner/Bailey/paper.ps`.

[5] Rodney J. Baxter. *Exactly solved models in statistical mechanics.* Academic Press, Inc. [Harcourt Brace Jovanovich, Publishers], London, 1989. Reprint of the 1982 original.

[6] Danièle Beauquier, Maurice Nivat, Éric Remila, and Mike Robson. Tiling figures of the plane with two bars. *Comput. Geom.*, 5(1):1–25, 1995.

[7] Anders Björner, Michel Las Vergnas, Bernd Sturmfels, Neil White, and Günter M. Ziegler. *Oriented matroids*, volume 46 of *Encyclopedia of Mathematics and its Applications*. Cambridge University Press, Cambridge, second edition, 1999.

[8] H. W. J. Blöte and H. J. Hilhorst. Roughening transitions and the zero-temperature triangular Ising antiferromagnet. *J. Phys. A*, 15:L631–L637, 1982.

[9] M. O. Blunt, J. C. Russell, M. del C. Giménez-López, J. P. Garrahan, X. Lin, M. Schröder, N. R. Champness, and P. H. Beton. Random tiling and topological defects in a two-dimensional molecular network. *Science*, 322:1077–1081, 2008.

[10] Béla Bollobás. *Modern graph theory*, volume 184 of *Graduate Texts in Mathematics*. Springer-Verlag, New York, 1998.

[11] Alexei Borodin and Vadim Gorin. Lectures on integrable probability. arXiv:1212.3351, 2012.

[12] Alexei Borodin, Vadim Gorin, and Eric M. Rains. $q$-distributions on boxed plane partitions. *Selecta Math. (N.S.)*, 16(4):731–789, 2010.

[13] Frédéric Bosio and Marc A. A. van Leeuwen. A bijection proving the Aztec diamond theorem by combing lattice paths. *Electron. J. Combin.*, 20(4):Paper 24, 30, 2013.

[14] Mireille Bousquet-Mélou, James Propp, and Julian West. Perfect matchings for the three-term Gale-Robinson sequences. *Electron. J. Combin.*, 16(1):Research Paper 125, 37, 2009.

[15] Jérémie Bouttier, Guillaume Chapuy, and Sylvie Corteel. From Aztec diamonds to pyramids: steep tilings. arXiv:1407.0665, 2014.

[16] David Bressoud and James Propp. How the alternating sign matrix conjecture was solved. *Notices Amer. Math. Soc.*, 46(6):637–646, 1999.

[17] Richard A. Brualdi and Stephen Kirkland. Aztec diamonds and digraphs, and Hankel determinants of Schröder numbers. *J. Combin. Theory Ser. B*, 94(2):334–351, 2005.

[18] Jin-Yi Cai and Aaron Gorenstein. Matchgates revisited. *Theory of Computing*, 10(7):167–197, 2014.

[19] L. Carlitz and Richard P. Stanley. Branchings and partitions. *Proc. Amer. Math. Soc.*, 53(1):246–249, 1975.

[20] Gabriel D. Carroll and David Speyer. The cube recurrence. *Electron. J. Combin.*, 11(1):Research Paper 73, 31 pp. (electronic), 2004.

[21] S.-C. Chang and R. Shrock. Some exact results on spanning trees on lattices. *J. Phys. A*, 39:5653–5658, 2006.

[22] Sunil Chhita and Benjamin Young. Coupling functions for domino tilings of Aztec diamonds. *Adv. Math.*, 259:173–251, 2014.

[23] Mihai Ciucu. A generalization of Kuo condensation. arXiv:1404.5003, 2014.

[24] Mihai Ciucu. Lozenge tilings with gaps in a 90° wedge domain with mixed boundary conditions. arXiv:1403.0925, 2014; in press at *Comm. Math. Phys.*

[25] Mihai Ciucu. Enumeration of perfect matchings in graphs with reflective symmetry. *J. Combin. Theory Ser. A*, 77(1):67–97, 1997.

[26] Mihai Ciucu. A complementation theorem for perfect matchings of graphs having a cellular completion. *J. Combin. Theory Ser. A*, 81(1):34–68, 1998.

[27] Mihai Ciucu. Perfect matchings and perfect powers. *J. Algebraic Combin.*, 17(3):335–375, 2003.

[28] Mihai Ciucu. A random tiling model for two dimensional electrostatics. *Mem. Amer. Math. Soc.*, 178(839):x+144, 2005.

[29] Mihai Ciucu. A visual proof of a result of Knuth on spanning trees of Aztec diamonds in the case of odd order. *Discrete Math.*, 307(15):1957–1960, 2007.

[30] Mihai Ciucu. Dimer packings with gaps and electrostatics. *Proc. Natl. Acad. Sci. USA*, 105(8):2766–2772, 2008.

[31] Mihai Ciucu. Symmetry classes of spanning trees of Aztec diamonds and perfect matchings of odd squares with a unit hole. *J. Algebraic Combin.*, 27(4):493–538, 2008.

[32] Mihai Ciucu. The scaling limit of the correlation of holes on the triangular lattice with periodic boundary conditions. *Mem. Amer. Math. Soc.*, 199(935):x+100, 2009.

[33] Mihai Ciucu. The Interaction of Collinear Gaps of Arbitrary Charge in a Two Dimensional Dimer System. *Comm. Math. Phys.*, 330(3):1115–1153, 2014.

[34] Mihai Ciucu and Ilse Fischer. Proof of two conjectures of Ciucu and Krattenthaler on the enumeration of lozenge tilings of hexagons with cut off corners. arXiv:1309.4640, 2013.

[35] Mihai Ciucu and Ilse Fischer. A triangular gap of side 2 in a sea of dimers in a 60° angle. *J. Phys. A*, 45(49):494011, 17, 2012.

[36] Mihai Ciucu and Christian Krattenthaler. A dual of MacMahon's theorem on plane partitions. *Proc. Natl. Acad. Sci. USA*, 110(12):4518–4523, 2013.

[37] Mihai Ciucu and Tri Lai. Proof of Blum's conjecture on hexagonal dungeons. *J. Combin. Theory Ser. A*, 125:273–305, 2014.

[38] Henry Cohn. 2-adic behavior of numbers of domino tilings. *Electron. J. Combin.*, 6:Research Paper 14, 7 pp. (electronic), 1999.

[39] Henry Cohn, Noam Elkies, and James Propp. Local statistics for random domino tilings of the Aztec diamond. *Duke Math. J.*, 85(1):117–166, 1996.

[40] Henry Cohn, Richard Kenyon, and James Propp. A variational principle for domino tilings. *J. Amer. Math. Soc.*, 14(2):297–346 (electronic), 2001.

[41] Henry Cohn, Michael Larsen, and James Propp. The shape of a typical boxed plane partition. *New York J. Math.*, 4:137–165 (electronic), 1998.

[42] F. Colomo and A. G. Pronko. The arctic curve of the domain-wall six-vertex model. *J. Stat. Phys.*, 138:662, 2010.

[43] F. Colomo and A. G. Pronko. The limit shape of large alternating sign matrices. *SIAM J. Discrete Math.*, 24(4):1558–1571, 2010.

[44] F. Colomo, A. G. Pronko, and P. Zinn-Justin. The arctic curve of the domain-wall six-vertex model in the anti-ferroelectric regime. *J. Stat. Mech.: Theor. Exp.*, page L03002, 2010.

[45] J. H. Conway and J. C. Lagarias. Tiling with polyominoes and combinatorial group theory. *J. Combin. Theory Ser. A*, 53(2):183–208, 1990.

[46] Cyndie Cottrell and Benjamin Young. Domino shuffling for the Del Pezzo 3 lattice. arXiv:1011.0045, 2010.

[47] S. J. Cyvin and I. Gutman. *Kekule Structures in Benzenoid Hydrocarbons*, volume 46 of *Lecture Notes in Chemistry*. Springer-Verlag, Berlin, 1988.

[48] N. Destainville, R. Mosseri, and F. Bailly. A formula for the number of tilings of an octagon by rhombi. *Theoret. Comput. Sci.*, 319(1-3):71–81, 2004.

[49] Nicolas Destainville and Suresh Govindarajan. Estimating the asymptotics of solid partitions. arXiv:1406.5605, 2014.

[50] Philippe Di Francesco. T-systems, networks and dimers. *Comm. Math. Phys.*, to appear, 2014. `http://xxx.lanl.gov/abs/1307.0095`.

[51] Tomislav Došlić. Perfect matchings in lattice animals and lattice paths with constraints. *Notices Amer. Math. Soc.*, 78(2):251–259, 2005.

[52] Julien Dubédat. Dimers and analytic torsion I. arXiv:1110.2808, 2014.

[53] Noam Elkies, Greg Kuperberg, Michael Larsen, and James Propp. Alternating-sign matrices and domino tilings. *J. Algebraic Combin.*, 1:111–132, 219–234, 1992.

[54] Serge Elnitsky. Rhombic tilings of polygons and classes of reduced words in Coxeter groups. *J. Combin. Theory Ser. A*, 77(2):193–221, 1997.

[55] Sen-Peng Eu and Tung-Shan Fu. A simple proof of the Aztec diamond theorem. *Electron. J. Combin.*, 12:Research Paper 18, 8 pp. (electronic), 2005.

[56] Chungpeng Fan and F. Y. Wu. General lattice model of phase transitions. *Phys. Rev. B*, 2:723–733, Aug 1970.

[57] P. Fendley, R. Moessner, and S. L. Sondhi. Classical dimers on the triangular lattice. *Phys. Rev. B*, 66:214513, Dec 2002.

[58] Patrik Ferrari. Dimers and orthogonal polynomials: connections with random matrices. arXiv:1004.3212, 2013.

[59] Patrik L. Ferrari, Michael Prähofer, and Herbert Spohn. Fluctuations of an atomic ledge bordering a crystalline facet. *Phys. Rev. E*, 69:035102, Mar 2004.

[60] Laura Florescu, Daniela Morar, David Perkinson, Nick Salter, and Tianyuan Xu. Sandpiles and dominos. arXiv:1406.0100, 2014.

[61] R. H. Fowler and G. S. Rushbrooke. An attempt to expand the statistical theory of perfect solutions. *Trans. Faraday Soc.*, 33:1272–1294, 1937.

[62] Markus Fulmek. Graphical condensation, overlapping Pfaffians and superpositions of matchings. *Electron. J. Combin.*, 17(1):Research Paper 83, 42, 2010.

[63] David Gale. Somos sequence update. *Math. Intelligencer*, 13(4):49–50, 1991.

[64] David Gale. The strange and surprising saga of the Somos sequences. *Math. Intelligencer*, 13(1):40–42, 1991.

[65] Ira Gessel and Gérard Viennot. Binomial determinants, paths, and hook length formulae. *Adv. in Math.*, 58(3):300–321, 1985.

[66] Alexander B. Goncharov and Richard Kenyon. Dimers and cluster integrable systems. *Ann. Sci. Éc. Norm. Supér. (4)*, 46(5):747–813, 2013.

[67] M. Gordon and W. H. T. Davison. Theory of resonance topology of fully aromatic hydrocarbons. I. *J. of Chem. Phys.*, 20:428–435, 1952.

[68] D. Grensing, I. Carlsen, and H.-Chr. Zapp. Some exact results for the dimer problem on plane lattices with non-standard boundaries. *Phil. Mag. A*, 41:777–781, 1980.

[69] Branko Grünbaum and G. C. Shephard. *Tilings and patterns*. W. H. Freeman and Company, New York, 1987.

[70] Heng Guo and Tyson Williams. The complexity of planar Boolean #CSP with complex weights. In *Automata, Languages, and Programming. Part I*, volume 7965 of *Lecture Notes in Comput. Sci.*, pages 516–527. Springer, Heidelberg, 2013.

[71] M. Hutchinson and M. Widom. Enumeration of octagonal tilings. arXiv:1306.5977, 2013.

[72] Kyung-Won Hwang, Naeem N. Sheikh, and Stephen G. Hartke. A note on divisibility of the number of matchings of a family of graphs. *Electron. J. Combin.*, 16(1):Note 10, 4, 2009.

[73] A. G. Izergin. Partition function of a six-vertex model in a finite volume. *Soviet Phys. Dokl.*, 32:878–879, 1987.

[74] É. Janvresse, T. de la Rue, and Y. Velenik. A note on domino shuffling. *Electron. J. Combin.*, 13(1):Research Paper 30, 19 pp. (electronic), 2006.

[75] William Jockusch. Perfect matchings and perfect squares. *J. Combin. Theory Ser. A*, 67(1):100–115, 1994.

[76] Kurt Johansson. Random matrices and determinantal processes. In *Mathematical Statistical Physics*, pages 1–55. Elsevier B. V., Amsterdam, 2006.

[77] P. John and H. Sachs. Wegesysteme und Linearfaktoren in hexagonalen und quadratischen Systemen (Path systems and linear factors in hexagonal and square systems). In *Graphen in Forschung und Unterricht (Festschrift K. Wagner)*, pages 85–100. Barbara Franzbecker Verlag Bad Salzdetfurth, 1985.

[78] P. W. Kasteleyn. The statistics of dimers on a lattice I: The number of dimer arrangements on a quadratic lattice. *Physica*, 27:1209–1225, 1961.

[79] P. W. Kasteleyn. Dimer statistics and phase transitions. *J. Math. Phys.*, 4:287–293, 1963.

[80] P. W. Kasteleyn. Graph theory and crystal physics. In *Graph Theory and Theoretical Physics*, pages 43–110. Academic Press, London, 1967.

[81] R. Kenyon. The Laplacian and Dirac operators on critical planar graphs. *Invent. Math.*, 150(2):409–439, 2002.

[82] Richard Kenyon. Talk given at the workshop on Random partitions and Calabi-Yau crystals, Amsterdam, 2005. Slides available at `http://www.math.brown.edu/~rkenyon/talks/pyramids.pdf`, 2005.

[83] Richard Kenyon. Lectures on dimers. In *Statistical Mechanics*, volume 16 of *IAS/Park City Math. Ser.*, pages 191–230. Amer. Math. Soc., Providence, RI, 2009.

[84] Richard Kenyon. Spanning forests and the vector bundle Laplacian. *Ann. Probab.*, 39(5):1983–2017, 2011.

[85] Richard Kenyon. Conformal invariance of loops in the double-dimer model. *Comm. Math. Phys.*, 326(2):477–497, 2014.

[86] Richard W. Kenyon, James G. Propp, and David B. Wilson. Trees and matchings. *Electron. J. Combin.*, 7:Research Paper 25, 34 pp. (electronic), 2000.

[87] Richard W. Kenyon and Scott Sheffield. Dimers, tilings and trees. *J. Combin. Theory Ser. B*, 92(2):295–317, 2004.

[88] Richard W. Kenyon and David B. Wilson. Double-dimer pairings and skew Young diagrams. *Electron. J. Combin.*, 18(1):Paper 130, 22, 2011.

[89] Jang Soo Kim, Karola Mészáros, Greta Panova, and David B. Wilson. Dyck tilings, increasing trees, descents, and inversions. *J. Combin. Theory Ser. A*, 122:9–27, 2014.

[90] K. Kokhas′. Domino tilings of Aztec diamonds and squares. *Zap. Nauchn. Sem. S.-Peterburg. Otdel. Mat. Inst. Steklov. (POMI)*, 360(Teoriya Predstavlenii, Dinamicheskie Sitemy, Kombinatornye Metody. XVI):180–230, 298, 2008.

[91] V. E. Korepin, N. M. Bogoliubov, and A. G. Izergin. *Quantum inverse scattering method and correlation functions*. Cambridge Monographs on Mathematical Physics. Cambridge University Press, Cambridge, England, 1993.

[92] Christoph Koutschan, Manuel Kauers, and Doron Zeilberger. Proof of George Andrews's and David Robbins's $q$-TSPP conjecture. *Proc. Natl. Acad. Sci. USA*, 108(6):2196–2199, 2011.

[93] C. Krattenthaler. Advanced determinant calculus. *Sém. Lothar. Combin.*, 42:Art. B42q, 67 pp. (electronic), 1999. The Andrews Festschrift (Maratea, 1998).

[94] C. Krattenthaler. Another involution principle-free bijective proof of Stanley's hook-content formula. *J. Combin. Theory Ser. A*, 88(1):66–92, 1999.

[95] C. Krattenthaler. Advanced determinant calculus: a complement. *Linear Algebra Appl.*, 411:68–166, 2005.

[96] Christian Krattenthaler. Watermelon configurations with wall interaction: exact and asymptotic results. *J. Physics Conf. Series*, 42:179–212, 2006.

[97] Atsuo Kuniba, Tomoki Nakanishi, and Junji Suzuki. $T$-systems and $Y$-systems in integrable systems. *J. Phys. A*, 44(10):103001, 146, 2011.

[98] Eric Kuo. Graphical condensation generalizations involving Pfaffians and determinants. arXiv:math/0605154, 2006.

[99] Eric Kuo. Applications of graphical condensation for enumerating matchings and tilings. *Theoret. Comput. Sci.*, 319(1-3):29–57, 2004.

[100] Greg Kuperberg. Self-complementary plane partitions by Proctor's minuscule method. *European J. Combin.*, 15(6):545–553, 1994.

[101] Greg Kuperberg. Symmetries of plane partitions and the permanent-determinant method. *J. Combin. Theory Ser. A*, 68(1):115–151, 1994.

[102] Greg Kuperberg. Another proof of the alternating-sign matrix conjecture. *Internat. Math. Res. Notices*, (3):139–150, 1996.

[103] Greg Kuperberg. Four symmetry classes of plane partitions under one roof. *J. Combin. Theory Ser. A*, 75(2):295–315, 1996.

[104] Greg Kuperberg. An exploration of the permanent-determinant method. *Electron. J. Combin.*, 5:Research Paper 46, 34 pp. (electronic), 1998.

[105] Greg Kuperberg. Kasteleyn cokernels. *Electron. J. Combin.*, 9(1):Research Paper 29, 30, 2002.

[106] Greg Kuperberg. Symmetry classes of alternating-sign matrices under one roof. *Ann. of Math. (2)*, 156(3):835–866, 2002.

[107] J. C. Lagarias and D. S. Romano. A polyomino tiling problem of Thurston and its configurational entropy. *J. Combin. Theory Ser. A*, 63(2):338–358, 1993.

[108] Tri Lai. New aspects of hexagonal dungeons. arXiv:1403.4481, 2014.

[109] Tri Lai. New aspects of regions whose tilings are enumerated by perfect powers. *Electron. J. Combin.*, 20(4):Paper 31, 47, 2013.

[110] L. S. Levitov. Equivalence of the dimer resonating-valence-bond problem to the quantum roughening problem. *Phys. Rev. Lett.*, 64:92–94, Jan 1990.

[111] Zhongyang Li. Conformal invariance of isoradial dimers. arXiv:1309.0151, 2013.

[112] Elliott H. Lieb. Residual entropy of square ice. *Phys. Rev.*, 162:162–172, Oct 1967.

[113] Bernt Lindström. On the vector representations of induced matroids. *Bull. London Math. Soc.*, 5:85–90, 1973.

[114] Fu Liu and Richard P. Stanley. A distributive lattice connected with arithmetic progressions of length three. arXiv:1103.5054, 2013.

[115] W. T. Lu and F. Y. Wu. Close-packed dimers on nonorientable surfaces. *Physics Letters A*, 293(5):235–246, 2002. Earlier preprint version at `http://arxiv.org/pdf/cond-mat/0110035.pdf`. An erratum is posted at `http://www.sciencedirect.com.libproxy.uml.edu/science/article/pii/S0375960102005182`.

[116] Michael Luby, Dana Randall, and Alistair Sinclair. Markov chain algorithms for planar lattice structures. *SIAM J. Comput.*, 31(1):167–192, 2001.

[117] Russell Lyons. Asymptotic enumeration of spanning trees. *Combin. Probab. Comput.*, 14(4):491–522, 2005.

[118] Russell Lyons. Identities and inequalities for tree entropy. *Combin. Probab. Comput.*, 19(2):303–313, 2010.

[119] Percy A. MacMahon. *Combinatory analysis. Vol. I, II (bound in one volume)*. Dover Phoenix Editions. Dover Publications, Inc., Mineola, NY, 2004. Reprint of An introduction to combinatory analysis (1920) and Combinatory analysis Vol. I, II (1915, 1916).

[120] Janice L. Malouf. An integer sequence from a rational recursion. *Discrete Math.*, 110(1-3):257–261, 1992.

[121] B. McCoy and T. T. Wu. *The two-dimensional Ising model*. Dover, 2nd edition, 2014.

[122] W. H. Mills, David P. Robbins, and Howard Rumsey, Jr. Alternating sign matrices and descending plane partitions. *J. Combin. Theory Ser. A*, 34(3):340–359, 1983.

[123] R. Moessner and S. L. Sondhi. On Ising and dimer models in two and three dimensions. *Phys. Rev. B*, 68:054405, Aug 2003.

[124] C. Moore and J. M. Robson. Hard tiling problems with simple tiles. *Discrete Comput. Geom.*, 26(4):573–590, 2001.

[125] Cristopher Moore and Stephan Mertens. *The nature of computation*. Oxford University Press, Oxford, 2011.

[126] Eric Nordenstam and Benjamin Young. Domino shuffling on Novak half-hexagons and Aztec half-diamonds. *Electron. J. Combin.*, 18(1):Paper 181, 22, 2011.

[127] Soichi Okada. Enumeration of symmetry classes of alternating sign matrices and characters of classical groups. *J. Algebraic Combin.*, 23(1):43–69, 2006.

[128] Lars Onsager. Crystal statistics. I. a two-dimensional model with an order-disorder transition. *Phys. Rev.*, 65:117–149, Feb 1944.

[129] Igor Pak and Jed Yang. The complexity of generalized domino tilings. *Electron. J. Combin.*, 20(4):Paper 12, 23, 2013.

[130] Igor Pak and Jed Yang. Tiling simply connected regions with rectangles. *J. Combin. Theory Ser. A*, 120(7):1804–1816, 2013.

[131] Jerome K. Percus. One more technique for the dimer problem. *J. Mathematical Phys.*, 10:1881–1888, 1969.

[132] Robert A. Proctor. Bruhat lattices, plane partition generating functions, and minuscule representations. *European J. Combin.*, 5(4):331–350, 1984.

[133] James Propp. Dimers and dominoes. arXiv:1405.2615, 2014.

[134] James Propp. Lattice structure for orientations of graphs. arXiv:math/0209005, 2002.

[135] James Propp. Lessons I learned from Richard Stanley. arXiv:1501.00719, 2014.

[136] James Propp. A new proof of a formula of Richard Stanley. *Poster presented at 1997 AMS annual meeting*, 1997. `http://jamespropp.org/san_diego.pdf`.

[137] James Propp. Enumeration of matchings: problems and progress. In *New Perspectives in Algebraic Combinatorics (Berkeley, CA, 1996–97)*, volume 38 of *Math. Sci. Res. Inst. Publ.*, pages 255–291. Cambridge Univ. Press, Cambridge, 1999. Updated version, dated 2015 and retitled "Problems and progress in enumeration of matchings", at `http://arxiv.org/abs/math/9904150`.

[138] James Propp. The many faces of alternating-sign matrices. In *Discrete Models: Combinatorics, Computation, and Geometry (Paris, 2001)*, Discrete Math. Theor. Comput. Sci. Proc., AA, pages 043–058 (electronic). Maison Inform. Math. Discrèt. (MIMD), Paris, 2001.

[139] James Propp. A reciprocity theorem for domino tilings. *Electron. J. Combin.*, 8(1):Research Paper 18, 9 pp. (electronic), 2001.

[140] James Propp. Generalized domino-shuffling. *Theoret. Comput. Sci.*, 303(2):267–301, 2003.

[141] James Propp. Lambda-determinants and domino-tilings. *Adv. in Appl. Math.*, 34(4):871–879, 2005.

[142] James Gary Propp and David Bruce Wilson. Exact sampling with coupled Markov chains and applications to statistical mechanics. In *Proceedings of the Seventh International Conference on Random Structures and Algorithms (Atlanta, GA, 1995)*, volume 9, pages 223–252, 1996.

[143] Kevin Purbhoo. Puzzles, tableaux, and mosaics. *J. Algebraic Combin.*, 28(4):461–480, 2008.

[144] A. V. Razumov and Yu. G. Stroganov. On refined enumerations of some symmetry classes of alternating-sign matrices. *Theor. Math. Phys.*, 141:1609–1630, 2004.

[145] A. V. Razumov and Yu. G. Stroganov. Enumeration of half-turn-symmetric alternating-sign matrices of odd order. *Theor. Math. Phys.*, 148:1174–1198, 2006.

[146] A. V. Razumov and Yu. G. Stroganov. Enumeration of quarter-turn symmetric alternating-sign matrices of odd order. *Theor. Math. Phys.*, 149:1639–1650, 2006.

[147] V. Reiner, D. Stanton, and D. White. The cyclic sieving phenomenon. *J. Combin. Theory Ser. A*, 108(1):17–50, 2004.

[148] Victor Reiner, Dennis Stanton, and Dennis White. What is . . . cyclic sieving? *Notices Amer. Math. Soc.*, 61(2):169–171, 2014.

[149] David P. Robbins and Howard Rumsey, Jr. Determinants and alternating sign matrices. *Adv. in Math.*, 62(2):169–184, 1986.

[150] Dan Romik. Connectivity patterns in loop percolation. arXiv:1303.6341, 2013.

[151] H. Sachs. Counting perfect matchings in lattice graphs. In *Topics in Combinatorics and Graph Theory (Oberwolfach, 1990)*, pages 577–584. Physica, Heidelberg, 1990.

[152] Bruce E. Sagan. The cyclic sieving phenomenon: a survey. In *Surveys in Combinatorics 2011*, volume 392 of *London Math. Soc. Lecture Note Ser.*, pages 183–233. Cambridge Univ. Press, Cambridge, 2011.

[153] Keiichi Shigechi and Paul Zinn-Justin. Path representation of maximal parabolic Kazhdan-Lusztig polynomials. *J. Pure Appl. Algebra*, 216(11):2533–2548, 2012.

[154] R. Shrock and F. Y. Wu. Spanning trees on graphs and lattices in $d$ dimensions. *Journal of Physics A*, 33:3881–3902, 2000.

[155] David E. Speyer. Perfect matchings and the octahedron recurrence. *J. Algebraic Combin.*, 25(3):309–348, 2007.

[156] Richard P. Stanley. Symmetries of plane partitions. *J. Combin. Theory Ser. A*, 43(1):103–113, 1986. with erratum in [157].

[157] Richard P. Stanley. Erratum: "Symmetries of plane partitions." *J. Combin. Theory Ser. A*, 44(2):310, 1987.

[158] Richard P. Stanley. *Enumerative Combinatorics. Volume 2*, volume 62 of *Cambridge Studies in Advanced Mathematics*. Cambridge University Press, Cambridge, 1999. With a foreword by Gian-Carlo Rota and appendix 1 by Sergey Fomin.

[159] Richard P. Stanley. *Enumerative Combinatorics. Volume 1*, volume 49 of *Cambridge Studies in Advanced Mathematics*. Cambridge University Press, Cambridge, second edition, 2012.

[160] John R. Stembridge. On minuscule representations, plane partitions and involutions in complex Lie groups. *Duke Math. J.*, 73(2):469–490, 1994.

[161] John R. Stembridge. Some hidden relations involving the ten symmetry classes of plane partitions. *J. Combin. Theory Ser. A*, 68(2):372–409, 1994.

[162] Volker Strehl. Counting domino tilings of rectangles via resultants. *Adv. in Appl. Math.*, 27(2-3):597–626, 2001. Special issue in honor of Dominique Foata's 65th birthday (Philadelphia, PA, 2000).

[163] Balázs Szendrői. Non-commutative Donaldson-Thomas invariants and the conifold. *Geom. Topol.*, 12(2):1171–1202, 2008.

[164] H. N. V. Temperley. Unnamed problem in problems section. In *Combinatorics: The Proceedings of the British Combinatorial Conference*, volume 13 of *London Mathematical Society Lecture Notes Series*, pages 202–204. University Press, 1974.

[165] H. N. V. Temperley and Michael E. Fisher. Dimer problem in statistical mechanics—an exact result. *Philos. Mag. (8)*, 6:1061–1063, 1961.

[166] William P. Thurston. Conway's tiling groups. *Amer. Math. Monthly*, 97(8):757–773, 1990.

[167] Forest Tong. On the number of perfect matchings of compound graphs. arXiv:1406.0669, 2014.

[168] L. G. Valiant. The complexity of computing the permanent. *Theoret. Comput. Sci.*, 8(2):189–201, 1979.

[169] D. J. A. Welsh. *Complexity: knots, colourings and counting*, volume 186 of *London Mathematical Society Lecture Note Series*. Cambridge University Press, Cambridge, 1993.

[170] David Bruce Wilson. Determinant algorithms for random planar structures. In *Proceedings of the Eighth Annual ACM-SIAM Symposium on Discrete Algorithms (New Orleans, LA, 1997)*, pages 258–267. ACM, New York, 1997.

[171] Weigen Yan, Yeong-Nan Yeh, and Fuji Zhang. Graphical condensation of plane graphs: a combinatorial approach. *Theoret. Comput. Sci.*, 349(3):452–461, 2005.

[172] Weigen Yan and Fuji Zhang. Graphical condensation for enumerating perfect matchings. *J. Combin. Theory Ser. A*, 110(1):113–125, 2005.

[173] Bo-Yin Yang. *Two enumeration problems about the Aztec diamonds*. ProQuest LLC, Ann Arbor, MI, 1991. Thesis (Ph.D.)–Massachusetts Institute of Technology.

[174] Ben Young. Computing a pyramid partition generating function with dimer shuffling. *J. Combin. Theory Ser. A*, 116(2):334–350, 2009.

[175] Benjamin Young. Generating functions for colored 3D Young diagrams and the Donaldson-Thomas invariants of orbifolds (with an appendix by Jim Bryan). *Duke Math. J.*, 152(1):115–153, 2010.

[176] Doron Zeilberger. Proof of the alternating sign matrix conjecture. *Electron. J. Combin.*, 3(2):Research Paper 13, approx. 84 pp. (electronic), 1996. The Foata Festschrift.

[177] Paul Zinn-Justin. Littlewood-Richardson coefficients and integrable tilings. *Electron. J. Combin.*, 16(1):Research Paper 12, 33, 2009.

# Chapter 10

## *Lattice Path Enumeration*

**Christian Krattenthaler**
*Universität Wien*

**CONTENTS**



## 10.1 Introduction

A **lattice path (path** for short) is what the name says: a path (walk) in a lattice in some $d$-dimensional Euclidean space. Formally, a lattice path $P$ is a sequence $P = (P_0, P_1, \ldots, P_l)$ of points $P_i$ in $\mathbb{Z}^d$. Figure 10.1 shows the lattice path $((0,0),(1,1),$

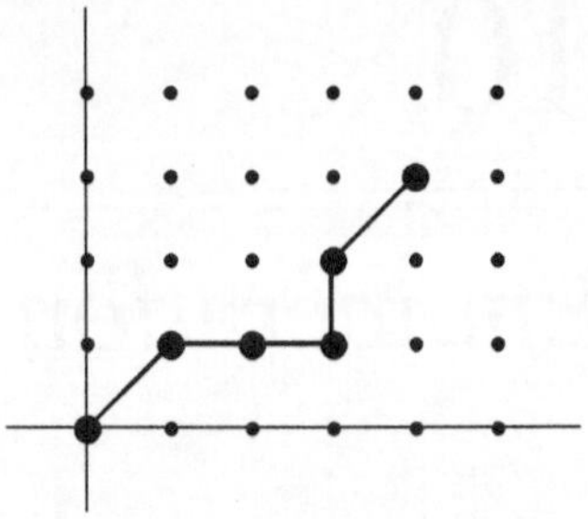

**Figure 10.1**
A lattice path.

$(2,1),(3,1),(3,2),(4,3))$. The point $P_0$ is called the **starting point** and $P_l$ is called the **end point** of $P$. The vectors $\overrightarrow{P_0P_1}, \overrightarrow{P_1P_2}, \dots, \overrightarrow{P_{l-1}P_l}$ are called the **steps** of $P$.

Lattice paths have been studied for a very long time, explicitly at least since the second half of the 19th century. At the beginning stand the investigations concerning the two-candidate ballot problem [8, 123] (see the paragraph below Corollary 10.3.2 in Section 10.3) and the gambler's ruin problem [65] (see [38, Ch. XIV, Sec. 2] and Example 10.11.3 in Section 10.11). Since then, lattice paths have penetrated many fields of mathematics, computer science, and physics. The reason for their ubiquity is, on the one hand, that they are well-suited to encode various (combinatorial) objects and their properties and, thus, problems in various fields can be solved by solving lattice path problems. On the other hand, since lattice paths are, at the outset, reasonably simple combinatorial objects, the study of physical, probabilistic, or statistical models is attractive in its own right. In particular, the importance of lattice path enumeration in non-parametric statistics seems to explain that the only books that are entirely devoted to lattice path combinatorics that I am aware of, namely [95] and [97], are written by statisticians.

The aim of this chapter is to provide an overview of results and methods in lattice path enumeration. Since, in view of the vast literature on the subject, comprehensiveness is hopeless, I have made a personal selection of topics that I consider of importance in the theory, the same applying to the methods I present here.

Clearly, when one talks of "enumeration," this comes in two different "flavors": exact and asymptotic. In this chapter, I only rarely touch asymptotics, but rather concentrate on exact enumeration results. In most cases, corresponding asymptotic results are easily derivable from the exact formulas by using standard methods from asymptotic analysis. See [43] for "the" standard text on asymptotic methods in combinatorial enumeration.

In many cases, I omit proofs. The proofs that are given are either reasonably short, or they serve to illustrate a key method or idea in lattice path enumeration. If one attempts to make a list of the important methods in lattice path enumeration, then this will include the following.

1. **Generating functions** (of course), in combination with the **Lagrange inversion formula** and/or **residue calculus** (see the second proof of Theo-

rem 10.4.5, the proof of Theorem 10.3.4, and the proof of Theorem 10.12.1 for examples);

2. **bijections** (they appear explicitly or implicitly at many places);
3. **reflection principle** (see the proof of Theorem 10.3.1 and Section 10.18);
4. **cycle lemma** (see Section 10.4);
5. **transfer matrix method** (see the proof of Theorem 10.11.1);
6. **kernel method** (see the proof of Theorem 10.12.2 and the paragraphs thereafter);
7. the **path switching involution for non-intersecting lattice paths** (see Section 10.13);
8. manipulation of **two-rowed arrays** for turn enumeration (see Section 10.14);
9. **orthogonal polynomials, continued fractions** (see Sections 10.9–10.11).

We start with some simple results on the enumeration of paths in the $d$-dimensional integer lattice in Section 10.2. The sections that follow, Sections 10.3–10.7, discuss so-called **simple lattice paths** in the plane integer lattice $\mathbb{Z}^2$; these are paths in $\mathbb{Z}^2$ consisting of horizontal and vertical unit steps in the positive direction. While still staying in the plane integer lattice, beginning from Section 10.8, we allow three kinds of steps: Changing the geometry slightly by a rotation about 45°, these are up-, down-, and level-steps. The case of Motzkin paths is intimately related to the theory of orthogonal polynomials and continued fractions. This link is explained in Sections 10.9–10.11. Section 10.12 provides a loose collection of further results for lattice paths in the plane integer lattice, with many pointers to the literature. The subsequent section, Section 10.13, is devoted to the theory of non-intersecting lattice paths, which is an extremely useful enumeration theory with many applications, particularly in the enumeration of tilings, plane partitions, and tableaux, but is also of great interest in its own right. Turn statistics are investigated in Section 10.14. Again, the original motivation comes from statistics, but more recent work, most importantly work on counting non-intersecting lattice paths by their number of turns, arose from problems in commutative algebra. Then we move into higher-dimensional space. Sections 10.15–10.17 present standard results for lattice paths in higher-dimensional lattices. How far one can go with the reflection principle is explained in Section 10.18. The brief Section 10.19 gives some glimpses of $q$-analogues, including pointers to the connections of lattice path enumeration with the Rogers–Ramanujan identities.

We conclude this introduction by fixing some notation that will be used consistently in this chapter. (It is in part inspired by standard probability notation.) Given lattice points $A$ and $E$, a set $\mathbb{S}$ of steps (vectors), a set of restrictions $R$, and a non-negative integer $m$, we write

$$L_m(A \to E; \mathbb{S} \mid R) \tag{10.1}$$

for the set of all lattice paths from $A$ to $E$ with $m$ steps, from $\mathbb{S}$, which obey the restrictions in $R$. The lattice itself in which these paths are considered will be always clear from the context and is therefore not included in the notation. For example, the path in Figure 10.1 is in

$$L_5\big((0,0) \to (4,3); \{(1,0),(0,1),(1,1)\} \mid x \geq y\big),$$

where $x \geq y$ indicates the restriction that the $x$-coordinate of any lattice point of the path is at least as large as its $y$-coordinate, or, equivalently, obeys the restriction to stay weakly below the diagonal $x = y$.

Parts in (10.1) may be left out if we do not intend to require the corresponding restriction, or if that restriction is clear from the context. For example, the set of lattice paths from $A$ to $E$ with horizontal and vertical unit steps in the positive direction without further restriction will be denoted by $L\big(A \to E; \{(1,0),(0,1)\}\big)$, or sometimes even shorter, if the step set is clear from the context, $L\big(A \to E\big)$.

When we consider **weighted counting**, then we shall also use a uniform notation. Given a set $\mathcal{M}$ and a weight function $w$ on $\mathcal{M}$, we denote by $GF(\mathcal{M};w)$ the **generating function** for $\mathcal{M}$ with respect to $w$, i.e.,

$$GF(\mathcal{M};w) := \sum_{x \in \mathcal{M}} w(x). \tag{10.2}$$

Finally, by convention, whenever we write a binomial coefficient $\binom{n}{k}$, it is assumed to be zero if $k$ is not an integer satisfying $0 \leq k \leq n$.

## 10.2 Lattice paths without restrictions

In this short section, we briefly cover the simplest enumeration problems for lattice paths. If we are given a set of steps $S$, then the number of paths starting from the origin and using $n$ steps from $S$ is $|S|^n$. If we are also fixing the end point, then we cannot expect a reasonable formula in this generality.

However, in the case of (positive) unit steps such formulae are available. Namely, the number of paths in the plane integer lattice $\mathbb{Z}^2$ from $(a,b)$ to $(c,d)$ consisting of horizontal and vertical unit steps in the positive direction is

$$\big|L\big((a,b) \to (c,d)\big)\big| = \binom{c+d-a-b}{c-a}, \tag{10.3}$$

since each path from $(a,b)$ to $(c,d)$ can be identified with a sequence of $(c-a)$ horizontal steps and $(d-b)$ vertical steps, the number of those sequences being given by the binomial coefficient in (10.3).

More generally, for the same reason, the number of paths in the $d$-dimensional integer lattice $\mathbb{Z}^d$ from $\mathbf{a} = (a_1, a_2, \dots, a_d)$ to $\mathbf{e} = (e_1, e_2, \dots, e_d)$ consisting of positive unit steps in the direction of some coordinate axis is given by a multinomial

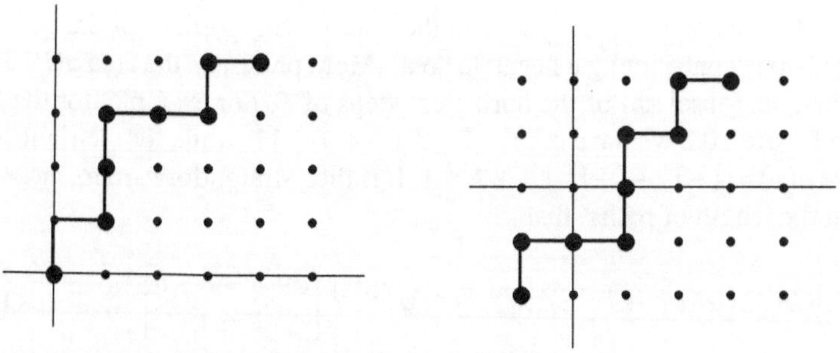

**Figure 10.2**
Two paths illustrating formula (10.7).

coefficient, namely

$$|L(\mathbf{a} \to \mathbf{e})| = \binom{\sum_{i=1}^{d}(e_i - a_i)}{e_1 - a_1, e_2 - a_2, \ldots, e_d - a_d} := \frac{\left(\sum_{i=1}^{d}(e_i - a_i)\right)!}{(e_1 - a_1)!\,(e_2 - a_2)! \cdots (e_d - a_d)!}. \tag{10.4}$$

There is another special case, in which one can write down a closed form expression for the number of paths between two given points with a fixed number of steps. Namely, the number of paths with $n$ horizontal and vertical unit steps (in the positive *or* negative direction) from $(a,b)$ to $(c,d)$ is given by

$$\left|L_n\big((a,b) \to (c,d); \{(\pm 1,0), (0,\pm 1)\}\big)\right| = \binom{n}{\frac{n+c+d-a-b}{2}} \binom{n}{\frac{n+c-d-a+b}{2}}. \tag{10.5}$$

See [60] and the references given there.

If one considers other step sets then it may often be possible to obtain (non-closed) formulae by "mixing" steps. A typical example is the case where we consider lattice paths in the plane allowing three types of steps, namely horizontal unit steps $(1,0)$, vertical unit steps $(0,1)$, and diagonal steps $(1,1)$. Let $S = \{(1,0),(0,1),(1,1)\}$ be this step set. If we want to know how many lattice paths there exist from $(a,b)$ to $(c,d)$ consisting of steps from $S$, then we find

$$\left|L\big((a,b) \to (c,d); S\big)\right| = \sum_{k=0}^{c-a} \binom{c+d-a-b-k}{k, c-a-k, d-b-k}, \tag{10.6}$$

since, if we fix the number of diagonal steps to $k$, then the number of ways to mix $k$ diagonal steps, $c-a-k$ horizontal steps, and $d-b-k$ vertical steps, is given by the multinomial coefficient, which represents the summand in (10.6). In the special case where $(a,b) = (0,0)$, the corresponding numbers are called **Delannoy numbers**, and, if $(c,d) = (n,n)$, **central Delannoy numbers.**

As a first excursion to weighted counting, we consider the generating function for lattice paths in $\mathbb{Z}^2$ from $A = (a,b)$ to $E = (c,d)$ consisting of horizontal and vertical unit steps in the positive direction, in which each path is weighted by $q^{a(P)}$, where

$a(P)$ denotes the area between the path and the $x$-axis (with portions of the path that lie below the $x$-axis contributing a negative area). More precisely, the area $a(P)$ is the sum of the heights (abscissa) of the horizontal steps of $P$. For example, for the left-hand path in Figure 10.2 we have $a(.) = 1+3+3+4 = 11$, while for the right-hand path we have $a(.) = (-1)+(-1)+1+2 = 1$. It is then straightforward to check (by induction on the length of paths) that

$$GF\left(L((a,b) \to (c,d)); q^{a(.)}\right) = q^{b(c-a)} \begin{bmatrix} c+d-a-b \\ c-a \end{bmatrix}_q, \tag{10.7}$$

where $\left[\begin{smallmatrix} c+d-a-b \\ c-a \end{smallmatrix}\right]_q$ denotes the **$q$-binomial coefficient** defined by

$$\begin{bmatrix} n \\ k \end{bmatrix}_q := \frac{(1-q^n)(1-q^{n-1})\cdots(1-q)}{(1-q^k)(1-q^{k-1})\cdots(1-q)(1-q^{n-k})(1-q^{n-k-1})\cdots(1-q)},$$

and $\left[\begin{smallmatrix} n \\ k \end{smallmatrix}\right]_q = 0$ if $k < 0$. This result connects lattice path enumeration with the theory of integer partitions. What we have computed in (10.7) is equivalent to the classical result that the generating function for integer partitions with at most $k$ parts, each of which is bounded above by $n$ is given by $\left[\begin{smallmatrix} n+k \\ k \end{smallmatrix}\right]_q$. We shall say a little bit more about $q$-counting in Section 10.19. The reader is referred to [2] for an excellent survey of the theory of partitions.

## 10.3 Linear boundaries of slope 1

Next we want to count paths from $(a,b)$ to $(c,d)$, where $a \geq b$ and $c \geq d$, which stay weakly below the main diagonal $y = x$. So, what we want to know is the number $\left|L((a,b) \to (c,d) \mid x \geq y)\right|$. This problem is most conveniently solved by the so-called **reflection principle** most often attributed to André [1]. However, while André did solve the ballot problem, he did not use the reflection principle. Its origin lies most likely in the method of images of electrostatics, see Sections 2.3–2.6 in [64].

**Theorem 10.3.1** *Let $a \geq b$ and $c \geq d$. The number of all paths from $(a,b)$ to $(c,d)$ staying weakly below the line $y = x$ is given by*

$$\left|L((a,b) \to (c,d) \mid x \geq y)\right| = \binom{c+d-a-b}{c-a} - \binom{c+d-a-b}{c-b+1}. \tag{10.8}$$

*Proof.* First we observe that the number in question is the number of all paths from $(a,b)$ to $(c,d)$ minus the number of those paths that cross the line $y = x$,

$$\begin{aligned}\left|L((a,b) \to (c,d) \mid x \geq y)\right| &= \left|L((a,b) \to (c,d))\right| \\ &\quad - \left|L((a,b) \to (c,d) \mid x \ngeq y \text{ at least once})\right|. \end{aligned} \tag{10.9}$$

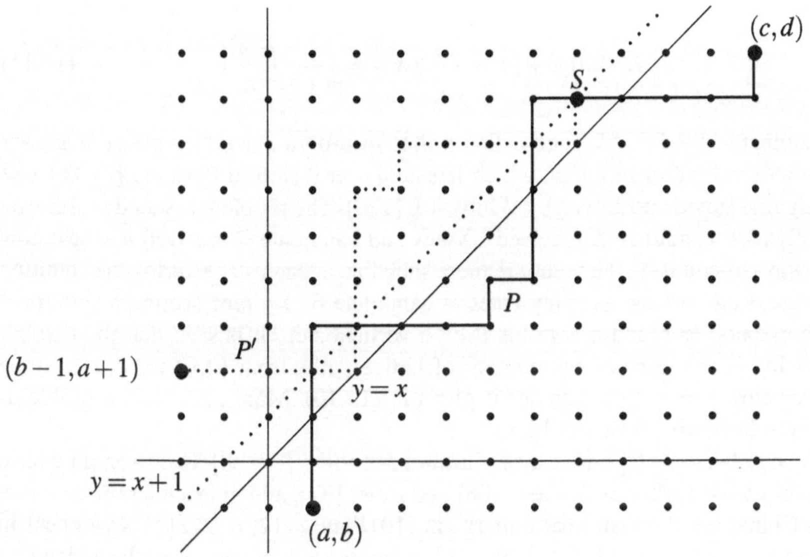


**Figure 10.3**
A path $P$ crossing the line $x = y$.

By (10.3) we already know $|L((a,b) \to (c,d))|$. The reflection principle shows that paths from $(a,b)$ to $(c,d)$ that cross $y = x$ are in bijection with paths from $(b-1, a+1)$ to $(c,d)$. This implies

$$|L((a,b) \to (c,d) \mid x \not\geq y \text{ at least once})| = |L((b-1,a+1) \to (c,d))|.$$

Hence, using (10.3) again, we establish (10.8).

The claimed bijection is obtained as follows. Consider a path $P$ from $(a,b)$ to $(c,d)$ crossing the line $y = x$. See Figure 10.3 for an example. Then $P$ must meet the line $y = x+1$. Among all the meeting points of $P$ and $y = x+1$, choose the right-most. Denote this point by $S$. Now reflect the portion of $P$ from $(a,b)$ to $S$ in the line $y = x+1$, leaving the portion from $S$ to $(c,d)$ invariant. Thus we obtain a new path $P'$ from $(b-1, a+1)$ to $(c,d)$. To construct the reverse mapping we only have to observe that any path from $(b-1, a+1)$ to $(c,d)$ must meet $y = x+1$ since $(b-1, a+1)$ and $(c,d)$ lie on different sides of $y = x+1$. Again we choose the right-most meeting point, denote it by $S$, and reflect the portion from $(b-1, a+1)$ to $S$ in the line $y = x+1$, thus obtaining a path from $(a,b)$ to $(c,d)$ that meets the line $y = x+1$ or, equivalently, crosses the line $y = x$. ∎

In particular, for $a = b = 0$ we obtain the following compact formula.

**Corollary 10.3.2** *If $c \geq d$ we have*

$$|L((0,0) \to (c,d) \mid x \geq y)| = \frac{c+1-d}{c+d+1}\binom{c+d+1}{d} \tag{10.10}$$

*and*

$$\left|L\big((0,0)\to(n,n)\mid x\geq y\big)\right| = \frac{1}{n+1}\binom{2n}{n}. \tag{10.11}$$

The numbers $\frac{c+1-d}{c+d+1}\binom{c+d+1}{d}$ are called **ballot numbers** since they give the answer to the classical ballot problem, which is usually attributed to Bertrand [8], but was actually first stated and solved by Whitworth [123]. The problem is stated as follows: In an election candidate $A$ received $c$ votes and candidate $B$ received $d$ votes; how many ways of counting the votes are there such that at each stage during the counting candidate $A$ has at least as many votes as candidate $B$? By representing a vote for $A$ by a horizontal step and a vote for $B$ by a vertical step, it is seen that the number in question is the same as the number of lattice paths from $(0,0)$ to $(c,d)$ staying weakly below $y=x$. This number is given in (10.10). More about ballot problems appears in Sections 10.12 and 10.18.

The numbers $\frac{1}{n+1}\binom{2n}{n}$ are called **Catalan numbers** [27, 28]. However, they have been considered earlier by Segner [106] and Euler [37], and independently even earlier in China; see the historical remarks in [101] and [112, p. 212]. They appear in numerous places; see [112, Ex. 6.19], with many more occurrences in the addendum [113].

An iterated reflection argument will give the number of paths between two diagonal lines.

**Theorem 10.3.3** *Let $a+t\geq b\geq a+s$ and $c+t\geq d\geq c+s$. The number of all paths from $(a,b)$ to $(c,d)$ staying weakly below the line $y=x+t$ and above the line $y=x+s$ is given by*

$$\begin{aligned}&\left|L\big((a,b)\to(c,d)\mid x+t\geq y\geq x+s\big)\right|\\&\quad=\sum_{k\in\mathbb{Z}}\left(\binom{c+d-a-b}{c-a-k(t-s+2)}-\binom{c+d-a-b}{c-b-k(t-s+2)+t+1}\right).\end{aligned} \tag{10.12}$$

Since this is (as well as Theorem 10.3.1) an instance of the general formula (10.145) for the number of paths staying in regions defined by hyperplanes, we omit the proof.

The formula in Theorem 10.3.3 is very convenient for computing the number of paths as long as the parameters are not too large. On the other hand, it is of no use if one is interested in asymptotic information, because the summands on the right-hand side of (10.12) alternate in sign so that there is considerable cancellation. However, with the help of little residue calculus, the formula can be transformed into a surprising formula featuring cosines and sines, from which asymptotic information can be easily read off.

**Theorem 10.3.4** *Let $a+t\geq b\geq a+s$ and $c+t\geq d\geq c+s$. The number of all paths from $(a,b)$ to $(c,d)$ staying weakly below the line $y=x+t$ and above the line $y=x+s$ is given by*

$$\left|L((a,b)\to(c,d)\mid x+t\geq y\geq x+s)\right|$$
$$=\sum_{k=1}^{\lfloor(t-s+1)/2\rfloor}\frac{4}{t-s+2}\left(2\cos\frac{\pi k}{t-s+2}\right)^{c+d-a-b}$$
$$\cdot\sin\left(\frac{\pi k(a-b+t+1)}{t-s+2}\right)\cdot\sin\left(\frac{\pi k(c-d+t+1)}{t-s+2}\right). \quad (10.13)$$

*Proof.* Trivially, the binomial coefficient $\binom{n}{k}$ is the coefficient of $z^{-1}$ in the Laurent series

$$\frac{(1+z)^n}{z^{k+1}}.$$

Thus, the sum (10.12) equals the coefficient of $z^{-1}$ in

$$\sum_{k=0}^{\infty}\left(\frac{(1+z)^{c+d-a-b}z^{k(t-s+2)}}{z^{c-a+(c+d-a-b)(t-s+2)+1}}-\frac{(1+z)^{c+d-a-b}z^{k(t-s+2)}}{z^{c-b+t+(c+d-a-b)(t-s+2)+2}}\right)$$
$$=\frac{(1+z)^{c+d-a-b}}{z^{c-a+(c+d-a-b)(t-s+2)+1}(1-z^{t-s+2})}-\frac{(1+z)^{c+d-a-b}}{z^{c-b+t+(c+d-a-b)(t-s+2)+2}(1-z^{t-s+2})}$$
$$=\frac{(1+z)^{c+d-a-b}\left(z^{(-c+d+a-b)/2}-z^{(-c+d-a+b)/2-t-1}\right)}{z^{(c+d-a-b)/2+(c+d-a-b)(t-s+2)+1}(1-z^{t-s+2})}. \quad (10.14)$$

(In the second line we used the formula for the geometric series. It can be either regarded as a summation in the formal sense, or else one must assume that $|z|<1$.) Equivalently, the sum (10.12) equals the residuum of the Laurent series (10.14) at $z=0$. Now consider the contour integral of (10.14) (with respect to $z$, of course) along a circle of radius $r$ around the origin. It is a standard fact that in the limit $r\to\infty$ this integral vanishes, because the integrand (10.14) is of the order $O(1/z^2)$. Therefore, by the theorem of residues, the sum of the residues of (10.14) must be 0, or, equivalently, the residuum at $z=0$, which we are interested in, equals the negative of the sum of the other residues. As the other poles of (10.14) are the $(t-s+2)$th roots of unity different from 1, we obtain

$$-\sum_{k=1}^{t-s+1}\frac{\left(1+e^{\frac{2\pi ik}{t-s+2}}\right)^{c+d-a-b}\left(e^{\frac{\pi ik}{t-s+2}(-c+d+a-b)}-e^{\frac{\pi ik}{t-s+2}(-c+d-a+b-2t-2)}\right)}{e^{\frac{2\pi ik}{t-s+2}(\frac{c+d-a-b}{2}+1)}\left(-(t-s+2)e^{\frac{2\pi ik}{t-s+2}(t-s+1)}\right)}$$
$$=\sum_{k=1}^{t-s+1}\frac{1}{t-s+2}\left(2\cos\frac{\pi k}{t-s+2}\right)^{c+d-a-b}e^{\frac{\pi ik}{t-s+2}(-c+d-t-1)}$$
$$\cdot\left(e^{\frac{\pi ik}{t-s+2}(a-b+t+1)}-e^{-\frac{\pi ik}{t-s+2}(a-b+t+1)}\right)$$

for the sum (10.12). Now, in the last line, we pair the $k$th and the $(t-s+2-k)$th summand. Thus, upon little manipulation, the above sum turns into

$$\sum_{k=1}^{\lfloor (t-s+1)/2 \rfloor} \frac{1}{t-s+2}\left(2\cos\frac{\pi k}{t-s+2}\right)^{c+d-a-b}$$
$$\cdot\left(e^{-\frac{\pi ik}{t-s+2}(c-d+t+1)} - e^{\frac{\pi ik}{t-s+2}(c-d+t+1)}\right)\left(e^{\frac{\pi ik}{t-s+2}(a-b+t+1)} - e^{-\frac{\pi ik}{t-s+2}(a-b+t+1)}\right).$$

Clearly, this formula is equivalent to (10.13). ∎

From the generating function formula given in Section 10.11 (see Example 10.11.2), one can see that this asymptotic formula comes from **Chebyshev polynomials.**

## 10.4 Simple paths with linear boundaries of rational slope, I

When we want to count simple lattice paths (recall the meaning of "simple" from the introduction) in the plane bounded by an arbitrary line $y = kx + d$, $k, d \in \mathbb{R}$, the reflection principle obviously does not help, since the reflection of a lattice path in a generic line does not necessarily give a lattice path. In fact, a solution in form of a determinant can be given when the boundary is viewed as a special case of a set of general boundaries (see Section 10.7, Theorem 10.7.1; another solution was proposed by Takács [118], which is of similar complexity as it involves the solution of a large system of linear equations). However, there are cases where simpler expressions can be obtained, and these are discussed in this section. All of them can be derived from a very basic combinatorial lemma, the so-called **cycle lemma**, which exists in several variations.

The first case we discuss is the enumeration of simple lattice paths from the origin to a lattice point $(r,s)$, with $r$ and $s$ relatively prime, which stay weakly below the line connecting the origin and $(r,s)$.

**Theorem 10.4.1** *Let r and s be relatively prime positive integers. The number of all paths from* $(0,0)$ *to* $(r,s)$ *staying weakly below the line* $ry = sx$ *is given by*

$$\left|L((0,0) \to (r,s) \mid sx \geq ry)\right| = \frac{1}{r+s}\binom{r+s}{r}. \tag{10.15}$$

**Remark 10.4.2** The numbers in (10.15) are nowadays called **rational Catalan numbers** (cf. [4]), the Catalan numbers being the special case where $r = n$ and $s = n+1$.

The above result follows easily from a form of the cycle lemma, which is known in the statistics literature as Spitzer's lemma [110].

**Lemma 10.4.3 (Spitzer's Lemma)** *Let* $a_1, a_2, \ldots, a_N$ *be real numbers with the property that* $a_1 + a_2 + \cdots + a_N = 0$ *and no other partial sum of consecutive* $a_i$*'s, read*

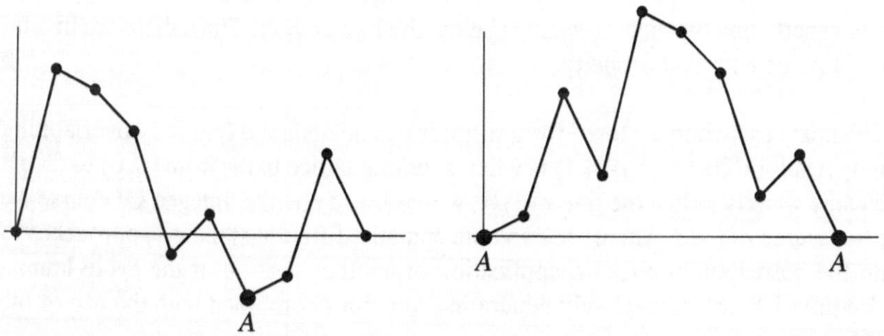


**Figure 10.4**
An instance of Spitzer's lemma.

*cyclically (by which we mean sums of the form* $a_j + a_{j+1} + \cdots + a_k$ *with* $j \le k$ *and* $k - j < N$*, where indices are interpreted modulo* $N$*), vanishes. Then there exists a unique cyclic permutation* $a_i, a_{i+1}, \ldots, a_N, a_1, \ldots, a_{i-1}$ *with the property that for all* $j = 1, 2, \ldots, N$ *the sum of the first* $j$ *letters of this permuted array is non-negative.*

**Remark 10.4.4** This lemma could be further generalized by weakening the above assumption to demand that $K$ partial sums of consecutive $a_i$'s, read cyclically, of **minimal length** vanish, with the conclusion that there be $K$ cyclic permutations with the above non-negativity property.

*Proof.* We interpret the real numbers $a_i$ as steps of a path (although not necessarily of a **lattice** path), by concatenating the steps $(1, a_1)$, $(1, a_2)$, ..., $(1.a_N)$ to a path starting at the origin. See the left half of Figure 10.4 for a typical example.

Since the sum of all $a_i$'s vanishes, the end point of the path lies on the $x$-axis. We identify this end point with the starting point (located at the origin), so that we consider this path as a cyclic object.

By the non-vanishing of cyclic subsums, there is a unique point of minimal height, $A$ say. (This may also be the starting/end point, which we identified.) In the figure this point is marked by a thick dot. Now "permute" the path cyclically, that is, take the portion of the path from $A$ to the end, and concatenate it with the initial portion of the path until $A$. See the right half of Figure 10.4 for the result in our example. Obviously, the new path always lies strictly above the $x$-axis, except at the beginning and at the end. This identifies the cyclic permutation of the $a_i$'s with the required property. ∎

*Proof of Theorem 10.4.1.* We consider *all* paths from $(0,0)$ to $(r,s)$. There are $\binom{r+s}{s}$ such paths. Given a path $P$ from $(0,0)$ to $(r,s)$, we consider the sequence $a_1, a_2, \ldots, a_{r+s}$, where $a_i = s$ if the $i$th step of the path is a horizontal step, and $a_i = -r$ if the $i$th step of the path is a vertical step. Since $r$ and $s$ are relatively prime, no cyclic subsum of the $a_i$'s, except the complete sum, can vanish. The cycle lemma in

Lemma 10.4.3 then implies that, out of the $r+s$ cyclic "permutations" of the path $P$, there is exactly one that stays (weakly) below the line $sx = ry$. Thus, there are in total $\frac{1}{r+s}\binom{r+s}{r}$ paths with that property. ■

The next case where a closed form formula can be obtained (partially overlapping with the result in Theorem 10.4.1) is when counting lattice paths from $(0,0)$ to $(c,d)$, which stay weakly below the line $x = \mu y$, where $\mu$ is a positive integer. Of course we have to assume $c \geq \mu d$. There are two conceptually different standard approaches to obtain the corresponding result: application of another version of the cycle lemma (see Lemma 10.4.6), respectively generating functions combined with the use of the Lagrange inversion formula.

**Theorem 10.4.5** *Let $\mu$ be a non-negative integer and $c \geq \mu d$. The number of all lattice paths from the origin to $(c,d)$ that lie weakly below $x = \mu y$ is given by*

$$\left|L((0,0) \to (c,d) \mid x \geq \mu y)\right| = \frac{c - \mu d + 1}{c + d + 1}\binom{c+d+1}{d}. \tag{10.16}$$

This result is essentially equivalent to the cycle lemma due to Dvoretzky and Motzkin [36]. It has been rediscovered many times; see [35] for a partial survey and many related references, as well as [112, Lemma 5.3.6 and Example 5,3,7].

**Lemma 10.4.6 (Cycle Lemma)** *Let $\mu$ be a non-negative integer. For any sequence $p_1p_2\cdots p_{m+n}$ of $m$ 1's and $n$ 2's, with $m \geq \mu n$, there exist exactly $m - \mu n$ cyclic permutations $p_ip_{i+1}\cdots p_{m+n}p_1\cdots p_{i-1}$, $1 \leq i \leq m+n$, that have the property that for all $j = 1,2,\ldots,m+n$ the first $j$ letters of this permutation contain more 1's than $\mu$ times the number of 2's.*

*Proof.* A sequence $p_1p_2\cdots p_{m+n}$ of $m$ 1's and $n$ 2's can be seen as a lattice path from $(0,0)$ to $(m,n)$ by interpreting the 1's as horizontal steps and the 2's as vertical steps. Cyclically permuting $p_1p_2\cdots p_{m+n}$ means to cut the corresponding lattice path into two pieces and put them together in exchanged order, thus obtaining a new lattice path from $(0,0)$ to $(m,n)$. Finally, the property that in each initial string of a sequence the number of 1's dominates (i.e., is larger than) $\mu$ times the number of 2's means that the corresponding lattice path stays strictly below the line $x = \mu y$, with the exception of the starting point $(0,0)$.

For the proof of the lemma interpret $p_1p_2\cdots p_{m+n}$ as a path, as described before, and join a shifted copy of this path at the end point $(m,n)$, another shifted copy at $(2m,2n)$, etc. Figure 10.5 shows an example with $\mu = 2$, $m = 9$, $n = 3$. The path $P$ corresponds to the sequence 121111122111. Cyclic permutations of $p_1p_2\cdots p_{m+n}$ correspond to cutting a piece of $m+n$ successive steps out of this lattice path structure. Then imagine a sun to be located in direction $(\mu,1)$ illuminating the lattice path structure. A cyclic permutation will satisfy the dominance property for each initial string if and only if the first step of the corresponding lattice path is illuminated. In Figure 10.5 the illuminated steps are indicated by thick lines. It is an easy matter of fact that of any $m+n$ successive steps there are exactly $m - \mu n$ illuminated steps.

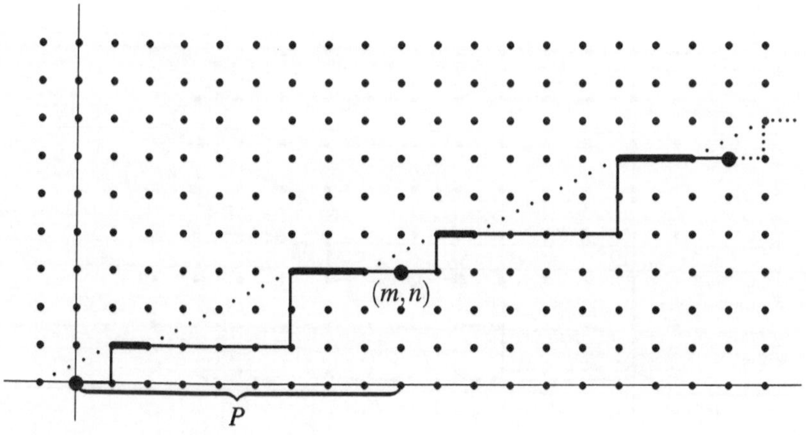


**Figure 10.5**
An instance of the cycle lemma with $\mu = 2$, $m = 9$, and $n = 3$.

Therefore out of the $m+n$ cyclic permutations of $p_1 p_2 \cdots p_{m+n}$ there are exactly $m - \mu n$ cyclic permutations having the dominance property for each initial string. ■

*First proof of Theorem 10.4.5.* We want to count paths from $(0,0)$ to $(c,d)$ staying weakly below $x = \mu y$. To fit with the cycle lemma we adjoin a horizontal step at the beginning and shift everything by one unit to the right. Thus we are now asking for the number of paths from $(0,0)$ to $(c+1,d)$ staying **strictly** below $x = \mu y$, except for the starting point $(0,0)$. Now one applies Lemma 10.4.6 with $m = c+1$ and $n = d$: given a path $P$ from $(0,0)$ to $(c+1,d)$, exactly $m - \mu n = c + 1 - \mu d$ of its cyclic "permutations" satisfy the property of staying **strictly** below $x = \mu y$, except for the starting point $(0,0)$. Thus, the total number of paths from $(0,0)$ to $(c,d)$ with that property is given by (10.16). ■

For instructional purposes, we also present the generating function proof.

*Second proof of Theorem 10.4.5.* The generating function proof works in two steps. First, an equation is found for the generating function of those paths that return in the end to the boundary $x = \mu y$. Then, in a second step, paths ending arbitrarily are decomposed into paths of the former type, leading to a generating function expression in terms of the earlier generating function to which the Lagrange inversion formula is applicable.

Let $P$ be a path in $L\big((0,0) \to (\mu d, d) \mid x \geq \mu y\big)$ (see Figure 10.6). For $l = 0, 1, \ldots, \mu - 1$, the path $P$ will meet the line $x = \mu y + l$ (which is parallel to our boundary $x = \mu y$) somewhere for the last time. Denote this point by $S_l$. Clearly, the path $P$ must leave $S_l$ by a horizontal step, which we denote by $s_h$ for short. This gives

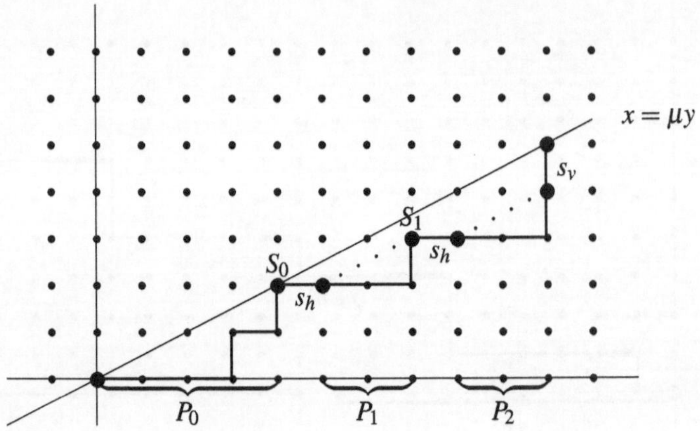


**Figure 10.6**
The path $P$ and the line $x = \mu y$.

us a unique decomposition of $P$ of the form

$$P = P_0 s_h P_1 s_h \dots s_h P_\mu s_v,$$

where $P_0$ is $P$'s portion from the origin up to $S_0$, $P_1$ is $P$'s portion from the point immediately following $S_0$ up to $S_1$, etc. By $s_v$ we denote the final vertical step. All the portions $P_i$ (when shifted appropriately) belong to $L((0,0) \to (\mu n, n) \mid x \geq \mu y)$ for some $n$. Let

$$\mathscr{L}_0 = \bigcup_{n \geq 0} L((0,0) \to (\mu n, n) \mid x \geq \mu y). \tag{10.17}$$

Then we have the following decomposition:

$$\mathscr{L}_0 = \{\varepsilon\} \cup \left((\mathscr{L}_0 s_h)^\mu \mathscr{L}_0 s_v\right).$$

Here, as always in the sequel, $\varepsilon$ denotes the empty path.

By elementary combinatorial principles, this immediately translates into a functional equation for the generating function

$$F_0(z) := \sum_{n \geq 0} \left| L((0,0) \to (\mu n, n) \mid x \geq \mu y) \right| z^n$$

for $\mathscr{L}_0$ (note that the summation index $n$ records the vertical height of the end point of paths), namely

$$F_0(z) = 1 + z F_0(z)^{\mu+1}. \tag{10.18}$$

If we write $F_0(z) = 1 + G_0(z)$, then Equation (10.18) in terms of the series $G_0(z)$ reads

$$\frac{G_0(z)}{(1 + G_0(z))^{\mu+1}} = z, \tag{10.19}$$

which simply says that $G_0(z)$ is the compositional inverse of $z/(1+z)^{\mu+1}$.

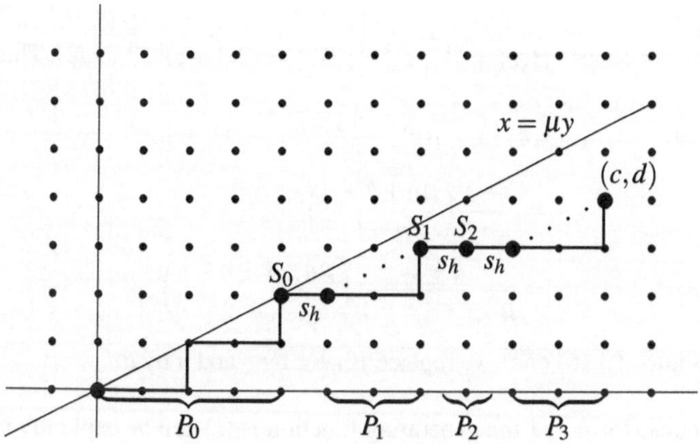


**Figure 10.7**
The more general problem.

Turning to the more general problem, consider a lattice path $P$ in $\left|L\big((0,0) \to (\mu d+k,d) \mid x \geq \mu y\big)\right|$ (see Figure 10.7). For $l = 0,1,\dots,k-1$ the path will meet the line $x = \mu y + l$ somewhere for the last time. Denote this point by $S_l$. Clearly, the path $P$ must leave $S_l$ by a horizontal step, for which we again write $s_h$. This gives us a decomposition of $P$ of the form

$$P = P_0 s_h P_1 s_h \dots s_h P_k,$$

where $P_0$ is the portion of $P$ from the origin up to $S_0$, $P_1$ is the portion of $P$ from the point immediately following $S_0$ up to $S_1$, and so on. Observe that again the portions $P_l$ belong to $\mathcal{L}_0$ (being defined in (10.17)). Let

$$\mathcal{L}_k = \bigcup_{n\geq 0} L\big((0,0) \to (\mu n+k,n) \mid x \geq \mu y\big).$$

Then we have the following decomposition:

$$\mathcal{L}_k = (\mathcal{L}_0 s_h)^k \mathcal{L}_0.$$

This translates again into an equation for the corresponding generating function

$$F_k(z) := \sum_{n\geq 0} \left|L\big((0,0) \to (\mu n+k,n) \mid x \geq \mu y\big)\right| z^n$$

for $\mathcal{L}_k$, namely into

$$F_k(z) = F_0(z)^{k+1} = (1+G_0(z))^{k+1}.$$

We noted above that $G_0(z)$ is the compositional inverse of $z/(1+z)^{\mu+1}$. Therefore, we may apply the Lagrange formula (see [112, Corollary 5.4.3]; for the current purpose, we have to choose $H(z) = (1+z)^{k+1}$, $f(z) = z/(1+z)^{\mu+1}$ there). This yields

$$\begin{aligned} L\big((0,0) \to (\mu n+k,n) \mid x \geq \mu y\big) &= \frac{1}{n}\langle z^{-1}\rangle (k+1)\,(1+z)^k \frac{(1+z)^{n(\mu+1)}}{z^n} \\ &= \frac{k+1}{n}\binom{\mu n+k+n}{n-1} \\ &= \frac{k+1}{\mu n+k+n+1}\binom{\mu n+k+n+1}{n}, \end{aligned}$$

which turns into (10.16) once we replace $\mu n+k$ by $c$ and $n$ by $d$. ■

In particular, for $\mu = 1$ the generating function $F_0(z)$ can be explicitly evaluated from solving the quadratic equation (10.18). In the case where the paths return to the boundary $x = y$, i.e., where $(c,d) = (n,n)$, this gives the familiar generating function for the Catalan numbers (compare with the second paragraph after Corollary 10.3.2)

$$\sum_{n\geq 0} C_n z^n = \frac{1-\sqrt{1-4z}}{2z}. \tag{10.20}$$

More generally, if $\mu$ is kept generic and $(c,d) = (\mu n, n)$ (that is, we consider again paths that return to the boundary), then the formula on the right-hand side of (10.16) becomes $\frac{1}{\mu n+1}\binom{(\mu+1)n}{n}$. These numbers are now commonly called **Fuß–Catalan numbers**, cf. [3, pp. 59–60] for more information on their significance and historical remarks.

So far we only counted paths bounded by $x = \mu y$ where the starting point lies on the boundary. If we drop this latter assumption and now want to enumerate all paths from $(a,b)$ to $(c,d)$ staying weakly below $x = \mu y$, there is still an answer, although only in terms of a sum. In fact, we can offer two different expressions. Which of these two is preferable depends on the particular situation, to be more precise, on which of the numbers $(a/\mu - b)$ or $(d - a/\mu)$ being larger (see Figure 10.8 for the pictorial significance of these numbers). Though the proof for the first expression is rather straightforward, the proof for the second expression is more difficult. The result below was first found by Korolyuk [73]. It is a special case of an even more general result of Niederhausen [98, Sec. 2.2] on the enumeration of simple paths with piecewise linear boundaries, which we will discuss in Section 10.6.

**Theorem 10.4.7** *Let $\mu$ be a non-negative integer, $a \geq \mu b$ and $c \geq \mu d$. The number of all lattice paths from $(a,b)$ to $(c,d)$ staying weakly below $x = \mu y$ is given by*

$$\begin{aligned} \big|L\big((a,b) \to (c,d) \mid x \geq \mu y\big)\big| &= \binom{c+d-a-b}{c-a} \\ &- \sum_{i=\lfloor a/\mu \rfloor+1}^{d} \binom{i(\mu+1)-a-b-1}{i-b} \frac{c-\mu d+1}{c+d-i(\mu+1)+1}\binom{c+d-i(\mu+1)+1}{d-i}, \end{aligned} \tag{10.21}$$

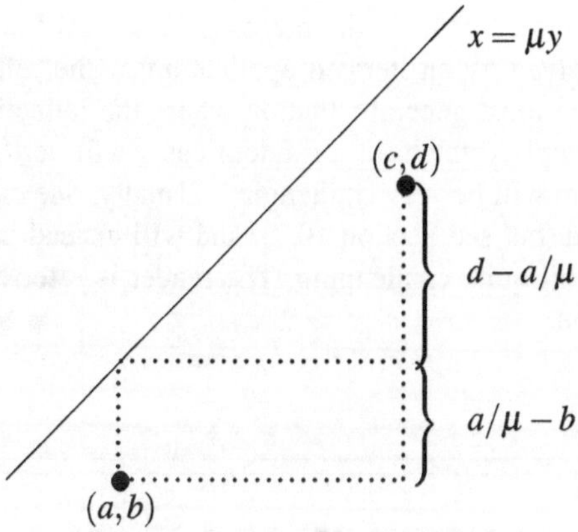


**Figure 10.8**
The significance of the numbers $(a/\mu - b)$ and $(d - a/\mu)$.

*and also*

$$\left|L\big((a,b) \to (c,d) \mid x \geq \mu y\big)\right| = \sum_{i=0}^{\lfloor a/\mu \rfloor - b} (-1)^i \binom{a - \mu(b+i)}{i} \times \frac{c - \mu d + 1}{c + d - (\mu+1)(b+i) + 1} \binom{c + d - (\mu+1)(b+i) + 1}{d - b - i}. \quad (10.22)$$

*Proof of* (10.21). The number of paths in question equals the number of all paths from $(a,b)$ to $(c,d)$ minus those paths that cross $x = \mu y$. To count the latter observe that any path crossing $x = \mu y$ must meet the line $x = \mu y - 1$, and for the last time in some point $(\mu i - 1, i)$ where $\lfloor a/\mu \rfloor + 1 \leq i \leq c$. Fix such an $i$, then the number of all these paths is

$$\left|L\big((a,b) \to (\mu i - 1, i)\big)\right| \cdot \left|L\big((\mu i, i) \to (c,d) \mid x \geq \mu y\big)\right|.$$

We already know the first number due to (10.3), and we also know the second number due to (10.16), since a shift in direction $(-\mu i, -i)$ shows that the second number equals $\left|L\big((0,0) \to (c - \mu i, d - i) \mid x \geq \mu y\big)\right|$. ∎

*Proof of* (10.22). This is the special case of Theorem 10.6.1 where $m = 2$, $\mu_1 = \nu_1 = 0$, $y_1 = \lfloor a/\mu \rfloor - b$, $\mu_2 = \mu$, $\nu_2 = \mu b - a$.

For a different, direct proof, in the sum in (10.21) replace the index $i$ by $i + b$; the new index then ranges from $\lfloor a/\mu \rfloor - b + 1$ to $d - b$; extend the sum to all $i$ between 0 and $d - b$, thereby adding a partial sum where $i$ ranges from 0 to $\lfloor a/\mu \rfloor - b$; the former sum can be evaluated by means of a convolution formula of the Hagen–Rothe type (cf. [54, Eq. (11)]), and the result is the binomial coefficient $\binom{c+d-a-b}{c-a}$. ∎

Enumeration of lattice paths in the presence of several linear boundaries can in the best cases be solved by an iterated application of the reflection principle; see Section 10.18 for the most general situation where the reflection principle applies. But, if it does not apply (which, in a random case, will certainly be so), then the enumeration problem will be very challenging. Usually, one cannot expect to find a useful exact formula (but see Section 10.7), and will instead investigate asymptotic behaviors. This is still quite challenging. The reader is referred to [22, 68, 69] for work in this direction.

## 10.5 Simple paths with linear boundaries with rational slope, II

In Section 10.4 we considered lattice paths bounded by a line $x = \mu y$, with $\mu$ a non-negative integer. Now we want to consider a more general linear boundary of the form $\nu x = \mu y$, where $\nu, \mu$ are non-negative integers. We describe a generating function approach, due to Sato [104], which works for a large class of cases. Alternative solutions, which work in all cases, in the form of a determinant, can be given as a special case of a set of general boundaries. These are discussed in Section 10.7, see in particular Theorem 10.7.1.

The problem that we want to attack here is to enumerate all lattice paths from an arbitrary starting point to an arbitrary end point staying weakly below the line $\nu x = \mu y$, where $\nu$ and $\mu$ are positive integers. A simple shift of the plane shows that this is equivalent to enumerating paths from the origin to an arbitrary end point staying weakly below $\nu x = \mu y - \rho$, for an appropriate $\rho$. Without loss of generality we may assume in the sequel that $\nu < \mu$. For the approach of Sato, this is the more convenient formulation of the problem. The idea is to introduce $\nu \times \nu$ matrices that contain the path numbers we are looking for. More precisely, define the $\nu \times \nu$ matrix $W(z;c,\rho)$ by

$$W(z;c,\rho) := \big(w(z;c+g,\rho+h)\big)_{0 \le g,h \le \nu-1}, \tag{10.23}$$

where

$$w(z;c,\rho) = \sum_{\mu n + c \equiv \rho \ (\mathrm{mod}\ \nu)} w(n;c,\rho)\, z^n, \tag{10.24}$$

with

$$w(n;c,\rho) = \begin{cases} \left|L\big((0,0) \to (\tfrac{\mu n + c - \rho}{\nu}, n) \mid \nu x \ge \mu y - \rho\big)\right|, & \mu n + c \equiv \rho \ (\mathrm{mod}\ \nu) \\ & \text{and } \mu n + c \ge \rho, \\ \binom{n + \frac{\mu n + c - \rho}{\nu}}{n}, & \mu n + c \equiv \rho \ (\mathrm{mod}\ \nu) \\ & \text{and } \mu n + c < \rho. \end{cases} \tag{10.25}$$

So, what the matrix $W(z;c,\rho)$ contains is generating functions of the path numbers

$w(n;c,\rho)$ that we want to know. The definition of $w(n;c,\rho)$ for $\mu n+c<\rho$ (in which case there cannot be any paths from $(0,0)$ to $(\frac{\mu n+c-\rho}{\nu},n))$ is just for technical convenience. Basically, the matrix $W(z;c,\rho)$ is

$$W(z;c,\rho)=\left(\sum_{\mu n+c+g-h\equiv\rho \pmod{\nu}}\left|L\left((0,0)\to\left(\frac{\mu n+c+g-\rho-h}{\nu},n\right)\mid \nu x\geq \mu y-\rho-h\right)\right|z^n\right)_{0\leq g,h\leq \nu-1}. \tag{10.26}$$

The following theorem of Sato [104, Theorem 1] tells how to compute $W(z;c,\rho)$.

**Theorem 10.5.1** *Let*

$$M=\left((-1)^{\nu-h-1}s_{(c+g+1,1^{\nu-h-1})}\left(u_0(z),\dots,u_{\nu-1}(z)\right)\right)_{0\leq g,h\leq \nu-1}, \tag{10.27}$$

*where*

$$s_{(\alpha,1^\beta)}(u_0,\dots,u_{\nu-1}) = \sum_{\nu-1\geq i_\alpha\geq i_{\alpha-1}\geq\dots i_1<j_1<\dots<j_\beta\leq\nu-1} u_{i_\alpha}(z)u_{i_{\alpha-1}}(z)\cdots u_{i_1}(z)u_{j_1}(z)\cdots u_{j_\beta}(z), \tag{10.28}$$

*$u_l(z)$ being defined by*

$$u_l(z)=e^{(2\pi i l/\nu)}\sum_{n\geq 0}\frac{1}{1+(\nu+\mu)n}\binom{\frac{1}{\nu}+(1+\frac{\mu}{\nu})n}{n}\left(ze^{2\pi i l\mu/\nu}\right)^n,$$
$$l=0,1,\dots,\nu-1. \tag{10.29}$$

*Furthermore, let*

$$\Phi(z;\rho)=\left(\sum_{\substack{\mu l\equiv\rho-g+h \pmod{\nu}\\ \mu l\leq\rho-g+h}}(-1)^l\binom{(\rho-g+h-\mu l)/\nu}{l}z^l\right)_{0\leq g,h\leq \nu-1}. \tag{10.30}$$

*Then, for any non negative integers $c$, $\rho$, $\nu$, $\mu$, with $\nu<\mu$, we have*

$$W(z;c,\rho)=M(z;c,\rho)\Phi(z;\rho). \tag{10.31}$$

**Note 10.5.2** *Note that $s_{(\alpha,1^\beta)}(u_0,\dots,u_{\nu-1})$ is a* **Schur function** *of* **hook shape** *(cf. [89, Ch. I, Sec. 3, Ex. 9]).*

It might be useful to discuss an example, in order to illustrate what this is all about.

**Example 10.5.3** *We take $\nu=2$, $\mu=3$. So, by (10.25), the quantity $w(n;c,\rho)$ represents the number of all lattice paths from $(0,0)$ to $((3n+c-\rho)/2,n)$ which stay weakly below the line $2x=3y-\rho$, where $c\equiv 3n-\rho \pmod 2$, i.e., $c\equiv n+\rho \pmod 2$.*

*By definition (10.23), we have*

$$W(z;c,\rho) = \big(w(z;c+g,\rho+h)\big)_{0\le g,h\le 1} = \begin{pmatrix} w(z;c,\rho) & w(z;c,\rho+1) \\ w(z;c+1,\rho) & w(z;c+1,\rho+1) \end{pmatrix}.$$

*Using (10.31), this can be written as*

$$W(z;c,\rho) = M(z;c,2)\,\Phi(z;\rho),$$

*where*

$$\Phi(z;\rho) = \begin{pmatrix} \sum_{\substack{\mu l\equiv\rho \ (\mathrm{mod}\ \nu)\\ \mu l\le\rho}} (-1)^l \binom{(\rho-\mu l)/\nu}{l} z^l & \sum_{\substack{\mu l\equiv\rho+1 \ (\mathrm{mod}\ \nu)\\ \mu l\le\rho+1}} (-1)^l \binom{(\rho+1-\mu l)/\nu}{l} z^l \\ \sum_{\substack{\mu l\equiv\rho-1 \ (\mathrm{mod}\ \nu)\\ \mu l\le\rho-1}} (-1)^l \binom{(\rho-1-\mu l)/\nu}{l} z^l & \sum_{\substack{\mu l\equiv\rho \ (\mathrm{mod}\ \nu)\\ \mu l\le\rho}} (-1)^l \binom{(\rho-\mu l)/\nu}{l} z^l \end{pmatrix}$$

*by (10.30), and*

$$M(z;c,2) = \begin{pmatrix} -s_{(c+1,1)}\big(u_0(z),u_1(z)\big) & s_{(c+1)}\big(u_0(z),u_1(z)\big) \\ -s_{(c+2,1)}\big(u_0(z),u_1(z)\big) & s_{(c+2)}\big(u_0(z),u_1(z)\big) \end{pmatrix}$$

*by (10.27), with $s_{(\alpha,1^\beta)}(u_0(z),u_1(z))$ being defined in (10.28), and*

$$u_l(z) = (-1)^l \sum_{n\ge 0} \frac{(-1)^{ln}}{1+5n}\binom{\frac{1}{2}+\frac{5}{2}n}{n} z^n,$$

*as given in (10.29).*

*So, in particular, in case that $c=\rho=0$ the matrix $\Phi(z;0)$ is the $2\times 2$ identity matrix, and so we have*

$$W(z;0,0) = \begin{pmatrix} w(z;0,0) & w(z;0,1) \\ w(z;1,0) & w(z;1,1) \end{pmatrix}$$
$$= M(z;0,2) = \begin{pmatrix} -u_0(z)u_1(z) & u_0(z)+u_1(z) \\ -u_0(z)u_1(z)(u_0(z)+u_1(z)) & u_0^2(z)+u_0(z)u_1(z)+u_1^2(z) \end{pmatrix}.$$

*Whence, for even $n$ the number of all lattice paths from $(0,0)$ to $(3n/2,n)$ which stay weakly below the line $2x=3y$, equals*

$$\left|L\big((0,0)\to(\tfrac{3n}{2},n)\mid 2x\ge 3y\big)\right| = \langle z^n\rangle w(z;0,0)$$
$$= \sum_{l=0}^{n}(-1)^l \frac{1}{1+5l}\binom{\frac{1}{2}+\frac{5}{2}l}{l}\cdot\frac{1}{1+5(n-l)}\binom{\frac{1}{2}+\frac{5}{2}(n-l)}{n-l},$$

*and for odd $n$ the number of all lattice paths from $(0,0)$ to $((3n-1)/2,n)$ which stay weakly below the line $2x=3y-1$, equals*

$$\left|L\big((0,0)\to(\tfrac{3n-1}{2},n)\mid 2x\ge 3y-1\big)\right| = \langle z^n\rangle w(z;0,1) = \frac{2}{1+5n}\binom{\frac{1}{2}+\frac{5}{2}n}{n}.$$

Sato [104] also derived a result of similar type for two parallel linear boundaries with rational slope. To be precise, we want to enumerate all lattice paths from an arbitrary starting point to an arbitrary end point staying weakly below a given line $\nu x = \mu y - \rho$ and above another given line $\nu x = \mu y + \sigma$, where $\mu, \nu, \rho, \sigma$ are non-negative integers. Again, without loss of generality we may assume that $\nu < \mu$ and that the starting point is the origin.

Following the approach we have taken earlier, we define the $\nu \times \nu$ matrix $T(z;c,\rho,\sigma)$ by

$$T(z;c,\rho,\sigma) := \big(t(z;c+g,\rho+h,\sigma-h)\big)_{0\le g,h\le \nu-1}, \tag{10.32}$$

where

$$t(z;c,\rho,\sigma) = \sum_{\mu n+c\equiv\rho \pmod{\nu}} t(n;c,\rho,\sigma)\, z^n, \tag{10.33}$$

with

$$t(n;c,\rho,\sigma) = \begin{cases} \left|L\big((0,0)\to(\frac{\mu n+c-\rho}{\nu},n)\mid \mu y+\sigma\ge \nu x\ge \mu y-\rho\big)\right|, & \\ & \mu n+c\equiv\rho \pmod{\nu} \\ & \text{and } \mu n+c\ge\rho, \\ \binom{n+\frac{\mu n+c-\rho}{\nu}}{n}, & \mu n+c\equiv\rho \pmod{\nu} \\ & \text{and } \mu n+c<\rho. \end{cases} \tag{10.34}$$

Similar to the one boundary case, what the matrix $T(z;c,\rho,\sigma)$ contains is generating functions of the path numbers $t(n;c,\rho,\sigma)$ that we want to compute. The definition of $t(n;c,\rho,\sigma)$ for $\mu n+c<\rho$ (in which case there cannot be any paths from $(0,0)$ to $(\frac{\mu n+c-\rho}{\nu},n)$) is just for technical convenience. Basically, the matrix $T(z;c,\rho,\sigma)$ is

$$T(z;c,\rho,\sigma) = \Bigg(\sum_{\mu n+c+g-h\equiv\rho \pmod{\nu}} \left|L\left((0,0)\to\left(\frac{\mu n+c+g-\rho-h}{\nu},n\right)\mid \mu y+\sigma-h\ge \nu x\ge \mu y-\rho-g\right)\right| z^n\Bigg)_{0\le g,h\le\nu-1}. \tag{10.35}$$

The following theorem of Sato [104, Theorem 4] tells us how to compute $T(z;c,\rho,\sigma)$.

**Theorem 10.5.4** *For any non-negative integers $c$, $\rho$, $\sigma$, $\nu$, $\mu$, with $\nu<\mu$, we have*

$$T(z;c,\rho,\sigma) = \Phi(z;\rho+\sigma+1-c-\mu)\,\Phi^{-1}(z;\rho+\sigma+1)\,\Phi(z;\rho), \tag{10.36}$$

*where $\Phi(z;\vartheta)$ is given by* (10.30).

**Example 10.5.5** *As an illustration, let us again consider the case $\nu=2$, $\mu=3$. By* (10.34)*, the quantity $t(n;c,\rho,\sigma)$ represents the number of all lattice paths from* $(0,0)$

*to* $((3n+c-\rho)/2, n)$ *which stay weakly below the line* $2x = 3y - \rho$ *and above the line* $2x = 3y + \sigma$*, where* $c \equiv 3n - \rho \pmod 2$*, i.e.,* $c \equiv n + \rho \pmod 2$*.*

*By definition* (10.32)*, we have*

$$T(z;c,\rho,\sigma) = \big(t(z;c+g,\rho+h,\sigma-h)\big)_{0\le g,h\le 1} = \begin{pmatrix} t(z;c,\rho,\sigma) & t(z;c,\rho+1,\sigma-1) \\ t(z;c+1,\rho,\sigma) & t(z;c+1,\rho+1,\sigma-1) \end{pmatrix}.$$

*By Theorem 10.5.4, this can be written as*

$$T(z;c,\rho,\sigma) = \begin{pmatrix} \phi(z;\rho+\sigma-c-1) & \phi(z;\rho+\sigma-c) \\ \phi(z;\rho+\sigma-c-2) & \phi(z;\rho+\sigma-c-1) \end{pmatrix} \times \begin{pmatrix} \phi(z;\rho+\sigma+1) & \phi(z;\rho+\sigma+2) \\ \phi(z;\rho+\sigma) & \phi(z;\rho+\sigma+1) \end{pmatrix}^{-1} \times \begin{pmatrix} \phi(z;\rho) & \phi(z;\rho+1) \\ \phi(z;\rho-1) & \phi(z;\rho) \end{pmatrix}, \tag{10.37}$$

*where*

$$\phi(z;a) = \sum_{\substack{\mu l \equiv a \pmod \nu \\ \mu l \le a}} (-1)^l \binom{(a-\mu l)/\nu}{l} z^l.$$

*In particular, if* $c = \sigma$ *and* $\rho = 0$*, so that we are, for example, interested in the number of all lattice paths from* $(0,0)$ *to* $((3n+c)/2, n)$ *that stay weakly below the line* $2x = 3y$ *and above the line* $2x = 3y + c$ *(the reader should observe that this means that the starting point is on the first line whereas the end point is on the second line), then our formula* (10.37) *reduces to*

$$T(z;c,\rho) = \begin{pmatrix} t(z;c,0,c) & t(z;c,1,c-1) \\ t(z;c+1,0,c) & t(z;c+1,1,c-1) \end{pmatrix} = \frac{1}{\phi^2(z;c+1) - \phi(z;c)\phi(z;c+2)} \begin{pmatrix} \phi(z;c) & \phi(z;c+1) \\ 0 & 0 \end{pmatrix}.$$

*Thus we obtain*

$$t(z;c,0,c) = \sum_{n \equiv c \pmod 2} \left| L\left((0,0) \to \left(\tfrac{3n+c}{2}, n\right) \mid 3y + c \ge 2x \ge 3y\right) \right| z^n = \frac{\phi(z;c)}{\phi^2(z;c+1) - \phi(z;c)\phi(z;c+2)}, \tag{10.38}$$

*with*

$$\phi(z;a) = \sum_{\substack{3l \equiv a \pmod 2 \\ 3l \le a}} (-1)^l \binom{(a-3l)/2}{l} z^l.$$

## 10.6 Simple paths with a piecewise linear boundary

In this section we generalize the one-sided linear boundary results in Corollary 10.3.2 and Theorems 10.4.5 and 10.4.7 to piecewise linear boundaries. To be more precise, we want to count lattice paths from the origin $(0,0)$ to $(c,d)$ staying weakly below the line segments

$$\{(x,y): x=\mu_1 y+\nu_1, 0=y_0\le y\le y_1\},\ \{(x,y): x=\mu_2 y+\nu_2, y_1<y\le y_2\},\\ \dots,\ \{(x,y): x=\mu_m y+\nu_m, y_{m-1}<y\le y_m=d\}, \tag{10.39}$$

for some sequence $0=y_0<y_1<\dots<y_m=d$ of non-negative integers, non-negative integers $\mu_1,\mu_2,\dots,\mu_m$, and integers $\nu_1,\nu_2,\dots,\nu_m$. Let us denote this piecewise linear restriction by $R_m$. See Figure 10.9 for an example. By an iteration argument it will be seen that the solution to this problem can be given in form of an $m$-fold sum.

The result below is due to Niederhausen [98, Sec. 2.2 in connection with (2.4) and (2.7)], but see also [96]. In order to understand the statement below, it is important to observe that the number of paths that we want to determine is a polynomial in $c$, while keeping all other variables fixed. We shall not provide a detailed argument here but, instead, refer to [98, Sec. 2.2]. For convenience, let us denote this polynomial by $L_{R_m,d}(c)$.

**Theorem 10.6.1** *The number of lattice paths from* $(0,0)$ *to* $(c,d)$ *staying weakly below the piecewise linear boundary* $R_m$ *given in* (10.39) *is equal to*

$$\left|L((0,0)\to(c,d)\mid R_m)\right| = \sum_{i=0}^{y_{m-1}} L_{R_{m-1},i}(\mu_m i+\nu_m-1)\\ \cdot\frac{c-\mu_m d-\nu_m+1}{c+d-i(\mu_m+1)-\nu_m+1}\binom{c+d-i(\mu_m+1)-\nu_m+1}{d-i}. \tag{10.40}$$

**Remark 10.6.2** Clearly, we may now apply Theorem 10.6.1 to $L_{R_{m-1},i}(\mu_m i+\nu_m-1)$ so that, iteratively, we obtain an $m$-fold sum. (In the last step, one applies (10.22).)

*Idea of proof of Theorem 10.6.1.* To begin with, let us assume that the piecewise linear boundary be convex. See Figure 10.9 for an example. Evidently, any path from $(0,0)$ to $(c,d)$ has to touch $x=\mu_m y+\nu_m$ for the last time, say in $(\mu_m i+\nu_m,i)$. In Figure 10.9 we have $m=3$, and the last touching point of the path $P$ with $x=\mu_m y+\nu_m$ is the point $(10,2)$, which is marked by a star. Then we utilize the same idea that led to the formula (10.21) to obtain the number in question being equal to

$$\left|L((0,0)\to(c,d)\mid R_m)\right|\\ =\sum_{i=0}^{y_{m-1}}\left|L((0,0)\to(\mu_m i+\nu_m-1,i)\mid R_m)\right|\\ \cdot\left|L((\mu_m i+\nu_m,i)\to(c,d)\mid x\ge\mu_m y+\nu_m)\right|$$

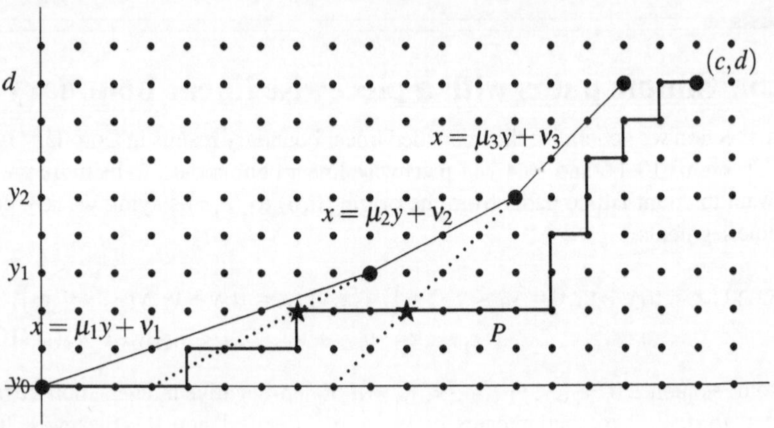


**Figure 10.9**
A piecewise linear boundary.

$$= \sum_{i=0}^{y_{m-1}} \left| L\big((0,0) \to (\mu_m i + \nu_m - 1, i) \mid R_{m-1}\big) \right| \cdot \frac{c - \mu_m d - \nu_m + 1}{c + d - i(\mu_m + 1) - \nu_m + 1} \binom{c + d - i(\mu_m + 1) - \nu_m + 1}{d - i},$$

by virtue of Theorem 10.4.5. In the summand, we were allowed to replace $R_m$ by $R_{m-1}$ since the summation ends at $i = y_{m-1}$, and thus the $m$th segment does not come into play.

Evidently, if the piecewise linear boundary should not be convex, then this argument breaks down. However, Niederhausen shows in [98, Sec. 2.2], using the polynomiality of the path numbers (and some results from umbral calculus), that the above formula continues to hold in that case also, even if the substitution of $\mu_m i + \nu_m - 1$ in the argument of the polynomial $L_{R_{m-1},i}(\,.\,)$ has no combinatorial meaning anymore. ∎

## 10.7 Simple paths with general boundaries

The most general problem to encounter is to count paths in a region that is bounded by nonlinear upper and lower boundaries as exemplified in Figure 10.10.

To have a convenient notation, let $a_1 \le a_2 \le \cdots \le a_n$ and $b_1 \le b_2 \le \cdots \le b_n$ be integers with $a_i \ge b_i$. We abbreviate $\mathbf{a} = (a_1, a_2, \ldots, a_n)$ and $\mathbf{b} = (b_1, b_2, \ldots, b_n)$. By $L\big((0,b_1) \to (n,a_n) \mid \mathbf{a} \ge \mathbf{y} \ge \mathbf{b}\big)$ we denote the set of all lattice paths from $(0,b_1)$ to $(n,a_n)$ that satisfy the property that for all $i = 1,2,\ldots,n$ the height $y_i$ of the $i$th horizontal step is in the interval $[b_i, a_i]$. If we also write $\mathbf{y}(P) = (y_1, y_2, \ldots, y_n)$ for the

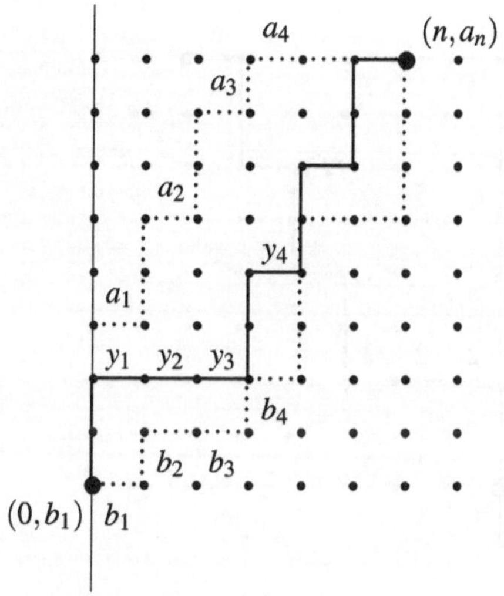


**Figure 10.10**
A path in a ladder-shaped region.

sequence of heights of horizontal steps of a path $P$, then the notation just introduced explains itself. Pictorially (see Figure 10.10), the described restriction means that we consider paths in a ladder-shaped region, the upper ladder being determined by $\mathbf{a}$, the lower ladder being determined by $\mathbf{b}$. See Figure 10.11, which displays an example with $n = 6$, $\mathbf{a} = (3,5,7,8,8,8)$, $\mathbf{b} = (0,1,1,2,5,5)$, $\mathbf{y}(P_0) = (2,2,2,4,6,8)$.

Originally, the result below was derived by Kreweras [85] using recurrence relations, but the most conceptual and most elegant way to attack this problem is by the method of non-intersecting lattice paths; see Section 10.13 and [115].

**Theorem 10.7.1** *Let* $\mathbf{a} = (a_1,a_2,\ldots,a_n)$ *and* $\mathbf{b} = (b_1,b_2,\ldots,b_n)$ *be integer sequences with* $a_1 \le a_2 \le \cdots \le a_n$, $b_1 \le b_2 \le \cdots \le b_n$, *and* $a_i \ge b_i$, $i = 1,2,\ldots,n$. *The number of all paths from* $(0,b_1)$ *to* $(n,a_n)$ *satisfying the property that, for all* $i = 1,2,\ldots,n$, *the height of the ith horizontal step is between* $b_i$ *and* $a_i$ *is given by*

$$\left|L((0,b_1) \to (n,a_n) \mid \mathbf{a} \ge \mathbf{y} \ge \mathbf{b})\right| = \det_{1\le i,j\le n}\left(\binom{a_i - b_j + 1}{j-i+1}\right). \tag{10.41}$$

*Proof.* We apply Theorem 10.13.3 with $\lambda = (1,1,\ldots,1)$ and $\mu = (0,0,\ldots,0)$, both vectors containing $n$ entries. This counts vectors $(\pi_1,\pi_2,\ldots,\pi_n)$ with $\pi_1 < \pi_2 < \cdots < \pi_n$ with a lower and an upper bound on each $\pi_i$. By replacing $\pi_i$ by $\pi_i - i$, this counting problem is translated into the counting problem we consider here, $\pi_i - i$ corresponding to the height of the $i$th horizontal step of a path. ∎

Of course, with increasing $n$ this formula will become less tractable. An alternative formula can be obtained by rotating the whole picture by 90° and applying

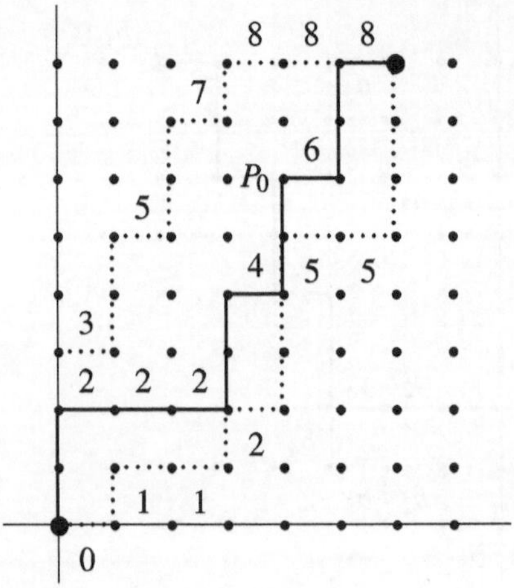


**Figure 10.11**
Lattice path with a general boundary.

formula (10.41) to the new situation. Now the size of the determinant is $a_n$, which is smaller than before if $a_n < n$, i.e., if the difference between the $y$-coordinates of end and starting point is less than the difference between the respective $x$-coordinates.

In some cases, a different type of formula might be preferable, which one may obtain by the so-called **dummy path technique**, as proposed in Krattenthaler and Mohanty [83]. Again, it comes from non-intersecting lattice paths. It is based on the following observation (see Stanley [111, Ex. 2.7.2]).

**Lemma 10.7.2** *Let $C_1, C_2, \dots, C_n$ be pairwise distinct points in $\mathbb{Z}^2$. Then the number of lattice paths from $(a,b)$ to $(c,d)$ that avoid $C_1, C_2, \dots, C_n$ is given by*

$$\det_{1 \le i,j \le n+1} \left( \left| L(A_j \to E_i) \right| \right), \tag{10.42}$$

*where $A_1 = (a,b)$, $A_2 = C_1$, $\dots$, $A_{n+1} = C_n$, $E_1 = (c,d)$, $E_2 = C_1$, $\dots$, $E_{n+1} = C_n$.*

*Proof.* We reformulate our counting problem in that we want to determine the number of families $(P_1, P_2, \dots, P_{n+1})$ of non-intersecting lattice paths, where $P_1$ runs from $A_1 = (a,b)$ to $E_1 = (c,d)$, and for $i = 1,2,\dots,n$ the "dummy path" $P_{i+1}$ runs from $A_{i+1} = C_i$ to $E_{i+1} = C_i$. By Theorem 10.13.1, with $G$ the directed graph with vertices $\mathbb{Z}^2$ and edges given by horizontal and vertical unit steps in the positive direction, all weights being 1, $A_i$ and $E_i$ as above, this number equals the determinant in (10.42). ∎

The idea now is that, given some (possibly two-sided) boundary, one "describes" this boundary by such "dummy points" (paths) and uses the above lemma to compute the number of paths that avoid these, thus avoiding the boundary. In some cases the boundary can be "described" by only very few "dummy points," which may lead to

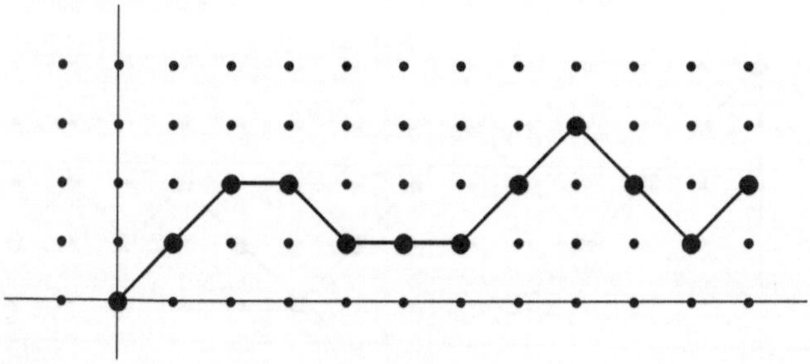

**Figure 10.12**
A Motzkin path.

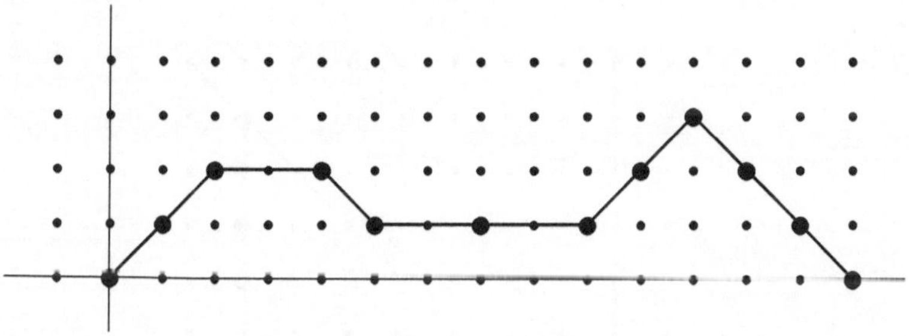

**Figure 10.13**
A Schröder path.

a useful formula. Several formulae that appear in the literature are instances of this idea (sometimes of minor variations), although it may not be stated there that way; see [95, Theorem 2 on p. 36] and [83] and the references given there.

## 10.8 Elementary results on Motzkin and Schröder paths

The subject of this section and the following three sections is lattice paths in $\mathbb{Z}^2$, which consist of **up-steps** $(1,1)$, **down-steps** $(1,-1)$, and **level-steps** $(1,0)$ or $(2,0)$ that do not pass below the $x$-axis. If the only allowed level-steps are unit steps $(1,0)$, then the corresponding paths are called **Motzkin paths**. If the only allowed level-steps are double steps $(2,0)$, then the corresponding paths are called **Schröder paths.** We call the special paths which consist of just up- and down-steps (but contain no level-steps) **Catalan paths**. In the special case, where these paths start and end on the $x$-axis, they are commonly called **Dyck paths**.

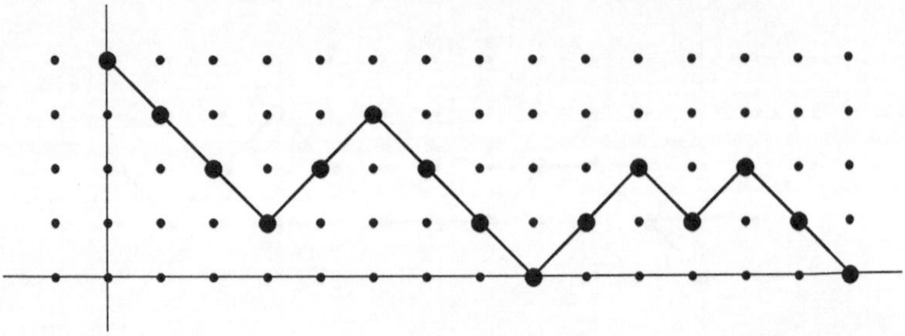

**Figure 10.14**
A Catalan path.

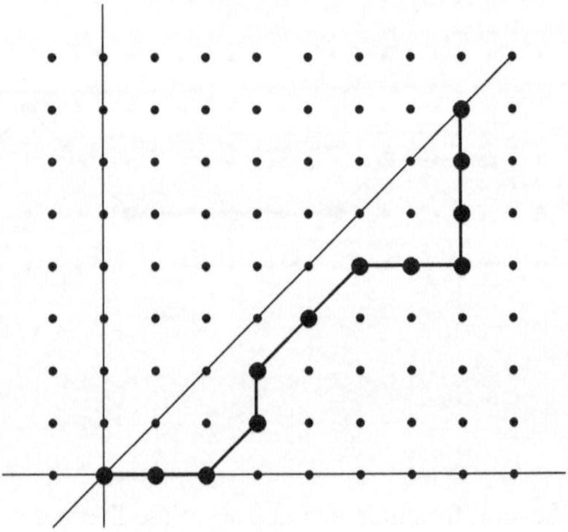

**Figure 10.15**
A Schröder path rotated-reflected.

Let $M = \{(1,1),(1,-1),(1,0)\}$ and $S = \{(1,1),(1,-1),(2,0)\}$, so that $M$ is the set of steps allowed in Motzkin paths (see Figure 10.12 for an example) and $S$ is the set of steps allowed in Schröder paths (see Figure 10.13 for an example).

A frequently used alternative way to view Schröder paths is by reflecting the picture with respect to the $x$-axis, rotating the result by 45°, and finally scaling everything by a factor of $1/\sqrt{2}$, so that the steps $(1,1),(1,-1),(2,0)$ are replaced by the steps $(1,0),(0,1),(1,1)$, in that order. Figure 10.15 shows the result of this translation when applied to the Schröder path in Figure 10.13. It translates Schröder paths into paths that consist of unit horizontal and vertical steps in the positive direction and of upward diagonal steps, which stay weakly below the main diagonal $y = x$. Without the diagonal restriction, the counting problem would be solved by the Delannoy numbers in (10.6).

Nevertheless, this translation, combined with Theorem 10.3.1, already tells us how to enumerate Motzkin and Schröder paths with given starting and end point.

**Theorem 10.8.1** *Let $b \geq 0$ and $d \geq 0$. The number of all paths from $(a,b)$ to $(c,d)$ that consist of steps out of $M = \{(1,1),(1,-1),(1,0)\}$ and do not pass below the x-axis* **(Motzkin paths)** *is given by*

$$\begin{aligned}&\left|L((a,b)\to(c,d);M \mid y\geq 0)\right|\\&\quad=\sum_{k=0}^{c-a}\binom{c-a}{k}\left(\binom{c-a-k}{(c+d-k-a-b)/2}-\binom{c-a-k}{(c+d-k-a+b+2)/2}\right),\end{aligned} \tag{10.43}$$

*where, by convention, a binomial coefficient is 0 if its bottom parameter is not an integer.*

*Furthermore, the number of all paths from $(a,b)$ to $(c,d)$ that consist of steps out of $S = \{(1,1),(1,-1),(2,0)\}$ and do not pass below the x-axis* **(Schröder paths)** *is given by*

$$\begin{aligned}\left|L((a,b)\to(c,d);S \mid y\geq 0)\right| &= \sum_{k=0}^{(c-a)/2}\binom{c-a-k}{k}\\&\cdot\left(\binom{c-a-2k}{(c+d-2k-a-b)/2}-\binom{c-a-2k}{(c+d-2k-a+b+2)/2}\right),\end{aligned} \tag{10.44}$$

*with the same convention for binomial coefficients.*

*Proof.* By the above described translation (reflection + rotation), a Motzkin path from $(a,b)$ to $(c,d)$ with exactly $k$ level-steps is translated into a path from $\left(\frac{a+b}{2},\frac{a-b}{2}\right)$ to $\left(\frac{c+d}{2},\frac{c-d}{2}\right)$, which consists of steps from $\{(1,0),(0,1),(\frac{1}{2},\frac{1}{2})\}$, among them exactly $k$ diagonal steps $(\frac{1}{2},\frac{1}{2})$, and which stays weakly below the main diagonal $y = x$. Clearly, if we remove the $k$ diagonal steps and concatenate the resulting path pieces, we obtain a simple path from $\left(\frac{a+b}{2},\frac{a-b}{2}\right)$ to $\left(\frac{c+d}{2}-\frac{k}{2},\frac{c-d}{2}-\frac{k}{2}\right)$ which stays weakly below $y = x$. The number of the latter paths was determined in Theorem 10.3.1. On the other hand, there are $\binom{c-a}{k}$ ways to reinsert the $k$ diagonal steps. Thus, Eq. (10.43) is established.

The proof of (10.44) is analogous. ∎

We will derive expressions for corresponding generating functions in Section 10.9, see Theorem 10.9.2.

It is worth stating the special case of Theorem 10.8.1 where the paths start and terminate on the $x$-axis separately.

**Corollary 10.8.2** *The number of Motzkin paths from $(0,0)$ to $(n,0)$ is given by*

$$\left|L((0,0)\to(n,0);M \mid y\geq 0)\right| = \sum_{k=0}^{\lfloor n/2\rfloor}\binom{n}{2k}\frac{1}{k+1}\binom{2k}{k}. \tag{10.45}$$

*If n is even, the number of Schröder paths from* $(0,0)$ *to* $(n,0)$ *is given by*

$$\left|L((0,0) \to (n,0); S \mid y \geq 0)\right| = \sum_{k=0}^{n/2} \binom{n/2+k}{2k} \frac{1}{k+1} \binom{2k}{k}. \qquad (10.46)$$

The numbers in (10.45) are called **Motzkin numbers**. The numbers in (10.46) are called **large Schröder numbers**. If $n \geq 1$, the latter are all divisible by 2 (which is easily seen by switching the first occurrence of a level-step with a pair consisting of an up-step and a down-step, and vice versa). Dividing these numbers by 2, we obtain the **little Schröder numbers**. Similarly to Catalan numbers, Motzkin and Schröder numbers also appear in numerous contexts; see [112, Ex. 6.38 and 6.39].

The summations in (10.43) and (10.44) do not simplify, not even in the special cases given in (10.45) and (10.46).

In concluding this section, we point out that Motzkin paths, or, more precisely, **decorated** Motzkin paths, are of utmost importance for the enumeration of many other combinatorial objects, most importantly for the enumeration of permutations and (set) partitions. A decorated Motzkin path (in the french literature: **"histoire"**) is a Motzkin path in which each step carries a certain label. In terms of enumeration, one may consider this as allowing several different steps of the same kind: for example, several different horizontal steps. In terms of generating functions, this labeling is reflected by appropriate weights of the steps. The importance of decorated Motzkin paths comes from the fact that several bijections have been constructed between them and permutations or partitions, which have the property that they "transfer" detailed information about permutations or partitions to the world of (decorated) Motzkin paths, allowing for very refined enumeration results for permutations and partitions. Such bijections have been constructed by Biane [9], Foata and Zeilberger [44], Françon and Viennot [46], Médicis and Viennot [94], and by Simion and Stanton [107]. See [33] for a unifying view.

## 10.9 A continued fraction for the weighted counting of Motzkin paths

We now assign a weight to each Motzkin path that starts and ends on the $x$-axis, and express the corresponding generating function in terms of a **continued fraction**. The corresponding result is due to Flajolet [42]. The weight is so general that the result also covers Schröder paths and Catalan paths.

Given a Motzkin path $P$, we define the weight $w(P)$ to be the product of the weights of all its steps, where the weight of an up-step is 1 (hence, does not contribute anything to the weight), the weight of a level-step at height $h$ is $b_h$, and the weight of a down-step from height $h$ to $h-1$ is $\lambda_h$. Figure 10.16 shows a Motzkin path with the steps labeled by their corresponding weights, so that the weight of the path is $b_2\lambda_2 b_1 b_1 \lambda_3 \lambda_2 \lambda_1 = b_1^2 b_2 \lambda_1 \lambda_2^2 \lambda_3$.

Then the following theorem is true.

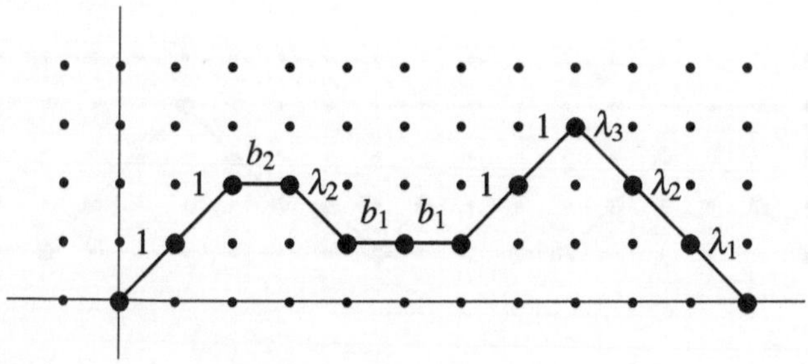


**Figure 10.16**
A labeled Motzkin path.

**Theorem 10.9.1** *With the weight w defined as above, the generating function for Motzkin paths running from the origin back to the x-axis, which stay weakly below the line $y = k$, is given by*

$$GF\big(L((0,0) \to (*,0); M \mid 0 \le y \le k); w\big) = \cfrac{1}{1 - b_0 - \cfrac{\lambda_1}{1 - b_1 - \cfrac{\lambda_2}{1 - b_2 - \cdots - \cfrac{\lambda_k}{1 - b_k}}}}. \tag{10.47}$$

*In particular, the generating function for* all *Motzkin paths running from the origin back to the x-axis is given by the infinite continued fraction*

$$GF\big(L((0,0) \to (*,0); M \mid 0 \le y); w\big) = \cfrac{1}{1 - b_0 - \cfrac{\lambda_1}{1 - b_1 - \cfrac{\lambda_2}{1 - b_2 - \cdots}}}. \tag{10.48}$$

*Proof.* Clearly, it suffices to prove (10.47). Equation (10.48) then follows upon letting $k \to \infty$.

We prove (10.47) by induction on $k$. For $k = 0$, Equation (10.47) is trivially true. Hence, let us assume the truth of (10.47) for $k$ replaced by $k - 1$. For accomplishing the induction step, we consider a Motzkin path starting at the origin, staying weakly below $y = k$, and finally returning to the $x$-axis, see Figure 10.17 for an example with $k = 3$.

Such a path can be uniquely decomposed into

$$l^{e_0} u P_1 d\, l^{e_1} u P_2 d\, l^{e_2} \ldots,$$

where $l$ denotes a level-step at height 0, $u$ an up-step, and $d$ a down-step, where $e_i$ are

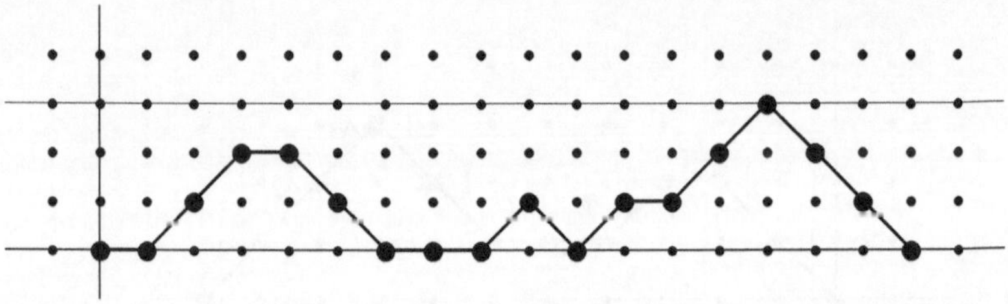

**Figure 10.17**
The path with decomposition $l^1uP_1d\,l^2uP_2d\,l^0uP_3d$.

non-negative integers, and where, for any $i$, $P_i$ is some path between the lines $y=1$ and $y=k$ that starts at and returns to the line $y=1$. For example, this decomposition applied to the path in Figure 10.17 yields

$$l^1uP_1d\,l^2uP_2d\,l^0uP_3d,$$

where $P_1 = uld$, $P_2$ is the empty path, and $P_3 = luudd$. This implies immediately the generating function equation

$$GF\big(L((0,0)\to(*,0);M\mid 0\le y\le k);w\big)$$
$$=\frac{1}{1-b_0-\lambda_1\cdot GF\big(L((0,1)\to(*,1);M\mid 1\le y\le k);w\big)}.$$

By induction, the generating function on the right-hand side is known: It is given by (10.47) with $k$ replaced by $k-1$, $b_i$ replaced by $b_{i+1}$, and $\lambda_i$ replaced by $\lambda_{i+1}$, for all $i$. This completes the induction step. ∎

This result has numerous consequences. First of all, it allows us to derive algebraic expressions for the generating functions $\sum_{n\ge0} M_n z^n$ and $\sum_{n\ge0} S_n z^n$, where $M_n$ denotes the number of all Motzkin paths from $(0,0)$ to $(n,0)$, and where $S_n$ denotes the number of all Schröder paths from $(0,0)$ to $(2n,0)$. By definition, $M_0 = S_0 = 1$. The numbers $M_n$ are called **Motzkin numbers**, while the numbers $S_n$ are called **large Schröder numbers**.

**Theorem 10.9.2** *We have*

$$\sum_{n\ge0} M_n z^n = \frac{1-z-\sqrt{1-2z-3z^2}}{2z^2} \tag{10.49}$$

*and*

$$\sum_{n\ge0} S_n z^n = \frac{1-z-\sqrt{1-6z+z^2}}{2z}. \tag{10.50}$$

*Proof.* By (10.48) with $b_i = z$ and $\lambda_i = z^2$ for all $i$ (the reader should note that for any down-step there is a corresponding up-step before), we have

$$\sum_{n\geq 0} M_n z^n = \cfrac{1}{1-z-\cfrac{z^2}{1-z-\cfrac{z^2}{1-z-\cdots}}}.$$

Thus, in particular, we have $M(z) = 1/(1-z-z^2M(z))$. The appropriate solution of this quadratic equation is exactly the right-hand side of (10.49).

Similarly, by setting $b_i = \lambda_i = z^2$ in (10.48) for all $i$, we obtain

$$\sum_{n\geq 0} S_n z^{2n} = \cfrac{1}{1-z^2-\cfrac{z^2}{1-z^2-\cfrac{z^2}{1-z^2-\cdots}}},$$

and eventually (10.50) after solving the analogous quadratic equation. ∎

In Section 10.11, we will express the continued fraction (10.47) in numerator/denominator form, the numerator and denominator being orthogonal polynomials.

We conclude this section with another continued fraction result, due to Roblet and Viennot [102]. We restrict our attention to Dyck paths, that is, to paths consisting of up- and down-steps, starting at the origin and returning to the $x$-axis, and never running below the $x$-axis. We refine the earlier defined weight $w$ in the following way, so that in addition it also takes into account peaks: Given a Dyck path $P$, we define the weight $\hat{w}(P)$ of $P$ to be the product of the weights of all its steps, where the weight of an up-step is 1, where the weight of a down-step from height $h$ to $h-1$, which follows immediately after an up-step (thus, together, forming a peak of the path), is $v_h$, and where the weight of a down-step from height $h$ to $h-1$, which follows after another down-step, is $\lambda_h$. Thus, the weight of the Dyck path in Figure 10.18 is $v_2v_4v_4\lambda_3v_3\lambda_2\lambda_1v_1 = v_1v_2v_3v_4^2\lambda_1\lambda_2\lambda_3$. With these definitions, the theorem of Roblet and Viennot [102, Prop. 1] reads as follows.

**Theorem 10.9.3** *With the weight $\hat{w}$ defined as above, the generating function $\sum_P \hat{w}(P)$, where the sum is over all Dyck paths starting at the origin and returning to the x-axis, is given by*

$$GF\big(L((0,0)\to(*,0);\{(1,1),(1,-1)\}\mid y\geq 0);\hat{w}\big)$$
$$= \cfrac{1}{1-(v_1-\lambda_1)-\cfrac{\lambda_1}{1-(v_2-\lambda_2)-\cfrac{\lambda_2}{1-(v_3-\lambda_3)-\cdots}}}. \qquad (10.51)$$

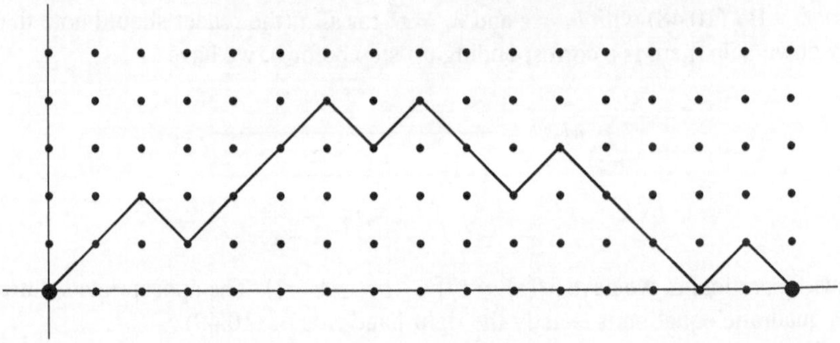

**Figure 10.18**
A Dyck path.

## 10.10 Lattice paths and orthogonal polynomials

Orthogonal polynomials play an important role in many different subject areas, whether they be pure or applied. The reader is referred to [117] for an in-depth introduction. It is well-known that the theory of orthogonal polynomials is intimately connected with Hankel determinants and continued fractions, which we just discussed in Section 10.9 from a combinatorial point of view. It is Viennot [119] who made the connection, and who showed that a large part of the theory of orthogonal polynomials is in fact combinatorics. The key objects in this **combinatorial theory of orthogonal polynomials** are **Motzkin paths**. If one is forced to, one may compress the interplay between the theory of orthogonal polynomials and path enumeration to two key facts: First, (generalized) moments of orthogonal polynomials are generating functions for Motzkin paths, see Theorem 10.10.3; second, generating functions for bounded Motzkin paths can be expressed in terms of orthogonal polynomials; see Theorem 10.11.1. But, of course, this combinatorial theory of orthogonal polynomials has much more to offer, of which we present an extract in this section, with a focus on path enumeration.

We call a sequence $(p_n(x))_{n\geq 0}$ of polynomials over $\mathbb{C}$, where $p_n(x)$ is of degree $n$. **orthogonal** if there exists a linear functional $L$ on polynomials over $\mathbb{C}$ (i.e., a linear map, which maps a polynomial to a complex number) such that

$$L(p_n(x)p_m(x)) = \begin{cases} 0, & \text{if } n \neq m, \\ \text{nonzero}, & \text{if } n = m. \end{cases} \tag{10.52}$$

We alert the reader that our definition deviates from the classical *analytic* definition in that we *do not* require $L(p_n(x)^2)$ to be *positive*. The above somewhat weaker notion of orthogonality is sometimes referred to as **formal orthogonality**. The term "formal" expresses the fact that the corresponding theory does not require any analytic tools, just formal, algebraic arguments. In fact, the formal theory could be equally well developed over any field $K$ of characteristic 0 (instead of over $\mathbb{C}$).

It is easy to see that it is not true that for every linear functional $L$ there is a corresponding sequence of orthogonal polynomials. Let us consider the example of the linear functional defined by $L(x^n) := 1$, $n = 0, 1, 2, \ldots$. Equivalently, this means that $L(p(x)) = p(1)$. In order to construct a corresponding sequence of orthogonal polynomials, we start with $p_0(x)$. This must be a polynomial of degree 0, but otherwise we are completely free. Without loss of generality we may choose $p_0(x) \equiv 1$. To determine $p_1(x)$, we use (10.52) with $m = 0$ and $n = 1$. Thus we obtain $p_1(x) = x - 1$. But then we have $L(p_1(x)^2) = L((x-1)^2) = 0$, which violates the requirement (10.52), with $m = n = 1$, that $L(p_1(x)^2)$ should be nonzero.

On the other hand, if we have a linear functional $L$ such that there exists a sequence of orthogonal polynomials, then it is easy to see that all other sequences are just linear multiples of the former sequence.

**Lemma 10.10.1** *Let $L$ be a linear functional on polynomials and $(p_n(x))_{n\geq 0}$ be a sequence of polynomials orthogonal with respect to $L$. If $(q_n(x))_{n\geq 0}$ is another sequence of polynomials orthogonal with respect to $L$, then there are nonzero numbers $a_n \in \mathbb{C}$ such that $q_n(x) = a_n\, p_n(x)$.*

Lemma 10.10.1 justifies that from now on we will restrict our attention to sequences of **monic** polynomials.

One of the key results in the theory of orthogonal polynomials is **Favard's Theorem**, which we state next.

**Theorem 10.10.2** *A sequence $(p_n(x))_{n\geq 0}$ of monic polynomials, $p_n(x)$ being of degree $n$, is orthogonal if and only if there exist sequences $(b_n)_{n\geq 0}$ and $(\lambda_n)_{n\geq 1}$, with $\lambda_n \neq 0$ for all $n \geq 1$, such that the three-term recurrence*

$$x p_n(x) = p_{n+1}(x) + b_n p_n(x) + \lambda_n p_{n-1}(x), \qquad \textit{for } n \geq 1, \tag{10.53}$$

*holds, with initial conditions $p_0(x) = 1$ and $p_1(x) = x - b_0$.*

In our context, more important than the statement of the theorem itself is its proof, which introduces Motzkin paths in a surprising way in (10.57), and in particular Theorem 10.10.3 below, which is the key ingredient in the proof, given after the proof of Theorem 10.10.3.

**Theorem 10.10.3** *Let the polynomials $p_n(x)$ be given by the three-term recurrence (10.53), and let $L$ be the linear functional defined by $L(1) = 1$ and $L(p_n(x)) = 0$ for $n \geq 1$. Then*

$$L(x^n p_k(x) p_l(x)) = \lambda_1 \cdots \lambda_l \cdot GF\big(L((0,k) \to (n,l); M \mid 0 \leq y); w\big), \tag{10.54}$$

*where $w$ is the weight on Motzkin paths defined in Section 10.9.*

*Proof.* We prove the assertion by induction on $n$.

If $n = 0$, then we have to show

$$L\big(p_k(x)\,p_l(x)\big) = \lambda_1 \cdots \lambda_l \cdot \delta_{k,l}, \tag{10.55}$$

where $\delta_{k,l}$ denotes the Kronecker delta. We establish this claim by induction on $k+l$. It is obviously true for $k = l = 0$. Without loss of generality, we assume $k \geq l$. Then, using the three-term recurrence (10.53) twice, together with the induction hypothesis, we have

$$\begin{aligned} L\big(p_k(x)\,p_l(x)\big) &= L\big(p_k(x)\,x p_{l-1}(x)\big) - b_{l-1}L\big(p_k(x)\,p_{l-1}(x)\big) - \lambda_{l-1}L\big(p_k(x)\,p_{l-2}(x)\big)\\ &= L\big(x p_k(x)\,p_{l-1}(x)\big)\\ &= L\big(p_{k+1}(x)\,p_{l-1}(x)\big) + b_k L\big(p_k(x)\,p_{l-1}(x)\big) + \lambda_k L\big(p_{k-1}(x)\,p_{l-1}(x)\big)\\ &= \lambda_k L\big(p_{k-1}(x)\,p_{l-1}(x)\big). \end{aligned}$$

Clearly, this achieves the induction step and thus establishes (10.55).

We may now continue with the induction on $n$. For the induction step, we apply (10.53) with $n = k$ on the left-hand side of (10.54). This leads to

$$\begin{aligned} &L\big(x^n\,p_k(x)\,p_l(x)\big)\\ &\quad = L\big(x^{n-1}\,p_{k+1}(x)\,p_l(x)\big) + b_k L\big(x^{n-1}\,p_k(x)\,p_l(x)\big) + \lambda_k L\big(x^{n-1}\,p_{k-1}(x)\,p_l(x)\big). \end{aligned}$$

By the induction hypothesis, we may interpret the right-hand side of this equality as a generating function for Motzkin paths, as described by (10.54) with $n$ replaced by $n-1$. It is then straightforward to see that this implies (10.54) itself. ∎

Now we have all the prerequisites available in order to prove Theorem 10.10.2.

*Proof of Theorem 10.10.2.* For showing the forward implication, let $(p_n(x))_{n\geq 0}$ be a sequence of monic polynomials, $p_n(x)$ of degree $n$, which is orthogonal with respect to the linear functional $L$. Then we can express $xp_n(x)$ in terms of a linear combination of the polynomials $p_{n+1}(x), p_n(x), \ldots, p_0(x)$,

$$xp_n(x) = p_{n+1}(x) + b_n p_n(x) + \lambda_n p_{n-1}(x) + \omega_{n,n-2}p_{n-2}(x) + \cdots + \omega_{n,0}p_0(x). \tag{10.56}$$

We have to show that in fact the first three terms on the right-hand side suffice, i.e., that all other terms are zero.

In order to do that, we multiply both sides of (10.56) by $p_i(x)$, for some $i < n-1$, and apply $L$ on both sides. Because of (10.52), on the right-hand side it is only the term $\omega_{n,i}L\big(p_i(x)^2\big)$ that survives. On the left-hand side we obtain $L\big(xp_i(x)p_n(x)\big)$. The polynomial $xp_i(x)$ of degree $i+1$ can be expressed as a linear combination of the polynomials $p_{i+1}(x), p_i(x), \ldots, p_0(x)$. Because of (10.52) and $i < n-1$, we therefore conclude that $L\big(xp_i(x)p_n(x)\big) = 0$. Hence, $\omega_{n,i}$ is indeed 0 for $i < n-1$. Similarly, we have

$$\lambda_n L\big(p_{n-1}(x)^2\big) = L\big(xp_{n-1}(x)p_n(x)\big) = L\big(p_n(x)^2\big),$$

which is nonzero because of (10.52). Hence, we have $\lambda_n \neq 0$, as desired.

For the proof of the backward implication, we must construct a linear functional $L$ such that (10.52) holds, given a sequence $(p_n(x))$ of polynomials, $p_n(x)$ of degree $n$, satisfying the three-term recurrence (10.53). We construct $L$ by defining $L(1)=1$ and $L\big(p_n(x)\big)=0$ for $n\geq 1$. Theorem 10.10.3 with $n=0$ immediately implies that $L\big(p_k(x)p_l(x)\big)=0$ if $k\neq l$, as there is no Motzkin path from $(0,k)$ to $(0,l)$, and that $L\big(p_k(x)^2\big)=\lambda_1\cdots\lambda_k\neq 0$. This completes the proof of the theorem. ∎

In the above proof, we have found a linear functional $L$ by defining (see Theorem 10.10.3) its moments $\mu_n := L(x^n)$ to be generating functions for Motzkin paths, namely

$$\mu_n = GF\big(L((0,0)\to(n,0);M\mid 0\leq y);w\big) = \sum_{\substack{P \text{ a Motzkin path}\\ \text{from } (0,0) \text{ to } (n,0)}} w(P), \qquad (10.57)$$

the weights of the paths carrying the coefficients in the three-term recurrence (10.53). This definition generates a linear functional with $\mu_0 = L(1) = 1$. It is easy to see that all other such linear functionals are constant nonzero multiples of the linear functional defined by (10.57). This justifies to restrict ourselves to linear functionals with first moment equal to 1.

In view of Theorem 10.9.1, the backward implication of Theorem 10.10.2 can also be phrased in the following way.

**Corollary 10.10.4** *Let $(p_n(x))_{n\geq 0}$ be a sequence of polynomials satisfying the three-term recurrence* (10.53) *with initial conditions $p_0(x)=1$ and $p_1(x)=x-b_0$. Then $(p_n(x))_{n\geq 0}$ is orthogonal with respect to the linear functional L, where the generating function of its moments $\mu_n = L(x^n)$ is given by*

$$\sum_{n\geq 0}\mu_n z^n = \cfrac{1}{1-b_0 z-\cfrac{\lambda_1 z^2}{1-b_1 z-\cfrac{\lambda_2 z^2}{1-b_2 z-\cdots}}}. \qquad (10.58)$$

*All other linear functionals with respect to which the sequence $(p_n(x))_{n\geq 0}$ is orthogonal are constant nonzero multiples of L.*

**Remark 10.10.5** A continued fraction of the type (10.58) is called a **Jacobi continued fraction** or **J-fraction**.

*Proof of Corollary 10.10.4.* Combine (10.57) and (10.48) with $b_i$ replaced by $b_i z$ and $\lambda_i$ replaced by $\lambda_i z$. ∎

Below, we illustrate what we have found so far by an example. The polynomials which appear in this example, the **Chebyshev polynomials**, are of particular importance for path counting.

**Example 10.10.6** *We choose $b_i = 0$ and $\lambda_i = 1$ for all i. Then the three-term recurrence* (10.53) *becomes*

$$xu_n(x) = u_{n+1}(x) + u_{n-1}(x), \qquad \text{for } n \geq 1, \tag{10.59}$$

*with initial values $u_0(x) = 1$ and $u_1(x) = x$. These polynomials are, up to reparametrization,* **Chebyshev polynomials of the second kind**. *To see that, recall that the latter are defined by*

$$U_n(\cos\vartheta) = \frac{\sin((n+1)\vartheta)}{\sin\vartheta},$$

*or, equivalently,*

$$U_n(x) = \frac{\sin((n+1)\arccos x)}{\sqrt{1-x^2}}.$$

*Because of the easily verified fact that*

$$\sin((n+1)\vartheta) + \sin((n-1)\vartheta) = 2\cos\vartheta \sin n\vartheta,$$

*the Chebyshev polynomials of the second kind satisfy the three-term recurrence*

$$2xU_n(x) = U_{n+1}(x) + U_{n-1}(x), \qquad \text{for } n \geq 1, \tag{10.60}$$

*with initial values $U_0(x) = 1$ and $U_1(x) = 2x$. Therefore we have*

$$U_n(x) = u_n(2x) \tag{10.61}$$

*for all n.*

*It is straightforward to verify*

$$U_n(x) = \sum_{k\geq 0} (-1)^k \binom{n-k}{k} (2x)^{n-2k}, \tag{10.62}$$

*whence, by* (10.61), *we have*

$$u_n(x) = \sum_{k\geq 0} (-1)^k \binom{n-k}{k} x^{n-2k}.$$

*Another well-known fact is*

$$\frac{2}{\pi}\int_0^{\pi} \sin((n+1)\vartheta)\sin((m+1)\vartheta)\,d\vartheta = \begin{cases} 1, & n = m, \\ 0, & n \neq m. \end{cases}$$

*Substitution of $x = \cos\vartheta$ then yields*

$$\frac{2}{\pi}\int_{-1}^{1} U_n(x)U_m(x)\sqrt{1-x^2}\,dx = \delta_{nm}. \tag{10.63}$$

*Thus the linear functional L for Chebyshev polynomials of the second kind is given by*

$$L(p(x)) = \frac{2}{\pi}\int_0^{\pi} p(x)\sqrt{1-x^2}\,dx.$$

*Using* (10.57) *we can now easily compute the corresponding moments. On the right-hand side of* (10.57) *all the terms corresponding to paths that contain a level-step vanish, because* $b_i = 0$ *for all* $i$. *Therefore, the right-hand side counts paths that contain only up-steps and down-steps (and never pass below the x-axis). Clearly, there cannot be such a path if* $n$ *is odd. If* $n$ *is even, then by* (10.11) *the number of these paths is the Catalan number* $\frac{1}{n/2+1}\binom{n}{n/2}$. *Hence, by also taking into account* (10.61), *we have shown that*

$$\frac{2}{\pi}\int_{-1}^{1} x^m\sqrt{1-x^2} = \begin{cases} \frac{1}{4^n}\frac{1}{n+1}\binom{2n}{n}, & m = 2n, \\ 0, & m = 2n+1. \end{cases}$$

*Chebyshev polynomials are not only tied to Catalan paths (Dyck paths), i.e., paths that consist of just up- and down-steps, but also to Motzkin paths. To see this, let us now choose* $b_i = \lambda_i = 1$ *for all* $i$. *Then the three-term recurrence* (10.53) *becomes*

$$xm_n(x) = m_{n+1}(x) + m_n(x) + m_{n-1}(x), \qquad \text{for } n \geq 1, \tag{10.64}$$

*with initial values* $m_0(x) = 1$ *and* $m_1(x) = x - 1$. *Comparison with* (10.60) *reveals that these polynomials are expressible by means of Chebyshev polynomials of the second kind as*

$$m_n(x) = U_n\left(\frac{x-1}{2}\right). \tag{10.65}$$

*We will take advantage of this relation in Section 10.11 to obtain further enumerative results on Motzkin paths.*

We now come back to the earlier observed fact that not all linear functionals allow for a corresponding sequence of orthogonal polynomials. Which linear functionals do is told by the following theorem. The criterion is given in terms of **Hankel determinants** of the moments of $L$. A Hankel determinant (or **persymmetric** or **Turánian determinant**) is a determinant of a matrix that has constant entries along antidiagonals, i.e., it is a determinant of the form $\det_{1\le i,j,\le n}(a_{i+j})$. We omit the proof here, but Viennot [119, Ch. IV, Cor. 6 and 7] has shown that it can be given by an elegant application of the main theorem on non-intersecting lattice paths, Theorem 10.13.1, by using the interpretation of moments in terms of generating functions for Motzkin paths as given in Theorem 10.10.3.

**Theorem 10.10.7** *Let* $L$ *be a linear functional on polynomials with nth moment* $\mu_n = L(x^n)$. *For any non-negative integer* $n$ *let*

$$\Delta_n = \det\begin{pmatrix} \mu_0 & \mu_1 & \mu_2 & \cdots & \mu_n \\ \mu_1 & \mu_2 & \cdots\cdots & & \mu_{n+1} \\ \mu_2 & \cdots\cdots\cdots & & & \mu_{n+2} \\ \vdots & & & & \vdots \\ \mu_n & \cdots\cdots\cdots & & & \mu_{2n} \end{pmatrix}$$

*and*

$$\chi_n = \det \begin{pmatrix} \mu_0 & \mu_1 & \cdots & \mu_{n-1} & \mu_n \\ \mu_1 & \mu_2 & \cdots & \mu_n & \mu_{n+1} \\ \vdots & \vdots & & \vdots & \vdots \\ \mu_{n-1} & \mu_n & \cdots & \mu_{2n-2} & \mu_{2n-1} \\ \mu_{n+1} & \mu_{n+2} & \cdots & \mu_{2n} & \mu_{2n+1} \end{pmatrix}.$$

*Let $(p_n(x))_{n\geq 0}$ be the sequence of monic polynomials that is orthogonal with respect to L. Then the polynomials satisfy the three-term recurrence* (10.53) *with*

$$\lambda_n = \frac{\Delta_n \Delta_{n-2}}{\Delta_{n-1}^2} \tag{10.66}$$

*and*

$$b_n = \frac{\chi_n}{\Delta_n} - \frac{\chi_{n-1}}{\Delta_{n-1}}. \tag{10.67}$$

*In particular, given a linear functional L on the set of polynomials, there exists a sequence of orthogonal polynomials that are orthogonal with respect to L if and only if all Hankel determinants $\Delta_n = \det_{0\leq i,j\leq n}(\mu_{i+j})$ of moments are nonzero.*

Implicit in (10.66) is the Hankel determinant evaluation

$$\Delta_n = \lambda_1^n \lambda_2^{n-1} \cdots \lambda_n^1, \tag{10.68}$$

which expresses the close interplay between Hankel determinants, moments of orthogonal polynomials, and Motzkin path enumeration (via Theorem 10.10.3).

We conclude this section with an explicit, determinantal formula for orthogonal polynomials, given the moments of the orthogonality functional. Again, Viennot [119, Ch. IV, §4] has given a beautiful combinatorial proof for this formula. using non-intersecting lattice paths.

**Theorem 10.10.8** *Let L be a linear functional defined on polynomials with moments $\mu_n = L(x^n)$. Then the corresponding sequence $(p_n(x))_{n\geq 0}$ of monic orthogonal polynomials is given by*

$$p_n(x) = \frac{1}{\Delta_{n-1}} \det \begin{pmatrix} \mu_0 & \mu_1 & \mu_2 & \cdots & \mu_n \\ \mu_1 & \mu_2 & \cdots & \mu_n & \mu_{n+1} \\ \mu_2 & \cdots & \mu_n & \mu_{n+1} & \mu_{n+2} \\ \cdots & \cdots & \cdots & \cdots & \cdots \\ \mu_{n-1} & \mu_n & \mu_{n+1} & \cdots & \mu_{2n-1} \\ 1 & x & \cdots & x^{n-1} & x^n \end{pmatrix}, \tag{10.69}$$

*where, again, $\Delta_{n-1} = \det_{0\leq i,j\leq n-1}(\mu_{i+j})$.*

*Proof.* It suffices to check that $L(x^m p_n(x)) = 0$ for $0 \le m < n$. Indeed, by (10.69) we have

$$L(x^m p_n(x)) = \frac{1}{\Delta_{n-1}} \det \begin{pmatrix} \mu_0 & \mu_1 & \mu_2 & \dots & \mu_n \\ \mu_1 & \mu_2 & \dots & \mu_n & \mu_{n+1} \\ \mu_2 & \dots & \mu_n & \mu_{n+1} & \mu_{n+2} \\ \dots & \dots & \dots & \dots & \dots \\ \mu_{n-1} & \mu_n & \mu_{n+1} & \dots & \mu_{2n-1} \\ \mu_m & \mu_{m+1} & \dots & \mu_{m+n-1} & \mu_{m+n} \end{pmatrix}.$$

Thus the result is zero, because for $0 \le m < n$ the $m$th and the last row in the above determinant are identical. ∎

In Section 10.11 we derive several further enumeration results on Motzkin paths that feature orthogonal polynomials.

We close this section by pointing out that Motzkin paths can be seen as so-called **heaps of pieces**. The corresponding theory has been developed by Viennot [120]. As a matter of fact, it is the combinatorial realization of the **Cartier–Foata monoid** [26].

For further intriguing work on the connections between lattice path counting, Hankel determinants, and continued fractions, the reader is referred to Gessel and Xin [52], and also Sulanke and Xin [116].

## 10.11 Motzkin paths in a strip

In Sections 10.8 and 10.9 we have derived enumeration results for Motzkin paths that start and terminate on the $x$-axis. In particular, Theorem 10.9.1 provided a continued fraction for the generating function with respect to a very general weight. This continued fraction can be compactly brought in numerator/denominator form, using orthogonal polynomials. In fact, more generally, a compact expression for the generating function of Motzkin paths that start and terminate at **arbitrary** points can be given, again using orthogonal polynomials.

In order to be able to state the corresponding result, we need two definitions. Recall that, given sequences $(b_n)_{n\ge 0}$ and $(\lambda_n)_{n\ge 1}$, with $\lambda_n \ne 0$ for all $n \ge 1$, the three-term recurrence (10.53),

$$xp_n(x) = p_{n+1}(x) + b_n p_n(x) + \lambda_n p_{n-1}(x), \qquad \text{for } n \ge 1, \tag{10.70}$$

with initial conditions $p_0(x) = 1$ and $p_1(x) = x - b_0$, produces a sequence $(p_n(x))_{n\ge 0}$ of orthogonal polynomials. We also need associated "shifted" polynomials (often simply called **associated orthogonal polynomials**), denoted by $(Sp_n(x))_{n\ge 0}$, which arise from the sequence $(p_n(x))$ by replacing $\lambda_i$ by $\lambda_{i+1}$ and $b_i$ by $b_{i+1}$, $i = 0, 1, 2, \dots$, everywhere in the three-term recurrence (10.70) and in the initial conditions. Furthermore, given a polynomial $p(x)$ of degree $n$, we denote the corresponding **reciprocal polynomial** $x^n p(1/x)$ by $p^*(x)$.

**Theorem 10.11.1** *With the weight $w$ defined as before Theorem 10.9.1, the generating function for Motzkin paths running from height $r$ to height $s$ that stay weakly below the line $y = k$ is given by*

$$\sum_{n\geq 0} GF\big(L((0,r)\to(n,s);M\mid 0\leq y\leq k);w\big)x^n$$

$$=\begin{cases} \dfrac{x^{s-r}p_r^*(x)S^{s+1}p_{k-s}^*(x)}{p_{k+1}^*(x)}, & \text{if } r\leq s,\\ \lambda_r\cdots\lambda_{s+1}\dfrac{x^{r-s}p_s^*(x)S^{r+1}p_{k-r}^*(x)}{p_{k+1}^*(x)}, & \text{if } r\geq s. \end{cases} \tag{10.71}$$

*In particular, the generating function for Motzkin paths running from the origin back to the x-axis that stay weakly below the line $y = k$ is given by*

$$\sum_{n\geq 0} GF\big(L((0,0)\to(n,0);M\mid 0\leq y\leq k);w\big)x^n = \frac{Sp_k^*(x)}{p_{k+1}^*(x)}. \tag{10.72}$$

*Proof.* Consider the directed graph, $P_{k+1}$ say, with vertices $v_0, v_1, \ldots v_k$, where for $h = 0,1,\ldots,k-1$ there is an arc from $v_h$ to $v_{h+1}$ as well as an arc from $v_{h+1}$ to $v_h$, and where there is a loop for each vertex $v_h$. Motzkin paths that never exceed height $k$ correspond in a one-to-one fashion to **walks** on $P_{k+1}$. In this correspondence, an up-step from height $h$ to $h+1$ in the Motzkin path corresponds to a step from vertex $v_h$ to vertex $v_{h+1}$ in the walk, and similarly for level- and down-steps. To make the correspondence also weight-preserving, we attach a weight of 1 to an arc from $v_h$ to $v_{h+1}$, $h = 0,1,\ldots,k-1$, a weight of $\lambda_h$ to an arc from $v_h$ to $v_{h-1}$, and a weight of $b_h$ to a loop at $v_h$.

By the **transfer matrix method** (see e.g. [111, Theorem 4.7.2]), the generating function for walks from $v_r$ to $v_s$ is given by

$$\frac{(-1)^{r+s}\det(I-xA;s,r)}{\det(I-xA)},$$

where $A$ is the (weighted) **adjacency matrix** of $P_{k+1}$, where $I$ is the $(k+1)\times(k+1)$ identity matrix, and where $\det(I-xA;s,r)$ is the minor of $(I-xA)$ with the $s$th row and $r$th column deleted.

Now, the (weighted) adjacency matrix of $P_{k+1}$ with the property that the weight of a particular walk would correspond to the weight $w$ of the corresponding Motzkin path is the tridiagonal matrix

$$A=\begin{pmatrix} b_0 & 1 & 0 & \ldots & & & \\ \lambda_1 & b_1 & 1 & 0 & \ldots & & \\ 0 & \lambda_2 & b_2 & 1 & 0 & \ldots & \\ \vdots & \ddots & \ddots & \ddots & \ddots & \ddots & \vdots \\ & \ldots & 0 & \lambda_{k-2} & b_{k-2} & 1 & 0 \\ & & \ldots & 0 & \lambda_{k-1} & b_{k-1} & 1 \\ & & & \ldots & 0 & \lambda_k & b_k \end{pmatrix}.$$

It is easily verified that, with this choice of $A$, we have $\det(I - xA) = p^*_{k+1}(x)$ (by expanding the determinant with respect to the last row and comparing with the three-term recurrence (10.53)), and, similarly, that the numerator in (10.71) agrees with $(-1)^{r+s}\det(I - xA; r, s)$. ∎

**Example 10.11.2** *We illustrate Theorem 10.11.1 for the special cases that were considered in Example 10.10.6.*

*Let first $b_i = 0$ and $\lambda_i = 1$ for all $i$. Combinatorially, we are talking about paths consisting of up- and down-steps, that is, Catalan paths (Dyck paths). Since for this choice of $b_i$'s and $\lambda_i$'s there is no difference between the orthogonal polynomials and the corresponding associated orthogonal polynomials arising from* (10.53)*, Example 10.10.6 tells us that*

$$p_n(x) = Sp_n(x) = U_n(x/2).$$

*From* (10.71)*, it then follows that*

$$\sum_{n\geq 0} \left|L((0,r) \to (n,s); \{(1,1),(1,-1)\} \mid 0 \leq y \leq k)\right| \cdot x^n$$
$$= \begin{cases} \dfrac{U_r(1/2x)\,U_{k-s}(1/2x)}{x\,U_{k+1}(1/2x)}, & \text{if } r \leq s,\\[2ex] \dfrac{U_s(1/2x)\,U_{k-r}(1/2x)}{x\,U_{k+1}(1/2x)}, & \text{if } r \geq s. \end{cases} \tag{10.73}$$

*Next let $b_i = \lambda_i = 1$ for all $i$. Combinatorially, we are talking about paths consisting of up-, down-, and level-steps, that is, Motzkin paths. Again, since for this choice of $b_i$'s and $\lambda_i$'s there is no difference between the orthogonal polynomials and the corresponding associated orthogonal polynomials arising from* (10.53)*, Example 10.10.6 tells us that*

$$p_n(x) = Sp_n(x) = U_n\left(\frac{x-1}{2}\right).$$

*From* (10.71)*, it then follows that*

$$\sum_{n\geq 0} \left|L((0,r) \to (n,s); M \mid 0 \leq y \leq k)\right| \cdot x^n$$
$$= \begin{cases} \dfrac{U_r\left(\frac{1-x}{2x}\right) U_{k-s}\left(\frac{1-x}{2x}\right)}{x\,U_{k+1}\left(\frac{1-x}{2x}\right)}, & \text{if } r \leq s,\\[2ex] \dfrac{U_s\left(\frac{1-x}{2x}\right) U_{k-r}\left(\frac{1-x}{2x}\right)}{x\,U_{k+1}\left(\frac{1-x}{2x}\right)}, & \text{if } r \geq s. \end{cases} \tag{10.74}$$

**Example 10.11.3** *The standard application of* (10.74) *concerns the* **gambler's ruin problem** *(see also [38, Ch. XIV]): two players A and B have initially a and $R - a$ dollars, respectively. They play several rounds, in each of which the probability that player A wins is $p_A$, the probability that player B wins is $p_B$, and the probability that there is a tie is $p_T = 1 - p_A - p_B$. If one player wins, (s)he takes a dollar from*

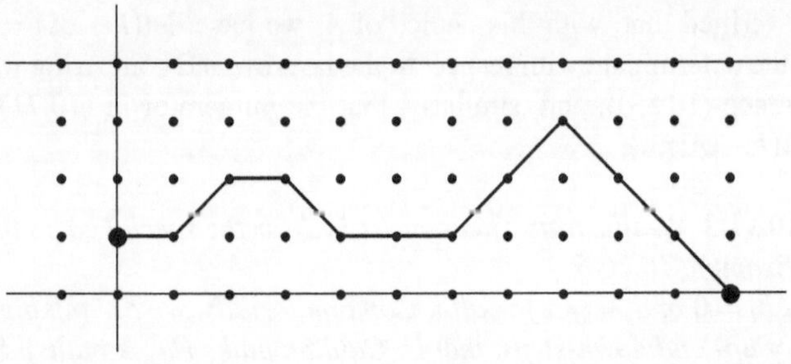


**Figure 10.19**
The gambler's ruin.

*the other. If there is a tie, nothing happens. The play stops when one of the players is bankrupt. What is the probability that player A, say, goes bankrupt after N rounds?*

*By disregarding the last round (which is necessarily a round in which B wins), this problem can be represented by a lattice path starting at $(0,a-1)$, ending at $(N-1,0)$, with steps $(1,1)$ (corresponding to player A to win a round), $(1,-1)$ (corresponding to player B to win a round), and $(1,0)$ (corresponding to a tie), which does not pass below the x-axis, and which does not pass above the horizontal line $y=R-2$. For example, the lattice path in Figure 10.19 corresponds to the play, where player A starts with two dollars, player B starts with four dollars, the outcome of the rounds is in turn TATBTTAABBBB (the letter A symbolizing a round where A won, with an analogous meaning of the letter B, and the letter T symbolizing a tie), so that A goes bankrupt after $N=12$ rounds (while B did not).*

*If we assign the* **weight** *$p_A$ to an up-step $(1,1)$, $p_B$ to a down-step $(1,-1)$, and $p_T$ to a level-step $(1,0)$, then the probability of this play is the product of the weights of all the steps of the path P times $p_B$ (corresponding to the last round where B wins and A goes bankrupt; in our example, it is $p_T p_A p_T p_B p_T p_T p_A p_A p_B p_B p_B p_B$). If we write $p(P)$ for the product of the weights of the steps of P, then, in order to solve the problem, we need to compute the sum $\sum_P p_B p(P)$, where the sum is over all the above described paths from $(0,a-1)$ to $(N-1,0)$.*

*Clearly,* (10.74) *with $r=a-1$ and $s=0$ provides the solution for the above problem, in terms of a generating function. Since the zeroes of the Chebyshev polynomials are explicitly known, one can apply partial fraction decomposition to obtain an explicit formula for the coefficients in the generating function. If this is carried out, then we get*

$$\left|L\big((0,r)\to(n,s);M\mid 0\le y\le k\big)\right|$$
$$=\frac{2}{k+2}\sum_{j=1}^{k+1}\left(2\cos\frac{\pi j}{k+2}+1\right)^n\cdot\sin\frac{\pi j(r+1)}{k+2}\cdot\sin\frac{\pi j(s+1)}{k+2}. \quad (10.75)$$

## 10.12 Further results for lattice paths in the plane

In this section we collect various further results on the enumeration of two-dimensional lattice paths, respectively pointers to further such results.

The first set of results that we describe concerns lattice paths in the plane integer lattice $\mathbb{Z}^2$ that consist of steps from a finite set $\mathbb{S}$ that contains steps of the form $(1,b)$. Here, $b$ is some integer. Say,

$$\mathbb{S} = \{(1,b_1),(1,b_2),\ldots,(1,b_m)\}. \tag{10.76}$$

We also assume that to each step $(1,b_j)$ there is associated a weight $w_j \in \mathbb{C}$.

Banderier and Flajolet [5] completely solved the exact and asymptotic enumeration of lattice paths consisting of steps from $\mathbb{S}$ obeying certain restrictions. We concentrate here on the exact enumeration results.

The key object in their theory is the **characteristic polynomial** of the step set $\mathbb{S}$,

$$P_{\mathbb{S}}(u) = \sum_{j=1}^{m} w_j u^{b_j}. \tag{10.77}$$

If we write $c = -\min_j b_j$ and $d = \max_j b_j$, then $P_{\mathbb{S}}(u)$ can be rewritten in the form

$$P_{\mathbb{S}}(u) = \sum_{j=-c}^{d} p_j u^j,$$

for appropriate coefficients $p_j$. Associated with the characteristic polynomial is the **characteristic equation**

$$1 - zP_{\mathbb{S}}(u) = 0, \tag{10.78}$$

or, equivalently,

$$u^c - zu^c P_{\mathbb{S}}(u) = u^c - z\sum_{j=0}^{c+d} p_{j-c}u^j = 0. \tag{10.79}$$

The form (10.79) has only non-negative powers in $u$, and it shows that, counting multiplicity, there are $c+d$ solutions to the characteristic equation when $u$ is expressed as a function in $z$. These $c+d$ solutions fall into two categories; There are $c$ "small branches" $u_1(z), u_2(z), \ldots, u_c(z)$ satisfying

$$u_j(z) \sim e^{2\pi i(j-1)/c} p_{-c}^{1/c} z^{1/c} \quad \text{as } z \to 0,$$

and $d$ "large branches" $u_{c+1}(z), u_{c+2}(z), \ldots, u_{c+d}(z)$ satisfying

$$u_j(z) \sim e^{2\pi i(c+1-j)/d} p_d^{-1/d} z^{-1/d} \quad \text{as } z \to 0.$$

One can show that there are functions $A(z)$ and $B(z)$ that are analytic and non-zero at 0 such that, in a neighborhood of 0,

$$u_j(z) = \omega^{j-1} z^{1/c} A(\omega^{j-1} z^{1/c}), \text{ with } \omega = e^{2\pi i/c}, \quad j = 1,2,\ldots,c, \tag{10.80}$$

$$u_j(z) = \varpi^{c+1-j} z^{-1/d} B(\varpi^{j-c-1} z^{1/d}), \text{ with } \varpi = e^{2\pi i/d}, \quad j = c+1, c+2, \ldots, c+d. \tag{10.81}$$

We are now in the position to state the enumeration results for lattice paths with steps from $\mathbb{S}$ without further restriction. In the formulation, we use $\ell(P)$ to denote the length of a path $P$, and $h(P)$ to denote the abscissa (height) of the end point of $P$.

**Theorem 10.12.1** *The generating function $\sum_P z^{\ell(P)} u^{h(P)}$ for lattice paths P that start at the origin and consist of steps from $\mathbb{S}$ as given in* (10.76) *equals*

$$GF\left(L((0,0) \to (*,*);\mathbb{S}); z^{\ell(\cdot)} u^{h(\cdot)}\right) = \frac{1}{1 - zP_{\mathbb{S}}(u)}, \tag{10.82}$$

*with $P_{\mathbb{S}}(u)$ the characteristic polynomial of $\mathbb{S}$ given in* (10.77). *Moreover, the generating function $\sum_P z^{\ell(P)}$ for those paths P that end at height* 0 *equals*

$$GF\left(L((0,0) \to (*,0);\mathbb{S}); z^{\ell(\cdot)}\right) = z \sum_{j=1}^{c} \frac{u_j'(z)}{u_j(z)} = z\frac{d}{dz}\left(u_1(z)u_2(z)\cdots u_c(z)\right), \tag{10.83}$$

*where $u_1(z), u_2(z), \ldots, u_c(z)$ are the small branches given in* (10.80). *Finally, for $k < c$ the generating function $\sum_P z^{\ell(P)}$ for those paths P that end at height k equals*

$$GF\left(L((0,0) \to (*,k);\mathbb{S}); z^{\ell(\cdot)}\right) = z \sum_{j=1}^{c} \frac{u_j'(z)}{u_j^{k+1}(z)} = -\frac{z}{k}\frac{d}{dz}\left(\sum_{j=1}^{c} u_j^{-k}(z)\right), \tag{10.84}$$

*where again $u_1(z), u_2(z), \ldots, u_c(z)$ are the small branches given in* (10.80), *while for $k > -d$ it equals*

$$GF\left(L((0,0) \to (*,k);\mathbb{S}); z^{\ell(\cdot)}\right) = -z \sum_{j=c+1}^{c+d} \frac{u_j'(z)}{u_j^{k+1}(z)} = \frac{z}{k}\frac{d}{dz}\left(\sum_{j=c+1}^{c+d} u_j^{-k}(z)\right), \tag{10.85}$$

*Proof.* By elementary combinatorial principles, the generating function $\sum_P z^{\ell(P)} u^{h(P)}$ for lattice paths $P$ that start at the origin and consist of steps from $\mathbb{S}$ is given by $\sum_{n\geq 0} z^n P_{\mathbb{S}}^n(u)$, which equals (10.82).

In order to determine the generating function $\sum_P z^{\ell(P)}$ for those lattice paths $P$ that end at height 0, we have to extract the coefficient of $u^0$ in (10.82). This can be achieved by computing the contour integral

$$\frac{1}{2\pi i}\int_C \frac{1}{1 - zP_{\mathbb{S}}(u)} \frac{du}{u}, \tag{10.86}$$

where $C$ is a contour encircling the origin in the positive direction. One has to choose $C$ so that, for sufficiently small $z$, the small branches lie within the contour, while the large branches lie outside. Then, by the residue theorem, only the small branches contribute to the integral (10.86). The residue at $u = u_j(z)$ equals (assuming that, in addition, we have chosen $z$ so that all small branches are different)

$$\operatorname*{Res}_{u=u_j(z)}\left(\frac{1}{u(1 - zP_{\mathbb{S}}(u))}\right) = -\frac{1}{zu_j(z)P_{\mathbb{S}}'(u_j(z))}.$$

The integral in (10.86) equals the sum of these residues. This sum simplifies to (10.83) since differentiation of both sides of the characteristic equation (10.78) shows that $P'_{\mathbb{S}}(u_j(z))^{-1} = -z^2 u'_j(z)$ for all small branches $u_j(z)$.

The arguments for establishing (10.84) and (10.85) are similar. ∎

The second set of results concerns lattice paths starting at the origin with steps from $\mathbb{S}$ that do not run below the $x$-axis.

**Theorem 10.12.2** *The generating function $\sum_P z^{\ell(P)} u^{h(P)}$ for lattice paths $P$ that start at the origin, consist of steps from $\mathbb{S}$ as given in* (10.76)*, and do not run below the $x$-axis, equals*

$$GF\big(L((0,0) \to (*,*);\mathbb{S} \mid y \geq 0); z^{\ell(\cdot)} u^{h(\cdot)}\big) = \frac{\prod_{j=1}^{c}(u - u_j(z))}{u^c(1 - zP_{\mathbb{S}}(u))}$$
$$= -\frac{1}{p_d z} \prod_{j=c+1}^{c+d} \frac{1}{(u - u_j(z))}, \tag{10.87}$$

*with $P_{\mathbb{S}}(u)$ the characteristic polynomial of $\mathbb{S}$ given in* (10.77)*, and $u_1(z), u_2(z), \ldots, u_c(z)$ and $u_{c+1}(z), u_{c+2}(z), \ldots, u_{c+d}(z)$ the small and large branches given in* (10.80) *and* (10.81)*. In particular, the generating function $\sum_P z^{\ell(P)}$ for those paths $P$ that end at height* 0 *equals*

$$GF\big(L((0,0) \to (*,0);\mathbb{S} \mid y \geq 0); z^{\ell(\cdot)}\big) = \frac{(-1)^{c-1}}{p_{-c} z} \prod_{j=1}^{c} u_j(z))$$
$$= \frac{(-1)^{d-1}}{p_d z} \prod_{j=c+1}^{c+d} \frac{1}{u_j(z))}. \tag{10.88}$$

*Proof.* Here, we use the so-called **kernel method** (cf. e.g. [18]). Let $F(z,u)$ denote the generating function on the left-hand side of (10.87). Then we have

$$F(z,u) = 1 + zP_{\mathbb{S}}(u)F(z,u) - z[u^{<0}]\big(P_{\mathbb{S}}(u)F(z,u)\big), \tag{10.89}$$

where $[u^{<0}]G(z,u)$ means that in the series $G(z,u)$ all monomials $z^n u^m$ with $m \geq 0$ are dropped, because any lattice path that is counted by $F(z,u)$ is either empty, or it consists of a step ($zP_{\mathbb{S}}(u)$ describes the possibilities) added to a path, except that the steps that would take the walk below level 0 are to be taken out (the operator $[u^{<0}]$ extracts the terms to be taken out). Since $P_{\mathbb{S}}(u)$ involves only a finite number of negative powers, we may rewrite (10.89) in the form

$$F(z,u)(1 - zP_{\mathbb{S}}(u)) = 1 - z \sum_{k=0}^{c-1} r_k(u) F_k(z), \tag{10.90}$$

for some Laurent polynomials $r_k(u)$ that can be computed from $P_{\mathbb{S}}(u)$ via (10.89),

$$r_k(u) = [u^{<0}](P_{\mathbb{S}}(u)u^k) = \sum_{j=-c}^{-k-1} p_j u^{j+k}.$$

Here, $F_k(z)$ is the generating function $\sum_P z^{\ell(P)}$ for those paths $P$ that end at height $k$.

In the current context, the factor $1-zP_{\mathbb{S}}(u)$ on the left-hand side of (10.90) (which is identical with the left-hand side of the characteristic equation (10.78)) is called the **kernel**. The idea of the kernel method is to substitute $u=u_j(z)$, $j=1,2,\dots,c$ (that is, the small branches) on both sides of (10.90) so that the kernel, and thus the left-hand side, vanishes. In this way, we arrive at the system of equations

$$u_j^c(z)-z\sum_{k=0}^{c-1}u_j^c(z)r_k(u_j(z))F_k(z)=0,\quad j=1,2,\dots,c.$$

This system of linear equations in the unknowns $F_0(z),F_1(z),\dots,F_{c-1}(z)$ could now be solved. Alternatively, we could observe that the expression

$$u^c-z\sum_{k=0}^{c-1}u^c r_k(u)F_k(z)$$

is a polynomial in $u$ of degree $c$ with leading monomial $u^c$. Its roots are exactly the small branches $u_j(z)$, $j=1,2,\dots,c$. Hence, it factorizes as

$$u^c-z\sum_{k=0}^{c-1}u^c r_k(u)F_k(z)=\prod_{j=1}^{c}(u-u_j(z)). \tag{10.91}$$

Extraction of the coefficient of $u^0$ on both sides immediately gives $F_0(z)$, the generating function for the paths that end at height 0. This leads directly to (10.88). The formula (10.87) follows from (10.90) and (10.91). ∎

Sometimes, the kernel method is also applicable if the set of steps $\mathbb{S}$ is infinite. This is. for instance, the case for **Łukasiewicz paths**, which are paths consisting of steps from $\mathbb{S}_L=\{(1,b):b\in\{-1,0,1,2,\dots\}\}$, starting at the origin, returning to the $x$-axis, and never running below it. In that case, the equation (10.90) for the generating function $\sum_P z^{\ell(P)}u^{h(P)}$ becomes

$$F(z,u)\left(1-\frac{z}{u(1-u)}\right)=1-zu^{-1}F_0(z), \tag{10.92}$$

where, as before, $F_0(z)$ is the generating function for those paths that end at height 0 (that is, return to the $x$-axis). Here, the kernel is

$$1-\frac{z}{u(1-u)},$$

and it vanishes for $u(z)=\frac{1-\sqrt{1-4z}}{2}$. If this is substituted in (10.92), then we obtain

$$GF\big(L((0,0)\to(*,0);\mathbb{S}_L\mid y\ge 0);z^{\ell(\cdot)}\big)=F_0(z)=\frac{1-\sqrt{1-4z}}{2z},$$

the Catalan number generating function (10.20). Hence, also Łukasiewicz paths of length $n$ are enumerated by the Catalan number $C_n=\frac{1}{n+1}\binom{2n}{n}$.

To conclude this topic, it must be mentioned that Banderier and Gittenberger [6] have extended the analyses of [5] to also include the area statistics.

A very cute problem, which arose in a probabilistic context around 2000, is the problem of counting paths (walks) in the **slit plane**. The slit plane is the integer lattice $\mathbb{Z}^2$ where one has taken out the half-axis $\{(k,0) : k \leq 0\}$. Investigation of this problem started with the conjecture that the number of paths in the slit plane that start at $(1,0)$ and do $2n+1$ horizontal or vertical unit steps (in the positive or in the negative direction) is given by the Catalan number $C_{2n+1}$. This conjecture was proved by Bousquet-Mélou and Schaeffer in [20], but they provide much stronger and more general results on the enumeration of lattice paths in the slit plane in that paper. When it is not possible to find exact formulas, then the focus is on the nature of the generating function, whether it be algebraic or not, D-finite or not, etc. Methods used are the cycle lemma and the kernel method.

An innocent looking three-candidate ballot problem stands at the beginning of another long line of investigation: Let $E_1, E_2, E_3$ be candidates in an election, $E_1$ receiving $e_1$ votes, $E_2$ receiving $e_2$ votes, and $E_3$ receiving $e_3$ votes, $e_1 \geq \max\{e_2, e_3\}$. How many ways of counting the votes are there such that at any stage during the counting candidate $E_1$ has at least as many votes as $E_2$ and at least as many votes as $E_3$? In lattice path formulation this means to count all simple lattice paths in $\mathbb{Z}^3$ from the origin to $(e_1, e_2, e_3)$ staying in the region $\{(x_1, x_2, x_3) : x_1 \geq x_2 \text{ and } x_1 \geq x_3\}$. * We state the result below. Solutions were given by Kreweras [85] and Niederhausen [99], see also Gessel [48]. This line of research was picked up later by Bousquet-Mélou [16] who showed, again with the help of the kernel method, that the generating function of these "Kreweras walks" is algebraic. It must be pointed out that this counting problem is a "non-example" for the reflection principle (see Section 10.18), that is, the reflection principle does not apply. The reason is that, if one tries to set it up for application of the reflection principle, then one realizes that the nice property that for permutations other than the identity permutation some hyperplane has to be touched would fail.

**Theorem 10.12.3** *Let $e_1 \geq \max\{e_2, e_3\}$. The number of all lattice paths in $\mathbb{Z}^3$ from $(0,0,0)$ to $(e_1, e_2, e_3)$ subject to $x_1 \geq x_2$ and $x_1 \geq x_3$ is given by*

$$\begin{aligned}|L((0,0,0) \to (e_1,e_2,e_3) \mid x_1 \geq \max\{x_2,x_3\})| &= \binom{e_1+e_2+e_3}{e_1,e_2,e_3} - \frac{e_2+e_3}{1+e_1}\binom{e_1+e_2+e_3}{e_1,e_2,e_3} \\ &+ \sum_{i,j\geq 1} (-1)^{i+j} \frac{(e_1+e_2+e_3)!\,(2i+2j-2)!\,(i+j-2)!}{i!\,(e_3-i)!\,j!\,(e_2-j)!\,(2i-1)!\,(2j-1)!\,(i+j+e_1)!}. \qquad (10.93)\end{aligned}$$

*In particular, if $e_1 = e_2$ this number simplifies to*

$$|L((0,0,0) \to (e_1,e_2,e_3) \mid x_1 \geq \max\{x_2,x_3\})| = 2^{2e_3+1}\left(\frac{e_1!}{(e_1-e_3)!}\right)^2 \frac{(2e_1-2e_3+1)!}{(2e_1+2)!}. \qquad (10.94)$$

*It seems that this is a non-planar lattice path problem, contradicting the title of the section. However, the problem can be translated into a two-dimensional problem, see [16].

We come to a relatively recent research field: the enumeration of walks in the quarter plane. The question that was posed is: Given a particular step set, can one find an explicit formula for the corresponding generating function, and, if not, is the generating function rational, algebraic, D-finite, or neither? For "small" step sets, the analysis is now complete, due to work by Bousquet-Mélou and Mishna [19], by Bostan and Kauers [12], and by Bostan, Kurkova, Raschel and Salvy [13, 14]. However, there is not yet a good understanding how, or whether at all, one can decide from the step set that the generating function has one of the above mentioned properties.

The last topic that I mention here is the connection between Dyck and Schröder path enumeration on the one hand, and Hilbert series for diagonal harmonics and Macdonald polynomials on the other hand. This topic would by itself require a whole chapter. We refer the reader to the survey [61] and the references therein. One of the most intriguing combinatorial problems originating from the investigations in this area is new statistics for Dyck paths, most prominently "bounce" and "dinv." It has been shown (algebraically) that the pair (bounce. area) is equally distributed as (area, bounce), and the same for area and dinv. However, although much effort has been put into it, so far nobody could come up with a direct combinatorial reason (in the best case: a bijection) why this symmetry holds.

## 10.13 Non-intersecting lattice paths

The technique of non-intersecting lattice paths is a powerful counting method. We have already seen its effectiveness in Section 10.7. Originally, non-intersecting paths arose in matroid theory, in the work of Lindström [88]. Lindström's result was rediscovered (not always in its most general form) in the 1980s at about the same time in three different communities, not knowing the work of each other at that time: in statistical physics by Fisher [41, Sec. 5.3] in order to apply it to the analysis of vicious walkers as a model of wetting and melting, in combinatorial chemistry by John and Sachs [70] and Gronau, Just, Schade, Scheffler and Wojciechowski [58] in order to compute Pauling's bond order in benzenoid hydrocarbon molecules, and in enumerative combinatorics by Gessel and Viennot [50, 51] in order to count tableaux and plane partitions. It must however be mentioned that in fact the same idea appeared even earlier in work by Karlin and McGregor [71, 72] in a probabilistic framework, as well as that the so-called "Slater determinant" in quantum mechanics (cf. [108] and [109, Ch. 11]) may qualify as an "ancestor" of the determinantal formula of Lindström. Since then, many more applications have been found, particularly in plane partition and rhombus tiling enumeration, see e.g. [23, 30, 40, 114] and Chapter 9 for more information on this topic.

We devote this section to developing the theory of non-intersecting lattice paths and give some sample applications. This will be continued in Section 10.14, where

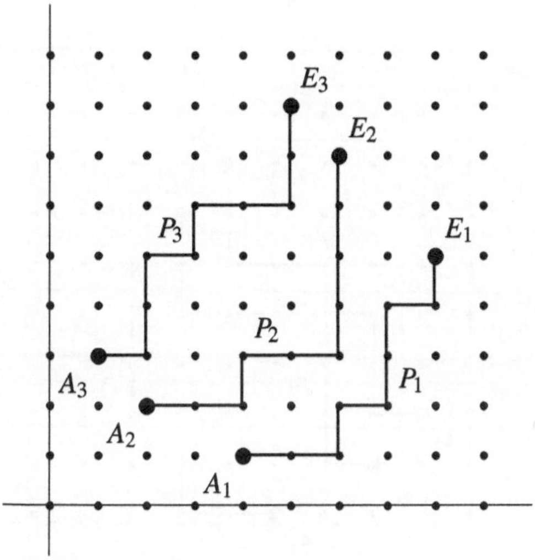


**Figure 10.20**
A family of non-intersecting paths.

we give results on the enumeration of non-intersecting lattice paths in the plane with respect to turns.

The most general version of the non-intersecting path theorem ([88, Lemma 1], [51, Theorem 1]) is formulated for paths in a **directed graph**. Let $G$ be a directed graph with vertices $V$ and (directed) edges $E$. A **path** (actually, the usual notion in graph theory is **walk**) in $G$ is a sequence $v_0, v_1, \ldots, v_m$ of vertices, for some $m$, such that there is an edge from $v_i$ to $v_{i+1}$, $i = 0, 1, \ldots, m-1$. We denote the set of all paths in $G$ from $A$ to $E$ by $L_G(A \to E)$. The directed graph $G$ is called **acyclic** if there is no non trivial closed path in $G$, i.e., if there is no path that starts and ends in the same vertex other than a zero-length path.

The central definition is that a family $\mathbf{P} = (P_1, P_2, \ldots, P_n)$ of paths $P_i$ in $G$ is called **non-intersecting** if no two paths of $\mathbf{P}$ have a vertex in common. Otherwise $\mathbf{P}$ is called **intersecting**. In the context of lattice path enumeration, the graph $G$ comes from a lattice. In many examples, the vertices of $G$ are the lattice points $\mathbb{Z}^2$ in the plane, and the edges of $G$ connect a point $(i, j)$ to $(i+1, j)$, respectively a point $(i, j)$ to $(i, j+1)$. Figure 10.20 displays a family of non-intersecting lattice paths in this sense, Figure 10.21 shows a family of intersecting lattice paths. (It is very important to note that, in the geometric realization of paths as piecewise linear trails, the corresponding trails may very well have common points, but never in starting and end points of steps, see Figure 10.22 for such an example. In particular, non-intersecting lattice paths may even cross each other in the geometric visualization.)

Returning to the general setup, we furthermore assume that to any edge $e$ in the graph $G$ there is assigned a weight $w(e)$ (an element in some commutative ring $\mathscr{R}$). The **weight** of a path $P$ is the product $w(P) = \prod_e w(e)$, where the product is over all edges $e$ of the path $P$. The weight $w(\mathbf{P})$ of a family $\mathbf{P} = (P_1, P_2, \ldots, P_n)$ of paths is defined as the product of all the weights of paths in the family, $w(\mathbf{P}) = w((P_1, P_2, \ldots, P_n)) = \prod_{i=1}^{n} w(P_i)$.

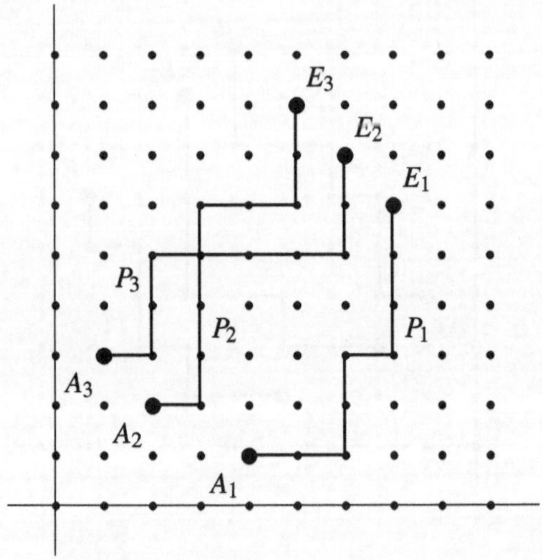


**Figure 10.21**
A family of intersecting paths.

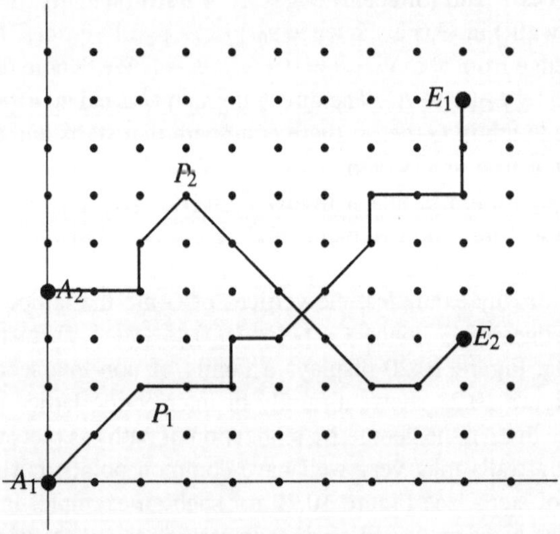


**Figure 10.22**
Non-intersecting lattice paths may even cross.

Given two sequences $\mathbf{A} = (A_1, A_2, \ldots, A_n)$ and $\mathbf{E} = (E_1, E_2, \ldots, E_n)$ of vertices of $G$, we write $L_G(\mathbf{A} \to \mathbf{E})$ for the set of all families $(P_1, P_2, \ldots, P_n)$ of paths, where $P_i$ runs from $A_i$ to $E_i$, $i = 1, 2, \ldots, n$, whereas $L_G(\mathbf{A} \to \mathbf{E} \mid \text{non-intersecting})$ denotes the subset of families of non-intersecting paths.

We need one more piece of notation. Given a permutation $\sigma \in \mathfrak{S}_n$ and a vector $\mathbf{v} = (v_1, v_2, \ldots, v_n)$, by $\mathbf{v}_\sigma$ we mean $(v_{\sigma(1)}, v_{\sigma(2)}, \ldots, v_{\sigma(n)})$. We are now in the position to state and prove the main theorem on non-intersecting paths, due to Lindström [88, Lemma 1].

**Theorem 10.13.1** *Let $G$ be a directed, acyclic graph, and let $\mathbf{A} = (A_1, A_2, \ldots, A_n)$ and $\mathbf{E} = (E_1, E_2, \ldots, E_n)$ be sequences of vertices in $G$. Then*

$$\sum_{\sigma \in \mathfrak{S}_n} (\operatorname{sgn} \sigma) \cdot GF\big(L_G(\mathbf{A}_\sigma \to \mathbf{E} \mid \text{non-intersecting}); w\big)$$
$$= \det_{1 \le i,j \le n} \big(GF\big(L_G(A_j \to E_i); w\big)\big). \quad (10.95)$$

*Proof.* By expanding the determinant on the right-hand side of (10.95), we obtain

$$\det_{1 \le i,j \le n} \big(GF\big(L_G(A_j \to E_i); w\big)\big) = \sum_{\sigma \in \mathfrak{S}_n} \operatorname{sgn} \sigma \prod_{i-1}^{n} GF\big(L_G(A_{\sigma(i)} \to E_i); w\big)$$
$$= \sum_{\substack{(\sigma, \mathbf{P}) \\ \sigma \in \mathfrak{S}_n, \mathbf{P} \in L_G(\mathbf{A}_\sigma \to \mathbf{E})}} \operatorname{sgn} \sigma \, w(\mathbf{P}). \quad (10.96)$$

The sum in (10.96) expresses the determinant in (10.95) as a generating function for pairs $(\sigma, \mathbf{P})$ in the set

$$\bigcup_{\sigma \in \mathfrak{S}_n} L_G(\mathbf{A}_\sigma \to \mathbf{E}). \quad (10.97)$$

We now define a sign-reversing, weight-preserving involution $\varphi$ on the set of all pairs $(\sigma, \mathbf{P})$ in (10.97) with the property that $\mathbf{P}$ is intersecting. Sign-reversing means that if $\varphi((\sigma, \mathbf{P})) = (\sigma_\varphi, \mathbf{P}_\varphi)$ then $\operatorname{sgn} \sigma = -\operatorname{sgn} \sigma_\varphi$, while weight-preserving means that $w(\mathbf{P}) = w(\mathbf{P}_\varphi)$. Suppose that we had already constructed such a $\varphi$. Then in the sum (10.96) all contributions of pairs $(\sigma, \mathbf{P})$ in (10.97) where $\mathbf{P}$ is intersecting would cancel. Only contributions of pairs $(\sigma, \mathbf{P})$ in (10.97) where $\mathbf{P}$ is non-intersecting would survive, establishing (10.95).

Next we construct the sign-reversing, weight-preserving involution $\varphi$. Let $(\sigma, \mathbf{P})$ be in $L_G(\mathbf{A}_\sigma \to \mathbf{E})$ where $\mathbf{P}$ is intersecting. In the left-hand picture of Figure 10.23 an example is shown with $G$ the directed graph corresponding to the integer lattice $\mathbb{Z}^2$, $n = 3$, and $\sigma = 213$. Among all pairs of paths with a common point, choose the lexicographically largest, say $(P_i, P_j)$, $i < j$, and among all common points of that pair choose the last along the paths. (It does not matter on which of the two paths of the pair we choose the last common point since the graph $G$ is acyclic.) Denote this common point by $M$. In our example, the common points between paths are $(3,2)$, $(3,3)$, $(4,5)$. The lexicographically largest pair of paths with common points is $(P_2, P_3)$. The last common point of this pair is $M = (4,5)$.

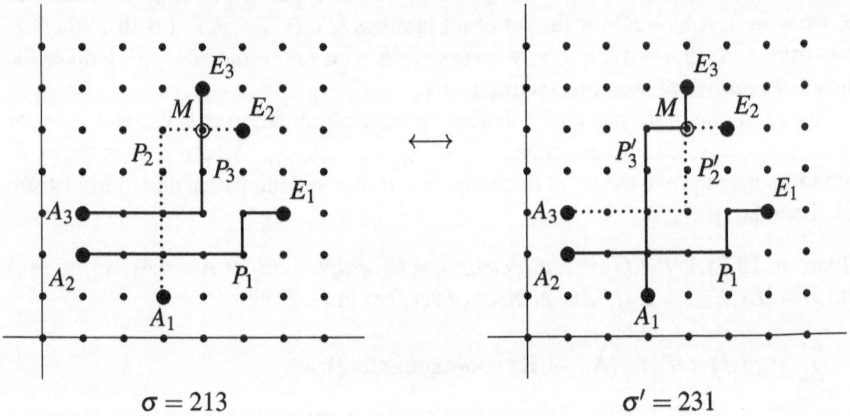


**Figure 10.23**
Lattice paths corresponding to 213 and 231.

Returning to the general construction of the involution $\varphi$, we now interchange the initial portions of $P_i$ and $P_j$ up to $M$. To be more precise, we form the new paths

$$P_i' = [\text{subpath of } P_j \text{ from } A_{\sigma(j)} \text{ to } M \text{ joined with subpath of } P_i \text{ from } M \text{ to } E_i]$$

and

$$P_j' = [\text{subpath of } P_i \text{ from } A_{\sigma(i)} \text{ to } M \text{ joined with subpath of } P_j \text{ from } M \text{ to } E_j].$$

Then we define

$$\begin{aligned}\varphi((\sigma,\mathbf{P})) &= \varphi((\sigma,(P_1,\ldots,P_i,\ldots,P_j,\ldots,P_n)))\\ &= (\sigma\circ(ij),(P_1,\ldots,P_i',\ldots,P_j',\ldots,P_n)),\end{aligned}$$

where $(i,j)$ denotes the transposition interchanging $i$ and $j$. The right-hand picture in Figure 10.23 shows what is obtained by this operation in our example. The image $\varphi((\sigma,\mathbf{P}))$ is again an element of the set in (10.97) since the new permutation of the starting points of $\mathbf{P}$ is exactly $\sigma\circ(ij)$. Moreover, $(P_1,\ldots,P_i',\ldots,P_j',\ldots,P_n)$ is intersecting since $P_i'$ and $P_j'$ are. From all this it is obvious that when $\varphi$ is applied to $\varphi((\sigma,\mathbf{P}))$ we arrive back at $(\sigma,\mathbf{P})$. Hence, $\varphi$ is an involution. Since $\sigma$ and $\sigma\circ(ij)$ differ in sign, $\varphi$ is sign-reversing. Finally, since the total (multi)set of edges in the path families does not change under application of $\varphi$, the map $\varphi$ is also weight-preserving. This finishes the proof. ∎

The most frequent situation in which the general result in Theorem 10.13.1 is applied is the one where non-intersecting paths can only occur if the starting and end points are connected via the identity permutation, that is, if $(P_1,P_2,\ldots,P_n)$ can only be non-intersecting if $P_i$ connects $A_i$ with $E_i$, $i = 1,2,\dots,n$. In that situation, Theorem 10.13.1 simplifies to the following result.

**Corollary 10.13.2** *Let $G$ be a directed, acyclic graph, and let $\mathbf{A} = (A_1,A_2,\dots,A_n)$ and $\mathbf{E} = (E_1,E_2,\dots,E_n)$ be sequences of vertices in $G$ such that the only permutation $\sigma$ that allows for a family $(P_1,P_2,\dots,P_n)$ of non-intersecting paths such that $P_i$ connects $A_{\sigma(i)}$ with $E_i$, $i = 1,2,\dots,n$, is the identity permutation. Then the generating function $\sum_{\mathbf{P}} w(\mathbf{P})$ for families $\mathbf{P} = (P_1,P_2,\dots,P_n)$ of non-intersecting paths, where $P_i$ is a path running from $A_i$ to $E_i$, $i = 1,2,\dots,n$, is given by*

$$GF\big(L_G(\mathbf{A} \to \mathbf{E} \mid \text{non-intersecting}); w\big) = \det_{1 \le i,j \le n} \big(GF\big(L_G(A_j \to E_i); w\big)\big). \quad (10.98)$$

The standard application of Corollary 10.13.2 concerns **semistandard tableaux**. These are important objects particularly in the representation theory of the general and the special linear groups, cf. [103].

Let $\lambda = (\lambda_1, \lambda_2, \dots, \lambda_n)$ and $\mu = (\mu_1, \mu_2, \dots, \mu_n)$ be $n$-tuples of integers such that

$$\lambda_1 \ge \lambda_2 \ge \lambda_3 \ge \cdots \ge \lambda_n, \quad (10.99a)$$
$$\mu_1 \ge \mu_2 \ge \mu_3 \ge \cdots \ge \mu_n, \quad (10.99b)$$
$$\text{and } \lambda_i \ge \mu_i \text{ for all } i. \quad (10.99c)$$

A semistandard tableau $T$ of shape $\lambda/\mu$ is an array of integers

$$\begin{array}{ccccccc} & & & \pi_{1,\mu_1+1} & \cdots\cdots\cdots & & \pi_{1,\lambda_1} \\ & \pi_{2,\mu_2+1} & \cdots & \pi_{2,\mu_1+1} & \cdots\cdots & \pi_{2,\lambda_2} & \\ & \cdot^{\cdot^{\cdot}} & & \vdots & & \cdot^{\cdot^{\cdot}} & \\ \pi_{n,\mu_n+1} & & \cdots\cdots\cdots\cdots & & \pi_{n,\lambda_n} & & \end{array} \quad (10.100)$$

such that entries along rows are weakly increasing and entries along columns are strictly increasing. A semistandard tableau of shape $(7,6,6,4)/(3,3,1,0)$ is shown in Figure 10.24(a). (The lower and upper bounds on the entries displayed to the left and right of the tableau should be ignored at this point.)

Let $\mathbf{a} = (a_1, a_2 \dots, a_n)$ and $\mathbf{b} = (b_1, b_2, \dots, b_n)$ be sequences of integers such that

$$a_1 \le a_2 \le \cdots \le a_n \quad (10.101a)$$
$$b_1 \le b_2 \le \cdots \le b_n, \quad (10.101b)$$
$$\text{and } a_i \ge b_i \text{ for all } i. \quad (10.101c)$$

We claim that *semistandard tableaux of shape $\lambda/\mu$ where the entries in row $i$ are at most $a_i$ and at least $b_i$ bijectively correspond to families $(P_1,P_2,\dots,P_n)$ of non-intersecting lattice paths $P_i$, where $P_i$ runs from $(\mu_i - i, b_i)$ to $(\lambda_i - i, a_i)$.*

This is seen as follows. Let $\pi$ be a semistandard tableau of shape $\lambda/\mu$ where the entries in row $i$ are at most $a_i$ and at least $b_i$. The semistandard tableau $\pi$ is

| | | | | | | | | |
|---|---|---|---|---|---|---|---|---|
| $-2 \le$ | | | | 1 | 1 | 4 | 5 | $\le 7$ |
| $-1 \le$ | | | | 5 | 5 | 7 | | $\le 8$ |
| $0 \le$ | | 2 | 5 | 6 | 9 | 9 | | $\le 9$ |
| $1 \le$ | 1 | 5 | 6 | 8 | | | | $\le 10$ |

(a) Semistandard tableau

$\updownarrow$

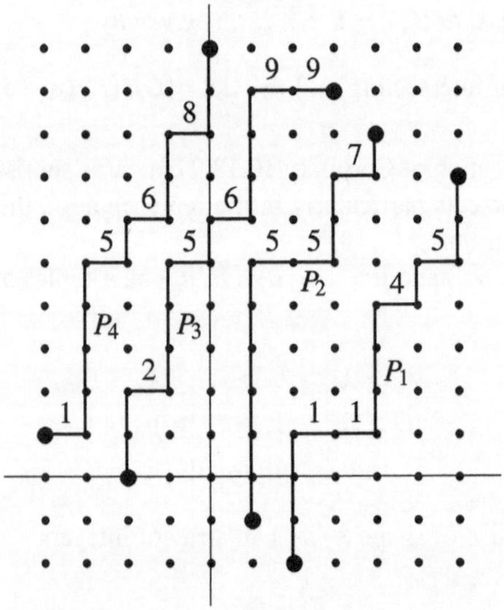


(b) Family of non-intersecting lattice paths

**Figure 10.24**
A column-strict reverse plane partition and the corresponding family of non-intersecting lattice paths.

mapped to a family of lattice paths by associating the $i$th row of $\pi$ with a path $P_i$ from $(\mu_i - i, b_i)$ to $(\lambda_i - i, a_i)$ where the entries in the $i$th row are interpreted as heights of the horizontal steps in the path $P_i$. Thus from $\pi$ we obtain the family $\mathbf{P} = (P_1, \ldots, P_n)$ of lattice paths. Figure 10.24(b) displays the family of lattice paths that in this way results from the array displayed in Figure 10.24(a).

Clearly, the property that the columns of $\pi$ are strictly increasing translates into the condition that $(P_1, P_2, \ldots, P_n)$ is non-intersecting.

By applying Corollary 10.13.2 to this situation, we obtain the following enumeration result.

**Theorem 10.13.3** *Let* $\lambda = (\lambda_1, \lambda_2, \ldots, \lambda_n)$ *and* $\mu = (\mu_1, \mu_2, \ldots, \mu_n)$ *be sequences of integers satisfying* (10.99). *Let* $\mathbf{a} = (a_1, a_2, \ldots, a_n)$ *and* $\mathbf{b} = (b_1, b_2, \ldots, b_n)$ *be sequences of integers satisfying* (10.101). *Then the number of all semistandard tableaux* $\pi$ *of shape* $\lambda/\mu$ *where the entries in row* $i$ *are at most* $a_i$ *and at least* $b_i$

*equals*

$$\det_{1\le i,j\le n}\left(\binom{a_i-b_j+\lambda_i-\mu_j-i+j}{\lambda_i-\mu_j-i+j}\right). \tag{10.102}$$

*More generally, the generating function $\sum_\pi q^{n(\pi)}$ for the same set of semistandard tableaux $\pi$, where $n(\pi)$ denotes the sum of all entries of $\pi$, equals*

$$\det_{1\le i,j\le n}\left(q^{b_j(\lambda_i-\mu_j-i+j)}\begin{bmatrix}a_i-b_j+\lambda_i-\mu_j-i+j\\ \lambda_i-\mu_j-i+j\end{bmatrix}_q\right). \tag{10.103}$$

If the shape $\lambda/\mu$ is a straight shape and the bounds $\mathbf{a}$ and $\mathbf{b}$ are constant, that is, if, say, $\mu=(0,0,\dots,0)$, $\mathbf{b}=(1,1,\dots,1)$, and $\mathbf{a}=(a,a,\dots,a)$, then the above determinants can be evaluated in closed form. Rewritten appropriately, the result is the **hook-content formula**. (We refer the reader to [103, Sec. 3.10] or [112, Cor. 7.21.6] for unexplained terminology).

**Theorem 10.13.4** *Let $\lambda=(\lambda_1,\lambda_2,\dots,\lambda_n)$ be a sequence of non-negative integers satisfying* (10.99). *Then the number of all semistandard tableaux $\pi$ of shape $\lambda$ with entries in row $i$ being positive integers at most $a$ equals*

$$\prod_\rho \frac{a+c(\rho)}{h(\rho)}, \tag{10.104}$$

*where the product is over all cells $\rho$ in the Ferrers diagram of the partition $\lambda$, $c(\rho)$ is the content of the cell $\rho$, and $h(\rho)$ is the hook-length of the cell $\rho$. More generally, the generating function $\sum_\pi q^{n(\pi)}$ for the same set of semistandard tableaux $\pi$ equals*

$$q^{\sum_{i=1}^n i\lambda_i}\prod_\rho \frac{1-q^{a+c(\rho)}}{1-q^{h(\rho)}}. \tag{10.105}$$

By introducing more general weights, in the same way one can provide combinatorial proofs for the Jacobi–Trudi-type identities for Schur functions (cf. [103, Sec. 4.5]), respectively formulas for so-called flagged Schur functions, originally introduced by Lascoux and Schützenberger [87], see also [121].

If in an array (10.100) one also introduces a relationship between the first row and the last row, then one is led to define so-called **cylindric partitions**, as was done by Gessel and Krattenthaler [49]. Also in that theory, non-intersecting paths play an essential role.

It may seem that the general result in Theorem 10.13.1 is a very artificial statement, perhaps being of no use. However, even this result does have several applications. For example, the most elegant proof of the determinant formula for higher-dimensional path counting under a general two-sided restriction ([115]; see Section 10.17, Theorem 10.17.1) fundamentally makes use of the full generality of Theorem 10.13.1. Further applications of the general formula (10.95) can be found in rhombus tiling enumeration (see [30, 40]), in combinatorial commutative algebra (see [79]), and in the combinatorial theory of orthogonal polynomials developed in Section 10.10 (see the proof of Theorem 10.10.8 as given in [119]).

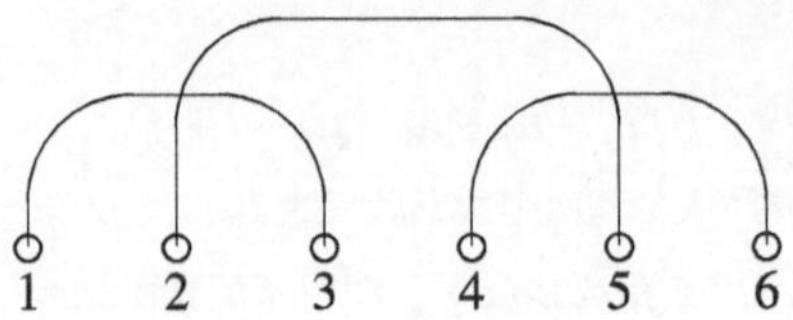


**Figure 10.25**
A perfect matching.

In several applications, one has to deal with the problem of enumerating non-intersecting lattice paths where the starting and end points are not fixed. Either it is only the starting points that are fixed and the end points are any points from a given set (or the other way round), or it is even that the starting points may come from one set and the end points from another. The solution to these counting problems comes from **Pfaffians**.

A Pfaffian is very similar to a determinant. Whereas in the definition of a determinant there appear *permutations*, in the definition of a Pfaffian there appear *perfect matchings*. A **perfect matching** of a set of objects, $\mathscr{A}$ say, is a pairing of the objects. For example, if $\mathscr{A} = \{1,2,3,4,5,6\}$, then $\{\{1,3\},\{2,5\},\{4,6\}\}$ is a matching of $\mathscr{A}$. A matching of $\{1,2,\dots,N\}$ can be realized geometrically by drawing points labeled $1,2,\dots,N$ along a line, and then connecting any two points whose labels are paired in the matching by a curve, so that there are no touching points between curves and no triple intersections. Figure 10.25 shows the geometric realization of $\{\{1,3\},\{2,5\},\{4,6\}\}$. Any two pairs $\{i,k\}$ and $\{j,l\}$ in a matching for which $i<j<k<l$ are called a **crossing** of the matching. The **sign** $\operatorname{sgn}\pi$ of a matching $\pi$ is $(-1)^c$, where $c$ is the number of crossings of $\pi$. Thus, the sign of $\{\{1,3\},\{2,5\},\{4,6\}\}$ is $(-1)^2=+1$. In the geometric realization of a matching, its sign can be read off as $(-1)^{c'}$, where $c'$ is the number of crossing points between two curves. (It is easily checked that it does not matter how we draw the curves.)

With these definitions, the **Pfaffian** $\operatorname{Pf}(A)$ of an upper triangular array $A = (a_{ij})_{1\le i<j\le 2n}$ is defined by

$$\operatorname{Pf}(A) := \sum_{\pi \text{ a perfect matching of } \{1,2,\dots,2n\}} \operatorname{sgn}\pi \prod_{\{i,j\}\in\pi} a_{ij}. \tag{10.106}$$

For example, for $n=2$ we have

$$\operatorname{Pf}\left((a_{ij})_{1\le i<j\le 4}\right) = a_{12}a_{34} - a_{13}a_{24} + a_{14}a_{23}.$$

Alternatively, the Pfaffian could be defined as the (appropriate) square root of a skew symmetric matrix. To be precise, if $A$ is a skew symmetric matrix, then

$$\operatorname{Pf}(A)^2 = \det(A), \tag{10.107}$$

where $\operatorname{Pf}(A)$ has to be interpreted as the Pfaffian of the upper triangular part of $A$. For a (combinatorial) proof of this fact see e.g. [114, Prop. 2.2].

Now let again $G$ be a directed, acyclic graph. Let $A_1, A_2, \ldots, A_n$ be vertices in $G$, and $\mathbf{E} = (\ldots, E_1, E_2, \ldots)$ be an ordered set of vertices. What we want to count is the number of **all families $\mathbf{P} = (P_1, P_2, \ldots, P_n)$ of non-intersecting paths, where $P_i$ runs from $A_i$ to some vertex in $\mathbf{E}$, $i = 1, 2, \ldots, n$.** The difference to the situation in Theorem 10.13.1 is that the end vertices of the paths are not fixed. Nevertheless, we adopt the earlier notation for this more general situation. To be precise, by $L_G(\mathbf{A} \to \mathbf{E} \mid \text{non-intersecting})$ we mean the set of all families $\mathbf{P} = (P_1, P_2, \ldots, P_n)$ of non-intersecting paths in $G$, where $P_i$ runs from $A_i$ to some vertex $E_{k_i}$ in $\mathbf{E}$, $i = 1, 2, \ldots, n$ and $k_1 < k_2 < \cdots < k_n$. The corresponding enumeration result is due to Okada [100, Theorem 3] and Stembridge [114, Theorem 3.1].

**Theorem 10.13.5** *Let $G$ be a directed, acyclic graph with a weight function $w$ on its edges. Let $\mathbf{A} = (A_1, A_2, \ldots, A_{2n})$ and $\mathbf{E} = (\ldots, E_1, E_2, \ldots)$ be sequences of vertices in $G$. Then*

$$\sum_{\sigma \in \mathfrak{S}_{2n}} (\operatorname{sgn} \sigma) \cdot GF\big(L_G(\mathbf{A}_\sigma \to \mathbf{E} \mid \text{non-intersecting}); w\big) = \operatorname{Pf}_{1 \le i < j \le 2n}\big(Q_G(i,j;w)\big), \tag{10.108}$$

*where $L_G(\mathbf{A}_\sigma \to \mathbf{E} \mid \text{non-intersecting})$ is the set of all families $(P_1, P_2, \ldots, P_{2n})$ of non-intersecting paths, $P_i$ connecting $A_{\sigma(i)}$ with $E_{k_i}$, $i = 1, 2, \ldots, 2n$ and $k_1 < k_2 < \ldots k_{2n}$, and $Q_G(i,j;w)$ is the generating function $\sum_{(P',P'')} w(P')w(P'')$ for all pairs $(P', P'')$ of non-intersecting lattice paths, where $P'$ connects $A_i$ with some $E_k$ and $P''$ connects $A_j$ with some $E_l$, with $k < l$.*

The proof uses the same involution idea as the proof of Theorem 10.13.1 does. See [114, Proof of Theorem 3.1].

Similarly to Theorem 10.13.1, the most frequent situation in which the general result in Theorem 10.13.5 is applied is the one where non-intersecting paths can only occur if the starting and end points are connected via the identity permutation, that is, if $(P_1, P_2, \ldots, P_{2n})$ can only be non-intersecting if $P_i$ connects $A_i$ with $E_{k_i}$, $i = 1, 2, \ldots, 2n$ and $k_1 < k_2 < \cdots < k_{2n}$. In that situation, Theorem 10.13.5 simplifies to the following result.

**Corollary 10.13.6** *Let $G$ be a directed, acyclic graph with a weight function $w$ on its edges. Let $\mathbf{A} = (A_1, A_2, \ldots, A_{2n})$ and $\mathbf{E} = (\ldots, E_1, E_2, \ldots)$ be sequences of vertices in $G$ such that the only permutation $\sigma$ that allows for a family $(P_1, P_2, \ldots, P_{2n})$ of non-intersecting paths such that $P_i$ connects $A_{\sigma(i)}$ with $E_{k_i}$, $i = 1, 2, \ldots, 2n$ and $k_1 < k_2 < \cdots < k_{2n}$, is the identity permutation. Then the generating function $\sum_{\mathbf{P}} w(\mathbf{P})$ for families $\mathbf{P} = (P_1, P_2, \ldots, P_{2n})$ of non-intersecting paths, where $P_i$ is a path running from $A_i$ to $E_{k_i}$, $i = 1, 2, \ldots, 2n$ and $k_1 < k_2 < \cdots < k_{2n}$, is given by*

$$GF\big(L_G(\mathbf{A} \to \mathbf{E} \mid \text{non-intersecting}); w\big) = \operatorname{Pf}_{1 \le i < j \le 2n}\big(Q_G(i,j;w)\big), \tag{10.109}$$

*where $Q_G(i,j;w)$ has the same meaning as in Theorem 10.13.5.*

Theorem 10.13.5 and Corollary 10.13.6 are only formulated for an even number of paths. However, a simple trick allows us to also use it for an odd number of paths:

One introduces a "phantom" vertex $X$ that cannot be reached by any other vertex (one can think of the vertex at infinity), and adjoins this point as a new starting point **and** as a new end point. A family of non-intersecting paths would necessarily contain the zero-length path from $X$ to $X$ as one of the paths, which therefore can be ignored. Theorem 10.13.5 applies, and yields the following corollary.

**Corollary 10.13.7** *Let $G$ be a directed, acyclic graph with a weight function $w$ on its edges. Let $\mathbf{A}=(A_1,A_2,\dots,A_{2n-1})$ and $\mathbf{E}=(\dots,E_1,E_2,\dots)$ be sequences of vertices in $G$. Then*

$$\sum_{\sigma\in\mathfrak{S}_{2n}} (\operatorname{sgn}\sigma)\cdot GF\big(L_G(\mathbf{A}_\sigma\to\mathbf{E}\mid \text{non-intersecting});w\big)=\operatorname{Pf}(Q_G(i,j;w))_{1\le i<j\le 2n}, \tag{10.110}$$

*where $L_G(\mathbf{A}_\sigma\to\mathbf{E}\mid$ non-intersecting$)$ has the same meaning as in Theorem 10.13.5, where for $j\le 2n-1$ the quantity $Q_G(i,j;w)$ has the same meaning as in Theorem 10.13.5, and where $Q_G(i,2n;w)$ is the generating function $\sum_P w(P)$ for all paths running from $A_i$ to some point of $\mathbf{E}$.*

*In particular, if the vertices $\mathbf{A}$ and $\mathbf{E}$ are such that the only permutation $\sigma$ that allows for a family $(P_1,P_2,\dots,P_{2n-1})$ of non-intersecting paths such that $P_i$ connects $A_{\sigma(i)}$ with $E_{k_i}$, $i=1,2,\dots,2n-1$ and $k_1<k_2<\cdots<k_{2n-1}$, is the identity permutation, then the generating function $\sum_{\mathbf{P}} w(\mathbf{P})$ for families $\mathbf{P}=(P_1,P_2,\dots,P_{2n-1})$ of non-intersecting paths, where $P_i$ is a path running from $A_i$ to $E_{k_i}$, $i=1,2,\dots,2n-1$ and $k_1<k_2<\cdots<k_{2n-1}$, is given by*

$$GF\big(L_G(\mathbf{A}\to\mathbf{E}\mid \text{non-intersecting});w\big)=\operatorname{Pf}(Q_G(i,j;w))_{1\le i<j\le 2n}. \tag{10.111}$$

For various applications of this theorem, see e.g. [77, 100, 114].

Next we consider a mixed case, in which the starting points of the paths are fixed, some end points are fixed, but some end points can be chosen from a given set. To be precise, let $m$ and $n$ be a positive integer with $m\le n$, let $\mathbf{A}=(A_1,A_2,\dots,A_n)$ and $\mathbf{E}=(E_1,E_2,\dots,E_m)$ be vertices in a given directed, acyclic graph $G$, let $\hat{\mathbf{E}}=(\dots,\hat{E}_1,\hat{E}_2,\dots)$ be an ordered set of (finitely many or infinitely many) vertices. What we want to determine is the generating function for *all families* $\mathbf{P}=(P_1,P_2,\dots,P_n)$ *of non-intersecting paths, where for $i=1,2,\dots,m$ the path $P_i$ runs from $A_i$ to $E_i$, and where for $i=m+1,m+2,\dots,n$ the path $P_i$ runs from $A_i$ to some point $\hat{E}_{k_i}$ in $\hat{\mathbf{E}}$, with* $k_{m+1}<k_{m+2}<\cdots<k_n$. We write $L_G(\mathbf{A}\to\mathbf{E}\cup\hat{\mathbf{E}}\mid$ non-intersecting$)$ for the set of these families of paths. The corresponding enumeration result is due to Stembridge [114, Theorem 3.2].

**Theorem 10.13.8** *Let $G$ be a directed, acyclic graph with a weight function $w$ on its edges, and let $m$ and $n$ be a positive integer such that $m\le n$ and $m+n$ is even. Let $\mathbf{A}=(A_1,A_2,\dots,A_n)$, $\mathbf{E}=(E_1,E_2,\dots,E_m)$ and $\hat{\mathbf{E}}=(\dots,\hat{E}_1,\hat{E}_2,\dots)$ be sequences of vertices in $G$. Then*

$$\sum_{\sigma\in\mathfrak{S}_n} (\operatorname{sgn}\sigma)\cdot GF\big(L_G(\mathbf{A}_\sigma\to\mathbf{E}\cup\hat{\mathbf{E}}\mid \text{non-intersecting});w\big)=\operatorname{Pf}\begin{pmatrix} Q & H\\ -H^t & 0\end{pmatrix}, \tag{10.112}$$

*where* $L_G(\mathbf{A}_\sigma \to \mathbf{E} \cup \hat{\mathbf{E}} \mid \text{non-intersecting})$ *is the set of all families* $(P_1, P_2, \dots, P_n)$ *of non-intersecting paths,* $P_i$ *connecting* $A_{\sigma(i)}$ *with* $E_i$, $i = 1, 2, \dots, m$, $P_i$ *connecting* $A_{\sigma(i)}$ *with* $\hat{E}_{k_i}$, $i = m+1, m+2, \dots, n$ *and* $k_{m+1} < k_{m+2} < \cdots < k_n$, *where* $Q = (Q_G(i,j;w))_{1 \le i,j \le n}$ *is a skew-symmetric matrix with* $Q_G(i,j;w)$ *denoting the generating function* $\sum_{(P',P'')} w(P')w(P'')$ *for all pairs* $(P',P'')$ *of non-intersecting lattice paths, where* $P'$ *connects* $A_i$ *with some* $\hat{E}_k$ *and* $P''$ *connects* $A_j$ *with some* $\hat{E}_l$, *with* $k < l$, *and where* $H = (H_G(i,j;w))_{1 \le i \le n, 1 \le j \le m}$ *is the rectangular matrix with* $H_G(i,j;w)$ *denoting the generating function* $\sum_P w(P)$ *for all paths* $P$ *from* $A_i$ *to* $E_j$. *The Pfaffian of a skew-symmetric matrix has to be interpreted according to the remark containing* (10.107).

*In particular, if the vertices* $\mathbf{A}$ *and* $\mathbf{E} \cup \hat{\mathbf{E}}$ *are such that the only permutation* $\sigma$ *that allows for a family* $(P_1, P_2, \dots, P_n)$ *of non-intersecting paths such that* $P_i$ *connects* $A_{\sigma(i)}$ *with* $E_i$, $i = 1, 2, \dots, m$, *and* $P_i$ *connects* $A_{\sigma(i)}$ *with* $\hat{E}_{k_i}$, $i = m+1, m+2, \dots, n$ *and* $k_{m+1} < k_{m+2} < \cdots < k_n$, *is the identity permutation, then the generating function* $\sum_{\mathbf{P}} w(\mathbf{P})$ *for all families* $(P_1, P_2, \dots, P_n)$ *of non-intersecting lattice paths, where for* $i = 1, 2, \dots, m$ *the path* $P_i$ *runs from* $A_i$ *to* $E_i$, *and where for* $i = m+1, m+2, \dots, n$ *the path* $P_i$ *runs from* $A_i$ *to some point* $\hat{E}_{k_i}$ *in* $\hat{\mathbf{E}}$, $k_{m+1} < k_{m+2} < \cdots < k_n$, *is given by*

$$GF\big(L_G(\mathbf{A} \to \mathbf{E} \cup \hat{\mathbf{E}} \mid \text{non-intersecting}); w\big) = (-1)^{\binom{m}{2}} \operatorname{Pf}\begin{pmatrix} Q & H \\ -H^t & 0 \end{pmatrix}. \quad (10.113)$$

Again, the proof uses the same involution idea as the proof of Theorem 10.13.1 does. See [114, Proof of Theorem 3.2]. Applications can for example be found in [31, 32, 114].

As a matter of fact, Theorem 10.13.8 is a special case of the so-called minor summation formula due to Ishikawa and Wakayama [66, Theorem 2].

**Theorem 10.13.9** *Let* $m$, $n$, $p$ *be integers such that* $n+m$ *is even and* $0 \le n-m \le p$. *Let* $M$ *be any* $n \times p$ *matrix,* $H$ *be any* $n \times m$ *matrix, and* $A = (a_{ij})_{1<i,j<p}$ *be any skew-symmetric matrix. Then we have*

$$\sum_K \operatorname{Pf}(A_K^K) \det(M_K \vdots H) = (-1)^{\binom{m}{2}} \operatorname{Pf}\begin{pmatrix} MAM^t & H \\ -H^t & 0 \end{pmatrix}.$$

*where* $K$ *runs over all* $(n-m)$*-element subsets of* $[1,p]$, $A_K^K$ *is the skew-symmetric matrix obtained by picking the rows and columns indexed by* $K$, *and* $M_K$ *is the submatrix of* $M$ *consisting of the columns corresponding to* $K$.

Theorem 10.13.8 results from the special case of Theorem 10.13.9 where $A$ is the $p \times p$ skew-symmetric matrix with all 1s above the diagonal, and $M$ and $H$ are matrices the entries of which are appropriately chosen path generating functions. This is based on the well-known fact (see e.g. [114, Prop. 2.3(c)]) that $\operatorname{Pf}(1)_{1 \le i < j \le 2N} = 1$ for all $N$.

The last theorem in this section addresses the case where starting **and** end points are chosen from given sets. To be precise, let $\mathbf{A} = (A_1, A_2, \dots, A_n)$ and $\mathbf{E} = (\dots, E_1, E_2, \dots)$ be ordered sets of vertices (finitely many or infinitely many in the case of $\mathbf{E}$). What we want to determine is the generating function for *all families $\mathbf{P} = (P_1, P_2, \dots, P_s)$ of non-intersecting paths, where for $i = 1, 2, \dots, s$ the path $P_i$ runs from some $A_{k_i}$ to some $E_{l_i}$*. The corresponding enumeration result is due to Okada [100, Theorem 4] and Stembridge [114, Theorem 4.1]. In the formulation below, by abuse of notation, $\mathbf{A}' \subseteq \mathbf{A}$ means that $\mathbf{A}'$ is a subsequence of $\mathbf{A}$, with an analogous meaning for $\mathbf{E}' \subseteq \mathbf{E}$.

**Theorem 10.13.10** *Let $G$ be a directed, acyclic graph with a weight function $w$ on its edges, and let $\mathbf{A} = (A_1, A_2, \dots, A_n)$ and $\mathbf{E} = (\dots, E_1, E_2, \dots)$ be sequences of vertices in $G$.*

*(a) If $n$ is even, then*

$$\sum_{s=0}^{n/2} t^s \sum_{\substack{\mathbf{A}' \subseteq \mathbf{A}, \mathbf{E}' \subseteq \mathbf{E} \\ |\mathbf{A}'| = |\mathbf{E}'| = 2s}} GF\left(L_G(\mathbf{A}' \to \mathbf{E}' \mid \text{non-intersecting}); w\right)$$
$$= \operatorname{Pf}_{1 \le i < j \le n}\left((-1)^{i+j-1} + tQ_G(i,j;w)\right), \quad (10.114)$$

*where $Q_G(i,j;w)$ has the same meaning as in Theorem 10.13.5.*

*(b) If $n$ is odd, then*

$$\sum_{s=0}^{n} t^s \sum_{\substack{\mathbf{A}' \subseteq \mathbf{A}, \mathbf{E}' \subseteq \mathbf{E} \\ |\mathbf{A}'| = |\mathbf{E}'| = s}} GF\left(L_G(\mathbf{A}' \to \mathbf{E}' \mid \text{non-intersecting}); w\right)$$
$$= \operatorname*{Pf}_{1 \le i < j \le n+1}\left((-1)^{i+j-1} + t^2 Q_G(i,j;w)\right), \quad (10.115)$$

*where for $j \le n$ the quantity $Q_G(i,j;w)$ has the same meaning as in Theorem 10.13.5, while $Q_G(i,n+1;w)$ equals the generating function $t^{-1}\sum_P w(P)$ for all paths $P$ from $A_i$ to some vertex in $\mathbf{E}$.*

*(c) If $n$ is even, then*

$$\sum_{s=0}^{n} t^s \sum_{\substack{\mathbf{A}' \subseteq \mathbf{A}, \mathbf{E}' \subseteq \mathbf{E} \\ |\mathbf{A}'| = |\mathbf{E}'| = s}} GF\left(L_G(\mathbf{A}' \to \mathbf{E}' \mid \text{non-intersecting}); w\right)$$
$$= \operatorname*{Pf}_{1 \le i < j \le n+2}\left((-1)^{i+j-1} + t^2 Q_G(i,j;w)\right), \quad (10.116)$$

*where for $j \le n+1$ the quantity $Q_G(i,j;w)$ has the same meaning as in (b), and $Q_G(i,n+2;w) = 0$.*

For applications of this theorem in plane partition enumeration see [100] and [114].

If one weakens the condition of non-intersection of lattice paths in the plane to the requirement that paths are allowed to touch each other in isolated points but not to change sides, then one arrives at the model of **osculating paths**. The motivation to consider this model comes from an observation of Bousquet-Mélou and Habsieger [17] that **alternating sign matrices** are in bijection with families of osculating paths with appropriate starting and end points. Alternating sign matrices are fascinating, but notoriously difficult to count, therefore it may be useful to investigate objects that are equivalent to them. So far, this point of view has not led to much, but recently Brak and Galleas [21] proved a constant term formula for families of osculating paths.

## 10.14 Lattice paths and their turns

In this section we consider **turns** of lattice paths. Literally, a **turn** of a lattice path is any vertex of a path where the direction of the path changes. The enumeration of lattice paths with a given number of turns is motivated by problems of correlated random walks, distribution of runs (cf. [95]), coefficients of Hilbert polynomials of determinantal and Pfaffian rings (cf. [84, 86]), and summations for Schur functions (cf. [76]).

The approach for the enumeration of simple plane lattice paths with respect to their number of turns that we present here is by encoding lattice paths in terms of two-rowed arrays, a point of view put forward in [78].

For simple lattice paths in the plane there are two types of turns. We call a vertex $T$ of a path a **North-East turn (NE-turn** for short) if $T$ is reached by a step towards north and left by a step towards east. We call a vertex $T$ of a path an **East-North turn (EN-turn** for short) if $T$ is reached by a step towards east and left by a step towards north. Thus, the NE-turns of the path $P_0$ in Figure 10.26 are $(1,1)$, $(2,3)$, and $(5,4)$, the EN-turns of $P_0$ are $(2,1)$, $(5,3)$, and $(6,4)$. We denote by $NE(P)$ the number of NE-turns of $P$ and by $EN(P)$ the number of EN-turns of $P$.

Now we describe the encoding of paths in terms of two-rowed arrays. Actually, we use two encodings, one corresponding to NE-turns, one corresponding to EN-turns. Let $(p_1,q_1)$, $(p_2,q_2)$, $\dots$, $(p_\ell,q_\ell)$ be the NE-turns of a path $P$. Then the **NE-turn representation** of $P$ is defined by the two-rowed array

$$\begin{matrix} p_1 & p_2 & \cdots & p_\ell \\ q_1 & q_2 & \cdots & q_\ell, \end{matrix} \tag{10.117}$$

which consists of two strictly increasing sequences. Clearly, if $P$ runs from $(a,b)$ to $(c,d)$ then $a \le p_1 < p_2 < \cdots < p_\ell \le c-1$ and $b+1 \le q_1 < q_2 < \cdots < q_\ell \le d$ are satisfied. If we wish to make this fact transparent, we write

$$\begin{matrix} a \le & p_1 & p_2 & \cdots & p_\ell & \le c-1 \\ b+1 \le & q_1 & q_2 & \cdots & q_\ell & \le d. \end{matrix} \tag{10.118}$$

For a given starting point and a given final point, by definition the empty array is

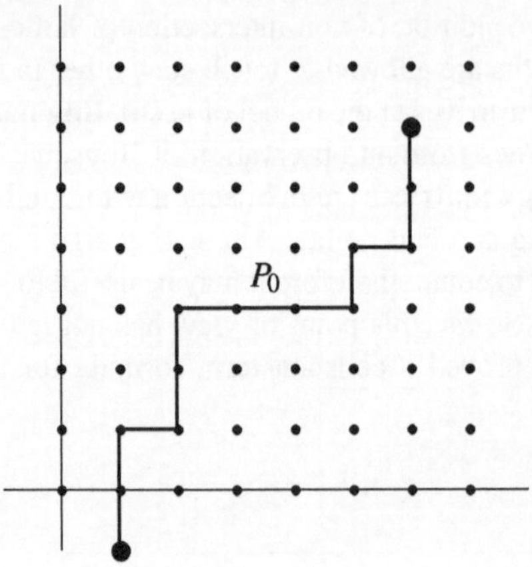


**Figure 10.26**
The path $P_0$ has NE turns $(1,1)$, $(2,3)$, and $(5,4)$, and it has EN turns $(2,1)$, $(5,3)$, and $(6,4)$.

the representation for the only path that has no NE-turn. For the path in our running example we obtain the NE-turn representation

$$\begin{matrix} 1 & 2 & 5 \\ 1 & 3 & 4 \end{matrix},$$

or, with bounds included,

$$\begin{matrix} 1\le & 1 & 2 & 5 & \le 5 \\ 0\le & 1 & 3 & 4 & \le 6 \end{matrix}.$$

Similarly, if $(p_1,q_1), (p_2,q_2), \ldots, (p_\ell,q_\ell)$ denote the EN-turns of a path $P$, then (10.117) is called the **EN-turn representation** of $P$. If $P$ runs from $(a,b)$ to $(c,d)$ then $a+1 \le p_1 < p_2 < \cdots < p_\ell \le c$ and $b \le q_1 < q_2 < \cdots < q_\ell \le d-1$ are satisfied. Again, as earlier, we write

$$\begin{matrix} a+1\le & p_1 & p_2 & \ldots & p_\ell & \le c \\ b\le & q_1 & q_2 & \ldots & q_\ell & \le d-1. \end{matrix} \tag{10.119}$$

For a given starting point and a given final point, by definition the empty array is the representation for the only path that has no EN-turn. For the path in our running example we obtain the EN-turn representation

$$\begin{matrix} 2 & 5 & 6 \\ 1 & 3 & 4 \end{matrix},$$

or, with bounds included,

$$\begin{matrix} 2\le & 2 & 5 & 6 & \le 6 \\ -1\le & 1 & 3 & 4 & \le 5 \end{matrix}.$$

Also two-rowed arrays with rows being of unequal length will be considered. These arrays will also have the property that the rows are strictly increasing. So, by convention, whenever we speak of two-rowed arrays, we mean two-rowed arrays with strictly increasing rows. For these arrays we will use a notation of the kind (10.118) or (10.119) as well. We shall frequently use the short notation $(\mathbf{a} \mid \mathbf{b})$ for two-rowed arrays, where $\mathbf{a}$ denotes the sequence $(a_i)$ of elements of the first row, and $\mathbf{b}$ denotes the sequence $(b_i)$ of elements of the second row.

From (10.118) we see at once that the number of all paths from $(a,b)$ to $(c,d)$ with exactly $\ell$ NE-turns is in fact equal to the number of $\ell$-element subsets of the set $\{a,a+1,\dots,c-1\}$ times the number of $\ell$-element subsets of $\{b+1,b+2,\dots,d\}$. A similar argument holds for EN-turns. Thus we have proved

$$\begin{aligned}\left|L\big((a,b)\to(c,d)\mid NE(.)=\ell\big)\right| &= \left|L\big((a,b)\to(c,d)\mid EN(.)=\ell\big)\right| \\ &= \binom{c-a}{\ell}\binom{d-b}{\ell}. \qquad (10.120)\end{aligned}$$

A lattice path statistic that is frequently used is the number of **runs** of a lattice path. A **run** in a path $P$ is a maximal subpath of $P$ consisting of steps of equal type. We write $\mathrm{run}(P)$ for the number of runs of $P$. The runs of the path $P_0$ in Figure 10.26 are the subpaths from $(1,-1)$ to $(1,1)$, from $(1,1)$ to $(2,1)$, from $(2,1)$ to $(2,3)$, from $(2,3)$ to $(5,3)$, from $(5,3)$ to $(5,4)$, from $(5,4)$ to $(6,4)$, and from $(6,4)$ to $(6,6)$. Thus we have $\mathrm{run}(P_0)=7$. Obviously, the number of runs of a path is exactly one more than the total number of turns (both NE-turns and EN-turns). Besides, there is also a close relation between NE-turns and runs, which allows us to translate any enumeration result for NE-turns into one for runs.

To avoid case-by-case formulations, depending on whether the number of runs is even or odd, we prefer to consider generating functions. Suppose we know the number of all paths from $A$ to $E$ satisfying some property $R$ and containing a given number of NE-turns. Then we know the generating function $GF(L(A\to E\mid R);x^{NE(.)})$. For brevity, let us denote it by $F(A\to E\mid R;x)$. We define four refinements of $F(A\to E\mid R;x)$. Let $F_{hv}(A\to E\mid R;x)$ be the generating function $\sum_P x^{NE(P)}$ where the sum is over all paths in $L(A\to E\mid R)$ that start with a horizontal step and end with a vertical step. The notations $F_{hh}(A\to E\mid R;x)$, $F_{vh}(A\to E\mid R;x)$, and $F_{vv}(A\to E\mid R;x)$ are defined analogously. The relation between enumeration by runs and enumeration by NE-turns is given by

$$\begin{aligned}GF(L(A\to E\mid R);x^{\mathrm{run}(.)}) &= xF_{hh}(A\to E\mid R;x^2)+x^2F_{hv}(A\to E\mid R;x^2)\\ &\quad + F_{vh}(A\to E\mid R;x^2)+xF_{vv}(A\to E\mid R;x^2). \qquad (10.121)\end{aligned}$$

All the four refinements of the NE-turn generating function can be expressed in terms of NE-turn generating functions. This is seen by setting up a few linear equations and solving them. Evidently, the following is true:

$$\begin{aligned}F(A\to E\mid R;x) &= F_{hh}(A\to E\mid R;x)+F_{hv}(A\to E\mid R;x)\\ &\quad + F_{vh}(A\to E\mid R;x)+F_{vv}(A\to E\mid R;x).\end{aligned}$$

Besides, if $E_1 = (1,0)$ and $E_2 = (0,1)$ denote the standard unit vectors, we have

$$\begin{aligned} F_{hh}(A \to E \mid R;x) + F_{hv}(A \to E \mid R;x) &= F(A+E_1 \to E \mid R;x) \\ F_{hv}(A \to E \mid R;x) + F_{vv}(A \to E \mid R;x) &= F(A \to E - E_2 \mid R;x) \\ F_{hv}(A \to E \mid R;x) &= F(A+E_1 \to E - E_2 \mid R;x). \end{aligned}$$

Solving for $F_{hh}$, $F_{hv}$, $F_{vh}$, $F_{vv}$ we get

$$F_{hh}(A \to E \mid R;x) = F(A+E_1 \to E \mid R;x) - F(A+E_1 \to E - E_2 \mid R;x) \quad (10.122a)$$

$$F_{hv}(A \to E \mid R;x) = F(A+E_1 \to E - E_2 \mid R;x) \quad (10.122b)$$

$$F_{vh}(A \to E \mid R;x) = F(A \to E \mid R;x) + (A+E_1 \to E - E_2 \mid R;x) \quad (10.122c)$$

$$- F(A+E_1 \to E \mid R;x) - F(A \to E - E_2 \mid R;x) \quad (10.122d)$$

$$F_{vv}(A \to E \mid R;x) = F(A \to E - E_2 \mid R;x) - F(A+E_1 \to E - E_2 \mid R;x). \quad (10.122e)$$

As we know from Section 10.3, counting paths restricted by $x = y$, or even by two lines $x = y+t$ and $x = y+s$, is effectively solved by the reflection principle. Of course, reflection by itself is useless for counting paths by turns, since the reflection of portions of paths does not take care of turns. It might introduce new turns or make turns disappear. However, there are "analogues" of reflection for two-rowed arrays, which are due to Krattenthaler and Mohanty [81].

**Theorem 10.14.1** *Let $a \geq b$ and $c \geq d$. The number of all paths from $(a,b)$ to $(c,d)$ staying weakly below $x = y$ with exactly $\ell$ NE-turns is given by*

$$\left|L\big((a,b) \to (c,d) \mid x \geq y, NE(.) = \ell\big)\right| = \binom{c-a}{\ell}\binom{d-b}{\ell} - \binom{c-b-1}{\ell-1}\binom{d-a+1}{\ell+1}, \quad (10.123)$$

*and with exactly $\ell$ EN-turns is given by*

$$\left|L\big((a,b) \to (c,d) \mid x \geq y, EN(.) = \ell\big)\right| = \binom{c-a}{\ell}\binom{d-b}{\ell} - \binom{c-b+1}{\ell}\binom{d-a-1}{\ell}. \quad (10.124)$$

*Proof.* We start with proving (10.123). By the NE-turn representation (10.118), the paths from $(a,b)$ to $(c,d)$ staying weakly below $x = y$ with exactly $\ell$ NE-turns can be represented by

$$\begin{array}{cccccc} a \leq & p_1 & p_2 & \dots & p_\ell & \leq c-1 \\ b+1 \leq & q_1 & q_2 & \dots & q_\ell & \leq d, \end{array} \quad (10.125a)$$

where

$$p_i \geq q_i, \qquad i = 1,2,\dots,\ell. \quad (10.125b)$$

Following the argument in the proof of Theorem 10.3.1, the number of these two-rowed arrays is the number of *all* two-rowed arrays of the type (10.125a) minus those two-rowed arrays of the type (10.125a) that violate (10.125b), i.e. where $p_i < q_i$ for some $i$ between 1 and $\ell$. We know the first number from (10.120).

Concerning the second number, we claim that two-rowed arrays of the type (10.125a) that violate (10.125b) are in one-to-one correspondence with two-rowed arrays of the type

$$\begin{array}{cccccccc} b+1\le & & & \bar p_2 & \cdots & \bar p_\ell & \le c-1 \\ a\le & \bar q_0 & \bar q_1 & \bar q_2 & \cdots & \bar q_\ell & \le d. \end{array} \tag{10.126}$$

The number of all these two-rowed arrays is $\binom{c-b-1}{\ell-1}\binom{d-a+1}{\ell+1}$, as desired. So it only remains to construct the one-to-one correspondence.

Take a two-rowed array $(\mathbf{p} \mid \mathbf{q})$ of the type (10.125a) such that $p_i < q_i$ for some $i$. Let $I$ be the largest integer such that $p_I < q_I$. Then map $(\mathbf{p} \mid \mathbf{q})$ to

$$\begin{array}{ccccccc} & q_1 & \cdots\cdots & q_{I-1} & p_{I+1} & \cdots & p_\ell \\ p_1 \quad p_2 & \cdots\cdots & p_I & q_I & q_{I+1} & \cdots & q_\ell \end{array}. \tag{10.127}$$

Note that both rows are strictly increasing because of $q_{I-1} < q_I < q_{I+1} \le p_{I+1}$ and $p_I < q_I$. By some case-by-case analysis it can be seen that (10.127) is of type (10.126). For example, if $I = \ell$ then we must check $q_{I-1} \le c-1$, among others. Clearly, this follows from the inequalities $q_{I-1} < q_I \le d \le c$.

The inverse of this map is defined in the same way. Let $(\bar{\mathbf{p}} \mid \bar{\mathbf{q}})$ be a two-rowed array of the type (10.126). Let $\bar I$ be the largest integer such that $\bar p_{\bar I} < \bar q_{\bar I}$. If there are none, take $\bar I = 2$. Then map $(\bar{\mathbf{p}} \mid \bar{\mathbf{q}})$ to

$$\begin{array}{ccccccc} \bar q_0 & \cdots\cdots & & \bar q_{\bar I-1} & \bar p_{\bar I+1} & \cdots & \bar p_\ell \\ \bar p_2 & \cdots & \bar p_{\bar I} & \bar q_{\bar I} & \cdots\cdots\cdots & & \bar q_\ell \end{array}. \tag{10.128}$$

It is not difficult to check that the mappings (10.127) and (10.128) are inverses of each other. This completes the proof of (10.123).

The second identity, (10.124), can be established similarly. ∎

**Remark 10.14.2** The above proof leads in fact to $q$-analogues; see [81].

A refinement of Theorem 10.3.3 taking into account turns may as well be derived in this way.

**Theorem 10.14.3** *Let $a+t \ge b \ge a+s$ and $c+t \ge d \ge c+s$. The number of all paths from $(a,b)$ to $(c,d)$ staying weakly below the line $y = x+t$ and above the line $y = x+s$ with exactly $\ell$ NE-turns is given by*

$$\begin{aligned} &\left|L((a,b)\to(c,d) \mid x+t\ge y\ge x+s, NE(.)=\ell)\right| \\ &\quad = \sum_{k\in\mathbb{Z}} \left( \binom{c-a-k(t-s)}{\ell+k}\binom{d-b+k(t-s)}{\ell-k} \right. \\ &\qquad \left. - \binom{c-b-k(t-s)+s-1}{\ell+k}\binom{d-a+k(t-s)-s+1}{\ell-k} \right). \end{aligned} \tag{10.129}$$

Some of the results in Section 10.4 allow also for refinements taking into account turns.

**Theorem 10.14.4** *Let $\mu$ be a positive integer and $c \geq \mu d$. The number of all lattice paths from the origin to $(c,d)$ that stay weakly below $x = \mu y$ with exactly $\ell$ NE-turns is given by*

$$\left|L((0,0) \to (c,d) \mid x \geq \mu y, NE(.) = \ell)\right| = \binom{c}{\ell}\binom{d}{\ell} - \mu\binom{c-1}{\ell-1}\binom{d+1}{\ell+1}, \tag{10.130}$$

*and with exactly $\ell$ EN-turns is given by*

$$\begin{aligned}&\left|L((0,0) \to (c,d) \mid x \geq \mu y, EN(.) = \ell)\right| \\ &= \frac{c-\mu d+1}{c+1}\binom{c+1}{\ell}\binom{d-1}{\ell-1} = \binom{c+1}{\ell}\binom{d-1}{\ell-1} - \mu\binom{c}{\ell-1}\binom{d}{\ell}. \end{aligned}\tag{10.131}$$

Two-rowed arrays may also be used to prove this result, see [78]. A very elegant alternative proof using a rotation operation on paths is given by Goulden and Serrano [55].

We conclude this section by stating results on the enumeration of families of **non-intersecting** lattice paths with respect to turns. This type of problem originally arose from the study of the Hilbert polynomial of certain determinantal and Pfaffian rings (cf. [84, 86]). The results are due to Krattenthaler [75]. We do not provide proofs. Suffice it to mention that they work by using two-rowed arrays.

Let $\mathbf{A} = (A_1, A_2, \ldots, A_n)$ and $\mathbf{E} = (E_1, E_2, \ldots, E_n)$ be points in $\mathbb{Z}^2$. How many families $\mathbf{P} = (P_1, P_2, \ldots, P_n)$ of non-intersecting lattice paths, where $P_i$ runs from $A_i$ to $E_i$, $i = 1, 2, \ldots, n$, are there such that the total number of NE-turns in $\mathbf{P}$ is some fixed number, $\ell$ say?

We give the following three theorems about the counting of non-intersecting lattice paths with a given number of turns. The first theorem concerns counting families of non-intersecting lattice paths with given starting and end points with a given number of NE-turns.

**Theorem 10.14.5** *Let $A_i = (a_1^{(i)}, a_2^{(i)})$ and $E_i = (e_1^{(i)}, e_2^{(i)})$ be lattice points satisfying*

$$a_1^{(1)} \leq a_1^{(2)} \leq \cdots \leq a_1^{(n)}, \quad a_2^{(1)} > a_2^{(2)} > \cdots > a_2^{(n)},$$

*and*

$$e_1^{(1)} < e_1^{(2)} < \cdots < e_1^{(n)}, \quad e_2^{(1)} \geq e_2^{(2)} \geq \cdots \geq e_2^{(n)}.$$

*The number of all families $\mathbf{P} = (P_1, P_2, \ldots, P_n)$ of non-intersecting lattice paths $P_i : A_i \to E_i$, such that the paths of $\mathbf{P}$ altogether contain exactly $\ell$ NE-turns, is*

$$\sum_{\ell_1+\cdots+\ell_n=\ell} \det_{1\leq i,j\leq n}\left(\binom{e_1^{(j)} - a_1^{(i)} + i - j}{\ell_i + i - j}\binom{e_2^{(j)} - a_2^{(i)} - i + j}{\ell_i}\right). \tag{10.132}$$

The second theorem concerns counting families of non-intersecting lattice paths staying weakly below $x = y$, with given starting and end points, by their number of NE-turns.

**Theorem 10.14.6** *Let $A_i = (a_1^{(i)}, a_2^{(i)})$ and $E_i = (e_1^{(i)}, e_2^{(i)})$ be lattice points satisfying*

$$a_1^{(1)} \le a_1^{(2)} \le \cdots \le a_1^{(n)}, \quad a_2^{(1)} > a_2^{(2)} > \cdots > a_2^{(n)},$$
$$e_1^{(1)} < e_1^{(2)} < \cdots < e_1^{(n)}, \quad e_2^{(1)} \ge e_2^{(2)} \ge \cdots \ge e_2^{(n)},$$

*and $a_1^{(i)} \ge a_2^{(i)}$, $e_1^{(i)} \ge e_2^{(i)}$, $i = 1,2,\dots,n$. The number of all families $\mathbf{P} = (P_1, P_2, \dots, P_n)$ of non-intersecting lattice paths $P_i : A_i \to E_i$, that stay weakly below the line $x = y$, and where the paths of $\mathbf{P}$ altogether contain exactly $\ell$ NE-turns, is*

$$\sum_{\ell_1+\cdots+\ell_n=\ell} \det_{1\le i,j\le n} \left( \binom{e_1^{(j)} - a_1^{(i)} + i - j}{\ell_i + i - j} \binom{e_2^{(j)} - a_2^{(i)} - i + j}{\ell_i} \right.$$
$$\left. - \binom{e_1^{(j)} - a_2^{(i)} - i - j + 1}{\ell_i - j} \binom{e_2^{(j)} - a_1^{(i)} + i + j - 1}{\ell_i + i} \right). \tag{10.133}$$

In the third theorem (basically) the same families of non-intersecting lattice paths as before are counted, but with respect to their number of EN-turns. By a rotation by 180° this could be translated into a result about counting families of non-intersecting lattice paths staying **above** $x = y$, with given starting and end points, with respect to **NE-turns**.

**Theorem 10.14.7** *Let $A_i = (a_1^{(i)}, a_2^{(i)})$ and $E_i = (e_1^{(i)}, e_2^{(i)})$ be lattice points satisfying*

$$a_1^{(1)} < a_1^{(2)} < \cdots < a_1^{(n)}, \quad a_2^{(1)} \ge a_2^{(2)} \ge \cdots \ge a_2^{(n)},$$
$$e_1^{(1)} \le e_1^{(2)} \le \cdots \le e_1^{(n)}, \quad e_2^{(1)} > e_2^{(2)} > \cdots > e_2^{(n)},$$

*and $a_1^{(i)} \ge a_2^{(i)}$, $e_1^{(i)} \ge e_2^{(i)}$, $i = 1,2,\dots,n$. The number of all families $\mathbf{P} = (P_1, P_2, \dots, P_n)$ of non-intersecting lattice paths $P_i : A_i \to E_i$, which stay weakly below the line $x = y$, and where the paths of $\mathbf{P}$ altogether contain exactly $\ell$ EN-turns, is*

$$\sum_{\ell_1+\cdots+\ell_n=\ell} \det_{1\le i,j\le n} \left( \binom{e_1^{(j)} - a_1^{(i)} + i - j}{\ell_i + i - j} \binom{e_2^{(j)} - a_2^{(i)} - i + j}{\ell_i} \right.$$
$$\left. - \binom{e_1^{(j)} - a_2^{(i)} - i - j + 3}{\ell_i - j + 1} \binom{e_2^{(j)} - a_1^{(i)} + i + j - 3}{\ell_i + i - 1} \right). \tag{10.134}$$

## 10.15 Multidimensional lattice paths

This section and the following three contain enumeration results for lattice paths in spaces of higher dimension. Most of the time, we shall be concerned with the $d$-dimensional lattice $\mathbb{Z}^d$. The coordinates in $d$-dimensional space will be denoted by $x_1, x_2, \dots, x_d$.

Obviously, as a basis to start with, we need the number of all simple paths in $\mathbb{Z}^d$ (that is, paths consisting of positive unit steps in the direction of some coordinate axis) from $(a_1,a_2,\dots,a_d)$ to $(e_1,e_2,\dots,e_d)$. Since these lattice paths can be seen as permutations of $e_1-a_1$ steps in $x_1$-direction, $e_2-a_2$ steps in $x_2$-direction, $\dots$, $e_d-a_d$ steps in $x_d$-direction, the answer is a multinomial coefficient,

$$\left|L\big((a_1,\dots,a_d)\to(e_1,\dots,e_d)\big)\right| = \binom{\sum_{i=1}^d (e_i-a_i)}{e_1-a_1,e_2-a_2,\dots,e_d-a_d}. \tag{10.135}$$

## 10.16 Multidimensional lattice paths bounded by a hyperplane

Here, we consider simple lattice paths in $\mathbb{Z}^{d+1}$ restricted by a hyperplane of the form $x_0=\sum_{i=1}^d \mu_i x_i$ where $\mu_i$, $i=0,1,\dots,d$, are nonnegative integers. It should be noted that the reflection principle does not apply because, in general, the set of steps is not invariant under reflection with respect to such a hyperplane (except of course when $\mu_i=1$ for all $i$, in which case the reflection principle does apply).

**Theorem 10.16.1** *Let $\mu_0,\mu_1,\dots,\mu_d$ be nonnegative integers and $c_0,c_1,\dots,c_d$ integers such that $c_0\ge\sum_{i=1}^d \mu_i c_i$. The number of all lattice paths from the origin to $(c_0,c_1,\dots,c_d)$ not crossing the hyperplane $x_0=\sum_{i=1}^d \mu_i x_i$ is given by*

$$\left|L\Big(0\to(c_0,c_1,\dots,c_d)\mid x_0\ge\sum_{i=1}^d \mu_i x_i\Big)\right| = \frac{c_0-\sum_{i=1}^d \mu_i c_i+1}{1+\sum_{i=0}^d c_i}\binom{1+\sum_{i=0}^d c_i}{c_0+1,c_1,c_2,\dots,c_d}. \tag{10.136}$$

We omit the proof. Both proofs of Theorem 10.4.5, the generating function proof and the proof by use of the cycle lemma, can be extended to proofs of the above theorem.

To conclude this section, we point out that Sato [105] has extended his generating function results for the number of paths in the plane integer lattice between two parallel lines that we presented in Section 10.5 to the multidimensional case. Similarly, the result of Niederhausen on the enumeration of paths in the plane integer lattice subject to a piece-wise linear boundary, which was presented in Section 10.6, has a multidimensional extension; see [98, Sec. 2.2].

## 10.17 Multidimensional paths with a general boundary

In this section we generalize the enumeration problem of Section 10.7 to arbitrary dimensions. Let $n_1,n_2,\dots,n_d$ be nonnegative integers and **a** and **b** be increasing integer functions defined on the box

$$[0,\mathbf{n}] := \prod_{i=1}^{d}\{0,1,\dots,n_i\}$$

such that $\mathbf{a} \geq \mathbf{b}$. $\mathbf{a}$ is increasing means that $\mathbf{a}(\mathbf{i}) \leq \mathbf{a}(\mathbf{j})$ whenever $\mathbf{i} \leq \mathbf{j}$ in the usual product order. We ask for the number of all paths in $\mathbb{Z}^{d+1}$ from $(0,\mathbf{b}(0))$ to $(\mathbf{n},\mathbf{a}(\mathbf{n}))$ that always stay in the region "that is bounded by $\mathbf{a}$ and $\mathbf{b}$," by which we mean the region

$$\{(\mathbf{i},y) : \mathbf{b}(\mathbf{i}) \leq y \leq \mathbf{a}(\mathbf{i})\}. \tag{10.137}$$

The generalization of Theorem 10.7.1, due to Handa and Mohanty [62], reads as follows.

**Theorem 10.17.1** *Let $n_1, n_2, \dots, n_d$ be nonnegative integers and $p = \prod_{i=1}^{d} n_i$. Assume that the points in the box $[0,\mathbf{n}]$ are $0 = \mathbf{u}_0, \mathbf{u}_1, \mathbf{u}_2, \dots, \mathbf{u}_p = \mathbf{n}$, ordered lexicographically. Then the number of all lattice paths in $\mathbb{Z}^{d+1}$ from $(0,\mathbf{b}(0))$ to $(\mathbf{n},\mathbf{a}(\mathbf{n}))$ that always stay in the region* (10.137) *equals*

$$\left|L\big((0,\mathbf{b}(0)) \to (\mathbf{n},\mathbf{a}(\mathbf{n})) \mid \mathbf{a} \geq \mathbf{y} \geq \mathbf{b}\big)\right|$$
$$= (-1)^{\sum_{i=1}^{d} n_i + \prod_{i=1}^{d} n_i} \det_{0 \leq i,j \leq \sum_{i=1}^{d} n_i - 1} \left( \binom{\mathbf{a}(\mathbf{u}_i) - \mathbf{b}(\mathbf{u}_{j+1}) + 1}{\mathbf{u}_{j+1} - \mathbf{u}_i} \right). \tag{10.138}$$

The most elegant and illuminating proof is by the use of non-intersecting lattice paths, see [115]. Sulanke proves in fact a $q$-analogue in [115].

## 10.18 The reflection principle in full generality

We have explained the reflection principle in the proof of Theorem 10.3.1 in Section 10.3, where it solved the problem of counting simple lattice paths in the plane bounded by the diagonal. Nothing prevents us from applying the same idea in a higher-dimensional setting. It is then natural to ask: How far can we go with the reflection principle? What is the most general situation where it applies? This question was raised and answered by Gessel and Zeilberger [53], and, independently, by Biane [10] in a more restricted setting; see also Grabiner and Magyar [57].

The standard example, which will serve as our running example, is the problem of counting all paths from $(a_1, a_2, \dots, a_d)$ to $(e_1, e_2, \dots, e_d)$ that always stay in the region $x_1 \geq x_2 \geq \cdots \geq x_d$. This problem is equivalent to several other enumeration problems, the most prominent being the ***d*-candidate ballot problem** (for the 2-candidate ballot problem see Section 10.3) and the problem of counting **standard Young tableaux** of a given shape.

In the ***d*-candidate ballot problem** there are $d$ candidates in an election, say $E_1, E_2, \dots, E_d$, $E_1$ receiving $e_1$ votes, $E_2$ receiving $e_2$ votes, ..., $E_d$ receiving $e_d$ votes.

How many ways of counting the votes are there, such that at any stage during the counting $E_1$ has at least as many votes as $E_2$, $E_2$ has at least as many votes as $E_3$, etc.? It is evident that by encoding each vote for candidate $E_i$ by a step in $x_i$-direction this ballot problem is transferred into counting paths from the origin to $(e_1, e_2, \dots, e_d)$ that are staying in the region $x_1 \geq x_2 \geq \cdots \geq x_d$.

A **standard Young tableaux** of skew shape $\lambda/\mu$, where $\lambda = (\lambda_1, \lambda_2, \dots, \lambda_d)$ and $\mu = (\mu_1, \mu_2, \dots, \mu_d)$ are $d$-tuples of non-negative integers that are in non-increasing order and satisfy $\lambda_i \geq \mu_i$ for all $i$, is an arrangement of the numbers $1, 2, \dots, \sum_{i=1}^{d}(\lambda_i - \mu_i)$ of the form

$$\begin{array}{ccccccc} & & & \pi_{1,\mu_1+1} & \cdots\cdots\cdots\cdots & \pi_{1,\lambda_1} \\ & \pi_{2,\mu_2+1} & \cdots & \pi_{2,\mu_1+1} & \cdots\cdots & \pi_{2,\lambda_2} & \\ \iddots & & & \vdots & \iddots & & \\ \pi_{d,\mu_d+1} & \cdots\cdots\cdots\cdots\cdots\cdots & & \pi_{d,\lambda_d} & & & \end{array}$$

such that numbers along rows and columns are increasing. See Chapter 14 for more information on these important combinatorial objects. By encoding an entry $i$ located in row $j$ of the tableau by a step in $x_j$-direction, $i = 1, 2, \dots, \sum_{i=1}^{d}(\lambda_i - \mu_i)$, it is easy to see that standard tableaux of shape $\lambda/\mu$ are in bijection with lattice paths from $\mu = (\mu_1, \mu_2, \dots, \mu_d)$ to $\lambda = (\lambda_1, \lambda_2, \dots, \lambda_d)$ consisting of positive unit steps in the direction of some coordinate axis, and that stay in the region $x_1 \geq x_2 \geq \cdots \geq x_d$.

It is a classical result due to MacMahon [90, p. 175] (see also [91, §103]) that the solution to the counting problem is given by a determinant, see e.g. [112, Prop. 7.10.3 combined with Cor. 7.16.3].

**Theorem 10.18.1** *Let* $A = (a_1, a_2, \dots, a_d)$ *and* $E = (e_1, e_2, \dots, e_d)$ *be points in* $\mathbb{Z}^d$ *with* $a_1 \geq a_2 \geq \cdots \geq a_d$ *and* $e_1 \geq e_2 \geq \cdots \geq e_d$*. The number of all lattice paths from* $A$ *to* $E$ *consisting of positive unit steps in the direction of some coordinate axis and staying in the region* $x_1 \geq x_2 \geq \cdots \geq x_d$ *equals*

$$|L(A \to E \mid x_1 \geq x_2 \geq \cdots \geq x_d)| = \left(\sum_{i=1}^{d}(e_i - a_i)\right)! \det_{1 \leq i,j \leq d}\left(\frac{1}{(e_i - a_j - i + j)!}\right). \tag{10.139}$$

If the starting point $A$ equals the origin, then the above determinant can be reduced to a Vandermonde determinant by elementary column operations, and thus it can be evaluated in closed form. If one rewrites the result appropriately, then one arrives at the celebrated **hook formula** due to Frame, Robinson and Thrall [45]. (We refer the reader to [103, Sec. 3.10] or [112, Cor. 7.21.6] for unexplained terminology).

**Theorem 10.18.2** *Let* $E = (e_1, e_2, \dots, e_d)$ *be a point in* $\mathbb{Z}^d$ *with* $e_1 \geq e_2 \geq \cdots \geq e_d \geq 0$*. The number of all lattice paths from the origin to* $E$ *consisting of positive unit steps in the direction of some coordinate axis and staying in the region* $x_1 \geq x_2 \geq \cdots \geq x_d$ *equals*

$$|L(A \to E \mid x_1 \geq x_2 \geq \cdots \geq x_d)| = \frac{\left(\sum_{i=1}^{d} e_i\right)!}{\prod_{\rho} h(\rho)}, \tag{10.140}$$

*where the product is over all cells $\rho$ in the Ferrers diagram of the partition $(e_1, e_2, \ldots, e_d)$, and $h(\rho)$ is the hook-length of the cell $\rho$.*

It was pointed out by Zeilberger [124] that the formula in (10.139) can be proved by means of the reflection principle. The natural environment for a "general reflection principle" is within the setting of **reflection groups**. A **reflection group** is a group that is generated by all reflections with respect to the hyperplanes $H$ in a given set $\mathscr{H}$ of hyperplanes (in some $\mathbb{R}^d$). We review the facts about reflection groups that are relevant for us below. For an excellent exposition of the subject see Humphreys [63]. As we already said, the situation of Theorem 10.18.1 will be our running example.

As above, let $\mathscr{H}$ be a (finite) set of hyperplanes in some $\mathbb{R}^d$. Let $W$ denote the group that is generated by the corresponding reflections. By definition, $W$ is a subgroup of $O(d)$. Some of the elements of $W$ happen to be reflections with respect to a hyperplane (not necessarily belonging to $\mathscr{H}$), and let $\mathscr{R}$ denote the collection of all these hyperplanes. Of course, $\mathscr{R}$ contains $\mathscr{H}$. In the example when $\mathscr{H}$ is the set of hyperplanes $H_i$ given by

$$H_i: \quad x_i - x_{i+1} = 0, \quad i = 1, 2, \ldots, d-1, \tag{10.141}$$

(these are the hyperplanes restricting the paths in Theorem 10.18.1), the group $W$ is the permutation group $\mathfrak{S}_d$, acting on $\mathbb{R}^d$ by permuting coordinates. All the reflections in this group are the interchanges of two coordinates $x_i$ and $x_j$, $1 \le i < j \le d$, corresponding to the transpositions $(i, j)$ in $\mathfrak{S}_d$. Hence, the corresponding set $\mathscr{R}$ of hyperplanes in this case is

$$\mathscr{R} = \{x_i - x_j = 0 : \ 1 \le i < j \le d\}. \tag{10.142}$$

The hyperplanes in $\mathscr{R}$ cut the space into different regions. The connected components of the complement of $\bigcup_{H \in \mathscr{R}} H$ in $\mathbb{R}^d$ are called **chambers**. Each chamber is enclosed by a set $\mathscr{R}_0$ of bordering hyperplanes. Clearly, $\mathscr{R}_0$ is a subset of $\mathscr{R}$. In our running example a typical chamber is the region

$$\{(x_1, x_2, \ldots, x_d) : \ x_1 > x_2 > \cdots > x_d\}, \tag{10.143}$$

bounded by the hyperplanes in (10.141). As a matter of fact, in this special case any chamber has the form

$$\{(x_1, x_2, \ldots, x_d) : \ x_{\sigma(1)} > x_{\sigma(2)} > \cdots > x_{\sigma(d)}\}, \tag{10.144}$$

where $\sigma$ is some permutation in $\mathfrak{S}_d$.

It can be shown that the reflections with respect to the hyperplanes in $\mathscr{R}_0$ generate the complete reflection group $W$. Another important fact is that, given one chamber $C$, all chambers are $w(C)$, where $w$ runs through the elements of the reflection group $W$, all $w(C)$'s being distinct.

Now we are in the position to formulate and prove Gessel and Zeilberger's result [53, Theorem 1]. The motivation for the technical conditions in the statement of the theorem involving $k_H$ and $r_H$ is that they make sure that it is not possible to "jump" over a hyperplane without touching it in a lattice point.

**Theorem 10.18.3** *Let $C$ be a chamber of some reflection group $W$, determined by the hyperplanes in the set $\mathscr{R}_0$. Let $\mathbb{S}$ be a set of steps that is invariant under $W$, i.e., $w(\mathbb{S}) = \mathbb{S}$, and with the property that for all hyperplanes $H \in \mathscr{R}_0$ and all steps $s \in \mathbb{S}$ the Euclidean inner product $(s, r_H)$ is either $0$ or $\pm k_H$, where $k_H$ is a fixed constant, $r_H$ is a fixed non-zero vector perpendicular to $H$, both depending only on the hyperplane $H$. Furthermore, let $A$ and $E$ be lattice points inside the chamber $C$ such that also $w(A)$ and $w(E)$ are lattice points for all $w \in W$, and such that for all hyperplanes $H \in \mathscr{R}_0$ the Euclidean inner product $(A, r_H)$ is an integral multiple of $k_H$.*

*Then the number of all lattice paths from $A$ to $E$, with exactly $m$ steps from $\mathbb{S}$, and staying strictly inside the chamber $C$, equals*

$$|L_m(A \to E; \mathbb{S} \mid \text{inside } C)| = \sum_{w \in W} (\operatorname{sgn} w)\, |L_m(w(A) \to E; \mathbb{S})|, \qquad (10.145)$$

*where $\operatorname{sgn} w = \det w$, considering $w$ as an orthogonal transformation of $\mathbb{R}^d$.*

**Remark 10.18.4** A weighted version of the above theorem in which steps carry weights such that images of steps under $W$ carry the same weight holds as well.

*Proof.* We may rewrite (10.145) in the form

$$|L_m(A \to E; \mathbb{S} \mid \text{inside } C)| = \sum_{(w,P)} \operatorname{sgn} w, \qquad (10.146)$$

where the sum is over all pairs $(w, P)$ with $w \in W$ and $P \in L_m(w(A) \to E; \mathbb{S})$. The proof of (10.146) is by a sign-reversing involution on the set of all such pairs $(w, P)$, where $P$ touches at least one of the hyperplanes in $\mathscr{H}_0$. Sign-reversing has to be understood with respect to $\operatorname{sgn} w$. Provided the existence of such an involution, the only contributions to the sum in (10.146) would be by pairs $(w, P)$ where $P$ does not touch any of the hyperplanes in $\mathscr{H}_0$. We claim that this can only be the case for $w = \text{id}$. In fact, as we already mentioned, it is one of the properties of a reflection group $W$ that, given one chamber $C$, all chambers are $w(C)$, $w \in W$, and all $w(C)$'s are distinct. Therefore, if $A$ is in $C$ and $w \neq \text{id}$, then $w(A)$ must be in a different chamber and so cannot be in $C$. Thus, evidently, any path from $w(A)$ to $E$, the point $E$ being inside $C$, must touch at least one of the bordering hyperplanes. This would prove (10.146) and hence the theorem.

Now we construct the promised involution. Fix some order of the hyperplanes in $\mathscr{H}_0$. Let $(w, P)$ be a pair with $w \in W$, $P \in L_m(w(A) \to E; \mathbb{S})$, and $P$ touching at least one of the hyperplanes in $\mathscr{H}_0$. Consider all meeting points of $P$ with hyperplanes in $\mathscr{H}_0$. Choose the last meeting point along the path $P$ and denote it by $M$. $M$ must be a lattice point because of the assumptions that involve the constants $k_H$. Let $H$ be the first hyperplane (in the chosen order) that meets $P$. Then we form the new path $P'$ by reflecting the portion of $P$ from the starting point $w(A)$ up to $M$ with respect to $H$ and leaving the portion from $M$ to $P$ invariant. By assumption, reflection of a step from $\mathbb{S}$ is again a step in $\mathbb{S}$. So, $P'$ also consists of steps from $\mathbb{S}$ only. Evidently, the starting point of $P'$ is $w_H w(A)$, where $w_H$ denotes the reflection with respect to $H$. Hence, $(w_H w, P')$ is a pair under consideration for the sum in (10.146), and $P'$ touches one of the hyperplanes in $\mathscr{H}_0$ (namely $H$). This mapping is an involution since nothing was changed after $M$. Moreover, we have $\operatorname{sgn} w_H w = -\operatorname{sgn} w$. Therefore it is also sign-reversing. This completes the proof of the theorem. ∎

In the case of our running example, $W$ is the group of permutations of coordinates, $C$ is given by (10.143), the set of hyperplanes is (10.141), the set of steps is $\{\varepsilon_1, \varepsilon_2, \ldots, \varepsilon_d\}$, with $\varepsilon_i$ denoting the positive unit vector in $x_i$-direction. If $H_i$ is the hyperplane $x_i - x_{i+1} = 0$, we may choose $r_{H_i} = \varepsilon_i - \varepsilon_{i+1}$, so that all constants $k_{H_i}$ are 1, $i = 1, 2, \ldots, d-1$. Since the number of lattice paths between two given lattice points is given by a multinomial coefficient (see (10.135)), it is then not difficult to see that (10.145) yields (10.139) in this special case.

Which other examples are covered by the general setup in Theorem 10.18.3? The answer is that all reflection groups that are "relevant" in our context are completely classified. The meaning of "relevant" is as follows. In order for formula (10.145) to make sense, the sum on the right-hand side of (10.145) should be finite. So, only reflection groups that are "discrete" and act "locally finite" will be of interest to us. It is exactly these reflection groups that are precisely known (see Humphreys [63, Sec. 4.10], Bourbaki [15, Ch. V, VI]).

The classification of all **finite** reflection groups says that any such finite reflection group decomposes into the direct product of irreducible reflection groups, all of which act on pairwise orthogonal subspaces. These irreducible reflection groups do not decompose further. There exist four infinite families of types $I_2(m)$ $(m = 1, 2, \ldots)$, $A_d$, $B_d = C_d$, $D_d$ $(d = 1, 2, \ldots)$ of such groups, and the seven exceptional groups of types $G_2$, $F_4$, $E_6$, $E_7$, $E_8$, and $H_3$, $H_4$. (The indices mark the dimension of the vector space on which they act faithfully.) In addition, for most of these irreducible finite reflection groups there exists an **affine** reflection group, which is infinite. The finite reflection group is generated by all reflections with respect to the hyperplanes that run through a given point (we assume that this is the origin). The affine reflection group is generated by a larger set of hyperplanes, which includes the aforementioned hyperplanes plus certain translates of them. The reflection groups corresponding to $G_2$ are the same as those for $I_2(6)$, therefore we need not consider $G_2$.

Grabiner and Magyar [57, p. 247] have determined all possible step sets (up to dilation) for each of the irreducible reflection groups such that the technical conditions of Theorem 10.18.3 are satisfied. Not for all types do there exist such step sets. It should be noted however that the "empty" step $(0, 0, \ldots, 0)$ can always be added to any possible step set. The following list describes all possible instances of Theorem 10.18.3 when applied to an irreducible finite or affine reflection group. The results for lattice paths in chambers of affine reflection groups have been made explicit by Grabiner [56].

- **Types $H_3, H_4, F_4, E_8, I_2(m)$:** There are no possible step sets.
- **Type $A_{d-1}$:** The set of reflecting hyperplanes is $\mathscr{R} = \{x_i - x_j = 0 : 1 \le i < j \le d\}$. Obviously, the reflection with respect to $x_i - x_j = 0$ acts by interchanging

the $i$th and $j$th coordinate. So, the associated finite reflection group is the group of permutations of the coordinates $x_1,x_2,\dots,x_d$, which is isomorphic to the permutation group $\mathfrak{S}_d$. A typical chamber is $C=\{(x_1,x_2,\dots,x_d):x_1>x_2>\cdots>x_d\}$.

Possible step sets are the sets

$$\mathbb{S}_k := \{w\cdot(1,\dots,1,0,\dots,0) : w\in\mathfrak{S}_d\}, \quad k=1,2,\dots,d$$

(with $k$ occurrences of 1), all compatible with each other, as well as

$$\mathbb{S}_k^{\pm} := \{w\cdot(\pm1,\dots,\pm1,0,\dots,0) : w\in\mathfrak{S}_d\}, \quad k=1 \text{ and } k=d$$

(with $k$ occurrences of $\pm1$), which can not be mixed together.

Theorem 10.18.1 is a direct consequence of Theorem 10.18.3 with $W=\mathfrak{S}_d$ and $\mathbb{S}=\mathbb{S}_1$.

The second standard application is the one for $\mathbb{S}=\mathbb{S}_d^{\pm}$.

**Theorem 10.18.5** *Let $A=(a_1,a_2,\dots,a_d)$ and $E=(e_1,e_2,\dots,e_d)$ be points in $\mathbb{Z}^d$, with all $a_i$'s of the same parity, all $e_i$'s of the same parity, $a_1>a_2>\cdots>a_d$ and $e_1>e_2>\cdots>e_d$. The number of all lattice paths from $A$ to $E$ consisting of $m$ steps from $\mathbb{S}_d^{\pm}$ and staying in the region $x_1>x_2>\cdots>x_d$ equals*

$$\left|L(A\to E;\mathbb{S}_d^{\pm}\mid x_1>x_2>\cdots>x_d)\right| = \det_{1\le i,j\le d}\left(\binom{m}{\frac{m+e_i-a_j}{2}}\right). \tag{10.147}$$

We should point out that the lattice paths in Theorem 10.18.5 are in bijection with configurations in the **lock step model**, a frequently studied **vicious walker model**. On the other hand, the lattice paths in Theorem 10.18.1 are in bijection with configurations in another popular vicious walker model, the so-called **random turns model**. We refer the reader to [80, Sec. 2] for more detailed comments on these connections.

The associated affine reflection group, the affine reflection group of type $\tilde{A}_{d-1}$, is generated by the reflections with respect to the hyperplanes $\mathscr{R}=\{x_i-x_j=k:\ 1\le i<j\le d,\ k\in\mathbb{Z}\}$. The elements of this group are called **affine permutations**. They act by permuting the coordinates $x_1,x_2,\dots,x_d$ and adding a vector $(k_1,k_2,\dots,k_d)$ with $k_1+k_2+\cdots+k_d=0$. A typical chamber * is $C=\{(x_1,x_2,\dots,x_d):\ x_1>x_2>\cdots>x_d>x_1-1\}$. For enumeration purposes, we inflate this chamber, see (10.148) below.

The probably first explicitly stated enumeration result for lattice paths in an affine chamber is the result below due to Filaseta [39], although it was not formulated in that way.

**Theorem 10.18.6** *Let $A=(a_1,a_2,\dots,a_d)$ and $E=(e_1,e_2,\dots,e_d)$ be points in $\mathbb{Z}^d$ with $a_1>a_2>\cdots>a_d$ and $e_1>e_2>\cdots>e_d$. The number of all paths from $A$ to $E$ consisting of steps from $\mathbb{S}_1$ and staying in the chamber*

$$\{(x_1,x_2,\dots,x_d) : x_1>x_2>\cdots>x_d>x_1-N\} \tag{10.148}$$

*Actually, the chambers of affine reflection groups are usually called **alcoves**.

*of type $\tilde{A}_{d-1}$, equals*

$$|L(A \to E; \mathbb{S}_1 \mid x_1 > x_2 > \cdots > x_d > x_1 - N)| = \Big(\sum_{i=1}^{d}(e_i - a_i)\Big)! \sum_{k_1+\cdots+k_d=0} \det_{1\le i,j\le d}\left(\frac{1}{(e_i - a_j + k_i N)!}\right). \quad (10.149)$$

See [82] for a $q$-analogue. It should be noted that Theorem 10.3.3 follows from the special case of the above theorem where $d = 2$.

For the step set $\mathbb{S}_1^{\pm}$ consisting of positive **and** negative unit steps in the direction of some coordinate axis, we obtain the following result.

**Theorem 10.18.7** *Let $m$ and $N$ be positive integers. Furthermore, let $A = (a_1, a_2, \ldots, a_d)$ and $E = (e_1, e_2, \ldots, e_d)$ be vectors of integers in the chamber* (10.148) *of type $\tilde{A}_{d-1}$. Then the number of lattice paths from $A$ to $E$ with exactly $m$ steps from $\mathbb{S}_1^{\pm}$, which stay in the alcove* (10.148), *is given by the coefficient of $x^m/m!$ in*

$$\sum_{k_1+\cdots+k_d=0} \det_{1\le i,j\le d}\left(I_{e_j - a_i + Nk_i}(2x)\right), \quad (10.150)$$

*where $I_\alpha(x)$ is the modified Bessel function of the first kind*

$$I_\alpha(x) = \sum_{j=0}^{\infty} \frac{(x/2)^{2j+\alpha}}{j!\,(j+\alpha)!}.$$

The last result for type $A_{d-1}$ which we state is the one for paths in an affine chamber of type $\tilde{A}_{d-1}$ with steps from $\mathbb{S}_d^{\pm}$.

**Theorem 10.18.8** [56, Eq. (35)] *Let $m$ and $N$ be positive integers. Furthermore, let $A = (a_1, a_2, \ldots, a_d)$ and $E = (e_1, e_2, \ldots, e_d)$ be vectors of integers in the chamber* (10.148) *of type $\tilde{A}_{d-1}$ such that all $a_i$'s have the same parity, and all $e_i$'s have the same parity. Then the number of lattice paths from $A$ to $E$ with exactly $m$ steps from $\mathbb{S}_d^{\pm}$, which stay in the chamber* (10.148), *is given by*

$$|L_m(A \to E; \mathbb{S}_d^{\pm} \mid x_1 > x_2 > \cdots > x_d > x_1 - N)| = \sum_{k_1+\cdots+k_d=0} \det_{1\le i,j\le d}\left(\binom{m}{\frac{m+e_i-a_j}{2} + Nk_j}\right). \quad (10.151)$$

- **Types $B_d$, $C_d$**: The finite reflection groups of types $B_d$ and $C_d$ are identical. The set of reflecting hyperplanes is $\mathscr{R} = \{\pm x_i \pm x_j = 0 : 1 \le i < j \le d\} \cup \{x_i = 0 : 1 \le i \le d\}$. Obviously, the reflection with respect to $x_i - x_j = 0$ acts by interchanging the $i$th and $j$th coordinate, the reflection with respect to $x_i + x_j = 0$ acts by interchanging the $i$th and $j$th coordinate and changing the sign of both, while the reflection with respect to $x_i = 0$ acts by changing the sign of the $i$th coordinate.

  Here, the possible step sets are only $\mathbb{S}_1^{\pm}$ and $\mathbb{S}_d^{\pm}$.

The associated finite reflection group is the group of **signed permutations** of the coordinates $x_1, x_2, \ldots, x_d$, which acts by permuting and changing signs of (some of) the coordinates $x_1, x_2, \ldots, x_d$. It is frequently called the **hyperoctahedral group** since it is the symmetry group of a $d$-dimensional octahedron. It is furthermore isomorphic to the semidirect product $(\mathbb{Z}/2\mathbb{Z})^d \rtimes \mathfrak{S}_d$. A typical chamber is $C = \{(x_1, x_2, \ldots, x_d) : x_1 > x_2 > \cdots > x_d > 0\}$.

We have the following enumeration result for lattice paths staying in this chamber.

**Theorem 10.18.9** *Let $A = (a_1, a_2, \ldots, a_d)$ and $E = (e_1, e_2, \ldots, e_d)$ be points in $\mathbb{Z}^d$, with all $a_i$'s of the same parity, all $e_i$'s of the same parity, $a_1 > a_2 > \cdots > a_d > 0$ and $e_1 > e_2 > \cdots > e_d > 0$. The number of all lattice paths from A to E consisting of m steps from $\mathbb{S}_d^{\pm}$ and staying in the region $x_1 > x_2 > \cdots > x_d > 0$ equals*

$$\left|L_m(A \to E; \mathbb{S}_d^{\pm} \mid x_1 > x_2 > \cdots > x_d)\right| = \det_{1 \le i,j \le d}\left(\binom{m}{\frac{m+e_i-a_j}{2}} - \binom{m}{\frac{m+e_i+a_j}{2}}\right). \tag{10.152}$$

The associated affine reflection group now comes in two flavors, types $\tilde{B}_d$ and $\tilde{C}_d$. A typical chamber of types $\tilde{C}_d$ is $C = \{(x_1, x_2, \ldots, x_d) : 1 > x_1 > x_2 > \cdots > x_d > 0\}$, while a typical chamber of type $\tilde{B}_d$ is $C = \{(x_1, x_2, \ldots, x_d) : x_1 > x_2 > \cdots > x_d > 0 \text{ and } x_1 + x_2 < 1\}$,

Next we quote the two results from [56] on the enumeration of lattice paths in chambers of type $\tilde{C}_d$.

**Theorem 10.18.10** [56, Eq. (23)] *Let m and N be positive integers. Furthermore, let $A = (a_1, a_2, \ldots, a_d)$ and $E = (e_1, e_2, \ldots, e_d)$ be vectors of integers in the chamber*

$$\{(x_1, x_2, \ldots, x_n) : N > x_1 > x_2 > \cdots > x_d > 0\} \tag{10.153}$$

*of type $\tilde{C}_d$. Then the number of lattice paths from A to E with exactly m steps from $\mathbb{S}_1^{\pm}$, which stay in the chamber* (10.153), *is given by the coefficient of $x^m/m!$ in*

$$\det_{1 \le i,j \le d}\left(\frac{1}{N}\sum_{r=0}^{2N-1} \sin\frac{\pi r e_i}{N} \cdot \sin\frac{\pi r a_j}{N} \cdot \exp\left(2x\cos\frac{\pi r}{N}\right)\right). \tag{10.154}$$

The result for lattice paths with steps from $\mathbb{S}_d^{\pm}$ is the following.

**Theorem 10.18.11** [56, Eq. (18)] *Let m and N be positive integers. Furthermore, let $A = (a_1, a_2, \ldots, a_d)$ and $E = (e_1, e_2, \ldots, e_d)$ be vectors of integers in the chamber* (10.153) *of type $\tilde{C}_d$ such that all $a_i$'s are of the same parity, and all $e_i$'s are of the same parity. Then the number of lattice paths from A to E with exactly m steps from $\mathbb{S}_d^{\pm}$, which stay in the chamber* (10.153), *is given by*

$$\det_{1 \le i,j \le d}\left(\frac{2^{m-1}}{N}\sum_{r=0}^{4N-1} \sin\frac{\pi r \lambda_t}{N} \cdot \sin\frac{\pi r \eta_h}{N} \cdot \cos^m\frac{\pi r}{2N}\right). \tag{10.155}$$

Enumeration results for lattice paths in a chamber of type $\tilde{B}_d$ can also be derived from Theorem 10.18.3. We omit to state them here, but instead refer to [56] and [80, Theorems 8 and 9].

- **Type** $D_d$: The set of reflecting hyperplanes is $\mathscr{R} = \{\pm x_i \pm x_j = 0 :\ 1 \le i < j \le d\}$. Obviously, $D_d$ is a subset of $B_d$ or $C_d$. The action of the reflection with respect to a hyperplane $\pm x_i \pm x_j = 0$ was already explained there.

  The associated finite reflection group is the group of signed permutations of the coordinates $x_1, x_2, \dots, x_d$ **with an even number of sign changes.** It acts by permuting the coordinates $x_1, x_2, \dots, x_d$ and changing an even number of signs thereof. A typical chamber is $C = \{(x_1, x_2, \dots, x_d) :\ x_1 > x_2 > \cdots > x_{d-1} > |x_d|\}$.

  The associated affine reflection group is generated by the reflections with respect to the hyperplanes $\mathscr{R} = \{\pm x_i \pm x_j = k :\ 1 \le i < j \le d,\ k \in \mathbb{Z}\}$. The elements of this group act by permuting the coordinates $x_1, x_2, \dots, x_d$, changing an even number of signs thereof, and adding a vector $(k_1, k_2, \dots, k_d)$ with $k_1 + k_2 + \cdots + k_d \equiv 0 \pmod 2$. A typical chamber is $C = \{(x_1, x_2, \dots, x_d) :\ x_1 > x_2 > \cdots > x_{d-1} > |x_d|, \text{ and } x_1 + x_2 < 1\}$.

We omit the explicit statement of enumeration results for types $D_d$ and $\tilde{D}_d$, which one may derive from Theorem 10.18.3, and instead refer to [56] and [80, Theorems 10 and 11].

- **Types** $E_6$ **and** $E_7$: There are possible step sets (see [57, p. 247]), but since this does not yield interesting enumeration results, we refrain from discussing these two cases further.

A non-example for the application of the reflection principle has been discussed in Section 10.12, see Theorem 10.12.3.

## 10.19 $q$-Counting of lattice paths and Rogers–Ramanujan identities

In this section, we discuss some $q$-analogues of earlier (plain) enumeration results, and we briefly present work showing the close link between lattice path enumeration and the celebrated Rogers–Ramanujan identities.

As we have already seen in the introduction, one source of $q$-analogues is area counting of lattice paths. This idea has also been used to construct a $q$-analogue of Catalan numbers. Given a Dyck path $P$ (see Section 10.8) from $(0,0)$ to $(2n,0)$, let $\tilde{a}(P) := \frac{1}{2}\left(a(P) - n\right)$. In other words, $\tilde{a}(P)$ is half of the area between $P$ and the "lowest" Dyck path from $(0,0)$ and $(2n,0)$, that is, the zig-zag path in which up-steps and down-steps alternate. Alternatively, $\tilde{a}(P)$ counts the squares with side length $\sqrt{2}$

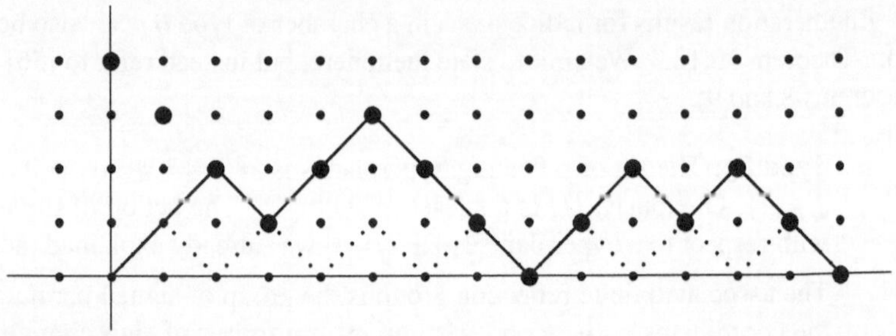

**Figure 10.27**
A Dyck path.

(rotated by 45°) that fit between $P$ and the zig-zag path. In Figure 10.27 this zig-zag path is indicated as the dotted path. For the Dyck path shown with full lines in the figure, we have $\tilde{a}(.) = 6$. This (modified) area statistics is now used to define the $q$-**Catalan number** $C_n(q)$ as the generating function for Dyck paths of length $2n$ with respect to the statistics $\tilde{a}(.)$,

$$C_n(q) = GF\big(L((0,0) \to (2n,0); \{(1,1), (1,-1)\}); q^{\tilde{a}(P)}\big). \tag{10.156}$$

By decomposing a given Dyck path $P$ uniquely into

$$P = s_u P_1 s_d P_2,$$

where $s_u$ denotes an up-step, $s_d$ denotes a down-step, and $P_1$ and $P_2$ are Dyck paths, one obtains the recurrence

$$C_n(q) = \sum_{k=0}^{n-1} q^k C_k(q) C_{n-k-1}(q), \quad n \geq 1, \tag{10.157}$$

with initial condition $C_0(q) = 1$. These $q$-Catalan numbers have been originally introduced by Carlitz and Riordan [25]. We shall say more about these further below.

A different statistics can be derived from turn enumeration (cf. Section 10.14). In the geometry we are considering here, turns are peaks and valleys of a Dyck path. For a peak at lattice point $S$, denote by $x(S)$ the number of steps along the path from the origin to $S$. (Equivalently, $x(S)$ is the ordinate of $S$.) In the Dyck path in Figure 10.27, the peaks are at $(2,2)$, $(5,3)$, $(10,2)$, and $(12,2)$. The ordinates are $x((2,2)) = 2$, $x((5,3)) = 5$, $x((10,2)) = 10$, $x((12,2)) = 12$, The **major index** of a Dyck path $P$, denoted by $\operatorname{maj}(P)$, is the sum of all values $x(S)$ over all peaks $S$ of $P$. For the Dyck path in Figure 10.27, we have $\operatorname{maj}(.) = 2+5+10+12 = 29$. Fürlinger and Hofbauer [47] used this statistic to define alternative $q$-Catalan numbers, namely

$$c_n(q) = GF\big(L((0,0) \to (2n,0); \{(1,1), (1,-1)\}); q^{\operatorname{maj}(P)}\big). \tag{10.158}$$

They showed that

$$c_n(q) = \frac{1-q}{1-q^{n+1}} \begin{bmatrix} 2n \\ n \end{bmatrix}_q, \tag{10.159}$$

the "natural" $q$-analogue of the Catalan number in view of its explicit formula $\frac{1}{n+1}\binom{2n}{n}$. More on these $q$-Catalan numbers and further work in this direction can be found in [47, 74, 81]. These ideas have been extended to Schröder paths and numbers by Bonin, Shapiro and Simion in [11].

Returning to the $q$-Catalan numbers of Carlitz and Riordan, we see that by the choice of $b_i = 0$, $i = 0, 1, \dots$, $\lambda_i = q^{i-1}z$, $i = 1, 2, \dots$ in Theorem 10.9.1, we obtain a continued fraction for the generating function of $q$-Catalan numbers $C_n(q)$, namely

$$\sum_{n=0}^{\infty} C_n(q) z^n = \cfrac{1}{1 - \cfrac{z}{1 - \cfrac{qz}{1 - \cfrac{q^2 z}{1 - \cfrac{q^3 z}{1 - \ddots}}}}}. \tag{10.160}$$

If one substitutes $z = -q$ in this continued fraction, then it becomes the reciprocal of the celebrated Ramanujan continued fraction (cf. [2, Ch. 7])

$$1 + \cfrac{q}{1 + \cfrac{q^2}{1 + \cfrac{q^3}{1 + \cfrac{q^4}{\ddots}}}} = \frac{\sum_{n=0}^{\infty} \frac{q^{n^2}}{(q;q)_n}}{\sum_{n=0}^{\infty} \frac{q^{n(n+1)}}{(q;q)_n}}, \tag{10.161}$$

where $(\alpha;q)_n := (1-\alpha)(1-\alpha q)\cdots(1-\alpha q^{n-1})$, $n \geq 1$, and $(\alpha;q)_0 := 1$.

Numerator and denominator on the right-hand side of this identity feature in the equally celebrated Rogers–Ramanujan identities (cf. also [2, Ch. 7])

$$\sum_{n=0}^{\infty} \frac{q^{n^2}}{(q;q)_n} = \frac{1}{(q;q^5)_\infty (q^4;q^5)_\infty} \tag{10.162}$$

and

$$\sum_{n=0}^{\infty} \frac{q^{n(n+1)}}{(q;q)_n} = \frac{1}{(q^2;q^5)_\infty (q^3;q^5)_\infty}. \tag{10.163}$$

The fact that we came across the left-hand sides of these identities by starting with lattice path counting problems may indicate that the Rogers–Ramanujan identities themselves may be linked with lattice path enumeration. Bressoud [24] was the first to actually set up such a link. Since then, this connection has been explored much further and extended in various directions, particularly so in the physics literature, see [7, 29, 34, 93, 122] and the references therein.

## 10.20 Self-avoiding walks

A path (walk) in a lattice in $d$-dimensional Euclidean space is called **self-avoiding** if it visits each point of the lattice at most once. One cannot expect useful formulas for the exact enumeration of self-avoiding paths (except in extremely simple lattices). This is the reason why, with a few exceptions, research in this area concentrates on **asymptotic** counting: how many self-avoiding walks are there in a particular lattice, consisting of $n$ steps from a given step set, asymptotically as $n$ tends to infinity? This is a notoriously difficult question, that has been investigated mainly in the physics and probability literature. In fact, the self-avoiding walk constitutes a fascinating, vast subject area, which would need a chapter by itself. We refer the reader to the standard book [92], and to the more recent volumes [59, 67] which contain more recent material on or relating to self-avoiding walks.

# References

[1] D. André. Solution directe du problème résolu par M. Bertrand. *Comptes Rendus Acad. Sci. Paris Sér. I*, 105:436–437, 1887.

[2] G. E. Andrews. *The theory of partitions*. Addison–Wesley, Reading, Encyclopedia of Math. and its Applications, vol. 2, 1976. Reprinted by Cambridge University Press, Cambridge, 1998.

[3] D. Armstrong. *Generalized noncrossing partitions and combinatorics of Coxeter groups*. Mem. Amer. Math. Soc., vol. 202, no. 949, Amer. Math. Soc., Providence, RI, 2009.

[4] D. Armstrong, B. Rhoades, and N. Williams. Rational associahedra and noncrossing partitions. *Electron. J. Combin.*, 20:Paper 54, 27 pp., 2013.

[5] C. Banderier and P. Flajolet. Basic analytic combinatorics of directed lattice paths. *Theoret. Computer Science*, 281:37–80, 2002.

[6] C. Banderier and B. Gittenberger. Analytic combinatorics of lattice paths: enumeration and asymptotics for the area. In *Fourth Colloquium on Mathematics and Computer Science Algorithms, Trees, Combinatorics and Probabilities*, pages 345–355, Assoc. Discrete Math. Theor. Comput. Sci., Nancy, 2006. Discrete Math. Theor. Comput. Sci. Proc., AG.

[7] A. Berkovich and P. Paule. Lattice paths, $q$-multinomials and two variants of the Andrews–Gordon identities. *Ramanujan J.*, 5:409–425, 2001.

[8] J. F. L. Bertrand. Solution d'un problème. *Comptes Rendus Acad. Sci. Paris Sér. I*, 105:369, 1887.

[9] P. Biane. Permutations suivant le type d'excedance et le nombre d'inversions et interprétation combinatoire d'une fraction continue de Heine. *European J. Combin.*, 14:277–284, 1993.

[10] P. Biane. Representations of symmetric groups and free probability. *Adv. Math.*, 138:126–181, 1998.

[11] J. Bonin, L. Shapiro, and R. Simion. Some $q$-analogues of the Schröder numbers arising from combinatorial statistics on lattice paths. *J. Statist. Plann. Inference*, 34:35–56, 1993.

[12] A. Bostan and M. Kauers. The complete generating function for Gessel walks is algebraic. *Proc. Amer. Math. Soc.*, 138:3063–3078, 2010.

[13] A. Bostan, I. Kurkova, and K. Raschel. A human proof of Gessel's lattice path conjecture. Preprint; arXiv:1309.1023, 2014.

[14] A. Bostan, K. Raschel, and B. Salvy. Non-D-finite excursions in the quarter plane. *J. Combin. Theory Ser. A*, 121:45–63, 2014.

[15] N. Bourbaki. *Groupes et algèbres de Lie*. Chapitres IV, V, VI, Éléments de Mathématiques, Fasc. 34, Hermann, Paris, 1968.

[16] M. Bousquet-Mélou. Walks in the quarter plane — Kreweras' algebraic model. *Ann. Appl. Probab.*, 15:1451–1491, 2005.

[17] M. Bousquet-Mélou and L. Habsieger. Sur les matrices à signes alternants. *Discrete Math.*, 139:57–72, 1995.

[18] M. Bousquet-Mélou and A. Jehanne. Polynomial equations with one catalytic variable, algebraic series and map enumeration. *J. Combin. Theory Ser. B*, 96:623–672, 2006.

[19] M. Bousquet-Mélou and M. Mishna. Walks with small steps in the quarter plane. In *Algorithmic probability and combinatorics*, pages 1–39. Contemp. Math., vol. 520, Amer. Math. Soc., Providence, R.I., 2010.

[20] M. Bousquet-Mélou and G. Schaeffer. Walks on the slit plane. *Probab. Theory Related Fields*, 124:305–344, 2002.

[21] R. Brak and W. Galleas. Constant term solution for an arbitrary number of osculating lattice paths. *Lett. Math. Phys.*, 103:1261–1272, 2013.

[22] R. Brak, A. L. Owczarek, A. Rechnitzer, and S. G. Whittington. A directed walk model of a long chain polymer in a slit with attractive walls. *J. Phys. A*, 38:4309–4325, 2005.

[23] F. Brenti. Combinatorics and total positivity. *J. Combin. Theory Ser. A*, 71:175–218, 1995.

[24] D. M. Bressoud. Lattice paths and the Rogers–Ramanujan identities. In *Number theory, Madras 1987*, pages 140–172, Berlin, 1989. Lecture Notes in Math., 1395, Springer.

[25] L. Carlitz and J. Riordan. Two element lattice path permutation numbers and their $q$-generalization. *Duke Math. J.*, 31:371–388, 1964.

[26] P. Cartier and D. Foata. *Problèmes combinatoires de commutation et réarrangements*. Lecture Notes in Mathematics, vol. 85. Springer–Verlag, Berlin, New York, 1969. Republished in the "books" section of the Séminaire Lotharingien de Combinatoire.

[27] E. C. Catalan. Note sur une équation aux différences finies. *J. Math. Pures Appl.*, 3:508–515, 1838.

[28] E. C. Catalan. Solution nouvelle de cette question : Un polygône étant donné, de combien de manières peut-on le partager en triangles au moyen de diagonales ? *J. Math. Pures Appl.*, 4:91–94, 1839.

[29] J. Cigler. Fibonacci–Zahlen, Gitterpunktwege und die Identitäten von Rogers–Ramanujan. *Math. Semesterber.*, 52:97–125, 2005.

[30] M. Ciucu, T. Eisenkölbl, C. Krattenthaler, and D. Zare. Enumeration of lozenge tilings of hexagons with a central triangular hole. *J. Combin. Theory Ser. A*, 95:251–334, 2001.

[31] M. Ciucu and C. Krattenthaler. The interaction of a gap with a free boundary in a two dimensional dimer system. *Commun. Math. Phys.*, 302:253–289, 2011.

[32] M. Ciucu and C. Krattenthaler. A factorization theorem for rhombus tilings of a hexagon with triangular holes. Preprint; arXiv:1403.3323, 2014.

[33] R. J. Clarke, E. Steingrímsson, and J. Zeng. New Euler–Mahonian statistics on permutations and words. *Adv. in Appl. Math.*, 18:237–270, 1997.

[34] S. Corteel and O. Mallet. Overpartitions, lattice paths, and Rogers–Ramanujan identities. *J. Combin. Theory Ser. A*, 114:1407–1437, 2007.

[35] N. Dershovitz and S. Zaks. The cycle lemma and some applications. *Europ. J. Combin.*, 11:35–40, 1990.

[36] A. Dvoretzky and T. Motzkin. A problem of arrangements. *Duke Math. J.*, 14:305–313, 1947.

[37] L. Euler. Summarium. *Novi Commentarii Academiae Scientiarum Petropolitanae*, 7:13–15, 1758/59. Reprinted in *Opera Omnia (1)* **26** (1953), xvi–xviii.

[38] W. Feller. *An introduction to probability theory and its applications*. vol. 1, 2nd ed., Wiley, New York, 1957.

[39] M. Filaseta. A new method for solving a class of ballot problem. *J. Combin. Theory Ser. A*, 39:102–111, 1985.

[40] I. Fischer. Enumeration of rhombus tilings of a hexagon which contain a fixed rhombus in the centre. *J. Combin. Theory Ser. A*, 96:31–88, 2001.

[41] M. E. Fisher. Walks, walls, wetting, and melting. *J. Statist. Phys.*, 34:667–729, 1984.

[42] P. Flajolet. Combinatorial aspects of continued fractions. *Discrete Math.*, 32:125–161, 1980.

[43] P. Flajolet and R. Sedgewick. *Analytic combinatorics*. Cambridge University Press, Cambridge, 2009.

[44] D. Foata and D. Zeilberger. Denert's permutation statistic is indeed Euler–Mahonian. *Stud. Appl. Math.*, 83:31–59, 1990.

[45] J. S. Frame, G. B. Robinson, and R. M. Thrall. The hook graphs of the symmetric group. *Canad. J. Math.*, 6:316–325, 1954.

[46] J. Françon and X. G. Viennot. Permutations selon leurs pics, creux, doubles montés et double descentes, nombres d'Euler et nombres de Genocchi. *Discrete Math.*, 28:21–35, 1979.

[47] J. Fürlinger and J. Hofbauer. $q$-Catalan numbers. *J. Combin. Theory A*, 40:248–264, 1985.

[48] I. M. Gessel. A probabilistic method for lattice path enumeration. *J. Statist. Plann. Inference*, 14:49–58, 1986.

[49] I. M. Gessel and C. Krattenthaler. Cylindric partitions. *Trans. Amer. Math. Soc.*, 349:429–479, 1997.

[50] I. M. Gessel and X. Viennot. Binomial determinants, paths, and hook length formulae. *Adv. Math.*, 58:300–321, 1985.

[51] I. M. Gessel and X. Viennot. Determinants, paths, and plane partitions. Preprint, 1989; available at `http://www.cs.brandeis.edu/~ira`.

[52] I. M. Gessel and G. Xin. The generating function of ternary trees and continued fractions. *Electron. J. Combin.*, 13:Research Paper 53, 48 pp., 2006.

[53] I. M. Gessel and D. Zeilberger. Random walk in a Weyl chamber. *Proc. Amer. Math. Soc.*, 115:27–31, 1992.

[54] H. W. Gould. Some generalizations of Vandermonde's convolution. *Amer. Math. Monthly*, 63:84–91, 1956.

[55] I. P. Goulden and L. G. Serrano. Maintaining the spirit of the reflection principle when the boundary has arbitrary integer slope. *J. Combin. Theory Ser. A*, 104:317–326, 2003.

[56] D. J. Grabiner. Random walk in an alcove of an affine Weyl group, and non-colliding random walks on an interval. *J. Combin. Theory Ser. A*, 97:285–306, 2002.

[57] D. J. Grabiner and P. Magyar. Random walks in Weyl chambers and the decomposition of tenbsor powers. *J. Alg. Combin*, 2:239–260, 1993.

[58] H.-D. O. F. Gronau, W. Just, W. Schade, P. Scheffler, and J. Wojciechowski. Path systems in acyclic directed graphs. *Zastos. Mat.*, 19:399–411, 1988.

[59] A. J. Guttmann, editor. *Polygons, polyominoes and polycubes*. Lecture Notes in Physics, vol. 775, Springer–Verlag, Dordrecht, 2009.

[60] R. Guy, C. Krattenthaler, and B. E. Sagan. Lattice paths, reflections, & dimension-changing bijections. *Ars Combin.*, 34:3–15, 1992.

[61] J. Haglund. *The $q,t$-Catalan numbers and the space of diagonal harmonics*. University Lecture Series, vol. 41, Amer. Math. Soc., Providence, R.I., 2008.

[62] B. R. Handa and S. G. Mohanty. Enumeration of higher-dimensional paths under restrictions. *Discrete Math.*, 26:119–128, 1979.

[63] J. E. Humphreys. *Reflection groups and Coxeter groups*. Cambridge University Press, Cambridge, 1990.

[64] K. Humphreys. A history and a survey of lattice path enumeration. *J. Statist. Plann. Inference*, 140:2237–2254, 2010.

[65] C. Huygens. De ratiociniis in ludo aleae. In van Schooten, editor, *Exercitationum Mathematicarum Libri quinque*. Amsterdam, 1657. Reprinted in Œuvres complètes de Christiaan Huygens, vol. 14, La Haye: Nijhoff.

[66] M. Ishikawa and M. Wakayama. Minor summation formula for pfaffians. *Linear and Multilinear Algebra*, 39:285–305, 1995.

[67] E. J. Janse van Rensburg. *The statistical mechanics of interacting walks, polygons, animals and vesicles*. Oxford Lecture Series in Mathematics and its Applications, vol. 18, Oxford University Press, Oxford, 2000.

[68] E. J. Janse van Rensburg, T. Prellberg, and A. Rechnitzer. Directed paths in a wedge. *J. Phys. A*, 40:14069–14084, 2007.

[69] E. J. Janse van Rensburg, T. Prellberg, and A. Rechnitzer. Partially directed paths in a wedge. *J. Combin. Theory Ser. A*, 115:623–650, 2008.

[70] P. John and H. Sachs. Wegesysteme und Linearfaktoren in hexagonalen und quadratischen Systemen. In *Graphs in research and teaching, (Kiel, 1985)*, pages 85–101, Bad Salzdetfurth, 1985. Franzbecker.

[71] S. Karlin and J. L. McGregor. Coincidence properties of birth-and-death processes. *Pacific J. Math.*, 9:1109–1140, 1959.

[72] S. Karlin and J. L. McGregor. Coincidence probabilities. *Pacific J. Math.*, 9:1141–1164, 1959.

[73] V. S. Korolyuk. On the discrepancy of empiric distributions for the case of two independent samples. *Izv. Akad. Nauk SSSR Ser. Math.*, 19:81–96, 1955. IMS & AMS Selected Translations Math. Statist. Prob. **4** (1963), 105–122.

[74] C. Krattenthaler. Counting lattice paths with a linear boundary, part 2: $q$-ballot and $q$-Catalan numbers. *Sitz.ber. Österr. Akad. Wiss., math.-naturwiss. Klasse*, 198:171–199, 1989.

[75] C. Krattenthaler. Counting nonintersecting lattice paths with turns. *Séminaire Lotharingien Combin.*, 34:paper B34i, 17 pp, 1995.

[76] C. Krattenthaler. *The major counting of nonintersecting lattice paths and generating functions for tableaux*. Mem. Amer. Math. Soc., vol. 115, no. 552, Providence, R.I., 1995.

[77] C. Krattenthaler. Determinant identities and a generalization of the number of totally symmetric self-complementary plane partitions. *Electron. J. Combin.*, 4(1):Article #R27, 62 pp, 1997.

[78] C. Krattenthaler. The enumeration of lattice paths with respect to their number of turns. In N. Balakrishnan, editor, *Advances in Combinatorial Methods and Applications to Probability and Statistics*, pages 29–58. Birkhäuser, Boston, 1997.

[79] C. Krattenthaler. On multiplicities of points on Schubert varieties in Grassmannians. *Séminaire Lotharingien Combin.*, 45:Article B45c, 11 pp, 2001.

[80] C. Krattenthaler. Asymptotics for random walks in alcoves of affine Weyl groups. *Séminaire Lotharingien Combin.*, 52:Article B52i, 72 pp, 2007.

[81] C. Krattenthaler and S. G. Mohanty. On lattice path counting by major and descents. *Europ. J. Combin.*, 14:43–51, 1993.

[82] C. Krattenthaler and S. G. Mohanty. $q$-generalization of a ballot problem. *Discrete Math.*, 126:195–208, 1994.

[83] C. Krattenthaler and S. G. Mohanty. Counting tableaux with row and column bounds. *Discrete Math.*, 136:273–286, 1995.

[84] C. Krattenthaler and M. Prohaska. A remarkable formula for counting nonintersecting lattice paths in a ladder with respect to turns. *Trans. Amer. Math. Soc.*, 351:1015–1042, 1999.

[85] G. Kreweras. Sur une classe de problèmes de dénombrement liés au treillis des partitions des entiers. *Cahiers du B.U.R.O. Univ. de Rech. Opér.*, 6:5–105, 1965.

[86] D. M. Kulkarni. Counting of paths and coefficients of Hilbert polynomial of a determinantal ideal. *Discrete Math.*, 154:141–151, 1996.

[87] A. Lascoux and M.-P. Schützenberger. Polynômes de Schubert. *C. R. Acad. Sci. Paris*, 294:447–450, 1982.

[88] B. Lindström. On the vector representations of induced matroids. *Bull. London Math. Soc.*, 5:85–90, 1973.

[89] I. G. Macdonald. *Symmetric functions and Hall polynomials*. second edition, Oxford University Press, New York/London, 1995.

[90] P. A. MacMahon. Memoir on the theory of the partitions of numbers — Part IV. *Phil. Trans. Royal Soc. London (A)*, 209:153–175, 1909.

[91] P. A. MacMahon. *Combinatory analysis*. vols. 1 and 2, Cambdridge University Press, Cambdridge, 1915, 1916. Reprinted by Chelsea, New York, 1955.

[92] N. Madras and G. Slade. *The self-avoiding walk*. Probability and its Applications, Birkhäuser Boston, Inc., Boston, MA, 1993.

[93] P. Mathieu. Paths and partitions: combinatorial descriptions of the parafermionic states. *J. Math. Phys.*, 50:095210, 43 pp., 2009.

[94] A. de Médicis and X. G. Viennot. Moments des $q$-polynômes de Laguerre et la bijection de Foata–Zeilberger. *Adv. Appl. Math.*, 15:262–304, 1994.

[95] S. G. Mohanty. *Lattice path counting and applications*. Academic Press, New York, 1979.

[96] S. G. Mohanty and T. V. Narayana. Some properties of compositions and their application to probability and statistics, I. *Biometrische Z.*, 3:252–258, 1963.

[97] T. V. Narayana. *Lattice path combinatorics with statistical applications*. Math. Expositions, no. 23, Univ. of Toronto Press, Toronto, 1979.

[98] H. Niederhausen. Linear recurrences under side conditions. *Europ. J. Combin.*, 1:353–368, 1980.

[99] H. Niederhausen. The ballot problem with three candidates. *Europ. J. Combin.*, 4:161–167, 1983.

[100] S. Okada. On the generating functions for certain classes of plane partitions. *J. Combin. Theory Ser. A*, 51:1–23, 1989.

[101] I. Pak. History of Catalan numbers. Preprint; arXiv:1408.5711, 2014. To appear as an appendix to Richard Stanley's book *Catalan Numbers*.

[102] E. Roblet and X. Viennot. Théorie combinatoire des $T$-fractions et approximants de Padé en deux points. *Discrete Math.*, 153:271–288, 1996.

[103] B. E. Sagan. *The symmetric group*. 2nd edition, Springer–Verlag, New York, 2001.

[104] M. Sato. Generating functions for the number of lattice paths between two parallel lines with a rational incline. *Math. Japonica*, 34:123–137, 1989.

[105] M. Sato. Generating functions for the number of lattice paths restricted by two parallel hyperplanes. *J. Statist. Plann. Inference*, 34:251–258, 1993.

[106] J. A. Segner. Enumeratio modorum, quibus figurae planaerectilineae per diagonales dividuntur in triangula. *Novi Commentarii Academiae Scientiarum Petropolitanae*, 7:203–209, 1758/59.

[107] R. Simion and D. Stanton. Octabasic Laguerre polynomials and permutation statistics. *J. Comput. Appl. Math.*, 68:297–329, 1996.

[108] J. C. Slater. The theory of complex spectra. *Phys. Rev.*, 34:1293–1322, 1929.

[109] J. C. Slater. *Quantum theory of matter*. 2nd ed., McGraw–Hill, New York, 1968.

[110] F. Spitzer. A combinatorial lemma and its application to probability theory. *Trans. Amer. Math. Soc.*, 82:323–339, 1956.

[111] R. P. Stanley. *Enumerative combinatorics*. vol. 1, Wadsworth & Brooks/Cole, Pacific Grove, California, 1986. Reprinted by Cambridge University Press, Cambridge, 1998.

[112] R. P. Stanley. *Enumerative combinatorics*. vol. 2, Cambridge University Press, Cambridge, 1999.

[113] R. P. Stanley. Catalan addendum. Continuation of Exercise 6.19 from [112]; available at `http://math.mit.edu/~rstan/ec/catadd.pdf`, 2013.

[114] J. R. Stembridge. Nonintersecting paths, pfaffians and plane partitions. *Adv. Math.*, 83:96–131, 1990.

[115] R. A. Sulanke. A determinant for $q$-counting lattice paths. *Discrete Math.*, 81:91–96, 1990.

[116] R. A. Sulanke and G. Xin. Hankel determinants for some common lattice paths. *Adv. in Appl. Math.*, 40:149–167, 2008.

[117] G. Szegő. *Orthogonal polynomials*. American Mathematical Society Colloquium Publications, vol. XXIII, New York, 1959.

[118] L. Takács. A generalization of the ballot problem and its application in the theory of queues. *J. Amer. Statist. Assoc.*, 57:327–337, 1962.

[119] X. Viennot. *Une théorie combinatoire des polynômes orthogonaux généraux*. UQAM, Montréal, Québec, 1983. Available at `www.xavierviennot.org/xavier/polynomes_orthogonaux.html`.

[120] X. G. Viennot. Heaps of pieces. I. Basic definitions and combinatorial lemmas. In *Combinatoire énumérative*, pages 321–350. Lecture Notes in Math., 1234, Springer, Berlin, 1986.

[121] M. L. Wachs. Flagged Schur functions, Schubert polynomials, and symmetrizing operators. *J. Combin. Theory Ser. A*, 40:276–289, 1985.

[122] T. A. Welsh. *Fermionic expressions for minimal model Virasoro characters*. Mem. Amer. Math. Soc., vol. 175, no. 827, 2005.

[123] W. A. Whitworth. Arrangements of $m$ things of one sort and $n$ things of another sort, under certain conditions of priority. *Messenger Math.*, 8:105–114, 1878.

[124] D. Zeilberger. André's reflection proof generalized to the many candidate ballot problem. *Discrete Math.*, 44:325–326, 1983.

# Chapter 11

# *Catalan Paths and $q,t$-enumeration*

**James Haglund**
*University of Pennsylvania*

## CONTENTS

## 11.1 Introduction

This chapter contains an account of a two-parameter version of the Catalan numbers, and corresponding two-parameter versions of related objects such as parking functions and Schröder paths, which have become important in algebraic combinatorics and other areas of mathematics as well. Although the original motivation for the definition of these objects was the study of Macdonald polynomials and the representation theory of diagonal harmonics, in this account we focus only on the combinatorics associated to their description in terms of lattice paths. Hence this chapter can be read by anyone with a modest background in combinatorics. In Section 11.2 we include basic facts involving $q$-analogues, permutation statistics, and symmetric functions, which we need in later sections. Sections 11.3, 11.4, and 11.5 contain the results on the $q,t$-versions of the Catalan numbers, parking functions, and Schröder paths, respectively. Section 11.6 contains a brief account of the recent exciting extensions of these objects that have arisen in the study of string theory, knot invariants, and the Hilbert scheme from algebraic geometry.

## 11.2 Introduction to $q$-analogues and Catalan numbers

### 11.2.1 Permutation statistics and Gaussian polynomials

In combinatorics a **$q$-analogue** of a counting function is typically a polynomial in $q$ that reduces to the function in question when $q=1$, and furthermore satisfies versions of some or all of the algebraic properties, such as recursions, of the function. We sometimes regard $q$ as a real parameter satisfying $0<q<1$. We define the $q$-analogue of the real number $x$, denoted $[x]$, as

$$[x] = \frac{1-q^x}{1-q}.$$

Work supported by NSF grant DMS-1200296. Much of this chapter is a condensed version of Chapters 1, 3, 4, and 5 of the author's book *The $q,t$-Catalan Numbers and the Space of Diagonal Harmonics: With an Appendix on the Combinatorics of Macdonald Polynomials*, 

By l'Hôpital's rule, $[x] \to x$ as $q \to 1^-$. Let $\mathbb{N}$ denote the nonnegative integers. For $n \in \mathbb{N}$, we define the $q$-analogue of $n!$, denoted $[n]!$, as

$$[n]! = \prod_{i=1}^{n}[i] = (1+q)(1+q+q^2)\cdots(1+q+\ldots+q^{n-1}).$$

We let $|S|$ denote the cardinality of a finite set $S$. By a **statistic** on a set $S$ we mean a combinatorial rule that associates an element of $\mathbb{N}$ to each element of $S$. A **permutation statistic** is a statistic on the symmetric group $S_n$. We use the one-line notation $\sigma_1\sigma_2\cdots\sigma_n$ for the element $\sigma = \begin{pmatrix} 1 & 2 & \ldots & n \\ \sigma_1 & \sigma_2 & \ldots & \sigma_n \end{pmatrix}$ of $S_n$. More generally, a **word** (or multiset permutation) $\sigma_1\sigma_2\cdots\sigma_n$ is a linear list of the elements of some multiset of nonnegative integers. (The reader may wish to consult [60, Chapter 1] for more background on multiset permutations.) An **inversion** of a word $\sigma$ is a pair $(i,j)$, $1 \le i < j \le n$ such that $\sigma_i > \sigma_j$. A **descent** of $\sigma$ is an integer $i$, $1 \le i \le n-1$, for which $\sigma_i > \sigma_{i+1}$. The set of such $i$ is known as the **descent set**, denoted $\mathrm{Des}(\sigma)$. We define the **inversion statistic** $\mathrm{inv}(\sigma)$ as the number of inversions of $\sigma$ and the **major index statistic** $\mathrm{maj}(\sigma)$ as the sum of the descents of $\sigma$, i.e.,

$$\mathrm{inv}(\sigma) = \sum_{\substack{i<j \\ \sigma_i>\sigma_j}} 1, \quad \mathrm{maj}(\sigma) = \sum_{\substack{i \\ \sigma_i>\sigma_{i+1}}} i.$$

For example, $\mathrm{inv}(613524) = 8$, while $\mathrm{Des}(613524) = \{1,4\}$ and $\mathrm{maj}(613524) = 5$.

A permutation statistic is said to be **Mahonian** if its distribution over $S_n$ is $[n]!$.

**Theorem 11.2.1** *Both inv and maj are Mahonian, i.e.,*

$$\sum_{\sigma\in S_n} q^{inv(\sigma)} = [n]! = \sum_{\sigma\in S_n} q^{maj(\sigma)}.$$

*Proof.* Given $\beta \in S_{n-1}$, let $\beta(k)$ denote the permutation in $S_n$ obtained by inserting $n$ between the $(k-1)$th and $k$th elements of $\beta$. For example, $2143(3) = 21543$. Clearly $\mathrm{inv}(\beta(k)) = \mathrm{inv}(\beta) + n - k$, so

$$\sum_{\sigma\in S_n} q^{\mathrm{inv}(\sigma)} = \sum_{\beta\in S_{n-1}} (1+q+q^2+\ldots+q^{n-1})q^{\mathrm{inv}(\beta)}$$

and thus by induction inv is Mahonian.

A modified version of this idea works for maj. Say the descents of $\beta \in S_{n-1}$ are at places $i_1 < i_2 < \cdots < i_s$. Then

$$\mathrm{maj}(\beta(n)) = \mathrm{maj}(\beta), \quad \mathrm{maj}(\beta(i_s+1)) = \mathrm{maj}(\beta)+1,$$
$$\ldots, \mathrm{maj}(\beta(i_1+1)) = \mathrm{maj}(\beta)+s, \quad \mathrm{maj}(\beta(1)) = s+1.$$

If the non-descents less than $n-1$ of $\beta$ are at places $\alpha_1 < \alpha_2 < \cdots < \alpha_{n-2-s}$, then

$$\mathrm{maj}(\beta(\alpha_1+1)) = \mathrm{maj}(\beta)+s-(\alpha_1-1)+\alpha_1+1 = \mathrm{maj}(\beta)+s+2.$$

To see why, note that $s-(\alpha_1-1)$ is the number of descents of $\beta$ to the right of $\alpha_1$, each of which will be shifted one place to the right by the insertion of $n$ just after $\beta_{\alpha_1}$. Also, we have a new descent at $\alpha_1+1$. By similar reasoning,

$$\operatorname{maj}(\beta(\alpha_2)) = \operatorname{maj}(\beta)+s-(\alpha_2-2)+\alpha_2+1 = \operatorname{maj}(\beta)+s+3,$$

$$\vdots$$

$$\begin{aligned}\operatorname{maj}(\beta(\alpha_{n-2-s})) &= \operatorname{maj}(\beta)+s-(\alpha_{n-2-s}-(n-2-s))+\alpha_{n-2-s}+1\\ &= \operatorname{maj}(\beta)+n-1.\end{aligned}$$

Thus

$$\sum_{\sigma\in S_n} q^{\operatorname{maj}(\sigma)} = \sum_{\beta\in S_{n-1}} (1+q+\ldots+q^s+q^{s+1}+\ldots+q^{n-1})q^{\operatorname{maj}(\beta)}$$

and again by induction maj is Mahonian. ■

Major P. MacMahon introduced the major-index statistic and proved it is Mahonian [54]. Foata [19] found a map $\Phi$ that sends a permutation $\sigma$ with a given major index to another $\phi(\sigma)$ with the same value for inv. Furthermore, if we denote the descent set of $\sigma^{-1}$ by $\operatorname{Ides}(\sigma)$, then $\operatorname{Ides}(\phi(\sigma)) = \operatorname{Ides}(\sigma)$. The image $\phi$ of the map $\Phi$ can be described as follows. If $n \le 2$, $\phi(\sigma)=\sigma$. If $n>2$, we add a number to $\phi$ one at a time: Begin by setting $\phi^{(1)}=\sigma_1$, $\phi^{(2)}=\sigma_1\sigma_2$ and $\phi^{(3)}=\sigma_1\sigma_2\sigma_3$. Then if $\sigma_2>\sigma_3$, draw a bar after each element of $\phi^{(3)}$ that is greater than $\sigma_3$, while if $\sigma_2<\sigma_3$, draw a bar after each element of $\phi^{(3)}$ that is less than $\sigma_3$. Also add a bar before $\phi_1^{(3)}$. For example, if $\sigma = 4137562$ we now have $\phi^{(3)} = |41|3$. Now regard the numbers between two consecutive bars as "blocks," and in each block, move the last element to the beginning, and finally remove all bars. We end up with $\phi^{(3)} = 143$.

Proceeding inductively, we begin by letting $\phi^{(i)}$ be the result of adding $\sigma_i$ to the end of $\phi^{(i-1)}$. Then if $\sigma_{i-1}>\sigma_i$, draw a bar after each element of $\phi^{(i)}$ that is greater than $\sigma_i$, while if $\sigma_{i-1}<\sigma_i$, draw a bar after each element of $\phi^{(i)}$ that is less than $\sigma_i$. Also draw a bar before $\phi_1^{(i)}$. Then in each block, move the last element to the beginning, and finally remove all bars. If $\sigma = 4137562$ the successive stages of the algorithm yield

$$\begin{aligned}\phi^{(3)} &= 143\\ \phi^{(4)} &= |1|4|3|7 \to 1437\\ \phi^{(5)} &= |1437|5 \to 71435\\ \phi^{(6)} &= |71|4|3|5|6 \to 174356\\ \phi^{(7)} &= |17|4|3|5|6|2 \to 7143562\end{aligned}$$

and so $\phi(4137562) = 7143562$.

**Proposition 11.2.2** *The equality $\operatorname{maj}(\sigma) = \operatorname{inv}(\phi(\sigma))$ holds. Furthermore, $\operatorname{Ides}(\sigma) = \operatorname{Ides}(\phi(\sigma))$, and also $\phi(\sigma)$ and $\sigma$ have the same last letter.*

*Proof.* We claim $\mathrm{inv}(\phi^{(k)}) = \mathrm{maj}(\sigma_1 \cdots \sigma_k)$ for $1 \le k \le n$. Clearly this is true for $k \le 2$. Assume it is true for $k < j$, where $2 < j \le n$. If $\sigma_{j-1} > \sigma_j$, $\mathrm{maj}(\sigma_1 \cdots \sigma_j) = \mathrm{maj}(\sigma_1 \cdots \sigma_{j-1}) + j - 1$. On the other hand, for each block arising in the procedure creating $\phi^{(j)}$, the last element is greater than $\sigma_j$, which creates a new inversion, and when it is moved to the beginning of the block, also creates a new inversion with each element in its block. It follows that $\mathrm{inv}(\phi^{(j)}) = \mathrm{inv}(\phi^{(j-1)}) + j - 1$. Similar remarks hold if $\sigma_{j-1} < \sigma_j$. In this case $\mathrm{maj}(\sigma_1 \cdots \sigma_{j-1}) = \mathrm{maj}(\sigma_1 \cdots \sigma_j)$. Also, each element of $\phi$ that is not the last element in its block is larger than $\sigma_j$, which creates a new inversion, but a corresponding inversion between this element and the last element in its block is lost when we cycle the last element to the beginning. Hence $\mathrm{inv}(\phi^{(j-1)}) = \mathrm{inv}(\phi^{(j)})$ and the claim follows.

Note that $\mathrm{Ides}(\sigma)$ equals the set of all $i$, $1 \le i < n$ such that $i+1$ occurs before $i$ in $\sigma$. In order for the $\phi$ map to change this set, at some stage, say when creating $\phi^{(j)}$, we must move $i$ from the end of a block to the beginning, passing $i-1$ or $i+1$ along the way. But this could only happen if $\sigma_j$ is strictly between $i$ and either $i-1$ or $i+1$, an impossibility. ∎

We now show that the map $\Phi$ is invertible by constructing the permutation $\beta = \Phi^{-1}(\sigma)$. Begin by setting $\beta^{(1)} = \sigma$. Then if $\sigma_n > \sigma_1$, draw a bar **before** each number in $\beta^{(1)}$ which is less than $\sigma_n$, and also before $\sigma_n$. If $\sigma_n < \sigma_1$, draw a bar before each number in $\beta^{(1)}$ which is greater than $\sigma_n$, and also before $\sigma_n$. Next move each number at the beginning of a block to the end of the block.

The last letter of $\beta$ is now fixed. Next set $\beta^{(2)} = \beta^{(1)}$, and compare the $(n-1)$th letter with the first, creating blocks as above, and draw an extra bar before the $(n-1)$th letter. For example, if $\sigma = 7143562$ the successive stages of the $\beta$ algorithm yield

$$\beta^{(1)} = |71|4|3|5|6|2 \to 1743562$$
$$\beta^{(2)} = |17|4|3|5|62 \to 7143562$$
$$\beta^{(3)} = |7143|562 \to 1437562$$
$$\beta^{(4)} = |1|4|3|7562 \to 1437562$$
$$\beta^{(5)} = |14|37562 \to 4137562$$
$$\beta^{(6)} = \beta^{(7)} = 4137562$$

and so $\Phi^{-1}(7143562) = 4137562$. Notice that at each stage we are reversing the steps of the $\Phi$ algorithm, and it is easy to see this holds in general.

An **involution** on a set $S$ is a bijective map from $S$ to $S$ whose square is the identity. Foata and Schützenberger [20] showed that the map $i\Phi i\Phi^{-1}i$, where $i$ is the inverse map on permutations, is an involution on $S_n$ that interchanges inv and maj.

For $n, k \in \mathbb{N}$ with $0 \le k \le n$, let

$$\begin{bmatrix} n \\ k \end{bmatrix} = \begin{bmatrix} n \\ k \end{bmatrix}_q = \frac{[n]!}{[k]![n-k]!} = \frac{(1-q^n)(1-q^{n-1})\cdots(1-q^{n-k+1})}{(1-q^k)(1-q^{k-1})\cdots(1-q)}$$

denote the **Gaussian polynomial**. These are special cases of more general objects known as $q$-**binomial coefficients**, which are defined for $x \in \mathbb{R}$ as

$$\begin{bmatrix} x \\ k \end{bmatrix} = \frac{(q^{x-k+1};q)_k}{(q;q)_k},$$

where $(a;q)_k = (a)_k = (1-a)(1-qa)\cdots(1-q^{k-1}a)$ is the "$q$-rising factorial."

A **partition** $\lambda$ is a nonincreasing finite sequence $\lambda_1 \geq \lambda_2 \geq \ldots$ of positive integers. $\lambda_i$ is called the $i$th **part** of $\lambda$. We let $\ell(\lambda)$ denote the number of parts, and $|\lambda| = \sum_i \lambda_i$ the sum of the parts. For various formulas it will be convenient to assume $\lambda_j = 0$ for $j > \ell(\lambda)$. The **Ferrers graph** of $\lambda$ is an array of unit squares, called **cells**, with $\lambda_i$ cells in the $i$th row, where the first cell in each row is left-justified. We often use $\lambda$ to refer to its Ferrers graph. We define the **conjugate partition**, $\lambda'$ as the partition whose Ferrers graph is obtained from $\lambda$ by reflecting across the diagonal $x = y$, as in Figure 11.1. Here $(i,j) \in \lambda$ refers to a cell with (column, row) coordinates $(i,j)$, with the lower-left-hand cell of $\lambda$ having coordinates $(1,1)$. The notation $x \in \lambda$ means $x$ is a cell in $\lambda$. For technical reasons we say that 0 has one partition, the empty set $\emptyset$, with $\ell(\emptyset) = 0 = |\emptyset|$.

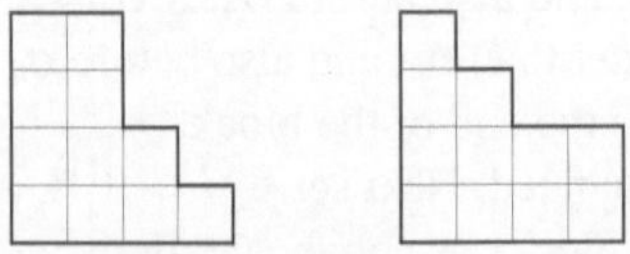

**Figure 11.1**
On the left, the Ferrers graph of the partition $(4,3,2,2)$, and on the right, that of its conjugate $(4,3,2,2)' = (4,4,2,1)$.

The following result shows the Gaussian polynomials are in fact polynomials in $q$, which is not obvious from their definition.

**Theorem 11.2.3** *For $n, k \in \mathbb{N}$,*

$$\begin{bmatrix} n+k \\ k \end{bmatrix} = \sum_{\lambda \subseteq n^k} q^{|\lambda|}, \tag{11.1}$$

*where the sum is over all partitions $\lambda$ whose Ferrers graph fits inside a $k \times n$ rectangle, i.e., for which $\lambda_1 \leq n$ and $\ell(\lambda) \leq k$.*

*Proof.* Let

$$P(n,k) = \sum_{\lambda \subseteq n^k} q^{|\lambda|}.$$

Clearly

$$P(n,k) = \sum_{\substack{\lambda \subseteq n^k \\ \lambda_1 = n}} q^{|\lambda|} + \sum_{\substack{\lambda \subseteq n^k \\ \lambda_1 \leq n-1}} q^{|\lambda|} = q^n P(n,k-1) + P(n-1,k).$$

On the other hand

$$\begin{aligned} q^n \begin{bmatrix} n+k-1 \\ k-1 \end{bmatrix} + \begin{bmatrix} n-1+k \\ k \end{bmatrix} &= q^n \frac{[n+k-1]!}{[k-1]![n]!} + \frac{[n-1+k]!}{[k]![n-1]!} \\ &= \frac{q^n [k][n+k-1]! + [n-1+k]![n]}{[k]![n]!} \\ &= \frac{[n+k-1]!}{[k]![n]!} (q^n(1+q+\ldots+q^{k-1}) + 1 + q + \ldots + q^{n-1}) \\ &= \frac{[n+k]!}{[k]![n]!}. \end{aligned}$$

Since $P(n,0) = P(0,k) = 1$, both sides of (11.1) thus satisfy the same recurrence and initial conditions. ■

Given $\alpha = (\alpha_0, \ldots, \alpha_s) \in \mathbb{N}^{s+1}$, let

$$\{0^{\alpha_0} 1^{\alpha_1} \cdots s^{\alpha_s}\}$$

denote the multiset with $\alpha_i$ copies of $i$, where $\alpha_0 + \ldots + \alpha_s = n$. We let $M_\alpha$ denote the set of all permutations of this multiset and refer to $\alpha$ as the **weight** of any given one of these words. Also let

$$\begin{bmatrix} n \\ \alpha_0, \ldots, \alpha_s \end{bmatrix} = \frac{[n]!}{[\alpha_0]! \cdots [\alpha_s]!}$$

denote the **$q$-multinomial coefficient**.

The following result is due to MacMahon [26].

**Theorem 11.2.4** *Both inv and maj are multiset Mahonian, i.e., given $\alpha \in \mathbb{N}^{s+1}$,*

$$\sum_{\sigma \in M_\alpha} q^{inv(\sigma)} = \begin{bmatrix} n \\ \alpha_0, \ldots, \alpha_s \end{bmatrix} = \sum_{\sigma \in M_\alpha} q^{maj(\sigma)}. \tag{11.2}$$

**Remark 11.2.5** Foata's map also proves Theorem 11.2.4 bijectively. To see why, given a multiset permutation $\sigma$ of $M(\beta)$ let $\sigma'$ denote the **standardization** of $\sigma$, defined to be the permutation obtained by replacing the $\beta_0$ 0's by the numbers 1 through $\beta_0$, in increasing order as we move left to right in $\sigma$, then replacing the $\beta_1$ 1's by the numbers $\beta_0 + 1$ through $\beta_0 + \beta_1$, again in increasing order as we move left to right in $\sigma$, etc. For example, the standardization of 31344221 is 51678342. Note that

$$\text{Ides}(\sigma') \subseteq \{\beta_1, \beta_1 + \beta_2, \ldots\} \tag{11.3}$$

and in fact standardization gives a bijection between elements of $M(\beta)$ and permutations satisfying (11.3). Since the map $\Phi$ fixes the inverse descent set, $\Phi$ maps $M(\beta)$ to itself bijectively, sending maj to inv.

**Exercise 11.2.6** *If $\sigma$ is a word of length $n$ define the co-major index of $\sigma$ as follows.*

$$comaj(\sigma) = \sum_{\substack{\sigma_i > \sigma_{i+1} \\ 1 \le i < n}} n - i. \tag{11.4}$$

*Show that Foata's map $\phi$ implies there is a bijective map $co\phi$ on words of fixed weight such that*

$$comaj(\sigma) = inv(co\phi(\sigma)).$$

### 11.2.2 The Catalan numbers and Dyck paths

A **lattice path** is a sequence of North $N(0,1)$ and East $E(1,0)$ steps in the first quadrant of the $xy$-plane, starting at the origin $(0,0)$ and ending at say $(m,n)$. We let $L_{m,n}$ denote the set of all such paths, and $L^+_{m,n}$ the subset of $L_{m,n}$ consisting of paths that never go below the line $y = \frac{n}{m}x$. A **iDyck path**, sometimes called a **Catalan path**, is an element of $L^+_{n,n}$ for some $n$.

Let $C_n = \frac{1}{n+1}\binom{2n}{n}$ denote the $n$th **Catalan number**, so

$$C_0, C_1, \ldots = 1, 1, 2, 5, 14, 42, \ldots.$$

There are now over 200 known combinatorial interpretations for the Catalan numbers. (See [58, Ex. 6.19, p. 219] for a list of 66 of these interpretations.) One of these is the number of elements of $L^+_{n,n}$. For $1 \le k \le n$, the number of Dyck paths from $(0,0)$ to $(k,k)$ that touch the line $y = x$ only at $(0,0)$ and $(k,k)$ is $C_{k-1}$, since such a path must begin with a $N$ step, end with an $E$ step, and never go below the line $y = x+1$ as it goes from $(0,1)$ to $(k-1,k)$. The number of ways to extend such a path to $(n,n)$ and still remain a Dyck path is clearly $C_{n-k}$. It follows that

$$C_n = \sum_{k=1}^{n} C_{k-1}C_{n-k}, \qquad n \ge 1. \tag{11.5}$$

There are two natural $q$-analogues of $C_n$. Given $\pi \in L_{n,m}$, let $\sigma(\pi)$ be the element of $M_{(m,n)}$ resulting from the following algorithm. First initialize $\sigma$ to the empty string. Next start at $(0,0)$, move along $\pi$ and add a 0 to the end of $\sigma(\pi)$ every time a $N$ step is encountered, and add a 1 to the end of $\sigma(\pi)$ every time an $E$ step is encountered. Similarly, given $\sigma \in M_{(m,n)}$, let $\pi(\sigma)$ be the element of $L_{n,m}$ obtained by inverting the above algorithm. We call the transformation of $\pi$ to $\sigma$ or its inverse the **coding** of $\pi$ or $\sigma$. For $\pi \in L^+_{n,n}$, let $a_i(\pi)$ denote the number of complete squares, in the $i$th row from the bottom of $\pi$, which are to the right of $\pi$ and to the left of the line $y = x$. We refer to $a_i(\pi)$ as the **length** of the $i$th row of $\pi$. Furthermore call $(a_1(\pi), a_2(\pi), \ldots, a_n(\pi))$

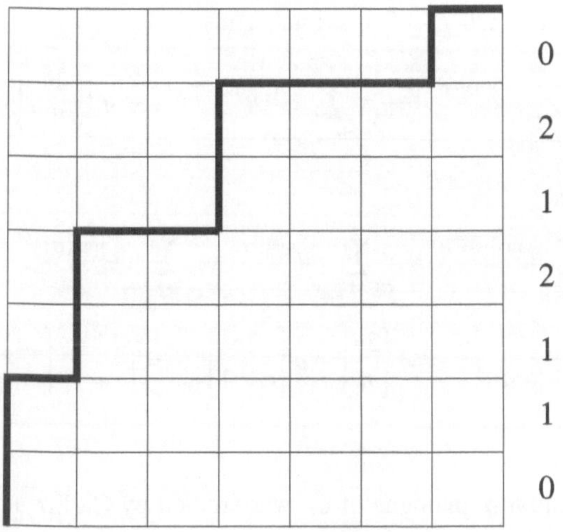


**Figure 11.2**
A Dyck path, with row lengths on the right. The area statistic is $1+1+2+1+2=7$.

the **area vector** of $\pi$, and set $\text{area}(\pi) = \sum_i a_i(\pi)$. For example, the path in Figure 11.2 has area vector $(0,1,1,2,1,2,0)$, and $\sigma(\pi) = 00100110011101$. By convention we say $L^+_{0,0}$ contains one path, the empty path $\emptyset$, with $\text{area}(\emptyset) = 0$.

Let $M^+_{(m,n)}$ denote the elements $\sigma$ of $M_{(m,n)}$ corresponding to paths in $L^+_{n,m}$. Words in $M^+_{n,n}$ are thus characterized by the property that in any initial segment there are at least as many 0's as 1's. The first $q$-analogue of $C_n$ is given by the following.

**Theorem 11.2.7 (MacMahon [54, p. 214])** *We have*

$$\sum_{\pi \in L^+_{n,n}} q^{maj(\sigma(\pi))} = \frac{1}{[n+1]} \begin{bmatrix} 2n \\ n \end{bmatrix}. \tag{11.6}$$

*Proof.* We give a *bijective* proof, taken from [22]. Let $M^-_{(m,n)} = M_{(m,n)} \setminus M^+_{(m,n)}$, and let $L^-_{n,m} = L_{n,m} \setminus L^+_{n,m}$ be the corresponding set of lattice paths. Given a path $\pi \in L^-_{n,n}$, let $P$ be the point with smallest $x$-coordinate among those lattice points $(i,j)$ in $\pi$ for which $j-i$ is maximal, i.e., whose distance from the line $y=x$ in a SE direction is maximal. (Since $\pi \in L^-_{n,n}$, this maximal value of $i-j$ is positive.) Let $P'$ be the lattice point on $\pi$ before $P$. There must be an east step connecting $P'$ to $P$ (preceded by another east step unless $P'$ is the origin). Change this east step into a north step and shift the remainder of the path after $P$ up one unit and left one unit. We now have a path $\phi(\pi)$ from $(0,0)$ to $(n-1,n+1)$, and moreover $\text{maj}(\sigma(\phi(\pi))) = \text{maj}(\sigma(\pi)) - 1$.

It is easy to see that this map is invertible. Given a lattice path $\pi'$ from $(0,0)$ to $(n-1,n+1)$, let $P'$ be the point with maximal $x$-coordinate among those lattice points $(i,j)$ in $\pi'$ for which $j-i$ is maximal. Thus

$$\sum_{\sigma\in M^-_{(n,n)}} q^{\mathrm{maj}(\sigma)} = \sum_{\sigma'\in M_{(n+1,n-1)}} q^{\mathrm{maj}(\sigma')+1} = q\begin{bmatrix}2n\\n+1\end{bmatrix}, \tag{11.7}$$

using (11.2). Hence

$$\sum_{\pi\in L^+_{n,n}} q^{\mathrm{maj}(\sigma(\pi))} = \sum_{\sigma\in M_{(n,n)}} q^{\mathrm{maj}(\sigma)} - \sum_{\sigma\in M^-_{(n,n)}} q^{\mathrm{maj}(\sigma)} \tag{11.8}$$

$$= \begin{bmatrix}2n\\n\end{bmatrix} - q\begin{bmatrix}2n\\n+1\end{bmatrix} = \frac{1}{[n+1]}\begin{bmatrix}2n\\n\end{bmatrix}. \tag{11.9}$$

■

The second natural $q$-analogue of $C_n$ was studied by Carlitz and Riordan [15]. They define

$$C_n(q) = \sum_{\pi\in L^+_{n,n}} q^{\mathrm{area}(\pi)}. \tag{11.10}$$

For example, the paths in $L^+_{3,3}$ given in Figure 11.3 have area, from left-to-right, $3,2,1,1,0$, so $C_3(q) = 1+2q+q^2+q^3$.

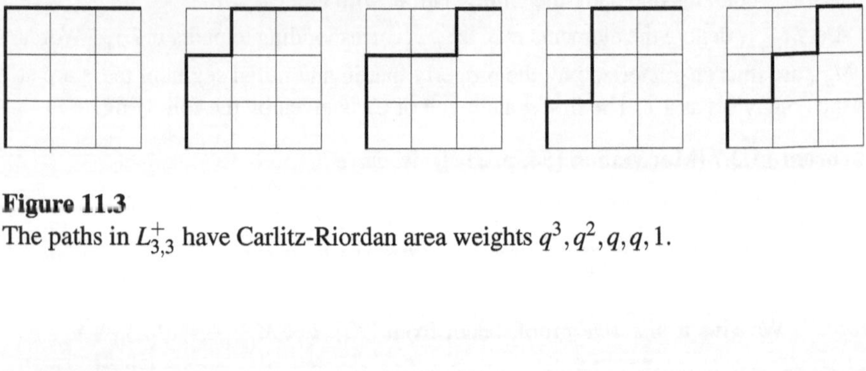

**Figure 11.3**
The paths in $L^+_{3,3}$ have Carlitz-Riordan area weights $q^3, q^2, q, q, 1$.

**Proposition 11.2.8** *The recurrence relation*

$$C_n(q) = \sum_{k=1}^{n} q^{k-1} C_{k-1}(q) C_{n-k}(q), \quad n\geq 1 \tag{11.11}$$

*holds.*

*Proof.* As in the proof of (11.5), we break up our path $\pi$ according to the "point of first return" to the line $y=x$. If this occurs at $(k,k)$, then the area of the part of $\pi$ from $(0,1)$ to $(k-1,k)$, when viewed as an element of $L^+_{k-1,k-1}$, is $k-1$ less than the area of this portion of $\pi$ when viewed as a path in $L^+_{n,n}$. ■

**Exercise 11.2.9** *Define a* **co-inversion** *of $\sigma$ to be a pair $(i,j)$ with $i<j$ and $\sigma_i<\sigma_j$. Show*

$$C_n(q) = \sum_{\pi \in L^+_{n,n}} q^{coinv(\sigma(\pi)) - \binom{n+1}{2}}, \tag{11.12}$$

*where $coinv(\sigma)$ is the number of co-inversions of $\sigma$.*

### 11.2.3 The $q$-Vandermonde convolution

Let

$$_{p+1}\phi_p\left(\begin{matrix} a_1, & a_2, & \dots, & a_{p+1} \\ & b_1, & \dots, & b_p \end{matrix};q;z\right) = \sum_{k=0}^{\infty} \frac{(a_1)_k \cdots (a_{p+1})_k}{(q)_k (b_1)_k \cdots (b_p)_k} z^k \tag{11.13}$$

denote the basic hypergeometric series. A good general reference for this subject is [28]. The following result is known as **Cauchy's $q$-binomial series.**

**Theorem 11.2.10** *The identity*

$$_1\phi_0\left(\begin{matrix} a \\ - \end{matrix};q;z\right) = \sum_{k=0}^{\infty} \frac{(a)_k}{(q)_k} z^k = \frac{(az)_\infty}{(z)_\infty}, \quad |z|<1, |q|<1 \tag{11.14}$$

*holds, where $(a;q)_\infty = (a)_\infty = \prod_{i=0}^{\infty}(1-aq^i)$.*

*Proof.* The following proof is based on the proof in [28, Chap. 1]. Let

$$F(a,z) = \sum_{k=0}^{\infty} \frac{(a)_k}{(q)_k} z^k.$$

Then

$$F(a,z) - F(a,qz) = (1-a)zF(aq,z) \tag{11.15}$$

and

$$F(a,z) - F(aq,z) = -azF(aq,z). \tag{11.16}$$

Eliminating $F(aq,z)$ from (11.15) and (11.16) we get

$$F(a,z) = \frac{(1-az)}{(1-z)} F(a,qz).$$

Iterating this $n$ times, then taking the limit as $n \to \infty$ we get

$$\begin{aligned} F(a,z) &= \lim_{n\to\infty} \frac{(az;q)_n}{(z;q)_n} F(a,q^n z) \\ &= \frac{(az;q)_\infty}{(z;q)_\infty} F(a,0) = \frac{(az;q)_\infty}{(z;q)_\infty}. \end{aligned} \tag{11.17}$$

■

**Corollary 11.2.11 (The $q$-binomial theorem)** *We have*

$$\sum_{k=0}^{n} q^{\binom{k}{2}} \begin{bmatrix} n \\ k \end{bmatrix} z^k = (-z;q)_n \tag{11.18}$$

*and*

$$\sum_{k=0}^{\infty} \begin{bmatrix} n+k \\ k \end{bmatrix} z^k = \frac{1}{(z;q)_{n+1}}. \tag{11.19}$$

*Proof.* To prove (11.18), set $a = q^{-n}$ and $z = -zq^n$ in (11.14) and simplify. To prove (11.19), let $a = q^{n+1}$ in (11.14) and simplify. ∎

For any function $f(z)$, let $f(z)|_{z^k}$ denote the coefficient of $z^k$ in the Maclaurin series for $f(z)$.

**Corollary 11.2.12** *The identity*

$$\sum_{k=0}^{h} q^{(n-k)(h-k)} \begin{bmatrix} n \\ k \end{bmatrix} \begin{bmatrix} m \\ h-k \end{bmatrix} = \begin{bmatrix} m+n \\ h \end{bmatrix} \text{ holds.} \tag{11.20}$$

*Proof.* By (11.18),

$$\begin{aligned} q^{\binom{h}{2}} \begin{bmatrix} m+n \\ h \end{bmatrix} &= \prod_{k=0}^{m+n-1} (1+zq^k)|_{z^h} \\ &= \prod_{k=0}^{n-1} (1+zq^k) \prod_{j=0}^{m-1} (1+zq^n q^j)|_{z^h} \\ &= (\sum_{k=0}^{n-1} q^{\binom{k}{2}} \begin{bmatrix} n \\ k \end{bmatrix} z^k)(\sum_{j=0}^{m-1} q^{\binom{j}{2}} \begin{bmatrix} m \\ j \end{bmatrix} (zq^n)^j)|_{z^h} \\ &= \sum_{k=0}^{h} q^{\binom{k}{2}} \begin{bmatrix} n \\ k \end{bmatrix} q^{\binom{h-k}{2}} \begin{bmatrix} m \\ h-k \end{bmatrix} (q^n)^{h-k}. \end{aligned}$$

The result now reduces to the identity

$$\binom{k}{2} + \binom{h-k}{2} + n(h-k) - \binom{h}{2} = (n-k)(h-k).$$

∎

**Corollary 11.2.13** *The identity*

$$\sum_{k=0}^{h} q^{(m+1)k} \begin{bmatrix} n-1+k \\ k \end{bmatrix} \begin{bmatrix} m+h-k \\ h-k \end{bmatrix} = \begin{bmatrix} m+n+h \\ h \end{bmatrix} \tag{11.21}$$

*holds.*

*Proof.* By (11.19),

$$\begin{aligned}\begin{bmatrix} m+n+h \\ h \end{bmatrix} &= \frac{1}{(z)_{m+n+1}}\Big|_{z^h} \\ &= \frac{1}{(z)_{m+1}}\frac{1}{(zq^{m+1})_n}\Big|_{z^h} \\ &= \left(\sum_{j=0}^{h} z^j \begin{bmatrix} m+j \\ j \end{bmatrix}_q\right)\left(\sum_{k=0}^{h} (zq^{m+1})^k \begin{bmatrix} n-1+k \\ k \end{bmatrix}_q\right)\Big|_{z^h} \\ &= \sum_{k=0}^{h} q^{(m+1)k} \begin{bmatrix} n-1+k \\ k \end{bmatrix}\begin{bmatrix} m+h-k \\ h-k \end{bmatrix}_q .\end{aligned}$$

■

We note that (11.20) and (11.21) have alternative, elementary proofs based on $q$-counting lattice paths. Both identities are special cases of the following result, known as the **$q$-Vandermonde convolution**. For a proof see [28, Chap. 1].

**Theorem 11.2.14** *Let $n \in \mathbb{N}$. Then*

$$ {}_2\phi_1\left(\begin{matrix} a, & q^{-n} \\ & c \end{matrix}; q; q\right) = \frac{(c/a)_n}{(c)_n} a^n. \tag{11.22}$$

**Exercise 11.2.15** *By reversing summation in (11.22), show that*

$$ {}_2\phi_1\left(\begin{matrix} a, & q^{-n} \\ & c \end{matrix}; q; cq^n/a\right) = \frac{(c/a)_n}{(c)_n}. \tag{11.23}$$

**Exercise 11.2.16** *Show Newton's binomial series*

$$\sum_{n=0}^{\infty} \frac{a(a+1)\cdots(a+n-1)}{n!} z^n = \frac{1}{(1-z)^a}, \quad |z|<1,\ a \in \mathbb{R} \tag{11.24}$$

*can be derived from (11.14) by replacing a by $q^a$ and letting $q \to 1^-$. For simplicity you can assume $a, z \in \mathbb{R}$.*

### 11.2.4 Symmetric functions

#### 11.2.4.1 The basics

Given $f(x_1,\ldots,x_n) \in K[x_1,x_2,\ldots,x_n]$ for some field $K$, and $\sigma \in S_n$, let

$$\sigma f = f(x_{\sigma_1},\ldots,x_{\sigma_n}). \tag{11.25}$$

We say $f$ is a **symmetric function** if $\sigma f = f$ for all $\sigma \in S_n$. It will be convenient to work with more general functions $f$ depending on countably many indeterminates $x_1,x_2,\ldots$, indicated by $f(x_1,x_2,\ldots)$, in which case we view $f$ as a formal power series in the $x_i$, and say it is a symmetric function if it is invariant under any permutation of the variables. The standard references on this topic are [58, Chap. 7] and [53]. We will often let $X_n$ and $X$ stand for the set of variables $\{x_1,\dots,x_n\}$ and $\{x_1,x_2,\dots\}$, respectively.

We let $\Lambda$ denote the ring of symmetric functions in $x_1,x_2,\dots$ and $\Lambda^n$ the vector subspace consisting of symmetric functions that are homogeneous of degree $n$. The most basic symmetric functions are the monomial symmetric functions, which depend on a partition $\lambda$ in addition to a set of variables. They are denoted by $m_\lambda(X) = m_\lambda(x_1,x_2,\dots)$. In a symmetric function it is typical to leave off explicit mention of the variables, with a set of variables being understood from context, so $m_\lambda = m_\lambda(X)$. We illustrate these first by means of examples. We let $\mathrm{Par}(n)$ denote the set of partitions of $n$, and use the notation $\lambda \vdash n$ as an abbreviation for $\lambda \in \mathrm{Par}(n)$.

**Example 11.2.17** *Let* $\lambda = (1,1)$. *Then we have*

$$\begin{aligned} m_{1,1} &= \sum_{i<j} x_i x_j \\ m_{2,1,1}(X_3) &= x_1^2x_2x_3 + x_1x_2^2x_3 + x_1x_2x_3^2 \\ m_2(X) &= \sum_i x_i^2. \end{aligned}$$

In general, $m_\lambda(X)$ is the sum of all distinct monomials in the $x_i$ whose multiset of exponents equals the multiset of parts of $\lambda$. Any element of $\Lambda$ can be expressed uniquely as a linear combination of the $m_\lambda$.

We let $1^n$ stand for the partition consisting of $n$ parts of size 1. The function $m_{1^n}$ is called the $n$th **elementary symmetric function**, which we denote by $e_n$. Then

$$\prod_{i=1}^{\infty}(1+zx_i) = \sum_{n=0}^{\infty} z^n e_n, \quad e_0 = 1. \tag{11.26}$$

Another important special case is $m_n = \sum_i x_i^n$, known as the **power-sum symmetric functions**, denoted by $p_n$. We also define the **complete homogeneous symmetric functions** $h_n$, by $h_n = \sum_{\lambda \vdash n} m_\lambda$, or equivalently

$$\frac{1}{\prod_{i=1}^{\infty}(1-zx_i)} = \sum_{n=0}^{\infty} z^n h_n, \quad h_0 = 1. \tag{11.27}$$

For $\lambda \vdash n$, we define $e_\lambda = \prod_i e_{\lambda_i}$, $p_\lambda = \prod_i p_{\lambda_i}$, and $h_\lambda = \prod_i h_{\lambda_i}$. For example,

$$\begin{aligned} e_{2,1} &= \sum_{i<j} x_i x_j \sum_k x_k = m_{2,1} + 3m_{1,1,1} \\ p_{2,1} &= \sum_i x_i^2 \sum_j x_j = m_3 + m_{2,1} \\ h_{2,1} &= (\sum_i x_i^2 + \sum_{i<j} x_i x_j) \sum_k x_k = m_3 + 2m_{2,1} + 3m_{1,1,1}. \end{aligned}$$

Assuming we have at least $n$ variables, it is known that $\{e_\lambda, \lambda \vdash n\}$ forms a basis for $\Lambda^n$, and so do $\{p_\lambda, \lambda \vdash n\}$ and $\{h_\lambda, \lambda \vdash n\}$.

**Definition 11.2.18** *Two simple functions on partitions we will often use are*

$$n(\lambda) = \sum_i (i-1)\lambda_i = \sum_i \binom{\lambda_i'}{2}$$
$$z_\lambda = \prod_i i^{m_i} m_i!,$$

*where $m_i = m_i(\lambda)$ is the number of parts of $\lambda$ equal to i. For example, $n(54331) = 4+6+9+4 = 23$, and $z(5433111) = 5*4*3^2*2!*3! = 2160$.*

**Exercise 11.2.19** *Use (11.26) and (11.27) to show that*

$$e_n = \sum_{\lambda \vdash n} \frac{(-1)^{n-\ell(\lambda)} p_\lambda}{z_\lambda},$$
$$h_n = \sum_{\lambda \vdash n} \frac{p_\lambda}{z_\lambda}.$$

We let $\omega$ denote the ring endomorphism $\omega : \Lambda \to \Lambda$ defined by

$$\omega(p_k) = (-1)^{k-1} p_k.$$

Thus $\omega$ is an involution with $\omega(p_\lambda) = (-1)^{|\lambda|-\ell(\lambda)} p_\lambda$, and by Exercise 11.2.19, $\omega(e_n) = h_n$, and more generally $\omega(e_\lambda) = h_\lambda$.

For $f \in \Lambda$, the special value $f(1,q,q^2,\ldots,q^{n-1})$ is known as the **principal specialization** (of order $n$) of $f$.

**Theorem 11.2.20** *The following identities hold.*

$$e_m(1,q,\ldots,q^{n-1}) = q^{\binom{m}{2}} \begin{bmatrix} n \\ m \end{bmatrix}_q$$
$$h_m(1,q,\ldots,q^{n-1}) = \begin{bmatrix} n-1+m \\ m \end{bmatrix}_q$$
$$p_m(1,q,\ldots,q^{n-1}) = \frac{1-q^{nm}}{1-q^m}.$$

*Proof.* The principal specializations for $e_m$ and $h_m$ follow directly from (11.18), (11.19), (11.26) and (11.27). ∎

**Remark 11.2.21** The principal specialization of $m_\lambda$ doesn't have a particularly simple description, although if $ps_n^1$ denotes the set of $n$ variables, each equal to 1, then [58, p. 303]

$$m_\lambda(ps_n^1) = \binom{n}{m_1, m_2, m_3, \ldots},$$

where again $m_i$ is the multiplicity of the number $i$ in the multiset of parts of $\lambda$.

**Remark 11.2.22** $\Lambda$ is known to be isomorphic to $K[p_1, p_2, \ldots]$. Hence, although identities like

$$h_{2,1} = m_3 + 2m_{2,1} + 3m_{1,1,1}$$

appear at first to depend on a set of variables, it is customary to view them as polynomial identities in the $p_\lambda$. Since the $p_k$ (in infinitely many variables) are algebraically independent, we can specialize them to whatever we please, forgetting about the original set of variables $X$.

We define the **Hall scalar product**, a bilinear form from $\Lambda \times \Lambda$ to $\mathbb{Q}$, by

$$\langle p_\lambda, p_\beta \rangle = z_\lambda \chi(\lambda = \beta),$$

where for any logical statement $L$

$$\chi(L) = \begin{cases} 1 & \text{if } L \text{ is true} \\ 0 & \text{if } L \text{ is false.} \end{cases}$$

Clearly $\langle f, g \rangle = \langle g, f \rangle$. Also, $\langle \omega f, \omega g \rangle = \langle f, g \rangle$, which follows from the definition if $f = p_\lambda, g = p_\beta$, and by bilinearity for general $f, g$ since the $p_\lambda$ form a basis for $\Lambda$.

**Theorem 11.2.23** *The $h_\lambda$ and the $m_\beta$ are dual with respect to the Hall scalar product, i.e.,*

$$\langle h_\lambda, m_\beta \rangle = \chi(\lambda = \beta). \tag{11.28}$$

*Proof.* See [53] or [58]. ■

For any $f \in \Lambda$, and any basis $\{b_\lambda, \lambda \in \text{Par}\}$ of $\Lambda$, let $f|_{b_\lambda}$ denote the coefficient of $b_\lambda$ when $f$ is expressed in terms of the $b_\lambda$. Then (11.28) implies

**Corollary 11.2.24**

$$\langle f, h_\lambda \rangle = f|_{m_\lambda}. \tag{11.29}$$

#### 11.2.4.2 Tableaux and Schur functions

Given $\lambda, \mu \in \text{Par}(n)$, a **semi-standard Young tableau** (or SSYT) of **shape** $\lambda$ and weight $\mu$ is a filling of the cells of the Ferrers graph of $\lambda$ with the elements of the multiset $\{1^{\mu_1} 2^{\mu_2} \cdots\}$, so that the numbers weakly increase across rows and strictly increase up columns. Let $SSYT(\lambda, \mu)$ denote the set of these fillings, and $K_{\lambda,\mu}$ the cardinality of this set. The $K_{\lambda,\mu}$ are known as the **Kostka numbers**. Our definition also makes sense if our weight is a **weak composition** of $n$ , i.e., any finite sequence of nonnegative integers whose sum is $n$. For example, $K_{(3,2),(2,2,1)} = K_{(3,2),(2,1,2)} = K_{(3,2),(1,2,2)} = 2$ as in Figure 11.4.

If the Ferrers graph of a partition $\beta$ is contained in the Ferrers graph of $\lambda$, denoted $\beta \subseteq \lambda$, let $\lambda/\beta$ refer to the subset of cells of $\lambda$ that are not in $\beta$. This is referred to

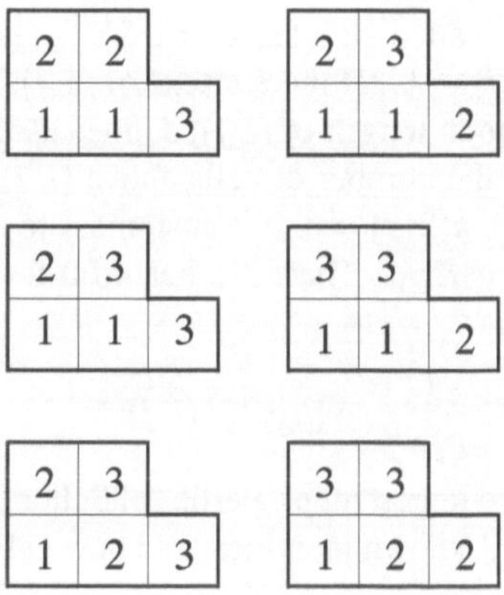

**Figure 11.4**
Some SSYT of shape $(3,2)$.

as a **skew shape**. Define a SSYT of shape $\lambda/\beta$ and weight $\nu$, where $|\nu| = |\lambda| - |\beta|$, to be a filling of the cells of $\lambda/\beta$ with elements of $\{1^{\nu_1}2^{\nu_2}\cdots\}$, again with weak increase across rows and strict increase up columns. The number of such tableaux is denoted $K_{\lambda/\beta,\nu}$.

Let wcomp$(\mu)$ denote the set of all weak compositions whose multiset of nonzero parts equals the multiset of parts of $\mu$. It follows easily from Figure 11.4 that $K_{(3,2),\alpha} = 2$ for all $\alpha \in \text{wcomp}(2,2,1)$. Hence

$$\sum_{\alpha,T} \prod_i x_i^{\alpha_i} = 2m_{(2,2,1)}, \tag{11.30}$$

where the sum is over all tableaux $T$ of shape $(3,2)$ and weight some element of wcomp$(2,2,1)$.

This is a special case of a more general phenomenon. For $\lambda \in \text{Par}(n)$, define

$$s_\lambda = \sum_{\alpha,T} \prod_i x_i^{\alpha_i},$$

where the sum is over all weak compositions $\alpha$ of $n$, and all possible tableaux $T$ of shape $\lambda$ and weight $\alpha$. Then

$$s_\lambda = \sum_{\mu \vdash n} K_{\lambda,\mu} m_\mu, \tag{11.31}$$

i.e., $K_{\lambda,\beta} = K_{\lambda,\alpha}$ for all compositions $\beta, \alpha$ whose multiset of parts is the same (we leave it to the interested reader to prove this fact bijectively). The $s_\lambda$ are called **Schur functions**, and are fundamental to the theory of symmetric functions. Two special cases of (11.31) are $s_n = h_n$ (since $K_{n,\mu} = 1$ for all $\mu \in \text{Par}(n)$) and $s_{1^n} = e_n$ (since $K_{1^n,\mu} = \chi(\mu = 1^n)$).

A SSYT of weight $1^n$ is called **standard**, or a SYT. The set of SYT of shape $\lambda$ is denoted $SYT(\lambda)$. For $(i,j) \in \lambda$, let the **content** of $(i,j)$, denoted $c(i,j)$, be $i-j$. Also, let $h(i,j)$ denote the **hook length** of $(i,j)$, defined as the number of cells to the right of $(i,j)$ in row $j$ plus the number of cells above $(i,j)$ in column $i$ plus 1. For example, if $\lambda = (5,5,3,3,1)$, $h(2,2) = 6$. It is customary to let $f^\lambda$ denote the number of SYT of shape $\lambda$, i.e., $f^\lambda = K_{\lambda,1^n}$. There is a beautiful formula for $f^\lambda$, namely

$$f^\lambda = \frac{n!}{\prod_{(i,j)\in\lambda} h(i,j)}. \tag{11.32}$$

Below we list some of the important properties of Schur functions. See [58, Chapter 7] for proofs of these well-known identities, and how (11.32) can be derived from (11.33).

**Theorem 11.2.25** *Let $\lambda, \mu \in Par$. Then*

1. *The Schur functions are orthonormal with respect to the Hall scalar product, i.e.,*

$$< s_\lambda, s_\mu >= \chi(\lambda = \mu).$$

*Thus for any $f \in \Lambda$,*

$$< f, s_\lambda >= f|_{s_\lambda}.$$

2. *Action by $\omega$:*

$$\omega(s_\lambda) = s_{\lambda'}.$$

3. *Principal Specialization: For $\lambda \in Par$,*

$$s_\lambda(1,q,q^2,\ldots,q^{n-1}) = q^{n(\lambda)} \prod_{(i,j)\in\lambda} \frac{[n+a-l]}{[a+l+1]}, \tag{11.33}$$

*where for a given square $(i,j) \in \lambda$, we define the coarm $a'$ and coleg $l'$ as in Figure 11.5 from Section 11.3.*

4. *Cauchy Identities: For any two alphabets of variables $X, Y$, let $XY$ denote the set of variables $\{x_i y_j\}$. Then*

$$e_n(XY) = \sum_{\lambda\in Par(n)} s_\lambda(X) s_{\lambda'}(Y) \tag{11.34}$$

$$h_n(XY) = \sum_{\lambda\in Par(n)} s_\lambda(X) s_\lambda(Y).$$

### 11.2.4.3 Statistics on tableaux

There is a $q$-analogue of the Kostka numbers, denoted by $K_{\lambda,\mu}(q)$, which has many applications in representation theory and the combinatorics of tableaux. Originally defined algebraically in an indirect fashion, the $K_{\lambda,\mu}(q)$ are polynomials in $q$ that satisfy $K_{\lambda,\mu}(1) = K_{\lambda,\mu}$. Foulkes [21] conjectured that there should be a statistic stat$(T)$ on SSYT $T$ of shape $\lambda$ and weight $\mu$ such that

$$K_{\lambda,\mu}(q) = \sum_{T \in SSYT(\lambda)} q^{\text{stat}(T)}.$$

This conjecture was resolved by Lascoux and Schützenberger [45], who found a statistic **charge** to generate these polynomials. Butler [14] provided a detailed account of their proof, filling in a lot of missing details. A short proof, based on the combinatorial formula for Macdonald polynomials, is contained in [39, Appendix A].

Assume we have a tableau $T \in SSYT(\lambda,\mu)$ where $\mu \in$ Par. It will be more convenient for us to describe a slight modification of charge$(T)$, called cocharge$(T)$, which is defined as $n(\mu)$ – charge. The **reading word** read$(T)$ of $T$ is obtained by reading the entries in $T$ from left to right in the top row of $T$, then continuing left to right in the second row from the top of $T$, etc. For example, the tableau in the upper-left of Figure 11.4 has reading word 22113. To calculate cocharge$(T)$, perform the following algorithm on read$(T)$. (Note: The algorithm below applies to any word, not just words that are the reading word of some SYT.)

**Algorithm 11.2.26**

1. *Start at the end of read$(T)$ and scan left until you encounter a* 1, *say this occurs at spot $i_1$, so read$(T)_{i_1} = 1$. Then start there and scan left until you encounter a* 2. *If you hit the end of read$(T)$ before finding a* 2, *loop around and continue searching left, starting at the end of read$(T)$. Say the first* 2 *you find equals read$(T)_{i_2}$. Now iterate, start at $i_2$ and search left until you find a* 3, *etc. Continue in this way until you have found* $4, 5, \ldots, \mu_1$, *with $\mu_1$ occurring at spot $i_{\mu_1}$. Then the first subword of textread$(T)$ is defined to be the elements of the set $\{read(T)_{i_1}, \ldots, read(T)_{i_{\mu_1}}\}$, listed in the order in which they occur in read$(T)$ if we start at the beginning of read$(T)$ and move left to right. For example, if read$(T)$ =* 21613244153 *then the first subword equals* 632415, *corresponding to places* $3, 5, 6, 8, 9, 10$ *of read$(T)$.*

   *Next remove the elements of the first subword from read$(T)$ and find the first subword of what's left. Call this the second subword. Remove this and find the first subword in what's left and call this the third subword of read$(T)$, etc. For the word* 21613244153, *the subwords are* 632415, 2143, 1.

2. *The value of charge$(T)$ will be the sum of the values of charge on each of the subwords of rw$(T)$. Thus it suffices to assume rw$(T) \in S_m$ for some m, in which case we set*

$$cocharge(rw(T)) = comaj(rw(T)^{-1}),$$

*where $read(T)^{-1}$ is the usual inverse in $S_m$, with comaj as in (11.4). (Another way of describing cocharge(read(T)) is the sum of $m-i$ over all $i$ for which $i+1$ occurs before $i$ in read(T).) For example, if $\sigma = 632415$, then $\sigma^{-1} = 532461$ and $cocharge(\sigma) = 5+4+1 = 10$, and finally*

$$cocharge(21613244153) = 10+4+0 = 14.$$

Note that to compute charge, we could create subwords in the same manner, and count $m-i$ for each $i$ with $i+1$ occurring to the right of $i$ instead of to the left. For $\lambda, \mu \in \mathrm{Par}(n)$ we set

$$\begin{aligned}\tilde{K}_{\lambda,\mu}(q) &= q^{n(\mu)}K_{\lambda,\mu}(1/q) \\ &= \sum_{T \in SSYT(\lambda,\mu)} q^{\mathrm{cocharge}(T)}.\end{aligned}$$

In addition to the cocharge statistic, there is a **major index statistic** on SYT that is often useful. Given a SYT $T$ of shape $\lambda$, define a **descent** of $T$ to be a value of $i$, $1 \le i < |\lambda|$, for which $i+1$ occurs in a row above $i$ in $T$. Let

$$\begin{aligned}\mathrm{maj}(T) &= \sum i \\ \mathrm{comaj}(T) &= \sum |\lambda| - i,\end{aligned}$$

where the sums are over the descents of $T$. Then [58, p.363]

$$\begin{aligned}s_\lambda(1,q,q^2,\ldots) &= \frac{1}{(q)_n} \sum_{T \in SYT(\lambda)} q^{\mathrm{maj}(T)} \\ &= \frac{1}{(q)_n} \sum_{T \in SYT(\lambda)} q^{\mathrm{comaj}(T)}.\end{aligned}$$

### 11.2.5 Representation theory

Let $G$ be a finite group. A (matrix) representation of $G$ is a group homomorphism from $G$ to $GL_n(\mathbb{C})$, the set of invertible $n \times n$ matrices with entries in $\mathbb{C}$. See [56] and [44] for a detailed discussion of the representation theory of the symmetric group and other finite groups. We include an informal discussion of some of the main ideas here, in order to motivate the combinatorial problems we will be discussing.

We will identify a representation with the image of a homomorphism from $G$ to $GL_n(\mathbb{C})$, namely the set of square invertible matrices $\{M(g), g \in G\}$ with the property that

$$M(g)M(h) = M(gh) \quad \text{for all } g,h \in G. \tag{11.35}$$

On the left-hand side of (11.35) we are using ordinary matrix multiplication, and on the right-hand side, to define $gh$, multiplication in $G$. The number of rows of a given $M(g)$ is called the dimension of the representation.

An **action** of $G$ on a set $S$ is a map from $G \times S$ to $S$, denoted by $g(s)$ for $g \in G$, $s \in S$, which satisfies

$$g(h(s)) = (gh)(s) \quad \text{for all } g,h \in G, s \in S,$$

with $e(s) = s$ for all $s \in S$, where $e$ is the identity in $G$. Let $V$ be a finite-dimensional $\mathbb{C}$-vector space, with basis $w_1, w_2, \ldots w_n$. Any linear action of $G$ on $V$ makes $V$ into a $\mathbb{C}G$ module. A module is called **irreducible** if it has no submodules other than $\{0\}$ and itself. Maschke's theorem [44] says that every nonzero **C**$G$-module $V$ can be expressed as a direct sum of irreducible submodules.

If we form a matrix $M(g)$ whose $i$th row consists of the coefficients of the $w_j$ when expanding $g(w_i)$ in the $w$ basis, then $\{M(g), g \in G\}$ is a representation of $G$. In general $\{M(g), g \in G\}$ will depend on the choice of basis, but the trace of the matrices will not. The trace of the matrix $M(g)$ is called the **character** of the module (under the given action), which we denote char$(V)$. If $V = \bigoplus_{j=1}^{d} V_j$, where each $V_j$ is irreducible, then an ordered basis of $V$ can be chosen so that the matrix $M$ will be in block-diagonal form, where the sizes of the blocks are the dimensions of the $V_j$. Clearly $\text{char}(V) = \sum_{j=1}^{d} \text{char}(V_j)$. It turns out that there are only a certain number of possible functions that occur as characters of irreducible modules, namely one for each conjugacy class of $G$. These are called the **irreducible characters** of $G$.

In the case $G = S_n$, the conjugacy classes are in one-to-one correspondence with partitions $\lambda \in \text{Par}(n)$, and the irreducible characters are denoted $\chi^\lambda$. The dimension of a given $V_\lambda$ with character $\chi^\lambda$ is known to be $f^\lambda$. The value of a given $\chi^\lambda(\sigma)$ depends only on the conjugacy class of $\sigma$. For the symmetric group the conjugacy class of an element is determined by rearranging the lengths of the disjoint cycles of $\sigma$ into nonincreasing order to form a partition called the **cycle type** of $\sigma$. Thus we can talk about $\chi^\lambda(\beta)$, which means the value of $\chi^\lambda$ at any permutation of cycle type $\beta$. For example, $\chi^{(n)}(\beta) = 1$ for all $\beta \vdash n$, so $\chi^{(n)}$ is called the **trivial character**. Also, $\chi^{1^n}(\beta) = (-1)^{n-\ell(\beta)}$ for all $\beta \vdash n$, so $\chi^{1^n}$ is called the **sign character**, since $(-1)^{n-\ell(\beta)}$ is the sign of any permutation of cycle type $\beta$.

One reason Schur functions are important in representation theory is the following (see [58, p. 347] and [53, Chapter 1]).

**Theorem 11.2.27** *When expanding the $s_\lambda$ into the $p_\lambda$ basis, the coefficients are the $\chi^\lambda$. To be exact*

$$p_\mu = \sum_{\lambda \vdash n} \chi^\lambda(\mu) s_\lambda$$

$$s_\lambda = \sum_{\mu \vdash n} z_\mu^{-1} \chi^\lambda(\mu) p_\mu.$$

Let $\mathbb{C}[X_n] = \mathbb{C}[x_1, \ldots, x_n]$. Given $f(x_1, \ldots, x_n) \in \mathbb{C}[X_n]$ and $\sigma \in S_n$, then

$$\sigma f = f(x_{\sigma_1}, \ldots, x_{\sigma_n})$$

defines an action of $S_n$ on $\mathbb{C}[X_n]$.

Assume $V$ is a homogeneous subspace of $\mathbb{C}[X_n]$ which can be decomposed as

$$V = \bigoplus_{i=0}^{\infty} V^{(i)},$$

where $V^{(i)}$ is the subspace consisting of all elements of $V$ that are homogeneous of degree $i$ in the $x_j$, and is finite-dimensional. This gives a **grading** of the space $V$, and we define the **Hilbert series** $\mathscr{H}(V;q)$ of $V$ to be the sum

$$\mathscr{H}(V;q) = \sum_{i=0}^{\infty} q^i \dim(V^{(i)}),$$

where dim indicates the dimension as a $\mathbb{C}$-vector space. If in addition $V$ is fixed by the $S_n$ action, we define the **Frobenius series** $\mathscr{F}(V;q)$ of $V$ to be the symmetric function

$$\sum_{i=0}^{\infty} q^i \sum_{\lambda \in \mathrm{Par}(i)} \mathrm{Mult}(\chi^{\lambda}, V^{(i)}) s_{\lambda},$$

where $\mathrm{Mult}(\chi^{\lambda}, V^{(i)})$ is the multiplicity of the irreducible character $\chi^{\lambda}$ in the character of $V^{(i)}$ under the action. In other words, if we decompose $V^{(i)}$ into irreducible $S_n$-submodules, $\mathrm{Mult}(\chi^{\lambda}, V^{(i)})$ is the number of these submodules whose trace equals $\chi^{\lambda}$.

A polynomial in $\mathbb{C}[X_n]$ is **alternating**, or an **alternant**, if

$$\sigma f = (-1)^{\mathrm{inv}(\sigma)} f \quad \text{for all } \sigma \in S_n.$$

The set of alternants in $V$ forms a subspace called the subspace of alternants, or **anti-symmetric elements**, denoted $V^{\varepsilon}$. This is also an $S_n$-submodule of $V$.

**Proposition 11.2.28** *The Hilbert series of $V^{\varepsilon}$ equals the coefficient of $s_{1^n}$ in the Frobenius series of $V$, i.e.,*

$$\mathscr{H}(V^{\varepsilon};q) = \langle \mathscr{F}(V;q), s_{1^n} \rangle.$$

*Proof.* Let $B$ be a basis for $V^{(i)}$ with the property that the matrices $M(\sigma)$ are in block form. Then $b \in B$ is also in $(V^{\varepsilon})^{(i)}$ if and only if the column of $M(\sigma)$ corresponding to $b$ has entries $(-1)^{\mathrm{inv}(\sigma)}$ on the diagonal and 0's elsewhere, i.e., is a block corresponding to $\chi^{1^n}$. Thus

$$\langle \mathscr{F}(V;q), s_{1^n} \rangle = \sum_{i=0}^{\infty} q^i \dim((V^{\varepsilon})^{(i)}) = \mathscr{H}(V^{\varepsilon};q).$$

■

**Remark 11.2.29** Since the dimension of the representation corresponding to $\chi^{\lambda}$ equals $f^{\lambda}$, which by (11.29) equals $< s_{\lambda}, h_{1^n} >$, we have

$$\langle \mathscr{F}(V;q), h_{1^n} \rangle = \mathscr{H}(V;q).$$

**Example 11.2.30** *Since a basis for $\mathbb{C}[X_n]$ can be obtained by taking all possible monomials in the $x_i$,*

$$\mathcal{H}(\mathbb{C}[X_n];q) = (1-q)^{-n}.$$

*Taking into account the $S_n$-action, it is known [40, Section 1.4] that*

$$\begin{aligned}\mathcal{F}(\mathbb{C}[X_n];q) &= \sum_{\lambda \in Par(n)} s_\lambda \frac{\sum_{T \in SYT(\lambda)} q^{maj(T)}}{(q)_n} \qquad (11.36)\\ &= \sum_{\lambda \in Par(n)} s_\lambda s_\lambda(1,q,q^2,\ldots) = \prod_{i,j} \frac{1}{(1-q^i x_j z)}|_{z^n}.\end{aligned}$$

#### 11.2.5.1 The ring of coinvariants and the space of diagonal harmonics

The set of symmetric polynomials in the $x_i$, denoted $\mathbb{C}[X_n]^{S_n}$, which is generated by $1, e_1, \ldots e_n$, is called the **ring of invariants**. The quotient ring $R_n = \mathbb{C}[x_1,\ldots,x_n]/ < e_1, e_2, \ldots, e_n >$, or equivalently $\mathbb{C}[x_1,\ldots,x_n]/ < p_1, p_2, \ldots, p_n >$, obtained by forming the quotient by the ideal generated by all symmetric polynomials of positive degree, is known as the **ring of coinvariants**. It is known that $R_n$ is finite dimensional as a $\mathbb{C}$-vector space, with $\dim(R_n) = n!$, and more generally that

$$\mathcal{H}(R_n;q) = [n]!.$$

E. Artin [10] derived a specific basis for $R_n$, namely the cosets of

$$\{\prod_{1 \le i \le n} x_i^{\alpha_i}, 0 \le \alpha_i \le i-1\}.$$

Also,

$$\mathcal{F}(R_n;q) = \sum_{\lambda \in \mathrm{Par}(n)} s_\lambda \sum_{T \in SYT(\lambda)} q^{\mathrm{maj}(T)}, \qquad (11.37)$$

a result that Stanley [57, 59] attributes to unpublished work of Lusztig. This shows the Frobenius series of $R_n$ is $(q)_n$ times the Frobenius series of $\mathbb{C}[X_n]$.

Let

$$V_n = \det \begin{bmatrix} 1 & x_1 & \ldots & x_1^{n-1} \\ 1 & x_2 & \ldots & x_2^{n-1} \\ & & \vdots & \\ 1 & x_n & \ldots & x_n^{n-1} \end{bmatrix} = \prod_{1 \le i < j \le n} (x_j - x_i)$$

be the Vandermonde determinant. The **space of harmonics** $H_n$ can be defined as the $\mathbb{C}$-vector space spanned by $V_n$ and its partial derivatives of all orders. Haiman [40] provides a detailed proof that $H_n$ is isomorphic to $R_n$ as an $S_n$ module, and notes that an explicit isomorphism $\alpha$ is obtained by letting $\alpha(h), h \in H_n$, be the element of

$\mathbb{C}[X_n]$ represented modulo $< e_1, \ldots, e_n >$ by $h$. Thus $\dim(H_n) = n!$ and moreover the character of $H_n$ under the $S_n$-action is given by (11.37). He also argues that (11.37) follows immediately from (11.36) and the fact that $H_n$ generates $\mathbb{C}[X_n]$ as a free module over $\mathbb{C}[X_n]^{S_n}$.

There is a natural extension of this construction to two sets of variables, which has a very rich algebraic and combinatorial structure. Let the ring of **diagonal coinvariants** $DR_n$ be defined as

$$DR_n = \mathbb{C}[X_n, Y_n] / \left\langle \sum_{i=1}^{n} x_i^h y_i^k, \forall h+k > 0 \right\rangle.$$

By analogy we also define the space of **diagonal harmonics** $DH_n$ by

$$DH_n = \left\{ f \in \mathbb{C}[X_n, Y_n] : \sum_{i=1}^{n} (\partial x_i)^h (\partial y_i)^k f = 0, \forall h+k > 0 \right\}.$$

Many of the properties of $H_n$ and $R_n$ carry over to two sets of variables. For example the symmetric group acts "diagonally" on $DR_n$ and $DH_n$ by permuting the $X$ and $Y$ variables in the same way, which turns $DH_n$ and $DR_n$ into finite-dimensional isomorphic $S_n$-modules. We can decompose $DH_n$ by homogeneous $X$ and $Y$ degree, and the $S_n$ action respects this bi-grading. Hence we can talk about the Hilbert series $\mathscr{H}(DH_n; q, t)$ and and the Frobenius series $\mathscr{F}(DH_n; q, t)$.

## 11.3 The $q,t$-Catalan numbers

Given a cell $x \in \lambda$, let the **arm** $a = a(x)$, **leg** $l = l(x)$, **coarm** $a' = a'(x)$, and **coleg** $l' = l'(x)$ be the number of cells strictly between $x$ and the border of $\lambda$ in the east, north, west, and south directions, respectively, as in Figure 11.5. Also, define

$$B_\mu = B_\mu(q,t) = \sum_{x \in \mu} q^{a'} t^{l'}, \quad \Pi_\mu = \Pi_\mu(q,t) = \prod_{x \in \mu}{}' (1 - q^{a'} t^{l'}),$$

where a prime symbol $'$ above a product or a sum over cells of a partition $\mu$ indicates we ignore the corner $(1,1)$ cell, and $B_\emptyset = 0$, $\Pi_\emptyset = 1$. For example, $B_{(2,2,1)} = 1 + q + t + qt + t^2$ and $\Pi_{(2,2,1)} = (1-q)(1-t)(1-qt)(1-t^2)$. Note that

$$n(\mu) = \sum_{x \in \mu} l' = \sum_{x \in \mu} l.$$

In 1996 Garsia and Haiman [27] introduced an amazing two-parameter Catalan sequence, $C_n(q,t)$, which they defined as the following sum of rational functions:

$$C_n(q,t) = \sum_{\mu \vdash n} \frac{T_\mu^2 M \Pi_\mu B_\mu}{w_\mu}, \tag{11.38}$$

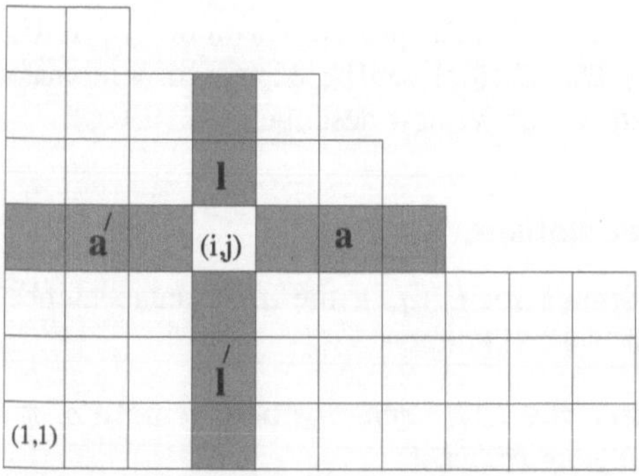


**Figure 11.5**
The arm $a$, coarm $a'$, leg $l$, and coleg $l'$ of a cell.

where

$$M = (1-q)(1-t), \quad T_\mu = t^{n(\mu)} q^{n(\mu')}, \quad w_\mu = \prod_{x \in \mu} (q^a - t^{l+1})(t^l - q^{a+1}).$$

The definition of $C_n(q,t)$ was motivated by ideas involving algebraic geometry, and Garsia and Haiman conjectured that $C_n(q,t) \in \mathbb{N}[q,t]$. More specifically, they conjectured that $C_n(q,t)$ is the sign character in the Frobenius series for $DH_n$, i.e.,

$$C_n(q,t) = \langle \mathscr{F}(DH_n; q,t), s_{1^n} \rangle. \tag{11.39}$$

Mark Haiman proved this conjecture in 2001 by obtaining the following expression for $\mathscr{F}(DH_n; q,t)$ in terms of Macdonald polynomials.

**Theorem 11.3.1** *[42] We have*

$$\mathscr{F}(DH_n; q,t) = \sum_{\mu \vdash n} \frac{T_\mu M \Pi_\mu B_\mu \tilde{H}_\mu(X;q,t)}{w_\mu}, \tag{11.40}$$

*where the $\tilde{H}_\mu(X;q,t)$ form the modified Macdonald polynomial symmetric function basis.*

Although we will not describe these polynomials explicitly here, the interested reader can find a combinatorial description of them in [36] or [39, Appendix A].

We mention that

$$\langle \tilde{H}_\mu(X;q,t), s_{1^n}(X) \rangle = T_\mu,$$

so (11.40) implies (11.39). The right-hand side of (11.40) can be expressed more compactly as $\nabla e_n(X)$, where $\nabla$ is the linear operator on symmetric functions satisfying

$$\nabla \tilde{H}_\mu(X;q,t) = T_\mu \tilde{H}_\mu(X;q,t).$$

Hence we can refer to $\mathscr{F}(DH_n; q,t)$ and $\nabla e_n$ interchangeably.

Around the same time Haiman proved Theorem 11.3.1, Garsia and Haglund proved independently that $C_n(q,t)$ can be expressed combinatorially in terms of statistics on Dyck paths which we now describe.

### 11.3.1 The bounce statistic

Our combinatorial formula for $C_n(q,t)$, the $q,t$-Catalan number, involves a new statistic on Dyck paths we call **bounce**.

**Definition 11.3.2** *Given $\pi \in L^+_{n,n}$, define the* **bounce path** *of $\pi$ to be the path described by the following algorithm.*

*Start at $(0,0)$ and travel North along $\pi$ until you encounter the beginning of an E step. Then turn East and travel straight until you hit the diagonal $y = x$. Then turn North and travel straight until you again encounter the beginning of an E step of $\pi$, then turn East and travel to the diagonal, etc. Continue in this way until you arrive at $(n,n)$.*

*We can think of our bounce path as describing the trail of a billiard ball shot North from $(0,0)$, which "bounces" right whenever it encounters a horizontal step and "bounces" up when it encounters the line $y = x$. The bouncing ball will strike the diagonal at places*

$$(0,0),(j_1,j_1),(j_2,j_2),\ldots,(j_{b-1},j_{b-1}),(j_b,j_b) = (n,n).$$

*We define the bounce statistic bounce$(\pi)$ to be the sum*

$$bounce(\pi) = \sum_{i=1}^{b-1} n - j_i,$$

*and we call $b$ the number of bounces, with $j_1$ the length of the first bounce, $j_2 - j_1$ the length of the second bounce, etc. The lattice points where the bouncing billiard ball switches from traveling North to East are called the* **peaks** *of $\pi$. The first peak is the peak with smallest $y$ coordinate, the second peak the one with next smallest $y$ coordinate, etc. For the path $\pi$ in Figure 11.6, there are five bounces of lengths $3,2,2,3,1$ and bounce$(\pi) = 19$. The first two peaks have coordinates $(0,3)$ and $(3,5)$.*

Let

$$F_n(q,t) = \sum_{\pi \in L^+_{n,n}} q^{\text{area}(\pi)} t^{\text{bounce}(\pi)}.$$

**Theorem 11.3.3** *The equality*

$$C_n(q,t) = F_n(q,t) \tag{11.41}$$

*holds.*

Combining Theorem 11.3.3 with Theorem 11.3.1 we have the following.

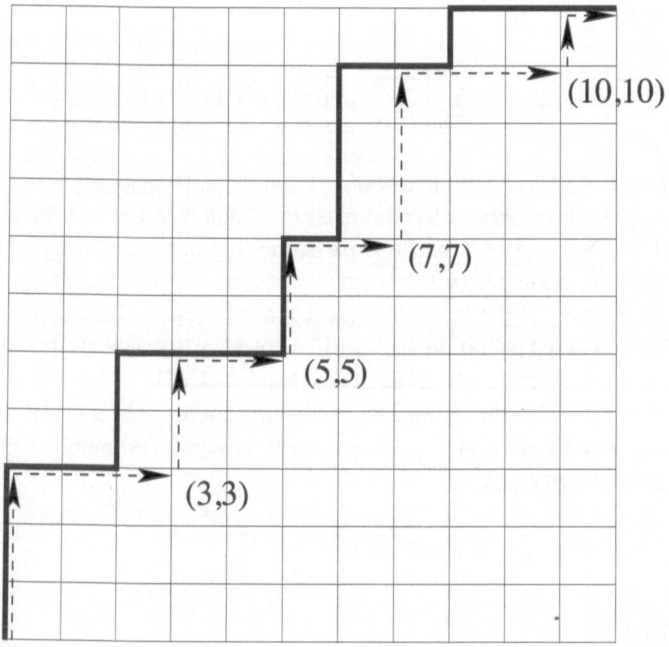


**Figure 11.6**
The bounce path (dotted line) of a Dyck path (solid line). The bounce statistic equals $11-3+11-5+11-7+11-10 = 8+6+4+1 = 19$.

**Corollary 11.3.4** *The identity*

$$\mathscr{H}(DH_n^{\varepsilon}; q,t) = \sum_{\pi \in L_{n,n}^+} q^{area(\pi)} t^{bounce(\pi)} \tag{11.42}$$

*holds.*

Theorem 11.3.3 was first conjectured by Haglund in 2000 [33] after a prolonged study of tables of $C_n(q,t)$. It was then proved by Garsia and Haglund [25, 26]. At the present time there is no known way of proving Corollary 11.3.4 without using both Theorems 11.3.3 and 11.3.1.

The proof of Theorem 11.3.3 is based on a recursive structure underlying $F_n(q,t)$.

**Definition 11.3.5** *Let $L_{n,n}^+(k)$ denote the set of all $\pi \in L_{n,n}^+$ which begin with exactly $k$ N steps followed by an E step. By convention $L_{0,0}^+(k)$ consists of the empty path if $k=0$ and is empty otherwise. Set*

$$F_{n,k}(q,t) = \sum_{\pi \in L_{n,n}^+(k)} q^{area(\pi)} t^{bounce(\pi)}, \quad F_{n,0} = \chi(n=0).$$

**Theorem 11.3.6** *[33]. For* $1 \le k \le n$,

$$F_{n,k}(q,t) = \sum_{r=0}^{n-k} \begin{bmatrix} r+k-1 \\ r \end{bmatrix}_q t^{n-k} q^{\binom{k}{2}} F_{n-k,r}(q,t). \tag{11.43}$$

*Proof.* Given $\beta \in L^+_{n,n}(k)$, with first bounce $k$ and second bounce say $r$, then $\beta$ must pass through the lattice points with coordinates $(1,k)$ and $(k,k+r)$ (the two large dots in Figure 11.7). Decompose $\beta$ into two parts, the first part being the portion of $\beta$ starting at $(0,0)$ and ending at $(k,k+r)$, and the second the portion starting at $(k,k+r)$ and ending at $(n,n)$. If we adjoin a sequence of $r$ $N$ steps to the beginning of the second part, we obtain a path $\beta'$ in $L^+_{n-k,n-k}(r)$. It is easy to check that bounce$(\beta) =$ bounce$(\beta') + n - k$. It remains to relate area$(\beta)$ and area$(\beta')$.

Clearly the area inside the triangle below the first bounce step is $\binom{k}{2}$. If we fix $\beta'$, and let $\beta$ vary over all paths in $L^+_{n,n}(k)$ that travel through $(1,k)$ and $(k,k+r)$, then the sum of $q^{\text{area}(\beta)}$ will equal

$$q^{\text{area}(\beta')} q^{\binom{k}{2}} \begin{bmatrix} k+r-1 \\ r \end{bmatrix}_q$$

by (11.1). Thus

$$\begin{aligned} F_{n,k}(q,t) &= \sum_{r=0}^{n-k} \sum_{\beta' \in L^+_{n-k,n-k}(r)} q^{\text{area}(\beta')} t^{\text{bounce}(\beta')} t^{n-k} q^{\binom{k}{2}} \begin{bmatrix} k+r-1 \\ r \end{bmatrix}_q \\ &= \sum_{r=0}^{n-k} \begin{bmatrix} r+k-1 \\ r \end{bmatrix}_q t^{n-k} q^{\binom{k}{2}} F_{n-k,r}(q,t). \end{aligned}$$

■

**Corollary 11.3.7** *We haves*

$$F_n(q,t) = \sum_{b=1}^{n} \sum_{\substack{\alpha_1+\alpha_2+\ldots+\alpha_b=n \\ \alpha_i>0}} t^{\alpha_2+2\alpha_3+\ldots+(b-1)\alpha_b} q^{\sum_{i=1}^{b}\binom{\alpha_i}{2}} \prod_{i=1}^{b-1} \begin{bmatrix} \alpha_i+\alpha_{i+1}-1 \\ \alpha_{i+1} \end{bmatrix}_q, \tag{11.44}$$

*where the inner sum is over all compositions* $\alpha$ *of* $n$ *into* $b$ *positive integers.*

*Proof.* This follows by iterating the recurrence in Theorem 11.3.6. The inner term in the sum over $b$ equals the sum of $q^{\text{area}} t^{\text{bounce}}$ over all paths $\pi$ whose bounce path has $b$ steps of lengths $\alpha_1, \ldots, \alpha_b$. For such a $\pi$, the contribution of the first bounce to bounce$(\pi)$ is $n - \alpha_1 = \alpha_2 + \ldots + \alpha_b$, the contribution of the second bounce is $n - \alpha_1 - \alpha_2 = \alpha_3 + \ldots + \alpha_b$, et cetera, so bounce$(\pi) = \alpha_2 + 2\alpha_3 + \ldots + (b-1)\alpha_b$. ■

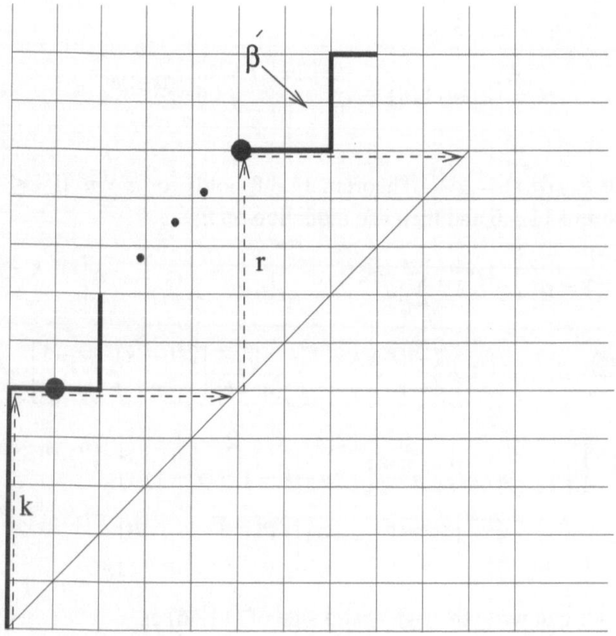


**Figure 11.7**
A path whose first two bounce steps are $k$ and $r$.

### 11.3.2 The special values $t = 1$ and $t = 1/q$

Garsia and Haiman proved that

$$C_n(q,1) = C_n(q)$$
$$q^{\binom{n}{2}}C_n(q,1/q) = \frac{1}{[n+1]}\begin{bmatrix}2n\\ n\end{bmatrix}_q, \tag{11.45}$$

which shows that both the Carlitz-Riordan and MacMahon $q$-Catalan numbers are special cases of $C_n(q,t)$. In this section we derive analogous results for $F_{n,k}(q,1)$ and $F_{n,k}(q,1/q)$.

By definition we have

$$F_n(q,1) = C_n(q).$$

It is perhaps worth mentioning that the $F_{n,k}(q,1)$ satisfy the simple recurrence

$$F_{n,k}(q,1) = \sum_{m=k}^{n} q^{m-1}F_{m-1,k-1}(q,1)F_{n-m}(q,1).$$

This follows by grouping paths in $L^+_{n,n}(k)$ according to the first time they return to the diagonal, at say $(m,m)$, then arguing as in the proof of Proposition 11.11.

The $F_{n,k}(q,1/q)$ satisfy the following refinement of (11.45).

**Theorem 11.3.8** *For $1 \le k \le n$,*

$$q^{\binom{n}{2}} F_{n,k}(q,1/q) = \frac{[k]}{[n]} \begin{bmatrix} 2n-k-1 \\ n-k \end{bmatrix}_q q^{(k-1)n}.$$

*Proof.* Since $F_{n,n}(q,t) = q^{\binom{n}{2}}$, Theorem 11.3.8 holds for $k = n$. If $1 \le k < n$, we start with Theorem 11.3.6 and then use induction on $n$:

$$\begin{aligned} q^{\binom{n}{2}} F_{n,k}(q,q^{-1}) &= q^{\binom{n}{2}} q^{-\binom{n-k}{2}} \sum_{r=1}^{n-k} q^{\binom{n-k}{2}} F_{n-k,r}(q,q^{-1}) q^{\binom{k}{2}-(n-k)} \begin{bmatrix} r+k-1 \\ r \end{bmatrix}_q \\ &= q^{\binom{n}{2}+\binom{k}{2}-(n-k)} q^{-\binom{n-k}{2}} \sum_{r=1}^{n-k} \begin{bmatrix} r+k-1 \\ r \end{bmatrix}_q \frac{[r]}{[n-k]} \begin{bmatrix} 2(n-k)-r-1 \\ n-k-r \end{bmatrix}_q q^{(r-1)(n-k)} \\ &= q^{(k-1)n} \sum_{r=1}^{n-k} \begin{bmatrix} r+k-1 \\ r \end{bmatrix}_q \frac{[r]}{[n-k]} \begin{bmatrix} 2(n-k)-2-(r-1) \\ n-k-1-(r-1) \end{bmatrix}_q q^{(r-1)(n-k)} \\ &= q^{(k-1)n} \frac{[k]}{[n-k]} \sum_{u=0}^{n-k-1} \begin{bmatrix} k+u \\ u \end{bmatrix}_q q^{u(n-k)} \begin{bmatrix} 2(n-k)-2-u \\ n-k-1-u \end{bmatrix}_q. \end{aligned} \tag{11.46}$$

Using (11.19) we can write the right-hand side of (11.46) as

$$\begin{aligned} q^{(k-1)n} \frac{[k]}{[n-k]} \frac{1}{(zq^{n-k})_{k+1}} \frac{1}{(z)_{n-k}} \Big|_{z^{n-k-1}} &= q^{(k-1)n} \frac{[k]}{[n-k]} \frac{1}{(z)_{n+1}} \Big|_{z^{n-k-1}} \\ &= q^{(k-1)n} \frac{[k]}{[n-k]} \begin{bmatrix} n+n-k-1 \\ n \end{bmatrix}_q. \end{aligned}$$

■

**Corollary 11.3.9** *The identity*

$$q^{\binom{n}{2}} F_n(q,1/q) = \frac{1}{[n+1]} \begin{bmatrix} 2n \\ n \end{bmatrix}_q$$

*holds.*

*Proof.* N. Loehr has pointed out that we can use

$$F_{n+1,1}(q,t) = t^n F_n(q,t), \tag{11.47}$$

which by Theorem 11.3.8 implies

$$\begin{aligned} q^{\binom{n+1}{2}} F_{n+1,1}(q,1/q) &= \frac{[1]}{[n+1]} \begin{bmatrix} 2(n+1)-2 \\ n+1-1 \end{bmatrix}_q = \frac{[1]}{[n+1]} \begin{bmatrix} 2n \\ n \end{bmatrix}_q \\ &= q^{\binom{n+1}{2}-n} F_n(q,1/q) = q^{\binom{n}{2}} F_n(q,1/q). \end{aligned}$$

■

### 11.3.3 The symmetry problem and the dinv statistic

From its definition, it is easy to show $C_n(q,t) = C_n(t,q)$, since the arm and leg values for $\mu$ equal the leg and arm values for $\mu'$, respectively, which implies $q$ and $t$ are interchanged when comparing terms in (11.38) corresponding to $\mu$ and $\mu'$. This also follows from the theorem that $C_n(q,t) = \mathscr{H}(DH_n^{\varepsilon}; q,t)$. Thus we have

$$\sum_{\pi \in L_{n,n}^+} q^{\text{area}(\pi)} t^{\text{bounce}(\pi)} = \sum_{\pi \in L_{n,n}^+} q^{\text{bounce}(\pi)} t^{\text{area}(\pi)}, \tag{11.48}$$

a surprising statement in view of the apparent dissimilarity of the area and bounce statistics. At present there is no other known way to prove (11.48) other than as a corollary of Theorem 11.3.3.

**Problem 11.3.10** *Prove (11.48) by exhibiting a bijection on Dyck paths that interchanges area and bounce.*

A solution to Problem 11.3.10 should lead to a deeper understanding of the combinatorics of $DH_n$. We now give a combinatorial proof from [33] of a very special case of (11.48), by showing that the marginal distributions of area and bounce are the same, i.e., $F_n(q,1) = F_n(1,q)$.

**Theorem 11.3.11** *The identity*

$$\sum_{\pi \in L_{n,n}^+} q^{area(\pi)} = \sum_{\pi \in L_{n,n}^+} q^{bounce(\pi)} \tag{11.49}$$

*holds.*

*Proof.* Given $\pi \in L_{n,n}^+$, let $a_1 a_2 \cdots a_n$ denote the sequence whose $i$th element is the $i$th coordinate of the area vector of $\pi$, i.e., the length of the $i$th row (from the bottom) of $\pi$. A moment's thought shows that such a sequence is characterized by the property that it begins with zero, consists of $n$ nonnegative integers, and has no 2-ascents, i.e., values of $i$ for which $a_{i+1} > a_i + 1$. To construct such a sequence we begin with an arbitrary multiset of row lengths, say $\{0^{\alpha_1} 1^{\alpha_2} \cdots (b-1)^{\alpha_b}\}$ and then choose a multiset permutation $\tau$ of $\{0^{\alpha_1} 1^{\alpha_2}\}$ that begins with 0 in $\binom{\alpha_1 - 1 + \alpha_2}{\alpha_2}$ ways. Next we will insert the $\alpha_3$ twos into $\tau$, the requirement of having no 2-ascents translating into having no consecutive 02 pairs. This means the number of ways to do this is $\binom{\alpha_2 - 1 + \alpha_3}{\alpha_3}$, independent of the choice of $\tau$. The formula

$$F_n(q,1) = \sum_{b=1}^{n} \sum_{\substack{\alpha_1 + \ldots + \alpha_b = n \\ \alpha_i > 0}} q^{\sum_{i=2}^{b} \alpha_i (i-1)} \prod_{i=1}^{b-1} \binom{\alpha_i - 1 + \alpha_{i+1}}{\alpha_{i+1}} \tag{11.50}$$

follows, since the product above counts the number of Dyck paths with a specified multiset of row lengths, and the power of $q$ is the common value of area for all these paths. Comparing (11.50) with the $q = 1, t = q$ case of formula (11.44) completes the proof. ∎

There is another pair of statistics for the $q,t$-Catalan discovered by M. Haiman [41]. It involves pairing area with a different statistic we call dinv, for **diagonal inversion** or **$d$-inversion**. It is defined, with $a_i$ the length of the $i$th row from the bottom, as follows.

**Definition 11.3.12** *Let $\pi \in L^+_{n,n}$. Let*

$$\begin{aligned} dinv(\pi) = &|\{(i,j) : 1 \le i < j \le n \quad a_i = a_j\}| \\ &+ |\{(i,j) : 1 \le i < j \le n \quad a_i = a_j + 1\}|. \end{aligned}$$

*In words, $dinv(\pi)$ is the number of pairs of rows of $\pi$ of the same length, or that differ by one in length, with the longer row below the shorter. For example, for the path on the left in Figure 11.8, with row lengths on the right, the inversion pairs $(i,j)$ are $(3,7),(4,7),(5,7),(6,8)$ (corresponding to rows that differ by one in length), and $(2,7),(3,4),(3,5),(3,8),(4,5),(4,8),(5,8)$ (corresponding to pairs of rows of the same length), thus dinv $= 11$. We call inversion pairs between rows of the same length "equal-length" inversions, and the other kind "offset-length" inversions.*

**Figure 11.8**
A path $\pi$ with row lengths to the right, and the image $\zeta(\pi)$.

**Theorem 11.3.13** *The identity*

$$\sum_{\pi \in L^+_{n,n}} q^{dinv(\pi)} t^{area(\pi)} = \sum_{\pi \in L^+_{n,n}} q^{area(\pi)} t^{bounce(\pi)} \tag{11.51}$$

*holds.*

*Proof.* We will describe a bijective map $\zeta$ on Dyck paths with the property that

$$\begin{aligned} \text{dinv}(\pi) &= \text{area}(\zeta(\pi)) \\ \text{area}(\pi) &= \text{bounce}(\zeta(\pi)). \end{aligned} \tag{11.52}$$

Say $b-1$ is the length of the longest row of $\pi$. The lengths of the bounce steps of $\zeta$ will be $\alpha_1,\ldots,\alpha_b$, where $\alpha_i$ is the number of rows of length $i-1$ in $\pi$. To construct the actual path $\zeta$, place a pen at the lattice point $(\alpha_1,\alpha_1+\alpha_2)$ (the second peak of the bounce path of $\zeta$). Start at the end of the area sequence and travel left. Whenever you encounter a 1, trace a South step with your pen. Whenever you encounter a 0, trace a West step. Skip over all other numbers. Your pen will end up at the top of the first peak of $\zeta$. Now go back to the end of the area sequence, and place your pen at the top of the third peak. Traverse the area sequence again from right to left, but this time whenever you encounter a 2 trace out a South step, and whenever you encounter a 1, trace out a West step. Skip over any other numbers. Your pen will end up at the top of the second peak of $\zeta$. Continue at the top of the fourth peak looking at how the rows of length 3 and 2 are interleaved, etc.. See Figure 11.8.

It is easy to see this map is a bijection, since given $\zeta$, from the bounce path we can determine the multiset of row lengths of $\pi$. We can then build up the area sequence of $\pi$ just as in the proof of Theorem 11.3.11. From the portion of the path between the first and second peaks we can see how to interleave the rows of lengths 0 and 1, and then we can insert the rows of length 2 into the area sequence, etc.

Note that when tracing out the part of $\zeta$ between the first and second peaks, whenever we encounter a 0 and trace out a West step, the number of area squares directly below this West step and above the bounce path of $\zeta$ equals the number of 1's to the left of this 0 in the area sequence, which is the number of offset-length inversion pairs involving the corresponding row of length 0. Since the area below the bounce path clearly counts the total number of equal-length inversions, it follows that $\operatorname{dinv}(\pi)=\operatorname{area}(\zeta(\pi))$.

Now by direct calculation,

$$\begin{aligned}\operatorname{bounce}(\zeta) &= n-\alpha_1+n-\alpha_1-\alpha_2+\ldots n-\alpha_1-\ldots-\alpha_{b-1} \\ &= (\alpha_2+\ldots+\alpha_b)+\ldots+(\alpha_b)=\sum_{i=1}^{b-1} i\alpha_{i+1}=\operatorname{area}(\pi).\end{aligned} \tag{11.53}$$

■

**Remark 11.3.14** The construction of the bounce path for a Dyck path occurs in an independent context, in work of Andrews, Krattenthaler, Orsina and Papi [5] on the enumeration of ad-nilpotent ideals of a Borel subalgebra of $\mathfrak{sl}(n+1,\mathbb{C})$. They prove the number of times a given nilpotent ideal needs to be bracketed with itself to become zero equals the number of bounces of the bounce path of a certain Dyck path associated to the ideal. Another of their results is a bijective map on Dyck paths that sends a path with $b$ bounces to a path whose longest row is of length $b-1$. The $\zeta$ map above is just the inverse of their map. Because they only considered the number of bounces, and not the bounce statistic per se, they did not notice any connection between $C_n(q,t)$ and their construction.

Theorem 11.3.3 now implies the following.

**Corollary 11.3.15** *We have*

$$C_n(q,t) = \sum_{\pi \in L^+_{n,n}} q^{dinv(\pi)} t^{area(\pi)}. \tag{11.54}$$

*We also have*

$$F_{n,k}(q,t) = \sum_{\substack{\pi \in L^+_{n,n} \\ \pi \text{ has exactly } k \text{ rows of length } 0}} q^{dinv(\pi)} t^{area(\pi)}, \tag{11.55}$$

*since under the $\zeta$ map, paths with k rows of length 0 correspond to paths whose first bounce step is of length k.*

**Remark 11.3.16** N. Loehr has noted that if one can find a map that fixes area and sends bounce to dinv, by combining this with the $\zeta$ map one would have a map that interchanges area and dinv, solving Problem 11.3.10.

**Exercise 11.3.17** *Let*

$$G_{n,k}(q,t) = \sum_{\substack{\pi \in L^+_{n,n} \\ \pi \text{ has exactly } k \text{ rows of length } 0}} q^{dinv(\pi)} t^{area(\pi)}. \tag{11.56}$$

*Without referencing any results on the bounce statistic, prove combinatorially that*

$$G_{n,k}(q,t) = t^{n-k} q^{\binom{k}{2}} \sum_{r=0}^{n-k} \begin{bmatrix} r+k-1 \\ r \end{bmatrix}_q G_{n-k,r}(q,t).$$

**Exercise 11.3.18** *Haiman's conjectured statistics for $C_n(q,t)$ actually involved a different description of dinv. Let $\lambda(\pi)$ denote the partition consisting of the $\binom{n}{2}$ − area($\pi$) squares above $\pi$ but inside the $n \times n$ square. (This is the Ferrers graph of a partition in the so-called English convention, which is obtained from the French convention of Figure 11.1 by reflecting the graph about the x-axis. In this convention, the leg $l(s)$ of a square s is defined as the number of squares of $\lambda$ below s in the column and above the lower border $\pi$, and the arm $a(s)$ as the number of squares of $\lambda$ to the right and in the row.) Then Haiman's original version of dinv was the number of cells s of $\lambda$ for which*

$$l(s) \le a(s) \le l(s) + 1.$$

*Prove this definition of dinv is equivalent to Definition 11.3.12.*

### 11.3.4 q-Lagrange inversion

$q$-Lagrange inversion is useful when analyzing the special case $t = 1$ of $\mathscr{F}(DH_n;q,t)$. In this section we derive a general $q$-Lagrange inversion theorem based on work of Garsia and Haiman. We will be working in the ring of formal power series, and we begin with a result of Garsia [23].

**Theorem 11.3.19** *If*

$$(F \circ_q G)(z) = \sum_n f_n G(z)G(qz)\cdots G(q^{n-1}z),$$

*where $F = \sum_n f_n z^n$, then*

$$F \circ_q G = z \text{ and } G \circ_{q^{-1}} F = z$$

*are equivalent to each other and also to*

$$(\Phi \circ_{q^{-1}} F) \circ_q G = \Phi = (\Phi \circ_q G) \circ_{q^{-1}} F \qquad \text{for all } \Phi.$$

Given $\pi \in L^+_{n,n}$, let $\beta(\pi) = \beta_1(\pi)\beta_2(\pi)\cdots$ denote the partition consisting of the vertical step lengths of $\pi$ (i.e., the lengths of the maximal blocks of consecutive 0's in $\sigma(\pi)$) arranged in nonincreasing order. For example, for the path on the left in Figure 11.8 we have $\beta = (3,2,2,1)$. By convention we set $\beta(\emptyset) = \emptyset$. Define $H(z)$ via the equation $1/H(-z) := \sum_{k=0}^{\infty} e_k z^k$. Using Theorem 11.3.19, Haiman [40, pp. 47-48] derived the following.

**Theorem 11.3.20** *There is a unique solution $h_n^*(q), n \geq 0$ to the equation*

$$\sum_{k=0}^{\infty} e_k z^k = \sum_{n=0}^{\infty} q^{-\binom{n}{2}} h_n^*(q) z^n H(-q^{-1}z)H(-q^{-2}z)\cdots H(-q^{-n}z), \qquad h_0^*(q) = 1.$$

*For $n > 0$, $h_n^*(q)$ has the explicit expression*

$$h_n^*(q) = \sum_{\pi \in L^+_{n,n}} q^{area(\pi)} e_{\beta(\pi)}.$$

*For example, we have*

$$h_3^*(q) = q^3 e_3 + q^2 e_{2,1} + 2q e_{2,1} + e_{1^3}.$$

We now derive a slight generalization of Theorem 11.3.20 that stratifies Dyck paths according to the length of their first bounce step.

**Theorem 11.3.21** *Let $c_k, k \geq 0$ be a set of variables. Define $h_n^*(\mathbf{c},q), n \geq 0$ via the equation*

$$\sum_{k=0}^{\infty} e_k c_k z^k = \sum_{n=0}^{\infty} q^{-\binom{n}{2}} h_n^*(\mathbf{c},q) z^n H(-q^{-1}z)H(-q^{-2}z)\cdots H(-q^{-n}z), \qquad h_0^*(\mathbf{c},q) = c_0. \tag{11.57}$$

*Then for $n \geq 0$, $h_n^*(\mathbf{c},q)$ has the explicit expression*

$$h_n^*(\mathbf{c},q) = \sum_{k=0}^{n} c_k \sum_{\pi \in L^+_{n,n}(k)} q^{area(\pi)} e_{\beta(\pi)}.$$

*For example, we have*

$$h_3^*(\mathbf{c},q) = q^3 e_3 c_3 + (q^2 e_{2,1} + q e_{2,1}) c_2 + (q e_{2,1} + e_{1^3}) c_1.$$

*Proof.* Our proof follows Haiman's proof of Theorem 11.3.20 closely. Set $H^*(z,\mathbf{c};q) := \sum_{n=0}^{\infty} h_n^*(\mathbf{c},q)z^n$, $H^*(z;q) := \sum_{n=0}^{\infty} h_n^*(q)z^n$, $\Phi = H^*(zq,\mathbf{c};q)$, $F = zH(-z)$, and $G = zH^*(qz;q)$. Replacing $z$ by $zq$ in (11.57) we see that Theorem 11.3.21 is equivalent to the statement

$$\begin{aligned}\sum_{k=0}^{\infty} e_k c_k q^k z^k &= \sum_{n=0}^{\infty} q^n h_n^*(\mathbf{c},q) zH(-z) zq^{-1}H(-q^{-1}z)\cdots zq^{1-n}H(-q^{1-n}z)\\ &= \Phi \circ_{q^{-1}} F.\end{aligned}$$

On the other hand, Theorem 11.3.20 can be expressed as

$$\frac{1}{H(-z)} = \sum_{n=0}^{\infty} q^{-\binom{n}{2}} h_n^*(q) z^n H(-z/q)\cdots H(-z/q^n),$$

or

$$\begin{aligned} z &= \sum_{n=0}^{\infty} q^{-\binom{n}{2}} h_n^*(q) z^{n+1} H(-z)H(-z/q)\cdots H(-z/q^n)\\ &= \sum_{n=0}^{\infty} q^n h_n^*(q)\,\{zH(-z)\}\left\{\frac{z}{q}H(-z/q)\right\}\cdots\left\{\frac{z}{q^n}H(-z/q^n)\right\}\\ &= G \circ_{q^{-1}} F.\end{aligned}$$

Thus assuming Theorem 11.3.21 we have, using Theorem 11.3.19,

$$\begin{aligned}\Phi &= (\Phi \circ_{q^{-1}} F)\circ_q G \qquad (11.58)\\ &= (\sum_{k=0}^{\infty} e_k \mu_k q^k z^k)\circ_q G.\end{aligned}$$

Comparing coefficients of $z^n$ in (11.58) and simplifying we see that Theorem 11.3.21 is equivalent to the statement

$$q^n h_n^*(\mathbf{c},q) = \sum_{k=0}^{n} q^{\binom{k}{2}+k} e_k c_k \sum_{\substack{n_1+\ldots+n_k=n-k\\ n_i\geq 0}} q^{n-k}\prod_{i=1}^{k} q^{(i-1)n_i} h_{n_i}^*(q). \qquad (11.59)$$

To prove (11.59) we use the "factorization of Dyck paths" as discussed in [40]. This can be be represented pictorially as in Figure 11.9. The terms multiplied by $c_k$ correspond to $\pi \in L_{n,n}^+(k)$. The area of the parallelogram whose left border is on the line $y = x+i-1$ is $(i-1)n_i$, where $n_i$ is the length of the left border of the parallelogram. Using the fact that the path to the left of this parallelogram is in $L_{n_i,n_i}^+$, (11.59) now becomes transparent. ■

Letting $e_k - 1$, $c_j = \chi(j=k)$ and replacing $q$ by $q^{-1}$ and $z$ by $z/q$ in Theorem 11.3.21 we get the following.

**Corollary 11.3.22** *For* $1 \leq k \leq n$,

$$z^k = \sum_{n\geq k} q^{\binom{n}{2}-n+k} F_{n,k}(q^{-1},1) z^n (z)_n.$$

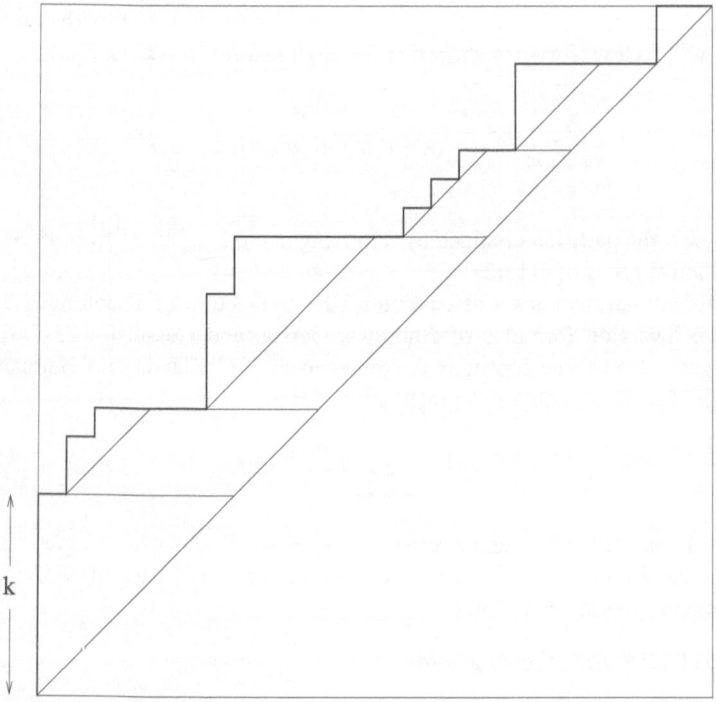


**Figure 11.9**
A Dyck path factored into smaller paths.

Theorem 11.3.21 is a $q$-analogue of the general Lagrange inversion formula [4, p.629]

$$f(x) = f(0) + \sum_{n=1}^{\infty} \frac{x^n}{n!\phi(x)^n} \left[ \frac{d^{n-1}}{dx^{n-1}} (f'(x)\phi^n(x)) \right]_{x=0}, \tag{11.60}$$

where $\phi$ and $f$ are analytic in a neighborhood of 0, with $\phi(0) \neq 0$. To see why, assume without loss of generality that $f(0) = 1$, and set $f(x) = \sum_{k=0}^{\infty} c_k x^k$ and $\phi = \frac{1}{H(-x)} = \sum_{k=0}^{\infty} e_k x^k$ in (11.60) to get

$$\begin{aligned} \sum_{k=1}^{\infty} c_k x^k &= \sum_{n=1}^{\infty} \frac{x^n}{n!} H(-x)^n \frac{d^{n-1}}{dx^{n-1}} \left( \sum_{k=1}^{\infty} k c_k x^{k-1} (\sum_{m=0}^{\infty} e_m x^m)^n \right) |_{x=0} \\ &= \sum_{n=1}^{\infty} \frac{x^n}{n!} H(-x)^n \sum_{k=1}^{n} k c_k (n-1)! (\sum_{m=0}^{\infty} e_m x^m)^n |_{x^{n-k}} \\ &= \sum_{n=1}^{\infty} x^n H(-x)^n \sum_{k=1}^{n} c_k \frac{k}{n} \sum_{\substack{j_1+j_2+\ldots+j_n=n-k \\ j_i \geq 0}} e_{j_1} e_{j_2} \cdots e_{j_n} \\ &= \sum_{n=1}^{\infty} x^n H(-x)^n \sum_{k=1}^{n} c_k \frac{k}{n} \sum_{\alpha \vdash n-k} e_\alpha \binom{n}{k-\ell(\alpha), n_1(\alpha), n_2(\alpha), \ldots}. \end{aligned} \tag{11.61}$$

The equivalence of (11.61) to the $q=1$ case of Theorem 11.3.21 (with $c_k$ replaced by $c_k/e_k$) will follow if we can show that for any fixed $\alpha \vdash n-k$,

$$\sum_{\substack{\pi \in L^+_{n,n}(k) \\ \beta(\pi)-k=\alpha}} 1 = \frac{k}{n}\binom{n}{k-\ell(\alpha), n_1(\alpha), n_2(\alpha), \ldots}, \tag{11.62}$$

where $\beta - k$ is the partition obtained by removing one part of size $k$ from $\beta$. See [33] for an inductive proof of (11.62).

In [40] Haiman includes a discussion of the connection of Theorem 11.3.20 to $q$-Lagrange inversion formulas of Andrews, Garsia, and Gessel [3, 23, 29]. Further background on these formulas is contained in [61]. Garsia and Haiman used $q$-Lagrange inversion to obtain the interesting identity

$$\nabla e_n|_{t=1} = \sum_{\pi \in L^+_{n,n}} q^{\mathrm{area}(\pi)} e_{\beta(\pi)} \tag{11.63}$$

for the $t=1$ case of the Frobenius series.

Garsia and Haiman were also able to obtain the $t = 1/q$ case of $\mathcal{F}(DH_n;q,t)$, which can be expressed as follows.

**Theorem 11.3.23** *The following holds:*

$$q^{\binom{n}{2}} \mathcal{F}(DH_n;q,1/q) = \frac{1}{[n+1]} e_n(XY), \tag{11.64}$$

*where $Y = \{1, q, \ldots, q^n\}$. Equivalently, by the Cauchy identity (11.34),*

$$\left\langle q^{\binom{n}{2}} \mathcal{F}(DH_n;q,1/q), s_\lambda \right\rangle = \frac{1}{[n+1]} s_{\lambda'}(1, q, \ldots, q^n). \tag{11.65}$$

Note that by Theorem 11.2.20, the special case $\lambda = 1^n$ of (11.65) reduces to (11.45), the formula for MacMahon's maj-statistic $q$-Catalan.

**Problem 11.3.24** *Find a $q,t$-version of the Lagrange inversion formula that will yield an identity for $\mathcal{F}(DH_n;q,t)$, and that reduces to Theorem 11.3.21 when $t=1$ and incorporates (11.64).*

## 11.4 Parking functions and the Hilbert series

### 11.4.1 Extension of the dinv statistic

Another beautiful corollary of Haiman's formula for $\mathcal{F}(DH_n;q,t)$ is that the dimension of $DH_n$ equals $(n+1)^{n-1}$, which is the number of **parking functions** on $n$ cars.

See the chapter in this volume on parking functions for more information on these important combinatorial objects. In this chapter we will view parking functions geometrically, as a Dyck path $\pi$ together with a placement of the numbers, or "cars," 1 through $n$ in the squares just to the right of $N$ steps of $\pi$, with strict decrease down columns. See Figure 11.10.

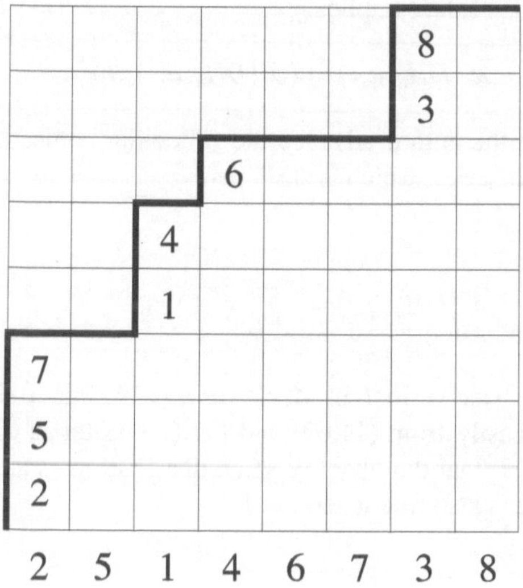


**Figure 11.10**
A parking function $P$.

We now describe an extension of the dinv statistic to parking functions. Let $\mathscr{P}_n$ denote the set of all parking functions on $n$ cars. Given $P \in \mathscr{P}_n$ with associated Dyck path $\pi = \pi(P)$, if car $i$ is in row $j$ we say occupant$(j) = i$. Let dinv$(P)$ be the number of pairs $(i, j)$, $1 \le i < j \le n$ such that

$$\begin{aligned}\operatorname{dinv}(P) = &|\{(i,j) : 1 \le i < j \le n, \quad a_i = a_j, \text{ and occupant}(i) < \text{occupant}(j)\}| \\ &+ |\{(i,j) : 1 \le i < j \le n, \quad a_i = a_j + 1, \text{ and occupant}(i) > \text{occupant}(j)\}|.\end{aligned}$$

Thus dinv$(P)$ is the number of pairs of rows of $P$ of the same length, with the row above containing the larger car, or which differ by one in length, with the longer row below the shorter, and the longer row containing the larger car. For example, for the parking function in Figure 11.10, the inversion pairs $(i, j)$ are $(1,7)$, $(2,7)$, $(2,8)$, $(3,4)$, $(4,8)$ and $(5,6)$, so dinv$(P) = 6$.

We define area$(P) =$ area$(\pi)$, and also define the **reading word** of $P$, denoted read$(P)$, to be the permutation obtained by reading the cars along diagonals in a southwest direction, starting with the diagonal farthest from the line $y = x$, then working inwards. For example, the parking function in Figure 11.10 has area 9 reading word 64781532.

**Remark 11.4.1** Note that $\text{read}(P) = n\cdots 21$ if and only if $\text{dinv}(P) = \text{dinv}(\pi)$. We call this parking function the Maxdinv parking function for $\pi$, which we denote by $\text{Maxdinv}(\pi)$.

Recall that Remark 11.2.29 implies

$$\mathcal{H}(DH_n;q,t) = \langle \mathcal{F}(DH_n;q,t), h_{1^n} \rangle . \tag{11.66}$$

In [38] N. Loehr and the author advance the following conjectured combinatorial formula for the Hilbert series, which is still open.

**Conjecture 11.4.2**

$$\mathcal{H}(DH_n;q,t) = \sum_{P\in\mathcal{P}_n} q^{dinv(P)} t^{area(P)}. \tag{11.67}$$

Conjecture 11.4.2 has been verified in Maple for $n \leq 11$. The truth of the conjecture when $q = 1$ follows easily from (11.63) and (11.66). Later in this section we will show (Corollary 11.4.9) that dinv has the same distribution as area over $\mathcal{P}_n$, which implies the conjecture is also true when $t = 1$.

### 11.4.2 An explicit formula

Given $\tau \in S_n$, with descents at places $i_1 < i_2 < \ldots < i_k$, we call the first $i_1$ letters of $\tau$ the first run of $\tau$, the next $i_2 - i_1$ letters of $\tau$ the second run of $\tau$, …, and the last $n - i_k$ letters of $\tau$ the $(k+1)$th run of $\tau$. For example, the runs of 58246137 are 58, 246 and 137. It will prove convenient to call element 0 the $(k+2)$th run of $\tau$. Let $\text{cars}(\tau)$ denote the set of parking functions whose cars in rows of length 0 consist of the elements of the $(k+1)$th run of $\tau$ (in any order), whose cars in rows of length 1 consist of the elements of the $k$th run of $\tau$ (in any order), …, and whose cars in rows of length $k$ consist of the elements of the first run of $\tau$ (in any order). For example, the elements of $\text{cars}(31254)$ are listed in Figure 11.11.

Let $\tau$ be as above, and let $i$ be a given integer satisfying $1 \leq i \leq n$. If $\tau_i$ is in the $j$th run of $\tau$, we define $w_i(\tau)$ to be the number of elements in the $j$th run that are larger than $\tau_i$, plus the number of elements in the $(j+1)$th run that are smaller than $\tau_i$. For example, if $\tau = 385924617$, then the values of $w_1, w_2, \ldots, w_9$ are $1,1,3,3,3,2,1,2,1$.

**Theorem 11.4.3** *Given* $\tau \in S_n$,

$$\sum_{P\in cars(\tau)} q^{dinv(P)} t^{area(P)} = t^{maj(\tau)} \prod_{i=1}^{n} [w_i(\tau)]_q. \tag{11.68}$$

*Proof.* We will build up elements of $\text{cars}(\tau)$ by starting at the end of $\tau$, where elements go in rows of length 0, and add elements right to left. We define a partial

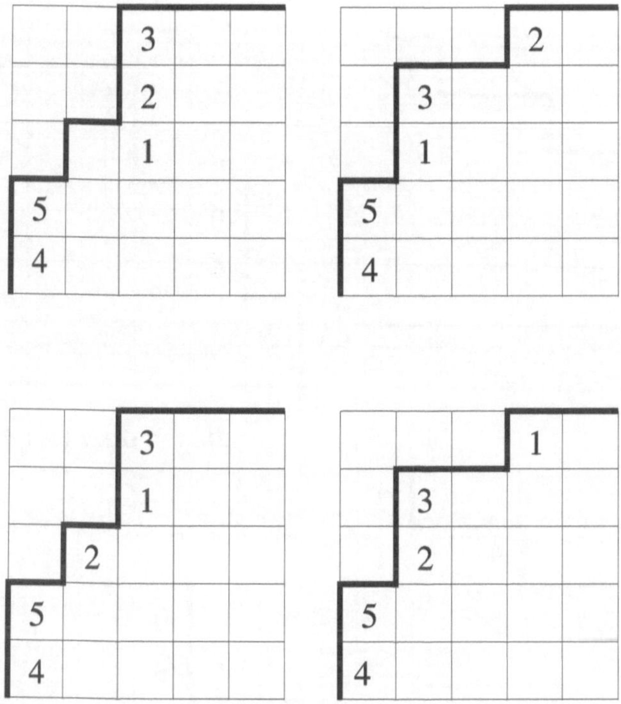


**Figure 11.11**
The elements of cars(31254).

parking function to be a Dyck path $\pi \in L^+_{m,m}$ for some $m$, together with a placement of $m$ distinct positive integers (not necessarily the integers 1 through $m$) to the right of the $N$ steps of $\pi$, with strict decrease down columns. Say $\tau = 385924617$ and we have just added car 9 to obtain a partial parking function $A$ with cars 1 and 7 in rows of length 0, cars 2, 4 and 6 in rows of length 1, and car 9 in a row of length 2, as in the upper-left grid of Figure 11.12. The rows with $*$'s to the right are rows above which we can insert a row of length 2 with car 5 in it and still have a partial parking function. Note the number of starred rows equals $w_3(\tau)$, and that in general $w_i(\tau)$ can be defined as the number of ways to insert a row containing car $\tau_i$ into a partial parking function containing cars $\tau_{i+1},\ldots,\tau_n$, in rows of the appropriate length, and still obtain a partial parking function.

Consider what happens to dinv as we insert the row with car 5 above a starred row to form $A'$. Pairs of rows that form inversions in $A$ will also form inversions in $A'$. Furthermore the rows of length 0 in $A$, or of length 1 with a car larger than 5, cannot form inversions with car 5 no matter where it is inserted. However, a starred row will form an inversion with car 5 if and only if car 5 is in a row below it. It follows that if we weight insertions by $q^{\text{dinv}}$, inserting car 5 at the various places will generate a

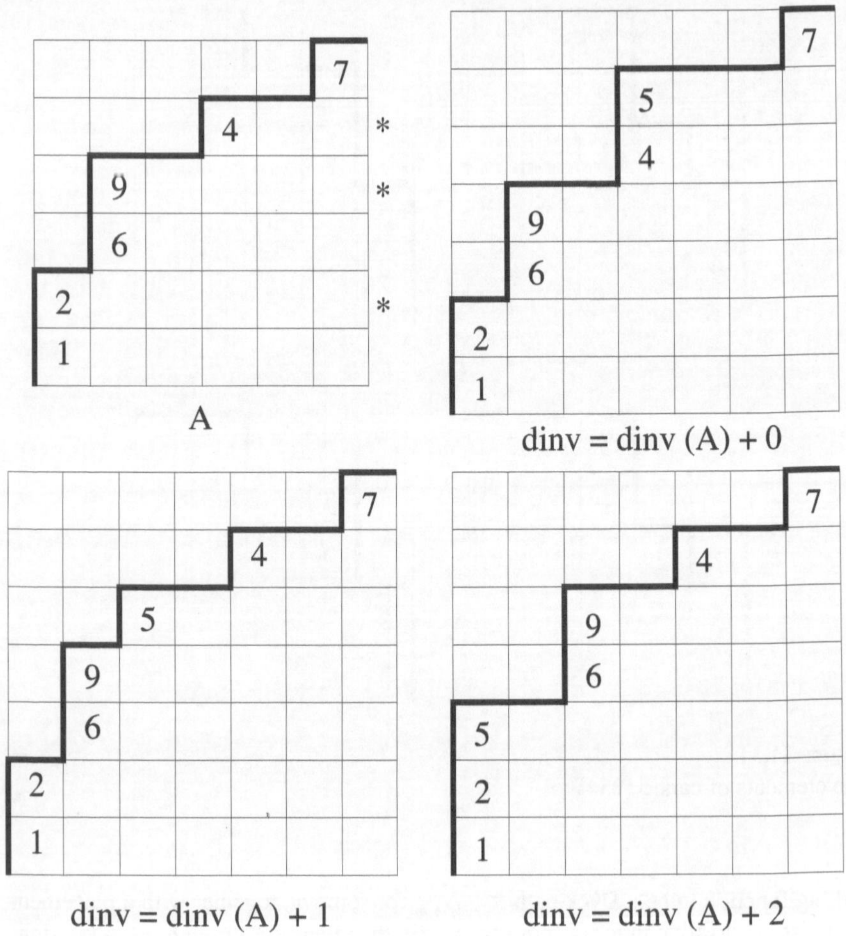


**Figure 11.12**
Partial parking functions occurring in the proof of Theorem 11.4.3.

factor of $[w_i(\tau)]$ times the weight of $A$, as in Figure 11.12. Finally note that for any path $\pi$ corresponding to an element of cars$(\tau)$, maj$(\tau) = $ area$(\pi)$. ■

By summing Theorem 11.4.3 over all $\tau \in S_n$ we get the following.

**Corollary 11.4.4** *The identity*

$$\sum_{P \in \mathscr{P}_n} q^{dinv(P)} t^{area(P)} = \sum_{\tau \in S_n} t^{maj(\tau)} \prod_{i=1}^{n} [w_i(\tau)]_q.$$

*holds.*

**Problem 11.4.5** *Prove*

$$\sum_{P \in \mathscr{P}_n} q^{dinv(P)} t^{area(P)} \tag{11.69}$$

*is symmetric in $q,t$.*

**Remark 11.4.6** Beyond merely proving the symmetry of (11.69), one could hope to find a bijective proof. It is interesting to note that by (11.68) the symmetry in $q,t$ when one variable equals 0 reduces to the fact that both inv and maj are Mahonian. Hence any bijective proof of symmetry may have to involve generalizing Foata's bijective transformation of maj into inv.

### 11.4.3 The statistic area′

By a **diagonal labeling** of a Dyck path $\pi \in L^+_{n,n}$ we mean a placement of the numbers 1 through $n$ in the squares on the main diagonal $y = x$ in such a way that for every consecutive $EN$ pair of steps of $\pi$, the number in the same column as the $E$ step is smaller than the number in the same row as the $N$ step. Let $\mathscr{A}_n$ denote the set of pairs $(A, \pi)$ where $A$ is a diagonal labeling of $\pi \in L^+_{n,n}$. Given such a pair $(A, \pi)$, we let $\text{area}'(A, \pi)$ denote the number of area squares $x$ of $\pi$ for which the number on the diagonal in the same column as $x$ is smaller than the number in the same row as $x$. Also define $\text{bounce}(A, \pi) = \text{bounce}(\pi)$.

The following result appears in [38].

**Theorem 11.4.7** *There is a bijection between $\mathscr{P}_n$ and $\mathscr{A}_n$ that sends $(dinv, area)$ to $(area', bounce)$.*

*Proof.* Given $P \in \mathscr{P}_n$ with associated path $\pi$, we begin to construct a pair $(A, \zeta(\pi))$ by first letting $\zeta(\pi)$ be the same path formed by the $\zeta$ map from the proof of Theorem 11.3.13. The length $\alpha_1$ of the first bounce of $\zeta$ is the number of rows of $\pi$ of length 0, etc.. Next place the cars that occur in $P$ in the rows of length 0 in the lowest $\alpha_1$ diagonal squares of $\zeta$, in such a way that the order in which they occur, reading top to bottom, in $P$ is the same as the order in which they occur, reading top to bottom, in $\zeta$. Then place the cars that occur in $P$ in the rows of length 1 in the next $\alpha_2$ diagonal squares of $\zeta$, in such a way that the order in which they occur, reading top to bottom, in $P$ is the same as the order in which they occur, reading top to bottom, in $\zeta$. Continue in this way until all the diagonal squares are filled, resulting in the pair $(A, \zeta)$. See Figure 11.13 for an example.

The properties of the $\zeta$ map immediately imply $\text{area}(\pi) = \text{bounce}(A, \zeta)$ and that $(A, \zeta) \in \mathscr{A}_n$. The reader will have no trouble showing that the equation $\text{dinv}(P) = \text{area}'(A, \zeta)$ is also implicit in the proof of Theorem 11.3.13. ∎

**Remark 11.4.8** Drew Armstrong [7] has found an interpretation for the area′ statistic, as well as the bounce and dinv statistics, in terms of hyperplane arrangements. See also [9].

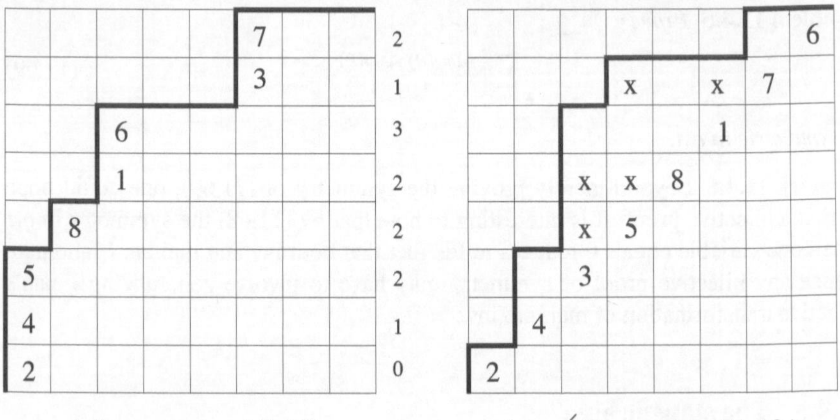


**Figure 11.13**
The map in the proof of Theorem 11.4.7. Squares contributing to area′ are marked with $x$'s.

### 11.4.4 The pmaj statistic

We now define a statistic on parking functions called pmaj, due to Loehr and Remmel [52, 50], which generalizes the bounce statistic. Given $P \in \mathscr{P}_n$, we define the **pmaj-parking order**, denoted $\beta(P)$, by the following procedure. Let $C_i = C_i(P)$ denote the set of cars in column $i$ of $P$, and let $\beta_1$ be the largest car in $C_1$. We begin by parking car $\beta_1$ in spot 1. Next we perform the "dragnet" operation, which takes all the cars in $C_1 \setminus \{\beta_1\}$ and combines them with $C_2$ to form $C_2'$. Let $\beta_2$ be the largest car in $C_2'$ that is smaller than $\beta_1$. If there is no such car, let $\beta_2$ be the largest car in $C_2'$. Park car $\beta_2$ in spot 2 and then iterate this procedure. Assuming we have just parked car $\beta_{i-1}$ in spot $i-1$, $3 \le i < n$, we let $C_i' = C_{i-1}' \setminus \{\beta_{i-1}\}$ and let $\beta_i$ be the largest car in $C_i'$ that is smaller than $\beta_{i-1}$, if any, while otherwise $\beta_i$ is the largest car in $C_i'$. For the example in Figure 11.14, we have $C_1 = \{5\}$, $C_2 = \{1,7\}$, $C_3 = \{\}$, etc. and $C_2' = \{1,7\}$, $C_3' = \{7\}$, $C_4' = \{2,4,6\}$, $C_5' = \{2,3,4\}$, etc., with $\beta = 51764328$.

Now let $\operatorname{rev}(\beta(P)) = (\beta_n, \beta_{n-1}, \ldots, \beta_1)$ and define $\operatorname{pmaj}(P) = \operatorname{maj}(\operatorname{rev}(\beta(P)))$. For example, for the parking function of Figure 11.14 we have $\operatorname{rev}(\beta) = 82346715$ and pmaj $= 1+6 = 7$. Given $\pi \in L_{n,n}^{+}$, it is easy to see that if $P$ is the parking function for $\pi$ obtained by placing car $i$ in row $i$ for $1 \le i \le n$, then $\operatorname{pmaj}(P) = \operatorname{bounce}(\pi)$. We call this parking function the **primary pmaj parking function** for $\pi$.

We now describe a bijection $\Gamma$ from $\mathscr{P}_n$ to $\mathscr{P}_n$ from [52] that sends (area, pmaj) $\to$ (dinv, area). The crucial observation behind it is this. Fix $\gamma \in S_n$ and consider the set of parking functions that satisfy $\operatorname{rev}(\beta(P)) = \gamma$. We can build up this set recursively by first forming a partial parking function consisting of car $\gamma_n$ in column 1. If $\gamma_{n-1} < \gamma_n$, then we can form a partial parking function consisting of two cars whose pmaj parking order is $\gamma_n\gamma_{n-1}$ in two ways. We can either have both cars $\gamma_n$

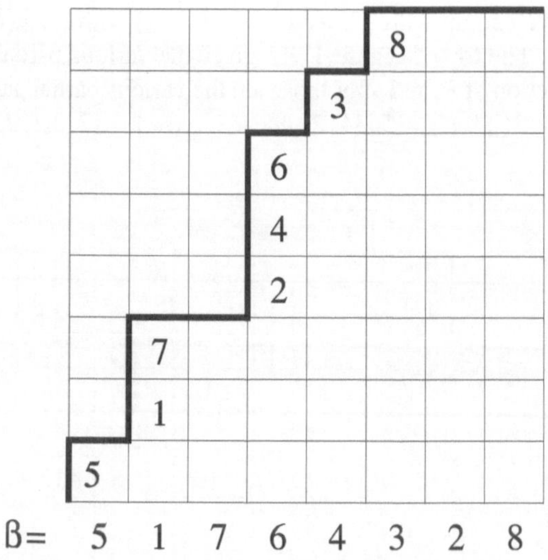


**Figure 11.14**
A parking function $P$ with pmaj parking order $\beta = 51764328$.

and $\gamma_{n-1}$ in column 1, or car $\gamma_n$ in column 1 and car $\gamma_{n-1}$ in column 2. If $\gamma_{n-1} > \gamma_n$, then we must have car $\gamma_n$ in column 1 and car $\gamma_{n-1}$ in column 2. In the case where $\gamma_{n-1} < \gamma_n$, there were two choices for columns to insert car $\gamma_{n-1}$ into, corresponding to the fact that $w_{n-1}(\gamma) = 2$. When $\gamma_{n-1} > \gamma_n$, there was only one choice for the column to insert $\gamma_{n-1}$ into, and correspondingly $w_{n-1}(\gamma) = 1$.

More generally, say we have a partial parking function consisting of cars in the set $\{\gamma_n, \ldots, \gamma_{i+1}\}$ whose pmaj parking order is $\gamma_n \cdots \gamma_{i+2}\gamma_{i+1}$. It is easy to see that the number of ways to insert car $\gamma_i$ into this so the new partial parking function has pmaj parking order $\gamma_n \cdots \gamma_{i+1}\gamma_i$ is exactly $w_i(\gamma)$. Furthermore, as you insert car $\gamma_i$ into columns $n-i+1, n-i, \ldots, n-i-w_i(\gamma)+2$ the area of the partial parking function increases by 1 each time. It follows that

$$\sum_{P \in \mathscr{P}_n} q^{\text{area}(P)} t^{\text{pmaj}(P)} = \sum_{\gamma \in S_n} t^{\text{maj}(\gamma)} \prod_{i=1}^{n-1} [w_i(\gamma)].$$

Moreover, we can identify the values of (area, pmaj) for individual parking functions by considering permutations $\gamma \in S_n$ and corresponding $n$-tuples $(u_1, \ldots, u_n)$ with $0 \leq u_i < w_i(\gamma)$ for $1 \leq i \leq n$. (Note $u_n$ always equals 0). Then $\text{maj}(\gamma) = \text{pmaj}(P)$, and $u_1 + \ldots + u_n = \text{area}(P)$. (For those familiar with the description of parking functions in terms of preference functions, we have $f(\beta_{n+1-i}) = n+1-i-u_i$ for $1 \leq i \leq n$.)

Now given such a pair $\gamma \in S_n$ and corresponding $n$-tuple $(u_1, \ldots, u_n)$, from the proof of Theorem 11.4.3 we can build up a parking function $Q$ recursively by inserting cars $\gamma_n, \gamma_{n-1}, \ldots$ one at a time, where for each $j$ the insertion of car $\gamma_j$ adds $u_j$

to $\mathrm{dinv}(Q)$. Thus we end up with a bijection $\Gamma : P \mapsto Q$ with $(\mathrm{area}(P), \mathrm{pmaj}(P)) = (\mathrm{dinv}(Q), \mathrm{area}(Q))$. The top of Figure 11.15 gives the various partial parking functions in the construction of $P$, and after those are the various partial parking functions in the construction of $Q$, for $\gamma = 563412$ and $(u_1, \ldots, u_6) = (2,0,1,0,1,0)$.

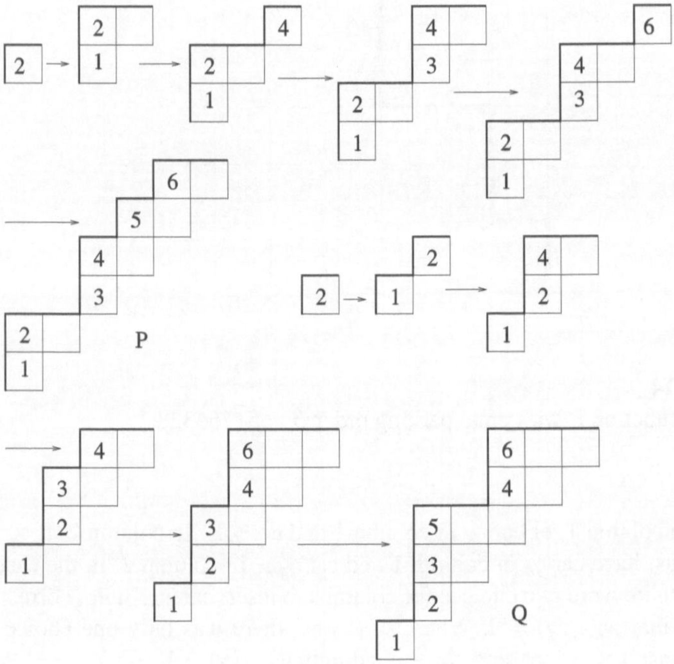


**Figure 11.15**
The recursive construction of the $P$ and $Q$ parking functions in the $\Gamma$ correspondence for $\gamma = 563412$ and $u = (2,0,1,0,1,0)$.

**Corollary 11.4.9** *The marginal distributions of pmaj, area, and dinv over $\mathscr{P}_n$ are all the same, i.e.,*

$$\sum_{P \in \mathscr{P}_n} q^{pmaj(P)} = \sum_{P \in \mathscr{P}_n} q^{area(P)} = \sum_{P \in \mathscr{P}_n} q^{dinv(P)}.$$

**Exercise 11.4.10** *Notice that in Figure 11.15, the final correspondence is between parking functions that equal the primary pmaj and Maxdinv parking functions for their respective paths. Show that this is true in general, i.e., that when P equals the primary pmaj parking function for $\pi$ then the bijection $\Gamma$ reduces to the inverse of the bijection $\zeta$ from the proof of Theorem 11.3.13.*

### 11.4.5 The cyclic-shift operation

Given $S \subseteq \{1,\ldots,n\}$, let $\mathscr{P}_{n,S}$ denote the set of parking functions for which $C_1(P) = S$. If $x \in \{1,2,\ldots,n\}$, define

$$CYC_n(x) = \begin{cases} x+1 & \text{if } x < n \\ 1 & \text{if } x = n. \end{cases}$$

For any set $S \subseteq \{1,2,\ldots,n\}$, let $CYC_n(S) = \{CYC_n(x) : x \in S\}$. Assume $S = \{s_1 < s_2 < \cdots < s_k\}$ with $s_k < n$. Given $P \in \mathscr{P}_{n,S}$, define the cyclic-shift of $P$, denoted $CYC_n(P)$, to be the parking function obtained by replacing $C_i$, the cars in column $i$ of $P$, with $CYC_n(C_i)$, for each $1 \leq i \leq n$. Note that the column of $P$ containing car $n$ will have to be sorted, with car 1 moved to the bottom of the column. The map $CYC_n(P)$ is undefined if car $n$ is in column 1. See the top portion of Figure 11.16 for an example.

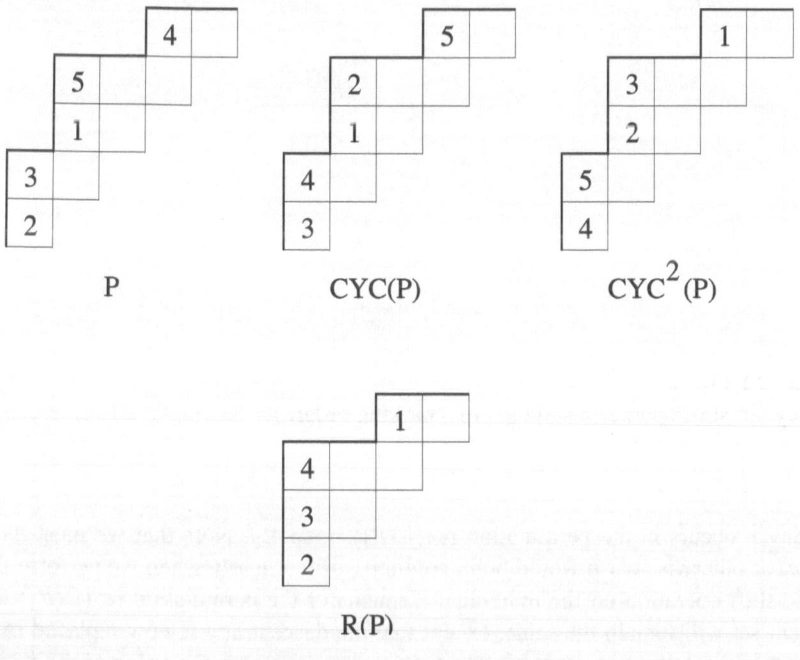


**Figure 11.16**
The map $R(P)$.

**Proposition 11.4.11** [50] *Suppose $P \in \mathscr{P}_{n,S}$ with $S = \{s_1 < s_2 < \cdots < s_k\}$, $s_k < n$. Then*

$$pmaj(P) = pmaj(CYC(P)) + 1. \tag{11.70}$$

*Proof.* Imagine adding a second coordinate to each car, with car $i$ initially represented by $(i,i)$. If we list the second coordinates of each car as they occur in the pmaj parking order for $P$, by definition we get the sequence $\beta_1(P)\beta_2(P)\cdots\beta_n(P)$. We now perform the cyclic-shift operation to obtain $CYC(P)$, but when doing so we operate only on the first coordinates of each car, leaving the second coordinates unchanged. The reader will have no trouble verifying that if we now list the second coordinates of each car as they occur in the pmaj parking order for $CYC(P)$, we again get the sequence $\beta_1(P)\beta_2(P)\cdots\beta_n(P)$. It follows that the pmaj parking order of $CYC(P)$ can be obtained by starting with the pmaj parking order $\beta$ of $P$ and performing the cyclic-shift operation on each element of $\beta$ individually. See Figure 11.17.

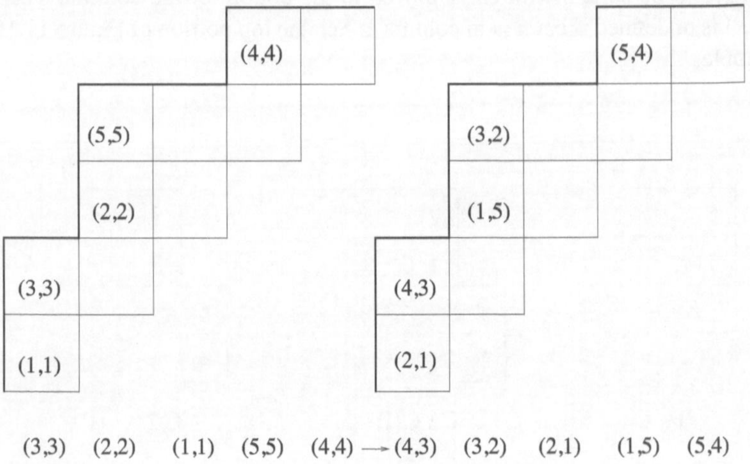


**Figure 11.17**
The cyclic-shift operation and the pmaj parking order.

Say $n$ occurs in the permutation $\mathrm{rev}(\beta(P))$ in spot $j$. Note that we must have $j < n$, or otherwise car $n$ would be in column 1 of $P$. Clearly when we perform the cyclic-shift operation on the individual elements of the permutation $\mathrm{rev}(\beta(P))$ the descent set will remain the same, except that the descent at $j$ is now replaced by a descent at $j-1$ if $j > 1$, or is removed if $j = 1$. In any case the value of the major index of $\mathrm{rev}(\beta(P))$ is decreased by 1. ∎

Using Proposition 11.4.11, Loehr derives the following recurrence.

**Theorem 11.4.12** *Let $n \geq 1$ and $S = \{s_1 < \cdots < s_k\} \subseteq \{1,\ldots,n\}$. Set*

$$P_{n,S}(q,t) = \sum_{P \in \mathcal{P}_{n,S}} q^{area(P)} t^{pmaj(P)}.$$

*Then*

$$P_{n,S}(q,t) = q^{k-1}t^{n-s_k} \sum_{T \subseteq \{1,\ldots,n\} \setminus S} P_{n-1,CYC_n^{n-s_k}(S \cup T/\{s_k\})}(q,t), \tag{11.71}$$

*with the initial conditions* $P_{n,\emptyset}(q,t) = 0$ *for all* $n$ *and* $P_{1,\{1\}}(q,t) = 1$.

*Proof.* For $P \in \mathscr{P}_{n,S}$, let $Q = CYC_n^{n-s_k}(P)$. Then

$$\begin{aligned} \mathrm{pmaj}(Q) + n - s_k &= \mathrm{pmaj}(P) \\ \mathrm{area}(Q) &= \mathrm{area}(P). \end{aligned}$$

Since car $n$ is in the first column of $Q$, in the pmaj parking order for $Q$ car $n$ is in spot 1. By definition, the dragnet operation will then combine the remaining cars in column 1 of $Q$ with the cars in column 2 of $Q$. Now car $n$ being in spot 1 translates into car $n$ being at the end of $\mathrm{rev}(\beta(Q))$, which means $n-1$ is not in the descent set of $\mathrm{rev}(\beta(Q))$. Thus if we define $R(P)$ to be the element of $\mathscr{P}_{n-1}$ obtained by parking car $n$ in spot 1, performing the dragnet, then truncating column 1 and spot 1 as in Figure 11.16, we have

$$\mathrm{pmaj}(R(P)) = \mathrm{pmaj}(P) - (n - s_k).$$

Furthermore, performing the dragnet leaves the number of area cells in columns $2,3,\ldots,n$ of $Q$ unchanged but eliminates the $k-1$ area cells in column 1 of $Q$. Thus

$$\mathrm{area}(R(P)) = \mathrm{area}(P) - k + 1$$

and the recursion now follows easily. ∎

Loehr also derives the following compact formula for $P_{n,S}$ when $t = 1/q$.

**Theorem 11.4.13** *For* $n \geq 0$ *and* $S = \{s_1 < \cdots < s_k\} \subseteq \{1,\ldots,n\}$,

$$q^{\binom{n}{2}} P_{n,S}(1/q,q) = q^{n-k}[n]^{n-k-1} \sum_{x \in S} q^{n-x}. \tag{11.72}$$

*Proof.* Our proof is, for the most part, taken from [50]. If $k = n$,

$$P_{n,S}(1/q,q) = P_{n,\{1,\ldots,n\}}(1/q,q) = q^{-\binom{n}{2}},$$

while the right-hand side of (11.72) equals $[n]^{-1}[n] = 1$. Thus (11.72) holds for $k = n$. It also holds trivially for $n = 0$ and $n = 1$. So assume $n > 1$ and $0 < k < n$. From (11.71),

$$\begin{aligned} P_{n,S}(1/q,q) &= q^{n+1-s_k-k} \sum_{T \subseteq \{1,\ldots,n\} \setminus S} P_{n-1,CYC_n^{n-s_k}(S \cup T \setminus \{s_k\})}(1/q,q) \qquad (11.73) \\ &= q^{n+1-s_k-k} \sum_{j=0}^{n-k} \sum_{\substack{T \subseteq \{1,\ldots,n\} \setminus S \\ |T| = j}} P_{n-1,CYC_n^{n-s_k}(S \cup T \setminus \{s_k\})}(1/q,q). \end{aligned}$$

The summand when $j = n-k$ equals

$$P_{n-1,\{1,\dots,n-1\}}(1/q,q) = q^{-\binom{n-1}{2}}.$$

For $0 \le j < n-k$, by induction the summand equals

$$q^{n-1-(j+k-1)-\binom{n-1}{2}}[n-1]^{n-1-(j+k-1)-1} \sum_{x \in CYC_n^{n-s_k}(S\cup T\setminus\{s_k\})} q^{n-1-x}, \qquad (11.74)$$

since $j+k-1 = |CYC_n^{n-s_k}(S\cup T\setminus\{s_k\})|$. Plugging (11.74) into (11.73) and reversing summation we now have

$$q^{-2n+s_k+k}q^{\binom{n}{2}}P_{n,S}(1/q,q) = 1 + \sum_{j=0}^{n-k-1} q^{n-k-j}[n-1]^{n-k-j-1}\sum_{x=1}^{n-1} q^{n-1-x} \qquad (11.75)$$
$$\times \sum_T \chi(T \subseteq \{1,\dots,n\}\setminus S, |T| = j, CYC_n^{s_k-n}(x) \in S\cup T\setminus\{s_k\}).$$

To compute the inner sum over $T$ above, we consider two cases.

1. $x = n-(s_k-s_i)$ for some $i \le k$. Since $x < n$, this implies $i < k$, and since $CYC_n^{s_k-n}(x) = s_i$, we have $CYC_n^{s_k-n}(x) \in S\cup T\setminus\{s_k\}$. Thus the inner sum above equals the number of $j$-element subsets of $\{1,\dots,n\}\setminus S$, or $\binom{n-k}{j}$.
2. $x \ne n-(s_k-s_i)$ for all $i \le k$. By Exercise 11.4.14 below, the inner sum over $T$ in (11.75) equals $\binom{n-k-1}{j-1}$.

Applying the above analysis to (11.75) we now have

$$q^{-2n+s_k+k}q^{\binom{n}{2}}P_{n,S}(1/q,q) = 1 + \sum_{j=0}^{n-k-1} q^{n-k-j}[n-1]^{n-k-j-1}$$
$$\times \sum_{x \text{ satisfies (1)}} \left[\binom{n-k-1}{j} + \binom{n-k-1}{j-1}\right] q^{n-1-x}$$
$$+ \sum_{j=0}^{n-k-1} q^{n-k-j}[n-1]^{n-k-j-1} \sum_{x \text{ satisfies (2)}} \binom{n-k-1}{j-1} q^{n-1-x}.$$

Now $x$ satisfies (1) if and only if $n-1-x = s_k - s_i - 1$ for some $i < k$, and so

$$q^{-2n+s_k+k}q^{\binom{n}{2}}P_{n,S}(1/q,q) = 1 + \sum_{j=0}^{n-k-1} q^{n-k-j}[n-1]^{n-k-j}\binom{n-k-1}{j-1}$$
$$+ \sum_{j=0}^{n-k-1} q^{n-k-j-1}[n-1]^{n-k-j-1}\binom{n-k-1}{j}\sum_{i=1}^{k-1} q^{s_k-s_i}$$
$$= \sum_{m=0\,(m=j-1)}^{n-k-1} \binom{n-k-1}{m}(q[n-1])^{n-k-m-1}$$

$$+\sum_{i=1}^{k-1} q^{s_k-s_i} \sum_{j=0}^{n-k-1} (q[n-1])^{n-k-j-1}\binom{n-k-1}{j}$$

$$= (1+q[n-1])^{n-k-1} + \sum_{i=1}^{k-1} q^{s_k-s_i}(1+q[n-1])^{n-k-1}$$

$$= [n]^{n-k-1}(1+\sum_{i=1}^{k-1} q^{s_k-s_i}).$$

Thus

$$q^{\binom{n}{2}} P_{n,S}(1/q,q) = q^{n-k}[n]^{n-k-1} \sum_{x\in S} q^{s_k-x+n-s_k}.$$

■

**Exercise 11.4.14** *Show that if $x \neq n-(s_k-s_i)$ for all $i \leq k$, the inner sum over $T$ in (11.75) equals $\binom{n-k-1}{j-1}$.*

As a corollary of his formula for $\mathscr{F}(DH_n;q,t)$, Haiman proves a conjecture he attributes in [40] to Stanley, namely that

$$q^{\binom{n}{2}} \mathscr{H}(DH_n;1/q,q) = [n+1]^{n-1}. \tag{11.76}$$

Theorem 11.4.13 and (11.76) together imply Conjecture 11.4.2 is true when $t = q$ and $q = 1/q$. To see why, first observe that

$$P_{n+1,\{n+1\}}(q,t) = \sum_{P\in\mathscr{P}_n} q^{\text{area}(P)} t^{\text{pmaj}(P)}.$$

Hence by Theorem 11.4.13,

$$q^{\binom{n}{2}} \sum_{P\in\mathscr{P}_n} q^{-\text{area}(P)} q^{\text{pmaj}(P)} = q^{\binom{n}{2}} P_{n+1,\{n+1\}}(1/q,q)$$

$$= q^{-n} q^{\binom{n+1}{2}} P_{n+1,\{n+1\}}(1/q,q)$$

$$= q^{-n} q^{n}[n+1]^{n-1} q^{0} = [n+1]^{n-1}.$$

The main impediment to proving Conjecture 11.4.2 seems to be the lack of a recursive decomposition of the Hilbert series along the lines of (11.43).

### 11.4.6 Tesler matrices

For an $n \times n$ upper triangular matrix, we define the **$j$th hook sum**, where $1 \leq j \leq n$, to be the sum of all the entries in the $j$th row of the matrix, minus the sum of all the entries in the $j$th column strictly above the diagonal. A **Tesler matrix** of order $n$ is an $n \times n$ upper-triangular matrix of nonnegative integers, such that all the hook sums equal 1. Let Tes$(n)$ denote the set of Tesler matrices of order $n$. For example, the elements of Tes$(3)$ are

$$\begin{bmatrix}1&0&0\\0&1&0\\0&0&1\end{bmatrix},\begin{bmatrix}1&0&0\\0&0&1\\0&0&2\end{bmatrix},\begin{bmatrix}0&1&0\\0&2&0\\0&0&1\end{bmatrix},\begin{bmatrix}0&1&0\\0&1&1\\0&0&2\end{bmatrix}, \tag{11.77}$$

$$\begin{bmatrix}0&1&0\\0&0&2\\0&0&3\end{bmatrix},\begin{bmatrix}0&0&1\\0&1&0\\0&0&2\end{bmatrix},\begin{bmatrix}0&0&1\\0&0&1\\0&0&3\end{bmatrix}.$$

Let $[k]_{q,t} = (t^k - q^k)/(t-q)$ denote the $q,t$-analog of the integer $k$, and recall that $M = (1-q)(1-t)$. To each Tesler matrix $C$ we associate the weight

$$\mathrm{wt}(C) = (-M)^{\mathrm{pos}(C)-n} \prod_{c_{ij}>0} [c_{ij}]_{q,t},$$

where pos$(C)$ is the number of positive entries in $C$. For example, the weight of

$$\begin{bmatrix}0&1&0\\0&1&1\\0&0&2\end{bmatrix}$$

is $(t+q)(-M) = -(t+q)(1-q)(1-t)$.

Using the theory of Macdonald polynomials and (11.40), Haglund [35] proved the following result, which gives a formula for $\mathscr{H}(DH_n;q,t)$ in terms of Tesler matrices.

**Theorem 11.4.15** *[35] The following equality holds:*

$$\mathscr{H}(DH_n;q,t) = \sum_{C\in Tes(n)} wt(C). \tag{11.78}$$

**Example 11.4.16** *When $n=3$, the terms in (11.77), with weights, give*

$$\mathscr{H}(DH_3;q,t) = 1+(t+q)+(t+q)-(1-q)(1-t)(t+q)+$$
$$(t+q)(t^2+tq+q^2)+(t+q)+(t^2+tq+q^2).$$

Note that Formula (11.78) is clearly a polynomial, and clearly symmetric in $q,t$. It gives a possible way of attacking Conjecture 11.4.2 without the use of symmetric function theory, since in principle one could figure out how to cancel the negative terms in the right-hand side of (11.78), leaving a positive expression as in Conjecture 11.4.2. P. Levande [47, 48] has shown how to do this cancellation when $t=1$ and when $t=0$.

# 11.5 The $q,t$-Schröder polynomial

## 11.5.1 The Schröder bounce and area statistics

In this section we develop the theory of the $q,t$-Schröder polynomial, which gives a combinatorial interpretation, in terms of statistics on Schröder lattice paths, for the coefficient of a Schur hook shape in $\mathscr{F}(DH_n;q,t)$. A **Schröder path** is a lattice path from $(0,0)$ to $(n,n)$ consisting of $N(0,1)$, $E(1,0)$ and diagonal $D(1,1)$ steps that never goes below the line $y=x$. We let $L^+_{n,n,d}$ denote the set of Schröder lattice paths consisting of $n-d$ $N$ steps, $n-d$ $E$ steps, and $d$ $D$ steps. We refer to a triangle whose vertex set is of the form $\{(i,j),(i+1,j),(i+1,j+1)\}$ for some $(i,j)$ as a **lower triangle**, and define the **area** of a Schröder path $\pi$ to be the number of lower triangles below $\pi$ and above the line $y=x$. Note that if $\pi$ has no $D$ steps, then the Schröder definition of area agrees with the definition of the area of a Catalan path. We let $a_i(\pi)$ denote the length of the $i$th row, i.e., the number of lower triangles between the path and the diagonal in the $i$th row from the bottom of $\pi$, so $\text{area}(\pi)=\sum_{i=1}^n a_i(\pi)$.

Given $\pi \in L^+_{n,n,d}$, let $\sigma(\pi)$ be the word of 0's, 1's and 2's obtained in the following way. Initialize $\sigma$ to be the empty string, then start at $(0,0)$ and travel along $\pi$ to $(n,n)$, adding a 0, 1, or 2 to the end of $\sigma(\pi)$ when we encounter a $N$, $D$, or $E$ step, respectively, of $\pi$. (If $\pi$ is a Dyck path, then this definition of $\sigma(\pi)$ is the same as the previous definition from Section 11.2, except that we end up with a word of 0's and 2's instead of 0's and 1's. Since all our applications involving $\sigma(\pi)$ depend only on the relative order of the elements of $\sigma(\pi)$, this change is only a superficial one. We define the statistic bounce$(\pi)$ by means of the following algorithm.

**Algorithm 11.5.1**

1. *First remove all D steps from $\pi$, and collapse to obtain a Catalan path $\Gamma(\pi)$. More precisely, let $\Gamma(\pi)$ be the Catalan path for which $\sigma(\Gamma(\pi))$ equals $\sigma(\pi)$ with all 1's removed. Recall the ith peak of $\Gamma(\pi)$ is the lattice point where the bounce path for $\Gamma(\pi)$ switches direction from N to E for the ith time. The lattice point at the beginning of the corresponding E step of $\pi$ is called the ith peak of $\pi$.*

2. *For each D step x of $\pi$, let nump(x) be the number of peaks of $\pi$ below x. Then define*

$$bounce(\pi) = bounce(\Gamma(\pi)) + \sum_x nump(x),$$

*where the sum is over all D steps of $\pi$. For example, if $\pi$ is the Schröder path on the left in Figure 11.18, with $\Gamma(\pi)$ on the right, then bounce$(\pi) = (3+1)+(0+1+1+2) = 8$. Note that if $\pi$ has no D steps, this definition of bounce$(\pi)$ agrees with the previous definition from Section 11.3.*

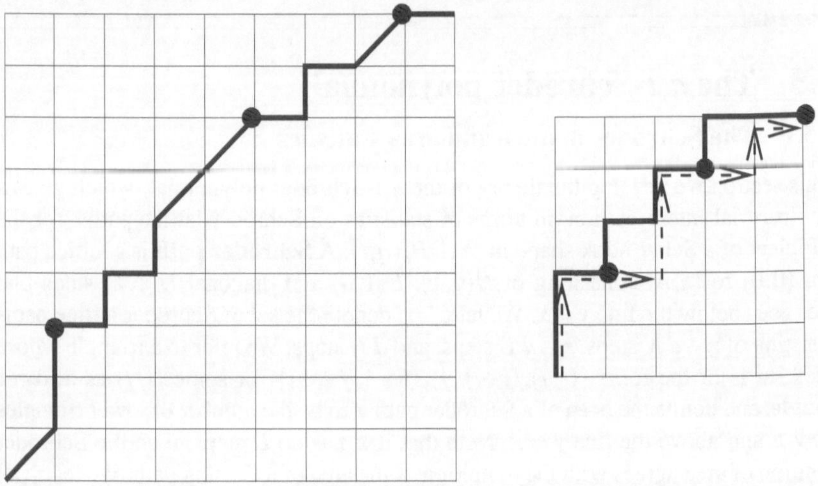

**Figure 11.18**
On the left, a Schröder path $\pi$ with the peaks marked by large dots. On the right is $\Gamma(\pi)$ and its bounce path and peaks.

We call the vector whose $i$th coordinate is the length of the $i$th bounce step of $\Gamma(\pi)$ the **bounce vector** of $\pi$. Say $\Gamma(\pi)$ has $b$ bounce steps, and call the set of rows of $\pi$ between peaks $i$ and $i+1$ **section** $i$ of $\pi$ for $1 \le i < b$. In addition we call **section** 0 the set of rows below peak 1, and **section** $b$ the set of rows above peak $b$. If $\pi$ has $\beta_i$ $D$ steps in section $i$, $0 \le i \le b$, we refer to $(\beta_0, \beta_1, \ldots, \beta_b)$ as the **shift vector** of $\pi$.

For example, the path on the left in Figure 11.18 has bounce vector $(2,2,1)$ and shift vector $(1,2,1,0)$. We refer to the portion of $\sigma(\pi)$ corresponding to the $i$th section of $\pi$ as the $i$th section of $\sigma(\pi)$.

Given $n, d \in \mathbb{N}$, we define the **$q,t$-Schröder polynomial** $S_{n,d}(q,t)$ as follows.

$$S_{n,d}(q,t) = \sum_{\pi \in L^+_{n,n,d}} q^{\mathrm{area}(\pi)} t^{\mathrm{bounce}(\pi)}.$$

These polynomials were introduced by Egge, Haglund, Killpatrick and Kremer [18]. They conjectured the following result, which was subsequently proved by Haglund using plethystic results involving Macdonald polynomials [34].

**Theorem 11.5.2** *For all* $0 \le d \le n$,

$$S_{n,d}(q,t) = \langle \mathscr{F}(DH_n; q,t), e_{n-d} h_d \rangle .$$

Since

$$S_{n,0}(q,t) = F_n(q,t) = C_n(q,t),$$

the $d = 0$ case of Theorem 11.5.2 reduces to Theorem 11.3.3.

Let $\tilde{L}^+_{n,n,d}$ denote the set of paths $\pi$ which are in $L^+_{n,n,d}$ and also have no $D$ step above the highest $N$ step, i.e., no 1's in $\sigma(\pi)$ after the rightmost 0. Define

$$\tilde{S}_{n,d}(q,t) = \sum_{\pi \in \tilde{L}^+_{n,n,d}} q^{\text{area}(\pi)} t^{\text{bounce}(\pi)}.$$

Then the following holds.

**Theorem 11.5.3** *Theorem 11.5.2 is equivalent to the statement that for all* $0 \le d \le n-1$,

$$\tilde{S}_{n,d}(q,t) = \left\langle \mathscr{F}(DH_n;q,t), s_{d+1,1^{n-d-1}} \right\rangle.$$

*Proof.* Given $\pi \in \tilde{L}^+_{n,n,d}$, we can map $\pi$ to a path $\alpha(\pi) \in L^+_{n,n,d+1}$ by replacing the highest $N$ step and the following $E$ step of $\pi$ by a $D$ step. By Exercise 11.5.4 below, this map leaves area and bounce unchanged. Conversely, if $\alpha \in L^+_{n,n,d}$ has a $D$ step above the highest $N$ step, we can map it to a path $\pi \in \tilde{L}^+_{n,n,d-1}$ in an area and bounce preserving fashion by changing the highest $D$ step to a $NE$ pair. It follows that for $1 \le d \le n$,

$$\begin{aligned} S_{n,d}(q,t) &= \sum_{\pi \in \tilde{L}^+_{n,n,d}} q^{\text{area}(\pi)} t^{\text{bounce}(\pi)} + \sum_{\pi \in \tilde{L}^+_{n,n,d-1}} q^{\text{area}(\pi)} t^{\text{bounce}(\pi)} \\ &= \tilde{S}_{n,d}(q,t) + \tilde{S}_{n,d-1}(q,t). \end{aligned}$$

Since $S_{n,0}(q,t) = \tilde{S}_{n,0}(q,t)$, $e_{n-d}h_d = s_{d+1,1^{n-d-1}} + s_{d,1^{n-d}}$ for $0 < d \le n-1$, and $S_{n,n}(q,t) = 1 = \tilde{S}_{n,n-1}(q,t)$, the result follows by a simple inductive argument. ∎

**Exercise 11.5.4** *Given* $\pi$ *and* $\alpha(\pi)$ *in the proof of Theorem 11.5.3, show that*

$$\begin{aligned} area(\pi) &= area(\alpha(\pi)) \\ bounce(\pi) &= bounce(\alpha(\pi)). \end{aligned}$$

Define $q,t$-analogues of the **big Schröder numbers** $r_n$ and **little Schröder numbers** $\tilde{r}_n$ as follows.

$$\begin{aligned} r_n(q,t) &= \sum_{d=0}^{n} S_{n,d}(q,t) \\ \tilde{r}_n(q,t) &= \sum_{d=0}^{n-1} \tilde{S}_{n,d}(q,t). \end{aligned}$$

The numbers $r_n(1,1)$ count the total number of Schröder paths from $(0,0)$ to $(n,n)$. The $\tilde{r}_n(1,1)$ are known to count many different objects [58, p.178], including the number of Schröder paths from $(0,0)$ to $(n,n)$ that have no $D$ steps on the line $y = x$. From our comments above we have the simple identity $r_n(q,t) = 2\tilde{r}_n(q,t)$, and using

Haiman's formula for $\mathscr{F}(DH_n;q,t)$ we get the polynomial identities

$$\sum_{d=0}^{n} w^d S_{n,d}(q,t) = \sum_{\mu\vdash n} \frac{T_\mu \prod_{x\in\mu}(w+q^{a'}t^{l'})M\Pi_\mu B_\mu}{w_\mu}$$

$$\sum_{d=0}^{n-1} w^d \tilde{S}_{n,d}(q,t) = \sum_{\mu\vdash n} \frac{T_\mu \prod_{x\in\mu, x\neq(0,0)}(w+q^{a'}t^{l'})M\Pi_\mu B_\mu}{w_\mu}. \qquad (11.79)$$

An interesting special case of (11.79) is

$$\tilde{r}_{n,d}(q,t) = \sum_{\mu\vdash n} \frac{T_\mu \prod'_{x\in\mu}(1-q^{2a'}t^{2l'})MB_\mu}{w_\mu}.$$

### 11.5.2 Recurrences and explicit formulae

We begin with a useful lemma about area and Schröder paths.

**Lemma 11.5.5 (The "boundary lemma")** *Given $a,b,c \in \mathbb{N}$, let boundary$(a,b,c)$ be the path whose $\sigma$ word is $2^c1^b0^a$. Then*

$$\sum_{\pi} q^{area'(\pi)} = \begin{bmatrix} a+b+c \\ a,b,c \end{bmatrix}_q,$$

*where the sum is over all paths $\pi$ from $(0,0)$ to $(c+b,a+b)$ consisting of a N steps, b D steps and c E steps, and area$'(\pi)$ is the number of lower triangles between $\pi$ and boundary$(a,b,c)$.*

*Proof.* Given $\pi$ as above, we claim the number of coinversions of $\sigma(\pi)$ equals area$'(\pi)$. To see why, start with $\pi$ as in Figure 11.19, and note that when consecutive *ND*, *DE*, or *NE* steps are interchanged, area$'$ decreases by 1. Thus area$'(\pi)$ equals the number of such interchanges needed to transform $\pi$ into boundary$(a,b,c)$, or equivalently to transform $\sigma(\pi)$ into $2^c1^b0^a$. But this is just coinv$(\sigma(\pi))$. Thus

$$\begin{aligned}\sum_{\pi} q^{\text{area}'(\pi)} &= \sum_{\sigma\in M_{(a,b,c)}} q^{\text{coinv}(\sigma)} \\ &= \sum_{\sigma\in M_{(a,b,c)}} q^{\text{inv}(\sigma)} \\ &= \begin{bmatrix} a+b+c \\ a,b,c \end{bmatrix}_q\end{aligned}$$

by (11.2). ■

Given $n,d,k \in \mathbb{N}$ with $1 \le k \le n$, let $L^+_{n,n,d}(k)$ denote the set of paths in $L^+_{n,n,d}$ that have $k$ total $D$ plus $N$ steps below the lowest $E$ step. We view $L^+_{n,n,n}(k)$ as containing

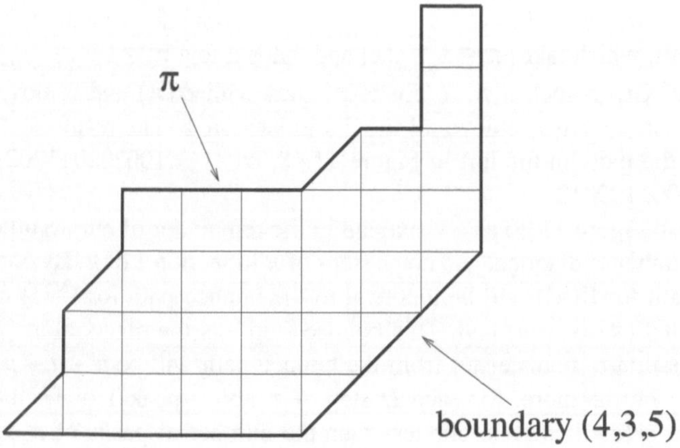


**Figure 11.19**
The region between a path $\pi$ and the corresponding boundary path. For this region area$' = 27$.

the path with $n$ $D$ steps if $k = n$ and $L^+_{n,n,n}(k)$ as being the empty set if $k < n$. Define

$$S_{n,d,k}(q,t) = \sum_{\pi \in L^+_{n,n,d}(k)} q^{\text{area}(\pi)} t^{\text{bounce}(\pi)}, \tag{11.80}$$

with $S_{n,n,k}(q,t) = \chi(k = n)$. There is a recursive structure underlying the $S_{n,d,k}(q,t)$ which extends that underlying the $F_{n,k}(q,t)$. The following result is derived in [34], and is similar to recurrence relations occurring in [18]. For any two integers $n, k$ we use the notation

$$\delta_{n,k} = \chi(n = k).$$

**Theorem 11.5.6** *Let $n, k, d \in \mathbb{N}$ with $1 \le k \le n$. Then*

$$S_{n,n,k}(q,t) = \delta_{n,k}, \tag{11.81}$$

*and for $0 \le d < n$,*

$$S_{n,d,k}(q,t) = t^{n-k} \sum_{p=\max(1,k-d)}^{\min(k,n-d)} \begin{bmatrix} k \\ p \end{bmatrix}_q q^{\binom{p}{2}} \sum_{j=0}^{n-k} \begin{bmatrix} p+j-1 \\ j \end{bmatrix}_q S_{n-k,d+p-k,j}(q,t), \tag{11.82}$$

*with the initial conditions*

$$S_{0,0,k} = \delta_{k,0}, \quad S_{n,d,0} = \delta_{n,0}\delta_{d,0}.$$

*Proof.* If $d = n$ then (11.81) follows directly from the definition. If $d < n$ then $\pi$ has at least one peak. Say $\pi$ has $p$ $N$ steps and $k - p$ $D$ steps in section 0. First assume $p < n-d$, i.e., $\Pi$ has at least two peaks. We now describe an operation we call **truncation**, which takes $\pi \in L^+_{n,n,d}(k)$ and maps it to a $\pi' \in L^+_{n-k,n-k,d-k+p}$ with one less peak. Given such a $\pi$, to create $\pi'$ start with $\sigma(\pi)$ and remove the first $k$ letters (section 0). Also remove all the 2's in section 1. The result is $\sigma(\pi')$. For example, for the path on the left in Figure 11.18, $\sigma(\pi) = 10020201120212$, $k = 3$ and $\sigma(\pi') = 001120212$.

We will use Figure 11.20 as a visual aid in the remainder of our argument. Let $j$ be the total number of diagonal and north steps of $\pi$ in section 1 of $\pi$. By construction the bounce path for $\Gamma(\pi')$ will be identical to the bounce path for $\Gamma(\Pi)$ except the first bounce of $\Gamma(\pi)$ is truncated. This bounce step hits the diagonal at $(p,p)$, and so the contribution to bounce$(\pi')$ from the bounce path will be $n-d-p$ less than to bounce$(\pi)$. Furthermore, for each $D$ step of $\pi$ above peak 1 of $\Pi$, the number of peaks of $\pi'$ below it will be one less than the number of peaks of $\pi$ below it. It follows that

$$\begin{aligned}\text{bounce}(\pi) &= \text{bounce}(\pi') + n - d - p + d - (k-p)\\ &= \text{bounce}(\pi') + n - k.\end{aligned}$$

Since the area below the triangle of side $p$ from Figure 11.20 is $\binom{p}{2}$,

$$\text{area}(\pi) = \text{area}(\pi') + \binom{p}{2} + \text{area0} + \text{area1},$$

where area0 is the area of section 0 of $\pi$, and area1 is the portion of the area of section 1 of $\pi$ not included in area$(\pi')$. When we sum $q^{\text{area0}(\pi)}$ over all $\pi \in L^+_{n,n,d}(k)$ that get mapped to $\pi'$ under truncation, we generate a factor of

$$\begin{bmatrix} k \\ p \end{bmatrix}_q$$

by the $c = 0$ case of the boundary lemma.

From the proof of the boundary lemma, area1 equals the number of coinversions of the first section of $\sigma(\pi)$ involving pairs of 0's and 2's or pairs of 1's and 2's. We need to consider the sum of $q$ to the number of such coinversions, summed over all $\pi$ that map to $\pi'$, or equivalently, summed over all ways to interleave the $p$ 2's into the fixed sequence of $j$ 0's and 1's in section 0 of $\pi'$. Taking into account the fact that such an interleaving must begin with a 2 but is otherwise unrestricted, we get a factor of

$$\begin{bmatrix} p-1+j \\ j \end{bmatrix}_q,$$

since each 2 will form a coinversion with each 0 and each 1 occurring before it. It is now clear how the various terms in (11.82) arise.

Finally, we consider the case when there is only one peak, so $p = n-d$. Since there are $d-(k-p) = d-k+n-d = n-k$ $D$ steps above peak 1 of $\pi$, we have

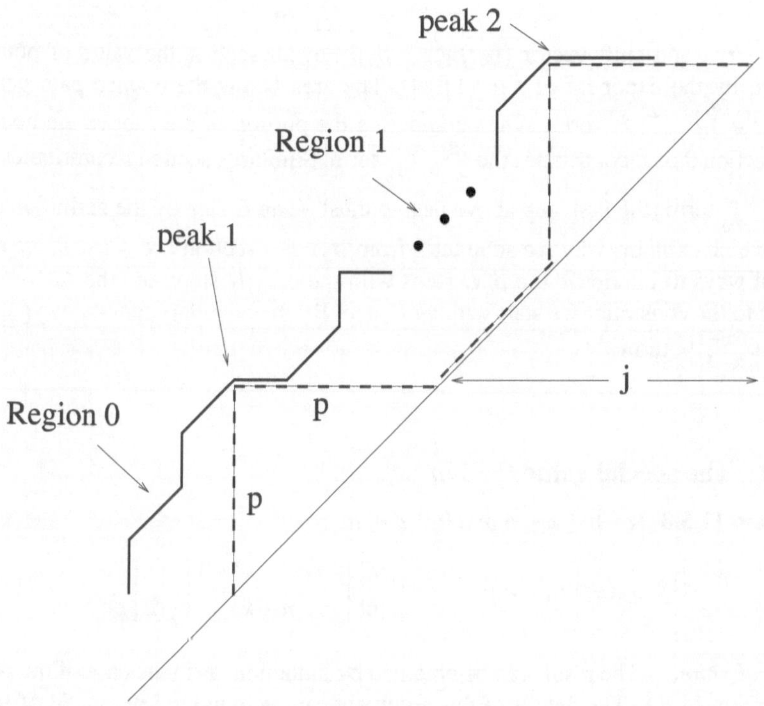


**Figure 11.20**
A path $\pi$ decomposed into various regions under truncation.

$\text{bounce}(\pi) = n-k$. Taking area into account, by the above analysis we get

$$S_{n,d,k}(q,t) = t^{n-k} q^{\binom{n-d}{2}} \begin{bmatrix} k \\ n-d \end{bmatrix}_q \begin{bmatrix} n-d-1+n-k \\ n-k \end{bmatrix}_q \tag{11.83}$$

which agrees with the $p = n-d$ term on the right-hand side of (11.82) since $S_{n-k,n-k,j}(q,t) = \delta_{j,n-k}$ from the initial conditions. ■

The following explicit formula for $S_{n,d}(q,t)$ was obtained in [18].

**Theorem 11.5.7** *For all* $0 \le d < n$,

$$S_{n,d}(q,t) = \sum_{b=1}^{n-d} \sum_{\substack{\alpha_1+\ldots+\alpha_b=n-d,\, \alpha_i>0 \\ \beta_0+\beta_1+\ldots+\beta_b=d,\, \beta_i\ge 0}} \begin{bmatrix} \beta_0+\alpha_1 \\ \beta_0 \end{bmatrix}_q \begin{bmatrix} \beta_b+\alpha_b-1 \\ \beta_b \end{bmatrix}_q q^{\binom{\alpha_1}{2}+\ldots+\binom{\alpha_b}{2}}$$

$$t^{\beta_1+2\beta_2+\ldots+b\beta_b+\alpha_2+2\alpha_3+\ldots+(b-1)\alpha_b} \prod_{i=1}^{b-1} \begin{bmatrix} \beta_i+\alpha_{i+1}+\alpha_i-1 \\ \beta_i, \alpha_{i+1}, \alpha_i-1 \end{bmatrix}_q . \tag{11.84}$$

*Proof.* Consider the sum of $q^{\text{area}}t^{\text{bounce}}$ over all $\pi$ that have bounce vector $(\alpha_1,\ldots,\alpha_b)$, and shift vector $(\beta_0,\beta_1,\ldots,\beta_b)$. For all such $\pi$ the value of bounce is given by the exponent of $t$ in (11.84). The area below the bounce path generates the $q^{\binom{\alpha_1}{2}+\cdots+\binom{\alpha_b}{2}}$ term. When computing the portion of area above the bounce path, section 0 of $\pi$ contributes the $\begin{bmatrix}\beta_0+\alpha_1\\ \beta_0\end{bmatrix}_q$ term. Similarly, section $b$ contributes the $\begin{bmatrix}\beta_b+\alpha_b-1\\ \beta_b\end{bmatrix}_q$ term (the first step above peak $b$ must be an $E$ step by the definition of a peak, which explains why we subtract 1 from $\alpha_b$). For section $i$, $1\le i<b$, we sum over all ways to interleave the $\beta_i$ $D$ steps with the $\alpha_{i+1}$ $N$ steps and the $\alpha_i$ $E$ steps, subject to the constraint we start with an $E$ step. By the boundary lemma, we get the $\begin{bmatrix}\beta_i+\alpha_{i+1}+\alpha_i-1\\ \beta_i,\alpha_{i+1},\alpha_i-1\end{bmatrix}_q$ term. ■

### 11.5.3 The special value $t=1/q$

**Theorem 11.5.8** *For $1\le k\le n$ and $0\le d\le n$,*

$$q^{\binom{n}{2}-\binom{d}{2}}S_{n,d,k}(q,1/q)=q^{(k-1)(n-d)}\frac{[k]}{[n]}\begin{bmatrix}2n-k-d-1\\ n-k\end{bmatrix}_q\begin{bmatrix}n\\ d\end{bmatrix}_q. \tag{11.85}$$

*Sketch of proof.* The result can be obtained by induction, as in the case of the proof of Theorem 11.3.8. The details of this argument can be found in [39, pp. 54-56]. ■

**Corollary 11.5.9** *For $0\le d\le n$,*

$$q^{\binom{n}{2}-\binom{d}{2}}S_{n,d}(q,1/q)=\frac{1}{[n-d+1]}\begin{bmatrix}2n-d\\ n-d,n-d,d\end{bmatrix}_q. \tag{11.86}$$

*Proof.* It is easy to see combinatorially that $S_{n+1,d,1}(q,t)=t^nS_{n,d}(q,t)$. Thus

$$\begin{aligned}q^{\binom{n}{2}-\binom{d}{2}}S_{n,d}(q,1/q)&=q^{\binom{n}{2}+n-\binom{d}{2}}S_{n+1,d,1}(q,1/q)\\ &=q^{\binom{n+1}{2}-\binom{d}{2}}S_{n+1,d,1}(q,1/q)\\ &=q^{(1-1)(n+1-d)}\frac{[1]}{[n+1]}\begin{bmatrix}2(n+1)-1-d-1\\ n\end{bmatrix}_q\begin{bmatrix}n+1\\ d\end{bmatrix}_q\\ &=\frac{1}{[n-d+1]}\begin{bmatrix}2n-d\\ n-d,n-d,d\end{bmatrix}_q.\end{aligned}$$

■

**Remark 11.5.10** Corollary 11.5.9 proves that $S_{n,d}(q,t)$ is symmetric in $q,t$ when $t=1/q$. For we have

$$S_{n,d}(q,1/q)=q^{-\binom{n}{2}+\binom{d}{2}}\frac{1}{[n-d+1]}\begin{bmatrix}2n-d\\ n-d,n-d,d\end{bmatrix}_q,$$

and replacing $q$ by $1/q$ we get

$$S_{n,d}(1/q,q) = q^{\binom{n}{2}-\binom{d}{2}} \frac{q^{n-d}}{[n-d+1]} \begin{bmatrix} 2n-d \\ n-d, n-d, d \end{bmatrix}_q q^{2\binom{n-d}{2}+\binom{d}{2}-\binom{2n-d}{2}},$$

since $[n]!_{1/q} = [n]!q^{-\binom{n}{2}}$. Now

$$\binom{n}{2} - \binom{d}{2} + n - d + 2\binom{n-d}{2} + \binom{d}{2} - \binom{2n-d}{2} = \binom{d}{2} - \binom{n}{2},$$

so $S_{n,d}(q,1/q) = S_{n,d}(1/q,q)$. It is of course an open problem to show $S_{n,d}(q,t) = S_{n,d}(t,q)$ bijectively, since the $d=0$ case is Open Problem 11.3.10.

### 11.5.4 A Schröder dinv-statistic

Let

$$C_n(q,t,w) = \sum_{d=0}^{n} w^d S_{n,d}(q,t),$$

and for $\pi \in L^+_{n,n}$, let $a'_i(\pi)$ equal the number of area squares in the $i$th column from the right. Also set $a'_0(\pi) = -1$ for all $\pi$. For example, for the path on the right in Figure 11.8, we have $(a'_0, a'_1, a'_2, \ldots, a'_8) = (-1,0,1,1,2,3,3,1,0)$. The $(q,t)$-Schröder can be expressed strictly in terms of Dyck paths as follows.

**Theorem 11.5.11** *The following equality holds:*

$$C_n(q,t,w) = \sum_{\pi \in L^+_{n,n}} q^{area(\pi)} t^{bounce(\pi)} \prod_{\substack{1 \le i \le n \\ a'_i > a'_{i-1}}} (1 + w/q^{a'_i}). \tag{11.87}$$

*Proof.* (sketch) Show the coefficient of $w^d$ in (11.87) satisfies the same recurrence as (11.82). ∎

Next define the **reading order** of the rows of $\pi$ to be the order in which the rows are listed by decreasing value of $a_i$, where if two rows have the same $a_i$-value, the row above is listed first. For example, for the path on the left in Figure 11.8, the reading order is

row 6, row 8, row 5, row 4, row 3, row 7, row 2, row 1.

Finally let $b_k = b_k(\pi)$ be the number of inversion pairs as in Definition 11.3.12 which involve the $k$th row in the reading order and rows before it in the reading order, and set $b_0 = -1$. For example, for the path on the left in Figure 11.8, we have

$$(b_0, b_1, \ldots, b_8) = (-1,0,1,1,2,3,3,1,0). \tag{11.88}$$

Note that in this example $a'_i(\zeta(\pi)) = b_i(\pi)$ for all $i$, and it is not hard to see that this is true in general. Hence we have

$$C_n(q,t,w) = \sum_{\pi \in L^+_{n,n}} q^{\mathrm{dinv}(\pi)} t^{\mathrm{area}(\pi)} \prod_{\substack{1 \le i \le n \\ b_i > b_{i-1}}} (1 + w/q^{b_i}). \tag{11.89}$$

It is also easy to see that $b_i > b_{i-1}$ if and only if row$_i$ contains a column top, i.e., the $N$ step of the path in that row is followed immediately by an $E$ step.

Now to get a term in (11.89) contributing to $S_{n,d}(q,t)$ we make a selection of $d$ column tops, and for the rows containing those column tops we subtract the corresponding $b_i$ contribution to dinv. If we start with the path $\pi$ and replace those column tops by $D$ steps, we get a Schröder path, and we can define dinv of this path as $\mathrm{dinv}(\pi)$ minus the sum of the chosen $b_i$.

### 11.5.5 The shuffle conjecture

In [37] a nice conjecture for the expansion of $\mathscr{F}(DH_n;q,t)$ into monomials is introduced. It is often referred to as the shuffle conjecture, since it can be phrased in the following simple way: decompose $\{1,2,\ldots,n\}$ into increasing sequences of consecutive integers $\alpha_1, \alpha_2, \ldots, \alpha_k$ of lengths $|\alpha_1|, |\alpha_2|, \ldots$ and decreasing sequences of consecutive integers $\beta_1, \ldots, \beta_s$ of lengths $|\beta_1|, |\beta_2|, \ldots$. For example, if $n = 8$ we might have

$$k = 1, s = 2, \alpha_1 = \{6,7,8\}, \beta_1 = \{5,4,3\}, \beta_2 = \{2,1\}. \tag{11.90}$$

Given such a decomposition, we say a permutation $\sigma \in S_n$ is an $\alpha, \beta$**-shuffle** if for each $i, j$ all the terms in $\alpha_i$ occur in increasing order in $\sigma$, and all the terms of $\beta_j$ occur in decreasing order in $\sigma$. For example, for $\alpha, \beta$ as in (11.90), the permutation 26571483 is an $\alpha, \beta$-shuffle.

**Conjecture 11.5.12 (The Shuffle Conjecture [37])**

$$\langle \mathscr{F}(DH_n;q,t), h_{|\alpha_1|} h_{|\alpha_2|} \cdots h_{|\alpha_k|} e_{|\beta_1|} e_{|\beta_2|} \cdots e_{|\beta_s|} \rangle = \sum_{\substack{P \in \mathscr{P}_n \\ read(P) \text{ is an } \alpha,\beta\text{-shuffle}}} q^{dinv(P)} t^{area(P)}. \tag{11.91}$$

If our decomposition of $\{1,2,\ldots,n\}$ is into $n$ sequences consisting of only one element each, then the set of $\alpha, \beta$-shuffles is the set of all parking functions. By Remark 11.2.29 the left-hand side of (11.91) gives $\mathscr{H}(DH_n;q,t)$, and so in this case the shuffle conjecture reduces to Conjecture 11.4.2. On the other hand, consider the case $k = s = 1$, $\alpha_1 = n-d+1, \ldots, n$, $\beta_1 = n-d, \ldots, 1$. The left-hand side of (11.91) reduces to $\langle \mathscr{F}(DH_n;q,t), h_d e_{n-d} \rangle$, the $q,t$-Schröder. If read$(P)$ is an $\alpha, \beta$-shuffle, we must have all the elements of $\alpha_1$ occurring at the top of columns. It is easy to see that none of these cars can form dinv-pairs with any other car that occurs before them in the reading order. We can identify columns containing elements of $\alpha_1$ at the top with selections in the right-hand side of (11.89) of the $d$-column tops to be regarded as

$D$ steps, and one finds in fact that the Schröder dinv-statistic in the right-hand side of (11.89) is the same as dinv$(P)$.

For a subset $D$ of $\{1,2,\ldots,n-1\}$, let

$$\mathcal{Q}_{n,D}(Z) = \sum_{\substack{a_1 \le a_2 \le \cdots \le a_n \\ a_i = a_{i+1} \implies i \notin D}} z_{a_1} z_{a_2} \cdots z_{a_n}$$

denote Gessel's **fundamental quasisymmetric function**. Another equivalent form of the shuffle conjecture appearing in [37] is the statement that

$$\mathcal{F}(DH_n;q,t) = \sum_{P \in \mathcal{P}_n} q^{\mathrm{dinv}(P)} t^{\mathrm{area}(P)} \mathcal{Q}_{n,\mathrm{Ides}(\mathrm{read}(P))}(x_1,\ldots,x_n), \tag{11.92}$$

where recall that for any permutation $\sigma \in S_n$, $\mathrm{Ides}(\sigma)$ is the set of all $i$ for which $i+1$ occurs before $i$ in $\sigma$. The authors show that the sum

$$A_\pi(x_1,\ldots,x_n;q) = \sum_{\substack{P \in \mathcal{P}_n \\ \pi \text{ fixed}}} q^{\mathrm{dinv}(P)} \mathcal{Q}_{n,\mathrm{Ides}(\mathrm{read}(P))}(x_1,\ldots,x_n), \tag{11.93}$$

obtained by restricting the right-hand side of (11.92) to a fixed path $\pi$, is a special case of a family of symmetric functions introduced by Lascoux, Leclerc, and Thibon [46] commonly called LLT polynomials. In [46] it is conjectured that these polynomials are always Schur positive, and two recent preprints claim to give independent proofs of this conjecture. One of these is by Grojnowski and Haiman [32], and uses the representation theory of Hecke algebras. The other, by S. Assaf [11], is a purely combinatorial construction involving objects called dual equivalence graphs.

In [39, Theorem 6.8, p. 98] it is shown that if $\pi$ is a path with the property that each non-empty column has its base on the diagonal $x = y$ (a so-called **balanced path**), then the Schur coefficients of $A_\pi(x_1,\ldots,x_n;q)$ can be expressed in terms of the charge statistic of Algorithm 11.2.26. Hence one could hope that for general LLT polynomials, or at least for the type of LLT's corresponding to Dyck paths, there is some similar formula. Note that the product $e_{\beta(\pi)}$ in (11.63) can be easily expanded in terms of Schur functions, and in fact the coefficient of $s_\lambda$ in $e_\beta$ is $K_{\lambda',\beta}$. Thus there may be a way of associating powers of $q$ with SSYT, depending on the shape of $\pi$ in some way, to generate the LLT polynomial $A_\pi(x_1,\ldots,x_n;q)$.

**Problem 11.5.13** *Find a nice formula, perhaps in terms of a generalized charge statistic, for the coefficients in the Schur expansion of the LLT polynomial corresponding to a given Dyck path, i.e., the right-hand side of (11.93).*

Assaf's construction in terms of dual equivalence graphs does yield a combinatorial formula of sorts for an arbitrary LLT polynomial, but her construction is rather involved, and it is not yet known how to reproduce explicit formulas, like the one for balanced paths in terms of charge, from her formula. Note that the shuffle conjecture gives a prediction for the coefficient of a monomial symmetric function in $\mathcal{F}(DH_n;q,t)$, but does not (except in the hook case) give a nice conjecture for the more desirable Schur coefficients.

## 11.6 Rational Catalan combinatorics

### 11.6.1 The superpolynomial invariant of a torus knot

Throughout this section $(m,n)$ is a fixed pair of relatively prime, positive integers. In the last few years an exciting generalization of the $q,t$-Schröder and the shuffle conjecture has been introduced. This generalization depends on an arbitrary pair $(m,n)$, and has an interpretation in terms of knot theory. By a **knot** we mean an embedding of a circle in $\mathbb{R}^3$, and by a **knot invariant** a polynomial that is the same on equivalent knots. Two classical knot invariants on a knot $K$ are the Jones polynomial $V_K(t)$ and the HOMFLY polynomial $P_K(a,q)$. Dunfield, Gukov, and Rasmussen [17] hypothesized the existence of a **superpolynomial** knot invariant $P_K(a,q,t)$, which would contain the HOMFLY and Jones polynomials as limiting cases, as well as having other desirable properties. Possible definitions of the superpolynomial for torus knots $T_{(m,n)}$ have recently been suggested by Angnanovic and Shakirov [1] (see also [2]), Cherednik [16], and Oblomkov, Rasmussen, and Shende [55]. All three methods seem to give the same polynomial, and in fact Gorsky and Negut [31] have proved the descriptions in [1] and [16] do in fact give the same polynomial. The description in [1] is in terms of Macdonald polynomials, and Gorsky first realized that when $m = n+1$, if we use the Cherednik parametrization, then the superpolynomial can be expressed as the function $C_n(q,t,-a)$ from (11.89), giving a completely new interpretation for the $q,t$-Schröder.

In [55] a conjectured combinatorial expression for the superpolynomial of $T(m,n)$ in terms of weighted lattice paths is given, which we now describe. Let $\text{Grid}(m,n)$ be the $n \times m$ grid of labeled squares whose upper-left-hand corner square is labeled with $(n-1)(m-1)-1$, and whose labels decrease by $m$ as you go down columns and by $n$ as you go across rows. For example,

$\text{Grid}(3,7) =$

| | | |
|---|---|---|
| 11 | 4 | −3 |
| 8 | 1 | −6 |
| 5 | −2 | −9 |
| 2 | −5 | −12 |
| −1 | −8 | −15 |
| −4 | −11 | −18 |
| −7 | −14 | −21 |

To the corners of the squares of $\text{Grid}(m,n)$ we associate Cartesian coordinates, where the lower-left-hand corner of the grid has coordinates $(0,0)$, and the upper-right-hand corner of the grid $(m,n)$. Let $L^+_{(m,n)}$ denote the set of lattice paths $\pi$ for which none of the squares with negative labels are above $\pi$. (This agrees with our definition of $L^+_{(m,n)}$ from Section 11.2.2 as paths that stay above the line $my = nx$). For a given $\pi$, we let $\text{area}(\pi)$ denote the number of squares in $\text{Grid}(m,n)$ with positive labels that are below $\pi$. Furthermore, let $\text{dinv}(\pi)$ denote the number of squares in $\text{Grid}(m,n)$

that are above $\pi$ and whose arm and leg lengths satisfy

$$\frac{a}{l+1} < \frac{m}{n} < \frac{a+1}{l}. \tag{11.94}$$

For example, if $(m,n) = (3,7)$ and $\pi = NNNNNEENNE$, then $\text{area}(\pi) = 2$ (corresponding to the squares with labels 2 and 5). Also, $\text{dinv}(\pi) = 2$; the squares with labels $11, 8, 4, 1$ have $a = l = 1, a = 1, l = 0, a = 0, l = 1, a = l = 0$, respectively, and so the squares with labels 8 and 11 do not satisfy (11.94), while the squares with labels 1 and 4 do.

Next we define a generalization of the formula (11.89) for general $(m,n)$. Given $\pi \in L^+_{(m,n)}$, let $R(\pi)$ denote the set of labels of squares that are at the top of some column of $\pi$. Say these labels occur in columns $c_1, c_2, \ldots, c_k$ as we move left to right. Then for $1 \le i \le k$, let $t_i$ denote the label of the square that is in the same row as the square at the top of column $c_i$, and also in column $c_{i+1}$, and set $T(\pi) = \{t_1, t_2, \ldots, t_{k-1}\}$. For example, if $\pi$ is the path on the left of Figure 11.22, then

$$\begin{aligned} R(\pi) &= \{-3, 1, 5\} \\ T(\pi) &= \{-6, -2\}. \end{aligned} \tag{11.95}$$

Now form a vector $\alpha(\pi) = (\alpha_1, \ldots, \alpha_k)$ consisting of the elements of $R(\pi)$ in decreasing order, and let $c_i(\pi)$ denote the number of elements of $R(\pi)$ that are larger than $\alpha_i$, minus the number of elements of $T(\pi)$ that are larger than $\alpha_i$. For the example of (11.95), we have $\alpha = (5, 1, -3)$, and so $c_1 = 0 - 0 = 0, c_2 = 1 - 0 = 1, c_3 = 2 - 1 = 1$. Furthermore set $c_0 = -1$.

**Conjecture 11.6.1** [55] *For any pair $(m,n)$ of positive, relatively prime integers,*

$$P_{T(m,n)}(-w,q,t) = \sum_{\pi \in L^+_{m,n}} q^{dinv(\pi)} t^{area(\pi)} \prod_{\substack{c_i > c_{i-1} \\ 1 \le i \le k}} (1 + w/q^{c_i}). \tag{11.96}$$

*Here $T(m,n)$ is the $(m,n)$ torus knot, which winds around the torus $m$ times in one direction and $n$ times in the other before returning to the starting point, and we use the parametrization of the superpolynomial occurring in [16, p. 18, eq. (2.12)].*

**Exercise 11.6.2** *Show that if $m = n+1$, (11.96) reduces to (11.89).*

There is also a version of the shuffle conjecture for any $(m,n)$. Let an $(m,n)$-parking function be a path $\pi \in L^+_{(m,n)}$ together with a placement of the integers 1 through $n$ (called cars) just to the right of the N steps of $\pi$, with strict decrease down columns. For such a pair $P$, we let $\text{rank}(j)$ be the label of the square that contains $j$, and we set

$$\text{tdinv}(P) = |\{(i,j) : 1 \le i < j \le n \text{ and } \text{rank}(i) < \text{rank}(j) < \text{rank}(i) + m\}|.$$

Furthermore we let the reading word $\text{read}(P)$ be the permutation obtained by listing the cars by decreasing order of their ranks. For the $(3,7)$-parking function of Figure

11.21, tdinv $= 3$, with inversion pairs formed by pairs of cars $(6,7)$, $(4,6)$, and $(2,4)$, and the reading word is 7642531. Let maxtdinv$(\pi)$ be tdinv of the parking function for $\pi$ whose reading word is the reverse of the identity, and for any parking function $P$ for $\pi$ set

$$\mathrm{dinv}(P) - \mathrm{dinv}(\pi) + \mathrm{tdinv}(P) - \mathrm{maxtdinv}(\pi).$$

Then the combinatorial side of the $(m,n)$ shuffle conjecture is the function

$$B_{(m,n)}(x_1,\ldots,x_n;q,t) = \sum_{(m,n)\text{ parking functions }P} q^{\mathrm{dinv}(P)} t^{\mathrm{area}(\pi)} \mathcal{Q}_{n,\mathrm{Ides}(\mathrm{read}(P))}(x_1,\ldots,x_n). \tag{11.97}$$

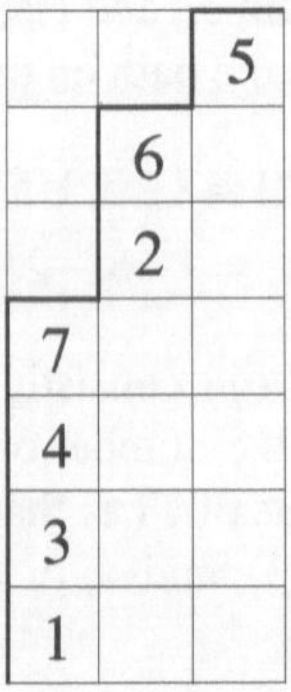


**Figure 11.21**
A $(3,7)$-parking function.

Gorsky and Negut [31] show how the results of Aganagic and Shakirov on torus knot invariants can be expressed in terms of Macdonald polynomials using advanced objects such as the Hilbert scheme. Bergeron, Garsia, Leven, and Xin [12, 13] have shown how this Macdonald polynomial construction can be done combinatorially with plethystic symmetric function operators, and in fact they define operators $Q_{(m,n)}$ for any relatively prime $(m,n)$ by a recursive procedure. The rational shuffle conjecture can then be phrased as $Q_{(m,n)}(-\mathbf{1})^{\mathbf{n}} = B_{(m,n)}(x_1,\ldots,x_n;q,t)$. The symmetric function $Q_{(n+1,n)}(-\mathbf{1})^{\mathbf{n}}$ reduces to $\nabla e_n$, and so the rational shuffle conjecture reduces to the original shuffle conjecture when $m = n+1$.

Gorsky and Negut also conjecture that $Q_{(m,n)}(-\mathbf{1})^{\mathbf{n}}$ is the Frobenius series of the unique finite-dimensional irreducible representation of the rational Cherednik algebra with parameter $m/n$, with respect to a certain bigrading. Hikita [43] has shown that $B_{(m,n)}(x_1,\ldots,x_n;q,t)$ is the bigraded Frobenius series of certain $S_n$-modules arising in the study of the homology of type $A$ affine Springer fibers, which gives another possible way of attacking the rational shuffle conjecture. We mention that the portion of the right-hand side of (11.97) corresponding to a fixed path $\pi$ is an LLT polynomial, and hence is Schur positive.

Many of the results in this chapter involving the special cases $t = 1$ and $t = 1/q$ have elegant extensions to general, relatively prime $(m,n)$. For example, Gorsky and Negut [31] prove that when $t = 1/q$ the $q,t$-Catalan for $(m,n)$, obtained by taking the coefficient of $s_{1^n}$ in $Q_{(m,n)}(-\mathbf{1})^{\mathbf{n}}$, reduces to

$$\frac{1}{[m]}\begin{bmatrix} m+n-1 \\ n \end{bmatrix}.$$

Another example is that the total number of $(m,n)$-parking functions is $m^{n-1}$.

### 11.6.2 Tesler matrices and the superpolynomial

There is a general formula for $Q_{(m,n)}(-\mathbf{1})^{\mathbf{n}}$ in terms of Tesler matrices.

**Theorem 11.6.3 (Gorsky, Negut** 2013) *For any pair of positive, relatively prime integers $(m,n)$,*

$$Q_{m,n}(-\mathbf{1})^{\mathbf{n}} = \sum_{C\in Tes(m,n)} \prod_{i=1}^{m} e_{c_{ii}} \prod_{\substack{1\leq i<m \\ c_{i,i+1}>0}} ([c_{i,i+1}+1]_{q,t} - [c_{i,i+1}]_{q,t}) \prod_{\substack{2\leq i+1<j\leq m \\ c_{i,j}>0}} (-M)[c_{i,j}]_{q,t}. \tag{11.98}$$

*Here $Tes(m,n)$ is the set of $m\times m$ upper-triangular matrices $C$ satisfying*

$$c_{i,i} + \sum_{j>i} c_{i,j} - \sum_{j<i} c_{j,i} = \left\lfloor \frac{in}{m} \right\rfloor - \left\lfloor \frac{(i-1)n}{m} \right\rfloor, \quad 1\leq i\leq m. \tag{11.99}$$

**Example 11.6.4** *Given a symmetric function $f$ expressed as a polynomial in the $e_k$, the coefficient of $s_{1^n}$ in $f$ can be found by simply replacing each $e_k$ by 1. Now when $m = n+1$ the conditions (11.99) reduce to the hook sums all equal 1 except for the first, which equals 0. Having a first hook sum equal to 0 forces the first row to be all zeros, and so $Tes(n+1,n)$ is really just the same as $Tes(n)$. Hence taking the weights from (11.98) when $n = 3$ for the matrices in (11.77), setting all $e_{c_{ii}} = 1$, we get*

$$\begin{aligned} C_3(q,t) &= 1 + ([2]-[1]) + ([2]-[1]) + ([2]-[1])^2 + \\ &\quad + ([3]-[2])([2]-[1]) - M - M([2]-[1]) \\ &= 1 + (q+t-1) + (q+t-1) + (q+t-1)^2 \\ &\quad + (q^2+qt+t^2-q-t))(q+t-1) \\ &\quad - (1-q)(1-t) - (1-q)(1-t)(q+t-1) \\ &= q^3 + q^2t + qt + qt^2 + t^3. \end{aligned}$$

Garsia and Haglund [24] independently obtained a Tesler matrix expression for $\nabla e_n$, although it is a bit more complicated to state than the $m = n+1$ case of (11.98).

The superpolynomial for the $(m,n)$ Torus knot can be defined analytically as

$$\mathscr{P}_{T(m,n)}(a,q,t) = \sum_{d\geq 0} (-a)^{n-d} \langle Q_{(m,n)}(-1)^{\mathbf{n}}, e_d h_{n-d} \rangle.$$

As a corollary of (11.98), Gorsky and Negut obtain the following.

**Corollary 11.6.5** *For any pair of positive, relatively prime integers $(m,n)$, we have*

$$\mathscr{P}_{T(m,n)}(a,q,t) = \sum_{C\in Tes(m,n)} \prod_{\substack{1\leq i\leq m\\ c_{i,i}>0}} (1-a) \prod_{1\leq i<m} ([c_{i,i+1}+1]_{q,t} - [c_{i,i+1}]_{q,t})$$

$$\cdot \prod_{\substack{2\leq i+1<j\leq m\\ c_{i,j}>0}} (-M)[c_{i,j}]_{q,t}.$$

There are a number of intriguing open problems involving the combinatorics of $(m,n)$-Catalan paths. For example, there is a candidate extension of the zeta map of Figure 11.8 that can be described as follows. Given a path $\pi \in L^+_{(m,n)}$, call the set of corners of grid squares that are touched by $\pi$ the "vertices" of $\pi$. Next define $S(\pi)$ to be the set consisting of the labels of those squares whose upper-left-hand corners are vertices of $\pi$. A given label in $S(\pi)$ is called an $N$ label if the vertex associated to it is the start of an $N$ step, otherwise it is called an $E$ label. For example, if $\pi$ is the path on the left in Figure 11.22, then

$$\pi = NNNNNENENENE \quad S(\pi) = \{-10,-7,-4,-1,2,5,-2,1,-6,-3\}. \quad (11.100)$$

We now define the "sweep map" of [6], denoted $\zeta$, from $L^+_{(m,n)}$ to $L^+_{(m,n)}$ as follows: Order the elements of $S(\pi)$ in increasing order to create a vector of labels $D(\pi) = (d_1, d_2, \ldots, d_{m+n})$. Then create a path $\phi(\pi)$ by defining the $i$th step of $\phi(\pi)$ to be an $N$ step if $d_i$ is an $N$ label, and an $E$ step if $d_i$ is an $E$ label. For the example in (11.100), we have

$$D(\pi) = (-10,-7,-6,-4,-3,-2,-1,1,2,5) \quad \phi(\pi) = NNNNENNNENE.$$

**Exercise 11.6.6** *Show that when $m = n+1$, paths in $L^+_{(m,n)}$ are in bijection with paths in $L^+_{(n,n)}$, and that the sweep map reduces to the $\zeta$ map of Figure 11.8.*

**Problem 11.6.7** *Prove that for general coprime $(m,n)$ the sweep map is a bijection from $L^+_{(m,n)} \to L^+_{(m,n)}$.*

This problem has been studied by Gorsky, Mazin, and Vazirani [30] and Armstrong, Loehr, and Warrington [6]. See also [8]. In [30] it is shown that the sweep map is a bijection whenever $m = kn+1$ or $m = kn-1$ for some positive integer $k$. We note that in the case $m = kn+1$ Loehr [49, 51] (see also [39, pp. 108-109]) has defined an extension of the bounce statistic, which when combined with area generates the $q,t$-Catalan for $m = kn+1$. For this "paths in a $n\times kn$ rectangle" case there is also an interpretation for the rational shuffle conjecture in terms of a generalization of diagonal harmonics (see [37]).

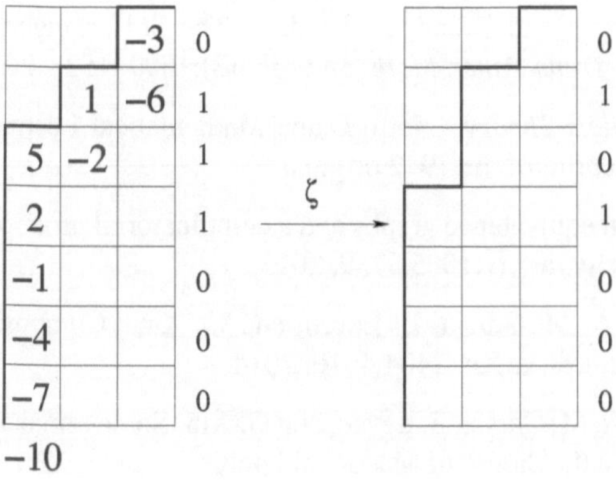


**Figure 11.22**
The sweep map.

# References


[1] M. Aganagic and S. Shakirov. Knot homology from refined Chern-Simons theory. arXiv:1105.5117, 2011.

[2] Mina Aganagic and Shamil Shakirov. Refined Chern-Simons theory and knot homology. In *String-Math 2011*, volume 85 of *Proc. Sympos. Pure Math.*, pages 3–31. Amer. Math. Soc., Providence, RI, 2012.

[3] G. E. Andrews. Identities in combinatorics, II: a $q$-analog of the Lagrange inversion theorem. *Proc. Amer. Math. Soc.*, 53:240–245, 1975.

[4] G. E. Andrews, R. Askey, and R. Roy. *Special Functions*, volume 71 of *Encyclopedia of Mathematics and its Applications*. Cambridge University Press, Cambridge, 1999.

[5] G. E. Andrews, C. Krattenthaler, L. Orsina, and P. Papi. *ad*-Nilpotent $b$-ideals in $\mathfrak{sl}(n)$ having a fixed class of nilpotence: combinatorics and enumeration. *Trans. Amer. Math. Soc.*, 354(10):3835–3853 (electronic), 2002.

[6] D. Armstrong, N. Loehr, and G. Warrington. Rational parking functions and Catalan numbers. arXiv:1403.1845, 2014.

[7] Drew Armstrong. Hyperplane arrangements and diagonal harmonics. *J. Comb.*, 4(2):157–190, 2013.

[8] Drew Armstrong, Christopher R. H. Hanusa, and Brant C. Jones. Results and conjectures on simultaneous core partitions. *European J. Combin.*, 41:205–220, 2014.

[9] Drew Armstrong and Brendon Rhoades. The Shi arrangement and the Ish arrangement. *Trans. Amer. Math. Soc.*, 364(3):1509–1528, 2012.

[10] E. Artin. *Galois Theory*. Notre Dame Mathematical Lectures. Notre Dame Ind., 1976. Reprint of the 1942 original.

[11] S. Assaf. Dual equivalence graphs and a combinatorial proof of LLT and Macdonald positivity. arXiv:1005.3759, 2013.

[12] F. Bergeron, A. M. Garsia, E. Leven, and G. Xin. Compositional $(km, kn)$-shuffle conjectures. arXiv:1404.4616, 2014.

[13] F. Bergeron, A. M. Garsia, E. Leven, and G. Xin. Some remakable new plethystic operators in the thoery of Macdonald polynomials. arXiv:1405.0316, 2014.

[14] L. M. Butler. Subgroup lattices and symmetric functions. *Mem. Amer. Math. Soc.*, 112(539):vi+160, 1994.

[15] L. Carlitz and J. Riordan. Two element lattice permutation numbers and their $q$-generalization. *Duke J. Math.*, 31:371–388, 1964.

[16] Ivan Cherednik. Jones polynomials of torus knots via DAHA. *Int. Math. Res. Not. IMRN*, (23):5366–5425, 2013.

[17] Nathan M. Dunfield, Sergei Gukov, and Jacob Rasmussen. The superpolynomial for knot homologies. *Experiment. Math.*, 15(2):129–159, 2006.

[18] E. Egge, J. Haglund, K. Killpatrick, and D. Kremer. A Schröder generalization of Haglund's statistic on Catalan paths. *Electron. J. Combin.*, 10:Research Paper 16, 21 pp. (electronic), 2003.

[19] D. Foata. On the Netto inversion number of a sequence. *Proc. Amer. Math. Soc.*, 19:236–240, 1968.

[20] D. Foata and M.-P. Schützenberger. Major index and inversion number of permutations. *Math. Nachr.*, 83:143–159, 1978.

[21] H. O. Foulkes. A survey of some combinatorial aspects of symmetric functions. In *Permutations (Actes Colloq., Univ. René-Descartes, Paris, 1972)*, pages 79–92. Gauthier-Villars, Paris, 1974.

[22] J. Fürlinger and J. Hofbauer. $q$-Catalan numbers. *J. Combin. Theory Ser. A*, 40:248–264, 1985.

[23] A. M. Garsia. A $q$-analogue of the Lagrange inversion formula. *Houston J. Math.*, 7:205–237, 1981.

[24] A. M. Garsia and J. Haglund. A polynomial expression for the character of diagonal harmonics. *Annals of Combinatorics*. to appear.

[25] A. M. Garsia and J. Haglund. A positivity result in the theory of Macdonald polynomials. *Proc. Nat. Acad. Sci. U.S.A.*, 98:4313–4316, 2001.

[26] A. M. Garsia and J. Haglund. A proof of the $q,t$-Catalan positivity conjecture. *Discrete Math.*, 256:677–717, 2002.

[27] A. M. Garsia and M. Haiman. A remarkable $q,t$-Catalan sequence and $q$-Lagrange inversion. *J. Algebraic Combin.*, 5(3):191–244, 1996.

[28] G. Gasper and M. Rahman. *Basic hypergeometric series, second ed.*, volume 96 of *Encyclopedia of Mathematics and its Applications*. Cambridge University Press, Cambridge, 2004. With a foreword by Richard Askey.

[29] I. Gessel. A noncommutative generalization and $q$-analog of the Lagrange inversion formula. *Trans. Amer. Math. Soc.*, 257:455–482, 1980.

[30] E. Gorsky, M. Mazin, and M. Vazirani. Affine permutations and rational slope parking functions. arXiv:1403.0303, 2014.

[31] E. Gorsky and A. Negut. Refined knot invariants and Hilbert schemes. arXiv:1304.3328, 2013.

[32] I. Grojnowski and M. Haiman. Affine Hecke algebras and positivity of LLT and Macdonald polynomials. 2006. Preprint at http://www.math.berkeley.edu/ mhaiman/.

[33] J. Haglund. Conjectured statistics for the $q,t$-Catalan numbers. *Adv. Math.*, 175(2):319–334, 2003.

[34] J. Haglund. A proof of the $q,t$-Schröder conjecture. *Int. Math. Res. Not.*, 11:525–560, 2004.

[35] J. Haglund. A polynomial expression for the Hilbert series of the quotient ring of diagonal coinvariants. *Adv. Math.*, 227:2092–2106, 2011.

[36] J. Haglund, M. Haiman, and N. Loehr. A combinatorial formula for Macdonald polynomials. *Jour. Amer. Math. Soc.*, 18:735–761, 2005.

[37] J. Haglund, M. Haiman, N. Loehr, J. B. Remmel, and A. Ulyanov. A combinatorial formula for the character of the diagonal coinvariants. *Duke Math. J.*, 126:195–232, 2005.

[38] J. Haglund and N. Loehr. A conjectured combinatorial formula for the Hilbert series for Diagonal Harmonics. *Discrete Math. (Proceedings of the FPSAC 2002 Conference held in Melbourne, Australia)*, 298:189–204, 2005.

[39] James Haglund. *The q,t-Catalan numbers and the space of diagonal harmonics*, volume 41 of *University Lecture Series*. American Mathematical Society, Providence, RI, 2008. With an appendix on the combinatorics of Macdonald polynomials.

[40] M. Haiman. Conjectures on the quotient ring by diagonal invariants. *J. Algebraic Combin.*, 3:17–76, 1994.

[41] M. Haiman. Private communication, 2000.

[42] M. Haiman. Vanishing theorems and character formulas for the Hilbert scheme of points in the plane. *Invent. Math.*, 149:371–407, 2002.

[43] Tatsuyuki Hikita. Affine Springer fibers of type $A$ and combinatorics of diagonal coinvariants. *Adv. Math.*, 263:88–122, 2014.

[44] G. James and M. Liebeck. *Representations and characters of groups*. Cambridge University Press, New York, second edition, 2001.

[45] A. Lascoux and M.-P. Schützenberger. Sur une conjecture de H. O. Foulkes. *C. R. Acad. Sci. Paris Sér. A-B*, 286:A323–A324, 1978.

[46] Alain Lascoux, Bernard Leclerc, and Jean-Yves Thibon. Ribbon tableaux, Hall-Littlewood functions, quantum affine algebras, and unipotent varieties. *J. Math. Phys.*, 38(2):1041–1068, 1997.

[47] P. Levande. Special cases of the parking functions conjecture and upper-triangular matrices. *DMTCS, Proceedings of the 23rd Internat. Conf. on Formal Power Series and Algebraic Combinatorics (FPSAC 2011)*, pages 635–644, 2011. See wwww.dmtcs.org/dmtcs-ojs/index.php/proceedings/issue/view/119.

[48] Paul Levande. *Combinatorial structures and generating functions of Fishburn numbers, parking functions, and Tesler matrices*. ProQuest LLC, Ann Arbor, MI, 2012. Thesis (Ph.D.)–University of Pennsylvania.

[49] N. Loehr. *Multivariate Analogues of Catalan Numbers, Parking Functions, and their Extensions*. Ph.D. Thesis, University of California at San Diego, 2003.

[50] Nicholas A. Loehr. Combinatorics of $q$, $t$-parking functions. *Adv. in Appl. Math.*, 34(2):408–425, 2005.

[51] Nicholas A. Loehr. Conjectured statistics for the higher $q,t$-Catalan sequences. *Electron. J. Combin.*, 12:Research Paper 9, 54 pp. (electronic), 2005.

[52] Nicholas A. Loehr and Jeffrey B. Remmel. Conjectured combinatorial models for the Hilbert series of generalized diagonal harmonics modules. *Electron. J. Combin.*, 11(1):Research Paper 68, 64 pp. (electronic), 2004.

[53] I. G. Macdonald. *Symmetric Functions and Hall Polynomials*. Oxford Mathematical Monographs, second ed., Oxford Science Publications. The Clarendon Press Oxford University Press, New York, 1995.

[54] P. A. MacMahon. *Combinatory Analysis*. Two volumes (bound as one). Chelsea Publishing Co., New York, 1960.

[55] Alexei Oblomkov, Jacob Rasmussen, and Vivek Shende. The Hilbert scheme of a plane curve singularity and the HOMFLY homology of its link. arXiv:1201.2115, 2012.

[56] Bruce E. Sagan. *The symmetric group*, volume 203 of *Graduate Texts in Mathematics*. Springer-Verlag, New York, second edition, 2001. Representations, combinatorial algorithms, and symmetric functions.

[57] R. P. Stanley. Invariants of finite groups and their applications to combinatorics. *Bull. Amer. Math. Soc. (new series)*, 1:475–511, 1979.

[58] R. P. Stanley. *Enumerative Combinatorics*, volume 2. Cambridge University Press, Cambridge, United Kingdom, 1999.

[59] R. P. Stanley. Recent progress in algebraic combinatorics. *Bull. Amer. Math. Soc.*, 40:55–68, 2003.

[60] Richard P. Stanley. *Enumerative combinatorics. Volume 1*, volume 49 of *Cambridge Studies in Advanced Mathematics*. Cambridge University Press, Cambridge, second edition, 2012.

[61] D. Stanton. Recent results for the $q$-Lagrange inversion formula. In *Ramanujan revisited (Urbana-Champaign, Ill., 1987)*, pages 525–536. Academic Press, Boston, MA, 1988.

# Chapter 12

# *Permutation Classes*

**Vincent Vatter**
*University of Florida*
*Department of Mathematics*
*Gainesville, Florida USA*

**CONTENTS**

## 12.1 Introduction

Hints of the study of patterns in permutations date back a century, to Volume I, Section III, Chapter V of MacMahon's 1915 magnum opus *Combinatory Analysis* [126]. In that work, MacMahon showed that the permutations that can be partitioned into two decreasing subsequences (in other words, the 123-avoiding permutations) are counted by the Catalan numbers. Twenty years later, Erdős and Szekeres [84] proved that every permutation of length at least $(k-1)(\ell-1)+1$ must contain either $12\cdots k$ or $\ell\cdots 21$. Roughly twenty-five years after Erdős and Szekeres, Schensted's famous paper [146] on increasing and decreasing subsequences was published.

Most, however, date the study of permutation classes to 1968, when Knuth published Volume 1 of *The Art of Computer Programming* [118]. In Section 2.2.1 of that book, Knuth introduced sorting with stacks and double-ended queues (deques), which leads naturally to the notion of permutation patterns. In particular, Knuth observed that a permutation can be sorted by a stack if and only if it avoids 231 and showed that these permutations are also counted by the Catalan numbers. He inspired many subsequent papers, including those of Even and Itai [85] in 1971, Tarjan [156] in 1972, Pratt [140] in 1973, Rotem [145] in 1975, and Rogers [144] in 1978.

Near the end of his paper, Pratt wrote that

> From an abstract point of view, the [containment order] on permutations is even more interesting than the networks we were characterizing. This relation seems to be the only partial order on permutations that arises in a simple and natural way, yet it has received essentially no attention to date.

Pratt's suggestion to study this order in the abstract was taken up a dozen years later by Simion and Schmidt in their seminal 1985 paper "Restricted permutations" [148]. The field has continually expanded since then, and is now the topic of the conference *Permutation Patterns*, held each year since its inauguration (by Albert and Atkinson) at the University of Otago in 2003.

Several overviews of the field have been published, including Kitaev's 494-page compendium *Patterns in Permutations and Words* [112], one chapter in Bóna's undergraduate textbook *A Walk Through Combinatorics* [54] and several in his monograph *Combinatorics of Permutations* [50], and Steingrímsson's survey article [155] for the 2013 *British Combinatorial Conference*. In addition, the proceedings of the conference *Permutation Patterns 2007* [125] contains surveys by Albert [2], Atkinson [22], Bóna [53], Brignall [62], Kitaev [111], Klazar [117], and Steingrímsson [154] on various aspects of the field.

This survey differs significantly from prior overviews. This is partly because there is a lot of new material to discuss. In particular, Section 12.2.5 presents Fox's results on growth rates of principal classes. After that, Section 12.3 is peppered with recent results, while Section 12.4 presents some new results improving on those published. More significantly, this survey differentiates itself from previous summaries of the area by its focus on permutation classes in general.

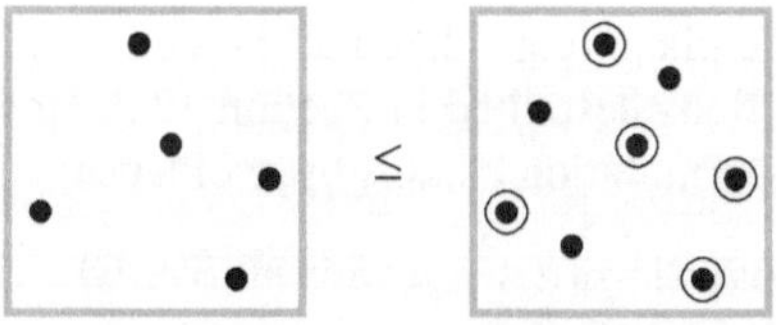

**Figure 12.1**
The containment order on permutations.

In order to maintain this focus, a great many beautiful results have been omitted. Thus, despite the impressive results of Elizalde [83], consecutive patterns will not be discussed. Nor will there be any discussion of mesh patterns, which began as a generalization of the "generalized" (now called vincular) patterns introduced by Babson and Steingrímsson [28] in their classification of Mahonian statistics but have since been shown to be worthy of study on their own via the wonderful Reciprocity Theorem of Brändén and Claesson [60]. We similarly neglect two questions raised by Wilf in [165]: packing densities (which Presutti and Stromquist [141] have shown can be incredibly interesting) and the topology of the poset of permutations (where McNamara and Steingrímsson [130] have established some significant results). Indeed, we even ignore the original application to sorting, despite the deep results of Albert and Bousquet-Mélou [10] and Pierrot and Rossin [137, 138]. Alas, even this list of omitted topics contains omissions, for which I apologize.

### 12.1.1 Basics

Throughout this survey we think of permutations in one-line notation, so a permutation of length $n$ is simply an ordering of the set $[1,n] = \{1,2,\ldots,n\}$ of integers. The permutation $\pi$ **contains** the permutation $\sigma$ of length $k$ if it has a subsequence of length $k$ that is **order isomorphic** to $\sigma$, i.e., that has the same pairwise comparisons as $\sigma$. For example, the subsequence 38514 is order isomorphic to 25413, so 25413 is contained in the permutation 36285714. Permutation containment is perhaps best seen by drawing the **plot** of a permutation, which is the set of points $\{(i,\pi(i))\}$ as shown on the right of Figure 12.1. Permutation containment is a partial order on the set of all finite permutations, so if $\sigma$ is contained in $\pi$ we write $\sigma \leq \pi$. If $\sigma \not\leq \pi$, we say that $\pi$ **avoids** $\sigma$.

The central objects of study in this survey are **permutation classes**, which are downsets (a.k.a., lower order ideals) of permutations under the containment order. Thus if $\mathcal{C}$ is a class containing the permutation $\pi$ and $\sigma \leq \pi$ then $\sigma$ must also lie in $\mathcal{C}$. Given any set $X$ of permutations, one way to obtain a permutation class is to take the **downward closure** of $X$,

$$\mathrm{Sub}(X) = \{\sigma : \sigma \leq \pi \text{ for some } \pi \in X\}.$$

There are many other ways to specify a permutation class, for example as the set of permutations sortable by a particular machine, as the set of permutations that can be "drawn on" a figure in the plane (considered at the end of this subsection), or by a number of other constructions described in Section 12.3. However, by far the most common way to define a permutation class is by avoidance:

$$\mathrm{Av}(B) = \{\pi : \pi \text{ avoids all } \beta \in B\}.$$

If one permutation of $B$ is contained in another then we may remove the larger one without changing the class. Thus we may take $B$ to be an **antichain**, meaning that no element of $B$ contains any others. In the case that $B$ is an antichain we call it the **basis** of this class. The case where $B$ is a singleton has received considerable attention; we call such classes **principal**.

The pictorial view shown in Figure 12.1 makes it clear that the containment order has the eight symmetries of the square, which (from the permutation viewpoint) are generated by inverse and reverse. For example, the classes Av(132), Av(213), Av(231), and Av(312) are all symmetric (isomorphic as partially ordered sets), as are the classes Av(123) and Av(321). This shows that there are only two essentially different principal classes avoiding a permutation of length 3. There are seven essentially different principal classes avoiding a permutation of length 4.

We are frequently interested in the enumeration of permutation classes. Thus letting $\mathscr{C}_n$ denote the set of permutations of length $n$ in the class $\mathscr{C}$, we wish to determine (either exactly or asymptotically) the behavior of the sequence $|\mathscr{C}_0|$, $|\mathscr{C}_1|$, ... (this sequence is called the **speed** of the class in some contexts). One way of doing this is to explicitly compute the generating function of the class,

$$\sum_{n\geq 0} |\mathscr{C}_n| x^n = \sum_{\pi\in\mathscr{C}} x^{|\pi|},$$

where here $|\pi|$ denotes the length of $\pi$. We are often interested in whether this generating function is **rational** (the quotient of two polynomials), **algebraic** (meaning that there is a polynomial $p(x,y) \in \mathbb{Q}[x,y]$ such that $p(x, f(x)) = 0$), or **D-finite** (if its derivatives span a finite dimensional vector space over $\mathbb{Q}(x)$).

In practice it is often quite difficult to compute generating functions of permutation classes, and thus we must content ourselves with the rough asymptotics of $|\mathscr{C}_n|$. To do so we define the **upper** and **lower growth rate** of the class $\mathscr{C}$ by

$$\overline{\mathrm{gr}}(\mathscr{C}) = \limsup_{n\to\infty} \sqrt[n]{|\mathscr{C}_n|} \quad \text{and} \quad \underline{\mathrm{gr}}(\mathscr{C}) = \liminf_{n\to\infty} \sqrt[n]{|\mathscr{C}_n|},$$

respectively. It is not known if these two quantities agree in general.

**Conjecture 12.1.1** *For every permutation class $\mathscr{C}$, $\overline{\mathrm{gr}}(\mathscr{C}) = \underline{\mathrm{gr}}(\mathscr{C})$.*

If the upper and lower growth rates of a class are equal, we refer to their common value as the **(proper) growth rate** of the class. In order to establish a sufficient condition for the existence of proper growth rates, we need two definitions. Pictorially, the **(direct) sum** of the permutations $\pi$ and $\sigma$, denoted by $\pi \oplus \sigma$, is shown on the left of Figure 12.2. If $\pi$ has length $k$ and $\sigma$ has length $\ell$, we can also define the sum of $\pi$ and $\sigma$ by

$$(\pi\oplus\sigma)(i) = \begin{cases} \pi(i) & \text{for } i \in [1,k], \\ \sigma(i-k)+k & \text{for } i \in [k+1, k+\ell]. \end{cases}$$

The analogous operation depicted on the right of Figure 12.2 is called the **skew sum**.

**Figure 12.2**
The sum and skew sum operations.

The permutation class $\mathscr{C}$ is said to be **sum closed** (respectively, **skew closed**) if $\pi \oplus \sigma \in \mathscr{C}$ (respectively, $\pi \ominus \sigma \in \mathscr{C}$) for every pair of permutations $\sigma, \pi \in \mathscr{C}$. The permutation $\pi$ is further said to be **sum** (respectively, **skew**) **decomposable** if it can be expressed as a nontrivial sum (respectively, skew sum) of permutations, and **sum** (respectively, **skew**) **indecomposable** otherwise. It is easy to establish that a class is sum (respectively, skew) closed if and only if all of its basis elements are sum (respectively, skew) indecomposable. By observing that a single permutation cannot be both sum and skew decomposable, we obtain the following.

**Observation 12.1.2** *Every principal permutation class is either sum or skew closed.*

The sequence $\{a_n\}$ is said to be **supermultiplicative** if $a_{m+n} \geq a_m a_n$ for all $m$ and $n$. Fekete's Lemma states that if the sequence $\{a_n\}$ is supermultiplicative then $\lim \sqrt[n]{a_n}$ exists and is equal to $\sup \sqrt[n]{a_n}$. A simple application of this lemma gives us the following result.

**Proposition 12.1.3 (Arratia [19])** *Every sum closed (or, by symmetry, skew closed) permutation class has a (possibly infinite) growth rate. In particular, this holds for every principal class.*

*Proof.* Suppose $\mathscr{C}$ is a sum closed permutation class, and thus $\pi \oplus \sigma \in \mathscr{C}_{m+n}$ for all $\pi \in \mathscr{C}_m$ and $\sigma \in \mathscr{C}_n$. Moreover, a given $\tau \in \mathscr{C}_{m+n}$ arises in this way from at most one such pair so $|\mathscr{C}_{m+n}| \geq |\mathscr{C}_m||\mathscr{C}_n|$. This shows that the sequence $\{|\mathscr{C}_n|\}$ is supermultiplicative, and thus $\lim \sqrt[n]{|\mathscr{C}_n|}$ exists by Fekete's Lemma. ∎

In particular, $\mathrm{gr}(\mathscr{C}) \geq \sqrt[n]{|\mathscr{C}_n|}$ for all sum or skew closed classes $\mathscr{C}$ and all integers $n$, which can be used to establish lower bounds for growth rates of these classes. (We appeal to this fact in the proofs of both Proposition 12.2.1 and Theorem 12.2.10.)

Continuing our exploration of the sum operation, for every permutation $\pi$ there are unique sum indecomposable permutations $\alpha_1, \ldots, \alpha_k$ (called the **sum components** of $\pi$) such that $\pi = \alpha_1 \oplus \cdots \oplus \alpha_k$. Therefore the permutations in a sum closed class can be viewed as sequences of sum indecomposable permutations. In particular, if a class is sum closed and the generating function for its nonempty sum indecomposable members is $s$, then the generating function for the class is $1/(1-s)$.

A permutation is **layered** if it is the sum of decreasing permutations. Clearly every decreasing permutation is sum indecomposable, so by our previous remarks the generating function for the class of layered permutations is

$$\frac{1}{1-\frac{x}{1-x}} = \frac{1-x}{1-2x}.$$

(We could have instead observed that the layered permutations are in bijection with integer compositions.) It is not difficult to show that the basis of the class of layered permutations is $\{231,312\}$.

A permutation is **separable** if it can be built from the permutation 1 by repeated sums and skew sums. For example, the permutation 576984132 is separable:

$$\begin{aligned} 765984132 &= 32154 \ominus 1 \ominus 132 \\ &= (321 \oplus 21) \ominus 1 \ominus (1 \oplus 21) \\ &= ((1 \ominus 1 \ominus 1) \oplus (1 \ominus 1)) \ominus 1 \ominus (1 \oplus (1 \ominus 1)). \end{aligned}$$

The term separable is due to Bose, Buss, and Lubiw [56], who proved that the separable permutations are Av(2413,3142), although these permutations first appeared in the much earlier work of Avis and Newborn [27].

Recall that the **little Schröder number** indexed by $n-1$ counts the number of ways to insert parenthesis into a sequence of $n$ symbols in such a way that every pair of parentheses surrounds at least two symbols or parenthesized groups and no parentheses surround the entire sequence (see Stanley [152, Exercise 6.39.a]), while the **large Schröder numbers** are twice the little Schröder numbers. It follows immediately from our decomposition above that the separable permutations are counted by the large Schröder numbers. For example, the permutation from our example can be encoded by the pair

$$\ominus, ((\bullet\bullet\bullet)(\bullet\bullet))\bullet(\bullet(\bullet\bullet)),$$

where $\ominus$ indicates that the outermost division is a skew sum and each pair of parentheses within the expression denotes the opposite type of decomposition as the pair enclosing it. These numbers begin 1, 2, 6, 22, 90, 394, 1806, ..., and have the algebraic generating function

$$\frac{3-x-\sqrt{1-6x+x^2}}{2}.$$

It follows that the growth rate of the separable permutations is approximately 5.83.

The pictorial perspective presented in Figure 12.1 leads naturally to a geometric treatment of the containment order, in which permutations are objects of a moduli space. From this viewpoint, the fundamental objects are **figures**, which are simply subsets of the plane. Given two figures $\Phi, \Psi \subseteq \mathbb{R}^2$, the figure $\Phi$ is **involved** in the figure $\Psi$, denoted $\Phi \leq \Psi$ if there are subsets $A, B \subseteq \mathbb{R}$ and increasing injections $\tau_x : A \to \mathbb{R}$ and $\tau_y : B \to \mathbb{R}$ such that

$$\Phi \subseteq A \times B \text{ and } \tau(\Phi) \subseteq \Psi,$$

where $\tau(\Phi) = \{(\tau_x(a), \tau_y(b)) : (a,b) \in \Phi\}$.

The involvement order is a **preorder** on the collection of all figures (it is reflexive and transitive but not necessarily antisymmetric). If $\Phi \leq \Psi$ and $\Psi \leq \Phi$, then we say that $\Phi$ and $\Psi$ are **equivalent** figures and write $\Phi \approx \Psi$. If two figures have only finitely many points, it can be shown that they are equivalent if and only if one can be transformed to the other by stretching and shrinking the axes. Figure 12.3 shows two examples of equivalent figures.

**Figure 12.3**
Two pairs of equivalent figures.

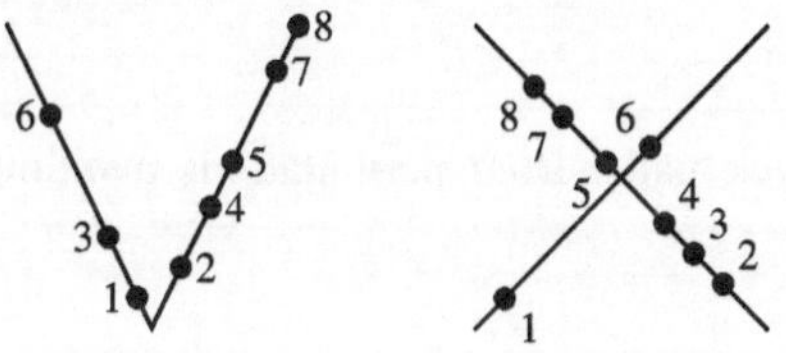


**Figure 12.4**
On the left, the permutation 63124578 can be drawn on a V. On the right, the permutation 18756432 can be drawn on an X.

We call a figure **independent** if no two points of the figure lie on a common horizontal or vertical line. It is clear that every finite independent figure is equivalent to the plot of a unique permutation (which we could call its **modulus** if adopting the moduli space perspective). Under this identification, the involvement order on equivalence classes of finite independent figures is isomorphic to the containment order on permutations. Every figure $\Phi \subseteq \mathbb{R}^2$ therefore defines a permutation class,

$$\mathrm{Sub}(\Phi) = \{\pi : \pi \text{ is equivalent to a finite independent figure involved in } \Phi\},$$

which we call a **figure class**. If $\pi \in \mathrm{Sub}(\Phi)$ then we say that $\pi$ can be **drawn** on the figure $\Phi$. Two examples of this are shown in Figure 12.4. For example, the class Sub(V) of permutations that can be drawn on the leftmost figure in Figure 12.3 contains all permutations that consist of a decreasing sequence followed by an increasing subsequence. It is not difficult to show that Sub(V) = Av(132, 231). The class Sub(X) was studied by Elizalde [82], who established a bijection between this class and a set of permutations studied by Knuth in Volume 3 of *The Art of Computer Programming* [119]. Both of these classes are examples of geometric grid classes, introduced in Section 12.4.2.

### 12.1.2 Avoiding a permutation of length three

Here we briefly consider the two simplest nontrivial principal classes: Av(231) and Av(321). As mentioned in the introduction, both of these classes are counted by the Catalan numbers (and thus have growth rates of 4), so there ought to be nice bijections between them and Dyck paths. Indeed there are, as shown in Figure 12.5, and it is almost the same bijection for both classes.

The bijection depicted in Figure 12.5 is originally due to Knuth [118, 119], though it was Krattenthaler [121] who gave the nonrecursive formulation we de-

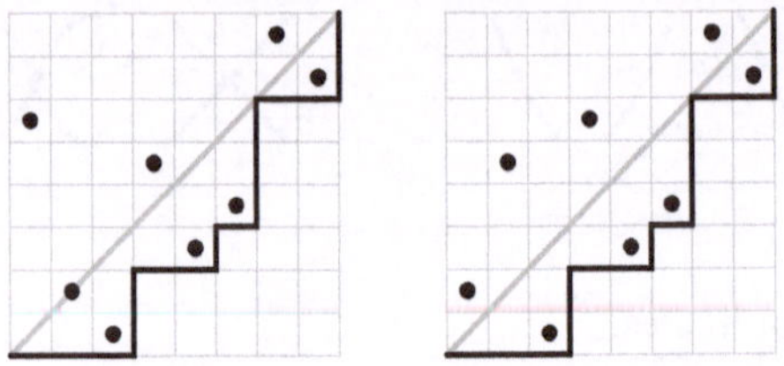

**Figure 12.5**
The construction of Dyck paths from permutations avoiding 231 (left) and 321 (right).

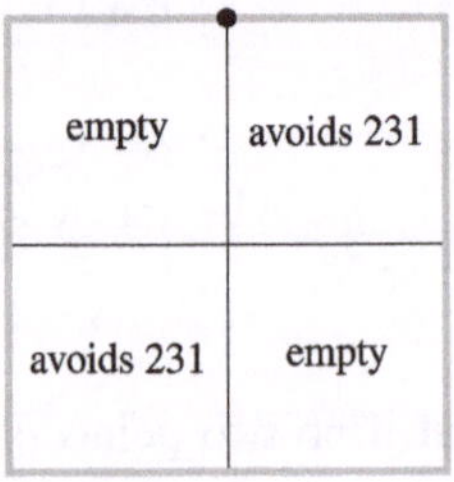


**Figure 12.6**
A 231-avoiding permutation.

scribe. In this bijection, we plot a permutation and then draw the Dyck path that lies just below its right-to-left minima (an entry is a **right-to-left minimum** if it is smaller than every entry to its right). It is then possible to show that the positions of the non-right-to-left minima can be determined from this Dyck path. There are at least eight more essentially different bijections between these two classes. For the rest of them we refer to Claesson and Kitaev's survey [77].

Despite the elegance of this bijection, Av(231) and Av(321) are *very* different classes. Indeed, Miner and Pak [131] make a compelling argument that there is no truly "ultimate" bijection between these two classes. As discussed in Section 12.3, Av(231) has only countably many subclasses, while Av(321) has uncountably many. Indeed, Albert and Atkinson [3] have proved that every proper subclass of Av(231) has a rational generating function, while a simple counting argument shows that Av(321) contains subclasses whose generating functions aren't even $D$-finite.

Another way to see the contrast between Av(231) and Av(321) is to consider the decomposition of a 231-permutation shown in Figure 12.6. As this figure shows, all entries to the left of the maximum in a 231-avoiding permutation must lie below all entries to the right of the maximum. Going in the other direction, every permutation constructed in this manner avoids 231. Thus if we let $f$ denote the generating function for the 231-avoiding permutations (including the empty permutation), we see immediately that

$$f = xf^2 + 1,$$

so $f$ is the generating function for the Catalan numbers.

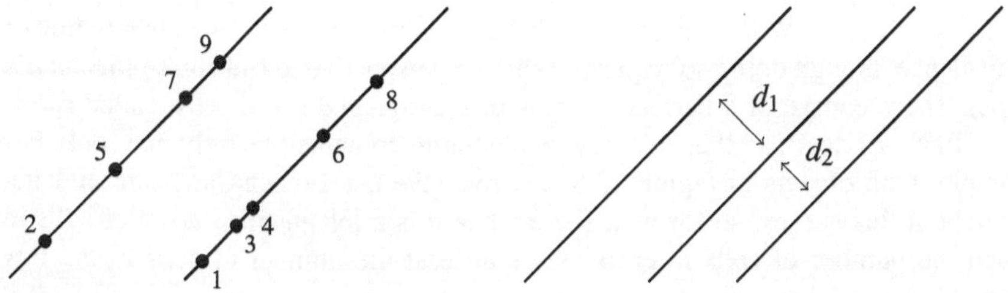


**Figure 12.7**
On the left, a drawing of the permutation 257193468 on two parallel lines. On the right, three parallel lines.

Despite the similarities of Figure 12.5, there is no analogue of Figure 12.6 for the 321-avoiding permutations. Perhaps the best structural description of this class one can give is the following.

**Proposition 12.1.4** *The class of permutations that can be expressed as the union of two increasing subsequences is* Av(321).

*Proof.* Label each entry of $\pi \in \text{Av}(321)$ by the length of the longest decreasing subsequence that ends at that entry. Clearly we have only used the labels 1 and 2, and for each value of $i$ the entries labeled by $i$ form an increasing subsequence. ∎

In his thesis, Waton proved the following geometric version of Proposition 12.1.4 (illustrated on the left of Figure 12.7).

**Proposition 12.1.5 (Waton [163])** *The class of permutations that can be drawn on any two parallel lines of positive slope is* Av(321).

Our proof of Proposition 12.1.4 generalizes easily to show that every $k\cdots 21$-avoiding permutation can be expressed as the union of $k-1$ increasing subsequences. However, Waton showed that Proposition 12.1.5 does not extend in the same way. Consider three parallel lines at distances $d_1$ and $d_2$ from each other as depicted on the right of Figure 12.7. Clearly every permutation that can be drawn on these lines avoids 4321, but Waton proved that every different choice of the ratio $d_1/d_2$ leads to a different proper subclass of Av(4321).

### 12.1.3 Wilf-equivalence

As we saw in the last subsection, the permutations 231 and 321 are equally avoided. Instead of being a coincidence of small numbers, this turns out to be a common phenomenon. The classes $\mathscr{C}$ and $\mathscr{D}$ are said to be **Wilf-equivalent** if they are equinumerous, i.e., if $|\mathscr{C}_n| = |\mathscr{D}_n|$ for every $n$. In this subsection we are primarily concerned with Wilf-equivalence of principal classes, so instead of saying that $\text{Av}(\beta)$ and $\text{Av}(\gamma)$ are Wilf-equivalent we simply say that $\beta$ and $\gamma$ are Wilf-equivalent.

Large swaths of Wilf-equivalences can be explained via a much stronger notion of equivalence, though doing so requires a shift of perspective to **full rook placements (frps)**. These consist of a Ferrers board with a designated set of cells, called **rooks** (here drawn as dots), so that each row and column contains precisely one rook. For example, both objects in Figure 12.8 are frps. (We use French/Cartesian indexing throughout this survey, so for us a Ferrers board is a left-justified array of cells in which the number of cells in each row is at least the number of cells in the row above.)

There is a natural partial order on the set of all frps: given frps $R$ and $S$, we say that $R$ is **contained** in $S$ if $R$ can be obtained from $S$ by deleting rows and columns. Furthermore, this partial order generalizes the containment order on permutations. To make this precise, let us call a frp **square** if the underlying Ferrers board is square and say that the permutation $\pi$ of length $n$ corresponds to the $n \times n$ frp with rooks in the cells $(i, \pi(i))$ for every $i$. When restricted to square frps, the containment order on frps is equivalent to the containment order on the corresponding permutations.

Because of this correspondence between permutations and square frps, we say that a frp **contains** the permutation $\sigma$ if it contains the square frp corresponding to $\sigma$ and otherwise say that it **avoids** $\sigma$. Observe that the entire square containing $\sigma$ must be contained within the frp; for example, the frp below contains 12 but avoids 21.

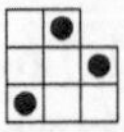

We say that the permutations $\beta$ and $\gamma$ are **shape-Wilf-equivalent** if given any shape $\lambda$, the number of $\beta$-avoiding frps of shape $\lambda$ is the same as the number of $\gamma$-avoiding frps of shape $\lambda$. Shape-Wilf-equivalence implies Wilf-equivalence by considering only square shapes, but it also implies quite a bit more.

**Proposition 12.1.6 (Babson and West [29])** *If $\beta$ and $\gamma$ are shape-Wilf-equivalent, then for every permutation $\delta$, $\beta \oplus \delta$ and $\gamma \oplus \delta$ are also shape-Wilf-equivalent.*

*Proof.* Suppose that there is a bijection between $\beta$-avoiding and $\gamma$-avoiding frps of every shape. Now fix a shape $\lambda$. We construct a bijection between $\beta \oplus \delta$-avoiding frps and $\gamma \oplus \delta$-avoiding frps of shape $\lambda$. Let $R$ be a $\beta \oplus \delta$-avoiding frp of shape $\lambda$.

We call a cell of $R$ **dangerous** if there is a copy of $\delta$ completely contained in the region above and to the right of the cell. The entire set of dangerous cells is called the **danger zone** (see Figure 12.8). The danger zone forms a (possibly empty) Ferrers board nestled in the bottom-left corner of $R$. Ignoring the rookless rows and columns of the danger zone, we thus obtain a $\beta$-avoiding frp. We may then use the bijection between $\beta$-avoiding and $\gamma$-avoiding frps of that shape to produce a $\gamma \oplus \delta$-avoiding frp of shape $\lambda$, as desired. ∎

Only two shape-Wilf-equivalence results are known. We sketch bijective proofs for both.

**Theorem 12.1.7 (Backelin, West, and Xin [30])** *For every value of k, the permutations $k \cdots 21$ and $12 \cdots k$ are shape-Wilf-equivalent.*

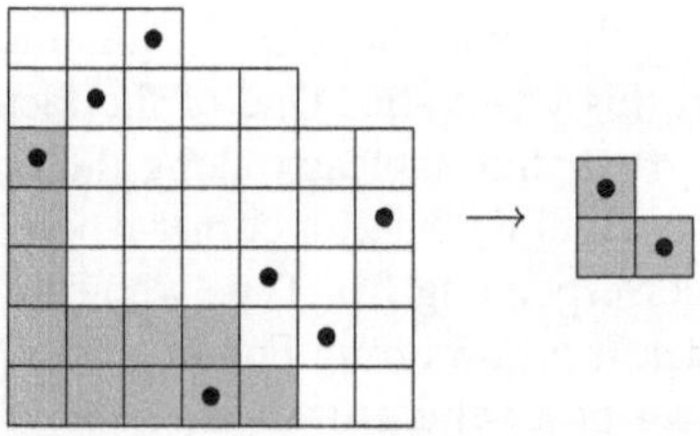

**Figure 12.8**
On the left, a frp with its danger zone for the permutation 12 shaded. On the right, the frp that the danger zone contracts to after removing rookless rows and columns.

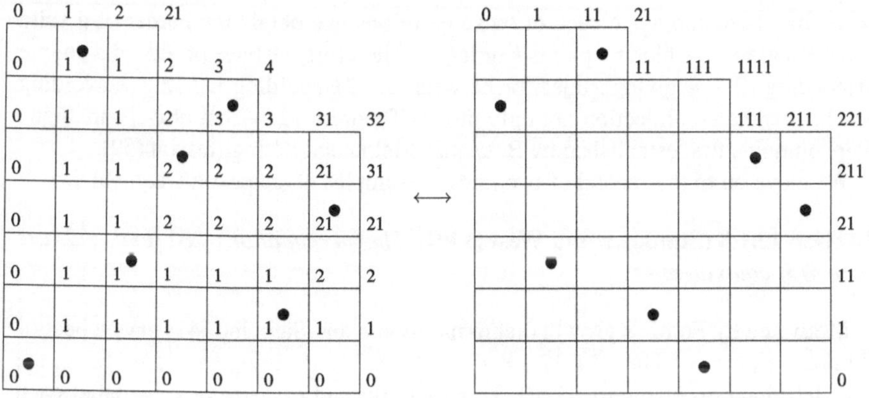

**Figure 12.9**
Krattenthaler's bijection, shown here converting a 321-avoiding frp to a 123-avoiding frp.

By Proposition 12.1.6 this implies that the permutations $k\cdots 21\oplus\beta$ and $12\cdots k\oplus\beta$ are Wilf-equivalent for every permutation $\beta$. Special cases of Theorem 12.1.7 had been established before the general case above; West proved the $k = 2$ case in his thesis [164], while the $k = 3$ case was shown by Babson and West [29]. Krattenthaler [122] was the first to give a truly nice bijection between $k\cdots 21$-avoiding and $12\cdots k$-avoiding frps, using the growth diagrams of Fomin [87].

Consider the labeled frp shown on the left of Figure 12.9. In this diagram every corner is labeled by an integer partition (the labels lie slightly above and to the right of the corners). This partition (written in shortened notation) represents the shape of the resulting standard Young tableaux when the partial permutation lying below and to the left of that corner is passed through the Robinson–Schensted correspondence. As one can imagine, there are a variety of restrictions on such growth diagrams. For example, the labels along the northwest-southeast boundary form an **oscillating tableaux**, meaning that each partition differs from the previous partition by precisely one box.

We will not concern ourselves with the other requirements here, except to call labeled tableaux formed in this way **valid**. One of the facts that Krattenthaler established is that given the oscillating tableaux along the border of a valid growth diagram, one can reconstruct all of the interior corner labels, and from these, the position of the rooks in the corresponding frp. Thus we need only concern ourselves with the labels of this border. It follows from Fomin's work that by conjugating the partitions labeling the border of a valid growth diagram one obtains another valid growth diagram.

Now recall that if a frp avoids $k \cdots 21$, its shape under Robinson–Schensted has fewer than $k$ parts (Schensted's Theorem [146]). Equivalently, all parts along the border of its valid growth diagram will have fewer than $k$ parts. As Krattenthaler observed, this combination of Fomin's growth diagrams and Schensted's Theorem shows that if we conjugate each of those partitions, we obtain the border of a valid growth diagram in which no part is $k$ or larger. Therefore we have produced a unique frp avoiding $12 \cdots k$, giving a bijection between $k \dots 21$-avoiding and $12 \cdots k$-avoiding frps. Krattenthaler's bijection not only proves Theorem 12.1.7 but also its analogue for involutions, first established by Bousquet-Mélou and Steingrímsson [59].

We move on to the second of two known examples of shape-Wilf-equivalence.

**Theorem 12.1.8 (Stankova and West [149])** *The permutations* 231 *and* 312 *are shape-Wilf-equivalent.*

Motivated by Fomin's growth diagrams, Bloom and Saracino [41] gave a beautiful bijective proof of Theorem 12.1.8. Consider a 231-avoiding frp, as shown in the upper-left diagram of Figure 12.10. This time, instead of partitions, we label each corner of the northwest-southeast border by the length of the longest increasing sequence in the partial permutation lying below and to the left of that corner. We then read these labels to construct a labeled Dyck path, as shown on the right of this figure.

We need a final term. A **weak tunnel** in a Dyck path is a horizontal segment between two vertices of the path that stays weakly below the path (the paths shown in Figure 12.10 each have a weak tunnel indicated in gray). Bloom and Saracino proved that these labeled Dyck paths satisfy three rules:

- **Monotone property:** Labels increase by at most 1 after an up step and decrease by at most 1 after a down step.
- **Zero property:** The 0 labels are precisely those on the $x$-axis.
- **231-avoiding tunnel property:** Given two vertices at the same height connected by a weak tunnel, the label of the leftmost vertex is at most the label of the rightmost vertex. (This was called the "diagonal property" in [40].)

They also proved that there is a bijection between labeled Dyck paths and 231-avoiding frps satisfying these three properties. Suppose we perform the same operation on 312-avoiding frps, as shown on the bottom of Figure 12.10. Bloom and Saracino showed that the resulting labeled Dyck paths have all of the same properties except for one; the 231-avoiding tunnel property becomes the

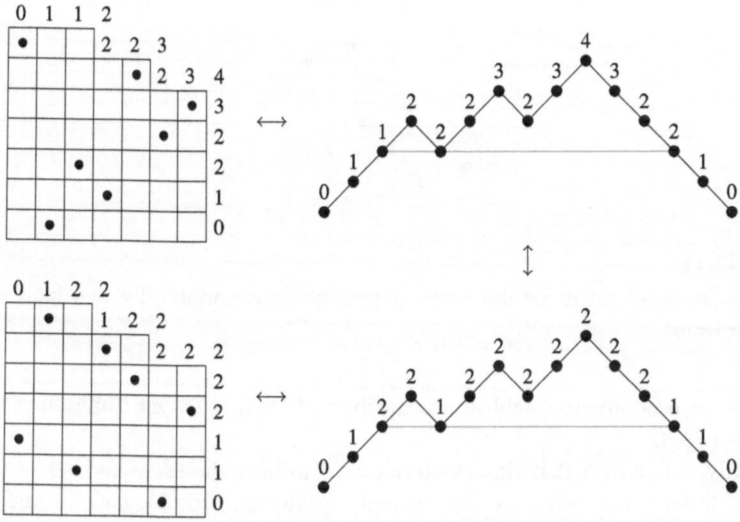


**Figure 12.10**
Bloom and Saracino's bijection between 231-avoiding (upper-left) and 312-avoiding frps (bottom-left).

- **312-avoiding tunnel property:** Given two vertices at the same height connected by a weak tunnel, the label of the rightmost vertex is at most the label of the leftmost vertex.

To prove Theorem 12.1.8, we merely need to construct a bijection between labeled Dyck paths satisfying the monotone, zero, and 231-avoiding tunnel properties and those that satisfy the monotone, zero, and 312-avoiding tunnel properties. This (as is illustrated on the right of Figure 12.10) is done by fixing the labels on the $x$-axis at zero, and then replacing the label $\ell$ in the labeled Dyck path of a 231-avoiding frp by $h-\ell+1$, where $h$ is the height of the corresponding vertex.

As no other shape-Wilf-equivalences are known, we must pose the following question.

**Question 12.1.9** *Do Theorems 12.1.7 and 12.1.8 (and their implications) constitute all of the shape-Wilf-equivalences?*

To finish mapping the known world of Wilf-equivalence of principal classes, we need one more result.

**Theorem 12.1.10 (Stankova [150])** *The permutations* 1342 *and* 2413 *are Wilf-equivalent, but not shape-Wilf-equivalent.*

Bloom [39] was the first to describe a nice bijection between these classes (more precisely, he constructed a bijection between 1423-avoiding permutations and 2413-avoiding permutations). By showing that his bijection preserves certain permutation

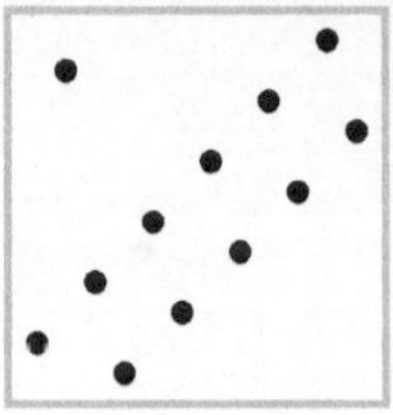

**Figure 12.11**
A typical basis element for the class of permutations sortable by two increasing stacks in series.

statistics, he was able to establish a conjecture of Dokos, Dwyer, Johnson, Sagan, and Selsor [80].

Finally, we remark that classes do not need to have the same number of basis elements to be Wilf-equivalent. For example, Atkinson, Murphy, and Ruškuc [24] showed that the class Av(1342) is Wilf-equivalent to the infinitely based class

$$\mathrm{Av}(\{2\ (2k-1)\ 4\ 1\ 6\ 3\ 8\ 5\ \cdots\ (2k)\ (2k-3)\ :\ k \geq 2\}),$$

which consists of those permutations that can be sorted by two increasing stacks in series (for a visualization of these basis elements, see Figure 12.11; a proof of this result using frps is given in Bloom and Vatter [42]). Burstein and Pantone [70] later showed that Av(1234) and Av(1324, 3416725) are Wilf-equivalent, along with other **unbalanced Wilf-equivalences.**

### 12.1.4 Avoiding a longer permutation

In the study of principal classes, there is a steep increase in difficulty when we increase the length of the basis element from 3 to 4. We begin this section with a routine bound, but end it with the easily-stated but notorious problem of computing the growth rate of Av(1324).

**Proposition 12.1.11** *The growth rate of* $\mathrm{Av}(k\cdots 21)$ *is at most* $(k-1)^2$.

*Proof.* Let $\pi$ be a $k\cdots 21$-avoiding permutation of length $n$. By the obvious generalization of Proposition 12.1.4 mentioned in Section 12.1.2, we can partition the entries of $\pi$ into $k-1$ increasing subsequences. Label these subsequences 1 to $k-1$ and extend this labeling to their entries. We can easily reconstruct $\pi$ if we know the positions and values of the entries of each of these subsequences. We can record this information in two lists, one reading left to right, the other reading bottom to top. Thus there are at most $(k-1)^{2n}$ permutations of length $n$ that avoid $k\cdots 21$, as desired. ∎

There does not seem to be an elementary way to establish a matching lower bound for Proposition 12.1.11. Bóna [51, Corollary 5.8] proved that for every per-

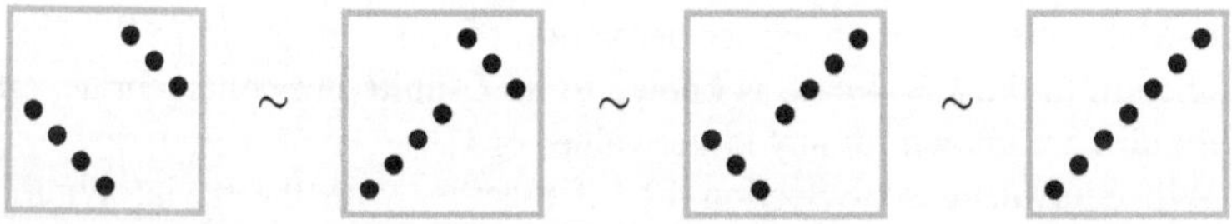

**Figure 12.12**
Demonstrating the Wilf-equivalence of a two-layer permutation and a monotone permutation of the same length. The first and last equivalences follow from Theorem 12.1.7, while the central equivalence is an application of symmetry.

mutation $\beta$,

$$\sqrt{\mathrm{gr}(\mathrm{Av}(\beta \ominus 21))} = 1 + \sqrt{\mathrm{gr}(\mathrm{Av}(\beta \ominus 1))},$$

which implies that $\mathrm{gr}(\mathrm{Av}(k\cdots 21)) = (k-1)^2$, but his proof is quite involved. The first derivation of $\mathrm{gr}(\mathrm{Av}(k\cdots 21))$ follows from a much more general result of Regev.

**Theorem 12.1.12 (Regev [142])** *The growth rate of* $\mathrm{Av}(k\cdots 21)$ *is* $(k-1)^2$.

This immediately implies the following, which will be useful later when we bound the growth rate of principal classes avoiding layered permutations (Theorem 12.3.8).

**Corollary 12.1.13** *For any layered permutation* $\beta$ *of length* $k$ *and consisting of at most two layers, the growth rate of* $\mathrm{Av}(\beta)$ *is* $(k-1)^2$.

*Proof.* The single layer case is Regev's Theorem 12.1.12, while the two-layer case follows by Theorem 12.1.7 and symmetry, as shown in Figure 12.12. ∎

Gessel [95] later gave an explicit formula for the generating functions of the classes $\mathrm{Av}(k\cdots 21)$ in terms of determinants. Stanley [153] expressed the significance of Gessel's result by writing that

> Gessel's theorem reduces the theorems of Baik, Deift, and Johansson to "just" analysis, viz., the Riemann-Hilbert problem in the theory of integrable systems, followed by the method of steepest descent to analyze the asymptotic behavior of integrable systems.

(The quote refers to Baik, Deift, and Johansson's proof [31] that, after appropriate rescaling, the distribution of the longest increasing subsequence statistic on permutations of length $n$ converges to the Tracy-Widom distribution.)

Gessel's result was later rederived by Bousquet-Mélou [58] using the kernel method. In the case $k = 4$, she expressed the enumeration as an explicit sum,

$$|\mathrm{Av}_n(4321)| = \frac{1}{(n+1)^2(n+2)} \sum_{i=0}^{n} \binom{2i}{i}\binom{n+1}{i+1}\binom{n+2}{i+2}.$$

(Bousquet-Mélou had given an earlier derivation of this formula in [57].) The generating function in the $k = 4$ case is known to be $D$-finite but nonalgebraic, and no "nice" formulas are known for any larger values of $k$.

The Wilf-equivalences of Section 12.1.3 together with the trivial symmetries show that (from an enumerative perspective) there are only two more cases of principal classes avoiding a permutation of length 4: $\mathrm{Av}(1342)$ and $\mathrm{Av}(1324)$. Bóna was the first to enumerate $\mathrm{Av}(1342)$, by constructing a bijection between the skew indecomposable 1342-avoiding permutations and $\beta(0,1)$-trees, which had been previously counted by Tutte [158]. (Because the class $\mathrm{Av}(1342)$ is skew closed, its generating function can be computed easily from the generating function for its skew indecomposables, as mentioned in Section 12.1.1.)

**Theorem 12.1.14 (Bóna [46])** *The generating function for* $\mathrm{Av}(1342)$ *is*

$$\frac{32x}{1+20x-8x^2-(1-8x)^{3/2}},$$

*and thus the growth rate of this class is* 8.

Bloom and Elizalde [40] have recently shown how to derive this generating function using the techniques of Bloom and Saracino [41] discussed in Section 12.1.3, which we briefly sketch.

First, instead of counting $\mathrm{Av}(1342)$ they count the symmetric class $\mathrm{Av}(3124)$. A frp is **board minimal** if its rooks do not lie in any smaller Ferrers board, or equivalently, if it has a rook in each of its upper-right corners. There is an obvious bijection, which we denote by $\chi$, between permutations and board minimal frps. It is also easy to see that were a frp from $\chi(\mathrm{Av}(3124))$ to contain 312, there would be a rook in an upper-right corner above and to the right of this copy of 312, a contradiction. Therefore these frps avoid 312.

Next label the border of these frps as in Section 12.1.3 and convert them to Dyck paths. The corresponding Dyck paths must satisfy the monotone, zero, and 312-avoiding tunnel properties, and in addition, because they arise from board minimal frps, they satisfy the **peak property**: The labels rounding any peak (an up step followed by a down step) are $\ell$, $\ell+1$, $\ell$ for some value of $\ell$.

Thus we are reduced to counting the family of labeled Dyck paths satisfying the monotone, zero, 312-avoiding tunnel, and peak properties. Bloom and Elizalde were able to find the generating function for these by introducing a catalytic variable and applying Tutte's quadratic method (which is described in Flajolet and Sedgewick [86, VII. 8.2]).

This leaves a single case, that of $\mathrm{Av}(1324)$. As reported in [81], at the conference *Permutation Patterns 2005*, Zeilberger made the colorful claim that

> Not even God knows $|\mathrm{Av}_{1000}(1324)|$.

Steingrímsson is less pessimistic, writing in [155] that

> I'm not sure how good Zeilberger's God is at math, but I believe that some humans will find this number in the not so distant future.

| | | | | | | | | | | | | | |
|---|---|---|---|---|---|---|---|---|---|---|---|---|---|
| 1 | | | | | | | | | | | | | |
| 1 | 1 | | | | | | | | | | | | |
| 1 | 2 | 2 | 1 | | | | | | | | | | |
| 1 | 2 | 5 | 6 | 5 | 3 | 1 | | | | | | | |
| 1 | 2 | 5 | 10 | 16 | 20 | 20 | 15 | 9 | 4 | 1 | | | |
| 1 | 2 | 5 | 10 | 20 | 32 | 51 | 67 | 79 | 80 | 68 | 49 | 29 | ... |
| 1 | 2 | 5 | 10 | 20 | 36 | 61 | 96 | 148 | 208 | 268 | 321 | 351 | ... |
| 1 | 2 | 5 | 10 | 20 | 36 | 65 | 106 | 171 | 262 | 397 | 568 | 784 | ... |
| 1 | 2 | 5 | 10 | 20 | 36 | 65 | 110 | 181 | 286 | 443 | 664 | 985 | ... |

**Figure 12.13**
The number of 1324-avoiding permutations, grouped into columns by their number of inversions.

Bóna currently holds the record for the lowest upper bound on the growth rate of Av(1324).

**Theorem 12.1.15 (Bóna [44])** *The growth rate of* Av(1324) *is at most* 13.74.

Bóna established Theorem 12.1.15 by refining the approach of Claesson, Jelínek, and Steingrímsson [76] presented in Section 12.3.1. Claesson, Jelínek, and Steingrímsson made a further conjecture which, if true, would give a better upper bound. Consider the table shown in Figure 12.13. In this table, the entry in column $k+1$ of row $n$ is the number of 1324-avoiding permutations of length $n$ with precisely $k$ inversions. As the gray cells are meant to indicate, it appears that the entries along a column are weakly increasing and eventually constant. They proved that the columns are eventually constant, but could not show that they are weakly increasing. If this conjecture could be proved, it would give the improved bound of $e^{\pi\sqrt{2/3}} \leq 13.01$ on the growth rate of Av(1324). Indeed this might be part of a much more general phenomenon; empirical evidence suggests that the analogous table for $\beta$-avoiding permutations has weakly increasing columns for all $\beta \neq 12\cdots k$.

Arratia [19] had conjectured that $\mathrm{gr}(\mathrm{Av}(\beta)) \leq (k-1)^2$ for all permutations $\beta$ of length $k$, and Bóna [51] had gone even further, conjecturing that equality held if and only if $\beta$ was layered. We know now (by Fox's results presented in Section 12.2.5) that these conjectures are far from the truth, but the first known counterexample was the class Av(1324). Using quite a bit of computation and the insertion encoding (introduced by Albert, Linton, and Ruškuc [15]), Albert, Elder, Rechnitzer, Westcott, and Zabrocki [13] showed that

$$\mathrm{gr}(\mathrm{Av}(1324)) \geq 9.47.$$

Extrapolating from the amount their lower bounds improved as they increased the accuracy of their approximation, they guessed that the true value lies between 11 and 12. Madras and Liu [127] later used Markov chain Monte Carlo methods to estimate that this growth rate lies between 10.71 and 11.83.

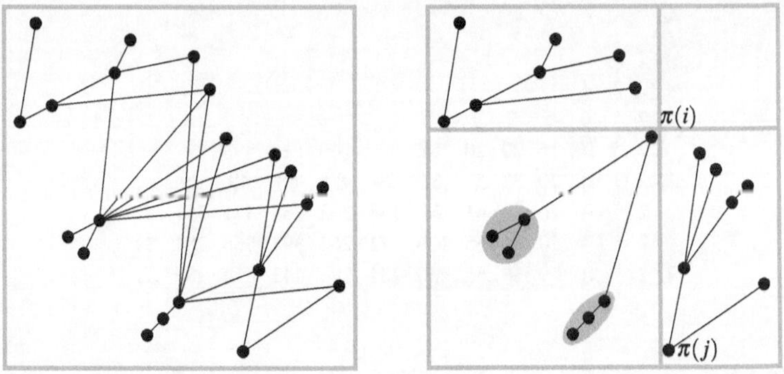


**Figure 12.14**
The drawing on the left shows the Hasse diagram of the poset of a 1324-avoiding permutation. On the right, this Hasse diagram has been divided into three trees.

At present, the best estimate seems to be due to Conway and Guttmann [78]. They extended the approach of Johansson and Nakamura [106] (who had computed the first 31 terms of the enumeration) to compute the first 36 terms of the enumeration of this class and then applied the methodology laid out in Guttmann [98] to approximate that

$$|\operatorname{Av}_n(1324)| \sim C\mu^n \nu^{\sqrt{n}} n^g,$$

where $C \approx 9.5$, $\nu \approx 0.04$, $g \approx -1.1$. and $\mu$—the growth rate—is approximately 11.60.

Bevan [36] has recently established that $\operatorname{gr}(\operatorname{Av}(1324)) > 9.81$ by constructing a large subset of this class with a particularly regular structure. Sketching his proof would require quite a detour, but we attempt to convey a hint of his novel approach.

Given a permutation $\pi$, define the poset $P_\pi$ on the points $\{(i, \pi(i))\}$ in which $(i, \pi(i)) \leq (j, \pi(j))$ if both $i < j$ and $\pi(i) < \pi(j)$, i.e., if this pair of entries forms a noninversion in $\pi$. The drawing on the left of Figure 12.14 shows the Hasse diagram of the poset of a 1324-avoiding permutation. Because this permutation avoids 1324, its Hasse diagram does not contain a diamond:

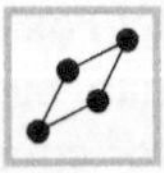

Therefore if we restrict the Hasse diagram to $\pi(1)$ and all entries lying above and to its right we see a rooted tree, as drawn on the right of Figure 12.14. Next let $\pi(i)$ denote the greatest point lying below $\pi(1)$. If we restrict the Hasse diagram to $\pi(i)$ and all entries lying below and to its left, we see another rooted tree. Finally, for this example, if we let $\pi(j)$ denote the leftmost entry to the right of $\pi(i)$ and restrict the Hasse diagram to this entry and all entries above and to its right, we see yet another rooted tree. Let us call rooted trees like the first and third **up trees** and trees like the second **down trees**.

Bevan's construction consists of alternating up trees and down trees, while interleaving their vertices. However, this alone is not sufficient to avoid 1324. For example, the following interleaving of a down tree and an up tree contains 1324.

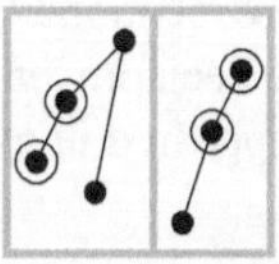

Thus we must be careful in performing this interleaving. In constructing the permutation shown in Figure 12.14, we avoided creating a copy of 1324 by interleaving the vertices of the up trees with the principal subtrees of the down tree, as indicated by the shading in Figure 12.14 (a **principal subtree** of a rooted tree is a component of the forest obtained after removing the root).

## 12.2 Growth rates of principal classes

In the 1980s, Stanley and Wilf independently conjectured that every proper permutation class (that is, every class except the class of all permutations) has a finite upper growth rate. Obviously, it suffices to prove this for principal classes. In this context, the growth rate of $\mathrm{Av}(\beta)$ is often (though not in this survey) called the **Stanley–Wilf limit** of $\beta$.

Two noteworthy partial results were proved in the 20th century. First Bóna [49] proved the Stanley–Wilf Conjecture for $\mathrm{Av}(\beta)$ when $\beta$ is a layered permutation (we prove a stronger result in Theorem 12.3.8). Then Alon and Friedgut [18] used a result of Klazar [113] about Davenport-Schinzel sequences to prove that for every $\beta$ there is a constant $c_\beta$ such that $|\mathrm{Av}_n(\beta)| < c_\beta^{n\gamma(n)}$ where $\gamma$ grows even more slowly than the inverse Ackermann function (the inverse Ackermann function itself is considered to be less than 5 for all "reasonable" values of $n$). This result caused Wilf's belief to waver; in 2002 he [165] wrote that

> This conjecture had been considered a "sure thing," but the results of Alon and Friedgut seem to make it somewhat less certain because a similar bound, involving the Ackermann function, in the Davenport-Schinzel theory turns out to be best possible.

In 2004, Marcus and Tardos [129] presented an elegant proof of the Stanley–Wilf Conjecture. About their proof, Zeilberger [168] wrote

> Once I thought that proofs of long-standing conjectures had to be difficult. After all didn't brilliant people make them, and didn't these people, and many other brilliant people, unsuccessfully try to prove them? Isn't that a meta-proof that the proof, if it exists, should be long, difficult, and ugly? And if not, shouldn't they, at least, contain entirely new ideas and/or techniques?

We present Marcus and Tardos' proof in Section 12.2.3. Of course, the proof of the Stanley–Wilf Conjecture only raises new questions. For example, what can be said about the relationship between the growth rate of $\mathrm{Av}(\beta)$ and the length of $\beta$? Many conjectures (some of which were mentioned in Section 12.1.4) have been made about this relationship, and most have been proven false. The first truly general result in this direction is due to Valtr, and appeared in a paper of Kaiser and Klazar [107]. We present a specialization.

**Proposition 12.2.1** *For all sufficiently large $k$ and all permutations $\beta$ of length $k$,* $\mathrm{gr}(\mathrm{Av}(\beta)) > k^2/27$.

*Proof.* Set $n = \lfloor k^2/9 \rfloor$ (the integer part of $k^2/9$) and let $\beta$ be any pattern of length $k$. Using linearity of expectation and Stirling's Formula, we see that the probability that a permutation of length $n$ chosen uniformly at random contains $\beta$ is at most

$$\frac{1}{k!}\binom{n}{k} < \frac{n^k}{(k!)^2} \leq \frac{k^{2k}}{9^k\,(k/e)^{2k}} < \left(\frac{e^2}{9}\right)^k.$$

This quantity tends to 0 as $k \to \infty$ so for sufficiently large $k$ at least half of the permutations of length $n$ avoid $\beta$. Now Proposition 12.1.3 and Stirling's Formula show that

$$\mathrm{gr}(\mathrm{Av}(\beta)) \geq \sqrt[n]{|\mathrm{Av}_n(\beta)|} \geq \sqrt[n]{n!/2} \geq \frac{\sqrt[n]{\lceil k^2/9 \rceil !}}{\sqrt[n]{n/2}} \geq \frac{k^2}{9e\sqrt[n]{n/2}} > k^2/27$$

for large enough $k$, as desired. ∎

Therefore growth rates of principal class grow at least quadratically in terms of the length of the avoided pattern, but the upper bound conjectured by Arratia is false (as we saw in Section 12.1.4). The fact that Arratia's Conjecture failed for a layered permutation only lent more credence to the long-standing conjecture that among all permutations $\beta$ of length $k$, the permutation that is easiest to avoid (thereby giving the greatest growth rate) is layered. In his survey for the 2013 *British Combinatorial Conference*, Steingrímsson [155] wrote that

> Although the evidence is strong in support of the conjecture that the most easily avoided pattern of any given length is a layered pattern, there is currently no general conjecture that fits all the known data about the particular layered patterns with the most avoiders. However, there are some ideas about what form the most avoided layered patterns ought to have, and specific conjectures that have not been shown to be false.

The oldest of the specific conjectures Steingrímsson mentions appeared in Burstein's 1998 thesis [69]. Bóna [52] made a competing conjecture in 2007 that the most easily avoided permutation of length $k$ was $1 \oplus 21 \oplus 21 \oplus \cdots \oplus 21$ for odd $k$ and $1 \oplus 21 \oplus 21 \oplus \cdots \oplus 21 \oplus 1$ for even $k$.

After these conjectures were made, Claesson, Jelínek, and Steingrímsson [76] established that for every layered permutation $\beta$ of length $k$ the growth rate of $\mathrm{Av}(\beta)$ is less than $4k^2$ (we present their argument in Section 12.3.1). Bóna [45] refined this approach to show that for his conjectured easiest-to-avoid permutations, the corresponding growth rates were at most $2.25k^2$. Because of this, the function

$$g(k) = \max\{\mathrm{gr}(\mathrm{Av}(\beta)) : |\beta| = k\}$$

was widely believed to grow quadratically. Thus it was quite a shock when Fox [88] showed in 2014 that $g(k)$ grows faster than any polynomial. We prove a specialization of Fox's Theorem in Section 12.2.5.

Before moving on, because this section concerns estimates we briefly define the pieces of big-oh notation we use. Let $f(n)$ and $g(n)$ be positive functions. These functions are **asymptotic**, written $f \sim g$, if $f(n)/g(n) \to 1$ as $n \to \infty$. We write $f = O(g)$ if there is a constant $C$ such that $f(n) \leq Cg(n)$ for all sufficiently large $n$. We similarly write $f = \Omega(g)$ if there is a constant $c$ such that $f(n) \geq cg(n)$ for all sufficiently large $n$. Finally, we write $f = \Theta(g)$ if both $f = O(g)$ and $f = \Omega(g)$.

### 12.2.1 Matrices and the interval minor order

In the context of principal classes, the most general results on growth rates have come from expanding the vista to the context of (zero/one) matrices. The study of patterns within matrices dates back to at least 1951, when Zarankiewicz [167] posed what is now known as the Zarankiewicz Problem. While his problem is often phrased in terms of bipartite graphs, what he actually asked was

> Soit $R_n$, où $n > 3$, un réseau plan formé de $n^2$ points rangés en $n$ lignes et $n$ colonnes. Trouver le plus petit nombre naturel $k_2(n)$ tel que tout sous-ensemble de $R_n$ formé de $k_2(n)$ points contienne 4 points situés simultanément dans 2 lignes et dans 2 colonnes de ce réseau.

Translated, the Zarankiewicz Problem asks for the minimum number of ones such that any way those ones are arranged in an $n \times n$ matrix there will be a $2 \times 2$ submatrix of all ones.

Let us introduce some notation to discuss this and more general problems. The **weight** of the matrix $M$ is the number of ones it contains, and we denote this quantity by $\mathrm{wt}(M)$. We also say that the matrix $P$ is a **submatrix** of $M$ if $P$ can be obtained from $M$ by deleting rows, columns, and ones (in this last case, changing them to zeros). A generalization of Zarankiewicz's problem is to determine the asymptotics of the function

$$\mathrm{ex}(n;P) = \max\{\mathrm{wt}(M) : M \text{ is an } n \times n \text{ matrix avoiding } P \text{ as a submatrix}\},$$

and the quantity Zarankiewicz asked about was $\mathrm{ex}(n;J_2)+1$, where we denote by $J_k$ the $k \times k$ all-one matrix.

Shortly after Zarankiewicz posed his problem, Kővári, Sós, and Turán [120] showed that the answer is asymptotic to $n^{3/2}$. We include a short proof below. The upper bound is from the original proof while the lower bound was presented by Reiman [143] in 1958.

**Theorem 12.2.2** *The function* $\mathrm{ex}(n;J_2)$ *is asymptotic to* $n^{3/2}$.

*Proof.* We begin with the upper bound. Take $M$ to be a $J_2$-avoiding $n\times n$ matrix of weight $\mathrm{ex}(n;J_2)$. For each $i$, let $d_i$ denote the number of ones in row $i$. Call two ones in the same row a **couple**, so row $i$ contains $\binom{d_i}{2}$ couples. Clearly $M$ can have at most $\binom{n}{2}$ total couples because the same two columns cannot participate as a couple in two different rows so

$$\binom{n}{2}\geq\sum_{i=1}^{n}\binom{d_i}{2}.$$

Next we use the Cauchy-Schwarz inequality to conclude that $\sum d_i^2\geq(\sum d_i)^2/n$. By noting that $\mathrm{wt}(M)=\sum d_i$, we obtain

$$n^2-n\geq\sum_{i=1}^{n}d_i^2-\sum_{i=1}^{n}d_i\geq\frac{(\mathrm{wt}(M))^2}{n}-\mathrm{wt}(M),$$

showing that $\mathrm{ex}(n;J_2)$ does not grow asymptotically faster than $n^{3/2}$.

For the lower bound, let $q$ be a prime power and suppose that $n=q^2+q+1$. It is known that there is a finite projective plane of order $q$, with $n$ points and $n$ lines. Label the points $p_1,\ldots,p_n$ and the lines $\ell_1,\ldots,\ell_n$ and define the incidence matrix $M$ by setting its $(i,j)$ entry equal to one if and only if the point $p_i$ lies on the line $\ell_j$. Clearly $M$ does not contain $J_2$, because any two points determine a unique line. Moreover, a basic counting argument shows that in a projective plane of order $q$ every point lies on $q+1$ lines. Thus we have

$$\mathrm{wt}(M)=(q+1)(q^2+q+1)\approx n^{3/2}.$$

The fact that $\mathrm{ex}(n;J_2)\sim n^{3/2}$ follows from this and basic results about the distribution of primes. ■

As Kővári, Sós, and Turán [120] noted in their original paper, the upper bound in Theorem 12.2.2 can be generalized to show that $\mathrm{ex}(n;J_k)=O(n^{2-1/k})$ for all $k$ and they conjectured that this is the true order of magnitude, i.e., that $\mathrm{ex}(n;J_k)=\Theta(n^{2-1/k})$. While this conjecture is widely believed to be true, it remains open. The best general lower bound shows that $\mathrm{ex}(n;J_k)=\Omega(n^{2-2/(k+1)})$.

Almost 40 years passed before the function $\mathrm{ex}(n;P)$ was investigated for matrices other than $J_k$, in two papers published around the same time and, oddly enough, concerning the same forbidden pattern (up to symmetry). Bienstock and Győri [38] established that

$$\mathrm{ex}\left(n;\begin{pmatrix}1&1&\\1&&1\end{pmatrix}\right)=O(n\log n).$$

(It is common practice in this field to suppress zeros.) Along with giving an independent proof of this result, Füredi [91] also presented a construction (set $M(i,j)=1$

if and only if $j-i$ is a power of 2) showing that this is the correct rate of growth. Both papers were motivated by problems in discrete geometry: Bienstock and Győri were investigating the complexity of an algorithm for computing obstacle-avoiding rectilinear paths in the plane, while Füredi used this extremal function to bound the number of unit distances between the vertices of a convex polygon.

The first systematic investigation of these extremal functions was performed by Füredi and Hajnal [92] in 1992. They found the extremal functions (up to constant factors) for several matrices. Some of these extremal functions are quite exotic; for example, using the work of Hart and Sharir [100] Füredi and Hajnal showed that

$$\mathrm{ex}\left(n; \begin{pmatrix} 1 & & 1 & \\ & 1 & & 1 \end{pmatrix}\right) = \Theta(n\alpha(n)),$$

where $\alpha(n)$ is the inverse Ackermann function. At the end of their article, they asked

is it true that the complexity of all permutation configurations are linear?

The statement that $\mathrm{ex}(n;M_\beta)$ is linear in $n$ for every permutation $\beta$ became known as the Füredi–Hajnal Conjecture; here $M_\beta$ denotes the **permutation matrix** of $\beta$, which is defined by $M_\beta(i,j)=1$ if and only if $\beta(i)=j$. Indeed, at the time Marcus and Tardos proved the Füredi–Hajnal Conjecture, they did not know about the Stanley–Wilf Conjecture, nor that Klazar had shown that the Füredi–Hajnal Conjecture implied the Stanley–Wilf Conjecture (we prove this in the next subsection).

Despite the rich history of these submatrix problems, later advances—especially those of Fox [88]—show that the permutation containment order is much more closely related to a different order on matrices, the **interval minor** order. In retrospect, this order has featured in the proofs of all of the main results presented in this section, before it had ever been formally defined. Such a formal definition requires a few preliminaries. The **contraction** of two adjacent rows of a matrix is obtained by replacing those two rows by a single row that has a one in a column if either of the original rows had a one in that column. We define the contraction of a column analogously. We further say that $P$ is a **contraction** of $M$ if $P$ can be obtained from $M$ by a sequence of contractions of adjacent rows or columns.

To give an alternative definition of contractions, given an $n\times n$ matrix $M$ and intervals $X,Y\subseteq[1,n]$, we denote by $M(X\times Y)$ the **block** of $M$ consisting of those entries $M(i,j)$ with $(i,j)\in X\times Y=\{(x,y) : x\in X, y\in Y\}$. The contraction $P$ of $M$ can be defined by a pair of sequences $1=c_1\le\cdots\le c_{t+1}=n+1$ (the column divisions) and $1=r_1\le\cdots\le r_{u+1}=n+1$ (the row divisions) where we set

$$P(i,j)=\begin{cases} 1 & \text{if the block } M([c_i,c_{i+1})\times[r_j,r_{j+1})) \text{ contains a one and} \\ 0 & \text{otherwise.}\end{cases}$$

(We continue to use Cartesian coordinates so that the permutation matrix of $\pi$ looks like its plot.)

Finally, the matrix $P$ is an **interval minor** of $M$ if $P$ is a submatrix of a contraction of $M$. In other words, the interval minor order allows for the contraction of adjacent rows or columns, and the deletion of ones. As we will see, at the coarsest

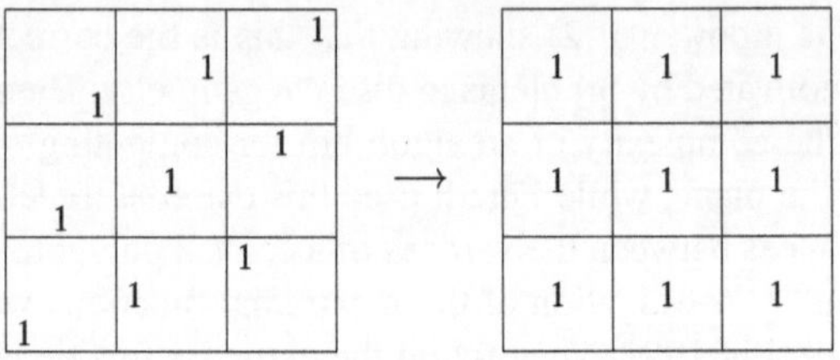

**Figure 12.15**
A $9 \times 9$ permutation matrix containing $J_3$ as an interval minor.

level, questions about growth rates of principal permutation classes are intimately connected with the following question:

> How much can an $n \times n$ matrix weigh if does not contain $J_k$ as an interval minor?

We present two answers, in Sections 12.2.3 (an upper bound) and 12.2.5 (a lower bound). The connection between this question and growth rates is explored in Sections 12.2.2 and 12.2.4. We begin with two elementary observations.

**Observation 12.2.3** *If the matrix $M$ contains the permutation matrix $M_\beta$ as an interval minor then it also contains $M_\beta$ as a submatrix.*

Due to this fact, when we say that a matrix contains a permutation matrix we needn't specify which order we are considering. Indeed, permutation matrices are essentially characterized by the condition in Observation 12.2.3—it is an easy exercise to show that if $P$ is such that every matrix containing $P$ as an interval minor also contains $P$ as a submatrix then, modulo all-zero rows and columns, $P$ is a permutation matrix.

Observation 12.2.3 implies that if a matrix contains $J_k$ as an interval minor then it contains *every* $k \times k$ permutation matrix. Thus to get an upper bound on $\mathrm{gr}(\mathrm{Av}(\beta))$ we can consider the set of matrices that avoid $J_{|\beta|}$ as an interval minor. In particular, the Füredi–Hajnal Conjecture will follow if we can establish that, for every $k$, there is a constant $c_k$ such that $\mathrm{wt}(M) \leq c_k n$ for every $n \times n$ matrix that does not contain $J_k$ as an interval minor.

The link between avoiding $J_k$ and lower bounds on growth rates comes via the following fact, whose truth is demonstrated by Figure 12.15.

**Observation 12.2.4** *For every nonnegative integer $k$, there is a permutation $\beta$ of length $k^2$ whose permutation matrix contains $J_k$ as an interval minor.*

Therefore if many permutation matrices avoid $J_k$ as an interval minor, they also avoid $M_\beta$ for some permutation $\beta$ of length $k^2$.

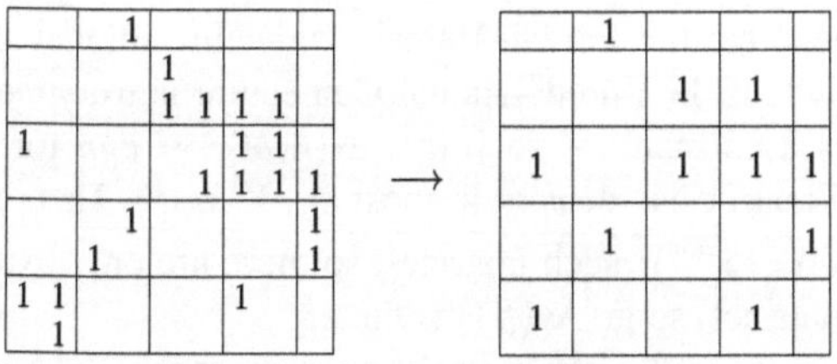

**Figure 12.16**
The interval minor $M^{/2}$ formed in the proof of Klazar's Theorem 12.2.5.

### 12.2.2 The number of light matrices

The main result of this subsection provides the link between the extremal Füredi–Hajnal Conjecture and the enumerative Stanley–Wilf Conjecture. We state it in a slightly different way than it was originally presented. Let us define the **density** of the matrix $M$ as

$$\delta(M) = \max\left\{\frac{\mathrm{wt}(P)}{m} : P \text{ is an } m \times m \text{ interval minor of } M\right\}.$$

Note that this definition measures the maximum average number of ones per row over all square interval minors of $M$. In particular, the density of a matrix can be greater than 1; for example, the permutation matrix shown on the left of Figure 12.15 has density at least 3, because it contains $J_3$ as a minor.

We seek to bound the number of matrices of small density.

**Theorem 12.2.5 (Klazar [114])** *Let $d \geq 0$ be a constant. There are fewer than $15^{2dn}$ matrices of size $n \times n$ and density at most $d$.*

*Proof.* We prove the theorem by induction on $n$. As the base case $n = 1$ is trivial, we may assume that $n \geq 2$ and that the claim holds for all lesser values of $n$.

Let $M$ be an $n \times n$ matrix of density at most $d$ and denote by $M^{/2}$ the interval minor of $M$ formed by using the row and column divisions $\{1, 3, 5, \ldots, n+1\}$, without deleting any ones. Thus $M^{/2}$ is formed by contracting $2 \times 2$ blocks of $M$ (or possibly smaller blocks on the top and right), as shown in Figure 12.16.

By the definition of density, $\mathrm{wt}(M^{/2}) \leq d\lfloor n/2 \rfloor$. Now we ask how many $n \times n$ matrices could condense to a given $\lceil n/2 \rceil \times \lceil n/2 \rceil$ matrix $M^{/2}$. Each nonzero entry of $M^{/2}$ can come from at most 15 possible blocks of $M$ (all blocks except for the all-zero block), while the zero entries of $M^{/2}$ can only come from an all-zero block of $M$. Since $M^{/2}$ has at most $d\lceil n/2 \rceil$ nonzero entries, the number of $n \times n$ matrices that could condense to it is at most $15^{d\lceil n/2 \rceil}$.

Therefore, the number of $n \times n$ matrices of density at most $d$ is at most $15^{d\lceil n/2 \rceil}$ times the number of possibilities for $M^{/2}$, which by induction is less than $15^{2d\lceil n/2 \rceil}$. In other words, the number of $n \times n$ matrices of density at most $d$ is less than

$$15^{d\lceil n/2 \rceil} \cdot 15^{2d\lceil n/2 \rceil} = 15^{3d\lceil n/2 \rceil} < 15^{2dn}$$

for $n \geq 2$, completing the proof of the theorem. ∎

This result shows that the Füredi–Hajnal Conjecture implies the Stanley–Wilf Conjecture as follows: if the Füredi–Hajnal Conjecture is true, then for every permutation $\beta$ there is a constant $c_\beta$ such that $\mathrm{ex}(n;M_\beta) \leq c_\beta n$ for all $n$. Therefore every $M_\beta$-avoiding matrix has density at most $c_\beta$. Klazar's Theorem 12.2.5 shows that there are at most $(15^{2c_\beta})^n$ such matrices, so there are only exponentially many such permutation matrices, so $\mathrm{gr}(\mathrm{Av}(\beta))$ is finite.

As we have presented it, the bound in Klazar's Theorem 12.2.5 is essentially best possible. To see this, let $M$ be any $n \times n$ matrix with $dn$ ones and density $d$. Then $M$ itself contains $2^{dn} = (2^d)^n$ submatrices of size $n \times n$ and density at most $d$. Thus to improve on the link between bounds in the Füredi–Hajnal Conjecture and those in the Stanley–Wilf Conjecture, one must restrict to permutation matrices. Via some quite delicate bounds, Cibulka [75] showed that the number of $n \times n$ permutation matrices of density at most $d$ is at most $(2.88d^2)^n$. Fox [88] gives a much simpler proof that the $n$th root of this number is $O(d^2)$.

### 12.2.3 Matrices avoiding $J_k$ are light

We now present Marcus and Tardos' proof of the Füredi–Hajnal Conjecture. The general approach is again to consider a specific contraction.

**Theorem 12.2.6 (Marcus and Tardos [129])** *If the $n \times n$ matrix $M$ does not contain $J_k$ as an interval minor then*

$$\delta(M) \leq 2k^4 \binom{k^2}{k}.$$

*Proof.* Let $f_k(n)$ denote the maximum weight of an $n \times n$ matrix that avoids $J_k$ as an interval minor. We apply induction on $n$, noting that the $n = 1$ case is trivial. Let $M$ be an $n \times n$ matrix of weight $f_k(n)$ not containing $J_k$ as an interval minor. We form the interval minor $M^{/q}$ by choosing our row and column divisions to be $\{1, q+1, 2q+1, \ldots n+1\}$; here $q$ is a parameter to be determined at the end of the proof. Thus we are contracting $M$ into $q \times q$ blocks (possibly including smaller blocks on the top and right). Of course, $M^{/q}$ also does not contain $J_k$ as an interval minor.

The cleverest part of the proof is the following definition. A $q \times q$ block of $M$ is called **tall** if it contains ones in at least $k$ different rows and/or **wide** if it contains ones in at least $k$ different columns. We begin by bounding the number of tall and/or wide blocks of $M$.

Suppose that more than $(k-1)\binom{q}{k}$ tall blocks of $M$ correspond to entries in the same row of $M^{/q}$. We contract each such block into a single column. This gives an interval minor of $M$ containing $q$ rows and $(k-1)\binom{q}{k}$ columns in which every column contains at least $k$ ones. By deleting extraneous ones as needed, we may assume that each column has precisely $k$ ones. However, that leaves us with only $\binom{q}{k}$ choices of columns for this interval minor, and thus (under our assumption about the number of tall blocks in this row) at least one column must occur at least $k$ times, showing that $M$ contains $J_k$ as an interval minor, a contradiction.

By symmetry, at most $(k-1)\binom{q}{k}$ entries of each column of $M^{/q}$ correspond to wide blocks in $M$. As $M^{/q}$ has $\lceil n/q \rceil$ rows and columns, we have the bound

$$\#\text{ of tall and/or wide entries of } M^{/q} \leq 2\left\lceil \frac{n}{q} \right\rceil (k-1)\binom{q}{k},$$

which is linear in $n$. For these entries of $M^{/q}$ we use the trivial bound that they correspond to blocks with at most $q^2$ nonzero entries. The remaining entries of $M^{/q}$, corresponding to blocks that are neither tall nor wide, can have at most $(k-1)^2$ nonzero entries. Putting these bounds together, we obtain by induction that

$$\text{wt}(M) \leq 2q^2 \left\lceil \frac{n}{q} \right\rceil (k-1)\binom{q}{k} + (k-1)^2 f_k\left(\left\lceil \frac{n}{q} \right\rceil\right).$$

In order to get a sense of the solution to this recurrence, we iterate it approximately:

$$\begin{aligned}
f_k(n) &\lesssim 2qn(k-1)\binom{q}{k} + (k-1)^2 f_k\left(\frac{n}{q}\right), \\
&\lesssim 2qn(k-1)\binom{q}{k} + (k-1)^2\left(2n(k-1)\binom{q}{k} + (k-1)^2 f_k\left(\frac{n}{q^2}\right)\right), \\
&\lesssim 2qn(k-1)\binom{q}{k} + 2n(k-1)^3\binom{q}{k} + 2\frac{n}{q}(k-1)^5\binom{q}{k} + \cdots.
\end{aligned}$$

We can further approximate this quantity by viewing it as a geometric series with ratio $(k-1)^2/q$:

$$f_k(n) \lesssim 2qn(k-1)\binom{q}{k}\sum_{i=0}^{\infty}\left(\frac{(k-1)^2}{q}\right)^i.$$

This series has a finite sum so long as we choose $q > (k-1)^2$, in which case we get a linear upper bound on $\text{wt}(M)$, completing the proof. In particular, setting $q = k^2$, we see that

$$\sum_{i=0}^{\infty}\left(\frac{(k-1)^2}{k^2}\right)^i = \frac{k^2}{2k-1} < k,$$

and the bound becomes

$$f_k(n) \lesssim 2k^4\binom{k^2}{k} n.$$

While we presented only estimates above in order to motivate the choice of $q$, with this bound on $f_k(n)$ now discovered it would only take a simple inductive argument to give a rigorous proof. ∎

We have proved quite a bit more than the Stanley–Wilf Conjecture. Every proper downset of matrices in the interval minor order must avoid some matrix $J_k$. Thus every proper downset of matrices in this order has bounded density by Marcus and Tardos' Theorem 12.2.6 and thus grows at most exponentially by Klazar's Theorem 12.2.5. Proper permutation classes are merely subsets of these downsets.

### 12.2.4 Dense matrices contain many permutations

Having established upper bounds on growth rates, we now turn our attention to lower bounds. In this subsection we prove a simplified version of a result of Cibulka [75], showing that dense matrices contain many permutation matrices. In the following subsection, we use this result to show that growth rates of principal classes can be very large. Our first result in this direction, below, gives a lower bound for how many permutation matrices a dense matrix with a certain amount of regularity must contain.

**Proposition 12.2.7 (Cibulka [75])** *Suppose that the matrix $M$ has $n$ columns and $m \geq n$ rows, and that each column contains at least $n$ ones. Then $M$ contains at least $n!/\binom{m}{n}$ permutation matrices of size $n \times n$.*

*Proof.* Proceeding from left to right, we can find at least $n!$ copies of $n \times n$ permutation matrices in $M$ by choosing a nonzero entry in a new row from each column. Of course, some permutation matrices may be contained in $M$ many times. However, $M$ cannot contain more than $\binom{m}{n}$ copies of a given $n \times n$ permutation matrix because that is the number of ways to select the rows from which the nonzero entries are then chosen. ∎

The bound we actually use follows from Proposition 12.2.7 via Stirling's Formula:

$$\frac{n!}{\binom{m}{n}} \geq \frac{n!}{m^n/n!} = \frac{(n!)^2}{m^n} \geq \left(\frac{n^2}{e^2 m}\right)^n.$$

The main result of this subsection establishes the link between the weight of a matrix and the number of permutation matrices contained in it. Note that the conclusion of this theorem is only interesting for large densities $\delta(M)$.

**Theorem 12.2.8 (based on Cibulka [75, Theorem 6])** *For every $n \times n$ matrix $M$, there is some value of $m \geq 1$ such that $M$ contains $\gamma^m$ or more $m \times m$ permutation matrices, where $\gamma = \delta(M)^{1/9}/16$.*

*Proof.* The argument resembles that used to prove Marcus and Tardos' Theorem 12.2.6, though with an additional clever idea and different parameters. In particular, instead of being given $k$ (formerly the size of the forbidden interval minor), here we choose $k$ to be the smallest perfect square such that $k \geq e^4\gamma^2$. Clearly we can satisfy this requirement for some $k$ at most $2e^4\gamma^2$.

Define $f_\gamma(n)$ to be the greatest possible weight of an $n \times n$ matrix that does not contain $\gamma^m$ or more $m \times m$ permutation matrices for any value of $m$. The theorem is equivalent to the fact that

$$f_\gamma(n) < (16\gamma)^9 n, \tag{12.1}$$

which we establish by induction on $n$. If $n \leq (16\gamma)^9$, then (12.1) holds trivially. Thus we may assume that $n > (16\gamma)^9$ and take $M$ to be an $n \times n$ matrix of weight $f_\gamma(n)$ that does not contain $\gamma^m$ or more $m \times m$ permutation matrices for any value of $m$.

Again we begin by forming the interval minor $M^{/q}$ by choosing our row and column divisions to be $\{1, q+1, 2q+1, \ldots n+1\}$, although this time we set $q = 4k^2$

instead of $k^2$. We have by induction that $\mathrm{wt}(M^{/q}) < (16\gamma)^9 \lceil n/q \rceil$. The definitions of tall and wide are unchanged: a $q \times q$ block of $M$ is tall (respectively, wide) if it contains ones in at least $k$ different rows (respectively, columns).

The next step is to bound the number of tall/wide blocks of $M$, but this time we obtain a different bound because of our assumption that $M$ contains few permutation matrices. We bound the number of tall blocks that correspond to a single row of $M^{/q}$, since the bound for wide blocks in a column will follow by symmetry.

Given a set of tall blocks corresponding to entries in the same row of $M^{/q}$, we group them into contiguous sets of $k$ tall blocks apiece. Our bound comes from considering two different types of groups. First, if the ones in all of the blocks of a group of tall blocks lie in only $k^{3/2}$ rows, we call the group **concentrated**. In this case, we form a minor by removing the all-zero rows of the block (this can be viewed as a contraction) and then contracting each block of the group to a single column. In doing so we obtain a matrix with $k$ columns and at most $k^{3/2}$ rows in which every column contains at least $k$ ones (because the original blocks were tall). Applying Proposition 12.2.7 to this interval minor shows that it (and thus also $M$) contains at least

$$\frac{k!}{\binom{k^{3/2}}{k}} \geq \left(\frac{k^2}{e^2 k^{3/2}}\right)^k = \left(\frac{\sqrt{k}}{e^2}\right)^k \geq \gamma^k$$

$k \times k$ permutation matrices. Thus $M$ cannot contain a concentrated group of tall blocks.

It remains to consider groups of tall blocks that are not concentrated. If we have a nonconcentrated group of tall blocks then these blocks can all be contracted into a single column with at least $k^{3/2}$ ones. Thus if $M$ contains $k^{3/2}$ nonconcentrated groups then by contracting each group into a single column we obtain a matrix with $k^{3/2}$ columns and $q = 2k^2$ rows in which each column contains at least $k^{3/2}$ ones. Applying Proposition 12.2.7 to this interval minor shows that it contains at least

$$\frac{(k^{3/2})!}{\binom{2k^2}{k^{3/2}}} \geq \left(\frac{k^3}{2e^2k^2}\right)^{k^{3/2}} = \left(\frac{k}{2e^2}\right)^{k^{3/2}} \geq \left(\frac{e^2\gamma^2}{2}\right)^{k^{3/2}} > \gamma^{k^{3/2}}$$

$k^{3/2} \times k^{3/2}$ permutation matrices. Therefore $M$ cannot contain $k^{3/2}$ nonconcentrated groups within the same row of blocks.

We now count the ones in $M$. A given row of $M^{/q}$ must correspond to fewer than $k^{5/2}$ tall blocks because when arranged in groups of $k$ tall blocks apiece, no group can be concentrated, and $M$ cannot contain $k^{3/2}$ nonconcentrated groups. Using the trivial bound that a $q \times q$ block can contain at most $q^2$ ones, we see that the combined weight of an entire row of tall blocks is at most $q^2 k^{5/2}$. As there are $\lceil n/q \rceil$ such rows of blocks and also $\lceil n/q \rceil$ columns of blocks that are subject to analogous constraints, the total weight of all tall and/or wide blocks in $M$ is less than $2\lceil n/q \rceil q^2 k^{5/2}$.

Next we must bound the weight of the blocks that are neither tall nor wide. Each such block has weight less than $k^2$. Moreover, the number of such blocks is at most $\mathrm{wt}(M^{/q}) < (16\gamma)^9 \lceil n/q \rceil$. Combining our bounds (and remembering that $q = 4k^2$

while $k \leq 2e^4\gamma^2$) shows that

$$\begin{aligned}\mathrm{wt}(M) &< 2\left\lceil\frac{n}{q}\right\rceil q^2k^{5/2} + (16\gamma)^9\left\lceil\frac{n}{q}\right\rceil k^2,\\ &< \left(4qk^{5/2} + 2\frac{(16\gamma)^9k^2}{q}\right)n,\\ &= \left(16k^{9/2} + \frac{(16\gamma)^9}{2}\right)n.\end{aligned}$$

It can be now be checked that

$$16k^{9/2} \leq 16\cdot 2^{9/2}e^{18}\gamma^9 < \frac{(16\gamma)^9}{2},$$

establishing (12.1) and completing the proof. ■

The clever new idea in the proof of Theorem 12.2.8 is that of concentration, which allows one to handle tall/wide blocks in either of two different ways. In Cibulka's original proof he also refined the notion of tall/wide blocks, distinguishing between tall, very tall, and ultratall blocks. With these additional considerations, he was able to lower the 9 in the statement of the theorem to 4.5.

### 12.2.5 Dense matrices avoiding $J_k$

Our final result of this section shows that the formerly widely held belief that $\mathrm{gr}(\mathrm{Av}(\beta))$ grows quadratically in $|\beta|$ is false. This result, proved by Fox [88], is established by means of an elegant construction.

Throughout this section we are concerned with a matrix $M$ of size $n \times n$ where $n$ is a power of 2, say $n = 2^r$. We divide the interval $[1, 2^r]$ into a number of **dyadic intervals**. The interval $[1, 2^r]$ is itself dyadic. Its two halves, $[1, 2^{r-1}]$ and $[2^{r-1}+1, 2^r]$ are also dyadic. In turn, the halves of each of those intervals are dyadic, and this process continues until we reach the singletons, which are all dyadic. More formally, a dyadic interval is any interval of the form $[a2^b + 1, (a+1)2^b]$ for nonnegative integers $a$ and $b$. A **dyadic rectangle** is then a product of two dyadic intervals (in the same sense as the products we used to define contraction).

The number of dyadic intervals in $[1, 2^r]$ is

$$\sum_{b=0}^{r} 2^b = 2^{r+1} - 1 = 2n - 1,$$

and every element of $[1, n]$ lies in precisely $r+1$ such dyadic intervals (one of each size). The final observation we need to make about dyadic intervals is their most important property, at least from the viewpoint of our upcoming construction. Suppose that $I_1$ and $I_2$ are subintervals of $[1, n]$, and for $i \in \{1, 2\}$ let $D_i$ denote the smallest dyadic interval containing $I_i$. If $D_1 = D_2$, then both $I_1$ and $I_2$ must contain elements of both halves of this dyadic interval. Therefore we can conclude that the smallest dyadic intervals containing two *disjoint* subintervals of $[1, n]$ are *distinct*.

We are now ready to describe the construction of $M$. Order the dyadic subintervals of $[1,n]$ as $D_1,\dots,D_{2n-1}$ arbitrarily and let $A$ be a $(2n-1)\times(2n-1)$ **auxiliary matrix**. We build $M$ from $A$ by setting

$$M(i,j)=\begin{cases} 1 & \text{if } A(k,\ell)=1 \text{ for every pair } k,\ell \text{ with } (i,j)\in D_k\times D_\ell,\\ 0 & \text{otherwise.}\end{cases}$$

The following result illustrates the usefulness of this construction.

**Proposition 12.2.9 (Fox [88])** *Suppose that A and M are related as above. If A avoids $J_k$ as a submatrix, then M avoids $J_k$ as an interval minor.*

*Proof.* Suppose that $M$ contains $J_k$ as an interval minor. Thus there are row and column divisions $1=r_1\le\cdots\le r_{k+1}=n+1$ and $1=c_1\le\cdots\le c_{k+1}=n+1$ such that for every $i$ and $j$, the block $M([c_i,c_{i+1})\times[r_j,r_{j+1}))$ contains a nonzero entry.

Now choose indices $\overline{c}_1,\dots,\overline{c}_k$ such that for every $i$, the smallest dyadic interval containing $[c_i,c_{i+1})$ is $D_{\overline{c}_i}$ and similarly choose indices $\overline{r}_1,\dots,\overline{r}_k$ such that for every $j$, the smallest dyadic interval containing $[r_j,r_{j+1})$ is $D_{\overline{r}_j}$. Our observation above shows that because the intervals we are considering are disjoint, these are sets of distinct indices. Thus it follows by the construction of $M$ that $A(\overline{c}_i,\overline{r}_j)=1$ for all indices $i$ and $j$, i.e., that $A$ contains $J_k$ as a submatrix, completing the proof. ■

As one might expect for an extremal result, the proof of Fox's Theorem is probabilistic. In order to find a dense matrix $M$ avoiding $J_k$ as an interval minor we choose the entries of the auxiliary matrix $A$ uniformly at random and then (essentially) construct $M$ as above. While this part of the proof is conceptually straight-forward, the reader may find the requisite inequalities a bit laborious.

**Theorem 12.2.10 (Fox [88])** *For all sufficiently large positive integers k such that $\sqrt{k}/8$ is an integer, there exists a matrix of density at least $2^{\sqrt{k}/16}$ that avoids $J_k$ as an interval minor.*

*Proof.* Suppose that $r=\sqrt{k}/8$ is an integer and set $n=2^r$ so that every integer in the interval $[1,n]$ lies in precisely $r+1$ dyadic intervals. The proof uses a parameter $q$, which is a probability that will be determined near the end and will satisfy $1/2>q\ge 4r/k$.

We begin by building the auxiliary matrix $A$ of size $(2n-1)\times(2n-1)$. Choose each entry of $A$ uniformly at random so that it is 1 with probability $1-q$. Let $X$ denote the random variable counting the number of copies of $J_k$ occurring as a submatrix in $A$. Clearly

$$\mathbb{E}[X]=\binom{2n-1}{k}^2(1-q)^{k^2}.$$

The standard bound $1-x\le e^{-x}$ shows that $(1-q)^{k^2}\le e^{-qk^2}$. For the binomial coefficient, we have

$$\binom{2n-1}{k}<\frac{(2n)^k}{k!}<\frac{(2n)^k}{2^k}=n^k,$$

where the second inequality follows because $k \geq 4$. Combining these bounds yields

$$\mathbb{E}[X] < n^{2k} e^{-qk^2}.$$

The right-hand side of this inequality is maximized when $q$ is as small as possible. As

$$q \geq \frac{4r}{k} = \frac{4\log_2 n}{k} > \frac{4\ln n}{k},$$

we see that

$$\mathbb{E}[X] < n^{2k} e^{-4k\ln n} = n^{-2k}.$$

In particular,

$$\Pr[A \text{ contains } J_k \text{ as a submatrix}] \leq \mathbb{E}[X] < n^{-2k},$$

so almost all such matrices avoid $J_k$ as a submatrix.

Now construct $M$ as in Proposition 12.2.9. The probability that an entry of $M$ is equal to 1 is thus $(1-q)^{(r+1)^2}$ because every entry lies in $(r+1)^2$ dyadic rectangles. It is worth remarking that the entries of $M$ are highly correlated (when viewed as $n^2$ random variables). In particular, the probability that $M$ is the zero matrix is at least $q$, because one of the entries of $A$ corresponds to the dyadic rectangle $[1,n] \times [1,n]$, and thus must be 1 if $M$ is to have any nonzero entries. Of course, this correlation does not prevent us from appealing to linearity of expectation, which shows that

$$\mathbb{E}[\mathrm{wt}(M)] = n^2(1-q)^{(r+1)^2}.$$

However, we must guarantee that $M$ avoids $J_k$ as an interval minor, and this construction does not provide such a guarantee (though it makes it exceedingly likely). Therefore we take $M$ as above if it avoids $J_k$ as an interval minor (i.e., if $A$ avoids $J_k$ as a submatrix), and otherwise set $M$ equal to the zero matrix. From our previous computation of $\mathbb{E}[X]$, we see that

$$\begin{aligned}
\mathbb{E}[\mathrm{wt}(M)] &\geq n^2(1-q)^{(r+1)^2} - n^2 \cdot \Pr[A \text{ contains } J_k \text{ as a submatrix}], \\
&> n^2(1-q)^{(r+1)^2} - n^{2-2k}, \\
&> n^2(1-q)^{(r+1)^2} - 1.
\end{aligned}$$

Thus there is an $n \times n$ matrix $M$ avoiding $J_k$ as an interval minor with weight at least this expectation.

To finish the proof we need to choose $q$ and then manipulate the inequalities to show that the density of $M$ is at least $2^{\sqrt{k}/16}$, which is equivalent to showing that $\mathrm{wt}(M) \geq n^{3/2}$. For the rest of the proof we assume that $k \geq 48^2 = 2308$ and set $q = 1/\sqrt{k}$. As promised at the beginning of the proof,

$$\frac{1}{2} > q = \frac{1}{\sqrt{k}} > \frac{4(\sqrt{k}/8)}{k} = \frac{4r}{k}.$$

We now claim that for this value of $q$ and the particular matrix $M$ we have selected,

$$\mathrm{wt}(M) \geq n^2(1-q)^{(r+1)^2} - 1 > n^2 2^{-3qr^2} - 1 > n^{3/2}. \tag{12.2}$$

Of these three inequalities, the first follows from our choice of $M$ and the second requires the most work. Canceling the 1 and the $n^2$, we want to show that $(1-q)^{(r+1)^2} > n^{-3qr}$. Taking logarithms of both sides, this is equivalent to

$$(r+1)^2 \log(1-q) > -(3\log 2)qr^2 \tag{12.3}$$

For $q < 1/2$, we have the bound $\log(1-q) > -3q/2$ so the left-hand side of (12.3) can be bounded by

$$(r+1)^2 \log(1-q) > -\frac{3}{2}(r+1)^2 q.$$

For $r \geq 6$ (which we have because $k \geq 48^2$),

$$-\frac{3}{2}(r+1)^2 > -(3\log 2)r^2 \; (\approx -2.08r^2),$$

verifying (12.3) and thus the second inequality of (12.2).

All that remains is to show the final inequality of (12.2). Recall that $n = 2^r$, so

$$n^2 2^{-3qr^2} - 1 = n^{2-3qr} - 1 = n^{2-3/8} - 1.$$

This quantity is greater than $n^{3/2} = 2^{\sqrt{k}/16} n$ for $n \geq 4$, completing the proof. ∎

It remains only to connect this result to growth rates of principal classes, which we do via our previous definition of

$$g(k) = \max\{\mathrm{gr}(\mathrm{Av}(\beta)) : |\beta| = k\}.$$

Choose $k$ large enough so that Theorem 12.2.10 holds. Recall by Observation 12.2.4 that there are permutations of length $k^2$ that contain a $J_k$ minor. Let $\beta$ be such a permutation and set $\ell = k^2$. Fox's Theorem shows that there is a matrix $M$ of density at least $2^{\ell^{1/4}/16}$ that avoids $J_k$ as a minor and thus also avoids the permutation matrix of $\beta$. Cibulka's Theorem 12.2.8 then shows that there is a value of $m$ such that $M$ contains $\gamma^m$ or more $m \times m$ permutation matrices, where

$$\gamma = \frac{\delta(M)^{1/9}}{16} = \frac{2^{\ell^{1/4}/144}}{16} = 2^{\ell^{1/4}/144-4}.$$

Every permutation matrix contained in $M$ also avoids the permutation matrix of $\beta$. From our observation in Section 12.1.1 about supermultiplicativity it follows that $\mathrm{gr}(\mathrm{Av}(\beta)) \geq \gamma$, showing that

$$g(k) = 2^{\Omega(k^{1/4})}.$$

Thus $g(k)$ grows faster than every polynomial. (There is no commonly agreed upon term for functions that grow like $2^{n^\alpha}$ for $0 < \alpha < 1$. Because such functions grow much more slowly than $2^n$, they are often called **subexponential** when found as upper bounds for algorithmic problems. However, this term does not accurately convey how quickly such functions do grow, so terms such as **stretched exponential** and **mildly exponential** are also used.)

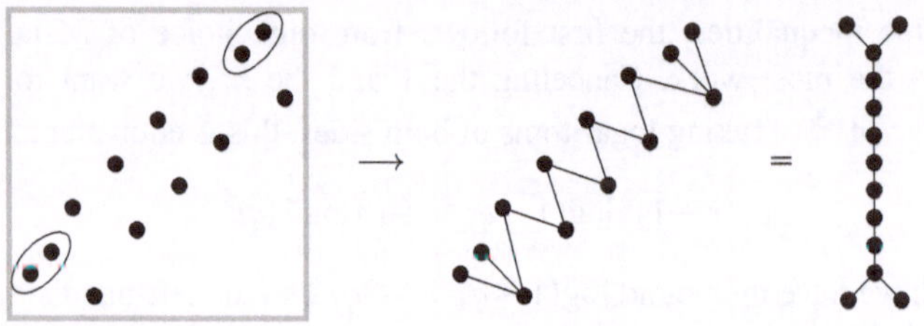

**Figure 12.17**
A member of the infinite antichain commonly denoted $U$.

## 12.3 Notions of structure

We saw in the previous section that proper permutation classes have finite upper growth rates. However, none of those techniques are of any use if we want to know the precise growth rate of a class. Indeed, even if we only want to estimate growth rates, it is hard to imagine those techniques providing much insight. Thus it seems that to know more about growth rates of permutation classes we must take a detailed look at their structure.

As we will see, there are a great many aspects of "structure" for permutation classes that have so far resisted unification. For a first aspect of structure, consider the permutation shown on the left of Figure 12.17. In the middle and right of this figure, two drawings of the **permutation graph**, $G_\pi$, of this permutation are presented. This is the graph on the vertices $\{(i,\pi(i))\}$ in which $(i,\pi(i))$ and $(j,\pi(j))$ are adjacent if they form an inversion, i.e., $i<j$ and $\pi(i)>\pi(j)$. The drawing on the right of this figure shows that the permutation graph of this permutation is a **split-end path**, that is, it is constructed from a path by adding four vertices, two adjacent to each leaf. Clearly we could modify this construction to build an infinite set of permutations whose permutation graphs are all split-end paths.

It is easy to see that if $\sigma \le \pi$ then $G_\sigma$ is an induced subgraph of $G_\pi$, though the converse need not hold (for example, $G_\pi$ and $G_{\pi^{-1}}$ are isomorphic, but of course unless $\pi$ is an involution, it is not contained in $\pi^{-1}$). Therefore, since no split-end path is an induced subgraph of another, it follows that the infinite set of permutations built this way is an infinite antichain (recall that an antichain is a set of pairwise incomparable permutations). The same approach shows that the family of permutations alluded to in Figure 12.11 also forms infinite antichain; the graph of the permutation from that figure is shown below.

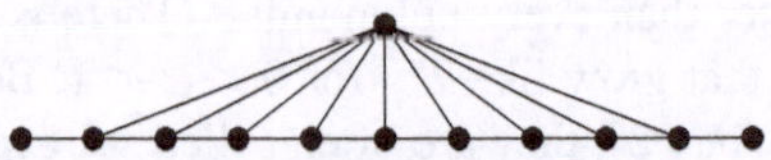

A **well-quasi-order (wqo)** is a **quasi-order** (a binary and transitive, but not necessarily antisymmetric, binary relation) that contains neither an infinite strictly descending sequence nor an infinite antichain. Because containment is a partial order on permutations, we use the term **well-partially-ordered (wpo)** to describe this property instead. Of course, permutation classes cannot contain infinite strictly decreasing sequences, so wpo is synonymous with the absence of infinite antichains in this context. Well-quasi/partial-orders have been studied quite extensively in a variety of contexts. Cherlin [74] surveys these investigations from a particularly general perspective.

The infinite antichain we constructed above from split-end paths lies in the class Av(321). Atkinson, Murphy, and Ruškuc [23] showed that the only wpo principal classes are Av(1), Av(21), Av(231), and their symmetries. This supports our statement in Section 12.1.2 that despite their Wilf-equivalence, the classes Av(231) and Av(321) are very different.

The following property of wpo classes can be tremendously useful.

**Proposition 12.3.1** *The subclasses of a wpo permutation class satisfy the* **descending chain condition**, *i.e., if $\mathscr{C}$ is a wpo class, there does not exist an infinite sequence $\mathscr{C} = \mathscr{C}^0 \supsetneq \mathscr{C}^1 \supsetneq \mathscr{C}^2 \supsetneq \cdots$ of permutation classes.*

*Proof.* Suppose to the contrary that the wpo class $\mathscr{C}$ were to contain an infinite strictly decreasing sequence of subclasses $\mathscr{C} = \mathscr{C}^0 \supsetneq \mathscr{C}^1 \supsetneq \mathscr{C}^2 \supsetneq \cdots$. For each $i \geq 1$, choose $\beta_i \in \mathscr{C}^{i-1} \setminus \mathscr{C}^i$. The set of minimal elements of $\{\beta_1, \beta_2 \ldots\}$ is an antichain and therefore finite, so there is an integer $m$ such that $\{\beta_1, \beta_2 \ldots, \beta_m\}$ contains these minimal elements. In particular, $\beta_{m+1} \geq \beta_i$ for some $1 \leq i \leq m$. However, we chose $\beta_{m+1} \in \mathscr{C}^m \setminus \mathscr{C}^{m+1}$, and because $\beta_{m+1}$ contains $\beta_i$, it does not lie in $\mathscr{C}^i$ and thus cannot lie in $\mathscr{C}^m$, a contradiction. ∎

In their 1996 paper, Noonan and Zeilberger [134] conjectured that every finitely based permutation class has a *D*-finite generating function. Clearly the finite basis hypothesis is necessary—because there are infinite antichains of permutations, there are uncountably many permutation classes with distinct generating functions, but only countably many *D*-finite generating functions with rational coefficients. The status of the Noonan-Zeilberger Conjecture is still unresolved, though Zeilberger no longer believes it to be true (as witnessed by his quote about Av(1324) presented in Section 12.1.4). Moreover, the work of Conway and Guttman [78] suggests that Av(1324) has a non-*D*-finite generating function, giving us a concrete potential counterexample.

One might also ask about the other direction. If a class has a "nice" generating function, does it have a good deal of structure? Indeed, at the same conference, Zeilberger asked for necessary and sufficient conditions for a class to have a rational/algebraic/*D*-finite/... generating function [81]. However, the work of Albert, Brignall, and Vatter [12] shows that this question is almost certainly intractable.

To briefly sketch this argument, let $\mathscr{C}$ be a proper permutation class, so Marcus and Tardos' Theorem 12.2.6 shows that there is some constant $\gamma$ such that $|\mathscr{C}_n| < \gamma^n$ for all $n$. By modifying the construction of the antichain in Figure 12.17, it can be shown that for every finite $\gamma$, there is an infinite antichain $A$ containing at least $\delta^n$ permutations of each sufficiently long length $n$ for some constant $\delta > \gamma$. Moreover, $A$ can be constructed in such a way that its downward closure, Sub($A$), has a rational generating function. Therefore if we start with the class $\mathscr{C} \cup \mathrm{Sub}(A)$ and remove $|\mathscr{C}_n|$ elements of length $n$ from $A$ for every sufficiently long length $n$, we obtain a permutation class containing $\mathscr{C}$ that has (up to the addition of a polynomial) the same rational generating function as $\mathrm{Sub}(A)$, proving the following.

**Theorem 12.3.2 (Albert, Brignall, and Vatter [12])** *Every permutation class except for the class of all permutations is contained in a class with a rational generating function.*

In light of Theorem 12.3.2, it is clearly hopeless to attempt to establish a structural characterization of classes with rational (or algebraic, or $D$-finite, ...) generating functions. A similar issue arises in the definition of perfect graphs. In that context, we know that the chromatic number of a graph, $\chi(G)$, is at least its clique number, $\omega(G)$ and we would like to ask which graphs achieve equality, but we're faced with the problem that given any graph, its union with a sufficiently large complete graph will satisfy $\chi = \omega$. Thus we say that a graph is **perfect** if $\chi = \omega$ for the graph and *all of its induced subgraphs*. In the permutation class context, we parallel this by saying that a permutation class is **strongly rational** (respectively, **strongly algebraic**) if it and all of its subclasses have rational (respectively, algebraic) generating functions. Our counting argument from before now yields the following implication.

**Proposition 12.3.3** *Every strongly algebraic permutation class is wpo.*

I have conjectured that this is actually the characterization of strongly algebraic permutation classes, though this conjecture has never before appeared in print.

**Conjecture 12.3.4** *A permutation class is strongly algebraic if and only if it is wpo.*

As of yet, very little has been established about strongly algebraic classes, but there are some results on strongly rational classes. Using the substitution decomposition (the topic of Section 12.3.2), Albert and Atkinson [3] proved that every proper subclass of $\mathrm{Av}(231)$ has a rational generating function. More generally, Albert, Atkinson, and Vatter [8] showed that in a strongly rational class, the sum indecomposable permutations also have a rational generating function. Other properties of strongly rational classes are discussed in Section 12.4.2.

In the rest of this section we present several more notions of structure and study their interactions. Most of the tools developed here are applied in the final section, where we ask about the set of all growth rates of permutation classes.

### 12.3.1 Merging and splitting

Here we consider a very coarse notion of structure. Given any two permutation classes $\mathscr{C}$ and $\mathscr{D}$, their **merge**, $\mathscr{C} \odot \mathscr{D}$, consists of those permutations whose entries can be partitioned into two subsequences, one order isomorphic to a permutation in $\mathscr{C}$ and the order isomorphic to a permutation in $\mathscr{D}$ (usually thought of as coloring the entries of the permutation red or blue). For example, Proposition 12.1.4 shows that

$$\mathrm{Av}(321) = \mathrm{Av}(21) \odot \mathrm{Av}(21),$$

and this generalizes to any class of the form $\mathrm{Av}(k\cdots 21)$. What if we change one of the $\mathrm{Av}(21)$ classes to $\mathrm{Av}(12)$? Then we obtain the class of permutations that can be expressed as the union of an increasing subsequence and a decreasing subsequence. This class was introduced by Stankova [150] who named it the class of **skew-merged** permutations and computed its basis:

$$\mathrm{Av}(21) \odot \mathrm{Av}(12) = \mathrm{Av}(2143, 3412).$$

Atkinson [20] was the first to compute the (algebraic) generating function of this class.

It should be noted that the merge operation preserves very few properties. The example of $\mathrm{Av}(321)$ shows that it does not preserve wpo or rational generating functions, while the example of $\mathrm{Av}(4321)$ shows that it does not preserve algebraic generating functions. Moreover, it is relatively easy to construct examples of finitely based classes whose merge is not finitely based, so merge does not preserve finite bases (though Kézdy, Snevily, and Wang [110] proved that the merge of $\mathrm{Av}(12\cdots k)$ with $\mathrm{Av}(\ell\cdots 21)$ always has a finite basis). However, we can get a bound on the growth rates of merges by the following result (which seems to have first appeared, though implicitly, in Albert [1]).

**Proposition 12.3.5** *For any two permutation classes $\mathscr{C}$ and $\mathscr{D}$,*

$$\overline{\mathrm{gr}}(\mathscr{C} \odot \mathscr{D}) \le \left(\sqrt{\overline{\mathrm{gr}}(\mathscr{C})} + \sqrt{\overline{\mathrm{gr}}(\mathscr{D})}\right)^2.$$

*Proof.* Given a permutation of length $k$ from $\mathscr{C}$ and another permutation of length $n-k$ from $\mathscr{D}$, there are $\binom{n}{k}^2$ ways to merge them to form a permutation of length $n$: choose $k$ positions to be occupied by the permutation from $\mathscr{C}$, then choose the $k$ values for this subsequence, and then the permutation is determined. Therefore

$$|(\mathscr{C} \odot \mathscr{D})_n| \le \sum_{k=0}^{n} \binom{n}{k}^2 |\mathscr{C}_k||\mathscr{D}_{n-k}| \le \left(\sum_{k=0}^{n} \binom{n}{k} \sqrt{|\mathscr{C}_k||\mathscr{D}_{n-k}|}\right)^2. \qquad (12.4)$$

To avoid introducing epsilon, we sketch the proof from this point, though it is not difficult to make it formal. Suppose that $|\mathscr{C}_n| \approx \gamma^n$ and $|\mathscr{D}_n| \approx \delta^n$. Then (12.4) becomes

$$|(\mathscr{C} \odot \mathscr{D})_n| \lesssim \left(\sum_{k=0}^{n} \binom{n}{k} \sqrt{\gamma^k \delta^{n-k}}\right)^2 = \left(\sqrt{\gamma} + \sqrt{\delta}\right)^{2n}.$$

Taking $n$th roots gives the desired inequality. ∎

Which classes can we obtain via merges? Let us say that the class $\mathscr{C}$ is **splittable** if $\mathscr{C}$ is contained in $\mathscr{D} \odot \mathbb{E}$ for proper subclasses $\mathscr{D}, \mathbb{E} \subsetneq \mathscr{C}$. Thus $\mathrm{Av}(k\ldots 21)$ is splittable for every $k$, as is the class of skew-merged permutations. But clearly the class $\mathrm{Av}(21)$ of increasing permutations is not splittable, as is (with a bit more thought) the class of layered permutations. In its fullest generality, this remains an open (and seemingly quite difficult) question.

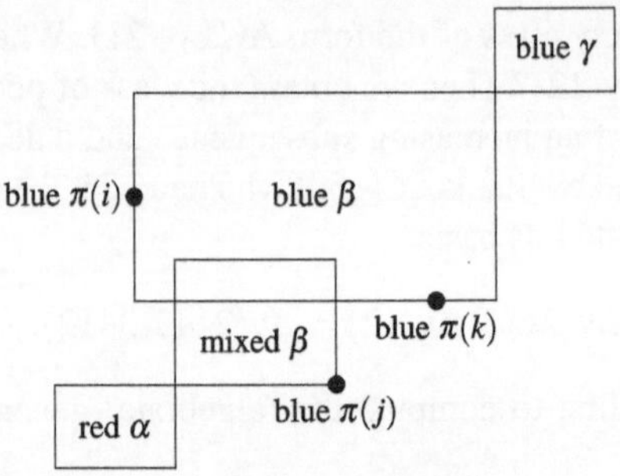


**Figure 12.18**
The final contradiction in the proof of Proposition 12.3.7.

**Question 12.3.6** *Which permutation classes are splittable?*

Question 12.3.6 remains open even for principal classes. The case of $\text{Av}(\beta)$ where $\beta$ is layered has been considered several times in the literature. Bóna [52] presented the first splittability result for such classes, though not in this language. His result was then generalized by Claesson, Jelínek, and Steingrímsson [76], which was in turn generalized by Jelínek and Valtr to the following.

**Proposition 12.3.7 (Jelínek and Valtr [105])** *For all nonempty permutations $\alpha$, $\beta$, and $\gamma$, we have*

$$\text{Av}(\alpha \oplus \beta \oplus \gamma) \subseteq \text{Av}(\alpha \oplus \beta) \odot \text{Av}(\beta \oplus \gamma).$$

*In particular, every principal class whose basis element is the sum of three (or more) nonempty permutations is splittable.*

*Proof.* Take $\pi \in \text{Av}(\alpha \oplus \beta \oplus \gamma)$. We seek to color the entries of $\pi$ red and blue so that the red entries avoid $\alpha \oplus \beta$ and the blue entries avoid $\beta \oplus \gamma$. We proceed from left to right, coloring the entry $\pi(i)$ red unless

(B1) doing so would create a red copy of $\alpha \oplus \beta$, or

(B2) there is already a blue entry below and to the left of $\pi(i)$.

If either of these conditions hold, we color $\pi(i)$ blue.

The red entries of $\pi$ avoid $\alpha \oplus \beta$ by definition, so it suffices to show that the blue entries avoid $\beta \oplus \gamma$. Suppose otherwise, and let $\pi(k)$ denote the bottommost entry of a blue copy of $\beta \oplus \gamma$ in $\pi$. If $\pi(k)$ were colored blue because of rule (B1), then $\pi$ would contain a copy of $\alpha \oplus \beta$ ending at $\pi(k)$, and thus this together with the blue copy of $\gamma$ shows that $\pi$ would contain $\alpha \oplus \beta \oplus \gamma$. As this leads to a contradiction, $\pi(k)$ must have been colored blue because of rule (B2).

Let $\pi(j)$ denote the leftmost blue entry below and to the left of $\pi(k)$. By this choice of $\pi(j)$, we know that it was colored blue because of rule (B1), and thus it forms the rightmost entry of an otherwise-red copy of $\alpha \oplus \beta$ in $\pi$. We now seek to establish the contradiction shown in Figure 12.18. Choose $\pi(i)$ to be the leftmost entry in the blue copy of $\beta \oplus \gamma$ which $\pi(k)$ lies in. Clearly if the red copy of $\alpha$ were to lie completely to the left of $\pi(i)$ then $\pi$ would contain $\alpha \oplus \beta \oplus \gamma$, so some entries of this copy of $\alpha$ must lie to the right of $\pi(i)$. However, this means that all of the entries of the mixed copy of $\beta$ must lie to the right of $\pi(i)$, and thus because $\pi(i)$ is blue, none of the red entries of this copy of $\beta$ may lie above $\pi(i)$. Finally, the blue entry $\pi(j)$ must also lie below $\pi(i)$ because it lies below $\pi(k)$. However, this implies that the mixed copy of $\beta$ must lie entirely below and to the left of the blue copy of $\gamma$, contradicting our assumption that $\pi$ avoids $\alpha \oplus \beta \oplus \gamma$. ∎

While the work of Jelínek and Valtr [105] focuses on the abstract notions of splittability, the earlier work of Claesson, Jelínek, and Steingrímsson [76] was concerned with the applications of Propositions 12.3.5 and (their version of) 12.3.7 to growth rates of principal classes with layered basis elements. In particular, we see that $\text{Av}(1324) \subseteq \text{Av}(132) \odot \text{Av}(213)$, and thus the growth rate of Av(1324) is at most 16. Bóna [44] has since refined this approach to give an upper bound of 13.74.

Our next result was promised in the beginning of Section 12.2.

**Theorem 12.3.8 (Claesson, Jelínek, and Steingrímsson [76])** *For every layered permutation $\beta$ of length k, the growth rate of* $\text{Av}(\beta)$ *is less than* $4k^2$.

*Proof.* Given a positive integer $\ell$, we denote by $\delta_\ell$ the permutation $\ell \cdots 21$. We prove the stronger inequality that for a sequence $\ell_1, \ldots, \ell_m$ of positive integers summing to $k$,

$$\text{gr}(\text{Av}(\delta_{\ell_1} \oplus \cdots \oplus \delta_{\ell_m})) \leq (2k - \ell_1 - \ell_m - m + 1)^2 < 4k^2.$$

The proof is by induction on $m$. The $m = 1$, 2 cases follow from Corollary 12.1.13, so we may assume that $m \geq 3$. By Propositions 12.3.5 and 12.3.7,

$$\begin{aligned} \sqrt{\text{gr}(\text{Av}(\delta_{\ell_1} \oplus \cdots \oplus \delta_{\ell_m}))} &< \sqrt{\text{gr}(\text{Av}(\delta_{\ell_1} \oplus \delta_{\ell_2}))} + \sqrt{\text{gr}(\text{Av}(\delta_{\ell_2} \oplus \cdots \oplus \delta_{\ell_m}))}, \\ &\leq (\ell_1 + \ell_2 - 1) + (2(k - \ell_1) - \ell_2 - \ell_m - (m-1) + 1), \\ &= 2k - \ell_1 - \ell_m - m - 1, \end{aligned}$$

as desired. ∎

We conclude this subsection by returning to a (very) special case of the splittability question. What about principal classes for which the basis element is the sum of just two permutations, i.e., those of the form $\text{Av}(\alpha \oplus \beta)$? Here we can apply Proposition 12.3.7 to see that

$$\text{Av}(\alpha \oplus \beta) \subseteq \text{Av}(\alpha \oplus 1 \oplus \beta) \subseteq \text{Av}(\alpha \oplus 1) \odot \text{Av}(1 \oplus \beta).$$

Thus so long as neither $\alpha \oplus 1$ nor $1 \oplus \beta$ is equal to $\alpha \oplus \beta$, such classes are splittable. By symmetry, this leaves only those classes of the form $\text{Av}(1 \oplus \beta)$. We have already remarked that Av(12) is not splittable. Jelínek and Valtr [105] showed that Av(132) is also not splittable. However, via an intricate argument, they were able to prove that every class of the form $\mathrm{Av}(1 \oplus \beta)$ with $|\beta| \geq 3$ is splittable, which is the final piece needed to obtain the following result.

**Theorem 12.3.9 (Jelínek and Valtr [105])** *For all sum (or skew) decomposable permutations $\beta$ of length at least four, the class $\mathrm{Av}(\beta)$ is splittable.*

### 12.3.2 The substitution decomposition

While the merge construction can provide somewhat reasonable upper bounds on growth rates on splittable classes, it sheds little light on the exact enumeration problem. In this subsection, we investigate the substitution decomposition, which has proved very useful for computing generating functions of classes, especially when combined with the techniques of Section 12.4.2. It should be noted that, as with most of the structural notions discussed in this section, the substitution decomposition has been studied for a wide variety of combinatorial objects. For a slightly outdated survey, we refer to Möhring and Radermacher [132]. This concept dates back to a 1953 talk of Fraïssé [89], although its first significant application was in Gallai's 1967 paper [93] (see [94] for a translation).

An **interval** in the permutation $\pi$ is a set of contiguous indices $I = [a, b]$ such that the set of values $\pi(I) = \{\pi(i) : i \in I\}$ is also contiguous (four intervals are indicated in the permutation on the left of Figure 12.19). Given a permutation $\sigma$ of length $m$ and nonempty permutations $\alpha_1, \ldots, \alpha_m$, the **inflation** of $\sigma$ by $\alpha_1, \ldots, \alpha_m$, denoted $\sigma[\alpha_1, \ldots, \alpha_m]$, is the permutation of length $|\alpha_1| + \cdots + |\alpha_m|$ obtained by replacing each entry $\sigma(i)$ by an interval that is order isomorphic to $\alpha_i$ in such a way that the intervals themselves are order isomorphic to $\sigma$. For example, the permutation shown in Figure 12.19 is

$$2413[1, 132, 321, 12] = 4\,798\,321\,56.$$

Every permutation of length $n \geq 1$ has **trivial** intervals of lengths 0, 1, and $n$; all other intervals are termed **proper**. We further say that the empty permutation and the permutation 1 are **trivial**. A nontrivial permutation is **simple** if it has no proper intervals. The shortest simple permutations are thus 12 and 21, there are no simple permutations of length three, and the simple permutations of length four are 2413 and 3142. Simple permutations and inflations are linked by the following result. Its proof follows in a straight-forward manner once one establishes the fact that the intersection of two intervals is itself an interval.

**Proposition 12.3.10 (Albert and Atkinson [3])** *Every nontrivial permutation $\pi$ is an inflation of a unique simple permutation $\sigma$. Moreover, if $\pi = \sigma[\alpha_1, \ldots, \alpha_m]$ for a simple permutation $\sigma$ of length $m \geq 4$, then each $\alpha_i$ is unique. If $\pi$ is sum decomposable, then there is a unique sequence of sum indecomposable permutations $\alpha_1, \ldots, \alpha_m$ such that $\pi = \alpha_1 \oplus \cdots \oplus \alpha_m$. The same holds, mutatis mutandis, with sum replaced by skew sum.*

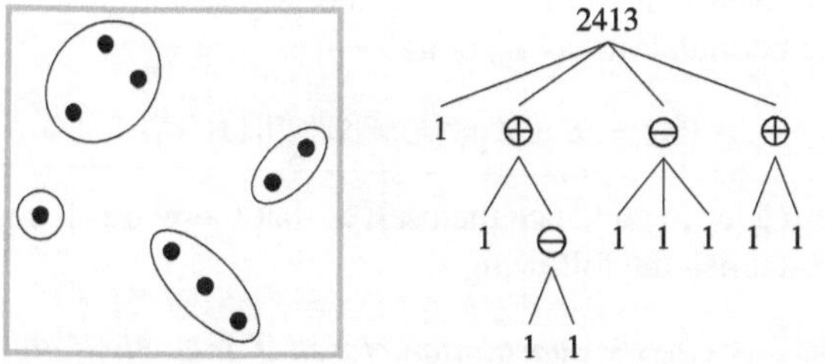


**Figure 12.19**
The plot of the permutation 479832156 and its substitution decomposition tree.

By recursively decomposing the permutation $\pi$ and its intervals as suggested by Proposition 12.3.10, we obtain a rooted tree called the **substitution decomposition tree** of $\pi$ (an example is shown on the right of Figure 12.19). The **substitution depth** of $\pi$ is the height of its substitution decomposition tree, so for example, the substitution depth of the permutation from Figure 12.19 is 3, while the substitution depth of every simple or monotone permutation is 1, and the substitution depth of every nonmonotone layered permutation is at most 2.

The main result of Albert and Atkinson's formative paper on the substitution decomposition is the following.

**Theorem 12.3.11 (Albert and Atkinson [3])** *Every permutation class with only finitely many simple permutations is strongly algebraic (and thus in particular, wpo). Moreover, every permutation class with only finitely many simple permutations and bounded substitution depth is strongly rational.*

For example, Theorem 12.3.11 shows that the class of separable permutations is strongly algebraic, because the only simple permutations in this class are $\{1, 12, 21\}$. However, there are (relatively speaking) few permutation classes with only finitely many simple permutations, so it may appear that the substitution decomposition is not as applicable as might be hoped. We remedy this when we present a generalization in Section 12.4.2 (Theorem 12.4.16).

A different generalization is due to Brignall, Huczynska, and Vatter [65]. They gave a proof of Theorem 12.3.11 using "query-complete sets of properties" showing that the conclusion holds not just for the permutation class itself but for a great many subsets of it. For example, if a permutation class contains only finitely many simple permutations then the generating function for the alternating (up-down) permutations, or the even permutations, or the involutions in the class is also algebraic.

Given two classes $\mathscr{D}$ and $\mathscr{U}$, the **inflation** of $\mathscr{D}$ by $\mathscr{U}$ is defined as

$$\mathscr{D}[\mathscr{U}] = \{\sigma[\alpha_1, \ldots, \alpha_m] : \sigma \in \mathscr{D}_m \text{ and } \alpha_1, \ldots, \alpha_m \in \mathscr{U}\}.$$

The class $\mathscr{C}$ is said to be **substitution closed** if $\mathscr{C}[\mathscr{C}] \subseteq \mathscr{C}$. The **substitution closure** $\langle\mathscr{C}\rangle$ of a class $\mathscr{C}$ can be defined in any number of ways, for example as the smallest substitution closed class containing $\mathscr{C}$, or as the largest permutation class with the same set of simple permutations as $\mathscr{C}$, or as

$$\langle\mathscr{C}\rangle = \mathscr{C} \cup \mathscr{C}[\mathscr{C}] \cup \mathscr{C}[\mathscr{C}[\mathscr{C}]] \cup \cdots.$$

We leave it to the reader to convince themselves that these are all equivalent definitions and then to establish the following.

**Proposition 12.3.12** *A class is substitution closed if and only if all of its basis elements are simple permutations.*

Atkinson, Ruškuc, and Smith [26] investigated bases of substitution closures of principal classes. In a great many cases these bases are infinite, but they were able to compute several bases in the finite cases. For example, they showed that

$$\langle \text{Av}(321)\rangle = \text{Av}(25314, 35142, 41352, 42513, 362514, 531642)$$

by bounding the length of potential basis elements of this class and then conducting an exhaustive search by computer.

If $\mathscr{C} \subseteq \langle\mathscr{D}\rangle$ for a proper subclass $\mathscr{D} \subsetneq \mathscr{C}$ then we say that $\mathscr{C}$ is **deflatable**. We then have the following analogue of the splittability question.

**Question 12.3.13** *Which permutation classes are deflatable?*

An equivalent definition is that the class $\mathscr{C}$ is deflatable if and only if one can find a permutation $\pi \in \mathscr{C}$ that is not contained in any simple permutation of $\mathscr{C}$. For example, the class of separable permutations is clearly deflatable, as it is equal to $\langle\{1, 12, 21\}\rangle$. It follows that Av(231), as a subclass of the separable permutations, is also deflatable. Neither of these classes is splittable, so there are classes that can be decomposed via inflations that can't be decomposed via the merge operation. In the other direction, Av(321) is trivially splittable, but it can be shown that it is not deflatable.

While using the substitution decomposition to enumerate deflatable classes has become a common technique (several applications of this approach are cited in Section 12.4.2), Question 12.3.13 has only recently attracted attention. Interestingly, the most comprehensive result on deflatability to date, quoted below, shows that most of the classes known to be splittable by Theorem 12.3.9 are *not* deflatable. Thus deflatability and splittability are in some sense good complements to each other.

**Theorem 12.3.14 (Albert, Atkinson, Homberger, and Pantone [6])** *For all choices of nonempty permutations $\alpha$, $\beta$, and $\gamma$, the class* $\text{Av}(\alpha \oplus \beta \oplus \gamma)$ *is not deflatable. Moreover, for $|\alpha|, |\beta| \geq 2$, the class* $\text{Av}(\alpha \oplus \beta)$ *is not deflatable.*

They also showed that all classes of the form $\text{Av}(1 \oplus \beta)$ with $|\beta| \leq 4$ are not deflatable, but the class $\text{Av}(1 \oplus 23541)$ is deflatable.

Atkinson posed a much different conjecture about the simple permutations in a class at the conference *Permutation Patterns 2007*. Let $\text{Si}_n(\mathscr{C})$ denote the set of all simple permutations in $\mathscr{C}$ of length $n$.

**Conjecture 12.3.15 (Atkinson)** *The ratio* $|\mathrm{Si}_n(\mathscr{C})|/|\mathscr{C}_n|$ *tends to* 0 *as* $n \to \infty$ *for every proper permutation class* $\mathscr{C}$.

What happens if we let $\mathscr{C}$ be the (nonproper) class of all permutations in Atkinson's Conjecture? It is relatively easy to show that the probability that a permutation of length $n$ contains an interval of size 3 or greater tends to 0 as $n \to \infty$, so (asymptotically) the only obstructions to simplicity are intervals of size 2. Let $X$ denote the number of intervals of size 2 in a permutation of length $n$ chosen uniformly at random. It is routine to compute that $\mathbb{E}[X] = 2$. Moreover, in the 1940s, Kaplansky [108] and Wolfowitz [166] showed (independently) that $X$ is asymptotically Poisson distributed, and thus $\mathbf{P}[X = 0] = 1/e^2$ (a more modern proof using what is known as the Chen-Stein Method appears in Corteel, Louchard, and Pemantle [79], and it is not difficult to establish this result via Brun's Sieve if one is so inclined).

Therefore the number of simple permutations of length $n$ is asymptotic to $n!/e^2$. Albert, Atkinson, and Klazar [7] gave refined asymptotics for this quantity, showing that it is

$$\frac{n!}{e^2}\left(1 - \frac{4}{n} + \frac{2}{n(n-1)} + O\left(\frac{1}{n^3}\right)\right).$$

They also established that the sequence counting simple permutations of length $n$ is not $D$-finite.

Our next result can be used to verify that a given permutation class contains only finitely many simple permutations.

**Theorem 12.3.16 (Schmerl and Trotter [147])** *Every simple permutation of length* $n$ *contains a simple permutation of length* $n-1$ *or* $n-2$.

It should be noted that Schmerl and Trotter proved their theorem in the more general context of binary, irreflexive relational structures, of which permutations are a special case. A streamlined proof of Schmerl and Trotter's theorem for the special case of permutations is given in Brignall and Vatter [67].

The simple permutations that do not contain a simple permutation with one fewer entry are quite rare. We say that an **alternation** is a permutation whose plot can be divided into two halves, by a single horizontal or vertical line, so that for every pair of entries from the same part there is an entry from the other part that **separates** them, i.e., there is an entry from the other part that lies either horizontally or vertically between them. A **parallel alternation** is an alternation in which these two sets of entries form monotone subsequences, either both increasing or both decreasing (the permutation on the left of Figure 12.20 is a parallel alternation). The statement of Schmerl and Trotter's Theorem can be refined to say that every simple permutation that is not a parallel alternation contains a simple permutation with one fewer entry.

A **wedge alternation** is an alternation in which the two halves of entries form monotone subsequences, one increasing and one decreasing. Wedge alternations can be made simple by the addition of a single entry in one of two positions, as shown in the second and third permutations of Figure 12.20. We call the resulting permutations **wedge simple permutations**. Brignall, Huczynska, and Vatter [64] proved a Ramsey-like result for simple permutations, which states that every "long enough"

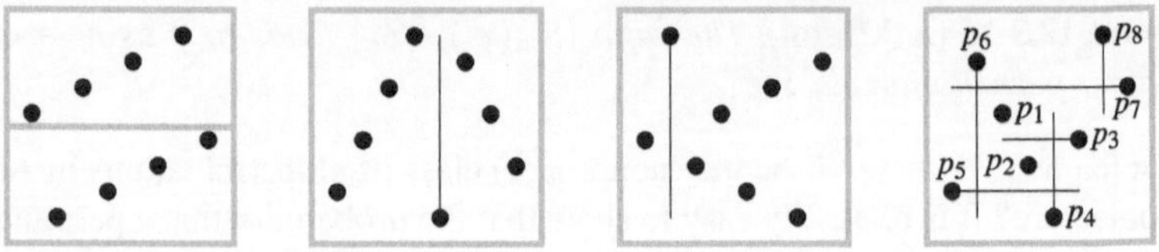


**Figure 12.20**
From left to right, a parallel alternation, two wedge simple permutations, and a proper pin sequence.

simple permutation contains a large simple parallel alternation, a large wedge simple permutation, or a large member of a third family of simple permutations, called **proper pin sequences**. This latter family (a member of which is shown on the far right of Figure 12.20) is defined inductively.

An **axes-parallel rectangle** is any rectangle of the form $X \times Y$ for intervals $X$ and $Y$. The **rectangular hull** of a set of points in the plane is defined as the smallest axes-parallel rectangle containing them. Given independent points $\{p_1, \ldots, p_i\}$ in the plane, a **proper pin** for these points is a point $p$ that lies outside their rectangular hull and separates $p_i$ from $\{p_1, \ldots, p_{i-1}\}$. A proper pin sequence is then constructed by starting with independent points $p_1$ and $p_2$, choosing $p_3$ to be a proper pin for $\{p_1, p_2\}$, then choosing $p_4$ to be a proper pin for $\{p_1, p_2, p_3\}$, and so on. It can be shown that every proper pin sequence is either simple or can be made simple by the removal of a single point. We can now state the theorem.

**Theorem 12.3.17 (Brignall, Huczynska, and Vatter [64])** *There is a function $f(k)$ such that every simple permutation of length at least $f(k)$ contains a simple permutation of length at least k that is either a parallel alternation, a wedge simple permutation, or a proper pin sequence.*

By considering each of the three types of simple permutations guaranteed by Theorem 12.3.17, this result implies that there is a function $g(k)$ such that every simple permutation of length at least $g(k)$ contains two simple permutations of length $k$ that are either disjoint or nearly so (they might share a single entry).

This corollary puts the results of Bóna [47, 48] under the substitution decomposition umbrella. He showed that for any value of $r$, the number of permutations with at most $r$ copies of 132 has an algebraic generating function. (Mansour and Vainshtein [128] later described a way to compute these generating functions.) Clearly the class of 132-avoiding permutations contains only finitely many simple permutations (it is a subclass of the separable permutations). To put this another way, all simple permutations of length at least 4 contain 132. Therefore, by the above corollary to Theorem 12.3.17, all simple permutations of length at least $g(4)$ contain 2 copies of 132. More generally, the class of permutations with at most $r$ copies of 132 does not contain any simple permutations of length $g^r(4)$ or longer, and thus has an algebraic generating function by Theorem 12.3.11. (A simpler derivation of this implication of Theorem 12.3.17 is given in Vatter [160].)

As shown in Brignall, Ruškuc, and Vatter [66], Theorem 12.3.17 can be used to devise a procedure to determine whether a class (specified by a finite basis) contains infinitely many simple permutations. Bassino, Bouvel, Pierrot, and Rossin [33] have since given a much more practical algorithm for this decision problem.

We conclude this subsection by returning to the splittability question. The following is a special case of the results of Jelínek and Valtr [105].

**Proposition 12.3.18** *Substitution closed classes are not splittable.*

*Proof.* Let $\mathscr{C}$ be a substitution closed class. If $12 \notin \mathscr{C}$ then $\mathscr{C}$ must be the class of all decreasing permutations, which is clearly not splittable, so we may assume that $12 \in \mathscr{C}$. Now take two proper subclasses $\mathscr{D}, \mathbb{E} \subsetneq \mathscr{C}$ and choose $\sigma \in \mathscr{C} \setminus \mathscr{D}$ and $\tau \in \mathscr{C} \setminus \mathbb{E}$. Because $12 \in \mathscr{C}$, the permutation $\rho = \sigma \oplus \tau = 12[\sigma, \tau]$ lies in $\mathscr{C}$ but in neither $\mathscr{D}$ nor $\mathbb{E}$.

Now consider $\pi = \rho[\rho, \ldots, \rho]$. By definition, $\pi \in \mathscr{C}$, but we claim that $\pi \notin \mathscr{D} \odot \mathbb{E}$. Indeed, if we tried to color the entries of $\pi$ red and blue so that the red subpermutation lied in $\mathscr{D}$ and the blue subpermutation lied in $\mathbb{E}$, we would have to use both colors in each interval order isomorphic to $\rho$ (because $\rho \notin \mathscr{D} \cup \mathbb{E}$), but then there would be monochromatic copy of $\rho$ containing one entry per interval. ∎

By Proposition 12.3.18 and our previous estimates on the number of simple permutations, we see that if $\beta$ is chosen uniformly at random from all permutations of length $k$, there is (asymptotically) a $1/e^2$ probability that it is simple, and thus that $\mathrm{Av}(\beta)$ is not splittable. As Jelínek and Valtr [105] note, it would be interesting to obtain a better understanding of the probability that $\mathrm{Av}(\beta)$ is not splittable.

### 12.3.3 Atomicity

To motivate the notion of atomicity, let us continue with the splittability question. If the class $\mathscr{C}$ can be expressed as the union of two of its proper subclasses, then it is trivially contained in the merge of these two classes. We say that $\mathscr{C}$ is **atomic** if it cannot be expressed as the union of two of its proper subclasses. Thus the splittability question is only interesting for atomic classes.

Atomicity was first used in the context of permutations by Atkinson in his 1999 paper "Restricted permutations" [21] (which anticipated many of the structural notions discussed in this section), where he showed that

$$\mathrm{Av}(321, 2143) = \mathrm{Av}(321, 2143, 2413) \cup \mathrm{Av}(321, 2143, 3142),$$

and used this to enumerate the former class. In the wider context of relational structures, the notion has a much longer history, dating back to a 1954 article of Fraïssé [90]. It is not difficult to show that the **joint embedding property** is a necessary and sufficient condition for the permutation class $\mathscr{C}$ to be atomic; this condition states that for all $\sigma, \tau \in \mathscr{C}$, there is a $\pi \in \mathscr{C}$ containing both $\sigma$ and $\tau$. Fraïssé established another necessary and sufficient condition for atomicity, which we describe only in the permutation context (Fraïssé proved his results in the context of arbitrary relational structures). Given two linearly ordered sets $A$ and $B$ and a bijection $f : A \to B$, every finite subset $\{a_1 < \cdots < a_n\} \subseteq A$ maps to a finite sequence $f(a_1), \ldots, f(a_n) \in B$ that is order isomorphic to a unique permutation. We say that this permutation is order isomorphic to $f(\{a_1, \ldots, a_n\})$ and call the set of all permutations that are order isomorphic to $f(X)$ for finite subsets $X \subseteq A$ the **age of** $f$. While this class is typically denoted $\mathrm{Age}(f : A \to B)$, given our previous definitions we choose instead to refer to it as $\mathrm{Sub}(f : A \to B)$.

**Theorem 12.3.19 (Fraïssé [90]; see also Hodges [101, Section 7.1])** *The following are equivalent for a permutation class $\mathscr{C}$:*

*(1) $\mathscr{C}$ is atomic,*

*(2) $\mathscr{C}$ satisfies the joint embedding property, and*

*(3) $\mathscr{C} = \mathrm{Sub}(f : A \to B)$ for a bijection $f$ and countable linear orders $A$ and $B$.*

*Proof.* We first show that (1) and (2) are equivalent. Suppose to the contrary that $\mathscr{C}$ satisfies the joint embedding property but $\mathscr{C} \subseteq \mathscr{D} \cup \mathbb{E}$ for proper subclasses $\mathscr{D}, \mathbb{E} \subsetneq \mathscr{C}$. Clearly we may assume that neither $\mathscr{D}$ nor $\mathbb{E}$ is a subset of the other, so there are permutations $\sigma \in \mathscr{D} \setminus \mathbb{E}$ and $\tau \in \mathbb{E} \setminus \mathscr{D}$. By the joint embedding property there is some $\pi \in \mathscr{C}$ containing both of these permutations, but $\pi$ cannot lie in either $\mathscr{D}$ or $\mathbb{E}$, a contradiction. Next suppose that $\mathscr{C}$ does not satisfy the joint embedding property, so there are permutations $\sigma, \tau \in \mathscr{C}$ such that no permutation in $\mathscr{C}$ contains both. Therefore every permutation in $\mathscr{C}$ avoids either $\sigma$ or $\tau$, so we see that $\mathscr{C}$ is contained in the union of its proper subclasses $\mathscr{C} \cap \mathrm{Av}(\sigma)$ and $\mathscr{C} \cap \mathrm{Av}(\tau)$.

Next we show that (2) and (3) are equivalent. Suppose that $\mathscr{C} = \mathrm{Sub}(f : A \to B)$ and take $\sigma, \tau \in \mathscr{C}$. Thus $\sigma$ is order isomorphic to $f(A_\sigma)$ for a subset $A_\sigma \subseteq A$ and $\tau$ is order isomorphic to $f(A_\tau)$ for a subset $A_\tau \subseteq A$, so the permutation that is order isomorphic to $f(A_\sigma \cup A_\tau)$ contains both $\sigma$ and $\tau$. Next suppose that $\mathscr{C}$ satisfies the joint embedding property and list the elements of $\mathscr{C}$ as $\sigma_1, \sigma_2, \ldots$. Define $\pi_0$ to be the empty permutation, and for $i \geq 1$, choose a permutation in $\mathscr{C}$ that contains both $\sigma_i$ and $\pi_{i-1}$ to be $\pi_i$. Thus every element of $\mathscr{C}$ is contained in some $\pi_i$ (and thus also all $\pi_j$ for $j \geq i$). Clearly we may choose $A_0$, $B_0$, and $f_0$ so that $\pi_0$ is order isomorphic to $f_0(A_0)$. Now for $i \geq 1$, we construct $A_i$, $B_i$, and $f_i$ so that $A_i \supseteq A_{i-1}$, $B_i \supseteq B_{i-1}$, $f_i(A_i)$ is order isomorphic to $\pi_i$, and such that $f_i(a) = f_{i-1}(a)$ for all $a \in A_{i-1}$. The desired bijection is then $f = \lim_{i\to\infty} f_i$, with $A = \bigcup A_i$ and $B = \bigcup B_i$. ∎

Recall that every principal permutation class is either sum or skew closed (Observation 12.1.2), so all principal classes satisfy the joint embedding property and are trivially atomic. Despite this, the notion of atomicity has proved to be quite useful in the study of permutation classes. In particular, we apply the following two propositions in our final subsection.

**Proposition 12.3.20** *Every wpo permutation class can be expressed as a finite union of atomic classes.*

*Proof.* Consider the binary tree whose root is the wpo class $\mathscr{C}$, all of whose leaves are atomic classes, and in which the children of the non-atomic class $\mathscr{D}$ are two proper subclasses $\mathscr{D}^1, \mathscr{D}^2 \subsetneq \mathscr{D}$ such that $\mathscr{D}^1 \cup \mathscr{D}^2 = \mathscr{D}$. Because $\mathscr{C}$ is wpo its subclasses satisfy the descending chain condition (Proposition 12.3.1), so this tree contains no infinite paths and thus is finite by König's Lemma; its leaves are the desired atomic classes. ■

Thanks to the following result, the problem of computing growth rates of wpo classes can be reduced to that of computing growth rates of atomic classes. We leave the proof as an easy exercise for the reader.

**Proposition 12.3.21** *The upper growth rate of a wpo permutation class is equal to the greatest upper growth rate of its atomic subclasses.*

The notion of representing permutation classes as ages raises several interesting questions. For instance, define $\mathfrak{T}(A,B)$ as the set of all permutation classes that can be expressed as $\mathrm{Sub}(f : A \to B)$. Given two linear orders $A$ and $B$, we might ask if we can characterize $\mathfrak{T}(A,B)$. Atkinson, Murphy, and Ruškuc [25] were the first to investigate this question; they characterized the classes of the form $\mathrm{Sub}(f : \mathbb{N} \to \mathbb{N})$, which they called **natural classes**. It is worth noting that every principal class is, up to symmetry, a natural class—sum closed classes are natural classes, while skew closed classes can be expressed as $\mathrm{Sub}(f : -\mathbb{N} \to \mathbb{N})$. Huczynska and Ruškuc [103] later studied classes of the form $\mathrm{Sub}(f : A \to \mathbb{N})$ for arbitrary linear orders $A$, which they called **supernatural classes**. Of the many interesting questions that remain open, we quote their "contiguity question":

**Question 12.3.22 (Huczynska and Ruškuc [103])** *If* $\mathscr{C} \in \mathfrak{T}(i\mathbb{N},\mathbb{N}) \cap \mathfrak{T}(k\mathbb{N},\mathbb{N})$, *is* $\mathscr{C} \in \mathfrak{T}(j\mathbb{N},\mathbb{N})$ *for all* $i \le j \le k$*?*

A stronger version of the joint embedding property, called amalgamation, has also been considered. Let $\mathscr{C}$ be a permutation class containing the permutation $\rho$. We say that $\mathscr{C}$ is **$\rho$-amalgamable** if given two permutations $\sigma, \tau \in \mathscr{C}$, each with a marked copy of $\rho$, we can find a permutation $\pi \in \mathscr{C}$ containing both $\sigma$ and $\tau$ such that the two marked copies of $\rho$ coincide. A class is called **homogeneous** if it is $\rho$-amalgamable for every $\rho \in \mathscr{C}$. Cameron [71] proved that there are precisely five infinite homogeneous permutation classes: $\mathrm{Av}(12)$, $\mathrm{Av}(21)$, the layered permutations, the reversed layered permutations, and the class of all permutations. As Cameron noted, the permutation case turned out to be far simpler than several noteworthy cases considered before, such as the undirected graph case (established in 1980 by Lachlan and Woodrow [124]), the tournament case (due to Lachlan [123]), and the directed graph case (Cherlin [73]).

We conclude this subsection by relating amalgamation and splittability.

**Proposition 12.3.23 (Jelínek and Valtr [105])** *Every permutation class that is not* 1*-amalgamable is splittable.*

*Proof.* Suppose that the class $\mathscr{C}$ is not 1-amalgamable, so there are permutations $\sigma, \tau \in \mathscr{C}$ with marked entries $\sigma(i)$ and $\tau(j)$ respectively such that no permutation in $\mathscr{C}$ contains copies of $\sigma$ and $\tau$ in which the entries corresponding to $\sigma(i)$ and $\tau(j)$ coincide. We claim that $\mathscr{C}$ is equal to the merge of its proper subclasses $\mathscr{C} \cap \text{Av}(\sigma)$ and $\mathscr{C} \cap \text{Av}(\tau)$. Let $\pi \in \mathscr{C}$ be arbitrary. Color an entry of $\pi$ red if it can play the role of $\sigma(i)$ in a copy of $\sigma$ in $\pi$, and blue otherwise. None of the red entries can play the role of $\tau(j)$ and none of the blue entries can play the role of $\sigma(i)$, showing that the red entries avoid $\tau$ and the blue entries avoid $\sigma$, proving the result. ∎

## 12.4 The set of all growth rates

The final topic of this survey, the set of growth rates of all permutation classes, relies on the rich toolbox of structural results collected in the previous section as well as the theory of the three different flavors of grid classes introduced in Sections 12.4.1–12.4.3. Like the ideas of the previous section, growth rates have been studied for a wide variety of combinatorial objects, for example, posets, set partitions, graphs, ordered graphs, tournaments, and ordered hypergraphs. For details of the work on growth rates in this more general context, we refer to Bollobás' survey for the 2007 *British Combinatorial Conference* [43].

Characterizing the set of growth rates of permutation classes is easy at the very low end of the spectrum. The number of permutations of length $n$ in a class must be an integer, and if it is ever 0 then the class is empty for all larger lengths. Thus there are no growth rates properly between 0 and 1. The next jump takes more work, and is referred to as the **Fibonacci Dichotomy**.

**Theorem 12.4.1 (Kaiser and Klazar [107])** *Suppose that $\mathscr{C}$ is a permutation class. If $|\mathscr{C}_n|$ is ever less than the nth Fibonacci number, then $|\mathscr{C}_n|$ is eventually polynomial.*

The Fibonacci numbers referred to above are the **combinatorial** Fibonacci numbers, which begin $1, 1, 2, 3, 5, \ldots$ for $n = 0, 1, 2, 3, 4, \ldots$. Classes with this enumeration certainly exist. For example, both $\bigoplus\{1, 21\}$ and $\bigominus\{1, 12\}$ are counted by the Fibonacci numbers. Huczynska and Vatter [104] gave a proof of Theorem 12.4.1 using monotone grid classes, which we sketch in Section 12.4.1. We note in passing that while versions of Theorem 12.4.1 (i.e., jumps from polynomial to superpolynomial growth) are known to exist for a variety of different combinatorial structures (see Klazar [116] for numerous examples), the most general possible form of such a result remains open, as discussed in Pouzet and Thiéry [139].

It follows that the first three growth rates of permutation classes are 0, 1, and $\varphi$, where $\varphi$ is the golden ratio. Next consider the classes $\bigoplus\text{Sub}(k \cdots 21)$. It is easy to see that these classes are counted by (one version of) the **k-generalized Fibonacci**

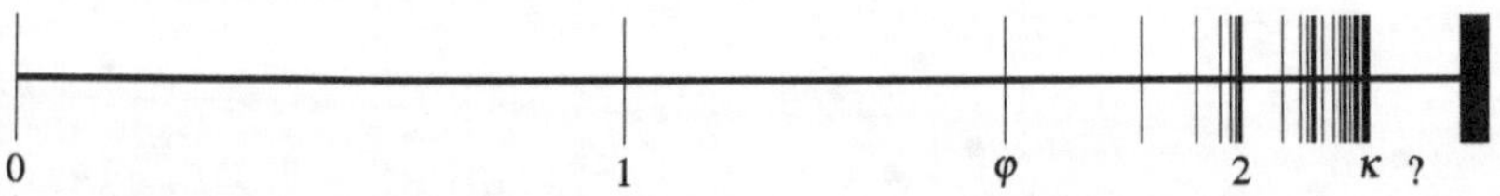


**Figure 12.21**
The set of all growth rates of permutation classes, as presently known.

**numbers**, i.e., that they have generating functions of the form

$$\frac{1}{1-x-x^2-\cdots-x^k}.$$

The growth rates of these classes are the largest roots of the polynomials $x^k - x^{k-1} - \cdots - 1$, or equivalently (by multiplying by $x-1$), the largest roots of the polynomials $x^{k+1} - 2x^k + 1$. These classes converge to the class of layered permutations, and thus their growth rates converge to 2. Kaiser and Klazar [107] showed that these are the only growth rates of permutation classes below 2, making 2 the least accumulation point of growth rates. Their work was later generalized by Balogh, Bollobás, and Morris [32], who showed that growth rates of ordered graph classes take on precisely the same values below 2 (every permutation class can be viewed as a class of ordered graphs, but not vice versa).

This result has since been extended in Vatter [161] where the growth rates up to

$$\kappa = \text{the unique real root of } x^3 - 2x^2 - 1 \approx 2.21$$

were characterized. One implication of this result is that growth rates of permutation classes and ordered graph classes diverge above 2. Sections 12.4.2 and 12.4.3 introduce the machinery (geometric and generalized grid classes) needed to study small permutation classes, while Section 12.4.4 sketches the proof of the following result.

**Theorem 12.4.2 (Vatter [161])** *The sub-$\kappa$ growth rates of permutation classes consist precisely of* 0, 1, 2, *and roots of the families of polynomials (for all nonnegative $k$ and $\ell$)*

- $x^{k+1} - 2x^k + 1$ *(the sub-2 growth rates, identified by Kaiser and Klazar [107]),*
- $(x^3 - 2x^2 - 1)x^{k+\ell} + x^\ell + 1$,
- $(x^3 - 2x^2 - 1)x^k + 1$ *(accumulation points of growth rates which themselves accumulate at $\kappa$),*
- $x^4 - x^3 - x^2 - 2x - 3$, $x^5 - x^4 - x^3 - 2x^2 - 3x - 1$, $x^3 - x^2 - x - 3$, *and* $x^4 - x^3 - x^2 - 3x - 1$.

Theorem 12.4.2 shows that $\kappa$ distinguishes itself on the number-line of growth rates by being the first accumulation point of accumulation points. It is also the

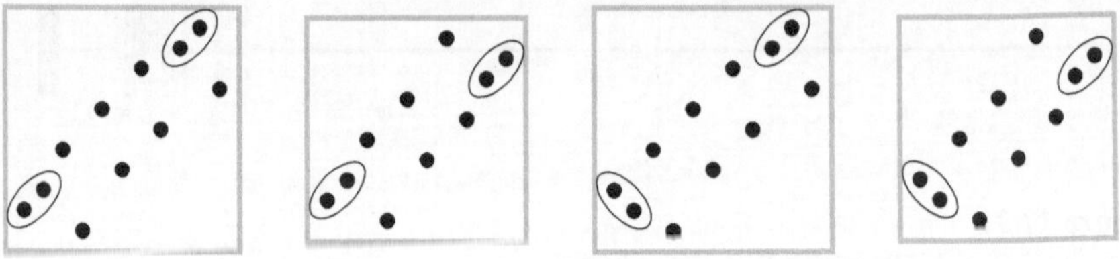

**Figure 12.22**
An infinite antichain used to make an interval of growth rates.

least growth rate at which one finds uncountably many permutation classes; indeed, Klazar [115] originally defined $\kappa$ as

$$\kappa = \inf\{\gamma : \text{uncountably many classes } \mathscr{C} \text{ satisfy } \overline{\mathrm{gr}}(\mathscr{C}) < \gamma\}.$$

One direction follows without too much work—the growth rate of the downward closure of the infinite antichain from Figure 12.17 is equal to $\kappa$, so there are uncountably many classes of growth rate $\kappa$. The other direction will be proved in Section 12.4.4. The phase transition at $\kappa$ also has ramifications for exact enumeration:

**Theorem 12.4.3 (Albert, Ruškuc, and Vatter [16])** *Every permutation class with growth rate less than $\kappa$ is strongly rational.*

For this variety of reasons we call classes of growth rate less than $\kappa$ **small**. At the other end of the spectrum, Albert and Linton [14] constructed an uncountable set of growth rates of permutation classes, and conjectured that at some point the set of growth rates of permutation classes contains all subsequent real numbers. Let us define

$$\lambda = \inf\{\gamma : \text{every real number } x > \gamma \text{ is the growth rate of a permutation class}\}.$$

By refining their techniques, Vatter [159] showed that $\lambda$ exists and is less than 2.49. Bevan [35] has since lowered this bound, showing that $\lambda < 2.36$. Together, this result and Theorem 12.4.2 establish the number-line of growth rates shown in Figure 12.21.

The results about $\lambda$ have been proved by constructing families of sum closed classes with a great deal of flexibility in their growth rates. For example, one of the constructions from [159] features the infinite antichain shown in Figure 12.22. Let $A$ denote the members of this antichain of length at least 5. Clearly $A$ contains two permutations of each of these lengths, both of which are sum indecomposable. We can compute the generating function for $\bigoplus\mathrm{Sub}(A)$ if we know how many sum indecomposable permutations of each length are properly contained in a member of $A$. This sequence can be shown to be

$$(r_n) = 1, 1, 3, 5, 6, 6, 6, 6, \ldots$$

(starting at $n = 1$). Therefore the sequence counting sum indecomposable permutations in $\bigoplus\mathrm{Sub}(A)$ is

$$(t_n) = 1, 1, 3, 5, 8, 8, 8, 8, \ldots.$$

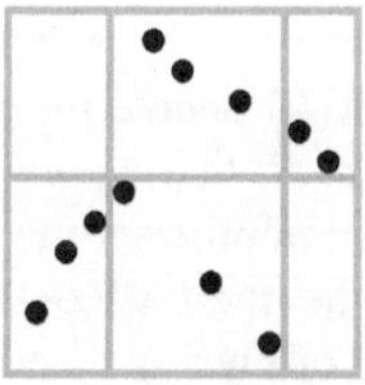

**Figure 12.23**
A $\begin{pmatrix} 0 & -1 & -1 \\ 1 & -1 & 0 \end{pmatrix}$-gridding of a permutation. Here the column divisions are given by $c_1 = 1$, $c_2 = 4$, $c_3 = 10$, and $c_4 = 12$, while the row divisions are $r_1 = 1$, $r_2 = 7$, and $r_3 = 12$.

If we want to construct a subclass of $\bigoplus \text{Sub}(A)$ we may choose any subset of $A$ together with all of the sum indecomposable permutations properly contained the members of $A$. Thus for any sequence $(s_n)$ which satisfies $r_n \leq s_n \leq t_n$ for all $n$ we can construct a permutation class with generating function $1/(1 - \sum s_n x^n)$. The growth rates of such classes can be shown to consist of the entire interval of real numbers between approximately 2.49 and 2.51.

### 12.4.1 Monotone grid classes

As mentioned in Section 12.3.1, Stankova [150] computed the basis of the skew-merged permutations,

$$\text{Av}(21) \odot \text{Av}(12) = \text{Av}(2143, 3412).$$

This class is the prototypical example of a monotone grid class, although we need some notation to introduce these in general.

The definition of monotone grid classes is quite similar to the definition of interval minors from Section 12.2.1, although here we are subdividing the plot of a permutation, rather than a matrix. Let $\pi$ be a permutation of length $n$ and choose intervals $X, Y \subseteq [1, n]$. We write $\pi(X \times Y)$ to denote the permutation that is order isomorphic to those entries with indices from $X$ and values from $Y$.

Given a $t \times u$ matrix $M$ consisting of 0, 1, and $-1$ entries, an $M$-gridding of the permutation $\pi$ of length $n$ is a choice of column divisions $1 = c_1 \leq \cdots \leq c_{t+1} = n+1$ and row divisions $1 = r_1 \leq \cdots \leq r_{u+1} = n+1$ such that for all $i$ and $j$, $\pi([c_i, c_{i+1}) \times [r_j, r_{j+1}))$ is increasing if $M(i, j) = 1$, decreasing if $M(i, j) = -1$, and empty if $M(i, j) = 0$. Figure 12.23 shows an example. The class of all permutations possessing $M$-griddings is called the **monotone grid class** of $M$, and denoted by $\text{Grid}(M)$.

Our first major result characterizes the classes contained in $\text{Grid}(M)$ for some finite $0/\pm1$ matrix $M$. We call such classes **monotonically griddable** (thus there are both monotone grid classes and monotonically griddable classes, and it is important to be recognizant of this distinction). First we need to establish an alternate characterization of monotone griddability. Recall from Section 12.3.2 that an axes-parallel rectangle is any rectangle of the form $X \times Y$ for intervals $X$ and $Y$. If $R = X \times Y$ is an axes-parallel rectangle, we let $\pi(R)$ denote the permutation which is order isomorphic to the entries of $\pi$ lying in $R$, that is, with indices in $X$ and values in $Y$. The axes-parallel rectangle $R$ is then **nonmonotone** for the permutation $\pi$ if $\pi(R)$ is nonmonotone. We further say that the line $L$ **slices** the rectangle $R$ if $L$ intersects the interior of $R$. Finally, a collection $\mathfrak{L}$ of lines and a collection $\mathfrak{R}$ of rectangles, we say that $\mathfrak{L}$ slices $\mathfrak{R}$ if every rectangle in $\mathfrak{R}$ is sliced by some line in $\mathfrak{L}$.

**Proposition 12.4.4** *The class $\mathscr{C}$ is monotonically griddable if and only if there is a constant $\ell$ such that for every permutation $\pi \in \mathscr{C}$, the collection of axes-parallel nonmonotone rectangles of $\pi$ can be sliced by a collection of $\ell$ vertical or horizontal lines.*

*Proof.* First, if $\mathscr{C} \subseteq \text{Grid}(M)$ for a $0/\pm1$ matrix $M$ of size $t \times u$, then every $\pi \in \mathscr{C}$ has an $M$-gridding with at most $t+u-2$ vertical and horizontal lines and by definition these lines must slice every nonmonotone rectangle of $\pi$.

For the other direction, suppose that there is a constant $\ell$ such that for every $\pi \in \mathscr{C}$ there is a collection $\mathfrak{L}_\pi$ of vertical and horizontal lines that slice every nonmonotone rectangle of $\pi$. These lines define a monotone gridding of $\pi$ of size $t \times u$ with $t+u \leq \ell+2$, i.e., they show that $\pi \in \text{Grid}(M)$ for some $0/\pm1$ matrix of this size. There are only finitely many such matrices, so letting $M^\oplus$ denote the direct sum of all such matrices we see that $\mathscr{C} \subseteq \text{Grid}(M^\oplus)$. ■

We can now state and prove the characterization of the monotonically griddable classes. One direction is clear. Using the notation

$$\oplus^a 21 = \underbrace{21 \oplus \cdots \oplus 21}_{a \text{ copies of } 21},$$

it follows that if $\oplus^a 21 \in \text{Grid}(M)$ then $M$ must have at least $a$ rows and $a$ columns. Analogously, $\ominus^b 12$ cannot lie in the grid class of a matrix that is smaller than $b \times b$. In other words, if $\mathscr{C}$ is monotonically griddable, it must avoid $\oplus^a 21$ and $\ominus^b 12$ for some values of $a$ and $b$. As we prove below, this is also a sufficient condition for monotone griddability. The proof we give relies on Stankova's computation of the basis of the skew-merged permutations.

**Theorem 12.4.5 (Huczynska and Vatter [104])** *A permutation class is monotonically griddable if and only if it does not contain arbitrarily long sums of* 21 *or skew sums of* 12.

*Proof.* As we have already observed the other direction of the proof, let $\mathscr{C}$ be a permutation class avoiding $\oplus^a 21$ and $\ominus^b 12$. We prove by induction on $a+b$ that there is a function $\ell(a,b)$ such that given any permutation $\pi \in \mathscr{C}$, the nonmonotone rectangles of $\pi$ can be sliced by a collection of $\ell(a,b)$ vertical or horizontal lines.

If either $a$ or $b$ equals 1 then $\mathscr{C}$ consists solely of monotone permutations and thus we may set $\ell(1,b) = \ell(a,1) = 0$. The next case to consider is $a = b = 2$, mean-

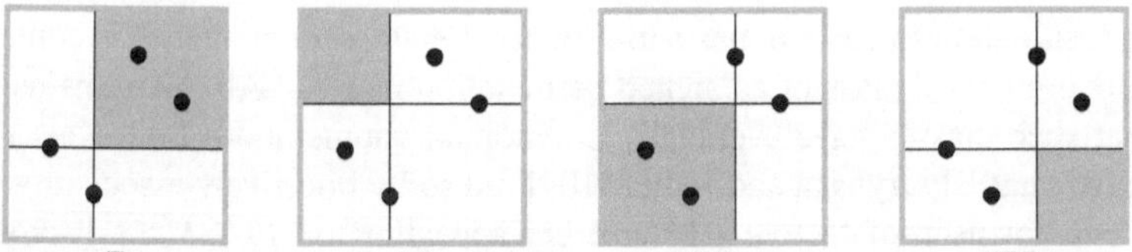


**Figure 12.24**
The four regions in the proof of Theorem 12.4.5.

ing that $\mathscr{C}$ avoids both 2143 and 3412. Thus $\mathscr{C}$ is a subclass of the skew-merged permutations, and so we may set $\ell(2,2) = 2$.

For the inductive step, we may assume by symmetry that $a \geq 3$ and $b \geq 2$. Take an arbitrary $\pi \in \mathscr{C}$. If $\pi$ avoids $\oplus^2 21 = 2143$ then its nonmonotone rectangles can be sliced by $\ell(2,b)$ vertical and horizontal lines by induction. Thus we may assume that $\pi$ contains 2143. We fix a specific copy of 2143 in $\pi$ and then partition the entries of $\pi$ into the four regions shown in Figure 12.24. Clearly every nonmonotone rectangle either lies completely in one of these regions or is sliced by one of the four lines bordering these regions. Now notice that the entries in regions (I) and (III) avoid $\oplus^{a-1}21$, while those in regions (II) and (IV) avoid $\ominus^{b-1}12$. This shows that we can set

$$\ell(a,b) = 2\ell(a-1,b) + 2\ell(a,b-1) + 4,$$

completing the proof. ∎

This result allows us to sketch the proof of the Fibonacci Dichotomy (Theorem 12.4.1). Suppose that $\mathscr{C}$ is a permutation class and $|\mathscr{C}_n|$ is less than the $n$th Fibonacci number. The classes $\bigoplus\{1,21\}$ and $\bigominus\{1,12\}$ are both counted by the Fibonacci numbers. Therefore $\mathscr{C}$ cannot contain either of these classes, and thus is monotonically griddable by Theorem 12.4.5.

Next we recall the definition of alternations, first encountered in Section 12.3.2. An alternation is a permutation whose plot can be divided into two parts, by a single horizontal or vertical line, so that for every pair of entries from the same part there is an entry from the other part that separates them. A parallel alternation is one in which the two halves of the alternation form monotone subsequences, either both increasing or both decreasing, while for a wedge alternation one of these is increasing and the other is decreasing. It follows from the Erdős-Szekeres Theorem that every sufficiently long alternation contains a long parallel or wedge alternation. If $\mathscr{C}$ were to contain arbitrarily long alternations of either type, its growth rate would be at least 2. Thus there is a bound on the length of alternations in $\mathscr{C}$. We now appeal to the following result.

**Proposition 12.4.6 (Huczynska and Vatter [104])** *Suppose that $\mathscr{C}$ is monotonically griddable and that the length of alternations in $\mathscr{C}$ is bounded. Then $\mathscr{C} \subseteq$ Grid$(M)$ for a $0/\pm1$ matrix $M$ in which no two nonzero entries share a row or column.*

This result shows that if $\mathscr{C}$ has sub-Fibonacci enumeration, then it is contained in the monotone grid class of a "signed permutation matrix." All that remains is to show that such classes have eventually polynomial enumeration. Both Kaiser and Klazar [107] and Huczynska and Vatter [104] did so by bijectively associating such classes with downsets of vectors in $\mathbb{N}^t$ and then appealing to a 1976 *Monthly* problem posed (and solved) by Stanley [151]. Homberger and Vatter [102] present a constructive proof that gives an algorithm for computing these polynomials.

Monotone grid classes were first introduced (under a different name) in Murphy and Vatter [133], where the focus was on which monotone grid classes are wpo. Let $M$ be a $0/\pm1$ matrix of size $t \times u$. The **cell graph** of $M$ is the graph on the vertices $\{(i,j) : M(i,j) \neq 0\}$ in which $(i,j)$ and $(k,\ell)$ are adjacent if the corresponding cells of $M$ share a row or a column and there are no nonzero entries between them in this row or column. We say that the matrix $M$ is a **forest** if its cell graph is a forest. Viewing the absolute value of $M$ as the adjacency matrix of a bipartite graph, we obtain a different graph, its **row-column graph**, which is the bipartite graph on the vertices $x_1, \ldots, x_t, y_1, \ldots, y_u$ where there is an edge between $x_i$ and $y_j$ if and only if $M(i,j) \neq 0$. It is not difficult to show that the cell graph of a matrix is a forest if and only if its row-column graph is also a forest (a formal proof is given in Vatter and Waton [162]).

**Theorem 12.4.7 (Murphy and Vatter [133])** *The grid class* Grid$(M)$ *is wpo if and only if the cell graph of $M$ is a forest.*

Theorem 12.4.7 has since been generalized in several directions. In the next subsection we will see a generalization of the wpo part of this result (Theorem 12.4.15). Brignall [63] has presented another generalization of Theorem 12.4.7 in the context of the generalized grid classes of Section 12.4.3.

Somewhat surprisingly, we know almost nothing about the bases of monotone grid classes. The skew-merged permutations have a finite basis, and Waton [163] showed that the monotone grid class of $J_2$ (the $2 \times 2$ all-one matrix) has a finite basis. Thus we remain far away from the following.

**Conjecture 12.4.8** *Every monotone grid class has a finite basis.*

We have used grid classes to establish the Fibonacci dichotomy, but what about growth rates of grid classes themselves? In exploring this question, Bevan [34] uncovered a beautiful connection between permutation patterns and algebraic graph theory. Recall that the **spectral radius** of a graph is the largest eigenvalue of its adjacency matrix.

**Theorem 12.4.9 (Bevan [34])** *The growth rate of* Grid$(M)$ *exists and is equal to the square of the spectral radius of the row-column graph of $M$.*

It is well beyond the scope of this survey to attempt to sketch Bevan's proof, but we nonetheless attempt to convey some essence of his approach. Every connected component in the row-column graph of $M$ corresponds to some submatrix of $M$, and we call this submatrix a **connected component** of $M$. It is not difficult to establish the following result, which was first used in Vatter [161].

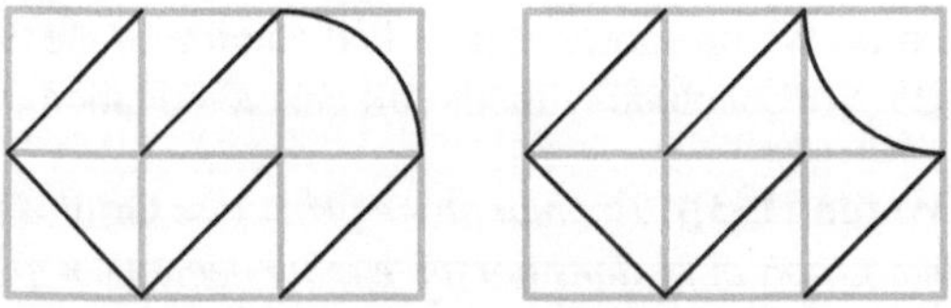

**Figure 12.25**
Waton proved that the monotone grid class of the matrix $\begin{pmatrix} 1 & 1 & -1 \\ -1 & 1 & 1 \end{pmatrix}$ is equal to the union of the two figure classes shown above.

**Proposition 12.4.10** *The upper growth rate of* Grid$(M)$ *is equal to the greatest upper growth rate of the monotone grid class of a connected component of* $M$.

Spectral radii follow the analogous pattern—the spectral radius of the graph $G$ is equal to the greatest spectral radius of a connected component of $G$. Therefore it suffices to consider grid classes with connected row-column graphs. A **tour** on a graph is a walk (a sequence of not-necessarily distinct vertices, each connected by an edge) that ends where it began. Bevan showed that if the row-column graph of $M$ is connected, then the number of permutations of length $n$ in Grid$(M)$ is, up to a polynomial factor, equal to the number of **balanced** tours of length $2n$ in the row-column graph of $G$, where a balanced tour is a tour that traverses every edge the same number of times in each direction. He then showed that this balance condition does not affect the asymptotics of tours, establishing the theorem.

As a consequence of Bevan's Theorem 12.4.9, we may appeal to the significant literature on algebraic graph theory to determine the possible growth rates of monotone grid classes. First, obviously every growth rate of a monotone grid class is an algebraic integer, because it is the square of an eigenvalue of a zero/one matrix. More strikingly, if the row-column graph of $M$ is a cycle, then the growth rate of Grid$(M)$ is equal to 4, no matter how long the cycle. Also, if the growth rate of a monotone grid class is less than this, then it is equal to $4\cos^2(\pi/k)$ for some integer $k \geq 3$. Finally, for every growth rate $\gamma \geq 2+\sqrt{5} \approx 4.24$, there is a monotone grid class with growth rate arbitrarily close to $\gamma$.

The final question we address in this subsection is whether monotone grid classes are atomic. Many monotone grid classes (such as the class of skew-merged permutations) are atomic. However, monotone grid classes are not, in general, atomic. In his thesis [163], Waton established necessary and sufficient conditions for a grid class to be atomic. Let $M$ be a $0/\pm 1$ matrix. Given a cycle in its row-column graph, we say that the **sign** of the cycle is the product of the entries corresponding to its edges. So, for example, the matrix

$$\begin{pmatrix} 1 & 1 & -1 \\ -1 & 1 & 1 \end{pmatrix}$$

contains a positive cycle (corresponding to columns 1 and 3) and two negative cycles (corresponding to columns 1 and 2 and columns 2 and 3). Indeed, the grid class of this matrix is not atomic; Waton showed that it is the union of the two figure classes shown in Figure 12.25. We conclude with his full characterization.

**Theorem 12.4.11 (Waton [163])** *The monotone grid class* Grid($M$) *is atomic if and only if the row-column graph of M does not have a connected component containing a negative cycle together with any other cycle.*

### 12.4.2 Geometric grid classes

In Section 12.1.1 we introduced the class of permutations that can be drawn on an X and called it Sub(X), though to avoid cumbersome notation we denote it by $\mathscr{X}$ here. We return to this class throughout this subsection because it is the prototypical example of a **geometric** grid class. Clearly the class $\mathscr{X}$ is a subclass of the skew-merged permutations, but it is not all of the skew-merged permutations. For example, the permutation 3142 cannot be drawn on an X because once we place the 3, 1, and 4 on the X, there is no place for the 2 to lie simultaneously above the 1 and to the right of the 4:

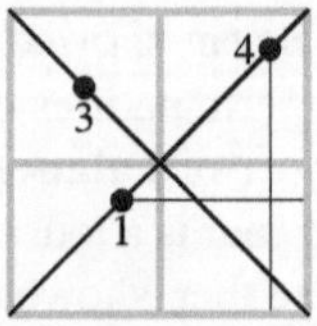


By symmetry, 2413 also cannot be drawn on an X, so $\mathscr{X}$ is also a subclass of the separable permutations. Indeed, it is not hard to see that $\mathscr{X}$ is the class of skew-merged separable permutations,

$$\mathscr{X} = \text{Av}(2143, 2413, 3142, 3412),$$

because every permutation drawn on an X must have some point that is at least as far away from the center of the X as every other point. Therefore every permutation in $\mathscr{X}$ is of the form $1 \oplus \pi$, $\pi \oplus 1$, $1 \ominus \pi$, or $\pi \ominus 1$ for some $\pi \in \mathscr{X}$. This leads quickly to the generating function of the class. From the above, we have

$$\mathscr{X} = \{1\} \cup (1 \oplus \mathscr{X}) \cup (\mathscr{X} \oplus 1) \cup (1 \ominus \mathscr{X}) \cup (\mathscr{X} \ominus 1).$$

Moreover, $(1 \oplus \mathscr{X}) \cap (\mathscr{X} \oplus 1) = (1 \oplus \mathscr{X} \oplus 1)$, so we see (by symmetry) that if $f$ is generating function for nonempty permutations in $\mathscr{X}$, $f = x + 4xf - 2x^2 f$. Thus the generating function for $\mathscr{X}$ is

$$\frac{x}{1 - 4x + 2x^2}.$$

It follows that the growth rate of $\mathscr{X}$ is $2 + \sqrt{2} \approx 3.41$. This is significantly lower than the growth rate of the skew-merged permutations (4, which follows from Atkinson [20] or Bevan's Theorem 12.4.9) or that of the separable permutations (approximately 5.83, as we saw in Section 12.1.1).

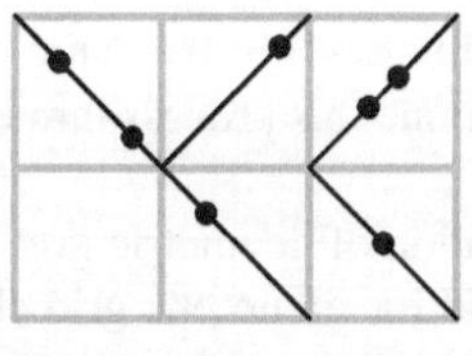

**Figure 12.26**
The permutation 6327415 lies in the geometric grid class of the matrix $\begin{pmatrix} -1 & 1 & 1 \\ 0 & -1 & -1 \end{pmatrix}$.

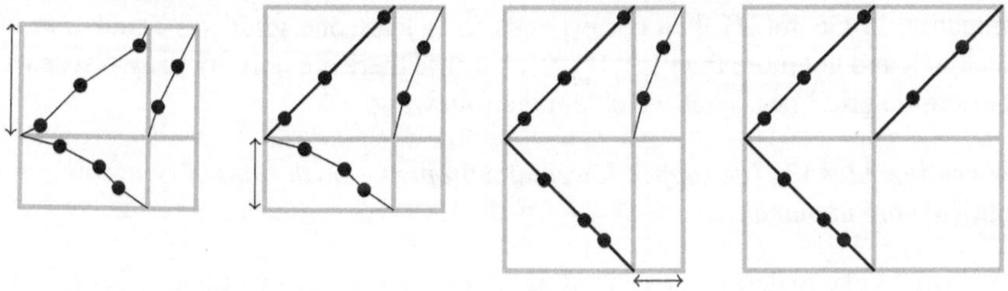

**Figure 12.27**
"Straightening" a member of a monotone grid class to show that it also lies in the corresponding geometric grid class.

Suppose that we have a $0/\pm1$ matrix $M$, as in the definition of a monotone grid class. The **standard figure** of $M$, denoted $\Lambda_M$, is the figure in $\mathbb{R}^2$ consisting of two types of line segments for every pair of indices $i, j$ such that $M(i,j) \neq 0$:

- the increasing open line segment from $(i-1, j-1)$ to $(i, j)$ if $M(i,j) = 1$ or
- the decreasing open line segment from $(i-1, j)$ to $(i, j-1)$ if $M(i,j) = -1$.

The **geometric grid class** of $M$, denoted $\text{Geom}(M)$, is the set of all permutations that can be drawn on $\Lambda_M$ (an example is shown in Figure 12.26). The inclusion $\text{Geom}(M) \subseteq \text{Grid}(M)$ always holds, but our example of $\mathscr{X}$ shows that the geometric grid class of $M$ may be a proper subclass of its monotone grid class. Indeed, this happens precisely when $M$ contains a cycle:

**Proposition 12.4.12** *The classes* $\text{Grid}(M)$ *and* $\text{Geom}(M)$ *are equal if and only if the cell graph of* $M$ *is a forest.*

Figure 12.27 gives a sense of how to prove Proposition 12.4.12. Suppose that the cell graph of $M$ is a tree (the forest case follows easily from the tree case) and choose one cell to be the root. Now take the plot of any gridded permutation $\pi \in \text{Grid}(M)$ and stretch (or shrink) this row vertically so that the points lie on a line of slope $\pm1$. Next, for every neighbor cell in the same row (there can be at most two), we stretch the $x$-axis, while for every neighbor cell in the same column we stretch the $y$-axis. Because $M$ is a tree, we can continue this process throughout all cells without having to revisit a cell.

Next we turn to the enumeration of geometric grid classes, where it turns out we can say a lot more than we could for monotone grid classes. First, though, we need a bit more precision. There are infinitely many ways to draw every permutation $\pi \in \mathrm{Geom}(M)$ on the standard figure $\Lambda_M$ because we can move the points by tiny amounts without changing the underlying permutation, but clearly we want to consider minute changes such as this as equivalent. Therefore we say that a **gridded permutation** is a (drawing of a) permutation together with grid lines corresponding to $M$, and we denote the set of all gridded permutations in $\mathrm{Geom}(M)$ by $\mathrm{Geom}^{\sharp}(M)$. Every permutation in $\mathrm{Geom}(M)$ then corresponds to at least one gridded permutation in $\mathrm{Geom}^{\sharp}(M)$, and not more than $\binom{n}{t-1}\binom{n}{u-1}$, because there are only so many places we can insert the grid lines. Thus we obtain the following.

**Observation 12.4.13** *The (upper, lower, and proper) growth rates of* $\mathrm{Geom}(M)$ *and* $\mathrm{Geom}^{\sharp}(M)$ *are identical.*

We would like to describe an encoding of the gridded permutations in $\mathrm{Geom}^{\sharp}(M)$ by words over a finite alphabet. First let us return to the permutations drawn on an X to see an easy example before presenting the general construction. Below is our drawing from Figure 12.4 with two changes: First, the points are labeled by what quadrant they lie in (we consider the center of the X to be the origin), and second, the line segments have been assigned orientations.

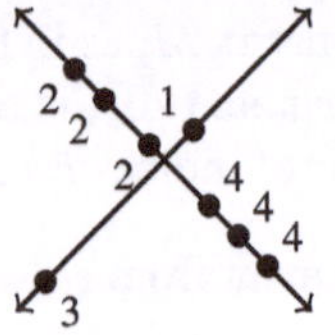


To encode this gridded permutation, we order these points according to their distance from the beginning of their line segment, which in this case is also their distance from the center of the figure, and record the labels of the points in this order. While it may take a ruler to verify it, in our example above the encoding is 21242443.

Suppose that we have an arbitrary $0/\pm 1$ matrix $M$. The important property of the orientation above is that it is **consistent**—in each column all lines are oriented either left or right, and in each row all lines are oriented either up or down. This is not possible for all matrices, for example, the standard figure of the matrix

$$\begin{pmatrix} 1 & 1 \\ 1 & -1 \end{pmatrix}$$

cannot be consistently oriented. But there is a way to remedy this. The grid lines of the standard figure consist of $x = i$ and $y = i$ for all (relevant) integers $i$. If we also add grid lines at half-integer values of $x$ and $y$, each cell is chopped into four but the geometric grid class itself is not changed.

In terms of the matrix we started with, this operation is equivalent to performing the substitutions

$$0 \leftarrow \begin{pmatrix} 0 & 0 \\ 0 & 0 \end{pmatrix}, \qquad 1 \leftarrow \begin{pmatrix} 0 & 1 \\ 1 & 0 \end{pmatrix}, \qquad -1 \leftarrow \begin{pmatrix} -1 & 0 \\ 0 & -1 \end{pmatrix},$$

and we call the resulting matrix, written $M^{\times 2}$, the **double refinement** of $M$. Thus the standard figure of $M^{\times 2}$ consists of $2 \times 2$ blocks that are all subfigures of the X, and thus we can use its consistent orientation to define a consistent orientation of $M^{\times 2}$.

For the rest of this discussion let us suppose that we have a consistent orientation for the standard figure of $M$ (replacing $M$ by $M^{\times 2}$ if necessary). Next we fix a **cell alphabet**, $\Sigma_M$, containing one letter corresponding to each nonempty cell of $M$. Finally, we encode a gridded drawing (of a permutation) by ordering its points by their distance from the beginning of their line segment (we can assume there are no ties by possibly moving points by a minuscule amount) and then recording the corresponding "cell letter" of each point in this order.

Formally, we have just defined a map

$$\varphi^\sharp : \Sigma_M^* \longrightarrow \mathrm{Geom}^\sharp(M).$$

This is in general a many-to-one map, because the letters of $\Sigma_M$ often "commute"; for example, in our encoding of permutations drawn on an X, interchanging adjacent occurrences of 1 and 3 does not change the image (the gridded permutation), and the same holds for 2 and 4. This is because the corresponding cells of the standard figure are **independent**, meaning that they share neither a row nor a column. For general matrices $M$, we write $v \equiv_M w$ if $w$ can be obtained from $v$ via a sequence of interchanges of adjacent occurrences of letters corresponding to independent cells. It is not difficult to see that $\varphi^\sharp$ is actually a bijection when restricted to equivalence classes of words modulo $\equiv_M$, i.e., that the map

$$\varphi^\sharp : \Sigma_M^* / \equiv_M \longrightarrow \mathrm{Geom}^\sharp(M)$$

is a bijection.

Objects such as $\Sigma^* / \equiv$ are known as **trace** (or **partially commutative**) **monoids**. They were first introduced by Cartier and Foata [72] in 1969 who used them to give a combinatorial proof of MacMahon's Master Theorem. Cartier and Foata showed via Möbius inversion that the generating function for equivalence classes of $\Sigma^* / \equiv$ (by length) is

$$\frac{1}{1 - (c_1 x - c_2 x^2 + c_3 x^3 - c_4 x^4 + \cdots)},$$

where $c_k$ is the number of $k$-elements subsets of $\Sigma$ that pairwise commute (see also Flajolet and Sedgewick [86, V.3.3]). In our example of permutations drawn on an X, we have that $c_0 = 1$, $c_1 = 4$, $c_2 = 2$, and $c_3 = c_4 = 0$. Thus the generating function for the corresponding trace monoid is

$$\frac{1}{1 - 4x + 2x^2}.$$

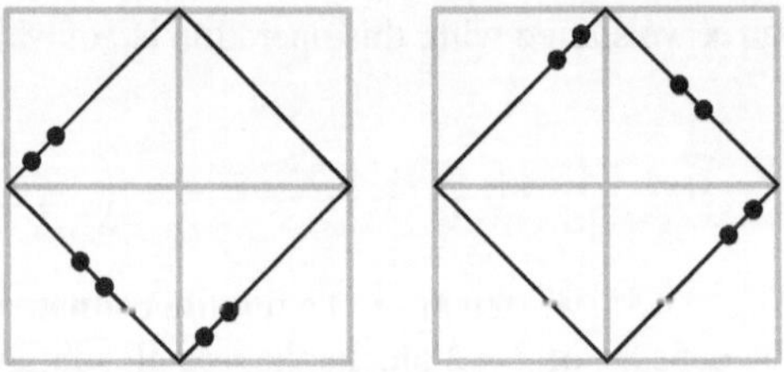

**Figure 12.28**
Two very different griddings of the permutation 564312.

As Observation 12.4.13 tells us should be the case, this generating function has the same growth rate as the class $\mathscr{X}$.

Growth rates of trace monoids were studied by Goldwurm and Santini [96]. Bevan [37] used their work to express the growth rate of Geom($M$) in terms of the row-column graph of $M$, giving a geometric analogue of his Theorem 12.4.9. A **k-matching** of a graph is a set of $k$ edges, no two incident with the same vertex. Letting $m_k(G)$ denote the number of $k$-matchings of a graph on $n$ vertices, its **matching polynomial** is defined by

$$\mu_G(x) = \sum_{k\geq 0} (-1)^k m_k(G) x^{n-2k}.$$

(No matching can contain over $\lfloor n/2 \rfloor$ edges, so this is indeed a polynomial.) We state Bevan's result below in terms of the double refinement of $M$ because this is necessary in order to guarantee that we have a consistent orientation.

**Theorem 12.4.14 (Bevan [37])** *The growth rate of* Geom($M$) *exists and is equal to the square of the largest root of the matching polynomial of the row-column graph of* $M^{\times 2}$.

In contrast to Theorem 12.4.9, this result shows that changing the sign of an entry of $M$ can change the growth rate of the corresponding geometric grid class. For example, Bevan [37] notes that

$$\mathrm{gr}\left(\mathrm{Geom}\begin{pmatrix} -1 & 0 & -1 \\ 1 & -1 & 1 \end{pmatrix}\right) = 4 < 3+\sqrt{2} = \mathrm{gr}\left(\mathrm{Geom}\begin{pmatrix} 1 & 0 & -1 \\ 1 & -1 & 1 \end{pmatrix}\right).$$

Having addressed the asymptotic enumeration of geometric grid classes, we move on to their exact enumeration. As in our study of monotonically griddable classes, we are interested not only in geometric grid classes themselves, but also in their subclasses. Thus we say that the class $\mathscr{C}$ is **geometrically griddable** if $\mathscr{C} \subseteq$ Geom($M$) for some finite $0/\pm 1$ matrix $M$. Unlike the case with monotonically griddable classes, there is no known characterization of geometrically griddable classes.

With exact enumeration we have a new type of problem. While our map $\varphi^\sharp$ restricts to a bijection between $\Sigma_M^*/\equiv_M$ and Geom$^\sharp(M)$, a given permutation may

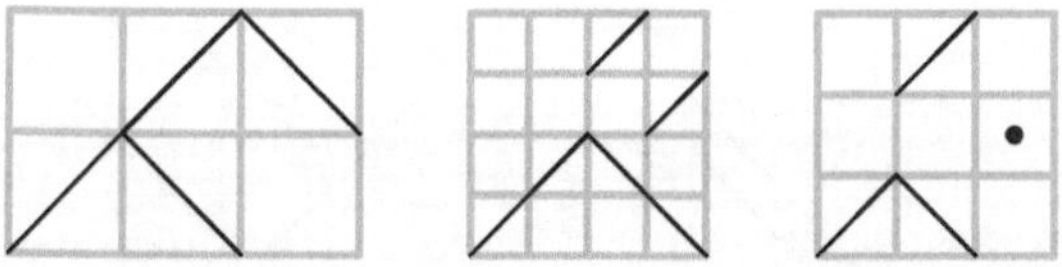

**Figure 12.29**
The simple permutations in Av(3124,4312) lie in the union of the two geometric grid classes shown on the left; the intersection of these two classes is the geometric grid class shown on the right.

have many different griddings, as demonstrated in Figure 12.28. In [4], this issue is addressed by imposing an order on all $\mathrm{Geom}^\sharp(M)$ griddings of a given permutation $\pi$. Among all of these griddings, we would like to select the minimal one, called the **preferred gridding**. By using a trick involving "marking entries" it can be shown that the set of all preferred griddings of permutations in a given geometrically griddable class is in bijection with a regular language, establishing the following result.

**Theorem 12.4.15 (Albert, Atkinson, Bouvel, Ruškuc, and Vatter [4])** *Every geometrically griddable class is strongly rational and finitely based.*

In particular, every geometrically griddable class is wpo (we have rewritten history a bit here; the wpo property plays an important role in the proof of Theorem 12.4.15, and thus must be established first). By Proposition 12.4.12, Theorem 12.4.15 thereby generalizes one direction of Theorem 12.4.7.

Continuing in this line of research, Albert, Ruškuc, and Vatter studied the inflations and substitution closures of geometrically griddable classes. Their main result is the following generalization of Albert and Atkinson's Theorem 12.3.11.

**Theorem 12.4.16 (Albert, Ruškuc, and Vatter [16])** *Let $\mathscr{C}$ be a geometrically griddable class. Its substitution closure, $\langle\mathscr{C}\rangle$, is strongly algebraic. Moreover, for every strongly rational class $\mathscr{U}$, the inflation $\mathscr{C}[\mathscr{U}]$ is strongly rational.*

It follows that all such classes are wpo. The proof of this result is well beyond the scope of this survey. Both parts of the theorem rely on the notion of "query-complete sets of properties" introduced by Brignall, Huczynska, and Vatter [65] in their generalization of Albert and Atkinson's Theorem 12.3.11. In addition, the proof of the second part of the theorem relies on the work of Brignall [61] on decompositions of permutations contained in inflations of the form $\mathscr{C}[\mathscr{U}]$ and the work of Albert, Atkinson, and Vatter [8] on strongly rational classes (in particular, that the sum indecomposable permutations in a strongly rational class themselves have a rational generating function).

Theorem 12.4.16 greatly expands the applicability of the substitution decomposition, and has found numerous applications. One example is the work of Pantone [135], who used this approach to enumerate the class Av(3124,4312). Pantone

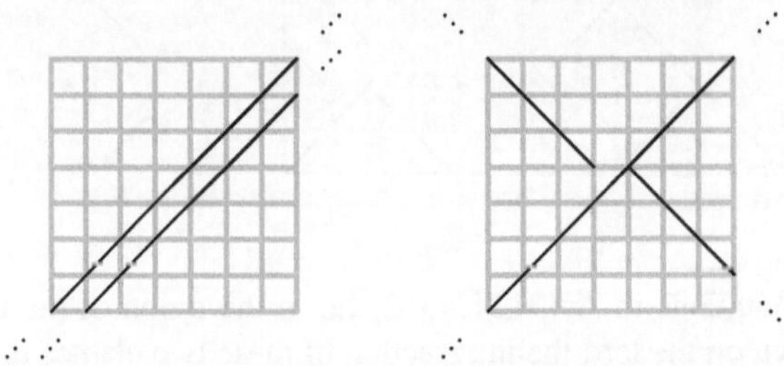

**Figure 12.30**
Geometric descriptions of the 321-avoiding permutations and the simple skew-merged permutations (up to symmetry).

showed that the simple permutations in this class lie in the union of the first and second geometric grid classes shown in Figure 12.29. He then counted the class of interest by computing the intersection of these two geometric grid classes (shown on the right of this figure; here the dot is a single point), constructing regular languages in bijection with the simple permutations in all three geometric grid classes, and then applying the machinery of the substitution decomposition. Three simpler examples in the same spirit are described in Albert, Atkinson, and Vatter [9].

Lately, though no general theory has been established, there has been some promising work done using geometric grid classes of *infinite* matrices. Recall Waton's Proposition 12.1.5, which shows that Av(321) is the class of permutations that can be drawn on two parallel lines. By taking these lines to have slope 1 and adding the grid lines $x = i$ and $y = i$ for all integers $i$, we see that Av(321) is the infinite geometric grid class shown on the left of Figure 12.30. This viewpoint, known as the **staircase decomposition** has been used by Guillemot and Vialette [97] in their study of the complexity of the permutation containment problem, by Albert, Atkinson, Brignall, Ruškuc, Smith, and West [5] in their study of growth rates of classes of the form $\text{Av}(321,\beta)$, and by Albert and Vatter [17] in their study of 321-avoiding simple permutations. This latter work has since been refined by Bóna, Homberger, Pantone, and Vatter [55] to enumerate the involutions in the classes Av(1342) and Av(2341).

Albert and Vatter [17] also gave a geometric interpretation of the skew-merged permutations. It can be shown that every simple skew-merged permutation is, up to symmetry, an element of the infinite geometric grid class shown on the right of Figure 12.30, and they used this fact to give a more structural derivation of the generating function for skew-merged permutations, originally due to Atkinson [20]. Albert and Brignall [11] studied the class of permutations whose corresponding Schubert varieties are defined by inclusions. They showed that the simple permutations of this class lie in an infinite sequence of geometric grid classes that they called **crenellations** (see Figure 12.31), and used this insight to enumerate the class.

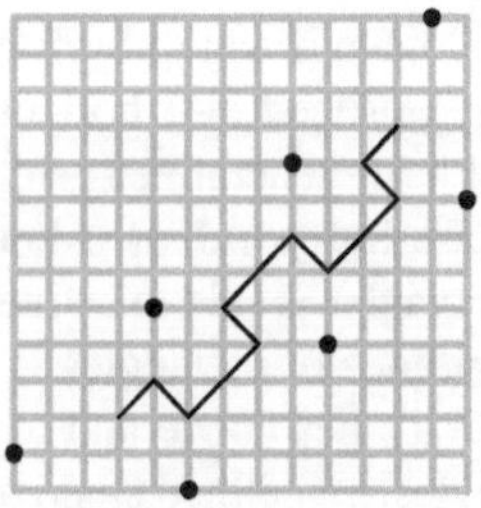

**Figure 12.31**
A crenellation.

### 12.4.3 Generalized grid classes

In a **generalized grid class** the cells are allowed to contain nonmonotone permutations. Formally, let $\mathscr{M}$ be a $t \times u$ matrix of permutation classes. An $\mathscr{M}$-gridding of the permutation $\pi$ of length $n$ is a choice of column divisions $1 = c_1 \leq \cdots \leq c_{t+1} = n+1$ and row divisions $1 = r_1 \leq \cdots \leq r_{u+1} = n+1$ such that for all $i$ and $j$, $\pi([c_i, c_{i+1}) \times [r_j, r_{j+1}))$ lies in the class $\mathscr{M}_{i,j}$.

The class of all permutations with $\mathscr{M}$-griddings is the grid class of $\mathscr{M}$, denoted Grid($\mathscr{M}$). In the context of monotone grid classes, we defined the class $\mathscr{C}$ to be monotonically griddable if $\mathscr{C} \subseteq \text{Grid}(M)$ for some finite $0/\pm 1$ matrix $M$. Now we define the class $\mathscr{C}$ to be **$\mathscr{G}$-griddable** if $\mathscr{C} \subseteq \text{Grid}(\mathscr{M})$ for some finite matrix $\mathscr{M}$ whose entries are all subclasses of $\mathscr{G}$ (for the purposes of this definition, we may take them all to be equal to $\mathscr{G}$). From this perspective, a class is monotonically griddable if and only if it is $(\text{Av}(21) \cup \text{Av}(12))$-griddable (though this fact takes a bit of thought).

As Theorem 12.4.5 characterizes monotonically griddable classes, our next result characterizes generalized grid classes.

**Theorem 12.4.17 (Vatter [161])** *The permutation class $\mathscr{C}$ is $\mathscr{G}$-griddable if and only if it does not contain arbitrarily long sums or skew sums of basis elements of $\mathscr{G}$, that is, if there exists a constant $m$ so that $\mathscr{C}$ does not contain $\beta_1 \oplus \cdots \oplus \beta_m$ or $\beta_1 \ominus \cdots \ominus \beta_m$ for any basis elements $\beta_1, \ldots, \beta_m$ of $\mathscr{G}$.*

We prove this result by appealing to a result of Gyárfás and Lehel from 1970, but a proof from first principles can be found in Vatter [161]. One direction is again clear—if $\mathscr{C}$ contains arbitrarily long sums or skew sums of basis elements of $\mathscr{G}$, then it is not $\mathscr{G}$-griddable.

Now suppose that $\mathscr{C}$ does not contain arbitrarily long sums or skew sums of basis elements of $\mathscr{G}$ and consider the set

$$\mathfrak{R}_\pi = \{\text{axes-parallel rectangles } R : \pi(R) \notin \mathscr{G}\}.$$

Thus for every $R \in \mathfrak{R}_\pi$, $\pi(R)$ contains a basis element of $\mathscr{G}$, so in any $\mathscr{G}$-gridding of $\pi$ every rectangle in $\mathfrak{R}_\pi$ must be sliced by a grid line. Clearly the converse is also true (if every rectangle in $\mathfrak{R}_\pi$ is sliced by a grid line, these grid lines define a

$\mathscr{G}$-gridding of $\pi$), so $\mathscr{C}$ is $\mathscr{G}$-griddable if and only if there is a constant $\ell$ such that, for every $\pi \in \mathscr{C}$, the set $\mathfrak{R}_\pi$ can be sliced by $\ell$ horizontal or vertical lines. (This is the analogue of Proposition 12.4.4 for generalized grid classes.)

We say that two rectangles are **independent** if both their $x$- and $y$-axis projections are disjoint, and a set of rectangles is said to be independent if they are pairwise independent. An **increasing sequence** of rectangles is a sequence $R_1, \ldots, R_m$ of independent rectangles such that $R_{i+1}$ lies above and to the right of $R_i$ for all $i \geq 0$. **Decreasing sequences** of rectangles are defined analogously. By our assumptions on $\mathscr{C}$, there is some constant $m$ such that for every $\pi \in \mathscr{C}$ the set $\mathfrak{R}_\pi$ contains neither an increasing nor a decreasing sequence of size $m$.

An independent set of rectangles corresponds to a permutation. Therefore, by the Erdős-Szekeres Theorem, $\mathfrak{R}_\pi$ cannot contain an independent set of size greater than $(m-1)^2+1$. The proof of Theorem 12.4.17 is therefore completed by the following result of Gyárfás and Lehel, which was later strengthened by Károlyi and Tardos [109].

**Theorem 12.4.18 (Gyárfás and Lehel [99])** *There is a function $f(m)$ such that for any collection $\mathfrak{R}$ of axes-parallel rectangles in the plane that has no independent set of size greater than m, there exists a set of $f(m)$ horizontal and vertical lines that slice every rectangle in $\mathfrak{R}$.*

Our next question is which permutation classes can be gridded by one of their proper subclasses. We call the class $\mathscr{C}$ **grid irreducible** if it is not $\mathscr{G}$-griddable for any proper subclass $\mathscr{G} \subsetneq \mathscr{C}$. Somewhat surprisingly, this question has a concrete answer.

**Proposition 12.4.19** *The class $\mathscr{C}$ is grid irreducible if and only if $\mathscr{C} = \{1\}$ or if for every $\pi \in \mathscr{C}$ either $\bigoplus \mathrm{Sub}(\pi) \subseteq \mathscr{C}$ or $\bigominus \mathrm{Sub}(\pi) \subseteq \mathscr{C}$.*

*Proof.* One direction is immediate from Theorem 12.4.17: if $\mathscr{C}$ contains arbitrarily long sums or skew sums of all of its elements, then it is grid irreducible. Moreover, the only finite grid irreducible class is $\{1\}$, which takes care of that part of the other direction.

Now suppose that $\mathscr{C}$ is an infinite permutation class. If there is some $\pi \in \mathscr{C}$ such that neither $\bigoplus \mathrm{Sub}(\pi)$ nor $\bigominus \mathrm{Sub}(\pi)$ is contained in $\mathscr{C}$ then $\mathscr{C}$ is $(\mathscr{C} \cap \mathrm{Av}(\pi))$-griddable by Theorem 12.4.17, and thus is not grid irreducible. ∎

As shown in the proof of this result, if neither $\bigoplus \mathrm{Sub}(\pi_1)$ nor $\bigominus \mathrm{Sub}(\pi_1)$ is contained in $\mathscr{C}$ then it is $(\mathscr{C} \cap \mathrm{Av}(\pi_1))$-griddable. We may then apply this to $\mathscr{C} \cap \mathrm{Av}(\pi_1)$ to see that if neither $\bigoplus \mathrm{Sub}(\pi_2)$ nor $\bigominus \mathrm{Sub}(\pi_2)$ is contained in $\mathscr{C} \cap \mathrm{Av}(\pi_1)$, it is $(\mathscr{C} \cap \mathrm{Av}(\pi_1, \pi_2))$-griddable. Continuing in this manner, we can construct a descending chain of classes

$$\begin{array}{ccccccc} \mathscr{C} & \supsetneq & \mathscr{C} \cap \mathrm{Av}(\pi_1) & \supsetneq & \mathscr{C} \cap \mathrm{Av}(\pi_1, \pi_2) & \supsetneq & \cdots \\ \| & & \| & & \| & & \\ \mathscr{C}^{(0)} & \supsetneq & \mathscr{C}^{(1)} & \supsetneq & \mathscr{C}^{(2)} & \supsetneq & \cdots \end{array}$$

such that $\mathscr{C}$ is $\mathscr{C}^{(i)}$-griddable for all $i$. Because there are infinite strictly descending chains of permutation classes, there is no guarantee that this process will terminate (see Vatter [161] for such a construction). However, if $\mathscr{C}$ is wpo then it satisfies the descending chain condition by Proposition 12.3.1 and thus this process must stop. This proves the following.

**Proposition 12.4.20** *If the class $\mathscr{C}$ is wpo then it is $\mathscr{G}$-griddable for the grid irreducible class*

$$\mathscr{G} = \{\pi : \textit{either} \bigoplus \mathrm{Sub}(\pi) \subseteq \mathscr{C} \textit{ or} \bigominus \mathrm{Sub}(\pi) \subseteq \mathscr{C}\}.$$

We need a final result about generalized grid classes, which shows that atomic grid irreducible classes are very constrained.

**Proposition 12.4.21** *Suppose that the class $\mathscr{C}$ is atomic. Then it is grid irreducible if and only if it is sum or skew closed.*

*Proof.* Theorem 12.4.17 shows that every sum or skew closed class is grid irreducible, so it suffices to prove the reverse direction. Suppose that $\mathscr{C}$ is atomic, but neither sum nor skew closed. Thus there are permutations $\pi, \sigma \in \mathscr{C}$ such that $\bigoplus \mathrm{Sub}(\pi), \bigominus \mathrm{Sub}(\tau) \not\subseteq \mathscr{C}$. Because $\mathscr{C}$ is atomic, we can find a permutation $\tau \in \mathscr{C}$ containing both $\pi$ and $\sigma$. Thus $\mathscr{C}$ does not contain arbitrarily long sums or skew sums of $\tau$, so it is $(\mathscr{C} \cap \mathrm{Av}(\tau))$-griddable (again by Theorem 12.4.17). This shows that $\mathscr{C}$ is not grid irreducible, completing the proof. ∎

We conclude by investigating a special case of generalized grid classes and their relation to the splittability question. We call the $2 \times 1$ generalized grid class $\mathrm{Grid}(\mathscr{D}\ \mathbb{E})$ the **horizontal juxtaposition** of the classes $\mathscr{D}$ and $\mathbb{E}$ (the obvious symmetry of this construction is called a **vertical juxtaposition**). Atkinson [21] introduced juxtapositions and established the structure of their basis elements.

Using our results on generalized grid classes we are able to completely characterize the classes which can be expressed as nontrivial juxtapositions. If $\mathscr{C}$ is not atomic then $\mathscr{C} \subseteq \mathscr{D} \cup \mathbb{E}$ for proper subclasses $\mathscr{D}, \mathbb{E} \subsetneq \mathscr{C}$, so $\mathscr{C}$ is also contained in the nontrivial juxtaposition $\mathrm{Grid}(\mathscr{D}\ \mathbb{E})$. Suppose instead that $\mathscr{C}$ is atomic. If $\mathscr{C}$ is either sum or skew closed then it is grid irreducible by Proposition 12.4.19 and thus not contained in a juxtaposition of proper subclasses. Otherwise, Proposition 12.4.21 shows that $\mathscr{C}$ is $\mathscr{D}$-griddable for a proper subclass $\mathscr{D} \subsetneq \mathscr{C}$. Choose $\mathscr{M}$ to be a matrix of minimal possible size such that all of its entries are equal to $\mathscr{D}$ and $\mathscr{C} \subseteq \mathrm{Grid}(\mathscr{M})$. We may suppose by symmetry that $\mathscr{M}$ has at least two columns, and thus by slicing it by a vertical line we see that $\mathscr{C}$ is contained in the horizontal juxtaposition of two proper subclasses, establishing the following.

**Proposition 12.4.22** *A permutation class is contained in the juxtaposition of two proper subclasses if and only if it is neither sum nor skew closed.*

In particular, if a class is contained in the juxtaposition of two proper subclasses then it is splittable. Thus the splittability question is only interesting for sum or skew closed classes that are not substitution closed.

### 12.4.4 Small permutation classes

Our goal in this subsection is to outline the proofs of Theorem 12.4.2, which specifies all possible growth rates of small permutations classes, and 12.4.3, which states that all small permutation classes are strongly rational, while glossing over some of the more technical details. Adopting a perspective that is slightly historically backward, we prove Theorem 12.4.3 first, by showing that every small permutation class is the inflation of a geometric grid class by a strongly rational class, and is thereby strongly rational itself. We then show that if $\mathscr{C}$ is a small permutation class, its growth rate is either equal to 0, 1, or 2, or is equal to the growth rate of the largest sum or skew closed class contained in it (thereby establishing that these classes have proper growth rates). The actual list of possible growth rates provided in Theorem 12.4.2 can then be produced by characterizing all possible enumerations of sum indecomposable permutations in classes with growth rates less than $\kappa$, though we do not go into that rather lengthy undertaking here.

For the moment, let $\gamma$ be an arbitrary real number and suppose we would like to grid all permutation classes of growth rate less than $\gamma$. That is, we would like to find a single **cell class** $\mathscr{G}$ such that every permutation of (upper) growth rate less than $\gamma$ is $\mathscr{G}$-griddable.

To this end, define

$$\mathscr{G}_\gamma = \{\pi : \text{either gr}\,(\bigoplus \text{Sub}(\pi)) < \gamma \text{ or gr}\,(\bigominus \text{Sub}(\pi)) < \gamma\}.$$

If $\pi \in \mathscr{G}_\gamma$, then at least one of $\bigoplus \text{Sub}(\pi)$ or $\bigominus \text{Sub}(\pi)$ has growth rate less than $\gamma$, so in order to grid all permutation classes of growth rate less than $\gamma$ we must have $\pi$ in our cell class. Our next result shows that this cell class can indeed be used to grid all classes of growth rate less than $\gamma$.

**Proposition 12.4.23** *For every real number $\gamma$, if the permutation class $\mathscr{C}$ satisfies $\overline{\text{gr}}(\mathscr{C}) < \gamma$ then it is $\mathscr{G}_\gamma$-griddable.*

*Proof.* Suppose to the contrary that $\mathscr{C}$ satisfies $\overline{\text{gr}}(\mathscr{C}) < \gamma$ but is not $\mathscr{G}_\gamma$-griddable, and let $f$ denote the generating function of $\mathscr{C}$. Now fix an arbitrary integer $m$. We seek to derive a contradiction by showing that $f(1/\gamma) > m$, which, because $m$ is arbitrary, will imply that $\overline{\text{gr}}(\mathscr{C}) \geq \gamma$. By Theorem 12.4.17 and symmetry, we may assume that there is a permutation of the form $\beta_1 \oplus \cdots \oplus \beta_m$ contained in $\mathscr{C}$ where each $\beta_i$ is a basis element of $\mathscr{G}_\gamma$. If $\text{gr}(\bigoplus \beta_i)$ were less than $\gamma$ then $\beta_i$ would lie in $\mathscr{G}_\gamma$, so we know that $\text{gr}(\bigoplus \beta_i) \geq \gamma$ for every index $i$.

Let $s_i$ denote the generating function for the nonempty sum indecomposable permutations contained in (or equal to) $\beta_i$, so that the generating function for $\bigoplus \beta_i$ is given by $1/(1-s_i)$. These generating functions have positive singularities that are less than or equal to $1/\gamma$, and because each $s_i$ is a polynomial this singularity must be a pole. Moreover, each $s_i$ has positive coefficients, which implies that the unique positive solution to $s_i(x) = 1$ occurs for $x \leq 1/\gamma$. In particular, $s_i(1/\gamma) \geq 1$ for all indices $i$.

Now consider the set of permutations of the form $\alpha_1 \oplus \cdots \oplus \alpha_k$ for some $k \leq m$, where for each $i$, $\alpha_i$ is a nonempty sum indecomposable permutation contained in

$\beta_i$. Clearly this is a subset (but likely not a subclass) of $\mathscr{C}$. Moreover, the generating function for this set of permutations is

$$s_1 + s_1 s_2 + \cdots + s_1 \cdots s_m,$$

which is at least $m$ when evaluated at $1/\gamma$. This implies that the generating function for $\mathscr{C}$ is also at least $m$ when evaluated at $1/\gamma$, completing the contradiction. ∎

Proposition 12.4.23, which was not used in the original proof of Theorem 12.4.2, gives us an easily computable membership test to determine whether $\pi \in \mathscr{G}_\gamma$. The small permutation classes are all $\mathscr{G}_{\kappa-\varepsilon}$-griddable for some $\varepsilon > 0$ so we begin by establishing some restrictions on $\mathscr{G}_\kappa$.

The reader may have noticed that both infinite antichains we have encountered in this survey (in Figures 12.11 and 12.17) consist of variations on a single theme. While that particular coincidence should not be taken too far (there are many more infinite antichains based on different motifs, for example, all those lying in monotone grid classes with cycles), the theme of those two antichains plays a significant role in the characterization of small permutation classes. We define the **increasing oscillating sequence** as

$$4,1,6,3,8,5,\ldots,2k+2,2k-1,\ldots.$$

(This sequence, itself listed as A065164 in the OEIS [157], also arises in the study of juggling and genomics, see Buhler, Eisenbud, Graham, and Wright [68] and Pevzner [136], respectively.)

An **increasing oscillation** is defined to be any sum indecomposable permutation that is order isomorphic to a subsequence of the increasing oscillating sequence. It is easily seen that there are only two increasing oscillations of each length, which are inverses of each other. Finally, a **decreasing oscillation** is the reverse of an increasing oscillation, and an **oscillation** is either an increasing or a decreasing oscillation.

Let $\mathscr{O}$ denote the downward closure of the set of all oscillations,

$$\mathscr{O} = \text{Sub}(\text{increasing and decreasing oscillations of all lengths}).$$

It can be shown that their substitution closure $\langle \mathscr{O} \rangle$ has a finite basis (Vatter [161, Proposition A.2]). From this, it can then be computed that for every basis element $\beta$ of $\langle \mathscr{O} \rangle$, both $\bigoplus \text{Sub}(\beta)$ and $\bigominus \text{Sub}(\beta)$ have growth rates greater than 2.24. Therefore Proposition 12.4.23 immediately gives us the following.

**Proposition 12.4.24** *The cell class $\mathscr{G}_\kappa$ is contained in $\langle \mathscr{O} \rangle$.*

This gives us a place to start, but we need to get some control on the structure of small permutation classes. In particular, we need an analogue of Proposition 12.4.6 that will allow us to "chop" the griddings of these classes. The direct analogue of Proposition 12.4.6 is below, while Figure 12.32 shows why we must impose these hypotheses.

**Theorem 12.4.25 (Vatter [161, Theorem 5.4])** *Suppose that the cell class $\mathscr{G}$ contains only finitely many simple permutations and has bounded substitution depth. If the class $\mathscr{C}$ is $\mathscr{G}$-griddable and the length of alternations in $\mathscr{C}$ is bounded, then*

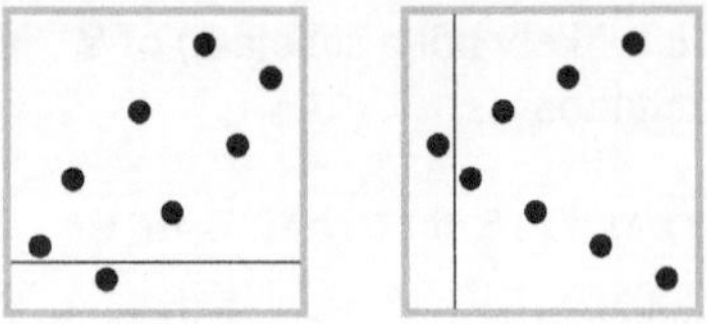

**Figure 12.32**
Two permutations that show why the hypotheses of Theorem 12.4.25 are necessary On the left is a long increasing oscillation, while on the right is a permutation with large substitution depth, each with an extra grid line. Both situations require a large number of additional grid lines in order to grid them into independent rectangles.

*$\mathscr{C} \subseteq \mathrm{Grid}(\mathscr{M})$ for a matrix $\mathscr{M}$ in which every nonempty entry is a subclass of $\mathscr{G}$ and no two nonempty entries share a row or a column.*

Clearly Theorem 12.4.25 is not enough for our purposes, since small permutation classes may contain arbitrarily long alternations. Still, the examples of Figure 12.32 demonstrate obstructions to chopping any sort of gridding, and the actual chopping result used for small permutation classes (Theorem 5.4 of [161]) has precisely the same conditions on $\mathscr{G}$. Thus we need to verify that our cell classes $\mathscr{G}_{\kappa-\varepsilon}$ have finitely many simple permutations and bounded substitution depth.

We tackle substitution depth first. It can be shown (Proposition 4.2 of [161]) that every permutation of substitution depth at least $8n$ contains a wedge alternation of length at least $n$. Furthermore, using the membership test provided by Proposition 12.4.23 one can prove that for every $\gamma < 1+\varphi \approx 2.62$, the cell class $\mathscr{G}_\gamma$ does not contain arbitrarily long wedge alternations. Thus there is some constant $d$ such that every permutation in $\mathscr{G}_\kappa$ has substitution depth at most $d$.

It is not difficult to establish that the growth rate of $\mathscr{O}$ is precisely $\kappa$. Moreover, using the fact that $\mathscr{O}$ is the union of two posets (corresponding to the increasing and decreasing oscillations) neither of which has an antichain containing more than three elements, it follows that for every $\varepsilon > 0$ we have $\mathscr{G}_{\kappa-\varepsilon} \subseteq \langle \mathscr{O}^k \rangle$ for some $k$, where

$$\mathscr{O}^k = \mathrm{Sub}(\text{oscillations of length at most } k).$$

With these two computations handled, we can then chop the griddings of small permutation classes to obtain the following result.

**Theorem 12.4.26 (Vatter [161, Theorem 5.4])** *Suppose that $\overline{\mathrm{gr}}(\mathscr{C}) < \kappa - \varepsilon$ for some $\varepsilon > 0$. Then $\mathscr{C} \subseteq \mathrm{Grid}(\mathscr{M})$ for a finite matrix $\mathscr{M}$ such that*

*(1) every entry is equal to $\mathscr{G}_{\kappa-\varepsilon}$, the class of monotone permutations, or is empty,*

*(2) every nonempty entry that shares a row or column with another nonempty entry is equal to the class of monotone permutations, and*

*(3) no nonempty entry shares a row or column with more than one other nonempty entry.*

Moreover, we have seen above that $\mathcal{G}_{\kappa-\varepsilon}$ is a subclass of $\langle \mathcal{O}^k \rangle$ and contains only permutations of substitution depth at most $d$. This gives us the following.

**Proposition 12.4.27** *For every $\varepsilon > 0$, the cell class $\mathcal{G}_{\kappa-\varepsilon}$ is strongly rational.*

*Proof.* Recall that by the second part of Theorem 12.4.15, the inflation of a geometrically griddable class by a strongly rational class is itself strongly rational. Choose $d$ so that every permutation in $\mathcal{G}_{\kappa-\varepsilon}$ has substitution depth at most $d$. Thus each of these permutations is either the inflation of a simple permutation by permutations of substitution depth at most $d-1$ or the sum or skew sum of such permutations. Thus if we define

$$\tilde{\mathcal{O}}^k = \mathcal{O}^k \cup \mathrm{Av}(21) \cup \mathrm{Av}(12),$$

we see that

$$\mathcal{G}_{\kappa-\varepsilon} \subseteq \underbrace{\tilde{\mathcal{O}}^k[\tilde{\mathcal{O}}^k[\cdots]]}_{d \text{ copies of } \tilde{\mathcal{O}}^k}.$$

Because $\tilde{\mathcal{O}}^k$ is the union of a finite class with the class of all monotone permutations, it is geometrically griddable, and thus the strong rationality of $\mathcal{G}_{\kappa-\varepsilon}$ follows by iteratively applying the second part of Theorem 12.4.15. ∎

It is possible to show the stronger result that $\mathcal{G}_{\kappa}$ is strongly rational, but this requires more work, and Proposition 12.4.27 is enough for our purposes. The first of these purposes is to prove Theorem 12.4.3, which states that small permutation classes are strongly rational. By Theorem 12.4.26, every small permutation class is contained in $\mathrm{Grid}(M)[\mathcal{G}_{\kappa-\varepsilon}]$ for a matrix $M$ in which no nonzero entry shares a row or column with more than one other nonzero entry. It follows that $M$ is a forest so $\mathrm{Grid}(M) = \mathrm{Geom}(M)$ by Proposition 12.4.12. The result now follows by the second part of Theorem 12.4.16.

Our last goal is the characterization of growth rates of small permutation classes (Theorem 12.4.2). We observed in Proposition 12.4.10 that the upper growth rate of $\mathrm{Grid}(M)$ is equal to the greatest upper growth rate of the monotone grid class of a connected component of $M$. A similar argument holds for generalized grid classes and their subclasses. Thus given a class $\mathcal{C} \subseteq \mathrm{Grid}(\mathcal{M})$, its upper growth rate is equal to the upper growth rate of the restriction of $\mathcal{C}$ to a connected component of $\mathcal{M}$. We can now sketch the reduction to sum closed classes.

**Theorem 12.4.28** *If the permutation class $\mathcal{C}$ satisfies $\overline{\mathrm{gr}}(\mathcal{C}) < \kappa$ then $\mathrm{gr}(\mathcal{C})$ exists and is equal to 0, 1, 2, or the growth rate of a subclass that is either sum or skew closed.*

*Proof.* Suppose that $\overline{\mathrm{gr}}(\mathcal{C}) < \kappa$ and set $\mathcal{C}^0 = \mathcal{C}$. By our previous work, we can find a cell class $\mathcal{G}^0 \subseteq \mathcal{C} \cap \mathcal{G}_{\kappa-\varepsilon}$ for some $\varepsilon > 0$ such that $\mathcal{C}$ is $\mathcal{G}^0$-griddable. Next we can find a matrix $\mathcal{M}^0$ with all entries equal to subclasses of $\mathcal{G}^0$ and satisfying the conclusions of Theorem 12.4.26 such that $\mathcal{C}^0 \subseteq \mathrm{Grid}(\mathcal{M}^0)$. By our remarks above, the upper growth rate of $\mathcal{C}^0$ is the upper growth rate of the restriction of $\mathcal{C}^0$ to a connected component of $\mathcal{M}^0$. This connected component is either a single cell (in which case we get a subclass of $\mathcal{G}^0$) or a pair of monotone cells. It can be shown that all subclasses of a $1 \times 2$ monotone grid class have growth rates 0, 1, or 2, so we are done if the latter situation holds. Thus we may assume the former situation holds and choose $\mathcal{C}^1 \subseteq \mathcal{G}^0$ such that $\overline{\mathrm{gr}}(\mathcal{C}^0) = \overline{\mathrm{gr}}(\mathcal{C}^1)$.

Importantly, Proposition 12.4.27 shows that $\mathcal{C}^1$ is strongly rational, and thus wpo. By Proposition 12.3.21, this implies that the upper growth rate of $\mathcal{C}^1$ is equal to that of one of its atomic subclasses, say $\mathcal{A}^1 \subseteq \mathcal{C}^1$. By Proposition 12.4.20, we may now choose a grid irreducible class $\mathcal{G}^1$ such that $\mathcal{A}^1$ is $\mathcal{G}^1$-griddable. Then we choose a matrix $\mathcal{M}^1$ with all entries equal to subclasses of $\mathcal{G}^1$, which satisfies the conclusions of Theorem 12.4.26 and with $\mathcal{A}^1 \subseteq \mathrm{Grid}(\mathcal{M}^1)$. The upper growth rate of $\mathcal{A}^1$ is then equal to that of a restriction of $\mathcal{A}^1$ to a connected component of $\mathcal{M}^1$. We are done as before if this component consists of two cells, so we may assume that $\overline{\mathrm{gr}}(\mathcal{A}^1) = \overline{\mathrm{gr}}(\mathcal{C}^2)$ for a subclass $\mathcal{C}^2 \subseteq \mathcal{G}^1$.

By repeating this process indefinitely, we either find that the upper growth rate of $\mathcal{C}$ is equal to 0, 1, or 2, or we construct an infinite descending chain

$$\mathcal{C} = \mathcal{C}^0 \supseteq \mathcal{G}^0 \supseteq \mathcal{C}^1 \supseteq \mathcal{A}^1 \supseteq \mathcal{G}^1 \supseteq \mathcal{C}^2 \supseteq \mathcal{A}^2 \supseteq \mathcal{G}^2 \supseteq \mathcal{C}^3 \supseteq \cdots,$$

all with identical upper growth rates. Moreover, because $\mathcal{G}^0$ is wpo, Proposition 12.3.1 shows that this chain must terminate. Thus there is some $i$ such that $\mathcal{C}^i = \mathcal{A}^i = \mathcal{G}^i$. This implies that the class $\mathcal{C}^i$ is both atomic and grid irreducible, and thus it is either sum or skew closed by Proposition 12.4.21, proving the theorem. ■

Thus we have reduced the characterization of growth rates of small permutation classes to the characterization of growth rates of small *sum closed* permutation classes, as promised. From this point Theorem 12.4.2 follows after a fairly technical analysis of sum indecomposable permutations.

**Acknowledgments.** This chapter has greatly benefited by the comments, corrections, and suggestions of the referee as well as those of Michael Albert, David Bevan, Jonathan Bloom, Robert Brignall, Cheyne Homberger, Vít Jelínek, and Jay Pantone.

# References

[1] Michael H. Albert. On the length of the longest subsequence avoiding an arbitrary pattern in a random permutation. *Random Structures Algorithms*, 31(2):227–238, 2007.

[2] Michael H. Albert. An introduction to structural methods in permutation patterns. In Steve Linton, Nik Ruškuc, and Vincent Vatter, editors, *Permutation*

*Patterns*, volume 376 of *London Mathematical Society Lecture Note Series*, pages 153–170. Cambridge University Press, 2010.

[3] Michael H. Albert and Mike D. Atkinson. Simple permutations and pattern restricted permutations. *Discrete Math.*, 300(1-3):1–15, 2005.

[4] Michael H. Albert, Mike D. Atkinson, Mathilde Bouvel, Nik Ruškuc, and Vincent Vatter. Geometric grid classes of permutations. *Trans. Amer. Math. Soc.*, 365:5859–5881, 2013.

[5] Michael H. Albert, Mike D. Atkinson, Robert Brignall, Nik Ruškuc, Rebecca Smith, and Julian West. Growth rates for subclasses of Av(321). *Electron. J. Combin.*, 17:Paper 141, 16 pp., 2010.

[6] Michael H. Albert, Mike D. Atkinson, Cheyne Homberger, and Jay Pantone. Deflatability of permutation classes. arXiv:1409.5296 [math.CO], 2014.

[7] Michael H. Albert, Mike D. Atkinson, and Martin Klazar. The enumeration of simple permutations. *J. Integer Seq.*, 6(4):Article 03.4.4, 18 pp., 2003.

[8] Michael H. Albert, Mike D. Atkinson, and Vincent Vatter. Subclasses of the separable permutations. *Bull. Lond. Math. Soc.*, 43:859–870, 2011.

[9] Michael H. Albert, Mike D. Atkinson, and Vincent Vatter. Inflations of geometric grid classes of permutations: three case studies. *Australas. J. Combin.*, 58(1):27–47, 2014.

[10] Michael H. Albert and Mireille Bousquet-Mélou. Permutations sortable by two stacks in parallel and quarter plane walks. *European J. Combin.*, 43:131–164, 2015.

[11] Michael H. Albert and Robert Brignall. Enumerating indices of schubert varieties defined by inclusions. *J. Combin. Theory Ser. A*, 123:154–168, 2014.

[12] Michael H. Albert, Robert Brignall, and Vincent Vatter. Large infinite antichains of permtutations. *Pure Math. Appl. (PU.M.A.)*, 24(2):47–57, 2013.

[13] Michael H. Albert, Murray Elder, Andrew Rechnitzer, P. Westcott, and Mike Zabrocki. On the Stanley–Wilf limit of 4231-avoiding permutations and a conjecture of Arratia. *Adv. in Appl. Math.*, 36(2):96–105, 2006.

[14] Michael H. Albert and Steve Linton. Growing at a perfect speed. *Combin. Probab. Comput.*, 18:301–308, 2009.

[15] Michael H. Albert, Steve Linton, and Nik Ruškuc. The insertion encoding of permutations. *Electron. J. Combin.*, 12(1):Paper 47, 31 pp., 2005.

[16] Michael H. Albert, Nik Ruškuc, and Vincent Vatter. Inflations of geometric grid classes of permutations. *Israel J. Math.*, to appear. arXiv:1202.1833 [math.CO], 2012.

[17] Michael H. Albert and Vincent Vatter. Generating and enumerating 321-avoiding and skew-merged simple permutations. *Electron. J. Combin.*, 20(2):Paper 44, 11 pp., 2013.

[18] Noga Alon and Ehud Friedgut. On the number of permutations avoiding a given pattern. *J. Combin. Theory Ser. A*, 89(1):133–140, 2000.

[19] Richard Arratia. On the Stanley–Wilf conjecture for the number of permutations avoiding a given pattern. *Electron. J. Combin.*, 6:Note 1, 4 pp., 1999.

[20] Mike D. Atkinson. Permutations which are the union of an increasing and a decreasing subsequence. *Electron. J. Combin.*, 5:Paper 6, 13 pp., 1998.

[21] Mike D. Atkinson. Restricted permutations. *Discrete Math.*, 195(1-3):27–38, 1999.

[22] Mike D. Atkinson. Permuting machines and permutation patterns. In Steve Linton, Nik Ruškuc, and Vincent Vatter, editors, *Permutation Patterns*, volume 376 of *London Mathematical Society Lecture Note Series*, pages 67–88. Cambridge University Press, 2010.

[23] Mike D. Atkinson, Maximillian M. Murphy, and Nik Ruškuc. Partially well-ordered closed sets of permutations. *Order*, 19(2):101–113, 2002.

[24] Mike D. Atkinson, Maximillian M. Murphy, and Nik Ruškuc. Sorting with two ordered stacks in series. *Theoret. Comput. Sci.*, 289(1):205–223, 2002.

[25] Mike D. Atkinson, Maximillian M. Murphy, and Nik Ruškuc. Pattern avoidance classes and subpermutations. *Electron. J. Combin.*, 12(1):Paper 60, 18 pp., 2005.

[26] Mike D. Atkinson, Nik Ruškuc, and Rebecca Smith. Substitution-closed pattern classes. *J. Combin. Theory Ser. A*, 118:317–340, 2011.

[27] David Avis and Monroe Newborn. On pop-stacks in series. *Utilitas Math.*, 19:129–140, 1981.

[28] Eric Babson and Einar Steingrímsson. Generalized permutation patterns and a classification of the Mahonian statistics. *Sém. Lothar. Combin.*, 44:Article B44b, 18 pp., 2000.

[29] Eric Babson and Julian West. The permutations $123p_4 \cdots p_m$ and $321p_4 \cdots p_m$ are Wilf-equivalent. *Graphs Combin.*, 16(4):373–380, 2000.

[30] Jörgen Backelin, Julian West, and Guoce Xin. Wilf-equivalence for singleton classes. *Adv. in Appl. Math.*, 38(2):133–148, 2007.

[31] Jinho Baik, Percy Deift, and Kurt Johansson. On the distribution of the length of the longest increasing subsequence of random permutations. *J. Amer. Math. Soc.*, 12(4):1119–1178, 1999.

[32] József Balogh, Béla Bollobás, and Robert Morris. Hereditary properties of ordered graphs. In M. Klazar, J. Kratochvíl, M. Loebl, J. Matoušek, R. Thomas, and P. Valtr, editors, *Topics in Discrete Mathematics*, volume 26 of *Algorithms Combin.*, pages 179–213. Springer, Berlin, 2006.

[33] F. Bassino, Mathilde Bouvel, Adeline Pierrot, and Dominique Rossin. An algorithm for deciding the finiteness of the number of simple permutations in permutation classes. arXiv:1307.2006 [math.CO], 2014.

[34] David Bevan. Growth rates of permutation grid classes, tours on graphs, and the spectral radius. *Trans. Amer. Math. Soc.*, to appear. arXiv:1302.2037 [math.CO].

[35] David Bevan. Intervals of permutation class growth rates. arXiv:1410.3679 [math.CO], 2014.

[36] David Bevan. Permutations avoiding 1324 and patterns in Łukasiewicz paths. arXiv:1406.2890 [math.CO], 2014.

[37] David Bevan. Growth rates of geometric grid classes of permutations. *Electron. J. Combin.*, 21(4):Paper 4.51, 17 pp., 2014.

[38] Dan Bienstock and Ervin Győri. An extremal problem on sparse 0-1 matrices. *SIAM J. Discrete Math.*, 4(1):17–27, 1991.

[39] Jonathan Bloom. A refinement of Wilf-equivalence for patterns of length 4. *J. Combin. Theory Ser. A*, 124:166–177, 2014.

[40] Jonathan Bloom and Sergi Elizalde. Pattern avoidance in matchings and partitions. *Electron. J. Combin.*, 20(2):Paper 5, 38 pp., 2013.

[41] Jonathan Bloom and Dan Saracino. A simple bijection between 231-avoiding and 312-avoiding placements. *J. Combin. Math. Combin. Comput.*, 89:23–32, 2014.

[42] Jonathan Bloom and Vincent Vatter. Two vignettes on full rook placements. *Australas. J. Combin.*, to appear. arXiv:1310.6073 [math.CO].

[43] Béla Bollobás. Hereditary and monotone properties of combinatorial structures. In A. Hilton and J. Talbot, editors, *Surveys in Combinatorics 2007*, volume 346 of *London Mathematical Society Lecture Note Series*, pages 1–39. Cambridge University Press, 2007.

[44] Miklós Bóna. A new record for 1324-avoiding permutations. *European J. Math.*, to appear. arXiv:1404.4033 [math.CO].

[45] Miklós Bóna. On the best upper bound for permutations avoiding a pattern of a given length. arXiv:1209.2404 [math.CO], 2012.

[46] Miklós Bóna. Exact enumeration of 1342-avoiding permutations: a close link with labeled trees and planar maps. *J. Combin. Theory Ser. A*, 80(2):257–272, 1997.

[47] Miklós Bóna. The number of permutations with exactly $r$ 132-subsequences is $P$-recursive in the size! *Adv. in Appl. Math.*, 18(4):510–522, 1997.

[48] Miklós Bóna. Permutations with one or two 132-subsequences. *Discrete Math.*, 181(1-3):267–274, 1998.

[49] Miklós Bóna. The solution of a conjecture of Stanley and Wilf for all layered patterns. *J. Combin. Theory Ser. A*, 85(1):96–104, 1999.

[50] Miklós Bóna. *Combinatorics of Permutations*. Discrete Mathematics and its Applications (Boca Raton). Chapman & Hall/CRC, Boca Raton, FL, 2004.

[51] Miklós Bóna. The limit of a Stanley–Wilf sequence is not always rational, and layered patterns beat monotone patterns. *J. Combin. Theory Ser. A*, 110(2):223–235, 2005.

[52] Miklós Bóna. New records in Stanley–Wilf limits. *European J. Combin.*, 28(1):75–85, 2007.

[53] Miklós Bóna. On three different notions of monotone subsequences. In Steve Linton, Nik Ruškuc, and Vincent Vatter, editors, *Permutation Patterns*, volume 376 of *London Mathematical Society Lecture Note Series*, pages 89–114. Cambridge University Press, 2010.

[54] Miklós Bóna. *A Walk Through Combinatorics*. World Scientific Publishing Co. Pte. Ltd., Hackensack, NJ, third edition, 2011.

[55] Miklós Bóna, Cheyne Homberger, Jay Pantone, and Vincent Vatter. Pattern-avoiding involutions: exact and asymptotic enumeration. arXiv:1310.7003 [math.CO], 2014.

[56] Prosenjit Bose, Jonathan F. Buss, and Anna Lubiw. Pattern matching for permutations. *Inform. Process. Lett.*, 65(5):277–283, 1998.

[57] Mireille Bousquet-Mélou. Four classes of pattern-avoiding permutations under one roof: generating trees with two labels. *Electron. J. Combin.*, 9(2):Paper 19, 31 pp., 2003.

[58] Mireille Bousquet-Mélou. Counting permutations with no long monotone subsequence via generating trees and the kernel method. *J. Algebraic Combin.*, 33(4):571–608, 2011.

[59] Mireille Bousquet-Mélou and Einar Steingrímsson. Decreasing subsequences in permutations and Wilf equivalence for involutions. *J. Algebraic Combin.*, 22(4):383–409, 2005.

[60] Petter Brändén and Anders Claesson. Mesh patterns and the expansion of permutation statistics as sums of permutation patterns. *Electron. J. Combin.*, 18(2):Paper 5, 14 pp., 2011.

[61] Robert Brignall. Wreath products of permutation classes. *Electron. J. Combin.*, 14(1):Paper 46, 15 pp., 2007.

[62] Robert Brignall. A survey of simple permutations. In Steve Linton, Nik Ruškuc, and Vincent Vatter, editors, *Permutation Patterns*, volume 376 of *London Mathematical Society Lecture Note Series*, pages 41–65. Cambridge University Press, 2010.

[63] Robert Brignall. Grid classes and partial well order. *J. Combin. Theory Ser. A*, 119:99–116, 2012.

[64] Robert Brignall, Sophie Huczynska, and Vincent Vatter. Decomposing simple permutations, with enumerative consequences. *Combinatorica*, 28:385–400, 2008.

[65] Robert Brignall, Sophie Huczynska, and Vincent Vatter. Simple permutations and algebraic generating functions. *J. Combin. Theory Ser. A*, 115(3):423–441, 2008.

[66] Robert Brignall, Nik Ruškuc, and Vincent Vatter. Simple permutations: decidability and unavoidable substructures. *Theoret. Comput. Sci.*, 391(1–2):150–163, 2008.

[67] Robert Brignall and Vincent Vatter. A simple proof of a theorem of Schmerl and Trotter for permutations. *J. Comb.*, to appear. arXiv:1409.4725 [math.CO].

[68] Joe Buhler, David Eisenbud, Ron Graham, and Colin Wright. Juggling drops and descents. *Amer. Math. Monthly*, 101(6):507–519, 1994.

[69] Alexander Burstein. *Enumeration of Words with Forbidden Patterns*. PhD thesis, University of Pennsylvania, 1998.

[70] Alexander Burstein and Jay Pantone. Two examples of unbalanced Wilf-equivalence. *J. Comb.*, to appear. arXiv:1402.3842 [math.CO].

[71] Peter J. Cameron. Homogeneous permutations. *Electron. J. Combin.*, 9(2):Paper 2, 9 pp., 2002/03.

[72] Pierre Cartier and Dominique Foata. *Problèmes combinatoires de commutation et réarrangements*. Lecture Notes in Mathematics, No. 85. Springer-Verlag, Berlin, 1969.

[73] Gregory L. Cherlin. The classification of countable homogeneous directed graphs and countable homogeneous $n$-tournaments. *Mem. Amer. Math. Soc.*, 131(621):xiv+161, 1998.

[74] Gregory L. Cherlin. Forbidden substructures and combinatorial dichotomies: WQO and universality. *Discrete Math.*, 311(15):1543–1584, 2011.

[75] Josef Cibulka. On constants in the Füredi–Hajnal and the Stanley–Wilf conjecture. *J. Combin. Theory Ser. A*, 116(2):290–302, 2009.

[76] Anders Claesson, Vít Jelínek, and Einar Steingrímsson. Upper bounds for the Stanley–Wilf limit of 1324 and other layered patterns. *J. Combin. Theory Ser. A*, 119:1680–1691, 2012.

[77] Anders Claesson and Sergey Kitaev. Classification of bijections between 321- and 132-avoiding permutations. *Sém. Lothar. Combin.*, 60:Art. B60d, 30 pp., 2008.

[78] Andrew R. Conway and Anthony J. Guttmann. On the growth rate of 1324-avoiding permutations. *Adv. in Appl. Math.*, 64:50–69, 2015.

[79] Sylvie Corteel, Guy Louchard, and Robin Pemantle. Common intervals of permutations. *Discrete Math. Theor. Comput. Sci.*, 8(1):189–214, 2006.

[80] Theodore Dokos, Tim Dwyer, Bryan P. Johnson, Bruce E. Sagan, and Kimberly Selsor. Permutation patterns and statistics. *Discrete Math.*, 312(18):2760–2775, 2012.

[81] Murray Elder and Vincent Vatter. Problems and conjectures presented at the Third International Conference on Permutation Patterns, University of Florida, March 7–11, 2005. arXiv:math/0505504 [math.CO].

[82] Sergi Elizalde. The $\mathscr{X}$-class and almost-increasing permutations. *Ann. Comb.*, 15:51–68, 2011.

[83] Sergi Elizalde. The most and the least avoided consecutive patterns. *Proc. Lond. Math. Soc.*, 106(5):957–979, 2013.

[84] Paul Erdős and George Szekeres. A combinatorial problem in geometry. *Compos. Math.*, 2:463–470, 1935.

[85] Shimon Even and Alon Itai. Queues, stacks, and graphs. In *Theory of Machines and Computations (Proc. Internat. Sympos., Technion, Haifa, 1971)*, pages 71–86. Academic Press, New York, 1971.

[86] Philippe Flajolet and Robert Sedgewick. *Analytic Combinatorics*. Cambridge University Press, Cambridge, 2009.

[87] Sergey Fomin. The generalized Robinson–Schensted–Knuth correspondence. *Zap. Nauchn. Sem. Leningrad. Otdel. Mat. Inst. Steklov. (LOMI)*, 155:156–175, 1986.

[88] Jacob Fox. Stanley–Wilf limits are typically exponential. arXiv:1310.8378 [math.CO], 2013.

[89] Roland Fraïssé. On a decomposition of relations which generalizes the sum of ordering relations. *Bull. Amer. Math. Soc.*, 59:389, 1953.

[90] Roland Fraïssé. Sur l'extension aux relations de quelques propriétés des ordres. *Ann. Sci. Ecole Norm. Sup. (3)*, 71:363–388, 1954.

[91] Zoltán Füredi. The maximum number of unit distances in a convex $n$-gon. *J. Combin. Theory Ser. A*, 55(2):316–320, 1990.

[92] Zoltán Füredi and Péter Hajnal. Davenport–Schinzel theory of matrices. *Discrete Math.*, 103(3):233–251, 1992.

[93] Tibor Gallai. Transitiv orientierbare Graphen. *Acta Math. Acad. Sci. Hungar*, 18:25–66, 1967.

[94] Tibor Gallai. *A translation of T. Gallai's paper: "Transitiv orientierbare Graphen"*, pages 25–66. Wiley-Intersci. Ser. Discrete Math. Optim. Wiley, Chichester, 2001.

[95] Ira M. Gessel. Symmetric functions and *P*-recursiveness. *J. Combin. Theory Ser. A*, 53(2):257–285, 1990.

[96] Massimiliano Goldwurm and Massimo Santini. Clique polynomials have a unique root of smallest modulus. *Inform. Process. Lett.*, 75(3):127–132, 2000.

[97] Sylvain Guillemot and Stéphane Vialette. Pattern matching for 321-avoiding permutations. In Yingfei Dong, Ding-Zhu Du, and Oscar Ibarra, editors, *Algorithms and Computation (20th International Symposium, ISAAC 2009, Honolulu, Hawaii, USA, December 16-18, 2009)*, volume 5878 of *Lecture Notes in Comput. Sci.*, pages 1064–1073, Berlin, 2009. Springer.

[98] Anthony J. Guttmann. Analysis of series expansions for non-algebraic singularities. arXiv:1405.5327 [math-ph], 2014.

[99] András Gyárfás and Jenő Lehel. A Helly-type problem in trees. In *Combinatorial Theory and its Applications, II (Proc. Colloq., Balatonfüred, 1969)*, pages 571–584. North-Holland, Amsterdam, 1970.

[100] Sergiu Hart and Micha Sharir. Nonlinearity of Davenport–Schinzel sequences and of generalized path compression schemes. *Combinatorica*, 6(2):151–177, 1986.

[101] Wilfrid Hodges. *Model Theory*, volume 42 of *Encyclopedia of Mathematics and its Applications*. Cambridge University Press, Cambridge, 1993.

[102] Cheyne Homberger and Vincent Vatter. On the effective and automatic enumeration of polynomial permutation classes. arXiv:1308.4946 [math.CO], 2013.

[103] Sophie Huczynska and Nik Ruškuc. Pattern classes of permutations via bijections between linearly ordered sets. *European J. Combin.*, 29(1):118–139, 2008.

[104] Sophie Huczynska and Vincent Vatter. Grid classes and the Fibonacci dichotomy for restricted permutations. *Electron. J. Combin.*, 13:R54, 14 pp., 2006.

[105] Vít Jelínek and Pavel Valtr. Splittings and Ramsey properties of permutation classes. *Adv. in Appl. Math.*, 63:41–67, 2015.

[106] Fredrik Johansson and Brian Nakamura. Using functional equations to enumerate 1324-avoiding permutations. *Adv. in Appl. Math.*, 56:20–34, 2014.

[107] Tomáš Kaiser and Martin Klazar. On growth rates of closed permutation classes. *Electron. J. Combin.*, 9(2):Paper 10, 20 pp., 2003.

[108] Irving Kaplansky. The asymptotic distribution of runs of consecutive elements. *Ann. Math. Statistics*, 16:200–203, 1945.

[109] Gyula Károlyi and Gábor Tardos. On point covers of multiple intervals and axis-parallel rectangles. *Combinatorica*, 16(2):213–222, 1996.

[110] André E. Kézdy, Hunter S. Snevily, and Chi Wang. Partitioning permutations into increasing and decreasing subsequences. *J. Combin. Theory Ser. A*, 73(2):353–359, 1996.

[111] Sergey Kitaev. A survey on partially ordered patterns. In Steve Linton, Nik Ruškuc, and Vincent Vatter, editors, *Permutation Patterns*, volume 376 of *London Mathematical Society Lecture Note Series*, pages 115–135. Cambridge University Press, 2010.

[112] Sergey Kitaev. *Patterns in Permutations and Words*. Monographs in Theoretical Computer Science. An EATCS Series. Springer, Heidelberg, 2011.

[113] Martin Klazar. A general upper bound in extremal theory of sequences. *Comment. Math. Univ. Carolin.*, 33(4):737–746, 1992.

[114] Martin Klazar. The Füredi–Hajnal conjecture implies the Stanley–Wilf conjecture. In *Formal Power Series and Algebraic Combinatorics (Moscow, 2000)*, pages 250–255. Springer, 2000.

[115] Martin Klazar. On the least exponential growth admitting uncountably many closed permutation classes. *Theoret. Comput. Sci.*, 321(2-3):271–281, 2004.

[116] Martin Klazar. On growth rates of permutations, set partitions, ordered graphs and other objects. *Electron. J. Combin.*, 15(1):Paper 75, 22 pp., 2008.

[117] Martin Klazar. Overview of some general results in combinatorial enumeration. In Steve Linton, Nik Ruškuc, and Vincent Vatter, editors, *Permutation Patterns*, volume 376 of *London Mathematical Society Lecture Note Series*, pages 3–40. Cambridge University Press, 2010.

[118] Donald E. Knuth. *The Art of Computer Programming. Volume 1*. Addison-Wesley Publishing Co., Reading, Mass., 1968.

[119] Donald E. Knuth. *The Art of Computer Programming. Volume 3*. Addison-Wesley Publishing Co., Reading, Mass., 1973.

[120] Thomas Kövari, Vera T. Sós, and Pál Turán. On a problem of K. Zarankiewicz. *Colloquium Math.*, 3:50–57, 1954.

[121] Christian Krattenthaler. Permutations with restricted patterns and Dyck paths. *Adv. in Appl. Math.*, 27(2-3):510–530, 2001.

[122] Christian Krattenthaler. Growth diagrams, and increasing and decreasing chains in fillings of Ferrers shapes. *Adv. in Appl. Math.*, 37(3):404–431, 2006.

[123] Alistair H. Lachlan. Countable homogeneous tournaments. *Trans. Amer. Math. Soc.*, 284(2):431–461, 1984.

[124] Alistair H. Lachlan and Robert E. Woodrow. Countable ultrahomogeneous undirected graphs. *Trans. Amer. Math. Soc.*, 262(1):51–94, 1980.

[125] Steve Linton, Nik Ruškuc, and Vincent Vatter, editors. *Permutation Patterns*, volume 376 of *London Mathematical Society Lecture Note Series*. Cambridge University Press, Cambridge, 2010.

[126] Percy A. MacMahon. *Combinatory Analysis*. Cambridge University Press, London, 1915/16.

[127] Neal Madras and Hailong Liu. Random pattern-avoiding permutations. In *Algorithmic Probability and Combinatorics*, volume 520 of *Contemp. Math.*, pages 173–194. Amer. Math. Soc., Providence, RI, 2010.

[128] Toufik Mansour and Alek Vainshtein. Counting occurrences of 132 in a permutation. *Adv. in Appl. Math.*, 28(2):185–195, 2002.

[129] Adam Marcus and Gábor Tardos. Excluded permutation matrices and the Stanley–Wilf conjecture. *J. Combin. Theory Ser. A*, 107(1):153–160, 2004.

[130] Peter McNamara and Einar Steingrímsson. On the topology of the permutation pattern poset. arXiv:1305.5569 [math.CO], 2013.

[131] Sam Miner and Igor Pak. The shape of random pattern-avoiding permutations. *Adv. in Appl. Math.*, 55:86–130, 2014.

[132] Rolf H. Möhring and Franz J. Radermacher. Substitution decomposition for discrete structures and connections with combinatorial optimization. In *Algebraic and Combinatorial Methods in Operations Research*, volume 95 of *North-Holland Math. Stud.*, pages 257–355. North-Holland, Amsterdam, 1984.

[133] Maximillian M. Murphy and Vincent Vatter. Profile classes and partial well-order for permutations. *Electron. J. Combin.*, 9(2):Paper 17, 30 pp., 2003.

[134] John Noonan and Doron Zeilberger. The enumeration of permutations with a prescribed number of "forbidden" patterns. *Adv. in Appl. Math.*, 17(4):381–407, 1996.

[135] Jay Pantone. The enumeration of permutations avoiding 3124 and 4312. arXiv:1309.0832 [math.CO], 2013.

[136] Pavel A. Pevzner. *Computational Molecular Biology*. Computational Molecular Biology. MIT Press, Cambridge, MA, 2000.

[137] Adeline Pierrot and Dominique Rossin. 2-stack pushall sortable permutations. arXiv:1303.4376 [cs.DM], 2013.

[138] Adeline Pierrot and Dominique Rossin. 2-stack sorting is polynomial. arXiv:1304.2860 [math.CO], 2013.

[139] Maurice Pouzet and Nicolas M. Thiéry. Some relational structures with polynomial growth and their associated algebras I: Quasi-polynomiality of the profile. *Electron. J. Combin.*, 20(2):Paper 1, 35 pp., 2013.

[140] Vaughan R. Pratt. Computing permutations with double-ended queues, parallel stacks and parallel queues. In *STOC '73: Proceedings of the Fifth Annual ACM Symposium on the Theory of Computing*, pages 268–277, New York, NY, USA, 1973. ACM Press.

[141] Cathleen Battiste Presutti and Walter Stromquist. Packing rates of measures and a conjecture for the packing density of 2413. In Steve Linton, Nik Ruškuc, and Vincent Vatter, editors, *Permutation Patterns*, volume 376 of *London Math. Soc. Lecture Note Ser.*, pages 287–316. Cambridge Univ. Press, Cambridge, 2010.

[142] Amitai Regev. Asymptotic values for degrees associated with strips of Young diagrams. *Adv. in Math.*, 41(2):115–136, 1981.

[143] István Reiman. Über ein Problem von K. Zarankiewicz. *Acta. Math. Acad. Sci. Hungar.*, 9:269–273, 1958.

[144] Douglas G. Rogers. Ascending sequences in permutations. *Discrete Math.*, 22(1):35–40, 1978.

[145] Doron Rotem. On a correspondence between binary trees and a certain type of permutation. *Inform. Process. Lett.*, 4(3):58–61, 1975/76.

[146] Craige E. Schensted. Longest increasing and decreasing subsequences. *Canad. J. Math.*, 13:179–191, 1961.

[147] James H. Schmerl and William T. Trotter. Critically indecomposable partially ordered sets, graphs, tournaments and other binary relational structures. *Discrete Math.*, 113(1-3):191–205, 1993.

[148] Rodica Simion and Frank W. Schmidt. Restricted permutations. *European J. Combin.*, 6(4):383–406, 1985.

[149] Zvezdelina Stankova and Julian West. A new class of Wilf-equivalent permutations. *J. Algebraic Combin.*, 15(3):271–290, 2002.

[150] Zvezdelina E. Stankova. Forbidden subsequences. *Discrete Math.*, 132(1-3):291–316, 1994.

[151] Richard P. Stanley. Solution to problem E2546. *Amer. Math. Monthly*, 83(10):813–814, 1976.

[152] Richard P. Stanley. *Enumerative combinatorics. Vol. 2*, volume 62 of *Cambridge Studies in Advanced Mathematics*. Cambridge University Press, Cambridge, 1999.

[153] Richard P. Stanley. Recent progress in algebraic combinatorics. *Bull. Amer. Math. Soc. (N.S.)*, 40(1):55–68, 2003.

[154] Einar Steingrímsson. Generalized permutation patterns—a short survey. In Steve Linton, Nik Ruškuc, and Vincent Vatter, editors, *Permutation Patterns*, volume 376 of *London Mathematical Society Lecture Note Series*, pages 137–152. Cambridge University Press, 2010.

[155] Einar Steingrímsson. Some open problems on permutation patterns. In Simon R. Blackburn, Stefanie Gerke, and Mark Wildon, editors, *Surveys in Combinatorics 2013*, volume 409 of *London Mathematical Society Lecture Note Series*, pages 239–263. Cambridge University Press, 2013.

[156] Robert Tarjan. Sorting using networks of queues and stacks. *J. Assoc. Comput. Mach.*, 19:341–346, 1972.

[157] The On-line Encyclopedia of Integer Sequences. Published electronically at `http://oeis.org/`.

[158] William T. Tutte. A census of planar maps. *Canad. J. Math.*, 15:249–271, 1963.

[159] Vincent Vatter. Permutation classes of every growth rate above 2.48188. *Mathematika*, 56:182–192, 2010.

[160] Vincent Vatter. Small configurations in simple permutations. *European J. Combin.*, 31:1781–1784, 2010.

[161] Vincent Vatter. Small permutation classes. *Proc. Lond. Math. Soc. (3)*, 103:879–921, 2011.

[162] Vincent Vatter and Steve Waton. On partial well-order for monotone grid classes of permutations. *Order*, 28:193–199, 2011.

[163] Steve Waton. *On Permutation Classes Defined by Token Passing Networks, Gridding Matrices and Pictures: Three Flavours of Involvement*. PhD thesis, Univ. of St Andrews, 2007.

[164] Julian West. *Permutations with forbidden subsequences and stack-sortable permutations*. PhD thesis, M.I.T., 1990.

[165] Herbert S. Wilf. The patterns of permutations. *Discrete Math.*, 257(2-3):575–583, 2002.

[166] Jacob Wolfowitz. Note on runs of consecutive elements. *Ann. Math. Statistics*, 15:97–98, 1944.

[167] Kazimierz Zarankiewicz. Problem P101. *Colloq. Math.*, 2(3–4):301, 1951.

[168] Doron Zeilberger. Opinion 58: 'I saw their proof now I'm a believer': The Marcus-Tardos proof of the Füredi–Hajnal and Stanley–Wilf conjectures. `http://www.math.rutgers.edu/~zeilberg/Opinion58.html`, Dec. 2003.

# Chapter 13

# *Parking Functions*

**Catherine H. Yan**
*Department of Mathematics, Texas A&M University, College Station, TX*

## CONTENTS

## 13.1 Introduction

The notion of parking functions was introduced by Konheim and Weiss [53] as a colorful way to describe their work on computer storage. The parking problem can be stated as follows. There are $n$ cars $C_1, \dots, C_n$ that want to park on a one-way street with ordered parking spaces $0, 1, \dots, n-1$. Each car $C_i$ has a preferred space $a_i$. The cars enter the street one at a time in the order $C_1, C_2, \dots, C_n$. A car tries to park in its preferred space. If that space is occupied, then it parks in the next available space. If there is no space then the car leaves the street. The sequence $(a_1, \dots, a_n)$ is called a **parking function of length** $n$ if all the cars can park, i.e., no car leaves the street.

It is easy to see that a sequence $(a_1, \dots, a_n)$ is a parking function if and only if it has at least $i$ terms less than $i$, for each $1 \le i \le n$. Let $a_{(1)} \le a_{(2)} \le \cdots \le a_{(n)}$ be the non-decreasing rearrangement of terms $a_i$'s, where $a_{(i)}$ is called the $i$th order statistics of $(a_1, \dots, a_n)$. Then the sequence $(a_1, \dots, a_n)$ is a parking function if and only if $0 \le a_{(i)} < i$. * Another equivalent definition is that a sequence $(a_1, \dots, a_n)$ is a parking function if and only if there is a permutation $\sigma \in \mathfrak{S}_n$ such that $0 \le a_{\sigma(i)} < i$.

The number of parking functions of length $n$ is $(n+1)^{n-1}$, a result obtained in the very first paper [53] on the subject by analytic method. A simple and elegant proof was given by Pollak, (see [71]), which can be described as follows: Add an additional space $n$ and arrange all $n+1$ spaces clockwise in a circle. Again assume that $n$ cars enter the street one at a time, each with a preferred space $a_i \in \{0, 1, \dots, n\}$. Preference $a_i = n$ is treated like any other preference: If space $n$ is occupied, car $C_i$ moves clockwise to the first unoccupied space. Every sequence of preferences leaves one space unoccupied, and because of symmetry the number of sequences leaving a given space, say $k$, unoccupied is the same for every $k$, $0 \le k \le n$. Hence the number with $k = n$, which is the number of parking functions, is $\frac{1}{n+1}$ of the total number of preference sequences. This gives $\dfrac{(n+1)^n}{n+1} = (n+1)^{n-1}$.

It is immediately noticed that $(n+1)^{n-1}$ is the number of labeled trees on $n+1$ vertices, by the famous Cayley's formula [18]. Many bijections between the set of parking functions and the set of labeled trees are constructed. They reveal deep connections between parking functions and other combinatorial structures, and lead to various generalizations and applications in different fields, notably in algebra, interpolation theory, probability and statistics, representation theory, and geometry. In this chapter we survey the basic results and developments on the combinatorics theory of parking functions in the last 20 years.

**Notation.** We write $[n]$ for the set $\{1, 2, \dots, n\}$, and $[n]_0$ for the set $\{0, 1, \dots, n\}$. We will often think of a function $f : [n] \to [n]$ as the sequence of its values $f(1), f(2), \dots, f(n)$. If $(h_i)$ is a sequence, we will use the boldface letter $\mathbf{h}$ to rep-

*In literature, some people assume that $a_i$ are positive integers and use the condition that $1 \le a_{(i)} \le i$. Throughout this chapter we always allow the value of 0 and require that $0 \le a_{(i)} < i$.

resent it. The letter **a** is reserved for a parking function of length $n$. We denote by $\mathscr{PK}_n$ the set of parking functions of length $n$,

A labeled tree on $[n]_0$ is a connected graph on the vertex set $[n]_0$ with no cycles. It is equivalent to a labeled rooted forest on $[n]$ obtained from the tree by deleting vertex 0 and replacing edges connecting 0 to vertex $i$ by a root at $i$. A labeled rooted forest on $[n]$ is also called an **acyclic function** on $[n]$. We denote by $\mathscr{T}_{n+1}$ the set of labeled trees with the vertex set $[n]_0$.

## 13.2 Parking functions and labeled trees

There are many bijections between the set of parking functions of length $n$ and the set of labeled trees on $[n]_0$, for example, see [72, 71, 51, 30, 31, 56, 65]. Here we introduce some elegant constructions, each of which reveals an intrinsic property of parking functions.

### 13.2.1 Labeled trees with Prüfer code

A Prüfer code, first introduced by Prüfer [70] to prove Cayley's formula, associates with each labeled tree with $n$ vertices a unique sequence of length $n-2$. Given a tree $T \in \mathscr{T}_{n+1}$, its Prüfer code is generated by iteratively removing vertices from the tree until only two vertices remain. Specifically, at step $i$, one removes the leaf with the smallest label and sets the $i$th element of the Prüfer sequence to be the label of this leaf's neighbor. For instance, the Prüfer code of the tree in Figure 13.1 is $(2,4,7,0,2,3)$.

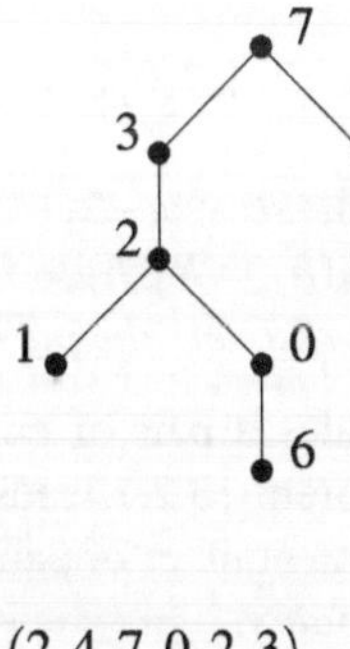


**Figure 13.1**
A labeled tree with Prüfer code $(2,4,7,0,2,3)$.

A bijection between $\mathscr{PK}_n$ and $\mathscr{T}_{n+1}$ was constructed by Pollak [71] using Prüfer code.

**Theorem 13.2.1** *For each parking function* $\mathbf{a} = (a_1, a_2, \ldots, a_n)$*, define the difference sequence* $(c_1, c_2, \ldots, c_{n-1})$ *by letting*

$$c_i = a_{i+1} - a_i \mod n+1. \tag{13.1}$$

*Let* $T(\mathbf{a})$ *be the labeled tree in* $\mathcal{T}_{n+1}$ *whose Prüfer code is* $(c_1, c_2, \ldots, c_{n-1})$*. Then the map* $\mathbf{a} \to T(\mathbf{a})$ *is a bijection from* $\mathcal{PK}_n$ *to* $([n]_0)^{n-1}$*.*

*Sketch of Proof.* The inverse equations of (13.1) can be written as

$$a_i = a_1 + c_1 + \cdots + c_{i-1}, \quad \mod n+1 \quad 2 \leq i \leq n. \tag{13.2}$$

That is, a parking function is determined by its difference sequence if $a_1$ is known. The following algorithm describes how to find an $a_1$ for each $(c_1, c_2, \ldots, c_{n-1}) \in \{0, 1, \ldots, n\}^{n-1}$ such that $(a_1, a_2, \ldots, a_n)$ determined by Equations (13.2) is a parking function.

**Algorithm 13.2.2 (Algorithm to determine** $a_1$**)** *Given* $(c_1, c_2, \ldots, c_{n-1}) \in \{0, 1, \ldots, n\}^{n-1}$,

1. *Let*

$$\begin{aligned} h_1 &= 0, \\ h_i &= c_1 + \cdots + c_{i-1} \mod n+1, \quad 2 \leq i \leq n. \end{aligned}$$

2. *Let* $\mathbf{r}(h) = (r_0, \ldots, r_n)$ *be the* **specification** *of* $h$*, i.e.,* $r_i = \text{card}\{j : h_j = i\}$*. Let* $R_j(h) = r_0 + \cdots r_j - j - 1$ *for* $0 \leq j \leq n$*.*
3. *Find* $d$ *to be the smallest index such that* $R_d(h) = \min\{R_j(h) : 0 \leq j \leq n\}$*. Then*

$$a_1 = n - d.$$

The uniqueness of $a_1$ follows from the fact that the number of parking functions of length $n$ is equal to the number of Prüfer codes of length $n-1$. ■

**Example 13.2.3** *For the Prüfer code* $(2,4,7,0,2,5)$*, we have* $h = (0,2,6,5,5,7,4)$ *and hence* $\mathbf{r}(h) = (1,0,1,0,1,2,1,1)$*.*

*The sequence* $R(h)$ *is given by* $(0,-1,-1,-2,-2,-1,-1,-1)$*, where the minimal value of* $R_j(h)$ *is reached at* $R_3(h) = R_4(h) = -2$*. The smallest index* $d$ *is 3. Hence we have* $a_1 = n - 3 = 4$*, which recovers the parking function* $(4,6,2,1,1,3,0)$*.*

A zero in the Prüfer code indicates a pair of consecutive like numbers in the parking function. Consequently Theorem 13.2.1 leads to an explicit formula for the enumerator of parking functions by the number of pairs of like consecutive numbers. Precisely, we have the following corollary.

**Corollary 13.2.4** *The equality*

$$\sum_{\alpha \in \mathcal{PK}_n} q^{|\{i \,:\, a_i = a_{i+1}\}|} = (q+n)^{n-1}.$$

*holds, where* $|S|$ *is the cardinality of a set* $S$*.*

### 13.2.2 Inversions of labeled trees

Let $\mathbf{a} = (a_1, \ldots, a_n)$ be a parking function of length $n$. Consider again the one-way street parking scenario, and assume that the car $C_i$ parked at space $p_i$, for $1 \le i \le n$. Then $p_1 p_2 \cdots p_n$ is a permutation on the letters $\{0, 1, \ldots, n-1\}$. Let $D(\mathbf{a})$ be the total **displacement**, i.e.

$$D(\mathbf{a}) = \sum_{i=1}^{n} (p_i - a_i) = \binom{n}{2} - \sum_{i=1}^{n} a_i.$$

Then $D(\mathbf{a})$ is the total number of failed trials before all the cars find their parking spaces. In the language of hashing in computer algorithms, $D(\mathbf{a})$ represents the number of linear probing. See Section 13.3.1 for more discussion.

Define the **displacement-enumerator of parking functions** as the polynomial

$$P_n(q) = \sum_{\mathbf{a} \in \mathscr{PK}_n} q^{D(\mathbf{a})} = q^{\binom{n}{2}} \sum_{\mathbf{a} \in \mathscr{PK}_n} q^{-(a_1 + \cdots + a_n)}.$$

The degree of $P_n(q)$ is $\binom{n}{2}$. This polynomial also enumerates labeled trees by some important tree statistics, one of which is the number of **inversions.**

Let $T \in \mathscr{T}_{n+1}$. View vertex 0 as the root of $T$. If $\{i, j\}$ is an edge of $T$ and $j$ lies on the unique path connecting 0 to $i$, we say that $i$ is the **predecessor** of $j$ and $j$ is a **successor** of $i$. In a rooted tree the degree of a vertex $i$ is the number of its successors. A leaf is a non-root vertex with no successor.

An **inversion** of the tree $T$ is a pair $(i, j)$ for which $i < j$ and $j$ lies on the unique path connecting 0 to $i$. Let $\mathrm{inv}(T)$ denote the number of inversions of $T$. The **inversion enumerator $I_n(q)$ of labeled trees on $n+1$ vertices** is defined as the polynomial

$$I_n(q) = \sum_{T \in \mathscr{T}_{n+1}} q^{\mathrm{inv}(T)},$$

which is also called the inversion enumerator of labeled forests on $[n]$.

Kreweras [56] studied polynomial systems satisfying a recurrence relation, and found several combinatorial interpretations, among which are the polynomials $I_n(q)$ and $P_n(q)$. In Kreweras paper, parking functions are called **suites majeures.**

**Theorem 13.2.5**

1. *The inversion enumerator $I_n(q)$ of labeled trees on $n+1$ vertices satisfies the recurrence relation*

$$\begin{aligned} I_1(q) &= 1, \\ I_{n+1}(q) &= \sum_{i=0}^{n} \binom{n}{i} (q^i + q^{i-1} + \cdots + 1) I_i(q) I_{n-i}(q). \end{aligned}$$

2. *The displacement enumerator of parking functions satisfies the recurrence relation*

$$\begin{aligned} P_1(q) &= 1, \\ P_{n+1}(q) &= \sum_{i=0}^{n} \binom{n}{i} (q^i + q^{i-1} + \cdots + 1) P_i(q) P_{n-i}(q). \end{aligned}$$

3. *Consequently,*

$$I_n(q) = P_n(q).$$

In fact, these polynomials are immediately related to connected graphs.

**Theorem 13.2.6** *Let $C_n(q)$ be the edge-enumerator of connected graphs. Precisely,*

$$C_n(q) = \sum_G q^{|E(G)|-n},$$

*where $G$ ranges over all connected graphs (without loops or multiple edges) on $n+1$ labeled vertices, and $E(G)$ is the set of edges of $G$. Then*

$$I_n(1+q) = P_n(1+q) = C_n(q).$$

Theorem 13.2.6 implies that the sum of $\binom{D(\mathbf{a})}{k}$ taken over all parking functions of length $n$, as well as the sum of $\binom{\mathrm{inv}(T)}{k}$ taken over all labeled trees on $n+1$ vertices, are equal to the total number of connected graphs with $n+k$ edges on $n+1$ labeled vertices. Analysis of these generating functions is essential in characterizing the evolution of random graphs, see Janson, Knuth, Łuczak and Pittel [48].

Using the exponential formula on graphs and connected graphs, one derives the following generating function identities [81].

**Theorem 13.2.7** *The following identities hold.*

1.

$$\sum_{n\geq 1} I_n(q)(q-1)^{n-1}\frac{x^n}{n!} = \sum_{n\geq 1} P_n(q)(q-1)^{n-1}\frac{x^n}{n!} = \log \sum_{n\geq 0} q^{\binom{n}{2}}\frac{x^n}{n!}.$$

2.

$$\sum_{n\geq 0} I_n(q)(q-1)^{n-1}\frac{x^n}{n!} = \sum_{n\geq 0} P_n(q)(q-1)^{n-1}\frac{x^n}{n!} = \frac{\sum_{n\geq 0} q^{\binom{n+1}{2}}\frac{x^n}{n!}}{\sum_{n\geq 0} q^{\binom{n}{2}}\frac{x^n}{n!}}.$$

Based on Theorem 13.2.5, Kreweras constructed recursively a bijection that carries the total displacement $D(\mathbf{a})$ of parking functions to the number of inversions of labeled trees. A more direct and elegant construction was given by Knuth [51], which we describe next.

**Algorithm 13.2.8 (A Bijection $\phi : PK_n \to \mathscr{T}_{n+1}$ such that $D(\mathbf{a}) = \mathrm{inv}(\phi(\mathbf{a}))$)** *Let $\mathbf{a} = (a_1,\ldots,a_n) \in \mathscr{PK}_n$ and $p_i$ be the space that car $C_i$ occupies. Let $p'_i = p_i + 1$, hence $\mathbf{p}' = p'_1p'_2\cdots p'_n$ is a permutation of length $n$. Let $\mathbf{q} = q_1q_2\cdots q_n$ be the inverse permutation of $\mathbf{p}'$, that is, car $C_{q_i}$ is parked at the space $i-1$.*

1. *Construct an auxiliary tree by letting the predecessor of vertex $k$ be the first element on the right of $k$ and larger than $k$ in the permutation $q_1q_2\cdots q_n$; if there is no such element, let the predecessor be 0.*

2. *Make a copy of the auxiliary tree. Then relabel the nonzero vertices of the new tree by proceeding as follows, in preorder (i.e., any vertex is processed before its successors): If the label of the current vertex was k in the auxiliary tree, swap its current label with the label that is currently $(p'_k - a_k)$th smallest in the subtree rooted at the current vertex. The final tree is $\phi(\mathbf{a})$.*

**Example 13.2.9** *Let* $\mathbf{a} = (4,0,1,0,3,6,4)$. *Then* $\mathbf{p} = p_1 \cdots p_n = 4\ 0\ 1\ 2\ 3\ 6\ 5$ *and* $\mathbf{p}' = p'_1 p'_2 \cdots p'_n = 5\ 1\ 2\ 3\ 4\ 7\ 6$. *Consequently, the inverse of the* $\mathbf{p}'$ *is* $\mathbf{q} = q_1 \cdots q_n = 2\ 3\ 4\ 5\ 1\ 7\ 6$. *Figure 13.2 shows the auxiliary tree and the final tree defined by Knuth's bijection. One checks that* $D(\mathbf{a}) = \mathrm{inv}(\phi(\mathbf{a})) = 3$.

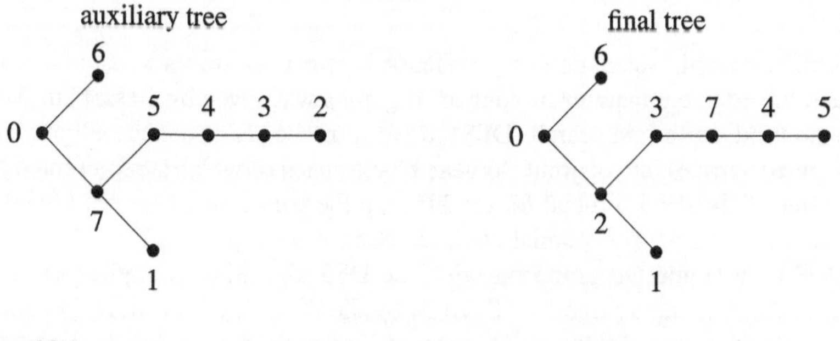


**Figure 13.2**
The auxiliary tree and the tree $\phi(\mathbf{a})$ corresponding to $\mathbf{a} = (4,0,1,0,3,6,4)$.

To reverse the procedure, we can reconstruct the auxiliary tree from a labeled tree $T \in \mathscr{T}_{n+1}$ by proceeding in preorder to swap the label of each vertex with the largest label currently in its subtree. Comparing the auxiliary tree and the final tree we could obtain the values of $p'_k - a_k$ for each $1 \le k \le n$. The permutation $q_1 \cdots q_n$ can be read from the auxiliary tree in postorder, which is recursively defined by

- If a tree $T$ is null, then the empty list is the postorder listing of $T$.
- If $T$ comprises a single node, that node itself is the postorder list of $T$.
- Otherwise, $T$ has root $r$ and the successors of $r$ are $t_1 > t_2 > \cdots > t_k$. Let $T_i$ be the subtree with root $t_i$. The postorder listing of $T$ is the nodes of $T_1$ in postorder, …, the nodes of $T_k$ in postorder, all followed by the root r.

For example, the postorder of the auxiliary tree in Figure 13.2 is 2 3 4 5 1 7 6 (0), which is $\mathbf{q} = q_1 \cdots q_n$. Taking the inverse of $\mathbf{q}$, one recovers $p'_1 p'_2 \cdots p'_n$, and hence $a_1 \cdots a_n$.

Knuth's bijection has another nice property: In the tree $\phi(\mathbf{a})$, if a vertex with label $t$ is labeled $k$ in the auxiliary tree, then there are $p_k - a_k$ nodes with smaller labels than $t$ in the subtree of $\phi(\mathbf{a})$ rooted at $t$. In a labeled tree $T$, call a vertex $v$ **leader** if it is the smallest among all the vertices of the subtree of $T$ rooted at $b$. In particular, every leaf is a leader. Denote by $\text{lead}(T)$ the number of leaders in a labeled tree $T$. On the other hand, for a parking function $\mathbf{a}$ we say a car $C_i$ is **lucky** if $p_i = a_i$, that is, $C_i$ is parked at its preferred space. Let $\text{lucky}(\mathbf{a})$ denote the number of lucky cars. Then Knuth's bijection $\phi$ satisfies $\text{lucky}(\mathbf{a}) = \text{lead}(\phi(\mathbf{a}))$.

The statistics "lucky($\mathbf{a}$)" for parking functions and "lead($T$)" for labeled trees were studied by Gessel and Seo [37], and Seo and Shin [73], respectively, where they gave an explicit formula for the corresponding enumerators as

$$\sum_{\mathbf{a}\in\mathscr{P}\mathscr{K}_n} u^{\text{lucky}(\mathbf{a})} = \sum_{T\in\mathscr{T}_{n+1}} u^{\text{lead}(T)} = u\prod_{i=1}^{n-1}(i+(n-i+1)u).$$

A combinatorial explanation of the relation between the inversion-enumerator of trees and the edge-enumerator of connected graphs was given by Gessel and Wang [38], who used depth-first search (DFS) to partition the set of connected graphs on $n+1$ labeled vertices into disjoint Boolean blocks, each of which is represented by a labeled tree. This idea is crucial for establishing the connection between $G$-parking functions and the Tutte polynomial of $G$, (cf. Section 13.5. )

Let $G$ be a connected graph on $[n]_0$. The DFS algorithm is applied to $G$ and returns a certain spanning tree $T = \text{DFS}(G)$ by the following procedure. We start at vertex 0 (which is viewed at the root of the tree), and at each step we go to the greatest adjacent unvisited vertex if there is one, otherwise, we backtrack. For example, from the graph on the left of Figure. 13.3, we get the spanning tree on the right.

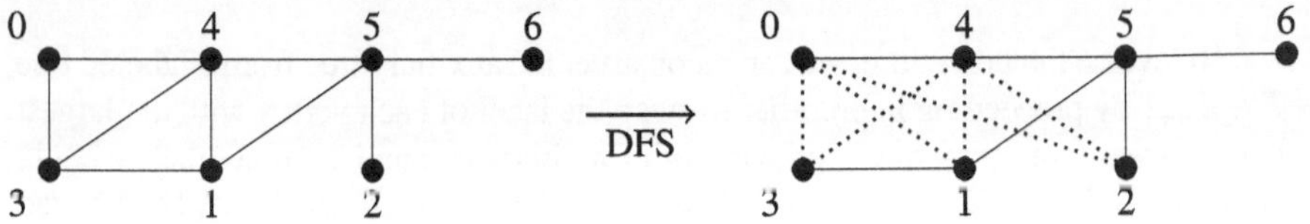


**Figure 13.3**
The spanning tree found by DFS on a connected graph.

Fix a labeled tree $T \in \mathscr{T}_{n+1}$. Let $\mathscr{G}(T)$ be the set of connected graphs $G$ for which $\text{DFS}(G) = T$. Define a set $\mathscr{E}(T)$ of edges not in $T$ whose elements are in one-to-one correspondence with the inversions of $T$. More precisely, with every inversion $(j,k)$, $(j > k > 0)$, associate the edge between $k$ and the predecessor of $j$. For the tree on the right of Figure 13.3, the edges of $\mathscr{E}(T)$ are indicated by dotted lines.

Gessel and Wang characterized the set of connected graphs in $\mathscr{G}(T)$.

**Theorem 13.2.10** *$\mathscr{G}(T)$ consists of those graphs obtained from $T$ by adjoining an arbitrary subset of edges in $\mathscr{E}(T)$.*

An immediate corollary of Theorem 13.2.10 is the equation

$$\sum_{G\in\mathscr{G}(T)} q^{|E(G|)-n} = (1+q)^{\mathrm{inv}(T)}.$$

Summing over all trees $T \in \mathscr{T}_{n+1}$, we have

$$C_n(q) = \sum_{G \text{ connected}} q^{|E(G)|-n} = \sum_{T\in\mathscr{T}_{n+1}} q^{\mathrm{inv}(T)} = I_n(1+q).$$

In the next subsection we will see that applying the breadth-first search on the connected graphs leads to the equation between $P_n(1+q)$ and $C_n(q)$.

### 13.2.3 Graph searching algorithms

For a sequence $\mathbf{a} = (a_1,\dots,a_n)$ with $0 \le a_i < n$, its **specification** is the vector $\vec{r}(\mathbf{a}) = (r_0,\dots,r_{n-1})$ where $r_i = |\{j : a_j = i\}|$. If **a** is a parking function, then the vector $\vec{r}(\mathbf{a})$ is **balanced**, that is,

$$\begin{aligned} r_0+\cdots+r_i-i-1 &\ge 0 \qquad \text{for } 0\le i<n, \\ r_0+\cdots+r_{n-1} &= n. \end{aligned}$$

In addition, define the **order permutation** $\sigma(\mathbf{a}) = \sigma_1\sigma_2\cdots\sigma_n$ of a parking function **a** by letting

$$\sigma_i = |\{j : a_j < a_i, \text{ or } a_j = a_i \text{ and } j \le i\}|.$$

In other words, $\sigma_i$ is the position of the entry $a_i$ in the non-decreasing rearrangement of **a**. For example, for $\mathbf{a} = (2\ 0\ 3\ 0\ 4\ 8\ 1\ 5\ 4)$, the specification is $\vec{r}(\mathbf{a}) = (2,1,1,1,4,1,0,0,1)$, and the order permutation is $\sigma(\alpha) = 4\,1\,5\,2\,6\,9\,3\,8\,7$. Clearly a parking function determines its specification and order permutation. Conversely, knowing the vector $\vec{r}(\mathbf{a})$ and the permutation $\sigma(\mathbf{a})$, we can easily recover **a** by replacing $i$ in $\sigma(\mathbf{a})$ by the $i$th smallest term in the list $0^{r_0}1^{r_1}\cdots(n-1)^{r_{n-1}}$.

Not every pair of a vector $\vec{r}$ and a permutation $\sigma$ can be the specification and the order permutation of a parking function. The vector and the permutation must be **compatible** with each other, that is, the terms in the inverse $\sigma^{-1}$ of $\sigma$ are increasing on every interval with the indices $\{1+\sum_{i=0}^{k-1} r_i, 2+\sum_{i=0}^{k-1} r_i, \dots, \sum_{i=0}^{k} r_i\}$ (if $r_k \ne 0$). Equivalently, the terms $1+\sum_{i=0}^{k-1} r_i, 2+\sum_{i=0}^{k-1} r_i, \dots, \sum_{i=0}^{k} r_i$ appear from left to right in $\sigma$.

Let $\mathscr{C}_n$ be the set of all pairs $(\vec{r},\sigma)$ with $\vec{r} \in \mathbb{N}^n$ and $\sigma$ a permutation of length $n$ compatible with $\vec{r}$.

**Theorem 13.2.11** *The map* $\rho : \mathbf{a} \to (\vec{r}(\mathbf{a}), \sigma(\mathbf{a}))$ *is a bijection from* $\mathscr{PK}_n$ *to* $\mathscr{C}_n$.

On the other hand, there are many ways to construct bijections between $\mathscr{C}_n$ and $\mathscr{T}_{n+1}$, the set of labeled trees with the vertex set $[n]_0$. Each such bijection, combined with Theorem 13.2.11, gives a bijection between $\mathscr{PK}_n$ and $\mathscr{T}_{n+1}$.

**Theorem 13.2.12** *The set $\mathscr{C}_n$ is in one-to-one correspondence with $\mathscr{T}_{n+1}$.*

Here we introduce a family of bijections between $\mathscr{C}_n$ and $\mathscr{T}_{n+1}$, each of which is determined by a choice function and corresponds to a searching algorithm on trees. Generally speaking, a searching algorithm checks the vertices of a tree one-by-one, starting with the root 0. At each step, we pick a new vertex that is connected to the "checked" vertices. The choice function would tell us which new vertex to pick.

Let $\Pi$ be the set of all ordered pairs $(F,W)$ such that $F$ is a tree whose vertex set $V(F)$ is a subset of $[n]_0$ containing the root 0, and $\emptyset \neq W \subseteq \text{Leaf}(F)$ where $\text{Leaf}(F)$ is the set of leaves of $F$. A **choice function** $\gamma$ is a function from $\Pi$ to $[n]$ such that $\gamma(F,W) \in W$. Some choice functions are described in Examples 13.2.14–13.2.17.

Fix a choice function $\gamma$. Given a tree $T \in \mathscr{T}_{n+1}$, we define a linear order of the vertices $V(\gamma) = v_0, v_1, v_2, \ldots, v_n$. First, set $v_0 = 0$. For each $0 < i < n$, assuming $v_0, \ldots, v_{i-1}$ are determined. Let $W_i = \{v$ : the predecessor of $v$ is in $\{v_0, \ldots, v_{i-1}\}\}$, and $F_i$ be the subtree obtained from $T$ by restricting to $W_i \cup \{v_0, \ldots, v_{i-1}\}$. Then let $v_i = \gamma(F_i, W_i)$.

For each $v_i$, order the successors of $v_i$ from small to large. Let $\pi_\gamma$ be obtained by reading the successors of $v_0$, followed by successors of $v_1$, then successors of $v_2$, and so on. Finally, let $\vec{r}_\gamma = (r_0, r_1, \ldots, r_{n-1})$ where $r_i$ is the number of successors of the vertex $v_i$.

**Theorem 13.2.13** *The map $\phi_\gamma : T \to (\vec{r}_\gamma, \pi_\gamma^{-1})$ described above is a bijection from $\mathscr{T}_{n+1}$ onto $\mathscr{C}_n$, where $\pi_\gamma^{-1}$ is the inverse of $\pi_\gamma$.*

We explain the bijections with some explicit examples of choice functions, and apply each bijection to the tree $T$ in Figure 13.4.

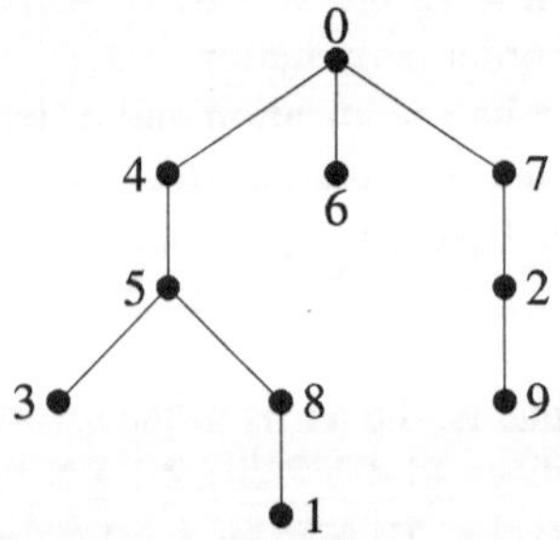


**Figure 13.4**
A tree in $\mathscr{T}_{10}$.

**Example 13.2.14 (The vertex-adding order)** *Let $\gamma(F,W)$ be the vertex in $W$ with minimal label. Then the vertex ordering is $V(\gamma) = 0\,4\,5\,3\,6\,7\,2\,8\,1\,9$. Hence*

$$\vec{r}_\gamma = (3,1,2,0,0,1,1,1,0), \quad \pi_\gamma = 467538291.$$

**Example 13.2.15 (Depth-first search order)** *Also known as the preorder traversal, the depth-first order is the list of vertices in the order that they were first visited by the DFS. Here we adapt the version of DFS described in the end of Section 13.2.2.*

*The choice function $\gamma_{df}$ is defined as $\gamma_{df}(F,W) = v$ where $v$ is the minimal element of $W$ under the depth-first search order. Then the vertex ordering defined by $\gamma_{df}$ is $V(df) = 0\,7\,2\,9\,6\,4\,5\,8\,1\,3$. Hence*

$$\vec{r}_{\gamma_{df}} = (3,1,1,0,0,1,2,1,0), \quad \pi_{\gamma_{df}} = 467295381.$$

**Example 13.2.16 (Breadth-first search order)** *Breadth-first search (BFS) is another commonly used graph searching algorithm, which begins at the root and explores all the neighboring nodes before going to the next node. The order that the vertices are visited under the BSF is called the BFS order, and denoted by $<_{bf}$. In a labeled tree $T \in \mathcal{T}_{n+1}$, let $\mathrm{level}(i)$ be the distance of a node $i$ to the root 0. A simple version of the BFS order of the vertices of $T$ is to let $i <_{bf} j$ if $\mathrm{level}(i) < \mathrm{level}(j)$, or if $\mathrm{level}(i) = \mathrm{level}(j)$ but $i < j$.*

*The choice function $\gamma_{bf}$ is defined as $\gamma_{bf}(F,W) = v$ where $v$ is the minimal element of $W$ under the breadth-first search order $<_{bf}$. Then for the tree in Figure 13.4, the vertex ordering defined by $\gamma_{bf}$ is $V(\gamma_{df}) = 0\,4\,6\,7\,2\,5\,3\,8\,9\,1$. Hence*

$$\vec{r}_{\gamma_{bf}} = (3,1,0,1,1,2,0,1,0), \quad \pi_{\gamma_{bf}} = 467529381.$$

**Example 13.2.17 (Breadth-first search order with a queue)** *This is a variation of the BFS order in Example 13.2.16. It corresponds to an implementation of BFS with a queue structure. It is a vertex ordering, denoted by $<_{bfq}$, which can be defined explicitly by letting $i <_{bfq} j$ if (1) $\mathrm{level}(i) < \mathrm{level}(j)$, or (2) $\mathrm{level}(i) = \mathrm{level}(j)$ and $\mathrm{pre}(i) <_{bfq} \mathrm{pre}(j)$, where $\mathrm{pre}(i)$ is the predecessor of $i$; or (3) $\mathrm{pre}(i) = \mathrm{pre}(j)$ and $i < j$.*

*The choice function $\gamma_{bfq}(F,W)$ always chooses the minimal element of $W$ under the order $<_{bfq}$. Under this choice function the vertex ordering is $0\,4\,6\,7\,5\,2\,3\,8\,9\,1$. And*

$$\vec{r}_{\gamma bf,q} = (3,1,0,1,2,1,0,1,0), \quad \pi_{\gamma_{bfq}} = 467523891.$$

*Note that in this case, the permutation $\pi_{\gamma_{bfq}}$ is the same as the vertex ordering (after removing 0).*

Theorem 13.2.11 and the bijections given in Theorem 13.2.13 have some interesting implications. For example, combining the map $\rho$ of Theorem 13.2.11 and any bijection of Theorem 13.2.13, we get that the number of parking functions with $k$ entries equal to 0 is equal to the number of labeled trees on $n+1$ vertices whose root vertex has degree $k$, which in turn equals the number of rooted forests on $[n]$ with $k$ tree components. The latter is well-known to be $\binom{n-1}{k-1} n^{n-k}$. Written in terms of generating functions, we have

**Corollary 13.2.18** *The enumerator of parking functions by the number of elements equal to 0 is $x(x+n)^{n-1}$, i.e.,*

$$\sum_{\mathbf{a} \in \mathcal{PK}_n} x^{|\{i\,:\,a_i=0\}|} = x(x+n)^{n-1}.$$

Next we describe a special implementation of the breadth-first search algorithm with a queue structure, which finds a particular spanning tree for any connected graph. The algorithm, combining with Theorems 13.2.11 and 13.2.13, gives the combinatorial explanation of the equation $P_n(1+q) = C_n(q)$.

Let $G$ be a connected graph on $[n]_0$. The BFS algorithm is described as a queue $Q$ that starts at vertex 0. At each stage we take the vertex $x$ at the head of $Q$, remove $x$ from $Q$, and add all unvisited neighbors $y$ of $x$ to $Q$ in numerical order. We will call that operation "processing $x$." We continue the above procedure until the queue is empty. The spanning tree $T = BFS(G)$ is obtained by adding all edges of the form $\{x,y\}$, where $x$ is the vertex being processed, and $y$ is an unvisited neighbor of $x$. For the connected graph $G$ in Figure 13.3, the BFS gives the spanning tree shown on the right-hand side of Figure 13.5.

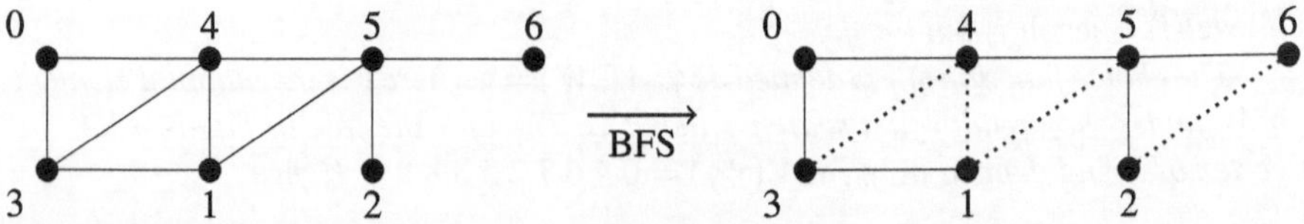


**Figure 13.5**
Spanning tree of Figure 13.3 found by BFS, where the dotted edges belong to $\mathscr{E}_1(T)$.

Let $r_t$ be the number of vertices found by the $t$th vertex processed, for $t = 0, \ldots, n$, which in our example, are $(r_0, r_1, \ldots, r_6) = (2,1,1,0,2,0,0)$. Note that $r_n$ is always 0. Let $q_t$ be the size of the queue after the $t$th vertex is processed. Then $q_0 = 1$ and $q_t = q_{t-1} + r_{t-1} - 1$, which, in our example, are 1, 2, 2, 2, 1, 2, 1, 0.

For the example in Figure 13.5, the queue $Q$, $r_t$ and $q_t$ at each stage $t$ are listed in the following table.

| t | 0 | 1 | 2 | 3 | 4 | 5 | 6 | 7 |
|---|---|---|---|---|---|---|---|---|
| Q | 0 | 3<br>4 | 4<br>1 | 1<br>5 | 5 | 2<br>6 | 6 | $\emptyset$ |
| $r_t$ | 2 | 1 | 1 | 0 | 2 | 0 | 0 | |
| $q_t$ | 1 | 2 | 2 | 2 | 1 | 2 | 1 | 0 |

A necessary and sufficient condition for the graph being connected is that the sequence $\{q_t\}$ satisfies

$$q_{n+1} = 0, \qquad \text{and} \qquad q_i > 0 \text{ for } i \leq n.$$

In terms of $r_t$, the above constraints are equivalent to

$$r_0 + \cdots r_{i-1} \geq i \text{ for } i < n, \qquad r_0 + \cdots + r_{n-1} = n.$$

In other words, $\vec{r} = (r_0, \ldots, r_{n-1})$ is balanced. Also note that the queue uniquely determines a permutation $\pi$ of length $n$, namely, the order that the vertices are processed in the queue. One can read it from the head of the queue. In our example, it is 341526. The vector $\vec{r}$ and the permutation $\pi^{-1}$ are compatible. Actually, they are uniquely determined by the spanning tree $T$ found by the BFS.

Let $\mathbf{a}(T)$ be the parking function that corresponds to the pair $(\vec{r}, \pi^{-1})$ by the bijection $\rho$ of Theorem 13.2.11, where $T = BFS(G)$. One easily computes that the displacement of $a(T)$ is $D(\mathbf{a}(T)) = \binom{n}{2} - \sum_{i=0}^{n-1} ir_i$. In our example, $\mathbf{a}(G) = (1,4,0,0,2,4)$, and $D(\mathbf{a}(G)) = 4$.

Let $\mathscr{G}_1(T)$ be the set of connected graphs for which the spanning tree found by the BFS is $T$. The following crucial observation is due to Spencer [76]: An edge $\{i,j\}$ can be added to $T$ without changing the spanning tree under the BFS if and only if in the queue, when the first of the two vertices is processed, the other is currently in the queue. Let $\mathscr{E}_1(T)$ be the set of all such edges. In our example, $\mathscr{E}_1(T) = \{\{3,4\}, \{4,1\}, \{1,5\}, \{2,6\}\}$. See Figure 13.5 to compare with the table of the queue.

It follows that

**Theorem 13.2.19** *$\mathscr{G}_1(T)$ consists of these graphs obtained from $T$ by adjoining an arbitrary subset of edges in $\mathscr{E}_1(T)$.*

Thus

$$\sum_{G \in \mathscr{G}_1(T)} q^{|E(G)|-n} = (1+q)^{|\mathscr{E}_1(T)|}.$$

Now we compute $|\mathscr{E}_1(T)|$. From the queue $Q$, we have

$$|\mathscr{E}_1(T)| = \sum_{i=1}^{n} (q_i - 1) = \sum_{i=1}^{n} (r_0 + \cdots + r_{i-1} - i) = \sum_{i=0}^{n-1} (n-i) r_i - \binom{n+1}{2}.$$

Since $\sum_{i=0}^{n-1} r_i = n$, the number of the last formula equals

$$n \sum_{i=0}^{n-1} r_i - \sum_{i=0}^{n-1} ir_i - \binom{n+1}{2} = \binom{n}{2} - \sum_{i=0}^{n-1} ir_i = D(\mathbf{a}(T)).$$

Thus

$$\sum_{G \in \mathscr{G}_1(T)} q^{|E(G)|-n} = (1+q)^{D(\mathbf{a}(T))}.$$

Summing over all trees $T$ on $[n]_0$, we obtain

$$C_n(q) = \sum_{G \text{ connected}} q^{|E(G)|-n} = \sum_{T \in \mathscr{T}_{n+1}} (1+q)^{D(\mathbf{a}(T))} = P_n(1+q).$$

### 13.2.4 External activity of labeled trees

Another combinatorial statistic of labeled trees that corresponds to the displacement of parking functions is the external activity, a notion proposed by Tutte in defining a bivariate polynomial, called **Tutte polynomial**, for undirected graphs. Tutte polynomial plays an important role in graph enumeration. The evaluations of Tutte polynomial at various points give the numbers of spanning trees, spanning forests, connected subgraphs, acyclic orientations, etc. Here we just recall the definition of external activity. In Section 13.5.3 we will present a much stronger relation between Tutte polynomial and general parking functions.

Consider the complete graph $K$ on $[n]_0$. Fix a total ordering of its edges. For a tree $T \in \mathscr{T}_{n+1}$, an edge $e \in K - T$ is **externally active** if it is the smallest edge in the unique cycle contained in $T \cup \{e\}$. The **external activity** $ea(T)$ is the number of externally active edges of $T$. Define the **external activity enumerator** of trees in $\mathscr{T}_{n+1}$ as

$$EA_n(q) = \sum_{T \in \mathscr{T}_{n+1}} q^{ea(T)}.$$

Then

$$EA_n(q) = I_n(q) = P_n(q).$$

The equation $EA_n(q) = I_n(q)$ was proved by Björner [13] using his results on shellability and homology in matroids. Beissinger [9] constructed a bijection from parking functions to labeled trees that carries the displacement of parking functions to the external activity of trees.

### 13.2.5 Sparse connected graphs

Theorem 13.2.6 has a nice application in graphical enumeration. Let $c(n+1,k)$ be the number of labeled connected graphs on $n+1$ vertices with exactly $n+k$ edges. For example, $c(n+1,0) = (n+1)^{n-1}$ by Cayley's formula. Theorem 13.2.6 implies that

$$c(n+1,k) = \sum_j p_j \binom{j}{k},$$

where $p_j$ is the number of parking functions that have displacement $D(\mathbf{a}) = j$. Let $F_k(n)$ be the $k$th falling factorial moment of $D(\mathbf{a})$, i.e.,

$$F_k(n) = \frac{1}{(n+1)^{n-1}} \sum_{\mathbf{a} \in \mathscr{PK}_n} (D(\mathbf{a}))_k = \frac{1}{(n+1)^{n-1}} \sum_{\mathbf{a} \in \mathscr{PK}_n} k! \binom{D(\mathbf{a})}{k},$$

where $(n)_k$ is the falling factorial $n(n-1)\cdots(n-k+1)$. It follows that

**Theorem 13.2.20**

$$c(n+1,k) = \frac{(n+1)^{n-1}}{k!} F_k(n).$$

Theorem 13.2.20 can also be obtained by using a result of Spencer [76] that

$$\frac{c(n+1,k)}{c(n+1,0)} = E\left[\binom{M}{k}\right]$$

where $M$ is a certain random variable defined on all labeled trees with $n+1$ vertices with uniform distribution. This formula, together with the bijection between labeled trees and parking functions, yields Theorem 13.2.20.

For $k = 1,2$, the formulas for $F_k(n)$ are computed in [36, 52] as

$$F_1(n) = \frac{1}{2}\sum_{i=2}^{n}\binom{n}{i}\frac{i!}{(1+n)^{i-1}},$$

$$F_2(n) = \frac{n(n-1)(n-2)}{24(n+1)^2}(15Q_3(n+1,n-3)+7Q_2(n+1,n-3)+2Q_1(n+1,n-3)),$$

where

$$Q_r(m,n) = \sum_{k\geq 0}\binom{r+k}{k}\frac{n(n-1)\cdots(n-k+1)}{m^k}. \tag{13.3}$$

Explicit formulas for higher moments are computed by Kung and Yan [58, Theorem 7.1] by solving a recursion based on a combinatorial decomposition of parking functions. The decomposition is the key idea in connecting parking functions to Gončarov polynomials, a system of polynomials from interpolation theory that provides a natural tool for studying the algebraic properties of parking functions.

In [84] Wright found the following asymptotic formula: for fixed $k$ and $n$ tending to infinity,

$$c(n+1,k) = \rho_{k-1}(n+1)^{n-1+3k/2}(1+O(n^{-1/2})).$$

The Wright constants $\rho_k$ are given by

$$\rho_k = \frac{\pi^{1/2}2^{(1-3k)/2}\sigma_k}{\Gamma((3k/2)+1)},$$

where $\sigma_k$ is defined by a second order recursion

$$\sigma_{k+1} = \frac{3(k+1)\sigma_k}{2} + \sum_{s=1}^{k-1}\sigma_s\sigma_{k-s} \qquad (k\geq 2),$$

with initial values $\sigma_0 = 1/4$, $\sigma_1 = 5/16$, and $\sigma_2 = 15/16$. The first several values of $\rho_k$ are $\rho_0 = \sqrt{2\pi}/4$, $\rho_1 = 5/24$, and $\rho_2 = 5\sqrt{2\pi}/2^8$. Using combinatorial analysis, Kung and Yan [58] gave another formula for the Wright constants $\rho_k$, which only involves a linear recurrence.

**Remark 13.2.21** Theorems 13.2.11 and 13.2.12 are due to Foata and Riordan [30], where they used the breadth-first search order of Example 13.2.17 to prove Theorem 13.2.12. A simpler version of choice functions first appeared in Françon [31] as "selection procedures" that lead to a family of bijections between parking functions and labeled trees, each corresponding to a searching algorithm. The idea was extended to parking functions associated with graphs by Chebikin and Pylyavskyy [21], and Kostic and Yan [54]. Spencer [76] used the implementation of BFS with a queue to develop an exact formula for the number of labeled connected graphs on $[n]$ with $n-1+k$ edges ($k$ fixed) in terms of appropriate expectations. Moving to asymptotics, Spencer showed that the expectations can be expressed in terms of a certain restricted Brownian motion. Spencer's approach was generalized by Kostic and Yan

[54] to establish the relation between parking functions associated to a graph $G$ and the Tutte polynomial of $G$, (cf. Section 13.5). Results of [54] also reveal the connections between external activities of the spanning trees of a graph $G$ and the parking functions associated to $G$.

## 13.3 Many faces of parking functions

The parking function is an object lying in the center of combinatorics and appearing in many discrete and algebraic structures. In this section we describe some of the most important examples and their implications.

### 13.3.1 Hashing and linear probing

An efficient method for storing and retrieving data in computer programming is known as **hashing** or **scatter storage technique**. It has a hash function $h$ that assigns a hash value $h(K)$ to each item $K$. However, two or more items may have the same hash value and hence cause a hash collision. **Linear probing and insertion** is a simple and basic algorithm for resolving hash collisions by sequentially searching the hash table for a free location. It inserts $n$ items in $m > n$ cells by the following rules. Begin with all the cells $(0,1,\ldots,m-1)$ empty. Then, for $1 \leq k \leq n$, insert the $k$th item into the first nonempty cell in the sequence $h_k, h_k+1, h_k+2, \ldots (\text{mod } m)$, where $h_k$ is the hash value of the $k$th item and is in the range $0 \leq h_k < m$. For a comprehensive description of hashing, as well as other storage and retrieval methods, see Section 6.4 of [51].

In the above description, the sequence $(h_1,\ldots,h_n)$ is called the **hash function**. It is **confined** if the linear probing with $h_1,\ldots,h_n$ will leave the cell $m-1$ empty. One notes immediately that when $m = n+1$, the confined hash functions are exactly parking functions of length $n$.

If the $k$th item is inserted into position $p_k$, then $D(\mathbf{h}) = \sum_{k=1}^{n} [(p_k - h_k) \mod m]$ is the number of linear probes, or the total displacement of the items from their hash addresses, which gives the name **displacement** of the corresponding statistic of parking functions. Let $D_{m,n}(q) = \sum_{\mathbf{h}} q^{D(\mathbf{h})}$, where $\mathbf{h}$ ranges over all $m^n$ possible hash functions $(h_1,\ldots,h_n)$ with $0 \leq h_k < m$, and let $F_{m,n}(q)$ be the same sum restricted to confined hash functions. Using Pollak's argument with $m$ cells, one notes that given $(h_1,\ldots,h_n)$, the $m$ hash sequences $\{(h_i+j) \mod m, 0 \leq j < m\}$ all have the same total displacement, and exactly $(m-n)/m$ of them are confined. Therefore $F_{mn}(q) = \frac{m-n}{m} D_{mn}(q)$, and the probability generating function for $D(\mathbf{h})$ satisfies

$$\frac{D_{mn}(q)}{D_{mn}(1)} = \frac{F_{mn}(q)}{F_{mn}(1)}. \tag{13.4}$$

Equation (13.4) allows us to reduce the computation of the probability distribution of $D(\mathbf{h})$ over random hash functions to that of confined hash functions, which is easier since the linear probing does not "wrap around" when the hash function is confined. In particular, we have $F_{n+1,n}(q) = P_n(q)$, the displacement enumerator of parking functions of length $n$. In general, for $m = n + r$, a confined hash function $\mathbf{h} = h_1, \ldots, h_n$ leaves the cells $n_1 < n_2 < \cdots < n_r$ empty if and only if $\mathbf{h}$ is obtained by merging $r$ parking functions of lengths $n_1, n_2 - n_1 - 1, \ldots, n_r - n_{r-1} - 1$. Let $s_1 = n_1$ and $s_i = n_i - n_{i-1} - 1$ for $2 \le i \le r$. Then $\sum_{i=1}^{r} s_i = n$, and

$$F_{n+r,n}(q) = \sum_{s_1+s_2+\cdots+s_r=n} \binom{n}{s_1, s_2, \ldots, s_r} P_{s_1}(q) P_{s_2}(q) \cdots P_{s_r}(q). \qquad (13.5)$$

Let

$$F(q,z) = \sum_{n \ge 0} P_n(q) \frac{z^n}{n!}.$$

It follows from Equation (13.5) that

$$\sum_{n \ge 0} F_{n+r,n}(q) \frac{z^n}{n!} = F(q,z)^r.$$

In other words, the distribution of total displacement for linear probing with random hash functions is determined by the exponential generating function of $P_n(q)$ for parking functions. Some identities related to $F(q,z)$ and its variations are given in Theorem 13.2.7.

In analyzing the performance of hashing as a storage method, one is interested in the expected value of $D(\mathbf{h})$ over all hash functions $h : [n] \to [m]$, assuming all are equally likely. This value is $D'_{m,n}(1)/D_{m,n}(1)$, which equals $F'_{m,n}(1)/F_{m,n}$. Using combinatorial analysis and the Lagrange inversion formula, one gets an explicit formula for the expected value of $D(\mathbf{h})$, (see e.g. [36, 52]),

**Theorem 13.3.1** *The expected value of linear probes $D(\mathbf{h})$ as $h$ varies over all hash functions from $[n]$ to $[m]$ $(n < m)$ is*

$$\frac{1}{2} \sum_{i=2}^{n} \binom{n}{i} i! m^{1-i} = \frac{1}{2} \left[ \frac{n(n-1)}{m} + \frac{n(n-1)(n-2)}{m^2} + \cdots \right] = \frac{n}{2} (Q_0(m, n-1) - 1),$$

*where $Q_r(m,n)$ is defined in* (13.3).

Knuth also computed the second factorial moments of $D(\mathbf{h})$, and obtained the following expected value of $D(\mathbf{h})(D(\mathbf{h}) - 1)$ in Formula (5.5) of [52].

$$\begin{aligned} & Exp[D(\mathbf{h})(D(\mathbf{h}) - 1)] \\ = \; & \frac{n(n-1)(n-2)}{12m^2} [15 Q_3(m, n-3) + (4 + 3m - 3n) Q_2(m, n-3) \\ & + (5 - 3m + 3n) Q_1(m, n-3)]. \end{aligned}$$

Moment analysis and characterizations of limit distributions for the linear probes are carried out by Flajolet, Poblete and Viola [29], and Janson [47].

### 13.3.2 Lattice of noncrossing partitions

A **set partition** of $[n]$ is a family of pairwise disjoint nonempty subsets $B_1, \ldots, B_k$ whose union is $[n]$. A partition $\{B_1, \ldots, B_k\}$ is **noncrossing** if there are no elements $a < b < c < d$ such that $a, c \in B_i$, $b, d \in B_j$ and $i \neq j$. The study of noncrossing partitions can be traced back to Becker [8] under the name "planar rhyme schemes." The systematic study of noncrossing partitions began with Kreweras [55] and Poupart [69], and became one of the central topics in contemporary combinatorics.

Let $\pi$ and $\pi'$ be two partitions of $[n]$. We say that $\pi$ is a **refinement** of $\pi'$ if every block of $\pi$ is contained in a block of $\pi'$. Refinement induces a partial order on the set of partitions: $\pi < \pi'$ if and only if $\pi$ is a refinement of $\pi'$. Under this order there is a unique maximal element, the partition with only one block $[n]$. In addition any two partitions $\pi$ and $\pi'$ have a greatest lower bound $\pi \wedge \pi'$ whose blocks are the non-empty intersections of one block of $\pi$ and one block of $\pi'$. It is easy to check that if both $\pi$ and $\pi'$ are noncrossing, then so is $\pi \wedge \pi'$. Hence the set of noncrossing partitions form a lattice, which is denoted $\mathrm{NC}_n$. The lattice of noncrossing partitions enjoys a number of remarkable properties, and plays a fundamental role in the combinatorics of Coxeter groups and free probability. See [1] and [66] for references. Next is a list of basic properties of lattice of non-crossing partitions.

1. It is well-known that the number of noncrossing partitions of $[n]$ is the $n$th Catalan number $C_n = \frac{1}{n+1}\binom{2n}{n}$.

2. The lattice $\mathrm{NC}_n$ is graded of rank $n-1$. The rank of a noncrossing partition $\pi$ is $n - |\pi|$, where $|\pi|$ is the number of blocks in $\pi$.

3. The number of noncrossing partitions with exactly $k$ blocks is given by the Narayana number $N(n,k)$, where
$$N(n,k) = \frac{1}{n}\binom{n}{k}\binom{n}{k-1}.$$

4. The lattice $\mathrm{NC}_n$ has the minimum $\hat{0} = \{1\}\{2\}\cdots\{n\}$ and the maximum $\hat{1} = [n]$.

5. $\mathrm{NC}_n$ is self-dual, i.e., there is an isomorphism $\xi : \mathrm{NC}_n \to \mathrm{NC}_n$ such that $\pi < \sigma$ if and only if $\xi(\sigma) < \xi(\pi)$. The Kreweras complement operation provides such an isomorphism, [55].

A surprising connection between parking functions and the lattice of noncrossing partitions was revealed by Stanley [80], who constructed an edge labeling of $\mathrm{NC}_{n+1}$ whose maximal chains are labeled by the set $\mathscr{PK}_n$, The labeling also leads to a local action of the symmetric group $\mathfrak{S}_n$ on $\mathrm{NC}_{n+1}$.

In a locally finite poset $P$, an edge of $P$ is a pair $(u,v)$ such that $v$ covers $u$, i.e., $u < v$ and there is no element $t$ satisfying $u < t < v$. For $\mathrm{NC}_{n+1}$, a partition $\pi'$ covers partition $\pi$ if and only if $\pi'$ is obtained from $\pi$ by merging two blocks into one. Let

$(\pi, \pi')$ be an edge of $\mathrm{NC}_{n+1}$ where two blocks $B$ and $B'$ of $\pi$ are merged to form one block of $\pi'$. Suppose that $\min B < \min B'$. Define

$$\Lambda(\pi, \pi') = \max\{i \in B \, : \, i < \min B'\}.$$

For instance, if $B = \{2,3,6,16\}$ and $B' = \{8,9,10,14\}$, then $\Lambda(\pi, \pi') = 6$. This labeling is well-defined since the set $\{i \in B \, : \, i < \min B'\}$ is nonempty.

Figure 13.6 shows the labeling $\Lambda$ on the lattice $\mathrm{NC}_3$.

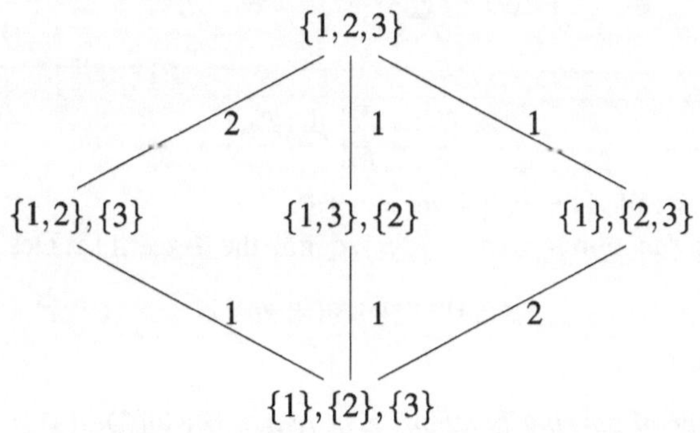


**Figure 13.6**
The labeling $\Lambda$ on $\mathrm{NC}_3$.

The labeling $\Lambda$ of the edges of $\mathrm{NC}_{n+1}$ extends in a natural way to a labeling of the maximal chains. Namely, given a maximal chain $\mathfrak{m} : \hat{0} = \pi_0 < \pi_1 < \cdots < \pi_n = \hat{1}$ where $\pi_i$ covers $\pi_{i-1}$, we set

$$\Lambda(\mathfrak{m}) = (\Lambda(\pi_0, \pi_1), \Lambda(\pi_1, \pi_2), \ldots, \Lambda(\pi_{n-1}, \pi_n)).$$

Note that $\Lambda(\pi_i, \pi_{i+1}) \in [n]$. Adjust the values by letting

$$\Lambda_1(\mathfrak{m}) = (\Lambda(\pi_0, \pi_1) - 1, \Lambda(\pi_1, \pi_2) - 1, \ldots, \Lambda(\pi_{n-1}, \pi_n) - 1).$$

**Theorem 13.3.2** *The label $\Lambda_1(\mathfrak{m})$ of the maximal chains of $\mathrm{NC}_{n+1}$ consists of the parking functions of length $n$, each occurring once.*

As an immediate consequence, we recover a result of Kreweras [55].

**Corollary 13.3.3** *The number of maximal chains of noncrossing partitions of $\{1,2,\ldots,n+1\}$ is $(n+1)^{n-1}$.*

The labeling $\Lambda$ induces an R-labeling by letting $\Lambda^*(\pi, \sigma) = |\pi| - \Lambda(\pi, \sigma)$, as in the sense of [78, Def. 3.13.1]. More precisely, for every interval $[\pi, \pi']$ of $NC_{n+1}$, there is a unique maximal chain $\mathfrak{m} : \pi = \pi_0 < \pi_1 < \cdots < \pi_j = \pi'$ such that

$$\Lambda^*(\pi_0, \pi_1) \leq \Lambda^*(\pi_1, \pi_2) \leq \cdots \leq \Lambda^*(\pi_{j-1}, \pi_j).$$

For any poset equipped with an R-labeling, there is a general theorem [78, Theorem 3.13.2] that allows us to enumerate the labeling of maximal chains with a given descent set in terms of the rank-selected Möbius invariants. Applying this theorem to $NC_{n+1}$ and parking functions yields the following result.

For a finite graded poset $P$ of rank $n$ with $\hat{0}$ and $\hat{1}$ and with rank function $\rho$, let $S$ be a subset of $[n-1]$ and denote $\alpha_P(S)$ to be the number of chains $\hat{0} = t_0 < t_1 < \cdots < t_s = \hat{1}$ of $P$ such that $\{\rho(t_1), \rho(t_2), \ldots, \rho(t_{s-1})\} = S$. The function $\alpha_P(S)$ is called the **flag $f$-vector** of $P$. Further define

$$\beta_P(S) = \sum_{T \subseteq S} (-1)^{|S-T|} \alpha_P(T),$$

or equivalently,

$$\alpha_P(S) = \sum_{T \subseteq S} \beta_P(T).$$

The function $\beta_P$ is called the **flag $h$-vector** of $P$.

For a parking function $\mathbf{a} = (a_1, \ldots, a_n)$, define the **descent set** $\mathrm{Des}(\mathbf{a})$ by letting

$$\mathrm{Des}(\mathbf{a}) = \{i \,:\, a_i > a_{i+1}\}.$$

**Theorem 13.3.4** *Let $S \subseteq [n-1]$.*

1. *The number of parking functions* $\mathbf{a}$ *of length $n$ with* $\mathrm{Des}(\mathbf{a}) = S$ *is equal to* $\beta_{\mathrm{NC}_{n+1}}([n-1]-S)$.

2. *The number of parking functions* $\mathbf{a}$ *of length $n$ satisfying* $S \subseteq \mathrm{Des}(\mathbf{a})$ *is equal to* $\alpha_{\mathrm{NC}_{n+1}}([n-1]-S)$.

The value of $\alpha_{\mathrm{NC}_{n+1}}(T)$ for $T \subseteq [n-1]$ is computed in Theorem 3.2 of [26]. Assume $T = \{t_1 < t_2 < \cdots < t_r\}$. Set $t_0 = 0$, $t_{r+1} = n$, and $\delta_i = t_i - t_{i-1}$ for $1 \le i \le r+1$. Then

$$\alpha_{\mathrm{NC}_{n+1}}(T) = \frac{1}{n} \prod_{i=1}^{r+1} \binom{n+1}{\delta_i}.$$

Theorem 13.3.2 is used by Kim and Seo [50] to study the minimal transitive factorizations for permutations of cycle type $(n)$ and $(1, n-1)$. Given a permutation $\sigma$ in $\mathfrak{S}_n$, the **minimal transitive factorizations** of $\sigma$ is a set of $n+\ell-2$ transpositions that generate the full symmetric group such that their product is $\sigma$, where $\ell$ is the number of cycles of $\sigma$. If the permutation $\sigma$ is of cycle type $\lambda = (\lambda_1, \ldots, \lambda_\ell)$, then the number of minimal transitive factorizations of $\sigma$ is

$$(n+\ell-2)!\,n^{\ell-3} \prod_{i=1}^{\ell} \frac{\lambda_i^{\lambda_i}}{(\lambda_i - 1)!},$$

a formula originally suggested by Hurwitz but proved by Goulden and Jackson [41], and Bousquet-Mélou and Schaeffer [16]. In particular, when $\sigma$ is of cycle type $(n)$ or $(1, n-1)$, Hurwitz's formula reduces to $n^{n-2}$ and $(n-1)^n$. Kim and Seo [50] presented combinatorial proofs for these two cases using an interplay between the minimal transitive factorizations, cycle chord diagrams, Stanley's labeling of maximal chains of $\mathrm{NC}_n$, and parking functions.

### 13.3.3 Hyperplane arrangements

A (real) hyperplane arrangement is a discrete set of affine hyperplanes in $\mathbb{R}^n$. An especially important arrangement is the **braid arrangement** $\mathscr{B}_n$, which consists of all the hyperplanes $x_i - x_j = 0$ for $1 \leq i < j \leq n$. For a hyperplane arrangement $\mathscr{A}$, if we remove the union of the hyperplanes from $\mathbb{R}^n$, then we obtain a disjoint union of open cells, called the **regions** of $\mathscr{A}$. Fix a region $R_0$ of $\mathscr{A}$ and call it the **base region**. Given a region $R$ of $\mathscr{A}$, let $d(R)$, the **distance** of $R$, be the number of hyperplanes $H$ in $\mathscr{A}$ that separate $R_0$ from $R$, i.e., $R_0$ and $R$ lie in different sides of $H$. Define the **distance enumerator** of $\mathscr{A}$ (with respect to $R_0$) to be the generating function

$$D_{\mathscr{A}}(q) = \sum_R q^{d(R)},$$

where $R$ ranges over all regions of $\mathscr{A}$. For finite $\mathscr{A}$, $D_{\mathscr{A}}(q)$ is a polynomial in $q$, and $D_{\mathscr{A}}(1)$ is the number of regions of the hyperplane arrangement $\mathscr{A}$.

As an example, consider the braid arrangement $\mathscr{B}_n$ that has $n!$ regions, each corresponding to a permutation of $\mathfrak{S}_n$. It is natural to let $R_0$ be the region defined by the conditions $x_1 > x_2 > \cdots > x_n$. There is an elegant way of labeling the regions of $\mathscr{B}_n$ by integer sequences that leads to the formula of $D_{\mathscr{B}_n}(q)$. Let $e_i \in \mathbb{N}^n$ be the vector with a 1 in the $i$th coordinate and 0's elsewhere. First label the base region $R_0$ by $\lambda(R_0) = (0, \ldots, 0)$. Suppose now that $R$ has been labeled, and that $R'$ is an unlabeled region that is separated from $R$ by a unique hyperplane $x_i = x_j$, where $i < j$. Then set

$$\lambda(R') = \lambda(R) + e_i.$$

It is easy to check that this labeling is well-defined, independent of the order in which the regions are labeled, and the labels of regions of $\mathscr{B}_n$ are the sequences $(c_1, \ldots, c_n)$ such that $0 \leq c_i \leq n - i$. Hence

$$D_{\mathscr{B}_n}(q) = (1+q)(1+q+q^2)\cdots(1+q+\cdots+q^{n-1}),$$

the standard $q$-analog of $n!$.

A deformation of the braid arrangement is the Shi arrangement $\mathscr{S}_n$ consisting of hyperplanes $x_i - x_j = 0, 1$ for $1 \leq i < j \leq n$. This deformation was first considered by Shi [74] in his investigation of the affine Weyl group $\tilde{A}_n$. Shi showed by grouptheoretic techniques that the number of regions of Shi arrangement is $(n+1)^{n-1}$. Pak and Stanley [79] gave a bijective proof by constructing a labeling of $\mathscr{S}_n$ that is analogous to that of $\mathscr{B}_n$.

**Algorithm 13.3.5 (A labeling of $\mathscr{S}_n$)** *Let the base region $R_0$ be defined by $x_1 > x_2 > \cdots > x_n$ and $x_1 - x_n < 1$. Note that this is the unique region of $\mathscr{S}_n$ contained between all pairs of parallel hyperplanes. Label $R_0$ by the n-tuple $(0, 0, \ldots, 0)$. Now suppose a region $R$ is labeled, and $R'$ is an unlabeled region that is separated from $R$ by a hyperplane $H$ of $\mathscr{S}_n$. Then define*

$$\lambda(R') = \begin{cases} \lambda(R) + e_i, & \text{if } H \text{ is given by } x_i - x_j = 0 \text{ with } i < j \\ \lambda(R) + e_j, & \text{if } H \text{ is given by } x_i - x_j = 1 \text{ with } i < j. \end{cases}$$

*The labeling $\lambda$ is independent of the order of the hyperplanes, hence well-defined. In addition, if $\lambda(R) = (a_1, \ldots, a_n)$, then $d(R) = a_1 + \cdots + a_n$, that is, there are $a_1 + \cdots + a_n$ hyperplanes of $\mathscr{S}_n$ that separate $R$ and $R_0$.*

Figure 13.7 shows the labeling of $\mathscr{S}_3$, where the hyperplanes are projected to $x_1 + x_2 + x_3 = 0$.

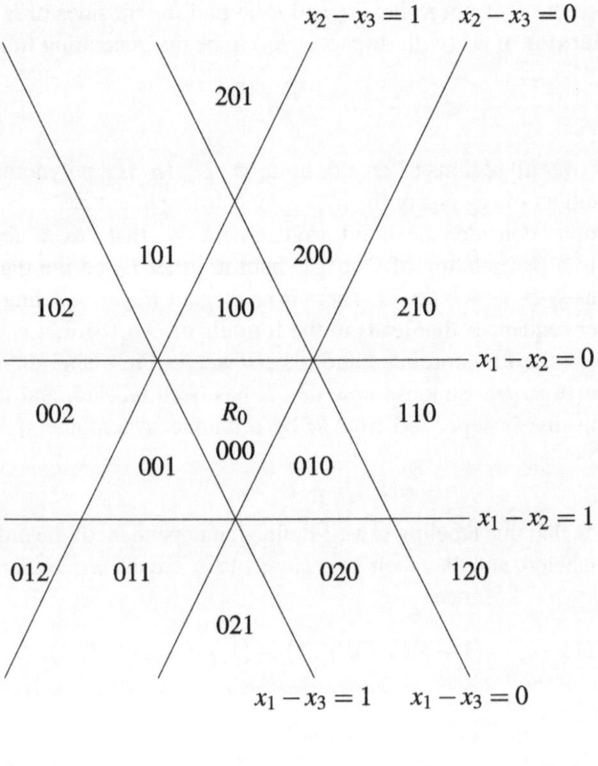


**Figure 13.7**
The labeling $\lambda(R)$ for the Shi arrangement $\mathscr{S}_3$.

**Theorem 13.3.6** *The labeling $\lambda$ defined above is a bijection from the regions of $\mathscr{S}_n$ to the set $\mathscr{PK}_n$ of all parking functions of length n. Consequently, the number of regions R for which i hyperplanes separate R from $R_0$ is equal to the number of parking functions of length n with displacement $D(\mathbf{a}) = \binom{n}{2} - i$.*

In addition to Theorem 13.3.6, Stanley [79] generalized the labeling of $\mathscr{S}_n$ to the extended Shi arrangement, and established connections between the extended Shi arrangement and other combinatorial subjects, in particular, generalized parking functions and rooted $k$-forests, (c.f. 13.4.4).

Athanasiadis and Linusson [5] constructed another simple bijection between the regions of the Shi arrangement and the set of parking functions. Their bijection can be generalized to any subarrangement of $\mathscr{S}_n$ containing the hyperplanes $x_i - x_j = 0$ and to the extended Shi arrangements. It also implies that the number of relatively bounded regions of $\mathscr{S}_n$ is $(n-1)^{n-1}$, where a region is relatively bounded if its intersection with the hyperplane $x_1 + \cdots + x_n = 0$ is bounded as a subset of Euclidean space. Athanasiadis and Linusson's bijection maps the relatively bounded regions of $\mathscr{S}_n$ to the **prime parking functions** of length $n$, a concept due to Gessel. Under our notation a **prime parking function** of length $n$ is a sequence $(a_1, \ldots, a_n)$ of non-negative integers such that for all $1 \leq j \leq n-1$, the cardinality of the set $\{a_i : a_i < j\}$ is at least $j+1$.

Armstrong [2] considered the regions of the Shi arrangement as antichains or ideals in the poset of positive roots, and mapped the regions bijectively to certain labeled Dyck paths. The same labeled Dyck paths are commonly used in the study of diagonal harmonics to encode parking functions, (c.f. Section 13.6). Combining these two results one gets essentially the same bijection of Athanasiadis and Linusson's. There is some further elaboration of this labeling scheme in [4], which reveals the relations between the Shi arrangement and other hyperplane arrangements, as well as the relations among various statistics of parking functions.

### 13.3.4 Allowable input-output pairs in a priority queue

A **priority queue** is an abstract data type equipped with the operations INSERT and DELETEMIN. Let $\sigma = \sigma_1\sigma_2\ldots\sigma_n$ be a permutation of $[n]$. Each INSERT operation will insert the next element of $\sigma$ into the queue, and each DELETEMIN operation will remove the minimal element of the queue and place it in the output stream. A sequence of $n$ INSERT and $n$ DELETEMIN is an allowable sequence if any initial subsequence contains at least as many INSERT's as DELETEMIN's. The application of an allowable sequence to a permutation $\sigma$ is called a priority queue computation. Assume the outcome is $\tau$. Then $(\sigma, \tau)$ is called an allowable pair of permutations in a priority queue.

For example, when $n = 3$, the allowable sequences are *IDIDID*, *IDIIDD*, *IIDIDD*, *IIDDID*, *IIIDDD*, where *I* stands for INSERT and *D* for DELETEMIN. There are 16 allowable pairs of permutations, as listed below.

| | | | |
|---|---|---|---|
| (123, 123) | (132, 132) | (132, 123) | (213, 213) |
| (213, 123) | (231, 231) | (231, 213) | (231, 123) |
| (312, 312) | (312, 123) | (312, 132) | (321, 321) |
| (321, 312) | (321, 213) | (321, 231) | (321, 123) |

Let $\mathscr{Q}_n$ be the set of all allowable pairs of permutations on $[n]$. Atkinson and Thiyagarajah [6] found that the number of allowable pairs in $\mathscr{Q}_n$ is $(n+1)^{n-1}$. Gilbey and Kalikow [39] constructed a bijection between $\mathscr{Q}_n$ and $\mathscr{PK}_n$ that for each allowable pair $(\sigma, \tau)$, the corresponding parking function has $\tau$ as output, that is, the car $C_{\tau_i}$ is parked at the space $i-1$ under the parking rules. In other words, $\tau$ is the permutation $\mathbf{q}$ as defined in Knuth's bijection $\phi$ described in Section 13.2.2. In addition, the bijection of Gilbey and Kalikow's has the extra property of preserving the set of breakpoints. For a parking function $\mathbf{a} = (a_1, \ldots, a_n)$, an integer $b \in \{0, 1, \ldots, n\}$ is a breakpoint if and only if $|\{i : p(i) \leq b-1| = b$, while for an allowable pair $(\sigma, \tau)$, $b \in [0, n]$ is a breakpoint if and only if $\{\sigma_1, \ldots, \sigma_b\} = \{\tau_1, \ldots, \tau_b\}$. Note that under these definitions, 0 and $n$ are always breakpoints of $\mathbf{a} \in \mathscr{P}\mathscr{K}_n$ and any $(\sigma, \tau) \in \mathscr{Q}_n$.

### 13.3.5 Two variations of parking functions

We conclude this section with two variations of parking functions.

The first variation is the defective parking functions considered by Cameron, Johannsen, Prellberg, and Schweitzer [17]. Suppose that there are $m$ drivers entering a one-way street with $n$ parking spaces, each with a preferred parking space. The parking procedure is the same as before. If $k$ drivers fail to park, the preference sequence is called a **defective parking function of defect** $k$, whose number is denoted by $\mathrm{cp}(n, m, k)$. The original parking functions correspond to the case that $m = n$ and $k = 0$.

For the case $m < n$ and $k = 0$, Pollak's argument can be adapted to give $\mathrm{cp}(n, m, 0) = (n+1-m)(n+1)^{m-1}$. The formula for general case is much harder. Cameron et al. established a recurrence relation for $\mathrm{cp}(n, m, k)$, and expressed it as an equation for a three-variable generating function. Solving the equation by using the kernel method, they obtained the following result.

**Theorem 13.3.7** *The number of defective parking functions of defect k is given by*

$$\mathrm{cp}(n, m, k) = S(n, m, k) - S(n, m, k+1),$$

*where $S(n, m, k)$ is the number of car parking preferences of m cars on n spaces, such that at least k cars do not find a parking space. The values of $S(n, m, k)$ can be computed explicitly by*

$$S(n, m, k) = \begin{cases} n^m & \text{if } k \leq m-n, \\ \sum_{i=0}^{m-k} \binom{m}{i} (n-m+k)(n-m+k+i)^{i-1}(m-k-i)^{m-i} & \text{otherwise.} \end{cases}$$

These formulas were also derived by Pitman and Stanley [67] using the volume polynomial of a polytope, and by Yan [86] with combinatorial means.

Cameron et al. investigated the asymptotic behavior of defective parking functions. They proved that if $m = n + \lfloor y\sqrt{n} \rfloor$, then the limiting probability of at most $\lfloor x\sqrt{n} \rfloor$ drivers failing to park is

$$\lim_{n \to \infty} \frac{1}{n^m} \sum_{k=0}^{\lfloor x\sqrt{n} \rfloor} \mathrm{cp}(n, m, k) = \begin{cases} 1 - e^{-2x(x-y)} & \text{if } x > y, \\ 0 & \text{otherwise.} \end{cases}$$

This limit distribution implies that if $m < n + k$ for a fixed constant $k$, then

$$\frac{\mathrm{cp}(n, m, k)}{n^m} \sim \frac{2}{n} \cdot (2k - m + n) \cdot e^{-2k(k+n-m)/n}.$$

Another problem for defective parking functions is to find the limiting probability that all parking spaces are occupied. For $m < n$ this probability is clearly 0. It turns out that when $m$ is linear in $n$, that is, $m = \lfloor cn \rfloor$ for a constant $c > 0$ and $k = m - n$,

$$\lim_{n\to\infty} \frac{\text{cp}(n,m,k)}{n^m} = \begin{cases} 0 & \text{if } c \leq 1, \\ 1 - e^{-c} \cdot \sum_{i\geq 1} \frac{(ci/e^c)^{i-1}}{i!} & \text{if } c > 1. \end{cases} \tag{13.6}$$

Formula (13.6) is proved in Theorem 10 of [17] and also in [77] by Spencer and Yan who used the Galton-Watson branching process to investigate the connections between parking functions and random labeled trees. The approach of [77] was further extended by Chassaing and Marckert [20] to obtain tight bounds for the moments of the width of rooted labeled trees.

The second variation was given by Zara [88] who described an interesting parking problem. Assume that in the middle of the well-known parking procedure, $n - k$ spaces are already taken, and the spaces $\mathbf{q} = \{q_1 < q_2 < \cdots < q_k\}$ are still available. Some other $k$ cars want to take these spaces; they enter the street and start advancing. When they are in front of spaces $\mathbf{p} = \{p_1 < p_2 < \cdots < p_k\}$, the lights go off. The cars can only advance towards the end of the street and park in an unoccupied space. Will all of them be able to find parking spaces?

It is clear that all the cars can park if and only if $p_i \leq q_i$ for all $1 \leq i \leq k$. If this condition is satisfied, we say that $\mathbf{p} \preceq \mathbf{q}$ and call $\mathbf{p}, \mathbf{q}$ an initial condition. Given an initial condition $\mathbf{p}, \mathbf{q}$, a possible final arrangement is given by a permutation $\tau \in \mathfrak{S}_k$: for each $i = 1, \ldots, k$, the car at $p_i$ takes the space $q_{\tau(i)}$. A permutation $\tau$ is attainable from $\mathbf{p}, \mathbf{q}$ if $\tau$ can appear as the permutation associated with a final arrangement when the initial condition is $\mathbf{p}, \mathbf{q}$. In particular, if everyone plays safe and takes the first available space, just as what is defined in the notion of parking functions, the resulting $\tau$ is called the **safe permutation** for $\mathbf{p}, \mathbf{q}$.

Zara gave a complete characterization of safe permutations: If $\tau$ is a safe permutation, then it avoids pattern 231. Conversely, a 231-avoiding permutation $\tau$ can be realized as the safe permutation for some initial conditions if and only if is$(\tau)$ is at least $2k - n$, where is$(\tau)$ is the maximal length of the increasing subsequences of $\tau$. There is a surprising connection between the safe permutations and the paths in the Johnson graph $J(n,k)$, which is the graph whose vertices are the $k$-element subsets of $[n]$, and whose edges are $(V_1, V_2)$ such that $V_1 \cap V_2$ has size $k - 1$.

## 13.4 Generalized parking functions

### 13.4.1 u-parking functions

For a finite sequence $(x_1, x_2, \ldots, x_n)$ of real numbers, rearrange the terms in non-decreasing order as $x_{(1)} \leq x_{(2)} \leq \cdots \leq x_{(n)}$. The term $x_{(i)}$ is the $i$th **order statistic** of the sequence $\mathbf{x}$. Let $\mathbf{u}$ be a non-decreasing sequence $(u_1, u_2, u_3, \ldots)$ of positive integers. A $\mathbf{u}$-parking function of length $n$ is a sequence $(x_1, x_2, \ldots, x_n)$ satisfying

$0 \leq x_{(i)} < u_i$. The parking functions studied in previous sections correspond to the case $\mathbf{u} = (1, 2, \ldots, n)$, and are usually referred as **ordinary parking functions** in enumeration literature, or as **classical parking functions** in diagonal harmonic research (c.f. Section 13.6). However, in this paper we shall reserve the words "classical parking functions" to the case that $\mathbf{u}$ is an arithmetic progression (c.f. Section 13.4.4).

Let $\mathscr{PK}_n(\mathbf{u})$ be the set of $\mathbf{u}$-parking functions of length $n$, and $PK_n(\mathbf{u})$ the number of elements in $\mathscr{PK}_n(\mathbf{u})$. Clearly $PK_n(\mathbf{u})$ is a function of $u_1, \ldots, u_n$. Less obvious is the fact that it is a homogeneous polynomial of $u_1, u_2, \ldots, u_n$. Furthermore, there is a determinant formula for $PK_n(\mathbf{u})$.

**Theorem 13.4.1** *The number $PK_n(\mathbf{u})$ of $\mathbf{u}$-parking functions of length $n$ equals $n!\det(D)$, where $D$ is the $n \times n$ matrix with $ij$th entry equal to*

$$\frac{u_i^{j-i+1}}{(j-i+1)!},$$

*if $j-i+1 \geq 0$ and 0 otherwise.*

Theorem 13.4.1 is the discrete analog of a formula of Steck [82] for the cumulative distribution function of the random vector for order statistics of $n$ independent random variables with uniform distribution on an interval. However, Steck's formula is for sequences of real values. To use it one needs to establish a connection between real sequences and integer sequences. One way to do it is to use the parking polytope introduced in Section 13.4.2. In Section 13.4.3 we provide another proof of Theorem 13.4.1 via the theory of Gončarov polynomials.

The determinant formula in Theorem 13.4.1, while giving a solution for $PK_n(\mathbf{u})$, is not easy to compute for a general sequence $\mathbf{u}$. An easy case known is when $\mathbf{u}$ is an arithmetic progression, i.e., $u_i = a + (i-1)b$ for some positive integers $a, b$. In that case $PK_n(a, a+b, \ldots, a+(n-1)b) = a(a+nb)^{n-1}$, and the corresponding parking functions are called **classical parking functions**. Many enumerative results and combinatorial correspondences described in the previous two sections can be extended to classical parking functions, which we will discuss in Section 13.4.4.

For a sequence $\mathbf{u}$, let $\Delta(\mathbf{u})$ be the difference sequence $(u_1, u_2 - u_1, u_3 - u_2, \ldots)$. Besides classical parking functions, the only $PK_n(\mathbf{u})$ that are computed explicitly are the following two cases.

**Theorem 13.4.2**

*1. If*

$$\Delta(\mathbf{u}) = (a, \overbrace{b, \ldots, b}^{n-m-1}, \overbrace{c, \ldots, c}^{m-1}, d),$$

*then*

$$PK_n(\mathbf{u}) = a \sum_{j=0}^{m} \binom{n}{j} (m+1-j)(c-b) \left[a + (n-j)\right]^{n-j-1}$$
$$\cdot \left[((m+1)c - (m+1-j)b)^{j-1} + j(d-c)(mc - (m+1-j)b)^{j-2}\right].$$

2. *If*

$$\Delta(\mathbf{u}) = (a, \overbrace{b, \ldots, b}^{n-m-1}, d, \overbrace{c, \ldots, c}^{m-1}),$$

*then*

$$PK_n(\mathbf{u}) = a \sum_{j=0}^{m} \binom{n}{j} [d + (m-j)c - (m+1-j)] \cdot [d + mc - (m+1-j)b]^{j-1} [a + (n-j)b]^{n-j-1}. \quad (13.7)$$

Theorem 13.4.2 was proved in [86] by a combinatorial decomposition that partitions every integer sequence into two subsequences: a "maximum" parking function and a subsequence consisting of terms of higher values. Formula (13.7) with $c = 0$ was also given in [67] by connecting it to the empirical cumulative distributive function based on a sample of $n$ independent uniform $(0,1)$ variables crossing an arbitrary line through the unit square. Note that when $\Delta(\mathbf{u}) = (n-(m-k)+1, 1, \ldots, 1, 0, \ldots, 0)$, $PK_n(\mathbf{u})$ is exactly the number of defective parking functions with $m$ drivers, $n$ parking spaces, and at most $k$ defects, as described in Section 13.3.5.

In Section 13.4.2 we introduce a combinatorial representation of a **u**-parking function of length $n$: it equals the volume of certain polytope in $\mathbb{R}^n$, which admits a number of interpretations, in terms of empirical distributions, plane partitions, and polytopal subdivisions. Then we present in Section 13.4.3 a special sequence of polynomials, namely, Gončarov polynomials, that form a natural basis for working with **u**-parking functions. The connection between Gončarov polynomials and **u**-parking functions was established by extending the decomposition of [86]. Many properties of **u**-parking functions can be derived from the theory of parking polytopes and the Gončarov polynomials. In particular, there are various formulas that allow us to compute the value of $PK_n(\mathbf{u})$ efficiently.

### 13.4.2 A parking polytope

In [67] Pitman and Stanley introduced an $n$-dimensional polytope $\Pi_n$ whose volume defines a polynomial that has many combinatorial interpretations. Actually the volume polynomial of $\Pi_n$ is a variant form of **u**-parking functions. We present some basic properties of the polytope $\Pi_n$ and its volume polynomial.

Let

$$\Pi_n(\mathbf{x}) := \{\mathbf{y} \in \mathbb{R}^n \;:\; y_i \geq 0 \text{ and } \sum_{i=1}^{j} y_i \leq \sum_{i=1}^{j} x_i \text{ for all } 1 \leq j \leq n\}$$

for arbitrary $\mathbf{x} = (x_1, \ldots, x_n)$ with $x_i > 0$ for all $i$. The $n$-dimensional volume

$$V_n(\mathbf{x}) = Vol(\Pi_n(\mathbf{x}))$$

is a homogeneous polynomial of degree $n$ in the variables $x_1, \ldots, x_n$, which is called the **volume polynomial**. For example, volume polynomials of small indices are

$$V_1(\mathbf{x}) \;=\; x_1,$$

$$V_2(\mathbf{x}) = x_1x_2 + \frac{1}{2}x_1^2,$$

$$V_3(\mathbf{x}) = x_1x_2x_3 + \frac{1}{2}x_1^2x_2 + \frac{1}{2}x_1x_2^2 + \frac{1}{2}x_1^2x_3 + \frac{1}{6}x_1^3.$$

The volume polynomial $V_n(\mathbf{x})$ is closely related to the order statistics of $n$ independent random variables. Let $U_1, U_2, \ldots, U_n$ be $n$ independent random variables uniformly distributed in $(0,1)$, and $\{U_{(i)} : i \le i \le n)$ be the order statistics of $U_1, U_2, \ldots, U_n$. Because the random vectors $(U_{(i)} : 1 \le i \le n)$ and $(1 - U_{(n+1-i)} : 1 \le i \le n)$ have the same uniform distribution with constant density $n!$ on the simplex

$$\{\mathbf{u} \in \mathbf{R}^n : 0 \le u_1 \le \cdots \le u_n \le 1\},$$

one obtains that for arbitrary vectors $\mathbf{u}$ and $\mathbf{r}$ in this simplex

$$\Pr(U_{(i)} \le u_j \text{ for all } 1 \le j \le n) = n! V_n(x_1, \ldots, x_n) \tag{13.8}$$

where $(x_1, \ldots, x_n) = \Delta(\mathbf{u})$, or equivalently, $u_i = \sum_{j=1}^{i} x_j$ , and

$$\Pr(U_{(i)} \ge r_i \text{ for all } 1 \le i \le n) = n! V_n(x_1, \ldots, x_n) \tag{13.9}$$

where $x_j = r_{n+2-j} - r_{n+1-j}$ (with the convention that $r_{n+1} = 1$). Formulas (13.8) and (13.9) allow us to compute $V_n(\mathbf{x})$ using the probability theory.

**Theorem 13.4.3** *For a positive integer n,*

$$V_n(\mathbf{x}) = \sum_{\mathbf{k} \in K_n} \prod_{i=1}^{n} \frac{x_i^{k_i}}{k_i!} = \frac{1}{n!} \sum_{\mathbf{k} \in K_n} \binom{n}{k_1, \ldots, k_n} x_1^{k_1} \cdots x_n^{k_n}, \tag{13.10}$$

*where $K_n$ is the set of balanced vectors of length n, that is,*

$$K_n = \{\mathbf{k} \in \mathbf{N}^n : \sum_{i=1}^{j} k_i \ge j \textit{ for all } 1 \le i \le n-1 \textit{ and } \sum_{i=1}^{n} k_i = n\}.$$

*Proof.* Let $u_i = \sum_{j=1}^{i} x_j$. By homogeneity of $V_n$, it suffices to prove Formula (13.10) with $u_n \le 1$. Fix $\mathbf{u}$ and consider the probability in (13.8). For $1 \le i \le n+1$ let $N_i$ be the number of $U_i$ that lying in the interval $(u_{i-1}, u_i]$, with the conventions that $u_0 = 0$ and $u_{n+1} = 1$.

Since $U_i$ are independent and uniformly distributed in $(0,1)$, the random vector $(N_i : 1 \le k \le n+1)$ has the **multinomial distribution** with parameters $n$ and $(x_1, \ldots, x_n, x_{n+1})$, where $x_{n+1} = 1 - u_n = 1 - \sum_{i=1}^{n} x_i$. In other words, for any sequence $(k_1, \ldots, k_{n+1})$ with $\sum_{i=1}^{n+1} = n$, we have

$$\Pr[(N_1, \ldots, N_{n+1}) = (k_1, \ldots, k_{n+1})] = \binom{n}{k_1, \ldots, k_{n+1}} \prod_{i=1}^{n+1} x_i^{k_i} = n! \prod_{i=1}^{n+1} \frac{x_i^{k_i}}{k_i!}.$$

Note that $N_i$ also equals the number of $U_{(i)}$'s that lie in $(u_{i-1}, u_i]$. Hence $U_{(i)} \le u_i$ if and only if $\sum_{i=1}^{i} N_j \ge i$. It follows

$$\begin{aligned}
& \Pr(U_{(i)} \le u_i \text{ for all } 1 \le i \le n) \\
= \; & \Pr(\sum_{i=1}^{i} N_j \ge i \text{ for all } 1 \le i \le n) \\
= \; & \sum_{\mathbf{k} \in K_n} \Pr((N_1, \ldots N_n, N_{n+1}) = (k_1, \ldots, k_n, 0)) \\
= \; & n! \sum_{\mathbf{k} \in K_n} \prod_{i=1}^{n} \frac{x_i^{k_i}}{k_i!}.
\end{aligned}$$

Combining the above equation with (13.8) we prove Theorem 13.4.3. ∎

From the above proof follow two more probabilistic interpretations of $V_n(\mathbf{x})$.

**Corollary 13.4.4** *Let $(N_i, 1 \le i \le n+1)$ be a random vector with multinomial distribution with parameters $n$ and $(p_1, \ldots, p_{n+1})$, as if $N_i$ is the number of times $i$ appears in a sequence of $n$ independent trials with probability $p_i$ of getting $i$ on each trial for $1 \le i \le n+1$, where $\sum_{i=1}^{n+1} p_i = 1$. Then*

$$\Pr\left(\sum_{j=1}^{i} N_j \ge i \textit{ for all } 1 \le i \le n\right) = n! V_n(p_1, p_2, \ldots, p_n),$$

*and*

$$\Pr\left(\sum_{j=1}^{i} N_j < i \textit{ for all } 1 \le i \le n\right) = n! V_n(p_{n+1}, p_n, \ldots, p_2).$$

The relation between $V_n(\mathbf{x})$ and the generalized parking functions is given by the following theorem.

**Theorem 13.4.5** *Assume that $\mathbf{x} = (x_1, \ldots, x_n) \in \mathbb{N}^n$. Let $u_j = \sum_{i=1}^{j} x_i$, i.e., $\mathbf{x} = \Delta(\mathbf{u})$. Then*

$$PK_n(\mathbf{u}) = \sum_{\mathbf{a} \in \mathscr{P}\mathscr{K}_n} x_{a_1+1} \cdots x_{a_n+1} = n! V_n(\mathbf{x}). \tag{13.11}$$

*Proof.* Given an ordinary parking function $\mathbf{a} = (a_1, \ldots, a_n)$, replace each $i-1$ by an integer in the set $\{x_1 + \cdots + x_{i-1}, \ldots, x_1 + \cdots + x_i - 1\}$. The number of ways to do this is given in the middle expression in (13.11), and every $\mathbf{u}$-parking function is obtained exactly once in this way. This gives the first equality. The second equality follows from Theorem 13.4.3 and 13.2.11, as every ordinary parking function $\mathbf{a}$ is obtained by a balanced vector $\mathbf{k} \in K_n$ which determines that the terms of $\mathbf{a}$ contain exactly $k_i$ $(i-1)$'s, and a compatible permutation that determines how the terms are arranged. Given a balanced vector $\mathbf{k} \in K_n$, it is clear that there are $\binom{n}{k_1, \ldots, k_n}$ permutations that are compatible with $\mathbf{k}$. ∎

In particular, Formula (13.10) gives an explicit way to compute $PK_n(\mathbf{u})$, which is much easier than the determinant formula in Theorem 13.4.1.

An interesting special case of Theorem 13.4.5 is when we take $x_i = q^{i-1}$. In this case we have

$$n!V_n(1,q,q^2,\dots,q^{n-1}) = \sum_{\alpha \in PK_n} q^{a_1+\cdots+a_n} = q^{\binom{n}{2}} P_n(1/q) = q^{\binom{n}{2}} I_n(1/q),$$

where $P_n(q)$ is the displacement enumerator of ordinary parking functions, and $I_n(q)$ is the inversion enumerator of labeled trees.

The polytope $\Pi_n(\mathbf{x})$ has many interesting properties. For example, it admits a subdivision into a collection of $n$-dimensional chambers, with the volume of each chamber corresponding to a term of the volume polynomial. It also relates to plane partitions, rooted binary trees, and another polytope called **associahedron**. More combinatorial properties of $\Pi_i(\mathbf{x})$ are discussed in [67].

### 13.4.3 Theory of Gončarov polynomials

In this section we describe a polynomial sequence that forms a natural basis for working with **u**-parking functions. The involved polynomials are called Gončarov polynomials, which arose from the Gončarov interpolation problem in numerical analysis.

> [**Gončarov Interpolation**] Given two sequences of real or complex numbers $a_0, a_1, \dots, a_n$ and $b_0, b_1, \dots, b_n$, find a polynomial $p(x)$ of degree $n$ such that for each $i, 0 \le i \le n$, the $i$th derivative $p^{(i)}(x)$ evaluated at $a_i$ equals $b_i$.

Gončarov polynomials are the basis of solutions to the Gončarov interpolation problem and correspond to the cases that all $b_i$ but one are zero. To state their properties, we start with a brief introduction of the theory of sequences of polynomial biorthogonal to a sequence of linear functionals. The details can be found in [60].

Let $\mathscr{P}$ be the vector space of all polynomials in the variable $x$ over a field $F$ of characteristic zero. Let $D : \mathscr{P} \to \mathscr{P}$ be the differentiation operator. For a scalar $a$ in the field $F$, let

$$\varepsilon(a) : \mathscr{P} \to F,\ p(x) \mapsto p(a)$$

be the linear functional that evaluates $p(x)$ at $a$.

Let $\varphi_s(D), s = 0,1,2,\dots$ be a sequence of linear operators on $\mathscr{P}$ of the form

$$\varphi_s(D) = D^s \sum_{r=0}^{\infty} b_{sr} D^r,$$

where the coefficients $b_{s0}$ are assumed to be non-zero. There exists a unique sequence $p_n(x), n = 0,1,2,\dots$ of polynomials such that $p_n(x)$ has degree $n$ and

$$\varepsilon(0)\varphi_s(D)p_n(x) = n!\delta_{sn},$$

where $\delta_{sn}$ is the Kronecker delta.

The polynomial sequence $p_n(x)$ is said to be **biorthogonal** to the sequence $\varphi_s(D)$ of operators, or, the sequence $\varepsilon(0)\varphi_s(D)$ of linear functionals. Using Cramer's rule to solve the linear system and Laplace's expansion to group the results, we can express $p_n(x)$ by the the following **determinantal formula**:

$$p_n(x) = \frac{n!}{b_{00}b_{10}\cdots b_{n0}} \begin{vmatrix} b_{00} & b_{01} & b_{02} & \cdots & b_{0,n-1} & b_{0n} \\ 0 & b_{10} & b_{11} & \cdots & b_{1,n-2} & b_{1,n-1} \\ 0 & 0 & b_{20} & \cdots & b_{2,n-3} & b_{2,n-2} \\ \vdots & \vdots & \vdots & \ddots & \vdots & \vdots \\ 0 & 0 & 0 & \cdots & b_{n-1,0} & b_{n-1,1} \\ 1 & x & x^2/2! & \cdots & x^{n-1}/(n-1)! & x^n/n! \end{vmatrix}.$$

Since $\{p_n(x)\}_{n=0}^{\infty}$ forms a basis of $\mathscr{P}$, any polynomial can be uniquely expressed as a linear combination of $p_n(x)$'s. Explicitly, we have the **expansion formula**: If $p(x)$ is a polynomial of degree $n$, then

$$p(x) = \sum_{i=0}^{n} \frac{d_i p_i(x)}{i!},$$

where $d_i = \varepsilon(0)\varphi_i(D)p(x)$. In particular,

$$x^n = \sum_{i=0}^{n} \frac{n! b_{i,n-i} p_i(x)}{i!},$$

which gives a **linear recursion** for $p_n(x)$. Equivalently, one can write the above equation in terms of formal power series equations, and obtain the **Appell relation**

$$e^{xt} = \sum_{n=0}^{\infty} \frac{p_n(x)\varphi_n(t)}{n!},$$

where $\varphi_n(t) = t^s \sum_{r=0}^{\infty} b_{sr} t^r$.

A special example of sequences of biorthogonal polynomials is the Gončarov polynomials. Let $(a_0, a_1, a_2, \ldots)$ be a sequence of numbers or variables called *nodes*. The sequence of **Gončarov polynomials**

$$g_n(x; a_0, a_1, \ldots, a_{n-1}),\ n = 0, 1, 2, \ldots$$

is the sequence of polynomials biorthogonal to the operators

$$\varphi_s(D) = D^s \sum_{r=0}^{\infty} \frac{a_s^r D^r}{r!} = \varepsilon(a_s) D^s.$$

As indicated by the notation, $g_n(x; a_0, a_1, \ldots, a_{n-1})$ depends only on the nodes $a_0, a_1, \ldots, a_{n-1}$. In particular, when all the $a_i$ equal $a$, we have

$$g_n(x; a, a, \ldots, a) = (x-a)^n$$

and Gončarov interpolation is just expansion as a power series at $x = a$. When $a_0, a_1, a_2, \ldots$ form an arithmetic progression $a, a+b, a+2b, \ldots$, we get Abel polynomials

$$g_n(x; a, a+b, a+2b, \ldots, a+(n-1)b) = (x-a)(x-a-nb)^{n-1}.$$

Gončarov polynomials have many nice algebraic and analytic properties, which make them very useful in analysis and combinatorics. Next we list some basic properties, of which the first four follow from the theory of biorthogonal polynomials.

**Theorem 13.4.6** *The Gončarov polynomials $g_n(x; a_0, a_1, \ldots, a_{n-1})$ have the following properties.*

1. *Determinant formula.*

$$g_n(x; a_0, a_1, \ldots, a_{n-1}) = n! \begin{vmatrix} 1 & a_0 & \frac{a_0^2}{2!} & \frac{a_0^3}{3!} & \cdots & \frac{a_0^{n-1}}{(n-1)!} & \frac{a_0^n}{n!} \\ 0 & 1 & a_1 & \frac{a_1^2}{2!} & \cdots & \frac{a_1^{n-2}}{(n-2)!} & \frac{a_1^{n-1}}{(n-1)!} \\ 0 & 0 & 1 & a_2 & \cdots & \frac{a_2^{n-3}}{(n-3)!} & \frac{a_2^{n-2}}{(n-2)!} \\ \vdots & \vdots & \vdots & \vdots & \ddots & \vdots & \vdots \\ 0 & 0 & 0 & 0 & \cdots & 1 & a_{n-1} \\ 1 & x & \frac{x^2}{2!} & \frac{x^3}{3!} & \cdots & \frac{x^{n-1}}{(n-1)!} & \frac{x^n}{n!} \end{vmatrix}.$$

2. *Expansion formula. If $p(x)$ is a polynomial of degree $n$, then*

$$p(x) = \sum_{i=0}^{n} \frac{\varepsilon(a_i)D^i p(x)}{i!} g_i(x; a_0, a_1, \ldots, a_{i-1}).$$

3. *Linear recurrence. Let $p(x) = x^n$ in the expansion formula, we have*

$$x^n = \sum_{i=0}^{n} \binom{n}{i} a_i^{n-i} g_i(x; a_0, a_1, \ldots, a_{i-1}). \tag{13.12}$$

4. *Appell relation.*

$$e^{xt} = \sum_{n=0}^{\infty} g_n(x; a_0, a_1, \ldots, a_{n-1}) \frac{t^n e^{a_n t}}{n!}.$$

5. *Differential relations. The Gončarov polynomials can be equivalently defined by the differential relations*

$$Dg_n(x; a_0, a_1, \ldots, a_{n-1}) = ng_{n-1}(x; a_1, a_2, \ldots, a_{n-1}),$$

*with initial conditions*

$$g_n(a_0; a_0, a_1, \ldots, a_{n-1}) = \delta_{0,n}.$$

6. *Integral relations.*

$$\begin{aligned} g_n(x;a_0,a_1,\ldots,a_{n-1}) &= n\int_{a_0}^{x} g_{n-1}(t;a_1,a_2,\ldots,a_{n-1})dt \\ &= n!\int_{a_0}^{x} dt_1 \int_{a_1}^{t_1} dt_2 \cdots \int_{a_{n-1}}^{t_{n-1}} dt_n. \end{aligned}$$

7. *Shift invariance formula.*

$$g_n(x+\xi;a_0+\xi,a_1+\xi,\ldots,a_{n-1}+\xi) = g_n(x;a_0,a_1,\ldots,a_{n-1}).$$

8. *Perturbation formula.*

$$\begin{aligned} &g_n(x;a_0,\ldots,a_{m-1},a_m+b_m,a_{m+1},\ldots,a_{n-1}) \\ &\quad = g_n(x;a_0,\ldots a_{m-1},a_m,a_{m+1},\ldots,a_{n-1}) \\ &\quad - \binom{n}{m} g_{n-m}(a_m+b_m;a_m,a_{m+1},\ldots,a_{n-1})g_m(x;a_0,a_1,\ldots,a_{m-1}). \end{aligned}$$

9. *Sheffer relation.*

$$g_n(x+y;a_0,\ldots,a_{n-1}) = \sum_{i=0}^{n} \binom{n}{i} g_{n-i}(y;a_i,\ldots,a_{n-1})x^i.$$

*In particular,*

$$g_n(x;a_0,\ldots,a_{n-1}) = \sum_{i=0}^{n} \binom{n}{i} g_{n-i}(0,a_i,\ldots,a_{n-1})x^i.$$

*That is, coefficients of Gončarov polynomials are constant terms of (shifted) Gončarov polynomials.*

It turns out that **u**-parking functions provide a combinatorial interpretation of Gončarov polynomials.

**Theorem 13.4.7** *Let* $\mathbf{u} = (u_1,u_2,\ldots)$ *be a sequence of non-decreasing positive integers. Then we have*

$$\begin{aligned} PK_n(\mathbf{u}) = PK_n(u_1,u_2,\ldots,u_n) &= g_n(x;x-u_1,x-u_2,\ldots,x-u_n) \\ &= g_n(0;-u_1,-u_2,\ldots,-u_n) \\ &= (-1)^n g_n(0;u_1,u_2,\ldots,u_n). \end{aligned}$$

*Proof.* Let $\mathbf{x} = (x_1,x_2,\ldots,x_n)$ be an integer sequence whose terms satisfy $0 \le x_i < x$ for a positive integer $x$, and $(x_{(1)},x_{(2)},\ldots,x_{(n)})$ be the order statistics of $\mathbf{x}$. Let $m$ be the maximum index such that

$$x_{(i)} < u_i \quad \text{for} \quad i = 1,2,\ldots,m. \qquad (13.13)$$

Then, the subsequence $(x_{i_1}, x_{i_2}, \ldots, x_{i_m})$ from which the sequence $(x_{(1)}, x_{(2)}, \ldots, x_{(m)})$ was obtained by rearrangement is a **u**-parking function of length $m$. Furthermore, $m$ is the maximum index satisfying condition (13.13) if and only if

$$x_{(n)} \geq x_{(n-1)} \geq \cdots \geq x_{(m+1)} \geq u_{m+1}.$$

Equivalently, the complementary subsequence $(x_{j_1}, x_{j_2}, \ldots, x_{j_{n-m}})$, obtained by deleting the subsequence $(x_{i_1}, x_{i_2}, \ldots, x_{i_m})$ from the original sequence, takes values in the interval $[u_{m+1}, x-1]$. Since the maximum index $m$ and the set $\{i_1, i_2, \ldots, i_m\}$ are uniquely determined by the sequence $(x_1, x_2, \ldots, x_n)$, and any pair of subsequences satisfying the conditions in the theorem can be reassembled into a sequence in $[0, x-1]^n$, this decomposition yields a bijection, which leads to the equation

$$x^n = \sum_{m=0}^{n} \binom{n}{m} (x - u_{m+1})^{n-m} PK_m(u_1, u_2, \ldots, u_m). \tag{13.14}$$

Comparing the recursion (13.14) with the linear recursion (13.12) for Gončarov polynomials, we obtain

$$PK_n(u_1, u_2, \ldots, u_n) = g_n(x; x-u_1, x-u_2, \ldots, x-u_n).$$

By the shift invariance formula,

$$g_n(x; x-u_1, x-u_2, \ldots, x-u_n) = g_n(0; -u_1, -u_2, \ldots, -u_n).$$

Since the Gončarov polynomial $g_n(x; a_0, a_1, \ldots, a_{n-1})$ is a homogeneous polynomial of total degree $n$ in $x, a_0, a_1, \ldots, a_{n-1}$, we have

$$g_n(0; -u_1, -u_2, \ldots, -u_n) = (-1)^n g_n(0; u_1, u_2, \ldots, u_n).$$

■

Any reasonable formula for Gončarov polynomials yields a reasonable formula for **u**-parking functions. For example, when $u_i = a + (i-1)b$, we obtain the following special case.

**Corollary 13.4.8**

$$PK_n(a, a+b, a+2b, \ldots, a+(n-1)b) = a(a+nb)^{n-1}.$$

The homogeneity of Gončarov polynomials implies

**Corollary 13.4.9**

$$PK_n(bu_1, bu_2, \ldots, bu_n) = b^n PK_n(u_1, u_2, \ldots, u_n).$$

Using the determinant formula of Theorem 13.4.6, we get a proof of Theorem 13.4.1, the discrete analogue of Steck's formula. The two formulas of Theorem 13.4.2 can be proved by using the perturbation formula of Gončarov polynomials. In addition, one could compute the sum enumerator of **u**-parking functions by applying the properties of Gončarov polynomials and the decomposition of Theorem 13.4.7.

The **sum enumerator** $S_n(q;\mathbf{u})$ of the set of **u**-parking functions is the polynomial in $q$ defined by

$$S_n(q;\mathbf{u}) = \sum_{(a_1,a_2,\ldots,a_n)} q^{a_1+a_2+\ldots+a_n}$$

where the sum ranges over all **u**-parking functions $(a_1,a_2,\ldots,a_n)$. The sum enumerator may be regarded as a "$q$-analogue" of $PK_n(\mathbf{u})$.

**Theorem 13.4.10** *The sum enumerator* $S_n(q;\mathbf{u})$ *satisfies the equation*

$$(1+q+q^2+\ldots+q^{x-1})^n = \sum_{m=0}^{n}\binom{n}{m}(q^{u_{m+1}}+q^{u_{m+1}+1}+\ldots+q^{x-1})^{n-m}S_m(q;\mathbf{u}).$$

*Proof.* Since sum enumerators are multiplicative, the sum enumerator of integer sequences $(x_1,\ldots,x_n)$ with $0 \le x_i < x$ is

$$(1+q+q^2+\ldots+q^{x-1})^n.$$

For the same reason, the sum enumerator of sequences that are decomposed into a **u**-parking function of length $m$ and a sequence in $[u_{m+1},x-1]^{n-m}$ is

$$(q^{u_{m+1}}+q^{u_{m+1}+1}+\ldots+q^{x-1})^{n-m}S_m(q,\mathbf{u}).$$

The recursion now follows. ■

Comparing the recursion in Theorem 13.4.10 with the linear recursion (13.12), we obtain

$$S_n(q;\mathbf{u}) = PK_n(1+q+\ldots+q^{u_1-1},1+q+\ldots+q^{u_2-1},\ldots,1+q+\ldots+q^{u_n-1}).$$

Using Theorem 13.4.7 and the shift invariance formula, we can express sum enumerators in terms of Gončarov polynomials, as

$$S_n(q;\mathbf{u}) = g_n\left(\frac{1}{1-q};\frac{q^{u_1}}{1-q},\frac{q^{u_2}}{1-q},\ldots,\frac{q^{u_n}}{1-q}\right).$$

By homogeneity of Gončarov polynomials,

$$(1-q)^n S_n(q;\mathbf{u}) = g_n(1;q^{u_1},q^{u_2},\ldots,q^{u_n}).$$

Hence, sum enumerators satisfy the simpler linear recursion

$$1 = \sum_{m=0}^{n}\binom{n}{m}q^{u_{m+1}(n-m)}(1-q)^m S_m(q;\mathbf{u}). \tag{13.15}$$

They also satisfy the following Appell relation

$$\exp(t) = \sum_{n=0}^{\infty}(1-q)^n S_n(q;\mathbf{u})\exp(q^{u_{n+1}}t)\frac{t^n}{n!}.$$

In the case of ordinary parking functions with $u_i = i$, we have

$$(1-q)^n S_n(q;1,2,\ldots,n) = g_n(1;q,q^2,\ldots,q^n),$$

and the sum enumerator is related to the displacement enumerator of ordinary parking functions by the equation

$$S_n(q;1,2,\ldots,n) = q^{\binom{n}{2}} P_n(1/q).$$

Hence (13.15) leads to a linear recurrence for $P_n(q)$:

$$1 = \sum_{m=0}^{n} q^{\binom{m+1}{2}-(m+1)n} \cdot (q-1)^m P_m(q).$$

More properties and computation of parking functions via Gončarov polynomials are given in [59, 60]. For example, [60] gives the generating functions for factorial moments of sums of **u**-parking functions, while the explicit formulas for the first and second factorial moments of sums of **u**-parking functions are given in [59]. Khare, Lorentz and Yan [49] studied multivariate Gončarov polynomials, and extended many algebraic and analytic properties of Gončarov polynomials to the multivariate case. They also established a connection between multivariate Gončarov polynomials and order statistics of integer sequences, which leads to a higher dimensional generalization of parking functions.

If one changes the differential operator $D$ to the backward different operator $\Delta$ that maps a polynomial $p(x)$ to $\Delta p(x) = p(x) - p(x-1)$, then one gets the **difference Gončarov polynomials**, whose combinatorial counterpart is the set of lattice paths with a given right boundary. Theory of difference Gončarov polynomials is presented in [57].

### 13.4.4 Classical parking functions

A special class of **u**-parking functions is the classical parking functions, for which the entries of the vector **u** form an arithmetic progression. In particular, if $u_i = a + (i-1)b$ for some positive integers $a$ and $b$, we will call the **u**-parking functions $(a,b)$**-parking functions**. Classical parking functions have a rich theory. Many combinatorial representations of ordinary parking functions can be generalized to this case, much in the same way as theory of Catalan structures (structures counted by the Catalan numbers $\frac{1}{n+1}\binom{2n}{n}$ ) generalizes to that of the Fuss-Catalan structures (structures counted by the Fuss-Catalan numbers, or $k$-Catalan numbers $\frac{1}{kn+1}\binom{nk+1}{n}$). In this section we give a brief summary of such generalizations.

**Basic enumeration.** It is well-known that the number of $(a,b)$-parking functions is $a(a+nb)^{n-1}$.

Let $E_k(n;a,b)$ be the expected value of the $k$th factorial moment of the sum of the terms in a uniform random $(a,b)$-parking function, that is,

$$E_k(n;a,b) = \frac{1}{a(a+nb)^{n-1}} \sum_{(x_1,x_2,\ldots,x_n)} (x_1+x_2+\ldots+x_n)_k,$$

where $(n)_k$ is the falling factorial $n(n-1)\cdots(n-k+1)$ and the sum ranges over all $(a,b)$-parking functions $(x_1,x_2,\ldots,x_n)$. Explicit formulas for $E_1(n;a,b)$ and $E_2(n;a,b)$ are computed by Kung and Yan in [58, 59].

**Theorem 13.4.11** *The expected sums of terms in a random $(a,b)$-parking function of length n is*

$$E_1(n;a,b) = \frac{n(a+nb+1)}{2} - \frac{1}{2}\sum_{j=1}^{n}\binom{n}{j}\frac{j!b^j}{(a+nb)^{j-1}}.$$

**Theorem 13.4.12** *The second factorial moment of the sum of terms of a random $(a,b)$-parking function of length n is*

$$\frac{1}{4}n(n-1)(a+nb+1)^2 + \frac{1}{3}n(a+nb+1)(a+nb-1)$$
$$-\frac{n(a+nb+1)}{2}\sum_{j=1}^{n}\binom{n}{j}\frac{j!b^j}{(a+nb)^{j-1}} + \sum_{j=1}^{n}\binom{n}{j}\frac{j!b^j}{(a+nb)^{j-1}}\left(\frac{b}{6}j^3 - \frac{a}{6}j + \frac{1}{2}\right).$$

Kung and Yan also presented a general form for the higher moments of the expected sum of $(a,b)$-parking functions, see Theorem 1.1 of [58].

**Rooted $b$-forests.** The correspondence between ordinary parking functions and labeled trees can be extended to $(a,b)$-parking functions, where the labeled trees are replaces by forests of labeled trees with edge colors. Precisely, define a **rooted $b$-forest on** $[n]$ to be a rooted forest on the vertex set $[n]$ whose edges are colored with the colors $0,1,\ldots,b-1$. There is no further restriction on the possible coloring of the edges. Let $\mathscr{F}_n(a,b)$ be the set of all sequences $(T_1,T_2,\ldots,T_a)$ of length $a$ such that

(1) each $T_i$ is a rooted $b$ -forest,

(2) $T_i$ and $T_j$ are disjoint if $i \neq j$, and

(3) the union of the vertex sets of $T_1,\ldots,T_a$ is $[n]$.

Denote by $\mathscr{PK}_n((a,b))$ the set of $(a,b)$-parking functions of length $n$. Then there is a one-to-one correspondence between $\mathscr{PK}_n((a,b))$ and $\mathscr{F}_n(a,b)$. In fact, both of them can be mapped bijectively to the set $\mathscr{C}_n(a,b)$ that consists of pairs $(\vec{r},\sigma)$ such that

(1) $\vec{r} = (r_0,r_1,\ldots,r_{a+(n-1)b-1}) \in \mathbb{N}^{a+(n-1)b}$ is $(a,b)$-balanced, that is,

$$r_0+r_1+\cdots+r_{a+ib}-1 \geq i+1 \quad \text{for } i=0,1,\ldots,n-2,$$
$$r_0+r_1+\cdots+r_{a+(n-1)b-1} = n.$$

(2) $\sigma \in \mathfrak{S}_n$ is compatible with $\vec{r}$, that is, the terms in the inverse $\sigma^{-1}$ of $\sigma$ is increasing on every interval of the indices $\{1+\sum_{i=1}^{k} r_i, 2+\sum_{i=1}^{k} r_i, \ldots, \sum_{i=1}^{k+1} r_i\}$ (if $r_{k+1} \neq 0$).

These results were proved in [87] by using the breadth-first search on rooted $b$-forests, which generalize Theorems 13.2.11 and 13.2.12. Eu, Fu and Lai [27, 28] applied the construction of [87] to enumerate $(a,b)$-parking functions by their leading terms, or with certain symmetric restrictions and periodic restrictions.

**Inversions of $b$-forests and multicolored graphs.** The notion of inversions can be extended to the set $\mathscr{F}_n(a,b)$ as follows: Let $F = (T_1, T_2, \ldots, T_a)$ be a sequence of rooted $b$-forests on $[n]$. Denote the color of an edge $e$ by $\kappa(e)$. Define the $(a,b)$-inversion $\mathrm{inv}^{(a,b)}(F)$ by letting

$$\mathrm{inv}^{(a,b)}(F) = \mathrm{inv}(F) + \sum_{i=1}^{a}(i-1)|T_i| + \sum_{x\in[n]}\sum_{e\in K(x)} \kappa(e),$$

where $\mathrm{inv}(F)$ is the number of inversions of $T_1 \cup T_2 \cdots \cup T_a$ as an ordinary rooted forest, $K(x)$ is the set of edges lying between the vertex $x$ and the root of the unique tree to which $x$ belongs. Define the *$(a,b)$-inversion enumerator* $I_n^{(a,b)}(q)$ by

$$I_n^{(a,b)}(q) = \sum_{F\in\mathscr{F}_n(a,b)} q^{\mathrm{inv}^{(a,b)}(F)}.$$

For $a = 1$, $I_n^{(a,b)}(q) = I_n^{(1,b)}(q)$ is the $b$-inversion enumerator studied in [81, 85], and $I_n^{(1,1)}(q) = I_n(q)$ is the ordinary inversion enumerator of labeled trees.

For an $(a,b)$-parking function $\mathbf{a} = (a_1, \ldots, a_n)$, the $(a,b)$-**displacement** $D^{(a,b)}(\mathbf{a})$ is defined as

$$D^{(a,b)}(\mathbf{a}) = b\binom{n}{2} + an - \sum_i a_i,$$

and the $(a,b)$-**displacement enumerator** is

$$P_n^{(a,b)}(q) = \sum_{\mathbf{a}\in\mathscr{PK}_n((a,b))} q^{D^{(a,b)}(\mathbf{a})}.$$

Then we have

**Theorem 13.4.13**

$$I_n^{(a,b)}(q) = P_n^{(a,b)}(q).$$

Both of the above polynomials are related to another polynomial $C_n^{(a,b)}(q)$ that counts the number of edges in connected multigraphs. To wit, define a **multicolored** $(a,b)$-**graph** on $[n]$ to be a graph $G$ with the vertex set $[n]$ such that

(1) The edges of $G$ are colored with colors $0, 1, \ldots, b-1$,

(2) There are no loops or multiple edges of the same color in $G$. But $G$ may have edges with the same endpoints but different colors, and

(3) every vertex $r$ is assigned with a subset $f(r)$ of $[a] = \{1, 2, \ldots, a\}$. We say that $r$ is a **root** of $G$ if $f(r) \neq \emptyset$.

(4) For any subgraph $H$ of $G$, define $R(H) = \sum_{r\in H}|f(r)|$ to be the number of roots in $H$, counting multiplicity. Every connected component $G'$ of $G$ has at least one root, i.e., $R(G') > 0$.

Denote by $e(G)$ the number of edges of $G$, and by $r(G) = \sum_r |f(r)|$ the number of roots of $G$. Let

$$C_n^{(a,b)}(q) = \sum_G q^{e(G)+r(G)-n}$$

where $G$ ranges over all multicolored $(a,b)$-graphs on $[n]$. Set $C_0^{(a,b)}(q) = 1$ for all $a,b \in \mathbb{N}$. Then

**Theorem 13.4.14**

$$I_n^{(a,b)}(1+q) = P_n^{(a,b)}(1+q) = C_n^{(a,b)}(q).$$

Theorem 13.4.14 is proved in [87] by applying the DFS and BFS to multicolored graphs. The case that $a = 1$ was proved earlier in [81, 85] by recurrence relations. In [75] Shin and Zeng constructed a bijective proof for the case that $a$ is a multiple of $b$, which gives a refinement of Theorem 13.4.13 in that case.

**$b$-divisible noncrossing partitions.** Theorems 13.3.2 and 13.3.4 have been generalized by Stanley [79] to the lattice of $b$-divisible noncrossing partitions. Fix an positive integer $b$. A **$b$-divisible noncrossing partition** is a noncrossing partition $\pi$ for which every block size is divisible by $b$. Thus $\pi$ is a noncrossing partition for a set $[bn]$ for some $n \geq 0$. Let $NC_n^{(b)}$ be the poset of all $b$-divisible noncrossing partitions of $[bn]$. The combinatorial properties of the poset $NC_n^{(b)}$ were first studied by Edelman [26]. In particular, if a pair $(\pi,\pi')$ is an edge of $NC_n^{(b)}$, then $(\pi,\pi')$ is an edge of $NC_{bn}$. Hence the edge labeling $\Lambda$ and $\Lambda_1$ of $NC_n$ defined in Section 13.3.2 restrict to edge-labelings of $NC_n^{(b)}$.

Theorem 13.3.2 is generalized as follows.

**Theorem 13.4.15** *The label $\Lambda_1(\mathfrak{m})$ of the maximal chains of $\mathrm{NC}_{n+1}^{(b)}$ consists of the $(b,b)$-parking functions of length $n$, each occurring once.*

The posets $NC_n^{(b)}$ do not have a $\hat{0}$ when $b > 1$. For these posets one regards the minimal elements as having rank 0, and defines $\alpha_{NC_{n+1}^{(b)}}(S)$ and $\beta_{NC_{n+1}^{(b)}}(S)$ for $S \subseteq [n-1]_0$. Theorem 13.3.4 is generalized as follows.

**Theorem 13.4.16** *Let $S \subseteq [n-1]$.*

1. *The number of $[b,b]$-parking functions $\mathbf{a}$ of length $n$ with $\mathrm{Des}(\mathbf{a}) = S$ is equal to*

$$\beta_{\mathrm{NC}_{n+1}^{(b)}}([n-1]-S) + \beta_{\mathrm{NC}_{n+1}^{(b)}}([n-1]_0 - S).$$

2. *The number of $[b,b]$-parking functions $\mathbf{a}$ of length $n$ satisfying $S \subseteq \mathrm{Des}(\mathbf{a})$ is equal to*

$$\alpha_{\mathrm{NC}_{n+1}^{(b)}}([n-1]_0 - S).$$

**Extended Shi arrangements.** For $b \geq 1$ the **extended Shi arrangement** $\mathscr{S}_n^b$ is the collection of hyperplanes

$$x_i - x_j = -b+1, -b+2, \ldots, b, \text{ for } 1 \leq i < j \leq n.$$

The labeling of the regions of the Shi arrangement defined in Section 13.3.3 can be extended to this case, as described by Stanley [81].

Define the base region $R_0$ of $\mathscr{S}_n^b$ by

$$R_0 : x_1 > x_2 > \cdots > x_n > x_1 - 1.$$

First label the region $R_0$ by $(0,0,\ldots,0) \in \mathbb{N}^n$. Suppose now that $R$ has been labeled, and that $R'$ is an unlabeled region that is separated from $R$ by a unique hyperplane $H : x_i - x_j = m$, where $i < j$. Then define

$$\lambda(R') = \begin{cases} \lambda(R) + e_i, & \text{If } H \text{ is given by } x_i - x_j = m \text{ with } i < j \text{ and } m > 0 \\ \lambda(R) + e_j, & \text{if } H \text{ is given by } x_i - x_j = m \text{ with } i < j \text{ and } m \leq 0. \end{cases}$$

Then

**Theorem 13.4.17** *The labels $\lambda(R)$ of the extended Shi arrangement $\mathscr{S}_n^{(b)}$ are just the $(1,b)$-parking functions of length n, each occurring exactly once. In addition, the distance enumerator of the extended Shi arrangement is equal to the sum enumerator of $(1,b)$-parking functions.*

## 13.5 Parking functions associated with graphs

### 13.5.1 *G*-parking functions

In 2004 Postnikov and Shapiro [68] proposed a new generalization of parking functions, namely, the $G$-parking functions associated with a general connected digraph $G$. Let $G$ be a directed graph with the vertex set $[n]_0$, where multiple edges and loops are allowed. We will view the vertex 0 as the **root**. As usual in graph theory, a directed edge is represented by a pair $(i,j)$ of vertices, where $i$ is the tail of the edge and $j$ is the head of the edge. For a vertex $i$, the **indegree** $\text{indeg}(i)$ is the number of edges with tail $i$, and the **outdegree** $\text{outdeg}(i)$ is the number of edges with head $i$. In addition, for any subset $U \subseteq [n]$ and vertex $i \in U$, we define $\text{outdeg}_U(i)$ to be the cardinality of the set $\{(i,j) \in E(G) \mid j \notin U\}$, where $E(G)$ is the set of edges of $G$.

**Definition 13.5.1** *A $G$-parking function is a function $f$ from $[n]$ to the set of nonnegative integers $\mathbb{N}$ satisfying the following condition: For each subset $U \subseteq [n]$ of vertices of $G$, there exists a vertex $i \in U$ such that $f(i) < \text{outdeg}_U(i)$.*

For the complete graph $K_{n+1}$ on $[n]_0$, such defined functions are exactly the ordinary parking functions if we view $K_{n+1}$ as the digraph with one directed edge $(i,j)$ for each $i \neq j$, and record the function $f$ as a sequence $(f(1), f(2), \ldots, f(n))$.

In a digraph $G$ an **oriented spanning tree** $T$ of $G$ is a subgraph $T$ in which (1) the root 0 has outdegree 0, (2) all other vertices have outdegree 1, and (3) there exists a unique directed path from any vertex $i$ to the root 0. The number of oriented spanning trees of $G$, denoted $\kappa(G)$, is sometimes called the **complexity** of $G$. The value of $\kappa(G)$ can be computed by the Matrix-Tree Theorem; see [78, Chap 5.6].

The first important result of $G$-parking functions is the following theorem that extends the relation between ordinary parking functions and labeled trees.

**Theorem 13.5.2** *The number of G-parking functions equals the number of oriented spanning trees of the digraph G.*

The motivation for Postnikov and Shapiro to work with $G$-parking functions is to study a general class of algebras formed by taking the quotient of the polynomial ring modulo monotone monomial ideals and their deformations. As a special example, for any digraph $G$ they define two algebras $\mathscr{A}_G$ and $\mathscr{B}_G$ and describe their monomial bases. The basis elements correspond to $G$-parking functions. Then they proved Theorem 13.5.2 by applying general formulas for the Hilbert series and dimensions of the algebras given by a monotone monomial ideal.

Shortly after [68] appeared, Chebikin and Pylyavskyy [21] constructed a family of bijections between the set of $G$-parking functions and the set of oriented spanning trees of the digraph $G$. Their bijections were built on "different total orders" on the vertices of subtrees of $G$, similar to the "selection procedure" of Françon's [31] in constructing bijections between ordinary parking functions and rooted labeled trees. We will describe these bijections in a slightly more general setting in Section 13.5.3.

### 13.5.2 Abelian sandpile model

There is an interesting relation between $G$-parking functions and the **abelian sandpile model**, one of the archetype models of self-organized criticality in physical systems. The sandpile model was originated from an automaton model by Bak, Tang, and Wiesenfeld [7] for evolutions of dynamical systems on $\mathbb{Z}^2$, and was first formulated by Dhar [25] on any finite connected graphs. The model was also considered by combinatorialists as a game on a graph called the **chip firing game** or the **dollar game**, e.g. [12, 14, 15].

To show its connection to Tutte polynomials, we describe a simple version of the sandpile model for undirected graphs with possible multiple edges but no loops. A review of the more general theory with a class of toppling matrices, as well as the history and references on the subject can be found in the appendix of [68].

Let $G$ be a connected graph on the vertex set $[n]_0$, where the vertex 0 is the root. A configuration (sandpile) is a collection of indistinguishable chips distributed among the vertices of $G$. More precisely, it is a function $u$ from $[n]$ to $\mathbb{N}$ indicating how many chips are at each vertex $i \neq 0$. A non-root vertex $i$ is called **unstable** if it has at least as many chips as its degree, and an unstable vertex can **topple** by sending chips to adjacent vertices, one along each incident edge. In formula, toppling at vertex $i$

changes a configuration $u$ to $u'$ by

$$u'(j) = \begin{cases} u(i) - \deg(i) & \text{if } j = i, \\ u(j) + e(i,j) & \text{otherwise,} \end{cases}$$

for $j \neq 0$, where $e(i,j)$ is the number of edges between $i$ and $j$.

In a configuration toppling one vertex may cause neighboring vertices to become unstable. A configuration is said to be **stable** if all the non-root vertices are stable. It is easy to see that for a connected graph $G$ any initial configuration can be transformed into a stable configuration by a sequence of topplings at non-root vertices. Moreover, the final stable configuration does not depend on the order of the topplings, leading to the name "abelian sandpile model."

The root 0 plays a special role, for which we set that $u(0) = -\sum_{i \neq 0} u(i)$. For a stable configuration $u$, we can let the root topple by increasing the chips at any neighbor $i$ of 0 by $e(0,i)$, where $e(0,i)$ is the number of edges connecting 0 to $i$, and deceasing the "chips" at the root by $\deg(0)$. By allowing the root to topple for stable configurations, the sandpile model can continue indefinitely, and produce an infinite number of stable configurations. A configuration is **recurrent** in an evolving system if it could be observed after a long period of the evolution of the system. For abelian sandpile model, the system is considered to evolve by adding some chips at a random vertex and then applying the toppling rules. This leads to the following definition.

**Definition 13.5.3** *A configuration $u$ is* **recurrent** *if it is stable and there exists a positive configuration $v \neq 0$ such that $u$ can be obtained by a sequence of topplings from $u + v$.*

Recurrent configurations are also called **critical configurations** in [12]. There are many characterizations of recurrent configurations, see, for example, [25, 12, 24, 22]. The following explicit characterization is due to Dhar. Let us say that a configuration $u$ is **allowed** if for any non-empty subset $I \subseteq [n]$ there exists $j \in I$ such that

$$u(j) \geq \sum_{i \in I \setminus \{j\}} e(i,j).$$

**Theorem 13.5.4** *A configuration is recurrent if and only if it is stable and allowed.*

Comparing Theorem 13.5.4 and the definition of $G$-parking functions, we see that recurrent configurations and $G$-parking functions are complementary to each other, as pointed out in [68]. The special case for complete graph $K_{n+1}$ was proved earlier by Cori and Rossin [24].

**Theorem 13.5.5** *Let $G$ be a connected graph on $[n]_0$. A configuration $u : [n] \to \mathbb{N}$ is recurrent if and only only if the function $u^\vee$ given by $u^\vee(i) = \deg(i) - 1 - u(i)$ is a $G$-parking function.*

Dhar described a simple recursive procedure, the **burning algorithm**, to determine if a given stable configuration is allowed. Translated to $G$-parking functions, it leads to a linear algorithm that tests whether a given function $u : [n] \to \mathbb{N}$ is a $G$-parking function: First we mark the root 0. At each iteration of the algorithm, we mark all vertices $i$ that have more marked neighbors than the value $u(i)$. Then $u$ is a $G$-parking function if and only if all the vertices are marked in the end.

The following example shows how the algorithm works. In Figure 13.8 there are two vertex-functions on the complete graph $K_5$, where each vertex $i \neq 0$ is labeled by $i/f(i)$, i.e., the first label is the index of the vertex, and the second label is the function value. The first one shows a $G$-parking function, as one can mark all the vertices in the order $0,1,4,2,3$, while the second is not: Once one marks vertices $0,1,4$, no more vertices can be marked.

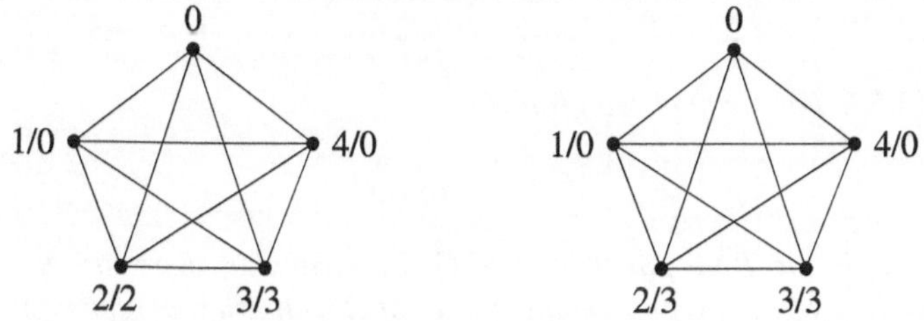


**Figure 13.8**
Dhar's burning algorithm.

The set of recurrent configurations of a sandpile model on $G$ forms an abelian group whose order is equal to the number of spanning trees of the graph. In particular, let the **level** of a recurrent configuration $u$ be given by

$$\text{level}(u) = \sum_{i\in[n]} u_i + \deg(0) - |E(G)|.$$

Written in terms of $G$-parking function, the statistic level$(u)$ induces a **weight** of $u^\vee$ as

$$\text{wt}(u^\vee) = |E(G)| - n - \sum_{i=1}^{n} u^\vee(i).$$

Let $P(G,y)$ be the level enumerator of the recurrent configurations of the sandpile model of $G$, or equivalently, the weight enumerator of $G$-parking functions, in formula,

$$P(G,y) = \sum_{u} y^{\text{level}(u)} = \sum_{u^\vee} y^{\text{wt}(u^\vee)},$$

where $u$ ranges over all recurrent configurations. The polynomial $P(G,y)$ is the natural generalization of the displacement enumerator of the ordinary parking functions, and is a specialization of the Tutte polynomial $T_G(x,y)$, which we recall next.

Suppose we are given a connected graph $G$ and a total ordering of its edges. Consider a spanning tree $T$ of $G$. An edge $e \in G-T$ is **externally active** if it is the smallest edge in the unique cycle contained in $T \cup e$. An edge $e \in T$ is **internally active** if it is the smallest edges in the unique cocycle contained in $(G-T)\cup\{e\}$ (a cocycle is a minimal edge cut). We let $\mathrm{ea}(T)$ be the number of externally active edges of $T$, and $\mathrm{ia}(T)$ the number of internally active edges in $T$. Tutte [83] then defined his polynomial as

$$T_G(x,y) = \sum_{T \subseteq G} x^{\mathrm{ia}(T)} y^{\mathrm{ea}(T)},$$

where the sum is over all spanning trees $T$ of $G$. Tutte showed that $T_G(x,y)$ is well-defined, i.e., independent of the total ordering of the edges of $G$.

The next theorem reveals the relation between $P(G,y)$ and the Tutte polynomial $T_G(x,y)$.

**Theorem 13.5.6** *For a connected graph G,*

$$P(G,y) = T_G(1,y)$$

*where $T_G(x,y)$ is the Tutte polynomial of G. In particular, there are as many G-parking functions of weight i as spanning trees of G with external activity i.*

Theorem 13.5.6 was conjectured by Biggs and proved by Merino López [64] by using a recursive characterization of Tutte polynomials. Cori and Le Borgne [23] constructed a bijection between recurrent configurations and spanning trees of $G$, which carries external activities of the spanning trees to the levels of the recurrent configurations of the sandpile model.

### 13.5.3 Multiparking functions, graph searching, and the Tutte polynomial

To get a complete expression of the bivariate Tutte polynomial, Kostic and Yan [54] introduced the notion of $G$-multiparking functions, which are in one-to-one correspondence with spanning forests of the graph $G$. They described a family of bijections between the spanning forests of a graph $G$ and the $G$-multiparking functions, each of which corresponds to a graph searching algorithm. In particular, the bijection induced by the breadth-first search leads to a new characterization of external activity, and hence a representation of the bivariate Tutte polynomial by statistics of multiparking functions. This generalizes Theorems 13.2.6 and 13.5.6, as well as the correspondence between ordinary parking functions and labeled trees.

For simplicity and clarity, in the following we assume that $G$ is a simple connected graph on $[n]$. The treatment of general directed graphs can be found in [54].

**Definition 13.5.7** *Let G be a simple graph with $V(G)=[n]$. A* **G-multiparking function** *is a function $f: V(G)=[n] \to \mathbb{N}\cup\{\infty\}$, such that for every $U \subseteq V(G)$ either* **(A)** *i is the vertex of smallest index in U and $f(i)=\infty$, or* **(B)** *there exists a vertex $i \in U$ such that $0 \le f(v_i) < \mathrm{outdeg}_U(i)$.*

The vertices that satisfy $f(i) = \infty$ in **(A)** will be called **roots of** $f$ and those that satisfy **(B)** (in $U$) are said to be **well-behaved** in $U$. The $G$-multiparking functions with only one root (which is necessarily vertex 1) are equivalent to $G$-parking functions defined by Postnikov and Shapiro.

An easy way to check whether a function is a $G$-multiparking function is to modify Dhar's burning algorithm to allow the root vertices.

**Proposition 13.5.8** *A vertex function is a G-multiparking function if and only if there exists an ordering $\pi(1), \pi(2), \ldots, \pi(n)$ of the vertices of the graph G such that for every j, $\pi(j)$ satisfies either condition* **(A)** *or condition* **(B)** *in $U_j := \{\pi(j), \ldots, \pi(n)\}$.*

Let $\mathscr{MP}_G$ be the set of $G$-multiparking functions and $\mathscr{F}_G$ the set of spanning forests of $G$.

**Theorem 13.5.9** *The set $\mathscr{MP}_G$ is in one-to-one correspondence with the set $\mathscr{F}_G$.*

Again there are many ways to construct bijections from $\mathscr{MP}_G$ to $\mathscr{F}_G$. Extending the idea of Section 13.2.3, we introduce a family of such bijections, each determined by a choice function and corresponding to a searching algorithm on trees/forests.

For a graph $G$, a **sub-forest** $F$ of $G$ is a subgraph of $G$ without cycles. (Here we don't assume any root in $F$). A leaf of $F$ is a vertex $v \in V(F)$ of degree 1 in $F$. Denote the set of leaves of $F$ by Leaf(F). Let $\prod$ be the set of all ordered pairs $(F, W)$ such that $F$ is a sub-forest of $G$, and $\emptyset \neq W \subseteq \text{Leaf(F)}$. As in Section 13.2.3 define a **choice function** $\gamma$ as a function from $\prod$ to $V(G)$ such that $\gamma(F, W) \in W$.

It is easier to describe the bijection from $\mathscr{F}_G$ to $\mathscr{MP}_G$, hence we present it first in the following algorithm. Fix a choice function $\gamma$. Let $G$ be a graph on $[n]$ with a spanning forest $F$. Let $T_1, \ldots, T_k$ be the trees of $F$ with respective minimal vertices $r_1 = 1 < r_2 < \cdots < r_k$.

**Algorithm 13.5.10**

- **Step 1. Determine a processing order $\pi$.** *Define a permutation $\pi = (\pi(1), \pi(2), \ldots, \pi(n)) = (v_1 v_2 \ldots v_n)$ on the vertices of G as follows. First, $v_1 = 1$. Assuming $v_1, v_2, \ldots, v_i$ are determined,*
  - *If there is no edge of F connecting vertices in $V_i = \{v_1, v_2, \ldots, v_i\}$ to vertices outside $V_i$, let $v_{i+1}$ be the vertex of smallest index not already in $V_i$;*
  - *Otherwise, let $W = \{v \notin V_i : v \text{ is adjacent to some vertices in } V_i\}$, and $F'$ be the forest obtained by restricting F to $V_i \cup W$. Let $v_{i+1} = \gamma(F', W)$.*
- **Step 2. Define a G-multiparking function** $f = f_F$. *Set $f(r_1) = f(r_2) = \cdots = f(r_k) = \infty$. For any other vertex v, let $r_v$ be the minimal vertex in the tree containing v, and $v, v^p, u_1, \ldots, u_t, r_v$ be the unique path from v to $r_v$. (That is, $v^p = \text{pre}(v)$. ) Set $f(v)$ to be the cardinality of the set*

$$\{v_j | \{v, v_j\} \in E(G),\ \pi^{-1}(v_j) < \pi^{-1}(v^p)\}.$$

The inverse bijection is quite complicated, which is achieved by the following algorithm. Given a $G$-multiparking function $f \in \mathcal{MP}_G$, Algorithm 13.5.11 finds a spanning forest $F \in \mathcal{F}_G$. Explicitly, we define quadruples $(\mathrm{val}_i, P_i, Q_i, F_i)$ recursively for $i = 0, 1, \ldots, n$, where $\mathrm{val}_i : V(G) \to \mathbb{Z}$ is the *value function*, $P_i$ is the set of **processed** vertices, $Q_i$ is the set of vertices *to be processed*, and $F_i$ is a sub-forest of $G$ with $V(F_i) = P_i \cup Q_i$, $Q_i \subseteq \mathrm{Leaf(F_i)}$ or $Q_i$ consists of an isolated vertex of $F_i$.

**Algorithm 13.5.11**

- **Step 1: initial condition.** *Let* $\mathrm{val}_0 = f$, $P_0$ *be empty, and* $F_0 = Q_0 = \{1\}$.

- **Step 2: choose a new vertex** $v$. *At time* $i \geq 1$, *let* $v = \gamma(F_{i-1}, Q_{i-1})$, *where* $\gamma$ *is the choice function.*

- **Step 3: process vertex** $v$. *For every vertex* $w$ *adjacent to* $v$ *and* $w \notin P_{i-1}$, *set* $\mathrm{val}_i(w) = \mathrm{val}_{i-1}(w) - 1$. *For any other vertex* $u$, *set* $\mathrm{val}_i(u) = \mathrm{val}_{i-1}(u)$. *Let* $N = \{w | \mathrm{val}_i(w) = -1, \mathrm{val}_{i-1}(w) \neq -1\}$. *Update* $P_i$, $Q_i$ *and* $F_i$ *by letting*

  1. $P_i = P_{i-1} \cup \{v\}$,
  2. $Q_i = Q_{i-1} \cup N \setminus \{v\}$ *if* $Q_{i-1} \cup N \setminus \{v\} \neq \emptyset$; *otherwise* $Q_i = \{u\}$ *where* $u$ *is the vertex of the lowest-index in* $[n] - P_i$.
  3. *Let* $F_i$ *be a graph on* $P_i \cup Q_i$ *whose edges are obtained from those of* $F_{i-1}$ *by joining edges* $\{w, v\}$ *for each* $w \in N$.

  *We say that the vertex* $v$ *is processed at time* $i$.

*Iterate steps 2–3 until* $i = n$. *We must have* $P_n = [n]$ *and* $Q_n = \emptyset$. *Define* $\Phi = \Phi_{\gamma,G} : \mathcal{MP}_G \to \mathcal{F}_G$ *by letting* $\Phi(f) = F_n$.

Note that the forest $F = \Phi(f)$ is built tree by tree by Algorithm 13.5.11 That is, if $T_i$ and $T_j$ are tree components of $F$ with roots $r_i$, $r_j$ and $r_i < r_j$, then any vertex of tree $T_i$ is processed before any vertex of $T_j$. The bijection $\Phi$ maps $G$-multiparking functions with $k$ roots to spanning forests with $k$ components.

We explain Algorithms 13.5.10 and 13.5.11 with an example where the choice function is given by the breadth-first order with a queue structure, i.e., the order $<_{bfq}$ described in Example 13.2.17, with additional rules comparing vertices in different trees. Explicitly, we view the minimal vertex of each tree component as the root of the tree. The order $<_{bfq}$ is defined as follows.

1. Any vertex of tree $T_i$ is less than any vertex of tree $T_j$ if the root of $T_i$ is less than the root of $T_j$

2. For vertices $i, j$ in the same tree component, let $i <_{bfq} j$ if

   (a) $\mathrm{level}(i) < \mathrm{level}(j)$, or

   (b) $\mathrm{level}(i) = \mathrm{level}(j)$ and $\mathrm{pre}(i) <_{bfq} \mathrm{pre}(j)$, or

   (c) $\mathrm{pre}(i) = \mathrm{pre}(j)$ and $i < j$.

**Example 13.5.12** *A graph $G$ and a multiparking function $f$ are given in Figure 13.9, in which each vertex is labeled by $i/f(i)$, i.e., the first label is the index of the vertex, and the second label is the function value.*

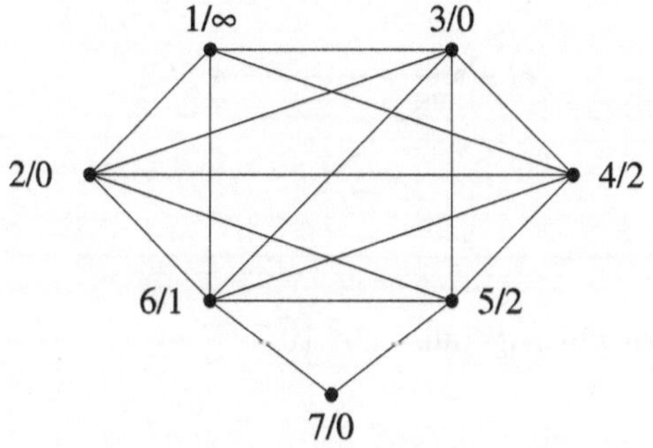


**Figure 13.9**
A graph and a multiparking function.

The spanning forest created by Algorithm 13.5.11 is indicated by the thick lines in Figure 13.10, while other edges are in dotted lines. In Figure 13.10, each vertex is labeled by $i/\text{val}_n(i)$. The $(P_i, Q_i, F_i)$ for this instance are as follows, where each $Q_i$ is an ordered set, and the first element in $Q_i$ is the next one to be processed. Since $P_i$ and $F_i$ are increasing sets, we just indicate when a new element is added to them.

| t | 0 | 1 | 2 | 3 | 4 | 5 | 6 | 7 |
|---|---|---|---|---|---|---|---|---|
| $Q_t$ | (1) | (2,3) | (3,6) | (6,4) | (4,5,7) | (5,7) | (7) | $\emptyset$ |
| new $i$ in $P_t$ | | 1 | 2 | 3 | 6 | 4 | 5 | 7 |
| new edges in $F_t$ | | $\{1,2\}, \{1,3\}$ | $\{2,6\}$ | $\{3,4\}$ | $\{6,5\}, \{6,7\}$ | | | |

Conversely, starting with a spanning tree, we can easily recover the $G$-multiparking function using Algorithm 13.5.10. For the spanning tree in Figure 13.10, the processing order is $1,2,3,6,4,5,7$. We have that $f(1) = \infty$ since it is the minimal element. Let $v$ be the vertex 5. Then $v^p = 6$ and $v$ is connected to two other vertices, namely, vertex 2 and 3, which are less than 6 in the processing order. Hence $f(5) = 2$. Similarly we can compute $f(i)$ for other vertices, and get

$$f(1) = \infty, \ f(2) = f(3) = f(7) = 0, \ f(4) = f(5) = 2, \ f(6) = 1.$$

To get a connection to the Tutte polynomial of $G$, we use another expression of $T_G(x,y)$. Let $H$ be a (spanning) subgraph of $G$. Denote by $c(H)$ the number of components of $H$. Define two invariants associated with $H$ as

$$\sigma(H) = c(H) - 1, \qquad \sigma^*(H) = |E(H)| - |V(G)| + c(H). \tag{13.16}$$

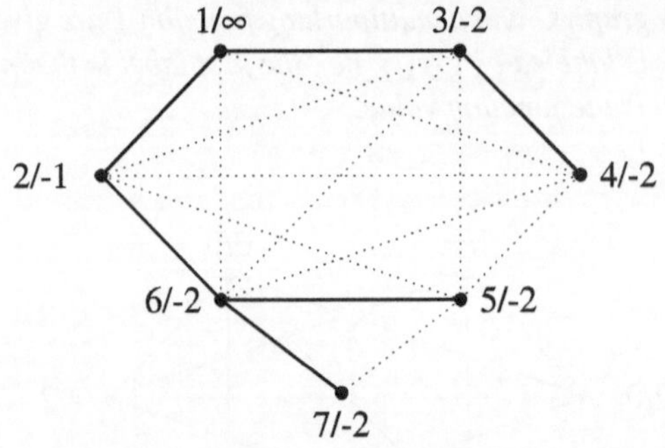


**Figure 13.10**
A spanning forest found by Algorithm 13.5.10.

The following alternative description of Tutte polynomials is well-known, for example, see [11].

**Theorem 13.5.13**

$$t_G(1+x,1+y) = \sum_{H \subseteq G} x^{\sigma(H)} y^{\sigma^*(H)}, \tag{13.17}$$

*where the sum is over all spanning subgraphs $H$ of $G$.*

For a subgraph $H$ of $G$, we apply the breadth-first search by using the implementation with a queue structure, as described in Section 13.2.3, to find a spanning forest. Since $H$ may not be connected, we modify the algorithm by adding the least unvisited vertex to the queue whenever the queue is empty but not all the vertices are processed. By the same argument as before, we partition all the subgraphs of $G$ into intervals, one for each spanning forest $F$. Each interval turns out to be a Boolean algebra consisting of all the ways to add some extra edges to its representing spanning forest. Expressing the Tutte polynomial in terms of sums over such intervals leads to an equation between the Tutte polynomial and a bivariate generating function of $G$-multiparking functions.

Given a $G$-multiparking function $f$, let $\pi = \pi_1\pi_2\ldots\pi_n$ be an ordering of the vertex set $[n]$ determined by applying the burning algorithm in a greedy way: Let $\pi_1 = 1$. After determining $\pi_1,\ldots,\pi_{i-1}$, if $V_i = V(G) - \{\pi_1,\ldots,\pi_{i-1}\}$ has a well-behaved vertex, that is, a vertex $v \in V_i$ such that $0 \le f(v) < \text{outdeg}_{V_i}(v)$, let $\pi_i$ be one of them; otherwise, let $\pi_i$ be the minimal vertex of $V_i$. In the latter case $\pi_i$ has to be a root of $f$.

Define the **total root record** $\text{Rec}(f)$ of $f$ as the cardinality of the set

$$\{\{v,w\} \in E(G) \mid v \text{ is a root of } f \text{ and } w \text{ appears in front of } v \text{ in the order } \pi\}.$$

It can be proved that the value $\text{Rec}(f)$ is independent of the order $\pi$ and hence well-defined. Let the weight of a $G$-multiparking function $f$ be

$$\mathrm{wt}(f) = |E(G)| + r(f) - n - \mathrm{Rec}(f) - \Big(\sum_{v \neq 0} f(v)\Big),$$

where $r(f)$ is the number of roots of $f$. Then we have

**Theorem 13.5.14** *For any connected graph G,*

$$T_G(1+x, y) = \sum_{f \in \mathscr{MP}(G)} x^{r(f)-1} y^{\mathrm{wt}(f)}.$$

*where $\mathscr{MP}(G)$ is the set of G-multiparking functions on G.*

Note that if $r(f) = 1$, the function $f$ can only have one root at vertex 1, therefore $\mathrm{Rec}(f) = 0$ and $\mathrm{wt}(f) = |E(G)| - n + 1 - \left(\sum_{v \neq 1} f(v)\right)$. Hence Theorem 13.5.6 is a special case of Theorem 13.5.14 by taking $x = 0$.

**Corollary 13.5.15** *For connected graph G, the number of G-multiparking functions is $t_G(2,1)$. Among them, those with an odd number of roots is counted by $\frac{1}{2}(t_G(2,1) + t_G(0,1))$, and those with an even number of roots is counted by $\frac{1}{2}(t_G(2,1) - t_G(0,1))$.*

*Proof.* The first sentence is obtained by taking $x = y = 1$ in Theorem 13.5.14. For the second sentence, one notices that $t_G(0,1) = \sum_f (-1)^{r(f)-1}$ is the difference between the number of $G$-multiparking functions with an odd number of roots, and those with an even number of roots. ∎

Theorem 13.5.14 also leads to a new expression of $t_{K_{n+1}}(x,y)$, the Tutte polynomial of complete graphs with $n+1$ vertices, in terms of ordinary parking functions. The exponent of the variable $x$ enumerates ordinary parking functions by the number of **critical left-to-right maxima**. Given an ordinary parking function $\mathbf{b} = (b_1, \ldots, b_n)$, we say that a term $b_i = j$ is **critical** if in $\mathbf{b}$ there are exactly $j$ terms less than $j$, and exactly $n-1-j$ terms larger than $j$. For example, in $\mathbf{b} = (3,0,0,2)$, the terms $b_1 = 3$ and $b_4 = 2$ are critical. Among them, only $b_1 = 3$ is also a left-to-right maximum.

Let $\alpha(\mathbf{a})$ be the number of critical left-to-right maxima in an ordinary parking function $\mathbf{a}$. We have

**Theorem 13.5.16**

$$t_{K_{n+1}}(x,y) = \sum_{\mathbf{a} \in \mathscr{PK}_n} x^{\alpha(\mathbf{a})} y^{\binom{n}{2} - \sum_i b_i},$$

*where $\mathscr{PK}_n$ is the set of ordinary parking functions of length n.*

An interesting question is how to express the bivariate Tutte polynomials in terms of $G$-parking functions only, avoiding introducing roots or using spanning forests. An answer was provided by Chang, Ma and Yeh [19], who proposed a notion of **critical bridge** for $G$-parking functions, and proved that

$$T_G(x,y) = \sum_f x^{\mathrm{cb}(f)} y^{\mathrm{wt}(f)},$$

where $\mathrm{cb}(f)$ is the number of critical bridges of $f$, and the sum ranges over all $G$-parking functions.

## 13.6 Final remarks

We close this survey by briefly mentioning a fast growing area, the combinatorial study of Macdonald polynomials, in which parking functions play an important role. Macdonald polynomials [63] is a family of multivariable orthogonal polynomials with applications to a wide variety of subjects, including algebraic combinatorics, harmonic analysis, Hilbert schemes, and representation theory. The combinatorics theory behind Macdonald polynomials was pioneered by Garsia, Haiman, and their school, and was developed dramatically in the last 20 years, with exciting new problems and progressions. Parking functions and their encoding as labeled Dyck paths provide powerful tools to study the intrinsic combinatorial structures. Conversely, the theory of Macdonald polynomials leads to many new generalizations and extensions that greatly enrich the theory of parking functions.

A comprehensive coverage on the combinatorics of Macdonald polynomials and the space of diagonal harmonics is given in the monograph [43] by Haglund. The enumerative results associated to statistics on parking functions and labeled lattice paths, as well as the two-parameter versions of related objects such as Catalan numbers and Schröder paths, are included in the chapter **Catalan Paths and $q,t$-enumeration** of this book. Here we give a gentle introduction of the role played by parking functions and the connection to previous sections.

Parking functions appear in the spaces of diagonal harmonics and diagonal coinvariants, which concern the action of the symmetric group $\mathfrak{S}_n$ on two sets $(x_1,\dots,x_n)$ and $(y_1,\dots,y_n)$ of $n$ variables. For a polynomial in $R_n = \mathbb{C}[x_1,\dots,x_n,y_1,\dots,y_n]$, let $\mathfrak{S}_n$ act diagonally, i.e.,

$$w\cdot x_i = x_{w(i)}, \qquad w\cdot y_i = y_{w(i)}.$$

It is known that

$$\{\sum_{i=1}^{n} x_i^h y_i^k, h+k\geq 0\} \tag{13.18}$$

generate $R_n^{\mathfrak{S}_n}$, the ring of invariants under the diagonal action. The **quotient ring** $DR_n$ **of diagonal coinvariants** is defined by $DR_n = R_n/I$, where $I$ is the ideal of $R_n$ generated by elements of (13.18) with zero constant term. By analogy the **space of diagonal harmonics** $DH_n$ is defined by

$$DH_n = \left\{ f\in R_n \ : \ \sum_{i=1}^{n} \frac{\partial^h}{x_i^h}\frac{\partial^k}{y_j^k} f = 0, \forall h+k>0 \right\}.$$

The spaces $DR_n$ and $DH_n$ are finite dimensional isomorphic vector spaces. A major theorem of $DH_n$ is the so-called $(n+1)^{n-1}$ conjecture, which was proposed by Garsia and Haiman [34] and later proved by Haiman [46].

**Theorem 13.6.1** *The dimension of $DH_n$ is $(n+1)^{n-1}$.*

Let $V_{h,k,n}$ be the set of polynomials $f \in R_n$ such that $f$ is homogeneous of degree $h$ in the $x$-variables and homogeneous of degree $k$ in the $y$-variables. By convention, the zero polynomial belongs to every $V_{h,k,n}$. We have the decomposition

$$R_n = \bigoplus_{h \geq 0} \bigoplus_{k \geq 0} V_{h,k,n},$$

which turns $R_n$ into a doubly graded vector space. One can write

$$DH_n = \bigoplus_{h \geq 0} \bigoplus_{k \geq 0} (DH_n \cap V_{h,k,n}).$$

The **Hilbert series of diagonal harmonics** is the bivariate polynomial

$$H_n(q,t) = \sum_{h \geq 0} \sum_{k \geq 0} \dim(DH_n \cap V_{h,k,n}) q^h t^k.$$

Similarly, one can turn $DR_n$ into a doubly graded vector space. Ignoring the ring structure of $DR_n$, the spaces $DR_n$ and $DH_n$ are isomorphic as doubly graded $\mathfrak{S}_n$-modules.

Haiman [46] proved that the Hilbert series $H_n(q,t)$ has the following properties.

(1) $H_n(1,1) = (n+1)^{n-1}$.

(2) $q^{n(n-1)/2} H_n(1/q, q) = (1+q+\cdots+q^n)^{n-1}$.

(3) $H_n(q,t) = H_n(t,q)$.

By the first property, $H_n(q,t)$ is a sum of $(n+1)^{n-1}$ monomials $q^h t^k$. This suggests that there could be a combinatorial interpretation of $H_n(q,t)$ as a joint enumerator of two statistics over parking functions. More precisely, one seeks two functions $\text{stat}_1$ and $\text{stat}_2$ from $\mathscr{PK}_n$ to $\mathbb{N}$ such that

$$H_n(q,t) = \sum_{\alpha \in \mathscr{PK}_n} q^{\text{stat}_1(\mathbf{a})} t^{\text{stat}_2(\mathbf{a})}.$$

The first conjectured combinatorial formula for $H_n(q,t)$ was proposed by Haglund and Loehr [45], and involved statistics "area" and "dinv." These two statistics are defined on labeled Dyck paths, a geometric encoding of parking functions.

A **Dyck path of order** $n$ is a lattice path from $(0,0)$ to $(n,n)$ consisting of steps East ($E = (1,0)$) and North ($N = (0,1)$), and never going below the line $y = x$. A **labeled Dyck path** is a Dyck path in which the $n$ north steps are labeled 1 to $n$ in such a way that the labels of consecutive north steps increase from bottom to top. See Figure 13.11 for an example, where one places each label in the lattice square to the right of the corresponding north step.

There is a simple bijection between labeled Dyck paths and parking functions. Let $P$ be a labeled Dyck path. For each $i \in [n]$, find the north step with label $i$ and let $a_i$ be the $x$-coordinate of this step. The fact that $P$ is a Dyck path implies that the sequence $(a_1, \ldots, a_n)$ is a parking function. Conversely, given a parking function, one places all cars that prefer the space $j$ in column $j+1$ in increasing order, such that there is exactly one car in each row, and cars preferring earlier spots appear in lower rows of the figure. The labels determine a labeled lattice path in the obvious way, which is easily seen to be a Dyck path. Example 13.6.2 shows this correspondence.

**Example 13.6.2** *For the parking function* $\mathbf{a} = (2,0,1,0,1,3,6,0)$, *Figure 13.11 shows the corresponding labeled Dyck path. To see how the labels are assigned, take the first column as an example. In* $\mathbf{a}$ *there are three 0s, which means that cars* $C_2$, $C_4$ *and* $C_8$ *prefer space 0. Hence there are three north steps in the Dyck path at* $x = 0$, *with labels* 2,4,8 *from bottom to top.*

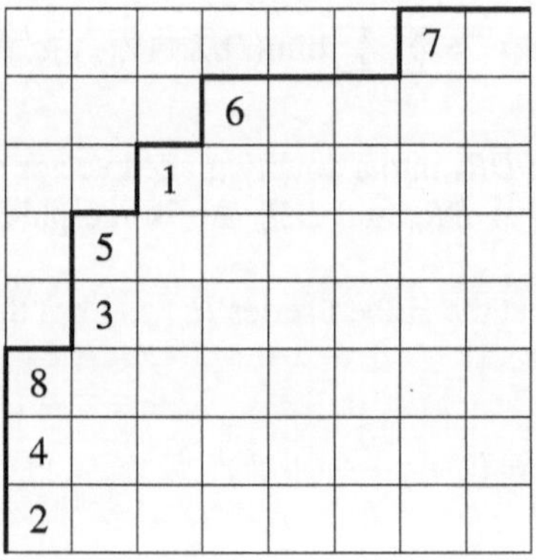


**Figure 13.11**
The parking function $\mathbf{a} = (2,0,1,0,1,3,6,0)$ and the corresponding labeled Dyck path.

The statistic area($\mathbf{a}$) of a parking function $\mathbf{a}$ is equivalent to the total displacement $D(\mathbf{a})$ defined in Section 2. The other statistic, dinv($\mathbf{a}$), is obtained by counting certain inversion pairs between rows of the labeled Dyck path. The exact definitions and detailed descriptions of these two statistics are given in the chapter **Catalan Paths and $q,t$-enumeration**, and hence omitted here. Let

$$CH_n(q,t) = \sum_{\mathbf{a} \in \mathscr{PK}_n} q^{\mathrm{dinv}(\mathbf{a})} t^{\mathrm{area}(\mathbf{a})}.$$

Haglund and Loehr [45] conjectured that $CH_n(q,t) = H_n(q,t)$. This conjecture has been verified in Maple for $n \leq 11$. The truth of the conjecture when $q = 1$ follows from results of Garsia and Haiman in [35]. It is also the reason that one considers parking functions but not other combinatorial objects counted by $(n+1)^{n-1}$ here: when $q=1$ there is a natural action of $\mathfrak{S}_n$ on the set $\mathscr{PK}_n$ that gives the Frobenius characteristic of the diagonal harmonics, up to the sign representation. When $t=1$, the conjecture follows from a result of Loehr and Remmel [62], who proved bijectively that area and dinv have the same distribution. In addition, Loehr [61] obtained a recursion characterizing the polynomial $CH_n(q,t)$, and proved the specialization

$$q^{n(n-1)/2}CH_n(1/q,q) = (1+q+\cdots+q^n)^{n-1}.$$

The joint symmetry of $CH_n(q,t)$, i.e., $CH_n(q,t) = CH_n(t,q)$, as well as Haglund and Loehr's conjecture are still open. They are part of a larger program on finding the combinatorial description for the character of the space of diagonal harmonics $DH_n$, which is the same as the character of diagonal coinvariants $DR_n$. It is known that the character of $DR_n$ as a doubly-graded $\mathfrak{S}_n$-module can be expressed using the Frobenius characteristic map as $\nabla e_n$, where $e_n$ is the $n$th elementary symmetric function, and $\nabla$ is an operator from the theory of Macdonald polynomials [46]. In 2005 Haglund, Haiman, Loehr, Remmel, and Ulyanov [44] proposed an explicit combinatorial formula of the expansion of $\nabla e_n$ into monomials. It is usually referred to as the "shuffle conjecture" since the formula can be described in terms of statistics on certain permutations associated to parking functions, which are shuffles of blocks of increasing and decreasing sequences. The shuffle conjecture contains a Garsia-Haglund formula for the $(q,t)$-Catalan number $C_n(q,t)$ [32, 33], the Haglund-Loehr conjecture for $H_n(q,t)$, and a formula for $(q,t)$-Schröder polynomials [42] as special cases.

In the literature on diagonal harmonics, ordinary parking functions enumerated by $(n+1)^{n-1}$ are called "classical parking functions," and it is standard to encode them as labeled Dyck paths. There are several extensions and generalizations of the shuffle conjecture, including an $m$-parameter extension [44], the **rational shuffle conjecture** [40], and the **compositional rational shuffle conjecture** [10]. The combinatorial foundations of these extensions are given by rational Dyck paths and rational parking functions. Explicitly, let $a,b$ be positive integers. An $(a,b)$-Dyck path is a lattice path from $(0,0)$ to $(b,a)$ consisting of $N$ and $E$ steps, and never going below the line $y=\frac{a}{b}x$. A rational $(a,b)$-parking function is an $(a,b)$-Dyck path together with a labeling of the north steps by the set $[a]$ such that labels increase in each column from bottom to top. Note that when $b=ka$ for some integer $k$, the rational $(a,b)$-parking functions are the same as the classical $(1,k)$-parking functions of length $a$ discussed in Section 13.4.4. For other cases, rational parking functions are not equivalent to **u**-parking functions with an arithmetic **u**. In particular, rational $(a,b)$-parking functions are counted by $b^{a-1}$ for co-prime positive integers $a,b$, where the ordinary case corresponds to $(a,b)=(n,n+1)$. The combinatorial theory of rational Dyck paths and rational parking functions, as well as their relations with the representation theory of diagonal coinvariants are discussed in [3].

**Acknowledgments.** First I would like to express my deepest gratitude to Gian-Carlo Rota and Richard Stanley for introducing me to the wonderful world of combinatorics, in particular, the subject of parking functions, and for their constant support and encouragement throughout my career. Many colleagues have contributed valuable ideas and comments, by either joint works or personal communications, for which I thank them: Drew Armstrong, Miklós Bóna, William Chen, Ira Gessel, Jim Haglund, Niraj Khare, Dimitrije Kostic, Joseph Kung, Rudolph Lorentz, Svetlana Poznanovik, Bruce Sagan, Joel Spencer, Xinyu Sun, Michelle Wachs, Yeong-Nan Yeh. I am also grateful to Hua Peng for his help on editing the manuscript, and to an anonymous referee for many helpful comments and suggestions.

This paper was written while I was taking a faculty development leave, and was supported by the National Science Foundation and the Association of Former Students at Texas A&M University.

# References

[1] D. Armstrong. *Generalized noncrossing partitions and combinatorics of Coxeter groups.* Mem. Amer. Math. Soc., vol. 202, no. 949, Amer. Math. Soc., Providence, RI, 2009.

[2] D. Armstrong. Hyperplane arrangements and diagonal harmonics. In *23rd International Conference on Formal Power Series and Algebraic Combinatorics (FPSAC 2011)*, Discrete Math. Theor. Comput. Sci. Proc., AO, pages 39–50. Assoc. Discrete Math. Theor. Comput. Sci., Nancy, 2011.

[3] D. Armstrong, N. Loehr, and G. S. Warrington. Rational parking functions and catalan numbers. arXiv:1403.1845, 2014.

[4] D. Armstrong and B. Rhoades. The Shi arrangement and the Ish arrangement. *Trans. Amer. Math. Soc.*, 364(3):1509–1528, 2012.

[5] C. A. Athanasiadis and S. Linusson. A simple bijection for the regions of the Shi arrangement of hyperplanes. *Discrete Math.*, 204(1-3):27–39, 1999.

[6] M. D. Atkinson and M. Thiyagarajah. The permutational power of a priority queue. *BIT*, 33(1):2–6, 1993.

[7] P. Bak, C. Tang, and K. Wiesenfeld. Self-organized criticality: an explanation of $1/f$ noise. *Phys. Rev. Letters*, 59(4):381–384, 1987.

[8] H. W. Becker. Planar rhyme schemes. *Bull. Amer. Math. Soc.*, 58:39, 1952.

[9] J. S. Beissinger and U. N. Peled. A note on major sequences and external activity in trees. *Electron. J. Combin.*, 4(2):Research Paper 4, 1997.

[10] F. Bergeron, A. Leven, A. Garsia, and G. Xin. Compositional $(km,kn)$shuffle conjectures. arXiv:1404.4616, 2014.

[11] N. L. Biggs. *Algebraic graph theory*. Cambridge Mathematical Library. Cambridge University Press, Cambridge, second edition, 1993.

[12] N. L. Biggs. Chip-firing and the critical group of a graph. *J. Algebraic Combin.*, 9(1):25–45, 1999.

[13] A. Björner. The homology and shellability of matroids and geometric lattices. In *Matroid applications*, volume 40 of *Encyclopedia Math. Appl.*, pages 226–283. Cambridge Univ. Press, Cambridge, 1992.

[14] A. Björner and L. Lovász. Chip-firing games on directed graphs. *J. Algebraic Combin.*, 1(4):305–328, 1992.

[15] A. Björner, L. Lovász, and P. W. Shor. Chip-firing games on graphs. *European J. Combin.*, 12(4):283–291, 1991.

[16] M. Bousquet-Mélou and G. Schaeffer. Enumeration of planar constellations. *Adv. in Appl. Math.*, 24(4):337–368, 2000.

[17] P. J. Cameron, D. Johannsen, T. Prellberg, and P. Schweitzer. Counting defective parking functions. *Electron. J. Combin.*, 15(1):Research Paper 92, 2008.

[18] A. Cayley. A theorem on trees. *Quart. J. Math.*, 23:376–378, 1889.

[19] H. Chang, J. Ma, and Y.-N. Yeh. Tutte polynomials and *G*-parking functions. *Adv. in Appl. Math.*, 44(3):231–242, 2010.

[20] P. Chassaing and J.-F. Marckert. Parking functions, empirical processes, and the width of rooted labeled trees. *Electron. J. Combin.*, 8(1):Research Paper 14, 2001.

[21] D. Chebikin and P. Pylyavskyy. A family of bijections between *G*-parking functions and spanning trees. *J. Combin. Theory Ser. A*, 110(1):31–41, 2005.

[22] R. Cori, A. Dartois, and D. Rossin. Avalanche polynomials of some families of graphs. In *Mathematics and computer science. III*, Trends Math., pages 81–94. Birkhäuser, Basel, 2004.

[23] R. Cori and Y. Le Borgne. The sand-pile model and Tutte polynomials. *Adv. in Appl. Math.*, 30(1-2):44–52, 2003.

[24] R. Cori and D. Rossin. On the sandpile group of dual graphs. *European J. Combin.*, 21(4):447–459, 2000.

[25] D. Dhar. Self-organized critical state of sandpile automaton models. *Phys. Rev. Lett.*, 64(14):1613–1616, 1990.

[26] P. H. Edelman. Chain enumeration and noncrossing partitions. *Discrete Math.*, 31(2):171–180, 1980.

[27] S-P. Eu, T-S. Fu, and C-J. Lai. On the enumeration of parking functions by leading terms. *Adv. in Appl. Math.*, 35(4):392–406, 2005.

[28] S-P. Eu, T-S. Fu, and C-J. Lai. Cycle lemma, parking functions and related multigraphs. *Graphs Combin.*, 26(3):345–360, 2010.

[29] P. Flajolet, P. Poblete, and A. Viola. On the analysis of linear probing hashing: Average-case analysis of algorithms. *Algorithmica*, 22(4):490–515, 1998.

[30] D. Foata and J. Riordan. Mappings of acyclic and parking functions. *Aequationes Math.*, 10:10–22, 1974.

[31] J. Françon. Acyclic and parking functions. *J. Combinatorial Theory Ser. A*, 18:27–35, 1975.

[32] A. M. Garsia and J. Haglund. A positivity result in the theory of Macdonald polynomials. *Proc. Natl. Acad. Sci. USA*, 98(8):4313–4316, 2001.

[33] A. M. Garsia and J. Haglund. A proof of the $q,t$-Catalan positivity conjecture. *Discrete Math.*, 256(3):677–717, 2002.

[34] A. M. Garsia and M. Haiman. A graded representation model for Macdonald's polynomials. *Proc. Nat. Acad. Sci. U.S.A.*, 90(8):3607–3610, 1993.

[35] A. M. Garsia and M. Haiman. A remarkable $q,t$-Catalan sequence and $q$-Lagrange inversion. *J. Algebraic Combin.*, 5(3):191–244, 1996.

[36] I. M. Gessel and B. E. Sagan. The Tutte polynomial of a graph, depth-first search, and simplicial complex partitions. *Electron. J. Combin.*, 3(2):Research Paper 9, 1996.

[37] I. M. Gessel and S. Seo. A refinement of Cayley's formula for trees. *Electron. J. Combin.*, 11(2):Research Paper 27, 2004/06.

[38] I. M. Gessel and D. L. Wang. Depth-first search as a combinatorial correspondence. *J. Combin. Theory Ser. A*, 26(3):308–313, 1979.

[39] J. D. Gilbey and L. H. Kalikow. Parking functions, valet functions and priority queues. *Discrete Math.*, 197/198:351–373, 1999.

[40] E Gorsky and A Negut. Refined knot invariants and Hilbert schemes. arXiv:1304.3328, 2013.

[41] I. P. Goulden and D. M. Jackson. Transitive factorisations into transpositions and holomorphic mappings on the sphere. *Proc. Amer. Math. Soc.*, 125(1):51–60, 1997.

[42] J. Haglund. A proof of the $q,t$-Schröder conjecture. *Int. Math. Res. Not.*, (11):525–560, 2004.

[43] J. Haglund. *The q,t-Catalan numbers and the space of diagonal harmonics*, volume 41 of *University Lecture Series*. American Mathematical Society, Providence, RI, 2008.

[44] J. Haglund, M. Haiman, N. Loehr, J. B. Remmel, and A. Ulyanov. A combinatorial formula for the character of the diagonal coinvariants. *Duke Math. J.*, 126(2):195–232, 2005.

[45] J. Haglund and N. Loehr. A conjectured combinatorial formula for the Hilbert series for diagonal harmonics. *Discrete Math.*, 298(1-3):189–204, 2005.

[46] M. Haiman. Vanishing theorems and character formulas for the Hilbert scheme of points in the plane. *Invent. Math.*, 149(2):371–407, 2002.

[47] S. Janson. Asymptotic distribution for the cost of linear probing hashing. *Random Structures Algorithms*, 19(3-4):438–471, 2001.

[48] S. Janson, D. E. Knuth, T. Łuczak, and B. Pittel. The birth of the giant component. *Random Structures Algorithms*, 4(3):231–358, 1993.

[49] N. Khare, R. Lorentz, and C. H. Yan. Bivariate gončarov polynomials and integer sequences. *Sci. China Math.*, 57(8):1561–1578, 2014.

[50] D. Kim and S. Seo. Transitive cycle factorizations and prime parking functions. *J. Combin. Theory Ser. A*, 104(1):125–135, 2003.

[51] D. E. Knuth. *The art of computer programming. Volume 3.* Addison-Wesley Publishing Co., Reading, Mass.-London-Don Mills, Ont., 1973. Sorting and searching, Addison-Wesley Series in Computer Science and Information Processing.

[52] D. E. Knuth. Linear probing and graphs. *Algorithmica*, 22(4):561–568, 1998.

[53] A. G. Konheim and B. Weiss. An occupancy discipline and applications. *SIAM J. Appl. Math.*, 14:12661274, 1966.

[54] D. Kostić and C. H. Yan. Multiparking functions, graph searching, and the Tutte polynomial. *Adv. in Appl. Math.*, 40(1):73–97, 2008.

[55] G. Kreweras. Sur les partitions non croisées d'un cycle. *Discrete Math.*, 1(4):333–350, 1972.

[56] G. Kreweras. Une famille de polynômes ayant plusieurs propriétés énumeratives. *Period. Math. Hungar.*, 11(4):309–320, 1980.

[57] J. P. S. Kung, X. Sun, and C. H. Yan. Goncarov-type polynomials and applications in combinatorics. *Preprint*, 2007. Available at http://www.math.tamu.edu/ cyan/Files/DGP.pdf.

[58] J. P. S. Kung and C. H. Yan. Exact formulas for moments of sums of classical parking functions. *Adv. in Appl. Math.*, 31(1):215–241, 2003.

[59] J. P. S. Kung and C. H. Yan. Expected sums of general parking functions. *Ann. Comb.*, 7(4):481–493, 2003.

[60] J. P. S. Kung and C. H. Yan. Gončarov polynomials and parking functions. *J. Combin. Theory Ser. A*, 102(1):16–37, 2003.

[61] N. A. Loehr. Combinatorics of $q$, $t$-parking functions. *Adv. in Appl. Math.*, 34(2):408–425, 2005.

[62] N. A. Loehr and J. B. Remmel. Conjectured combinatorial models for the Hilbert series of generalized diagonal harmonics modules. *Electron. J. Combin.*, 11(1):Research Paper 68, 2004.

[63] I. G. Macdonald. *Symmetric functions and Hall polynomials*. Oxford Mathematical Monographs. The Clarendon Press, Oxford University Press, New York, second edition, 1995.

[64] C. Merino López. Chip firing and the Tutte polynomial. *Ann. Comb.*, 1(3):253–259, 1997.

[65] P. Moszkowski. Arbres et suites majeures. *Period. Math. Hungar.*, 20(2):147–154, 1989.

[66] A. Nica and R. Speicher. *Lectures on the combinatorics of free probability*, volume 335 of *London Mathematical Society Lecture Note Series*. Cambridge University Press, Cambridge, 2006.

[67] J. Pitman and R. P. Stanley. A polytope related to empirical distributions, plane trees, parking functions, and the associahedron. *Discrete Comput. Geom.*, 27(4):603–634, 2002.

[68] A. Postnikov and B. Shapiro. Trees, parking functions, syzygies, and deformations of monomial ideals. *Trans. Amer. Math. Soc.*, 356(8):3109–3142, 2004.

[69] Y. Poupard. Étude et dénombrement parallèles des partitions non-croisées d'un cycle et des découpages d'un polygone convexe. *Discrete Math.*, 2(3):279–288, 1972.

[70] H. Prüfer. Neuer beweis eines satzes ber permutationen. *Arch. Math. Phys.*, 27:742–744, 1918.

[71] J. Riordan. Ballots and trees. *J. Combinatorial Theory*, 6:408–411, 1969.

[72] M. P. Schützenberger. On an enumeration problem. *J. Combinatorial Theory*, 4:219–221, 1968.

[73] S. Seo and H. Shin. A generalized enumeration of labeled trees and reverse Prüfer algorithm. *J. Combin. Theory Ser. A*, 114(7):1357–1361, 2007.

[74] J. Y. Shi. *The Kazhdan-Lusztig cells in certain affine Weyl groups*, volume 1179 of *Lecture Notes in Mathematics*. Springer-Verlag, Berlin, 1986.

[75] H. Shin and J. Zeng. A further correspondence between $(bc,\bar{b})$-parking functions and $(bc,\bar{b})$-forests. In *21st International Conference on Formal Power Series and Algebraic Combinatorics (FPSAC 2009)*, Discrete Math. Theor. Comput. Sci. Proc., AK, pages 793–804. Assoc. Discrete Math. Theor. Comput. Sci., Nancy, 2009.

[76] J. Spencer. Enumerating graphs and Brownian motion. *Comm. Pure Appl. Math.*, 50(3):291–294, 1997.

[77] J. Spencer and C. H. Yan. An enumeration problem and branching processes. *Preprint*, 1999. Available at http://www.math.tamu.edu/ cyan/Files/Ball.pdf.

[78] R. P. Stanley. *Enumerative combinatorics. Vol. I.* The Wadsworth & Brooks/Cole Mathematics Series. Wadsworth & Brooks/Cole Advanced Books & Software, Monterey, CA, 1986.

[79] R. P. Stanley. Hyperplane arrangements, interval orders, and trees. *Proc. Nat. Acad. Sci. U.S.A.*, 93(6):2620–2625, 1996.

[80] R. P. Stanley. Parking functions and noncrossing partitions. *Electron. J. Combin.*, 4(2):Research Paper 20, 1997.

[81] R. P. Stanley. Hyperplane arrangements, parking functions and tree inversions. In *Mathematical essays in honor of Gian-Carlo Rota (Cambridge, MA, 1996)*, volume 161 of *Progr. Math.*, pages 359–375. Birkhäuser Boston, Boston, MA, 1998.

[82] G. P. Steck. The Smirnov two sample tests as rank tests. *Ann. Math. Statist.*, 40:1449–1466, 1969.

[83] W. T. Tutte. A contribution to the theory of chromatic polynomials. *Canadian J. Math.*, 6:80–91, 1954.

[84] E. M. Wright. The number of connected sparsely edged graphs. *J. Graph Theory*, 1(4):317–330, 1977.

[85] C. H. Yan. Generalized tree inversions and $k$-parking functions. *J. Combin. Theory Ser. A*, 79(2):268–280, 1997.

[86] C. H. Yan. On the enumeration of generalized parking functions. In *Proceedings of the Thirty-first Southeastern International Conference on Combinatorics, Graph Theory and Computing (Boca Raton, FL, 2000)*, volume 147, pages 201–209, 2000.

[87] C. H. Yan. Generalized parking functions, tree inversions, and multicolored graphs. *Adv. in Appl. Math.*, 27(2-3):641–670, 2001.

[88] C. Zara. Parking functions, stack-sortable permutations, and spaces of paths in the Johnson graph. *Electron. J. Combin.*, 9(2):Research paper 11, 2002/03.

# Chapter 14

# *Standard Young Tableaux*

**Ron Adin and Yuval Roichman**
*Bar-Ilan University, Ramat-Gan, Israel*

## CONTENTS

## 14.1 Introduction

### 14.1.1 Appetizer

Consider throwing balls labeled $1,2,\ldots,n$ into a V-shaped bin with perpendicular sides.

**Question 14.1.1** *What is the total number of resulting configurations? How many configurations are there of any particular shape?*

In order to answer these questions, at least partially, recall the symmetric group $\mathscr{S}_n$ of all permutations of the numbers $1, \dots, n$. An **involution** is a permutation $\pi \in \mathscr{S}_n$ such that $\pi^2$ is the identity permutation.

**Theorem 14.1.2** *The total number of configurations of $n$ balls is equal to the number of involutions in the symmetric group $\mathscr{S}_n$.*

Theorem 14.1.2 may be traced back to Frobenius and Schur. A combinatorial proof will be outlined in Section 14.4 (see Corollary 14.4.14).

**Example 14.1.3** *There are four configurations on three balls. Indeed,*

$$\{\pi \in \mathscr{S}_3 : \pi^2 = 1\} = \{123, 132, 213, 321\}.$$

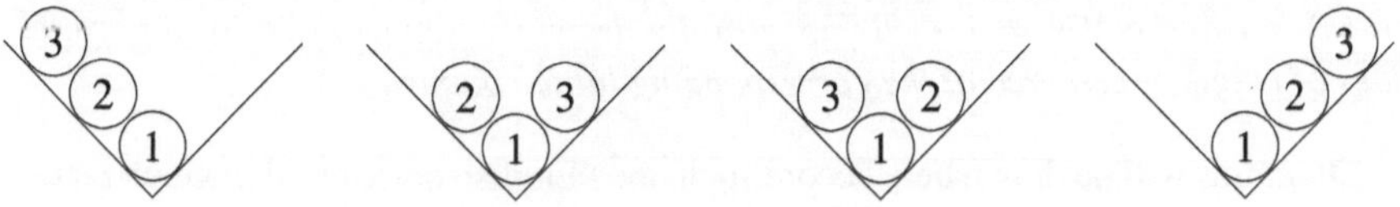

The **inversion number** of a permutation $\pi$ is defined by

$$\operatorname{inv}(\pi) := \#\{i < j : \pi(i) > \pi(j)\}.$$

The **left weak order** on $\mathscr{S}_n$ is defined by

$$\pi \leq \sigma \Longleftrightarrow \operatorname{inv}(\pi) + \operatorname{inv}(\sigma\pi^{-1}) = \operatorname{inv}(\sigma).$$

The following surprising result was first proved by Stanley [112].

**Theorem 14.1.4** *The number of configurations of $\binom{n}{2}$ balls that completely fill $n-1$ levels in the bin is equal to the number of maximal chains in the weak order on $\mathscr{S}_n$.*

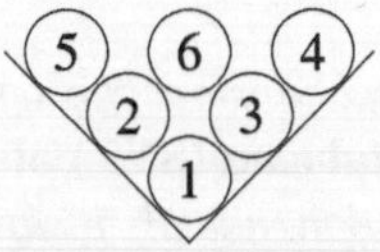

The configurations of balls in a bin are called **standard Young tableaux**. We shall survey in this chapter results related to Question 14.1.1 and its refinements. Variants and extensions of Theorem 14.1.2 will be described in Section 14.4. Variants and extensions of Theorem 14.1.4 will be described in Section 14.11.

### 14.1.2 General

This chapter is devoted to the enumeration of standard Young tableaux of various shapes, both classical and modern, and to closely related topics. Of course, there is a limit as to how far afield one can go. We chose to include here, for instance, $r$-tableaux and $q$-enumeration, but many interesting related topics were left out. Here are some of them, with a minimal list of relevant references for the interested reader: Semi-standard Young tableaux [69, 114], (reverse) plane partitions [114], solid (3-dimensional) standard Young tableaux [25], symplectic and orthogonal tableaux [55, 20, 127, 12, 128], oscillating tableaux [70, 103, 96, 22, 79], cylindric (and toric) tableaux [86].

## 14.2 Preliminaries

### 14.2.1 Diagrams and tableaux

**Definition 14.2.1** *A* **diagram** *is a finite subset $D$ of the two-dimensional integer lattice $\mathbb{Z}^2$. A point $c = (i, j) \in D$ is also called the* **cell** *in row $i$ and column $j$ of $D$; write* $\mathrm{row}(c) = i$ *and* $\mathrm{col}(c) = j$. *Cells are usually drawn as squares with axis-parallel sides of length* 1, *centered at the corresponding lattice points.*

Diagrams will be drawn here according to the "English notation," by which $i$ enumerates rows and increases downwards, while $j$ enumerates columns and increases from left to right:

| | | |
|---|---|---|
| (1,1) | (1,2) | (1,3) |
| (2,1) | (2,2) | |

For alternative conventions see Section 14.2.4.

**Definition 14.2.2** *Each diagram $D$ has a natural component-wise partial order, inherited from $\mathbb{Z}^2$:*

$$(i, j) \leq_D (i', j') \iff i \leq i' \text{ and } j \leq j'.$$

*As usual, $c <_D c'$ means $c \leq_D c'$ but $c \neq c'$.*

**Definition 14.2.3** *Let $n := |D|$, and consider the set $[n] := \{1, \ldots, n\}$ with its usual linear order. A* **standard Young tableau (SYT) of shape** *$D$ is a map $T : D \to [n]$ which is an order-preserving bijection, namely it satisfies*

$$c \neq c' \Longrightarrow T(c) \neq T(c')$$

*as well as*

$$c \leq_D c' \Longrightarrow T(c) \leq T(c').$$

Geometrically, a standard Young tableau $T$ is a filling of the $n$ cells of $D$ by the numbers $1,\dots,n$ such that each number appears once, and numbers increase in each row (as the column index increases) and in each column (as the row index increases). Write $\mathrm{sh}(T) = D$. Examples will be given below.

Let $\mathrm{SYT}(D)$ be the set of all standard Young tableaux of shape $D$, and denote its size by

$$f^D := |\mathrm{SYT}(D)|.$$

The evaluation of $f^D$ (and some of its refinements) for various diagrams $D$ is the main concern of the current chapter.

### 14.2.2 Connectedness and convexity

We now introduce two distinct notions of connectedness for diagrams, and one notion of convexity; for another notion of convexity see Observation 14.2.13.

**Definition 14.2.4** *Two distinct cells in* $\mathbb{Z}^2$ *are* **adjacent** *if they share a horizontal or vertical side; the cells adjacent to* $c = (i, j)$ *are* $(i \pm 1, j)$ *and* $(i, j \pm 1)$. *A diagram* $D$ *is* **path-connected** *if any two cells in it can be connected by a* **path**, *which is a finite sequence of cells in* $D$ *such that any two consecutive cells are adjacent. The maximal path-connected subsets of a nonempty diagram* $D$ *are its* **path-connected components**.

For example, the following diagram has two path-connected components:

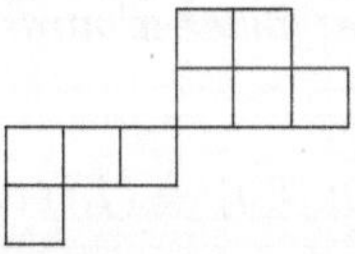

**Definition 14.2.5** *The* **graph** *of a diagram* $D$ *has all the cells of* $D$ *as vertices, with two distinct cells* $c, c' \in D$ *connected by an (undirected) edge if either* $c <_D c'$ *or* $c' <_D c$. *The diagram* $D$ *is* **order-connected** *if its graph is connected. In any case, the* **order-connected components** *of* $D$ *are the subsets of* $D$ *forming connected components of its graph.*

For example, the following diagram (in English notation) is order-connected:

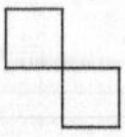

while the following diagram has two order-connected components, with cells marked 1 and 2, respectively:

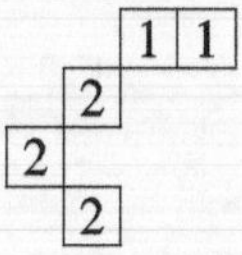


Of course, every path-connected diagram is also order-connected, so that every order-connected component is a disjoint union of path-connected components.

**Observation 14.2.6** *If $D_1,\ldots,D_k$ are the order-connected components of a diagram $D$, then*

$$f^D = \binom{|D|}{|D_1|,\ldots,|D_k|}\prod_{i=1}^{k} f^{D_i} = |D|!\cdot\prod_{i=1}^{k}\frac{f^{D_i}}{|D_i|!}.$$

**Definition 14.2.7** *A diagram $D$ is* **line-convex** *if its intersection with every axis-parallel line is either empty or convex, namely if each of its rows $\{j\in\mathbb{Z}\,|\,(i,j)\in D\}$ (for $i\in\mathbb{Z}$) and columns $\{i\in\mathbb{Z}\,|\,(i,j)\in D\}$ (for $j\in\mathbb{Z}$) is either empty or an interval $[p,q]=\{p,p+1,\ldots,q\}\subseteq\mathbb{Z}$.*

For example, the following diagram is path-connected but not line-convex:

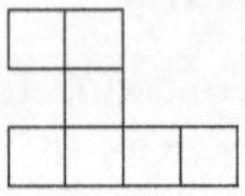

### 14.2.3 Invariance under symmetry

The number of SYT of shape $D$ is invariant under some of the geometric operations (isometries of $\mathbb{Z}^2$) that transform $D$. It is clearly invariant under arbitrary translations $(i,j)\mapsto(i+a,j+b)$. The group of isometries of $\mathbb{Z}^2$ that fix a point, say $(0,0)$, is the dihedral group of order 8. $f^D$ is invariant under a subgroup of order 4.

**Observation 14.2.8** *$f^D$ is invariant under arbitrary translations of $\mathbb{Z}^2$, as well as under*

- *reflection in a diagonal line: $(i,j)\mapsto(j,i)$ or $(i,j)\mapsto(-j,-i)$; and*
- *reflection in the origin (rotation by $180°$): $(i,j)\mapsto(-i,-j)$.*

Note that $f^D$ is not invariant, in general, under reflections in a vertical or horizontal line ($(i,j)\mapsto(i,-j)$ or $(i,j)\mapsto(-i,j)$) or rotations by $90°$ ($(i,j)\mapsto(-j,i)$ or $(i,j)\mapsto(j,-i)$). Thus, for example, each of the following diagrams, interpreted according to the English convention (see Section 14.2.4),

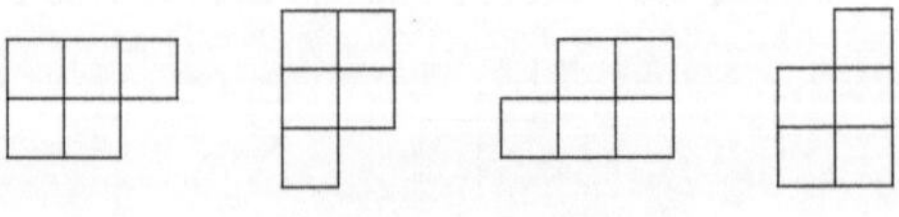

has $f^D=5$, whereas each of the following diagrams

has $f^D=2$.

### 14.2.4 Ordinary, skew and shifted shapes

The best known and most useful diagrams are, by far, the ordinary ones. They correspond to partitions.

**Definition 14.2.9** *A* **partition** *is a weakly decreasing sequence of positive integers: $\lambda = (\lambda_1, \ldots, \lambda_t)$, where $t \geq 0$ and $\lambda_1 \geq \ldots \geq \lambda_t > 0$. We say that $\lambda$ is a partition of* **size** *$n = |\lambda| := \sum_{i=1}^{t} \lambda_i$ and* **length** *$\ell(\lambda) := t$, and write $\lambda \vdash n$. The empty partition $\lambda = ()$ has size and length both equal to zero.*

**Definition 14.2.10** *Let $\lambda = (\lambda_1, \ldots, \lambda_t)$ be a partition. The* **ordinary** *(or* **straight**, *or* **left-justified**, *or* **Young**, *or* **Ferrers**) **diagram of shape** *$\lambda$ is the set*

$$D = [\lambda] := \{(i,j) \mid 1 \leq i \leq t, 1 \leq j \leq \lambda_i\}.$$

*We say that $[\lambda]$ is a diagram of* **height** *$\ell(\lambda) = t$.*

We shall adopt here the "English" convention for drawing diagrams, by which row indices increase from top to bottom and column indices increase from left to right. For example, in this notation the diagram of shape $\lambda = (4,3,1)$ is

$[\lambda] =$ 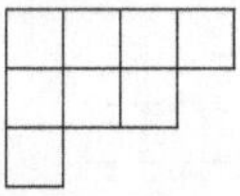 (English notation).

An alternative convention is the "French" one, by which row indices increase from bottom to top (and column indices increase from left to right):

$[\lambda] =$ 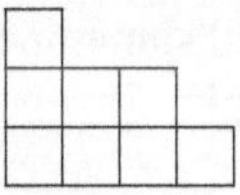 (French notation).

Note that the term "Young tableau" itself mixes English and French influences. There is also a "Russian" convention, rotated 45°:

$[\lambda] =$ 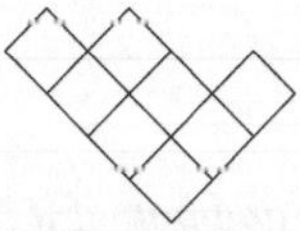 (Russian notation).

This notation leads naturally to the "gravitational" setting used to introduce SYT at the beginning of Section 14.1.

A partition $\lambda$ may also be described as an infinite sequence, by adding trailing zeros: $\lambda_i := 0$ for $i > t$. The partition $\lambda'$ **conjugate** to $\lambda$ is then defined by

$$\lambda'_j := |\{i \mid \lambda_i \geq j\}| \qquad (\forall j \geq 1).$$

The diagram $[\lambda']$ is obtained from the diagram $[\lambda]$ by interchanging rows and columns. For the example above, $\lambda' = (3,2,2,1)$ and

$[\lambda'] =$ 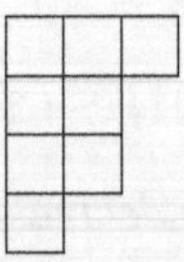

An ordinary diagram is clearly path-connected and line-convex. If $D = [\lambda]$ is an ordinary diagram of shape $\lambda$ we shall sometimes write $\mathrm{SYT}(\lambda)$ instead of $\mathrm{SYT}(D)$ and $f^\lambda$ instead of $f^D$.

**Example 14.2.11**

$T =$

| | | | |
|---|---|---|---|
| 1 | 2 | 5 | 8 |
| 3 | 4 | 6 | |
| 7 | | | |

$\in \mathrm{SYT}(4,3,1)$.

Note that, by Observation 14.2.8, $f^\lambda = f^{\lambda'}$.

**Definition 14.2.12** *If $\lambda$ and $\mu$ are partitions such that $[\mu] \subseteq [\lambda]$, namely $\mu_i \le \lambda_i$ $(\forall i)$, then the* **skew diagram of shape** $\lambda/\mu$ *is the set difference*

$$D = [\lambda/\mu] := [\lambda] \setminus [\mu] = \{(i,j) \in [\lambda] \,:\, \mu_i + 1 \le j \le \lambda_i\}$$

*of two ordinary shapes.*

For example,

$$[(6,4,3,1)/(4,2,1)] =$$

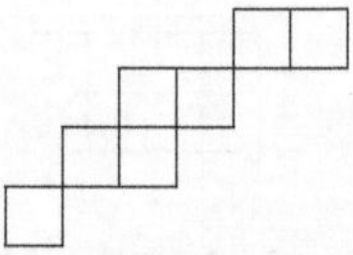

A skew diagram is line-convex, but not necessarily path-connected. In fact, its path-connected components coincide with its order-connected components. If $D = [\lambda/\mu]$ is a skew diagram of shape $\lambda/\mu$ we shall sometimes write $\mathrm{SYT}(\lambda/\mu)$ instead of $\mathrm{SYT}(D)$ and $f^{\lambda/\mu}$ instead of $f^D$. For example,

$T =$

| | | | | | |
|---|---|---|---|---|---|
| | | | | 1 | 4 |
| | | 3 | 7 | | |
| | 5 | 6 | | | |
| 2 | | | | | |

$\in \mathrm{SYT}((6,4,3,1)/(4,2,1))$.

Skew diagrams have an intrinsic characterization.

**Observation 14.2.13** *A diagram $D$ is skew if and only if it is* **order-convex**, *namely:*

$$c, c'' \in D,\ c' \in \mathbb{Z}^2,\ c \le c' \le c'' \Longrightarrow c' \in D,$$

*where $\le$ is the natural partial order in $\mathbb{Z}^2$, as in Definition 14.2.2.*

Another important class is that of shifted shapes, corresponding to strict partitions.

**Definition 14.2.14** *A partition $\lambda = (\lambda_1, \ldots, \lambda_t)$ $(t \ge 0)$ is* **strict** *if the part sizes $\lambda_i$ are strictly decreasing: $\lambda_1 > \ldots > \lambda_t > 0$. The* **shifted diagram of shape** $\lambda$ *is the set*

$$D = [\lambda^*] := \{(i,j) \,|\, 1 \le i \le t,\ i \le j \le \lambda_i + i - 1\}.$$

*Note that $(\lambda_i + i - 1)_{i=1}^t$ is a weakly decreasing sequence of positive integers.*

For example, the shifted diagram of shape $\lambda = (4,3,1)$ is

$[\lambda^*] =$ 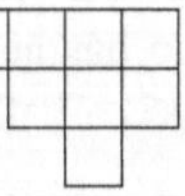

A shifted diagram is always path-connected and line-convex. If $D = [\lambda^*]$ is a shifted diagram of shape $\lambda$ we shall sometimes write $\mathrm{SYT}(\lambda^*)$ instead of $\mathrm{SYT}(D)$ and $g^\lambda$ instead of $f^D$. For example,

$$T = \begin{array}{|c|c|c|c|} \hline 1 & 2 & 4 & 6 \\ \hline \multicolumn{1}{c|}{} & 3 & 5 & 8 \\ \cline{2-4} \multicolumn{1}{c}{} & \multicolumn{1}{c|}{} & 7 & \multicolumn{1}{c}{} \\ \cline{3-3} \end{array} \in \mathrm{SYT}((4,3,1)^*).$$

## 14.2.5 Interpretations

There are various interpretations of a standard Young tableau, in addition to the interpretation (in Definition 14.2.3) as a linear extension of a partial order. Some of these interpretations play a key role in enumeration.

### 14.2.5.1 The Young lattice

A standard Young tableau of ordinary shape describes a **growth process** of diagrams of ordinary shapes, starting from the empty shape. For example, the tableau $T$ in Example 14.2.11 corresponds to the process

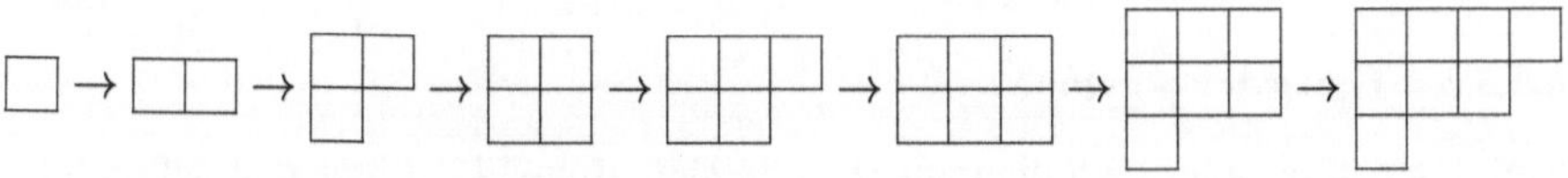

Consider the **Young lattice** whose elements are all partitions, ordered by inclusion (of the corresponding diagrams). By the above, a SYT of ordinary shape $\lambda$ is a maximal chain, in the Young lattice, from the empty partition to $\lambda$. The number of such maximal chains is therefore $f^\lambda$. More generally, a SYT of skew shape $\lambda/\mu$ is a maximal chain from $\mu$ to $\lambda$ in the Young lattice.

A SYT of shifted shape can be similarly interpreted as a maximal chain in the **shifted Young lattice**, whose elements are strict partitions ordered by inclusion.

### 14.2.5.2 Ballot sequences and lattice paths

**Definition 14.2.15** *A sequence $(a_1,\ldots,a_n)$ of positive integers is a* **ballot sequence**, *or* **lattice permutation**, *if for any integers $1 \le k \le n$ and $r \ge 1$,*

$$\#\{1 \le i \le k \mid a_i = r\} \ge \#\{1 \le i \le k \mid a_i = r+1\},$$

*namely: In any initial subsequence $(a_1,\ldots,a_k)$, the number of entries equal to $r$ is not less than the number of entries equal to $r+1$.*

A ballot sequence describes the sequence of votes in an election process with several candidates (and one ballot), assuming that at any given time candidate 1 has at least as many votes as candidate 2, who has at least as many votes as candidate 3, etc. For example, $(1,1,2,3,2,1,4,2,3)$ is a ballot sequence for an election process with 9 voters and 4 candidates.

For a partition $\lambda$ of $n$, denote by $\mathrm{BS}(\lambda)$ the set of ballot sequences $(a_1,\ldots,a_n)$ with $\#\{i \mid a_i = r\} = \lambda_r$ $(\forall r)$.

**Observation 14.2.16** *The map* $\phi : \mathrm{SYT}(\lambda) \to \mathrm{BS}(\lambda)$ *defined by*

$$\phi(T)_i := \mathrm{row}(T^{-1}(i)) \qquad (1 \le i \le n)$$

*is a bijection.*

For example, if $T$ is the SYT in Example 14.2.11 then $\phi(T) = (1,1,2,2,1,2,3,1)$.

Clearly, a ballot sequence $a = (a_1,\ldots,a_n) \in \mathrm{BS}(\lambda)$ with maximal entry $t$ corresponds to a **lattice path** in $\mathbb{R}^t$, from the origin 0 to the point $\lambda$, where in step $i$ of the path coordinate $a_i$ of the point increases by 1. In fact, $\mathrm{BS}(\lambda)$ is in bijection with the set of all lattice paths from 0 to $\lambda$ that lie entirely in the cone

$$\{(x_1,\ldots,x_t) \in \mathbb{R}^t \mid x_1 \ge \ldots \ge x_t \ge 0\}.$$

A SYT of skew shape $\lambda/\mu$ corresponds to a lattice path in this cone from $\mu$ to $\lambda$.

A SYT of shifted shape corresponds to a **strict** ballot sequence, describing a lattice path within the cone

$$\{(x_1,\ldots,x_t) \in \mathbb{R}^t \mid x_1 > \ldots > x_s > x_{s+1} = \ldots = x_t = 0 \text{ for some } 0 \le s \le t\}.$$

#### 14.2.5.3 The order polytope

Using the partial order on a diagram $D$, as in Definition 14.2.2, one can define the corresponding **order polytope**

$$P(D) := \{f : D \to [0,1] \mid c \le_D c' \Longrightarrow f(c) \le f(c')\ (\forall c, c' \in D)\}.$$

The order polytope is a closed convex subset of the unit cube $[0,1]^D$. Denoting $n := |D|$, each linear extension $c_1 < \ldots < c_n$ of the partial order $\le_D$ corresponds to a simplex

$$\{f : D \to [0,1] \mid f(c_1) \le \ldots \le f(c_n)\}$$

of volume $1/n!$ inside $P(D)$. The union of these simplices is $P(D)$, and this union is almost disjoint: Any intersection of two or more simplices is contained in a hyperplane, and thus has volume 0. Each simplex, or linear extension, corresponds to a SYT of shape $D$. Therefore

**Observation 14.2.17** *If* $P(D)$ *is the order polytope of a diagram* $D$ *then*

$$\mathrm{vol}\, P(D) = \frac{f^D}{|D|!}.$$

#### 14.2.5.4 Other interpretations

The number of SYT of ordinary shape can be interpreted as a coefficient in a power series, or as the constant term in a Laurent series; see Remark 14.6.14.

SYT of certain ordinary, skew and shifted shapes may be interpreted as **reduced words** for suitable elements in the symmetric group. This interpretation will be developed and explained in Section 14.11.

SYT may be interpreted as **permutations**; $f^\lambda$ then measures the size of certain subsets of the symmetric group, such as descent classes, sets of involutions, Knuth classes and pattern avoidance classes. Examples will be given in Sections 14.4 and 14.10.

The Young lattice has a vast generalization to the concept of **differential posets** [113, 42].

There are other deep algebraic and geometric interpretations. The interested reader is encouraged to start with the beautiful survey [9] and the excellent textbooks [53, 114, 99, 37, 69]. In the current survey we will focus on combinatorial aspects and mostly ignore the algebraic and geometric approaches.

### 14.2.6 Miscellanea

The concepts to be defined here are not directly related to standard Young tableaux, but will be used later in this survey.

A **composition** of a nonnegative integer $n$ is a sequence $(\mu_1, \ldots, \mu_t)$ of positive integers such that $\mu_1 + \ldots + \mu_t = n$. The components $\mu_i$ are *not* required to be weakly decreasing; in fact, every composition may be re-ordered to form a (unique) partition. $n = 0$ has a unique (empty) composition.

A permutation $\sigma \in \mathcal{S}_n$ **avoids a pattern** $\pi \in \mathcal{S}_k$ if the sequence $(\sigma(1), \ldots, \sigma(n))$ does not contain a subsequence $(\sigma(t_1), \ldots, \sigma(t_k))$ (with $t_1 < \ldots < t_k$) that is order-isomorphic to $\pi$, namely: $\sigma(t_i) < \sigma(t_j) \iff \pi(i) < \pi(j)$. For example, $21354 \in \mathcal{S}_5$ is 312-avoiding, but 52134 is not (since 523 is order-isomorphic to 312).

## 14.3 Formulas for thin shapes

### 14.3.1 Hook shapes

A **hook shape** is an ordinary shape that is the union of one row and one column. For example,

$[(6, 1^3)] =$ 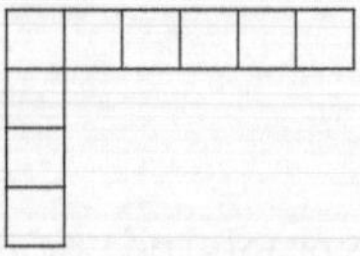

One of the simplest enumerative formulas is the following.

**Observation 14.3.1** *For every $n \geq 1$ and $0 \leq k \leq n-1$,*

$$f^{(n-k,1^k)} = \binom{n-1}{k}.$$

*Proof.* The letter 1 must be in the corner cell. The SYT is uniquely determined by the choice of the other $k$ letters in the first column. ∎

Note that, in a hook shape $(n+1-k, 1^k)$, the letter $n+1$ must be in the last cell of either the first row or the first column. Thus

$$f^{(n+1-k,1^k)} = f^{(n-k,1^k)} + f^{(n+1-k,1^{k-1})}.$$

By Observation 14.3.1, this is equivalent to Pascal's identity

$$\binom{n}{k} = \binom{n-1}{k} + \binom{n-1}{k-1} \qquad (1 \leq k \leq n-1).$$

**Observation 14.3.2** *The total number of hook shaped SYT of size $n$ is $2^{n-1}$.*

*Proof.* There is a bijection between hook shaped SYT of size $n$ and subsets of $\{1,\ldots,n\}$ containing 1: Assign to each SYT the set of entries in its first row.

Alternatively, a hook shaped SYT of size $n \geq 2$ is uniquely constructed by adding a cell containing $n$ at the end of either the first row or the first column of a hook shaped SYT of size $n-1$, thus recursively multiplying the number of SYT by 2.

Of course, the claim also follows from Observation 14.3.1. ∎

### 14.3.2 Two-rowed shapes

Consider now ordinary shapes with at most two rows.

**Proposition 14.3.3** *For every $n \geq 0$ and $0 \leq k \leq n/2$*

$$f^{(n-k,k)} = \binom{n}{k} - \binom{n}{k-1},$$

*where $\binom{n}{-1} = 0$ by convention. In particular,*

$$f^{(m,m)} = f^{(m,m-1)} = C_m = \frac{1}{m+1}\binom{2m}{m},$$

*the $m$-th Catalan number.*

*Proof.* We shall outline two proofs, one by induction and one combinatorial.

For a proof by induction on $n$ note first that $f^{(0,0)} = f^{(1,0)} = 1$.

If $0 < k < n/2$ then there are two options for the location of the letter $n$: at the end of the first row or at the end of the second. Hence

$$f^{(n-k,k)} = f^{(n-k-1,k)} + f^{(n-k,k-1)} \qquad (0 < k < n/2).$$

Thus, by the induction hypothesis and Pascal's identity,

$$\begin{aligned} f^{(n-k,k)} &= \binom{n-1}{k} - \binom{n-1}{k-1} + \binom{n-1}{k-1} - \binom{n-1}{k-2} \\ &= \binom{n-1}{k} - \binom{n-1}{k-2} \\ &= \binom{n}{k} - \binom{n}{k-1}. \end{aligned}$$

The cases $k = 0$ and $k = n/2$ are left to the reader.

For a combinatorial proof, recall (from Section 14.2.5.2) the lattice path interpretation of a SYT and use André's reflection trick: A SYT of shape $(n-k,k)$ corresponds to a lattice path from $(0,0)$ to $(n-k,k)$ which stays within the cone $\{(x,y) \in \mathbb{R}^2 \,|\, x \geq y \geq 0\}$, namely does not touch the line $y = x+1$. The number of *all* lattice paths from $(0,0)$ to $(n-k,k)$ is $\binom{n}{k}$. If such a path touches the line $y = x+1$, reflect its "tail" (starting from the first touch point) in this line to get a path from $(0,0)$ to the reflected endpoint $(k-1, n-k+1)$. The reflection defines a bijection between all the "touching" paths to $(n-k,k)$ and all the (necessarily "touching") paths to $(k-1, n-k+1)$, whose number is clearly $\binom{n}{k-1}$. ∎

**Corollary 14.3.4** *The total number of SYT of size n and at most* 2 *rows is* $\binom{n}{\lfloor n/2 \rfloor}$.

*Proof.* By Proposition 14.3.3,

$$\sum_{k=0}^{\lfloor n/2 \rfloor} f^{(n-k,k)} = \sum_{k=0}^{\lfloor n/2 \rfloor} \left( \binom{n}{k} - \binom{n}{k-1} \right) = \binom{n}{\lfloor n/2 \rfloor}.$$

∎

### 14.3.3 Zigzag shapes

A **zigzag shape** is a path-connected skew shape that does not contain a $2 \times 2$ square. For example, every hook shape is zigzag. Here is an example of a zigzag shape of size 11:

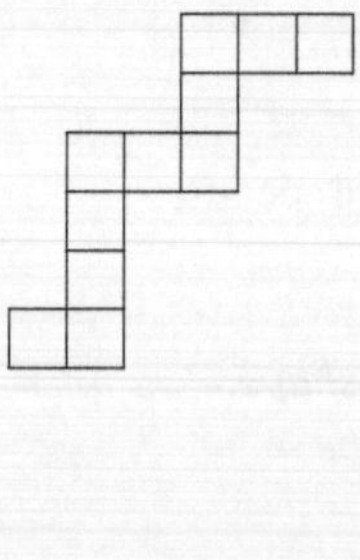

The number of SYT of a specific zigzag shape has an interesting formula, to be presented in Section 14.7.1. The total number of SYT of given size and various zigzag shapes is given by the following folklore statement, to be refined later (Proposition 14.10.12).

**Proposition 14.3.5** *The total number of zigzag shaped SYT of size $n$ is $n!$.*

*Proof.* Define a map from zigzag shaped SYT of size $n$ to permutations in $\mathscr{S}_n$ by simply listing the entries of the SYT, starting from the southwestern corner and moving along the shape. This map is a bijection, since an obvious inverse map builds a SYT from a permutation $\sigma = (\sigma_1, \ldots, \sigma_n) \in \mathscr{S}_n$ by attaching a cell containing $\sigma_{i+1}$ to the right of the cell containing $\sigma_i$ if $\sigma_{i+1} > \sigma_i$, and above this cell otherwise. ∎

## 14.4 Jeu de taquin and the RS correspondence

### 14.4.1 Jeu de taquin

Jeu de taquin is a very powerful combinatorial algorithm, introduced by Schützenberger [106]. It provides a unified approach to many enumerative results. In general, it transforms a SYT of skew shape into some other SYT of skew shape, using a sequence of **slides**. We shall describe here a version of it, using only **forward slides**, which transforms a SYT of skew shape into a (unique) SYT of ordinary shape. Our description follows [99].

**Definition 14.4.1** *Let $D$ be a nonempty diagram of skew shape. An* **inner corner** *for $D$ is a cell $c \notin D$ such that*

1. *$D \sqcup \{c\}$ is a skew shape, and*
2. *there exists a cell $c' \in D$ such that $c \leq c'$ (in the natural partial order of $\mathbb{Z}^2$, as in Definition 14.2.2).*

**Example 14.4.2** *Here is a skew shape with marked inner corners:*

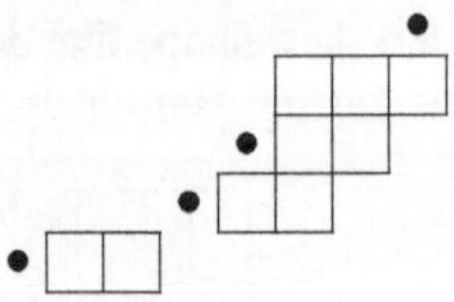

Here is the main jeu de taquin procedure:

**Algorithm 14.4.3**
***Input:*** *$T$, a SYT of arbitrary skew shape.*
***Output:*** *$T'$, a SYT of ordinary shape.*

**procedure** JDT($T$)
  $D \leftarrow \mathrm{sh}(T)$
  *Choose a cell* $c_0 = (i_0, j_0)$ *such that* $D \subseteq (c_0)_+ := \{(i,j) \in \mathbb{Z}^2 : i \geq i_0,\ j \geq j_0\}$
  **while** $c_0 \notin D$ **do**
    *Choose* $c = (i,j) \in (c_0)_+ \setminus D$ *which is an inner corner for* $D$
    $T \leftarrow$ FORWARDSLIDE$(T, c)$
    $D \leftarrow \mathrm{sh}(T)$
  **end while**
  **return** $T$    ▷ *Now* $c_0 \in D \subseteq (c_0)_+$, *so* $D$ *has ordinary shape*
**end procedure**

And here is the procedure FORWARDSLIDE:

**Algorithm 14.4.4**
***Input:*** $(T_{in}, c_{in})$, *where* $T_{in}$ *is a SYT of skew shape* $D_{in}$ *and* $c_{in}$ *is an inner corner for* $D_{in}$.
***Output:*** $T_{out}$, *a SYT of skew shape* $D_{out} = D_{in} \cup \{c_{in}\} \setminus \{c'\}$ *for some* $c' \in D_{in}$.

**procedure** FORWARDSLIDE($T_{in}, c_{in}$)
  $T \leftarrow T_{in}$, $c \leftarrow c_{in}$
  $D \leftarrow \mathrm{sh}(T)$
  **if** $c = (i,j)$ **then**
    $c_1 \leftarrow (i+1, j)$
    $c_2 \leftarrow (i, j+1)$
  **end if**
  **while** *at least one of* $c_1$ *and* $c_2$ *is in* $D$ **do**
    $$c' \leftarrow \begin{cases} c_1, & \text{if } c_1 \in D \text{ but } c_2 \notin D, \text{ or } c_1, c_2 \in D \text{ and } T(c_1) < T(c_2) \\ c_2, & \text{if } c_2 \in D \text{ but } c_1 \notin D, \text{ or } c_1, c_2 \in D \text{ and } T(c_2) < T(c_1) \end{cases}$$
    $D' \leftarrow D \cup \{c\} \setminus \{c'\}$    ▷ $c \notin D$, $c' \in D$
    *Define* $T' \in \mathrm{SYT}(D')$ *by:* $T' = T$ *on* $D \setminus \{c'\}$ *and* $T'(c) := T(c')$
    $D \leftarrow D'$, $T \leftarrow T'$, $c \leftarrow c'$
    **if** $c = (i,j)$ **then**
      $c_1 \leftarrow (i+1, j)$
      $c_2 \leftarrow (i, j+1)$
    **end if**
  **end while**
  **return** $T$
**end procedure**

The JDT algorithm employs certain random choices, but actually the following holds.

**Proposition 14.4.5** [106, 130, 131] *For any SYT* $T$ *of skew shape, the resulting SYT* JDT$(T)$ *of ordinary shape is independent of the choices made during the computation.*

**Example 14.4.6** *Here is an example of a forward slide, with the initial $c_{in}$ and the intermediate cells $c$ marked:*

$$T_{in} = \begin{array}{cccc} & \bullet & 3 & 6 \\ & 1 & 4 & 7 \\ 2 & 5 & 8 & \end{array} \to \begin{array}{cccc} & 1 & 3 & 6 \\ & \bullet & 4 & 7 \\ 2 & 5 & 8 & \end{array} \to \begin{array}{cccc} & 1 & 3 & 6 \\ & 4 & \bullet & 7 \\ 2 & 5 & 8 & \end{array} \to \begin{array}{cccc} & 1 & 3 & 6 \\ & 4 & 7 & \bullet \\ 2 & 5 & 8 & \end{array} = T_{out},$$

*and here is an example of a full jeu de taquin (where each step is a forward slide):*

$$T = \begin{array}{cccc} & \bullet & 3 & 6 \\ & 1 & 4 & 7 \\ 2 & 5 & 8 & \end{array} \to \begin{array}{cccc} & 1 & 3 & 6 \\ \bullet & 4 & 7 & \\ 2 & 5 & 8 & \end{array} \to \begin{array}{cccc} \bullet & 1 & 3 & 6 \\ 2 & 4 & 7 & \\ 5 & 8 & & \end{array} \to \begin{array}{ccc} 1 & 3 & 6 \\ 2 & 4 & 7 \\ 5 & 8 & \end{array} = \text{JDT}(T).$$

### 14.4.2 The Robinson-Schensted correspondence

The Robinson-Schensted (RS) correspondence is a bijection from permutations in $\mathscr{S}_n$ to pairs of SYT of size $n$ and the same ordinary shape. Its original motivation was the study of the distribution of longest increasing subsequences in a permutation. For a detailed description see, e.g., the textbooks [114], [99], and [37]. We will use the jeu de taquin algorithm to give an alternative description.

**Definition 14.4.7** *Denote $\delta_n := [(n, n-1, n-2, \ldots, 1)]$. For a permutation $\pi \in \mathscr{S}_n$ let $T_\pi$ the skew SYT of antidiagonal shape $\delta_n/\delta_{n-1}$ in which the entry in the i-th column from the left is $\pi(i)$.*

**Example 14.4.8**

$$\pi = 53412 \implies T_\pi = \begin{array}{ccccc} & & & & 2 \\ & & & 1 & \\ & & 4 & & \\ & 3 & & & \\ 5 & & & & \end{array}.$$

**Definition 14.4.9 (The Robinson-Schensted (RS) correspondence)** *For a permutation $\pi \in \mathscr{S}_n$ let*

$$P_\pi := \text{JDT}(T_\pi) \quad \textit{and} \quad Q_\pi := \text{JDT}(T_{\pi^{-1}}).$$

**Example 14.4.10**

$$\pi = 2413 \implies T_\pi = \begin{array}{cccc} & & & 3 \\ & & 1 & \\ & 4 & & \\ 2 & & & \end{array}, \quad T_{\pi^{-1}} = \begin{array}{cccc} & & & 2 \\ & & 4 & \\ & 1 & & \\ 3 & & & \end{array}.$$

*Then*

$$T_\pi = \begin{array}{cccc} & & & 3 \\ & \bullet & 1 & \\ & 4 & & \\ 2 & & & \end{array} \to \begin{array}{cccc} & & \bullet & 3 \\ & 1 & & \\ & 4 & & \\ 2 & & & \end{array} \to \begin{array}{ccc} & & 3 \\ & 1 & \\ \bullet & 4 & \\ 2 & & \end{array} \to \begin{array}{ccc} & & 3 \\ \bullet & 1 & \\ 2 & 4 & \end{array} \to \begin{array}{ccc} & \bullet & 3 \\ 1 & 4 & \\ 2 & & \end{array} \to \begin{array}{cc} \bullet & 3 \\ 1 & 4 \\ 2 & \end{array}$$

$$\to \begin{array}{cc} 1 & 3 \\ 2 & 4 \end{array} = P_\pi$$

*and*

$T_{\pi^{-1}} =$ 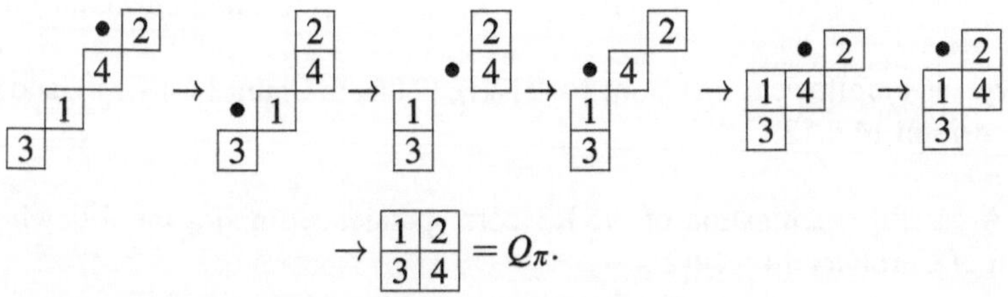 $= Q_\pi.$

**Theorem 14.4.11** *The RS correspondence is a bijection from all permutations in $\mathcal{S}_n$ to all pairs of SYT of size n and the same ordinary shape.*

Thus

**Claim 14.4.12** *For every permutation $\pi \in \mathcal{S}_n$,*

*(i)* $\mathrm{sh}(P_\pi) = \mathrm{sh}(Q_\pi)$.

*(ii)* $\pi \leftrightarrow (P,Q) \Longrightarrow \pi^{-1} \leftrightarrow (Q,P)$.

A very fundamental property of the RS correspondence is the following.

**Proposition 14.4.13** [104] *The height of* $\mathrm{sh}(\pi)$ *is equal to the size of the longest decreasing subsequence in $\pi$. The width of* $\mathrm{sh}(\pi)$ *is equal to the size of the longest increasing subsequence in $\pi$.*

A version of the RS correspondence for shifted shapes was given, initially, by Sagan [100]. An improved algorithm was found, independently, by Worley [135] and Sagan [102]. See also [48].

### 14.4.3 Enumerative applications

In this section we list just a few applications of the above combinatorial algorithms.

**Corollary 14.4.14**

*(1)* *The total number of pairs of SYT of the same shape is n!. Thus*

$$\sum_\lambda (f^\lambda)^2 = n!$$

*(2)* *The total number of SYT of size n is equal to the number of involutions in* $\mathcal{S}_n$ [110, A000085]. *Thus*

$$\sum_{\lambda \vdash n} f^\lambda = \sum_{k=0}^{\lfloor n/2 \rfloor} \binom{n}{2k} (2k-1)!!,$$

*where* $(2k-1)!! := 1 \cdot 3 \cdot \ldots \cdot (2k-1)$.

(3) *Furthermore, for every positive integer k, the total number of SYT of height* $< k$ *is equal to the number of* $[k,k-1,\ldots,1]$*-avoiding involutions in* $\mathcal{S}_n$.

*Proof.* (1) follows from Claim 14.4.12(i), (2) from Claim 14.4.12(ii), and (3) from Proposition 14.4.13. ■

A careful examination of the RS correspondence implies the following refinement of Corollary 14.4.14(2).

**Theorem 14.4.15** *The total number of SYT of size n with* $n-2k$ *odd rows is equal to* $\binom{n}{2k}(2k-1)!!$, *the number of involutions in* $\mathcal{S}_n$ *with* $n-2k$ *fixed points.*

**Corollary 14.4.16** *The total number of SYT of size* $2n$ *and all rows even is equal to* $(2n-1)!!$, *the number of fixed point free involutions in* $\mathcal{S}_{2n}$.

For further refinements see, e.g., [120, Ex. 45–46, 85].

Recalling the simple formula for the number of two-rowed SYT (Corollary 14.3.4), it is tempting to look for the total number of SYT of shapes with more rows.

**Theorem 14.4.17** [87] *The total number of SYT of size n and at most* 3 *rows is*

$$M_n = \sum_{k=0}^{\lfloor n/2 \rfloor} \binom{n}{2k} C_k,$$

*the nth Motzkin number* [110, A001006].

*Proof.* By Observation 14.2.6 together with Proposition 14.3.3, the number of SYT of skew shape $(n-k,k,k)/(k)$ is equal to

$$\binom{n}{2k} C_k,$$

where $C_k$ is $k$th Catalan number. On the other hand, by careful examination of the jeu de taquin algorithm, one can verify that it induces a bijection from the set of all SYT of skew shape $(n-k,k,k)/(k)$ to the set of all SYT of shapes $(n-k-j,k,j)$ for $0 \le j \le \min(k,n-2k)$. Thus

$$\sum_{\substack{\lambda \vdash n \\ \ell(\lambda) \le 3}} f^{\lambda} = \sum_{k=0}^{\lfloor n/2 \rfloor} \sum_{j} f^{(n-k-j,k,j)} = \sum_{k=0}^{\lfloor n/2 \rfloor} \binom{n}{2k} C_k,$$

completing the proof. ■

See [28] for a bijective proof of Theorem 14.4.17 via a map from SYT of height at most 3 to Motzkin paths.

The $n$th Motzkin number also counts non-crossing involutions in $\mathcal{S}_n$. It follows that

**Corollary 14.4.18** *The total number of SYT of height at most* 3 *is equal to the number of non-crossing involutions in* $\mathscr{S}_n$.

Somewhat more complicated formulas have been found for shapes with more rows.

**Theorem 14.4.19** [45]

1. *The total number of SYT of size* $n$ *and at most* 4 *rows is equal to* $C_{\lfloor (n+1)/2 \rfloor} C_{\lceil (n+1)/2 \rceil}$.
2. *The total number of SYT of size* $n$ *and at most* 5 *rows is equal to* $6 \sum_{k=0}^{\lfloor n/2 \rfloor} \binom{n}{2k} C_k \frac{(2k+2)!}{(k+2)!(k+3)!}$.

The following shifted analogue of Corollary 14.4.14(1) was proved by Schur [105], more than a hundred years ago, in a representation theoretical setting. A combinatorial proof, using the shifted RS correspondence, was given by Sagan [100]. An improved shifted RS algorithm was found, independently, by Worley [135] and Sagan [102]. See the end of Section 14.2.4 for the notation $g^\lambda$.

**Theorem 14.4.20**

$$\sum_{\text{strict } \lambda \vdash n} 2^{n-\ell(\lambda)} (g^\lambda)^2 = n!$$

## 14.5 Formulas for classical shapes

There is an explicit formula for the number of SYT of each classical shape—ordinary, skew or shifted. In fact, there are several equivalent fomulas, all unusually elegant. These formulas, with proofs, will be given in this section. Additional proof approaches (mostly for ordinary shapes) will be described in Section 14.6.

### 14.5.1 Ordinary shapes

In this subsection we consider ordinary shapes $D = [\lambda]$, corresponding to partitions $\lambda$. Recall the notation $f^\lambda := |\mathrm{SYT}(\lambda)|$ for the number of standard Young tableaux of shape $\lambda$. Several explicit formulas are known for this number—a product formula, a hook length formula and a determinantal formula.

Historically, ordinary tableaux were introduced by Young in 1900 [136]. The first explicit formula for the number of SYT of ordinary shape was the product formula. It was obtained in 1900 by Frobenius [36, eqn. 6] in an algebraic context, as the degree of an irreducible character $\chi^\lambda$ of $\mathscr{S}_n$. Independently, MacMahon [71, p. 175] in 1909 (see also [72, §103]) obtained the same formula for the number of ballot sequences (see Definition 14.2.15 above), which are equinumerous with SYT. In 1927

Young [137, pp. 260–261] showed that $\deg(\chi^\lambda)$ is actually equal to the number of SYT of shape $\lambda$, and also provided his own proof [137, Theorem II] of MacMahon's result.

**Theorem 14.5.1 (Ordinary product formula)** *For a partition $\lambda = (\lambda_1, \ldots, \lambda_t)$, let $\ell_i := \lambda_i + t - i$ $(1 \le i \le t)$. Then*

$$f^\lambda = \frac{|\lambda|!}{\prod_{i=1}^{t} \ell_i!} \cdot \prod_{(i,j):i<j} (\ell_i - \ell_j).$$

The best known and most influential of the explicit formulas is doubtlessly the Frame-Robinson-Thrall hook length formula, published in 1954 [34]. The story of its discovery is quite amazing [99]: Frame was led to conjecture the formula while discussing the work of Staal, one of Robinson's students, during Robinson's visit to him in May 1953. Robinson could not believe, at first, that such a simple formula exists, but became convinced after trying some examples, and together they proved it. A few days later, Robinson gave a lecture followed by a presentation of the new result by Frame. Thrall, who was in the audience, was very surprised because he had just proved the same result on the same day!

**Definition 14.5.2** *For a cell $c = (i, j) \in [\lambda]$ let*

$$H_c := [\lambda] \cap (\{(i,j)\} \cup \{(i,j') \mid j' > j\} \cup \{(i',j) \mid i' > i\})$$

*be the corresponding* **hook**, *and let*

$$h_c := |H_c| = \lambda_i + \lambda'_j - i - j + 1.$$

*be the corresponding* **hook length**.

For example, in the following diagram the cells of the hook $H_{(1,2)}$ are marked:

| | | | |
|---|---|---|---|
|  | • | • | • |
|  | • |  | |
|  | | | |

and in the following diagram each cell is labeled by the corresponding hook length:

| | | | |
|---|---|---|---|
| 6 | 4 | 3 | 1 |
| 4 | 2 | 1 | |
| 1 | | | |

**Theorem 14.5.3 (Ordinary hook length formula)** *For any partition $\lambda = (\lambda_1, \ldots, \lambda_t)$,*

$$f^\lambda = \frac{|\lambda|!}{\prod_{c \in [\lambda]} h_c}.$$

Last, but not least, is the determinantal formula. Remarkably, it also has a generalization to the skew case; see the next subsection.

**Theorem 14.5.4 (Ordinary determinantal formula)** *For any partition $\lambda = (\lambda_1, \dots, \lambda_t)$,*

$$f^\lambda = |\lambda|! \cdot \det \left[ \frac{1}{(\lambda_i - i + j)!} \right]_{i,j=1}^{t},$$

*using the convention $1/k! := 0$ for negative integers $k$.*

We shall now show that all these formulas are equivalent. Their validity will then follow from a forthcoming proof of Theorem 14.5.6, which is a generalization of Theorem 14.5.4. Other proof approaches will be described in Section 14.6.

**Claim 14.5.5** *The formulas in Theorems 14.5.1, 14.5.3 and 14.5.4 are equivalent.*

*Proof.* To prove the equivalence of the product formula (Theorem 14.5.1) and the hook length formula (Theorem 14.5.3), it suffices to show that

$$\prod_{c \in [\lambda]} h_c = \frac{\prod_{i=1}^{t} (\lambda_i + t - i)!}{\prod_{(i,j):i<j} (\lambda_i - \lambda_j - i + j)}.$$

This follows by induction on the number of columns, once we show that the product of hook lengths for all the cells in the first column of $[\lambda]$ satisfies

$$\prod_{i=1}^{t} h_{(i,1)} = \prod_{i=1}^{t} (\lambda_i + t - i);$$

and this readily follows from the obvious

$$h_{(i,1)} = \lambda_i + t - i \qquad (\forall i).$$

Actually, one also needs to show that the ordinary product formula is valid even when the partition $\lambda$ has trailing zeros (so that $t$ in the formula may be larger than the number of nonzero parts in $\lambda$). This is not difficult, since adding one zero part $\lambda_{t+1} = 0$ (and replacing $t$ by $t+1$) amounts, in the product formula, to replacing each $\ell_i = \lambda_i + t - i$ by $\ell_i + 1$ $(1 \le i \le t)$ and adding $\ell_{t+1} = 0$, which multiplies the RHS of the formula by

$$\frac{1}{\prod_{i=1}^{t} (\ell_i + 1) \cdot \ell_{t+1}!} \cdot \prod_{i=1}^{t} (\ell_i + 1 - \ell_{t+1}) = 1.$$

To prove equivalence of the product formula (Theorem 14.5.1) and the determinantal formula (Theorem 14.5.4), it suffices to show that

$$\det \left[ \frac{1}{(\ell_i - t + j)!} \right]_{i,j=1}^{t} = \frac{1}{\prod_{i=1}^{t} \ell_i!} \cdot \prod_{(i,j):i<j} (\ell_i - \ell_j) \quad ,$$

where

$$\ell_i := \lambda_i + t - i \qquad (1 \le i \le t)$$

as in Theorem 14.5.1. Using the **falling factorial** notation

$$(a)_n := \prod_{i=1}^{n} (a+1-i) \qquad (n \geq 0),$$

this claim is equivalent to

$$\det\left[(\ell_i)_{t-j}\right]_{i,j=1}^{t} = \prod_{(i,j):i<j} (\ell_i - \ell_j)$$

which, in turn, is equivalent (under suitable column operations) to the well-known evaluation of the Vandermonde determinant

$$\det\left[\ell_i^{t-j}\right]_{i,j=1}^{t} = \prod_{(i,j):i<j} (\ell_i - \ell_j).$$

See [99, pp. 132–133] for an inductive proof avoiding explicit use of the Vandermonde. ∎

### 14.5.2 Skew shapes

The determinantal formula for the number of SYT of an ordinary shape can be extended to apply to a general skew shape. The formula is due to Aitken [5, p. 310], and was rediscovered by Feit [29]. No product or hook length formula is known in this generality (but a product formula for a staircase minus a rectangle has been found by DeWitt [21]; see also [61]). Specific classes of skew shapes, such as zigzags and strips of constant width, have interesting special formulas; see Section 14.7.

**Theorem 14.5.6 (Skew determinantal formula [5, 29, 114])** *The number of SYT of skew shape $\lambda/\mu$, for partitions $\lambda = (\lambda_1, \ldots, \lambda_t)$ and $\mu = (\mu_1, \ldots, \mu_s)$ with $\mu_i \leq \lambda_i$ $(\forall i)$, is*

$$f^{\lambda/\mu} = |\lambda/\mu|! \cdot \det\left[\frac{1}{(\lambda_i - \mu_j - i + j)!}\right]_{i,j=1}^{t},$$

*with the conventions $\mu_j := 0$ for $j > s$ and $1/k! := 0$ for negative integers $k$.*

The following proof is inductive. There is another approach that uses the Jacobi-Trudi identity.

*Proof.* (Adapted from [29].) By induction on the size $n := |\lambda/\mu|$. Denote

$$a_{ij} := \frac{1}{(\lambda_i - \mu_j - i + j)!}.$$

For $n = 0$, $\lambda_i = \mu_i$ $(\forall i)$. Thus

$$i = j \Longrightarrow \lambda_i - \mu_i - i + i = 0 \Longrightarrow a_{ii} = 1$$

and

$$i > j \Longrightarrow \lambda_i - \mu_j - i + j < \lambda_i - \mu_j = \lambda_i - \lambda_j \leq 0 \Longrightarrow a_{ij} = 0.$$

Hence the matrix $(a_{ij})$ is upper triangular with diagonal entries 1, and $f^{\lambda/\mu} = 1 = 0!\det(a_{ij})$.

For the induction step assume that the claim holds for all skew shapes of size $n-1$, and consider a shape $\lambda/\mu$ of size $n$ with $t$ rows. The cell containing $n$ must be the last cell in its row and column. Therefore

$$f^{\lambda/\mu} = \sum_{i'} f^{(\lambda/\mu)_{i'}}$$

where $(\lambda/\mu)_{i'}$ is the shape $\lambda/\mu$ minus the last cell in row $i'$, and summation is over all the rows $i'$ that are nonempty and whose last cell is also last in its column. Explicitly, summation is over all $i'$ such that $\lambda_{i'} > \mu_{i'}$ as well as $\lambda_{i'} > \lambda_{i'+1}$. By the induction hypothesis,

$$f^{\lambda/\mu} = (n-1)! \sum_{i'} \det(a_{ij}^{(i')})$$

where $a_{ij}^{(i')}$ is the analogue of $a_{ij}$ for the shape $(\lambda/\mu)_{i'}$ and summation is over the above values of $i'$. In fact,

$$a_{ij}^{(i')} = \begin{cases} a_{ij}, & \text{if } i \neq i'; \\ (\lambda_i - \mu_j - i + j) \cdot a_{ij}, & \text{if } i = i'. \end{cases}$$

This holds for all values (positive, zero or negative) of $\lambda_i - \mu_j - i + j$. The rest of the proof consists of two steps.

**Step 1:** The above formula for $f^{\lambda/\mu}$ holds with summation extending over all $1 \leq i' \leq t$. Indeed, it suffices to show that

$$\lambda_{i'} = \mu_{i'} \text{ or } \lambda_{i'} = \lambda_{i'+1} \Longrightarrow \det(a_{ij}^{(i')}) = 0.$$

If $\lambda_{i'} = \lambda_{i'+1}$ then

$$\lambda_{i'+1} - \mu_j - (i'+1) + j = (\lambda_{i'} - 1) - \mu_j - i' + j \qquad (\forall j),$$

so that the matrix $(a_{ij}^{(i')})$ has two equal rows and hence its determinant is 0. If $\lambda_{i'} = \mu_{i'}$ then

$$j \leq i' < i \Longrightarrow \lambda_i - \mu_j - i + j < \lambda_i - \mu_j \leq \lambda_i - \mu_{i'} \leq \lambda_{i'} - \mu_{i'} = 0$$

and

$$j \leq i' = i \Longrightarrow (\lambda_{i'} - 1) - \mu_j - i' + j < \lambda_{i'} - \mu_j \leq \lambda_{i'} - \mu_{i'} = 0.$$

Thus the matrix $(a_{ij}^{(i')})$ has a zero submatrix corresponding to $j \leq i' \leq i$, which again implies that its determinant is zero, for example, by considering the determinant as a sum over permutations $\sigma \in \mathcal{S}_t$ and noting that, by the pigeon hole principle, there is no permutation satisfying $j = \sigma(i) > i'$ for all $i \geq i'$.

**Step 2:** Let $A_{ij}$ be the $(i,j)$-cofactor of the matrix $A = (a_{ij})$, so that

$$\det A = \sum_j a_{ij} A_{ij} \qquad (\forall i)$$

and also

$$\det A = \sum_i a_{ij} A_{ij} \qquad (\forall j).$$

Then, expanding along row $i'$,

$$\det(a_{ij}^{i'}) = \sum_j a_{i'j}^{(i')} A_{i'j} = \sum_j (c_{i'} - d_j) a_{i'j} A_{i'j}$$

where $c_i := \lambda_i - i$ and $d_j := \mu_j - j$. Thus

$$\begin{aligned}
\frac{f^{\lambda/\mu}}{(n-1)!} &= \sum_{i'=1}^{t} \det(a_{ij}^{(i')}) = \sum_{i'}\sum_j (c_{i'} - d_j) a_{i'j} A_{i'j} \\
&= \sum_{i'}\sum_j c_{i'} a_{i'j} A_{i'j} - \sum_{i'}\sum_j d_j a_{i'j} A_{i'j} \\
&= \sum_{i'} c_{i'} \det A - \sum_j d_j \det A = \left( \sum_{i'} c_{i'} - \sum_j d_j \right) \det A \\
&= \left( \sum_{i'} \lambda_{i'} - \sum_j \mu_j \right) \det A = |\lambda/\mu| \det A = n \det A
\end{aligned}$$

which completes the proof. ■

### 14.5.3 Shifted shapes

For a strict partition $\lambda$, let $g^\lambda := |\mathrm{SYT}(\lambda^*)|$ be the number of standard Young tableaux of shifted shape $\lambda$. Like ordinary shapes, shifted shapes have three types of formulas, namely product, hook length and determinantal. The product formula was proved by Schur [105], using representation theory, and then by Thrall [132], using recursion and combinatorial arguments.

**Theorem 14.5.7 (Schur's shifted product formula [105, 132], [69, p. 267, eq. (2)])** *For any strict partition $\lambda = (\lambda_1, \ldots, \lambda_t)$,*

$$g^\lambda = \frac{|\lambda|!}{\prod_{i=1}^{t} \lambda_i!} \cdot \prod_{(i,j):i<j} \frac{\lambda_i - \lambda_j}{\lambda_i + \lambda_j}.$$

**Definition 14.5.8** *For a cell $c = (i,j) \in [\lambda^*]$ let*

$$H_c^* := [\lambda^*] \cap \left( \{(i,j)\} \cup \{(i,j') \mid j' > j\} \cup \{(i',j) \mid i' > i\} \cup \{(j+1,j') \mid j' \geq j+1\} \right)$$

*be the correponding* **shifted hook***; note that the last set is relevant only for $j < t$. Let*

$$h_c^* := |H_c^*| = \begin{cases} \lambda_i + \lambda_{j+1}, & \text{if } j < t; \\ \lambda_i - j + |\{i' \mid i' \geq i,\, \lambda_{i'} + i' \geq j+1\}|, & \text{if } j \geq t. \end{cases}$$

*be the corresponding* **shifted hook length**.

For example, in the following diagram the cells in the shifted hook $H^*_{(1,2)}$ are marked

| | | | | |
|---|---|---|---|---|
| | • | • | • | • |
| | • | | | |
| | | • | • | |

and in the following diagram each cell is labeled by the corresponding shifted hook length.

| | | | | |
|---|---|---|---|---|
| 9 | 7 | 5 | 4 | 2 |
| | 6 | 4 | 3 | 1 |
| | | 2 | 1 | |

**Theorem 14.5.9 (Shifted hook length formula [69, p. 267, eq. (1)])** *For any strict partition $\lambda = (\lambda_1, \ldots, \lambda_t)$,*

$$g^\lambda = \frac{|\lambda|!}{\prod_{c \in [\lambda^*]} h^*_c}.$$

**Theorem 14.5.10 (Shifted determinantal formula)** *For any strict partition $\lambda = (\lambda_1, \ldots, \lambda_t)$,*

$$g^\lambda = \frac{|\lambda|!}{\prod_{(i,j):i<j}(\lambda_i + \lambda_j)} \cdot \det\left[\frac{1}{(\lambda_i - t + j)!}\right]_{i,j=1}^{t},$$

*using the convention $1/k! := 0$ for negative integers $k$.*

The formulas in Theorems 14.5.7, 14.5.9 and 14.5.10 can be shown to be equivalent in much the same way as was done for ordinary shapes in Section 14.5. Note that the factors of the first denominator in the determinantal formula (Theorem 14.5.10) are precisely the shifted hook lengths $h^*_c$ for cells $c = (i, j)$ in the region $j < t$.

## 14.6 More proofs of the hook length formula

### 14.6.1 A probabilistic proof

Probabilistic proofs rely on procedures for a random choice of an object from a set. The key observation is that a uniform distribution implies an exact evaluation and "almost uniform" distributions yield good bounds.

A seminal example is the Greene-Nijenhuis-Wilf probabilistic proof of the ordinary hook length formula, to be described here. Our outline follows Sagan's description, in the first edition of [99], of the original proof of Greene, Nijenhuis and Wilf [46].

We start with a procedure that generates a random SYT of a given ordinary shape $D$. Recall from Definition 14.5.2 the notions of **hook** $H_c$ and **hook length** $h_c$ corresponding to a cell $c \in D$. A **corner** of $D$ is a cell that is last in its row as well as in its column (equivalently, has hook length 1).

**Algorithm 14.6.1**

***Input:*** *$D$, a diagram of ordinary shape.*
***Output:*** *A random $T \in \mathrm{SYT}(D)$.*

1: ***procedure*** RANDOMSYT($D$)
2:     ***while*** *$D$ is not empty* ***do***
3:         $n \leftarrow |D|$
4:         *Choose randomly a cell $c \in D$*    ▷ *with uniform probability $1/n$*
5:         ***while*** *$c$ is not a corner of $D$* ***do***
6:             *Choose randomly a cell $c' \in H_c \setminus \{c\}$*    ▷ *with uniform probability $1/(h_c - 1)$*
7:             $c \leftarrow c'$
8:         ***end while***
9:         $T(c) \leftarrow n$
10:         $D \leftarrow D \setminus \{c\}$
11:     ***end while***
12:     ***return*** *$T$*
13: ***end procedure***

We claim that this procedure produces each SYT of shape $D$ with the same probability. Explicitly,

**Lemma 14.6.2** *The procedure* RANDOMSYT *produces each SYT of shape $D$ with probability*

$$p = \frac{1}{|D|!} \prod_{c \in D} h_c .$$

*Proof.* By induction on $n := |D|$. The claim clearly holds for $n = 0, 1$.

Suppose that the claim holds for all shapes of size $n-1$, and let $D$ be an ordinary shape of size $n$. Let $T \in \mathrm{SYT}(D)$, and assume that $T(v) = n$ for some corner $v = (\alpha, \beta)$. Denote $D' := D \setminus \{v\}$, and let $T' \in \mathrm{SYT}(D')$ be the restriction of $T$ to $D'$.

In order to produce $T$, the algorithm needs to first produce $v$ (in rows 4–8, given $D$), and then move on to produce $T'$ from $D'$. By the induction hypothesis, it suffices to show that the probability that rows 4–8 produce the corner $v = (\alpha, \beta)$ is

$$\frac{\prod_{c \in D} h_c / n!}{\prod_{c \in D'} h'_c / (n-1)!} = \frac{1}{n} \prod_{c \in D'} \frac{h_c}{h'_c} = \frac{1}{n} \prod_{i=1}^{\alpha-1} \frac{h_{i,\beta}}{h_{i,\beta} - 1} \prod_{j=1}^{\beta-1} \frac{h_{\alpha,j}}{h_{\alpha,j} - 1},$$

where $h'_c$ denotes hook length in $D'$. This is equal to

$$\frac{1}{n} \prod_{i=1}^{\alpha-1} \left(1 + \frac{1}{h_{i,\beta} - 1}\right) \prod_{j=1}^{\beta-1} \left(1 + \frac{1}{h_{\alpha,j} - 1}\right) = \frac{1}{n} \sum_{\substack{A \subseteq [\alpha-1] \\ B \subseteq [\beta-1]}} \prod_{i \in A} \frac{1}{h_{i,\beta} - 1} \prod_{j \in B} \frac{1}{h_{\alpha,j} - 1} .$$

Following Sagan [99] we call any possible sequence of cells of $D$ obtained by lines 4–8 of the procedure (starting at a random $c$ and ending at the given corner $v$) a **trial**. For each trial $\tau$, let

$$A(\tau) := \{i < \alpha : \exists j \text{ s.t. } (i, j) \text{ is a cell in the trial } \tau\} \subseteq [1, \alpha - 1]$$

be its **horizontal projection** and let

$$B(\tau) := \{j < \beta : \exists i \text{ s.t. } (i,j) \text{ is a cell in the trial } \tau\} \subseteq [1, \beta - 1]$$

be its **vertical projection.**

It then suffices to show that for any given $A \subseteq [1, \alpha - 1]$ and $B \subseteq [1, \beta - 1]$, the sum of probabilities of all trials $\tau$ ending at $v = (\alpha, \beta)$ such that $A(\tau) = A$ and $B(\tau) = B$ is

$$\frac{1}{n} \prod_{i \in A} \frac{1}{h_{i,\beta} - 1} \prod_{j \in B} \frac{1}{h_{\alpha,j} - 1}.$$

This may be proved by induction on $|A \cup B|$. ∎

Lemma 14.6.2 says that the algorithm produces each $T \in \mathrm{SYT}(D)$ with the same probability $p$. The number of SYT of shape $D$ is therefore $1/p$, proving the hook length formula (Theorem 14.5.3).

For a fully detailed proof see [46] or the first edition of [99]. A similar method was applied in [101] to prove the hook length formula for shifted shapes (Theorem 14.5.9 above).

### 14.6.2 Bijective proofs

There are several bijective proofs of the (ordinary) hook length formula. Franzblau and Zeilberger [35] gave a bijection that is rather simple to describe, but breaks the row-column symmetry of hooks. Remmel [95] used the Garsia-Milne involution principle [40] to produce a composition of maps, "bijectivizing" recurrence relations. Zeilberger [140] then gave a bijective version of the probabilistic proof of Greene, Nijenhuis and Wilf [46] (described in the previous subsection). Krattenthaler [58] combined the Hillman-Grassl algorithm [51] and Stanley's $(P, \omega)$-partition theorem with the involution principle. Novelli, Pak and Stoyanovskii [76] gave a complete proof of a bijective algorithm previously outlined by Pak and Stoyanovskii [80]. A generalization of their method was given by Krattenthaler [59].

Bijective proofs for the shifted hook length formula were given by Krattenthaler [58] and Fischer [30].

A bijective proof of the ordinary determinantal formula was given by Zeilberger [139]; see also [66] and [57].

We shall briefly describe here the bijections of Franzblau-Zeilberger and of Novelli-Pak-Stoyanovskii. Only the algorithms (for the map in one direction) will be specified; the interested reader is referred to the original papers (or to [99]) for more complete descriptions and proofs.

The basic setting for both bijections is the following.

**Definition 14.6.3** *Let $\lambda$ be a partition of $n$ and $A$ a set of positive integers such that $|A| = |\lambda|$. A* **Young tableaux** *of (ordinary) shape $\lambda$ and image $A$ is a bijection $R : [\lambda] \to A$, not required to be order-preserving. A* **pointer tableau** *(or* **hook function***) of shape $\lambda$ is a function $P : [\lambda] \to \mathbb{Z}$ that assigns to each cell $c \in [\lambda]$ a pointer $p(c')$*

*which encodes some cell $c'$ in the hook $H_c$ of $c$ (see Definition 14.5.2). The pointer corresponding to $c' \in H_c$ is defined as follows:*

$$p(c') := \begin{cases} j, & \text{if } c' \text{ is } j \text{ columns to the right of } c, \text{ in the same row;} \\ 0, & \text{if } c' = c; \\ -i, & \text{if } c' \text{ is } i \text{ rows below } c, \text{ in the same column.} \end{cases}$$

*Let* $\mathrm{YT}(\lambda, A)$ *denote the set of all Young tableaux of shape* $\lambda$ *and image* $A$, $\mathrm{PT}(\lambda)$ *the set of all pointer tableaux of shape* $\lambda$, *and* $\mathrm{SPT}(\lambda, A)$ *the set of all pairs* $(T,P)$ *("standard and pointer tableaux") where* $T \in \mathrm{SYT}(\lambda, A)$ *and* $P \in \mathrm{PT}(\lambda)$. $\mathrm{YT}(\lambda)$ *is a shorthand for* $\mathrm{YT}(\lambda, [n])$ *where* $n = |\lambda|$, *and* $\mathrm{SPT}(\lambda)$ *a shorthand for* $\mathrm{SPT}(\lambda, [n])$.

**Example 14.6.4** *A typical hook, with each cell marked by its pointer:*

$$\begin{array}{|c|c|c|c|c|} \hline 0 & 1 & 2 & 3 & 4 \\ \hline -1 \\ \cline{1-1} -2 \\ \cline{1-1} \end{array}$$

The hook length formula that we want to prove may be written as

$$n! = f^{\lambda} \cdot \prod_{c \in [\lambda]} h_c \, .$$

The LHS of this formula is the size of $\mathrm{YT}(\lambda)$, while the RHS is the size of $\mathrm{SPT}(\lambda)$. Any explicit bijection $f : \mathrm{YT}(\lambda) \to \mathrm{SPT}(\lambda)$ will prove the hook length formula. As promised, we shall present algorithms for two such bijections.

**The Franzblau-Zeilberger algorithm [35].** The main procedure, that we will call FZ-SORTTABLEAU, "sorts" a YT $R$ of ordinary shape, column by column from right to left, to produce a SPT $(T,P)$ of the same shape. The pointer tableau $P$ records each step of the sorting, keeping just enough information to enable reversal of the procedure. Note that $\varnothing$ denotes the empty tableau.

**Algorithm 14.6.5**
***Input:*** $R \in \mathrm{YT}(\lambda)$.
***Output:*** $(T,P) \in \mathrm{SPT}(\lambda)$.

```
procedure FZ-SORTTABLEAU(R)
    (T,P) ← (∅,∅)                                  ▷ Initialize
    m ← number of columns of R
    for j ← m downto 1 do                          ▷ Add columns from right to left
        c ← column j of R
        (T,P) ← INSERTCOLUMN(T,P,c)
    end for
    return (T,P)
end procedure
```

The algorithm makes repeated use of the following procedure INSERTCOLUMN:

**Algorithm 14.6.6**

***Input:*** $(T,P,c)$, *where* $(T,P) \in \mathrm{SPT}(\mu,A)$ *for some ordinary shape* $\mu$ *and some set* $A$ *of positive numbers of size* $|A| = |\mu|$ *such that all the rows of* $T$ *are increasing, and* $c = (c_1,\ldots,c_m)$ *is a vector of distinct positive integers* $c_i \notin A$ *whose length* $m \geq \ell(\mu)$.

***Output:*** $(T',P') \in \mathrm{SPT}(\mu',A')$, *where* $A' = A \cup \{c_1,\ldots,c_m\}$ *and* $\mu'$ *is obtained from* $\mu$ *by attaching a new first column of length* $m$.

**procedure** INSERTCOLUMN$(T,P,c)$
    **for** $i \leftarrow 1$ **to** $m$ **do**
        $T \leftarrow$ INSERT$(T,i,c_i)$ ▷ *Insert* $c_i$ *into row* $i$ *of* $T$, *keeping the row entries increasing*
        $d_i \leftarrow$ (*new column index of* $c_i$) $- 1$ ▷ *Initialize the pointer* $d_i$
    **end for**
    **while** *T is not a Standard Young Tableau* **do**
        $(k,x) \leftarrow T^{-1}(\min\{T(i,j) \mid T(i-1,j) > T(i,j)\})$ ▷ *The smallest entry out of order*
        $y \leftarrow d_{k-1} + 1$ ▷ *Claim:* $y > 0$
        $T \leftarrow$ EXCHANGE$(T,(k,x),(k-1,y))$
        $y' \leftarrow$ *new column index of the old* $T(k-1,y)$ ▷ *The new row index is* $k$
        ▷ *Update the pointers* $d_{k-1}$ *and* $d_k$

$$d_{k-1} \leftarrow \begin{cases} v, & \text{if } d_k = v \geq 0,\ v \neq x-1; \\ -1, & \text{if } d_k = x-1; \\ -(u+1), & \text{if } d_k = -u < 0. \end{cases}$$

        $d_k \leftarrow y' - 1$
    **end while**
    $P \leftarrow$ ATTACH$(d,P)$ ▷ *Attach* $d$ *to* $P$ *as a first column*
    **return** $(T,P)$ ▷ *T is now a SYT*
**end procedure**

This procedure makes use of some elementary operations, which may be described as follows:

- INSERT$(T,i,c_i)$ inserts the entry $c_i$ into row $i$ of $T$, reordering this row to keep it increasing.
- EXCHANGE$(T,(k,x),(\ell,y))$ exchanges the entries in cells $(k,x)$ and $(\ell,y)$ of $T$ and then reorders rows $k$ and $\ell$ to keep them increasing.
- ATTACH$(d,P)$ attaches the vector $d$ to the pointer tableau $P$ as a new first column.

**Example 14.6.7** *An instance of* INSERTCOLUMN$(T,P,c)$ *with*

$$T = \begin{array}{|c|c|} \hline 1 & 8 \\ \hline 4 \\ \cline{1-1} 7 \\ \cline{1-1} \end{array}, \quad P = \begin{array}{|c|c|} \hline 1 & 0 \\ \hline 0 \\ \cline{1-1} 0 \\ \cline{1-1} \end{array}, \quad c = \begin{array}{|c|} \hline 12 \\ \hline 5 \\ \hline 3 \\ \hline 6 \\ \hline \end{array}$$

*proceeds as follows (with the smallest entry out of order set in boldface):*

$$(T,d) = \begin{array}{|c|c|c|}\hline 1 & 8 & 12 \\ \hline 4 & 5 \\ \cline{1-2} \mathbf{3} & 7 \\ \cline{1-2} 6 \\ \cline{1-1}\end{array} \quad \begin{array}{|c|}\hline 2 \\ \hline 1 \\ \hline 0 \\ \hline 0 \\ \hline\end{array} \rightarrow \begin{array}{|c|c|c|}\hline 1 & 8 & 12 \\ \hline 3 & \mathbf{4} \\ \cline{1-2} 5 & 7 \\ \cline{1-2} 6 \\ \cline{1-1}\end{array} \quad \begin{array}{|c|}\hline 2 \\ \hline -1 \\ \hline 0 \\ \hline 0 \\ \hline\end{array} \rightarrow \begin{array}{|c|c|c|}\hline 1 & 4 & 8 \\ \hline 3 & 12 \\ \cline{1-2} 5 & \mathbf{7} \\ \cline{1-2} 6 \\ \cline{1-1}\end{array} \quad \begin{array}{|c|}\hline -2 \\ \hline 1 \\ \hline 0 \\ \hline 0 \\ \hline\end{array}$$

$$\rightarrow \begin{array}{|c|c|c|}\hline 1 & 4 & 8 \\ \hline 3 & 7 \\ \cline{1-2} 5 & 12 \\ \cline{1-2} 6 \\ \cline{1-1}\end{array} \quad \begin{array}{|c|}\hline -2 \\ \hline 0 \\ \hline 1 \\ \hline 0 \\ \hline\end{array}$$

*and yields*

$$T = \begin{array}{|c|c|c|}\hline 1 & 4 & 8 \\ \hline 3 & 7 \\ \cline{1-2} 5 & 12 \\ \cline{1-2} 6 \\ \cline{1-1}\end{array} \;, \quad P = \begin{array}{|c|c|c|}\hline -2 & 1 & 0 \\ \hline 0 & 0 \\ \cline{1-2} 1 & 0 \\ \cline{1-2} 0 \\ \cline{1-1}\end{array} \;.$$

*An instance of* FZ-SORTTABLEAU($R$) *with*

$$R = \begin{array}{|c|c|c|c|}\hline 9 & 12 & 8 & 1 \\ \hline 2 & 5 & 4 \\ \cline{1-3} 11 & 3 & 7 \\ \cline{1-3} 10 & 6 \\ \cline{1-2}\end{array}$$

*proceeds as follows (the second step being the instance above):*

$$\begin{array}{|c|}\hline 1 \\ \hline\end{array} \quad \begin{array}{|c|}\hline 0 \\ \hline\end{array} \rightarrow \begin{array}{|c|c|}\hline 1 & 8 \\ \hline 4 \\ \cline{1-1} 7 \\ \cline{1-1}\end{array} \quad \begin{array}{|c|c|}\hline 1 & 0 \\ \hline 0 \\ \cline{1-1} 0 \\ \cline{1-1}\end{array} \rightarrow \begin{array}{|c|c|c|}\hline 1 & 4 & 8 \\ \hline 3 & 7 \\ \cline{1-2} 5 & 12 \\ \cline{1-2} 6 \\ \cline{1-1}\end{array} \quad \begin{array}{|c|c|c|}\hline -2 & 1 & 0 \\ \hline 0 & 0 \\ \cline{1-2} 1 & 0 \\ \cline{1-2} 0 \\ \cline{1-1}\end{array}$$

$$\rightarrow \begin{array}{|c|c|c|c|}\hline 1 & 3 & 4 & 8 \\ \hline 2 & 7 & 9 \\ \cline{1-3} 5 & 10 & 12 \\ \cline{1-3} 6 & 11 \\ \cline{1-2}\end{array} \quad \begin{array}{|c|c|c|c|}\hline 0 & -2 & 1 & 0 \\ \hline 2 & 0 & 0 \\ \cline{1-3} -1 & 1 & 0 \\ \cline{1-3} 1 & 0 \\ \cline{1-2}\end{array} = (T,P) \;.$$

**The Novelli-Pak-Stoyanovskii algorithm [76].** Again, we prove the hook length formula

$$n! = f^\lambda \cdot \prod_{c \in [\lambda]} h_c$$

by building an explicit bijection $f : \mathrm{YT}(\lambda) \to \mathrm{SPT}(\lambda)$. However, instead of building the tableaux column by column, we shall use a modified jeu de taquin to unscramble the entries of $R \in \mathrm{YT}(\lambda)$ so that rows and columns increase. Again, a pointer tableau will keep track of the process so as to make it invertible. Our description will essentially follow [99].

First, define a linear (total) order on the cells of a diagram $D$ of ordinary shape by defining

$$(i,j) \trianglelefteq (i',j') \iff \text{either } j > j' \text{ or } j = j' \text{ and } i \geq i'.$$

For example, the cells of the following diagram are labeled 1 to 7 according to this linear order:

$$\begin{array}{|c|c|c|}\hline 7 & 4 & 2 \\ \hline 6 & 3 & 1 \\ \hline 5 \\ \cline{1-1}\end{array}$$

If $R \in \mathrm{YT}(\lambda)$ and $c \in D := [\lambda]$, let $R^{\trianglelefteq c}$ (respectively $R^{\triangleleft c}$) be the tableau consisting of all cells $b \in D$ with $b \trianglelefteq c$ (respectively, $\triangleleft c$).

Define a procedure MFORWARDSLIDE, which is the procedure FORWARDSLIDE from the description of jeu de taquin, with the following two modifications:

1. Its input is $(T, c)$ with $T \in \mathrm{YT}$ rather than $T \in \mathrm{SYT}$.
2. Its output is $(T, c)$ (see there), rather than just $T$.

**Algorithm 14.6.8**
***Input:*** $R \in \mathrm{YT}(\lambda)$.
***Output:*** $(T, P) \in SPT(\lambda)$.

```
procedure NPS(R)
    T ← R
    D ← sh(T)
    P ← 0 ∈ PT(D)                          ▷ A pointer tableau of shape D filled with zeros
    while T is not standard do
        c ← the ◁-maximal cell such that T^{◁c} is standard
        (T′, c′) ← MFORWARDSLIDE(T^{◁c}, c)
        for b ∈ D do          ▷ Replace T^{⊴c} by T′, except that T(c′) ← the old T(c)
                     ⎧ T(b),   if b ▷ c;
            T(b) ←   ⎨ T′(b),  if b ⊴ c and b ≠ c′;
                     ⎩ T(c),   if b = c′.
        end for
        Let c = (i_0, j_0) and c′ = (i_1, j_1)      ▷ Necessarily i_0 ≤ i_1 and j_0 ≤ j_1
        for i from i_0 to i_1 − 1 do
            P(i, j_0) ← P(i+1, j_0) − 1
        end for
        P(i_1, j_0) ← j_1 − j_0
    end while
    return T                            ▷ Now c_0 ∈ D ⊆ (c_0)_+, so D has ordinary shape
end procedure
```

**Example 14.6.9** *For*

$$R = \begin{array}{|c|c|}\hline 6 & 2 \\ \hline 4 & 3 \\ \hline 5 & 1 \\ \hline\end{array}\ ,$$

*here is the sequence of pairs* $(T, P)$ *produced during the computation of* NPS$(R)$ *(with* $c$ *in boldface):*

$$\begin{array}{|c|c|}\hline 6 & 2 \\ \hline 4 & \mathbf{3} \\ \hline 5 & 1 \\ \hline\end{array}\ \begin{array}{|c|c|}\hline 0 & 0 \\ \hline 0 & 0 \\ \hline 0 & 0 \\ \hline\end{array} \ \to\ \begin{array}{|c|c|}\hline 6 & \mathbf{2} \\ \hline 4 & 1 \\ \hline 5 & 3 \\ \hline\end{array}\ \begin{array}{|c|c|}\hline 0 & 0 \\ \hline 0 & -1 \\ \hline 0 & 0 \\ \hline\end{array} \ \to\ \begin{array}{|c|c|}\hline 6 & 1 \\ \hline 4 & 2 \\ \hline \mathbf{5} & 3 \\ \hline\end{array}\ \begin{array}{|c|c|}\hline 0 & -2 \\ \hline 0 & 0 \\ \hline 0 & 0 \\ \hline\end{array}$$

$$\rightarrow \begin{array}{|c|c|}\hline 6&1\\\hline \mathbf{4}&2\\\hline 3&5\\\hline\end{array}\ \begin{array}{|c|c|}\hline 0&-2\\\hline 0&0\\\hline 1&0\\\hline\end{array} \rightarrow \begin{array}{|c|c|}\hline \mathbf{6}&1\\\hline 2&4\\\hline 3&5\\\hline\end{array}\ \begin{array}{|c|c|}\hline 0&-2\\\hline 1&0\\\hline 1&0\\\hline\end{array} \rightarrow \begin{array}{|c|c|}\hline 1&4\\\hline 2&5\\\hline 3&6\\\hline\end{array}\ \begin{array}{|c|c|}\hline 0&-2\\\hline 0&0\\\hline 1&0\\\hline\end{array}.$$

### 14.6.3 Partial difference operators

MacMahon [72] has originally used partial difference equations, also known as recurrence relations, to solve various enumeration problems, among them the enumeration of ballot sequences, or equivalently SYT of an ordinary shape (see Section 14.2.5.2). Zeilberger [138] improved on MacMahon's proof by extending the domain of definition of the enumerating functions, thus simplifying the boundary conditions: In PDE terminology, a Neumann boundary condition (zero normal derivatives) was replaced by a Dirichlet boundary condition (zero function values). He also made explicit use of the algebra of partial difference operators; we shall present here a variant of his approach.

Consider, for example, the two dimensional ballot problem, namely finding the number $F(m_1,m_2)$ of lattice paths from $(0,0)$ to $(m_1,m_2)$ that stay in the region $m_1 \geq m_2 \geq 0$. MacMahon [72, p. 127] has set the partial difference equation

$$F(m_1,m_2) = F(m_1-1,m_2) + F(m_1,m_2-1) \qquad (m_1 > m_2 > 0)$$

with the boundary conditions

$$F(m_1,m_2) = F(m_1,m_2-1) \qquad (m_1 = m_2 > 0)$$

and

$$F(m_1,0) = 1 \qquad (m_1 \geq 0).$$

By extending $F$ to the region $m_1 \geq m_2 - 1$, the recursion can be required to hold for almost all $m_1 \geq m_2 \geq 0$:

$$F(m_1,m_2) = F(m_1-1,m_2) + F(m_1,m_2-1) \qquad (m_1 \geq m_2 \geq 0,\ (m_1,m_2) \neq (0,0))$$

with

$$F(m_1,m_2) = 0 \qquad (m_1 = m_2 - 1 \text{ or } m_2 = -1)$$

and

$$F(0,0) = 1.$$

In general, consider functions $f : \mathbb{Z}^n \to \mathbb{C}$ and define the fundamental **shift operators** $X_1, \ldots, X_n$ by

$$(X_i f)(m_1, \ldots, m_n) := f(m_1, \ldots, m_i + 1, \ldots, m_n) \qquad (1 \leq i \leq n).$$

For $\alpha \in \mathbb{Z}^n$ write $X^\alpha = X_1^{\alpha_1} \cdots X_n^{\alpha_n}$, so that $(X^\alpha f)(m) = f(m+\alpha)$. A typical linear partial difference operator with constant coefficients has the form

$$P = p(X_1^{-1}, \ldots, X_n^{-1}) = \sum_{\alpha \geq 0} a_\alpha X^{-\alpha},$$

for some polynomial $p(z)$ with complex coefficients, so that $a_\alpha \in \mathbb{C}$ for each $\alpha = (\alpha_1, \ldots, \alpha_n) \in \mathbb{N}^n$ (where $\mathbb{N}$ is the set of nonnegative integers) and $a_\alpha \neq 0$ for only finitely many values of $\alpha$. We also assume that $p(0) = a_0 = 1$.

**Definition 14.6.10** *Define the* **discrete delta function** $\delta : \mathbb{Z}^n \to \mathbb{C}$ *by*

$$\delta(m) = \begin{cases} 1, & \text{if } m = 0; \\ 0, & \text{otherwise.} \end{cases}$$

*A function $f : \mathbb{Z}^n \to \mathbb{C}$ satisfying $Pf = \delta$ is called a* **fundamental solution** *corresponding to the operator P. If f is supported in $\mathbb{N}^n$, it is called a* **canonical** *fundamental solution.*

It is clear that each operator $P$ as above has a unique canonical fundamental solution.

In the following theorem we consider a slightly more general type of operators, which can be written as $X^\alpha p(X_1^{-1}, \ldots, X_n^{-1})$ for a polynomial $p$ and some $\alpha \geq 0$.

**Theorem 14.6.11 (A variation on [138, Theorem 2])** *Let $F_n = F_n(m_1, \ldots, m_n)$ be the canonical fundamental solution corresponding to an operator $P = p(X_1^{-1}, \ldots, X_n^{-1})$, where $p(z)$ is a symmetric polynomial with $p(0) = 1$. Denote*

$$\Delta_n := \prod_{(i,j):i<j} (I - X_i X_j^{-1}).$$

*Then $G_n = \Delta_n F_n$ is the unique solution of the equation $Pg = 0$ in the region*

$$\{(m_1, \ldots, m_n) \in \mathbb{Z}^n \mid m_1 \geq \ldots \geq m_n \geq 0\} \setminus \{(0, \ldots, 0)\}$$

*subject to the boundary conditions*

$$(\exists i)\, m_i = m_{i+1} - 1 \Longrightarrow g(m_1, \ldots, m_n) = 0,$$

$$m_n = -1 \Longrightarrow g(m_1, \ldots, m_n) = 0$$

*and*

$$g(0, \ldots, 0) = 1.$$

*Proof.* Since each $X_i$ commutes with the operator $P$, so does $\Delta_n$. Since $m_1 + \ldots + m_n$ is invariant under $X_i X_j^{-1}$ and $F_n$ is a solution of $Pf = 0$ in the complement of the hyperplane $m_1 + \ldots + m_n = 0$, so is $G_n$. It remains to verify that $G_n$ satisfies the prescribed boundary conditions. Now, by definition,

$$G_n(m) = (I - X_1 X_2^{-1}) A_{1,2} F_n(m)$$

where the operator $A_{1,2}$ is symmetric with respect to $X_1$, $X_2$. Since $F_n(m)$ is a symmetric function, we can write

$$G_n(m) = (I - X_1 X_2^{-1}) H(m)$$

where $H$ is symmetric with respect to $m_1$ and $m_2$. Suppressing the dependence on $m_3,\ldots,m_n$,

$$G_n(m_1,m_2) = (I - X_1X_2^{-1})H(m_1,m_2) = H(m_1,m_2) - H(m_1+1,m_2-1) = 0$$

whenever $m_1 = m_2 - 1$, by the symmetry of $H$. Similarly, for $i = 1,\ldots,n-1$, $G_n(m) = 0$ on $m_i = m_{i+1} - 1$. On $m_n = -1$,

$$G_n(m_1,\ldots,m_{n-1},-1) = \prod_{(i,j):1\le i<j\le n-1} (I - X_iX_j^{-1}) \cdot \prod_{1\le i\le n-1} (I - X_iX_n^{-1})F_n(m_1,\ldots,m_{n-1},-1).$$

Since $F_n(m) = 0$ for $m_n < 0$,

$$G_n(m_1,\ldots,m_{n-1},-1) = 0.$$

Finally, $F_n(m_1,\ldots,m_n) = 0$ on all of the hyperplane $m_1 + \ldots + m_n = 0$ except the origin $0 = (0,\ldots,0)$. Therefore

$$G_n(0) = \Delta_n F_n(0) = F_n(0) = 1. \qquad \blacksquare$$

For every function $f : \mathbb{Z}^n \to \mathbb{C}$ whose support is contained in a translate of $\mathbb{N}^n$ (i.e., such that there exists $N \in \mathbb{Z}$ such that $f(m_1,\ldots,m_n) = 0$ whenever $m_i < N$ for some $i$) there is a corresponding generating function (formal Laurent series)

$$\mathrm{gf}(f) := \sum_m f(m)z^m \in \mathbb{C}((z_1,\ldots,z_n)).$$

Let $p(z_1,\ldots,z_n)$ be a polynomial with complex coefficients and $p(0) = 1$, and let $P = p(X_1^{-1},\ldots,X_n^{-1})$ be the corresponding operator. Since $\mathrm{gf}(X^{-\alpha} f) = z^\alpha\,\mathrm{gf}(f)$ we have

$$\mathrm{gf}(Pf) = p(z_1,\ldots,z_n)\cdot \mathrm{gf}(f),$$

and therefore $Pf = \delta$ implies $\mathrm{gf}(f) = 1/p(z_1,\ldots,z_n)$.

**Definition 14.6.12 (MacMahon [72])** *Let $A \subseteq \mathbb{Z}^n$ and $f : A \to \mathbb{C}$. A formal Laurent series $\sum_m a(m)z^m$ is a* **redundant generating function for** $f$ **on** $A$ *if $f(m) = a(m)$ for all $m \in A$.*

**Theorem 14.6.13 (MacMahon [72, p. 133])** *Let $g(m)$ be the number of lattice paths from 0 to $m$, where travel is restricted to the region*

$$A = \{(m_1,\ldots,m_n) \in \mathbb{Z}^n \,|\, m_1 \ge m_2 \ge \ldots \ge m_n \ge 0\}.$$

*Then*

$$\prod_{(i,j):i<j} \left(1 - \frac{z_j}{z_i}\right) \cdot \frac{1}{1 - z_1 - \ldots - z_n}$$

*is a redundant generating function for $g$ on $A$ and therefore*

$$g(m) = \frac{(m_1+\ldots+m_n)!}{(m_1+n-1)!\cdots m_n!} \prod_{(i,j):i<j} (m_i - m_j + j - i).$$

This gives, of course, the ordinary product formula (Theorem 14.5.1).

*Proof.* Apply Theorem 14.6.11 with $P = I - X_1^{-1} - \ldots - X_n^{-1}$. The canonical fundamental solution of $Pf = \delta$ is easily seen to be the multinomial coefficient

$$F_n(m_1,\ldots,m_n) = \begin{cases} \frac{(m_1+\ldots+m_n)!}{m_1!\cdots m_n!}, & \text{if } m_i \geq 0\,(\forall i); \\ 0, & \text{otherwise,} \end{cases}$$

with generating function $(1 - z_1 - \ldots - z_n)^{-1}$.

The number of lattice paths in the statement of Theorem 14.6.13 clearly satisfies the conditions on $g$ in Theorem 14.6.11, and therefore

$$g = G_n = \Delta_n F_n \qquad (\text{on } A).$$

This implies the claimed redundant generating function for $g$ on $A$.

To get an explicit expression for $g(m)$ note that $(m_1 + \ldots + m_n)!$ is invariant under $X_i X_j^{-1}$, so that

$$\begin{aligned} g(m_1,\ldots,m_n) &= \prod_{(i,j):i<j} (I - X_i X_j^{-1}) \left[\frac{(m_1+\ldots+m_n)!}{m_1!\cdots m_n!}\right] \\ &= (m_1+\ldots+m_n)! \cdot \prod_{(i,j):i<j} (I - X_i X_j^{-1}) \left[\frac{1}{m_1!\cdots m_n!}\right]. \end{aligned}$$

Consider

$$\begin{aligned} H(m_1,\ldots,m_n) &:= \prod_{(i,j):i<j} (I - X_i X_j^{-1}) \left[\frac{1}{m_1!\cdots m_n!}\right] \\ &= \prod_{(i,j):i<j} (X_i^{-1} - X_j^{-1}) \cdot \prod_i X_i^{n-i} \left[\frac{1}{m_1!\cdots m_n!}\right] \\ &= \prod_{(i,j):i<j} (X_i^{-1} - X_j^{-1}) \left[\frac{1}{\ell_1!\ell_2!\cdots \ell_n!}\right], \end{aligned}$$

where $\ell_i := m_i + n - i$ $(1 \leq i \leq n)$. Clearly $H$ is an alternating (anti-symmetric) function of $\ell_1,\ldots,\ell_n$, which means that

$$H(m_1,\ldots,m_n) = \frac{Q(\ell_1,\ell_2,\ldots,\ell_n)}{\ell_1!\ell_2!\cdots\ell_n!},$$

where $Q$ is an alternating polynomial of degree $n-1$ in each of its variables. $g$, and therefore also $H$ and $Q$, vanish on each of the hyperplanes $m_i = m_{i+1} - 1$, namely $\ell_i = \ell_{i+1}$ $(1 \leq i \leq n-1)$. Hence $\ell_i - \ell_{i+1}$, and by symmetry also $\ell_i - \ell_j$ for each $i \neq j$, divide $Q$. Hence

$$Q(\ell) = c \prod_{(i,j):i<j} (\ell_i - \ell_j)$$

and

$$g(m) = c\frac{(m_1+\ldots+m_n)!}{(m_1+n-1)!\cdots m_n!}\prod_{(i,j):i<j}(m_i-m_j-i+j)$$

for a suitable constant $c$, which is easily found to be 1 by evaluating $g(0)$. ■

**Remark 14.6.14** Theorem 14.6.13 gives an expression of the number of SYT of ordinary shape $\lambda = (\lambda_1,\ldots,\lambda_t)$ as the coefficient of $z^\ell$ (where $\ell_i = \lambda_i + t - i$) in the power series

$$\prod_{(i,j):i<j}(z_i-z_j)\cdot\frac{1}{1-z_1-\ldots-z_t},$$

or as the constant term in the Laurent series

$$\prod_i z_i^{-\ell_i}\prod_{(i,j):i<j}(z_i-z_j)\cdot\frac{1}{1-z_1-\ldots-z_t}.$$

## 14.7 Formulas for skew strips

We focus our attention now on two important families of skew shapes, which are of special interest: Zigzag shapes and skew strips of constant width.

### 14.7.1 Zigzag shapes

Recall (from Section 14.3.3) that a **zigzag shape** is a path-connected skew shape that does not contain a $2\times 2$ square.

**Definition 14.7.1** *For any subset $S\subseteq[n-1] := \{1,\ldots,n-1\}$ define a zigzag shape $D = \text{zigzag}_n(S)$, with cells labeled $1,\ldots,n$, as follows: Start with an initial cell labeled 1. For each $1\le i\le n-1$, given the cell labeled i, add an adjacent cell labeled $i+1$ above cell i if $i\in S$, and to the right of cell i otherwise.*

**Example 14.7.2**

$n=9, S=\{1,3,5,6\}$ $\longrightarrow$

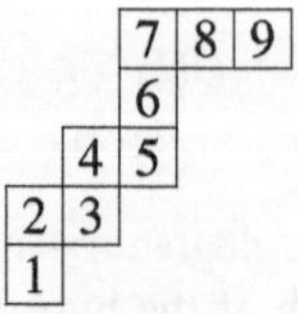


$\longrightarrow$ $\text{zigzag}_9(S) =$

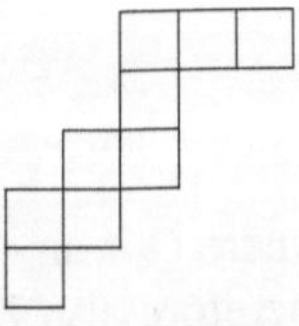

This defines a bijection between the set of all subsets of $[n-1]$ and the set of all zigzag shapes of size $n$ (up to translation). The set $S$ consists of the labels of all the cells in the shape $\text{zigzag}_n(S)$ such that the cell directly above is also in the shape. These are exactly the last (rightmost) cells in all the rows except the top row.

Recording the lengths of all the rows in the zigzag shape, from bottom up, it follows that zigzag shapes of size $n$ are also in bijection with all the **compositions** of $n$. In fact, given a subset $S = \{s_1, \ldots, s_k\} \subseteq [n-1]$ (with $s_1 < \ldots < s_k$), the composition corresponding to $\text{zigzag}_n(S)$ is simply $(s_1, s_2 - s_1, \ldots, s_k - s_{k-1}, n - s_k)$.

**Theorem 14.7.3** [72, 119] *Let $S = \{s_1, \ldots, s_k\} \subseteq [n-1]$ $(s_1 < \ldots < s_k)$ and set $s_0 :=$ 0 and $s_{k+1} := n$. Then*

$$f^{\text{zigzag}_n(S)} = n! \cdot \det\left[\frac{1}{(s_{j+1} - s_i)!}\right]_{i,j=0}^{k} = \det\left[\binom{n - s_i}{s_{j+1} - s_i}\right]_{i,j=0}^{k} = \det\left[\binom{s_{j+1}}{s_{j+1} - s_i}\right].$$

For example, the zigzag shape

$[\lambda/\mu] =$ 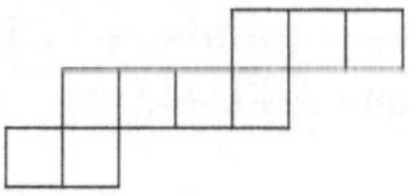

corresponds to $n = 9$ and $S = \{2, 6\}$, and therefore

$$f^{\lambda/\mu} = 9! \cdot \det\begin{bmatrix} \frac{1}{2!} & \frac{1}{6!} & \frac{1}{9!} \\ 1 & \frac{1}{4!} & \frac{1}{7!} \\ 0 & 1 & \frac{1}{3!} \end{bmatrix} = \det\begin{bmatrix} \binom{9}{2} & \binom{9}{6} & \binom{9}{9} \\ \binom{7}{0} & \binom{7}{4} & \binom{7}{7} \\ 0 & \binom{3}{0} & \binom{3}{3} \end{bmatrix} = \det\begin{bmatrix} \binom{2}{2} & \binom{6}{6} & \binom{9}{9} \\ \binom{2}{0} & \binom{6}{4} & \binom{9}{7} \\ 0 & \binom{6}{0} & \binom{9}{3} \end{bmatrix}$$

Theorem 14.7.3 is a special case of the determinantal formula for skew shapes (Theorem 14.5.6). We shall now consider a specific family of examples.

**Example 14.7.4** *Consider, for each nonnegative integer n, one special zigzag shape $D_n$. For n even it has all rows of length 2:*

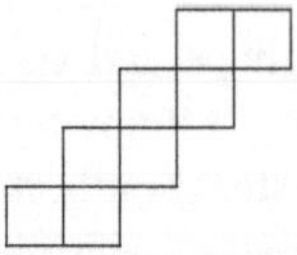

*and for n odd it has a top row of length 1 and all others of length 2:*

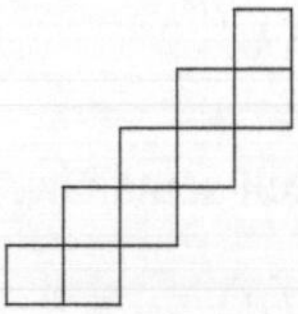

*Clearly, by Definition 14.7.1, $D_n = \text{zigzag}_n(S)$ for $S = \{2, 4, 6, \ldots\} \subseteq [n-1]$.*

**Definition 14.7.5** *A permutation $\sigma \in S_n$ is* **up-down** *(or* **alternating**, *or* **zigzag**) *if*

$$\sigma(1) < \sigma(2) > \sigma(3) < \sigma(4) > \ldots \quad .$$

**Observation 14.7.6** *If we label the cells of $D_n$ as in Definition 14.7.1 (see Example 14.7.2), then clearly each standard Young tableau $T : D_n \to [n]$ becomes an up-down permutation, and vice versa. (For an extension of this phenomenon see Proposition 14.10.12.)*

Up-down permutations were already studied by André [6, 7] in the nineteenth century. He showed that their number $A_n$ [110, A000111] satisfies

**Proposition 14.7.7**

$$\sum_{n=0}^{\infty} \frac{A_n x^n}{n!} = \sec x + \tan x.$$

They are therefore directly related to the **secant** (or **zig**, or **Euler**) **numbers** $E_n$ [110, A000364], the **tangent** (or **zag**) **numbers** $T_n$ [110, A000182] and the **Bernoulli numbers** $B_n$ [110, A000367 and A002445] by

$$A_{2n} = (-1)^n E_{2n} \qquad (n \geq 0)$$

and

$$A_{2n-1} = T_n = \frac{(-1)^{n-1} 2^{2n} (2^{2n} - 1)}{2n} B_{2n} \qquad (n \geq 1).$$

Note that there is an alternative convention for denoting Euler numbers, by which $E_n = A_n$ for all $n$.

**Proposition 14.7.8**

$$2A_{n+1} = \sum_{k=0}^{n} \binom{n}{k} A_k A_{n-k} \qquad (n \geq 1)$$

*with* $A_0 = A_1 = 1$.

*Proof.* In a SYT of the required shape and size $n+1$, the cell containing $n+1$ must be the last in its row and column. Removing this cell leaves at most two path-connected components, with the western/southern one necessarily of odd size (for $n \geq 1$). It follows that

$$A_{n+1} = \sum_{\substack{k=0 \\ k \text{ odd}}}^{n} \binom{n}{k} A_k A_{n-k} \qquad (n \geq 1).$$

Applying a similar argument to the cell containing 1 gives

$$A_{n+1} = \sum_{\substack{k=0 \\ k \text{ even}}}^{n} \binom{n}{k} A_k A_{n-k} \qquad (n \geq 0),$$

and adding the two formulas gives the required recursion. ∎

Indeed, the recursion for $A_n$ (Proposition 14.7.8) can be seen to be equivalent to the generating function (Proposition 14.7.7) since $f(x) = \sec x + \tan x$ satisfies the differential equation

$$2f'(x) = 1 + f(x)^2$$

with $f(0) = 1$.

Proposition 14.7.3 thus gives the determinantal formulas

$$(-1)^n E_{2n} = (2n)! \cdot \det \begin{pmatrix} \frac{1}{2!} & \frac{1}{4!} & \frac{1}{6!} & \cdot & \cdot & \frac{1}{(2n-4)!} & \frac{1}{(2n-2)!} & \frac{1}{(2n)!} \\ 1 & \frac{1}{2!} & \frac{1}{4!} & \cdot & \cdot & \cdot & \frac{1}{(2n-4)!} & \frac{1}{(2n-2)!} \\ 0 & 1 & \frac{1}{2!} & \cdot & \cdot & \cdot & \cdot & \frac{1}{(2n-4)!} \\ \vdots & \vdots & \vdots & \cdot & \cdot & \cdot & \vdots & \vdots \\ 0 & 0 & 0 & \cdot & \cdot & \cdot & 1 & \frac{1}{2!} \end{pmatrix}$$

and

$$T_n = (2n-1)! \cdot \det \begin{pmatrix} \frac{1}{2!} & \frac{1}{4!} & \frac{1}{6!} & \cdot & \cdot & \frac{1}{(2n-4)!} & \frac{1}{(2n-2)!} & \frac{1}{(2n-1)!} \\ 1 & \frac{1}{2!} & \frac{1}{4!} & \cdot & \cdot & \cdot & \frac{1}{(2n-4)!} & \frac{1}{(2n-3)!} \\ 0 & 1 & \frac{1}{2!} & \cdot & \cdot & \cdot & \cdot & \frac{1}{(2n-5)!} \\ \vdots & \vdots & \vdots & \cdot & \cdot & \cdot & \vdots & \vdots \\ 0 & 0 & 0 & \cdot & \cdot & \cdot & 1 & \frac{1}{1!} \end{pmatrix}$$

### 14.7.2 Skew strips of constant width

The **basic skew strip of width $m$ and height $n$** is the skew diagram

$$D_{m,n} = [(n+m-1, n+m-2, \ldots, m+1, m)/(n-1, n-2, \ldots, 1, 0)].$$

It has $n$ rows, of length $m$ each, with each row shifted one cell to the left with respect to the row just above it. For example, $D_{2,n}$ is $D_{2n}$ from Example 14.7.4 above while

$D_{3,5} =$ 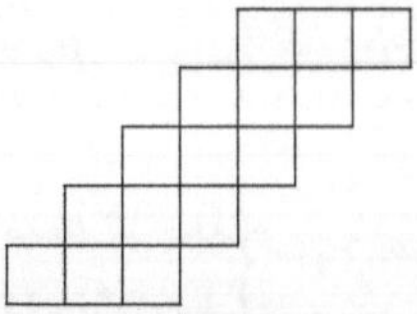

The **general skew strip of width $m$ and height $n$** (**$m$-strip**, for short), $D_{m,n,\lambda,\mu}$, has arbitrary partitions $\lambda$ and $\mu$, each of height at most $k := \lfloor m/2 \rfloor$, as "head" (northeastern tip) and "tail" (southwestern tip), respectively, instead of the basic partitions $\lambda = \mu = (k, k-1, \ldots, 1)$. For example,

$D_{6,7,(4,2,1),(3,3,1)} =$ 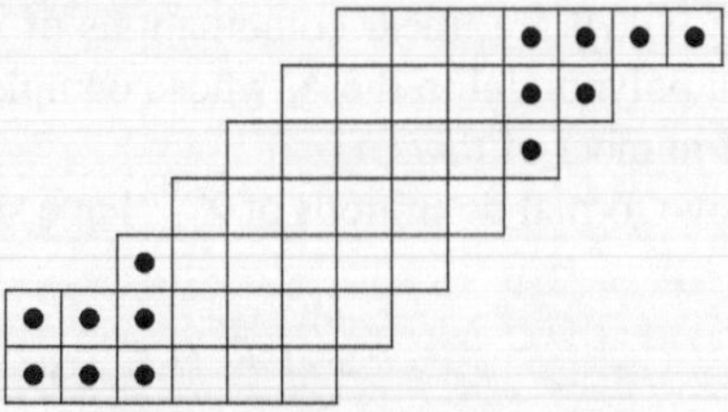

where $m = 6$, $n = 7$, $k = 3$ and the head and tail have marked cells.

The determinantal formula for skew shapes (Theorem 14.5.6) expresses $f^D$ as an explicit determinant of order $n$, the number of rows. Baryshnikov and Romik [11], developing an idea of Elkies [27], gave an alternative determinant of order $k$, half the length of a typical row. This is a considerable improvement if $m$, $k$, $\lambda$ and $\mu$ are fixed while $n$ is allowed to grow.

The general statement needs a bit of notation. Denote, for a non-negative integer $n$,

$$A'_n := \frac{A_n}{n!}, \qquad A''_n := \frac{A'_n}{2^{n+1}-1}, \qquad A'''_n := \frac{(2^n-1)A''_n}{2^n},$$

where $A_n$ are André's numbers (as in the previous subsection); and let

$$\varepsilon(n) := \begin{cases} (-1)^{n/2}, & \text{if } n \text{ is even,} \\ 0, & \text{if } n \text{ is odd.} \end{cases}$$

Define, for nonnegative integers $N$, $p$ and $q$,

$$\begin{aligned} X_N^{(0)}(p,q) \quad := \quad & \sum_{i=0}^{\lfloor p/2 \rfloor} \sum_{j=0}^{\lfloor q/2 \rfloor} \frac{(-1)^{i+j} A'_{N+2i+2j+1}}{(p-2i)!(q-2j)!} + \varepsilon(p+1) \sum_{j=0}^{\lfloor q/2 \rfloor} \frac{(-1)^j A'_{N+p+2j+1}}{(q-2j)!} \\ & + \varepsilon(q+1) \sum_{i=0}^{\lfloor p/2 \rfloor} \frac{(-1)^i A'_{N+2i+q+1}}{(p-2i)!} + \varepsilon(p+1)\varepsilon(q+1) A'_{N+p+q+1} \end{aligned}$$

and

$$\begin{aligned} X_N^{(1)}(p,q) \quad := \quad & \sum_{i=0}^{\lfloor p/2 \rfloor} \sum_{j=0}^{\lfloor q/2 \rfloor} \frac{(-1)^{i+j} A'''_{N+2i+2j+1}}{(p-2i)!(q-2j)!} + \varepsilon(p) \sum_{j=0}^{\lfloor q/2 \rfloor} \frac{(-1)^j A''_{N+p+2j+1}}{(q-2j)!} \\ & + \varepsilon(q) \sum_{i=0}^{\lfloor p/2 \rfloor} \frac{(-1)^i A''_{N+2i+q+1}}{(p-2i)!} + \varepsilon(p)\varepsilon(q) A'''_{N+p+q+1} \quad . \end{aligned}$$

**Theorem 14.7.9** [11, Theorem 4] *Let $D = D_{m,n,\lambda,\mu}$ be an $m$-strip with head and tail partitions $\lambda = (\lambda_1, \dots, \lambda_k)$ and $\mu = (\mu_1, \dots, \mu_k)$, where $k := \lfloor m/2 \rfloor$. For $1 \le i \le k$ define $L_i := \lambda_i + k - i$ and $M_i := \mu_i + k - i$, and denote*

$$m\%2 := \begin{cases} 0, & \text{if } m \text{ is even;} \\ 1, & \text{if } m \text{ is odd.} \end{cases}$$

*Then*

$$f^D = (-1)^{\binom{k}{2}} |D|! \cdot \det \left[ X_{2n-m+1}^{(m\%2)}(L_i, M_j) \right]_{i,j=1}^{k}.$$

Note that $X_N^{(\varepsilon)}(p,q)$ are linear combinations of $A_{N+1}, \dots, A_{N+p+q+1}$, so that $f^D$ is expressed as a polynomial in the $A_i$ whose complexity depends on the row length $m$ but not on the number of rows $n$.

The impressive formal definitions of $X_N^{(0)}$ have simple geometric interpretations:

$$X_{2n-1}^{(0)}(p,q) = \frac{f^D}{|D|!}$$

where

$$D = \text{zigzag}_{2n+p+q}(\{p+2, p+4, \ldots, p+2n-2\}) =$$

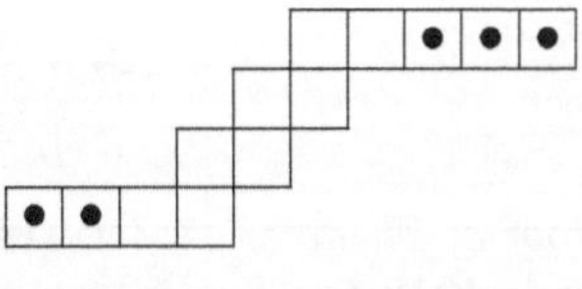

($p$ marked southwestern cells in a row, $2n$ unmarked cells, and $q$ marked northeastern cells in a row), and

$$X_{2n}^{(0)}(p,q) = \frac{f^D}{|D|!}$$

where

$$D = \text{zigzag}_{2n+p+q+1}(\{p+2, p+4, \ldots, p+2n, p+2n+1, \ldots, p+2n+q\})$$

$$=$$

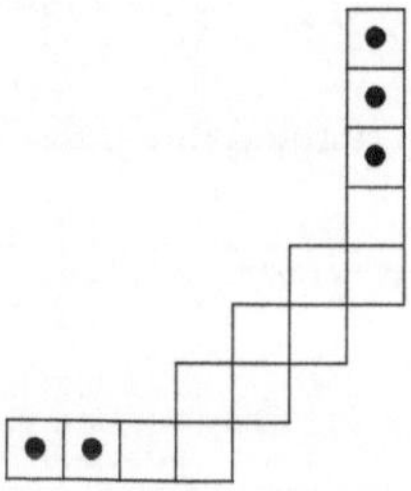

($p$ marked southwestern cells in a row, $2n+1$ unmarked cells, and $q$ marked northeastern cells in a column). These are 2-strips, i.e. zigzag shapes. It is possible to define $X_N^{(1)}(p,q)$ similarly in terms of 3-strips, a task left as an exercise to the reader [11].

Here are some interesting special cases.

**Corollary 14.7.10** [11, Theorem 1] *3-strips:*

$$f^{D_{3,n,(),()}} = \frac{(3n-2)!\,T_n}{(2n-1)!\,2^{2n-2}},$$

$$f^{D_{3,n,(1),()}} = \frac{(3n-1)!\,T_n}{(2n-1)!\,2^{2n-1}},$$

$$f^{D_{3,n}} = f^{D_{3,n,(1),(1)}} = \frac{(3n)!\,(2^{2n-1}-1)T_n}{(2n-1)!\,2^{2n-1}(2^{2n}-1)}.$$

**Corollary 14.7.11** [11, Theorem 2] *4-strips:*

$$f^{D_{4,n,(),()}} = (4n-2)!\left(\frac{T_n^2}{(2n-1)!^2} + \frac{E_{2n-2}E_{2n}}{(2n-2)!(2n)!}\right),$$

$$f^{D_{4,n}} = f^{D_{4,n,(1),(1)}} = (4n)!\left(\frac{E_{2n}^2}{(2n)!^2} - \frac{E_{2n-2}E_{2n+2}}{(2n-2)!(2n+2)!}\right).$$

**Corollary 14.7.12** [11, Theorem 3] 5-*strip:*

$$f^{D_{5,n,(),()}} = \frac{(5n-6)!\,T_{n-1}^2}{(2n-3)!2^{2^{4n-6}}(2^{2n-2}-1)}.$$

*Proof of Theorem 14.7.9 (sketch).* The proof uses transfer operators, following Elkies [27]. Elkies considered, essentially, the zigzag shapes (2-strips) $D_n$ from Example 14.7.4, whose SYT correspond to alternating (up-down) permutations. Recall from Section 14.2.5.3 the definition of the order polytope $P(D_n)$, whose volume is, by Observation 14.2.17,

$$\operatorname{vol} P(D_n) = \frac{f^{D_n}}{n!}.$$

This polytope can be written as

$$P(D_n) = \{(x_1,\ldots,x_n) \in [0,1]^n : x_1 \le x_2 \ge x_3 \le x_4 \ge \ldots\},$$

and therefore its volume can also be computed by an iterated integral:

$$\operatorname{vol} P(D_n) = \int_0^1 dx_1 \int_{x_1}^1 dx_2 \int_0^{x_2} dx_3 \int_{x_3}^1 dx_4 \cdots.$$

Some manipulations now lead to the expression

$$\operatorname{vol} P(D_n) = \langle T^{n-1}(\mathbf{1}), \mathbf{1} \rangle$$

where $\mathbf{1} \in L^2[0,1]$ is the function with constant value 1, $\langle \cdot, \cdot \rangle$ is the standard inner product on $L^2[0,1]$, and $T : L^2[0,1] \to L^2[0,1]$ is the compact self-adjoint operator defined by

$$(Tf)(x) := \int_0^{1-x} f(y)dy \qquad (\forall f \in L^2[0,1]).$$

The eigenvalues $\lambda_k$ and corresponding orthonormal eigenfunctions $\phi_k$ of $T$ can be computed explicitly, leading to the explicit formula

$$\operatorname{vol} P(D_n) = \sum_k \lambda_k^{n-1} \langle \mathbf{1}, \phi_k \rangle^2 = \frac{2^{n+2}}{\pi^{n+1}} \sum_{k=-\infty}^{\infty} \frac{1}{(4k+1)^{n+1}} \qquad (n \ge 1)$$

which gives a corresponding expression for

$$A_n = f^{D_n} = n! \operatorname{vol} P(D_n).$$

Baryshnikov and Romik extended this treatment of a 2-strip to general $m$-strips, using additional ingredients. For instance, the iterated integral for a 3-strip

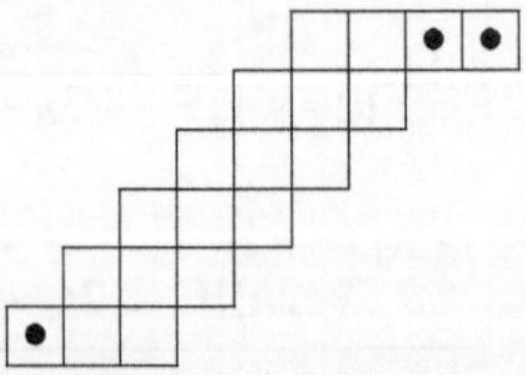

gives

$$\operatorname{vol} P(D_{3,n,\lambda,\mu}) = \langle (BA)^{n-1} T_\mu(\mathbf{1}), T_\lambda(\mathbf{1}) \rangle$$

where $\Omega := \{(u,v) \in [0,1]^2 : u \le v\}$, the transfer operators $A : L^2[0,1] \to L^2(\Omega)$ and $B : L^2(\Omega) \to L^2[0,1]$ are defined by

$$(Af)(u,v) := \int_u^v f(x)\,dx$$

and

$$(Bg)(x) := \int_0^x \int_x^1 g(u,v)\,dv\,du,$$

and $T_\lambda$, $T_\mu$ are operators corresponding to the "head" and "tail" partitions $\lambda$ and $\mu$. ■

In a slightly different direction, Stanley [118] defines $\hat{D}_{a,b,c,n}$ to be the skew shape whose diagram has $n$ rows, with $a$ cells in the top row and $b$ cells in each other row, and such that each row is shifted $c-1$ columns to the left with respect to the preceding (higher) row. For example,

$\hat{D}_{5,4,3,4} =$ 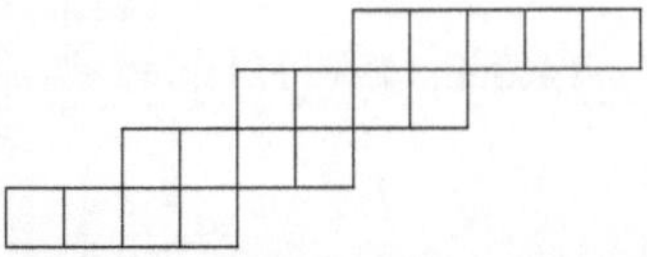

**Theorem 14.7.13** [118, Corollary 2.5] *For $a$, $b$ and $c$ with $c \le b < 2c$,*

$$\sum_{n \ge 0} f^{\hat{D}_{a,b,c,n+1}} \frac{x^{n+1}}{(a+nb)!} = \frac{x \sum_{n\ge 0} \frac{(-x)^n}{(b+nc)!}}{(b-c)! - x \sum_{n \ge 0} \frac{(-x)^n}{(a+nc)!}}.$$

Two special cases deserve special attention: $a = b$ and $b = c$.

For $a = b$ all the rows are of the same length.

**Corollary 14.7.14** *For $a$ and $c$ with $c \le a < 2c$,*

$$1 + \sum_{n \ge 1} f^{\hat{D}_{a,a,c,n}} \frac{x^n}{(na)!} = \left(1 - \frac{x}{(a-c)!} \sum_{n \ge 0} \frac{(-x)^n}{(a+nc)!}\right)^{-1}.$$

In particular, for $a = b = 3$ and $c = 2$, $\hat{D}_{3,3,2,n} = D_{3,n}$ as in Theorem 14.7.10:

$$\sum_{n \ge 0} f^{D_{3,n}} \frac{x^{2n}}{(3n)!} = \left(\sum_{n \ge 0} \frac{(-x^2)^n}{(2n+1)!}\right)^{-1} = \frac{x}{\sin x}.$$

This result was already known to Gessel and Viennot [43].

For $b = c$ one obtains a zigzag shape: $\hat{D}_{a,c,c,n+1} = \text{zigzag}_{a+nc}(S)$ for $S = \{c, 2c, \ldots, nc\}$.

**Corollary 14.7.15** *For any positive $a$ and $c$,*

$$\sum_{n \ge 0} f^{\text{zigzag}_{a+nc}(\{c,2c,\ldots,nc\})} \frac{x^{n+1}}{(a+nc)!} = \frac{x \sum_{n \ge 0} \frac{(-x)^n}{(c+nc)!}}{1 - x \sum_{n \ge 0} \frac{(-x)^n}{(a+nc)!}}.$$

## 14.8 Truncated and other non-classical shapes

**Definition 14.8.1** *A diagram of* **truncated shape** *is a line-convex diagram obtained from a diagram of ordinary or shifted shape by deleting cells from the northeastern corner (in the English notation, where row lengths decrease from top to bottom).*

For example, here are diagrams of a truncated ordinary shape

$$[(4,4,2,1)\setminus(1)] =$$

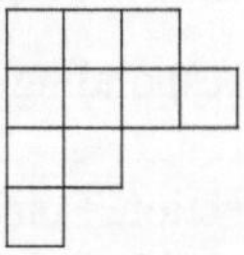

and a truncated shifted shape:

$$[(4,3,2,1)^*\setminus(1,1)] =$$

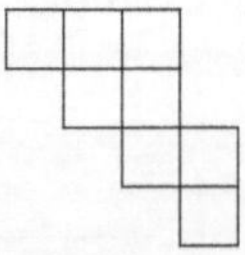

Modules associated to truncated shapes were introduced and studied in [54, 94]. Interest in the enumeration of SYT of truncated shapes was recently enhanced by a new interpretation [4]: The number of geodesics between distinguished pairs of antipodes in the flip graph of inner-triangle-free triangulations is twice the number of SYT of a corresponding truncated shifted staircase shape. Motivated by this result, extensive computations were carried out for the number of SYT of these and other truncated shapes. It was found that, in some cases, these numbers are unusually "smooth," i.e., all their prime factors are relatively very small. This makes it reasonable to expect a product formula. Subsequently, such formulas were conjectured and proved for rectangular and shifted staircase shapes truncated by a square, or nearly a square, and for rectangular shapes truncated by a staircase; see [1, 81, 125, 123].

### 14.8.1 Truncated shifted staircase shape

In this subsection, $\lambda = (\lambda_1, \ldots, \lambda_t)$ (with $\lambda_1 > \ldots > \lambda_t > 0$) will be a strict partition, with $g^\lambda$ denoting the number of SYT of shifted shape $\lambda$.

For any nonnegative integer $n$, let $\delta_n := (n, n-1, \ldots, 1)$ be the corresponding shifted staircase shape. By Schur's formula for shifted shapes (Theorem 14.5.7),

**Corollary 14.8.2** *The number of SYT of shifted staircase shape $\delta_n$ is*

$$g^{\delta_n} = N! \cdot \prod_{i=0}^{n-1} \frac{i!}{(2i+1)!},$$

*where $N := |\delta_n| = \binom{n+1}{2}$.*

The following enumeration problem was actually the original motivation for the study of truncated shapes, because of its combinatorial interpretation, as explained in [4].

**Theorem 14.8.3** [1, 81] *The number of SYT of truncated shifted staircase shape* $\delta_n \setminus (1)$ *is equal to*

$$g^{\delta_n} \frac{C_n C_{n-2}}{2C_{2n-3}},$$

*where* $C_n = \frac{1}{n+1}\binom{2n}{n}$ *is the n-th Catalan number.*

**Example 14.8.4** *There are* $g^{\delta_4} = 12$ *SYT of shape* $\delta_4$*, but only* 4 *SYT of truncated shape* $\delta_4 \setminus (1)$*:*

$$\begin{array}{cccc} 1 & 2 & 3 & \\ & 4 & 5 & 6 \\ & & 7 & 8 \\ & & & 9 \end{array}, \quad \begin{array}{cccc} 1 & 2 & 4 & \\ & 3 & 5 & 6 \\ & & 7 & 8 \\ & & & 9 \end{array}, \quad \begin{array}{cccc} 1 & 2 & 3 & \\ & 4 & 5 & 7 \\ & & 6 & 8 \\ & & & 9 \end{array}, \quad \begin{array}{cccc} 1 & 2 & 4 & \\ & 3 & 5 & 7 \\ & & 6 & 8 \\ & & & 9 \end{array}.$$

Theorem 14.8.3 may be generalized to a truncation of a $(k-1) \times (k-1)$ square from the northeastern corner of a shifted staircase shape $\delta_{m+2k}$.

**Example 14.8.5** *For* $m = 1$ *and* $k = 3$*, the truncated shape is*

$[\delta_5 \setminus (2^2)] =$ 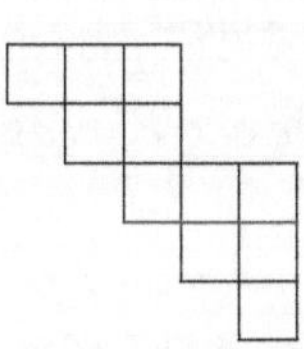

**Theorem 14.8.6** [1, Corollary 4.8] *The number of SYT of truncated shifted staircase shape* $\delta_{m+2k} \setminus ((k-1)^{k-1})$ *is*

$$g^{(m+k+1,\ldots,m+3,m+1,\ldots,1)} g^{(m+k+1,\ldots,m+3,m+1)} \cdot \frac{N!M!}{(N-M-1)!(2M+1)!},$$

*where* $N = \binom{m+2k+1}{2} - (k-1)^2$ *is the size of the shape and* $M = k(2m+k+3)/2 - 1$.

Similar results were obtained in [1] for truncation by "almost squares," namely by $(k^{k-1}, k-1)$.

### 14.8.2 Truncated rectangular shapes

In this section, $\lambda = (\lambda_1, \ldots, \lambda_m)$ (with $\lambda_1 \geq \ldots \geq \lambda_m \geq 0$) will be a partition with (at most) $m$ parts. Two partitions that differ only in trailing zeros will be considered equal.

For any nonnegative integers $m$ and $n$, let $(n^m) := (n, \ldots, n)$ ($m$ times) be the corresponding rectangular shape. The Frobenius-Young formula (Theorem 14.5.1) implies the following.

**Observation 14.8.7** *The number of SYT of rectangular shape $(n^m)$ is*

$$f^{(n^m)} = (mn)! \cdot \frac{F_m F_n}{F_{m+n}},$$

*where*

$$F_m := \prod_{i=0}^{m-1} i!.$$

Consider truncating a $(k-1) \times (k-1)$ square from the northeastern corner of a rectangular shape $((n+k-1)^{m+k-1})$.

**Example 14.8.8** *Let $n = 3$, $m = 2$ and $k = 3$. Then*

$$[(5^4) \setminus (2^2)] =$$ 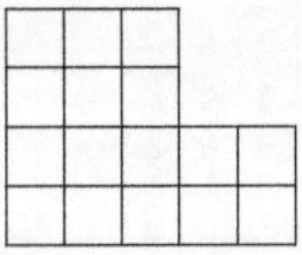

**Theorem 14.8.9** [1, Corollary 5.7] *The number of SYT of truncated rectangular shape $((n+k-1)^{m+k-1}) \setminus ((k-1)^{k-1})$ is*

$$\frac{N!(mk-1)!(nk-1)!(m+n-1)!k}{(mk+nk-1)!} \cdot \frac{F_{m-1}F_{n-1}F_{k-1}}{F_{m+n+k-1}},$$

*where $N$ is the size of the shape and $F_n$ is as in Observation 14.8.7.*

In particular,

**Corollary 14.8.10** *The number of SYT of truncated rectangular shape $((n+1)^{m+1}) \setminus (1)$ is*

$$\frac{N!(2m-1)!(2n-1)! \cdot 2}{(2m+2n-1)!(m+n)} \cdot \frac{F_{m-1}F_{n-1}}{F_{m+n}},$$

*where $N = (m+1)(n+1) - 1$ is the size of the shape and $F_n$ is as in Observation 14.8.7.*

Similar results were obtained in [1, 81] for truncation by almost squares $\kappa = (k^{k-1}, k-1)$.

Not much is known for truncation by rectangles. The following formula was conjectured in [1] and proved by Sun [124] using Selberg-type integrals.

**Proposition 14.8.11** [124] *For $n \geq 2$*

$$f^{(n^n) \setminus (2)} = \frac{(n^2-2)!(3n-4)!^2 \cdot 6}{(6n-8)!(2n-2)!(n-2)!^2} \cdot \frac{F_{n-2}^2}{F_{2n-4}},$$

*where $F_n$ is as in Observation 14.8.7.*

The following result was proved by Snow [111].

**Proposition 14.8.12** [111] *For $n \geq 2$ and $k \geq 0$*

$$f^{(n^{k+1})\setminus(n-2)} = \frac{(kn-k)!(kn+n)!}{(kn+n-k)!} \cdot \frac{F_k F_n}{F_{n+k}},$$

*where $F_n$ is as in Observation 14.8.7.*

A different method to derive product formulas, for other families of truncated shapes, has been developed by Panova [81]. Consider a rectangular shape truncated by a staircase shape.

**Example 14.8.13**

$$[(4^5)\setminus\delta_2] =$$

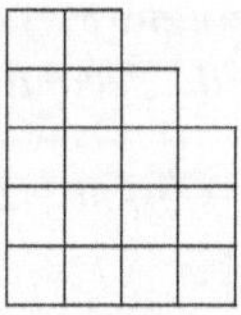

**Theorem 14.8.14** [81, Theorem 2] *Let $m \geq n \geq k$ be positive integers. The number of SYT of truncated shape $(n^m)\setminus\delta_k$ is*

$$\binom{N}{m(n-k-1)} f^{(n-k-1)^m} g^{(m,m-1,\ldots,m-k)} \frac{E(k+1,m,n-k-1)}{E(k+1,m,0)},$$

*where $N = mn - \binom{k+1}{2}$ is the size of the shape and*

$$E(r,p,s) = \begin{cases} \prod\limits_{r<l<2p-r+2} \frac{1}{(l+2s)^{r/2}} \prod\limits_{2\leq l\leq r} \frac{1}{((l+2s)(2p-l+2s+2))^{\lfloor l/2 \rfloor}}, & \text{if } r \text{ is even;} \\ \frac{((r-1)/2+s)!}{(p-(r-1)/2+s)!} E(r-1,p,s), & \text{if } r \text{ is odd.} \end{cases}$$

### 14.8.3 Other truncated shapes

The following elegant result regarding **shifted strips** was recently proved by Sun.

**Theorem 14.8.15** [126, §4.2] *The number of SYT of truncated shifted shape with n rows and 4 cells in each row*

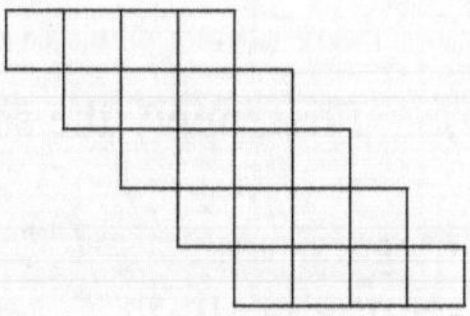

*is the $(2n-1)$th Pell number* [110, A000129]

$$\frac{1}{2\sqrt{2}}\left((1+\sqrt{2})^{2n-1} - (1-\sqrt{2})^{2n-1}\right).$$

Sun applied a probabilistic version of computations of volumes of order polytopes to enumerate SYT of truncated and other exotic shapes. In [123] he obtained product formulas for the number of SYT of certain truncated skew shapes. This includes the shape $((n+k)^{r+1}, n^{m-1})/(n-1)^r$ truncated by a rectangle or an "almost rectangle," the truncated shape $((n+1)^3, n^{m-2})/(n-2) \setminus (2^2)$, and the truncated shape $(n+1)^2/n^{m-2} \setminus (2)$.

Modules associated with non-line-convex shapes were considered in [54]. The enumeration of SYT of such shapes is a very recent subject of study. Special non-line-convex shapes with one box removed at the end or middle of a row were considered in [125]. For example,

**Proposition 14.8.16** [125, Theorem 5.2] *For $m \geq 0$, the number of SYT of shape $(m+3,3,3)$ with middle box in the second row removed, is*

$$\frac{m+5}{10}\binom{m+2}{2}\binom{m+9}{2}.$$

There are very few known results in this direction; problems in this area are wide open.

### 14.8.4 Proof approaches for truncated shapes

Different approaches were applied to prove the above product formulas. We will sketch one method and remark on another.

The **pivoting approach** of [1] is based on a combination of two different bijections from SYT to pairs of smaller SYT:

1. Choose a pivot cell $C$ in the northeastern boundary of a truncated shape $\zeta$ and subdivide the entries of a given SYT $T$ into those that are less than the entry $T(c)$ and those that are greater.

2. Choose a letter $t$ and subdivide the entries in a SYT $T$ into those that are less than or equal to $t$ and those that are greater than $t$.

Proofs are obtained by combining applications of the first bijection to truncated shapes and the second bijection to corresponding non-truncated ones. Here is a typical example.

*Proof of Theorem 14.8.3 (sketch).* First, apply the second bijection to a SYT of a shifted staircase shape.

Let $n$ and $t$ be nonnegative integers, with $t \leq \binom{n+1}{2}$. Let $T$ be a SYT of shifted staircase shape $\delta_n$, let $T_1$ be the set of all cells in $T$ with values at most $t$, and let $T_2$ be obtained from $T \setminus T_1$ by transposing the shape (reflecting in an anti-diagonal) and replacing each entry $i$ by $N-i+1$, where $N = |\delta_n| = \binom{n+1}{2}$. Clearly $T_1$ and $T_2$ are shifted SYT.

Here is an example with $n = 4$ and $t = 5$.

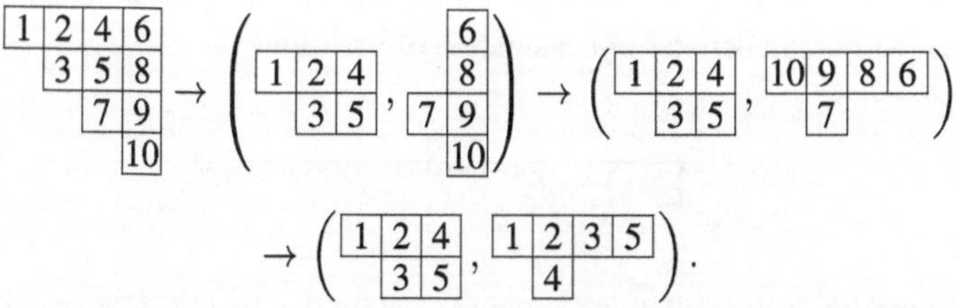

Notice that, treating strict partitions as sets, $\delta_4 = (4,3,2,1)$ is the disjoint union of $\mathrm{sh}(T_1) = (3,2)$ and $\mathrm{sh}(T_2) = (4,1)$. This is not a coincidence.

**Claim 14.8.17** *Treating strict partitions as sets, $\delta_n$ is the disjoint union of the shape of $T_1$ and the shape of $T_2$.*

In order to prove the claim notice that the borderline between $T_1$ and $T \setminus T_1$ is a lattice path of length exactly $n$, starting at the northeastern corner of the staircase shape $\delta_n$, and using only S (south) and W (west) steps, and ending at the southwestern boundary. If the first step is S then the first part of $\mathrm{sh}(T_1)$ is $n$, and the rest corresponds to a lattice path in $\delta_{n-1}$. Similarly, if the first step is W then the first part of $\mathrm{sh}(T_2)$ is $n$, and the rest corresponds to a lattice path in $\delta_{n-1}$. Thus exactly one of the shapes of $T_1$ and $T_2$ has a part equal to $n$. The claim follows by induction on $n$.

We deduce that, for any nonnegative integers $n$ and $t$ with $t \leq \binom{n+1}{2}$,

$$\sum_{\substack{\lambda \subseteq \delta_n \\ |\lambda| = t}} g^{\lambda} g^{\lambda^c} = g^{\delta_n}. \tag{14.1}$$

Here summation is over all strict partitions $\lambda$ with the prescribed restrictions, and $\lambda^c$ is the complement of $\lambda$ in $\delta_n = \{1, \ldots, n\}$ (where strict partitions are treated as sets). In particular, the LHS is independent of $t$.

Next apply the first bijection to SYT of truncated staircase shape $\delta_n \setminus (1)$. Choose as a pivot the cell $c = (2, n-1)$, just southwest of the missing northeastern corner. The entry $t = T(c)$ satisfies $2n-3 \leq t \leq \binom{n}{2} - 2n+2$. Let $T$ be a SYT of shape $\delta_n \setminus (1)$ with $T(c) = t$. One subdivides the other entries of $T$ into those that are (strictly) less than $t$ and those that are greater than $t$. The entries less than $t$ constitute $T_1$. To obtain $T_2$, replace each entry $i > t$ of $T$ by $N - i + 1$, where $N$ is the total number of entries in $T$, and suitably transpose the resulting array. It is easy to see that both $T_1$ and $T_2$ are shifted SYT.

**Example 14.8.18**

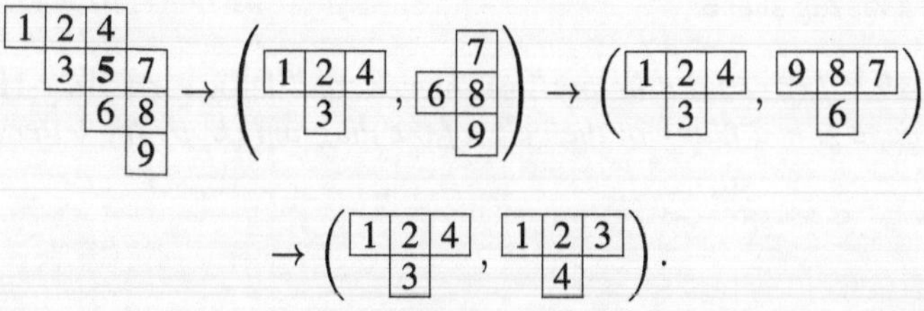

Next notice that the shape of $T_1$ is $(n-1,n-3)\cup\lambda$ while the shape of $T_2$ is $(n-1,n-3)\cup\lambda^c$, where $\lambda$ is a strict partition contained in $\delta_{n-4}$, $\lambda^c$ is its complement in $\delta_{n-4}$, and union denotes concatenation of partitions.

We deduce that

$$g^{\delta_n\setminus(1)} = \sum_t \sum_{\substack{\lambda\subseteq\delta_{n-4}\\|\lambda|=t}} g^{(n-1,n-3)\cup\lambda} g^{(n-1,n-3)\cup\lambda^c}. \tag{14.2}$$

Here summation is over all strict partitions $\lambda$ with the prescribed restrictions.

Finally, by Schur's formula (Theorem 14.5.7), for any strict partitions $\lambda$, $\lambda^c$ and $\mu=(\mu_1,\ldots,\mu_k)$ with $\mu_1>\ldots>\mu_k>m$ and $\lambda\cup\lambda^c=\delta_m$, the following holds.

$$g^{\mu\cup\lambda} g^{\mu\cup\lambda^c} = c(\mu,|\lambda|,|\lambda^c|)\cdot g^{\lambda} g^{\lambda^c}, \tag{14.3}$$

where

$$c(\mu,|\lambda|,|\lambda^c|) = \frac{g^{\mu\cup\delta_m} g^{\mu}}{g^{\delta_m}} \cdot \frac{|\delta_m|!(|\mu|+|\lambda|)!(|\mu|+|\lambda^c|)!}{(|\mu|+|\delta_m|)!|\mu|!|\lambda|!|\lambda^c|!}$$

depends only on the sizes $|\lambda|$ and $|\lambda^c|$ and not on the actual partitions $\lambda$ and $\lambda^c$.

Combining Equations (14.2), (14.1) and (14.3) together with some binomial identities completes the proof. ∎

For a detailed proof and applications of the method to other truncated shapes see [1].

A different proof was presented by Panova [81]. Panova's approach is sophisticated and involved and will just be outlined. The proof relies on a bijection from SYT of the given truncated shape to semi-standard Young tableaux of certain skew shapes. This translates the enumeration problem to the evaluation of sums of specializations of Schur functions. These evaluations are then reduced to computations of complex integrals, which are carried out by a comparison to another translation of the original enumerative problem to a volume of an associated order polytope.

## 14.9 Rim hook and domino tableaux

### 14.9.1 Definitions

The following concept generalizes the notion of SYT. Recall from Section 14.3.3 the definition of a zigzag shape.

**Definition 14.9.1** *Let $r$ and $n$ be positive integers and let $\lambda\vdash rn$. An $r$-***rim hook tableau** *of shape $\lambda$ is a filling of the cells of the diagram $[\lambda]$ by the letters $1,\ldots,n$ such that*

1. *each letter i fills exactly r cells, which form a zigzag shape called the i-th* **rim hook** *(or* **border strip***); and*

2. *for each* $1 \leq k \leq n$*, the union of the i-th rim hooks for* $1 \leq i \leq k$ *is a diagram of ordinary shape.*

*Denote by* $f_r^\lambda$ *the number of r-rim hook tableaux of shape* $\lambda \vdash rn$*.*

The $n$th rim hook forms a path-connected subset of the **rim** (southeastern boundary) of the diagram $[\lambda]$, and removing it leads inductively to a similar description for the other rim hooks.

Note that 1-rim hook tableaux are ordinary SYT; 2-rim hook tableaux are also called **domino tableaux.**

**Example 14.9.2** *Here is a domino tableau of shape* $(5,5,4)$*:*

| | | | | |
|---|---|---|---|---|
| 1 | 1 | 3 | 3 | 6 |
| 2 | 4 | 5 | 5 | 6 |
| 2 | 4 | 7 | 7 | |

*and here is a 3-rim hook tableau of shape* $(5,4,3)$*:*

| | | | | |
|---|---|---|---|---|
| 1 | 1 | 3 | 3 | 3 |
| 1 | 2 | 4 | 4 | |
| 2 | 2 | 4 | | |

**Definition 14.9.3** *An* **r-partition** *of n is a sequence* $\lambda = (\lambda^0, \ldots, \lambda^{r-1})$ *of partitions of total size* $|\lambda^0| + \ldots + |\lambda^{r-1}| = n$*. The corresponding* **r-diagram** $[\lambda^0, \ldots, \lambda^{r-1}]$ *is the sequence* $([\lambda^0], \ldots, [\lambda^{r-1}])$ *of ordinary diagrams. It is sometimes drawn as a skew diagram, with* $[\lambda^i]$ *lying directly southwest of* $[\lambda^{i-1}]$ *for every* $1 \leq i \leq r-1$*.*

**Example 14.9.4** *The 2-diagram of shape* $(\lambda^0, \lambda^1) = ((3,1),(2,2))$ *is*

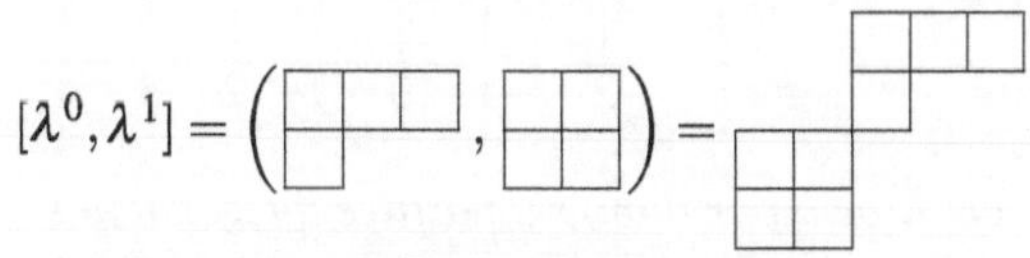

**Definition 14.9.5** *A* **standard Young** ***r*****-tableau (*****r*****-SYT)** $T = (T^0, \ldots, T^{r-1})$ *of shape* $\lambda = (\lambda^0, \ldots, \lambda^{r-1})$ *and total size n is obtained by inserting the integers* $1, 2, \ldots, n$ *as entries into the cells of the diagram* $[\lambda]$ *such that the entries increase along rows and columns.*

### 14.9.2 The *r*-quotient and *r*-core

**Definition 14.9.6** *Let* $\lambda$ *be a partition, and* $D = [\lambda]$ *the corresponding ordinary diagram. The* **boundary sequence** *of* $\lambda$ *is the* $0/1$ *sequence* $\partial(\lambda)$ *constructed as follows: Start at the southwestern corner of the diagram and proceed along the edges of the southeastern boundary up to the northeastern corner. Each horizontal (eastbound) step is encoded by* $1$*, and each vertical (north-bound) step by* $0$*.*

**Example 14.9.7**

$$\lambda = (3,1) \quad \rightarrow \quad D = [\lambda] =$$ 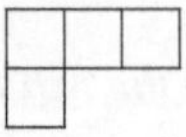 $$\rightarrow \quad \partial(\lambda) = (1,0,1,1,0)$$

The boundary sequence starts with 1 and ends with 0, unless $\lambda$ is the empty partition, for which $\partial(\lambda)$ is the empty sequence.

**Definition 14.9.8** *The* **extended boundary sequence** $\partial_*(\lambda)$ *of* $\lambda$ *is the doubly-infinite sequence obtained from* $\partial(\lambda)$ *by prepending to it the sequence* $(\ldots,0,0)$ *and appending to it the sequence* $(1,1,\ldots)$.

Geometrically, these additions represent a vertical ray and a horizontal ray, respectively, so that the tour of the boundary of $[\lambda]$ actually "starts" at the far south and "ends" at the far east.

**Example 14.9.9** *If* $\lambda = (3,1)$ *then* $\partial_*(\lambda) = (\ldots,0,0,1,0,1,1,0,1,1,\ldots)$, *and if* $\lambda$ *is empty then* $\partial_*(\lambda) = (\ldots,0,0,1,1,\ldots)$.

$\partial_*$ is clearly a bijection from the set of all partitions to the set of all doubly-infinite 0/1 sequences with initially only 0-s and eventually only 1-s.

**Definition 14.9.10** *There is a* **natural indexing** *of any (extended) boundary sequence, as follows: The index k of an element is equal to the number of* 1*-s weakly to its left minus the number of* 0*-s strictly to its right.*

**Example 14.9.11**

| | | | | | | | | | | | |
|---|---|---|---|---|---|---|---|---|---|---|---|
| 0/1 sequence: | … | 0 | 0 | 1 | 0 | 1 | 1 | 0 | 1 | 1 | … |
| | | ↑ | ↑ | ↑ | ↑ | ↑ | ↑ | ↑ | ↑ | ↑ | |
| Indexing: | … | −3 | −2 | −1 | 0 | 1 | 2 | 3 | 4 | 5 | … |

**Definition 14.9.12** *Let* $\lambda$ *be a partition,* $r$ *a positive integer, and* $s := \partial_*(\lambda)$.

1. *The* $r$**-quotient** $q_r(\lambda)$ *is a sequence of* $r$ *partitions obtained as follows: For each* $0 \le i \le r-1$ *let* $s^i$ *be the subsequence of* $s$ *corresponding to the indices that are congruent to* $i \pmod r$, *and let* $\lambda^i := \partial_*^{-1}(s^i)$. *Then* $q_r(\lambda) := (\lambda^0,\ldots,\lambda^{r-1})$.

2. *The* $r$**-core** *(or* $r$**-residue***)* $c_r(\lambda)$ *is the partition* $\lambda' = \partial_*^{-1}(s')$, *where* $s'$ *is obtained from* $s$ *by a sequence of moves that interchange a* 1 *in position* $i$ *with a* 0 *in position* $i+r$ *(for some* $i$*), as long as such a move is still possible.*

Denote $|q_r(\lambda)| := |\lambda^0| + \ldots + |\lambda^{r-1}|$.

**Theorem 14.9.13** *The equality* $|\lambda| = r \cdot |q_r(\lambda)| + |c_r(\lambda)|$ *holds.*

**Example 14.9.14** *For $\lambda = (6,4,2,2,2,1)$ and $r = 2$,*

$$s = \partial_*(\lambda) = (\ldots,0,0,0,1,0,1,0,0,\hat{0},1,1,0,1,1,0,1,1,1,\ldots)$$

*with a hat over the entry indexed 0. It follows that*

$$s^0 = (\ldots,0,0,0,0,0,1,1,0,1,\ldots) \quad \textit{and} \quad s^1 = (\ldots,0,1,1,0,1,0,1,1,1,\ldots).$$

*The 2-quotient is therefore $q_2(\lambda) = ((2),(3,2))$. The 2-core is*

$$c_2(\lambda) = \partial_*^{-1}(s') = \partial_*^{-1}(\ldots,0,0,0,0,0,0,0,1,\hat{0},1,0,1,1,1,1,1,1,1,\ldots) = (2,1).$$

*Indeed, $|\lambda| = 17 = 2\cdot 7 + 3 = 2\cdot|q_2(\lambda)| + |c_2(\lambda)|$.*

It is easy to see that, in this example, there are no 2-rim hook tableaux of shape $\lambda$.

**Theorem 14.9.15** *The following holds: $f_r^\lambda \neq 0 \iff$* the $r$-core $c_r(\lambda)$ is empty.

**Example 14.9.16** *Let $\lambda = (4,2)$, $n = 3$ and $r = 2$. Then*

$$s = \partial_*(\lambda) = (\ldots,0,0,0,1,\hat{1},0,1,1,0,1,1,1,\ldots),$$

*so that the 2-core*

$$s' = \partial_*^{-1}(\ldots,0,0,0,0,\hat{0},1,1,1,1,1,1,1,\ldots)$$

*is empty and the 2-quotient is*

$$q_2(\lambda) = (\partial_*^{-1}(\ldots,0,0,1,1,0,1,\ldots),\partial_*^{-1}(\ldots,0,1,0,1,1,1,\ldots)) = ((2),(1)).$$

*Of course, here $|\lambda| = 6 = 2\cdot 3 = r\cdot|q_2(\lambda)|$.*

In this example there are three domino tableaux of shape $(4,2)$, and also three 2-SYT of shape $((2),(1))$:

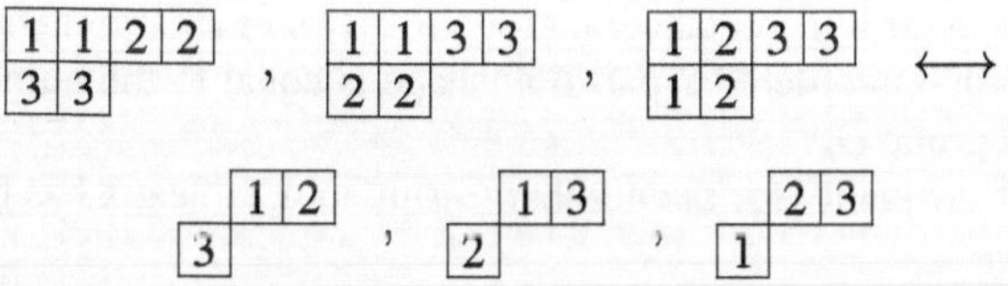

This is not a coincidence, as the following theorem shows.

**Theorem 14.9.17** *Let $\lambda$ be a partition with empty r-core, and let $q_r(\lambda)$ be its r-quotient. Then*

$$f_r^\lambda = f^{q_r(\lambda)}.$$

Theorem 14.9.17 may be combined with the hook length formula for ordinary shapes (Theorem 14.5.3) to obtain the following.

**Theorem 14.9.18** [53, p. 84] *If $f_r^\lambda \neq 0$ then*

$$f_r^\lambda = \frac{(|\lambda|/r)!}{\prod_{c\in[\lambda]: r|h_c} h_c/r}.$$

*Proof.* Note that $\lambda$ has an empty $r$-core; let $q_r(\lambda) = (\lambda^0, \ldots, \lambda^{r-1})$ be its $r$-quotient. By the hook length formula for ordinary shapes (Theorem 14.5.3) together with Observation 14.2.6,

$$f^{(\lambda^0,\ldots,\lambda^{r-1})} = \binom{|\lambda|/r}{|\lambda^0|,\ldots,|\lambda^{r-1}|}\prod_{i=0}^{r-1} f^{\lambda^i} = \frac{(|\lambda|/r)!}{\prod_{c\in[\lambda^0,\ldots,\lambda^{r-1}]} h_c}.$$

A careful examination of the $r$-quotient shows that it induces a bijection from the cells in $\lambda$ with hook length divisible by $r$ to all the cells in $(\lambda^0, \ldots, \lambda^{r-1})$, such that every cell $c \in [\lambda^0, \ldots, \lambda^{r-1}]$ with hook length $h_c$ corresponds to a cell $c' \in [\lambda]$ with hook length $h_{c'} = rh_c$. This completes the proof. ■

Stanton and White [121] generalized the RS correspondence to a bijection from $r$-colored permutations (i.e., elements of the wreath product $\mathbb{Z}_r \wr \mathcal{S}_n$) to pairs of $r$-rim hook tableaux of the same shape. The Stanton-White bijection, together with Theorem 14.9.17, implies the following generalization of Corollary 14.4.14.

**Theorem 14.9.19** *The following hold:*

(1) $$\sum_{\lambda\vdash rn} (f_r^\lambda)^2 = r^n n!$$

*and*

(2) $$\sum_{\lambda\vdash rn} f_r^\lambda = \sum_{k=0}^{\lfloor n/2\rfloor} \binom{n}{2k}(2k-1)!!r^{n-k}.$$

In particular, the total number of domino tableaux of size $2n$ is equal to the number of involutions in the hyperoctahedral group $B_n$. By similar arguments, the number of SYT of size $n$ and unordered 2-partition shape is equal to the number of involutions in the Weyl group $D_n$.

An important inequality for the number of rim hook tableaux has been found by Fomin and Lulov.

**Theorem 14.9.20** [33] *For any $\lambda \vdash rn$,*

$$f_r^\lambda \leq r^n n! \left(\frac{f^\lambda}{(rn)!}\right)^{1/r}.$$

See also [68, 97, 65].

## 14.10 $q$-Enumeration

This section deals primarily with three classical combinatorial parameters—the inversion number, the descent number and the major index. These parameters were originally studied in the context of permutations (and, more generally, words). The major index, for example, was introduced by MacMahon [72]. These permutation statistics were studied extensively by Foata and Schützenberger [31, 32], Garsia and Gessel [39], and others. Only later were these concepts defined and studied for standard Young tableaux.

### 14.10.1 Permutation statistics

We start with definitions of the main permutation statistics.

**Definition 14.10.1** *The* **descent set** *of a permutation* $\pi \in \mathscr{S}_n$ *is*

$$\mathrm{Des}(\pi) := \{i : \pi(i) > \pi(i+1)\},$$

*the* **descent number** *of* $\pi$ *is*

$$\mathrm{des}(\pi) := |\mathrm{Des}(\pi)|,$$

*and the* **major index** *of* $\pi$ *is*

$$\mathrm{maj}(\pi) := \sum_{i \in \mathrm{Des}(\pi)} i.$$

*The* **inversion set** *of* $\pi$ *is*

$$\mathrm{Inv}(\pi) := \{(i,j) : 1 \le i < j \le n, \pi(i) > \pi(j)\},$$

*and the* **inversion number** *of* $\pi$ *is*

$$\mathrm{inv}(\pi) := |\mathrm{Inv}(\pi)|.$$

We also use standard $q$-notation: For a nonnegative integer $n$ and nonnegative integers $k_1, \ldots, k_t$ with sum $n$, we write

$$[n]_q := \frac{q^n - 1}{q - 1}, \quad [n]_q! := \prod_{i=1}^{n} [i]_q, \quad \begin{bmatrix} n \\ k_1, \ldots, k_t \end{bmatrix}_q := \frac{[n]_q!}{\prod_{i=1}^{t} [k_i]_q!}.$$

Note that $[0]_q! = 1$.

**Theorem 14.10.2 (MacMahon's fundamental equidistribution theorem [72])** *For every positive integer* $n$

$$\sum_{\pi \in \mathscr{S}_n} q^{\mathrm{maj}(\pi)} = \sum_{\pi \in \mathscr{S}_n} q^{\mathrm{inv}(\pi)} = [n]_q!$$

A bijective proof was given in the classical paper of Foata [31]. Refinements and generalizations were given by many. In particular, Foata's bijection was applied to show that the major index and inversion number are equidistributed over inverse descent classes [32]. A different approach was suggested by Garsia and Gessel, who proved the following.

**Theorem 14.10.3** [39] *For every subset* $S = \{s_1, \ldots, s_k\} \subseteq [n-1]$ $(s_1 < \ldots < s_k)$

$$\sum_{\substack{\pi\in\mathcal{S}_n \\ \mathrm{Des}(\pi^{-1})\subseteq S}} q^{\mathrm{maj}(\pi)} = \sum_{\substack{\pi\in\mathcal{S}_n \\ \mathrm{Des}(\pi^{-1})\subseteq S}} q^{\mathrm{inv}(\pi)} = \begin{bmatrix} n \\ s_1, s_2 - s_1, s_3 - s_2, \ldots, n - s_k \end{bmatrix}_q .$$

The following determinantal formula [119, Example 2.2.5] follows by the inclusion-exclusion principle.

$$\sum_{\substack{\pi\in\mathcal{S}_n \\ \mathrm{Des}(\pi^{-1})= S}} q^{\mathrm{maj}(\pi)} = \sum_{\substack{\pi\in\mathcal{S}_n \\ \mathrm{Des}(\pi^{-1})= S}} q^{\mathrm{inv}(\pi)} = [n]!_q \det\left(\frac{1}{[s_{j+1}-s_i]!_q}\right)_{i,j=0}^{k}$$

$$= \det\left(\begin{bmatrix} n - s_i \\ s_{j+1} - s_i \end{bmatrix}_q\right)_{i,j=0}^{k} .$$

## 14.10.2 Statistics on tableaux

We start with definitions of descent statistics for SYT. Let $T$ be a standard Young tableau of shape $D$ and size $n = |D|$. For each entry $1 \le t \le n$ let $\mathrm{row}(T^{-1}(t))$ denote the index of the row containing the cell $T^{-1}(t)$.

**Definition 14.10.4** *The* **descent set** *of* $T$ *is*

$$\mathrm{Des}(T) := \{1 \le i \le n-1 \mid \mathrm{row}(T^{-1}(i)) < \mathrm{row}(T^{-1}(i+1))\},$$

*the* **descent number** *of* $T$ *is*

$$\mathrm{des}(T) := |\mathrm{Des}(T)|,$$

*and the* **major index** *of* $T$ *is*

$$\mathrm{maj}(T) := \sum_{i\in\mathrm{Des}(T)} i.$$

**Example 14.10.5** *Let*

$$T = \begin{array}{|c|c|c|c|} \hline 1 & 2 & 5 & 8 \\ \hline 3 & 4 & 6 \\ \cline{1-3} 7 \\ \cline{1-1} \end{array} .$$

*Then* $\mathrm{Des}(T) = \{2,5,6\}$, $\mathrm{des}(T) = 3$ *and* $\mathrm{maj}(T) = 2+5+6 = 13$.

For a permutation $\pi \in \mathcal{S}_n$, recall from Section 14.4 the notation $T_\pi$ for the skew (anti-diagonal) SYT that corresponds to $\pi$ and the notation $(P_\pi, Q_\pi)$ for the pair of ordinary SYT that corresponds to $\pi$ under the Robinson-Schensted correspondence. By definition,

$$\mathrm{Des}(T_\pi) = \mathrm{Des}(\pi^{-1}).$$

The jeu de taquin algorithm preserves the descent set of a SYT, and therefore

**Proposition 14.10.6** *For every permutation $\pi \in \mathcal{S}_n$,*

$$\mathrm{Des}(P_\pi) = \mathrm{Des}(\pi^{-1}) \qquad \text{and} \qquad \mathrm{Des}(Q_\pi) = \mathrm{Des}(\pi).$$

When it comes to inversion number, there is more than one possible definition for SYT.

**Definition 14.10.7** *An* **inversion** *in $T$ is a pair $(i, j)$ such that $1 \le i < j \le n$ and the entry for $j$ appears strictly south and strictly west of the entry for $i$:*

$$\mathrm{row}(T^{-1}(i)) < \mathrm{row}(T^{-1}(j)) \qquad \text{and} \qquad \mathrm{col}(T^{-1}(i)) > \mathrm{col}(T^{-1}(j)).$$

*The* **inversion set** *of $T$, $\mathrm{Inv}(T)$, consists of all the inversions in $T$, and the* **inversion number** *of $T$ is*

$$\mathrm{inv}(T) := |\mathrm{Inv}(T)|.$$

*The* **sign** *of $T$ is*

$$\mathrm{sign}(T) := (-1)^{\mathrm{inv}(T)}.$$

**Definition 14.10.8** *A* **weak inversion** *in $T$ is a pair $(i, j)$ such that $1 \le i < j \le n$ and the entry for $j$ appears strictly south and weakly west of the entry for $i$:*

$$\mathrm{row}(T^{-1}(i)) < \mathrm{row}(T^{-1}(j)) \qquad \text{and} \qquad \mathrm{col}(T^{-1}(i)) \ge \mathrm{col}(T^{-1}(j)).$$

*The* **weak inversion set** *of $T$, $\mathrm{Winv}(T)$, consists of all the weak inversions in $T$, and the* **weak inversion number** *of $T$ is*

$$\mathrm{winv}(T) := |\mathrm{Winv}(T)|.$$

**Observation 14.10.9** *For every standard Young tableaux $T$ of ordinary shape $\lambda$,*

$$\mathrm{winv}(T) = \mathrm{inv}(T) + \sum_j \binom{\lambda'_j}{2}.$$

*Here $\lambda'_j$ is the length of the $j$-th column of the diagram $[\lambda]$.*

**Example 14.10.10** *For $T$ as in Example 14.10.5, we have*

$$\mathrm{Inv}(T) = \{(2,3), (2,7), (4,7), (5,7), (6,7)\},$$

*$\mathrm{inv}(T) = 5$ and $\mathrm{sign}(T) = -1$. The weak inversion set consists of the inversion set plus all pairs of entries in the same column. Thus $\mathrm{winv}(T) = \mathrm{inv}(T) + \binom{3}{2} + \binom{2}{2} + \binom{2}{2} + \binom{1}{2} = 4 + 3 + 1 + 1 = 9$.*

For another (more complicated) inversion number on SYT see [47].

### 14.10.3 Thin shapes

We begin with refinements and $q$-analogues of results from Section 14.3.

#### 14.10.3.1 Hook shapes

It is easy to verify that

**Observation 14.10.11** *For any* $1 \le k \le n-1$

$$\sum_{\mathrm{sh}(T)=(n-k,1^k)} \mathbf{x}^{\mathrm{Des}(T)} = e_k,$$

*where*

$$e_k := \sum_{1 \le i_1 < i_2 < \ldots < i_k \le n-1} x_{i_1} \cdots x_{i_k}$$

*are the elementary symmetric functions.*

*Proof.* Let $T$ be a SYT of hook shape. Then for every $1 \le i \le n-1$, $i$ is a descent in $T$ if and only if the letter $i+1$ lies in the first column. ∎

Thus

$$\sum_{\mathrm{sh}(T)=\text{hook of size } n} \mathbf{x}^{\mathrm{Des}(T)} = \prod_{i=1}^{n-1}(1+x_i).$$

It follows that

$$\sum_{\mathrm{sh}(T)=\text{hook of size } n} t^{\mathrm{des}(T)} q^{\mathrm{maj}(T)} = \prod_{i=1}^{n-1}(1+tq^i). \tag{14.4}$$

Notice that for a $T$ of hook shape and size $n$, $\mathrm{sh}(T) = (n-k, 1^k)$ if and only if $\mathrm{des}(T) = k$. Combining this observation with Equation (14.4), the $q$-binomial theorem implies that

$$\sum_{\mathrm{sh}(T)=(n-k,1^k)} q^{\mathrm{maj}(T)} = q^{\binom{k}{2}} \begin{bmatrix} n-1 \\ k \end{bmatrix}_q .$$

Finally, the statistics winv and maj are equal on all SYT of hook shape. For most Young tableaux of non-hook zigzag shapes these statistics are not equal. However, the equidistribution phenomenon may be generalized to all zigzags. This will be shown below.

#### 14.10.3.2 Zigzag shapes

Recall from Section 14.7.1 that each subset $S \subseteq [n-1]$ defines a zigzag shape $\mathrm{zigzag}_n(S)$ of size $n$. The following statement refines Proposition 14.3.5.

**Proposition 14.10.12** *For any* $S \subseteq [n-1]$,

$$f^{\mathrm{zigzag}_n(S)} = \#\{\pi \in \mathscr{S}_n : \mathrm{Des}(\pi) = S\}.$$

*Proof.* Standard Young tableaux of the zigzag shape encoded by $S$ are in bijection with permutations in $S_n$ whose descent set is exactly $S$. The bijection converts such a tableau into a permutation by reading the cell entries starting from the southwestern corner. For example, for $S = \{1,3,5,6\}$,

$$T = \begin{array}{ccccc} & & 2 & 6 & 8 \\ & & 5 & & \\ & 3 & 7 & & \\ 1 & 4 & & & \\ 9 & & & & \end{array} \quad \mapsto \pi = [914375268].$$

■

Notice that in this example, $\pi^{-1} = [274368591]$ and $\mathrm{Des}(T) = \mathrm{Des}(\pi^{-1}) = \{2,3,6,8\}$. Also, $\mathrm{Winv}(T) = \mathrm{Inv}(\pi^{-1})$. This is a general phenomenon. Indeed,

**Observation 14.10.13** *Let $T$ be a SYT of shape $\mathrm{zigzag}_n(S)$ and let $\pi$ be its image under the bijection described in the proof of Proposition 14.10.12. Then*

$$S = \mathrm{Des}(\pi), \qquad \mathrm{Des}(T) = \mathrm{Des}(\pi^{-1}), \qquad \mathrm{Winv}(T) = \mathrm{Inv}(\pi^{-1}).$$

By Observation 14.10.13, there is a maj-winv preserving bijection from SYT of a given zigzag shape to permutations in the corresponding descent class. Combining this with Theorem 14.10.3 one obtains

**Proposition 14.10.14** *For a subset $S = \{s_1, \ldots, s_k\} \subseteq [n-1]$ $(s_1 < \ldots < s_k)$, set $s_0 := 0$ and $s_{k+1} := n$. Then*

$$\sum_{\mathrm{sh}(T)=\mathrm{zigzag}_n(S)} q^{\mathrm{maj}(T)} = \sum_{\mathrm{sh}(T)=\mathrm{zigzag}_n(S)} q^{\mathrm{winv}(T)} = [n]_q! \cdot \det\left(\frac{1}{[s_{j+1}-s_i]_q!}\right)_{i,j=0}^{k}.$$

#### 14.10.3.3 Two-rowed shapes

The major index and (weak) inversion number are not equidistributed over SYT of two-rowed shapes. However, the $q$-enumerations of both are nice. Two different $q$-Catalan numbers appear in the scene. First, consider enumeration by major index.

**Proposition 14.10.15** *For every $n \in \mathbb{N}$ and $0 \le k \le n/2$*

$$\sum_{\mathrm{sh}(T)=(n-k,k)} q^{\mathrm{maj}(T)} = \begin{bmatrix} n \\ k \end{bmatrix}_q - \begin{bmatrix} n \\ k-1 \end{bmatrix}_q.$$

*In particular,*

$$\sum_{\mathrm{sh}(T)=(m,m)} q^{\mathrm{maj}(T)} = q^m C_m(q)$$

*where*

$$C_m(q) = \frac{1}{[m+1]_q} \begin{bmatrix} 2m \\ m \end{bmatrix}_q$$

*is the m-th Fürlinger-Hofbauer q-Catalan number [38].*

Hence we have the following corollary.

**Corollary 14.10.16** *The equality*

$$\sum_{\text{height}(T)\leq 2} q^{\text{maj}(T)} = \begin{bmatrix} n \\ \lfloor n/2 \rfloor \end{bmatrix}_q.$$

*holds.*

For a bijective proof and refinements see [10].

The descent set is invariant under jeu de taquin. Hence the proof of Theorem 14.4.17 may be lifted to a $q$-analogue. Here is a $q$-analogue of Theorem 14.4.17.

**Theorem 14.10.17** *The major index generating function over SYT of size n and height* $\leq 3$ *is equal to*

$$m_n(q) = \sum_{k=0}^{\lfloor n/2 \rfloor} q^k \begin{bmatrix} n \\ 2k \end{bmatrix}_q C_k(q).$$

Furthermore, the following strengthened version of Corollary 14.4.14(3) holds.

**Corollary 14.10.18** *For every positive integer k*

$$\sum_{\{T\in \text{SYT}_n:\ \text{height}(T)<k\}} \mathbf{x}^{\text{Des}(T)} = \sum_{\{\pi\in \text{Avoid}_n(\sigma_k):\ \pi^2=id\}} \mathbf{x}^{\text{Des}(\pi)},$$

*where* $\text{Avoid}_n(\sigma_k)$ *is the subset of all permutations in* $\mathscr{S}_n$ *that avoid the pattern* $\sigma_k := [k, k-1, \ldots, 1]$.

*Proof.* Combine Theorem 14.4.13 with Proposition 14.10.6. ∎

Counting by inversions is associated with another $q$-Catalan number.

**Definition 14.10.19** [19] *Define the* **Carlitz-Riordan $q$-Catalan number** $\widetilde{C}_n(q)$ *by the recursion*

$$\widetilde{C}_{n+1}(q) := \sum_{k=0}^{n} q^k \widetilde{C}_k(q)\widetilde{C}_{n-k}(q)$$

*with* $\widetilde{C}_0(q) := 1$.

These polynomials are, essentially, generating functions for the area under Dyck paths of order $n$.

**Proposition 14.10.20** [108] *The equality*

$$\sum_{\text{sh}(T)=(m,m)} q^{\text{inv}(T)} = \widetilde{C}_m(q).$$

*holds.*

**Proposition 14.10.21** [108] *For* $0 \le k \le n/2$ *denote* $G_k(q) := \sum_{\mathrm{sh}(T)=(n-k,k)} q^{\mathrm{inv}(T)}$. *Then*

$$\sum_{k=0}^{\lfloor n/2 \rfloor} q^{\binom{n-2k}{2}} G_k(q)^2 = \widetilde{C}_n(q).$$

Enumeration of two-rowed SYT by descent number was studied by Barahovski.

**Proposition 14.10.22** [8] *For* $m \ge k \ge 1$,

$$\sum_{\mathrm{sh}(T)=(m,k)} t^{\mathrm{des}(T)} = \sum_{d=1}^{k} \frac{m-k+1}{k} \binom{k}{d} \binom{m}{d-1} t^d.$$

## 14.10.4 The general case

### 14.10.4.1 Counting by descents

There is a nice formula, due to Gessel, for the number of SYT of a given shape $\lambda$ with a given descent set. Since it involves a scalar product of symmetric functions, which are out of the scope of the current survey, we refer the reader to the original paper [41, Theorem 7].

There is also a rather complicated formula of Kreweras [62, 63] for the generating function of descent number on SYT of a given shape. However, the first moments of the distribution of this statistic may be calculated quite easily.

**Proposition 14.10.23** *For every partition* $\lambda \vdash n$ *and* $1 \le k \le n-1$

$$\#\{T \in \mathrm{SYT}(\lambda) : k \in \mathrm{Des}(T)\} = \left( \frac{1}{2} - \frac{\sum_i \binom{\lambda_i}{2} - \sum_j \binom{\lambda'_j}{2}}{n(n-1)} \right) f^\lambda.$$

*Here* $\lambda_i$ *is the length of the i-th row in the diagram* $[\lambda]$, *and* $\lambda'_j$ *is the length of the j-th column.*

For proofs see [50, 3].

One deduces that $\#\{T \in \mathrm{SYT}(\lambda) : k \in \mathrm{Des}(T)\}$ is independent of $k$. This phenomenon may be generalized as follows: For any composition $\mu = (\mu_1, \ldots, \mu_t)$ of $n$ let

$$S_\mu := \{\mu_1, \mu_1 + \mu_2, \ldots, \mu_1 + \ldots + \mu_{t-1}\} \subseteq [1, n-1].$$

The underlying partition of $\mu$ is obtained by reordering the parts in a weakly decreasing order.

**Theorem 14.10.24** *For every partition* $\lambda \vdash n$ *and any two compositions* $\mu$ *and* $\nu$ *of* $n$ *with the same underlying partition,*

$$\sum_{\mathrm{sh}(T)=\lambda} \mathbf{x}^{\mathrm{Des}(T) \setminus S_\mu} = \sum_{\mathrm{sh}(T)=\lambda} \mathbf{x}^{\mathrm{Des}(T) \setminus S_\nu}.$$

Proposition 14.10.23 implies that

**Corollary 14.10.25** *The expected descent number of a random SYT of shape $\lambda \vdash n$ is equal to*

$$\frac{n-1}{2} - \frac{1}{n}\left(\sum_i \binom{\lambda_i}{2} - \sum_j \binom{\lambda'_j}{2}\right).$$

The variance of descent number was computed in [3, 50], implying a concentration around the mean phenomenon. The proofs in [3] involve character theory, while those in [50] follow from a careful examination of the hook length bijection of Novelli, Pak and Stoyanovskii [76], described in Section 14.6.2 above.

#### 14.10.4.2 Counting by major index

Counting SYT of general ordinary shape by descents is difficult. Surprisingly, it was discovered by Stanley that counting by major index leads to a natural and beautiful $q$-analogue of the ordinary hook length formula (Proposition 14.5.3).

**Theorem 14.10.26 ($q$-Hook Length Formula [114, Corollary 7.21.5])** *For every partition $\lambda \vdash n$*

$$\sum_{T \in \mathrm{SYT}(\lambda)} q^{\mathrm{maj}(T)} = q^{\sum_i \binom{\lambda'_i}{2}} \frac{[n]_q!}{\prod_{c \in [\lambda]} [h_c]_q}.$$

This result follows from a more general identity, showing that the major index generating function for SYT of a skew shape is essentially the corresponding skew Schur function [114, Proposition 7.19.11]. If $|\lambda/\mu| = n$ then

$$s_{\lambda/\mu}(1, q, q^2, \ldots) = \frac{\sum_{T \in \mathrm{SYT}(\lambda/\mu)} q^{\mathrm{maj}(T)}}{(1-q)(1-q^2)\cdots(1-q^n)}.$$

An elegant $q$-analogue of Schur's shifted product formula (Proposition 14.5.7) was found by Stembridge.

**Theorem 14.10.27** [122, Corollary 5.2] *For every strict partition $\lambda = (\lambda_1, \ldots, \lambda_t) \vdash n$*

$$\sum_{T \in \mathrm{SYT}(\lambda^*)} q^{n \cdot \mathrm{des}(T) - \mathrm{maj}(T)} = \frac{[n]_q!}{\prod_{i=1}^t [\lambda_i]_q!} \cdot \prod_{(i,j): i<j} \frac{q^{\lambda_j} - q^{\lambda_i}}{1 - q^{\lambda_i + \lambda_j}}.$$

Theorem 14.10.26 may be easily generalized to $r$-tableaux.

**Corollary 14.10.28** *For every $r$-partition $\lambda = (\lambda^1, \ldots, \lambda^r)$ of total size $n$*

$$\sum_{T \in \mathrm{SYT}(\lambda)} q^{\mathrm{maj}(T)} = q^{\sum_i \binom{\lambda'_i}{2}} \cdot \frac{[n]_q!}{\prod_{c \in [\lambda]} [h_c]_q}.$$

The proof relies on a combination of Theorem 14.10.3 with the Stanton-White bijection for colored permutations.

#### 14.10.4.3 Counting by inversions

Unlike descent statistics, not much is known about enumeration by inversion statistics in the general case. The following result was conjectured by Stanley [116] and proved, independently, by Lam [64], Reifegerste [92] and Sjöstrand [109].

**Theorem 14.10.29** *The equality*

$$\sum_{\lambda \vdash n} \sum_{T \in \mathrm{SYT}(\lambda)} \mathrm{sign}(T) = 2^{\lfloor n/2 \rfloor}.$$

*holds.*

A generalization of Foata's bijective proof of MacMahon's fundamental equidistribution theorem (Theorem 14.10.2) to SYT of any given shape was described in [47], using a more involved concept of inversion number for SYT.

## 14.11 Counting reduced words

An interpretation of SYT as reduced words is presented in this section. This interpretation is based on Stanley's seminal paper [112] and follow ups. For further reading see [15, §7.4-7.5] and [16, §7].

### 14.11.1 Coxeter generators and reduced words

Recall that the symmetric group $\mathscr{S}_n$ is generated by the set $S := \{s_i : 1 \le i < n\}$ subject to the defining Coxeter relations:

$$s_i^2 = 1 \ (1 \le i < n); \quad s_i s_j = s_j s_i \ (|j-i| > 1); \quad s_i s_{i+1} s_i = s_{i+1} s_i s_{i+1} \ (1 \le i < n-1).$$

The elements of $S$ are called simple reflections and may be identified with the adjacent transposition in $\mathscr{S}_n$, where $s_i = (i, i+1)$.

The Coxeter length of a permutation $\pi \in \mathscr{S}_n$ is

$$\ell(\pi) := \min\{t : s_{i_1} \cdots s_{i_t} = \pi\},$$

the minimal length of an expression of $\pi$ as a product of simple reflections.

**Claim 14.11.1** *For every* $\pi \in \mathscr{S}_n$

$$\ell(\pi) = \mathrm{inv}(\pi).$$

A series of Coxeter generators $(s_{i_1}, \ldots, s_{i_t})$ is a reduced word of $\pi \in \mathscr{S}_n$ if the resulting product is a factorization of minimal length of $\pi$, that is $s_{i_1} \cdots s_{i_t} = \pi$ and $t = \ell(\pi)$. In this section, the enumeration of reduced words will be reduced to enumeration of SYT.

### 14.11.2 Ordinary and skew shapes

A **shuffle** of the two sequences $(1,2,\dots,k)$ and $(k+1,k+2,\dots,\ell)$ is a permutation $\pi \in \mathscr{S}_\ell$ in which both sequences appear as subsequences. For example ($k=3, \ell=7$): 4516237.

**Proposition 14.11.2** [26] *There exists a bijection $\lambda \mapsto \pi_\lambda$ from the set of all partitions to the set of all fixed point free shuffles such that:*

1. *If $\lambda$ has height $k$, width (length of first row) $\ell - k$ and size $n$ then $\pi_\lambda$ is a fixed point free shuffle of $(1,2,\dots,k)$ and $(k+1,k+2,\dots,\ell)$ with $\mathrm{inv}(\pi_\lambda) = n$.*

2. *The number of SYT of shape $\lambda$ is equal to the number of reduced words of $\pi_\lambda$.*

*Proof Sketch.* For the first claim, read the permutation from the shape as follows: Encode the rows by $1,2,\dots,k$ from bottom to top, and the columns by $k+1, k+2,\dots,\ell$ from left to right. Then walk along the southeastern boundary from bottom to top. If the $i$th step is horizontal, set $\pi_\lambda(i)$ to be its column encoding; otherwise set $\pi_\lambda(i)$ to be its row encoding.

**Example 14.11.3** *The shape*

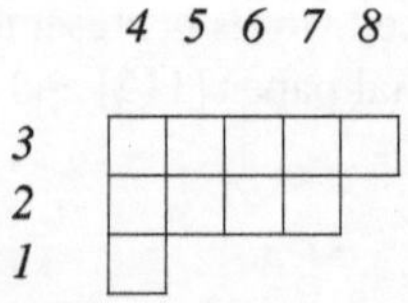


*corresponds to the shuffle permutation*

$$\pi = 41567283.$$

For the second claim, read the reduced word from the SYT as follows: If the letter $1 \le j \le n$ lies on the $i$th diagonal (from the left), set the $j$th letter in the word (from left to right) to be $s_i$.

**Example 14.11.4** *The SYT*

| | 4 | 5 | 6 | 7 | 8 |
|---|---|---|---|---|---|
| 3 | 1 | 2 | 3 | 6 | 8 |
| 2 | 4 | 5 | 9 | 10 | |
| 1 | 7 | | | | |

*corresponds to the reduced word (in adjacent transpositions)*

$$s_3 s_4 s_5 s_2 s_3 s_6 s_1 s_7 s_4 s_5 = 41567283.$$

The proof that this map is a bijection from all SYT of shape $\lambda$ to all reduced words of $\pi_\lambda$ is obtained by induction on the size of $\lambda$. ∎

**Corollary 14.11.5** *For every pair of positive integers $1 \le k \le \ell$, the number of reduced words of the permutation $[k+1, k+2, \ldots, \ell, 1, 2, \ldots, k]$ is equal to the number of SYT of rectangular shape $(k^{\ell-k})$.*

**Proposition 14.11.6** *There exists an injection from the set of all 321-avoiding permutations to the set of all skew shapes, which satisfies the following property: For every 321-avoiding permutation $\pi$ there exists a skew shape $\lambda/\mu$ such that the number of reduced words of $\pi$ is equal to the number of SYT of shape $\lambda/\mu$.*

The following theorem was conjectured and first proved by Stanley using symmetric functions [112]. A bijective proof was given later by Edelman and Greene [24].

**Theorem 14.11.7** [112, Corollary 4.3] *The number of reduced words (in adjacent transpositions) of the longest permutation $w_0 := [n, n-1, ..., 1]$ is equal to the number of SYT of staircase shape $\delta_{n-1} = (n-1, n-2, ..., 1)$.*

Corollary 14.11.5 and Theorem 14.11.7 are special instances of the following remarkable result.

**Theorem 14.11.8** [112, 24]

1. *For every permutation $\pi \in \mathcal{S}_n$, the number of reduced words of $\pi$ can be expressed as*

$$\sum_{\lambda \vdash \mathrm{inv}(\pi)} m_\lambda f^\lambda$$

*where $m_\lambda$ are nonnegative integers canonically determined by $\pi$.*

2. *The above sum is a unique $f^{\lambda_0}$ (i.e., $m_\lambda = \delta_{\lambda,\lambda_0}$) if and only if $\pi$ is 2143-avoiding.*

Reiner [93] applied Theorem 14.11.7 to show that the expected number of subwords of types $s_i s_{i+1} s_i$ and $s_{i+1} s_i s_{i+1}$ in a random reduced word of the longest permutation is exactly one. He conjectured that the distribution of their number is Poisson. For some recent progress see [129].

A generalization of Theorem 14.11.7 to type $B$ involves square shapes. The following theorem was conjectured by Stanley and proved by Haiman.

**Theorem 14.11.9** [49] *The number of reduced words (in the alphabet of Coxeter generators) of the longest signed permutation $w_0 := [-1, -2, ..., -n]$ in $B_n$ is equal to the number of SYT of square shape $(n^n)$.*

For a recent application see [82].

### 14.11.3 Shifted shapes

An interpretation of the number of SYT of a shifted shape was given by Edelman. Recall the left weak order from Section 14.1.1, and recall that $\sigma$ covers $\pi$ in this order if $\sigma = s_i\pi$ and $\ell(\sigma) = \ell(\pi)+1$. Edelman considered a modification of this order in which we further require that the letter moved to the left be larger than all letters that precede it.

**Theorem 14.11.10** [23, Theorem 3.2] *The number of maximal chains in the modified weak order is equal to the number of SYT of shifted staircase shape.*

A related interpretation of SYT of shifted shapes was given in [26]. A permutation $\pi \in \mathcal{S}_n$ is **unimodal** if $\mathrm{Des}(\pi^{-1}) = \{1,\ldots,j\}$ for some $0 \le j \le n-1$. Consider $U_n$, the set of all unimodal permutations in $\mathcal{S}_n$, as a poset under the left weak order induced from $\mathcal{S}_n$.

**Proposition 14.11.11** [26] *There exists a bijection $\lambda \mapsto \pi_\lambda$ from the set of all shifted shapes contained in the shifted staircase $\delta_{n-1} = (n-1,n-2,\ldots,1)$ to the set $U_n$ of all unimodal permutations in $\mathcal{S}_n$ such that:*

1. $|\lambda| = \mathrm{inv}(\pi_\lambda)$.
2. *The number of SYT of shifted shape $\lambda$ is equal to the number of maximal chains in the interval $[id,\pi_\lambda]$ in $U_n$.*

*Proof Sketch.* Construct the permutation $\pi_\lambda$ from the shape $\lambda$ as follows: Encode the rows of $\lambda$ by $1,2,\ldots$ from top to bottom, and the columns by $2,3,\ldots$ from left to right. Then walk along the southeastern boundary from bottom to top. If the $i$th step is horizontal, set $\pi_\lambda(i)$ to be its column encoding; otherwise set $\pi_\lambda(i)$ to be its row encoding.

**Example 14.11.12** *The shifted shape*

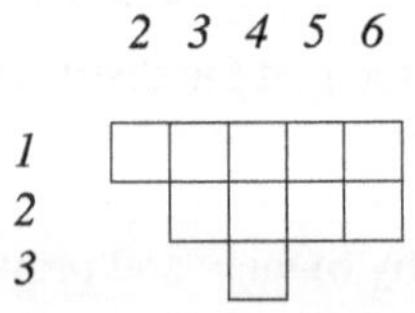


*corresponds to the permutation*

$$\pi = 435621.$$

Now construct the reduced word from the SYT $T$ as follows: If the letter $j$ lies in the $i$th diagonal (from left to right) of $T$ then set the $j$th letter in the word (from left to right) to be $s_i$.

**Example 14.11.13** *The SYT*

| | *2* | *3* | *4* | *5* | *6* |
|---|---|---|---|---|---|
| *1* | 1 | 2 | 3 | 6 | 8 |
| *2* | | 4 | 5 | 9 | 10 |
| *3* | | | 7 | | |

*corresponds to the reduced word (in adjacent transpositions)*

$$s_1 s_2 s_3 s_1 s_2 s_4 s_1 s_5 s_3 s_4 = 435621.$$

■

## 14.12 Appendix 1: Representation theoretic aspects

Representation theory may be considered as the birthplace of SYT; in fact, one cannot imagine group representations without the presence of SYT. Representation theory has been intimately related to combinatorics since its early days. The pioneering work of Frobenius, Schur and Young made essential use of integer partitions and tableaux. In particular, formulas for restriction, induction and decomposition of representations, as well as many character formulas, involve SYT. On the other hand, it is well known that many enumerative problems may be solved using representations. In this survey we restricted the discussion to combinatorial approaches. It should be noted that most results have representation theoretic proofs, and in many cases the discovery of the enumerative results was motivated by representation theoretic problems.

In this section we briefly point out several connections, assuming basic knowledge in non-commutative algebra, and give a very short sample of applications.

### 14.12.1 Degrees and enumeration

A SYT $T$ of shape $\lambda$ has an associated group algebra element $y_T \in \mathbb{C}[\mathcal{S}_n]$, called the **Young symmetrizer.** $y_T$ has a key role: It is an idempotent, and its principal right ideal

$$y_T\mathbb{C}[\mathcal{S}_n]$$

is an irreducible module of $\mathcal{S}_n$. All irreducible modules are generated, up to isomorphism, by Young symmetrizers, and two modules, which are generated by the Young symmetrizers of two SYT, are isomorphic if and only if these SYT have same shape. The irreducible characters of the symmetric group $\mathcal{S}_n$ over $\mathbb{C}$ are, thus, parameterized by the integer partitions of $n$.

**Proposition 14.12.1** *The degree of the character indexed by $\lambda \vdash n$ is equal to $f^\lambda$, the number of SYT of the ordinary shape $\lambda$.*

This phenomenon extends to skew and shifted shapes. The number of SYT of skew shape $\lambda/\mu$, $f^{\lambda/\mu}$, is equal to the degree of the decomposable module generated by a Young symmetrizer of a SYT of shape $\lambda/\mu$. Projective representations are indexed by shifted shapes; the number of SYT of shifted shape $\lambda$, $g^\lambda$, is equal to the degree of the associated projective representation.

Most of the results in this survey have representation theoretic proofs or interpretations. A few examples will be given here.

*Proof sketch of Proposition 14.3.3.* The symmetric group $\mathcal{S}_n$ acts naturally on subsets of size $k$. The associated character, $\mu^{(n-k,k)}$, is multiplicity free; its decomposition into irreducibles is

$$\mu^{(n-k,k)} = \sum_{i=0}^{k} \chi^{(n-i,i)}. \tag{14.5}$$

Hence

$$\chi^{(n-k,k)} = \mu^{(n-k,k)} - \mu^{(n-k+1,k-1)}.$$

The degrees thus satisfy

$$f^{(n-k,k)} = \chi^{(n-k,k)}(1) = \mu^{(n-k,k)}(1) - \mu^{(n-k+1,k-1)}(1) = \binom{n}{k} - \binom{n}{k-1}.$$

■

This argumentation may be generalized to prove Theorem 14.5.4. First, notice that (14.5) is a special case of the Young rule for decomposing permutation modules. The Young rule implies the determinantal Jacobi-Trudi formula for expressing an irreducible module as an alternating sum of permutation modules, see e.g. [53]. Evaluation of the characters at the identity permutation implies Theorem 14.5.4.

Next proceed to identities that involve sums of $f^\lambda$-s.

*Proof of Corollary 14.4.14(1).* Recall that for every finite group, the sum of squares of the degrees of the irreducibles is equal to the size of the group. This fact together with the interpretation of the $f^\lambda$-s as degrees of the irreducibles of the symmetric group $\mathcal{S}_n$ (Proposition 14.12.1) completes the proof. ■

The same proof yields Theorem 14.9.19(1).

The Frobenius-Schur indicator theorem implies that for every finite group, which may be represented over $\mathbb{R}$, the sum of degrees of the irreducibles is equal to the number of involutions in the group, implying Corollary 14.4.14(2). The proof of Theorem 14.9.19(2) is similar, see e.g. [18, 2].

*Proof sketch of Corollary 14.4.16.* The permutation module defined by the action of $\mathcal{S}_{2n}$ on the cosets of $B_n = \mathbb{Z}_2 \wr \mathcal{S}_n$ is isomorphic to a multiplicity free sum of all $\mathcal{S}_{2n}$-irreducible modules indexed by partitions with all parts even [69, §VII (2.4)]. Comparison of the dimensions completes the proof. ■

### 14.12.2 Characters and $q$-enumeration

The Murnaghan-Nakayama rule is a formula for computing values of irreducible $\mathscr{S}_n$-characters as signed enumerations of rim hook tableaux. Here is an example of special interest.

**Proposition 14.12.2** *For every $\lambda \vdash rn$, the value of the irreducible character $\chi^\lambda$ at a conjugacy class of cycle type $r^n$ is equal to the number of r-rim hook tableaux; namely,*

$$\chi^\lambda_{(r,\ldots,r)} = f^\lambda_r.$$

Another interpretation of $f^\lambda_r$ is as the degree of an irreducible module of the wreath product $\mathbb{Z}_r \wr \mathscr{S}_n$.

An equivalent formula for the irreducible character values is by weighted counts of all SYT of a given shape by their descents; $q$-enumeration then amounts to a computation of the corresponding Hecke algebra characters.

These character formulas may be applied to counting SYT by descents. Here is a simple example.

*Proof sketch of Proposition 14.10.23.* By the Murnaghan-Nakayama rule, the value of the character $\chi^\lambda$ at a transposition $s_i = (i, i+1)$ is equal to

$$\begin{aligned}|\{T \in \mathrm{SYT}(\lambda) : i \notin \mathrm{Des}(T)\}| - |\{T \in \mathrm{SYT}(\lambda) : i \in \mathrm{Des}(T)\}| \\ = f^\lambda - 2|\{T \in \mathrm{SYT}(\lambda) : i \in \mathrm{Des}(T)\}|.\end{aligned}$$

Combining this with the explicit formula for this character [52]

$$\chi^\lambda_{(2,1^{n-2})} = \frac{\sum_i \binom{\lambda_i}{2} - \sum_j \binom{\lambda'_j}{2}}{\binom{n}{2}} f^\lambda.$$

completes the proof. ∎

Finally, we quote two classical results, which apply enumeration by major index.

**Theorem 14.12.3 (Kraśkiewicz-Weyman, in a widely circulated manuscript finally published as [56])** *Let $\omega$ be a primitive* 1*-dimensional character on the cyclic group $C_n$ of order n. Then, for any partition $\lambda$ of n, the multiplicity of $\chi^\lambda$ in the induced character $Ind_{C_n}^{S_n}\omega$, which is also the character of the $S_n$ action on the multilinear part of the free Lie algebra on n generators (and of many other actions on combinatorial objects) is equal to the number of $T \in \mathrm{SYT}(\lambda)$ with* $\mathrm{maj}(T) \equiv 1 \pmod{n}$.

This result may actually be deduced from the following one.

**Theorem 14.12.4 (Lusztig-Stanley)** *For any partition $\lambda$ of n and any $0 \le k \le \binom{n}{2}$, the multiplicity of $\chi^\lambda$ in the character of the $S_n$ action on the k-th homogeneous component of the coinvariant algebra is equal to the number of $T \in \mathrm{SYT}(\lambda)$ with* $\mathrm{maj}(T) = k$.

A parallel powerful language is that of symmetric functions. The interested reader is referred to the excellent textbooks [114, Ch. 7] and [69].

## 14.13 Appendix 2: Asymptotics and probabilistic aspects

An asymptotic formula is sometimes available when a simple explicit formula is not known. Sometimes, such formulas do lead to the discovery of surprising explicit formulas. A number of important asymptotic results will be given in this appendix.

Recall the exact formulas (Corollary 14.3.4, Theorem 14.4.17 and Theorem 14.4.19) for the total number of SYT of ordinary shapes with small height. An asymptotic formula for the total number of SYT of bounded height was given by Regev [87]; see also [13, 117, 89].

**Theorem 14.13.1** [87] *Fix a positive integer k and a positive real number $\alpha$. Then, asymptotically as $n \to \infty$,*

$$F_{k,\alpha}(n) := \sum_{\substack{\lambda \vdash n \\ \ell(\lambda) \le k}} (f^\lambda)^{2\alpha} \sim k^{2\alpha n} \cdot n^{-\frac{1}{2}(k-1)(\alpha k+2\alpha-1)} \cdot c(k,\alpha),$$

*where*

$$c(k,\alpha) := k^{\frac{1}{2}k(\alpha k+\alpha-1)} (2\alpha)^{-\frac{1}{2}(k-1)(\alpha k+1)} (2\pi)^{-\frac{1}{2}(k-1)(2\alpha-1)} \prod_{i=1}^{k} \frac{\Gamma(i\alpha)}{\Gamma(\alpha)}.$$

In particular, $\lim_{n\to\infty} F_{k,\alpha}(n)^{1/n} = k^{2\alpha}$. Important special cases are: $\alpha = 1$, which gives (by Theorem 14.4.11 and Proposition 14.4.13) the asymptotics for the number of permutations in $\mathcal{S}_n$ that do not contain a decreasing subsequence of length $k+1$; and $\alpha = 1/2$, which gives (by Corollary 14.4.14(3)) the asymptotics for the number of involutions in $\mathcal{S}_n$ with the same property. See [87] for many other applications.

The proof of Theorem 14.13.1 uses the hook length formula for $f^\lambda$, factoring out the dominant terms from the sum and interpreting what remains (in the limit $n \to \infty$) as a $k$ dimensional integral. An explicit evaluation of this integral, conjectured by Mehta and Dyson [75, 74], has been proved using Selberg's integral formula [107].

Okounkov and Olshanski [77] introduced and studied a non-homogeneous analogue of Schur functions, the **shifted Schur function**. As a combinatorial application, they gave an explicit formula for the number of SYT of skew shape $\lambda/\mu$. Stanley [115] proved a formula for $f^{\lambda/\mu}$ in terms of values of symmetric group characters. By applying the Vershik-Kerov $\mathcal{S}_\infty$-theory together with the Okounkov-Olshanski theory of shifted Schur functions, he used that formula to deduce the asymptotics of $f^{\lambda/\mu}$. See also [118]

Asymptotic methods were applied to show that certain distinct ordinary shapes have the same multiset of hook lengths [90]. Bijective and other purely combinatorial proofs were given later [91, 14, 60, 44].

In two seminal papers, Logan and Shepp [67], and independently Vershik and Kerov [133], studied the problem of the **limit shape** of the pair of SYT that correspond, under the RS correspondence, to a permutation chosen uniformly at random from $\mathcal{S}_n$. In other words, choose each partition $\lambda$ of $n$ with probability $\mu_n(\lambda) = (f^\lambda)^2/n!$. This probability measure on the set of all partitions of $n$ is called **Plancherel measure.**

It was shown in [67, 133] that, under Plancherel measure, probability concentrates near one asymptotic shape. See also [17].

**Theorem 14.13.2** [67, 133] *Draw a random ordinary diagram of size n in Russian notation (see Section 14.2.4) and scale it down by a factor of $n^{1/2}$. Then, as n tends to infinity, the shape converges in probability, under Plancherel measure, to the following limit shape:*

$$f(x) = \begin{cases} \frac{2}{\pi}(x \arcsin \frac{x}{2} + \sqrt{4-x^2}), & \text{if } |x| \le 2; \\ |x|, & \text{if } |x| > 2. \end{cases}$$

This deep result had significant impact on mathematics in recent decades [98].

A closely related problem is to find the shape that maximizes $f^\lambda$. First, notice that Corollary 14.4.14 implies that

$$\sqrt{\frac{n!}{p(n)}} \le \max\{f^\lambda : \lambda \vdash n\} \le \sqrt{n!},$$

where $p(n)$ is the number of partitions of $n$.

**Theorem 14.13.3** [134]

*(1) There exist constants $c_1 > c_0 > 0$ such that*

$$e^{-c_1\sqrt{n}}\sqrt{n!} \le \max\{f^\lambda : \lambda \vdash n\} \le e^{-c_0\sqrt{n}}\sqrt{n!}.$$

*(2) There exists constants $c_1' > c_0' > 0$ such that*

$$\lim_{n\to\infty} \mu_n \left\{ \lambda \vdash n : c_0' < -\frac{1}{\sqrt{n}} \ln \frac{f^\lambda}{\sqrt{n!}} < c_1' \right\} = 1.$$

Similar phenomena occur when Plancherel measure is replaced by other measures. For the uniform measure see, e.g., [83, 84].

Motivated by the limit shape result, Pittel and Romik proved that there exists a limit shape to the two-dimensional surface defined by a uniform random SYT of rectangular shape [85].

Consider a fixed $(i, j) \in \mathbb{Z}_+^2$ and a SYT $T$ chosen according to some probability distribution on the SYT of size $n$. A natural task is to estimate the probability that $T(i, j)$ has a prescribed value. Regev was the first to give an asymptotic answer to this problem, for some probability measures, using $\mathcal{S}_\infty$-theory [88]; see also [78]. A combinatorial approach was suggested later by McKay, Morse and Wilf [73]. Here is an interesting special case.

**Proposition 14.13.4** [88, 73] *For a random SYT $T$ of order $n$ and a positive integer $k > 1$*

$$Prob(T(2,1) = k) \sim \frac{k-1}{k!} + O(n^{-3/2}).$$

In [73], Proposition 14.13.4 was deduced from the following theorem.

**Theorem 14.13.5** [73] *Let $\mu \vdash k$ be a fixed partition and let $T$ be a fixed SYT of shape $\mu$. Let $n \geq k$ and let $N(n;T)$ denote the number of SYT with $n$ cells that contain $T$. Then*

$$N(n;T) \sim \frac{t_n f^\lambda}{k!},$$

*where $t_n$ denotes the number of involutions in the symmetric group $\mathscr{S}_n$.*

It follows that

$$\sum_{\lambda \vdash n} f^{\lambda/\mu} \sim \frac{t_n f^\lambda}{k!}.$$

Stanley [115], applying techniques of symmetric functions, deduced precise formulas for $N(n;T)$ in the form of finite linear combinations of the $t_n$s.

**Acknowledgments.** Many people contributed comments and valuable information to this chapter. We especially thank Christos Athanasiadis, Tomer Bauer, Sergi Elizalde, Dominique Foata, Avital Frumkin, Curtis Greene, Ira Gessel, Christian Krattenthaler, Igor Pak, Arun Ram, Amitai Regev, Vic Reiner, Dan Romik, Bruce Sagan, Richard Stanley and Doron Zeilberger. We used Ryan Reich's package `ytableau` for drawing diagrams and tableaux, and the package `algorithmicx` by János Szász for typesetting algorithms in pseudocode.

# References

[1] R. M. Adin, R. C. King, and Y. Roichman. Enumeration of standard Young tableaux of certain truncated shapes. *Electron. J. Combin.*, 18:Paper 20, 14 pp., 2011.

[2] R. M. Adin, A. Postnikov, and Y. Roichman. A Gelfand model for wreath products. *Israel J. Math.*, 179:381–402, 2010.

[3] R. M. Adin and Y. Roichman. Descent functions and random Young tableaux. *Combin. Probab. Comput.*, 10:187–201, 2001.

[4] R. M. Adin and Y. Roichman. Triangle-free triangulations, hyperplane arrangements and shifted tableaux. *Electron. J. Combin.*, 19:Paper 32, 19 pp., 2012.

[5] A. C. Aitken. The monomial expansion of determinantal symmetric functions. *Proc. Royal Soc. Edinburgh (A)*, 61:300–310, 1943.

[6] D. André. Developpements de sec $x$ et de tang $x$. *Comptes Rendus Acad. Sci. Paris*, 88:965–967, 1879.

[7] D. André. Mèmoire sur les permutations alternées. *J. Math.*, 7:167–184, 1881.

[8] I. Barahovski. Counting descents in standard Young tableaux. arXiv:0810.4701, 2008.

[9] H. Barcelo and A. Ram. Combinatorial representation theory. In *New perspectives in algebraic combinatorics (Berkeley, CA, 1996–97)*, number 38 in Math. Sci. Res. Inst. Publ., pages 23–90. Cambridge Univ. Press, 1999.

[10] M. Barnabei, F. Bonetti, S. Elizalde, and M. Silimbani. Descent sets on 321-avoiding involutions and hook decompositions of partitions. arXiv:1401.3011, 2014.

[11] Y. Baryshnikov and D. Romik. Enumeration formulas for Young tableaux in a diagonal strip. *Israel J. Math.*, 178:157–186, 2010.

[12] A. Berele. A Schensted-type correspondence for the symplectic group. *J. Combin. Theory Ser. A*, 43:320–328, 1986.

[13] A. Berele and A. Regev. Hook Young diagrams with applications to combinatorics and to representations of Lie superalgebras. *Adv. in Math.*, 64:118–175, 1987.

[14] C. Bessenrodt. On hooks of skew Young diagrams and bars in shifted diagrams. *Ann. Comb.*, 5:37–49, 2001.

[15] A. Björner and F. Brenti. *Combinatorics of Coxeter groups*, volume 231 of *Graduate Texts in Mathematics*. Springer, 2005.

[16] A. Björner and R. P. Stanley. *A combinatorial miscellany*, volume 42 of *Monographies de L'Enseignement Mathématique*. L'Enseignement Mathématique, 2010.

[17] A. Borodin, A. Okounkov, and G. Olshanski. Asymptotics of Plancherel measures for symmetric groups. *J. Amer. Math. Soc.*, 13:481–515, 2000.

[18] D. Bump and D. Ginzburg. Generalized Frobenius-Schur numbers. *J. Algebra*, 278:294–313, 2004.

[19] L. Carlitz and J. Riordan. Two element lattice permutation numbers and their $q$-generalization. *Duke J. Math.*, 31:371–388, 1964.

[20] C. De Concini. Symplectic standard tableaux. *Adv. in Math.*, 34:1–27, 1979.

[21] E. DeWitt. *Identities relating Schur S-functions and Q-functions*. PhD thesis, University of Michigan, Ann Arbor, 2012. `http://deepblue.lib.umich.edu/handle/2027.42/93841`.

[22] S. Dulucq and B. E. Sagan. La correspondance de Robinson-Schensted pour les tableaux oscillants gauches. *Discrete Math.*, 139:129–142, 1995.

[23] P. Edelman. Tableaux and chains in a new partial order of $S_n$. *J. Combin. Theory Ser. A*, 51:181–204, 1989.

[24] P. Edelman and C. Greene. Balanced tableaux. *Adv. in Math.*, 63:42–99, 1987.

[25] S. B. Ekhad and D. Zeilberger. Computational and theoretical challenges on counting solid standard Young tableaux. `http://www.math.rutgers.edu/~zeilberg/mamarim/mamarimhtml/ssyt.html`, 2012.

[26] S. Elizalde and Y. Roichman. Arc permutations. *J. Algebraic Combin.*, 39:301–334, 2014.

[27] N. D. Elkies. On the sums $\sum_{k=-\infty}^{\infty}(4k+1)^{-n}$. *Amer. Math. Monthly*, 110:561–573, 2003.

[28] S.-P. Eu. Skew-standard tableaux with three rows. *Adv. in Appl. Math.*, 45:463–469, 2010.

[29] W. Feit. The degree formula for the skew-representations of the symmetric group. *Proc. Amer. Math. Soc.*, 4:740–744, 1953.

[30] I. Fischer. A bijective proof of the hook-length formula for shifted standard tableaux. arXiv:0112261, 2001.

[31] D. Foata. On the Netto inversion number of a sequence. *Proc. Amer. Math. Soc.*, 19:236–240, 1968.

[32] D. Foata and M.-P. Schützenberger. Major index and inversion number of permutations. *Math. Nachr.*, 83:143–159, 1978.

[33] S. Fomin and N. Lulov. On the number of rim hook tableaux. *J. Math. Sci.*, 87:4118–4123, 1997. Original (in Russian): Zap. Nauchn. Sem. S.-Peterburg. Otdel. Mat. Inst. Steklov. (POMI) **223** (1995), Teor. Predstav. Din. Sistemy, Kombin. i Algoritm. Metody. I, 219–226, 340.

[34] J. S. Frame, G. de B. Robinson, and R. M. Thrall. The hook graphs of the symmetric group. *Canad. J. Math.*, 6:316–324, 1954.

[35] D. S. Franzblau and D. Zeilberger. A bijective proof of the hook-length formula. *J. Algorithms*, 3:317–343, 1982.

[36] F. G. Frobenius. Über die charaktere der symmetrischer gruppe. *Sitzungsber. Kön. Preuss. Akad. Wissen. Berlin*, pages 516–534, 1900. reprinted in: Ges. Abh. III (J.-P. Serre, ed.), Springer-Verlag, Berlin, 1968, pp. 148–166.

[37] W. Fulton. *Young tableaux, with applications to representation theory and geometry*, volume 35 of *London Math. Society Student Texts*. Cambridge University Press, 1997.

[38] J. Fürlinger and J. Hofbauer. $q$-Catalan numbers. *J. Combin. Theory Ser. A*, 40:248–264, 1985.

[39] A. M. Garsia and I. Gessel. Permutation statistics and partitions. *Adv. in Math.*, 31:288–305, 1979.

[40] A. M. Garsia and S. C. Milne. Method for constructing bijections for classical partition identities. *Proc. Nat. Acad. Sci. U.S.A.*, 78:2026–2028, 1981.

[41] I. M. Gessel. Multipartite *P*-partitions and inner products of skew Schur functions. In *Combinatorics and Algebra (Boulder, Colo., 1983)*, number 34 in Contemp. Math., pages 289–317. Amer. Math. Soc., Providence, RI, 1984.

[42] I. M. Gessel. Counting paths in Young's lattice. *J. Statist. Plann. Inference*, 34:125–134, 1993.

[43] I. M. Gessel and X. G. Viennot. Determinants, paths, and plane partitions. preprint, 1989.

[44] I. Goulden and A. Yong. Dyck paths and a bijection for multisets of hook numbers. *Discrete Math.*, 254:153–164, 2002.

[45] D. Gouyou-Beauchamps. Standard Young tableaux of height 4 and 5. *European J. Combin.*, 10:69–82, 1989.

[46] C. Greene, A. Nijenhuis, and H. S. Wilf. A probabilistic proof of a formula for the number of Young tableaux of a given shape. *Adv. in Math.*, 31:104–109, 1979.

[47] J. Haglund and L. Stevens. An extension of the Foata map to standard Young tableaux. *Sém. Lothar. Combin.*, 56:Art. B56c, 15 pp., 2006/7.

[48] M. D. Haiman. On mixed insertion, symmetry, and shifted Young tableaux. *J. Combin. Theory Ser. A*, 50:196–225, 1989.

[49] M. D. Haiman. Dual equivalence with applications, including a conjecture of Proctor. *Discrete Math.*, 99:79–113, 1992.

[50] P. A. Hästö. On descents in standard Young tableaux. *Electron. J. Combin.*, 7:Research Paper 59, 13 pp., 2000.

[51] A. P. Hillman and R. M. Grassl. Reverse plane partitions and tableau hook numbers. *J. Combin. Theory Ser. A*, 21:216–221, 1976.

[52] R. E. Ingram. Some characters of the symmetric group. *Proc. Amer. Math. Soc.*, 1:358–369, 1950.

[53] G. D. James and A. Kerber. *The representation theory of the symmetric group*, volume 16 of *Encyclopedia of Math. and its Applications*. Addison-Wesley Publishing Co., 1981.

[54] G. D. James and M. H. Peel. Specht series for skew representations of symmetric groups. *J. Algebra*, 56:343–364, 1979.

[55] R. C. King. Modification rules and products of irreducible representations of the unitary, orthogonal, and symplectic groups. *J. Mathematical Phys.*, 12:1588–1598, 1971.

[56] W. Kraśkiewicz and J. Weyman. Algebra of coinvariants and the action of a Coxeter element. *Bayreuth. Math. Schr.*, 63:265–284, 2001.

[57] C. Krattenthaler. Enumeration of lattice paths and generating functions for skew plane partitions. *Manuscripta Math.*, 63:129–155, 1989.

[58] C. Krattenthaler. Bijective proofs of the hook formulas for the number of standard Young tableaux, ordinary and shifted. *Electron. J. Combin.*, 2:Research Paper 13, 9 pp., 1995.

[59] C. Krattenthaler. An involution principle-free bijective proof of Stanley's hook-content formula. *Discrete Math. Theor. Comput. Sci.*, 3:11–32, 1998/99.

[60] C. Krattenthaler. Bijections for hook pair identities. *Electron. J. Combin.*, 7:Research Paper 27, 13 pp., 2000.

[61] C. Krattenthaler and M. Schlosser. The major index generating function of standard Young tableaux of shapes of the form "staircase minus rectangle". *Contemporary Math.*, to appear. arXiv:1402.4538.

[62] G. Kreweras. Sur une extension du problème dit "de Simon Newcomb". *C. R. Acad. Sci. Paris Sér. A–B*, 263:A43–A45, 1966.

[63] G. Kreweras. Traitement simultané du "problème de Young" et du "problème de Simon Newcomb". *Cahiers de B.U.R.O.*, 10:23–31, 1967.

[64] T. Lam. Growth diagrams, domino insertion and sign-imbalance. *J. Combin. Theory Ser. A*, 107:87–115, 2004.

[65] M. Larsen and A. Shalev. Characters of symmetric groups: sharp bounds and applications. *Invent. Math.*, 174:645–687, 2008.

[66] N. Linial. A new derivation of the counting formula for Young tableaux. *J. Combin. Theory Ser. A*, 33:340–342, 1982.

[67] B. F. Logan and L. A. Shepp. A variational problem for random Young tableaux. *Adv. in Math.*, 26:206–222, 1977.

[68] N. Lulov and I. Pak. Rapidly mixing random walks and bounds on characters of the symmetric group. *J. Algebraic Combin.*, 16:151–163, 2002.

[69] I. G. Macdonald. *Symmetric Functions and Hall Polynomials*. Oxford Math. Monographs. Oxford Univ. Press, second edition, 1995.

[70] T. J. MacLarnan. *Tableau recursions and symmetric Schensted correspondences for ordinary, shifted and oscillating tableaux.* PhD thesis, UCSD, 1986.

[71] P. A. MacMahon. Memoir on the theory of the partitions of numbers – Part IV. *Phil. Trans. Royal Soc. London (A)*, 209:153–175, 1909.

[72] P. A. MacMahon. *Combinatory Analysis, vols. 1 and 2*. Cambridge Univ. Press, 1915, 1916.

[73] B. D. McKay, J. Morse, and H. S. Wilf. The distributions of the entries of Young tableaux. *J. Combin. Theory Ser. A*, 97:117–128, 2002.

[74] M. L. Mehta. *Random matrices and the statistical theory of energy levels.* Academic Press, 1967.

[75] M. L. Mehta and F. J. Dyson. Statistical theory of the energy levels of complex systems V. *J. Mathematical Phys.*, 4:713–719, 1963.

[76] J.-C. Novelli, I. Pak, and A. V. Stoyanovskii. A direct bijective proof of the hook-length formula. *Discrete Math. and Theor. Comp. Sci.*, 1:53–67, 1997.

[77] A. Okounkov and G. Olshanski. Shifted Schur functions. *St. Petersburg Math. J.*, 9:239–300, 1998. original (in Russian): Algebr. i Analiz **9** (1997), 73–146.

[78] G. Olshanski and A. Regev. Random Young tableaux and combinatorial identities. *Sém. Lothar. Combin.*, 46:Art. B46e, 30 pp., 2001/02.

[79] I. Pak and A. Postnikov. Oscillating tableaux, $S_p \times S_q$-modules, and Robinson-Schensted-Knuth correspondence. In *8th International Conference on Formal Power Series and Algebraic Combinatorics*, University of Minnesota, 1996. `http://www-igm.univ-mlv.fr/~fpsac/FPSAC96/articles.html`.

[80] I. M. Pak and A. V. Stoyanovskii. Bijective proof of the hook formula and its analogues. *Funct. Anal. Appl.*, 26:216–218, 1992. original (in Russian): Funktsional. Anal. i Prilozhen. **26** (1992), 80–82.

[81] G. Panova. Tableaux and plane partitions of truncated shapes. *Adv. in Appl. Math.*, 49:196–217, 2012.

[82] T. K. Petersen and L. Serrano. Cyclic sieving for longest reduced words in the hyperoctahedral group. *Electron. J. Combin.*, 17:Research Paper 67, 12 pp., 2010.

[83] B. Pittel. On a likely shape of the random Ferrers diagram. *Adv. in Appl. Math.*, 18:432–488, 1997.

[84] B. Pittel. On the distribution of the number of Young tableaux for a uniformly random diagram. *Adv. in Appl. Math.*, 29:185–214, 2002.

[85] B. Pittel and D. Romik. Limit shapes for random square Young tableaux. *Adv. in Appl. Math.*, 38:164–209, 2007.

[86] A. Postnikov. Affine approach to quantum Schubert calculus. *Duke Math. J.*, 128:473–509, 2005.

[87] A. Regev. Asymptotic values for degrees associated with strips of Young diagrams. *Adv. Math.*, 41:115–136, 1981.

[88] A. Regev. $S_\infty$ representations and combinatorial identities. *Trans. Amer. Math. Soc.*, 353:4371–4404, 2001.

[89] A. Regev. Asymptotics of Young tableaux in the strip, the $d$-sums. arXiv:1004.4476, 2010.

[90] A. Regev and A. M. Vershik. Asymptotics of Young diagrams and hook numbers. *Electron. J. Combin.*, 4:Paper 22, 12 pp., 1997.

[91] A. Regev and D. Zeilberger. Proof of a conjecture on multisets of hook numbers. *Ann. Comb.*, 1:391–394, 1997.

[92] A. Reifegerste. Permutation sign under the Robinson-Schensted correspondence. *Ann. Comb.*, 8:103–112, 2004.

[93] V. Reiner. Note on the expected number of Yang-Baxter moves applicable to reduced decompositions. *European J. Combin.*, 26:1019–1021, 2005.

[94] V. Reiner and M. Shimozono. Specht series for column-convex diagrams. *J. Algebra*, 174:489–522, 1995.

[95] J. B. Remmel. Bijective proofs of formulae for the number of standard Young tableaux. *Linear and Multilinear Algebra*, 11:45–100, 1982.

[96] T. Roby. The connection between the Robinson-Schensted correspondence for skew oscillating tableaux and graded graphs. *Discrete Math.*, 139:481–485, 1995.

[97] Y. Roichman. Upper bound on the characters of the symmetric groups. *Invent. Math.*, 125:451–485, 1996.

[98] D. Romik. *The Surprising Mathematics of Longest Increasing Subsequences*. Cambridge Univ. Press, 2014.

[99] B. E. Sagan. *The symmetric group: Representations, combinatorial algorithms, and symmetric functions*. Graduate Texts in Mathematics, Second edition. Springer Verlag, 2001.

[100] B. E. Sagan. An analog of Schensted's algorithm for shifted Young tableaux. *J. Combin. Theory Ser. A*, 27:10–18, 1979.

[101] B. E. Sagan. On selecting a random shifted Young tableau. *J. Algorithms*, 1:213–234, 1980.

[102] B. E. Sagan. Shifted tableaux, Schur $Q$-functions, and a conjecture of R. Stanley. *J. Combin. Theory Ser. A*, 45:62–103, 1987.

[103] B. E. Sagan. The ubiquitous Young tableau. In *Invariant theory and tableaux (Minneapolis, MN, 1988)*, number 19 in IMA Vol. Math. Appl., pages 262–298. Springer, New York, 1990.

[104] C. Schensted. Longest increasing and decreasing subsequences. *Canad. J. Math.*, 13:179–191, 1961.

[105] I. Schur. On the representation of the symmetric and alternating groups by fractional linear substitutions. *J. Reine Angew. Math.*, 139:155–250, 1911.

[106] M.-P. Schützenberger. La correspondance de Robinson. In D. Foata, editor, *Combinatoire et représentation du groupe symétrique*, number 579 in Lecture Notes in Math., pages 59–135. Springer-Verlag, New York, 1977.

[107] A. Selberg. Remarks on a multiple integral. *Norsk Mat. Tidsskr.*, 26:71–78, 1944. (Norwegian).

[108] M. Shynar. On inversions in standard Young tableaux. In *Proc. 16th Ann. Conf. Formal Power Series and Algebraic Combinatorics, Vancouver*, pages 239–246, 2004.

[109] J. Sjöstrand. On the sign-imbalance of partition shapes. *J. Combin. Theory Ser. A*, 111:190–203, 2005.

[110] N. J. A. Sloane. The on-line encyclopedia of integer sequences. `http://www.oeis.org/`.

[111] E. Snow. Enumeration of standard Young tableaux of certain truncated shapes. preprint, 2012.

[112] R. P. Stanley. On the number of reduced decompositions of elements of Coxeter groups. *European J. Combin.*, 5:359–372, 1984.

[113] R. P. Stanley. Differential posets. *J. Amer. Math. Soc.*, 1:919–961, 1988.

[114] R. P. Stanley. *Enumerative Combinatorics*, volume 2. Cambridge Univ. Press, Cambridge/New York, 1999.

[115] R. P. Stanley. On the enumeration of skew Young tableaux. *Adv. in Appl. Math.*, 30:283–294, 2003. Formal Power Series and Algebraic Combinatorics (Scottsdale, AZ, 2001).

[116] R. P. Stanley. Some remarks on sign-balanced and maj-balanced posets. *Adv. in Appl. Math.*, 34:880–902, 2005.

[117] R. P. Stanley. Increasing and decreasing subsequences and their variants. In *International Congress of Mathematicians 2006*, volume I, pages 545–579. Eur. Math. Soc., Zürich, 2007.

[118] R. P. Stanley. Two remarks on skew tableaux. *Electron. J. Combin.*, 18:Paper 16, 8 pp., 2011.

[119] R. P. Stanley. *Enumerative Combinatorics*, vol. 1, second edition. Cambridge Studies in Adv. Math., vol. 49. Cambridge Univ. Press, Cambridge, 2012.

[120] R. P. Stanley. Supplementary exercises for Chapter 7 of Enumerative Combinatorics, vol. 2, version of 24 December 2013. `http://math.mit.edu/~rstan/ec/ch7supp.pdf`.

[121] D. W. Stanton and D. E. White. A Schensted algorithm for rim hook tableaux. *J. Combin. Theory Ser. A*, 40:211–247, 1985.

[122] J. Stembridge. Nonintersecting paths, Pfaffians, and plane partitions. *Adv. in Math.*, 83:96–131, 1990.

[123] P. Sun. Evaluating the numbers of some skew standard Young tableaux of truncated shapes. *Electronic J. Combin.*, 22(1):P1.2, 2015.

[124] P. Sun. Note of the enumeration formula of SYT of truncated shape $n^m\setminus(2)$. *Europ. J. Combin.*, 46:126–133, 2015.

[125] P. Sun. A probabilistic approach for enumeration of certain Young tableaux. arXiv:1302.0628, 2013.

[126] P. Sun. Some enumeration formulas of SYTs. Preprint, 2014.

[127] S. Sundaram. *On the combinatorics of representations of $Sp(2n;\mathbb{C})$*. PhD thesis, MIT, 1978.

[128] S. Sundaram. Orthogonal tableaux and an insertion algorithm for $SO(2n+1)$. *J. Combin. Theory Ser. A*, 53:239–256, 1990.

[129] B. E. Tenner. On the expected number of commutations in reduced words. arXiv:1407.5636, 2014.

[130] G. P. Thomas. *Baxter algebras and Schur functions*. PhD thesis, Univ. College of Swansea, 1974.

[131] G. P. Thomas. On a construction of Schützenberger. *Discrete Math.*, 17:107–118, 1977.

[132] R. M. Thrall. A combinatorial problem. *Michigan Math. J.*, 1:81–88, 1952.

[133] A. M. Vershik and S. V. Kerov. Asymptotic behavior of the Plancherel measure of the symmetric group and the limit form of Young tableaux. *Dokl. Akad. Nauk SSSR*, 233:1024–1027, 1977. (Russian).

[134] A. M. Vershik and S. V. Kerov. Asymptotic of the largest and the typical dimensions of irreducible representations of a symmetric group. *Funct. Anal. Appl.*, 19:21–31, 1985.

[135] D. Worley. *A Theory of Shifted Young Tableaux*. PhD thesis, MIT, 1984.

[136] A. Young. On quantitative substitutional analysis (first paper). *Proc. London Math. Soc., Ser. 1*, 33:97–145, 1900. reprinted in: Collected Papers, Mathematical Expositions, no. 21, Univ. of Toronto Press, Toronto, 1977.

[137] A. Young. On quantitative substitutional analysis (third paper). *Proc. London Math. Soc., Ser. 2*, 28:255–292, 1928.

[138] D. Zeilberger. Partial difference equations in $m_1 \geq m_2 \geq \ldots \geq m_n \geq 0$ and their applications to combinatorics. *Discrete Math.*, 31:65–77, 1980.

[139] D. Zeilberger. André's reflection proof generalized to the many-candidate ballot problem. *Discrete Math.*, 44:325–326, 1983.

[140] D. Zeilberger. A short hook-lengths bijection inspired by the Greene-Nijenhuis-Wilf proof. *Discrete Math.*, 51:101–108, 1984.

# Chapter 15

# *Computer Algebra*

**Manuel Kauers**
*Research Institute for Symbolic Computation*
*Johannes Kepler University, Linz, Austria*

## CONTENTS

## 15.1 Introduction

The computer is an indispensable tool for a modern mathematician. It is not only useful for checking emails, or for writing articles, or for retrieving information from web services like Wikipedia, MathWorld, or Sloane's celebrated online encyclopedia of integer sequences (OEIS [170]). The computer is also useful for computing. People working in applied mathematics seem to be more aware of this than people working in pure mathematics, perhaps because of the immediate impact numerical computations have in natural sciences and engineering. While numerical computations typically do not lead to rigorous proofs, symbolic computations do have proof quality. Symbolic computation and computer algebra can therefore be a valuable tool for people working in pure mathematics. And more than many other areas of pure mathematics, combinatorics can benefit from the advances that have been made in computer algebra in the past few decades. One reason may be that computer algebra

The author was supported by the Austrian Science Fund FWF grants Y464-N18 and F050-04.

algorithms, just like all other algorithms, are applicable only to finite objects, and combinatorialists, unlike most other mathematicians, primarily study finite objects. Another reason may be that computer algebra algorithms developed during the past few decades are especially good at handling precisely those kinds of expressions, which tend to arise in the context of combinatorial problems.

The purpose of this chapter is to sustain this claim. In Section 15.2 we give a short general overview of the fundamental concepts of computer algebra, with some facts that are of independent interest as well as some background that is used in the subsequent sections. Section 15.3 is devoted to techniques for calculating information about generating functions. We give only a very limited selection of methods that have been invented for certain special combinatorial problems and focus on more general techniques. Symbolic summation is discussed in Section 15.4. With combinatorial problems leading invariably to all sorts of summation problems, the importance of computer algebra algorithms for simplifying sums is obvious. Many sums arising in combinatorics can nowadays be routinely tackled by a computer algebra system, even some that are well beyond the scope of any reasonable hand calculation.

The growing power of computer algebra in general, and especially its value for combinatorics, should not be underestimated. However, it should not be overestimated either. It is not realistic to expect that a serious research problem could somehow be solved autonomously by the computer after typing a few commands into a computer algebra system. Instead, solving a problem is most often an iterative process in which computer and user have to "collaborate." This sometimes requires bringing the problem at hand into a form that is better suited for some computational means, but possibly less useful for traditional paper-and-pencil reasoning. It sometimes also requires that several different computational techniques be applied together. One general and successful paradigm for solving problems by computer algebra is the guess-and-prove paradigm outlined in Section 15.5.

## 15.2 Computer algebra essentials

The characteristic feature of computer algebra algorithms is that they operate on exact representations of mathematical objects, as opposed to numerical algorithms, which accept approximate input and only produce approximate output. The advantage of working with exact representations is that the calculations are rigorous and the results have proof quality. The disadvantage is that exact computations may take much more time than approximate ones, and that some mathematical objects do not even admit a finite representation.

In the present section, we give an overview over the most important domains in which exact computations are possible, and introduce the most important algorithms available for solving problems in them. Obviously, this section cannot be a thorough introduction to computer algebra, only a collection of some basic facts that are essential for the later parts of the chapter. Further information can be found in the classical textbooks on the subject [134, 78, 196, 184].

## 15.2.1 Numbers

### 15.2.1.1 Integers and rational numbers

Perhaps the most natural domain in which exact computations can be done is the ring of **integers.** Obviously, every integer can be uniquely represented by a finite string of digits. This representation can be used for doing calculations (addition, multiplication, etc.) in the usual way. **Rational numbers** can be represented as pairs of integers (numerator and denominator), and we can do exact computations with those numbers as well.

Integers may have many digits, and may hence consume a lot of memory on a computer. Longer integers also need more computation time. The most common operations on integers can be performed in quasi-linear time. This means there is some fixed integer $k$ so that the operation can be performed on input with $n$ digits in time $\mathrm{O}(n\log(n)^k)$. In particular, addition, multiplication, division with remainder, greatest common divisor, least common multiple, and radix conversion can be performed in quasi-linear time [184, 42]. Factorization of integers is much harder. At the time of writing (2014), the fastest known algorithm requires $\exp(\mathrm{O}(n^{1/3}\log(n)^{2/3}))$ time to factor an integer with $n$ digits. In practice this means that integers with fifty thousand decimal digits cannot be factored in any reasonable amount of time, but all the other operations mentioned before need virtually no time at all.

### 15.2.1.2 Algebraic numbers

An object $\alpha$ is called **algebraic** over a field $F$ if it is the root of some univariate polynomial with coefficients in $F$. If $F = \mathbb{Q}$ we call $\alpha$ an **algebraic number.** For example, $\sqrt{2}$ is an algebraic number, as it is a root of $x^2 - 2$.

An algebraic number can be represented exactly by a polynomial $p$ of which it is a root (see Section 15.2.2 for more about polynomials) plus an approximate value that is sufficiently accurate to distinguish it from the other roots of $p$. For example, a possible representation of $\sqrt{2}$ would be $[x^2 - 2, 1]$ because 1 is closer to $\sqrt{2}$ than to the second root $-\sqrt{2}$. Root isolation algorithms [134, 151, 178, e.g.] may be used to determine the accuracy needed to distinguish the roots of a polynomial from each other, and to increase the accuracy of a given approximation if necessary.

The set $\overline{\mathbb{Q}}$ of all algebraic numbers forms a field, the **algebraic closure** of $\mathbb{Q}$. To compute the sum or the product of two elements $\alpha, \beta \in \overline{\mathbb{Q}}$ means that we know defining polynomials and approximate values for both of these values, and we want to know a defining polynomial and an approximate value for $\alpha + \beta$ or $\alpha\beta$. Adding or multiplying the two approximate values is of course easy. Defining polynomials can be constructed using Gröbner bases or resultants. The use of Gröbner bases for this purpose is explained in Example 15.2.7 below. For information on using resultants, see [36]. The construction of a defining polynomial for the multiplicative inverse is easy: If $p_0 + p_1 x + \cdots + p_d x^d$ is the defining polynomial for $\alpha$, then $p_d + p_{d-1}x + \cdots + p_1 x^{d-1} + p_0 x^d$ is a defining polynomial for $1/\alpha$.

The field $K = \mathbb{Q}(\alpha)$ generated by a single algebraic number $\alpha$ is completely determined by the (unique) monic polynomial $m$ of minimal degree of which $\alpha$ is a root. If $d$ is the degree of this polynomial, then $K$ is a vector space of dimension $d$ over $\mathbb{Q}$, and a basis is given by $1, \alpha, \alpha^2, \ldots, \alpha^{d-1}$. Consequently, the elements of $K$ are precisely the elements $p(\alpha)$ where $p$ is some polynomial of degree less than $d$. The (unique) polynomial $p$ of degree less $d$ is called the canonical representation of the element $p(\alpha) \in K$. If $p, q$ are the canonical representations of two elements of $K$, then the canonical representation of their sum is given by $p + q$ and the canonical representation of their product is given by $\text{rem}(pq, m)$, where $m$ is the defining polynomial of $K$. For computing the canonical representation of $1/p(\alpha)$, use the extended Euclidean algorithm for polynomials (see again Section 15.2.2) to find two polynomials $s, t$ with $ps + mt = 1$. Then $\text{rem}(s, m)$ is a canonical representation for $1/p(\alpha)$.

There are algorithms for computing the Galois group of a given field $\mathbb{Q}(\alpha)$, and hence for analyzing the subfield structure of a given field. In particular, it is possible to decide algorithmically whether a given algebraic number can be expressed in terms of radicals. Details can be found in [64].

#### 15.2.1.3 Real and complex numbers

There is no way to do exact computations in $\mathbb{R}$ or $\mathbb{C}$. For exact computations to be possible in some domain, it is necessary that there is a way of representing each of its elements in a finite way, i.e., as words of finite length over an alphabet with finitely many symbols. As there are only countably many words but uncountably many real or complex numbers, it follows that there is no way to exactly represent real or complex numbers, and hence there is also no way to do exact computations with them.

When using approximate representations, the question is how accurate the output is. The design of algorithms for doing numerical computations aims at minimizing both the error (i.e. the distance between output and actual solution) as well as the runtime. Algorithms in this area typically rely on some iteration that produces a sequence of approximate solutions that converges in the limit to the actual solution. The iteration is continued until the result is within the desired tolerance. The computational cost is determined by the convergence rate and the cost of a single iteration.

The design and analysis of such algorithms does not belong to the area of computer algebra but constitutes the area of numerical analysis. The two areas intersect at algorithms that take as input exact data and a target accuracy $N \in \mathbb{N}$ and return output that is guaranteed to lie in a ball of radius $1/N$ around the true solution. Such techniques are known as **arbitrary precision arithmetic,** and they are used for computing mathematical constants such as $\pi$ to billions of digits [29, 25, 42]. Note that with arbitrary precision arithmetic, it is always possible to recognize in a finite number of steps that two numbers are distinct, but it is impossible to recognize that they are equal.

Arbitrary precision arithmetic is relevant in combinatorial applications mainly in connection with asymptotic series.

#### 15.2.1.4 Finite fields and modular arithmetic

Numbers with more than a few thousand digits will seldom arise directly in combinatorial applications. However, it must be taken into account that numbers arising in intermediate results of a calculation can be much longer than those in the input or the output. This effect is known as **expression swell.**

There is no expression swell in finite fields. If $m$ is a positive integer, then the set $m\mathbb{Z} = \{\dots, -2m, -m, 0, m, 2m, \dots\} \subseteq \mathbb{Z}$ is an ideal in the ring $\mathbb{Z}$, and so $\mathbb{Z}_m := \mathbb{Z}/m\mathbb{Z}$ has a natural ring structure. The elements of this ring are equivalence classes of integers subject to the equivalence relation $a \sim b$ if and only if $m \mid a - b$. A set of representatives is given by $\{0, \dots, m-1\}$. If $p$ is a prime, then $\mathbb{Z}_p$ is a field. To find the multiplicative inverse of a nonzero element of $\mathbb{Z}_p$, represented by an integer $a$, use the extended Euclidean algorithm for integers to compute $s, t \in \mathbb{Z}$ with $as + pt = 1$. Then the equivalence class of $s$ is the multiplicative inverse of the equivalence class of $a$.

Finite fields are used to avoid expression swell. Instead of solving the problem in $\mathbb{Z}$, we solve it in some finite field $\mathbb{Z}_p$. We then know the equivalence class modulo $p$ to which the answer belongs. If we know furthermore that the integer answer is bounded by $M$, then choosing a prime $p > 2M$ will assure that every equivalence class contains only one element with absolute value at most $M$. This element must then be the solution to the integer problem. Applications of this technique are abundant. We illustrate it here only with some toy examples from linear algebra. Further applications can be found in the literature on computer algebra.

**Example 15.2.1 (Computing in modular images)** *Consider the linear system $Ax = b$ where*

$$A = \begin{pmatrix} 1570 & -440 & 8048 \\ 1775 & 1414 & 6723 \\ 1085 & 3436 & 621 \end{pmatrix}, \qquad b = \begin{pmatrix} -1328 \\ -1037 \\ 283 \end{pmatrix}.$$

*Suppose that for some reason we know that the system has a solution $x = (x_1, x_2, x_3) \in \mathbb{Z}^3$ whose components are bounded by $M = 10$ in absolute value. Choose $p = 23 > 20 = 2M$, interpret the entries of $A$ and $b$ as elements of $\mathbb{Z}_p$, and solve the linear system there. Then the coordinates of the solution vector also live in $\mathbb{Z}_p$. Select for each coordinate the unique representative whose absolute value is at most $M$. This gives the vector $x = (4, -1, -1) \in \mathbb{Z}^3$, which indeed satisfies $Ax = b$ in $\mathbb{Z}$.*

In general the solution of a linear system involves rational numbers. Also in this case, expression swell can be avoided by doing the computation modulo a prime. A rational number $a/b$ with $|a|, |b| < M$ can be reconstructed from its modular image $ab^{-1} \bmod p$ in $\mathbb{Z}_p$ provided that $p > 2M^2$. This is known as **rational reconstruction.** The classical algorithm for rational reconstruction is based on the Euclidean algorithm and explained in Section 5.10 of [184]. An alternative approach based on lattice reduction is given in Example 15.2.5 below.

**Example 15.2.2 (Rational Reconstruction)** *Consider the linear system $Ax = c$ with $A$ as in Example 15.2.1 and $c = (296, 817, 1051) \in \mathbb{Z}^3$. Suppose that for some reason we know that the system has a solution $x = (x_1, x_2, x_3) \in \mathbb{Q}^3$ whose components have numerators and denominators that are bounded by $M = 10$ in absolute value. Choose $p = 211 > 200 = 2M^2$, interpret the entries of $A$ and $c$ as elements of $\mathbb{Z}_p$, and solve the linear system there. The solution vector, expressed in terms of the canonical representatives, is $(69, 71, 141)$. Applying rational reconstruction to each coordinate gives $x = (-\frac{4}{3}, \frac{2}{3}, \frac{1}{3}) \in \mathbb{Q}^3$, which indeed satisfies $Ax = c$ in $\mathbb{Q}$.*

If we have the modular image of a number $x$ with respect to several (coprime) moduli $m_1, \ldots, m_\ell$, there is a way to merge them into a single modular image with respect to their product $m_1 m_2 \cdots m_\ell$. This is called **Chinese remaindering** (see Section 5.4 of [184] or Section 5.6 of [78] for how this works). The use of several small moduli instead of one large modulus has several advantages. First, for hardware reasons arithmetic with integers smaller than $2^{64}$ is significantly faster than arithmetic with larger integers. Second, the computation for different primes can be easily run in parallel on different machines. Third, in situations where an a priori bound of the solution is not available, we can incrementally increase the modulus by including more and more small primes.

**Example 15.2.3 (Chinese Remaindering)** *Consider the linear system $Ax = d$ with $A$ as in Example 15.2.1 and $d = (296, 817, 1050) \in \mathbb{Z}^3$. We seek the solution $x = (x_1, x_2, x_3) \in \mathbb{Q}^3$. First choose $p = 2147483647$. Modulo this prime, the solution of the linear system, expressed in terms of canonical representatives, is $(1146158679, 266089807, 564241799)$. Rational reconstruction applied to these numbers yields $(-\frac{32579}{14590}, -\frac{30069}{5722}, -\frac{33918}{16849})$, which is not a solution of the system. This indicates that the modulus was chosen too small. Next choose $q = 2147483629$. Modulo this prime, the solution reads $(1645758289, 1234352967, 769474034)$. Using the Chinese remainder algorithm, we find that the solution modulo $pq$ can be expressed as*

$$(572014109117159101, 1140337400488501161, 793099489171625462).$$

*Rational reconstruction applied to this modular image yields*

$$x = \left(-\frac{295758998}{219130875}, \frac{58807909}{87652350}, \frac{14758774}{43826175}\right),$$

*which indeed satisfies $Ax = d$ in $\mathbb{Q}$.*

#### 15.2.1.5 Lattice reduction

Lattice reduction was originally invented as a subprocedure for the first polynomial time algorithm for factoring univariate polynomials with rational coefficients [120, 184]. In combinatorics, it is mainly interesting for investigating possible relations among real numbers.

A lattice $L$ is defined as a subset of $\mathbb{Z}^m$ that is closed under addition and under multiplication by integers. A set $\{b_1,\dots,b_d\} \subseteq \mathbb{Z}^m$ is called a basis for $L$ if

$$L = \mathbb{Z}b_1 + \cdots + \mathbb{Z}b_d = \{x_1b_1 + \cdots + x_db_d : x_1,\dots,x_d \in \mathbb{Z}\}.$$

A lattice reduction algorithm takes as input some basis of a lattice $L$, and produces as output another basis for $L$ with the feature that its elements are short with respect to the Euclid norm. For example, given the basis $\{(123,234),(345,456)\}$ the output may be $\{(99,-12),(24,246)\}$.

The computation of the shortest vector in a lattice is a computationally hard problem that is not believed to be solvable by an efficient algorithm. However, there are algorithms that can efficiently construct a basis whose elements are perhaps not as short as can be but are at least guaranteed to be not much longer than the shortest ones, in a sense that can be made precise. The classical lattice reduction algorithm is known as **LLL** (named after the three inventors Lenstra, Lenstra, and Lovász). A comprehensive account on the application of this algorithm as well as the underlying algorithmic details can be found in the monograph [137]. Here we only give two example applications that may be relevant in combinatorial applications.

**Example 15.2.4 (Integer relations)** *Let $x_1,\dots,x_n$ be some real numbers. An* **integer relation** *for $x_1,\dots,x_n$ is a vector $(e_1,\dots,e_n) \in \mathbb{Z}^n$ such that $e_1x_1 + \cdots + e_nx_n = 0$. For example, $(33,-32,16)$ is an integer relation for the three real numbers*

$$x_1 = \sum_{n=1}^{\infty} \frac{1}{n^6}, \quad x_2 = \left(\sum_{n=1}^{\infty} \frac{1}{n^3}\right)^2, \quad x_3 = \sum_{n=1}^{\infty} \left(\frac{1}{n+1} \sum_{k=1}^{n} \frac{1}{k}\right)^3.$$

*Integer relations can be discovered using lattice reduction. In the example, suppose we are able to compute the approximations*

$$\begin{aligned} x_1 &\approx \xi_1 := 1.01734306198445, \\ x_2 &\approx \xi_2 := 1.44494079843363, \\ x_3 &\approx \xi_3 := 0.791611531524342. \end{aligned}$$

*Choose a big integer $w$, roughly of the reciprocal order of the approximation accuracy, for instance $w = 10^{15}$ in our example. Consider the lattice $L$ generated by*

$$\{\, (1,0,0,\lfloor w\xi_1 \rfloor),\ (0,1,0,\lfloor w\xi_2 \rfloor),\ (0,0,1,\lfloor w\xi_3 \rfloor) \,\}.$$

*A lattice reduction algorithm may turn this basis into*

$$\begin{aligned} \{\, &(33,-32,16,1),\ (1898474,452503,-3265793,4083522), \\ &(-2042705,-310957,3192786,6374529) \,\}. \end{aligned}$$

*The fact that the first vector in this basis is considerably shorter than the others is a strong indication (but of course not a proof) that its first components form an integer relation for $x_1,x_2,x_3$. It can be shown that every integer relation must eventually* *appear in a reduced lattice basis if the procedure is repeated with higher and higher accuracy. Also, if there is no relation, lattice reduction can be used to prove statements of the form "if an integer relation exists, at least one of its coefficients $e_i$ must exceed...." Results of this kind and further applications can be found in [30, 28, 15]. An alternative, more efficient but less general algorithm for finding integer relations is PSLQ [72, 73].*

**Example 15.2.5 (Rational and Algebraic Reconstruction)** *Suppose $x \in \mathbb{Z}$ is the representative of an equivalence class modulo $m \in \mathbb{Z}$ and we want to find $u, v \in \mathbb{Z}$ such that $x$ and $\frac{u}{v}$ have the same image in $\mathbb{Z}_m$. To solve this problem via lattice reduction, compute a short vector in the lattice generated by $\{\ (x,1),\ (m,0)\ \}$. If $u$ and $v$ are small compared to $m$, it is likely that $(u,v)$ will show up in a reduced basis. For example, for $x = 9876540$ and $m = 12345678$, applying lattice reduction to $\{\ (9876540,1),\ (12345678,0)\ \}$ we may obtain $\{\ (-12,5),\ (365262,876614)\ \}$. We can take $(u,v) = (-12,5)$.*

*More generally, we can also use lattice reduction to recover an algebraic number from its modular image. Let $\alpha$ be an algebraic number with a minimal polynomial of degree $d$, and let $x \in \mathbb{Q}(\alpha)$, say $x = \frac{u_0+u_1\alpha+\cdots+u_{d-1}\alpha^{d-1}}{v}$ for $u_0,\ldots,u_{d-1},v \in \mathbb{Z}$. Choose a large modulus $m$. An element $\bar{\alpha}$ of $\mathbb{Z}_m$ is a suitable image of $\alpha$ if it satisfies the minimal polynomial of $\alpha$ modulo $m$. Such an element may or may not exist. If it exists, then we can also find an image $\bar{x}$ of $x$ in $\mathbb{Z}_p$. To recover $x$ from $\bar{x}$, compute a short vector in the lattice generated by*

$$\{\ (\bar{x},0,\ldots,0,1),\ (0,\bar{x}/\bar{\alpha},0,\ldots,0,1),\ \ldots,\ (0,\ldots,0,\bar{x}/\bar{\alpha}^{d-1},1),$$
$$(m,0,\ldots,0,0),(0,m,0,\ldots,0,0),\ldots\ldots,\ (0,\ldots,0,m,0)\ \} \subseteq \mathbb{Z}^{d+1}.$$

*Here, by $\bar{x}/\bar{\alpha}^i$ we mean an arbitrary integer representative of the respective element of $\mathbb{Z}_m$. If $m$ is sufficiently large compared to $v$ and the $u_i$, a reduced basis of this lattice will contain the vector $(u_0,\ldots,u_{d-1},v)$.*

### 15.2.2 Polynomials

For the remainder of this chapter, let $R$ be a computable ring and $K$ be a computable field. To be computable means that the elements admit a finite representation and algorithms are available for performing arithmetic operations and for deciding whether two elements are equal.

By $R[x_1,\ldots,x_m]$ we denote the ring of polynomials in the variables $x_1,\ldots,x_m$ with coefficients in $R$. This is a computable ring if $R$ is. As a polynomial by definition has only finitely many terms, we can use a list of its monomials (coefficients and exponents) as a data structure.

By $K(x_1,\ldots,x_m)$ we denote the fraction field of $K[x_1,\ldots,x_m]$. Its elements are called **rational functions** in $x_1,\ldots,x_m$.

#### 15.2.2.1 Arithmetic and factorization

Polynomials with coefficients in a ring can be added, multiplied, composed, differentiated, and compared to zero. When polynomials are represented as lists of terms, all these operations require $O(n^m \log(n^m)\log(\log(n^m)))$ operations in the coefficient domain, where $n$ is the degree of the output and $m$ the number of variables.

Univariate polynomials with coefficients in a field form a Euclidean domain: For every $a,b \in K[x]$, $b \neq 0$ there exist (and we can compute) $q,r \in K[x]$ with $a = qb + r$ and $r = 0$ or $\deg(r) < \deg(b)$. We call $\operatorname{quo}(a,b) := q$ the quotient and $\operatorname{rem}(a,b) := r$ the remainder of $a$ and $b$. The number of field operations to perform division with remainder is roughly proportional to $\deg(a)$.

In a Euclidean domain, we can compute the greatest common divisor $g$ of two elements $a$ and $b$ by the Euclidean algorithm. With the extended Euclidean algorithm we can furthermore compute $s,t$ such that $g = as + tb$. Again, the number of required field operations is roughly proportional to the size of the input.

Factorization algorithms for $K[x]$ are available for various fields, in particular for $K = \mathbb{Z}_p$ ($p$ prime), for $K = \mathbb{Q}$ and for $K = \mathbb{Q}(\alpha)$ ($\alpha$ algebraic over $\mathbb{Q}$). Factorization of polynomials over any of these fields is considerably faster than factorization of integers. Observe that the factorization of a polynomial in general is different, depending on the assumed ground field.

**Example 15.2.6 (Polynomial Factorization)** *For $p = x^4 - x^2 - 2 \in K[x]$ we have the following factorizations:*

$$p = \begin{cases} (x^2+1)(x^2-2) & \text{if } K = \mathbb{Q} \\ (x+9)(x+17)(x+24)(x+32) & \text{if } K = \mathbb{Z}_{41} \\ (x^2+1)^2 & \text{if } K = \mathbb{Z}_3 \\ (x - \frac{1}{6}(\alpha^3+\alpha))(x+\frac{1}{6}(\alpha^3+\alpha)) & \\ \quad \times (x - \frac{1}{6}(\alpha^3-5\alpha))(x+\frac{1}{6}(\alpha^3-5\alpha)) & \text{if } K = \mathbb{Q}(\alpha) \text{ with } \alpha = \mathrm{i} + \sqrt{2} \\ (x^2+1)(x-1.414\ldots)(x+1.414\ldots) & \text{if } K = \mathbb{R} \end{cases}$$

Multivariate polynomials over a field do not form a Euclidean ring. There is nevertheless a notion of greatest common divisor, and an algorithm for computing it, but in general it is not possible to write the greatest common divisor $g$ of two multivariate polynomials $a$ and $b$ as $g = sa + tb$ for some other multivariate polynomials $s$ and $t$.

In some situations it is possible to reduce a computation for multivariate polynomials to one for univariate polynomials by selecting one of the variables and replacing all the others by suitable powers of it. For example, consider $a,b \in K[x,y]$ and suppose that $\deg_x(a) = \deg_x(b) = \deg_y(a) = \deg_y(b) = n$. Then we can find the greatest common divisor of $a$ and $b$ by replacing $y$ by $x^{n+1}$ in the input, obtaining univariate polynomials of degree $n(n+1)$, computing their greatest common divisor by the Euclidean algorithm, and then replacing each power $x^k$ by $x^{\operatorname{rem}(k,n+1)} y^{\operatorname{quo}(k,n+1)}$ in the result. This technique is known as **Kronecker substitution.**

#### 15.2.2.2 Evaluation and interpolation

Polynomials can be evaluated at elements of the coefficient domain, the result being an element of the coefficient domain. Evaluating a given univariate polynomial of degree $n$ at a given element of the ground ring costs $O(n)$ arithmetic operations in the ground ring (using Horner's rule). There is also an algorithm that evaluates the polynomial simultaneously at $n$ given elements, using not much more than a linear number of operations in the ground ring [184, Chapter 10].

Conversely, given the values of a polynomial of degree $n$ over a field at $n+1$ distinct elements, we can find the coefficients of the polynomial by interpolation. This can be done using a number of operations that is quasi-linear in $n$.

Evaluation and interpolation are used to avoid expression swell in $K[x]$, very much like modular images and Chinese remaindering are used to avoid expression swell in $\mathbb{Z}$. The value of a polynomial $p \in K[x]$ at an element $e \in K$ is the image of $p$ under the map $p \mapsto p \bmod (x-e)$. Computing the interpolating polynomial for some given values $(v_1, v_2, \ldots, v_n)$ at some given elements $(e_1, e_2, \ldots, e_n)$ means to merge several images $v_i$ valid modulo linear polynomials $x - e_i$ into a joint image $p$ valid modulo $(x-e_1)(x-e_2)\cdots(x-e_n)$.

Rational reconstruction is also available for polynomials. A rational function $a/b$ with $\deg(a), \deg(b) \le M$ can be reconstructed from its image modulo $p \in K[x]$ provided that $\deg(p) > 2M$.

#### 15.2.2.3 Gröbner bases

If $\alpha$ is a common root of two polynomials $a$ and $b$, then it is also a root of $a+b$ and a root of $ca$ for every other polynomial $c$. The set of polynomials that have $\alpha$ as a root is therefore an ideal of the polynomial ring $K[x]$. In the case of one variable, every ideal is generated by a single polynomial, the greatest common divisor of all the elements of the ideal. This is no longer true in the multivariate case. Here we only have Hilbert's basis theorem, which says that every ideal $I \subseteq K[x_1, \ldots, x_n]$ is generated by a finite number of elements, i.e., there exist $b_1, \ldots, b_m \in K[x_1, \ldots, x_n]$ such that $p \in I$ if and only if there exist $p_1, \ldots, p_m \in K[x_1, \ldots, x_n]$ such that $p = p_1b_1 + \cdots + p_mb_m$. We write $\langle b_1, \ldots, b_m \rangle := I$ for the ideal generated by $b_1, \ldots, b_m$.

The basis of an ideal is not uniquely determined, and it is in general not easy to see whether two given bases generate the same ideal or not. A **Gröbner basis** [49] is a distinguished ideal basis. In a sense, Gröbner bases play the role among ideal bases that matrices in reduced echelon form play among matrices. Given an arbitrary basis of an ideal $I$, a Gröbner basis for $I$ can be computed using Buchberger's algorithm.

Space limitations prevent us from giving a thorough introduction to Gröbner bases at this point. The interested reader is referred to the classical textbooks on the subject [66, 21, 7] or to the introductory articles [50, 51]. We list here only some of the problems that can be solved computationally using Gröbner bases. Given polynomials $p, b_1, \ldots, b_m, c_1, \ldots, c_\ell \in K[x_1, \ldots, x_n]$, we can do the following:

- Decide whether $p \in \langle b_1, \ldots, b_m \rangle$, and in the affirmative case, find cofactors ("certificates") $p_1, \ldots, p_m \in K[x_1, \ldots, x_n]$ with $p = p_1b_1 + \cdots + p_mb_m$.

- Decide whether there exists a positive integer $k$ such that $p^k \in \langle b_1,\dots,b_m\rangle$, and in the affirmative case, find such a $k$.
- Decide whether $\langle b_1,\dots,b_m\rangle = \langle c_1,\dots,c_\ell\rangle$.
- For any given $k \in \{1,\dots,n-1\}$, compute a basis for the ideal $\langle b_1,\dots,b_m\rangle \cap K[x_1,\dots,x_k]$ in the smaller ring $K[x_1,\dots,x_k]$.
- Compute a basis for the ideal $\langle b_1,\dots,b_m\rangle \cap \langle c_1,\dots,c_\ell\rangle \subseteq K[x_1,\dots,x_n]$.
- Compute a basis of the ideal $\ker(\phi)$, where $\phi\colon K[y_1,\dots,y_m] \to K[x_1,\dots,x_n]$ is the ring homomorphism that leaves $K$ fixed and sends each $y_i$ to $b_i$.
- Decide whether $p \in \operatorname{img}(\phi) = K[b_1,\dots,b_m]$, and in the affirmative case, find a polynomial $q \in K[y_1,\dots,y_m]$ such that $p = q(b_1,\dots,b_m)$.
- Compute a basis of the so-called syzygy-module consisting of all vectors $(p_1,\dots,p_m) \in K[x_1,\dots,x_n]^m$ with the property $p_1b_1 + \cdots + p_mb_m = 0$.
- Decide whether the number of points $(\alpha_1,\dots,\alpha_n) \in \overline{K}^n$ for which all the $b_i$ are simultaneously zero is finite, and if yes, list all these points.
- Determine the Hilbert dimension of the ideal $\langle b_1,\dots,b_m\rangle$.

**Example 15.2.7 (Algebraic number arithmetic)** *Let $\alpha,\beta \in \overline{\mathbb{Q}}$, and let $p,q \in \mathbb{Q}[x]$ be polynomials with $p(\alpha) = q(\beta) = 0$. Suppose we want to find a polynomial $r \in \mathbb{Q}[x]$ with $r(\gamma) = 0$, where $\gamma = \alpha\beta$. This can be done using Gröbner bases, as follows. Consider polynomials in $\mathbb{Q}[x,y,z]$ in which the variables $x,y,z$ are meant to represent the algebraic numbers $\alpha,\beta,\gamma$, respectively. For every polynomial $u \in \langle p(x), q(y), z-xy\rangle \subseteq \mathbb{Q}[x,y,z]$ we have $u(\alpha,\beta,\gamma) = 0$, because this is true for the generators by definition, and the definition of ideals implies that it is also true for every other element. Use Gröbner bases to compute a basis of the ideal $\langle p(x), q(y), z-xy\rangle \cap \mathbb{Q}[z]$. This basis will consist of a single polynomial $r \in \mathbb{Q}[z]$ with the property $r(\gamma) = 0$.*

*More generally, if $\gamma = g(\alpha,\beta)$ where $g = \frac{u}{v} \in \mathbb{Q}(x,y)$ is a rational function with $v(\alpha,\beta) \neq 0$, then we can find an annihilating polynomial for $\gamma$ by computing a basis of the ideal $\langle p(x), q(y), v(x,y)z - u(x,y)\rangle \cap \mathbb{Q}[z]$.*

**Example 15.2.8 (Equational Reasoning)** *Suppose we want to prove the formula*

$$\forall\, x,y,z \in \mathbb{C}: \Big( (xy+yz+zx = 0 \wedge x^2+y^2+z^2-4 = 0) \\ \Rightarrow (y^2+yz+z^2-2y-2z = 0 \vee y^2+yz+z^2+2y+2z = 0) \Big).$$

*Translated into geometry, the claim is that the set of common roots in $\mathbb{C}^3$ of the two polynomials in the first line is contained in the set of all points $\mathbb{C}^3$ where at least one of the two polynomials in the second line vanishes. The latter set is identical to the set of all roots in $\mathbb{C}^3$ of the product $p := (y^2+yz+z^2-2y-2z)(y^2+yz+z^2+2y+2z)$. By Hilbert's Nullstellensatz [66, Section 4.1], the claim is true if and only if there exists a positive integer $k$ with $p^k \in \langle xy+yz+zx, x^2+y^2+z^2-4\rangle \subseteq \mathbb{C}[x,y,z]$. It can be checked by a Gröbner basis computation that $k = 1$ does the job. Gröbner bases can also be used to compute the certificate relation*

$$p = (yz-xy-xz)(xy+yz+zx) + (y^2+2yz+z^2)(x^2+y^2+z^2-4).$$

### 15.2.2.4 Cylindrical algebraic decomposition

Gröbner bases are useful for studying the solution sets of systems of polynomial equations. Such solution sets are called **algebraic sets.** In a field with an order relation $\leq$, we can also consider polynomial inequalities. The solution set of a system of polynomial equations and inequalities is called a **semi-algebraic set.** Problems related to such sets can be solved using Cylindrical Algebraic Decomposition (CAD) [65, 53], also known as Collins's algorithm.

Again, space is too short to give a thorough introduction into this technique. We only mention the most relevant problems that can be solved by CAD and give two examples. Readers interested in using this technique are referred to the tutorial paper [98]. Algorithmic details can be found in the textbook [18]. If semi-algebraic sets are represented through a defining system of polynomial equations and inequalities, CAD can be used to:

- Decide whether or not a given semi-algebraic set is empty, finite, open, closed, connected, or bounded.
- Decide whether or not a given semi-algebraic set is contained in another one.
- Determine the (topological) dimension of a given semi-algebraic set.
- Determine a sample point of a given nonempty semi-algebraic set.
- Determine the number of points of a given finite semi-algebraic set.
- Determine a tight bounding box of a given bounded semi-algebraic set.
- Determine the connected components of a given semi-algebraic set.
- Determine the boundary, closure, or interior of a given semi-algebraic set.
- Determine the projection of a given semi-algebraic set in $\mathbb{R}^n$ to a coordinate subspace $\mathbb{R}^k$ $(k < n)$.
- Determine the semi-algebraic set consisting of all points $(x_1,\ldots,x_{n-1}) \in \mathbb{R}^{n-1}$ such that for all $x_n \in \mathbb{R}$ the given semi-algebraic set contains $(x_1,\ldots,x_{n-1},x_n)$.

CAD supports quantifier elimination. This means that if $\Phi$ is a formula built from polynomials in a finite number of variables $x_1,\ldots,x_n$ and $y_1,\ldots,y_m$, comparison symbols $=, \geq, \leq, >, <$, logical connectives $\wedge, \vee$, etc., and quantifiers $\forall, \exists$ according to the usual syntactic rules, and if $y_1,\ldots,y_m$ are the variables bound by the quantifiers and $x_1,\ldots,x_n$ are the free variables, then there exists (and CAD can compute) another formula $\Psi$ of polynomial inequalities in $x_1,\ldots,x_n$ only with the property that

$$\forall\, x_1,\ldots,x_n \in \mathbb{R} : \Phi(x_1,\ldots,x_n) \iff \Psi(x_1,\ldots,x_n).$$

**Example 15.2.9 (Quantifier Elimination)** *Applied to the quantified formula*

$$\Phi := \forall\, a \in \mathbb{R}\ \exists\, b \in \mathbb{R} : (a-2)(b-2) > c \Rightarrow a^2 + b^2 - c^2 \leq 0,$$

*CAD produces the equivalent formula* $\Psi := c \leq -2 \vee c \geq 0.$

**Example 15.2.10 (A Fibonacci Inequality)** *Sometimes it is possible to use CAD for proving inequalities about quantities that are not polynomials. As an example, consider the following inequality, where $F_n$ denotes the nth Fibonacci number defined by $F_0 = 0$, $F_1 = 1$, $F_{n+2} = F_{n+1} + F_n$ $(n \geq 0)$:*

$$\sum_{k=1}^{n} \frac{(2F_{k+1} - F_k)^2}{F_k} \geq \frac{(3F_{n+1} + F_n - 3)^2}{F_{n+1} + F_n - 1} \qquad (n \geq 2).$$

*We prove this inequality by induction on $n$. The validity for $n = 2$ is easily checked. Now assume the inequality holds for some $n \geq 2$. We have to show it also holds for $n+1$ in place of $n$, i.e.,*

$$\sum_{k=1}^{n+1} \frac{(2F_{k+1} - F_k)^2}{F_k} \geq \frac{(3F_{n+2} + F_{n+1} - 3)^2}{F_{n+2} + F_{n+1} - 1}$$

*which, using $\sum_{k=1}^{n+1} a_k = \sum_{k=1}^{n} a_k + a_{n+1}$ and the Fibonacci recurrence, can be rewritten to*

$$\sum_{k=1}^{n} \frac{(2F_{k+1} - F_k)^2}{F_k} + \frac{(F_{n+1} + 2F_n)^2}{F_{n+1}} \geq \frac{(4F_{n+1} + 3F_n - 3)^2}{2F_{n+1} + F_n - 1}.$$

*Now in order to complete the proof of the induction step, it suffices to prove the formula*

$$\forall x, y, z \in \mathbb{R}: \left(x \geq 0 \land y \geq z \geq 1 \land x \geq \frac{(3y + z - 3)^2}{y + z - 1}\right) \Rightarrow x + \frac{(y + 2z)^2}{y} \geq \frac{(4y + 3z - 3)^2}{2y + z - 1},$$

*in which the variables $x, y, z$ are used as representations of the sum $\sum_{k=1}^{n} \frac{(2F_{k+1} - F_k)^2}{F_k}$, the number $F_{n+1}$ and the number $F_n$, respectively. CAD has no trouble proving this formula (as an equivalent quantifier free formula it returns "true"). Further applications in this direction are described in [80, 96].*

### 15.2.3 Formal power series

The ring $R[[x]]$ of formal power series in $x$ is defined as the ring of all infinite sequences $(a_n)_{n=0}^{\infty}$ equipped with termwise addition as sum and with the convolution $(a_n)_{n=0}^{\infty}(b_n)_{n=0}^{\infty} := \left(\sum_{k=0}^{n} a_k b_{n-k}\right)_{n=0}^{\infty}$ as multiplication. It is customary to use the notation $\sum_{n=0}^{\infty} a_n x^n$ instead of $(a_n)_{n=0}^{\infty}$.

The ring $R[[x]]$ is not computable, even if $R$ is, for the same reason for which $\mathbb{R}$ is not computable: Since any ring $R$ has at least two elements, there are uncountably many sequences in $R$, viz. uncountably many formal power series, but there are only countably many words over a finite alphabet at our disposal for their representations. Computations in $R[[x]]$ are therefore necessarily limited.

#### 15.2.3.1 Truncated power series

The first possibility for computing with formal power series is to restrict the attention to a certain fixed number, $N$, of terms, and discard the remaining ones. A formal power series may then be written as $a_0 + a_1x + \cdots + a_{N-1}x^{N-1} + \mathrm{O}(x^N)$, where the symbol $\mathrm{O}(\cdot)$ does not have any analytic meaning but only indicates the truncation order. A formal power series of which only the first $N$ terms are known, for some finite $N \in \mathbb{N}$, is called a **truncated power series.**

It follows straight from the definition that if we know the first $N$ terms of two formal power series $a$ and $b$, then we can calculate the first $N$ terms of their sum and their product. From the first $N$ terms of $a$ we can also compute the first $N-1$ terms of the (term-wise) derivative $a'$ of $a$ and, if $\mathbb{Q} \subseteq R$, the first $N+1$ terms of the (term-wise) integral $\int a$ of $a$.

If $a$ is a formal power series whose zeroeth term admits a multiplicative inverse in $R$, then $a$ itself admits a multiplicative inverse in $R[[x]]$, and we can compute its first $N$ terms if we know the first $N$ terms of $a$. Furthermore, if $b$ is a formal power series whose zeroeth term is 0, then the composition $b \circ a = a(b(x)) \in R[[x]]$ is well-defined, and we can compute its first $N$ terms if we know the first $N$ terms of both $a$ and $b$. See, for example, [43, 42] for classical as well as more advanced algorithms.

The arithmetic for truncated power series has some similarities with arithmetic for polynomials. Note however that for a truncated power series, the terms beyond the truncation order $N$ are **unknown,** whereas for a polynomial of degree $N$ the higher order terms are **zero.** Algebraically, truncated power series of order $N$ can be identified with elements of the quotient ring $R[x]/\langle x^N \rangle$.

#### 15.2.3.2 Lazy power series

Truncated power series are in $R[[x]]$ what fixed precision floating point numbers are in $\mathbb{R}$. In the same sense, lazy power series arithmetic is in $R[[x]]$ what arbitrary precision arithmetic is in $\mathbb{R}$. The general idea is that a formal power series is represented by an algorithm which can calculate the $n$th coefficient of the series, for any given $n$. This representation is called **lazy** because it is in principle capable of computing any coefficient of a series but does not do so until explicitly asked for it.

It is clear that we can do arithmetic using the lazy representation. For example, when we know lazy representations for two series $a$ and $b$, the sum $a+b$ can be represented by an algorithm that computes the $n$th term of $a+b$ by first computing the $n$th term of $a$ using the lazy representation of $a$, then computing the $n$th term of $b$ using the lazy representation of $b$, and then returning their sum. Likewise for the other operations.

Lazy representations allow for recognizing in a finite number of steps that a given formal power series is different from zero: we only need to enumerate all its coefficients, one after the other, until we encounter a nonzero coefficient. This procedure will run forever if and only if the series at hand is the zero series. In contrast, there is no way to tell whether a formal power series represented by a truncated series of order $N$ whose first $N$ coefficients are zero is nonzero or not.

A common design pattern for lazy representations is to store a functional equation for the series in question (e.g., a differential equation satisfied by the series, or a recurrence equation for the coefficient sequence), plus possibly some initial values that together with the equation uniquely determine the whole object and allow for computing any desired number of additional terms. For example, the formal power series $a := \sqrt{1-x-x^2} = 1 - \frac{1}{2}x - \frac{5}{8}x^2 + \cdots \in \mathbb{Q}[[x]]$ is uniquely determined by the equation $a^2 - (1-x-x^2) = 0$ together with the initial value $a_0 = 1$, as follows for instance from the implicit function theorem for formal power series [171, 102].

#### 15.2.3.3 Generalized series

If $R$ is an integral domain, then so is $R[[x]]$. However, $R[[x]]$ is not a field, even when $R$ is, because the element $x$ does not have a multiplicative inverse in $R[[x]]$. For a field $K$, the quotient field of the ring $K[[x]]$ is denoted $K((x))$. Its elements can be identified with formal series of the form $\sum_{n=n_0}^{\infty} a_n x^n$ with $n_0 \in \mathbb{Z}$ some (possibly negative) integer. These series are called formal Laurent series. If $a$ is a formal Laurent series, then $[x^{\geq 0}]a$ is defined as the formal power series obtained from $a$ by discarding all (finitely many) terms with negative exponent.

Puiseux series are series of the form $\sum_{n=n_0}^{\infty} a_n x^{n/r}$ where $n_0 \in \mathbb{Z}$ and $r$ is a positive integer, called the ramification index of the series. When $K$ is algebraically closed, then the set of all Puiseux series, sometimes denoted by $K\langle\langle x \rangle\rangle$, is again algebraically closed [185]. Puiseux series arise as expansions of algebraic functions, i.e., roots of univariate polynomials over a rational function field.

Other types of infinite series may arise as solutions of linear differential equations with polynomial coefficients. So-called Fuchsian equations [88] have series of the form $x^{\alpha}a$ where $\alpha$ is some constant and $a$ is a formal power series. More generally, solutions of differential equations admit generalized series solutions of the form $e\,a$, where $e$ is a certain closed form expression and $a$ is a formal power series. See the remarks on solutions in Section 15.2.4 for a precise description of the closed form part.

Laurent series, Puiseux series, and some other types of generalized series have in common that they are determined by a formal power series plus some finite amount of additional data. Therefore, for their computational treatment similar remarks apply as for formal power series. Even more general series arise as asymptotic expansions of elementary functions or solutions of certain transcendental equations. For definitions and algorithmic techniques, see [153, 165, 179] and the references given there.

#### 15.2.3.4 The multivariate case

The ring $R[[x_1,\dots,x_n]]$ of formal power series in $n$ variables is defined in the obvious way. Concerning computations, similar remarks apply as for univariate power series. In particular, a power series admits a multiplicative inverse in $R[[x_1,\dots,x_n]]$ if and only if the coefficient of $x_1^0 x_2^0 \cdots x_n^0$ admits a multiplicative inverse in $R$. Furthermore, a multivariate power series $b \in R[[x_1,\dots,x_n]]$ may be substituted for $y_1$ into a power series $a \in R[[y_1,\dots,y_m]]$, giving a power series in $R[[x_1,\dots,x_n,y_2,\dots,y_m]]$, provided that the coefficient of $x_1^0 x_2^0 \cdots x_n^0$ in $b$ is zero.

As in the univariate case, there is in general no way to represent a formal power series in finite terms, but we can always do computations with truncated series, and under suitable circumstances there may be a way to define lazy series. Instead of a single truncation order, we may now have individual truncation orders for each variable. Alternatively, one can select a fixed tuple $(w_1,\dots,w_n)$ of positive real numbers, called a weight vector, and say that truncating a power series at order $N$ means to list all the coefficients $a_{i_1,\dots,i_n}$ with $0 \le w_1 i_1 + \cdots + w_n i_n \le N$. Note that for every fixed $N$, there are only finitely many.

In order to allow also negative exponents, we can proceed as follows. Let $C \subseteq \mathbb{R}^n$ be a pointed cone, i.e., a set with the properties $0 \in C$ and for all $x, y \in C$ and all $\lambda, \mu > 0$ we also have $\lambda x + \mu y \in C$ and $x \in C \wedge -x \in C \Rightarrow x = 0$. It can be shown that the set $R_C[[x_1,\dots,x_n]]$ consisting of all formal infinite sums $\sum_{(i_1,\dots,i_n)\in C\cap\mathbb{Z}^n} a_{i_1,\dots,i_n} x_1^{i_1} x_2^{i_2} \cdots x_n^{i_n}$ together with the natural addition and multiplication forms a ring.

**Example 15.2.11 (Series Expansions of Multivariate Rational Functions)** *For every rational function $r \in \mathbb{Q}(x_1,\dots,x_n)$ it is possible to find a cone $C$ and a vector $(e_1,\dots,e_n) \in \mathbb{Z}^n$ so that $r$ admits a series expansion in*

$$x_1^{e_1} x_2^{e_2} \cdots x_n^{e_n} \mathbb{Q}_C[[x_1,\dots,x_n]] = \left\{ \sum_{(\nu_1,\dots,\nu_n)\in C} a_{\nu_1,\dots,\nu_n} x_1^{\nu_1+e_1} \cdots x_n^{\nu_n+e_n} : a_{\nu_1,\dots,\nu_n} \in \mathbb{Q} \right\}.$$

*The cone $C$ is usually not unique, and depending on the choice of $C$, the coefficients in the expansion may be different. For example, consider $r = \frac{1}{x+y} \in \mathbb{Q}(x,y)$. Taking $C = \{(\lambda,\mu) \in \mathbb{R}^2 : \lambda \ge 0, \mu \le 0\}$ and $C' = \{(\lambda,\mu) \in \mathbb{R}^2 : \lambda \le 0, \mu \ge 0\}$, we have the two expansions*

$$\begin{aligned} \frac{1}{x+y} &= y^{-1} - xy^{-2} + x^2y^{-3} - x^3y^{-4} + \cdots && \in \mathbb{Q}_C[[x,y]] \\ &= x^{-1} - x^{-2}y + x^{-3}y^2 - x^{-4}y^3 + \cdots && \in \mathbb{Q}_{C'}[[x,y]]. \end{aligned}$$

For a series $a$ in some ring $R_C[[x_1,\dots,x_n]]$, the positive part $[x_1^{\geq} x_2^{\geq} \cdots x_n^{\geq}]a$ is defined as the formal power series obtained from $a$ by discarding all the terms $cx_1^{e_1} \cdots x_n^{e_n}$ where not all the exponents $e_1,\dots,e_n$ are positive. The positive part always belongs to the usual multivariate polynomial ring $R[[x_1,\dots,x_n]]$.

A detailed introduction to rings of multivariate formal series involving negative exponents is given in [13].

### 15.2.4 Operators

Just as algebraic numbers are represented exactly in a computer by the polynomial whose root they are, together with an approximation that is accurate enough to distinguish the number in question from the other roots of the polynomial, we can represent certain functions and infinite sequences by the differential equations or recurrence relations whose solutions they are, together with a suitable number of initial values. Linear operators that act on functions or sequences form rings which resemble the usual polynomial rings.

#### 15.2.4.1 Ore algebras

An Ore algebra is a ring of polynomials with a noncommutative multiplication that is defined in such a way that the product of two elements of the algebra applied to some function is the same as applying the two elements one after the other to this function. The general definition makes use of the notion of $\sigma$-derivation. If $R$ is a ring and $\sigma\colon R \to R$ is a ring homomorphism, then $\delta\colon R \to R$ is called a $\sigma$-derivation if we have $\delta(a+b) = \delta(a)+\delta(b)$ and $\delta(ab) = \delta(a)b+\sigma(a)\delta(b)$ for all $a,b \in R$.

Ore [140] introduces the following definition. Let $R$ be a ring, let $\sigma_1,\dots,\sigma_n\colon R \to R$ be endomorphisms, and let $\delta_1,\dots,\delta_n\colon R \to R$ be such that $\delta_i$ is a $\sigma_i$-derivation $(i = 1,\dots,n)$. The ring $R[\partial_1,\dots,\partial_n]$ of all (multivariate) polynomials in $\partial_1,\dots,\partial_n$ with coefficients in $R$, together with the usual (commutative) addition and the unique multiplication satisfying the **commutation rules** $\partial_i\partial_j = \partial_j\partial_i$ for all $i, j$ and $\partial_i a = \sigma_i(a)\partial_i + \delta_i(a)$ for $i = 1,\dots,n$ and all $a \in R$ is called an **Ore algebra** over $R$.

**Example 15.2.12 (Recurrence, $q$-Recurrence, and Differential Operators)** *1. The elements of the Ore algebra $K(x)[\partial]$ with $\sigma\colon K(x) \to K(x)$ defined via $\sigma(x) = x+1$ and $\sigma(c) = c$ for all $c \in K$, and with $\delta = 0$ can be interpreted as* **recurrence operators.** *In this case, we usually prefer to use the symbol $n$ instead of $x$ and the symbol $S_n$ instead of $\partial$.*

2. *Let $q \in K\setminus\{0\}$. The elements of the Ore algebra $K(x)[\partial]$ with $\sigma(x) = qx$ and $\sigma(c) = c$ for all $c \in K$, and with $\delta = 0$ are called* **$q$-recurrence operators.**

3. *The elements of the Ore algebra $K(x)[\partial]$ with $\sigma$ being the identity on $K(x)$ and $\delta = \frac{d}{dx}$ can be interpreted as* **differential operators.** *In this case, we usually write $D_x$ instead of $\partial$.*

Note that by repeated application of the commutation rules, we can write every element of an Ore algebra as a finite sum of monomials $u\partial_1^{e_1}\cdots\partial_n^{e_n}$ with $u$ in the coefficient domain and $e_1,\dots,e_n \in \mathbb{N}$.

If the coefficient domain is a field $K$ and there is only one generator $\partial$, as in the examples above, the Ore algebra is a left Euclidean domain. This means that for any given $A,B \in K[\partial]$ we can find two other operators $Q,R \in K[\partial]$ with $A = QB+R$, with $R$ being zero or having smaller order than $B$. Also the greatest common right divisor $\operatorname{gcrd}(A,B)$ and the least common left multiple $\operatorname{lclm}(A,B)$ of $A,B$ can be computed. See [48, 5] for an introduction into the classical algorithms, and [89, 34] for some more recent developments.

If the coefficient domain is not a field or if there is more than one generator, then the Ore algebra is in general no longer a left Euclidean domain, but there is still a notion of Gröbner bases for left ideals in such algebras, and there are algorithms for computing them. In particular, every left ideal in an Ore algebra over a ring $R$ is generated by a finite basis (provided that this is the case for ideals in the coefficient domain $R$). For details, see [63, 110].

#### 15.2.4.2 Actions and solutions

Elements of Ore algebras are used to represent functions that they map to zero. To make this precise, fix an Ore algebra $A = R[\partial_1, \dots, \partial_n]$. Let $M$ be some $R$-module, and for each $\partial_i$ choose a function $d_i \colon M \to M$ that satisfies certain compatibility conditions so that the action $A \times M \to M$, $(a, f) \mapsto a \cdot f$ defined by $a \cdot f = af$ for $a \in R$ and $f \in M$ and $\partial_i \cdot f = d_i(f)$ for $i = 1, \dots, n$ and $f \in M$ satisfies the laws $(a+b) \cdot f = (a \cdot f) + (b \cdot f)$ and $(ab) \cdot f = a \cdot (b \cdot f)$ for all $a, b \in A$ and $f \in M$.

**Example 15.2.13 (Some Natural Actions)** *1. Consider the Ore algebra $A = \mathbb{Z}[n][S_n]$, and let $M = \mathbb{C}^{\mathbb{N}}$ be the set of all infinite sequences $(a_n)_{n=0}^{\infty}$ of complex numbers. With the natural action of $A$ on $M$, an operator $L = \ell_0(n) + \ell_1(n)S_n + \cdots + \ell_r(n)S_n^r$ applied to a sequence $(a_n)_{n=0}^{\infty}$ in $M$ yields the sequence $\big(\ell_0(n)a_n + \ell_1(n)a_{n+1} + \cdots + \ell_r(n)a_{n+r}\big)_{n=0}^{\infty}$ in $M$.*

*2. For $K = \mathbb{Q}(q)$, let $A = K(x)[\partial]$ be the Ore algebra from Example 15.2.12.(2). Let $M = K^{\mathbb{N}}$ be the set of all infinite sequences $(a_n)_{n=0}^{\infty}$ of rational functions in $q$. The algebra $A$ acts on $M$ if we let $x \cdot (a_n)_{n=0}^{\infty} := (q^n a_n)_{n=0}^{\infty}$ and $\partial \cdot (a_n)_{n=0}^{\infty} := (a_{n+1})_{n=0}^{\infty}$.*

*3. The Ore algebra $A = \mathbb{Z}[x][D_x]$ acts in a natural way on the ring $M = \mathbb{C}[[x]]$ of formal power series with complex coefficients. It also acts in a natural way on the field $\mathscr{M}$ of meromorphic functions.*

*4. The Ore algebra $A = \mathbb{Q}[x, n][D_x, S_n]$ acts in a natural way on the $\mathbb{Q}$-vector space of all functions $f \colon \mathbb{C} \times \mathbb{N} \to \mathbb{C}$ that are analytic with respect to the first variable.*

An element $f \in M$ is called a **solution** of an operator $a \in A$ if $a \cdot f = 0$. For a fixed operator $a \in A$, we call $V(a) := \{ f \in M \mid a \cdot f = 0 \}$ the **solution space** of $a$ in $M$. If $C \subseteq R$ denotes the ring of all elements that are invariant under all the $\sigma_i$ and mapped to zero by all the $\delta_i$ that define the Ore algebra $A$, then $V(a)$ is a $C$-submodule of $M$. For a fixed element $f \in M$, we call $\operatorname{ann}(f) := \{ a \in A \mid a \cdot f = 0 \}$ the **annihilator** of $f$ in $A$. This is a left ideal of $A$.

If $A = K[\partial]$ for some field $K$, then we have $V(\operatorname{gcrd}(a,b)) = V(a) \cap V(b)$ and $V(\operatorname{lclm}(a,b)) \subseteq V(a) + V(b)$. The latter inclusion becomes an equality if $M$ is sufficiently rich.

There are various algorithms for finding "closed form" solutions of a given operator $a \in K(x)[\partial]$, where $K$ is a field of characteristic zero left fixed by $\sigma$ and mapped to 0 by $\delta$. In particular, we can find a basis of the $K$-vector space $V(a) \cap K(x)$ of all rational functions $f \in K(x)$ with $a \cdot f = 0$ [1, 2, 182, 45]. Moreover, for every $a \in K(x)[D_x]$ we can find a basis of the $K$-vector space $V(a) \cap E$, where $E$ is the field of all Liouvillian functions in $x$ [181, 168, 68]. For a recurrence operator $a \in K(x)[S_x]$, there are algorithms for finding all its hypergeometric solutions (see Section 15.4 below) or d'Alembertian solutions (see Section 15.4.2 below). There are also algorithms for solving recurrence equations in terms of hypergeometric series [54], or for solving differential equations in terms of special functions [47, 52].

Only in very special circumstances will an Ore operator have closed form solutions. More often, there will be no better way to express the solutions of an operator than the operator itself, plus possibly some finite amount of additional data. For example, a sequence solution $(f_n)_{n=0}^{\infty}$ of a recurrence operator $a \in K[x][S_x]$ is uniquely determined by the terms $f_0, f_1, \dots, f_{r-1}, f_{\xi_1}, f_{\xi_2}, \dots, f_{\xi_d}$, where $r$ is the order of the operator and $\xi_1 - r, \dots, \xi_d - r$ are the positive integer roots of the leading coefficient of $a$, which is an element of $K[x]$. In general, when there are positive integer roots, not every choice of initial values gives rise to a sequence solution. On the other hand, if there are no positive integer roots, then the sequence solutions $(f_n)_{n=0}^{\infty}$ of $a$ are in one-to-one-correspondence with their truncated versions $(f_n)_{n=0}^{r-1}$.

**Example 15.2.14 (Initial Values)** *1. Consider the recurrence operator* $a := (n+2)S_n^2 - (2n+3)S_n + (n+1) \in \mathbb{Q}[n][S_n]$. *In this case, the leading coefficient* $(n+2) \in \mathbb{Q}[n]$ *has no positive integer roots. Therefore the solution space* $V(a) \subseteq \mathbb{Q}^{\mathbb{N}}$ *has dimension two. For any two numbers* $u, v \in \mathbb{Q}$ *there exists precisely one sequence* $(f_n)_{n=0}^{\infty} \in V(a)$ *such that* $f_0 = u$ *and* $f_1 = v$. *For example, for the choice* $u = 0, v = 1$ *we get the harmonic numbers* $H_n = \sum_{k=1}^{n} \frac{1}{k}$.

2. *Consider the recurrence operator* $a := (n-6)(n-2)(n^2-10n+29)S_n^2 - (n^2 - 9n+24)(n^2-5n-2)S_n + 2(n-5)(n^2-8n+20) \in \mathbb{Q}[n][S_n]$. *This operator has order* $r = 2$ *and its leading coefficient has* $d = 2$ *positive integer roots. It follows that* $r = 2 \leq \dim_{\mathbb{Q}} V(a) \leq 4 = r+d$. *In fact, the dimension is 3, and the sequence solutions starting as follows form a basis:*

$$1,\ \tfrac{5}{7},\ \tfrac{10}{21},\ \tfrac{2}{7},\ 0,\ -\tfrac{5}{21},\ -\tfrac{2}{7},\ -\tfrac{4}{21},\ 0,\ \tfrac{26}{105},\ \tfrac{176}{315},\ \dots$$

$$0,\ 0,\ 0,\ 0,\ 1,\ 2,\ 2,\ \tfrac{4}{3},\ 0,\ -\tfrac{26}{15},\ -\tfrac{176}{45},\ \dots$$

$$0,\ 0,\ 0,\ 0,\ 0,\ 0,\ 0,\ 0,\ 1,\ 3,\ 6,\ \dots$$

We can let differential operators $a \in K[x][D_x]$ act on the ring $M = K[[x]]$ of formal power series. If $a$ has order $r$ and the coefficient of $D_x^r$ in $a$ (which is an element of $K[x]$) does not contain $x$ as a factor, then $a$ has $r$ linearly independent solutions in $M$, and there is a basis of the solution space $V(a)$ that has the form

$$\begin{aligned}
&1+0x+0x^2+\cdots+0x^{r-2}+0x^{r-1}+\square x^r+\square x^{r+1}+\cdots\\
&0+1x+0x^2+\cdots+0x^{r-2}+0x^{r-1}+\square x^r+\square x^{r+1}+\cdots\\
&\vdots\\
&0+0x+0x^2+\cdots+0x^{r-2}+1x^{r-1}+\square x^r+\square x^{r+1}+\cdots
\end{aligned}$$

where the symbols $\square$ stand for certain elements of $K$ that are determined by the operator $a$.

In general, if $x$ is a factor of the leading coefficient of $a$, the dimension of $V(a)$ in $K[[x]]$ may be smaller than the order $r$. In any case, it is possible to construct $r$ linearly independent generalized series solutions of the form

$$\exp(p(x^{-1/s}))x^{\alpha}q(x^{1/s}, \log(x))$$

where $s \in \mathbb{N} \setminus \{0\}$, $p$ is a polynomial of degree $< s$ with $p(0) = 0$ and coefficients in $\overline{K}$, the algebraic closure of $K$, $\alpha \in \overline{K}$, and $q = q(x,y) \in \overline{K}[[x]][y]$ is a polynomial in $y$ with coefficients that are formal power series in $x$. See [88, 181, 102] for details on this construction. An analogous result for the shift case says that any operator $a = a_0 + \cdots + a_r S_n^r \in K[n][S_n]$ with $a_0 \neq 0 \neq a_r$ admits $r$ linearly independent generalized series solutions of the form

$$\left(\frac{n}{\mathrm{e}}\right)^{\gamma n} \phi^n \exp(p(n^{1/s}))n^{\alpha} q(n^{-1/s}, \log(n))$$

where $\mathrm{e} = \exp(1)$, $\gamma \in \mathbb{Q}$, $\phi \in \overline{K}$, $p$ is a polynomial of degree $< s$ with $p(0) = 0$ and coefficients in $\overline{K}$, $\alpha \in \overline{K}$, and $q = q(x,y) \in \overline{K}[[x]][y]$. This construction is explained, for example, in [187].

In the first place, generalized series solutions are only formal solutions. However, in typical examples arising in combinatorics, they can be interpreted as asymptotic expansions of function or sequence solutions, as illustrated in the following example.

**Example 15.2.15 (Asymptotic Solutions)** *The operator $(n+2)S_n^2 - (4n+2)S_n + (n+1) \in \mathbb{Q}[n][S_n]$ has a unique sequence solution $(f_n)_{n=0}^{\infty}$ with $f_0 = 1$, $f_1 = 2$. It also has two generalized series solutions, the first terms of which are*

$$(2+\sqrt{3})^n n^{-1/2}\left(1 + \tfrac{-3+\sqrt{3}}{12}n^{-1} + \tfrac{2-\sqrt{3}}{16}n^{-2} + \tfrac{-15+10\sqrt{3}}{192}n^{-3} + \tfrac{133-70\sqrt{3}}{1536}n^{-4} + \cdots\right)$$
$$(2-\sqrt{3})^n n^{-1/2}\left(1 + \tfrac{-3-\sqrt{3}}{12}n^{-1} + \tfrac{2+\sqrt{3}}{16}n^{-2} + \tfrac{-15-10\sqrt{3}}{192}n^{-3} + \tfrac{133+70\sqrt{3}}{1536}n^{-4} + \cdots\right).$$

*We can expect the asymptotic behavior of the sequence solution $(f_n)_{n=0}^{\infty}$ to be described by a certain linear combination of these two series solutions. Indeed, because of $|2+\sqrt{3}| > |2-\sqrt{3}|$, the first series dominates the second, so it is plausible to expect that we have*

$$f_n = c(2+\sqrt{3})^n n^{-1/2}\left(1 + \tfrac{-3+\sqrt{3}}{12}n^{-1} + \tfrac{2-\sqrt{3}}{16}n^{-2} + \tfrac{-15+10\sqrt{3}}{192}n^{-3} + \mathrm{O}(n^{-4})\right)$$

*as $n \to \infty$, for a certain constant $c \in \mathbb{R}$. In order to get a feeling for $c$, consider the quotient sequence*

$$u_n = \frac{f_n}{(2+\sqrt{3})^n n^{-1/2}\left(1 + \tfrac{-3+\sqrt{3}}{12}n^{-1} + \tfrac{2-\sqrt{3}}{16}n^{-2} + \tfrac{-15+10\sqrt{3}}{192}n^{-3}\right)}.$$

*By direct calculation, we find that*

$$u_{10} \approx 0.58560385580375675584$$
$$u_{100} \approx 0.58560331421465401061$$
$$u_{1000} \approx 0.58560331416909798511$$
$$u_{10000} \approx 0.58560331416909349614.$$

*These figures provide convincing evidence (though not a formal proof) that the limit*

*$c = \lim_{n\to\infty} u_n$ exists, and that its value is approximately* 0.585603314169093496. *Using lattice reduction as explained in Example 15.2.4, we may be able to derive from this approximation a plausible conjecture for an exact formula for c. In this example, we may find* $c \stackrel{?}{=} \sqrt{\frac{3+2\sqrt{3}}{6\pi}}$.

## 15.3 Counting algorithms

Combinatorial algorithms can be divided into three categories. Algorithms from the first category are used for *generating* combinatorial objects, either all the objects from a particular class, or a random element according to some prescribed probability distribution. Algorithms from the second category are used for *searching* objects with special properties within a class of objects, for instance an element for which a certain objective function attains an optimal value. The third category are algorithms for *counting* the objects of a certain class, typically as a function of a parameter measuring the "size" of the objects in the class.

In this chapter we mostly discuss counting algorithms, as these are more closely related to computer algebra than the other two types. For encyclopedic overviews on generation and search algorithms, see [138, 116, 107] or the appropriate other chapters of this handbook.

### 15.3.1 Special purpose algorithms

When there is an algorithm for generating all the objects from a certain family, say of a given size $n$, then there is obviously also an algorithm for counting them. But such algorithms are usually prohibitively expensive: Their runtime is at least proportional to the number $a_n$ of objects, which is typically at least exponential in $n$. In this section we give an (unsystematic and incomplete) selection of examples for how faster algorithms can be obtained. The algorithms discussed in this section are only elementary applications of computer algebra using no more than arbitrary precision arithmetic or computations with truncated formal power series. Sometimes the construction of such algorithms is straightforward.

**Example 15.3.1 (Stirling Numbers)** *Let $S_2(n,k)$ be the Stirling numbers of the second kind, defined as the number of partitions of a set of size n into k nonempty disjoint subsets. It follows easily from the combinatorial definition that $S_2(n,k)$ satisfies the recurrence equation*

$$S_2(n,k) = S_2(n-1,k-1) + kS_2(n-1,k)$$

*for all $n,k \geq 1$. Together with the initial values $S_2(0,0) = 1$ and $S_2(i,0) = S_2(0,i) = 0$ for $i > 0$, this recurrence gives rise to an algorithm for computing $S_2(n,k)$ for any given $n,k \geq 1$. This algorithm will be reasonably efficient if it is implemented in such* *a way that no term $S_2(i,j)$ is computed more than once. For example, to compute the term $S_2(n,k)$, set up an array $S$ of size $(n+1)\times(k+1)$, initialize $S[0,0]=1$, $S[i,0]=0$ for $i=1,\dots,n$, $S[0,j]=0$ for $j=1,\dots,k$, and then for $i=1,\dots,n$ and $j=1,\dots,k$ set $S[i,j]=S[i-1,j-1]+jS[i-1,j]$. Finally return the value $S[n,k]$. This routine obviously requires only $nk$ operations in $\mathbb{Z}$ while a naive recursive implementation of the recurrence would run in exponential time because of excessive recalculation.*

It is typical that the number of objects in some combinatorial family can be computed using some sort of recursion. The recursion is however not always as simple, or as easy to see, as in the example above. Sometimes the key to an efficient counting algorithm is a recurrence for an associated object, from which the number of interest can be deduced in a second step.

**Example 15.3.2 (Restricted Lattice Walks)** *Consider the lattice walks in $\mathbb{N}^2\subseteq\mathbb{Z}^2$ starting at the origin $(0,0)$, consisting of exactly $n$ steps, each of which is one from $\{(0,-1),(-1,0),(1,1)\}$. Such walks are called* **Kreweras walks** *[117, 40, 41]. How many Kreweras walks with $n$ steps are there? The combinatorial description gives rise to a multivariate recurrence for the numbers $a_{n,i,j}$ counting the number of Kreweras walks with $n$ steps and end point $(i,j)$:*

$$a_{n,i,j}=a_{n-1,i+1,j}+a_{n-1,i,j+1}+a_{n-1,i-1,j-1}.$$

*Together with the initial values $a_{0,0,0}=1$ and $a_{0,i,j}=0$ if $i\neq 0$ or $j\neq 0$ and the boundary conditions $a_{n,-1,0}=a_{n,0,-1}=0$ for all $n$, the recurrence determines the numbers $a_{n,i,j}$ $(n,i,j\in\mathbb{N})$ uniquely and can be used to compute them. For example, we find $a_{9,4,5}=160$. The number $a_n$ of all Kreweras walks with $n$ steps (and arbitrary end point) is then obtained via $a_n=\sum_{i=0}^{\infty}\sum_{j=0}^{\infty}a_{n,i,j}=\sum_{i=0}^{n}\sum_{j=0}^{n}a_{n,i,j}$. The sequence starts with 1, 1, 3, 7, 17, 47, 125, 333, 939, 2597, 7183, … (A151265 in the OEIS [170]).*

Recurrence equations are not always as nice and natural as in the previous two examples, and they do not always give rise to efficient algorithms. An example for a family of combinatorial objects for which even the construction of "slow" recurrences is nontrivial are 1324-avoiding permutations [130, 91]. Due to the lack of better counting algorithms, only the first 31 terms of the counting sequence are known so far.

Instead of a recurrence for a counting sequence $(a_n)_{n=0}^{\infty}$ it is sometimes easier to come up with a functional equation for its **(ordinary) generating function** $a(x):=\sum_{n=0}^{\infty}a_nx^n\in\mathbb{Q}[[x]]$. Such functional equations can also give rise to efficient algorithms for calculating $a_n$.

**Example 15.3.3 (Unlabeled Rooted Trees)** *Let $a_n$ be the number of unlabeled rooted trees. There is no obvious recurrence for $a_n$, but it can be shown (e.g., [107, Section 7.2.1.6]) that the generating function $a(x)=\sum_{n=1}^{\infty}a_nx^n$ satisfies the functional equation*

$$a(x)=x\exp\big(a(x)+\tfrac{1}{2}a(x^2)+\tfrac{1}{3}a(x^3)+\cdots\big).$$

*This equation has a unique formal power series solution, and its coefficients can be calculated recursively. For, if the coefficients $a_1, a_2, \ldots, a_{n-1}$ are known, then we can calculate the nth coefficient $a_n = [x^n]a(x)$ using the formula that is obtained by picking the coefficient of $x^n$ from the right-hand side:*

$$\begin{aligned} a_n &= [x^n]\, x \exp\big(a(x) + \tfrac{1}{2}a(x^2) + \cdots\big) \\ &= [x^{n-1}] \exp\big(a(x) + \tfrac{1}{2}a(x^2) + \cdots\big) \\ &= [x^{n-1}] \sum_{m=0}^{\infty} \frac{1}{m!} \Big( \sum_{k=1}^{\infty} \frac{1}{k} \sum_{i=1}^{\infty} a_i x^{ik} \Big)^m \\ &= [x^{n-1}] \sum_{m=0}^{n-1} \frac{1}{m!} \Big( \sum_{i,k \in \mathbb{N}: 1 \le ikm < n} \frac{a_i}{k} x^{ik} \Big)^m . \end{aligned}$$

*Note that the truncation in the last step is correct because $[x^0]a(x) = 0$ implies that none of the discarded terms has a chance to contribute to the coefficient of $x^{n-1}$. Note also that for every fixed n, the final expression involves only a finite number of operations, and only depends on the terms $a_1, a_2, \ldots, a_{n-1}$ computed earlier.*

*The resulting sequence starting with* 1, 1, 2, 4, 9, 20, 48, 115, 286, … *(A000081).*

Another fruitful source of counting algorithms is the transfer matrix method, which can be used to derive formulas for generating functions. We give here only an example application. A general introduction can be found in [172, Section 4.7]; for implementations and additional examples see [131, 133, 193].

**Example 15.3.4 (Tilings)** *We want to tile a rectangle of size $n \times 3$ without holes using stones of shape $1 \times 1$ and $1 \times 2$ (possibly rotated by 90 degrees). In order to determine the number of different tilings, consider the following matrix, that encodes the possible transitions from one partial tiling to another by placing a stone at the lowest and leftmost free position. The variables $x_1$ and $x_2$ correspond to placing a stone of size 1 or 2, respectively, and the variable z keeps track of the total height of the construction. For example, the entry $x_2z$ at position $(3,6)$ means that a partial tiling with a "skyline" of type can be transformed into a partial tiling with a skyline of type by placing a stone of size 2 (vertically) in the middle position. This enlarges the overall height of the tower by one, which is reflected by the multiplication by z.*

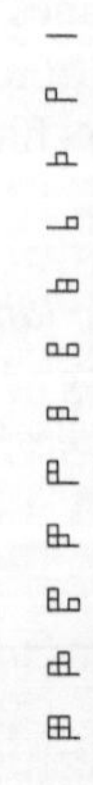

$$\begin{pmatrix} 0 & x_2 & 0 & x_2 & x_1 & x_1 & x_1 & 0 & 0 & 0 & 0 & x_2 \\ x_1z & 0 & 0 & 0 & x_2z & 0 & 0 & x_2 & x_1 & x_1 & 0 & 0 \\ 0 & 0 & 0 & 0 & 0 & x_2z & 0 & 0 & 0 & 0 & x_1 & 0 \\ 0 & 0 & 0 & 0 & 0 & 0 & x_2z & 0 & 0 & 0 & 0 & 0 \\ 0 & 0 & 0 & 0 & 0 & 0 & 0 & 0 & 0 & 0 & x_2 & 0 \\ 0 & 0 & 0 & x_1 & 0 & 0 & 0 & 0 & x_2 & 0 & 0 & 0 \\ x_2z & x_1 & x_1 & 0 & 0 & 0 & 0 & 0 & 0 & x_2 & 0 & x_1 \\ x_2z^2 & 0 & 0 & 0 & 0 & 0 & 0 & 0 & 0 & 0 & 0 & 0 \\ 0 & 0 & x_2z & 0 & 0 & 0 & 0 & x_1 & 0 & 0 & 0 & 0 \\ 0 & 0 & 0 & x_2z & 0 & 0 & 0 & 0 & 0 & 0 & 0 & 0 \\ 0 & x_2z & 0 & 0 & 0 & 0 & 0 & 0 & 0 & 0 & 0 & 0 \\ 0 & 0 & 0 & 0 & 0 & 0 & 0 & x_2 & 0 & 0 & 0 & 0 \end{pmatrix}$$

*If A denotes this matrix, then the entry at $(1,1)$ of the matrix power $A^k$ is a polynomial in $x_1, x_2, z$ that encodes all the possibilities to tile a rectangle with exactly k stones. A monomial $c x_1^{e_1} x_2^{e_2} z^n$ in this polynomial means that there are c different ways to fill a $3 \times n$ rectangle with exactly $e_1$ stones of size 1 and exactly $e_2 = k - e_1$ stones of size 2. The generating function for the number of all tilings (without regard to the number or type of stones) is given by $\sum_{k=0}^{\infty} (A^k)_{1,1}|_{x_1=1, x_2=1}$. Its first terms are* 1, 3, 22, 131, 823, 5096, ... *(A033506).*

Finally, there are situations where a counting algorithm can be obtained from a result about the asymptotic behavior of the counting sequence.

**Example 15.3.5 (Partition Numbers)** *Let $p(n)$ be the number of ways to write the integer n as a sum of positive integers (regardless of the order of the summands). The sequence starts with* 1, 1, 2, 3, 5, 7, 11, 15, 22, 30, 42, ... *(A000041)*

*One way to compute these numbers is via the pentagonal number theorem of Euler, which states that*

$$p(n) = \sum_k (-1)^{k-1} p(n - \tfrac{1}{2}k(3k-1)) \qquad (n \geq 1).$$

*This recurrence, together with the initial values $p(0) = 1$ and $p(n) = 0$ for $n < 0$, can be used to calculate $p(n)$ for any positive integer n.*

*A more efficient algorithm can be derived from the Hardy-Ramanujan-Rademacher formula, which in its simplest form amounts to the asymptotic estimate $p(n) \sim \frac{1}{4\sqrt{3}n} \exp(\pi\sqrt{2n/3})$ $(n \to \infty)$ and in its full form is a convergent series with an explicit error term that is so accurate that it can be used to approximate the value of $p(n)$ to a guaranteed error of less than $1/2$. Then rounding to the nearest integer gives the exact value of $p(n)$. See [90] for details.*

The examples in this section have in common that algorithms for counting certain combinatorial objects were obtained as a result of applying some counting theorem about the objects. It would be easily possible to give many further examples of this type, as nearly every theorem in combinatorics can be rephrased as an algorithm for counting objects, even if the theorems themselves were derived by hand. We will now discuss how to apply computer algebra in a more essential way, using it not only for implementing combinatorial theorems but for also for their derivation.

### 15.3.2 Combinatorial species

Recursive definitions of combinatorial objects can often be translated systematically into recurrence equations for the corresponding counting functions. The theory of combinatorial species [92, 23] provides a uniform and systematic framework for such constructions, and gives rise to algorithms for automatically translating recursive specifications of classes of combinatorial objects into functional equations for the corresponding generating functions, which can then be processed further by other algorithms.

#### 15.3.2.1 Formal definition and associated series

A combinatorial species is simply defined as a functor from the category of finite sets with bijections into itself. For example, the species DAG of directed acyclic graphs maps every finite set $V$ to the set $\mathrm{DAG}(V)$ of all directed acyclic graphs with vertex set $V$, and it maps a bijection $\sigma\colon V \to W$ into the bijection $\mathrm{DAG}(\sigma)\colon \mathrm{DAG}(V) \to \mathrm{DAG}(W)$ that relabels the vertices according to $\sigma$. Observe that DAG can be regarded as a formal description of the concept of directed acyclic graphs that abstracts away from the type of objects used as vertices.

If $F$ is a combinatorial species, then the cardinality of $F(V)$ only depends on the cardinality of $V$, not on the particular elements of $V$. We can therefore ask in general for the number $a_n$ of different objects $F$ produces when applied to a set $V$ of cardinality $n$. We define $\bar{a}_F(x) := \sum_{n=0}^{\infty} |F(\{1,\ldots,n\})| \frac{x^n}{n!}$ as the **exponential generating function** for $F$, so that, for example $[x^5]\bar{a}_{\mathrm{DAG}}(x) = \frac{29281}{5!}$ means that there are 29281 different directed acyclic graphs with five distinct nodes.

For counting unlabeled structures, define an equivalence relation on $F(V)$ by setting $p \sim q$ for $p,q \in F(V)$ if there exists a bijection $\sigma\colon V \to V$ such that $F(\sigma)(p) = q$. The equivalence classes of $\sim$ can be interpreted as the unlabeled counterparts of the objects in $F(V)$. We define the **(ordinary) generating function** for $F$ as $a_F(x) := \sum_{n=0}^{\infty} |F(\{1,\ldots,n\})/\!\sim| x^n$. For example, $[x^5]a_{\mathrm{DAG}}(x) = 302$ means that there are 302 different directed acyclic graphs with five indistinguishable nodes.

We are primarily interested in the ordinary and the exponential generating functions associated to a species $F$, but it turns out that an additional series is needed for computing them. Every bijection $\sigma\colon V \to V$ is a permutation of the elements of $V$, which can be written uniquely in terms of cycles. Write $\sigma_i$ for the number of cycles of length $i$ in this representation. For example, $\sigma_1$ is the number of fixed points of $\sigma$. Then the **cycle index series** of $F$ is defined as

$$Z_F(x_1,x_2,\ldots) := \sum_{n=0}^{\infty} \frac{1}{n!}\left(\sum_{\pi\in S_n} F(\pi)_1 x_1^{\pi_1} x_2^{\pi_2} \cdots\right) \in \mathbb{Q}[[x_1,x_2,\ldots]],$$

where $S_n$ denotes the symmetric group, i.e., the set of bijections of $\{1,\ldots,n\}$, and $F(\pi)_1$ is the number of fixed points of the bijection $F(\pi)$ from $F(\{1,\ldots,n\})$ to itself. We have $\bar{a}_F(x) = Z_F(x,0,0,\ldots)$ and $a_F(x) = Z_F(x,x^2,x^3,\ldots)$. See [23] for examples and a more elaborate discussion.

#### 15.3.2.2 Standard constructions

In general it is not obvious how to determine the cycle index series for a species $F$. However, many species of interest (and the corresponding series) can be constructed starting from a small number of elementary species (and the corresponding series) using a small number of operations on species. The most important constructions are summarized below.

- *Sum.* Disjoint union of $F$-objects and $G$-objects:

$$(F+G)(V) := F(V) \,\dot\cup\, G(V) = (F(V)\times\{\bullet\}) \cup (G(V)\times\{\circ\}).$$

- *Product.* All pairs $(f,g)$ where $f$ is an $F$-object and $g$ is a $G$-object:

$$(FG)(V) := \bigcup_{U \subseteq V} (F(U) \times G(V \setminus U))$$

One part of the elements of $V$ is used for labeling $f$, the remaining elements are used for labeling $g$.

- *Substitution.* $F$-objects labeled by $G$-objects:

$$(F \circ G)(V) := \bigcup_{V_1, V_2, \ldots, V_m} F(\{1, \ldots, m\}) \times G(V_1) \times G(V_2) \times \cdots \times G(V_m),$$

where the union is taken over all set partitions of $V$ into non-empty disjoint subsets $V = V_1 \dot{\cup} V_2 \dot{\cup} \cdots \dot{\cup} V_m$.

- *Pointing.* All $F$-objects in which one of the labels is singled out: $\Theta F(V) := F(V) \times V$. For example, if $T$ is the species of trees, then $\Theta T$ is the species of rooted trees.
- *Restriction.* For a species $F$ and an integer $k$, define $F_{=k}$ by $F_{=k}(U) := U$ if $|U| = k$ and $F_{=k}(U) = \emptyset$ otherwise. The restrictions $F_{\leq k}$ and $F_{\geq k}$ are defined analogously.

Typical starting points for constructions are simple species such as the 0-species with $0(V) = \emptyset$ for all $V$, the 1-species with $1(\emptyset) = \{\emptyset\}$ and $1(V) = \emptyset$ for all $V \neq \emptyset$, or the singleton species $X$ with $X(V) = V$ for all $V$ with $|V| = 1$ and $X(V) = \emptyset$ otherwise.

Further useful building blocks of a construction are the species of sets, defined by $\text{SET}(V) = \{V\}$ for all $V$, the power set species $\text{PSET}(V) := \{U : U \subseteq V\}$, and the species of cycles, defined as

$$\text{CYC}(\{v_1, \ldots, v_m\}) = \{(v_{\pi(1)}, \ldots, v_{\pi(m)}) : \pi \in S_m\}/_{\sim},$$

where $S_m$ is the symmetric group and $\sim$ the equivalence relation that identifies two tuples if and only if they are cyclic permutations of one another.

The cycle index series for these series are known explicitly, and it is known how each of the operations above affects the cycle index series. This knowledge can be encoded into algorithms that produce for a given species specification a lazy series representation of the corresponding generating functions.

**Example 15.3.6 (Direct Constructions)** *The singleton species $X$ can be viewed as representing an atomic object, e.g., ○. For fixed positive $k \in \mathbb{N}$, the objects in the species $X^k := XX \cdots X$ ($k$ factors) can be viewed as chains of exactly $k$ atoms, i.e., something like ○–○– ⋯ –○. The objects in $X + X^2 + X^3 + X^4$ are then chains consisting of at least one and at most four nodes. The pointing operation singles out one of the nodes as special, so an object in $\Theta(X + X^2 + X^3 + X^4)$ can be viewed as something like ○–●–○–○ or ○–○–●–○, for example. Next, consider the species $S = \text{CYC} \circ \Theta(X + X^2 + X^3 + X^4)$. Its objects can be thought of as graphs that are constructed by cyclically linking together a finite number of chains with at most four nodes, the link being attached to one of the nodes, for example*

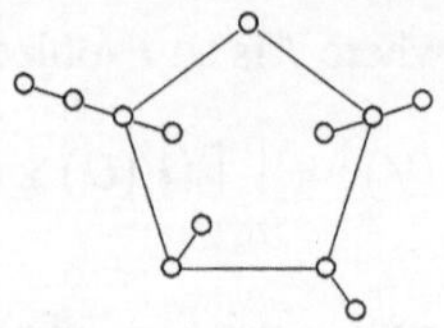

*The first terms of the generating functions $s(x)$ and $\bar{s}(x)$ for $S$ can be computed directly from the defining expression $S$ stated above, the result being*

$$s(x) = x + 3x^2 + 8x^3 + 21x^4 + 50x^5 + 128x^6 + \cdots$$
$$\text{and } \bar{s}(x) = x + 4x^2 + 10x^3 + 26x^4 + 54x^5 + 144x^6 + \cdots,$$

*respectively.*

#### 15.3.2.3 Recursive specifications

Recursive definitions of combinatorial structures can be formulated as equations for combinatorial species in terms of the constructions introduced above. For example, a (finite) sequence is by definition either the empty sequence or an object followed by a sequence. In the language of species this definition can be formulated as equation $\text{SEQ} = 1 + X\,\text{SEQ}$. The (unique) species solution SEQ of this equation is the species of finite sequences.

More generally, we can consider systems of equations, consisting of several equations and several unknown species. In general it is not obvious whether a given system has a solution, or whether its solution is unique. For systems composed of sum, product, substitution, restriction, 1, $X$, SET, SEQ, and CYC, Pivoteau, Salvy and Soria [148] introduce the notion of well-foundedness, and they prove that every well-founded system has a unique solution. This result generalizes the implicit species theorem of Joyal [92, 23], which in turn generalizes the implicit function theorem for formal power series (Section 15.2.3). It can be decided algorithmically whether a system is well-founded, and in this case, there are algorithms for efficiently computing the first terms of the generating functions, for drawing uniform random elements of prescribed size, and for numerically determining the asymptotic behavior of the counting sequences [148, 146, 147, 69, 118, 119].

**Example 15.3.7 (Funny Trees)** *Consider the (artificial) family of trees in which nodes are either black or white, the order of the subtrees of a white node matters but the order of the subtrees of a black node does not matter, and white nodes cannot have white children. The species $T$ of such trees may be defined by the following system, with two auxiliary variables $W$ and $B$ representing the species of trees with white or black roots, respectively, and $X_W$ and $X_B$ two distinguishable singleton species:*

$$T = W + B, \quad W = X_W\,(\text{SEQ} \circ B), \quad B = X_B\,(\text{SET} \circ T).$$

*Starting from this specification, the first terms of the counting sequence can be determined automatically, the result being* 0, 2, 3, 9, 30, 110, 423,...

More can be done for more restricted classes of systems. Of particular interest are systems composed only from sum, product, 1, $X$, SEQ, $\text{SEQ}_{=k}$, $\text{SEQ}_{\leq k}$ and $\text{SEQ}_{\geq k}$ ($k$ a fixed integer). Flajolet and Sedgewick [75, Sect. I.5.4.] show that the solution species of such systems always have algebraic generating functions. (Their "symbolic method" is a variant of the theory of species, see [148] for a detailed comparison.) The system of species equations can be automatically translated into a system of polynomial equations for the generating functions corresponding to the species variables, and Gröbner bases (Section 15.2.2) can be used to obtain from this system annihilating polynomials for the generating functions of each of the individual species.

**Example 15.3.8 (Planar Trees)** *The species* TREE *of nonempty planar trees is the solution of the equation* $\text{TREE} = X\,(\text{SEQ} \circ \text{TREE})$. *Taking this equation as input, it can be found algorithmically that the ordinary generating function* $T(x)$ *of nonempty planar trees satisfies the equation* $T(x)^2 - T(x) + x = 0$.

### 15.3.3 Partition analysis

The sequence $p(n)$ counting all the integer partitions was already mentioned in Example 15.3.5. In this section we discuss algorithms for counting generalized classes of partitions, or partitions subject to certain constraints. It turns out that the counting sequences of many interesting classes of partitions are much simpler than $p(n)$. Simple expressions for their generating functions can often be obtained algorithmically.

#### 15.3.3.1 Ω-Calculus

At the end of the 19th century, MacMahon [125, 126, 127] considered classes of partitions whose generating functions can be expressed as the positive part (cf. Section 15.2.3) of a simple power series. For example, the generating function counting all the partitions of $n$ into at most three parts can be written in the form

$$\left( [\lambda_1^{\geq}\lambda_2^{\geq}] \sum_{i_1,i_2,i_3=0}^{\infty} \lambda_1^{i_1-i_2}\lambda_2^{i_2-i_3}x^{i_1+i_2+i_3} \right)\Bigg|_{\lambda_1=\lambda_2=1} \in \mathbb{Q}[[x]].$$

The series to which the operator $[\lambda_1^{\geq}\lambda_2^{\geq}]$ is applied is a multivariate Laurent series expansion of the rational function

$$\frac{1}{(1-\lambda_1 x)(1-\lambda_2\lambda_1^{-1}x)(1-\lambda_2^{-1}x)} \in \mathbb{Q}(\lambda_1,\lambda_2,x).$$

MacMahon reserves the names $\lambda_1, \lambda_2, \ldots$ for variables to be eliminated by positive part extraction, and he uses the notation $\underset{\geq}{\Omega}$ for the operator $([\lambda_1^{\geq}\lambda_2^{\geq}\ldots]\cdot)|_{\lambda_1=\lambda_2=\cdots=1}$. The variables $\lambda_i$ are called **slack variables.**

Expressions involving the Omega operator can be simplified using general rewrite rules such as

$$\underset{\geq}{\Omega}\frac{C}{(1-A\lambda)(1-B/\lambda)} = \frac{C}{(1-A)(1-AB)}$$

where $A, B, C$ are free of $\lambda$. By applying this rule twice (once for eliminating $\lambda_1$, and then once more for eliminating also $\lambda_2$), we find

$$\underset{\geq}{\Omega} \frac{1}{(1-\lambda_1 x)(1-\lambda_2\lambda_1^{-1}x)(1-\lambda_2^{-1}x)} = \underset{\geq}{\Omega} \frac{1}{(1-\lambda_2^{-1}x)(1-x)(1-\lambda_2 x^2)} = \frac{1}{(1-x)(1-x^2)(1-x^3)}.$$

In a series of articles [10, 12, 11, 9, etc.], Andrews, Paule, and Riese have turned this approach into an algorithm. The algorithm is applicable to multivariate rational functions whose denominator factors into binomials $1-\lambda_1^{u_1}\cdots\lambda_m^{u_m}x_1^{v_1}\cdots x_k^{v_k}$ for certain fixed integers $u_1,\ldots,u_m,v_1,\ldots,v_k$. The result is a rational function in $x_1,\ldots,x_k$. An alternative algorithm for solving the same problem, also based on MacMahon's ideas, was proposed by Xin [188].

Rational functions of the required form arise naturally from geometric series in $\lambda_1,\ldots,\lambda_m$ and $x_1,\ldots,x_k$ in which the exponents are integer-linear combinations of the summation variables. The exponents of the $x_i$ carry the integers whose partitions are to be counted, and the exponents of the $\lambda_i$ express inequality constraints by which the type of partitions to be counted is defined. The approach is limited to problems that can be formulated in terms of a finite number of $\lambda_i$ and $x_i$.

**Example 15.3.9 (Plane Partitions)** *A plane partition of n is a matrix of nonnegative integers whose entries sum to n and where the entries are non-increasing along rows (from left to right) and columns (from top to bottom). Let $a_n$ be the number of plane partitions of n of size $2\times 2$, and let $a(x)=\sum_{n=0}^{\infty}a_n x^n$ be its generating function. In order to obtain a rational expression for $a(x)$, apply the partition analysis algorithm to*

$$\sum_{i_{11},i_{12},i_{21},i_{22}=0}^{\infty} \lambda_1^{i_{11}-i_{12}}\lambda_2^{i_{11}-i_{21}}\lambda_3^{i_{12}-i_{22}}\lambda_4^{i_{21}-i_{22}}x^{i_{11}+i_{12}+i_{21}+i_{22}} = \frac{1}{(1-\lambda_1\lambda_2 x)(1-\lambda_1^{-1}\lambda_3 x)(1-\lambda_2^{-1}\lambda_4 x)(1-\lambda_3^{-1}\lambda_4^{-1}x)}.$$

*The result is* $\frac{1}{(1-x)(1-x^2)^2(1-x^3)}$.

*It is also possible to compute a generating function that actually "generates" all the plane partitions of size $2\times 2$, by applying the algorithm to the series in $\lambda_1,\lambda_2,\lambda_3,\lambda_4,x_{11},x_{12},x_{21},x_{22}$ obtained from the series above by replacing $x^{i_{11}+i_{12}+i_{21}+i_{22}}$ by $x_{11}^{i_{11}}x_{12}^{i_{12}}x_{21}^{i_{21}}x_{22}^{i_{22}}$. The resulting rational function is then*

$$\frac{1-x_{11}^2x_{12}x_{21}}{(1-x_{11})(1-x_{11}x_{12})(1-x_{11}x_{21})(1-x_{11}x_{12}x_{21})(1-x_{11}x_{12}x_{21}x_{22})}.$$

*Its multivariate series expansion is the sum of all terms $x_{11}^{i_{11}}x_{12}^{i_{12}}x_{21}^{i_{21}}x_{22}^{i_{22}}$ for which $\binom{i_{11}\ i_{12}}{i_{21}\ i_{22}}$ is a plane partition of size $2\times 2$.*

Not only inequalities but also equational constraints can be handled by the Omega calculus. Because $A = 0$ if and only if $A \geq 0 \wedge A \leq 0$, such constraints can be encoded using two slack variables whose exponents differ by a factor of $-1$. While this is fine in theory, implementations may handle equational constraints in a more efficient way. MacMahon writes $\underset{=}{\Omega}$ for the operator $[\lambda_1^0 \lambda_2^0 \cdots]$.

For extracting the positive part or the constant coefficient from a series that is not of the form required for Omega calculus but is still holonomic, see Section 15.4.3.

### 15.3.3.2 Ehrhart theory

Partition analysis is closely related to discrete geometry. The objects of interest in discrete geometry are areas in a finite-dimensional real space defined as intersections of finitely many half spaces (viz., solutions of systems of finitely many linear inequalities). One of the questions of interest is to count the number of points with integer coordinates in such areas.

For a bounded set $P \subseteq \mathbb{R}^n$, the **Ehrhart function** [71] is defined as

$$\mathrm{Ehr}_P\colon \mathbb{N} \to \mathbb{N}, \qquad \mathrm{Ehr}_P(d) := |dP \cap \mathbb{Z}^n|,$$

where $dP := \{dx : x \in P\}$ refers to the $d$-fold dilate of $P$. Observe that $\mathrm{Ehr}_P(1)$ is the number of integer points in $P$.

A set $P \subseteq \mathbb{R}^n$ is called a **rational polytope** if it is the convex hull of finitely many points in $\mathbb{Q}^n$. This is the case if and only if $P$ is bounded and there is a matrix $A \in \mathbb{Z}^{m\times n}$ and a vector $b \in \mathbb{Z}^m$ such that $P = \{x \in \mathbb{R}^n : Ax \leq b\}$, where "$\leq$" is meant component-wise. In this case, $\mathrm{Ehr}_P(d)$ is a quasi-polynomial, i.e., its generating function $\sum_{d=0}^{\infty} \mathrm{Ehr}_P(d)x^d$ is a rational function whose denominator is a product of cyclotomic polynomials. Among the remarkable features of the Ehrhart function for rational polytopes is the Ehrhart-Macdonald reciprocity [124, 20], which states that $\mathrm{Ehr}_{P^\circ}(d) = (-1)^n \mathrm{Ehr}_P(-d)$, where $P^\circ$ denotes the interior of $P$. (Note that $P$ itself is a closed set.)

A rational polytope $P$ is called a **lattice polytope** if it is the convex hull of finitely many points in $\mathbb{Z}^n$. For such polytopes, the Ehrhart function is a polynomial in $d$, the **Ehrhart polynomial** of the polytope.

**Example 15.3.10 (Birkhoff Polytopes)** *For fixed $n \in \mathbb{N}$, let $B_n \subseteq \mathbb{R}^{n\times n}$ be the set of all matrices $A = ((a_{i,j}))_{i,j=1}^n$ with $a_{i,j} \geq 0$ and $\sum_{i=1}^n a_{i,j} = 1$ for all $j$ and $\sum_{j=1}^n a_{i,j} = 1$ for all $i$. Reading matrices as points in $\mathbb{R}^{n^2}$, the set $B_n$ is a polytope, called the $n$ th Birkhoff polytope. Beck and Pixton [19] have determined the Ehrhart polynomials of $B_n$ for $n = 1,\ldots,9$. While it is not hard to find $\mathrm{Ehr}_{B_1}(d) = 1$, $\mathrm{Ehr}_{B_2}(d) = d+1$, and perhaps $\mathrm{Ehr}_{B_3}(d) = \frac{1}{8}d^4 + \frac{3}{4}d^3 + \frac{15}{8}d^2 + \frac{9}{4}d + 1$, the computational cost increases rapidly with $n$. For the calculation of $\mathrm{Ehr}_{B_9}(d)$, they report a computation time of 324 CPU days on a 1.0 GHz Linux machine. Its degree is 64, and the leading coefficient is a rational number with a numerator consisting of 45 decimal digits and a denominator consisting of 85 decimal digits.*

Let $P \subseteq \mathbb{R}^n$ be a rational polytope with vertices $v_1, \ldots, v_m \in \mathbb{Q}^n$. Its dilates $dP$ ($d \geq 0$) can be viewed as slices of the cone $C \subseteq \mathbb{R}^{n+1}$ generated by $\{v_1, \ldots, v_m\} \times \{1\}$, i.e., $C = \{\lambda_1 \bar{v}_1 + \cdots + \lambda_m \bar{v}_m : \lambda_1, \ldots, \lambda_m \geq 0\}$ where the generator vectors $\bar{v}_i$ are obtained from the $v_i$ by appending a coordinate 1. The complete generating function $a$ of such a cone $C$ is defined as the formal sum of all the monomials $x_1^{e_1} \cdots x_n^{e_n} x_{n+1}^{e_{n+1}}$ with $(e_1, \ldots, e_{n+1}) \in C \cap \mathbb{Z}^{n+1}$. It lives in the ring $\mathbb{Z}_C[[x_1, \ldots, x_{n+1}]]$ (cf. Section 15.2.3; note that $C$ is a pointed cone). Setting $x_1 = \cdots = x_n = 1$ and $x_{n+1} = x$ in $a$ yields the generating function $\sum_{d=0}^{\infty} \mathrm{Ehr}_P(d) x^d$ of the Ehrhart polynomial for $P$.

The generating function $a$ is a multivariate series expansion of a certain rational function in $x_1, \ldots, x_{n+1}$. It is possible to compute the numerator and denominator of this rational function given as input the basis vectors $\bar{v}_1, \ldots, \bar{v}_m$. After scaling the $\bar{v}_i$ by an appropriate positive integer, if necessary, we may assume that all the $\bar{v}_i$ belong to $\mathbb{Z}^{n+1}$. For simplicity, let us further assume that the $\bar{v}_i$ are linearly independent in $\mathbb{R}^{n+1}$ (see [20] for a discussion of the general case). Writing $\bar{v}_i = (\bar{v}_{i,1}, \ldots, \bar{v}_{i,n+1})$, we then have

$$a = \frac{\sum_{(e_1, \ldots, e_{n+1}) \in \Pi(C) \cap \mathbb{Z}^{n+1}} x_1^{e_1} \cdots x_n^{e_n} x_{n+1}^{e_{n+1}}}{\prod_{i=1}^{m} (1 - x_1^{\bar{v}_{i,1}} \cdots x_{n+1}^{\bar{v}_{i,n+1}})},$$

where $\Pi(C) := \{\lambda_1 \bar{v}_1 + \cdots + \lambda_m \bar{v}_m : 0 \leq \lambda_1, \ldots, \lambda_m < 1\}$ is the **fundamental parallelepiped** of $C$. As $\Pi(C)$ is a bounded set, it contains at most finitely many integer points.

The numerator in this expression is problematic because the number of monomials may be quite large. Indeed, writing down the numerator expression may consume exponentially more space than writing down the input basis $\bar{v}_1, \ldots, \bar{v}_m$, even if we fix the dimension $n$ of the ambient space. It is therefore impossible to compute this rational function representation efficiently. However, there do exist cheaper representations: Barvinok [16] showed that for every cone $C \subseteq \mathbb{R}^{n+1}$ generated by linearly independent vectors $\bar{v}_1, \ldots, \bar{v}_m \in \mathbb{Z}^{n+1}$ there exist integers $L$, $e_1, \ldots, e_L$ and vectors $u_\ell = (u_{\ell,1}, \ldots, u_{\ell,n+1})$ and $w_{\ell,i} = (w_{\ell,i,1}, \ldots, w_{\ell,i,n+1})$ ($\ell = 1, \ldots, L$, $i = 1, \ldots, n+1$) such that

$$a = \sum_{\ell=1}^{L} \frac{(-1)^{e_\ell} x_1^{u_{\ell,1}} \cdots x_{n+1}^{u_{\ell,n+1}}}{\prod_{i=1}^{n+1} (1 - x_1^{w_{\ell,i,1}} \cdots x_{n+1}^{w_{\ell,i,n+1}})}$$

and the space needed to store this representation grows at most polynomially with the space needed to store the basis $\bar{v}_1, \ldots, \bar{v}_m$, provided that $n$ is fixed. He also provides a polynomial time algorithm for computing such a representation.

Barvinok's idea is thus to write the large rational function $a$ as a linear combination of small rational functions. Bringing his representation on a common denominator is not a good idea: it will lead to the exponentially longer expression stated earlier. Instead, one should try to extract the desired information directly from his representation. For example, to obtain the number of integer points in the rational polytope $P \subseteq \mathbb{R}^n$, compute the Barvinok representation of the complete generating function $a$ of the associated pointed cone $C \subseteq \mathbb{R}^{n+1}$, set $x_1 = \cdots = x_n = 1$ and $x_{n+1} = x$ in this expression, extract the coefficient of $x^1$ of the series expansions of each summand individually, and add up the results. See [17, 122, 44] for further information.

### 15.3.4 Computational group theory

Groups arise in combinatorics for various reasons. First, they may themselves be a source of counting or enumeration questions (How many elements does a group have? How many subgroups? How many groups of a particular type are there?, etc.). Secondly, and perhaps more importantly, groups are used for describing symmetries in combinatorial objects. For example, counting the number of ways to color the faces of a cube with three different colors requires some knowledge about the symmetry group of the cube. Similarly, a functional equation for the generating function of ternary trees can be obtained easily from a functional equation for the generating function of planar ternary trees by letting the symmetric group $S_3$ act on the subtrees of each internal node.

Details about such applications of groups in combinatorics are explained in other chapters of this handbook. In this section, we only give a short summary on how groups may be represented in a computer, and some algorithms for answering questions about groups. For further reading, we recommend the lecture notes [87] and the textbooks [167, 164, 86].

#### 15.3.4.1 Permutation groups

If $G$ is a group, a set $B \subseteq G$ is called a **generating set** if every element $g \in G$ can be written (not necessarily uniquely) as a finite product of elements of $B$. Trivially, $G$ itself is always a generating set for $G$, but most groups $G$ admit generating sets $B$ that are much smaller than $G$. Every finite group $G$ is isomorphic to a subgroup of a symmetric group $S_n$, and therefore, every finite group $G$ can be represented by a set $B$ of permutations.

The main algorithm for subgroups $G$ of $S_n$ specified via a generating set $B$ is the **Schreier-Sims algorithm** [166, 164, 87]. It constructs an alternative representation of $G$, called a **base and strong generating set** (BSGS), which makes it easy to answer a number of fundamental questions about $G$. From an application point of view, this is similar in spirit to the use of Gröbner bases for polynomial ideals, with $S_n$ playing the role of the polynomial ring, $G$ the role of an ideal, $B$ the role of a given basis, and the BSGS for $G$ playing the role of a Gröbner basis of the ideal. Computationally however, the algorithms are not related.

Some of the problems that can be solved using the Schreier-Sims algorithm (or related algorithms) for a group $G \leq S_n$ given via a generating set $B \subseteq G$ are the following:

- Compute the group order $|G|$.
- Construct a bijection between $\{1,\dots,|G|\}$ and $G$.
- Given $x \in \{1,\dots,n\}$, compute its stabilizer $G_x := \{g \in G : g \cdot x = x\} \leq G$.
- Given $x \in \{1,\dots,n\}$, compute its orbit $Gx = \{g \cdot x : g \in G\} \subseteq \{1,\dots,n\}$.
- Given $h \in S_n$, decide whether $h \in G$.
- Given $h_1, h_2 \in S_n$, decide whether there exists $g \in G$ with $h_1 = g^{-1}h_2 g$.
- Given $H \leq G$, compute the smallest normal subgroup of $G$ containing $H$.
- Given $H \leq G$, construct a set of representatives of $G/H$.

- Decide whether $G$ is solvable, or nilpotent.
- Construct a vector $B=(b_1,\dots,b_m)$ of pairwise different numbers $b_i\in\{1,\dots,n\}$ such that the only element of $G$ fixing each $b_i$ is the identity, and a set $S\subseteq G$ such that for all $i=1,\dots,m$ we have $\langle S\cap G^{(i)}\rangle=G^{(i)}$, where

$$G^{(i)}:=\{g\in G\mid\forall j\leq i:g\cdot b_j=b_j\}\leq G.$$

The last item contains the definition of a **base** (the vector $B$) and **strong generating set** (the set $S$).

**Example 15.3.11 (Schreier-Sims Algorithm)** *Consider the group*

$$G=\langle(1,2)(3,4),\ (1,2,3)(8,10,9),\ (1,6,7,5)(8,10)\rangle\leq S_{10}.$$

*A possible output of the Schreier-Sims algorithm applied to $G$ is the base $B=(10,5,2,3,4,7,1)$ and the strong generating set*

$$S=\{(6,7)(8,9),\ (4,7,6),\ (3,4)(6,7),\ (2,3,5)(8,9,10),\ (2,5)(3,4),\ (2,7)(3,4),\ (1,6,7,5)(8,10),\ (1,6)(8,9)\}.$$

*Once this data is known, it requires only some inexpensive calculations to derive that $|G|=15120$, that the two orbits of the natural action of $G$ on $\{1,\dots,10\}$ are $\{1,\dots,7\}$ and $\{8,9,10\}$, that the stabilizer of $x=8$ is the group $\langle(1,5)(9,10),\ (1,6)(9,10),\ (1,4,6),\ (1,6)(3,4),\ (2,6)(3,4),\ (2,7)(3,4)\rangle$, that the permutation $(1,2,3,4,5)$ belongs to $G$ and the permutation $(1,2,3,4)$ does not, that the smallest normal subgroup of $G$ containing $(1,2,3,4,5)$ is the group $\langle(1,2,3,4,5),\ (1,4,3,5,2),\ (1,3,6),\ (2,3,6,7,4)\rangle$, and so on.*

#### 15.3.4.2 Finitely presented groups

Let $\Omega$ be a set. A **word** over $\Omega$ is an element of $\Omega^*:=\bigcup_{n=0}^{\infty}\Omega^n$. The set $\Omega^*$ together with the **concatenation** $(a_1,\dots,a_n)(b_1,\dots,b_m):=(a_1,\dots,a_n,b_1,\dots,b_m)$ is a monoid, called the **free monoid** over $\Omega$. The neutral element is the **empty word** $\varepsilon:=()\in\Omega^0\subseteq\Omega^*$. It is customary to write a word $(a_1,a_2,\dots,a_n)$ simply as $a_1a_2\cdots a_n$.

Let $\Omega^{-1}:=\{\alpha^{-1}:\alpha\in\Omega\}$ be a disjoint copy of $\Omega$, and let $F\subseteq(\Omega\cup\Omega^{-1})^*$ be the set of all words that contain no subwords of the form $\alpha^{-1}\alpha$ or $\alpha\alpha^{-1}$. The operation $\circ\colon F\times F\to F$ is defined as concatenation followed by repeated removal of subwords of the form $\alpha^{-1}\alpha$ or $\alpha\alpha^{-1}$ until no such subwords remain. Then $(F,\circ)$ is a group, called the **free group** over $\Omega$.

Let now $R\subseteq F$, and write $\langle R\rangle_F$ for the smallest normal subgroup of $F$ containing all the elements of $R$. Then $G=F/\langle R\rangle_F$ is a group. If $\Omega$ and $R$ are finite, we call $G$ a **finitely presented group.** The common notation is $G=\langle\Omega\mid R\rangle$. For example, we have $\langle a\mid a^5\rangle\cong\mathbb{Z}/5\mathbb{Z}$ (easy), $\langle a,b\mid a^2,b^2,(ab)^3\rangle\cong S_3$ (less obvious), and $\langle a,b\mid a^{-1}b^{-1}ab\rangle\cong\mathbb{Z}^2$ (easy). Every finite group is finitely presented but not every finitely presented group is finite.

Finite presentations are a powerful tool for representing complicated groups by a finite (and often small) amount of data. However, there are not too many algorithms for solving problems about groups specified in this way. Many interesting problems turn out to be undecidable. In particular, it can be shown [139, 27] that there does not even exist an algorithm that would decide for given $\Omega$ and $R$ whether $\langle \Omega \mid R \rangle \cong \{1\}$. From this result it follows directly that it is also undecidable whether two given finitely presented groups are isomorphic, or whether two given words represent the same element of a given finitely presented group.

Most problems can therefore only be solved heuristically or by special purpose algorithms applicable only to finitely presented groups satisfying certain additional properties, or by semi-decision procedures (methods that terminate if the answer is "yes" but run forever if the answer is "no"). Two methods are of particular importance. The first is called **Knuth-Bendix completion** [108, 14]. This is a procedure that interprets the elements of $R$ as simplification rules and constructs an enlarged set $R' \supseteq R$ with the following two properties: (1) every word $\omega \in (\Omega \cup \Omega^{-1})^*$ can be brought by finitely many applications of the simplification rules in $R'$ to a unique word $\bar{\omega}$ that cannot be simplified any further, and (2) two words $\omega_1, \omega_2 \in (\Omega \cup \Omega^{-1})^*$ represent the same element of $\langle \Omega \mid R \rangle$ if and only if $\bar{\omega}_1 = \bar{\omega}_2$. The Knuth-Bendix procedure terminates if and only if there exists a finite set $R'$ with these properties. Computationally, the Knuth-Bendix procedure works similarly as Buchberger's algorithm for computing the Gröbner basis of a polynomial ideal, with the set $R'$ playing the role for the normal subgroup $\langle R \rangle_F = \langle R' \rangle_F \subseteq F$ that Gröbner bases play for polynomial ideals. In particular, once a finite $R'$ is available, many problems that are in general undecidable admit simple algorithmic solutions.

The second important method is the **Todd-Coxeter algorithm** [177, 167]. This method takes as input a finitely presented group $G = \langle \Omega \mid R \rangle$ and a finite subset $B \subseteq G$, and it produces as output a representation of the set $G/H$ of left cosets of $H$ in $G$. More precisely, when $n = |G/H|$ is finite, the algorithm produces as output a matrix $M \in \{1, \ldots, n\}^{n \times 2|\Omega|}$ in which each number $1, \ldots, n$ represents one of the cosets and the matrix entry $m_{i,j}$ is the coset obtained by multiplying the coset $i$ by the $j$th element of $\Omega \cup \Omega^{-1}$. If $G/H$ is infinite, the algorithm does not terminate. By convention, the index 1 is typically assigned to the coset $H = H \cdot 1$. The matrix $M$ is called a **coset table** for $G$ and $H$. The coset table can also be read as the description of a group homomorphism $\phi \colon G \to S_n$ with $\phi(H) = \{g \in \phi(G) : g \cdot 1 = 1\}$. For improvements and applications of the Todd-Coxeter algorithm, see [167, 87] and the references given there.

**Example 15.3.12 (Todd-Coxeter Algorithm)** *Let*

$$G = \langle a,\, b,\, c \mid a^3,\, b^3,\, c^4,\, (ab)^2,\, (ac)^2,\, (bc)^3 \rangle$$

*and let $H$ be the subgroup generated by $a$ and $bc$ in $G$. The Todd-Coxeter algorithm finds that there are eight cosets of $H$ in $G$. A coset table is*

| | $a$ | $a^{-1}$ | $b$ | $b^{-1}$ | $c$ | $c^{-1}$ |
|---|---|---|---|---|---|---|
| *1* | *1* | *1* | *2* | *3* | *4* | *2* |
| *2* | *3* | *4* | *3* | *1* | *1* | *5* |
| *3* | *4* | *2* | *1* | *2* | *6* | *7* |
| *4* | *2* | *3* | *4* | *4* | *5* | *1* |
| *5* | *7* | *6* | *7* | *8* | *2* | *4* |
| *6* | *5* | *7* | *6* | *6* | *8* | *3* |
| *7* | *6* | *5* | *8* | *5* | *3* | *8* |
| *8* | *8* | *8* | *5* | *7* | *7* | *6* |

*This means, for example, that multiplying the coset with index 5 by the element $b^{-1}$ from the right yields the coset with index 8. The coset with index 1 is H itself. Note that for the termination of the method it does not matter that G is an infinite group.*

*The homomorphism $\phi: G \to S_8$ sends a to $(2,3,4)(5,7,6)$, b to $(1,2,3)(5,7,8)$, and c to $(1,4,5,2)(3,6,8,7)$. Note that $\phi(a)$ and $\phi(bc) = (1,2,3)(5,7,8) \circ (1,4,5,2)(3,6,8,7) = (2,6,8)(3,4,5)$ both have 1 as fixed point.*

**Example 15.3.13 (Coxeter Groups)** *Coxeter groups [67, 26] are finitely presented groups $\langle \Omega \mid R \rangle$ where R is of a very special form: All elements of R must be of the form $(ab)^k$, where $a, b \in \Omega$ and k is a positive integer, with $k = 1$ if $a = b$. Due to this special structure, many problems are decidable for Coxeter groups even though they are undecidable for arbitrary finitely presented groups. In particular, there is a way to decide whether two words represent the same element of a given Coxeter group, and it is possible to determine whether a Coxeter group is finite or not (and to determine its size if it is finite).*

### 15.3.5 Software

Computer algebra packages for combinatorial species have already been developed long ago [22, 195, 70]. More recent implementations include the Maple package combstruct as well as packages for MuPad [85], Aldor [84], and Sage [152]. There is also a species library for the functional programming language Haskell [189]. These packages contain implementations of counting algorithms, enumeration procedures, and random samplers.

Andrews, Paule, and Riese have implemented their algorithms in form of a Mathematica package [10]. The algorithm of Xin is available in a Maple package [188]. For discrete geometry there are special purpose software systems such as polymake [77] and LattE [123], which contain implementations of Barvinok's algorithm as well as a number of other features not discussed here.

Most general purpose computer algebra systems provide some functionality for computing with groups. Typically these packages include implementations of the classical algorithms for computational group theory. Some offer useful tools for exploring and visualizing the structure of a given group, e.g., the subgroup lattice of a given permutation group. A description of the GroupTheory package for Maple 17 can be found in [149].

More advanced operations and more efficient implementations are provided by the special purpose system GAP [76]. At the time of writing, this system defines the state of the art in software for computational group theory. The functionality of GAP is also available in Sage [174].

## 15.4 Symbolic summation

Symbolic summation refers to algorithms for simplifying expressions involving the summation sign. In view of the ubiquity of sums in all parts of combinatorics, such algorithms are obviously of considerable importance.

### 15.4.1 Classical algorithms

#### 15.4.1.1 Hypergeometric terms

Classical summation algorithms deal with sums over hypergeometric terms. Let $C$ be a computable field and let $A = C(k)[S_k]$ be the Ore algebra of recurrence operators (see Section 15.2.4). A nonzero element $h$ of a function space $M$ on which $A$ acts is called **hypergeometric** (over $C(k)$) if it is the solution of some operator $a \in A$ of the form $a = S_k - r$, with $r \in C(k) \setminus \{0\}$.

In other words, $h$ is hypergeometric if the **shift quotient** $(S_k \cdot h)/h$ can be identified with a rational function $r \in C(k) \setminus \{0\}$. In yet other words, $h$ is hypergeometric if it can be written as a product $h = \prod_{i=0}^{k-1} r(i)$ for some rational function $r \in C(k) \setminus \{0\}$.

For example, $h = \frac{k!}{(2k+1)2^k}$ is a hypergeometric term because we have

$$S_k \cdot h = \frac{(k+1)!}{(2(k+1)+1)2^{k+1}} = \frac{(k+1)k!}{\frac{2k+3}{2k+1}(2k+1)2\,2^k} = \frac{(2k+1)(k+1)}{2(2k+3)}h.$$

Hypergeometric terms together with multiplication form a group. The sum $h_1 + h_2$ of two hypergeometric terms $h_1, h_2$ is hypergeometric if and only if $h_2/h_1$ can be identified with a rational function. In this case, we call $h_1$ and $h_2$ **similar.** By definition, every hypergeometric term $h$ is similar to its shift $S_k \cdot h$.

For the case of more variables, let $A = C(k_1, \dots, k_m)[S_{k_1}, \dots, S_{k_m}]$. A nonzero element $h$ from a function space $M$ on which $A$ acts is called hypergeometric if for every $i = 1, \dots, m$ there exists some operator $a \in A$ of the form $a = S_{k_i} - r_i$ for some $r_i \in C(k_1, \dots, k_m) \setminus \{0\}$ such that $a \cdot h = 0$. Again, another way of saying this is that all the shift quotients $(S_{k_i} \cdot h)/h$ $(i = 1, \dots, m)$ must agree with some rational functions. For example, $h = \binom{n}{k}$ is hypergeometric over $\mathbb{Q}(n,k)$, because we have

$$\frac{S_k \cdot h}{h} = \frac{\binom{n}{k+1}}{\binom{n}{k}} = \frac{n-k}{k+1} \in \mathbb{Q}(n,k) \quad \text{and} \quad \frac{S_n \cdot h}{h} = \frac{\binom{n+1}{k}}{\binom{n}{k}} = \frac{n+1}{n-k+1} \in \mathbb{Q}(n,k).$$

A hypergeometric term in $n,k$ is called **proper** if it can be written in the form

$$px^n y^k \prod_{i=1}^{M} \Gamma(a_i n + b_i k + c_i)^{e_i}$$

where $p \in K[n,k]$ is a polynomial, $x, y$ are constants, $M$ is a fixed integer, the $a_i, b_i, e_i$ are fixed integers, the $c_i$ are constants, and the shift acts as expected, e.g.,

$$S_k \cdot \Gamma(2n-3k+3) = \frac{1}{(2n-3k)(2n-3k+1)(2n-3k+2)} \Gamma(2n-3k+3).$$

For example, $h = \binom{n}{k}$ is proper hypergeometric because we can write it as $h = \frac{\Gamma(n+1)}{\Gamma(k+1)\Gamma(n-k+1)}$, which is of the required form. On the other hand, it can be shown that the hypergeometric term $h = \frac{1}{n^2+k^2}$ is not proper [186].

The notion of hypergeometric can be generalized to the $q$-case. Here, we let $C$ be a field of characteristic zero, $q \in C \setminus \{0\}$ some distinguished element, typically assumed not to be a root of unity, and we consider the Ore algebra $A = C(x)[\partial]$ defined by $\sigma(x) = qx$ (compare Examples 15.2.12.(2) and 15.2.13.(2)). A nonzero element $h$ of some function space $M$ on which $A$ acts is called $q$-hypergeometric if it is the solution of some operator $a \in A$ of the form $a = \partial - r$ for some $r \in C(x) \setminus \{0\}$. For example, $q^{\binom{k}{2}}$ is $q$-hypergeometric because it is annihilated by the operator $\partial - x$. Multivariate (proper) $q$-hypergeometric terms are defined analogously, see [186, 142] for details.

In most of the symbolic summation literature, hypergeometric terms are regarded as formal objects living in some extension field of the field of rational functions (see also Section 15.4.2 below for a more precise description). In combinatorial applications, it is usually necessary to interpret hypergeometric terms as sequences. Users of symbolic summation algorithms should be aware that this subtle difference can lead to misunderstandings about what a certain computation has or has not proved. See Chapter 5 of [102] for a discussion of hypergeometric summation algorithms that emphasize these issues.

#### 15.4.1.2 Gosper's algorithm

An expression of the form $\sum_{k=0}^{n} h_k$ is called **indefinite sum** if the upper summation bound is a variable, and this variable does not appear in the summand expression $h_k$. Finding a closed form expression for an indefinite sum is equivalent to finding a closed form solution $g_n$ of the so-called telescoping equation $g_{n+1} - g_n = h_n$. For if $g_n$ is such a solution, then we have $\sum_{k=0}^{n} h_k = g_{n+1} - g_0$.

Gosper's algorithm [81, 82, 141, 145, 109, 102] solves the indefinite summation problem for the case when the summand expression $h_k$ is a hypergeometric term. That is, given a hypergeometric term $h$, it decides whether there exists another hypergeometric term $g$ such that $\sigma(g) \quad g - h$, and if yes, it constructs such a term. A hypergeometric solution $g$ of the telescoping equation must necessarily have the form $g = ch$ for some rational function $c$. Gosper's algorithm can therefore also be considered as an algorithm for deciding whether there exists a rational solution $c$ to an equation of the form $r\sigma(c) - c = 1$, where $r$ is a known rational function.

**Example 15.4.1 (Indefinite Hypergeometric Summation)**

1. *Gosper's algorithm finds that* $\sum_{k=0}^{n} \frac{1}{4^k}\binom{2k}{k}$ *can be simplified to* $(1+2n)\frac{1}{4^n}\binom{2n}{n}$.

2. *It also proves that* $\sum_{k=1}^{n} \frac{1}{k}$ *does not have a hypergeometric closed form.*

#### 15.4.1.3 Zeilberger's algorithm

A sum that is not indefinite is called **definite.** In particular, sums whose upper summation bound is not a variable (as, e.g., in $\sum_{k=-\infty}^{\infty}\binom{n}{k}$), and sums whose upper summation bound is a variable that also appears in the summand expression (as, e.g., in $\sum_{k=0}^{n}\binom{n}{k}$) are definite.

Zeilberger's algorithm [190, 192, 145, 109, 102] solves the following problem: If $h_{n,k}$ is bivariate hypergeometric over $C(n,k)$, it searches for a nonzero operator $P \in C(n)[S_n]$ and a rational function $Q \in C(n,k)$ such that $(P-(S_k-1)Q)\cdot h_{n,k}=0$. Such a pair $(P,Q)$ is called a **creative telescoping relation**. $P$ is called a **telescoper** for $h_{n,k}$, and $Q$ a **certificate** for $P$ and $h_{n,k}$.

A creative telescoping relation for $h_{n,k}$ gives rise to a recurrence equation for the definite sum $F(n)=\sum_{k=0}^{n} h_{n,k}$. Writing $(P-(S_k-1)Q)\cdot h_{n,k}=0$ in the form

$$p_0(n)h_{n,k}+p_1(n)h_{n+1,k}+\cdots+p_r(n)h_{n+r,k}=q(n,k+1)h_{n,k+1}-q(n,k)h_{n,k}$$

and summing this equation over $k=0,\ldots,n$, we get

$$p_0(n)F(n)+p_1(n)\big(F(n+1)-h_{n+1,n+1}\big)+\cdots+p_r(n)\left(F(n+r)-\sum_{i=1}^{r} h_{n+r,n+i}\right)$$
$$=q(n,n+1)h_{n,n+1}-q(n,0)h_{n,0},$$

which can be rearranged to

$$p_0(n)F(n)+p_1(n)F(n+1)+\cdots+p_r(n)F(n+r)=\mathrm{rhs}(n)$$

for some explicit right-hand side. In summation problems arising in applications, the right-hand side usually turns out to be zero.

Not every hypergeometric term admits a creative telescoping relation. Zeilberger's algorithm finds such a relation if and only if one exists. A sufficient condition for the existence of a creative telescoping relation is that the summand be a proper hypergeometric term. This covers most situations arising in applications. It is however not necessary. A necessary and sufficient condition was given by Abramov [3].

**Example 15.4.2 (Definite Hypergeometric Summation)**

1. *For* $h_{n,k}=\binom{n}{k}^2$, *Zeilberger's algorithm finds the telescoper* $P=(n+1)S_n-(4n+2)$ *and the certificate* $Q=\frac{k^2(3n-2k+3)}{(n-k+1)^2}$. *For the sum* $F(n)=\sum_{k=-\infty}^{\infty}\binom{n}{k}^2=\sum_{k=0}^{n}\binom{n}{k}^2$, *it follows that* $(n+1)F(n+1)=(4n+2)F(n)$. *Together with the initial value* $F(0)=1$, *this implies* $F(n)=\prod_{k=0}^{n-1}\frac{4k+2}{k+1}=\binom{2n}{n}$.

2. *[143, Section 4.3] For $h_{n,k} = (-1)^k \binom{n}{k}\binom{3k}{n}$, Zeilberger's algorithm finds the telescoper $P = (4n+6)S_n^2 + (15n+21)S_n + (9n+9)$ and the certificate $Q = \frac{(n-3k)(n-3k+1)(n-3k+2)(2n+3)}{(n+2)(n-k+1)(n-k+2)}$. For the sum $F(n) = \sum_{k=0}^{n} h_{n,k}$, we obtain the recurrence $P \cdot F(n) = 0$. Among all telescopers for a given hypergeometric term, Zeilberger's algorithm always finds one that has smallest possible order. However, the sum may still satisfy a lower order recurrence. In the present example, we have $F(n) = (-3)^n$, which satisfies the first order recurrence $F(n+1) + 3F(n) = 0$. Such examples are rare.*

3. *Not all definite hypergeometric sums admit a hypergeometric closed form. As an example, the term $h_{n,k} = \binom{n}{k}^2\binom{n+k}{k}^2$ has the telescoper $P = (n+1)^3 - (2n+3)(17n^2+51n+39)S_n + (n+2)^3 S_n^2$. For the sum $F(n) = \sum_{k=0}^{n}\binom{n}{k}^2\binom{n+k}{k}^2$ we have $P \cdot F(n) = 0$, but there is no first order operator that annihilates $F(n)$, i.e., $F(n)$ is not hypergeometric (see Example 15.4.3.(2) below for a proof). This sum plays a role in Apéry's proof [180] of the irrationality of $\zeta(3)$.*

4. *For $h_{n,k} = \binom{2n}{k}$, Zeilberger's algorithm finds the telescoper $P = S_n - 4$ and the certificate $\frac{k(6n-2k+5)}{(2n-k+2)(2n-k+1)}$. For the sum $F(n) = \sum_{k=0}^{n}\binom{2n}{k}$ we therefore get the inhomogeneous recurrence $F(n+1) - 4F(n) = -\frac{1}{n+1}\binom{2n}{n}$.*

#### 15.4.1.4 Petkovšek's algorithm

In general, Zeilberger's algorithm does not deliver a closed form for a given definite hypergeometric sum, but only a recurrence that has the sum among its solutions. In the lucky case where this recurrence happens to be homogeneous and of first order, the recurrence can easily be rephrased as a hypergeometric closed form for the sum. In the general case, this is not so easy.

Petkovšek's algorithm [144, 145, 109, 102] solves this problem: Given a recurrence operator $P \in C(n)[S_n]$, it decides whether $P$ has hypergeometric solutions, i.e., whether there exist hypergeometric terms $h$ over $C(n)$ with $P \cdot h = 0$. In the affirmative case, it finds them all.

By definition $h$ is hypergeometric over $C(n)$ if there exists an operator $H = uS_n + v \in C(n)[S_n] \setminus \{0\}$ with $H \cdot h = 0$. If $P$ is any operator, then $h$ is a solution of $P$ if and only if $H$ is a **right factor** of $P$, i.e., we can write $P = UH$ for some operator $U \in C(n)[S_n]$. We can therefore say that Petkovšek's algorithm solves the problem of finding all the first order right factors of a given operator $P$.

Factoring Ore polynomials is computationally much harder than factoring commutative polynomials. The runtime of Petkovšek's algorithm is exponential in the number of irreducible factors of the leading and the trailing coefficient of the input operator. Better algorithms are known (see [183]), but they also require exponential time.

**Example 15.4.3 (Hypergeometric Solutions)**

1. *Suppose that for a certain definite sum $F(n)$, Zeilberger's algorithm produced the recurrence*

$$8(2n+5)(6n+11)F(n+3) - 6(24n^2+102n+101)F(n+2) \\ +3(18n^2+73n+64)F(n+1) - (n+1)(6n+17)F(n) = 0.$$

*Petkovšek's algorithm finds that this recurrence has two hypergeometric solutions: $\binom{2n}{n}$ and $\frac{1}{4^n}$. Being of third order, the recurrence has another solution, linearly independent of those, which is however not hypergeometric.*

*The sum $F(n)$ admits a closed form in terms of hypergeometric terms if and only if $F(n) = \alpha\binom{2n}{n} + \beta\frac{1}{4^n}$. To determine the constants $\alpha$ and $\beta$, if they exist, consider the system of linear equations obtained from this ansatz by setting $n = 1$ and $n = 2$. If we have $F(1) = \frac{23}{4}$, $F(2) = \frac{287}{16}$, $F(3) = \frac{3839}{64}$, then we find $\alpha = 3, \beta = -1$ and thus $F(n) = 3\binom{2n}{n} - \frac{1}{4^n}$ for all $n$. If instead we had $F(1) = \frac{1}{2}$, $F(2) = \frac{1}{4}$, $F(3) = \frac{1}{8}$, there would be no solution $\alpha, \beta$, and therefore no hypergeometric closed form of $F(n)$.*

2. *We have seen in Example 15.4.2.(3) that the sum $F(n) = \sum_{k=0}^{n}\binom{n}{k}^2\binom{n+k}{k}^2$ satisfies a certain second order recurrence. According to Petkovšek's algorithm, there is no hypergeometric solution of this recurrence. In particular, the solution $F(n)$ cannot be hypergeometric, viz. it does not satisfy a first order recurrence. Consequently, the recurrence found by Zeilberger's algorithm for $F(n)$ has minimal order in this example.*

### 15.4.2 ΠΣ-Theory

The classical algorithms for hypergeometric summation as outlined in the previous section have been generalized in various directions. The two most striking lines of development have led to sophisticated algorithms based on difference fields, as discussed in the present section, and based on operator ideals, as discussed in the following section.

#### 15.4.2.1 ΠΣ-Expressions and difference fields

A difference field is a field $K$ with a distinguished field automorphism $\sigma\colon K \to K$. The elements of a difference field $K$ are thought of as (formalizations of) sequences, and the automorphism $\sigma$ is meant to act on elements of $K$ like the shift operator acts on the corresponding sequences. Elements $c \in K$ with $\sigma(c) = c$ are called constants. The set of all constants of $K$ forms a subfield of $K$.

ΠΣ-fields are special difference fields introduced in 1981 by Karr [93, 94]. They are designed to formalize expressions that can be formed (with certain restrictions) from numbers, variables, arithmetic operations, summation sign and product sign according to the usual syntactic rules. ΠΣ-fields are constructed recursively from simpler $\Pi\Sigma$-fields by "adjoining a sum or a product." The construction starts with an arbitrary constant field.

More precisely, if $K$ is a $\Pi\Sigma$-field with constant field $C \subseteq K$ and automorphism $\sigma\colon K \to K$, then the field $E = K(t)$ of rational functions in $t$ with coefficients in $K$ is turned into a difference field extending $\sigma$ to $E$ via $\sigma(t) = \alpha t + \beta$ for some $\alpha \in K \setminus \{0\}$ and $\beta \in K$. Observe that when $\alpha = 1$ then the new element $t$ represents an indefinite sum over an element of $K$, and when $\beta = 0$ then $t$ represents an indefinite product over an element of $K$. If $\alpha$ and $\beta$ are such that all elements of $E$ left fixed by $\sigma$ are in $C$, then $E$ is again a $\Pi\Sigma$-field.

**Example 15.4.4 ($\Pi\Sigma$-fields)** *Expressions such as*

$$\sum_{k=1}^{n} \frac{k! + \sum_{i=1}^{k} \frac{1}{i}}{2^k + k}$$

*can easily be modelled as elements of $\Pi\Sigma$-fields. In order to construct a suitable field, start with the constant field $C = \mathbb{Q}$. Going through the expression from inside to outside, adjoin a new transcendental element for every sum or product encountered. Note that $n = \sum_{i=1}^{n} 1$ can be viewed as a shorthand notation for a sum, and $2^n$ and $n!$ are obvious shorthand notations for products. Thus we obtain the $\Pi\Sigma$-field $K = \mathbb{Q}(t_1, t_2, t_3, t_4, t_5)$ with $\sigma$ defined via*

$$\begin{aligned}
\sigma(t_1) &= t_1 + 1 && \text{“ } t_1 \approx n \text{ ”} \\
\sigma(t_2) &= 2t_2 && \text{“ } t_2 \approx 2^n \text{ ”} \\
\sigma(t_3) &= t_3 + \tfrac{1}{t_1+1} && \text{“ } t_3 \approx \textstyle\sum_{k=1}^{n} \frac{1}{k} \text{ ”} \\
\sigma(t_4) &= (t_1+1)t_4 && \text{“ } t_4 \approx n! \text{ ”} \\
\sigma(t_5) &= t_5 + \tfrac{1+(t_1+1)t_3+(t_1+1)^2 t_4}{(t_1+1)(t_1+1+2t_2)} && \text{“ } t_5 \approx \textstyle\sum_{k=1}^{n} \frac{k!+\sum_{i=1}^{k}\frac{1}{i}}{2^k+k} \text{ ”}
\end{aligned}$$

*Observe that there is some freedom in the arrangement of the field extension, but it is not completely arbitrary. For example, we may exchange $t_1$ and $t_2$, but not $t_1$ and $t_3$, because $\sigma(t_3)$ involves $t_1$.*

It is not obvious (but true) that all the constants of the field $K$ constructed in the example above belong to $\mathbb{Q}$, as is required for $K$ to be a $\Pi\Sigma$-field. As an example for a problematic construction, suppose we wanted to construct a $\Pi\Sigma$-field for the expression $(-1)^n$. A first attempt is to take $\mathbb{Q}(t)$ with $\sigma(t) = -t$, but then $\sigma(t^2) = \sigma(t)^2 = (-t)^2 = t^2$, so $t^2 \in \mathbb{Q}(t) \setminus \mathbb{Q}$ would be a new constant. Such new constants are forbidden in a $\Pi\Sigma$-field. A second idea might be to factor out the algebraic relation $((-1)^n)^2 - 1 = 0$, taking $\mathbb{Q}[t]/\langle t^2 - 1\rangle$ instead of $\mathbb{Q}(t)$, but since $t^2 - 1 = (t-1)(t+1)$ is not irreducible, this is not a field. In fact, Karr's original theory excludes sums involving the alternating sign. However, more recent versions of the theory developed by Schneider [155] also cover alternating signs.

For an expression involving sums and products to admit a natural translation into a $\Pi\Sigma$-field, it is necessary that the nesting is indefinite. This means that every subexpression must contain at most one free variable. For example, all the boxed subexpressions of

$$\boxed{\sum_{k=1}^{n} \boxed{\frac{1}{k}} \boxed{\sum_{i=1}^{k} \boxed{\frac{1}{i}}}}$$

contain only one free variable (and possibly several other variables that all are bound by a summation sign), so this expression is admissible. On the other hand, the boxed subexpression in

$$\sum_{k=1}^{n} \boxed{\frac{1}{n+k+1}}$$

contains two free variables $n, k$, so it is not admissible. We call expressions that are admissible in this sense **ΠΣ-expressions.** There is an algorithm that takes a ΠΣ-expression as input and produces a ΠΣ-field containing a corresponding element.

Some care must be applied when ΠΣ-expressions are interpreted as sequences. For example, the expression $n(n-1)(n-2)\prod_{k=1}^{n}(k^2-7k+12)$ represents the zero sequence but it may correspond to a nonzero element of a ΠΣ-field. If such an expression appears in the denominator of another expression, this may lead to results that are only correct in a certain algebraic sense but give no meaningful information about sequences. Fortunately, such issues rarely arise in practice, and for a large class of ΠΣ-expressions it can even be proven that they won't arise [161].

#### 15.4.2.2 Indefinite summation

Karr's algorithm solves the indefinite summation problem for ΠΣ-fields: Given a ΠΣ-field $K$ and an element $f \in K$, it decides whether there exists an element $g \in K$ such that $\sigma(g) - g = f$, and if so, it finds such a $g$.

**Example 15.4.5 (In-Field Summation)** *Let $H_n = \sum_{k=1}^{n} \frac{1}{k}$. Then Karr's algorithm can find the identities*

$$\sum_{k=1}^{n} H_k = (n+1)H_n - n, \qquad \sum_{k=1}^{n} H_k^2 = 2n - (2n+1)H_n + (n+1)H_n^2$$

*in the sense that given the left-hand side, it can calculate the right-hand side. The fact that the algorithm does not find any simplification for the sum $\sum_{k=1}^{n} H_k^3$ is a proof that this sum cannot be written as a rational function in $n$ and $H_n$.*

Karr's algorithm includes Gosper's algorithm as a special case. Indeed, if $h$ is hypergeometric over $\mathbb{Q}(n)$, then $\mathbb{Q}(n)(h)$ is a ΠΣ-field, and if there exists another hypergeometric term $g$ with $\sigma(g) - g = h$, then this hypergeometric term must belong to $\mathbb{Q}(n)(h)$, so Karr's algorithm will find it.

Karr's algorithm for simplifying ΠΣ-expressions can be viewed as a difference analogue of the Risch algorithm [150, 46] for deciding whether the integral of a given Liouvillian function (i.e. an expression composed of numbers, a variable $x$, and the functions exp and log) is again Liouvillian. In this analogy, log-expressions correspond to sums and exp-expressions correspond to products.

There is an extension of Karr's algorithm due to Schneider [160] that finds simplifications of a given sum that do not necessarily belong to the $\Pi\Sigma$-field in which the summand lives. Define the depth of an element of a $\Pi\Sigma$-field $K = C(t_1, \ldots, t_n)$ recursively by setting the depth of the elements of $C$ to zero, the depth of other elements to the maximal depth of all the generators $t_i$ appearing in it, and the depth of a generator $t_i$ satisfying $\sigma(t_i) = \alpha t_i + \beta$ as 1 more than the maximum depth of $\alpha$ and $\beta$.

The depth of an element of $K$ reflects the nesting depth of the corresponding $\Pi\Sigma$-expression. For example, if $K = \mathbb{Q}(t_1, t_2, t_3, t_4, t_5)$ is the $\Pi\Sigma$-field from Example 15.4.4, then $t_1$ and $t_2$ have depth 1, $t_3$ and $t_4$ have depth 2, and $t_5$ has depth 3. Schneider's extension of Karr's algorithm takes as input a $\Pi\Sigma$-field $K$ and an element $f \in K$, and it constructs as output an extension field $E$ of $K$ and an element $g \in E$ such that $\sigma(g) - g = f$ and $g$ has minimal possible depth.

**Example 15.4.6 (Nesting Depth Reduction)**

1. *We have seen in the previous example that the sum $\sum_{k=1}^{n} H_k^3$ cannot be expressed as a rational function in $n$ and $H_n$. However, Schneider's algorithm finds the representation*

$$\sum_{k=1}^{n} H_k^3 = -6n + \tfrac{3}{2}(2n+1)(2H_n - H_n^2) + (n+1)H_n^3 + \tfrac{1}{2}\sum_{k=1}^{n} \frac{1}{k^2}.$$

   *Observe that the sum in the last term on the right does not appear as a subexpression of the summand. Note also that the right-hand side is simpler than the left-hand side in the sense that $\sum_{k=1}^{n} H_k^3 = \sum_{k=1}^{n} \big(\sum_{i=1}^{k} \frac{1}{\sum_{j=1}^{i} 1}\big)^3$ has depth 3 while the right-hand side has only depth 2.*

2. *Schneider's algorithm proves that $\sum_{k=1}^{n} H_k^4$ does not admit any representation in terms of single sums.*

3. *Also the five-fold sum*

$$\sum_{k=1}^{n} \frac{\sum_{l=1}^{k} \frac{\sum_{m=1}^{l} \frac{\sum_{i=1}^{m} \frac{\sum_{j=1}^{i} \frac{1}{j}}{i}}{m^2}}{l}}{k}$$

   *does not admit a representation in terms of single sums. But it does have the following representation in terms of double sums, where we write $H_n^{(r)} = \sum_{k=1}^{n} \frac{1}{k^r}$ for the harmonic numbers of order $r$:*

$$\frac{1}{4}\Big(\frac{1}{3}(H_n^{(2)})^3+(H_n^{(4)}+\sum_{k=1}^{n}\frac{H_k^2}{k^2})H_n^{(2)}+\frac{2}{3}H_k^{(6)}-\sum_{k=1}^{n}\frac{H_k^{(4)}H_k}{k}-\sum_{k=1}^{n}\frac{(H_k^{(2)})^2H_k}{k}$$
$$+2\sum_{k=1}^{n}\frac{H_k^2}{k^4}+\sum_{k=1}^{n}\frac{H_k^4}{k^2}+H_n^2\sum_{k=1}^{n}\frac{H_k^2}{k^2}-\sum_{k=1}^{n}\frac{H_k^{(2)}H_k^2}{k^2}-2\sum_{k=1}^{n}\frac{H_k^3}{k^3}$$
$$+H_n\big(\sum_{k=1}^{n}\frac{H_k^{(4)}}{k}+\sum_{k=1}^{n}\frac{(H_k^{(2)})^2}{k}+2\sum_{k=1}^{n}\frac{H_k^2}{k^3}-2\sum_{k=1}^{n}\frac{H_k^3}{k^2}\big)\Big).$$

#### 15.4.2.3 Creative telescoping

When two elements $f_1, f_2$ of a $\Pi\Sigma$-field $K$ are not summable in $K$, i.e., there is no $g \in K$ with $\sigma(g) - g = f_1$ and also no $g \in K$ with $\sigma(g) - g = f_2$, it may still be the case that there exist constants $c_1, c_2$ in $K$ such that the linear combination $c_1 f_1 + c_2 f_2$ is summable. Given $f_0, \dots, f_r \in K$, the set

$$V := \{(g, c_0, \dots, c_r) \in K \times C^{r+1} : \sigma(g) - g = c_0 f_0 + \cdots + c_r f_r\}$$

is a vector space over the constant field $C$ of $K$. Given $K$ and $f_0, \dots, f_r$, Karr's algorithm can compute a basis for this vector space. This is known as **parameterized telescoping.** Note that it contains the usual telescoping problem for $r = 0$.

The parameterized telescoping problem arises as a subproblem of Karr's algorithm for solving the telescoping equation. Schneider [155] observed that it can be used to do creative telescoping in $\Pi\Sigma$-fields, thus generalizing Zeilberger's algorithm.

The procedure applies to any expression in two variables $n, k$ which is a $\Pi\Sigma$-expression with respect to $k$ when $n$ is regarded as a constant, and which maintains this property when $n$ is replaced by $n + i$ for any fixed $i \in \mathbb{N}$. If $f(n,k)$ denotes such an expression, we choose some $r \geq 0$ and set $f_0, \dots, f_r$ to the elements corresponding to $f(n,k), \dots, f(n+r,k)$ in a suitable $\Pi\Sigma$-field $K$. Then we use parameterized telescoping to find, if possible, a nontrivial linear combination $c_0 f_0 + \cdots + c_r f_r$ with constants $c_i$ (that may well involve $n$, but not $k$) that is equal to $\sigma(g) - g$ for some $g \in K$.

If no such relation is found, increase $r$ and try again. The procedure will find a creative telescoping relation after finitely many steps if and only if one exists. If there is no such relation, it will run forever. There are no results guaranteeing the existence of creative telescoping relations for (a reasonably large subclass of) the expressions to which the method applies, but the experience is that they do exist for all sums arising in combinatorics.

**Example 15.4.7 (Apéry's second sum)** *[180, 158] Apéry's proof of the irrationality of $\zeta(3)$ involves the sum*

$$F(n) = \sum_{k=0}^{n} \binom{n}{k}^2 \binom{n+k}{k}^2 \Big(\sum_{i=1}^{n} \frac{1}{i^3} + \sum_{i=1}^{k} \frac{(-1)^{i+1}}{2i^3 \binom{n}{i}\binom{n+i}{i}}\Big).$$

*Write $f(n,k)$ for the summand expression. When $n$ is fixed, this expression is a $\Pi\Sigma$-*

*expression with respect to the variable k. Also the two shifted expressions*

$$f(n+1,k) = \frac{(n+k+1)^2}{(n-k+1)^2}\binom{n}{k}^2\binom{n+k}{k}^2 \times\left(\sum_{i=1}^{n}\frac{1}{i^3}+\frac{1}{(n+1)^3}+\sum_{i=1}^{k}\frac{(n-i+1)(-1)^{i+1}}{2i^3(n+i+1)\binom{n}{i}\binom{n+i}{i}}\right)$$

$$f(n+2,k) = \frac{(n+k+2)^2(n+k+1)^2}{(n-k+2)^2(n-k+1)^2}\binom{n}{k}^2\binom{n+k}{k}^2\left(\sum_{i=1}^{n}\frac{1}{i^3}+\frac{1}{(n+1)^3} +\frac{1}{(n+2)^3}+\sum_{i=1}^{k}\frac{(n-i+2)(n-i+1)(-1)^{i+1}}{2i^3(n+i+2)(n+i+1)\binom{n}{i}\binom{n+i}{i}}\right)$$

*are $\Pi\Sigma$-expressions with respect to k when n is regarded as a constant. We can construct a $\Pi\Sigma$-field K with constant field $C=\mathbb{Q}(n,c)$ (where c is used for representing the constant sum $\sum_{i=1}^{n}\frac{1}{i^3}$), such that K contains elements $f_0, f_1, f_2$ corresponding to $f(n,k)$, $f(n+1,k)$, $f(n+2,k)$, respectively. Parameterized telescoping applied to $f_0, f_1, f_2 \in K$ gives rise to a creative telescoping relation for $f(n,k)$, which in turn gives rise to the recurrence*

$$(n+1)^3F(n)-(2n+3)(17n^2+51n+39)F(n+1)+(n+2)^3F(n+2)=0$$

*for the original definite sum. This is the same recurrence as in Example 15.4.2.(3).*

#### 15.4.2.4 D'Alembertian solutions

Given a linear recurrence equation as produced, for example, by creative telescoping, we may wish to determine whether it admits solutions that can be represented by suitable elements of $\Pi\Sigma$-fields. In general, this is not the case, but for sums arising in applications, it often is.

Consider an operator $P \in C(n)[S_n]$ with rational function coefficients. The first order right-hand factors of $P$ can be found using Petkovšek's algorithm (see Section 15.4.1). They correspond exactly to the hypergeometric solutions of $P$. Suppose $(S_n - u)$ for some $u \in C(n)$ is a right-hand factor of $P$, and write $P = U(S_n - u)$ for some $U \in C(n)[S_n]$. One solution of $P$ is then given by the product $h_n = \prod_{k=0}^{n} u(k-1)$, which evidently can be viewed as an element of a $\Pi\Sigma$-field. Suppose also $U$ has some first order right-hand factor $(S_n - v)$ with $v \in C(n)$, so that $U = V(S_n - v)$ for some $V \in C(n)[S_n]$, and $P = V(S_n - v)(S_n - u)$. Then $f_n = \prod_{k=1}^{n} v(k-1)$ is a solution of $U$, and $g_n := \left(\prod_{k=1}^{n} u(k-1)\right)\sum_{k=1}^{n}\prod_{i=1}^{k}\frac{v(i-2)}{u(i-1)}$ is a solution of $P$ because $(S_n - u)\cdot g_n = f_n$. Also this solution can evidently be viewed as an element of a $\Pi\Sigma$-field. Next, any hypergeometric solution of $V$ translates into a single sum solution of $U$, which in turn translates into a double sum solution of $P$, and so on.

In the best case, this procedure leads to $r$ linearly independent solutions of $P$ as nested sums and products. Such solutions are called **d'Alembertian** [6]. D'Alembertian solutions can be computed more generally for linear recurrence equations whose coefficients belong to an arbitrary $\Pi\Sigma$-field instead of just $C(n)$, and for inhomogeneous recurrences [4].

### 15.4.2.5 Nested definite sums

ΠΣ-expressions are composed of nested sums and products, but all sums and products arising in a ΠΣ-expression are indefinite in the sense that the upper bound of every summation or product sign is a variable that does not occur in the corresponding summand or factor expression. In the context of ΠΣ-theory, these expressions play the role of "closed forms." A typical application scenario is that a combinatorial problem leads to a definite sum whose summand is a ΠΣ-expression, and we wish to know whether the definite sum can be simplified to a ΠΣ-expression. Definite single sums are simplified by the following steps:

1. Using creative telescoping, compute a linear recurrence for the sum.
2. Compute the d'Alembertian solutions of this recurrence.
3. If possible, write the original sum as a linear combination of these solutions.
4. Compute an equivalent ΠΣ-expression of minimal depth.

Nested definite sums are simplified by applying these steps repeatedly, starting from the innermost definite sum and working outwards:

$$\boxed{\sum_{i=0}^{n}\boxed{\sum_{j=0}^{n}\boxed{\begin{array}{l}\Pi\Sigma\text{-expression w.r.t. } j\\ \text{with } n,i \text{ as parameters}\end{array}}}} \to \sum_{i=0}^{n}\boxed{\begin{array}{l}\Pi\Sigma\text{-expression w.r.t. } i\\ \text{with } n \text{ as parameter}\end{array}} \to \cdots$$

See the survey articles [156, 163, 162] for details and examples.

## 15.4.3 The holonomic systems approach

There are quantities that do not admit a closed form representation in terms of sums and products or other standard objects. Some of these can still be expressed as solutions of certain equations. In a sense, this is analogous to the choices for representing numbers (cf. Sect. 15.2.1): The numbers that can be expressed in terms of radicals form a proper subset of the algebraic numbers, and those in turn form a proper subset of all numbers. All algebraic numbers, even those that can not be written as radicals, can be described by the polynomial equation of which they are a root.

The situation is the same for power series. A formal power series $a \in K[[x]]$ is called **algebraic** if there exists a nonzero polynomial $p \in K[x,y]$ such that $p(x,a(x)) = 0$. The polynomial $p$ together with a finite number of terms of $a$ uniquely determine the infinite object $a$ by a finite amount of data. We can therefore do rigorous computations in the broad class of algebraic power series. An even broader class, discussed in the present section, is the class of D-finite functions. In the simplest case, these are power series $a \in K[[x]]$ that satisfy linear differential equations with polynomial coefficients, or sequences $(a_n)_{n=0}^{\infty}$ that satisfy a linear recurrence equation with polynomial coefficients. Again, the defining equation together with a suitable number of initial values uniquely determines such an infinite object by a finite amount of data. In this section we discuss algorithms operating on such representations.

#### 15.4.3.1 D-finite and holonomic functions

In its most general form, the notion of D-finiteness extends to objects in several variables that live in some vector space $M$ on which an Ore algebra $A = K[\partial_1,\dots,\partial_n]$ acts. We then say that an element $f \in M$ is called **D-finite** (with respect to the action of $A$ on $M$) if for all $i \in \{1,\dots,n\}$ we have $\operatorname{ann}(f) \cap K[\partial_i] \neq \{0\}$.

For sequences and power series in a single variable, to be D-finite just means that there is one nonzero operator (in $K(n)[S_n]$ or in $K(x)[D_x]$, respectively) which annihilates it. For objects in several variables, say some object $a(x_1,\dots,x_u,n_1,\dots,n_v)$ depending on $u$ "continuous" variables $x_1,\dots,x_u$ and $v$ "discrete" variables $n_1,\dots,n_v$, the definition says that it is D-finite if and only if it satisfies for each $i = 1,\dots,u$ a linear differential equation with polynomial coefficients involving no shifts and only derivatives with respect to $x_i$, and for each $j = 1,\dots,v$ a linear recurrence equation with polynomial coefficients involving no derivatives and only shifts with respect to $n_j$. There is no restriction on the variables that may show up in the polynomial coefficients.

Regardless of whether or not it is the annihilator of a function, an ideal $\mathfrak{a}$ is called D-finite if it has the property $\mathfrak{a} \cap K[\partial_i] \neq \{0\}$ for all $i$. In typical applications, a basis for the whole annihilator of a function $f$ may not be easy to obtain. To cover also these situations, we say that an ideal $\mathfrak{a}$ is a **D-finite description** for $f$ if $\mathfrak{a}$ is D-finite and $\operatorname{ann}(f) \subseteq \mathfrak{a}$. In most cases, multivariate D-finite objects can be specified by a D-finite description plus a finite number of initial values. However, the situation is not always as straightforward as in the univariate case.

**Example 15.4.8 (Initial Values in the Multivariate Case)**

1. *The sequence $f(n,k) = 3^n + (3nk+1)2^k$ is D-finite. A D-finite description is given by the ideal $\mathfrak{a} = \langle a_1, a_2 \rangle \subseteq \mathbb{Q}(n,k)[S_n,S_k]$ where*

$$a_1 = (6kn - 3k + 2)S_n^2 - (24kn - 6k + 8)S_n + (18kn + 9k + 6),$$
$$a_2 = (3kn + 6n + 1)S_k^2 - (9kn - 24n - 3)S_k + (6kn + 18n + 2).$$

*Note that $\mathfrak{a} \cap \mathbb{Q}(n,k)[S_n] = \langle a_1 \rangle \neq \{0\}$ and $\mathfrak{a} \cap \mathbb{Q}(n,k)[S_k] = \langle a_2 \rangle \neq \{0\}$. Using these two recurrences, every term $f(n,k)$ for $n,k \in \mathbb{N}$ can be computed recursively from the four initial values $f(0,0), f(1,0), f(0,1), f(1,1)$.*

2. *For doing computations it is not necessary (and not advisable!) to represent the ideal $\mathfrak{a}$ by a basis $a_1,\dots,a_n$ such that $\mathfrak{a} \cap K[\partial_i] = \langle a_i \rangle$. The definition only requires that such elements $a_i$ exist and not that we explicitly know them. For example, the ideal $\mathfrak{a}$ from part 1 can also be generated by the operators $a_1$ and*

$$a_3 = (6kn - 3k + 2)S_k + (3kn + 6n + 1)S_n - (15kn + 18n - 3k + 5).$$

*Note that with this basis all the terms $f(n,k)$ can be computed recursively from only two initial values $f(0,0), f(1,0)$. In general, it is preferable to work with bases for which the number of terms needed to identify a particular solution is as small as possible. The number of initial values is minimal when the basis is a Gröbner basis.*

We have seen in Example 15.2.14 for the univariate case that when a recurrence has integer singularities, we may need to specify a finite number of additional initial values. Singularities in the multivariate case are integer roots of multivariate polynomials. This is more troublesome because (a) there may be infinitely many, and (b) there is no algorithmic way to find them all [132].

If $f$ is D-finite with respect to the action of the algebra $K(x_1,\dots,x_u)[D_{x_1},\dots,D_{x_u}]$ of differential operators, then we also say that $f$ is **holonomic.** For a multivariate sequence, or more generally, for an object $f = f(x_1,\dots,x_u,n_1,\dots,n_v)$ depending on $u$ continuous variables $x_1,\dots,x_u$ and $v$ discrete variables $n_1,\dots,n_v$ on which the algebra $K(x_1,\dots,x_u,n_1,\dots,n_v)[D_{x_1},\dots,D_{x_u},S_{n_1},\dots,S_{n_v}]$ acts, we say that $f$ is **holonomic** if its generating function

$$\sum_{n_1=0}^{\infty}\cdots\sum_{n_v=0}^{\infty} f(x_1,\dots,x_u,n_1,\dots,n_v)z_1^{n_1}z_2^{n_2}\cdots z_v^{n_v}$$

is holonomic (or D-finite) with respect to the action of the algebra of differential operators $K(x_1,\dots,x_u,z_1,\dots,z_v)[D_{x_1},\dots,D_{x_u},D_{z_1},\dots,D_{z_v}]$.

This distinction between D-finite and holonomic is subtle, and it plays hardly any role in practice. Typically, an object is either D-finite and holonomic or it is neither holonomic nor D-finite. However, none of the two notions implies the other. For example, the sequence $1/(n^2+k^2)$ is D-finite but not holonomic, and the Kronecker-delta function $\delta_{n,k}$ viewed as a bivariate sequence is holonomic but not D-finite.

#### 15.4.3.2 Closure properties

In applications, a function $f$ will usually not be given in terms of a D-finite description but by some sort of expression. It is therefore important to recognize whether a given expression in fact represents a D-finite object, and if so, to construct a D-finite description for it.

This is done using the concept of closure properties. These are theorems that state that D-finiteness is preserved under various operations. By means of these theorems, it can often be seen "by inspection" that some complicated expression is D-finite, by looking at the innermost subexpressions first, which often are D-finite by definition, and then observing that the expression is composed of these subexpressions only through operations that preserve D-finiteness.

Of course, if the D-finiteness of a quantity cannot be established through closure properties, it may still be D-finite for other reasons. In fact, it can be very difficult to prove that some object is not D-finite. See [79, 74] and the references given there for some results in this direction.

The most important closure properties are the following. See [191, 173, 63, 110, 102] for proofs and further details.

1. Let $A$ be an Ore algebra acting on a function space $M$. If $f,g \in M$ are D-finite, then also $f+g$ is D-finite. If $f \in M$ is D-finite and $L \in A$, then $L \cdot f$ is D-finite. If $M$ is also a ring, and $f,g \in M$ are D-finite, then also $fg$ is D-finite.

2. Let $A = K(x)[D_x]$ act on $M = K[[x]]$, where $K$ is a field of characteristic zero. Let $f = \sum_{n=0}^{\infty} f_n x^n \in M$ be D-finite. Then also $\int f := \sum_{n=1}^{\infty} \frac{f_{n-1}}{n} x^n$ is D-finite with respect to $A$, and the coefficient sequence $(f_n)_{n=0}^{\infty}$ in $K^{\mathbb{N}}$ is D-finite with respect to $K(n)[S_n]$. Furthermore, if $g \in K[[x]]$ is algebraic with $g(0) = 0$, then also $f \circ g \in K[[x]]$ is D-finite with respect to $A$.

3. Let $A = K(n)[S_n]$ act on $M = K^{\mathbb{N}}$, where $K$ is a field of characteristic zero. Let $(f_n)_{n=0}^{\infty} \in M$ be D-finite. Then also $\sum f := (\sum_{i=0}^{n} f_i)_{n=0}^{\infty}$ is D-finite with respect to $A$, and the generating function $f = \sum_{n=0}^{\infty} f_n x^n \in K[[x]]$ is D-finite with respect to $K(x)[D_x]$. Furthermore, if $u, v \in \mathbb{N}$ are fixed, then $(f_{un+v})_{n=0}^{\infty} \in M$ is D-finite with respect to $A$.

These closure properties are constructive. This means that if D-finite descriptions $\mathfrak{a}$ and $\mathfrak{b}$ for two D-finite objects $f$ and $g$ are known, then we can compute a D-finite description of $f+g$, $fg$ and the other quantities. In the case of a single variable, if $f$ and $g$ are annihilated by operators $a$ and $b$, respectively, then $f+g$ is annihilated by the **least common left multiple** $\operatorname{lclm}(a,b)$. The corresponding operation for multiplication is called the **symmetric product:** if $c$ is the symmetric product of $a$ and $b$, then $c \cdot (fg) = 0$.

Closure properties are useful for recognizing expressions as D-finite objects, but can sometimes also be used to construct algorithmic proofs of identities among D-finite objects, as illustrated in the following example.

**Example 15.4.9 (Proving Identities)** *This example is borrowed from [99]. Consider the following identity for Hermite polynomials. We regard it as a (formal) power series with respect to t, where x and y are viewed as constant parameters. In the first term on the left, the expression $H_n(x)H_n(y)\frac{1}{n!}$ is regarded as a sequence in the discrete variable n, with x and y as parameters. Apply the closure properties algorithms as indicated by the braces in order to obtain a linear recurrence for the coefficients in the series expansion of the whole left-hand side as follows.*

$$\underbrace{\underbrace{\sum_{n=0}^{\infty} \underbrace{\underbrace{H_n(x)}_{\substack{\text{rec. of}\\ \text{ord. 2}}}\underbrace{H_n(y)}_{\substack{\text{rec. of}\\ \text{ord. 2}}}\underbrace{\frac{1}{n!}}_{\substack{\text{rec. of}\\ \text{ord. 1}}}}_{\text{rec. of order 4}} t^n}_{\text{differential equation of order 5}} - \underbrace{\underbrace{\underbrace{\frac{1}{\sqrt{1-4t^2}}}_{\substack{\text{alg.eq.}\\ \text{of deg. 2}}}}_{\substack{\text{diff.eq.}\\ \text{of ord. 1}}} \underbrace{\underbrace{\exp}_{\substack{\text{diff.eq.}\\ \text{of ord. 1}}}\Bigl(\underbrace{\frac{4t(xy - t(x^2+y^2))}{1-4t^2}}_{\text{alg.eq. of degree 1}}\Bigr)}_{\text{differential equation of order 1}}}_{\text{differential equation of order 1}}}_{\substack{\text{differential equation of order 5}\\ \rightsquigarrow\ \text{recurrence equation of order 4}}} = 0.$$

*If $c_n$ denotes the coefficient of $t^n$ in the series expansion of the entire left-hand side, we obtain that the sequence $(c_n)_{n=0}^{\infty}$ is annihilated by the recurrence operator*

$$(n+4)S_n^4 - 4xyS_n^3 - 4(2n - 2x^2 - 2y^2 + 5)S_n^2 - 16xyS_n + 16(n+1).$$

*Direct calculation confirms that $c_0 = c_1 = c_2 = c_3 = 0$, which together with the recurrence implies inductively that $c_n = 0$ for all $n \geq 0$. This proves the identity.*

### 15.4.3.3 Summation and integration

The concept of creative telescoping extends to holonomic functions. It is used for obtaining a D-finite description for a sum or an integral given a D-finite description for the summand or integrand. In the summation case, suppose we know a D-finite description $\mathfrak{a} \subseteq C(k, x_1, \ldots, x_v)[S_k, \partial_1, \ldots, \partial_v]$ for the summand $f(k, x_1, \ldots, x_v)$, where $k$ is the summation variable and $x_1, \ldots, x_v$ are other variables, on which the $\partial_i$ act in some way. Then a nonzero operator $P \in C(x_1, \ldots, x_v)[\partial_1, \ldots, \partial_v]$ free of $k$ and $S_k$ is called a **telescoper** for $f$ if there exists another operator $Q$, called a **certificate** for $P$, such that $P - (S_k - 1)Q \in \mathfrak{a}$.

From an ideal $\mathfrak{p} \subseteq C(x_1, \ldots, x_v)[\partial_1, \ldots, \partial_v]$ of telescopers with the property $\mathfrak{p} \cap C(x_1, \ldots, x_v)[\partial_i] \neq \{0\}$ we can obtain a D-finite description for the sum $\sum_{k=a}^{b} f(k, x_1, \ldots, x_v)$. In the simplest and most common case, when the sum has "natural boundaries," $\mathfrak{p}$ itself is already a description of the sum. To have natural boundaries means that $[Q \cdot f]_{k=a}^{b+1} = 0$ for all operators $Q$. This is the case, for example, when the variables $x_1, \ldots, x_v$ are discrete and for every specific point $(x_1, \ldots, x_v) \in \mathbb{Z}^v$ there are only finitely many values $k$ such that $f(k, x_1, \ldots, x_v) \neq 0$. In this situation, we also say that "$f$ has finite support."

When $f$ does not have natural boundaries, creative telescoping still gives rise to inhomogeneous functional equations of the form $P \cdot F = u$, where $F$ is the sum over $f$ and $u = [Q \cdot f]_{k=a}^{b+1}$. In this case, if we know telescopers $P_1, \ldots, P_m$ with corresponding certificates $Q_1, \ldots, Q_m$, and if we know (or can compute) for each $i$ an ideal $\mathfrak{q}_i$ of annihilating operators for the functions $[Q_i \cdot f]_{k=a}^{b+1}$, then the product ideal $(\mathfrak{q}_1 \cap \cdots \cap \mathfrak{q}_m)\langle P_1, \ldots, P_m \rangle$ is an ideal of annihilating operators for the sum $F$. It will be D-finite provided that $\langle P_1, \ldots, P_m \rangle$ and all the $\mathfrak{q}_i$ are D-finite.

Several algorithms are available for finding telescopers for holonomic functions. Zeilberger's "slow" algorithm [191, 63] uses a brute force elimination approach, Takayama's algorithm [175, 176, 63] reduces the problem to a Gröbner basis computation in an Ore algebra, Chyzak's algorithm [61] is a generalization of Zeilberger's algorithm for hypergeometric summation (Section 15.4.1), and the algorithm by Chen, Kauers and Koutschan [57, 111] is a generalization of an idea proposed by Apagodu and Zeilberger [136] for the hypergeometric case. For holonomic functions, it is guaranteed that these algorithms will find enough telescopers to generate a D-finite ideal.

**Example 15.4.10 (Apéry's second sum once more)** *Consider once more the sum*

$$F(n) = \sum_{k=0}^{n} \binom{n}{k}^2 \binom{n+k}{k}^2 \left( \sum_{i=1}^{n} \frac{1}{i^3} + \sum_{i=1}^{k} \frac{(-1)^{i+1}}{2i^3 \binom{n}{i}\binom{n+i}{i}} \right).$$

*In order to find a recurrence for this sum, process the sum from inside to outside.*

*The inner summation $F_1(n) = \sum_{i=1}^{n} \frac{1}{i^3}$ is univariate and can therefore be handled by the closure properties algorithms discussed in the previous section. Using that $(i+1)^3 S_i - i^3$ annihilates the summand, the result is that*

$$A = (n+2)^3 S_n^2 - (2n+3)(n^2+3n+3)S_n + (n+1)^3$$

*annihilates the sum $F_1(n)$. If we regard this sum as a bivariate object depending on $n$ and $k$, a D-finite description is given by the ideal $\mathfrak{a}_1 = \langle A, S_k - 1\rangle \subseteq \mathbb{Q}(n,k)[S_n,S_k]$.*

*The inner summation $F_2(n,k) = \sum_{i=1}^{k} \frac{(-1)^{i+1}}{2i^3\binom{n}{i}\binom{n+i}{i}}$ has a hypergeometric summand with annihilator $\langle (i+1)(i-n)(i+n+1)S_i - i^3,\ (i+n+1)S_n + (i-n-1)\rangle \subseteq \mathbb{Q}(n,i)[S_n,S_i]$. Given this information, a summation algorithm can find a D-finite description $\mathfrak{a}_2 \subseteq \mathbb{Q}(n,k)[S_n,S_k]$ of the form*

$$\begin{aligned}\mathfrak{a}_2 = \big\langle &(k+2)(n-k-1)(n+k+2)S_k^2 + (\ldots)S_k + (\ldots),\\ &(n+k+2)S_nS_k + (\ldots)S_n + (\ldots)S_k + (\ldots),\\ &(n+2)^3(n+k+2)S_n^2 + (\ldots)S_n + (\ldots)S_k + (\ldots)\big\rangle\end{aligned}$$

*for the sum $F_2(n,k)$. Now closure properties algorithms can turn $\mathfrak{a}_1$, $\mathfrak{a}_2$ and a description for $\binom{n}{k}^2\binom{n+k}{k}^2$ into a D-finite description $\mathfrak{a}$ for the summand of $F(n)$, and from this description $\mathfrak{a}$, a summation algorithm finds a messy fourth order recurrence for the sum $F(n)$.*

*Why does it not find the nice second order recurrence that we found in Example 15.4.7? The reason is that we lost a relation when we combined the D-finite descriptions $\mathfrak{a}_1$ and $\mathfrak{a}_2$ to a D-finite description for $F_1(n) + F_2(n,k)$. The usual algorithm for doing this yields in fact a D-finite description that is valid for every linear combination $\alpha F_1(n) + \beta F_2(n,k)$ ($\alpha, \beta \in \mathbb{Q}$), but it so happens that in the present example there is an additional annihilating operator for $\alpha = \beta = 1$. It has the form $(\ldots)S_k + (\ldots)S_n + (\ldots)$. If we apply creative telescoping to the ideal obtained by adding this operator to $\mathfrak{a}$, the result is the recurrence from Example 15.4.7, and the computation also goes much faster. See Example 15.5.3 for how such extra operators can be found.*

#### 15.4.3.4 Nested sums and integrals

Since the sum/integral over a holonomic function with respect to a discrete/continuous variable is a holonomic function in the remaining variables, it is clear that nested sums or integrals, and even mixed expressions involving summations as well as integrations, can be handled by applying creative telescoping, possibly in combination with algorithms for closure properties, repeatedly.

An alternative to repeated creative telescoping is multivariate creative telescoping, which is defined as follows. For notational simplicity, we formulate the definition only for the differential case. Let $A := C(x_1,\ldots,x_u,y_1,\ldots,y_v)[D_{x_1},\ldots,D_{x_u},\partial_1,\ldots,\partial_v]$ and let $\mathfrak{a} \subseteq A$ be a D-finite ideal. A telescoper for $\mathfrak{a}$ with respect to the variables $x_1,\ldots,x_u$ is an operator $P \in C(y_1,\ldots,y_v)[\partial_1,\ldots,\partial_v] \setminus \{0\}$ such that there exist

$Q_1, \ldots, Q_u \in A$ with

$$P - D_{x_1} Q_1 - D_{x_2} Q_2 - \cdots - D_{x_u} Q_u \in \mathfrak{a}.$$

Applications of this version of creative telescoping include the following.

- **Integrals.** If the integral $F = \int_\Omega f d(x_1, \ldots, x_u)$ has natural boundaries, then every telescoper for the integrand $f$ is an annihilating operator for the integral $F$.

- **Residues.** If $f = \sum_{i_1,\ldots,i_u} a_{i_1,\ldots,i_u}(y_1, \ldots, y_v) x_1^{i_1} \cdots x_u^{i_u}$ is a multivariate Laurent series in the sense of Section 15.2.3, define the residue of $f$ with respect to $x_1, \ldots, x_u$ as $F = [x_1^{-1} \ldots x_u^{-1}] f = a_{-1,\ldots,-1}(y_1, \ldots, y_v)$. Then every telescoper for $f$ with respect to $x_1, \ldots, x_u$ is an annihilating operator for $F$.

- **Constant Terms.** From the previous remark it follows immediately that a telescoper for $x_1^{-1} \cdots x_u^{-1} f$ is an annihilating operator for the constant term $F = [x_1^0 \ldots x_u^0] f$. More generally, when $(m_1, \ldots, m_u) \in \mathbb{Z}^u$ is fixed then every telescoper for $x_1^{-(m_1+1)} \cdots x_u^{-(m_u+1)} f$ with respect to $x_1, \ldots, x_u$ is an annihilating operator for $F = [x_1^{m_1} \ldots x_u^{m_u}] f = a_{m_1,\ldots,m_u}(y_1, \ldots, y_v)$.

- **Diagonals.** [121, 35] The diagonal of a multivariate power series $f(x_1, \ldots, x_u) = \sum_{i_1=0}^\infty \cdots \sum_{i_u=0}^\infty a_{i_1,\ldots,i_u} x_1^{i_1} \cdots x_u^{i_u}$ is defined as the univariate power series $d(y) := \sum_{n=0}^\infty a_{n,n,\ldots,n} y^n$. Every telescoper for the transformed series $\frac{1}{x_1 \cdots x_{u-1}} f(x_1, \ldots, x_{u-1}, \frac{y}{x_1 \ldots x_{u-1}})$ with respect to $x_1, \ldots, x_{u-1}$ is an annihilating operator for $d(y)$.

- **Positive Parts.** [33] If $f(x_1, \ldots, x_u)$ is a multivariate formal Laurent series in the sense of Section 15.2.3 whose support is contained in a half space $H \subseteq \mathbb{Z}^u$ that contains the principal orthant $\mathbb{N}^u$, then every telescoper for $\frac{1}{(1-x_1 y_1) \ldots (1-x_u y_u)} f(\frac{1}{y_1}, \ldots, \frac{1}{y_u})$ with respect to $y_1, \ldots, y_u$ is an annihilating operator for the positive part $\frac{1}{x_1 \cdots x_u} [x_1^> \ldots x_u^>] f$ of $f$.

### 15.4.4 Implementations and current research topics

The classical algorithms described in Section 15.4.1 were implemented in the 1990s and are now included in the standard libraries of many general purpose computer algebra systems. Hypergeometric single sums arising in combinatorics are usually no challenge for these implementations.

The advanced summation technology described in Sections 15.4.2 and 15.4.3 is not yet as widely available. The only known implementation of ΠΣ-theory is Schneider's Mathematica package Sigma [159]. For holonomic functions, there is Chyzak's package mgfun [60] for Maple (part of the Algolib library [8]), and a Mathematica package by Koutschan [110, 112]. For holonomic functions in a single variable, see [154, 129, 100].

There are some even more advanced summation algorithms. For example, there are extensions of the ΠΣ-theory towards sums involving special functions [157], unspecified summands [104, 103], or radical expressions [105], and the holonomic systems approach has been generalized to Abel type sums [128], sums involving Bernoulli numbers [59], Stirling numbers [95] and other non-holonomic quantities [62].

Most of the research work during the first decade of the 21st century was on extending the applicability of the classical summation algorithms to larger and larger classes. In recent years, the focus is moving more towards complexity analysis and the construction of more efficient algorithms. First results in this direction are various bounds on the sizes of telescopers [31, 55, 56] as well as alternative creative telescoping algorithms that construct telescopers without constructing corresponding certificates [38, 32, 58]. Since certificates tend to be much larger than the corresponding telescopers, algorithms that avoid their calculation have a good chance to be much more efficient. Note that for many summation and integration problems the certificate is not needed.

## 15.5 The guess-and-prove paradigm

The algorithms discussed in this chapter are useful for delegating boring and tedious tasks that often arise in the context of research in combinatorics to a machine. Even if a sum is simple enough that we may have a chance to simplify it by hand using traditional paper and pencil calculations, it may be a good idea to let the computer do it, and to devote our own precious time and energy to more important matters. Many applications of computer algebra are of this sort.

With the algorithms discussed in this chapter it is sometimes also possible to solve cutting edge research problems in combinatorics that are not accessible by any other means. It is however naive to expect that such problems can simply be solved by entering a command into a computer algebra system and waiting a minute or two for the solution. More realistically, in order to obtain nontrivial results using computer algebra, it is necessary to design an individual argument that reduces the problem at hand to one or more computational problems that can then be solved by the computer, possibly using enormous time and memory resources.

It is thus a common misconception that computer algebra would eventually make human creativity obsolete. Instead, the creativity that would otherwise be used to come up with a proof in the traditional way will now be needed to come up with a logical reduction of the problem under consideration to subproblems to which computer algebra techniques are applicable. Obviously, there is no systematic way of breaking a research problem into computable pieces, but a general guideline that often leads to success is to proceed along the following three steps: (1) automated generation of data, (2) automated extraction of conjectures from the data ("guessing"), (3) automated proof of the guessed conjectures.

Of course, a guessed conjecture may be false (although this virtually never happens in practice), and in this case, step (3) cannot succeed. When this happens, try again with some more data. Increasing the amount of data supplied to the guessing method will make it harder for false conjectures to show up in the output, so that eventually only true (and provable) conjectures will remain.

We illustrate the guess-and-prove paradigm on a collection of examples, starting with two artificial toy examples.

**Example 15.5.1 (Solving Transcendental Equations)** *Suppose we want to find a real number solution, other than $x = 1$, of the transcendental equation*

$$28(x-1)\log(3x-1) + (7x+2)\log(2x^2-2x+1) = 0.$$

*No algorithms are known for solving such equations directly. However, we can proceed as follows.*

1. *(generate data) Using a numerical algorithm, we can find that there is a solution $x$ with $x \approx 0.5714285714$.*
2. *(guess) Using lattice reduction (Section 15.2.1) or table lookup [169], we can find that this number is close to $x = 4/7$.*
3. *(prove) Simplifying the expression obtained by plugging the guessed solution $x = 4/7$ into the equation, we can check that it is indeed a solution.*

**Example 15.5.2 (Nonlinear Recurrences)** *Let $f(n)$ be defined recursively by $f(1) = 0$, $f(2) = 21$, $f(3) = 136$, and*

$$(4n^2-3)f(n+1)f(n-1) = (4n^2-19)f(n)^2 + 108n^4 - 106n^2 + 19.$$

*Suppose we want to know whether $f(n)$ is a polynomial sequence. No algorithms are known for deciding this directly. However, we can proceed as follows.*

1. *(generate data) Use the recurrence to compute $f(n)$ for $n = 1, \ldots, 20$, say.*
2. *(guess) Compute the interpolating polynomial of this data. It is $2x^4 - 3x^2 + 1$. The fact that the degree is significantly smaller than the number of interpolation points is already a good indication that this polynomial is a good guess.*
3. *(prove) For getting a rigorous proof, plug the polynomial into the equation and check that both sides are identical after simplification. This proves that we have indeed $f(n) = 2n^4 - 3n^2 + 1$ for all $n \in \mathbb{N}$.*

More generally, instead of finding a polynomial expression by interpolating data, we can also compute recurrence equations that match a given finite array of data. This technique enjoys a great popularity in experimental mathematics and it is often used as a synonym of "guessing," although it should better be only viewed as a particular technique for solving a particular type of guessing problems. The idea is easily explained. Suppose we know the first $N+1$ terms $c_0, \ldots, c_N$ of an infinite sequence $(c_n)_{n=0}^\infty$ and we are interested in operators $L \in K[n][S_n]$ that possibly annihilate $(c_n)_{n=0}^\infty$. Since most terms of the sequence are unknown, there is no chance to ensure that a particular operator does annihilate the whole sequence, but we can use as a necessary condition that any operator that is valid for all terms must in particular be valid for the finitely many terms we know explicitly. We can therefore **make an ansatz** with undetermined coefficients $\ell_{i,j}$ for an operator

$$\begin{aligned} L = {} & (\ell_{0,0} + \ell_{0,1}n + \ell_{0,2}n^2 + \cdots + \ell_{0,d}n^d) \\ & + (\ell_{1,0} + \ell_{1,1}n + \ell_{1,2}n^2 + \cdots + \ell_{1,d}n^d)S_n \\ & + \vdots \\ & + (\ell_{r,0} + \ell_{r,1}n + \ell_{r,2}n^2 + \cdots + \ell_{r,d}n^d)S_n^r \end{aligned}$$

of order $r$ and degree $d$, for some suitably chosen $r, d \in \mathbb{N}$, and require that the first $N-r$ terms of the sequence $L \cdot (c_n)_{n=0}^\infty$ be zero. This gives a system of $N-r$ homogeneous linear equations for the $(r+1)(d+1)$ many unknown coefficients $\ell_{i,j}$, which we can solve. Typically, when $N-r > (r+1)(d+1)$, such a system won't have any nontrivial solution, and on the other hand, every true operator must belong to the solution space. Therefore, if there is no solution, this proves that there is no operator of order $\leq r$ and degree $\leq d$ that annihilates $(c_n)_{n=0}^\infty$, and if there is a solution, then it is reasonable to "guess" that the corresponding operator will annihilate the entire sequence. The confidence of the guess can be increased by increasing the difference $N-r-(r+1)(d+1)$, but without using further information about the problem from which the sequence originates no finite amount of data will ever provide a rigorous proof that the operator is correct.

The idea to make an ansatz for an equation and to match it against known data is very versatile. It can also be used to search for linear or nonlinear algebraic or differential equations satisfied by a formal power series of which the first terms are known, and even multivariate recurrence equations satisfied by sequences depending on several indices or multivariate functional equations satisfied by power series in several variables. For the most common cases of guessing univariate recurrence or differential operators efficient algorithms are known whose performance is so much better than the naive ansatz sketched above that in applications it is not unusual that the computation time needed for guessing is negligible compared to the time needed for generating data. For details on such guessing algorithms and their implementations for various computer algebra systems, see [24, 154, 115, 97, 37, 83, 99].

We conclude this chapter with three non-trivial applications of the guess-and-prove paradigm.

**Example 15.5.3 (Annihilator Completion)** *In Example 15.4.10 we showed that the bivariate sequence*

$$f(n,k) = \sum_{i=1}^{n} \frac{1}{i^3} + \sum_{i=1}^{k} \frac{(-1)^{i+1}}{2i^3\binom{n}{i}\binom{n+i}{i}}$$

*has an annihilating ideal of the form*

$$\mathfrak{a} = \langle (\ldots)S_k^2 + (\ldots)S_k + (\ldots),$$

$$(\ldots)S_nS_k + (\ldots)S_n + (\ldots)S_k + (\ldots),$$
$$(\ldots)S_n^2 + (\ldots)S_n + (\ldots)S_k + (\ldots)\rangle \subseteq \mathbb{Q}(n,k)[S_n, S_k].$$

*This is, however, not the complete annihilator. There are annihilating operators of $f(n,k)$ that do not belong to $\mathfrak{a}$.*

*To find such an operator, use the known operators (or the original sum representation) to compute the terms $f(n,k)$ for, say, $n,k = 1,\ldots,20$, and use this data to guess an operator involving only terms $S_n^i S_k^j$ that are not divisible by the leading terms of the generators of $\mathfrak{a}$. This gives the conjectured operator*

$$L = (n+1)^2 S_n + 2(k-n)(k+1)S_k - (n^2 - 2kn + 2k^2 + 2k + 1).$$

*By construction, we then have $(L \cdot f)(n,k) = 0$ for $n,k = 0,\ldots,19$.*

*To prove that $L$ really annihilates $f$, use D-finite closure properties (Section 15.4.3) to compute an annihilating ideal $\mathfrak{b}$ for $g := L \cdot f$. We find $\mathfrak{b} = \langle (n+2)S_n - (n+1), S_k - 1\rangle$. By checking that $g(1,1) = 0$, the generators of $\mathfrak{b}$ imply that $g(n,k) = 0$ for all $n,k$, thus proving that $L$ annihilates $f$.*

**Example 15.5.4 (Restricted Lattice Walks II)** *Let $f(n,i,j)$ be the number of Kreweras walks with endpoint $(i,j)$, introduced in Example 15.3.2 above. Kreweras [117] showed that the generating function $F(t,x,y) = \sum_{n,i,j=0}^{\infty} f(n,i,j)x^i y^j t^n$ is algebraic. This result can also be obtained using the guess-and-proof paradigm, after making some preparatory observations.*

*From the combinatorial definition we have the recurrence equation*

$$f(n+1,i,j) = f(n,i+1,j+1) + f(n,i-1,j) + f(n,i,j-1)$$

*for $n \geq 0$, the boundary conditions $f(0,0,0) = 1$ and $f(n,i,-1) = f(n,-1,j) = 0$ for all $n,i,j$, and the symmetry $f(n,i,j) = f(n,j,i)$. For the generating function $F(t,x,y)$, these facts imply the functional equation*

$$\Big((x+y+x^2y^2) - \frac{xy}{t}\Big)F(t,x,y) = xF(t,x,0) + yF(t,y,0) - \frac{xy}{t}.$$

*If $Y(x,t) \in \mathbb{Q}(x)[[t]]$ is a root of the polynomial in front of $F(t,x,y)$, then this equation implies that the series $U = F(t,x,0)$ is a solution of the functional equation*

$$U(t,x) = \frac{1}{t}Y(t,x) - \frac{1}{x}Y(t,x)U(t,Y(t,x)).$$

*It is not difficult to see that $F(t,x,0)$ is in fact the only power series solution of this equation. Therefore, if we can show that the equation has an algebraic power series solution $U$, then this solution must be identical to $F(t,x,0)$, and thus $F(t,x,0)$ must be algebraic. It then follows from closure properties for algebraic functions that also $F(t,x,y)$ is algebraic.*

*To see that the functional equation for $U$ has an algebraic power series solution, first compute a number of terms of $F(t,x,0)$ using the recurrence for $f(n,i,j)$ stated above and the fact $[t^n]F(t,x,0) = \sum_{i=0}^{n} f(n,i,0)x^i$. Use this data to guess an algebraic* *equation possibly satisfied by $F(t,x,0)$. Check that this algebraic equation does have indeed a power series solution, and prove, using closure properties for algebraic functions, that this power series solution satisfies the functional equation.*

*A detailed description of all the steps of this computation can be found in [39]. A different method, also following the guess-and-proof paradigm, is explained in [106, 101]. This latter method can be used to derive and prove formulas for the counting sequence $f(n,0,0)$ of walks returning to the origin.*

**Example 15.5.5 (Determinant Identities)** *Zeilberger [194] points out that an approach by Mills et al. [135] for proving certain types of determinant identities can be executed in a guess-and-prove fashion. The q-TSPP conjecture [113] was proven in this way, as well as some other conjectured determinant identities [114].*

*The approach is applicable to identities for $n \times n$ determinants, where $n$ is a variable, and the $(i,j)$-th entry of the $n$-th matrix is given by a bivariate holonomic sequence $(a_{i,j})_{i,j=1}^{\infty}$, so that for $n = 1,2,3,4,\ldots$ we get the determinants*

$$a_{1,1},\quad \begin{vmatrix} a_{1,1} & a_{1,2} \\ a_{2,1} & a_{2,2} \end{vmatrix},\quad \begin{vmatrix} a_{1,1} & a_{1,2} & a_{1,3} \\ a_{2,1} & a_{2,2} & a_{2,3} \\ a_{3,1} & a_{3,2} & a_{3,3} \end{vmatrix},\quad \begin{vmatrix} a_{1,1} & a_{1,2} & a_{1,3} & a_{1,4} \\ a_{2,1} & a_{2,2} & a_{2,3} & a_{2,4} \\ a_{3,1} & a_{3,2} & a_{3,3} & a_{3,4} \\ a_{4,1} & a_{4,2} & a_{4,3} & a_{4,4} \end{vmatrix},\quad \ldots.$$

*Let it be conjectured that the $n$-th determinant in this sequence admits a closed form representation of the form $\prod_{k=1}^{n} b_k$, where $b_k$ is some holonomic sequence with $b_k \neq 0$ for all $k$.*

*Write $A_n$ for the nth determinant in the sequence, and let $A_n^{(j)}$ be the cofactors appearing in the Laplace expansion along the nth row, so that $A_n = \sum_{j=1}^{n} a_{n,j} A_n^{(j)}$ for all $n$. It is then easy to see that for the sequence $c_{n,j} := A_n^{(j)}/A_n$ we have the two identities*

$$\sum_{j=1}^{n} a_{i,j} c_{n,j} = 0 \quad (i = 1,\ldots,n-1), \qquad \text{and} \qquad c_{n,n} = 1.$$

*It is also easy to see that the sequence $c_{n,j}$ is uniquely determined by these two identities.*

*Now compute the values for $c_{n,j}$ for some $1 \leq n, j \leq N$, for some sufficiently large $N$, and use this data to guess a holonomic description of $c_{n,j}$. In theory, there is no reason why the sequence $(c_{n,j})_{n,j=1}^{\infty}$ should be holonomic, and if it isn't, the approach fails. But if a (guessed) holonomic description is discovered, then it can be decided using a summation algorithm (Section 15.4.3) whether every solution $\tilde{c}_{n,j}$ of the guessed holonomic system satisfies the two identities stated above, which, by uniqueness, implies that $c_{n,j} = \tilde{c}_{n,j}$, and hence that $c_{n,j}$ is holonomic. Finally, to conclude the proof, observe that the conjectured determinant identity is equivalent to $b_n = \sum_{j=1}^{n} a_{n,j} c_{n,j}$, which, using the guessed-and-proved holonomic description of $c_{n,j}$, can be proven using again a summation algorithm.*

**Acknowledgments.** I am grateful to Fredrik Johansson, Christoph Koutschan, Christian Krattenthaler, and Rika Yatchak for valuable remarks on an earlier version of this chapter.

# References

[1] S. A. Abramov and K.Yu. Kvashenko. Fast algorithms to search for the rational solutions of linear differential equations with polynomial coefficients. In *Proceedings of ISSAC'91*, pages 267–270, 1991.

[2] Sergei A. Abramov. Rational solutions of linear difference and $q$-difference equations with polynomial coefficients. In *Proceedings of ISSAC'95*, pages 285–289, July 1995.

[3] Sergei A. Abramov. When does Zeilberger's algorithm succeed? *Advances in Applied Mathematics*, 30(3):424–441, 2003.

[4] Sergei A. Abramov, Manuel Bronstein, Marko Petkovšek, and Carsten Schneider. In preparation.

[5] Sergei A. Abramov, Ha Q. Le, and Ziming Li. Univariate Ore polynomial rings in computer algebra. *Journal of the Mathematical Sciences*, 131:5885–5903, 2005.

[6] Sergei A. Abramov and Marko Petkovšek. D'Alembertian solutions of linear differential and difference equations. In *Proceedings of ISSAC'94*, pages 169–174, 1994.

[7] William Adams and Philippe Loustaunau. *An Introduction to Gröbner Bases*, volume 3 of *Graduate Studies in Mathematics*. American Mathematical Society, 1994.

[8] The Algo team. Algolib: The algorithms project's library and other packages of the algorithms project. `http://algo.inria.fr/libraries/`, 1990–2013.

[9] George E. Andrews and Peter Paule. MacMahon's dream. In *Partitions, q-series, and Modular Forms*, volume 23 of *Developments in Mathematics*, pages 1–12, 2012.

[10] George E. Andrews, Peter Paule, and Axel Riese. MacMahon's partition analysis III: The Omega package. *Electronic Journal of Combinatorics*, 22:887–904, 2001.

[11] George E. Andrews, Peter Paule, and Axel Riese. MacMahon's partition analysis VI: A new reduction algorithm. *Annals of Combinatorics*, 5:251–270, 2001.

[12] George E. Andrews, Peter Paule, Axel Riese, and Volker Strehl. MacMahon's partition analysis V: Bijections, recursions, and magic squares. In *Algebraic Combinatorics and Applications*, pages 1–39. Springer, 2001.

[13] Ainhoa Aparicio Monforte and Manuel Kauers. Formal Laurent series in several variables. *Expositiones Mathematicae*, 31(4):350–367, 2013.

[14] Franz Baader and Tobias Nipkow. *Term Rewriting and All That*. Cambridge University Press, 1999.

[15] David H. Bailey, Jonathan M. Borwein, Neil J. Calkin, Roland Girgensohn, D. Russel Luke, and Victor Moll. *Experimental Mathematics in Action*. A K Peters, 2007.

[16] Alexander I. Barvinok. A polynomial time algorithm for counting integral points in polyhedra when the dimension is fixed. *Mathematics of Operations Research*, 19(4):769–779, 1994.

[17] Alexander I. Barvinok. *Integer Points in Polyhedra*. European Mathematical Society, 2008.

[18] Saugata Basu, Richard Pollack, and Marie-Françoise Roy. *Algorithms in Real Algebraic Geometry*, volume 10 of *Algorithms and Computation in Mathematics*. Springer, 2nd edition, 2006.

[19] Matthias Beck and Dennis Pixton. The Ehrhart polynomial of the Birkhoff polytope. *Discrete and Computational Geometry*, 30:623–637, 2003.

[20] Matthias Beck and Robins Sinai. *Computing the Continuous Discretely, Integer-point enumeration in polyhedra*. Springer, 2007.

[21] Thomas Becker, Volker Weispfenning, and Heinz Kredel. *Gröbner Bases*. Springer, 1993.

[22] F. Bergeron and G. Cartier. Darwin: Computer algebra and enumerative combinatorics. In *Proceedings of STACS'88*, volume 294 of *Lecture Notes in Computer Science*, pages 393–394, 1988.

[23] Francois Bergeron, Gilbert Labelle, and Pierre Leroux. *Combinatorial Species and Tree-like Structures*. Cambridge University Press, 1998.

[24] Francois Bergeron and Simon Plouffe. Computing the generating function of a series given its first terms. *Experimental Mathematics*, 1(4):307–312, 1992.

[25] Lennart Berggren, Jonathan Borwein, and Peter Borwein. *Pi: A Source Book*. Springer, 2003.

[26] Anders Björner and Francesco Brenti. *Combinatorics of Coxeter Groups*. Springer, 2005.

[27] William Boone. Certain simple unsolvable problems of group theory iv. *Nederl. Akad. Wet., Proc., Ser. A*, 60:227–232, 1957.

[28] Jonathan Borwein and David Bailey. *Mathematics by Experiment: Plausible Reasoning in the 21st Century*. A K Peters, 2003.

[29] Jonathan M. Borwein and Peter B. Borwein. *Pi and the AGM: A Study in Analytic Number Theory and Computational Complexity*. Canadian Mathematical Society, 1998.

[30] Jonathan M. Borwein and Petr Lisoněk. Applications of integer relation algorithms. *Discrete Mathematics*, 23:65–82, 2000.

[31] Alin Bostan, Shaoshi Chen, Frédéric Chyzak, and Ziming Li. Complexity of creative telescoping for bivariate rational functions. In *Proceedings of ISSAC'10*, pages 203–210, 2010.

[32] Alin Bostan, Shaoshi Chen, Frédéric Chyzak, Ziming Li, and Guoce Xin. Hermite reduction and creative telescoping for hyperexponential functions. In *Proceedings of ISSAC'13*, pages 77–84, 2013.

[33] Alin Bostan, Frédéric Chyzak, Manuel Kauers, Lucien Pech, and Mark van Hoeij. Explicit differentiably finite generating functions of walks with small steps in the quarter plane. In preparation.

[34] Alin Bostan, Frédéric Chyzak, Bruno Salvy, and Ziming Li. Fast computation of common left multiples of linear ordinary differential operators. In *Proceedings of ISSAC'12*, pages 99–106, 2012.

[35] Alin Bostan, Frédéric Chyzak, Mark van Hoeij, and Lucien Pech. Explicit formula for the generating series of diagonal 3D rook paths. *Séminaire Lotharingien de Combinatoire*, 66:1–27, 2011.

[36] Alin Bostan, Philippe Flajolet, Bruno Salvy, and Eric Schost. Fast computation of special resultants. *Journal of Symbolic Computation*, 41(1):1–29, 2006.

[37] Alin Bostan and Manuel Kauers. Automatic classification of restricted lattice walks. In *Proceedings of FPSAC'09*, pages 201–215, 2009.

[38] Alin Bostan, Pierre Lairez, and Bruno Salvy. Creative telescoping for rational functions using the Griffiths-Dwork method. In *Proceedings of ISSAC'13*, pages 93–100, 2013.

[39] Alin Bostan and Manuel Kauers with an appendix by Mark van Hoeij. The complete generating function for Gessel walks is algebraic. *Proceedings of the AMS*, 138(9):3063–3078, 2010.

[40] Mireille Bousquet-Mélou. Walks in the quarter plane: Kreweras' algebraic model. *The Annals of Applied Probability*, 15(2):1451–1491, 2005.

[41] Mireille Bousquet-Mélou and Marni Mishna. Walks with small steps in the quarter plane. *Contemporary Mathematics*, 520:1–40, 2010.

[42] Richard Brent and Paul Zimmermann. *Modern Computer Arithmetic*. Cambridge University Press, 2011.

[43] R.P. Brent and H.T. Kung. Fast algorithms for manipulating formal power series. *Journal of the ACM*, 4:581–595, 1978.

[44] Felix Breuer. An invitation to Ehrhart theory: Polyhedral geometry and its applications in enumerative combinatorics. In *Computer Algebra and Polynomials*, Lecture Notes in Computer Science. Springer, 2014. to appear.

[45] Manuel Bronstein. On solutions of linear ordinary difference equations in their coefficient field. *Journal of Symbolic Computation*, 29:841–877, 2000.

[46] Manuel Bronstein. *Symbolic Integration I*, volume 1 of *Algorithms and Computation in Mathematics*. Springer, 2nd edition, 2005.

[47] Manuel Bronstein and Sébastien Lafaille. Solutions of linear ordinary differential equations in terms of special functions. In *Proceedings of ISSAC'02*, pages 23–28, 2002.

[48] Manuel Bronstein and Marko Petkovšek. An introduction to pseudo-linear algebra. *Theoretical Computer Science*, 157(1):3–33, 1996.

[49] Bruno Buchberger. *Ein Algorithmus zum Auffinden der Basiselemente des Restklassenrings nach einem nulldimensionalen Polynomideal*. PhD thesis, Universität Innsbruck, 1965.

[50] Bruno Buchberger and Manuel Kauers. Gröbner basis. *Scholarpedia*, 5(10):7763, 2010.

[51] Bruno Buchberger and Manuel Kauers. Buchberger's algorithm. *Scholarpedia*, 6(10):7764, 2011.

[52] Reinhold Burger, George Labahn, and Mark van Hoeij. Closed form solutions of linear ODEs having elliptic function coefficients. In *Proceedings of ISSAC'04*, pages 58–64, 2004.

[53] Bob F. Caviness and Jeremy R. Johnson, editors. *Quantifier Elimination and Cylindrical Algebraic Decomposition*, Texts and Monographs in Symbolic Computation. Springer, 1998.

[54] Yongjae Cha, Mark van Hoeij, and Giles Levy. Solving recurrence relations using local invariants. In *Proceedings of ISSAC'10*, pages 303–310, 2010.

[55] Shaoshi Chen and Manuel Kauers. Order-degree curves for hypergeometric creative telescoping. In *Proceedings of ISSAC'12*, pages 122–129, 2012.

[56] Shaoshi Chen and Manuel Kauers. Trading order for degree in creative telescoping. *Journal of Symbolic Computation*, 47(8):968–995, 2012.

[57] Shaoshi Chen, Manuel Kauers, and Christoph Koutschan. A generalized Apagodu-Zeilberger algorithm. In K. Nabeshima, editor, *Proceedings of ISSAC'14*, pages 107–114, 2014.

[58] Shaoshi Chen and Michael F. Singer. Residues and telescopers for bivariate rational functions. *Advances in Applied Mathematics*, 49(2):111–133, 2012.

[59] William Y. C. Chen and Lisa H. Sun. Extended Zeilberger's algorithm for identities on Bernoulli and Euler polynomials. arXiv:0810.0438.v1, 2008.

[60] Frédéric Chyzak. *Fonctions holonomes en calcul formel.* PhD thesis, INRIA Rocquencourt, 1998.

[61] Frédéric Chyzak. An extension of Zeilberger's fast algorithm to general holonomic functions. *Discrete Mathematics*, 217:115–134, 2000.

[62] Frédéric Chyzak, Manuel Kauers, and Bruno Salvy. A non-holonomic systems approach to special function identities. In John May, editor, *Proceedings of ISSAC'09*, pages 111–118, 2009.

[63] Frédéric Chyzak and Bruno Salvy. Non-commutative elimination in Ore algebras proves multivariate identities. *Journal of Symbolic Computation*, 26:187–227, 1998.

[64] Henri Cohen. *A Course in Computational Algebraic Number Theory.* Springer, 1993.

[65] George E. Collins. Quantifier elimination for the elementary theory of real closed fields by cylindrical algebraic decomposition. *Lecture Notes in Computer Science*, 33:134–183, 1975.

[66] David Cox, John Little, and Donal O'Shea. *Ideals, Varieties, and Algorithms.* Springer, 1992.

[67] H.S.M. Coxeter. Discrete groups generated by reflections. *Annals of Mathematics*, 35(3):588–621, 1934.

[68] James H. Davenport and Michael F. Singer. Elementary and Liouvillian solutions of linear differential equations. *Journal of Symbolic Computation*, 2(3):237–260, 1986.

[69] H. Decoste, G. Labelle, and P. Leroux. Une approche combinatoire pour l'itération de Newton-Raphson. *Advances in Applied Mathematics*, 3:407–416, 1982.

[70] Alain Denise, Isabelle Dutour, and Paul Zimmermann. CS: a MuPad package for counting and randomly generating combinatorial structures. In *Proceedings of FPSAC'98*, pages 195–204, 1998.

[71] Eugène Ehrhart. Sur les polyèdres rationelles homothétiques à $n$ dimensions. *C. R. Acad. Sci. Paris*, 254:616–618, 1962.

[72] H. Ferguson and D. Bailey. A polynomial time, numerically stable integer relation algorithm. Technical Report RNR-91-032, SRC-TR-92-066, NASA Ames Research Center, 1992.

[73] H. Ferguson, D. Bailey, and S. Arno. Analysis of PSLQ, an integer relation algorithm. *Mathematics of Computation*, 68(225):351–369, 1999.

[74] Philippe Flajolet, Stefan Gerhold, and Bruno Salvy. On the non-holonomic character of logarithms, powers, and the $n$-th prime function. *The Electronic Journal of Combinatorics*, 11(2):A2, 2005.

[75] Philippe Flajolet and Robert Sedgewick. *Analytic Combinatorics*. Cambridge University Press, 2009.

[76] GAP – groups, algorithms, programming. `http://www.gap-system.org/gap.html`, 1986–2014.

[77] Ewgenij Gawrilow and Michael Joswig. polymake: a framework for analyzing convex polytopes. In Gil Kalai and Günter M. Ziegler, editors, *Polytopes — Combinatorics and Computation*, pages 43–74. Birkhäuser, 2000.

[78] Keith O. Geddes, Stephen R. Czapor, and George Labahn. *Algorithms for Computer Algebra*. Kluwer, 1992.

[79] Stefan Gerhold. On some non-holonomic sequences. *Electronic Journal of Combinatorics*, 11(1):1–8, 2004.

[80] Stefan Gerhold and Manuel Kauers. A procedure for proving special function inequalities involving a discrete parameter. In Manuel Kauers, editor, *Proceedings of ISSAC'05*, pages 156–162, 2005.

[81] William Gosper. Decision procedure for indefinite hypergeometric summation. *Proceedings of the National Academy of Sciences of the United States of America*, 75:40–42, 1978.

[82] Ronald L. Graham, Donald E. Knuth, and Oren Patashnik. *Concrete Mathematics*. Addison-Wesley, second edition, 1994.

[83] Waldemar Hebisch and Martin Rubey. Extended Rate, more GFUN. *Journal of Symbolic Computation*, 46(8):889–903, 2011.

[84] Ralf Hemmecke and Martin Rubey. Aldor-combinat: An implementation of combinatorial species, 2006. `http://www.jku.at/people/hemmecke/aldor/combinat/`.

[85] Florent Hivert and Nicolas M. Thiery. MuPAD-combinat, an open-source package for research in algebraic combinatorics. *Séminaire Lotharingien de Combinatoire*, B51z, 2004.

[86] Derek Holt, Bettina Eick, and Eamonn O'Brian. *Handbook of Computational Group Theory*. Chapman & Hall/CRC, 2005.

[87] Alexander Hulpke. Computational group theory. `http://www.math.colostate.edu/~hulpke/CGT/cgtnotes.pdf`, 2010.

[88] E. L. Ince. *Ordinary Differential Equations*. Dover, 1926.

[89] Maximilian Jaroschek. Improved polynomial remainder sequences for Ore polynomials. *Journal of Symbolic Computation*, 58:64–76, 2013.

[90] Fredrik Johansson. Efficient implementation of the Hardy-Ramanujan-Rademacher formula. *LMS Journal of Computation and Mathematics*, 15:341–359, 2012.

[91] Fredrik Johansson and Brian Nakamura. Using functional equations to enumerate 1324-avoiding permutations. Technical Report 1309.7117, ArXiv, 2013.

[92] André Joyal. Une théorie combinatoire des séries formelles. *Advances in Mathematics*, 42:1–82, 1981.

[93] Michael Karr. Summation in finite terms. *Journal of the ACM*, 28:305–350, 1981.

[94] Michael Karr. Theory of summation in finite terms. *Journal of Symbolic Computation*, 1(3):303–315, 1985.

[95] Manuel Kauers. Summation algorithms for Stirling number identities. *Journal of Symbolic Computation*, 42(11):948–970, 2007.

[96] Manuel Kauers. Computer algebra and special function inequalities. In Tewodros Amdeberhan and Victor H. Moll, editors, *Tapas in Experimental Mathematics*, volume 457 of *Contemporary Mathematics*, pages 215–235. AMS, 2008.

[97] Manuel Kauers. Guessing handbook. Technical Report 09-07, RISC-Linz, 2009.

[98] Manuel Kauers. How to use cylindrical algebraic decomposition. *Séminaire Lotharingien de Combinatoire*, 65(B65a), 2011.

[99] Manuel Kauers. The holonomic toolkit. In Johannes Blümlein and Carsten Schneider, editors, *Computer Algebra in Quantum Field Theory: Integration, Summation and Special Functions*, Texts and Monographs in Symbolic Computation, pages 119–144. Springer, 2013.

[100] Manuel Kauers, Maximilian Jaroschek, and Fredrik Johansson. Ore polynomials in Sage. In *Computer Algebra and Polynomials*, Lecture Notes in Computer Science. Springer, 2014. to appear.

[101] Manuel Kauers, Christoph Koutschan, and Doron Zeilberger. Proof of Ira Gessel's lattice path conjecture. *Proceedings of the National Academy of Sciences*, 106(28):11502–11505, July 2009.

[102] Manuel Kauers and Peter Paule. *The Concrete Tetrahedron*. Springer, 2011.

[103] Manuel Kauers and Carsten Schneider. Application of unspecified sequences in symbolic summation. In Jean-Guillaume Dumas, editor, *Proceedings of ISSAC'06*, pages 177–183, 2006.

[104] Manuel Kauers and Carsten Schneider. Indefinite summation with unspecified sequences. *Discrete Mathematics*, 306(17):2073–2083, 2006.

[105] Manuel Kauers and Carsten Schneider. Symbolic summation with radical expressions. In Chris W. Brown, editor, *Proceedings of ISSAC'07*, pages 219–226, 2007.

[106] Manuel Kauers and Doron Zeilberger. The quasi-holonomic ansatz and restricted lattice walks. *Journal of Difference Equations and Applications*, 14(10):1119–1126, 2008.

[107] Donald E. Knuth. *The Art of Computer Programming 4A: Combinatorial Algorithms, Part 1*. Addison-Wesley, 2011.

[108] Donald E. Knuth and Peter Bendix. Simple word problems in universal algebras. In John Leech, editor, *Computational Problems in Abstract Algebra*, pages 263–297. Pergamon press, 1970.

[109] Wolfram Koepf. *Hypergeometric Summation*. Vieweg, 1998.

[110] Christoph Koutschan. *Advanced Applications of the Holonomic Systems Approach*. PhD thesis, RISC-Linz, Johannes Kepler Universität Linz, 2009.

[111] Christoph Koutschan. A fast approach to creative telescoping. *Mathematics in Computer Science*, 4(2–3):259–266, 2010.

[112] Christoph Koutschan. HolonomicFunctions (User's Guide). Technical Report 10-01, RISC Report Series, University of Linz, Austria, January 2010.

[113] Christoph Koutschan, Manuel Kauers, and Doron Zeilberger. Proof of George Andrews' and David Robbins' $q$-TSPP-conjecture. *Proceedings of the National Academy of Sciences*, 108(6):2196–2199, 2011.

[114] Christoph Koutschan and Thotsaporn Thanatipanonda. Advanced computer algebra for determinants. *Annals of Combinatorics*, 17(3):509–523, 2013.

[115] Christian Krattenthaler. RATE: A Mathematica guessing machine. `http://mat.univie.ac.at/~kratt/rate/rate.html`, 1997.

[116] D.L. Kreher and D.R. Stinson. *Combinatorial Algorithms: Generation, Enumeration, and Search*. Boca Raton, Florida, 1998.

[117] Germain Kreweras. Sur une classe de problèmes liés au treillis des partitions d'entiers. *Cahiers du B.U.R.O.*, 6:5–105, 1965.

[118] G. Labelle. On combinatorial differential equations. *Journal of Mathematical Analysis and Applications*, 113(2):344–381, 1986.

[119] G. Labelle. Some new computational methods in the theory of species. In *Combinatoire énumérative*, volume 1234 of *Lecture Notes in Computer Science*, pages 192–209. Springer, 1986.

[120] A. K. Lenstra, H. W. Lenstra, and L. Lovász. Factoring polynomials with rational coefficients. *Annals of Mathematics*, 126:515–534, 1982.

[121] L. Lipshitz. The diagonal of a D-finite power series is D-finite. *Journal of Algebra*, 113:373–378, 1988.

[122] Jesús A. De Loera, Raymond Hemmecke, and Matthias Köppe. *Algebraic and geometric ideas in the theory of discrete optimization*, volume 14 of *MPS-SIAM Series on Optimization*. Society for Industrial and Applied Mathematics, 2012.

[123] Jesús A. De Loera, Raymond Hemmecke, Jeremiah Tauzer, and Ruriko Yoshida. Effective lattice point counting in rational convex polytopes. *Journal of Symbolic Computation*, 38(4):1273–1302, 2004.

[124] I.G. Macdonald. Polynomials associated to finite cell complexes. *Journal of the London Mathematical Society*, 2(4):181–192, 1971.

[125] P. A. MacMahon. Memoir on the theory of the partition of numbers – part i. *Phil. Trans.*, 187:619–673, 1897.

[126] P. A. MacMahon. Memoir on the theory of the partition of numbers – part ii. *Phil. Trans.*, 192:351–401, 1899.

[127] P. A. MacMahon. Memoir on the theory of the partition of numbers – part iii. *Phil. Trans.*, 205:35–58, 1906.

[128] J. E. Majewicz. WZ-style certification and sister Celine's technique for Abel-type identities. *J. Difference Eqs and Appl.*, 2:55–65, 1996.

[129] Christian Mallinger. Algorithmic manipulations and transformations of univariate holonomic functions and sequences. Master's thesis, J. Kepler University, Linz, August 1996.

[130] Darko Marinov and Rados Radoicic. Counting 1324-avoiding permutations. *Electronic Journal of Combinatorics*, 9(2):R13, 2002.

[131] L. Martinelli, D. Merlini, R. Rocchini, A. Romoli, and C. Verri. A software to solve strip tiling problems. In *Proceedings of FPSAC 2000*, Lecture Notes in Computer Science, pages 801–806, 2000.

[132] Yuri V. Matiyasevich. *Hilbert's Tenth Problem*. MIT Press, 1993.

[133] D. Merlini, R. Sprugnoli, and M. C. Verri. A strip-like tiling algorithm. *Theoretical Computer Science*, 282(2):337–352, 2002.

[134] Maurice Mignotte. *Mathematics for Computer Algebra*. Springer, 1992.

[135] W. H. Mills, D. H. Robbins, and H. Rumsey. Proof of the MacDonald conjecture. *Inventiones Mathematicae*, 66:73–87, 1982.

[136] Mohamud Mohammed and Doron Zeilberger. Sharp upper bounds for the orders of the recurrences outputted by the Zeilberger and $q$-Zeilberger algorithms. *Journal of Symbolic Computation*, 39(2):201–207, 2005.

[137] Phong Q. Nguyen and Brigitte Vallée. *The LLL Algorithm*. Springer, 2010.

[138] Albert Nijenhuis and Herbert S. Wilf. *Combinatorial Algorithms*. Academic Press, 1978.

[139] P. S. Novikov. On the algorithmic unsolvability of the word problem in group theory (russian). *Trudy Mat. Inst. im. Steklov. no 44, Izdat. Akad. Nauk SSSR, Moskow*, 1955.

[140] O. Ore. Theory of non-commutative polynomials. *Annals of Mathematics*, 34:480–508, 1933.

[141] Peter Paule. Greatest factorial factorization and symbolic summation. *Journal of Symbolic Computation*, 20:235–268, 1995.

[142] Peter Paule and Axel Riese. A Mathematica $q$-Analogue of Zeilberger's Algorithm Based on an Algebraically Motivated Approach to $q$-Hypergeometric Telescoping. In M.E.H. Ismail and M. Rahman, editors, *Special Functions, q-Series and Related Topics*, volume 14 of *Fields Inst. Commun.*, pages 179–210. AMS, 1997.

[143] Peter Paule and Markus Schorn. A Mathematica version of Zeilberger's algorithm for proving binomial coefficient identities. *Journal of Symbolic Computation*, 20(5–6):673–698, 1995.

[144] Marko Petkovšek. Hypergeometric solutions of linear recurrences with polynomial coefficients. *Journal of Symbolic Computation*, 14(2–3):243–264, 1992.

[145] Marko Petkovšek, Herbert Wilf, and Doron Zeilberger. $A = B$. A K Peters, Ltd., 1997.

[146] Carine Pivoteau. *Génération aléatoire de structures combinatoires: méthode de Boltzmann effective*. PhD thesis, Universite Paris IV, 2008.

[147] Carine Pivoteau, Bruno Salvy, and Michèle Soria. Boltzmann oracle for combinatorial systems. *Discrete Mathematics and Theoretical Computer Science*, pages 475–488, 2008.

[148] Carine Pivoteau, Bruno Salvy, and Michèle Soria. Algorithms for combinatorial structures: Well-founded systems and Newton iterations. *Journal of Combinatorial Theory, Series A*, 119(8):1711–1773, 2012.

[149] E. J. Postma. The new group theory package in Maple 17. *Communications in Computer Algebra*, 47(3–4):174–178, 2013.

[150] Robert H. Risch. The solution of the problem of integration in finite terms. *Bulletin of the American Mathematical Society*, 79:605–608, 1970.

[151] Fabrice Rouiller and Paul Zimmermann. Efficient isolation of polynomial's real roots. *Journal of Computational and Applied Mathematics*, 162:33–50, 2004.

[152] The Sage-Combinat community. Sage-Combinat: enhancing Sage as a toolbox for computer exploration in algebraic combinatorics, 2008. `http://combinat.sagemath.org`.

[153] Bruno Salvy. Examples of automatic asymptotic expansions. *ACM Sigsam Bulletin*, 25(2):4–17, 1991.

[154] Bruno Salvy and Paul Zimmermann. Gfun: a Maple package for the manipulation of generating and holonomic functions in one variable. *ACM Transactions on Mathematical Software*, 20(2):163–177, 1994.

[155] Carsten Schneider. *Symbolic Summation in Difference Fields*. PhD thesis, RISC-Linz, Johannes Kepler Universität Linz, 2001.

[156] Carsten Schneider. The summation package Sigma: Underlying principles and a rhombus tiling application. *Discrete Mathematics and Theoretical Computer Science*, 6(2):365–386, 2004.

[157] Carsten Schneider. A new Sigma approach to multi-summation. *Advances in Applied Mathematics*, 34(4):740–767, 2005.

[158] Carsten Schneider. Apéry's double sum is plain sailing indeed. *Electronic Journal of Combinatorics*, 14(N5), 2007.

[159] Carsten Schneider. Symbolic summation assists combinatorics. *Sem. Lothar. Combin.*, B56b, 2007.

[160] Carsten Schneider. A refined difference field theory for symbolic summation. *Journal of Symbolic Computation*, 43:611–644, 2008.

[161] Carsten Schneider. Parameterized telescoping proves algebraic independence of sums. *Annals of Combinatorics*, 14:533–552, 2010.

[162] Carsten Schneider. Simplifying multiple sums in difference fields. In Johannes Blümlein and Carsten Schneider, editors, *Computer Algebra in Quantum Field Theory: Integration, Summation and Special Functions*, Texts and Monographs in Symbolic Computation, pages 325–360. Springer, 2013.

[163] Carsten Schneider. Fast algorithms for refined parameterized telescoping in difference fields. In *Proc. of the Workshop on Computer Algebra and Polynomials at the RICAM Special Semester 2013 Applications of Algebra and Number Theory*, Lecture Notes in Computer Science, 2014. to appear.

[164] Akos Seress. *Permutation Group Algorithms*. Cambridge University Press, 2003.

[165] John Shackell. *Symbolic Asymptotics*. Number 12 in Algorithms and Computation in Mathematics. Springer, 2004.

[166] Charles C. Sims. Computational methods in the study of permutation groups. In John Leech, editor, *Computational Problems in Abstract Algebra*, pages 169–183. Pergamon press, 1970.

[167] Charles C. Sims. *Computation with finitely presented groups*. Cambridge University Press, 1994.

[168] Michael F. Singer. Liouvillian solutions of $n$th order homogeneous linear differential equations. *American Journal of Mathematics*, 103:661–681, 1981.

[169] Nathan Singer, Andrew Shouldice, Lingyun Ye, Tomas Daske, Peter Dobscanyi, Dante Manna, O-Yeat Chan, and Jon Borwein. The inverse symbolic calculator. `http://isc.carma.newcastle.edu.au/`, 2013.

[170] Neil Sloane et al. The on-line encyclopedia of integer sequences. `http://www.oeis.org/`.

[171] Alan D. Sokal. A ridiculously simple and explicit implicit function theorem. *Séminaire Lotharingien de Combinatoire*, 61A, 2009.

[172] Richard P. Stanley. *Enumerative Combinatorics, Volume 1*. Cambridge Studies in Advanced Mathematics 49. Cambridge University Press, 1997. second edition: 2012.

[173] Richard P. Stanley. *Enumerative Combinatorics, Volume 2*. Cambridge Studies in Advanced Mathematics 62. Cambridge University Press, 1999.

[174] W. A. Stein et al. *Sage Mathematics Software*. The Sage Development Team, 2013. `http://www.sagemath.org`.

[175] Nobuki Takayama. An algorithm of constructing the integral of a module. In *Proceedings of ISSAC'90*, pages 206–211, 1990.

[176] Nobuki Takayama. Gröbner basis, integration and transcendental functions. In *Proceedings of ISSAC'90*, pages 152–156, 1990.

[177] J. A. Todd and H. S. M. Coxeter. A practical method for enumerating cosets of a finite abstract group. *Proceedings of the Edinburgh Mathematical Society Series II*, 5:26–34, 1936.

[178] Elias P. Tsigaridas and Ioannis Z. Emiris. On the complexity of real root isolation using continued fractions. *Theoretical Computer Science*, 392(1–3):158–173, 2008.

[179] Joris van der Hoeven. *Transseries and Real Differential Algebra*, volume 1888 of *Lecture Notes in Computer Science*. Springer, 2006.

[180] Alfred van der Poorten. A proof that Euler missed... — Apéry's proof of the irrationality for $\zeta(3)$. *The Mathematical Intelligencer*, 1:195–203, 1979.

[181] Marius van der Put and Michael Singer. *Galois Theory of Linear Differential Equations*. Springer, 2003.

[182] Mark van Hoeij. Rational solutions of linear difference equations. In *Proceedings of ISSAC'98*, pages 120–123, 1998.

[183] Mark van Hoeij. Finite singularities and hypergeometric solutions of linear recurrence equations. *Journal of Pure and Applied Algebra*, 139:109–131, 1999.

[184] Joachim von zur Gathen and Jürgen Gerhard. *Modern Computer Algebra.* Cambridge University Press, 1999.

[185] Robert J. Walker. *Algebraic Curves.* Princeton University Press, 1950.

[186] Herb S. Wilf and Doron Zeilberger. An algorithmic proof theory for hypergeometric (ordinary and $q$) multisum/integral identities. *Inventiones mathematicae*, 108:575–633, 1992.

[187] Jet Wimp and Doron Zeilberger. Resurrecting the asymptotics of linear recurrences. *Journal of Mathematical Analysis and Applications*, 111:162–176, 1985.

[188] Guoce Xin. A fast algorithm for MacMahon's partition analysis. *Electronic Journal of Combinatorics*, 11(R58), 2004.

[189] Brent Yorgey. A DSL for describing and computing with combinatorial species, 2012. `http://hackage.haskell.org/package/species`.

[190] Doron Zeilberger. A fast algorithm for proving terminating hypergeometric identities. *Discrete Mathematics*, 80:207–211, 1990.

[191] Doron Zeilberger. A holonomic systems approach to special function identities. *Journal of Computational and Applied Mathematics*, 32:321–368, 1990.

[192] Doron Zeilberger. The method of creative telescoping. *Journal of Symbolic Computation*, 11:195–204, 1991.

[193] Doron Zeilberger. Automatic counTilings. *The Personal Journal of Shalosh B. Ekhad and Doron Zeilberger*, 2006. `http://www.math.rutgers.edu/~zeilberg/mamarim/mamarimhtml/tilings.html`.

[194] Doron Zeilberger. The holonomic ansatz II: Automatic discovery(!) and proof(!!) of holonomic determinant evaluations. *Annals of Combinatorics*, 11(2):241–247, 2007.

[195] Paul Zimmermann. Gaia: a package for the random generation of combinatorial structures. *Maple Technical Newsletter*, 1:38–46, 1994.

[196] Richard Zippel. *Effective Polynomial Computation.* Springer, 1993.

[illegible] Cambridge University Press, [illegible]

[illegible] Vazirani. [illegible] [illegible]

[illegible]

[illegible] Zelikovsky. [illegible] *Journal of* [illegible] 195.

[illegible]

[illegible]

[illegible]

[illegible]

[illegible]

[illegible] Zelikovsky. Automatic [illegible]

[illegible] Zelikovsky. [illegible]

[illegible]

# Index